D0710485

THESAURUS

REVISED, 2D EDITION

ENGINEERING INFORMATION INC.
CASTLE POINT ON THE HUDSON
HOBOKEN, NJ 07030 USA

Engineering Information Thesaurus

Revised, 2nd Edition

Edited by Jessica L. Milstead

DIRECTOR, EDITORIAL SERVICES
Mary C. Berger

TECHNICAL STAFF
Robert Kreppel
Leah Kaufman
Mauro Pittaro, Jr.
Yakov Zeltser

PRODUCTION
Sally Foss
Kay Osborn

INTERNATIONAL STANDARD BOOK NUMBER ISBN 0-87394-144-6

First printing, January 1995.

Table of Contents

ENGINEERING INFORMATION THESAURUS

1995 Edition

BACKGROUND

The Ei Thesaurus was developed in response to requests from users of Engineering Information databases for an access tool that would make searching easier and more reliable. This second edition contains over 8,300 descriptors (preferred terms) and about 9,100 entry (non-preferred) terms, for a total of approximately 17,400 access points. 585 new terms have been added in this edition. They include 315 descriptors and 270 entry terms.[1] New descriptors are indicated by an entry date of January 1995 (data element DT in descriptor record).

A major feature of the second edition is that the appropriate classification code(s) have been added next to descriptors, so that the controlled vocabulary and classification systems are linked (see "Classification Codes" Page ix)

The Thesaurus replaces the SHE (Subject Headings for Engineering) section of the 1990 Ei Vocabulary and, effective January 1, 1993, serves as the indexing tool for the Engineering Index and the database Ei COMPENDEX*Plus and other index products. The new descriptors are in use for 1995 Ei products. The Thesaurus is not just a new edition of the Ei indexing vocabulary — it reflects a complete re-thinking of indexing policies, and is designed to provide new terminology both to meet the needs of searchers using electronic forms of the Ei databases, and to facilitate rapid search of the printed products as well. Users will observe several key changes:

- **The former heading-subheading organization has been abandoned.** Each index term stands alone. When a subheading represented a special aspect of the subject, it has been combined into a single term or divided into two separate terms, depending on the amount of material covered. Thus, "Electromagnetic waves—Absorption" has become "Electromagnetic wave absorption", but "Electron tubes—Reliability" is now indexed under two independent terms, "Electron tubes" and "Reliability". When a subheading refers to a specific type of thing represented by a main heading, the new term reverses the heading-subheading order. For example, "Light—Coherent" is now "Coherent Light".

- **Concepts previously expressed using inverted terms are now expressed using natural word order.** For example, "Electron tubes, Tetrode" is now "Tetrodes", " Electronic circuits, Frequency dividing" is now "Frequency dividing circuits" and "Furnaces, Metallurgical" is now "Metallurgical furnaces".

- **A full structure of cross references has been provided.** This greatly simplifies the tasks of determining the term(s) used to index a particular desired concept, and of locating related terms that might also be useful.

- **Provision has been made to permit searchers to determine the Ei Vocabulary equivalent of all new terms.** Users may also locate the Thesaurus equivalent of an Ei Vocabulary term. (See "Relationship to Ei Vocabulary" Page vii).

---

1 The term counts in the first edition of the Thesaurus were provided by the software used to construct the Thesaurus, and were inaccurate. The counting method in the current version of the software provides more accurate statistics, which are reflected in the numbers above.

ORGANIZATION OF THE THESAURUS

The Thesaurus is arranged alphabetically, word by word. For example, "Automobile windshields" precedes "Automobiles". Two kinds of terms may be found—descriptors and entry terms. Descriptors are terms actually used in indexing, while entry terms are synonymous or closely connected terms which, although not used for indexing documents, are given as cross references to guide the user to the descriptors.

Descriptors

Each descriptor record contains information about its relationships and, for terms that were not changed, the entry date (DT) on which the term was first used in indexing. Scope notes (SN) are provided when the meaning or usage of a term may not be clear from the term itself or from its relationships. In the second edition, most descriptors have the appropriate classification codes in the record (See "Classification Codes", Page ix).

Term Relationships

Four kinds of relationships may be provided with a descriptor. BT (Broader Term), NT (Narrower Term), RT (Related Term), and UF (Used For). BTs and NTs have a hierarchical relationship to the term. These hierarchical relationships are shown only one level up and down so, in some cases, it may be wise to check the term record for the broader or narrower terms to locate other relationships that might be useful.

RTs may indicate any non-hierarchical type of relationship. They suggest other terms where related concepts are to be found, and their appropriateness for a search should always be considered.

A UF relationship indicates a synonym or closely related term (entry term) which is not used in indexing. In some cases it may also assist in determining the way in which the descriptor is used in indexing.

An example of a descriptor record is shown below. (Note that descriptors appear in boldface.)

Well pumps 618.2 (446.1)(512.1.2)(512.2.2)
SN: Scope formerly limited to water well pumps
DT: Predates 1975
UF: Deep well pumps
BT: Pumps
NT: Oil well pumps
Water well pumps
RT: Submersible pumps
Wells

Entry Terms

The only information given for entry terms is the descriptor where that concept is indexed. Refer to the descriptor record for additional guidance. Two examples of entry terms are given below. (Note that entry terms appear in italics).

Abrasion
USE: Wear of materials

Crating
USE: Packaging

For every entry term that has a USE reference, there will be a reciprocal UF (Used For) reference. Reciprocal references for the entry terms shown above are:

Wear of materials
UF: Abrasion

Packaging
UF: Crating

Whenever an entry term leads to two different descriptors, a USE AND reference is provided:

Inflatable domes
USE: Domes AND
Inflatable structures

The reciprocal entries are:

Domes
UF: Inflatable domes

Inflatable structures
UF: Inflatable domes

Relationship To Ei Vocabulary

Former Ei Vocabulary terms (both heading and heading---subheading combinations) are designated by an asterisk * following the term or phrase. For each former term, the new thesaurus term is given as a USE: reference. Reciprocal relationships are also shown, through a UF: (Used For) reference. Former vocabulary terms should be used to search materials added to the database before January 1993. There are a number of important types of these references. Examples are shown below.

Simple heading:

*Microscopes, Electron**
USE: Electron microscopes

The reciprocal entry is:

Electron microscopes
UF: Microscopes, Electron*

Heading—subheading:

*Flow of fluids—Diffusers**
USE: Diffusers (fluid)

The reciprocal of this would be:

Diffusers (fluid)
UF: Flow of fluids—Diffusers*
Sugar factories—Diffusers*
Turbomachinery—Diffusers*

{Note that there were actually three former vocabulary terms that are now indexed under the term "Diffusers (fluid)"}

Heading or heading-subheading combination divided into two or more descriptors, either of which should be used to index the concept:

*Copper and alloys**
USE: Copper OR
Copper alloys

*Information science—Language translation and linguistics**
USE: Linguistics OR
Translation (languages)

Reciprocals are:

Copper
UF: Building materials—Copper*
Cast iron—Copper content*
Copper and alloys*
Cu
Springs—Copper*

Copper alloys
UF: Copper and alloys*

Linguistics
UF: Information science—Language translation and linguistics*

Translation (languages)
UF: Information science—Language translation and linguistics*

CLASSIFICATION CODES

Classification Codes comprise a numeric classification scheme that segments the literature in Ei databases into broad technical subject areas. Taken as a whole, the codes provide an overview of the scope of all of Ei's information products and services.

The codes are organized into six categories which are divided into thirty-eight subject series. These series are further subdivided into 182 specific subject areas whose codes are applied to articles Ei indexes.

As of January 1, 1993, these specific subject areas remain essentially unchanged but are further subdivided, resulting in over 800 individual classes. The expanded codes provide up to two additional levels of subject specificity. For example, Code 723 of the 720 series covers the broad area of Computer Software, Data Handling and Applications. An additional level of specificity, indicated by a decimal point, provides subdivisions under the original code. Some of these are further subdivided, indicated by another decimal point. For example:

Level 1: 720 Computers and Data Processing

Level 2: 723 Computer Software, Data Handling and Applications
Level 3: 723.1 Computer Programming
Level 4: 723.1.1 Computer Programming Languages

Up to four levels of increasing specificity are provided. Note that Level 1 Codes are not used in indexing, but instead are used to group like concepts together. In some cases a three-digit code (Level 2) is assigned to a descriptor indicating that the term is broader in meaning than the more specific codes, or that one or more of the specific codes is applicable depending on the use of the descriptor for indexing a document.

Each article is assigned at least one classification code. The number and level of the codes depend on the important concepts and applications presented in the article.

In this edition, the relationship between descriptors and the classification codes is presented for most descriptors. (Some descriptors such as "Analysis" will not have any codes indicated). An example of a descriptor/classification code line from a record is shown below:

Welding shops 538.2 (402.1)

Note that the codes may appear with or without parentheses. Those with parentheses are called optional codes to be used when appropriate for the specific document being indexed. In the above, the optional code would be used when the subject is being treated as a building.

Acknowledgement:

The Ei Thesaurus was produced using the Liu-Palmer Thesaurus Construction System, Release 2.1

ENGINEERING INFORMATION

THESAURUS

Abandoned mines 502.1 (504) (505.2)
DT: January 1993
UF: Mines and mining—Abandoned*
BT: Mines

Abatement (dust)
USE: Dust abatement

Aberrations (741) (941.3) (941.4)
SN: In optics and particle optics
DT: January 1993
UF: Optical aberrations
RT: Lenses
Optical instruments
Optics
Particle optics

Ability testing 912.4
DT: January 1993
UF: Personnel—Ability testing*
BT: Personnel testing
RT: Human engineering
Personnel

Ablation 641.2
DT: Predates 1975
NT: Laser ablation
RT: Ablative materials
Erosion
Heat shielding
Melting
Reentry
Sublimation
Vaporization
Wear of materials

Ablative materials (641.2)
DT: Predates 1975
BT: Materials
RT: Ablation

Abrasion
USE: Wear of materials

*Abrasion resistance**
USE: Wear resistance

Abrasive belts 606.2 (535.2)
DT: January 1993
UF: Belts—Abrasive*
BT: Belts
RT: Abrasives

Abrasive coatings 813.2, 606.1
DT: January 1993
UF: Abrasives—Coated products*
BT: Coatings
RT: Abrasives
Coated materials

Abrasive cutting 604.1, 606.1
DT: January 1993
UF: Metal cutting—Abrasive*
BT: Cutting
NT: Grinding (machining)
RT: Abrasives

Abrasive grinding
USE: Grinding (machining)

Abrasive wheels
USE: Grinding wheels

Abrasives 606.1
DT: Predates 1975
UF: Pumice
BT: Materials
RT: Abrasive belts
Abrasive coatings
Abrasive cutting
Corundum
Cubic boron nitride
Grinding wheels
Industrial diamonds

*Abrasives—Coated products**
USE: Abrasive coatings

ABS resins 815.1.1
DT: Predates 1975
UF: Acrylonitrile-butadiene-styrene resins
BT: Aromatic polymers
Terpolymers

Absolute temperature
USE: Temperature

Absorption 802.3 (931.2)
DT: Predates 1975
UF: Absorption (of matter)
BT: Sorption
NT: Gas absorption
Water absorption
RT: Adsorption
Chemisorption
Desorption

Absorption (of energy)
USE: Energy absorption

Absorption (of matter)
USE: Absorption

Absorption spectroscopy
(741) (941.3) (941.4) (801)
DT: January 1993
UF: Spectroscopy, Absorption*
BT: Spectroscopy

* Former Ei Vocabulary term

Abstracting 903.1
DT: January 1993
UF: Information science—Abstracting*
BT: Information analysis

Abutments (bridge) 401.1
DT: January 1993
UF: Bridge abutments
Bridges—Abutments*
BT: Bridge components

Ac
USE: Actinium

AC generator motors 705.2.1, 705.3.1
SN: Use for single alternating current machines capable of operating either as motors or as generators
DT: January 1993
UF: Electric generator motors, AC*
Motor generators (AC)
BT: AC generators
AC motors
RT: Pumped storage power plants

AC generators 705.2.1
DT: January 1993
UF: Electric generators, AC*
BT: AC machinery
Electric generators
NT: AC generator motors
Asynchronous generators
Synchronous generators
RT: Electric windings
Machine windings
Rotors (windings)
Stators

AC machinery 705.1 (705.2.1) (705.3.1)
DT: January 1993
UF: Alternating current machinery
Electric machinery, AC*
BT: Electric machinery
NT: AC generators
AC motors
Asynchronous machinery
Synchronous machinery
RT: Electric windings
Machine windings
Rotors (windings)
Stators

AC motors 705.3.1
DT: January 1993
UF: Electric motors, AC*
Repulsion motors
BT: AC machinery
Electric motors
NT: AC generator motors
Induction motors
Synchronous motors
RT: Machine windings
Rotors (windings)
Stators
Variable speed drives

AC network analyzers
USE: Electric network analyzers

Acceleration (931.1)
DT: Predates 1975
RT: Acceleration measurement
Accelerometers
Deceleration
Velocity

Acceleration control 731.3 (931.1)
DT: January 1993
UF: Control, Mechanical variables—Acceleration*
BT: Mechanical variables control

Acceleration measurement 943.2
DT: January 1993
UF: Mechanical variables measurement—Acceleration*
BT: Mechanical variables measurement
RT: Acceleration
Accelerometers

*Acceleration waves**
USE: Waves

Accelerator magnets 932.1.1 (701.1) (708.4)
DT: January 1993
UF: Accelerators—Magnets*
BT: Magnets
Particle accelerators

Accelerator shielding 932.1.1
DT: January 1993
UF: Accelerators—Shields*
BT: Particle accelerators
Radiation shielding

*Accelerators**
USE: Particle accelerators

*Accelerators, Betatron**
USE: Betatrons

*Accelerators, Cyclotron**
USE: Cyclotrons

*Accelerators, Electron ring**
USE: Electron ring accelerators

*Accelerators, Electrostatic**
USE: Electrostatic accelerators

*Accelerators, Linear**
USE: Linear accelerators

*Accelerators, Microtron**
USE: Microtrons

*Accelerators, Synchrocyclotron**
USE: Synchrocyclotrons

*Accelerators, Synchrotron**
USE: Synchrotrons

*Accelerators, Van de Graaff**
USE: Van de Graaff accelerators

*Accelerators—Accessories**
USE: Particle accelerator accessories

* Former Ei Vocabulary term

*Accelerators—Beam dynamics**
USE: Particle beam dynamics

*Accelerators—Beam extraction**
USE: Particle beam extraction

*Accelerators—Beam injection**
USE: Particle beam injection

*Accelerators—Colliding beam**
USE: Colliding beam accelerators

*Accelerators—Magnets**
USE: Accelerator magnets

*Accelerators—Power supply**
USE: Electric power utilization

*Accelerators—Shields**
USE: Accelerator shielding

*Accelerators—Storage rings**
USE: Storage rings

*Accelerators—Targets**
USE: Targets

Accelerometers 943.1
DT: Predates 1975
BT: Instruments
RT: Acceleration
Acceleration measurement
Gravimeters
Velocimeters

Accessories
DT: January 1993
UF: Industrial trucks—Attachments*
NT: Ant*enna accessories
Laser accessories
Machine tool attachments
Particle accelerator accessories
Photographic accessories
RT: Components

Accident prevention 914.1
DT: Predates 1975
UF: Railroad accident prevention
Safety
Safety engineering
BT: Prevention
NT: Blowout prevention
Laser safety
Pedestrian safety
RT: Accidents
Aircraft accidents
Codes (standards)
Collision avoidance
Crashworthiness
Enclosures
Environmental engineering
Escape ramps
Eye protection
Highway accidents
Industrial plants
Leak detection
Product liability
Protection
Safe handling
Safety devices
Safety factor
Skid resistance

Accidents 914.1
DT: January 1977
NT: Aircraft accidents
Blowouts
Electric accidents
Highway accidents
Nuclear reactor accidents
Railroad accidents
RT: Accident prevention
Automobile air bags
Avalanches (snowslides)
Contamination
Explosions
Failure (mechanical)
Fallout
Fire hazards
Fires
Hazardous materials spills
Hazards
Health hazards
Leak detection
Meteor impacts
Salvaging
Tornado generated missiles
Toxicity

Accounting (nuclear fuel)
USE: Nuclear fuel accounting

Acetal resins 815.1.1
DT: Predates 1975
UF: Polyacetals
BT: Polyethers
RT: Formaldehyde

Acetic acid 804.1
DT: Predates 1985
UF: Ethanoic acid
BT: Fatty acids

Acetone 804.1
DT: Predates 1975
BT: Ketones
RT: Solvents

Acetonitrile 804.1
DT: June 1990
BT: Organic compounds

Acetylation 802.2
DT: January 1993
BT: Chemical reactions

Acetylene 804.1
DT: Predates 1975
BT: Hydrocarbons

Acid rain 451.1
DT: January 1993
UF: Air pollution—Acid rain*
BT: Rain
RT: Air pollution

Acid resistance 802.2 (539.1)
DT: January 1993
UF: Acid resisting*
BT: Chemical resistance
RT: Acids
Corrosion resistance

*Acid resisting**
USE: Acid resistance

Acidity
USE: pH

Acidization 802.2
SN: Improvement of the permeability of petroleum reservoir rock by treatment with acid
DT: January 1993
UF: Oil wells—Acid treatment*
BT: Production
RT: Enhanced recovery
Natural gas deposits
Petroleum deposits
Well stimulation

Acids 803, 804
DT: October 1975
BT: Chemical compounds
NT: Inorganic acids
Organic acids
RT: Acid resistance

*Acids—Inorganic**
USE: Inorganic acids

*Acids—Organic**
USE: Organic acids

Acoustic arrays 752.1
SN: Systems for sound transmission or reception, designed to provide required directional characteristics
DT: January 1993
UF: Acoustic radiator arrays
Acoustic radiators—Arrays*
BT: Acoustic equipment
Arrays
RT: Acoustic generators
Acoustic radiators
Acoustic receivers

Acoustic bulk wave devices 752.1
DT: January 1986
UF: Bulk acoustic wave devices
Volume acoustic wave devices
BT: Acoustic devices

Acoustic delay lines 752.1
SN: Devices in which acoustic signals are propagated in a medium, to obtain a time delay for the signals
DT: Predates 1975
UF: Sonic delay lines
BT: Acoustic devices
NT: Ultrasonic delay lines

Acoustic devices 752.1
DT: Predates 1975
BT: Acoustic equipment
NT: Acoustic bulk wave devices
Acoustic delay lines
Acoustic surface wave devices
Acoustic transducers
Acoustooptical devices
Microwave acoustic devices
Ultrasonic devices
RT: Acoustics
Audio equipment
Audio systems
Instruments
Sound reproduction

*Acoustic devices—Microwave frequencies**
USE: Microwave acoustic devices

Acoustic dispersion 751.2
DT: January 1993
BT: Dispersion (waves)
NT: Ultrasonic dispersion
RT: Acoustic wave diffraction
Acoustic wave propagation
Acoustic wave refraction
Acoustic wave scattering
Acoustic wave velocity
Acoustic waves

Acoustic distortion 751.2
DT: January 1993
UF: Acoustic waves—Distortion*
BT: Distortion (waves)
NT: Ultrasonic distortion
RT: Acoustic waves

Acoustic emission testing 751.2 (422.2)
DT: January 1977
BT: Nondestructive examination
RT: Acoustic emissions

Acoustic emissions 751.2
SN: Generation of transient elastic waves, resulting from rapid release of strain energy caused by a change in the structure of a solid
DT: January 1993
UF: Emissions (acoustic)
Stress wave emissions
RT: Acoustic emission testing
Acoustics

Acoustic equipment 752.1
DT: Predates 1975
BT: Equipment
NT: Acoustic arrays
Acoustic devices
Acoustic generators
Acoustic microscopes
Acoustic receivers
Acoustic resonators
Sonar
Sounding apparatus
Ultrasonic equipment
RT: Acoustics
Audio equipment
Audio systems
Sound reproduction

* Former Ei Vocabulary term

Acoustic field measurement 751.2, 941.2
DT: January 1993
UF: Acoustic variables measurement—Acoustic field*
BT: Acoustic variables measurement
RT: Acoustic intensity measurement

Acoustic generators 752.4
SN: Generators of acoustic waves
DT: Predates 1975
UF: Generators (acoustic)
Sound generators
BT: Acoustic equipment
NT: Acoustic radiators
Bells
Loudspeakers
Musical instruments
Sirens
RT: Acoustic arrays
Photoacoustic effect
Signal generators
Transmitters
Vibrators

Acoustic holography 743.1.2 (751.2)
SN: Use of a phase interference pattern formed by acoustic beams that, when light interacts with the pattern, forms an image of an object placed in one of the beams
DT: January 1977
BT: Holography
RT: Acoustic imaging
Acoustics

Acoustic imaging 751.2
SN: Production of real-time images of the internal structure of opaque objects
DT: Predates 1975
UF: Acoustic lenses
BT: Imaging techniques
NT: Ultrasonic imaging
RT: Acoustic holography
Acoustics
Diagnosis
Lenses

Acoustic impedance 751.2
DT: January 1993
UF: Impedance (acoustic)
BT: Acoustic properties
RT: Acoustic impedance measurement
Acoustic wave velocity
Impedance matching (acoustic)
Natural frequencies

Acoustic impedance measurement 941.2 (751.2)
DT: January 1993
UF: Acoustic variables measurement—Impedance*
BT: Acoustic variables measurement
RT: Acoustic impedance
Acoustic wave velocity measurement

Acoustic insulating materials
USE: Sound insulating materials

Acoustic insulation
USE: Sound insulation

Acoustic intensity 751.2
DT: January 1993
UF: Sound intensity
BT: Acoustic properties
RT: Acoustic intensity measurement
Acoustic noise

Acoustic intensity measurement 941.2 (751.2)
DT: January 1993
UF: Acoustic variables measurement—Intensity*
BT: Acoustic variables measurement
RT: Acoustic field measurement
Acoustic intensity

Acoustic lenses
USE: Acoustic imaging

Acoustic logging 941.2, 751.2 (512.1.2) (512.2.2)
SN: Technique for measurement of the porosity of drill holes; the depth is compared against the travel time of a sonic impulse through a given portion of the formation; the rate depends upon the composition of the fluids and the rock in the formation
DT: January 1993
UF: Oil well logging—Acoustic*
BT: Well logging
RT: Natural gas well logging
Oil bearing formations
Oil well logging
Petroleum geology
Rocks

Acoustic masking
USE: Speech intelligibility

Acoustic measuring instruments 941.1
DT: January 1993
UF: Sound measuring instruments*
BT: Instruments
RT: Acoustic variables measurement
Audiometers

Acoustic microscopes 752.1, 941.1
SN: Instruments which use microwave acoustic radiation to visualize the microscopic detail of objects
DT: January 1993
UF: Microscopes, Acoustic*
BT: Acoustic equipment
Microscopes
RT: Imaging techniques

Acoustic mode (crystal)
USE: Lattice vibrations

Acoustic noise 751.4
DT: January 1993
UF: Air conditioning—Noise*
Automobiles—Noise*
Bearings—Noise*
Cams—Noise*
Cars—Noise*
Combustion—Noise*
Conveyors—Noise*

Acoustic noise *(continued)*
Cutting tools—Noise*
Fuel burners—Noise*
Gears—Noise*
Industrial plants—Noise*
Instruments—Noise*
Internal combustion engines—Noise*
Locomotives—Noise*
Machinery—Noise*
Mines and mining—Noise*
Noise (acoustic)
Noise, Acoustic*
Nuclear reactors—Noise*
Oil burners—Noise*
Power plants—Noise*
Refrigerators—Noise*
Tires—Noise*
Transportation—Noise*
Vehicles—Noise*
Ventilation—Noise*
BT: Acoustic waves
NT: White acoustic noise
RT: Acoustic intensity
Acoustic noise measurement
Acoustic wave absorption
Acoustic wave interference
Architectural acoustics
Audition
Human engineering
Mufflers
Noise abatement
Noise pollution
Shock waves
Vibrations (mechanical)

Acoustic noise measurement 751.4, 941.2
DT: January 1993
BT: Acoustic variables measurement
RT: Acoustic noise
Noise abatement

Acoustic properties 751.2
DT: January 1993
BT: Physical properties
NT: Acoustic impedance
Acoustic intensity
RT: Acoustic variables control
Acoustic variables measurement
Acoustics

Acoustic radiator arrays
USE: Acoustic arrays

Acoustic radiators 752.1, 752.4
SN: Surfaces that produce acoustic waves by means of their vibrations
DT: Predates 1975
UF: Radiators (acoustic)
BT: Acoustic generators
RT: Acoustic arrays
Doppler effect

*Acoustic radiators—Arrays**
USE: Acoustic arrays

Acoustic receivers 752.1 (716.3)
SN: Equipment which receives radio waves and converts them into sound waves
DT: Predates 1975
BT: Acoustic equipment
Signal receivers
RT: Acoustic arrays
Microphones

Acoustic resonators 752.1 (752.4)
SN: Enclosures that produce resonance of acoustic waves at particular frequencies
DT: Predates 1975
BT: Acoustic equipment
Cavity resonators
RT: Acoustics

Acoustic signal processing 751
SN: Extraction of information from signals which were propagated in the presence of acoustic noise
DT: June 1990
BT: Signal processing
RT: Acoustics
Speech analysis

Acoustic spectroscopy 751.2 (801) (941.2)
DT: January 1993
UF: Spectroscopy, Acoustic*
BT: Spectroscopy
NT: Photoacoustic spectroscopy
RT: Acoustics

Acoustic streaming 751.2
SN: Unidirectional currents of fluid flow resulting from the presence of sound waves
DT: January 1993
UF: Acoustics—Streaming*
Streaming (acoustic)
RT: Acoustic wave propagation
Acoustics

Acoustic surface wave devices 752.1
DT: October 1975
UF: Surface acoustic wave devices
BT: Acoustic devices
NT: Acoustic surface wave filters

Acoustic surface wave filters 752.1
DT: January 1981
UF: Surface acoustic wave filters
BT: Acoustic surface wave devices
Wave filters
RT: Crystal filters
Passive filters

Acoustic transducers 752.1 (752.4)
SN: Devices, such as phonograph pickups and microphones, that convert acoustical energy to electrical or mechanical energy
DT: Predates 1975
BT: Acoustic devices
Transducers
NT: Electroacoustic transducers
Pickups
Ultrasonic transducers
RT: Acoustooptical effects

* Former Ei Vocabulary term

Acoustic variables control 731.3 (751.2)
DT: January 1993
UF: Control, Acoustic variables*
Noise control
BT: Control
NT: Noise pollution control
RT: Acoustic properties
Acoustics

Acoustic variables measurement 941.2 (751.2)
DT: Predates 1975
UF: Sound measurement
BT: Measurements
NT: Acoustic field measurement
Acoustic impedance measurement
Acoustic intensity measurement
Acoustic noise measurement
Acoustic wave velocity measurement
Ultrasonic measurement
RT: Acoustic measuring instruments
Acoustic properties
Acoustics
Anechoic chambers
Magnetostrictive devices
Piezoelectric devices
Seismic prospecting
Seismographs
Vibration measurement

*Acoustic variables measurement—Acoustic field**
USE: Acoustic field measurement

*Acoustic variables measurement—Impedance**
USE: Acoustic impedance measurement

*Acoustic variables measurement—Intensity**
USE: Acoustic intensity measurement

*Acoustic variables measurement—Velocity**
USE: Acoustic wave velocity measurement

Acoustic velocity
USE: Acoustic wave velocity

Acoustic velocity measurement
USE: Acoustic wave velocity measurement

Acoustic wave absorption 751.1 (751.2)
DT: January 1993
UF: Acoustic waves—Absorption*
BT: Energy absorption
NT: Ultrasonic absorption
RT: Acoustic noise
Noise abatement
Sound insulating materials
Sound insulation

Acoustic wave backscattering 751.1 (751.2)
DT: January 1993
UF: Acoustic waves—Backscattering*
BT: Backscattering
RT: Acoustic wave propagation

Acoustic wave diffraction 751.1 (751.2)
DT: January 1993
UF: Acoustic waves—Diffraction*
BT: Diffraction
NT: Ultrasonic diffraction
RT: Acoustic dispersion
Acoustic wave propagation
Acoustic waves

Acoustic wave effects 751.1 (751.2)
DT: Predates 1975
BT: Wave effects
NT: Acoustoelectric effects
Acoustooptical effects
Sonoluminescence
Ultrasonic effects
RT: Acoustic wave propagation
Acoustic waves
Human engineering
Magnetoacoustic effects

Acoustic wave interference 751.1 (751.2)
DT: January 1993
UF: Acoustic waves—Interference*
BT: Wave interference
RT: Acoustic noise
Acoustic wave propagation

Acoustic wave propagation 751.1 (751.2) (931.2)
SN: In gases. For propagation in liquids and solids, use ACOUSTIC WAVE TRANSMISSION
DT: January 1993
UF: Acoustic waves—Propagation*
BT: Wave propagation
NT: Ultrasonic propagation
RT: Acoustic dispersion
Acoustic streaming
Acoustic wave backscattering
Acoustic wave diffraction
Acoustic wave effects
Acoustic wave interference
Acoustic wave reflection
Acoustic wave refraction
Acoustic wave scattering
Acoustic wave velocity
Acoustic waves
Atmospheric acoustics
Audition
Doppler effect
Shock waves
Speech

Acoustic wave reflection 751.1 (751.2)
DT: January 1993
UF: Acoustic waves—Reflection*
BT: Reflection
NT: Ultrasonic reflection
RT: Acoustic wave propagation
Acoustic waves
Architectural acoustics
Reverberation

Acoustic wave refraction 751.1 (751.2)
DT: January 1993
UF: Acoustic waves—Refraction*
BT: Refraction
NT: Ultrasonic refraction
RT: Acoustic dispersion
Acoustic wave propagation
Acoustic waves

* Former Ei Vocabulary term

Acoustic wave scattering 751.1 (751.2)
DT: January 1993
UF: Acoustic waves—Scattering*
BT: Scattering
NT: Ultrasonic scattering
RT: Acoustic dispersion
Acoustic wave propagation
Acoustic waves

Acoustic wave transmission 751.1 (751.2) (931.2)
SN: In liquids and solids. For transmission in gases, use ACOUSTIC WAVE PROPAGATION
DT: January 1993
UF: Acoustic waves—Transmission*
BT: Wave transmission
NT: Ultrasonic transmission
RT: Seismic waves
Sonar
Underwater acoustics

Acoustic wave velocity 751.1 (751.2)
DT: January 1993
UF: Acoustic velocity
Acoustic waves—Velocity*
Sound velocity
Speed of sound
Velocity of sound
BT: Velocity
NT: Ultrasonic velocity
RT: Acoustic dispersion
Acoustic impedance
Acoustic wave propagation
Acoustic wave velocity measurement
Acoustic waves
Shock waves

Acoustic wave velocity measurement 941.2, 751.1 (751.2)
DT: January 1993
UF: Acoustic variables measurement—Velocity*
Acoustic velocity measurement
BT: Acoustic variables measurement
RT: Acoustic impedance measurement
Acoustic wave velocity

Acoustic waves 751.1
SN: Includes infrasonic and sonic waves
DT: Predates 1975
UF: Sound
Sound waves
BT: Elastic waves
NT: Acoustic noise
Ion acoustic waves
Ultrasonic waves
RT: Acoustic dispersion
Acoustic distortion
Acoustic wave diffraction
Acoustic wave effects
Acoustic wave propagation
Acoustic wave reflection
Acoustic wave refraction
Acoustic wave scattering
Acoustic wave velocity
Acoustics
Backscattering
Doppler effect
Lattice vibrations
Noise pollution
Phonons
Shock waves
Vibrations (mechanical)

*Acoustic waves—Absorption**
USE: Acoustic wave absorption

*Acoustic waves—Backscattering**
USE: Acoustic wave backscattering

*Acoustic waves—Diffraction**
USE: Acoustic wave diffraction

*Acoustic waves—Distortion**
USE: Acoustic distortion

*Acoustic waves—Interference**
USE: Acoustic wave interference

*Acoustic waves—Propagation**
USE: Acoustic wave propagation

*Acoustic waves—Reflection**
USE: Acoustic wave reflection

*Acoustic waves—Refraction**
USE: Acoustic wave refraction

*Acoustic waves—Scattering**
USE: Acoustic wave scattering

*Acoustic waves—Transmission**
USE: Acoustic wave transmission

*Acoustic waves—Velocity**
USE: Acoustic wave velocity

Acoustics 751
DT: Predates 1975
BT: Physics
NT: Architectural acoustics
Atmospheric acoustics
Audio acoustics
Microwave acoustics
Ultrasonics
Underwater acoustics
RT: Acoustic devices
Acoustic emissions
Acoustic equipment
Acoustic holography
Acoustic imaging
Acoustic properties
Acoustic resonators
Acoustic signal processing
Acoustic spectroscopy
Acoustic streaming
Acoustic variables control
Acoustic variables measurement
Acoustic waves
Acoustics laboratories
Acoustooptical devices
Photoacoustic effect
Reverberation
Shock waves
Sound reproduction
Speech recognition
Vibrations (mechanical)

Acoustics laboratories (752) (751)
DT: Predates 1975
BT: Laboratories
NT: Anechoic chambers
RT: Acoustics

*Acoustics, Underwater**
USE: Underwater acoustics

*Acoustics—Reverberation**
USE: Reverberation

*Acoustics—Streaming**
USE: Acoustic streaming

Acoustoelasticity
USE: Elasticity

Acoustoelectric effects 751.2, 701.1 (751.1)
SN: Development of a direct-current voltage in a semiconductor or metal by an acoustic wave traveling parallel to the surface of the material
DT: January 1977
UF: Electroacoustic effects
BT: Acoustic wave effects
Electric field effects
RT: Semiconductor materials

Acoustoelectric testing
USE: Electroacoustic testing

Acoustoelectric transducers
USE: Electroacoustic transducers

Acoustoluminescence
USE: Sonoluminescence

Acoustomagnetic effects
USE: Magnetoacoustic effects

Acoustooptic modulators
USE: Bragg cells

Acoustooptical devices 752.1, 741.3
DT: January 1977
BT: Acoustic devices
NT: Bragg cells
RT: Acoustics
Acoustooptical effects
Optical instruments
Optics
Optoelectronic devices

Acoustooptical effects 751.2, 741.1
DT: January 1993
UF: Acoustooptics
Light—Acoustooptical effects*
BT: Acoustic wave effects
Optical properties
RT: Acoustic transducers
Acoustooptical devices
Coherent light
Light
Light modulation
Light modulators
Optoelectronic devices
Ultrasonics

Acoustooptics
USE: Acoustooptical effects

Acrylic monomers 804.1
DT: January 1986
BT: Monomers
Organic compounds
RT: Acrylics

Acrylics 815.1.1
DT: Predates 1975
BT: Thermoplastics
NT: Polyacrylates
Polyacrylonitriles
Polymethyl methacrylates
RT: Acrylic monomers
Synthetic fibers

Acrylonitrile-butadiene-styrene resins
USE: ABS resins

Actinides 547, 622.1
DT: Predates 1975
BT: Nonferrous metals
Radioactive elements
NT: Actinium
Protactinium
Thorium
Transuranium elements
Uranium

Actinium 547, 622.1
DT: January 1981
UF: Ac
BT: Actinides

Activated alumina 803, 804.2
DT: January 1993
UF: Alumina—Activated*
BT: Alumina
RT: Adsorbents
Catalysts

Activated carbon 804
DT: January 1993
UF: Activated charcoal
Carbon—Activated*
BT: Carbon
RT: Activated carbon treatment
Carbonaceous adsorbents

Activated carbon treatment (452.2) (802.3) (445.1) (803)
DT: January 1993
UF: Water treatment—Activated carbon*
BT: Water treatment
RT: Activated carbon

Activated charcoal
USE: Activated carbon

Activated silica treatment (452.2) (802.3) (445.1) (803)
DT: January 1993
UF: Water treatment—Activated silica*
BT: Water treatment

Activated sludge process 452.2 (802.2)
DT: January 1993
UF: Sewage treatment—Activated sludge*
BT: Sewage treatment

* Former Ei Vocabulary term

Activation analysis (801)
DT: January 1993
UF: Chemical analysis—Activation*
BT: Chemical analysis
NT: Neutron activation analysis
Radioactivation analysis

Activation energy (801.4) (931.3)
DT: January 1995
RT: Chemical activation
Physical chemistry

Activation reactions
USE: Chemical activation

*Activation**
USE: Chemical activation

Active filters 703.2
DT: January 1993
UF: Biquadratic filters
Electric filters, Active*
BT: Active networks
Electric filters

Active networks 703.1 (703.2)
DT: January 1993
UF: Electric networks, Active*
BT: Networks (circuits)
NT: Active filters
Negative impedance converters
RT: Amplifiers (electronic)
Oscillators (electronic)

Active solar buildings 402, 643.1, 657.1
DT: January 1993
UF: Buildings—Solar, Active*
BT: Solar buildings

Activity (catalysts)
USE: Catalyst activity

Activity (thermodynamics)
USE: Thermodynamic properties

Actuators 732.1 (731.2) (602.1) (632)
DT: Predates 1975
RT: Cams
Clutches
Electromagnets
Final control devices
Hydraulic rams
Jacks
Servomechanisms
Servomotors
Starters
Starting
Transducers
Valves (mechanical)

Acupuncture 461.6 (462.1)
DT: January 1987
BT: Patient treatment
NT: Electroacupuncture
RT: Biomedical engineering

Acylation 802.2
DT: January 1993
BT: Chemical reactions

Ada (programming language) 723.1.1
DT: January 1993
UF: Computer programming languages—Ada*
BT: Procedure oriented languages
RT: Pascal (programming language)

Adaptive algorithms (723) (921)
DT: January 1995
BT: Algorithms
RT: Image processing

Adaptive control systems 731.1
DT: January 1993
UF: Control systems, Adaptive*
BT: Adaptive systems
Control systems
NT: Self adjusting control systems
Self tuning control systems
RT: Intelligent control

Adaptive delta modulation
USE: Delta modulation

Adaptive filtering (731.1)
DT: January 1995
BT: Signal filtering and prediction
RT: Adaptive systems
Digital filters
Radar systems

Adaptive optics 741.1 (731.1)
DT: January 1993
BT: Optics
RT: Adaptive systems
Optical phase conjugation
Optical systems

Adaptive robots
USE: Intelligent robots

Adaptive systems (723.4) (731.1)
DT: January 1993
UF: Systems science and cybernetics—Adaptive systems*
BT: Cybernetics
NT: Adaptive control systems
RT: Adaptive filtering
Adaptive optics
Artificial intelligence
Automata theory
Cognitive systems
Control systems
Learning systems

Adcock direction finders 716.3
DT: January 1993
UF: Direction finding systems—Adcock*
BT: Radio direction finding systems

Adders (721.3) (713.5) (722.4)
DT: January 1993
UF: Binary adders
Carry circuits
Computers—Adders*
Half adders
BT: Digital circuits
RT: Carry logic
Counting circuits

* Former Ei Vocabulary term

Adders *(continued)*
Digital arithmetic
Logic circuits
Shift registers
Summing circuits

Addition reactions 802.2
DT: January 1993
BT: Chemical reactions

Additives 803 (804)
DT: January 1993
BT: Chemicals
NT: Cement additives
Concrete additives
Doping (additives)
Ferroalloys
Food additives
Fragrances
Fuel additives
Opacifiers
Petroleum additives
Rare earth additions
Rubber additions
RT: Antioxidants
Fillers
Flame retardants
Lubricants
Pigments
Plasticizers
Solvents

Adenosinetriphosphate 804.1 (461.2)
DT: January 1987
UF: ATP
BT: Coenzymes
Esters
RT: Muscle
Phosphates

Adhesion 801 (931.2)
DT: Predates 1975
UF: Concrete—Plaster adherence*
Tackiness
NT: Bond (masonry)
RT: Adhesive joints
Adhesives
Bond strength (materials)
Bonding
Composite materials
Curing
Laminating
Peeling
Sealing (closing)
Seizing
Stiction
Surface phenomena
Wetting

Adhesive joints (817.1) (818.5)
DT: January 1993
UF: Joints, Adhesive*
BT: Joints (structural components)
RT: Adhesion
Adhesives
Bonding
Gluing
Plastic adhesives
Rubber base adhesives
Wood products
Wooden construction

Adhesives (804.1) (817.1) (818.5) (804.2)
DT: Predates 1975
UF: Glues
Pastes (adhesives)
BT: Materials
NT: Bone cement
Hot melt adhesives
Plastic adhesives
Rubber base adhesives
RT: Adhesion
Adhesive joints
Binders
Bonding
Composite materials
Curing
Elastomers
Gluing
Laminating
Peeling
Plastics
Sealants
Seals

*Adhesives—Applicators**
USE: Applicators

*Adhesives—Bone cement**
USE: Bone cement

*Adhesives—Hot melt**
USE: Hot melt adhesives

*Adhesives—Peeling**
USE: Peeling

*Adhesives—Spreaders**
USE: Spreaders

Adiabatic engines 612.1
SN: Heat engines or thermodynamic systems which operate without gain or loss of heat
DT: January 1993
UF: Internal combustion engines—Adiabatic*
BT: Engines
RT: Internal combustion engines
Thermodynamics

Adjustable resistors
USE: Potentiometers (resistors)

ADM
USE: Delta modulation

Administration
USE: Management

Administrative data processing 723.2 (912.2)
DT: January 1993
UF: ADP
Business data processing
Data processing, Business*
BT: Data processing
NT: Financial data processing

Administrative data processing *(continued)*
Government data processing
Management information systems
RT: Electronic mail
Industrial management
Management
Management science
Office automation
Reservation systems
Spreadsheets
Teleconferencing
Word processing

Admittance (electric)
USE: Electric impedance

ADP
USE: Administrative data processing

Adsorbents 803 (804.1) (804.2)
DT: January 1987
BT: Materials
NT: Carbonaceous adsorbents
Fullers earth
Molecular sieves
RT: Activated alumina
Adsorption
Adsorption towers
Charcoal
Silica gel
Zeolites

*Adsorbents—Carbonaceous**
USE: Carbonaceous adsorbents

Adsorption 802.3
DT: Predates 1975
BT: Sorption
NT: Gas adsorption
RT: Absorption
Adsorbents
Adsorption isotherms
Adsorption towers
Chemisorption
Concentration (process)
Desorption
Wetting

Adsorption isotherms (641.1) (801.4) (931.2)
DT: January 1995
BT: Isotherms
RT: Adsorption

Adsorption towers 802.1
DT: January 1993
UF: Chemical equipment—Adsorption towers*
BT: Chemical equipment
RT: Adsorbents
Adsorption

Advertising
USE: Marketing

Aeration (sewage)
USE: Sewage aeration

Aeration (water)
USE: Water aeration

Aerial photographic surveys
USE: Photogrammetry

Aerial photography 742.1 (405.3)
DT: Predates 1975
BT: Photography
RT: Photogrammetry
Photomapping

Aerials
USE: Antennas

Aerodynamic flow
USE: Aerodynamics

Aerodynamic heating 651.1
DT: January 1993
UF: Aerodynamics—Aerodynamic heating*
BT: Heating
RT: Aerodynamics
Heat problems
Reentry

Aerodynamic loads 651.1
DT: January 1993
UF: Aerodynamics—Loads*
BT: Dynamic loads
RT: Aerodynamics

Aerodynamics 651.1
DT: Predates 1975
UF: Aerodynamic flow
Aeroelasticity
BT: Gas dynamics
NT: Flight dynamics
Flutter (aerodynamics)
Ground effect
Hypersonic aerodynamics
Lift
Subsonic aerodynamics
Supersonic aerodynamics
Transonic aerodynamics
RT: Aerodynamic heating
Aerodynamic loads
Aircraft
Airfoils
Atmospheric turbulence
Boundary layers
Buffeting
Compressible flow
Drag
Flow of fluids
Gases
Heat transfer
Hydrodynamics
Jets
Reentry
Shock tubes
Shock waves
Subsonic flow
Supersonic flow
Transonic flow
Turbulence
Turbulent flow
Unsteady flow

* Former Ei Vocabulary term

Aerodynamics *(continued)*
Viscous flow
Wind tunnels
Wings

*Aerodynamics—Aerodynamic heating**
USE: Aerodynamic heating

*Aerodynamics—Boundary layer**
USE: Boundary layers

*Aerodynamics—Buffeting**
USE: Buffeting

*Aerodynamics—Cascades**
USE: Cascades (fluid mechanics)

*Aerodynamics—Drag**
USE: Drag

*Aerodynamics—Flutter**
USE: Flutter (aerodynamics)

*Aerodynamics—Ground effect**
USE: Ground effect

*Aerodynamics—Hypersonic**
USE: Hypersonic aerodynamics

*Aerodynamics—Lift**
USE: Lift

*Aerodynamics—Loads**
USE: Aerodynamic loads

*Aerodynamics—Subsonic**
USE: Subsonic aerodynamics

*Aerodynamics—Supersonic**
USE: Supersonic aerodynamics

*Aerodynamics—Transonic**
USE: Transonic aerodynamics

*Aerodynamics—Wings and airfoils**
USE: Wings OR
Airfoils

Aeroelasticity
USE: Aerodynamics

Aeronautical engineering
USE: Aerospace engineering

Aeronautical flutter
USE: Flutter (aerodynamics)

Aeronautics
USE: Aviation

Aeroplanes
USE: Aircraft

Aerosols (804) (801.3)
DT: Predates 1975
BT: Colloids
NT: Atmospheric aerosols
RT: Air pollution
Atomization
Dust
Fallout
Flow visualization
Fly ash
Foams
Freons
Fume control
Particles (particulate matter)
Smoke
Sols
Turbidimeters

*Aerosols—Atmospheric**
USE: Atmospheric aerosols

Aerospace applications (652) (655)
DT: January 1993
UF: Television applications—Aerospace*
BT: Applications
RT: Aerospace engineering
Space applications

Aerospace engineering 658
DT: October 1975
UF: Aeronautical engineering
Aviation engineering
Space engineering
BT: Engineering
RT: Aerospace applications

Aerospace ground support 658
DT: Predates 1975
UF: Ground support (aerospace)
NT: Satellite ground stations
RT: Air traffic control
Aircraft communication
Aircraft landing systems
Missile silos
Rockets
Spacecraft
Telemetering
Telemetering equipment
Telemetering systems
Tracking (position)

*Aerospace vehicle tracking**
USE: Tracking (position)

Aerospace vehicles (652.1) (655.1)
SN: Vehicles capable of flight both within and outside the sensible atmosphere
DT: January 1986
BT: Aircraft
RT: Spacecraft
Tracking (position)

Affinity chromatography 801
DT: January 1993
UF: Chromatographic analysis—Affinity*
BT: Chromatography

Aflatoxins 461.6 (804.2)
DT: January 1993
UF: Carcinogens—Aflatoxins*
BT: Carcinogens
Hazardous materials
RT: Fungi

Afterburners (engine) 612.1.1
DT: January 1993
UF: Engine afterburners
Internal combustion engines—Afterburners*

* Former Ei Vocabulary term

Afterburners (engine)*(continued)*
BT: Internal combustion engines
RT: Exhaust systems (engine)

Afterburners (oven) 642.2
DT: January 1993
UF: Oven afterburners
Ovens, Industrial—Afterburners*
BT: Industrial ovens

Age hardening 531 (537)
DT: January 1993
UF: Metals and alloys—Age hardening*
Precipitation hardening
BT: Aging of materials
Hardening
Heat treatment
RT: Strain hardening

Agents 803 (804.1) (804.2)
DT: January 1993
BT: Chemicals
NT: Antioxidants
Antistatic agents
Blowing agents
Catalysts
Contrast media
Corrosion inhibitors
Deoxidants
Flame retardants
Flotation agents
Initiators (chemical)
Mold release agents
Mutagens
Plasticizers
Stabilizers (agents)
Surface active agents
Vulcanization agents
RT: Fluxes
Solvents

Agglomeration 802.3
DT: Predates 1975
UF: Aggregation
BT: Chemical operations
NT: Flocculation
RT: Coagulation
Concentration (process)
Densification
Granulation
Pelletizing
Precipitation (chemical)
Sedimentation

Aggregates 412.2, 406
DT: January 1993
UF: Roadbuilding materials—Aggregates*
BT: Materials
NT: Concrete aggregates
RT: Porous materials
Roadbuilding materials

Aggregation
USE: Agglomeration

Aging of materials (421)
DT: Predates 1975
UF: Aging*
NT: Age hardening
RT: Degradation
Heat treatment
Metallurgy
Phase transitions
Solid solutions
Weathering

*Aging**
USE: Aging of materials

Agreements
USE: Contracts

*Agricultural applications**
USE: Agriculture

Agricultural engineering 821
DT: Predates 1975
BT: Engineering
RT: Agriculture

Agricultural implements 821.1
DT: January 1993
UF: Agricultural machinery—Implements*
Tractors—Implements*
BT: Equipment
RT: Agricultural machinery
Tractors (agricultural)

Agricultural machinery 821.1
DT: Predates 1975
UF: Farm equipment
BT: Machinery
NT: Combines
Harvesters
Milking machines
Tractors (agricultural)
RT: Agricultural implements
Agriculture
Automatic guidance (agricultural machinery)
Farms
Lawn mowers

*Agricultural machinery—Automatic guidance**
USE: Automatic guidance (agricultural machinery)

*Agricultural machinery—Combines**
USE: Combines

*Agricultural machinery—Harvesters**
USE: Harvesters

*Agricultural machinery—Hydraulic equipment**
USE: Hydraulic equipment

*Agricultural machinery—Implements**
USE: Agricultural implements

*Agricultural machinery—Transmissions**
USE: Transmissions

Agricultural products 821.4
DT: January 1983
NT: Cotton
Dairy products
Fruits

* Former Ei Vocabulary term

Agricultural products *(continued)*
Grain (agricultural product)
Hemp
Seed
Sugar beets
Sugar cane
Tobacco
Wool
RT: Crops
Food products

*Agricultural products—Seed**
USE: Seed

Agricultural residues
USE: Agricultural wastes

Agricultural runoff 453.1, 821.5
DT: January 1993
UF: Water pollution—Agricultural runoffs*
BT: Runoff
RT: Agriculture
Soil pollution
Water pollution

Agricultural wastes 821.5
DT: January 1981
UF: Agricultural residues
Gas manufacture—Agricultural wastes*
BT: Wastes
NT: Hulls (seed coverings)
Manures
Straw
RT: Bagasse
Biogas

*Agricultural wastes—Hulls**
USE: Hulls (seed coverings)

*Agricultural wastes—Manures**
USE: Manures

*Agricultural wastes—Straw**
USE: Straw

Agriculture 821
DT: January 1983
UF: Agricultural applications*
Limestone—Agricultural applications*
BT: Industry
NT: Agronomy
Cultivation
Harvesting
Irrigation
RT: Agricultural engineering
Agricultural machinery
Agricultural runoff
Aquaculture
Crops
Farms
Fertilizers
Forestry
Irrigation canals
Nitrogen fertilizers
Orchards
Rural areas
Soil conservation

Agronomy 821.3
DT: January 1986
BT: Agriculture

AI
USE: Artificial intelligence

Aids for handicapped persons
USE: Human rehabilitation equipment

Air 804
SN: Use for the specific gaseous mixture. For the atmosphere, use EARTH ATMOSPHERE
DT: Predates 1975
BT: Fluids
NT: Compressed air
RT: Air purification
Air quality
Compressors
Earth atmosphere
Gases

Air bags (automobile)
USE: Automobile air bags

Air blast circuit breakers 704.2 (706.2)
DT: January 1993
UF: Electric circuit breakers, Air blast*
BT: Electric circuit breakers

Air brakes 602.1, 632.4
SN: Pneumatically actuated mechanical brakes
DT: January 1993
UF: Brakes, Air*
BT: Brakes
Pneumatic equipment
RT: Nonelectric final control devices

Air cleaners 451.2 (651.2)
DT: January 1993
UF: Air purifiers
Submarines—Air purifiers*
Wind tunnels—Air cleaners*
BT: Equipment
NT: Air filters
RT: Air pollution
Air purification
Air quality
Indoor air pollution
Purification
Submarines

Air compressors
USE: Compressors

Air conditioning 643.3
DT: Predates 1975
UF: Air conditioning—Hydronic systems*
Hydronic air conditioning systems
Underground air conditioning systems
BT: Climate control
NT: Gas fueled air conditioning
RT: Air conditioning ducts
Air curtains
Automobile cooling systems
Compressors
Cooling
Electric power utilization

* Former Ei Vocabulary term

Air conditioning *(continued)*
Environmental engineering
Evaporative cooling systems
Heating
Humidity control
Protective atmospheres
Refrigerants
Refrigerating machinery
Temperature
Thermal load
Thermoelectric equipment
Tropical buildings
Tropical engineering
Ventilation

Air conditioning ducts 643.4
DT: January 1993
UF: Air conditioning—Ducts*
BT: Ducts
RT: Air conditioning

*Air conditioning—Ducts**
USE: Air conditioning ducts

*Air conditioning—Evaporative cooling**
USE: Evaporative cooling systems

*Air conditioning—Gas fuel**
USE: Gas fueled air conditioning

*Air conditioning—Hydronic systems**
USE: Air conditioning

*Air conditioning—Noise**
USE: Acoustic noise

*Air conditioning—Solar energy systems**
USE: Solar equipment

*Air conditioning—Thermal load**
USE: Thermal load

*Air conditioning—Thermoelectric systems**
USE: Thermoelectric equipment

*Air conditioning—Underground**
USE: Underground equipment

Air curtains 643.5 (402)
DT: January 1993
UF: Doors—Air curtains*
RT: Air conditioning
Climate control
Doors

Air cushion vehicles (674.1)
DT: October 1975
UF: GEM (ground effect machines)
Ground effect vehicles
Hovercraft
Vehicles—Ground effect*
BT: Vehicles
RT: Boats
Ground effect
Hulls (ship)
Hydrofoil boats
Hydrofoils

Air cushioning 682.2
SN: Mechanical devices on railroad cars that use trapped air to permit smooth stops
DT: January 1993
UF: Cars—Air cushioning*
BT: Railroad car equipment

Air ejectors 618.3
SN: Devices that use fluid jets to remove air or other gases from an enclosed space
DT: Predates 1975
UF: Ejectors (air)
BT: Equipment
RT: Jets

Air engines 632.4, 641
DT: Predates 1975
BT: Gas engines
RT: Compressors
Stirling cycle

Air entrainment 802.3
DT: January 1993
UF: Entrainment (air)
BT: Processing

Air filters 451.2 (802.1)
DT: Predates 1975
BT: Air cleaners
Filters (for fluids)
RT: Filtration
Purification
Ventilation

Air lubrication 607.2
DT: January 1993
UF: Bearings—Air lubricated*
BT: Lubrication
RT: Bearings (machine parts)

Air navigation 431.5
DT: Predates 1975
BT: Aviation
Navigation
RT: Air traffic control
Aircraft
Aircraft communication
Inertial navigation systems

*Air navigation—Air traffic control**
USE: Air traffic control

*Air navigation—Inertial systems**
USE: Inertial navigation systems

Air permeability 931.2 (819.5)
DT: January 1993
UF: Textiles—Air permeability*
BT: Mechanical permeability

Air pollution 451
DT: Predates 1975
BT: Pollution
NT: Indoor air pollution
RT: Acid rain
Aerosols
Air cleaners
Air pollution control

Air pollution *(continued)*
Air purification
Air quality
Atmospheric aerosols
Atmospheric composition
Carbon dioxide
Carbon monoxide
Climate change
Coal dust
Earth atmosphere
Environmental engineering
Fallout
Fumigation
Gaseous effluents
Greenhouse effect
Industrial wastes
Mine dust
Mutagens
Nitrogen oxides
Ozone
Particulate emissions
Pollution induced corrosion
Smoke
Soot
Thermal plumes
Thermal pollution
Volatile organic compounds

Air pollution control 451.2
DT: January 1993
UF: Air pollution—Control*
BT: Pollution control
NT: Dust abatement
Fume control
Smoke abatement
RT: Air pollution
Air quality
Regional planning

Air pollution control equipment 451.2
DT: January 1993
UF: Air pollution—Control equipment*
BT: Pollution control equipment
NT: Dust collectors
RT: Catalytic converters
Scrubbers

*Air pollution—Acid rain**
USE: Acid rain

*Air pollution—Air quality**
USE: Air quality

*Air pollution—Control equipment**
USE: Air pollution control equipment

*Air pollution—Control**
USE: Air pollution control

*Air pollution—Fumigation**
USE: Fumigation

*Air pollution—Gaseous effluents**
USE: Gaseous effluents

*Air pollution—Indoor**
USE: Indoor air pollution

*Air pollution—Nitrogen oxides**
USE: Nitrogen oxides

*Air pollution—Particulate emissions**
USE: Particulate emissions

Air preheaters 616.1
DT: Predates 1975
BT: Heat exchangers
RT: Boilers

Air purification 451.2, 643 (672)
DT: January 1993
UF: Submarines—Air purification*
BT: Purification
RT: Air
Air cleaners
Air pollution
Submarines
Ventilation

Air purifiers
USE: Air cleaners

Air quality 443.1, 451.2
DT: January 1993
UF: Air pollution—Air quality*
RT: Air
Air cleaners
Air pollution
Air pollution control
Atmospheric composition
Earth atmosphere
Environmental engineering

Air supported structures
USE: Inflatable structures

Air terminals
USE: Airports

Air traffic control 431.5 (723.5)
DT: January 1993
UF: Air navigation—Air traffic control*
BT: Traffic control
RT: Aerospace ground support
Air navigation
Air transportation
Aircraft communication
Aircraft landing systems
Aircraft landing
Aviation
Collision avoidance
Control towers
Takeoff

Air transportation 431.1
DT: Predates 1975
UF: Airlines
Airways
BT: Transportation
NT: Helicopter services
VTOL/STOL services
RT: Air traffic control
Airports
Reservation systems
Transportation routes

* Former Ei Vocabulary term

Air turbines
USE: Compressed air motors

Airborne telescopes 741.3
DT: January 1993
UF: Telescopes—Airborne installations*
BT: Telescopes
RT: Astronomical satellites
Satellite observatories
Space optics

Airborne television transmitters 716.4
DT: January 1993
UF: Television transmitters—Airborne*
BT: Television transmitters
RT: Communication satellites

Aircraft 652.1
SN: Very general term; prefer specific type of aircraft
DT: Predates 1975
UF: Aeroplanes
Airplanes
NT: Aerospace vehicles
Aircraft parts and equipment
Airships
Amphibious aircraft
Balloons
Cargo aircraft
Deck landing aircraft
Delta wing aircraft
Freewing aircraft
Gliders
Helicopters
Jet aircraft
Man powered aircraft
Military aircraft
Personal aircraft
Propulsive wing aircraft
Reconnaissance aircraft
Research aircraft
Seaplanes
Supersonic aircraft
Training aircraft
Transport aircraft
VTOL/STOL aircraft
RT: Aerodynamics
Air navigation
Aircraft communication
Aircraft exhibitions
Aircraft fueling
Aircraft fuels
Aircraft landing
Aircraft landing systems
Aircraft manufacture
Aircraft models
Aircraft plants
Aircraft propulsion
Auxiliary power systems
Aviation
Flight dynamics
Flight simulators
Maneuverability
Meteorological problems
Navigation systems
Oxygen supply
Parachutes
Salvaging
Takeoff

Aircraft accidents 652.1, 914.1
DT: January 1995
UF: Aviation accidents
BT: Accidents
RT: Accident prevention
Aircraft emergency exits

Aircraft cabins
USE: Cabins (aircraft)

Aircraft carriers 672.1 (652.1.2)
DT: January 1993
BT: Warships
NT: Catapults (aircraft launchers)
RT: Arresting devices (aircraft)
Deck landing aircraft
Military aircraft

*Aircraft carriers—Catapults**
USE: Catapults (aircraft launchers)

*Aircraft carriers—Deck cooling**
USE: Cooling AND
Decks (ship)

Aircraft cockpits
USE: Cockpits (aircraft)

Aircraft communication 652.1, 716.3
DT: Predates 1975
BT: Telecommunication
RT: Aerospace ground support
Air navigation
Air traffic control
Aircraft
Aircraft radar equipment
Aviation
Helicopters
Mobile telecommunication systems

Aircraft emergency exits 652.1, 914.1
DT: January 1993
UF: Aircraft—Emergency exits*
Emergency exits (aircraft)
BT: Aircraft parts and equipment
RT: Aircraft accidents
Aircraft escape devices

Aircraft engine manufacture 653.1
DT: Predates 1975
BT: Aircraft manufacture
RT: Aircraft engines
Aircraft plants

Aircraft engines 653.1
DT: Predates 1975
BT: Aircraft parts and equipment
NT: Ramjet engines
RT: Aircraft engine manufacture
Compound engines
Ingestion (engines)
Internal combustion engines
Jet engines

* Former Ei Vocabulary term

*Aircraft engines, Jet and turbine**
USE: Jet engines OR
Turbojet engines

*Aircraft engines, Jet and turbine—Ducted fan**
USE: Ducted fan engines

*Aircraft engines, Jet and Turbine—Ingestion**
USE: Ingestion (engines)

*Aircraft engines, Jet and turbine—Nuclear propulsion**
USE: Nuclear propulsion

*Aircraft engines, Jet and turbine—Ramjet**
USE: Ramjet engines

*Aircraft engines, Jet and turbine—Regenerators**
USE: Regenerators

*Aircraft engines, Jet and turbine—Transmissions**
USE: Vehicle transmissions

*Aircraft engines, Jet and turbine—Turbine rings**
USE: Turbine rings

*Aircraft engines, Jet and turbine—Turbofan**
USE: Turbofan engines

*Aircraft engines, Jet and turbine—Turboprop**
USE: Turboprop engines

Aircraft escape devices 652.1, 914.1
DT: January 1993
UF: Aircraft—Escape devices*
Escape devices (aircraft)
BT: Aircraft parts and equipment
RT: Aircraft emergency exits

Aircraft exhibitions 652.1, 911.4
DT: Predates 1975
BT: Exhibitions
RT: Aircraft
Machinery exhibitions

Aircraft fabrics 652.2, 819.5
DT: January 1993
UF: Aircraft materials—Textiles*
BT: Fabrics
RT: Aircraft parts and equipment

Aircraft fuel tanks 652.3 (523)
DT: January 1993
UF: Aircraft—Fuel tanks*
BT: Fuel tanks

Aircraft fueling 523, 652.1
DT: January 1993
UF: Aircraft—Fueling*
BT: Fueling
RT: Aircraft
Aircraft fueling equipment
Aircraft fuels

Aircraft fueling equipment 523, 652.3
DT: January 1993
UF: Airport fueling equipment
Airports—Fueling equipment*
Fueling equipment (aircraft)
BT: Airport ground equipment
RT: Aircraft fueling
Aircraft fuels

Aircraft fuels 523, 652.1
DT: Predates 1975
BT: Fuels
RT: Aircraft
Aircraft fueling equipment
Aircraft fueling
Automotive fuels
Liquid fuels

Aircraft instruments 652.3
DT: Predates 1975
BT: Aircraft parts and equipment
Instruments
NT: Angle of attack indicators
Automatic pilots
Avionics
Bombsights
Proximity indicators
Stall indicators
Takeoff indicators
RT: Aircraft landing systems
Aircraft radar equipment
Aneroid altimeters
Fuel gages
Gyroscopes
Oxygen regulators
Radio altimeters
Radio direction finding systems
Radio telephone
Sextants
Tachometers

*Aircraft instruments—Angle of attack indicators**
USE: Angle of attack indicators

*Aircraft instruments—Automatic pilots**
USE: Automatic pilots

*Aircraft instruments—Fuel gages**
USE: Fuel gages

*Aircraft instruments—Oxygen regulators**
USE: Oxygen regulators

*Aircraft instruments—Proximity indicators**
USE: Proximity indicators

*Aircraft instruments—Sextants**
USE: Sextants

*Aircraft instruments—Stall indicators**
USE: Stall indicators

*Aircraft instruments—Takeoff indicators**
USE: Takeoff indicators

Aircraft landing 652.1 (731.2)
DT: January 1993
UF: Aircraft—Landing*
BT: Landing
RT: Air traffic control
Aircraft
Airports
Landing mats

* Former Ei Vocabulary term

*Aircraft landing gear**
USE: Landing gear (aircraft)

*Aircraft landing gear—Arresting devices**
USE: Arresting devices (aircraft)

*Aircraft landing gear—Retracting devices**
USE: Landing gear retracting devices

Aircraft landing systems 652.3
DT: January 1981
UF: Landing systems (aircraft)
RT: Aerospace ground support
Air traffic control
Aircraft
Aircraft instruments
Landing

Aircraft launchers
USE: Catapults (aircraft launchers)

Aircraft manufacture 652.1
DT: Predates 1975
BT: Manufacture
NT: Aircraft engine manufacture
RT: Aircraft
Aircraft plants

*Aircraft manufacture—Drafting practice**
USE: Drafting practice

*Aircraft manufacture—Finishing**
USE: Finishing

*Aircraft manufacture—Sandwich construction**
USE: Sandwich structures

Aircraft materials 652.2
SN: Also use the material if appropriate, e.g., COMPOSITE MATERIALS or MARAGING STEEL
DT: Predates 1975
BT: Materials
RT: Aircraft parts and equipment
Light metals

*Aircraft materials—Composite materials**
USE: Composite materials

*Aircraft materials—Graphite**
USE: Graphite

*Aircraft materials—Light metals**
USE: Light metals

*Aircraft materials—Maraging steel**
USE: Maraging steel

*Aircraft materials—Plastics**
USE: Plastics applications

*Aircraft materials—Powder metals**
USE: Powder metals

*Aircraft materials—Rubber**
USE: Rubber applications

*Aircraft materials—Sealants**
USE: Sealants

*Aircraft materials—Steel**
USE: Steel

*Aircraft materials—Textiles**
USE: Aircraft fabrics

*Aircraft materials—Tubing**
USE: Tubing

Aircraft mockups 652.1
DT: January 1993
UF: Aircraft—Models*
BT: Mockups
RT: Aircraft models

Aircraft models 652.1
SN: Reduced-scale models; for full-size models, use AIRCRAFT MOCKUPS
DT: Predates 1975
UF: Model aircraft
BT: Models
RT: Aircraft
Aircraft mockups

Aircraft parts and equipment 652.3
DT: January 1993
UF: Aircraft—Equipment*
Public address systems—Airborne*
BT: Aircraft
Vehicle parts and equipment
NT: Aircraft emergency exits
Aircraft engines
Aircraft escape devices
Aircraft instruments
Aircraft propellers
Aircraft radar equipment
Aircraft seats
Aircraft signal lights
Aircraft windshields
Airframes
Cabins (aircraft)
Cockpits (aircraft)
Empennages
Fuselages
Landing gear (aircraft)
Wings
RT: Aircraft fabrics
Aircraft materials
Airfoils
Navigation systems
Parachutes
Windshield wipers

Aircraft plants 652.1 (402.1)
DT: Predates 1975
BT: Industrial plants
RT: Aircraft
Aircraft engine manufacture
Aircraft manufacture

Aircraft propellers 652.3
DT: January 1993
UF: Aircraft—Jettisoned stores*
Aircraft—Propellers*
BT: Aircraft parts and equipment
Propellers
RT: Aircraft propulsion

Aircraft propulsion 653.1
DT: January 1993
UF: Propulsion—Aerospace applications*
BT: Propulsion
RT: Aircraft
Aircraft propellers
Propulsive wing aircraft

Aircraft radar equipment 716.2, 652.3
DT: January 1993
UF: Aircraft—Radar equipment*
BT: Aircraft parts and equipment
Radar equipment
RT: Aircraft communication
Aircraft instruments

Aircraft seats 652.3
DT: January 1993
UF: Aircraft—Seats*
BT: Aircraft parts and equipment
Seats

Aircraft signal lights 652.3 (707.2)
DT: January 1993
UF: Aircraft—Signal lights*
Signal lights (Aircraft)
BT: Aircraft parts and equipment
Electric lamps

Aircraft simulators
USE: Flight simulators

Aircraft windshields 652.3
DT: January 1993
UF: Aircraft—Windshields*
BT: Aircraft parts and equipment
Windshields
RT: Windshield wipers

Aircraft wings
USE: Wings

*Aircraft, Amphibian**
USE: Amphibious aircraft

*Aircraft, Military**
USE: Military aircraft

*Aircraft, Personal**
USE: Personal aircraft

*Aircraft, Research**
USE: Research aircraft

*Aircraft, Training**
USE: Training aircraft

*Aircraft, Transport**
USE: Transport aircraft

*Aircraft, VTOL/STOL**
USE: VTOL/STOL aircraft

*Aircraft—Airframes**
USE: Airframes

*Aircraft—Antisubmarine**
USE: Antisubmarine aircraft

*Aircraft—Arresting gear**
USE: Arresting devices (aircraft)

*Aircraft—Auxiliary equipment**
USE: Auxiliary equipment

*Aircraft—Auxiliary power**
USE: Auxiliary power systems

*Aircraft—Bomber**
USE: Bombers

*Aircraft—Cabins**
USE: Cabins (aircraft)

*Aircraft—Cockpits**
USE: Cockpits (aircraft)

*Aircraft—Crashworthiness**
USE: Crashworthiness

*Aircraft—Deck landing**
USE: Deck landing aircraft

*Aircraft—Delta wing**
USE: Delta wing aircraft

*Aircraft—Emergency exits**
USE: Aircraft emergency exits

*Aircraft—Empennage**
USE: Empennages

*Aircraft—Engine mounting**
USE: Engine mountings

*Aircraft—Equipment**
USE: Aircraft parts and equipment

*Aircraft—Escape devices**
USE: Aircraft escape devices

*Aircraft—Fighter**
USE: Fighter aircraft

*Aircraft—Flight dynamics**
USE: Flight dynamics

*Aircraft—Flight simulators**
USE: Flight simulators

*Aircraft—Forestry applications**
USE: Forestry

*Aircraft—Freewing**
USE: Freewing aircraft

*Aircraft—Freight**
USE: Cargo aircraft

*Aircraft—Fuel tanks**
USE: Aircraft fuel tanks

*Aircraft—Fueling**
USE: Aircraft fueling

*Aircraft—Fuselage**
USE: Fuselages

*Aircraft—Hydraulic equipment**
USE: Hydraulic equipment

* Former Ei Vocabulary term

*Aircraft—Jet flap**
USE: Jet flaps

*Aircraft—Jet propelled**
USE: Jet aircraft

*Aircraft—Jettisoned stores**
USE: Aircraft propellers

*Aircraft—Landing**
USE: Aircraft landing

*Aircraft—Man powered**
USE: Man powered aircraft

*Aircraft—Meteorological problems**
USE: Meteorological problems

*Aircraft—Models**
USE: Aircraft mockups

*Aircraft—Navigation systems**
USE: Navigation systems

*Aircraft—Oxygen supply**
USE: Oxygen supply

*Aircraft—Pneumatic equipment**
USE: Pneumatic equipment

*Aircraft—Propellers**
USE: Aircraft propellers

*Aircraft—Propulsive wing**
USE: Propulsive wing aircraft

*Aircraft—Radar equipment**
USE: Aircraft radar equipment

*Aircraft—Radiation hazards**
USE: Radiation hazards

*Aircraft—Radomes**
USE: Radomes

*Aircraft—Rain erosion**
USE: Erosion

*Aircraft—Reconnaissance**
USE: Reconnaissance aircraft

*Aircraft—Safety factor**
USE: Safety factor

*Aircraft—Salvaging**
USE: Salvaging

*Aircraft—Seats**
USE: Aircraft seats

*Aircraft—Shelters**
USE: Hangars

*Aircraft—Signal lights**
USE: Aircraft signal lights

*Aircraft—Supersonic speeds**
USE: Supersonic aircraft

*Aircraft—Takeoff**
USE: Takeoff

*Aircraft—Target drones**
USE: Target drones

*Aircraft—Windshields**
USE: Aircraft windshields

*Aircraft—Wings**
USE: Wings

Airfoils 652.1
DT: January 1993
UF: Aerodynamics—Wings and airfoils*
BT: Components
NT: Jet flaps
Wings
RT: Aerodynamics
Aircraft parts and equipment
Cascades (fluid mechanics)
Hydrofoil boats
Hydrofoils
Lift

Airframes 652.3
DT: January 1993
UF: Aircraft—Airframes*
BT: Aircraft parts and equipment
RT: Wings

Airlines
USE: Air transportation

Airplanes
USE: Aircraft

Airport buildings 402.2, 431.4
DT: January 1993
UF: Airports—Buildings*
BT: Airports
Buildings
NT: Control towers
Hangars

Airport fueling equipment
USE: Aircraft fueling equipment

Airport ground equipment 431.4
DT: January 1993
UF: Airports—Ground equipment*
BT: Airports
Equipment
NT: Aircraft fueling equipment
RT: Airport passenger transportation

Airport passenger transportation 431.2
DT: January 1993
UF: Airports—Passenger transportation*
BT: Transportation
RT: Airport ground equipment
Airport vehicular traffic

Airport runways 431.4
DT: Predates 1975
UF: Bituminous runways
Concrete runways
Runways (airport)
BT: Airports
NT: Emergency runways
Temporary runways
RT: Ice control
Landing
Landing mats

* Former Ei Vocabulary term

Airport runways *(continued)*
Runway foaming
Runway markings
Snow and ice removal

*Airport runways—Arresting devices**
USE: Arresting devices (aircraft)

*Airport runways—Bituminous**
USE: Bituminous paving materials

*Airport runways—Concrete**
USE: Concrete pavements

*Airport runways—Emergency**
USE: Emergency runways

*Airport runways—Foaming**
USE: Runway foaming

*Airport runways—Ice control**
USE: Ice control

*Airport runways—Landing mats**
USE: Landing mats

*Airport runways—Markings**
USE: Runway markings

*Airport runways—Snow and ice removal**
USE: Snow and ice removal

*Airport runways—Stabilization**
USE: Stabilization

*Airport runways—Temporary**
USE: Temporary runways

Airport vehicular traffic 431.4, 432
DT: January 1993
UF: Airports—Vehicular traffic*
Ground traffic (airports)
Vehicular traffic (airports)
BT: Transportation
RT: Airport passenger transportation
Airports
Highway systems

Airports 431.4
DT: Predates 1975
UF: Air terminals
BT: Facilities
NT: Airport buildings
Airport ground equipment
Airport runways
Heliports
Military airports
RT: Air transportation
Aircraft landing
Airport vehicular traffic
Baggage handling
Fog dispersal
Takeoff

*Airports—Baggage handling**
USE: Baggage handling

*Airports—Buildings**
USE: Airport buildings

*Airports—Control towers**
USE: Control towers

*Airports—Fog dispersal**
USE: Fog dispersal

*Airports—Fueling equipment**
USE: Aircraft fueling equipment

*Airports—Ground equipment**
USE: Airport ground equipment

*Airports—Helicopter services**
USE: Helicopter services

*Airports—Military**
USE: Military airports

*Airports—Passenger transportation**
USE: Airport passenger transportation

*Airports—Power supply**
USE: Electric power utilization

*Airports—Snow and ice removal**
USE: Snow and ice removal

*Airports—Vehicular traffic**
USE: Airport vehicular traffic

*Airports—VTOL/STOL services**
USE: VTOL/STOL services

Airships 652.5
SN: Lighter-than-air craft
DT: Predates 1975
UF: Blimps
Dirigibles
BT: Aircraft

Airways
USE: Air transportation

Al
USE: Aluminum

Alarm systems (914.1) (914.2)
DT: Predates 1975
UF: Warning systems
BT: Security systems
Signal systems
NT: Fire alarm systems
RT: Detectors
Electric equipment protection
Indicators (instruments)
Monitoring
Protection
Radar warning systems
Remote control
Safety devices
Signaling
Sirens
Smoke detectors

Alcohol fuels 523, 804.1
DT: January 1993
BT: Synthetic fuels
NT: Ethanol fuels
Methanol fuels
RT: Alcohols
Automotive fuels

Alcohol fuels *(continued)*
Diesel fuels
Gasohol

Alcohols 804.1
DT: Predates 1975
BT: Organic compounds
NT: Ethanol
Glycerol
Glycols
Methanol
RT: Alcohol fuels
Gasohol
Phenols

Aldehydes 804.1
DT: January 1986
BT: Organic compounds
NT: Formaldehyde
Xylose

Algae (453.1) (461.2) (471.5)
DT: January 1987
UF: Food products—Algae*
Reservoirs—Algae*
BT: Plants (botany)
NT: Seaweed
RT: Algae control
Marine biology
Microorganisms
Reservoirs (water)

Algae control 453.2
DT: January 1993
UF: Water treatment—Algae control*
BT: Pest control
RT: Algae
Herbicides
Water treatment

Algebra 921.1
DT: January 1993
UF: Mathematical techniques—Algebra*
BT: Mathematical techniques
NT: Boolean algebra
Linear algebra
Polynomials

ALGOL (programming language) 723.1.1
DT: January 1993
UF: Computer programming languages—ALGOL*
BT: Algorithmic languages

Algorithmic languages 723.1.1
DT: January 1993
UF: Computer metatheory—Algorithmic languages*
BT: Formal languages
High level languages
NT: ALGOL (programming language)
RT: Algorithms
Computation theory

Algorithms (723) (921)
DT: January 1993
UF: Computer algorithms
Computer programming—Algorithms*
Control algorithms
Mathematical techniques—Algorithms*
BT: Mathematical techniques
NT: Adaptive algorithms
Genetic algorithms
Learning algorithms
Parallel algorithms
RT: Algorithmic languages
Computer programming languages
Subroutines

Alignment 601.1 (603.1)
DT: January 1993
UF: Machine tools—Alignment*
Machinery—Alignment*
Shafts and shafting—Alignment*
RT: Leveling (machinery)
Machine tools
Machinery
Maintenance
Optical instruments
Shaft displacement
Straightening

Alkali metal alloys 549.1
DT: January 1993
BT: Alloys
NT: Cesium alloys
Lithium alloys
Potassium alloys
Rubidium alloys
Sodium alloys
RT: Alkali metal compounds
Alkali metals

Alkali metal compounds (804.1) (804.2)
DT: Predates 1975
BT: Metallic compounds
NT: Cesium compounds
Lithium compounds
Potassium compounds
Rubidium compounds
Sodium compounds
RT: Alkali metal alloys
Alkali metals

Alkali metals 549.1
DT: Predates 1975
BT: Nonferrous metals
NT: Cesium
Francium
Lithium
Potassium
Rubidium
Sodium
RT: Alkali metal alloys
Alkali metal compounds

Alkaline earth metal alloys 549.2
DT: January 1993
BT: Alloys
NT: Barium alloys
Beryllium alloys
Calcium alloys
Magnesium alloys
Strontium alloys

* Former Ei Vocabulary term

Alkaline earth metal alloys *(continued)*
RT: Alkaline earth metal compounds
Alkaline earth metals

Alkaline earth metal compounds (804.1) (804.2)
DT: June 1990
BT: Metallic compounds
NT: Barium compounds
Beryllium compounds
Calcium compounds
Magnesium compounds
Strontium compounds
RT: Alkaline earth metal alloys
Alkaline earth metals

Alkaline earth metals 549.2
DT: Predates 1975
BT: Nonferrous metals
NT: Barium
Beryllium
Calcium
Magnesium
Radium
Strontium
RT: Alkaline earth metal alloys
Alkaline earth metal compounds

Alkalinity
USE: pH

Alkanes
USE: Paraffins

Alkanoic acids
USE: Carboxylic acids

Alkenes
USE: Olefins

Alkenoic acids
USE: Carboxylic acids

Alkyd resins 815.1.1
DT: January 1987
BT: Organic compounds
Thermosets

Alkylation 802.2
DT: January 1993
BT: Chemical reactions
RT: Petroleum refining

All wheel drive vehicles (662) (663)
DT: January 1993
UF: Vehicles—All-wheel drive*
BT: Ground vehicles

Allergies 461.6, 461.9.1
DT: January 1993
UF: Biomedical engineering—Allergies*
RT: Antigen-antibody reactions
Biomedical engineering
Diseases
Immunology
Medicine

Alloying 531.1
DT: January 1993
NT: Gas alloying
Mechanical alloying
RT: Alloying elements
Alloys
Ferroalloys
Metallurgy
Rare earth additions
Solid solutions

Alloying elements 531.1
SN: Also use the element, if appropriate
DT: January 1993
BT: Chemical elements
RT: Alloying
Alloys
Ferroalloys
Rare earth additions

Alloys 531.1
DT: January 1993
UF: Metals and alloys*
BT: Materials
NT: Alkali metal alloys
Alkaline earth metal alloys
Aluminum alloys
Amorphous alloys
Binary alloys
Corrosion resistant alloys
Dental alloys
Ferroalloys
Gallium alloys
Germanium alloys
Heavy metal alloys
Indium alloys
Intermetallics
Plutonium alloys
Polonium alloys
Rare earth alloys
Silicon alloys
Superalloys
Thorium alloys
Transition metal alloys
Uranium alloys
RT: Alloying
Alloying elements
Composite materials
Corrosion
Machinability
Metal analysis
Metal cladding
Metal detectors
Metal extrusion
Metal finishing
Metal forming
Metal refining
Metallic compounds
Metallic glass
Metallic matrix composites
Metallography
Metalloids
Metals
Percolation (solid state)
Plating
Solid solutions
Welding

* Former Ei Vocabulary term

Allyl resins 815.1.1
DT: January 1987
BT: Thermosets

Alpha particle spectrometers (932.1) (941.3)
DT: January 1993
UF: Spectrometers, Alpha particle*
BT: Particle spectrometers
RT: Alpha particles

Alpha particles 931.3 (932.2)
DT: June 1990
BT: Charged particles
RT: Alpha particle spectrometers
Ions

Alpha rhythms
USE: Bioelectric potentials

Alternate fuels
USE: Synthetic fuels

Alternating current machinery
USE: AC machinery

Alternative fuels
USE: Synthetic fuels

Alternators (generators)
USE: Synchronous generators

Altimeters (aneroid)
USE: Aneroid altimeters

Altimeters (radio)
USE: Radio altimeters

*Altimeters**
USE: Aneroid altimeters

Alumina 804.2 (812.1)
DT: Predates 1975
UF: Cement—Alumina*
Powder metal products—Alumina*
BT: Aluminum compounds
Oxides
NT: Activated alumina
Corundum
Sintered alumina
RT: Aluminous refractories
Aluminum
Corundum deposits
Hydrated alumina
Minerals
Powder metal products
Powder metals

Alumina trihydrate
USE: Hydrated alumina

*Alumina—Activated**
USE: Activated alumina

*Alumina—Hydrated**
USE: Hydrated alumina

*Alumina—Sintered**
USE: Sintered alumina

Aluminothermic welding
USE: Thermit welding

Aluminous refractories 812.1
DT: January 1993
UF: Refractory materials—Alumina*
BT: Refractory materials
RT: Alumina
Aluminum compounds

Aluminum 541.1
DT: January 1993
UF: Al
Aluminum and alloys*
Aluminum recovery*
Boats—Aluminum applications*
Building materials—Aluminum*
Buildings—Aluminum applications*
Cast iron—Aluminum content*
Footbridges—Aluminum*
Packaging materials—Aluminum*
Slags—Aluminum recovery*
Water tanks and towers—Aluminum*
Windows—Aluminum*
BT: Light metals
RT: Alumina
Aluminum alloys
Aluminum beams and girders
Aluminum bridges
Aluminum castings
Aluminum cladding
Aluminum coated steel
Aluminum compounds
Aluminum containers
Aluminum corrosion
Aluminum foil
Aluminum foundry practice
Aluminum metallurgy
Aluminum metallography
Aluminum pipe
Aluminum plating
Aluminum sheet
Bauxite deposits
Bauxite mines
Bauxite ore treatment

Aluminum alloys 541.2
DT: January 1993
UF: Aluminum and alloys*
BT: Alloys
NT: Aluminum copper alloys
RT: Aluminum
Aluminum castings
Aluminum compounds

*Aluminum and alloys**
USE: Aluminum OR
Aluminum alloys

*Aluminum and alloys—Corrosion**
USE: Aluminum corrosion

Aluminum beams and girders 408.2, 541.1 (931.1)
DT: January 1993
UF: Beams and girders—Aluminum*
BT: Beams and girders
RT: Aluminum

 * Former Ei Vocabulary term

Aluminum bridges 401.1, 541.1
DT: January 1993
UF: Bridges, Aluminum*
BT: Bridges
RT: Aluminum

Aluminum castings 534, 541.1
DT: Predates 1975
BT: Metal castings
RT: Aluminum
Aluminum alloys
Metal casting

*Aluminum castings—Finishing**
USE: Metal finishing

Aluminum cladding 541.1 (535) (538)
DT: January 1993
UF: Oil tanks—Aluminum cladding*
BT: Metal cladding
RT: Aluminum
Oil tanks

Aluminum coated steel 545.3, 541.1
DT: January 1993
UF: Steel—Aluminum coating*
BT: Coated materials
Steel
RT: Aluminum

Aluminum compounds (804.1) (804.2)
DT: Predates 1975
BT: Metallic compounds
NT: Alumina
Chlorite minerals
Feldspar
Hydrated alumina
Semiconducting aluminum compounds
Zeolites
RT: Aluminous refractories
Aluminum
Aluminum alloys

Aluminum containers 541.1 (691.1) (694.1)
DT: January 1993
UF: Containers—Aluminum*
Tanks—Aluminum*
BT: Containers
RT: Aluminum

Aluminum copper alloys 541.2, 544.2
DT: January 1993
UF: Duralumin
BT: Aluminum alloys
Copper alloys

Aluminum corrosion 539.1, 541.1
DT: January 1993
UF: Aluminum and alloys—Corrosion*
BT: Corrosion
RT: Aluminum

*Aluminum deposits**
USE: Bauxite deposits

Aluminum foil 535.1, 541.1
DT: Predates 1975
BT: Aluminum sheet
Metal foil
RT: Aluminum

Aluminum foundry practice 534.2, 541.1
DT: Predates 1975
BT: Aluminum metallurgy
Foundry practice
RT: Aluminum
Aluminum plants

Aluminum hydrate
USE: Hydrated alumina

Aluminum hydroxide
USE: Hydrated alumina

Aluminum metallography 531.2, 541.1
DT: Predates 1975
BT: Metallography
RT: Aluminum
Aluminum metallurgy

Aluminum metallurgy 531.1, 541.1
DT: Predates 1975
BT: Metallurgy
NT: Aluminum foundry practice
Aluminum powder metallurgy
Aluminum refining
RT: Aluminum
Aluminum metallography
Aluminum plants
Bauxite ore treatment

Aluminum pipe 619.1, 541.1
DT: January 1993
UF: Aluminum pipelines
Pipe, Aluminum*
Pipelines, Aluminum*
BT: Pipe
RT: Aluminum

Aluminum pipelines
USE: Aluminum pipe AND
Pipelines

Aluminum plants 533, 541.1 (402.1)
DT: Predates 1975
BT: Industrial plants
NT: Aluminum rolling mills
RT: Aluminum foundry practice
Aluminum metallurgy

Aluminum plating 539.3, 541.1
DT: Predates 1975
BT: Plating
RT: Aluminum

Aluminum powder metallurgy 536, 541.1
DT: January 1993
UF: Powder metallurgy—Aluminum copper*
Powder metallurgy—Aluminum*
BT: Aluminum metallurgy
Powder metallurgy

*Aluminum recovery**
USE: Aluminum

Aluminum refining 533.2, 541.1
DT: Predates 1975
BT: Aluminum metallurgy
Metal refining

* Former Ei Vocabulary term

Aluminum rolling mills 535.1.1, 541.1
DT: Predates 1975
BT: Aluminum plants
Rolling mills

Aluminum sheet 535.1, 541.1
DT: Predates 1975
BT: Sheet metal
NT: Aluminum foil
RT: Aluminum

Am
USE: Americium

Amber 482.2.1, 815.1.1
DT: Predates 1975
BT: Natural polymers
RT: Gems

Ambient temperature
USE: Temperature

Ambulance cars 462.1, 682.1.1
DT: January 1993
UF: Cars—Ambulance*
BT: Railroad cars
RT: Ambulances
Health care

Ambulances 462.1, 662.1
DT: Predates 1975
BT: Emergency vehicles
RT: Ambulance cars
Health care

Americium 622.1
DT: Predates 1975
UF: Am
BT: Transuranium elements

Amination 802.2
DT: January 1993
BT: Chemical reactions
RT: Amines

Amines 804.1
DT: January 1986
BT: Nitrogen compounds
Organic compounds
NT: Chitin
Nitrosamines
RT: Amination

Amino acids 804.1 (461)
DT: January 1977
BT: Biological materials
Organic acids
RT: Polypeptides
Proteins

Amino resins (melamine)
USE: Melamine formaldehyde resins

Amino resins (urea)
USE: Urea formaldehyde resins

Ammeters 942.1
DT: Predates 1975
BT: Electric measuring instruments
RT: Electric current measurement
Ohmmeters
Voltmeters

Ammonia 804.2
DT: Predates 1975
UF: Refrigerants—Ammonia*
BT: Nitrogen compounds
RT: Ammonium compounds
Refrigerants

*Ammonia—Corrosive properties**
USE: Corrosive effects

Ammonium compounds (804.1) (804.2)
DT: Predates 1975
BT: Nitrogen compounds
RT: Ammonia
Salts

Ammunition 404.1
DT: January 1993
UF: Warships—Ammunition handling*
BT: Ordnance
RT: Projectiles

Amorphisation
USE: Amorphization

Amorphization 802.3 (933.2)
DT: January 1995
UF: Amorphisation
Crystalline-amorphous transformations
RT: Amorphous materials
Vitrification

Amorphous alloys 531 (933.2)
DT: January 1995
BT: Alloys
Amorphous materials

Amorphous films 933.2
DT: January 1993
UF: Films—Amorphous*
BT: Amorphous materials
Films

Amorphous materials 933.2
DT: January 1993
UF: Amorphous*
Materials—Amorphous*
BT: Materials
NT: Amorphous alloys
Amorphous films
Amorphous silicon
Glass
Metallic glass
RT: Amorphization

Amorphous silicon 549.3 (933.2)
DT: January 1995
BT: Amorphous materials
Silicon

*Amorphous**
USE: Amorphous materials

Amperometric sensors 942.1 (732.2) (801)
DT: January 1993
UF: Sensors—Amperometric measurements*

* Former Ei Vocabulary term

Amperometric sensors *(continued)*
BT: Sensors
RT: Electric current measurement
Glucose sensors

Amphibious aircraft 652.1, 674
DT: January 1993
UF: Aircraft, Amphibian*
BT: Aircraft
Amphibious vehicles

Amphibious automobiles 662.1, 674
DT: January 1993
UF: Automobiles—Amphibious*
BT: Amphibious vehicles
Automobiles

Amphibious vehicles 674.1
DT: January 1993
UF: Vehicles—Amphibious*
BT: Vehicles
NT: Amphibious aircraft
Amphibious automobiles
RT: Semisubmersibles
Submarines
Submersibles

Amplidynes
USE: Rotating magnetic amplifiers

Amplification (713.1) (741.1)
DT: January 1993
RT: Amplifiers (electronic)
Bode diagrams
Gain control
Gain measurement
Light amplifiers

Amplification measurement
USE: Gain measurement

Amplifiers (electronic) 713.1
DT: January 1993
UF: Amplifiers*
Television amplifiers
BT: Electronic equipment
NT: Audio frequency amplifiers
Bandpass amplifiers
Broadband amplifiers
Buffer amplifiers
Cathode followers
Chopper amplifiers
Dielectric amplifiers
Differential amplifiers
Direct coupled amplifiers
Feedback amplifiers
High frequency amplifiers
Intermediate frequency amplifiers
Logarithmic amplifiers
Microwave amplifiers
Operational amplifiers
Parametric amplifiers
Power amplifiers
Pulse amplifiers
Radio frequency amplifiers
Servo amplifiers
Tunnel diode amplifiers
Ultrahigh frequency amplifiers
Video amplifiers
RT: Active networks
Amplification
Circuit theory
Electric transformers
Gain control
Gain measurement
Networks (circuits)
Radio equipment
Radio receivers
Signal distortion
Signal receivers
Telecommunication repeaters
Telephone equipment
Transponders

Amplifiers (fluidic)
USE: Fluidic amplifiers

Amplifiers (light)
USE: Light amplifiers

*Amplifiers**
USE: Amplifiers (electronic)

*Amplifiers, Audio frequency**
USE: Audio frequency amplifiers

*Amplifiers, Bandpass**
USE: Bandpass amplifiers

*Amplifiers, Broadband**
USE: Broadband amplifiers

*Amplifiers, Buffer**
USE: Buffer amplifiers

*Amplifiers, Cathode follower**
USE: Cathode followers

*Amplifiers, Chopper**
USE: Chopper amplifiers

*Amplifiers, Dielectric**
USE: Dielectric amplifiers

*Amplifiers, Differential**
USE: Differential amplifiers

*Amplifiers, Diode**
USE: Diode amplifiers

*Amplifiers, Direct coupled**
USE: Direct coupled amplifiers

*Amplifiers, Feedback**
USE: Feedback amplifiers

*Amplifiers, High frequency**
USE: High frequency amplifiers

*Amplifiers, Intermediate frequency**
USE: Intermediate frequency amplifiers

*Amplifiers, Logarithmic**
USE: Logarithmic amplifiers

*Amplifiers, Magnetic**
USE: Magnetic amplifiers

*Amplifiers, Microwave**
USE: Microwave amplifiers

*Amplifiers, Operational**
USE: Operational amplifiers

*Amplifiers, Parametric**
USE: Parametric amplifiers

*Amplifiers, Power type**
USE: Power amplifiers

*Amplifiers, Pulse signal**
USE: Pulse amplifiers

*Amplifiers, Radio frequency**
USE: Radio frequency amplifiers

*Amplifiers, Rotary magnetic**
USE: Rotating magnetic amplifiers

*Amplifiers, Servo**
USE: Servo amplifiers

*Amplifiers, Tunnel diode**
USE: Tunnel diode amplifiers

*Amplifiers, Ultrahigh frequency**
USE: Ultrahigh frequency amplifiers

*Amplifiers, Video**
USE: Video amplifiers

*Amplifiers—Cascade connection**
USE: Cascade connections

Amplitude limiting circuits
USE: Limiters

Amplitude modulation (716) (717.1)
DT: Predates 1975
BT: Modulation
RT: Demodulation
Demodulators
Modulators

Amplitude selectors
USE: Limiters

Anaerobic digestion (452.4) (461.9) (802.2)
DT: January 1995
BT: Waste treatment
RT: Biochemical oxygen demand
Biological sewage treatment
Odor control

Analog computers 722.5
DT: January 1993
UF: Analogue computers
Computers, Analog*
BT: Computers
NT: Analog differential analyzers
Electric network analyzers
RT: Analog storage
Analog to digital conversion
Direct analogs
Error compensation
Function generators
Hybrid computers
Multiplying circuits

Analog data storage
USE: Analog storage

Analog differential analyzers 722.5
DT: January 1993
UF: Analogue differential analysers
Analyzers (differential)
Computers—Differential analyzers*
Differential analyzers (analog)
BT: Analog computers

Analog digital computers
USE: Hybrid computers

Analog digital conversion
USE: Analog to digital conversion

Analog storage 722.1 (722.5)
DT: January 1993
UF: Analog data storage
Analogue storage
Data storage, Analog*
BT: Data storage equipment
RT: Analog computers

Analog to digital conversion (722.4) (722.5)
DT: January 1993
UF: Analog digital conversion
Analogue digital conversion
Analogue to digital conversion
Converters (analog to digital)
Data conversion, Analog to digital*
Digitization
BT: Data processing
RT: Analog computers
Comparator circuits
Counting circuits
Hybrid computers
Image processing

Analogies
USE: Simulation

Analogue computers
USE: Analog computers

Analogue differential analysers
USE: Analog differential analyzers

Analogue digital conversion
USE: Analog to digital conversion

Analogue storage
USE: Analog storage

Analogue to digital conversion
USE: Analog to digital conversion

Analysis
SN: Very broad term; prefer specific type if available; otherwise also use the thing or process being analyzed
DT: January 1993
UF: Electric power systems—Load flow analysis*
Mineralogy—Analytical methods*
Transportation—Route analysis*
NT: Chemical analysis
Computer aided analysis
Failure analysis

* Former Ei Vocabulary term

Analysis *(continued)*
Image analysis
Lunar surface analysis
Particle size analysis
Stress analysis
Telephone traffic analysis
Waveform analysis
RT: Inspection
Sampling
Synthesis (chemical)
Testing

Analytic equipment 801, 802.1
SN: For chemical analysis
DT: January 1993
UF: Analytic instruments
Chemical analysis—Apparatus*
BT: Chemical equipment
RT: Chemical analysis
Colorimeters
Fluorometers

Analytic instruments
USE: Analytic equipment

Analytical chemistry
USE: Chemical analysis

Analytical geochemistry 481.2, 801
DT: January 1993
UF: Geochemistry—Analytical*
BT: Chemical analysis
Geochemistry

Analyzers (differential)
USE: Analog differential analyzers

Anchor cables 671.2
DT: January 1993
UF: Ship equipment—Anchor cables*
BT: Cables
Ship equipment
RT: Anchors

Anchorages (concrete construction) 408.2, 412
DT: January 1993
UF: Concrete construction—Anchorages*
BT: Construction equipment
RT: Concrete construction

Anchorages (foundations) 408.2, 483.2
DT: January 1993
UF: Foundations—Anchorages*
Structural design—Anchorages*
BT: Foundations
RT: Structural design

Anchors 671.2
DT: January 1993
UF: Ship equipment—Anchors*
BT: Ship equipment
RT: Anchor cables

Ancillary signal generators (715) (716)
SN: Instruments which generate signals for use in associated equipment
DT: January 1993
UF: Signal generators, Ancillary*
BT: Signal generators
RT: Electric clocks

Anechoic chambers 752.1
DT: June 1990
BT: Acoustics laboratories
RT: Acoustic variables measurement
Architectural acoustics
Reverberation
Test facilities

Anelastic relaxation (421) (931.2)
DT: January 1993
UF: Mechanical relaxation
BT: Relaxation processes
RT: Creep
Elasticity
Internal friction
Stress relaxation

Anemometers 944.3 (652.3)
DT: Predates 1975
BT: Flow measuring instruments
Meteorological instruments
Speed indicators
RT: Flow measurement
Laser Doppler velocimeters
Velocimeters
Wind

Aneroid altimeters 944.3 (652.3)
DT: January 1993
UF: Altimeters (aneroid)
Altimeters*
Barometric altimeters
BT: Range finders
RT: Aircraft instruments
Surveying instruments

Anesthesiology 461.6
DT: January 1993
UF: Biomedical engineering—Anesthesia*
BT: Medicine
RT: Anesthetics
Biomedical engineering

Anesthetics (461.6) (804.1) (804.2)
DT: January 1993
UF: Drug products—Anesthetics*
BT: Drug products
RT: Anesthesiology

Angiocardiography 461.6
DT: January 1993
UF: Biomedical engineering—Angiocardiography*
BT: Angiography
RT: Biomedical engineering
Cardiology
Contrast media

Angiography 461.6
DT: January 1993
UF: Biomedical engineering—Angiography*
BT: Diagnostic radiography
NT: Angiocardiography
RT: Biomedical engineering
Contrast media

* Former Ei Vocabulary term

Angle measurement 943.2
DT: January 1993
UF: Angular measurement
Mechanical variables measurement—Angles*
BT: Spatial variables measurement
RT: Micrometers

Angle of attack indicators 652.3
DT: January 1993
UF: Aircraft instruments—Angle of attack indicators*
BT: Aircraft instruments

Angle of contact
USE: Contact angle

Angular measurement
USE: Angle measurement

Animal cell culture (461.2) (801.2)
DT: January 1993
UF: Cell culture—Animal*
BT: Cell culture

Animal communication
USE: Biocommunications

Animation (723.5) (742.1)
DT: January 1993
UF: Computer graphics—Animation*
Motion pictures—Animation*
RT: Computer graphics
Motion pictures

Anionic polymerization 815.2
DT: January 1993
UF: Polymerization—Anionic polymerization*
BT: Polymerization
RT: Polycondensation
Ring opening polymerization

Anisotropy 931.2
DT: January 1993
BT: Physical properties
NT: Magnetic anisotropy
RT: Formability
Metallography
Optical Kerr effect
Textures

Annealing 537.1
SN: Heat treatment; for the algorithmic procedure use SIMULATED ANNEALING
DT: January 1993
UF: Glass manufacture—Annealing*
Heat treatment—Annealing*
Rapid thermal annealing
BT: Heat treatment
RT: Glass manufacture
Stress relaxation
Stress relief

Annealing (simulated)
USE: Simulated annealing

Anodes 714.1 (802.1) (704.1)
DT: January 1981
BT: Electrodes
RT: Anodic oxidation
Anodic polarization
Anodic protection
Electron tube components

Anodic oxidation 539.2.1 (802.2)
DT: January 1993
UF: Anodizing
Metals and alloys—Anodic oxidation*
BT: Oxidation
RT: Anodes
Coating techniques

Anodic polarization 801.4.1
DT: January 1995
BT: Polarization
RT: Anodes
Electrolysis

Anodic protection 539.2.1
DT: January 1993
UF: Corrosion protection, Anodic*
BT: Corrosion protection
RT: Anodes

Anodizing
USE: Anodic oxidation

Antarctic vehicles
USE: Arctic vehicles

Antenna accessories (716)
DT: January 1993
UF: Antennas—Accessories*
BT: Accessories
NT: Antenna feeders
Antenna grounds
Antenna reflectors
Radomes
RT: Antennas

Antenna arrays (716)
DT: January 1993
UF: Antennas—Arrays*
BT: Antennas
Arrays
NT: Antenna phased arrays
RT: Radio telescopes

Antenna feed systems
USE: Antenna feeders

Antenna feeders (716)
DT: January 1993
UF: Antenna feed systems
Antennas—Feed systems*
Feeds (antenna)
BT: Antenna accessories
RT: Waveguides

Antenna grounds (716)
DT: January 1993
UF: Antennas—Grounds*
Grounds (antenna)
BT: Antenna accessories
RT: Electric grounding

* Former Ei Vocabulary term

Antenna lobe patterns
USE: Directional patterns (antenna)

Antenna patterns
USE: Directional patterns (antenna)

Antenna phased arrays (716)
DT: January 1993
UF: Antennas—Phased arrays*
Phased arrays (antenna)
BT: Antenna arrays

Antenna radiation (711) (716)
DT: January 1993
UF: Antennas—Radiation*
BT: Electromagnetic waves
Radiation
NT: Directional patterns (antenna)
RT: Antennas

Antenna radiation patterns
USE: Directional patterns (antenna)

Antenna reflectors (716)
DT: January 1993
UF: Antennas—Reflectors*
BT: Antenna accessories
RT: Directional patterns (antenna)

Antennas (716)
DT: Predates 1975
UF: Aerials
BT: Telecommunication equipment
NT: Antenna arrays
Broadcasting antennas
Cylindrical antennas
Dipole antennas
Directive antennas
Helical antennas
Horn antennas
Microstrip antennas
Microwave antennas
Mobile antennas
Parabolic antennas
Radar antennas
Receiving antennas
Scanning antennas
Slot antennas
Television antennas
Traveling wave antennas
RT: Antenna accessories
Antenna radiation
Directional patterns (antenna)
Electric exciters
Radio equipment
Radio telescopes
Radio transmitters
Signal receivers
Telescopes
Television equipment
Transmitters
Waveguides

*Antennas, Cylindrical**
USE: Cylindrical antennas

*Antennas, Dipole**
USE: Dipole antennas

*Antennas, Directive**
USE: Directive antennas

*Antennas, Helical**
USE: Helical antennas

*Antennas, Horn**
USE: Horn antennas

*Antennas, Loop**
USE: Loop antennas

*Antennas, Microstrip**
USE: Microstrip antennas

*Antennas, Mobile**
USE: Mobile antennas

*Antennas, Parabolic**
USE: Parabolic antennas

*Antennas, Scanning**
USE: Scanning antennas

*Antennas, Slot**
USE: Slot antennas

*Antennas—Accessories**
USE: Antenna accessories

*Antennas—Arrays**
USE: Antenna arrays

*Antennas—Broadcasting**
USE: Broadcasting antennas

*Antennas—Directional patterns**
USE: Directional patterns (antenna)

*Antennas—Feed systems**
USE: Antenna feeders

*Antennas—Grounds**
USE: Antenna grounds

*Antennas—Microwave**
USE: Microwave antennas

*Antennas—Phased arrays**
USE: Antenna phased arrays

*Antennas—Radar**
USE: Radar antennas

*Antennas—Radiation**
USE: Antenna radiation

*Antennas—Receiving**
USE: Receiving antennas

*Antennas—Reflectors**
USE: Antenna reflectors

*Antennas—Television**
USE: Television antennas

*Antennas—Traveling wave**
USE: Traveling wave antennas

Anthracite 524
DT: January 1993
UF: Coal—Anthracite*

* Former Ei Vocabulary term

Anthracite *(continued)*
Hard coal
BT: Coal

Anthropometry 461.3
DT: January 1993
UF: Biomechanics—Anthropometry*
BT: Measurements
RT: Biomechanics

Anthropomorphic robots 731.5 (461.3)
DT: January 1993
UF: Robots, Industrial—Anthropomorphic*
BT: Robots
RT: Industrial robots

Anti-aliasing (723.5)
SN: Smoothing technique used in computer graphics
DT: January 1993
UF: Computer graphics—Anti-aliasing*
BT: Image processing
RT: Computer graphics

Antibiotics 461.6 (804.1)
DT: January 1993
UF: Drug products—Antibiotics*
BT: Drug products
RT: Fermentation
Microorganisms
Mutagens

Antibodies 461.9.1
DT: January 1987
BT: Proteins
NT: Monoclonal antibodies
RT: Antigen-antibody reactions
Antigens
Biocompatibility
Immunization
Immunology

Antiferromagnetic materials 708.4
DT: January 1993
UF: Magnetic materials—Antiferromagnetism*
BT: Magnetic materials
RT: Antiferromagnetism
Magnetic semiconductors

Antiferromagnetism 701.2 (708.4)
DT: January 1993
UF: Magnetism—Antiferromagnetism*
BT: Magnetism
RT: Antiferromagnetic materials
Metamagnetism

Antifouling paint 813.2
DT: January 1993
UF: Paint—Antifouling*
BT: Paint
Protective coatings
RT: Marine borers

Antifreeze solutions 803
DT: January 1987
BT: Solutions
RT: Chemicals

*Antifreeze solutions—Corrosive properties**
USE: Corrosive effects

Antifriction bearings 601.2 (931.1)
DT: January 1993
UF: Bearings—Antifriction*
BT: Bearings (machine parts)
NT: Ball bearings
Gas lubricated bearings
Roller bearings
Self lubricating bearings
RT: Journal bearings
Thrust bearings

Antigen-antibody reactions 461.9.1
DT: January 1993
UF: Immune reactions
Immunology—Antigen-antibody reactions*
RT: Allergies
Antibodies
Antigens
Biocompatibility
Graft vs. host reactions
Immunology

Antigens 461.9.1
DT: January 1987
BT: Biological materials
RT: Antibodies
Antigen-antibody reactions
Biocompatibility
Immunology
Proteins

Antiknock compounds 803, 523 (804.1) (804.2)
DT: Predates 1975
BT: Fuel additives
RT: Antiknock rating
Combustion knock

Antiknock rating 523
DT: January 1993
UF: Cetane number
Fuels—Antiknock rating*
Octane number
BT: Rating
RT: Antiknock compounds
Combustion knock
Fuels
Gasoline

Antimony 546.4
DT: January 1993
UF: Antimony and alloys*
Sb
BT: Heavy metals
NT: Semiconducting antimony
RT: Antimony alloys
Antimony compounds
Antimony deposits
Antimony metallurgy
Antimony mines
Antimony ore treatment

Antimony alloys 546.4
DT: January 1993
UF: Antimony and alloys*

* Former Ei Vocabulary term

Antimony alloys *(continued)*
BT: Heavy metal alloys
RT: Antimony
Antimony compounds
Pewter

*Antimony and alloys**
USE: Antimony OR
Antimony alloys

Antimony compounds (804.1) (804.2)
DT: Predates 1975
BT: Heavy metal compounds
NT: Semiconducting antimony compounds
RT: Antimony
Antimony alloys

Antimony deposits 504.3, 546.4
DT: Predates 1975
BT: Ore deposits
RT: Antimony
Antimony mines

Antimony metallurgy 531.1, 546.4
DT: Predates 1975
BT: Metallurgy
RT: Antimony
Antimony ore treatment

Antimony mines 504.3, 546.4
DT: January 1993
UF: Antimony mines and mining*
BT: Mines
RT: Antimony
Antimony deposits

*Antimony mines and mining**
USE: Antimony mines

Antimony ore treatment 533.1, 546.4
DT: Predates 1975
BT: Ore treatment
RT: Antimony
Antimony metallurgy

Antioxidants 803 (804.1) (804.2)
DT: January 1981
BT: Agents
RT: Additives
Oxidation
Stabilizers (agents)

Antireflection coatings 813.2 (741.3)
DT: January 1993
UF: Coatings—Antireflection coatings*
BT: Optical coatings
RT: Optical films

Antistatic agents
803 (811.1) (816.1) (819.5) (804.1) (804.2)
DT: January 1993
UF: Paper—Antistatic agents*
Plastics—Antistatic agents*
Textiles—Antistatic agents*
BT: Agents
RT: Paper
Plastics
Textiles

Antisubmarine aircraft 652.1.2
DT: January 1993
UF: Aircraft—Antisubmarine*
BT: Military aircraft

Antivibration mountings (601.2) (603.2) (612.1.1)
DT: January 1993
UF: Internal combustion engines—Antivibration mountings*
Machine tools—Antivibration mountings*
Machinery—Antivibration mountings*
BT: Mountings
RT: Engine mountings
Internal combustion engines
Machine tools
Machine vibrations
Machinery
Shock absorbers

Apartment houses 402.3
DT: Predates 1975
BT: Housing
RT: Houses
Intelligent buildings

APL (programming language) 723.1.1
DT: January 1993
UF: Computer programming languages—APL*
BT: High level languages

Appliances (domestic)
USE: Domestic appliances

Appliances (electric)
USE: Electric appliances

Application specific integrated circuits 714.2
DT: January 1995
UF: ASIC
Custom integrated circuits
Semicustom integrated circuits
BT: Monolithic integrated circuits
RT: Digital integrated circuits
Logic design
VLSI circuits

Applications
SN: Very general term; prefer specific type of application; also use Treatment Code A
DT: January 1993
NT: Aerospace applications
Computer applications
Ferrite applications
High temperature applications
Industrial applications
Laser applications
Marine applications
Medical applications
Military applications
Plasma applications
Plastics applications
Quartz applications
Robot applications
Rubber applications
Space applications
Television applications
Ultrasonic applications
Vacuum applications

* Former Ei Vocabulary term

Applicators 601
DT: January 1993
UF: Adhesives—Applicators*
Fungicides—Applicators*
Herbicides—Applicators*
Insecticides—Applicators*
Pesticides—Applicators*
BT: Equipment
NT: Spray guns
Spreaders

Applied mechanics
USE: Mechanics

Apprentices 912.4
DT: Predates 1975
BT: Personnel
RT: Personnel training

Approximation theory 921.6
DT: January 1993
UF: Mathematical techniques—Approximation theory*
BT: Mathematical techniques
Theory
NT: Chebyshev approximation
Least squares approximations
Linearization
RT: Boundary element method
Finite element method
Interpolation

Aquaculture 821.3 (471.5)
DT: January 1986
BT: Industry
RT: Agriculture
Biological filter beds
Fish ponds
Fisheries

*Aquaculture—Biological filter beds**
USE: Biological filter beds

*Aquaculture—Fish ponds**
USE: Fish ponds

Aqueous solutions
USE: Solutions

Aquifers 444.2
DT: January 1977
RT: Geology
Groundwater
Groundwater resources
Landforms
Recharging (underground waters)
Rocks
Water supply

*Aquifers—Recharging**
USE: Recharging (underground waters)

Ar
USE: Argon

Arc lamps 707.2
DT: January 1993
UF: Electric lamps, Arc*
BT: Discharge lamps
RT: Electric arcs

Arc plasma welding
USE: Plasma welding

Arc torches (plasma)
USE: Plasma torches

Arc welding
USE: Electric arc welding

Arch bridges 401.1
DT: January 1993
UF: Bridges, Arch*
BT: Bridges
RT: Arches

Arch dams 441.1
DT: January 1993
UF: Dams, Arch*
BT: Dams

Arches 408.2 (401.1)
DT: Predates 1975
BT: Structural members
RT: Arch bridges

Architectural acoustics 751.3 (402)
DT: Predates 1975
BT: Acoustics
Architecture
RT: Acoustic noise
Acoustic wave reflection
Anechoic chambers
Audio acoustics
Buildings
Reverberation

Architectural design 402, 408.1
DT: January 1987
BT: Architecture
Design
RT: Structural design

*Architectural design—Facades**
USE: Facades

*Architectural design—Interiors**
USE: Interiors (building)

Architecture 402
DT: Predates 1975
NT: Architectural acoustics
Architectural design
RT: Buildings
Construction
Model buildings
Structures (built objects)

Architecture (of computers)
USE: Computer architecture

Arctic buildings 402, 443.1
DT: January 1993
UF: Buildings—Arctic*
Cold region buildings
BT: Buildings
RT: Arctic engineering
Climate control
Cold effects
Undersnow structures

Arctic engineering (405.2) (409) (443) (454)
DT: June 1990
UF: Cold region engineering
BT: Engineering
RT: Arctic buildings
Arctic vehicles
Glaciers
Low temperature engineering
Permafrost
Undersnow structures

Arctic vehicles (432.1) (443.1)
DT: January 1993
UF: Antarctic vehicles
Cold region vehicles
Tractors—Antarctic expedition*
Vehicles—Antarctic expedition*
BT: Vehicles
RT: Arctic engineering
Undersnow structures

Area measuring instruments
USE: Planimeters

Argon 804
DT: Predates 1975
UF: Ar
BT: Inert gases

Arid regions 443, 444
DT: January 1993
UF: Water resources—Arid regions*
BT: Geographical regions
RT: Water resources
Water resources exploration

Arithmetic (digital)
USE: Digital arithmetic

Armor 404.1
DT: January 1993
UF: Military equipment—Armor*
BT: Ordnance
RT: Plate metal
Steel

Aromatic acids
USE: Carboxylic acids

Aromatic compounds 804.1
DT: January 1986
BT: Organic compounds
NT: Aromatic hydrocarbons
Aromatic polymers
Phenols
Styrene

Aromatic hydrocarbons 804.1
DT: Predates 1975
UF: Hydrocarbons—Aromatic*
BT: Aromatic compounds
Hydrocarbons
NT: Benzene
Naphthalene
Toluene
Xylene
RT: Aromatization

Aromatic polymers 815.1.1
DT: January 1977
UF: Polyacenes*
Polyacenic materials
BT: Aromatic compounds
Organic polymers
NT: ABS resins
Polyamideimides
Polyethylene terephthalates
Polystyrenes

Aromatization 802.2
DT: January 1993
BT: Chemical reactions
RT: Aromatic hydrocarbons
Petroleum refining

Arousal 461.4
DT: January 1993
UF: Human engineering—Arousal*
RT: Human engineering
Sensory perception

Array processors
USE: Parallel processing systems

Arrays (702.3) (716) (752.1)
DT: January 1993
NT: Acoustic arrays
Antenna arrays
Solar cell arrays
RT: Sensors

Arresting devices (aircraft) 652.3
DT: January 1993
UF: Aircraft landing gear—Arresting devices*
Aircraft—Arresting gear*
Airport runways—Arresting devices*
BT: Landing gear (aircraft)
RT: Aircraft carriers

Arsenals 404.1
DT: January 1993
UF: Military equipment—Arsenals*
BT: Military equipment

Arsenate minerals 482.2
DT: January 1993
UF: Mineralogy—Arsenates*
BT: Arsenic compounds
Minerals
RT: Arsenic

Arsenic 804
DT: Predates 1975
UF: As
BT: Metalloids
RT: Arsenate minerals
Arsenic compounds

Arsenic compounds (804.1) (804.2)
DT: Predates 1975
BT: Chemical compounds
NT: Arsenate minerals
RT: Arsenic

Articulation (speech)
USE: Speech

* Former Ei Vocabulary term

Artificial blood
USE: Blood substitutes

Artificial blood vessels
USE: Blood vessel prostheses

Artificial foods
USE: Synthetic foods

Artificial graphite 804.2
DT: January 1993
UF: Graphite—Artificial*
Synthetic graphite
BT: Graphite

Artificial hearts
USE: Artificial organs

Artificial hip joints
USE: Hip prostheses

Artificial intelligence 723.4
DT: January 1986
UF: AI
BT: Cybernetics
NT: Knowledge engineering
RT: Adaptive systems
Backpropagation
Brain models
Cognitive systems
Expert systems
Formal logic
Heuristic programming
Learning systems
Logic programming
Motion planning
Neural networks
Problem solving
Robotics
Robots
System theory
Theorem proving

Artificial limbs 462.4
DT: Predates 1975
UF: Prosthetics—Artificial limbs*
BT: Prosthetics
RT: Artificial organs
Human rehabilitation engineering
Human rehabilitation equipment
Patient rehabilitation

Artificial neural networks
USE: Neural networks

Artificial organs 462.4
DT: January 1993
UF: Artificial hearts
Prosthetics—Artificial organs*
BT: Prosthetics
RT: Artificial limbs

Artificial rain
USE: Cloud seeding

Artificial reality
USE: Virtual reality

Artificial satellites
USE: Satellites

Artificial vision
USE: Computer vision

As
USE: Arsenic

Asbestos 413.2, 482.2 (819.1)
DT: Predates 1975
UF: Heat insulating materials—Asbestos*
BT: Materials
RT: Asbestos cement
Asbestos deposits
Asbestos mines
Asbestos ore treatment
Hazardous materials
Textile fibers
Thermal insulating materials

Asbestos cement 412.1
DT: Predates 1975
UF: Cement asbestos
Water pipelines—Cement asbestos*
BT: Cements
RT: Asbestos
Asbestos cement pipe

Asbestos cement pipe 619.1, 412.1
DT: January 1993
UF: Pipe, Asbestos cement*
BT: Pipe
RT: Asbestos cement

Asbestos deposits 505.1
DT: Predates 1975
BT: Mineral resources
RT: Asbestos
Asbestos mines

Asbestos mines 505.1
DT: January 1993
UF: Asbestos mines and mining*
BT: Mines
RT: Asbestos
Asbestos deposits

*Asbestos mines and mining**
USE: Asbestos mines

Asbestos ore treatment 533.1
DT: Predates 1975
BT: Ore treatment
RT: Asbestos

Asdic
USE: Sonar

Ash (coal)
USE: Coal ash

Ash handling 691.1 (452.4)
DT: Predates 1975
BT: Materials handling
RT: Wastes

ASIC
USE: Application specific integrated circuits

* Former Ei Vocabulary term

Aspect ratio (943)
DT: January 1995
RT: Spatial variables measurement

Asphalt 411.1
DT: Predates 1975
BT: Bituminous materials
RT: Asphalt plants
Bituminous paving materials
Pavements
Tar

Asphalt pavements 406.2, 411.1
DT: January 1993
UF: Pavements—Asphalt*
BT: Pavements

Asphalt plants 411.1 (402.1) (513.2)
DT: Predates 1975
BT: Industrial plants
RT: Asphalt

Aspherics 741.1 (741.3)
DT: January 1993
BT: Optics
RT: Lenses

Assays 801
DT: January 1993
UF: Chemical analysis—Assays*
BT: Chemical analysis
NT: Bioassay
RT: Sampling

Assemblers (program)
USE: Program assemblers

Assembly (913.1)
DT: January 1993
BT: Production
NT: Robotic assembly
RT: Assembly machines
Clean rooms
Electronic equipment manufacture
Fabrication
Installation
Materials handling
Printed circuit manufacture
Surface mount technology

Assembly machines 601.1
DT: Predates 1975
UF: Industrial plants—Assembly machines*
BT: Machinery
RT: Assembly
Industrial plants
Robotic assembly

Assembly programs
USE: Program assemblers

Association reactions 802.2
DT: January 1993
UF: Association*
BT: Chemical reactions

*Association**
USE: Association reactions

Associative processing 723.2
DT: January 1993
BT: Data processing
RT: Associative storage

Associative storage 722.1
DT: January 1993
UF: Content addressable memories
Content addressable storage
Data storage, Digital—Associative*
BT: Digital storage
RT: Associative processing
Self organizing storage

Astatine 804 (622.1) (801.4.2)
DT: January 1981
UF: At
BT: Halogen elements
Radioactive elements

Astrodynamics
USE: Space flight

Astronautics
USE: Space flight

Astronomical satellites 655.2
DT: January 1993
UF: Satellites—Astronomical*
BT: Satellite observatories
RT: Airborne telescopes
Astronomy

Astronomy 657.2
DT: Predates 1975
BT: Natural sciences
NT: Astrophysics
Radar astronomy
Radio astronomy
RT: Astronomical satellites
Galaxies
Moon
Observatories
Planetariums
Relativity
Solar system
Space research
Telescopes

Astrophysics 657.2 (931.1)
DT: Predates 1975
BT: Astronomy
Physics
RT: Galaxies
Orbits
Relativity
Space research
Trajectories

Asymptotic stability (921.6)
DT: January 1995
BT: Stability
RT: Differential equations
Perturbation techniques

Asynchronous generators 705.2.1
DT: January 1993
UF: Electric generators, Asynchronous*

* Former Ei Vocabulary term

Asynchronous generators *(continued)*
Induction generators
BT: AC generators
Asynchronous machinery
RT: Electric windings
Machine windings

Asynchronous machinery 705.1 (705.2.1) (705.3.1)
DT: January 1993
UF: Electric machinery, Asynchronous*
Induction machinery
BT: AC machinery
NT: Asynchronous generators
Induction motors
RT: Machine windings

Asynchronous motors
USE: Induction motors

Asynchronous sequential logic 721.1 (721.3)
DT: January 1993
UF: Switching theory—Asynchronous sequential logic*
BT: Sequential circuits
RT: Sequential switching
Switching
Switching theory

Asynchronous transfer mode (716) (717) (718)
DT: January 1995
UF: ATM networks
BT: Time division multiplexing
RT: Broadband networks
Packet switching
Voice/data communication systems

At
USE: Astatine

Athletic facilities
USE: Gymnasiums

ATM
USE: Automatic teller machines

ATM networks
USE: Asynchronous transfer mode

Atmosphere (earth)
USE: Earth atmosphere

Atmospheres (extraterrestrial)
USE: Extraterrestrial atmospheres

*Atmospheres**
USE: Protective atmospheres

Atmospheric acoustics 443.1, 751 (481)
SN: Use for acoustics of terrestrial or extraterrestrial atmospheres
DT: Predates 1975
BT: Acoustics
RT: Acoustic wave propagation
Earth atmosphere
Extraterrestrial atmospheres

Atmospheric aerosols 443.1 (451.1) (801.3) (804.1) (804.2)
DT: January 1993
UF: Aerosols—Atmospheric*
BT: Aerosols
RT: Air pollution
Earth atmosphere

Atmospheric chemistry 443.1, 801.1
SN: Use for chemistry of terrestrial or extraterrestrial atmospheres
DT: January 1995
BT: Chemistry
RT: Atmospheric composition
Earth atmosphere
Extraterrestrial atmospheres
Meteorology
Photochemical reactions

Atmospheric composition 443.1, 801
SN: Use for composition of terrestrial or extraterrestrial atmospheres
DT: Predates 1975
RT: Air pollution
Air quality
Atmospheric chemistry
Atmospheric structure
Carbon dioxide
Climate change
Earth atmosphere
Extraterrestrial atmospheres
Meteorology

Atmospheric corrosion 539.1, 443.1 (802.2)
DT: January 1993
UF: Corrosion—Atmospheric*
BT: Corrosion
RT: Meteorology

Atmospheric density 443.1 (931.2)
SN: Use for density of terrestrial or extraterrestrial atmospheres
DT: January 1993
UF: Atmospheric pressure and density*
BT: Density of gases
RT: Atmospheric movements
Atmospheric pressure
Earth atmosphere
Extraterrestrial atmospheres
Meteorology

Atmospheric electricity 443.1, 701.1
SN: Use for electrical phenomena in terrestrial or extraterrestrial atmospheres
DT: Predates 1975
BT: Electricity
NT: Atmospherics
Lightning
RT: Atmospheric ionization
Electrostatics
Extraterrestrial atmospheres
Ionosphere
Meteorology
Thunderstorms

Atmospheric humidity 443.1
SN: Use for humidity of terrestrial or extraterrestrial atmospheres
DT: Predates 1975

* Former Ei Vocabulary term

Atmospheric humidity *(continued)*
UF: Humidity
Relative humidity
RT: Clouds
Earth atmosphere
Extraterrestrial atmospheres
Humidity control
Meteorology
Moisture
Psychrometers

Atmospheric interference
USE: Atmospherics

Atmospheric ionization 443.1, 802.2
SN: Use for ionization of terrestrial or extraterrestrial atmospheres
DT: Predates 1975
BT: Ionization of gases
RT: Atmospheric electricity
Earth atmosphere
Ionosphere
Meteorology
Photoionization

Atmospheric movements 443.1
SN: Use for movements in terrestrial or extraterrestrial atmospheres
DT: Predates 1975
NT: Atmospheric turbulence
Wind
RT: Atmospheric density
Atmospheric pressure
Atmospheric thermodynamics
Earth atmosphere
Extraterrestrial atmospheres
Hydrodynamics
Meteorology

Atmospheric optics 443.1, 741.1
SN: Use for optical properties and effects in terrestrial or extraterrestrial atmospheres
DT: Predates 1975
BT: Optics
RT: Atmospheric spectra
Earth atmosphere
Extraterrestrial atmospheres
Light absorption
Light reflection
Light scattering
Meteorology
Optical properties
Visibility

Atmospheric pressure 443.1
SN: Use for pressure of terrestrial or extraterrestrial atmospheres
DT: January 1993
UF: Atmospheric pressure and density*
BT: Pressure
RT: Atmospheric density
Atmospheric movements
Earth atmosphere
Extraterrestrial atmospheres
Meteorology

*Atmospheric pressure and density**
USE: Atmospheric density OR
Atmospheric pressure

Atmospheric radiation 443.1
SN: Use for thermal radiation from terrestrial or extraterrestrial atmospheres
DT: Predates 1975
BT: Heat radiation
RT: Atmospheric thermodynamics
Earth atmosphere
Extraterrestrial atmospheres
Infrared radiation
Meteorology

Atmospheric radioactivity 443.1, 622.1
SN: Use for radioactivity of terrestrial or extraterrestrial atmospheres
DT: Predates 1975
BT: Radioactivity
RT: Earth atmosphere
Extraterrestrial atmospheres

Atmospheric spectra 443.1, 741.1
SN: Use for spectra of terrestrial or extraterrestrial atmospheres
DT: Predates 1975
UF: Spectra (atmospheric)
RT: Atmospheric optics
Earth atmosphere
Extraterrestrial atmospheres

Atmospheric structure 443.1
SN: Use for structure of terrestrial or extraterrestrial atmospheres
DT: Predates 1975
RT: Atmospheric composition
Earth atmosphere
Extraterrestrial atmospheres
Meteorology

Atmospheric temperature 443.1
SN: Use for temperature of terrestrial or extraterrestrial atmospheres
DT: Predates 1975
BT: Temperature
RT: Atmospheric thermodynamics
Climate change
Earth atmosphere
Extraterrestrial atmospheres
Greenhouse effect
Meteorology
Thermal stratification

Atmospheric thermodynamics 443.1, 641.1
SN: Use for thermodynamics of terrestrial or extraterrestrial atmospheres
DT: Predates 1975
BT: Thermodynamics
RT: Atmospheric movements
Atmospheric radiation
Atmospheric temperature
Earth atmosphere
Extraterrestrial atmospheres
Meteorology

* Former Ei Vocabulary term

Atmospheric turbulence 443.1, 631.1 (931.1)
SN: Use for turbulence in terrestrial or extraterrestrial atmospheres
DT: January 1977
BT: Atmospheric movements
Turbulence
RT: Aerodynamics
Buffeting
Meteorology

Atmospherics 443.1, 701.1
DT: Predates 1975
UF: Atmospheric interference
Sferics
Static (atmospherics)
Strays (atmospheric)
BT: Atmospheric electricity
Electric discharges
RT: Lightning
Radio interference
Radio transmission
Signal interference

Atomic beams 931.3 (932.1)
DT: January 1986
BT: Particle beams
RT: Particle optics

Atomic clocks 943.3 (931.3)
DT: January 1993
UF: Clocks, Atomic*
BT: Clocks
RT: Electronic equipment
Frequency meters
Masers
Time measurement

Atomic energy
USE: Nuclear energy

Atomic force microscopy 741.3 (931.3)
DT: January 1995
BT: Microscopic examination
RT: Crystal atomic structure
Electron microscopy
Scanning electron microscopy
Scanning tunneling microscopy
Surface structure
Transmission electron microscopy

Atomic physics 931.3
DT: January 1993
UF: Physics—Atomic*
BT: Physics
RT: Atoms
Elementary particle sources
Elementary particles
Molecular physics
Nuclear physics
Research reactors

Atomic scattering factors
USE: Lattice vibrations

Atomic spectroscopy 931.3 (801)
DT: January 1993
BT: Spectroscopy
RT: Atoms
Electromagnetic waves

Atomic structure (crystals)
USE: Crystal atomic structure

Atomization 802.3
DT: January 1993
UF: Fuels—Atomization*
Liquids—Atomization*
RT: Aerosols
Atomizers
Fuels
Granulation
Liquid fuels
Liquids

Atomizers 802.1 (631.1.1)
DT: Predates 1975
BT: Chemical equipment
RT: Atomization

Atoms 931.3
DT: January 1986
RT: Atomic physics
Atomic spectroscopy
Conformations
Free radicals
Ions
Isotopes
Molecules

ATP
USE: Adenosinetriphosphate

Attenuation (711.1) (751.1)
DT: January 1993
NT: Electromagnetic wave attenuation
RT: Attenuation equalizers
Electric attenuators
Electric losses
Electromagnetic dispersion
Energy absorption
Internal friction
Scattering
Wave propagation
Wave transmission
Waves

Attenuation equalizers 714
DT: January 1993
UF: Equalizers, Attenuation*
BT: Equalizers
RT: Attenuation

Attenuators (electric)
USE: Electric attenuators

Attenuators (waveguide)
USE: Waveguide attenuators

Audio acoustics 751.1
SN: Use for the application of acoustics to an environment as related to electronic recording or reproduction of sound. For more general acoustics of buildings, use ARCHITECTURAL ACOUSTICS
DT: Predates 1975
BT: Acoustics

* Former Ei Vocabulary term

Audio acoustics *(continued)*
RT: Architectural acoustics
Audio equipment
Audio studios
Audio systems
Radio studios
Sound recording
Sound reproduction

Audio equipment (752.2.1) (752.3.1)
SN: Use for general subject of sound recording and reproducing equipment; prefer terms for specific equipment
DT: Predates 1975
BT: Electronic equipment
NT: Audio frequency amplifiers
Headphones
Loudspeakers
Microphones
Phonographs
Pickups
Public address systems
Tape recorders
RT: Acoustic devices
Acoustic equipment
Audio acoustics
Audio studios
Audio systems
Compact disk players
Phonograph records
Stereophonic recordings

*Audio equipment—Studios**
USE: Audio studios

Audio frequency amplifiers 713.1
DT: January 1993
UF: Amplifiers, Audio frequency*
BT: Amplifiers (electronic)
Audio equipment
RT: Phonographs

Audio recordings
USE: Stereophonic recordings

Audio studios (752.2)
DT: January 1993
UF: Audio equipment—Studios*
BT: Studios
RT: Audio acoustics
Audio equipment

Audio systems (751) (752.1)
SN: Use for the general subject only; prefer more specific terms
DT: Predates 1975
RT: Acoustic devices
Acoustic equipment
Audio acoustics
Audio equipment
Phonograph records
Sound recording
Sound reproduction
Stereophonic recordings

Audiometers 941
SN: For testing sensitivity of hearing
DT: Predates 1975
BT: Biomedical equipment
Instruments
RT: Acoustic measuring instruments
Audition

Audiotape recordings
USE: Stereophonic recordings

Audition 461.4, 941.1 (751.1)
SN: Ability to hear
DT: Predates 1975
UF: Hearing
BT: Sensory perception
NT: Underwater audition
RT: Acoustic noise
Acoustic wave propagation
Audiometers
Biocommunications
Ear protectors
Hearing aids
z Physiology
Speech
Speech intelligibility
Speech recognition

*Audition—Underwater**
USE: Underwater audition

Auditoriums 402.2 (752)
DT: Predates 1975
BT: Facilities
RT: Opera houses
Stages
Theaters (legitimate)

Auditory evoked potentials
USE: Bioelectric potentials

Auger conveyors
USE: Screw conveyors

Auger electron spectroscopy (801) (931.2) (941.4)
DT: January 1993
UF: Spectroscopy, Auger electron*
BT: Electron spectroscopy

Augers 502.2
DT: January 1993
UF: Mines and mining—Augers*
BT: Boring tools
RT: Mining drills

Austenite 531.2 (545.3)
DT: January 1993
UF: Iron and steel metallography—Austenite*
Metallography—Austenite*
BT: Metallographic phases
RT: Austenitic transformations
Steel metallography

Austenitic transformations 531.2 (545.3)
DT: January 1993
BT: Phase transitions
RT: Austenite

* Former Ei Vocabulary term

Autoclaves (462.1) (802.1)
DT: Predates 1975
BT: Containers
RT: Biomedical equipment
Chemical reactors
Curing
Hospitals
Pressure vessels
Sterilizers

Automata theory 721.1 (723)
DT: Predates 1975
BT: Computation theory
NT: Finite automata
Self reproducing automata
Sequential machines
Turing machines
RT: Adaptive systems
Cognitive systems
Computability and decidability
Computational grammars
Equivalence classes
Formal logic
Heuristic programming
Learning systems
Petri nets
Probabilistic logics
Robotics
Robots

*Automata theory—Computability and decidability**
USE: Computability and decidability

*Automata theory—Computational linguistics**
USE: Computational linguistics

*Automata theory—Context free grammars**
USE: Context free grammars

*Automata theory—Context free languages**
USE: Context free languages

*Automata theory—Context sensitive grammars**
USE: Context sensitive grammars

*Automata theory—Context sensitive languages**
USE: Context sensitive languages

*Automata theory—Finite automata**
USE: Finite automata

*Automata theory—Formal languages**
USE: Formal languages

*Automata theory—Grammars**
USE: Computational grammars

*Automata theory—Recursive functions**
USE: Recursive functions

*Automata theory—Self reproducing automata**
USE: Self reproducing automata

*Automata theory—Sequential machines**
USE: Sequential machines

*Automata theory—Theorem proving**
USE: Theorem proving

*Automata theory—Turing machines**
USE: Turing machines

Automatic control
USE: Control

Automatic guidance (agricultural machinery) 821.1, 731.2
DT: January 1993
UF: Agricultural machinery—Automatic guidance*
BT: Control
RT: Agricultural machinery
Automation

*Automatic operation**
USE: Automation

Automatic pilots 652.3 (731.2)
DT: January 1993
UF: Aircraft instruments—Automatic pilots*
BT: Aircraft instruments
RT: Automation

Automatic telephone exchanges 718.1 (731.2) (732)
DT: January 1993
UF: Electronic telephone exchanges
Telephone exchanges, Automatic*
BT: Automatic telephone systems
Telephone exchanges

Automatic telephone systems 718.1 (731.2) (732)
DT: January 1993
UF: Telephone systems, Automatic*
BT: Telephone systems
NT: Automatic telephone exchanges
RT: Automation
Switching systems
Telephone equipment

Automatic teller machines (723.2) (912.2)
DT: January 1993
UF: ATM
Business machines—Automated teller machines*
BT: Business machines
RT: Automation
Computers

Automatic testing (422) (423)
DT: January 1977
BT: Testing
RT: Automation
Equipment testing
Materials testing

Automatic train control 731.2, 682.1.2 (732)
DT: January 1993
UF: Railroads—Automatic train control*
BT: Control
RT: Automation
Railroads

Automation 731 (732)
DT: January 1993
UF: Automatic operation*

Automation *(continued)*
NT: Factory automation
Office automation
RT: Automatic guidance (agricultural machinery)
Automatic pilots
Automatic telephone systems
Automatic teller machines
Automatic testing
Automatic train control
Control
Control systems
Intelligent buildings
Mechanization
Process control
Robotics

Automobile air bags 662.1, 914.1
DT: January 1995
UF: Air bags (automobile)
BT: Automobile safety devices
RT: Accidents

Automobile bodies 662.4 (663.2)
DT: January 1993
UF: Automobiles—Bodies*
BT: Automobile parts and equipment
NT: Automobile frames

Automobile bumpers 662.4 (663.2)
DT: January 1993
UF: Automobiles—Bumpers*
Bumpers (automobile)
BT: Automobile parts and equipment

Automobile cooling systems 662.4 (663.2)
SN: For cooling the vehicle; for engine cooling use AUTOMOBILE ENGINES
DT: January 1993
BT: Automobile parts and equipment
Cooling systems
NT: Automobile radiators
RT: Air conditioning

Automobile door locks 605, 662.4 (663.2)
DT: January 1993
UF: Automobiles—Door locks*
BT: Automobile hardware
Locks (fasteners)

Automobile driver simulators 432, 912.4 (461.4)
DT: January 1993
UF: Automobile drivers—Simulators/Physical mock-ups*
Dummies (automobile crash testing)
BT: Simulators
RT: Crashworthiness
Human form models

Automobile drivers 432, 912.4
DT: Predates 1975
UF: Chauffeurs
Drivers (automobile)
BT: Transportation personnel
RT: Automobiles
Driver licensing
Driver training

*Automobile drivers—Licensing**
USE: Driver licensing

*Automobile drivers—Simulators/Physical mock-ups**
USE: Automobile driver simulators

Automobile electric equipment 704.2, 662.4 (715.2) (663.2)
DT: January 1993
UF: Automobiles—Electric equipment*
BT: Automobile parts and equipment
Electric equipment
RT: Automobile electronic equipment

Automobile electronic equipment 715.2, 662.4 (704.2) (663.2)
DT: January 1993
UF: Automobiles—Electronic equipment*
BT: Automobile parts and equipment
Electronic equipment
NT: Automobile radio equipment
Automobile television equipment
RT: Automobile electric equipment

Automobile engine manifolds 661.2
DT: January 1993
UF: Automobile engines—Manifolds*
Manifolds (automobile engines)
BT: Automobile engines

Automobile engine manufacture 661.1
DT: Predates 1975
BT: Automobile manufacture
RT: Automobile engines
Automobile plants

Automobile engineering
USE: Automotive engineering

Automobile engines 661.1
DT: Predates 1975
UF: Gas turbines—Automotive applications*
Turbomachinery—Automotive applications*
BT: Automobile parts and equipment
Internal combustion engines
NT: Automobile engine manifolds
Racing automobile engines
Small automobile engines
RT: Automobile engine manufacture
Nuclear powered vehicles
Running in

*Automobile engines—Braking**
USE: Braking

*Automobile engines—Intake silencers**
USE: Intake silencers

*Automobile engines—Light metals**
USE: Light metals

*Automobile engines—Manifolds**
USE: Automobile engine manifolds

*Automobile engines—Preignition**
USE: Preignition

*Automobile engines—Racing applications**
USE: Racing automobile engines

* Former Ei Vocabulary term

*Automobile engines—Running in**
USE: Running in

*Automobile engines—Small**
USE: Small automobile engines

*Automobile engines—Starting**
USE: Starting

Automobile exhibitions 662 (911.4)
DT: Predates 1975
BT: Exhibitions
RT: Automobiles
Machinery exhibitions

Automobile fabrics 819.5, 662.3 (663.2)
DT: January 1993
UF: Automobiles—Textiles*
BT: Fabrics
RT: Automobile parts and equipment

Automobile ferries
USE: Ferry boats

Automobile frames 662.4 (663.2)
DT: January 1993
UF: Automobiles—Frames*
Frames (automobile)
BT: Automobile bodies

Automobile fuel tanks 662.4 (663.2)
DT: January 1993
UF: Automobiles—Fuel tanks*
BT: Automobile parts and equipment
Fuel tanks

Automobile fuels
USE: Automotive fuels

Automobile furnishings 662.3 (663.2)
DT: January 1993
UF: Automobiles—Furnishings*
BT: Automobile parts and equipment

Automobile hardware 662.4 (663.2)
DT: January 1993
UF: Automobiles—Hardware*
BT: Automobile parts and equipment
Hardware
NT: Automobile door locks

Automobile headlights
USE: Headlights

Automobile heaters 662.4, 643.2 (663.2)
DT: January 1993
BT: Automobile parts and equipment
Heating equipment
RT: Automobile radiators

Automobile hydraulic equipment
USE: Automobile parts and equipment AND
Hydraulic equipment

Automobile instrument panels
662.4 (942) (943) (663.2)
DT: January 1993
UF: Automobiles—Instrument panels*
BT: Automobile parts and equipment
Instrument panels
RT: Automobile instruments

Automobile instruments
662.4 (942) (943) (663.2)
DT: January 1993
BT: Automobile parts and equipment
RT: Automobile instrument panels

Automobile manufacture 662.1
DT: Predates 1975
BT: Manufacture
NT: Automobile engine manufacture
RT: Automobile plants
Automobile proving grounds
Automobiles
Automotive engineering

*Automobile manufacture—Finishing**
USE: Finishing

Automobile materials 662.3
DT: Predates 1975
BT: Materials
RT: Automobile parts and equipment
Automobiles
Light metals
Tubing

*Automobile materials—Cast iron**
USE: Cast iron

*Automobile materials—Composite materials**
USE: Composite materials

*Automobile materials—Light metals**
USE: Light metals

*Automobile materials—Plastics**
USE: Plastics applications

*Automobile materials—Powder metals**
USE: Powder metals

*Automobile materials—Rubber**
USE: Rubber applications

*Automobile materials—Silicones**
USE: Silicones

*Automobile materials—Steel**
USE: Steel

*Automobile materials—Tubing**
USE: Tubing

Automobile mufflers 662.4 (663.2)
DT: January 1993
UF: Automobiles—Mufflers*
BT: Automobile parts and equipment
Mufflers

Automobile parts and equipment 662.4 (663.2)
DT: January 1993
UF: Automobile hydraulic equipment
Automobiles—Equipment*
BT: Automobiles
Ground vehicle parts and equipment
NT: Automobile bodies
Automobile bumpers
Automobile cooling systems
Automobile electric equipment

Automobile parts and equipment *(continued)*
Automobile electronic equipment
Automobile engines
Automobile fuel tanks
Automobile furnishings
Automobile hardware
Automobile heaters
Automobile instrument panels
Automobile instruments
Automobile mufflers
Automobile safety devices
Automobile seats
Automobile shock absorbers
Automobile springs
Automobile steering equipment
Automobile suspensions
Automobile transmissions
Automobile windshield wipers
Automobile windshields
RT: Automobile fabrics
Automobile materials
Automobile plants
Axles
Headlights
Windshield wipers

Automobile plants 662 (663) (911) (912) (402.1)
DT: Predates 1975
BT: Industrial plants
RT: Automobile engine manufacture
Automobile manufacture
Automobile parts and equipment
Automobiles

Automobile proving grounds 662 (663)
DT: January 1993
UF: Automobiles—Proving grounds*
Proving grounds (automobile)
BT: Facilities
RT: Automobile manufacture
Automobile testing
Automobiles

Automobile radiators 643.2, 662.4 (663.2) (616.1)
DT: January 1993
UF: Automobiles—Radiators*
BT: Automobile cooling systems
Radiators
RT: Automobile heaters

Automobile radio equipment 716.3, 662.4 (663.2)
DT: January 1993
BT: Automobile electronic equipment
Radio equipment

Automobile safety devices 662.4, 914.1 (663.2)
DT: January 1993
UF: Automobiles—Safety devices*
BT: Automobile parts and equipment
Safety devices
NT: Automobile air bags
Automobile seat belts

Automobile seat belts 662.4, 914.1 (663.2)
DT: January 1993
UF: Automobiles—Seat belts*
Seat belts (automobile)
BT: Automobile safety devices

Automobile seats 662.4 (663.2)
DT: January 1993
UF: Automobiles—Seats*
BT: Automobile parts and equipment
Seats
RT: Riding qualities

Automobile shock absorbers 662.4 (632.1) (663.2)
DT: January 1993
UF: Automobiles—Shock absorbers*
BT: Automobile parts and equipment
Shock absorbers

Automobile simulators 662 (912.4) (663)
SN: For training use
DT: January 1993
UF: Automobiles—Simulators*
BT: Simulators
RT: Personnel training

Automobile springs 662.4 (663.2)
DT: January 1993
UF: Automobiles—Springs and suspensions*
BT: Automobile parts and equipment
Vehicle springs

Automobile steering equipment 662.4 (663.2)
DT: January 1993
UF: Automobiles—Steering systems*
Steering equipment (automobile)
BT: Automobile parts and equipment
RT: Steering

Automobile suspensions 662.4 (632.1) (663.2)
DT: January 1993
UF: Automobiles—Springs and suspensions*
BT: Automobile parts and equipment
Vehicle suspensions

Automobile television equipment 716.4, 662.4 (663.2)
DT: January 1993
UF: Automobiles—Television equipment*
BT: Automobile electronic equipment
Television equipment

Automobile testing 662 (663)
SN: Also use the specific factor being tested, e.g., FUEL ECONOMY
DT: January 1993
BT: Equipment testing
RT: Automobile proving grounds
Automobiles

Automobile tires
USE: Tires

Automobile transmissions 661.2
DT: Predates 1975
BT: Automobile parts and equipment
Vehicle transmissions
RT: Power transmission

* Former Ei Vocabulary term

Automobile windshield wipers 662.4 (663.2)
DT: January 1993
BT: Automobile parts and equipment
RT: Automobile windshields

Automobile windshields 812.3, 662.4 (663.2)
DT: January 1993
UF: Automobiles—Windshields*
BT: Automobile parts and equipment
Windshields
RT: Automobile windshield wipers
Safety glass
Windshield wipers

Automobiles 662.1
SN: General subject, as well as private passenger automobiles
DT: Predates 1975
BT: Ground vehicles
NT: Amphibious automobiles
Automobile parts and equipment
Electric automobiles
Front wheel drive automobiles
Miniature automobiles
Racing automobiles
Rear engine automobiles
Taxicabs
RT: Automobile drivers
Automobile exhibitions
Automobile manufacture
Automobile materials
Automobile plants
Automobile proving grounds
Automobile testing
Automotive engineering
Automotive fuels
Fueling
Intelligent vehicle highway systems
License plates (automobile)
Military vehicles
Model automobiles
Riding qualities
Skidding
Stratified charge engines

*Automobiles—Amphibious**
USE: Amphibious automobiles

*Automobiles—Axles**
USE: Axles

*Automobiles—Bodies**
USE: Automobile bodies

*Automobiles—Bumpers**
USE: Automobile bumpers

*Automobiles—Crashworthiness**
USE: Crashworthiness

*Automobiles—Door locks**
USE: Automobile door locks

*Automobiles—Electric equipment**
USE: Automobile electric equipment

*Automobiles—Electric**
USE: Electric automobiles

*Automobiles—Electronic equipment**
USE: Automobile electronic equipment

*Automobiles—Engine mounting**
USE: Engine mountings

*Automobiles—Equipment**
USE: Automobile parts and equipment

*Automobiles—Frames**
USE: Automobile frames

*Automobiles—Front wheel drive**
USE: Front wheel drive automobiles

*Automobiles—Fuel tanks**
USE: Automobile fuel tanks

*Automobiles—Furnishings**
USE: Automobile furnishings

*Automobiles—Hardware**
USE: Automobile hardware

*Automobiles—Headlights**
USE: Headlights

*Automobiles—Heaters**
USE: Heating equipment

*Automobiles—Hydraulic equipment**
USE: Hydraulic equipment

*Automobiles—Instrument panels**
USE: Automobile instrument panels

*Automobiles—License plates**
USE: License plates (automobile)

*Automobiles—Lighting**
USE: Lighting

*Automobiles—Miniature**
USE: Miniature automobiles

*Automobiles—Mufflers**
USE: Automobile mufflers

*Automobiles—Noise**
USE: Acoustic noise

*Automobiles—Nuclear power**
USE: Nuclear powered vehicles

*Automobiles—Off highway**
USE: Off road vehicles

*Automobiles—Pneumatic equipment**
USE: Pneumatic equipment

*Automobiles—Proving grounds**
USE: Automobile proving grounds

*Automobiles—Racing**
USE: Racing automobiles

*Automobiles—Radiators**
USE: Automobile radiators

*Automobiles—Radio interference**
USE: Radio interference

*Automobiles—Rear engine**
USE: Rear engine automobiles

* Former Ei Vocabulary term

*Automobiles—Riding qualities**
USE: Riding qualities

*Automobiles—Safety devices**
USE: Automobile safety devices

*Automobiles—Safety factor**
USE: Safety factor

*Automobiles—Seat belts**
USE: Automobile seat belts

*Automobiles—Seats**
USE: Automobile seats

*Automobiles—Shock absorbers**
USE: Automobile shock absorbers

*Automobiles—Simulators**
USE: Automobile simulators

*Automobiles—Skidding**
USE: Skidding

*Automobiles—Springs and suspensions**
USE: Automobile springs OR
Automobile suspensions

*Automobiles—Steering systems**
USE: Automobile steering equipment

*Automobiles—Television equipment**
USE: Automobile television equipment

*Automobiles—Textiles**
USE: Automobile fabrics

*Automobiles—Weight control**
USE: Weight control

*Automobiles—Windshield wipers**
USE: Windshield wipers

*Automobiles—Windshields**
USE: Automobile windshields

Automotive engineering 664
DT: Predates 1975
UF: Automobile engineering
BT: Engineering
RT: Automobile manufacture
Automobiles

Automotive fuels (522) (523)
DT: Predates 1975
UF: Automobile fuels
BT: Fuels
RT: Aircraft fuels
Alcohol fuels
Automobiles
Diesel fuels
Ethanol fuels
Gasohol
Gasoline
Hydrogen fuels
Liquid fuels
Methanol fuels
Stratified charge engines

*Automotive fuels—Gasohol**
USE: Gasohol

Autonomous robots
USE: Robots

Auxiliary equipment
DT: January 1993
UF: Aircraft—Auxiliary equipment*
Industrial plants—Auxiliary equipment*
Nuclear reactors—Auxiliary equipment*
Power plants—Auxiliary equipment*
Ships—Auxiliary machinery*
BT: Equipment
NT: Auxiliary power systems

Auxiliary power systems (704) (652.3) (654.1)
DT: January 1993
UF: Aircraft—Auxiliary power*
Rockets and missiles—Auxiliary power systems*
BT: Auxiliary equipment
RT: Aircraft

Availability (913.1) (913.5)
DT: January 1995
RT: Failure analysis
Maintainability
Performance
Reliability

Avalanche diodes 714.1
DT: January 1993
UF: Semiconductor diodes, Avalanche*
BT: Semiconductor diodes
NT: IMPATT diodes
RT: Electric breakdown
Transit time devices

Avalanches (snowslides) 443.3
DT: January 1993
UF: Snow and snowfall—Avalanches and slides*
Snowslides
RT: Accidents
Landslides
Snow

Aviation 431.1
DT: Predates 1975
UF: Aeronautics
NT: Air navigation
Military aviation
RT: Air traffic control
Aircraft
Aircraft communication
Aviation medicine
Aviators
Landing
Takeoff

Aviation accidents
USE: Aircraft accidents

Aviation engineering
USE: Aerospace engineering

Aviation medicine 431.1, 461.6
DT: January 1993
BT: Medicine
RT: Aviation
Weightlessness

* Former Ei Vocabulary term

*Aviation, Military**
USE: Military aviation

*Aviation, Military—Fire control systems**
USE: Fire control systems

*Aviation, Military—Ground operations**
USE: Ground operations

*Aviation—Medical problems**
USE: Medical problems

*Aviation—Meteorology**
USE: Meteorology

Aviators 431.1, 912.4
DT: Predates 1975
BT: Transportation personnel
RT: Aviation
Training aircraft

*Aviators—Medical problems**
USE: Medical problems

Avionics 652.3, 715
DT: January 1977
BT: Aircraft instruments
RT: Electronics engineering

Awnings
USE: Sun hoods

Axial flow 631.1
DT: January 1993
UF: Flow of fluids—Axial*
BT: Flow of fluids
RT: Axial flow turbomachinery

Axial flow turbomachinery (612.3) (617) (618) (631.1)
DT: January 1993
UF: Turbomachinery—Axial flow*
BT: Turbomachinery
RT: Axial flow

Axles 662.4, 663.2, 682.1.1
DT: January 1993
UF: Automobiles—Axles*
Cars—Axles*
Earthmoving machinery—Axles*
Locomotives—Axles*
Railroad rolling stock—Axles*
BT: Components
RT: Automobile parts and equipment
Bogies (railroad rolling stock)
Earthmoving machinery
Locomotives
Railroad cars
Railroad rolling stock
Supports
Vehicle wheels
Wheels

Azeotropes 802.3
DT: January 1993
UF: Mixtures—Azeotrope*
BT: Mixtures

Azo dyes 803, 804.1
DT: January 1993
UF: Dyes and dyeing—Azo dyes*
BT: Dyes
Organic compounds

B
USE: Boron

Ba
USE: Barium

Backpropagation 723.4
DT: January 1995
RT: Artificial intelligence
Learning systems
Neural networks

Backscattering (711) (791.1)
DT: January 1993
BT: Scattering
NT: Acoustic wave backscattering
Electromagnetic wave backscattering
RT: Acoustic waves
Radar
Radar cross section
Radio transmission
Radiowave propagation effects
Rutherford backscattering spectroscopy

Backward wave tubes 714.1
DT: January 1993
UF: Electron tubes, Backward wave*
BT: Traveling wave tubes

Bacteria (461.9) (801.2)
DT: June 1990
BT: Microorganisms
NT: Coliform bacteria
RT: Bacteriology
Bacteriophages
Biological materials
Methanogens
Nitrification
Nitrogen fixation

Bacterial viruses
USE: Bacteriophages

Bacteriology (461.9) (801.2)
DT: October 1975
BT: Microbiology
NT: Sewage bacteriology
Water bacteriology
RT: Bacteria
Biological materials
Microorganisms

*Bacteriology—Coliforms**
USE: Coliform bacteria

Bacteriophages (461.9.1) (801.2)
DT: January 1993
UF: Bacterial viruses
Phage
Viruses—Bacterial*
Viruses—Bacteriophage*
BT: Viruses
RT: Bacteria

Bagasse 821.5 (524) (811)
DT: Predates 1975
UF: Pulp materials—Bagasse*
BT: Materials
RT: Agricultural wastes
Low grade fuel firing
Pulp materials

Baggage cars 682.1.1
DT: January 1993
UF: Cars—Baggage*
BT: Freight cars

Baggage handling 691 (431.4)
DT: January 1993
UF: Airports—Baggage handling*
BT: Materials handling
RT: Airports

Bainite 531.2
DT: January 1993
UF: Iron and steel metallography—Bainite*
Metallography—Bainite*
BT: Metallographic phases
RT: Bainitic transformations
Steel metallography

Bainitic transformations 531.2
DT: January 1993
BT: Phase transitions
RT: Bainite

Bakeries 822.1 (402.1)
DT: Predates 1975
BT: Food products plants
RT: Industrial ovens

*Bakeries—Ovens**
USE: Industrial ovens

Balances
USE: Precision balances

Balancing 601
SN: Use for the process of correcting unbalanced static and/or dynamic forces. For balancing to determine weight, use WEIGHING
DT: Predates 1975
RT: Balancing machines
Vibration control
Vibrations (mechanical)

Balancing machines 601
DT: Predates 1975
BT: Machinery
RT: Balancing

Ball bearings 601.2
DT: January 1993
UF: Bearings—Ball*
BT: Antifriction bearings
RT: Roller bearings
Thrust bearings

Ball milling 802.3 (533.1)
DT: January 1993
UF: Grinding mills—Ball milling*
BT: Grinding (machining)
RT: Ball mills
Mechanical alloying

Ball mills 802.1 (533.1)
DT: January 1993
BT: Grinding mills
RT: Ball milling
Granulators

Ballast (railroad track) 681.1
DT: January 1993
UF: Railroad plant and structures—Ballast*
BT: Materials
RT: Railroad tracks
Rock products

Ballast tanks 671.2 (672)
DT: January 1993
UF: Ships—Ballast tanks*
BT: Ship equipment
Tanks (containers)

Ballasts (lamp) 707.2
DT: January 1993
UF: Electric lamps—Ballasts*
BT: Discharge lamps

Ballistics 404.1, 931.1
DT: Predates 1975
BT: Kinematics
NT: Underwater ballistics
RT: Electromagnetic launchers
Military engineering
Military equipment
Projectiles

*Ballistics—Projectiles**
USE: Projectiles

*Ballistics—Underwater**
USE: Underwater ballistics

Balloons 652.5 (443.2)
DT: Predates 1975
BT: Aircraft
NT: Meteorological balloons
Radio balloons
RT: Inflatable structures

*Balloons—Meteorology**
USE: Meteorological balloons

*Balloons—Radio communication**
USE: Radio balloons

Bamboo 811.1, 821.4
DT: January 1993
UF: Pulp materials—Bamboo*
BT: Plants (botany)
RT: Pulp materials

Band elimination filters
USE: Notch filters

Band pass amplifiers
USE: Bandpass amplifiers

Band pass filters
USE: Bandpass filters

Band rejection filters
USE: Notch filters

* Former Ei Vocabulary term

Band structure 933
SN: Band theory of solids; electron energy bands
DT: Predates 1975
UF: Energy bands
BT: Electron energy levels
NT: Energy gap
RT: Electronic density of states
Electrons
Fermi surface
Semiconductor materials
Solids

Bandpass amplifiers 713.1
DT: January 1993
UF: Amplifiers, Bandpass*
Band pass amplifiers
BT: Amplifiers (electronic)
RT: Bandpass filters

Bandpass filters 703.2
DT: January 1993
UF: Band pass filters
Electric filters, Bandpass*
Parallel resonator filters
BT: Electric filters
RT: Bandpass amplifiers
Notch filters

Bandstop filters
USE: Notch filters

Bandwidth 716.1
DT: January 1995
RT: Bandwidth compression
Channel capacity
Natural frequencies

Bandwidth compression 716.1
DT: January 1993
UF: Information theory—Bandwidth compression*
BT: Telecommunication
RT: Bandwidth
Information theory
Television systems

Bang bang control systems 731.1
DT: January 1993
UF: Control systems, Bang bang*
BT: Control systems
RT: Control nonlinearities
Nonlinear control systems
Nonlinear systems
On-off control systems

Bank protection 407.2, 407.3 (914.1)
SN: Protection of river banks and lake shores. For seacoasts, use SHORE PROTECTION
DT: January 1993
UF: Coastal engineering—Bank protection*
Inland waterways—Bank protection*
BT: Protection
RT: Banks (bodies of water)
Coastal engineering
Environmental engineering
Erosion
Inland waterways
River control
Shore protection
Slope protection
Soil conservation

Banks (bodies of water) 407.2, 444.1
DT: January 1993
RT: Bank protection
Lakes
Rivers
Surface waters

Bar codes 723.2
DT: January 1987
BT: Codes (symbols)
RT: Marking machines

Barges (434.3) (674)
DT: Predates 1975
BT: Boats
NT: Pipe laying barges

*Barges—Constructing and outfitting**
USE: Outfitting (water craft) OR
Shipbuilding

*Barges—Pipe laying**
USE: Pipe laying barges

Barite 482.2 (505)
DT: Predates 1975
BT: Ores
RT: Barium

Barium 549.2
DT: January 1993
UF: Ba
Barium and alloys*
BT: Alkaline earth metals
RT: Barite
Barium alloys
Barium compounds
Barium metallography

Barium alloys 549.2
DT: January 1993
UF: Barium and alloys*
BT: Alkaline earth metal alloys
RT: Barium
Barium compounds

*Barium and alloys**
USE: Barium OR
Barium alloys

Barium compounds (804.1) (804.2)
DT: January 1993
BT: Alkaline earth metal compounds
NT: Barium titanate
RT: Barium
Barium alloys
Contrast media
Oxide superconductors

Barium metallography 531.2, 549.2
DT: June 1990
BT: Metallography
RT: Barium

Barium titanate 804.2, 812.1
DT: Predates 1975
BT: Barium compounds
Titanium compounds
RT: Ceramic materials
Ferroelectric materials

Bark stripping 811.2
DT: January 1993
UF: Stripping (bark)
Wood—Bark stripping*
BT: Stripping (removal)
RT: Lumber
Wood

Barometers 443.2, 944.3 (944.3)
DT: Predates 1975
BT: Meteorological instruments
Pressure transducers
RT: Pressure measurement

Barometric altimeters
USE: Aneroid altimeters

Barreling (539.3) (604.2) (606.2)
DT: January 1993
UF: Metal finishing—Barrel*
Metal finishing—Tumbling*
Tumbling
BT: Finishing

Bars (metal) 535.1.2
DT: January 1993
UF: Bars*
Rolling mill practice—Bars*
RT: Billets (metal bars)
Blooms (metal)
Rolling mill practice
Rolling mills

*Bars**
USE: Bars (metal)

Basalt (482.2) (414.2) (505)
DT: Predates 1975
BT: Volcanic rocks

Bascule bridges 401.1
DT: January 1993
UF: Bridges, Bascule*
BT: Cantilever bridges
Movable bridges

BASIC (programming language) 723.1.1
DT: January 1993
UF: Computer programming languages—BASIC*
BT: Procedure oriented languages

Basic converters
USE: Basic oxygen converters

Basic oxygen converters 532.1, 545.3
DT: January 1977
UF: Basic converters
Basic oxygen furnaces
Converters (metal refining)
BT: Metallurgical furnaces
RT: Basic oxygen process

Basic oxygen furnaces
USE: Basic oxygen converters

Basic oxygen process 531.1, 545.3
DT: January 1993
UF: Oxygen steelmaking
Steelmaking—Basic oxygen process*
BT: Steelmaking
NT: Spray steelmaking
RT: Basic oxygen converters

Basicity
USE: pH

Basins (settling)
USE: Settling tanks

Bast fibers 819.1
DT: January 1993
BT: Natural fibers
NT: Hemp fibers
Jute fibers
Kenaf fibers
RT: Textile fibers

Batch cell culture (461.2) (801.2)
DT: January 1993
UF: Cell culture—Batch*
BT: Cell culture

Batch data processing 723.2
DT: January 1993
UF: Batch processing (data)
Data processing—Batch processing*
BT: Data processing

Batch processing (data)
USE: Batch data processing

Bathymetry 471.3, 943.2
DT: January 1993
UF: Depth measurement (ocean)
Oceanography—Bathymetry*
BT: Measurements
RT: Oceanography

Bathyscaphes
USE: Submersibles

Batteries (electric)
USE: Electric batteries

Batteries (fuel cells)
USE: Fuel cells

Batteries (solar)
USE: Solar cells

Battery charging
USE: Charging (batteries)

Bauxite deposits 504.1, 541.1
DT: Predates 1975
UF: Aluminum deposits*
BT: Ore deposits
RT: Aluminum
Bauxite mines

Bauxite mines 504.1, 541.1
DT: January 1993
UF: Bauxite mines and mining*

Bauxite mines *(continued)*
BT: Mines
RT: Aluminum
Bauxite deposits

*Bauxite mines and mining**
USE: Bauxite mines

Bauxite ore treatment 533.1, 541.1
DT: Predates 1975
BT: Ore treatment
RT: Aluminum
Aluminum metallurgy

Be
USE: Beryllium

Beaches 407.3
DT: Predates 1975
BT: Landforms
RT: Coastal engineering
Erosion
Lakes
Oceanography
Shore protection

*Beaches—Sanitation**
USE: Sanitation

Beam lead devices 714.2
DT: January 1993
UF: Integrated circuits—Beam lead construction*
BT: Semiconductor devices
RT: Hybrid integrated circuits
Integrated circuits

Beam pens
USE: Light pens

Beam plasma interactions 932.3
DT: January 1993
UF: Interactions (plasmas)
Plasma beam interactions
Plasmas—Beam-plasma interactions*
BT: Plasma interactions
RT: Particle beams
Plasmas

Beam splitters (optical)
USE: Optical beam splitters

*Beam tracking**
USE: Particle beam tracking

Beams (light)
USE: Light

Beams (particle)
USE: Particle beams

Beams and girders 408.2 (931.1)
DT: Predates 1975
UF: Girders
BT: Structural members
NT: Aluminum beams and girders
Composite beams and girders
Concrete beams and girders
Curved beams and girders
Prestressed beams and girders
Steel beams and girders
Wooden beams and girders
RT: Bending (deformation)
Box girder bridges
Columns (structural)
Plate girder bridges
Plates (structural components)
Structural frames
Struts
Trusses

*Beams and girders—Aluminum**
USE: Aluminum beams and girders

*Beams and girders—Bending**
USE: Bending (forming)

*Beams and girders—Buckling**
USE: Buckling

*Beams and girders—Composite**
USE: Composite beams and girders

*Beams and girders—Concrete**
USE: Concrete beams and girders

*Beams and girders—Curved**
USE: Curved beams and girders

*Beams and girders—Deflection**
USE: Deflection (structures)

*Beams and girders—Prestressed**
USE: Prestressed beams and girders

*Beams and girders—Steel**
USE: Steel beams and girders

*Beams and girders—Wood**
USE: Wooden beams and girders

Beamsplitters (optical)
USE: Optical beam splitters

Bearing capacity (421)
SN: The load per unit area which can be safely supported
DT: January 1993
BT: Mechanical properties
RT: Dynamic loads
Load limits
Loads (forces)
Structural loads
Yield stress

Bearing mountings 601.2
DT: January 1993
UF: Bearings—Mountings*
BT: Mountings
RT: Bearings (machine parts)

Bearing pads 601.2
DT: January 1993
UF: Bearings—Pads*
Pads (bearing)
BT: Bearings (machine parts)

Bearings (machine parts) 601.2
DT: January 1993
UF: Bearings*
Low temperature bearings
Paper bearings

* Former Ei Vocabulary term

Bearings (machine parts) *(continued)*
BT: Machine components
NT: Antifriction bearings
Bearing pads
Foil bearings
Hydrostatic bearings
Jewel bearings
Journal bearings
Magnetic bearings
Miniature bearings
Nonmetallic bearings
Railroad bearings
Thrust bearings
RT: Air lubrication
Bearing mountings
Lubrication
Packing

Bearings (structural) 401.1, 408.2
DT: January 1993
UF: Bearings*
Bearings—Railroad*
BT: Structural members
NT: Bridge bearings
RT: Bushings
Supports

*Bearings**
USE: Bearings (machine parts) OR
Bearings (structural)

*Bearings—Air lubricated**
USE: Air lubrication

*Bearings—Antifriction**
USE: Antifriction bearings

*Bearings—Ball**
USE: Ball bearings

*Bearings—Foil**
USE: Foil bearings

*Bearings—Gas lubrication**
USE: Gas lubricated bearings

*Bearings—High temperature**
USE: High temperature applications

*Bearings—Hydrostatic**
USE: Hydrostatic bearings

*Bearings—Jewels**
USE: Jewel bearings

*Bearings—Journal**
USE: Journal bearings

*Bearings—Low temperature**
USE: Cryogenic equipment

*Bearings—Magnetic**
USE: Magnetic bearings

*Bearings—Miniature**
USE: Miniature bearings

*Bearings—Mountings**
USE: Bearing mountings

*Bearings—Noise**
USE: Acoustic noise

*Bearings—Nonmetallic materials**
USE: Nonmetallic bearings

*Bearings—Pads**
USE: Bearing pads

*Bearings—Paper**
USE: Paper products

*Bearings—Powder metal**
USE: Powder metal products

*Bearings—Railroad**
USE: Railroad bearings AND
Bearings (structural)

*Bearings—Roller**
USE: Roller bearings

*Bearings—Seizure**
USE: Seizing

*Bearings—Self lubricating**
USE: Self lubricating bearings

*Bearings—Shims**
USE: Shims

*Bearings—Temperature**
USE: Temperature

*Bearings—Thrust**
USE: Thrust bearings

Beating (pulp)
USE: Pulp beating

Beds (hospital)
USE: Hospital beds

Beepers
USE: Paging systems

Behavioral research 461.4
DT: January 1993
UF: Human engineering—Behavioral research*
BT: Research
Social sciences
RT: Economic and social effects
Human engineering

Behavioral science computing
USE: Social sciences computing

Bellows 601.2
DT: Predates 1975
BT: Components

Bells 752.1
DT: Predates 1975
UF: Ringers
BT: Acoustic generators
RT: Musical instruments

Belt conveyors 692.1
DT: January 1993
UF: Conveyors—Belt*
BT: Conveyors
RT: Belts

* Former Ei Vocabulary term

Belt drives
USE: Mechanical drives

Belts 602.1
DT: Predates 1975
UF: Oil field equipment—Belts*
BT: Machine components
NT: Abrasive belts
Wire belts
RT: Belt conveyors
Mechanical drives

*Belts—Abrasive**
USE: Abrasive belts

*Belts—Wire**
USE: Wire belts

Bendability
USE: Formability

Bending (deformation) (421) (422.2)
SN: Unintentional bending
DT: January 1993
BT: Deformation
RT: Beams and girders
Bending (forming)
Bending strength
Glass
Strain
Stress analysis
Torsional stress

Bending (forming) 535.2 (812.3) (619.1)
DT: January 1993
UF: Beams and girders—Bending*
Bowing
Glass—Bending*
Metal forming—Bending*
Tubes—Bending*
BT: Forming
RT: Bending (deformation)
Bending dies
Bending machines
Bending tools
Formability
Metal forming
Straightening
Tubes (components)

Bending brakes 603.1
DT: January 1993
UF: Brakes (bending)
Brakes/Bending
BT: Bending machines

Bending dies 603.1
DT: January 1993
UF: Dies—Bending*
BT: Dies
RT: Bending (forming)
Bending machines
Bending tools

Bending machines 603.1
DT: Predates 1975
BT: Metal forming machines
NT: Bending brakes
RT: Bending (forming)
Bending dies
Bending tools

Bending strength (421)
DT: January 1995
UF: Flexural strength
BT: Strength of materials
RT: Bending (deformation)

Bending tools 603.1 (604.2)
DT: January 1993
UF: Tools, jigs and fixtures—Bending*
BT: Tools
RT: Bending (forming)
Bending dies
Bending machines

Bends (in pipelines)
USE: Pipeline bends

Beneficiation 533.1
DT: January 1993
UF: Dressing (mineral)
Mineral dressing
Ore treatment—Beneficiation*
BT: Ore treatment
RT: Dewatering
Flotation
Metal refining
Purification
Refining
Segregation (metallography)
Size separation

Bentonite 482.2
DT: Predates 1975
BT: Clay
Silicate minerals
RT: Foundry sand

Benzene 804.1
SN: For benzine, use GASOLINE
DT: Predates 1975
UF: Benzol
BT: Aromatic hydrocarbons
RT: Benzene refining
Organic solvents
Solvents

Benzene refining 802.3, 804.1
DT: January 1993
UF: Benzene—Refining*
BT: Hydrocarbon refining
RT: Benzene

*Benzene—Refining**
USE: Benzene refining

Benzines
USE: Gasoline

Benzol
USE: Benzene

Berkelium 622.1
DT: January 1993
BT: Transuranium elements

Beryllia 804.2, 812.2
DT: January 1993
UF: Beryllium oxide
Refractory materials—Beryllia*
BT: Beryllium compounds
Oxides
RT: Refractory materials

Beryllium 542.1, 549
DT: January 1993
UF: Be
Beryllium and alloys*
BT: Alkaline earth metals
Light metals
RT: Beryllium alloys
Beryllium compounds
Beryllium deposits
Beryllium metallurgy
Beryllium metallography
Beryllium minerals
Beryllium ore treatment
Moderators

Beryllium alloys 542.1
DT: January 1993
UF: Beryllium and alloys*
BT: Alkaline earth metal alloys
RT: Beryllium
Beryllium compounds
Moderators

*Beryllium and alloys**
USE: Beryllium OR
Beryllium alloys

Beryllium compounds (804.1) (804.2)
DT: Predates 1975
BT: Alkaline earth metal compounds
NT: Beryllia
Beryllium minerals
RT: Beryllium
Beryllium alloys

Beryllium deposits 504.1, 542.1
DT: Predates 1975
BT: Ore deposits
RT: Beryllium

Beryllium metallography 531.2, 542.2
DT: Predates 1975
BT: Metallography
RT: Beryllium
Beryllium metallurgy

Beryllium metallurgy 531.1, 542.2
DT: Predates 1975
BT: Metallurgy
NT: Beryllium powder metallurgy
RT: Beryllium
Beryllium metallography
Beryllium ore treatment

Beryllium minerals 482.2
DT: January 1993
UF: Mineralogy—Beryllium materials*
BT: Beryllium compounds
Minerals
RT: Beryllium

Beryllium ore treatment 533.1, 542.1
DT: Predates 1975
BT: Ore treatment
RT: Beryllium
Beryllium metallurgy

Beryllium oxide
USE: Beryllia

Beryllium powder metallurgy 536.1, 542.1
DT: January 1993
UF: Powder metallurgy—Beryllium copper*
Powder metallurgy—Beryllium*
BT: Beryllium metallurgy
Powder metallurgy

Bessemer converters
USE: Metallurgical furnaces

Bessemer process 545.3
DT: January 1993
UF: Steelmaking—Bessemer process*
BT: Steelmaking

Beta ray spectrometers (931) (932) (941.4)
DT: January 1993
UF: Spectrometers, Beta ray*
BT: Particle spectrometers
RT: Electron optics

Betatrons 932.1.1
DT: January 1993
UF: Accelerators, Betatron*
BT: Particle accelerators
RT: Electron beams
Electron optics

Bevel gears 601.2
DT: January 1993
UF: Gears—Bevel*
BT: Gears

Beverages 822.3
DT: January 1987
BT: Food products
NT: Fruit juices
Wine
RT: Bottling plants
Breweries
Dairy products

Bi
USE: Bismuth

Bibliographic retrieval systems 903.3
DT: January 1993
UF: Information retrieval systems—Bibliographic*
BT: Information retrieval systems

Bibliographies (903)
DT: January 1993
RT: Information dissemination

Bicycles 432.2 (662.2)
DT: Predates 1975
BT: Ground vehicles
RT: Exercise equipment
Nonmotorized transportation

* Former Ei Vocabulary term

Bifurcation (mathematics) (921)
DT: January 1995
BT: Mathematical techniques
RT: Nonlinear equations

Bilaminate membranes (408.2) (631.1) (802.3) (817.1)
DT: January 1993
UF: Membranes—Bilaminate*
BT: Membranes
RT: Laminates

Billet mills 535.1.1
DT: January 1993
UF: Rolling mill practice—Billet*
BT: Hot rolling mills
RT: Blooming mills
Slab mills

Billets (metal bars) 535.1.2
DT: January 1993
NT: Blooms (metal)
RT: Bars (metal)
Ingots
Metal finishing
Plate metal
Rolling mill practice
Rolling mills

Bimetals 531
DT: January 1993
UF: Metals and alloys—Bimetals*
BT: Materials
RT: Skin effect
Thermometers
Thermostats

Binary adders
USE: Adders

Binary alloys (531.1) (531.2)
DT: January 1995
BT: Alloys
RT: Eutectics
Phase diagrams
Solid solutions
Systems (metallurgical)

Binary codes 723.1
DT: January 1995
BT: Codes (symbols)
RT: Bit error rate
Coding errors
Encoding (symbols)
Error correction
Error detection

Binary mixtures 804
DT: January 1995
BT: Mixtures

Binary sequences 721.1 (921.6)
DT: January 1993
UF: Computer metatheory—Binary sequences*
BT: Computation theory
RT: Convolutional codes

Binders 803 (534.2)
DT: January 1993
UF: Sand, Foundry—Binders*
BT: Materials
RT: Adhesives
Clay
Foundry sand
Sizing (finishing operation)

Binding energy 801.4
DT: January 1995
UF: TBE
Total binding energy
RT: Chemical bonds
Electron energy levels

Binocular vision 741.2
DT: January 1993
UF: Vision—Binocular effect*
BT: Vision

Binoculars 741.3
DT: Predates 1975
BT: Optical devices

Bins 694.4
DT: Predates 1975
BT: Containers

Bioassay 461.6 (801)
DT: January 1993
UF: Drug products—Bioassay*
BT: Assays
RT: Biological materials

Biocatalysts 801.2, 803 (461)
DT: January 1987
BT: Catalysts
NT: Enzymes
RT: Proteins

Biochemical engineering 805.1.1
DT: October 1975
BT: Chemical engineering
RT: Biochemistry

Biochemical oxygen demand (452.2) (452.4) (453)
DT: January 1993
UF: BOD
Oxygen demand (biochemical)
Sewage treatment—Biochemical oxygen demand*
BT: Chemical oxygen demand
RT: Anaerobic digestion
Sewage treatment

Biochemistry 801.2
DT: January 1983
BT: Biology
Chemistry
RT: Biochemical engineering
Bioconversion
Biodegradation
Biomarkers
Blood gas analysis
Body fluids
Calcification (biochemistry)

* Former Ei Vocabulary term

Biochemistry *(continued)*
DNA sequences
Metabolism
Mutagens
Nitrification
Nucleic acid sequences

*Biochemistry—Blood gas analysis**
USE: Blood gas analysis

*Biochemistry—Calcification**
USE: Calcification (biochemistry)

*Biochemistry—DNA sequences**
USE: DNA sequences

*Biochemistry—Glucose in body fluids**
USE: Glucose AND
Body fluids

*Biochemistry—Hemoglobin oxygen saturation**
USE: Hemoglobin oxygen saturation

*Biochemistry—Metabolism**
USE: Metabolism

*Biochemistry—Minerals in the body fluids**
USE: Body fluids AND
Minerals

*Biochemistry—Nucleic acid sequences**
USE: Nucleic acid sequences

Biocides (461.6) (803)
SN: Very general term; prefer a term for the specific type.
DT: January 1993
BT: Chemicals
NT: Disinfectants
Pesticides
RT: Fumigation

Biocommunications 461.1 (716.1) (721.1) (723)
DT: January 1993
UF: Animal communication
Systems science and cybernetics—Biocommunications*
BT: Communication
RT: Audition
Biocontrol
Biomedical engineering
Cybernetics
Learning systems
Speech

Biocompatibility 461.9.1 (462.5)
DT: January 1993
RT: Antibodies
Antigen-antibody reactions
Antigens
Biology
Biomaterials
Immunology
Vaccines

Biocontrol 461.1, 462.4, 731.1
DT: January 1993
UF: Systems science and cybernetics—Biocontrol*
BT: Cybernetics
RT: Biocommunications
Biology
Biomedical engineering
Prosthetics

Bioconversion 801.2, 802.2
DT: January 1993
UF: Conversion (biological)
BT: Chemical reactions
NT: Biodegradation
RT: Biochemistry
Biogas
Biomass
Biosynthesis
Biotechnology
Nitrogen fixation

Biodegradation 461.8, 801.2
DT: January 1993
BT: Bioconversion
RT: Biochemistry
Biotechnology

Bioelectric phenomena 461.1 (701.1)
DT: January 1993
UF: Bioelectricity
Biomedical engineering—Bioelectrical phenomena*
BT: Electricity
NT: Bioelectric potentials
RT: Biology
Biomedical engineering
Electrophysiology

Bioelectric potentials 461.1 (701.1)
DT: January 1993
UF: Alpha rhythms
Auditory evoked potentials
Biomedical engineering—Bioelectric potentials*
Evoked potentials
Potentials (bioelectric)
BT: Bioelectric phenomena
RT: Biomedical engineering
Electrocardiography
Electroencephalography
Electromyography

Bioelectricity
USE: Bioelectric phenomena

Bioengineering (biomedical)
USE: Biomedical engineering

Bioengineering (human)
USE: Human engineering

Bioengineering (systems science)
USE: Systems science

Biofeedback 461.4 (731.1)
DT: January 1993
UF: Human engineering—Biofeedback*
RT: Human engineering
Sensory perception

* Former Ei Vocabulary term

Biofilms 462.5
DT: January 1993
UF: Films—Biofilms*
BT: Films

Biogas 522, 821.5
DT: January 1986
BT: Hydrocarbons
RT: Agricultural wastes
Bioconversion
Fermentation
Gas fuel manufacture
Methane
Synthetic fuels

Biographies 903
DT: January 1993
UF: Engineers—Biographies*
RT: Engineers
History

Bioindicators
USE: Biomarkers

Biological filter beds 461.8, 821.3
DT: January 1993
UF: Aquaculture—Biological filter beds*
Filter beds
BT: Filters (for fluids)
RT: Aquaculture

Biological materials 461.2
SN: Components of living organisms. For materials employed in contact with living tissue, use BIOMATERIALS
DT: January 1977
BT: Materials
NT: Amino acids
Antigens
Biological membranes
Biopolymers
Body fluids
Cells
Cholesterol
Enzymes
Metabolites
Nucleic acids
Skin
Tissue
Tooth enamel
RT: Bacteria
Bacteriology
Bioassay
Biological materials preservation
Microorganisms
Viruses

Biological materials preservation 461.2
DT: January 1993
UF: Biological materials—Preservation*
Preservation (biological materials)
RT: Biological materials

*Biological materials—Blood vessels**
USE: Blood vessels

*Biological materials—Blood**
USE: Blood

*Biological materials—Bone**
USE: Bone

*Biological materials—Cardiovascular system**
USE: Cardiovascular system

*Biological materials—Cartilage**
USE: Cartilage

*Biological materials—Cells**
USE: Cells

*Biological materials—Cholesterol**
USE: Cholesterol

*Biological materials—Chromosomes**
USE: Chromosomes

*Biological materials—DNA**
USE: DNA

*Biological materials—Genes**
USE: Genes

*Biological materials—Ligaments**
USE: Ligaments

*Biological materials—Muscle**
USE: Muscle

*Biological materials—Preservation**
USE: Biological materials preservation

*Biological materials—RNA**
USE: RNA

*Biological materials—Skin**
USE: Skin

*Biological materials—Tendons**
USE: Tendons

*Biological materials—Tissue**
USE: Tissue

*Biological materials—Tooth enamel**
USE: Tooth enamel

Biological membranes 461.2
DT: January 1993
UF: Membranes—Biological*
BT: Biological materials
Membranes
NT: Cell membranes

Biological sewage treatment 452.2
DT: January 1993
UF: Sewage treatment—Biological treatment*
BT: Sewage treatment
RT: Anaerobic digestion

Biological water treatment
445.1 (801.2) (805.1.1)
DT: January 1993
UF: Water treatment—Biological treatment*
BT: Water treatment

Biology 461.9
DT: January 1993
BT: Natural sciences

Biology *(continued)*
NT: Biochemistry
Cytology
Immunology
Marine biology
Microbiology
Physiology
RT: Biocompatibility
Biocontrol
Bioelectric phenomena
Biomedical engineering
Biotechnology
Ecology
Growth kinetics
Plants (botany)
Vision

Biomarkers (481.1) (801.2) (805.1.1)
DT: January 1995
UF: Bioindicators
RT: Biochemistry
Geology

Biomass (805.1.1) (525.1) (821.5)
DT: January 1981
BT: Materials
Renewable energy resources
RT: Bioconversion
Cellulose
Lignin
Plants (botany)
Wood
Wood fuels

Biomaterials 462.5
SN: Natural or synthetic materials used in contact with living tissue, and biological fluids for prosthetic, diagnostic, therapeutic or storage applications. For components of living organisms use BIOLOGICAL MATERIALS.
DT: January 1987
BT: Materials
NT: Blood substitutes
Blood vessel prostheses
Bone cement
Dental materials
Heart valve prostheses
Implants (surgical)
Intraocular lenses
RT: Biocompatibility
Grafts
Hip prostheses
Joint prostheses
Myoelectrically controlled prosthetics
Pacemakers
Prosthetics

*Biomaterials—Blood substitutes**
USE: Blood substitutes

Biomechanics 461.3 (931.1)
DT: October 1975
BT: Mechanics
NT: Biped locomotion
Mastication
Respiratory mechanics
RT: Anthropometry
Biomedical engineering
Cadaveric experiments
Joints (anatomy)
Muscle
Musculoskeletal system
Sports medicine

*Biomechanics—Anthropometry**
USE: Anthropometry

*Biomechanics—Biped locomotion**
USE: Biped locomotion

*Biomechanics—Cadaveric experiments**
USE: Cadaveric experiments

*Biomechanics—Gait analysis**
USE: Gait analysis

*Biomechanics—Joints**
USE: Joints (anatomy)

*Biomechanics—Mastication**
USE: Mastication

*Biomechanics—Musculoskeletal systems**
USE: Musculoskeletal system

*Biomechanics—Respiratory mechanics**
USE: Respiratory mechanics

*Biomechanics—Sports**
USE: Sports medicine

Biomedical engineering 461.1
SN: Application of engineering to biomedical practice and clinical research
DT: Predates 1975
UF: Bioengineering (biomedical)
Biomedical optics
BT: Engineering
RT: Acupuncture
Allergies
Anesthesiology
Angiocardiography
Angiography
Biocommunications
Biocontrol
Bioelectric phenomena
Bioelectric potentials
Biology
Biomechanics
Biomedical equipment
Bionics
Biotechnology
Cadaveric experiments
Cardiology
Cardiovascular surgery
Computer aided diagnosis
Computerized tomography
Cryosurgery
Cryotherapy
Cytology
Dentistry
Diagnosis
Disease control
Drug therapy

Biomedical engineering *(continued)*
Echocardiography
Electrocardiography
Electroencephalography
Electromyography
Electronic medical equipment
Electrophysiology
Electrosurgery
Electrotherapeutics
Endocrinology
Endoscopy
Environmental engineering
Fetal monitoring
Fracture fixation
Gastroenterology
Gynecology
Hemodynamics
Human engineering
Hyperthermia therapy
Hypothermia
Implants (surgical)
Interferons
Living systems studies
Magnetic resonance imaging
Malaria control
Medical applications
Medicine
Microcirculation
Neonatal monitoring
Neurology
Neurophysiology
Neurosurgery
Noninvasive medical procedures
Nuclear medicine
Obstetrics
Oncology
Ophthalmology
Orthopedics
Patient monitoring
Patient rehabilitation
Patient treatment
Pediatrics
Physical therapy
Physiological models
Prosthetics
Respiratory therapy
Sterilization (cleaning)
Surgery
Thermography (imaging)
Transplantation (surgical)
Urology
X ray laboratories

*Biomedical engineering—Allergies**
USE: Allergies

*Biomedical engineering—Anesthesia**
USE: Anesthesiology

*Biomedical engineering—Angiocardiography**
USE: Angiocardiography

*Biomedical engineering—Angiography**
USE: Angiography

*Biomedical engineering—Bioelectric potentials**
USE: Bioelectric potentials

*Biomedical engineering—Bioelectrical phenomena**
USE: Bioelectric phenomena

*Biomedical engineering—Cadaveric experiments**
USE: Cadaveric experiments

*Biomedical engineering—Cardiology**
USE: Cardiology

*Biomedical engineering—Cardiovascular surgery**
USE: Cardiovascular surgery

*Biomedical engineering—Computer aided diagnosis**
USE: Computer aided diagnosis

*Biomedical engineering—Cryosurgery**
USE: Cryosurgery

*Biomedical engineering—Cryotherapy**
USE: Cryotherapy

*Biomedical engineering—Cytology**
USE: Cytology

*Biomedical engineering—Dentistry**
USE: Dentistry

*Biomedical engineering—Diagnosis**
USE: Diagnosis

*Biomedical engineering—Drug therapy**
USE: Drug therapy

*Biomedical engineering—Echocardiography**
USE: Echocardiography

*Biomedical engineering—Electrocardiography**
USE: Electrocardiography

*Biomedical engineering—Electroencephalography**
USE: Electroencephalography

*Biomedical engineering—Electromyography**
USE: Electromyography

*Biomedical engineering—Electronics**
USE: Electronic medical equipment

*Biomedical engineering—Electrophysiology**
USE: Electrophysiology

*Biomedical engineering—Electrosurgery**
USE: Electrosurgery

*Biomedical engineering—Endocrinology**
USE: Endocrinology

*Biomedical engineering—Endoscopy**
USE: Endoscopy

*Biomedical engineering—Fetal monitoring**
USE: Fetal monitoring

* Former Ei Vocabulary term

*Biomedical engineering—Fracture fixation**
USE: Fracture fixation

*Biomedical engineering—Gastroenterology**
USE: Gastroenterology

*Biomedical engineering—Gynecology**
USE: Gynecology

*Biomedical engineering—Hemodynamics**
USE: Hemodynamics

*Biomedical engineering—Hyperthermia therapy**
USE: Hyperthermia therapy

*Biomedical engineering—Hypothermia**
USE: Hypothermia

*Biomedical engineering—Living systems studies**
USE: Living systems studies

*Biomedical engineering—Microcirculation**
USE: Microcirculation

*Biomedical engineering—Neonatal monitoring**
USE: Neonatal monitoring

*Biomedical engineering—Neurology**
USE: Neurology

*Biomedical engineering—Neurophysiology**
USE: Neurophysiology

*Biomedical engineering—Neurosurgery**
USE: Neurosurgery

*Biomedical engineering—Noninvasive procedures**
USE: Noninvasive medical procedures

*Biomedical engineering—Obstetrics**
USE: Obstetrics

*Biomedical engineering—Oncology**
USE: Oncology

*Biomedical engineering—Ophthalmology**
USE: Ophthalmology

*Biomedical engineering—Orthopedics**
USE: Orthopedics

*Biomedical engineering—Patient monitoring**
USE: Patient monitoring

*Biomedical engineering—Patient rehabilitation**
USE: Patient rehabilitation

*Biomedical engineering—Patient treatment**
USE: Patient treatment

*Biomedical engineering—Pediatrics**
USE: Pediatrics

*Biomedical engineering—Physiological models**
USE: Physiological models

*Biomedical engineering—Plethysmography**
USE: Plethysmography

*Biomedical engineering—Positron emission tomography**
USE: Positron emission tomography

*Biomedical engineering—Radiotherapy**
USE: Radiotherapy

*Biomedical engineering—Respiratory therapy**
USE: Respiratory therapy

*Biomedical engineering—Surgery applications**
USE: Surgery

*Biomedical engineering—Surgical implants**
USE: Implants (surgical)

*Biomedical engineering—Urology**
USE: Urology

Biomedical equipment 462.1
DT: Predates 1975
UF: Medical equipment
Medical supplies
Surgical equipment
BT: Equipment
NT: Audiometers
Biosensors
Catheters
Defibrillators
Dental equipment
Disposable biomedical equipment
Electronic medical equipment
Exercise equipment
Hemodialyzers
Hospital beds
Human rehabilitation equipment
Hyperbaric chambers
Optometers
Orthotics
Oximeters
Oxygenators
Prosthetics
Respirators
Sensory aids
Sterilizers
Stretchers
Syringes
RT: Autoclaves
Biomedical engineering
Diagnosis
Disinfectants
Medicine
Microelectrodes
Operating rooms
Patient monitoring
Patient rehabilitation
Patient treatment
Positron emission tomography
Radioisotopes
Telemetering equipment
X ray apparatus

*Biomedical equipment—Catheters**
USE: Catheters

* Former Ei Vocabulary term

*Biomedical equipment—Defibrillators**
USE: Defibrillators

*Biomedical equipment—Disposable**
USE: Disposable biomedical equipment

*Biomedical equipment—Exercisers**
USE: Exercise equipment

*Biomedical equipment—Hemodialyzers**
USE: Hemodialyzers

*Biomedical equipment—Microelectrodes**
USE: Microelectrodes

*Biomedical equipment—Microspheres**
USE: Controlled drug delivery

*Biomedical equipment—Oximeters**
USE: Oximeters

*Biomedical equipment—Oxygenators**
USE: Oxygenators

*Biomedical equipment—Plastics**
USE: Plastics applications

*Biomedical equipment—Radionuclides**
USE: Radioisotopes

*Biomedical equipment—Respirators**
USE: Respirators

*Biomedical equipment—Rubber**
USE: Rubber products

*Biomedical equipment—Syringes**
USE: Syringes

*Biomedical equipment—Wheelchairs**
USE: Wheelchairs

Biomedical optics
USE: Biomedical engineering AND Optics

Bionics 461.1 (731.1)
DT: January 1993
UF: Systems science and cybernetics—Bionics*
BT: Cybernetics
RT: Biomedical engineering

Biopolymers 461.2, 815.1
DT: January 1987
BT: Biological materials
Natural polymers
NT: Hormones
Nucleic acids
Proteins

Biopulping 811.1.1
SN: Pulp manufacturing by biological means
DT: January 1993
UF: Pulp manufacture—Biopulping*
BT: Pulp manufacture
RT: Biotechnology

Bioreactors 802.1, 461.8 (462.1)
SN: For use in such applications as fermentation, biochemical processes, and genetic engineering
DT: January 1987
BT: Chemical reactors
NT: Chemostats
Fermenters
RT: Biotechnology

*Bioreactors—Chemostats**
USE: Chemostats

*Bioreactors—Fermenters**
USE: Fermenters

Biosensors (801.2) (732) (461.8) (462.1)
DT: January 1987
BT: Biomedical equipment
Sensors
NT: Catheter tip sensors
Enzyme sensors
Glucose sensors
Immunosensors
Ionic drug sensors
Microbial electrodes
RT: Biotechnology
Dissolved oxygen sensors
Microelectrodes
Oximeters
Oxygen sensors

*Biosensors—Catheter tip sensors**
USE: Catheter tip sensors

*Biosensors—Ionic drug sensors**
USE: Ionic drug sensors

*Biosensors—Microbial electrodes**
USE: Microbial electrodes

Biosynthesis 461.8, 801.2, 802.2 (805.1.1)
DT: January 1987
BT: Synthesis (chemical)
RT: Bioconversion
Biotechnology
Photosynthesis

Biotechnology 461.8
SN: Use for application of biochemistry, microbiology, and biochemical and chemical engineering to the processing of materials by microorganisms, enzymes, and cells
DT: January 1986
BT: Technology
NT: Genetic engineering
RT: Bioconversion
Biodegradation
Biology
Biomedical engineering
Biopulping
Bioreactors
Biosensors
Biosynthesis
Enzyme immobilization
Fermentation
Monoclonal antibodies
Nitrification

Biped locomotion 461.3 (931.1)
DT: January 1993
UF: Biomechanics—Biped locomotion*
Locomotion (biped)
BT: Biomechanics
NT: Gait analysis

* Former Ei Vocabulary term

Bipolar integrated circuits 714.2
DT: January 1993
BT: Integrated circuits
RT: Bipolar transistors
Heterojunction bipolar transistors

Bipolar semiconductor devices 714.2
DT: January 1993
UF: Semiconductor devices, Bipolar*
BT: Semiconductor devices
NT: Bipolar transistors

Bipolar transistors 714.2
DT: January 1993
UF: Transistors, Bipolar*
BT: Bipolar semiconductor devices
Transistors
NT: Heterojunction bipolar transistors
RT: Bipolar integrated circuits
Field effect transistors

Biquadratic filters
USE: Active filters

Birefringence 741.1
DT: January 1993
UF: Double refraction
Extraordinary ray
Light—Birefringence*
BT: Optical properties
RT: Kerr electrooptical effect
Light
Light polarization
Magnetooptical effects
Optical rotation
Photoelasticity
Refraction
Refractive index

Bismuth 549.3
DT: January 1993
UF: Bi
Bismuth and alloys*
BT: Heavy metals
RT: Bismuth alloys
Bismuth compounds
Bismuth metallography
Bismuth metallurgy
Bismuth mines
Bismuth ore treatment
Bismuth plating

Bismuth alloys 549.3
DT: January 1993
UF: Bismuth and alloys*
BT: Heavy metal alloys
RT: Bismuth
Bismuth compounds

*Bismuth and alloys**
USE: Bismuth OR
Bismuth alloys

Bismuth compounds (804.1) (804.2)
DT: Predates 1975
BT: Heavy metal compounds
NT: Semiconducting bismuth compounds
RT: Bismuth
Bismuth alloys
Oxide superconductors

Bismuth metallography 531.2, 549.3
DT: June 1990
BT: Metallography
RT: Bismuth
Bismuth metallurgy

Bismuth metallurgy 531.1, 549.3
DT: Predates 1975
BT: Metallurgy
RT: Bismuth
Bismuth metallography
Bismuth ore treatment

Bismuth mines 504.3, 549.3
DT: January 1993
UF: Bismuth mines and mining*
BT: Mines
RT: Bismuth

*Bismuth mines and mining**
USE: Bismuth mines

Bismuth ore treatment 533.1, 549.3
DT: Predates 1975
BT: Ore treatment
RT: Bismuth
Bismuth metallurgy

Bismuth plating 539.3, 549.3
DT: Predates 1975
BT: Plating
RT: Bismuth

Bistability (optical)
USE: Optical bistability

Bistable multivibrators
USE: Flip flop circuits

Bit error rate 723.1
DT: January 1995
RT: Binary codes
Error analysis
Error correction
Error detection

Bit holders 603.2
DT: January 1993
UF: Cutting tools—Bits and holders*
BT: Cutting tools

Bits 603.2
DT: January 1993
UF: Cutting tools—Bits and holders*
Drill bits
Oil well drilling—Bits*
BT: Cutting tools
RT: Drills

Bituminized fibers 411, 819
DT: January 1993
UF: Sewers—Bituminized fiber*
BT: Bituminous materials
Synthetic fibers

* Former Ei Vocabulary term

Bituminous coal 524 (503)
DT: January 1993
UF: Soft coal
BT: Coal

Bituminous coatings 411, 813.2
DT: January 1993
UF: Protective coatings—Bituminous*
BT: Bituminous materials
Coatings

Bituminous materials 411
DT: Predates 1975
BT: Materials
NT: Asphalt
Bituminized fibers
Bituminous coatings
Bituminous paving materials
Kerogen
RT: Rubber additions
Tar

*Bituminous materials—Rubber additions**
USE: Rubber additions

Bituminous paving materials 406, 411
DT: January 1993
UF: Airport runways—Bituminous*
Bituminous runways
Roadbuilding materials—Bituminous*
Roads and streets—Bituminous*
BT: Bituminous materials
RT: Asphalt
Roadbuilding materials

Bituminous runways
USE: Bituminous paving materials AND
Airport runways

Black coatings 813.2 (741.3) (657.1)
DT: January 1993
UF: Coatings—Black coatings*
BT: Coatings
RT: Solar absorbers

Blackboards (artificial intelligence)
USE: Expert systems

Blades (turbomachinery)
USE: Turbomachine blades

Blast cleaning 606.2 (539)
DT: January 1993
UF: Metal cleaning—Blast*
BT: Metal cleaning
RT: Blast finishing

Blast cupolas
USE: Hot blast cupolas

Blast enrichment (532.2) (534.2) (533.2)
DT: January 1993
UF: Blast furnace practice—Blast enrichment*
Enrichment (blast)
BT: Blast furnace practice

Blast finishing 606.2 (539)
DT: January 1993
UF: Metal finishing—Blast*
BT: Metal finishing
RT: Blast cleaning

Blast furnace practice 532.2, 534.2
DT: Predates 1975
BT: Iron ore treatment
NT: Blast enrichment
Burden (metallurgy)
RT: Blast furnaces
Fuel injection
Slags
Tapping (furnace)

*Blast furnace practice—Blast enrichment**
USE: Blast enrichment

*Blast furnace practice—Burden**
USE: Burden (metallurgy)

*Blast furnace practice—Fuel injection**
USE: Fuel injection

Blast furnace slag
USE: Slags

Blast furnaces 532.2
DT: Predates 1975
BT: Metallurgical furnaces
NT: Low shaft blast furnaces
RT: Blast furnace practice
Cupolas
Oxygen blast enrichment

*Blast furnaces—Blowers**
USE: Blowers

*Blast furnaces—Charging equipment**
USE: Charging equipment (furnaces)

*Blast furnaces—Low shaft**
USE: Low shaft blast furnaces

*Blast furnaces—Stoves**
USE: Stoves

Blast resistance 408.1 (412) (421)
DT: January 1993
UF: Concrete construction—Blast resistance*
Structural design—Blast resistance*
BT: Mechanical properties
RT: Blasting
Explosions
Structural design

Blasting (405.2) (502.1)
SN: Detonation of explosives
DT: Predates 1975
RT: Blast resistance
Construction
Dredging
Explosions
Mining
Trenching

Bleached pulp 811.1
DT: January 1993
UF: Pulp—Bleached*
BT: Pulp
RT: Bleaching

* Former Ei Vocabulary term

Bleaching 811.1.1 (802.2)
DT: January 1993
UF: Papermaking—Bleaching*
Pulp manufacture—Bleaching*
Textiles—Bleaching*
BT: Processing
RT: Bleached pulp
Papermaking
Pulp manufacture
Textile finishing
Textile processing

Bleeding systems 619.2.1
DT: January 1993
UF: Oil tanks—Bleeding systems*
BT: Oil tanks

Blending 802.3
DT: January 1993
UF: Coal preparation—Blending*
Petroleum products—Blending*
Plastics—Blending*
BT: Mixing
RT: Coal preparation
Petroleum products
Plastics
Polymer blends
Textile blends

Blends (fibers)
USE: Textile blends

Blends (plastic)
USE: Polymer blends

Blimps
USE: Airships

Block codes 723.1
DT: January 1995
BT: Codes (symbols)
RT: Coding errors
Encoding (symbols)
Error correction
Error detection

Block copolymers 815.1
DT: January 1977
UF: Block polymers
Diblock copolymers
BT: Copolymers
RT: Terpolymers

Block gages
USE: Gage blocks

Block polymers
USE: Block copolymers

Blood 461.2
DT: January 1993
UF: Biological materials—Blood*
BT: Body fluids
Cardiovascular system
RT: Blood gas analysis
Blood substitutes
Hemoglobin
Physiology

Blood dialysis equipment
USE: Hemodialyzers

Blood gas analysis 461.2, 801.2
DT: January 1993
UF: Biochemistry—Blood gas analysis*
BT: Chemical analysis
RT: Biochemistry
Blood
Hemoglobin oxygen saturation
Partial pressure sensors
Sensors

Blood substitutes 462.5
DT: January 1993
UF: Artificial blood
Biomaterials—Blood substitutes*
Fluorocarbon based blood substitutes
Hemoglobin based blood substitutes
Synthetic blood
BT: Biomaterials
RT: Blood

Blood vessel prostheses 462.4
DT: January 1993
UF: Artificial blood vessels
Prosthetics—Blood vessel prostheses*
Synthetic blood vessels
BT: Biomaterials
Prosthetics
RT: Blood vessels

Blood vessels 461.2
DT: January 1993
UF: Biological materials—Blood vessels*
BT: Cardiovascular system
RT: Blood vessel prostheses

Blooming mills 535.1.1
DT: January 1993
UF: Rolling mill practice—Bloom*
BT: Hot rolling mills
RT: Billet mills
Slab mills

Blooms (metal) 535.1.2
DT: January 1993
BT: Billets (metal bars)
RT: Bars (metal)
Ingots
Plate metal
Rolling mill practice
Rolling mills

Blow molding 816.1
DT: January 1993
UF: Blowing
Plastics—Blow molding*
BT: Plastics molding

Blowers 618.3
DT: Predates 1975
UF: Blast furnaces—Blowers*
Furnaces—Blowers*
Power plants—Blowers*
Wind tunnels—Blowers*
BT: Fans

* Former Ei Vocabulary term

Blowers *(continued)*
NT: Soot blowers
RT: Compressors
Power plants
Refrigerating machinery
Superchargers
Turbomachinery
Ventilation
Wind tunnels

Blowing
USE: Blow molding

Blowing agents 803, 816.1, 818.3.1
DT: January 1993
UF: Foaming agents
Plastics, Foamed—Blowing agents*
Plastics—Blowing agents*
BT: Agents
RT: Foamed plastics
Foamed rubber
Plastics
Rubber

Blowout preventers 914.1 (511.2) (512.2)
DT: January 1993
UF: Oil field equipment—Blowout preventers*
BT: Oil field equipment
RT: Blowout prevention
Blowouts
Safety devices

Blowout prevention 914.1 (511.2) (512.2)
DT: January 1993
UF: Oil well drilling—Blowout prevention*
BT: Accident prevention
RT: Blowout preventers
Blowouts
Oil well drilling

Blowouts 914.1 (511.2) (512.2)
DT: January 1993
UF: Natural gas wells—Blowout*
BT: Accidents
RT: Blowout preventers
Blowout prevention
Natural gas wells
Oil wells

Blueprints 745
DT: January 1993
RT: Copying
Drafting practice
Photocopying

Boards (printed circuits)
USE: Printed circuit boards

Boat building
USE: Shipbuilding

Boat construction
USE: Shipbuilding

Boat equipment 674.1
DT: January 1993
UF: Boats—Equipment*
BT: Boats
Vehicle parts and equipment
NT: Boat instruments
Fenders (boat)
Masts (boat)
RT: Ship equipment

Boat fenders
USE: Fenders (boat)

Boat instruments 674.1
DT: January 1993
BT: Boat equipment
Instruments

Boat lifts 407.2, 434.1
DT: January 1993
UF: Canals—Boat lifts*
BT: Equipment
RT: Canals

Boat outfitting
USE: Outfitting (water craft)

Boat stabilizers
USE: Stabilizers (marine vessel)

Boatbuilding
USE: Shipbuilding

Boats 674.1
DT: Predates 1975
UF: Inflatable boats
BT: Water craft
NT: Barges
Boat equipment
Ferry boats
Fire boats
Fishing vessels
Lifeboats
Motor boats
Pilot boats
Submersibles
Tugboats
Yachts
RT: Air cushion vehicles
Docking
Hulls (ship)
Marine engineering
Marine insurance
Mooring
Naval architecture
Production platforms
Sailing vessels
Ships
Single point mooring
Waterway transportation

*Boats—Aluminum applications**
USE: Aluminum

*Boats—Communication equipment**
USE: Telecommunication equipment

*Boats—Constructing and outfitting**
USE: Outfitting (water craft) OR
Shipbuilding

*Boats—Diesel**
USE: Diesel engines

* Former Ei Vocabulary term

*Boats—Docking**
USE: Docking

*Boats—Engines**
USE: Marine engines

*Boats—Equipment**
USE: Boat equipment

*Boats—Fenders**
USE: Fenders (boat)

*Boats—Inflatable**
USE: Inflatable equipment

*Boats—Masts**
USE: Masts (boat)

*Boats—Stabilizers**
USE: Stabilizers (marine vessel)

*Boats—Steam**
USE: Steamships

*Boats—Unloading**
USE: Unloading

*Boats—Winches**
USE: Winches

Bobbins 819.6
DT: January 1993
UF: Textile machinery—Bobbins*
BT: Textile machinery

BOD
USE: Biochemical oxygen demand

Bode diagrams 731.1
DT: January 1993
UF: Control systems—Bode diagrams*
RT: Amplification
Control system analysis
Control system synthesis
Frequency response
Graphic methods
System stability

Bodies of revolution 931.1 (631.1)
DT: January 1977
UF: Flow of fluids—Bodies of revolution*
BT: Geometry
NT: Cylinders (shapes)
Spheres
RT: Cones
Flow of fluids

Body fluids 461.2 (801.2)
DT: January 1993
UF: Biochemistry—Glucose in body fluids*
Biochemistry—Minerals in the body fluids*
Minerals in body fluids
BT: Biological materials
NT: Blood
RT: Biochemistry

Bogeys
USE: Bogies (railroad rolling stock)

Bogies (railroad rolling stock) 682.1 (682.1.1) (682.1.2)
DT: January 1993
UF: Bogeys
Cars—Bogies*
Locomotives—Bogies*
Railroad rolling stock—Bogies*
BT: Railroad rolling stock
RT: Axles
Locomotives
Wheels

Boiler circulation 614.2
DT: January 1993
UF: Boilers—Circulation*
Circulation (boiler)

Boiler codes 614.1, 902.2
DT: Predates 1975
BT: Codes (standards)
RT: Boilers
Laws and legislation

Boiler control 614.1, 731.2
DT: Predates 1975
BT: Control
RT: Boiler control instruments

Boiler control instruments 614.1, 732.2
DT: January 1993
UF: Boiler control—Instruments*
BT: Boilers
Control equipment
RT: Boiler control

*Boiler control—Combustion**
USE: Combustion

*Boiler control—Instruments**
USE: Boiler control instruments

*Boiler control—Water level**
USE: Water levels AND
Level control

Boiler corrosion 539.1, 614.2
DT: January 1993
UF: Boiler corrosion and deposits*
BT: Corrosion
RT: Boiler deposits
Boilers
Slags

*Boiler corrosion and deposits**
USE: Boiler corrosion OR
Boiler deposits

*Boiler corrosion and deposits—Slag**
USE: Slags

Boiler deposits 539.1, 614.2
DT: January 1993
UF: Boiler corrosion and deposits*
BT: Deposits
RT: Boiler corrosion
Boilers

* Former Ei Vocabulary term

Boiler firing 614.2
DT: Predates 1975
UF: Firing (boilers)
NT: Low grade fuel firing
Multiple fuel firing
RT: Combustion
Fueling
Pressurization
Stokers

*Boiler firing—Coal**
USE: Coal fired boilers

*Boiler firing—Coke**
USE: Coke fired boilers

*Boiler firing—Low grade fuel**
USE: Low grade fuel firing

*Boiler firing—Multiple fuel**
USE: Multiple fuel firing

*Boiler firing—Pressurization**
USE: Pressurization

*Boiler firing—Pulverized fuel**
USE: Pulverized fuel fired boilers

Boilers 614
SN: Use for steam-generating units heated by combustion of fuel
DT: Predates 1975
UF: Reboilers
Water boilers
BT: Steam generators
NT: Boiler control instruments
Coal fired boilers
Coke fired boilers
Cyclone boilers
Fire tube boilers
Gas fired boilers
High pressure boilers
Marine boilers
Oil fired boilers
Packaged boilers
Pulverized fuel fired boilers
Steam accumulators
Water tube boilers
RT: Air preheaters
Boiler codes
Boiler corrosion
Boiler deposits
Feedwater analysis
Feedwater heaters
Feedwater pumps
Feedwater treatment
Furnaces
Heating
Power plants
Pressurization
Safety valves
Soot blowers
Steam
Steam condensate
Steam condensers
Steam engines
Steam turbines
Stokers
Superheaters
Water levels

*Boilers—Circulation**
USE: Boiler circulation

*Boilers—Cyclone**
USE: Cyclone boilers

*Boilers—Electric heating**
USE: Electric heating

*Boilers—Fire tube**
USE: Fire tube boilers

*Boilers—Gas firing**
USE: Gas fired boilers

*Boilers—High pressure**
USE: High pressure boilers

*Boilers—Marine**
USE: Marine boilers

*Boilers—Packaged**
USE: Packaged boilers

*Boilers—Slag tap**
USE: Slags

*Boilers—Waste heat**
USE: Waste heat

*Boilers—Water tube**
USE: Water tube boilers

Boiling
USE: Boiling liquids

Boiling liquids 641.2 (631.1.1)
DT: January 1993
UF: Boiling
Flow of fluids—Boiling fluids*
Heat transfer—Boiling liquids*
BT: Liquids
NT: Nucleate boiling
RT: Heat transfer
Vaporization

Boiling water reactors 621.1 (932.2)
DT: January 1993
UF: BWR reactors
Nuclear reactors, Boiling water*
BT: Water cooled reactors

Bolometers 944.7
SN: Used to measure thermal radiation
DT: Predates 1975
BT: Radiometers
RT: Heat radiation
Infrared detectors
Standing wave meters
Temperature measurement
Temperature measuring instruments
Thermal variables measurement

Bolt tightening 605
DT: January 1993
UF: Bolts and nuts—Tightening*
Tightening (bolts)
RT: Bolts

* Former Ei Vocabulary term

Bolted joints 408.2
DT: January 1993
UF: Joints—Bolted*
BT: Joints (structural components)
RT: Bolts

Bolts 605
DT: January 1993
UF: Bolts and nuts*
BT: Fasteners
RT: Bolt tightening
Bolted joints
Cold heading
Nuts (fasteners)
Washers

*Bolts and nuts**
USE: Bolts OR
Nuts (fasteners)

*Bolts and nuts—Cold heading**
USE: Cold heading

*Bolts and nuts—Tightening**
USE: Bolt tightening

Bolus alba
USE: Kaolin

Bombers 404.1, 652.1.2
DT: January 1993
UF: Aircraft—Bomber*
BT: Military aircraft
NT: Bombsights
RT: Bombing

Bombing 404.1 (652.1.2)
DT: January 1993
UF: Bombs and bombing*
BT: Military operations
RT: Bombers
Bombs (ordnance)

Bombs (ordnance) 404
DT: January 1993
UF: Bombs and bombing*
BT: Ordnance
RT: Bombing

*Bombs and bombing**
USE: Bombing OR
Bombs (ordnance)

*Bombs and bombing—Sights**
USE: Bombsights

Bombsights 404.1
DT: January 1993
UF: Bombs and bombing—Sights*
BT: Aircraft instruments
Bombers
RT: Range finders

Bond (masonry) 405.2, 412
DT: January 1993
UF: Concrete reinforcements—Bond*
BT: Adhesion
RT: Bonding

Bond strength (chemical) 801.4
DT: January 1995
RT: Chemical bonds

Bond strength (materials) (421)
DT: January 1995
BT: Strength of materials
RT: Adhesion
Bonding

Bonding (538.1) (802.3) (813)
DT: Predates 1975
NT: Explosive bonding
Glass bonding
Gluing
Laminating
Roll bonding
Soldering
RT: Adhesion
Adhesive joints
Adhesives
Bond (masonry)
Bond strength (materials)
Cementing (shafts)
Cladding (coating)
Curing
Grouting
Joints (structural components)
Sealing (closing)
Well cementing

*Bonding—Dissimilar materials**
USE: Dissimilar materials

Bonds (chemical)
USE: Chemical bonds

Bone 461.2
DT: January 1993
UF: Biological materials—Bone*
BT: Musculoskeletal system
Tissue
RT: Orthopedics

Bone cement 462.5, 803
DT: January 1993
UF: Adhesives—Bone cement*
Cement (bone)
Dental equipment and supplies—Bone cement*
BT: Adhesives
Biomaterials
RT: Dental materials

Boolean algebra 721.1, 921.1
DT: January 1993
UF: Boolean lattices
Computer metatheory—Boolean algebra*
BT: Algebra
NT: Boolean functions
RT: Computation theory
Formal logic
Set theory

Boolean functions 721.1, 921.1
DT: January 1993
UF: Computer metatheory—Boolean functions*

* Former Ei Vocabulary term

Boolean functions *(continued)*
BT: Boolean algebra
Formal logic
Functions
RT: Function evaluation
Switching functions

Boolean lattices
USE: Boolean algebra

Booms (spill retention)
USE: Oil booms

Booster rockets
USE: Boosters (rocket)

Boosters (electric generators)
USE: Electric exciters

Boosters (rocket) 654.2
DT: January 1993
UF: Booster rockets
Rocket boosters
Rocket engines—Boosters*
BT: Rocket engines

Borate minerals 482.2
DT: January 1993
UF: Mineralogy—Borates*
BT: Boron compounds
Minerals
RT: Boron

Boreholes (501.1) (511.1)
DT: Predates 1975
NT: Deflected boreholes
Exploratory boreholes
Offshore boreholes
RT: Excavation
Fishing (oil wells)
Mine shafts
Well drilling
Well logging
Wells

*Boreholes—Circulating media**
USE: Circulating media

*Boreholes—Deflected**
USE: Deflected boreholes

*Boreholes—Diamond drilling**
USE: Diamond drilling

*Boreholes—Exploratory**
USE: Exploratory boreholes

*Boreholes—Fishing**
USE: Fishing (oil wells)

*Boreholes—Logging**
USE: Well logging

*Boreholes—Offshore**
USE: Offshore boreholes

Boride coatings 813.2 (812.1)
DT: January 1993
BT: Coatings
RT: Borides

Borides 804.2, 812.1 (539.2.2)
DT: Predates 1975
UF: Protective coatings—Borides*
BT: Boron compounds
RT: Boride coatings
Boron

Boriding 537.1
DT: January 1993
UF: Heat treatment—Boriding*
BT: Hardening
Heat treatment
RT: Case hardening
Spark hardening

Boring 604.2
DT: Predates 1975
RT: Boring tools
Cutting
Excavation
Machining
Perforating
Piercing
Reaming
Turning

Boring machines (earth boring)
USE: Earth boring machines

Boring machines (machine tools) 603.1
DT: January 1993
BT: Machine tools
RT: Drilling machines (machine tools)
Lathes

Boring tools 603.1
DT: Predates 1975
BT: Cutting tools
NT: Augers
RT: Boring
Machine tools

Boron 549.3
DT: Predates 1975
UF: B
BT: Metalloids
NT: Semiconducting boron
RT: Borate minerals
Borides
Boron deposits
Borophosphate glass
Borosilicate glass

Boron carbide 804.2, 812.1
DT: Predates 1975
BT: Boron compounds
Carbides

Boron compounds (804.1) (804.2)
DT: Predates 1975
BT: Chemical compounds
NT: Borate minerals
Borides
Boron carbide
Cubic boron nitride
RT: Borophosphate glass
Borosilicate glass

* Former Ei Vocabulary term

Boron deposits 504.1
DT: Predates 1975
BT: Ore deposits
RT: Boron

Borophosphate glass 812.3
DT: January 1993
UF: Glass—Borophosphate*
BT: Glass
RT: Boron
Boron compounds
Phosphates

Borosilicate glass 812.3
DT: January 1993
UF: Glass—Borosilicate*
BT: Glass
RT: Boron
Boron compounds
Silicates

Bottle washing machines 694.3
DT: Predates 1975
BT: Washing machines
RT: Bottles

Bottles 694.2
DT: Predates 1975
BT: Containers
NT: Dewars
Glass bottles
Plastic bottles
RT: Bottle washing machines
Bottling plants

*Bottles—Glass**
USE: Glass bottles

*Bottles—Plastics**
USE: Plastic bottles

Bottling plants 822.1(402.1)
DT: Predates 1975
BT: Food products plants
RT: Beverages
Bottles

Bottom hole pressure 512
DT: January 1993
UF: Oil wells—Bottom hole pressure*
BT: Pressure
RT: Oil wells

Bottoming cycle systems 614
DT: January 1993
UF: Cogeneration plants—Bottoming systems*
BT: Cogeneration plants

Boundaries (metal insulator)
USE: Metal insulator boundaries

Boundaries (semiconductor metal)
USE: Semiconductor metal boundaries

Boundaries (semiconductor)
USE: Semiconductor insulator boundaries

Boundary conditions (921)
DT: January 1995
RT: Differential equations

Boundary element method 921.6
DT: January 1993
UF: Mathematical techniques—Boundary element method*
BT: Numerical analysis
RT: Approximation theory
Boundary value problems
Difference equations
Differential equations
Finite element method

Boundary layer flow 631.1 (931.1)
DT: January 1993
UF: Flow of fluids—Boundary layer*
BT: Flow of fluids
RT: Boundary layers
Wall flow

Boundary layers (631.1) (641.2) (651.1)
DT: January 1993
UF: Aerodynamics—Boundary layer*
Heat transfer—Boundary layer*
RT: Aerodynamics
Boundary layer flow
Heat transfer

Boundary value problems (921.6)
DT: January 1993
UF: Mathematical techniques—Boundary value problems*
BT: Differential equations
RT: Boundary element method
Finite element method
Green's function
Numerical analysis
Numerical methods
Partial differential equations

Bow thrusters 674.1
SN: Rudders in the bow of a ship
DT: January 1993
UF: Ships—Bow thrusters*
BT: Ship steering equipment

Bowing
USE: Bending (forming)

Box girder bridges 401.1
DT: January 1993
UF: Bridges, Box girder*
BT: Bridges
RT: Beams and girders

Boxing (packaging)
USE: Packaging

Br
USE: Bromine

Braces (dental)
USE: Dental braces

Braces (for limbs and joints) 462.4
DT: January 1993
UF: Orthotics—Braces*
BT: Orthotics

Bragg cells 741.3, 752.1
DT: January 1993
UF: Acoustooptic modulators
BT: Acoustooptical devices
RT: Ultrasonic devices

Braille keyboards
USE: Computer keyboard substitutes (handicapped aid)

Brain 461.1
DT: January 1993
RT: Brain models
Electroencephalography
Neurology
Neurophysiology
Neurosurgery

Brain models 461.1 (723.4)
DT: January 1993
UF: Systems science and cybernetics—Brain models*
BT: Cybernetics
RT: Artificial intelligence
Brain
Learning systems
Neural networks

Brake hoses 602, 619.1
DT: January 1993
BT: Brakes
Hose

Brake linings 602
DT: January 1993
UF: Brakes—Lining*
BT: Brakes
Linings

Brakes 602
DT: Predates 1975
BT: Equipment
NT: Air brakes
Brake hoses
Brake linings
Electric brakes
Hydraulic brakes
RT: Braking
Clutches
Final control devices
Friction materials
Wheels

Brakes (bending)
USE: Bending brakes

*Brakes, Air**
USE: Air brakes

*Brakes, Electric**
USE: Electric brakes

*Brakes, Hydraulic**
USE: Hydraulic brakes

*Brakes—Hose couplings**
USE: Hose fittings

*Brakes—Lining**
USE: Brake linings

Brakes/Bending
USE: Bending brakes

Braking 602
DT: January 1993
UF: Automobile engines—Braking*
Cranes—Braking*
Electric machinery—Braking*
RT: Brakes
Control
Thrust reversal

Brass 544.2, 546.3
DT: Predates 1975
UF: Welding—Brass*
BT: Copper alloys
Zinc alloys
RT: Brass plating

Brass foundry practice
USE: Copper foundry practice

Brass plating 539.3, 544.2, 546.3
DT: Predates 1975
BT: Plating
RT: Brass

Brayton cycle 641.1
DT: January 1993
UF: Thermodynamics—Brayton cycle*
BT: Thermodynamics
RT: Gas turbines

Brazing 538.1.1
DT: Predates 1975
BT: Soldering
NT: Electric brazing
Electron beam brazing
Vacuum brazing
RT: Brazing filler metals
Fluxes
Welding

Brazing filler metals 538.1.1
DT: January 1993
UF: Brazing—Filler metals*
Brazing—Silver alloy fillers*
BT: Filler metals
RT: Brazing

*Brazing—Dissimilar base metals**
USE: Dissimilar metals

*Brazing—Electric**
USE: Electric brazing

*Brazing—Electron beam**
USE: Electron beam brazing

*Brazing—Filler metals**
USE: Brazing filler metals

*Brazing—Silver alloy fillers**
USE: Brazing filler metals AND
Silver alloys

*Brazing—Vacuum**
USE: Vacuum brazing

Breakage (coal)
USE: Coal breakage

Breakdown voltage
USE: Electric breakdown

Breakup (liquid drops)
USE: Drop breakup

Breakwaters 407.1
DT: Predates 1975
BT: Hydraulic structures
NT: Floating breakwaters
RT: Coastal engineering
Jetties
Oceanography
Shore protection

*Breakwaters—Floating**
USE: Floating breakwaters

*Breakwaters—Pneumatic**
USE: Pneumatic equipment

Breath controlled devices 461.5
DT: January 1993
UF: Human rehabilitation engineering—Breath-controlled devices*
BT: Human rehabilitation equipment
RT: Human rehabilitation engineering

Breeder reactors 621.1 (932.2)
DT: January 1993
UF: Nuclear reactors, Breeder*
BT: Nuclear reactors
RT: Breeding blankets

Breeding blankets 621.1 (932.2)
DT: January 1993
UF: Nuclear reactors—Breeding blankets*
BT: Nuclear reactors
RT: Breeder reactors

Breweries 822.1 (402.1)
DT: January 1986
BT: Food products plants
RT: Beverages
Distilleries

Brick 414.2
DT: Predates 1975
BT: Masonry materials
NT: Lime brick
Silica brick
RT: Brick buildings
Brick construction
Brickmaking
Refractory materials

Brick buildings 402, 414
DT: January 1993
UF: Buildings—Brick*
BT: Buildings
RT: Brick

Brick construction 405.2, 414.2
DT: Predates 1975
BT: Masonry construction
RT: Brick

*Brick—Efflorescence**
USE: Efflorescence

*Brick—Permeability**
USE: Mechanical permeability

*Brick—Waterproofing**
USE: Waterproofing

Brickmaking 414.1
DT: Predates 1975
BT: Manufacture
RT: Brick
Fly ash
Kilns

*Brickmaking—Fly ash**
USE: Fly ash

Bridge abutments
USE: Abutments (bridge)

Bridge approaches 401.1, 403
DT: January 1993
UF: Bridges—Approaches*
BT: Bridge components

Bridge bearings 401.1, 408.2
DT: January 1993
UF: Bridges—Bearings*
BT: Bearings (structural)
Bridge components
RT: Supports

Bridge cables 401.1, 408.2
DT: January 1993
UF: Bridges—Cables*
BT: Bridge components
Cables
RT: Cable stayed bridges

Bridge circuits (703.1) (713.5)
DT: January 1993
UF: Electronic circuits, Bridge*
BT: Networks (circuits)

Bridge clearances 401.1
DT: January 1993
UF: Bridges—Clearances*
Clearances (bridge)
RT: Bridges

Bridge components 401.1, 408.2
DT: January 1993
BT: Bridges
Components
NT: Abutments (bridge)
Bridge approaches
Bridge bearings
Bridge cables
Bridge decks
Bridge piers

Bridge conveyors 692.1
DT: January 1993
UF: Conveyors—Bridge*
BT: Conveyors

* Former Ei Vocabulary term

Bridge cranes 405.1, 693.1
DT: January 1993
UF: Cranes—Bridge*
BT: Cranes

Bridge decks 401.1
DT: January 1993
UF: Bridge floors
Bridges—Decks*
Decks (bridge)
Floors (bridge)
BT: Bridge components

Bridge floors
USE: Bridge decks

Bridge piers 401.1
DT: Predates 1975
UF: Piers (bridges)
BT: Bridge components
RT: Supports

*Bridge piers—Scour**
USE: Scour

Bridges 401.1
DT: Predates 1975
BT: Structures (built objects)
NT: Aluminum bridges
Arch bridges
Box girder bridges
Bridge components
Cantilever bridges
Composite bridges
Footbridges
Highway bridges
Military bridges
Movable bridges
Plate girder bridges
Prefabricated bridges
Railroad bridges
Steel bridges
Suspension bridges
Temporary bridges
Toll bridges
Wooden bridges
RT: Bridge clearances
Causeways
Culverts
Hydraulic structures
Load limits
Railings
Transportation
Urban planning
Widening (transportation arteries)

Bridges (electric measurement)
USE: Electric measuring bridges

*Bridges, Aluminum**
USE: Aluminum bridges

*Bridges, Arch**
USE: Arch bridges

*Bridges, Bascule**
USE: Bascule bridges

*Bridges, Box girder**
USE: Box girder bridges

*Bridges, Cable stayed**
USE: Cable stayed bridges

*Bridges, Cantilever**
USE: Cantilever bridges

*Bridges, Composite**
USE: Composite bridges

*Bridges, Concrete**
USE: Concrete bridges

*Bridges, Highway**
USE: Highway bridges

*Bridges, Lift**
USE: Lift bridges

*Bridges, Masonry**
USE: Masonry bridges

*Bridges, Military**
USE: Military bridges

*Bridges, Movable**
USE: Movable bridges

*Bridges, Plate girder**
USE: Plate girder bridges

*Bridges, Railroad**
USE: Railroad bridges

*Bridges, Steel**
USE: Steel bridges

*Bridges, Suspension**
USE: Suspension bridges

*Bridges, Swing**
USE: Swing bridges

*Bridges, Wood**
USE: Wooden bridges

*Bridges—Abutments**
USE: Abutments (bridge)

*Bridges—Approaches**
USE: Bridge approaches

*Bridges—Bearings**
USE: Bridge bearings

*Bridges—Cables**
USE: Bridge cables

*Bridges—Clearances**
USE: Bridge clearances

*Bridges—Decks**
USE: Bridge decks

*Bridges—Expansion joints**
USE: Expansion joints

*Bridges—Inspection equipment**
USE: Inspection equipment

*Bridges—Joints**
USE: Joints (structural components)

*Bridges—Load limits**
USE: Load limits

*Bridges—Luminous railings**
USE: Luminous railings

*Bridges—Pipeline crossings**
USE: Crossings (pipe and cable) AND
Pipelines

*Bridges—Prefabricated**
USE: Prefabricated bridges

*Bridges—Prestressed materials**
USE: Prestressed materials

*Bridges—Raising**
USE: Movable bridges

*Bridges—Reconstruction**
USE: Reconstruction (structural)

*Bridges—Supports**
USE: Supports

*Bridges—Temporary**
USE: Temporary bridges

*Bridges—Widening**
USE: Widening (transportation arteries)

Brillouin scattering 741.1
DT: January 1993
UF: Brillouin spectra
Light—Brillouin scattering*
Mandelstam-Brillouin scattering
BT: Light scattering

Brillouin spectra
USE: Brillouin scattering

Brines (444) (804.2)
DT: January 1987
UF: Geothermal wells—Brines*
Oil fields—Brines*
BT: Saline water
RT: Geothermal wells
Oil fields
Water wells

Briquets 524
SN: Also use material, e.g., COKE, if appropriate
DT: January 1993
UF: Coal briquets and briquetting*
Lignite—Briquets*
BT: Materials
RT: Briquetting
Fuels
Lignite
Sintered alumina

Briquetting 802.3, 524
DT: January 1993
UF: Coal briquets and briquetting*
Coke—Briquetting*
Peat—Briquetting*
BT: Compaction
RT: Briquets
Densification
Pelletizing
Preforming
Pressing (forming)

Brittle fracture (421)
DT: January 1995
BT: Fracture
RT: Brittleness
Ductility

Brittle materials
USE: Brittleness

Brittleness (421)
DT: January 1995
UF: Brittle materials
BT: Mechanical properties
RT: Brittle fracture
Ductility
Embrittlement
Hardness
Plasticity
Porosity
Toughness

Broaches 603.2
DT: Predates 1975
BT: Cutting tools
RT: Broaching
Broaching machines

Broaching 604.2
DT: Predates 1975
BT: Cutting
RT: Broaches
Broaching machines
Machining
Metal cutting
Piercing

Broaching machines 603.1
DT: Predates 1975
BT: Machine tools
RT: Broaches
Broaching

Broadband amplifiers 713.1
DT: January 1993
UF: Amplifiers, Broadband*
BT: Amplifiers (electronic)

Broadband ISDN
USE: Broadband networks AND
Voice/data communication systems

Broadband networks (716) (717) (718)
DT: January 1995
UF: Broadband ISDN
Wideband networks
BT: Telecommunication networks
RT: Asynchronous transfer mode
Local area networks
Visual communication

Broadcasting (716.3) (716.4)
SN: Very general term; prefer specific type of broadcasting
DT: Predates 1975
UF: Rediffusion*

* Former Ei Vocabulary term

Broadcasting *(continued)*
BT: Telecommunication services
NT: Radio broadcasting
Stereophonic broadcasting
Television broadcasting
RT: Broadcasting antennas

Broadcasting antennas (716.3) (716.4)
DT: January 1993
UF: Antennas—Broadcasting*
BT: Antennas
RT: Broadcasting

Broadcasting stations
USE: Radio stations

Bromine 804.2
DT: Predates 1975
UF: Br
BT: Halogen elements
RT: Bromine compounds

Bromine compounds (804.1) (804.2)
DT: January 1986
BT: Halogen compounds
RT: Bromine

Bronze 544.2, 546.2
DT: Predates 1975
UF: Shipbuilding materials—Bronze*
BT: Copper alloys
Tin alloys
RT: Bronze foundry practice
Bronze plating

Bronze foundry practice 534.2, 544.2, 546.2
DT: Predates 1975
BT: Foundry practice
RT: Bronze

Bronze plating 539.3, 544.2, 546.2
DT: Predates 1975
BT: Plating
RT: Bronze

Brown coal
USE: Lignite

Brownian motion
USE: Brownian movement

Brownian movement 801.3 (931)
DT: January 1993
UF: Brownian motion
Colloids—Brownian movement*
BT: Random processes
RT: Colloids
Dispersions
Kinetic theory of gases
Suspensions (fluids)

Brush conductors 705.1 (704)
DT: January 1993
UF: Electric conductors, Brush*
BT: Electric current collectors
RT: Electric commutators
Electric contacts
Electric machinery components

Brushes 605
SN: For brushes as electric current collectors, use BRUSH CONDUCTORS
DT: Predates 1975
BT: Tools

BTU content
USE: Calorific value

Bubble chambers 944.7 (932)
DT: Predates 1975
BT: Particle detectors
RT: Ionization chambers
Spark chambers

Bubble columns 802.1
DT: January 1993
UF: Chemical equipment—Bubble columns*
BT: Chemical equipment
RT: Bubbles (in fluids)

Bubble formation 631.1.2 (931.1)
DT: January 1993
UF: Flow of fluids—Bubble formation*
Liquids—Bubble formation*
BT: Fluid mechanics
RT: Bubbles (in fluids)
Flow of fluids
Liquids

Bubble memories
USE: Magnetic bubble memories

Bubbles (in fluids) 631.1.2 (931.1)
DT: January 1993
RT: Bubble columns
Bubble formation
Cavitation
Flow visualization
Fluid mechanics
Liquids

Bubbles (magnetic)
USE: Magnetic bubbles

Buchholz relays
USE: Relay protection

Buckling (408.2) (421) (931)
DT: January 1993
UF: Beams and girders—Buckling*
BT: Failure (mechanical)
RT: Deformation
High temperature testing
Strain
Torsional stress

Buckminsterfullerene
USE: Fullerenes

Buckyballs
USE: Fullerenes

Budget control (911)
DT: Predates 1975
UF: Budgeting
BT: Control
Management
RT: Industrial management

Budgeting
USE: Budget control

Buffer amplifiers 713.1
DT: January 1993
UF: Amplifiers, Buffer*
BT: Amplifiers (electronic)
RT: Buffer circuits

Buffer circuits (703.1) (713.1)
DT: January 1993
UF: Electronic circuits, Buffer*
BT: Networks (circuits)
RT: Buffer amplifiers

Buffer storage 722.1
DT: January 1995
UF: Cache memories
RT: Random access storage

Buffers (railroad cars)
USE: Railroad car buffers

Buffeting 651.1
DT: January 1993
UF: Aerodynamics—Buffeting*
RT: Aerodynamics
Atmospheric turbulence

Buffing 604.2 (539)
DT: January 1993
BT: Finishing
RT: Buffing machines
Grinding (machining)
Metal finishing
Polishing

Buffing machines (603.1) (605.1)
DT: Predates 1975
BT: Machine tools
RT: Buffing
Hand tools
Polishing machines

Building codes 402, 902.2, 902.3 (403)
DT: Predates 1975
BT: Codes (standards)
RT: Electric codes
Elevator codes
Laws and legislation
Urban planning
Ventilation codes
Welding codes

Building components 408.2 (402)
DT: January 1993
BT: Buildings
NT: Building signal systems
Building wiring
Ceilings
Chimneys
Doors
Facades
Facings
Interiors (building)
Partitions (building)
Stairs
Sun hoods
Walls (structural partitions)
Window screens
Windows
RT: Insulation
Lighting
Railings
Roofs
Structural members

Building materials (411) (412) (413) (414) (415)
SN: Materials used in construction of all types of structures and works, not just buildings
DT: Predates 1975
UF: Construction materials
Structural materials
BT: Materials
NT: Cements
Concrete blocks
Light weight building materials
Masonry materials
Particle board
Plastic building materials
Prestressed materials
Roadbuilding materials
Shipbuilding materials
Structural ceramics
Structural metals
Stucco
Tile
RT: Buildings
Gravel
Marble
Plaster
Plywood
Sealants
Slate
Structural design
Structures (built objects)
Synthetic marble
Wood laminates

*Building materials—Aluminum**
USE: Aluminum

*Building materials—Composite materials**
USE: Composite materials

*Building materials—Copper**
USE: Copper

*Building materials—Finishing**
USE: Finishing

*Building materials—Flammability tests**
USE: Flammability testing

*Building materials—Light weight**
USE: Light weight building materials

*Building materials—Plastics**
USE: Plastic building materials

*Building materials—Rubber**
USE: Rubber applications

*Building materials—Sealants**
USE: Sealants

* Former Ei Vocabulary term

Building moving 402
DT: January 1993
UF: Buildings—Moving*
BT: Moving
RT: Buildings
Construction

Building signal systems 402 (715.2) (914.1)
DT: January 1993
UF: Buildings—Signal systems*
BT: Building components
Signal systems

Building wiring 402 (704) (715)
DT: January 1993
UF: Electric wiring, Buildings*
BT: Building components
Electric wiring
RT: Electric codes

Buildings 402
DT: Predates 1975
BT: Structures (built objects)
NT: Airport buildings
Arctic buildings
Brick buildings
Building components
Concrete buildings
Exhibition buildings
Farm buildings
Filling stations
Fire houses
Grain elevators
Hospitals
Housing
Intelligent buildings
Lighthouses
Office buildings
Religious buildings
School buildings
Solar buildings
Store buildings
Tall buildings
Tropical buildings
Underground buildings
Wooden buildings
RT: Architectural acoustics
Architecture
Building materials
Building moving
Climate control
Construction
Demolition
Facilities
Hurricane effects
Hurricane resistance
Industrial plants
Loads (forces)
Model buildings
Modernization
Settlement of structures
Site selection
Structural design
Underpinnings
Wind stress

*Buildings—Aluminum applications**
USE: Aluminum

*Buildings—Arctic**
USE: Arctic buildings

*Buildings—Brick**
USE: Brick buildings

*Buildings—Ceilings**
USE: Ceilings

*Buildings—Climate control**
USE: Climate control

*Buildings—Concrete**
USE: Concrete buildings

*Buildings—Demolition**
USE: Demolition

*Buildings—Depreciation**
USE: Depreciation

*Buildings—Doors**
USE: Doors

*Buildings—Drive-in facilities**
USE: Drive-in facilities

*Buildings—Dynamic loads**
USE: Dynamic loads

*Buildings—Facings**
USE: Facings

*Buildings—Floors**
USE: Floors

*Buildings—Gas supply**
USE: Gas supply

*Buildings—Hurricane effects**
USE: Hurricane effects

*Buildings—Lift slab construction**
USE: Lift slab construction

*Buildings—Lighting**
USE: Lighting

*Buildings—Modernization**
USE: Modernization

*Buildings—Moving**
USE: Building moving

*Buildings—Partitions**
USE: Partitions (building)

*Buildings—Prefabricated**
USE: Prefabricated construction

*Buildings—Railings**
USE: Railings

*Buildings—Refuse incinerators**
USE: Refuse incinerators

*Buildings—Roofs**
USE: Roofs

*Buildings—Settlement**
USE: Settlement of structures

* Former Ei Vocabulary term

*Buildings—Shear walls**
USE: Shear walls

*Buildings—Signal systems**
USE: Building signal systems

*Buildings—Solar, Active**
USE: Active solar buildings

*Buildings—Solar, Passive**
USE: Passive solar buildings

*Buildings—Sound insulation**
USE: Sound insulation

*Buildings—Stairs**
USE: Stairs

*Buildings—Standby power systems**
USE: Standby power systems

*Buildings—Sun hoods**
USE: Sun hoods

*Buildings—Tall**
USE: Tall buildings

*Buildings—Tropics**
USE: Tropical buildings

*Buildings—Underground**
USE: Underground buildings

*Buildings—Underpinning**
USE: Underpinnings

*Buildings—Walls**
USE: Walls (structural partitions)

*Buildings—Waterproofing**
USE: Waterproofing

*Buildings—Welded steel**
USE: Welded steel structures

*Buildings—Wind stresses**
USE: Wind stress

*Buildings—Wood**
USE: Wooden buildings

Bulb turbines 632.2
DT: January 1993
UF: Hydraulic turbines—Bulb*
BT: Hydraulic turbines

Bulk acoustic wave devices
USE: Acoustic bulk wave devices

Bulk modulus
USE: Elastic moduli

Bulkheads (retaining walls) 407.1
DT: January 1993
UF: Port structures—Bulkheads*
BT: Port structures

Bumpers (automobile)
USE: Automobile bumpers

Bundled conductors 706.2
DT: January 1993
UF: Electric lines—Bundled conductors*
BT: Electric conductors
RT: Electric lines

Buoyancy 931.2 (631)
DT: January 1995
RT: Density (specific gravity)
Fluid mechanics
Porosity

Buoys (434) (471) (472)
DT: Predates 1975
BT: Marine signal systems

Burden (metallurgy) 532.2, 534.2
DT: January 1993
UF: Blast furnace practice—Burden*
BT: Blast furnace practice

Burners
USE: Fuel burners

Burning
USE: Combustion

Burnishing (539) (604.2) (606.2)
DT: January 1993
UF: Metal finishing—Burnishing*
BT: Finishing
RT: Metal finishing

Burst disks
USE: Rupture disks

Bursting disks
USE: Rupture disks

Bus conductors
USE: Busbars

Bus drivers 432.2, 912.4
DT: January 1993
UF: Drivers (bus)
Motor bus drivers*
BT: Transportation personnel
RT: Bus transportation
Buses
Driver training

Bus fleet operation
USE: Bus transportation

Bus garages 402.1, 432.2
DT: January 1993
UF: Garages—Motor bus*
BT: Garages (parking)
RT: Bus terminals
Bus transportation

* Former Ei Vocabulary term

Bus reactors
USE: Current limiting reactors

Bus terminals 402.1, 432.2
DT: January 1993
UF: Motor bus terminals*
Terminals (bus)
BT: Facilities
RT: Bus garages
Bus transportation
Buses

Bus transportation 432.2
DT: January 1993
UF: Bus fleet operation
Motor bus transportation*
BT: Motor transportation
RT: Bus drivers
Bus garages
Bus terminals
Buses

Busbars (704) (706.2)
DT: January 1993
UF: Bus conductors
Electric busbars
Electric conductors, Busbar*
BT: Electric conductors

Buses 663.1
DT: January 1993
UF: Motor buses*
BT: Ground vehicles
RT: Bus drivers
Bus terminals
Bus transportation
Trackless trolleys

Bushings 704.1
DT: January 1993
UF: Electric insulators—Bushings*
BT: Components
RT: Bearings (structural)
Electric insulators
Shafts (machine components)
Supports

Business data processing
USE: Administrative data processing

Business machines (722) (912.2)
DT: Predates 1975
BT: Machinery
NT: Automatic teller machines
Cash registers
Dictating machines
Point of sale terminals
Typewriters
RT: Computers
Data processing
Office equipment

*Business machines—Automated teller machines**
USE: Automatic teller machines

*Business machines—Cash registers**
USE: Cash registers

*Business machines—Point of sale terminals**
USE: Point of sale terminals

Butadiene 804.1 (818.2)
DT: Predates 1975
BT: Hydrocarbons
RT: Monomers
Olefins
Petrochemicals
Synthetic rubber

Butane 804.1 (521)
DT: January 1986
BT: Paraffins
RT: Fuels
Gas condensates
Liquefied petroleum gas
Natural gas

Butenes 804.1 (815.1.1)
DT: January 1986
UF: Butylenes
BT: Olefins

Butt welding 538.2.1
DT: January 1993
UF: Welding—Butt*
BT: Welding

Buttress dams 441.1
DT: January 1993
UF: Dams, Buttress*
BT: Dams

Butylenes
USE: Butenes

BWR reactors
USE: Boiling water reactors

By-products
USE: Byproducts

Byproducts 804
DT: January 1993
UF: By-products
BT: Materials
NT: Coal byproducts
RT: Industrial plants

C (element)
USE: Carbon

C (programming language) 723.1.1
DT: January 1993
UF: Computer programming languages—C*
BT: High level languages
RT: Computer operating systems
UNIX

* Former Ei Vocabulary term

C steel
USE: Carbon steel

C3 systems
USE: Command and control systems

Ca
USE: Calcium

Cabinetry
USE: Woodworking

Cabins (aircraft) 652.1
DT: January 1993
UF: Aircraft cabins
Aircraft—Cabins*
BT: Aircraft parts and equipment

Cable cores (535)
DT: January 1993
UF: Cores (cable)
Wire rope—Cores*
BT: Cables
RT: Wire rope

Cable ducts
USE: Electric conduits

Cable jointing 706.2
SN: Electric cables
DT: January 1993
UF: Cable splicing
Electric cables—Jointing*
Jointing (cables)
Splicing (cable)
BT: Joining
RT: Electric cables
Electric conductors
Electric connectors
Optical cables

Cable laying
USE: Electric cable laying

Cable sheathing 706.2
DT: January 1993
UF: Electric cables—Sheathing*
Sheathing (cable)
RT: Cable shielding
Electric cables
Electric conduits

Cable shielding 706.2
DT: January 1993
UF: Electric cables—Shielding*
BT: Electric shielding
RT: Cable sheathing
Electric cables

Cable ships 671, 706.2
DT: January 1993
BT: Ships
RT: Electric cable laying
Submarine cables

Cable splicing
USE: Cable jointing

Cable stayed bridges 401.1
DT: January 1993
UF: Bridges, Cable stayed*
BT: Cantilever bridges
RT: Bridge cables

Cable supported roofs 402, 408.2
DT: January 1993
UF: Roofs—Cable supported*
BT: Roofs
RT: Cables

Cable suspended roofs 402, 408.2
DT: January 1993
UF: Roofs—Cable suspended*
BT: Roofs
RT: Cables

Cable taping 706.2
DT: January 1993
UF: Electric cables—Taping*
Taping (cable)
RT: Electric cables
Electric tapes

Cable television systems 716.4
DT: January 1993
UF: CATV
Television systems, Cable*
BT: Television systems
RT: Coaxial cables
Subscription television
Television
Television equipment
Television networks

Cable terminations
USE: Electric connectors

Cables (535) (716.2)
DT: January 1993
BT: Wire products
NT: Anchor cables
Bridge cables
Cable cores
Electric cables
Elevator cables
Mooring cables
Tetherlines
Wire rope
RT: Cable supported roofs
Cable suspended roofs
Rope
Wire

* Former Ei Vocabulary term

Cableways 692.2
SN: Transporting devices. For devices used to transport electric conductors use ELECTRIC CONDUITS
DT: Predates 1975
BT: Materials handling equipment

Cabooses 682.1.1
DT: January 1993
UF: Cars—Caboose*
BT: Railroad cars

Cabs (truck) 663.1
DT: January 1993
UF: Motor trucks—Cabs*
BT: Trucks

Cache memories
USE: Buffer storage

CAD
USE: Computer aided design

Cadaveric experiments 461.1
DT: January 1993
UF: Biomechanics—Cadaveric experiments*
Biomedical engineering—Cadaveric experiments*
BT: Experiments
RT: Biomechanics
Biomedical engineering

Cadmium 549.3
DT: January 1993
UF: Cadmium and alloys*
Cd (element)
BT: Heavy metals
Transition metals
RT: Cadmium alloys
Cadmium compounds
Cadmium metallography
Cadmium metallurgy
Cadmium ore treatment
Cadmium plating

Cadmium alloys 549.3
DT: January 1993
UF: Cadmium and alloys*
BT: Heavy metal alloys
Transition metal alloys
RT: Cadmium
Cadmium compounds

*Cadmium and alloys**
USE: Cadmium OR
Cadmium alloys

Cadmium compounds (804.1) (804.2)
DT: Predates 1975
BT: Heavy metal compounds
Transition metal compounds
NT: Semiconducting cadmium compounds
RT: Cadmium
Cadmium alloys

Cadmium metallography 531.2, 549.3
DT: Predates 1975
BT: Metallography
RT: Cadmium
Cadmium metallurgy

Cadmium metallurgy 531.1, 549.3
DT: Predates 1975
BT: Metallurgy
RT: Cadmium
Cadmium metallography
Cadmium ore treatment

Cadmium ore treatment 533.1, 549.3
DT: Predates 1975
BT: Ore treatment
RT: Cadmium
Cadmium metallurgy

Cadmium plating 539.3, 549.3
DT: Predates 1975
BT: Plating
RT: Cadmium

Cadmium sulfide solar cells
702.3 (657.1) (714.2)
DT: January 1993
UF: Solar cells—Cadmium sulfide*
BT: Solar cells
RT: Semiconducting cadmium compounds

CAE
USE: Computer aided engineering

Cages (mine hoists) 502.1, 693
DT: January 1993
UF: Mine hoists—Cages*
RT: Materials handling equipment
Mine hoists

CAI
USE: Computer aided instruction

Caissons 405.1 (401.2)
SN: Watertight chambers used to protect underwater construction workers from water pressure
DT: Predates 1975
BT: Construction equipment
Hydraulic structures
RT: Tunnels
Underwater construction

* Former Ei Vocabulary term

Calcification (biochemistry) 801.2, 802.2
DT: January 1993
UF: Biochemistry—Calcification*
BT: Chemical reactions
RT: Biochemistry
Calcium

Calcination 802.3
DT: January 1993
BT: Chemical operations
RT: Pyrolysis

Calcite 482.2, 804.2 (414.2)
DT: Predates 1975
BT: Calcium compounds
Carbonates
Minerals

Calcium 549.2
DT: January 1993
UF: Ca
Calcium and alloys*
BT: Alkaline earth metals
RT: Calcification (biochemistry)
Calcium alloys
Calcium ion sensors

Calcium alloys 549.2
DT: January 1993
UF: Calcium and alloys*
BT: Alkaline earth metal alloys
RT: Calcium
Calcium compounds

*Calcium and alloys**
USE: Calcium OR
Calcium alloys

Calcium carbide 804.2, 812.1
DT: Predates 1975
BT: Calcium compounds
Carbides

Calcium compounds (804.1) (804.2)
DT: Predates 1975
BT: Alkaline earth metal compounds
NT: Calcite
Calcium carbide
Fluorspar
Hydrated lime
Lime
Perovskite
RT: Calcium alloys

Calcium hydroxide
USE: Hydrated lime

Calcium ion sensors 462.1, 801.2
DT: January 1993
UF: Sensors—Calcium ion sensors*
BT: Chemical sensors
RT: Calcium

Calcium oxide
USE: Lime

Calculating
USE: Calculations

Calculations 921 (721) (723)
DT: January 1993
UF: Calculating
Computation
NT: Transmission network calculations
RT: Mathematical techniques

Calendering 811.1.1, 816.1
DT: January 1993
UF: Paper—Calendering*
Plastics—Calendering*
BT: Finishing
RT: Calenders
Papermaking
Plastics machinery
Pressing (forming)

Calenders 811.1.2, 816.2
DT: January 1993
UF: Papermaking machinery—Calenders*
Plastics machinery—Calenders*
BT: Machinery
RT: Calendering
Papermaking machinery
Plastics machinery

Calibration (902.2) (941) (942) (943) (944)
DT: January 1993
BT: Standardization
RT: Frequency standards
Inspection
Instrument errors
Instruments
Measurements
Standards

Californium 622.1
DT: June 1990
UF: Cf
BT: Transuranium elements

Calorific value (521.1) (522) (523) (524)
DT: January 1993
UF: BTU content
Fuels—Calorific value*
BT: Physical properties
Thermodynamic properties
RT: Calorimetry
Combustion
Fuels
High temperature testing

Calorimeters 944.5 (801)
SN: For measurement of the quantity of heat
DT: Predates 1975
BT: Instruments
RT: Calorimetry
Temperature measuring instruments

*Calorimeters—Solar**
USE: Solar equipment

Calorimetry 944.6
DT: January 1981
BT: Measurements
NT: Differential scanning calorimetry
RT: Calorific value

* Former Ei Vocabulary term

Calorimetry *(continued)*
Calorimeters
Thermoanalysis

CAM
USE: Computer aided manufacturing

Camera films
USE: Photographic films

Camera lenses 742.2
DT: January 1993
UF: Cameras—Lenses*
Photographic lenses
BT: Cameras
Optical instrument lenses

Camera shutters 742.2
DT: January 1993
UF: Cameras—Shutters*
Shutters (camera)
BT: Cameras
Optical shutters

Camera tubes (television)
USE: Television camera tubes

Cameras 742.2
DT: Predates 1975
UF: Photogrammetry—Cameras*
Rockets and missiles—Cameras*
BT: Optical instruments
Photographic equipment
NT: Camera lenses
Camera shutters
Exposure controls
High speed cameras
Motion picture cameras
Streak cameras
Temperature indicating cameras
Ultrasonic cameras
Underground cameras
Underwater cameras
Video cameras
X ray cameras
RT: Exposure meters
Focusing
Optical filters
Optical shutters
Photogrammetry
Photography
Range finders

*Cameras—Attachments**
USE: Photographic accessories

*Cameras—Exposure control**
USE: Exposure controls

*Cameras—High speed**
USE: High speed cameras

*Cameras—Lenses**
USE: Camera lenses

*Cameras—Shutters**
USE: Camera shutters

*Cameras—Temperature indicating**
USE: Temperature indicating cameras

*Cameras—Ultrasonic**
USE: Ultrasonic cameras

*Cameras—Underground**
USE: Underground cameras

*Cameras—Underwater**
USE: Underwater cameras

Camouflage 404.1
DT: January 1993
UF: Military engineering—Camouflage*
RT: Military engineering
Military equipment
Military operations

Camping trailers
USE: Light trailers

Cams 601.3
DT: Predates 1975
BT: Machine components
RT: Actuators
Camshafts
Nonelectric final control devices

*Cams—Noise**
USE: Acoustic noise

Camshafts 601.2
DT: January 1993
UF: Internal combustion engines—Camshafts*
BT: Machine components
RT: Cams

Canal gates 407.2, 434.1
DT: January 1993
UF: Canals—Gates*
Gates (canal)
BT: Hydraulic structures
RT: Canals
Hydraulic gates

Canal linings 407.2 (405) (434)
DT: January 1993
UF: Canals—Lining*
BT: Canals
Linings

Canals 407.2, 434.1
DT: Predates 1975
BT: Inland waterways
NT: Canal linings
Irrigation canals
RT: Boat lifts
Canal gates
Locks (on waterways)
Waterway transportation

*Canals—Boat lifts**
USE: Boat lifts

*Canals—Gates**
USE: Canal gates

*Canals—Lining**
USE: Canal linings

* Former Ei Vocabulary term

Cancer treatment
USE: Oncology

Cancer-causing agents
USE: Carcinogens

Cancer-causing viruses
USE: Oncogenic viruses

*Cane sugar**
USE: Sugar (sucrose)

Canning 822.2
DT: January 1993
UF: Food products—Canning*
BT: Food processing
RT: Containers

Cantilever bridges 401.1
DT: January 1993
UF: Bridges, Cantilever*
BT: Bridges
NT: Bascule bridges
Cable stayed bridges

Capacitance 701.1 (703.1)
DT: January 1993
UF: Electric capacity
BT: Electric properties
RT: Capacitance measurement
Capacitors
Deep level transient spectroscopy
Electric charge
Electric power factor
Electrostatics
Permittivity

Capacitance measurement 942.2
DT: January 1993
UF: Electric measurements—Capacitance*
BT: Electric variables measurement
RT: Capacitance
Capacitors
Electric measuring bridges
Electric reactance measurement

Capacitor storage 704.1 (701.1)
DT: January 1993
BT: Electric energy storage
RT: Capacitors

Capacitors 704.1
DT: Predates 1975
UF: Condensers (electric)
Electric capacitors
Electric condensers
BT: Dielectric devices
Electric reactors
Electrostatic devices
NT: Ceramic capacitors
Electrolytic capacitors
Paper capacitors
Varactors
RT: Capacitance
Capacitance measurement
Capacitor storage
Electrets
Electric charge
Preferred numbers
Radio equipment

*Capacitors, Ceramic**
USE: Ceramic capacitors

*Capacitors, Electrolytic**
USE: Electrolytic capacitors

*Capacitors, Paper**
USE: Paper capacitors

Capillarity 631.1 (931.2)
DT: Predates 1975
BT: Surface phenomena
RT: Capillary flow
Capillary tubes
Contact angle
Fluid mechanics
Percolation (fluids)
Porosity
Rheology
Surface tension
Wetting

Capillary flow 631.1
DT: January 1993
UF: Flow of fluids—Capillaries*
BT: Flow of fluids
RT: Capillarity
Capillary tubes
Hemodynamics
Microcirculation

Capillary tubes 619.1 (644.3)
DT: Predates 1975
UF: Refrigerating machinery—Capillary tubes*
BT: Tubes (components)
RT: Capillarity
Capillary flow
Refrigerating machinery

Caps (container)
USE: Container closures

Car ferries
USE: Ferry boats

Carbamide
USE: Urea

Carbide cutting tools 603.2, 605.1
DT: January 1993
UF: Cutting tools—Carbide*
BT: Carbide tools
Cutting tools
RT: Carbide dies
Carbides

Carbide dies 603.2
DT: January 1993
UF: Dies—Carbide*
BT: Carbide tools
Dies
RT: Carbide cutting tools
Carbides

* Former Ei Vocabulary term

Carbide minerals 482.2
DT: January 1993
UF: Mineralogy—Carbides*
BT: Carbides
Minerals

Carbide tools 603.2
DT: January 1993
UF: Tools, jigs and fixtures—Carbide*
BT: Tools
NT: Carbide cutting tools
Carbide dies
RT: Carbides
Milling cutters
Taps

Carbides 804.2, 812.1
DT: Predates 1975
BT: Carbon inorganic compounds
NT: Boron carbide
Calcium carbide
Carbide minerals
Silicon carbide
Sintered carbides
Tantalum carbide
Titanium carbide
Tungsten carbide
Uranium carbide
RT: Carbide cutting tools
Carbide dies
Carbide tools

*Carbides—Sintered**
USE: Sintered carbides

Carbohydrates 804.1
DT: Predates 1975
BT: Organic compounds
NT: Polysaccharides
Sugars
RT: Food products

*Carbohydrates—Glucose**
USE: Glucose

Carbon 804
DT: Predates 1975
UF: C (element)
BT: Nonmetals
NT: Activated carbon
Carbon black
Carbon fibers
Charcoal
Diamonds
Fullerenes
Graphite
RT: Carbon carbon composites
Carbon fiber reinforced plastics
Carbon inorganic compounds
Carbon steel
Carbonaceous adsorbents
Carbonaceous refractories
Carbonaceous shale
Carbonate minerals
Carbonization
Coal
Coke
Decarbonization
Decarburization
Organic compounds
Soot

Carbon bisulfide
USE: Carbon disulfide

Carbon black 803 (818.3.1)
DT: Predates 1975
BT: Carbon
RT: Soot

Carbon carbon composites 415.4
DT: June 1990
BT: Composite materials
RT: Carbon
Carbon fibers

Carbon dioxide 804.2 (644.2)
DT: Predates 1975
UF: Refrigerants—Carbon dioxide*
BT: Carbon inorganic compounds
Oxides
RT: Air pollution
Atmospheric composition
Carbon dioxide process
Carbon dioxide recorders
Carbon dioxide arc welding
Carbon dioxide lasers
Carbonation
Greenhouse effect
Refrigerants

Carbon dioxide arc welding 538.2.1
DT: January 1993
UF: Gas metal arc welding
Welding—Carbon dioxide*
BT: Electric arc welding
RT: Carbon dioxide

Carbon dioxide gas lasers
USE: Carbon dioxide lasers

Carbon dioxide lasers 744.2
DT: January 1993
UF: Carbon dioxide gas lasers
Lasers, Carbon dioxide*
BT: Gas lasers
RT: Carbon dioxide

Carbon dioxide process 534.2
DT: January 1993
UF: Foundry practice—Carbon dioxide process*
BT: Metal casting
RT: Carbon dioxide
Foundry practice

Carbon dioxide recorders 943.3
DT: Predates 1975
BT: Recording instruments
RT: Carbon dioxide

*Carbon dioxide—Partial pressure sensors**
USE: Partial pressure sensors

Carbon disulfide 804.2
DT: Predates 1975
UF: Carbon bisulfide
BT: Organic compounds
Sulfur compounds
RT: Solvents

Carbon fiber reinforced plastics 817.1 (415.2)
DT: January 1993
UF: CFRP
Plastics, Reinforced—Carbon fiber*
BT: Fiber reinforced plastics
NT: Graphite fiber reinforced plastics
RT: Carbon
Carbon fibers

*Carbon fiber**
USE: Carbon fibers

Carbon fibers 804 (817.1)
DT: January 1993
UF: Carbon fiber*
BT: Carbon
Synthetic fibers
NT: Graphite fibers
RT: Carbon carbon composites
Carbon fiber reinforced plastics

Carbon inorganic compounds 804.2
DT: January 1993
BT: Inorganic compounds
NT: Carbides
Carbon dioxide
Carbon monoxide
Carbonates
Cyanides
RT: Carbon
Carbonaceous refractories

Carbon monoxide 804.2
DT: Predates 1975
BT: Carbon inorganic compounds
Oxides
RT: Air pollution
Synthesis gas

Carbon organic compounds
USE: Organic compounds

Carbon steel 545.3
DT: Predates 1975
UF: C steel
Eutectoid steel
Graphitic steel
Mild steel
BT: Steel
RT: Carbon

Carbon tetrachloride 803, 804.2
DT: Predates 1975
BT: Chlorine compounds
Organic compounds
RT: Solvents

*Carbon—Activated**
USE: Activated carbon

Carbonaceous adsorbents 803, 804.2 (505)
DT: January 1993
UF: Adsorbents—Carbonaceous*
BT: Adsorbents
RT: Activated carbon
Carbon

Carbonaceous refractories 812.2
DT: January 1993
UF: Refractory materials—Carbon*
BT: Refractory materials
RT: Carbon
Carbon inorganic compounds
Graphite

Carbonaceous shale 482.2, 483 (505)
DT: January 1993
UF: Shale—Carbonaceous*
BT: Shale
NT: Oil shale
RT: Carbon

Carbonate minerals 482.2
DT: January 1993
UF: Mineralogy—Carbonates*
BT: Carbonates
Minerals
NT: Magnesite
RT: Carbon
Potash

Carbonates 804.2
DT: January 1993
BT: Carbon inorganic compounds
NT: Calcite
Carbonate minerals
Potash

Carbonation 802.2
DT: January 1993
BT: Chemical reactions
RT: Carbon dioxide

Carbonitriding 537.1 (802.2)
DT: January 1993
UF: Gas cyaniding
Heat treatment—Carbonitriding*
Nicarbing
Nitrocarburizing
BT: Case hardening
RT: Nitriding

Carbonization 802.2 (445.1)
DT: Predates 1975
UF: Water treatment—Carbonization*
BT: Chemical reactions
NT: Coal carbonization
Coking
RT: Carbon
Decarbonization

Carbonylation 802.2
DT: January 1993
BT: Chemical reactions

Carboxylation 802.2
DT: January 1993
BT: Chemical reactions

* Former Ei Vocabulary term

Carboxylic acids 804.1
DT: January 1993
UF: Alkanoic acids
Alkenoic acids
Aromatic acids
BT: Organic acids
NT: Fatty acids

Carburetors 612.1.1
DT: January 1993
UF: Internal combustion engines—Carburetors*
BT: Internal combustion engines

Carburizing 537.1
SN: Excludes chemical reactions, except for metallic surface reactions
DT: January 1993
UF: Heat treatment—Carburizing*
Steel heat treatment—Carburizing*
BT: Case hardening
Heat treatment

Carcinogens 461, 804.1
DT: January 1977
UF: Cancer-causing agents
Oncogenic materials
BT: Chemicals
NT: Aflatoxins
Nitrosamines
RT: Hazardous materials
Mutagens
Oncogenic viruses
Oncology

*Carcinogens—Aflatoxins**
USE: Aflatoxins

*Carcinogens—Nitrosamines**
USE: Nitrosamines

Carding (textiles)
USE: Textile carding

Cardiology 461.6
DT: January 1993
UF: Biomedical engineering—Cardiology*
BT: Medicine
RT: Angiocardiography
Biomedical engineering
Cardiovascular system
Cardiovascular surgery
Defibrillators
Echocardiography
Electrocardiography
Pacemakers
Plethysmography

Cardiovascular surgery 461.6
DT: January 1993
UF: Biomedical engineering—Cardiovascular surgery*
BT: Surgery
RT: Biomedical engineering
Cardiology
Cardiovascular system
Pacemakers

Cardiovascular system 461.2
DT: January 1993
UF: Biological materials—Cardiovascular system*
NT: Blood vessels
Blood
RT: Cardiology
Cardiovascular surgery
Hemodynamics
Pacemakers
Plethysmography

Cargo aircraft 652.1.1 (431.3)
DT: January 1993
UF: Aircraft—Freight*
Freight aircraft
BT: Aircraft
RT: Cargo handling
Freight transportation
Transport aircraft

Cargo handling (431) (432) (433) (434) (691)
DT: January 1993
UF: Freight handling*
BT: Materials handling
NT: Door to door service (freight)
Piggyback freight handling
Refrigerated freight handling
RT: Cargo aircraft
Freight transportation

Carnot cycle 641.1
DT: January 1993
UF: Thermodynamics—Carnot cycle*
BT: Thermodynamics
RT: Stirling cycle

Carpet manufacture (819.4)
DT: January 1993
BT: Manufacture
RT: Fabrics
Linoleum
Textile processing

Carrier communication (716) (718)
DT: January 1993
UF: Transmission (carrier)
BT: Telecommunication
NT: Carrier telegraph
Carrier telephone
Carrier transmission on power lines
Diplex transmission
RT: Electric lines
Frequency hopping

Carrier concentration 701.1 (712.1) (931.3)
DT: January 1995
UF: Carrier density
Electron density
RT: Semiconductor materials

Carrier density
USE: Carrier concentration

Carrier telegraph 718.2
DT: January 1993
UF: Telegraph, Carrier*
BT: Carrier communication

* Former Ei Vocabulary term

Carrier telegraph *(continued)*
Telegraph
RT: Carrier transmission on power lines

Carrier telephone 718.1
DT: January 1993
UF: Telephone, Carrier*
BT: Carrier communication
Telephone
RT: Carrier transmission on power lines
Telephone lines

Carrier transmission on power lines 706.2 (718)
DT: January 1993
UF: Electric lines—Carrier transmission*
BT: Carrier communication
RT: Carrier telegraph
Carrier telephone
Electric lines

Carriers (charge)
USE: Charge carriers

Carry circuits
USE: Adders

Carry logic 721.2 (721.1)
DT: January 1993
UF: Computers—Carry logic*
BT: Digital arithmetic
RT: Adders
Logic circuits
Logic design

Carrying capacity (electric cables)
USE: Electric current carrying capacity (cables)

*Cars**
USE: Railroad cars

*Cars, Rail motor**
USE: Rail motor cars

*Cars—Air cushioning**
USE: Air cushioning

*Cars—Ambulance**
USE: Ambulance cars

*Cars—Axles**
USE: Axles

*Cars—Baggage**
USE: Baggage cars

*Cars—Bogies**
USE: Bogies (railroad rolling stock)

*Cars—Buffers**
USE: Railroad car buffers

*Cars—Caboose**
USE: Cabooses

*Cars—Couplings**
USE: Railroad car couplings

*Cars—Dining**
USE: Dining cars

*Cars—Dumpers**
USE: Dump cars

*Cars—Floors**
USE: Floors

*Cars—Freight**
USE: Freight cars

*Cars—Light weight**
USE: Light weight rolling stock

*Cars—Monorail**
USE: Monorail cars

*Cars—Noise**
USE: Acoustic noise

*Cars—Passenger**
USE: Passenger cars

*Cars—Postal**
USE: Postal cars

*Cars—Refrigerator**
USE: Refrigerator cars

*Cars—Riding qualities**
USE: Riding qualities

*Cars—Shock absorbers**
USE: Shock absorbers

*Cars—Springs and suspensions**
USE: Vehicle springs OR
Vehicle suspensions

*Cars—Street railroad**
USE: Trolley cars

*Cars—Subway**
USE: Subway cars

*Cars—Tank**
USE: Tank cars

*Cars—Track testing**
USE: Track test cars

*Cars—Traction**
USE: Traction (friction)

Cartilage 461.2
DT: January 1993
UF: Biological materials—Cartilage*
BT: Musculoskeletal system
Tissue

Cartography
USE: Mapping

Cascade connections 713.1
DT: January 1993
UF: Amplifiers—Cascade connection*
Tandem connections
BT: Electronic equipment

Cascade control systems 731.1
DT: January 1993
UF: Control systems, Cascade*
BT: Control systems
RT: Feedback

Cascades (fluid mechanics) 631.1
DT: January 1993
UF: Aerodynamics—Cascades*

* Former Ei Vocabulary term

Cascades (fluid mechanics) *(continued)*
Flow of fluids—Cascades*
Turbomachinery—Cascades*
BT: Fluid mechanics
RT: Airfoils
Fluid dynamics
Turbomachinery

CASE
USE: Computer aided software engineering

Case hardening 537.1 (545.3)
DT: January 1993
UF: Heat treatment—Case hardening*
Steel heat treatment—Case hardening*
BT: Hardening
Heat treatment
NT: Carbonitriding
Carburizing
RT: Boriding
Flame hardening
Nitriding
Spark hardening
Steel heat treatment

CASE tools
USE: Computer aided software engineering

Casein 804.1, 822.3
DT: January 1993
UF: Dairy products—Casein*
BT: Dairy products
Proteins
RT: Casein paint
Paint

Casein paint 804.1, 813.2
DT: January 1993
UF: Paint—Casein*
BT: Paint
RT: Casein

Cash registers (723.2)
DT: January 1993
UF: Business machines—Cash registers*
BT: Business machines
RT: Point of sale terminals

Casings (oil well)
USE: Oil well casings

Cast iron 534.2, 545.2
DT: Predates 1975
UF: Automobile materials—Cast iron*
Iron castings
Meehanite
Rolls—Cast iron*
Shipbuilding materials—Cast iron*
Welding—Iron castings*
BT: Iron alloys
Metal castings
NT: Malleable iron castings
Nodular iron
RT: Cast iron pipe
Graphitization
Iron
Metal casting

Cast iron pipe 534.2, 545.2, 619.1
DT: January 1993
UF: Cast iron pipelines
Pipe, Cast iron*
Pipelines, Cast iron*
BT: Pipe
RT: Cast iron

Cast iron pipelines
USE: Pipelines AND
Cast iron pipe

*Cast iron—Aluminum content**
USE: Aluminum

*Cast iron—Copper content**
USE: Copper

*Cast iron—Corrosion resisting**
USE: Corrosion resistant alloys

*Cast iron—Graphitization**
USE: Graphitization

*Cast iron—Hydrogen content**
USE: Hydrogen

*Cast iron—Lead content**
USE: Lead

*Cast iron—Nodular**
USE: Nodular iron

*Cast iron—Silicon content**
USE: Silicon

*Cast iron—Sulfur content**
USE: Sulfur

*Cast iron—Tin content**
USE: Tin

*Cast iron—Titanium content**
USE: Titanium

*Cast iron—Tungsten content**
USE: Tungsten

*Cast iron—Vanadium content**
USE: Vanadium

Casting 534.2
DT: January 1993
UF: Foundry practice—Casting*
Materials casting
Materials—Casting*
Refractory materials—Casting*
BT: Forming
NT: Centrifugal casting
Full mold process
Melt spinning
Metal casting
Plastics casting
RT: Foundry practice
Materials
Metal castings
Molding
Patternmaking
Rapid solidification
Tapping (furnace)

* Former Ei Vocabulary term

Castings 534.2
SN: Scope changed to castings in general from metal castings, for which use METAL CASTINGS
DT: Predates 1975
NT: Metal castings
Plastic castings

*Castings**
USE: Metal castings

CAT scans
USE: Computerized tomography

Catalysis 802.2
DT: Predates 1975
UF: Heterogeneous catalysis
Molecular catalysis
BT: Chemical reactions
NT: Fischer-Tropsch synthesis
RT: Catalysts
Catalytic converters
Catalytic cracking
Electropolymerization
Friedel-Crafts reaction
Petroleum refining
Reaction kinetics
Thermodynamics

Catalyst activity 803, 804 (802.2)
DT: January 1993
UF: Activity (catalysts)
Catalysts—Activity*
RT: Catalyst deactivation
Catalyst poisoning
Catalyst regeneration
Catalyst selectivity
Catalysts

Catalyst deactivation 803, 804 (802.2)
DT: January 1993
UF: Catalysts—Deactivation*
RT: Catalyst activity
Catalyst poisoning
Catalyst regeneration
Catalysts

Catalyst poisoning 803, 804 (802.2)
DT: January 1993
UF: Catalysts—Poisoning*
Poisoning (catalysts)
RT: Catalyst activity
Catalyst deactivation
Catalyst regeneration
Catalysts

Catalyst regeneration 803, 804 (802.2)
DT: January 1993
UF: Catalysts—Regeneration*
RT: Catalyst activity
Catalyst deactivation
Catalyst poisoning
Catalysts

Catalyst selectivity 803, 804 (802.2)
DT: January 1993
UF: Catalysts—Selectivity*
Selectivity (catalyst)
RT: Catalyst activity
Catalysts

Catalyst supports 803, 804
DT: January 1993
UF: Catalysts—Supported*
Supports (catalyst)
RT: Catalysts

Catalysts 803, 804
DT: Predates 1975
UF: Powder metal products—Catalytic*
BT: Agents
NT: Biocatalysts
RT: Activated alumina
Catalysis
Catalyst activity
Catalyst deactivation
Catalyst poisoning
Catalyst regeneration
Catalyst selectivity
Catalyst supports
Catalytic converters
Gas fuel manufacture
Initiators (chemical)
Powder metals
Reaction kinetics

*Catalysts—Activity**
USE: Catalyst activity

*Catalysts—Deactivation**
USE: Catalyst deactivation

*Catalysts—Poisoning**
USE: Catalyst poisoning

*Catalysts—Regeneration**
USE: Catalyst regeneration

*Catalysts—Selectivity**
USE: Catalyst selectivity

*Catalysts—Supported**
USE: Catalyst supports

Catalytic converters 612.1.1 (451.2)
DT: January 1993
UF: Internal combustion engines—Catalytic converters*
BT: Internal combustion engines
RT: Air pollution control equipment
Catalysis
Catalysts

Catalytic cracking 802.2 (513.1) (522)
DT: January 1993
UF: Gas manufacture—Catalytic cracking process*
BT: Cracking (chemical)
NT: Fluid catalytic cracking
Hydrocracking
RT: Catalysis
Gas fuel manufacture
Petroleum refining
Reforming reactions

Cataphoresis
USE: Electrophoresis

Catapults (aircraft launchers) 672.2 (404.1) (652.1.2)
DT: January 1993
UF: Aircraft carriers—Catapults*
Aircraft launchers
Launchers (aircraft)
BT: Aircraft carriers

Catheter tip sensors 462.1 (801)
DT: January 1993
UF: Biosensors—Catheter tip sensors*
BT: Biosensors
Catheters

Catheters 462.1
DT: January 1993
UF: Biomedical equipment—Catheters*
BT: Biomedical equipment
NT: Catheter tip sensors

Cathode followers 713.1
DT: January 1993
UF: Amplifiers, Cathode follower*
BT: Amplifiers (electronic)

Cathode ray oscilloscopes 942.2 (715)
DT: January 1993
UF: CRO
Digital storage oscilloscopes
Oscilloscopes (cathode ray)
Oscilloscopes, Cathode ray*
BT: Electric measuring instruments
RT: Cathode ray tubes
Cathode rays
Display devices
Oscillographs

Cathode ray tube screens
USE: Fluorescent screens

Cathode ray tubes 714.1
DT: January 1993
UF: CRT
Electron tubes, Cathode ray*
BT: Thermionic tubes
NT: Image converters
Image storage tubes
Television camera tubes
Television picture tubes
RT: Cathode ray oscilloscopes
Cathode rays
Display devices
Electron beams
Electron guns
Electron optics
Fluorescent screens
Magnetic lenses
Radar displays
Television equipment

Cathode rays (701.1) (741.1) (931.3) (932.1)
DT: Predates 1975
RT: Cathode ray oscilloscopes
Cathode ray tubes
Electron beams
Electron optics
Electrons
Fluorescent screens
Focusing

Cathodes (704.1) (714.1)
DT: Predates 1975
BT: Electrodes
NT: Field emission cathodes
Photocathodes
Thermionic cathodes
RT: Cathodic protection
Cathodoluminescence
Electron emission
Electron tube components

*Cathodes, Field emission**
USE: Field emission cathodes

*Cathodes, Thermionic**
USE: Thermionic cathodes

Cathodic protection 539.2 (704)
DT: January 1993
UF: Corrosion protection, Cathodic*
Electrolytic protection
BT: Corrosion protection
RT: Cathodes
Corrosion

Cathodic sputtering
USE: Sputtering

Cathodoluminescence 701.1, 741.1
DT: Predates 1975
UF: Cathodophosphorescence
Cathodothermoluminescence
BT: Luminescence
Radiation effects
RT: Cathodes

Cathodophosphorescence
USE: Cathodoluminescence

Cathodothermoluminescence
USE: Cathodoluminescence

Cationic polymerization 815.2
DT: January 1993
UF: Polymerization—Cationic polymerization*
BT: Polymerization
RT: Ring opening polymerization

CATV
USE: Cable television systems

Cauer filters
USE: Passive filters

Causeways 401.1
DT: Predates 1975
BT: Structures (built objects)
RT: Bridges
Culverts
Hydraulic structures

Caustic soda 804.2
DT: Predates 1975
BT: Hydrogen inorganic compounds
Sodium compounds

Causticity
USE: pH

* Former Ei Vocabulary term

Caves 481.1
DT: January 1993
UF: Geology—Caves*
BT: Landforms
RT: Subsidence

Cavitation 631.1.1
DT: Predates 1975
BT: Fluid mechanics
RT: Bubbles (in fluids)
Cavitation corrosion
Erosion
Stresses
Turbomachinery
Turbulence
Turbulent flow
Vortex flow
Wakes

Cavitation corrosion 539.1 (631.1.1)
DT: Predates 1975
UF: Cavitation erosion
BT: Corrosion
Erosion
RT: Cavitation

Cavitation erosion
USE: Cavitation corrosion

Cavity resonators (714.1) (714.3)
DT: January 1993
UF: Resonant cavities
Resonators, Cavity*
BT: Resonators
NT: Acoustic resonators
RT: Klystrons
Magnetrons
Masers
Microwave filters
Natural frequencies
Oscillators (electronic)
Resonance

Cavityless casting
USE: Full mold process

CBE
USE: Chemical beam epitaxy

CCD
USE: Charge coupled devices

CCTV
USE: Closed circuit television systems

Cd (element)
USE: Cadmium

CD players
USE: Compact disk players

CD-ROM 722.1
SN: In 1993 and 1994 CD/ROM was used
DT: January 1995
UF: CD/ROM
BT: Compact disks
Optical disk storage
RT: Compact disk players

CD/ROM
USE: CD-ROM

Ce
USE: Cerium

Ceilings 402
DT: January 1993
UF: Buildings—Ceilings*
BT: Building components

Cell culture (461.8) (801.2) (805.1.1)
DT: January 1987
UF: Culture (cells)
NT: Animal cell culture
Batch cell culture
Continuous cell culture
Plant cell culture
Tissue culture
RT: Cell immobilization
Cells
Growth kinetics

*Cell culture—Animal**
USE: Animal cell culture

*Cell culture—Batch**
USE: Batch cell culture

*Cell culture—Continuous**
USE: Continuous cell culture

*Cell culture—Growth kinetics**
USE: Growth kinetics

*Cell culture—Immobilization**
USE: Cell immobilization

*Cell culture—Plant**
USE: Plant cell culture

*Cell culture—Tissue**
USE: Tissue culture

Cell immobilization (461.8) (801.2) (805.1.1)
DT: January 1993
UF: Cell culture—Immobilization*
Immobilization (cell cultures)
RT: Cell culture

Cell membranes 461.2
DT: January 1993
UF: Membranes—Cell membrane*
BT: Biological membranes
Cells

Cellophane 811.3, 817.1
DT: January 1993
BT: Plastic films
RT: Cellulose

Cells 461.2
DT: January 1993
UF: Biological materials—Cells*
BT: Biological materials
NT: Cell membranes
Chromosomes
RT: Cell culture
Cytology

* Former Ei Vocabulary term

Cellular arrays 722.1
DT: January 1993
UF: Data storage, Digital—Cellular arrays*
BT: Digital integrated circuits
NT: Systolic arrays
RT: PROM

Cellular communications
USE: Cellular radio systems

Cellular neural networks 723.4
DT: January 1995
BT: Neural networks

Cellular plastics
USE: Foamed plastics

Cellular radio systems 716.3
DT: January 1993
UF: Cellular communications
Radio systems, Mobile—Cellular technology*
BT: Mobile radio systems
NT: Cellular telephone systems
RT: Radio telephone

Cellular telephone systems 718.1
DT: January 1993
UF: Personal communications
Telephone—Personal signaling*
BT: Cellular radio systems
Telephone systems

Cellulose 811.3, 815.1.1
DT: Predates 1975
BT: Polysaccharides
NT: Expansive cellulose
RT: Biomass
Cellophane
Cellulose derivatives
Cellulose films
Cotton
Lignin
Saccharification

Cellulose derivatives 811.3
DT: Predates 1975
BT: Derivatives
NT: Cellulosic resins
Nitrocellulose
RT: Cellulose
Protective colloids
Textile fibers

Cellulose films 811.3
DT: January 1993
BT: Films
RT: Cellulose

Cellulose nitrate
USE: Nitrocellulose

*Cellulose—Cotton**
USE: Cotton

*Cellulose—Expansive**
USE: Expansive cellulose

Cellulosic resins 811.3, 815.1.1
DT: Predates 1975
BT: Cellulose derivatives

Cement (bone)
USE: Bone cement

Cement (dental)
USE: Dental cement

Cement additives 412.1
DT: January 1993
UF: Cement—Additives*
BT: Additives
Cements

Cement asbestos
USE: Asbestos cement

*Cement handling**
USE: Materials handling AND
Cements

Cement industry 412.1
DT: Predates 1975
BT: Industry
NT: Cement manufacture
RT: Cement plants
Cements
Concrete industry
Construction industry

Cement manufacture 412.1
DT: January 1993
BT: Cement industry
Manufacture
RT: Cement plants

Cement plants 412.1 (402.1)
DT: Predates 1975
BT: Industrial plants
RT: Cement industry
Cement manufacture

*Cement**
USE: Cements

*Cement—Additives**
USE: Cement additives

*Cement—Alumina**
USE: Alumina

*Cement—Corrosion resisting**
USE: Chemical resistant materials

*Cement—Pozzolan**
USE: Pozzolan

*Cement—Setting**
USE: Setting

*Cement—Slag**
USE: Slag cement

*Cement—Waterproof**
USE: Waterproof cement

Cementing (shafts) (405.2) (502)
DT: January 1993
UF: Shaft sinking—Cementing*
BT: Shaft sinking
NT: Well cementing

* Former Ei Vocabulary term

RT: Bonding
Cements
Grouting

Cements 412.1
SN: Construction materials; for adhesive materials, use ADHESIVES
DT: January 1993
UF: Cement handling*
Cement*
BT: Building materials
NT: Asbestos cement
Cement additives
Portland cement
Pozzolan
Putty
Slag cement
Soil cement
Waterproof cement
RT: Cement industry
Cementing (shafts)
Concretes
Grouting
Mortar
Plaster
Sealing (closing)
Setting
Stucco
Well cementing

Centerless grinding 604.2
DT: January 1993
UF: Grinding—Centerless*
BT: Grinding (machining)

Centralized plant control
USE: Integrated control

Centralized signal control 681.3 (731.2)
DT: January 1993
UF: Railroad signals and signaling—Centralized control*
BT: Control
RT: Railroad signal systems
Signal systems

Centrifugal casting 534.2
DT: January 1993
UF: Foundry practice—Centrifugal casting*
BT: Casting
RT: Foundry practice
Investment casting
Permanent mold casting
Shell mold casting

Centrifugal pumps 618.2
DT: January 1993
UF: Pumps, Centrifugal*
BT: Rotary pumps
RT: Gear pumps
Turbine pumps
Vane pumps

Centrifugal separators
USE: Centrifuges

Centrifugation 802.3
DT: January 1993
BT: Separation
RT: Centrifuges
Dewatering
Extraction
Filtration

Centrifuges 802.1
DT: Predates 1975
UF: Centrifugal separators
Separators—Centrifugal*
BT: Separators
RT: Centrifugation
Space simulators

Ceramals
USE: Cermets

Ceramets
USE: Cermets

Ceramic capacitors 704.1 (812.1)
DT: January 1993
UF: Capacitors, Ceramic*
BT: Capacitors
Ceramic products

Ceramic coatings 812.1, 813.2
DT: January 1993
UF: Protective coatings—Ceramics*
BT: Coatings
RT: Ceramic products

Ceramic cutting tools 812.1 (603) (605)
DT: January 1993
UF: Cutting tools—Ceramic*
BT: Ceramic tools
Cutting tools
RT: Ceramic dies

Ceramic dies 603.2, 812.1
DT: January 1993
UF: Dies—Ceramic*
BT: Ceramic products
Dies
RT: Ceramic cutting tools

Ceramic fibers 812.1
DT: January 1993
UF: Ceramic materials—Fibers*
Refractory fibers
Vitreous fibers
BT: Ceramic materials
Synthetic fibers
RT: Ceramic matrix composites
Glass fibers

Ceramic foams 812.1
DT: January 1993
UF: Ceramic materials—Foams*
BT: Ceramic materials
Foams

Ceramic materials 812.1
DT: Predates 1973
UF: Ceramics
Electric insulating materials—Ceramic*

Ceramic materials *(continued)*
BT: Materials
NT: Ceramic fibers
Ceramic foams
Porcelain
Structural ceramics
RT: Barium titanate
Ceramic matrix composites
Ceramic products
Cermets
Clay
Cubic boron nitride
Dielectric materials
Electrets
Ferroelectric materials
Firing (of materials)
Hot pressing
Hydrostatic pressing
Mixed oxide fuels
Nepheline syenite
Oxides
Piezoelectric materials
Plaster
Refractory materials
Silicon nitride
Tile
Titanium carbide

*Ceramic materials—Efflorescence**
USE: Efflorescence

*Ceramic materials—Ferroelectric**
USE: Ferroelectric materials

*Ceramic materials—Fibers**
USE: Ceramic fibers

*Ceramic materials—Foams**
USE: Ceramic foams

*Ceramic materials—Hot pressing**
USE: Hot pressing

*Ceramic materials—Hydrostatic pressing**
USE: Hydrostatic pressing

*Ceramic materials—Piezoelectric**
USE: Piezoelectric materials

*Ceramic materials—Silicon nitride**
USE: Silicon nitride

Ceramic matrix composites 812.1 (415.4)
DT: January 1995
BT: Composite materials
RT: Ceramic fibers
Ceramic materials
Cermets
Reinforcement
Silicon nitride
Titanium carbide

Ceramic plants 812.1 (402)
DT: January 1993
BT: Industrial plants

Ceramic products 812.1
DT: Predates 1975
UF: Chinaware
NT: Ceramic capacitors
Ceramic dies
Ceramic tools
RT: Ceramic coatings
Ceramic materials
Dielectric devices
Enamels
Glass
Tile

*Ceramic products—Radioactivity**
USE: Radioactivity

Ceramic tools 812.1 (603) (605)
DT: January 1993
UF: Tools, jigs and fixtures—Ceramic*
BT: Ceramic products
Tools
NT: Ceramic cutting tools
RT: Milling cutters

Ceramics
USE: Ceramic materials

Cerenkov counters 944.7 (932.1) (932.2)
DT: Predates 1975
UF: Cherenkov counters
Counters (Cerenkov)
BT: Radiation counters

Cerium 547.2
DT: January 1993
UF: Ce
Cerium and alloys*
BT: Rare earth elements
RT: Cerium alloys
Cerium compounds
Cerium ore treatment

Cerium alloys 547.2
DT: January 1993
UF: Cerium and alloys*
BT: Rare earth alloys
RT: Cerium
Cerium compounds

*Cerium and alloys**
USE: Cerium OR
Cerium alloys

Cerium compounds (804.1) (804.2)
DT: Predates 1975
BT: Rare earth compounds
RT: Cerium
Cerium alloys

Cerium ore treatment 533.1, 547.2
DT: Predates 1975
BT: Ore treatment
RT: Cerium

Cermets 531, 812.1
DT: June 1990
UF: Ceramals
Ceramets
Metal ceramics
BT: Composite materials
RT: Ceramic materials

* Former Ei Vocabulary term

Cermets *(continued)*
Ceramic matrix composites
Powder metallurgy
Refractory materials
Titanium nitride

Certification of engineers
USE: Registration of engineers

Cesium 549.1
DT: January 1993
UF: Cesium and alloys*
Cs
BT: Alkali metals
RT: Cesium alloys
Cesium compounds

Cesium alloys 549.1
DT: January 1993
UF: Cesium and alloys*
BT: Alkali metal alloys
RT: Cesium

*Cesium and alloys**
USE: Cesium OR
Cesium alloys

Cesium compounds (804.1) (804.2)
DT: Predates 1975
BT: Alkali metal compounds
RT: Cesium

Cetane number
USE: Antiknock rating

Cf
USE: Californium

CFC
USE: Chlorofluorocarbons

CFD
USE: Computational fluid dynamics

CFRP
USE: Carbon fiber reinforced plastics

Chain conveyors 692.1
DT: January 1993
UF: Conveyors—Chain*
BT: Conveyors

Chain drives
USE: Mechanical drives

Chains 602.1
DT: Predates 1975
BT: Components
RT: Fasteners
Mechanical drives

Changeover (pipelines)
USE: Pipeline changeover

Changes
USE: Modification

Changes of state
USE: Phase transitions

Channel capacity 716.1
DT: January 1993
UF: Information capacity
Information theory—Channel capacity*
RT: Bandwidth
Information theory
Telecommunication

Channel flow 631.1
DT: January 1993
UF: Flow of fluids—Channel flow*
Heat transfer—Channel flow*
BT: Confined flow
NT: Open channel flow
RT: Flow patterns
Heat transfer
Pipe flow

Chaos
USE: Chaos theory

Chaos theory 921, 922 (931.1)
DT: June 1990
UF: Chaos
Optical chaos
BT: Theory
RT: Mathematical models
Probability
Quantum optics
Random processes
Statistical methods

Chapmanizing
USE: Nitriding

Character recognition (723)
DT: Predates 1975
BT: Pattern recognition
NT: Optical character recognition
RT: Character recognition equipment
Character sets
Pattern recognition systems
Scanning

Character recognition equipment 722.2
DT: Predates 1975
BT: Computer peripheral equipment
RT: Character recognition
Detectors

*Character recognition, Optical**
USE: Optical character recognition

Character sets 722.2
DT: January 1993
UF: Data processing—Character sets*
Printing—Character sets*
RT: Character recognition
Data processing
Printing

Characterization
DT: January 1993
RT: Composition

Charcoal 524 (811.2)
DT: Predates 1975
BT: Carbon

* Former Ei Vocabulary term

Charcoal *(continued)*
Wood products
RT: Adsorbents
Fuels
Wood fuels

Charge (electric)
USE: Electric charge

Charge carriers 701.1 (714.2)
DT: January 1993
UF: Carriers (charge)
Semiconductor materials—Charge carriers*
BT: Semiconductor materials
NT: Hot carriers

Charge coupled devices 714.2
DT: January 1993
UF: CCD
Charge-coupled devices
Semiconductor devices, Charge coupled*
BT: Charge transfer devices

Charge transfer 802.2
DT: January 1995
RT: Chemical reactions
Ionization

Charge transfer devices 714.2
DT: January 1993
UF: Semiconductor devices, Charge transfer*
BT: MIS devices
NT: Charge coupled devices

Charge-coupled devices
USE: Charge coupled devices

Charged particles (701.1) (931.3) (932.1.1)
DT: Predates 1975
BT: Elementary particles
NT: Alpha particles
Electrons
Ions
Protons
RT: Rutherford backscattering spectroscopy

Charges (fees)
USE: Fees and charges

Charging (batteries) 702.1.2
DT: January 1993
UF: Battery charging
Electric batteries—Charging*
RT: Electric batteries
Rectifying circuits
Secondary batteries

Charging (furnace) 534.2 (532)
DT: January 1993
UF: Cupola charging
Cupola practice—Charging*
Furnaces—Charging*
RT: Charging equipment (furnaces)
Filling
Furnaces
Loading

Charging equipment (furnaces) 532.2 (534.2)
DT: January 1993
UF: Blast furnaces—Charging equipment*
BT: Furnaces
RT: Charging (furnace)

Chauffeurs
USE: Automobile drivers

Chebyshev approximation 921
DT: January 1993
UF: Mathematical techniques—Chebyshev approximation*
BT: Approximation theory
Numerical analysis

Chelating
USE: Chelation

Chelation 802.2
DT: January 1993
UF: Chelating
BT: Chemical reactions
RT: Complexation
Ions
Organometallics

CHEMFET
USE: Chemically sensitive field effect transistors

Chemical activation 802.2, 804
DT: January 1993
UF: Activation reactions
Activation*
BT: Chemical reactions
RT: Activation energy
Thermodynamics

Chemical analysis (801) (804)
DT: Predates 1975
UF: Analytical chemistry
BT: Analysis
NT: Activation analysis
Analytical geochemistry
Assays
Blood gas analysis
Chromatographic analysis
Colorimetric analysis
Electrolytic analysis
Gas fuel analysis
Gravimetric analysis
Metal analysis
Microanalysis
Ore analysis
Petroleum analysis
Polarographic analysis
Sewage analysis
Spectroscopic analysis
Sulfur determination
Thermoanalysis
Trace analysis
Volumetric analysis
Water analysis
RT: Analytic equipment
Chemistry
Composition
Impurities
Indicators (chemical)
Ion selective electrodes

* Former Ei Vocabulary term

Chemical analysis *(continued)*
Lunar surface analysis
Measurements
Mutagens
Photometry
Physical chemistry
Polarimeters
Polariscopes
Sampling
Turbidimeters
X ray analysis
X ray diffraction analysis

*Chemical analysis—Activation**
USE: Activation analysis

*Chemical analysis—Apparatus**
USE: Analytic equipment

*Chemical analysis—Assays**
USE: Assays

*Chemical analysis—Indicators**
USE: Indicators (chemical)

*Chemical analysis—Titration**
USE: Titration

Chemical attack 802.2, 804
DT: January 1993
RT: Chemical resistance
Corrosion
Ozone resistance
Pitting

Chemical beam epitaxy 933.1.2 (802.2)
DT: January 1995
UF: CBE
BT: Vapor phase epitaxy
RT: Molecular beam epitaxy

Chemical bonds 801.4
DT: January 1995
UF: Bonds (chemical)
NT: Hydrogen bonds
RT: Binding energy
Bond strength (chemical)
Conformations
Crystal structure
Molecular structure

Chemical cleaning 802.3, 804 (539)
DT: January 1993
UF: Metal cleaning—Chemical*
BT: Cleaning
NT: Electrolytic cleaning
Pickling
RT: Chemical finishing
Chemical polishing
Ultrasonic cleaning

Chemical composition
USE: Composition

Chemical compounds (804.1) (804.2)
SN: Very broad term; prefer specific type of compound
DT: January 1993
NT: Acids
Arsenic compounds
Boron compounds
Derivatives
Halogen compounds
Hydrates
Inorganic compounds
Intercalation compounds
Metallic compounds
Minerals
Monomers
Nitrogen compounds
Organic compounds
Oxides
Phosphorus compounds
Polonium compounds
Polymers
Salts
Selenium compounds
Silicon compounds
Sulfur compounds
Tellurium compounds
RT: Chemistry

Chemical derivatives
USE: Derivatives

Chemical elements 804
DT: January 1993
NT: Alloying elements
Metals
Nonmetals
Radioactive elements
Trace elements

Chemical engineering 805.1
DT: Predates 1975
BT: Engineering
NT: Biochemical engineering
RT: Chemical exhibitions
Chemical industry
Chemical operations
Chemistry
Physical chemistry

Chemical equipment 802.1
DT: Predates 1975
BT: Equipment
NT: Adsorption towers
Analytic equipment
Atomizers
Bubble columns
Chemical reactors
Condensers (liquefiers)
Crystallizers
Distillation equipment
Electrolytic cells
Evaporators
Fluidized beds
Packed beds
Pulp digesters
RT: Chemical laboratories
Chemical lasers
Chemical reactions
Chemistry
Physical chemistry

*Chemical equipment—Adsorption towers**
USE: Adsorption towers

*Chemical equipment—Bubble columns**
USE: Bubble columns

*Chemical equipment—Condensers**
USE: Condensers (liquefiers)

*Chemical equipment—Crystallizers**
USE: Crystallizers

*Chemical equipment—Cyclones**
USE: Cyclone separators

*Chemical equipment—Distillation columns**
USE: Distillation columns

*Chemical equipment—Elastomers**
USE: Elastomers

*Chemical equipment—Fluidized beds**
USE: Fluidized beds

*Chemical equipment—Light sources**
USE: Light sources

*Chemical equipment—Lining**
USE: Linings

*Chemical equipment—Packed beds**
USE: Packed beds

*Chemical equipment—Plastics**
USE: Plastics applications

Chemical exhibitions 805, 911.4
DT: Predates 1975
BT: Exhibitions
RT: Chemical engineering
Chemistry

Chemical finishing 539, 802.3, 804
DT: January 1993
BT: Finishing
NT: Chemical polishing
Textile chemical treatment
RT: Chemical cleaning
Etching

Chemical indicators
USE: Indicators (chemical)

Chemical industry 805 (912.1)
DT: Predates 1975
BT: Industry
RT: Chemical engineering
Chemical plants
Chemicals
Petroleum industry
Plastics industry
Rubber industry

Chemical kinetics
USE: Reaction kinetics

Chemical laboratories 802.1 (402.1) (801)
DT: Predates 1975
BT: Laboratories
RT: Chemical equipment
Chemistry
Clinical laboratories

*Chemical laboratories—Dust control**
USE: Dust control

Chemical lasers (744.2) (802.1)
DT: January 1993
UF: Lasers, Chemical*
BT: Lasers
RT: Chemical equipment
Gas lasers

Chemical modification (802.2) (802.3)
DT: January 1993
UF: Electrochemical electrodes—Chemical modification*
BT: Chemical operations
Modification
RT: Electrochemical electrodes
Electrochemical sensors

Chemical operations 802.3
SN: Physical operations used in chemical processing. Very general term; prefer specific operation. Also use the application if appropriate.
DT: Predates 1975
UF: Unit operations (chemical)
NT: Agglomeration
Calcination
Chemical modification
Coagulation
Compounding (chemical)
Desulfurization
Dissolution
Emulsification
Fluidization
Gelation
Leaching
Mixing
Ozonization
Pelletizing
Separation
Sorption
RT: Chemical engineering
Chemical variables control
Chemistry
Crystallization
Drying
Heating equipment
Textile chemical treatment

*Chemical operations—Heaters**
USE: Heating equipment

Chemical oxygen demand (452.2) (452.4) (453)
DT: January 1993
UF: Sewage treatment—Chemical oxygen demand*
NT: Biochemical oxygen demand
RT: Sewage treatment

Chemical plants 802.1 (402.1)
DT: Predates 1975
BT: Industrial plants
NT: Petrochemical plants
RT: Chemical industry
Chemicals

Chemical polishing 604.2, 802.3
DT: January 1993
UF: Polishing—Chemical*
BT: Chemical finishing
Polishing
NT: Electrolytic polishing
RT: Chemical cleaning

Chemical pulp 811.1 (802.2)
DT: January 1993
UF: Pulp—Chemical*
BT: Pulp

Chemical reactions 802.2
DT: Predates 1975
UF: Reaction mechanisms
Unit processes
NT: Acetylation
Acylation
Addition reactions
Alkylation
Amination
Aromatization
Association reactions
Bioconversion
Calcification (biochemistry)
Carbonation
Carbonization
Carbonylation
Carboxylation
Catalysis
Chelation
Chemical activation
Combustion
Complexation
Condensation reactions
Coordination reactions
Crosslinking
Curing
Decarbonization
Decomposition
Degradation
Dehalogenation
Dehydration
Dehydrogenation
Denitrification
Dissociation
Electrolysis
Esterification
Fermentation
Free radical reactions
Grafting (chemical)
Halogenation
Hydration
Hydrogenation
Hydroxylation
Ion exchange
Ionization
Isomerization
Methanation
Nitration
Nitrification
Nitrogen fixation
Oxidation
Photochemical reactions
Polymerization
Redox reactions
Reduction
Reforming reactions
Saponification
Substitution reactions
Sulfonation
Synthesis (chemical)
Vitrification
Vulcanization
RT: Charge transfer
Chemical equipment
Chemical reactors
Chemical relaxation
Chemisorption
Chemistry
Differential scanning calorimetry
Drying
Ions
Physical chemistry
Radiation chemistry
Reaction kinetics
Separation
Stoichiometry
Thermodynamics

*Chemical reactions—Electrolytic**
USE: Electrolysis

*Chemical reactions—Nitrification**
USE: Nitrification

*Chemical reactions—Reaction kinetics**
USE: Reaction kinetics

*Chemical reactions—Redox**
USE: Redox reactions

Chemical reactors 802.1
DT: June 1990
BT: Chemical equipment
NT: Bioreactors
RT: Autoclaves
Chemical reactions
Chemistry

Chemical relaxation 802.2
DT: January 1993
BT: Relaxation processes
RT: Chemical reactions
Chemistry

Chemical resistance 802.2
DT: January 1993
NT: Acid resistance
Oil resistance
Ozone resistance
RT: Chemical attack
Chemical resistant materials
Corrosion resistance

Chemical resistant materials 802.2
DT: January 1993
UF: Cement—Corrosion resisting*
BT: Materials
RT: Chemical resistance
Corrosion resistant alloys

* Former Ei Vocabulary term

Chemical sensors 801
SN: In 1993 and 1994, GAS SENSORS was used
DT: January 1995
UF: Gas sensors
BT: Sensors
NT: Calcium ion sensors
Electrochemical sensors
Oxygen sensors
pH sensors

Chemical structure
USE: Structure (composition)

Chemical vapor deposition 802.2 (813.1)
DT: January 1993
UF: CVD
OMCVD
BT: Vapor deposition
NT: Metallorganic chemical vapor deposition
RT: Coatings
Films
Vapor phase epitaxy

Chemical variables control 802.3, 731.2
DT: January 1993
UF: Control, Chemical variables*
BT: Control
RT: Chemical operations
Moisture control

Chemical warfare 404.1, 803
DT: January 1993
UF: Gas warfare
Military engineering—Chemical warfare*
Poison gas warfare
BT: Military operations

Chemical wastes 452.3, 803, 804
DT: January 1993
UF: Industrial wastes—Chemicals*
BT: Industrial wastes
RT: Chemicals
Hazardous materials

Chemical water treatment 445.1, 803
DT: January 1993
UF: Water treatment—Chemicals*
BT: Water treatment

Chemically sensitive field effect transistors 714.1 (802.2)
DT: January 1993
UF: CHEMFET
Transistors, Field effect—Chemically sensitive*
BT: Field effect transistors
RT: Sensors

Chemicals 803, 804
DT: Predates 1975
BT: Materials
NT: Additives
Agents
Biocides
Carcinogens
Coolants
Dyes
Fillers
Indicators (chemical)
Inorganic chemicals
Organic chemicals
Solvents
Wood chemicals
RT: Antifreeze solutions
Chemical industry
Chemical plants
Chemical wastes
Hazardous materials

Chemicals removal (water treatment) 452.3, 803, 804
DT: January 1993
UF: Water treatment—Chemicals removal*
BT: Removal
Separation
Water treatment
NT: Hydrogen sulfide removal (water treatment)
Iron removal (water treatment)
Lead removal (water treatment)
Manganese removal (water treatment)
Radioisotope removal (water treatment)
RT: Color removal (water treatment)
Taste control (water treatment)

*Chemicals—Corrosive properties**
USE: Corrosive effects

*Chemicals—Vapor pressure**
USE: Vapor pressure

Chemiluminescence 741.1, 802.2
DT: Predates 1975
BT: Luminescence
RT: Photochemical reactions

Chemisorption 802.2, 802.3
DT: January 1993
BT: Sorption
RT: Absorption
Adsorption
Chemical reactions

Chemistry 801
DT: January 1986
BT: Natural sciences
NT: Atmospheric chemistry
Biochemistry
Colloid chemistry
Geochemistry
Petroleum chemistry
Physical chemistry
RT: Chemical analysis
Chemical compounds
Chemical engineering
Chemical equipment
Chemical exhibitions
Chemical laboratories
Chemical operations
Chemical reactions
Chemical reactors
Chemical relaxation

Chemostats 802.1, 461.8 (462.1)
DT: January 1993
UF: Bioreactors—Chemostats*
BT: Bioreactors

* Former Ei Vocabulary term

Chemotherapy
USE: Drug therapy

Cherenkov counters
USE: Cerenkov counters

Chewing
USE: Mastication

Chimney linings 402.1, 812.2 (614.2)
DT: January 1993
UF: Chimneys—Lining*
BT: Chimneys
Linings

Chimneys 402.1 (614.1)
DT: Predates 1975
BT: Building components
NT: Chimney linings
Flues
RT: Furnaces

*Chimneys—Demolition**
USE: Demolition

*Chimneys—Lining**
USE: Chimney linings

Chinaware
USE: Ceramic products

Chips (microprocessor)
USE: Microprocessor chips

Chitin 804.1
DT: January 1987
BT: Amines
Polysaccharides
Tissue

Chloride minerals 482.2
DT: January 1993
UF: Mineralogy—Chlorides*
BT: Chlorine compounds
Halide minerals

Chloridization
USE: Chlorination

Chlorinated rubber paint 813.2, 818.5
DT: January 1993
UF: Paint—Chlorinated rubber*
BT: Paint
RT: Chlorination
Rubber

Chlorination 802.2
DT: January 1993
UF: Chloridization
Water treatment—Chlorination*
BT: Halogenation
RT: Chlorinated rubber paint
Sewage treatment
Water treatment

Chlorine 804
DT: Predates 1975
UF: Cl
BT: Halogen elements
RT: Chlorine compounds

Chlorine compounds (804.1) (804.2)
DT: Predates 1975
BT: Halogen compounds
NT: Carbon tetrachloride
Chloride minerals
Chlorine containing polymers
Sodium chloride
RT: Chlorine

Chlorine containing polymers 815.1
DT: January 1977
BT: Chlorine compounds
Organic polymers

Chlorite minerals 482.2
DT: January 1993
UF: Mineralogy—Chlorites*
BT: Aluminum compounds
Iron compounds
Magnesium compounds
Silicate minerals

Chlorofluorocarbons 804
DT: January 1995
UF: CFC
BT: Halogen compounds
RT: Freons

Chlorophyll 804.1 (801.2)
DT: January 1987
BT: Nitrogen compounds
Pigments
RT: Photosynthesis

Chokes
USE: Electric inductors

Cholesteric liquid crystals 804
DT: January 1993
UF: Crystals, Liquid—Cholesteric*
BT: Liquid crystals

Cholesterol 461.2, 804.1
DT: January 1993
UF: Biological materials—Cholesterol*
BT: Biological materials
RT: Health hazards
Lipids

Chopper amplifiers 713.1
DT: January 1993
UF: Amplifiers, Chopper*
BT: Amplifiers (electronic)
RT: Choppers (circuits)

Choppers (circuits) (703.1) (713.1)
DT: January 1993
UF: Choppers*
BT: Switching circuits
RT: Chopper amplifiers
Electric inverters
Optical devices
Power converters

*Choppers**
USE: Choppers (circuits)

* Former Ei Vocabulary term

Christmas trees (wellheads) 511.2
DT: January 1993
UF: Oil field equipment—Christmas trees*
BT: Oil field equipment

Chromate coatings 813.2, 804.2
DT: January 1993
UF: Protective coatings—Chromate*
BT: Coatings
RT: Chromates

Chromates 804.2
DT: January 1993
BT: Chromium compounds
RT: Chromate coatings

Chromatographic analysis (801)
DT: Predates 1975
BT: Chemical analysis
RT: Chromatography
Separation
Sorption

*Chromatographic analysis—Affinity**
USE: Affinity chromatography

*Chromatographic analysis—Column**
USE: Column chromatography

*Chromatographic analysis—Gas**
USE: Gas chromatography

*Chromatographic analysis—Gel permeation**
USE: Gel permeation chromatography

*Chromatographic analysis—High performance liquid**
USE: High performance liquid chromatography

*Chromatographic analysis—High pressure liquid**
USE: High pressure liquid chromatography

*Chromatographic analysis—Hydrophobic**
USE: Hydrophobic chromatography

*Chromatographic analysis—Liquid**
USE: Liquid chromatography

*Chromatographic analysis—Molecular sieve**
USE: Molecular sieves

*Chromatographic analysis—Size exclusion**
USE: Size exclusion chromatography

*Chromatographic analysis—Thin layer**
USE: Thin layer chromatography

Chromatography (801) (802.3)
DT: January 1993
NT: Affinity chromatography
Column chromatography
Gas chromatography
Hydrophobic chromatography
Liquid chromatography
Size exclusion chromatography
Thin layer chromatography
RT: Chromatographic analysis
Separation

Chromite (812.2) (482.2) (504.3)
DT: Predates 1975
UF: Chromium iron ore
BT: Chromium compounds
Iron compounds
Ores
RT: Chromite deposits
Chromium mines
Chromium ore treatment

Chromite deposits 504.3
DT: Predates 1975
UF: Chromium deposits*
BT: Ore deposits
RT: Chromite
Chromium
Chromium mines

*Chromite mines and mining**
USE: Chromium mines

*Chromite ore treatment**
USE: Chromium ore treatment

Chromium 543.1
DT: January 1993
UF: Chromium and alloys*
Cr
BT: Refractory metals
RT: Chromite deposits
Chromium alloys
Chromium compounds
Chromium metallurgy
Chromium metallography
Chromium mines
Chromium ore treatment
Chromium plating
Strategic materials

Chromium alloys 543.1
DT: January 1993
UF: Chromium and alloys*
BT: Refractory alloys
NT: Stellite
RT: Chromium
Chromium compounds
Stainless steel

*Chromium and alloys**
USE: Chromium OR
Chromium alloys

Chromium compounds (804.1) (804.2)
DT: Predates 1975
BT: Refractory metal compounds
NT: Chromates
Chromite
RT: Chromium
Chromium alloys

*Chromium deposits**
USE: Chromite deposits

Chromium iron ore
USE: Chromite

Chromium metallography 531.2, 543.1
DT: Predates 1975
BT: Metallography
RT: Chromium
Chromium metallurgy

Chromium metallurgy 531.1, 543.1
DT: Predates 1975
BT: Metallurgy
NT: Chromium powder metallurgy
RT: Chromium
Chromium metallography
Chromium ore treatment

Chromium mines 504.3
DT: January 1993
UF: Chromite mines and mining*
Chromium mines and mining*
BT: Mines
RT: Chromite
Chromite deposits
Chromium

*Chromium mines and mining**
USE: Chromium mines

Chromium ore treatment 533.1, 543.1
DT: Predates 1975
UF: Chromite ore treatment*
BT: Ore treatment
RT: Chromite
Chromium
Chromium metallurgy

Chromium plating 539.3, 543.1
DT: Predates 1975
BT: Plating
RT: Chromium

Chromium powder metallurgy 536.1, 543.1
DT: January 1993
UF: Powder metallurgy—Chromium*
BT: Chromium metallurgy
Powder metallurgy

Chromogenics 741.1
DT: January 1993
BT: Optics
RT: Color
Electrochromism

Chromometers
USE: Colorimeters

Chromosomes 461.2
DT: January 1993
UF: Biological materials—Chromosomes*
BT: Cells
NT: Genes

Chronometers 943.3
DT: Predates 1975
BT: Mechanical clocks
RT: Navigation
Time measurement

Chucking machines 603.1
DT: Predates 1975
BT: Lathes
RT: Chucks
Turret lathes

Chucks (603) (605)
DT: Predates 1975
BT: Tools
RT: Chucking machines
Clamping devices
Guides (mechanical)

*Churches**
USE: Religious buildings

Cigarette filters (451.1) (804) (461.7)
DT: January 1993
UF: Cigarette manufacture—Smoke filters*
Smoke filters (cigarette)
BT: Filters (for fluids)
RT: Cigarette manufacture
Smoke

Cigarette manufacture (913.4)
DT: Predates 1975
BT: Manufacture
RT: Cigarette filters
Tobacco

*Cigarette manufacture—Smoke filters**
USE: Cigarette filters

Cinematography
USE: Photography AND
Motion pictures

Circuit analysis
USE: Electric network analysis

Circuit breakers
USE: Electric circuit breakers

Circuit breaking arcs (701.1) (704)
DT: January 1993
UF: Electric arcs, Circuit breaking*
BT: Electric arcs
RT: Electric circuit breakers
Electric sparks

Circuit coupling
USE: Coupled circuits

Circuit layout
USE: Integrated circuit layout

Circuit oscillations 703.1 (703.1.1) (703.1.2)
DT: January 1993
UF: Electronic circuits—Oscillations*
BT: Oscillations
RT: Networks (circuits)

Circuit resonance 703.1 (703.1.1) (703.1.2)
DT: January 1993
UF: Electronic circuits—Resonance*
BT: Resonance
RT: Networks (circuits)
Oscillations
Resonant circuits

Circuit schematics
USE: Schematic diagrams

* Former Ei Vocabulary term

Circuit theory 703.1 (703.1.1) (703.1.2)
DT: January 1993
UF: Electric networks*
Electric networks—Theory*
Network theory
BT: Theory
NT: Electric network analysis
Electric network parameters
Electric network synthesis
Electric network topology
Switching theory
Transmission line theory
RT: Amplifiers (electronic)
Coupled circuits
Distributed parameter networks
Electric filters
Laplace transforms
Lumped parameter networks
Networks (circuits)
Time varying networks
Z transforms

Circuit topology
USE: Electric network topology

Circuit tracking
USE: Electronic circuit tracking

Circuits
USE: Networks (circuits)

*Circuits, Coincidence**
USE: Coincidence circuits

Circular waveguides 714.3
DT: January 1993
UF: Cylindrical waveguides
Waveguides, Circular*
BT: Waveguides
RT: Helical waveguides

Circulating media (511) (512)
SN: Used in drilling
DT: January 1993
UF: Boreholes—Circulating media*
Oil well drilling—Circulating media*
Rock drilling—Circulating media*
BT: Drilling fluids
RT: Oil well drilling
Rock drilling

Circulation (blood)
USE: Hemodynamics

Circulation (boiler)
USE: Boiler circulation

Circulators (microwave)
USE: Microwave circulators

Circulators (waveguide)
USE: Waveguide circulators

City planning
USE: Urban planning

Civil defense 404.2
DT: Predates 1975
RT: Military engineering
Military operations
Nuclear explosions
Shelters (civil defense)

*Civil defense—Communications**
USE: Communication systems

*Civil defense—Food supply**
USE: Food supply

*Civil defense—Shelters**
USE: Shelters (civil defense)

Civil engineering 409
DT: Predates 1985
BT: Engineering
NT: Coastal engineering
Highway engineering
Sanitary engineering
RT: Municipal engineering
Structural design
Structures (built objects)
Urban planning

Cl
USE: Chlorine

Clad metals 535.1, 538.1
DT: January 1977
UF: Clad wire
Welding—Clad metals*
BT: Coated materials
NT: Copper clad steel
RT: Metal cladding
Metals

Clad wire
USE: Wire AND
Clad metals

Cladding (coating) (535.1) (538.1) (622.1)
DT: January 1993
UF: Cladding*
BT: Coating techniques
NT: Metal cladding
Nuclear fuel cladding
RT: Bonding
Coextrusion
Hard facing
Laminating
Metallizing
Plating
Wire

*Cladding**
USE: Cladding (coating)

Clamping devices (603.2) (605)
DT: Predates 1975
UF: Clamps
BT: Tools
RT: Chucks
Fixtures (tooling)

Clamps
USE: Clamping devices

* Former Ei Vocabulary term

Clarification 802.3 (822.2)
DT: January 1993
UF: Sugar manufacture—Clarification*
BT: Separation
RT: Clarifiers
Sedimentation
Sugar manufacture

Clarifiers 802.1 (452.2) (822.2)
DT: Predates 1975
UF: Sewage treatment plants—Clarifiers*
BT: Separators
RT: Clarification
Sewage treatment plants

Clark sensors
USE: Dissolved oxygen sensors

Classical field theory
USE: Continuum mechanics

Classical mechanics of continuous media
USE: Continuum mechanics

Classification (of information) 716.1
DT: January 1993
UF: Classification*
BT: Information analysis
RT: Indexing (of information)

Classification yards 681.2
DT: January 1993
UF: Railroad yards and terminals—Classification yards*
BT: Railroad yards and terminals

*Classification**
USE: Classification (of information)

Classifiers 802.1
DT: Predates 1975
UF: Coal preparation—Classifiers*
Ore treatment—Classifiers*
BT: Separators
NT: Screens (sizing)
RT: Ore treatment
Size determination
Size separation

Clay 483.1 (414.2) (812.1)
DT: Predates 1975
BT: Soils
NT: Bentonite
RT: Binders
Ceramic materials
Clay deposits
Clay minerals
Clay pipe
Clay products
Fullers earth
Ion exchangers
Minerals

Clay alteration (504) (481.2)
DT: January 1993
UF: Ore deposits—Clay alteration*
RT: Mineralogy
Ore deposits

Clay deposits 505.1
DT: Predates 1975
BT: Mineral resources
RT: Clay

Clay minerals 482.2
DT: January 1986
BT: Silicate minerals
NT: Kaolin
RT: Clay
Ion exchangers

Clay pipe 812.1, 619.1
DT: January 1993
UF: Pipe, Clay*
BT: Clay products
Pipe
RT: Clay

Clay products 812.1
DT: Predates 1975
NT: Clay pipe
RT: Clay

*Clay products—Efflorescence**
USE: Efflorescence

*Clay—Screening**
USE: Screening

Clean rooms 402.1 (461.1) (713) (714.2) (913)
DT: June 1990
BT: Facilities
RT: Assembly
Contamination
Environmental engineering
Integrated circuit manufacture
Semiconductor device manufacture

Cleaning 802.3
DT: January 1993
NT: Chemical cleaning
Decontamination
Disinfection
Dry cleaning
Purging
Snow and ice removal
Sterilization (cleaning)
Street cleaning
Surface cleaning
Textile scouring
Washing
RT: Coal preparation
Descaling
Detergents
Disinfectants
Finishing
Maintenance
Peeling
Purification
Refining
Sanitation
Scavenging
Soil resistance (textiles)
Surface treatment
Vacuum cleaners
Washing machines

* Former Ei Vocabulary term

Clearances (bridge)
USE: Bridge clearances

Climate change 443.1 (451)
DT: January 1995
NT: Global warming
Greenhouse effect
RT: Air pollution
Atmospheric composition
Atmospheric temperature
Climatology

Climate control 402, 731.2
DT: January 1993
UF: Buildings—Climate control*
BT: Control
NT: Air conditioning
Space heating
Ventilation
RT: Air curtains
Arctic buildings
Buildings
Cooling
Heating
Temperature control
Tropical buildings

Climatology 443
DT: January 1993
UF: Meteorology—Climatology*
BT: Meteorology
RT: Climate change
Drought
Earth atmosphere
Greenhouse effect
Precipitation (meteorology)

Clinical imaging
USE: Medical imaging

Clinical laboratories 462.2 (402) (801.2)
DT: January 1993
UF: Hospitals—Clinical laboratories*
BT: Laboratories
RT: Chemical laboratories
Hospitals

Clipping circuits
USE: Limiters

Clocks 943.3
DT: January 1993
BT: Timing devices
NT: Atomic clocks
Electric clocks
Mechanical clocks
RT: Time measurement
Watches

*Clocks, Atomic**
USE: Atomic clocks

*Clocks, Electric**
USE: Electric clocks

*Clocks, Mechanical**
USE: Mechanical clocks

Cloning 461.8.1 (801.2)
DT: January 1987
BT: Genetic engineering

Closed circuit television systems 716.4
DT: January 1993
UF: CCTV
Television systems, Closed circuit*
BT: Television systems
RT: Teleconferencing
Television
Television equipment

Closed cycle machinery (612) (617) (618) (644)
SN: Also use the type of equipment, e.g., Gas turbines
DT: January 1993
UF: Gas turbines—Closed cycle*
Refrigerators—Closed cycle*
Turbomachinery—Closed cycle*
BT: Machinery
RT: Gas turbines
Refrigerators
Turbomachinery

Closed loop control systems 731.1
DT: January 1995
UF: Negative feedback control systems
BT: Control systems
RT: Feedback

Closings (industrial plants)
USE: Plant shutdowns

Closures (containers)
USE: Container closeres

Cloth
USE: Fabrics

Clothing manufacture
USE: Garment manufacture

Cloud chambers 932, 944.7
DT: Predates 1975
BT: Particle detectors
RT: Ionization chambers

Cloud seeding 443.3
DT: January 1993
UF: Artificial rain
Rain and rainfall—Cloud seeding*
Seeding (clouds)
BT: Weather modification
RT: Clouds
Meteorology
Nucleation
Rain

Clouds 443
DT: January 1993
UF: Meteorology—Clouds*
BT: Earth atmosphere
RT: Atmospheric humidity
Cloud seeding
Meteorology
Precipitation (meteorology)
Storms
Thunderstorms

Clutches 602.2
DT: Predates 1975
BT: Equipment
NT: Electric clutches
RT: Actuators
Brakes
Couplings
Drives
Final control devices
Machine components

*Clutches, Electric**
USE: Electric clutches

Clutter (information theory) 716.1
DT: January 1993
UF: Information theory—Clutter*
BT: Information theory

Clutter (radar)
USE: Radar clutter

Cm
USE: Curium

CML
USE: Emitter coupled logic circuits

CMOS integrated circuits 714.2
DT: January 1993
UF: Complementary metal oxide semiconductor integrated circuits
Integrated circuits, CMOS*
BT: Monolithic integrated circuits

Co
USE: Cobalt

Co-ax cables
USE: Coaxial cables

Coagulation 802.3
DT: January 1993
BT: Chemical operations
RT: Agglomeration
Concentration (process)
Emulsions
Precipitation (chemical)
Sedimentation
Water treatment

Coal 524
DT: Predates 1975
UF: Fuel burners—Pulverized coal*
Geology—Coal*
BT: Fossil fuels
NT: Anthracite
Bituminous coal
Lignite
RT: Carbon
Coal ash
Coal byproducts
Coal carbonization
Coal combustion
Coal deposits
Coal fueled furnaces
Coal fueled gas turbines
Coal gas
Coal gasification
Coal handling
Coal hydrogenation
Coal industry
Coal liquefaction
Coal mines
Coal oil mixtures
Coal preparation
Coal reclamation
Coal research
Coal slurries
Coal storage
Coal tailings
Coal tar
Coal water mixtures
Coke
Coking properties
Furnace fuels
Geology
Petrography
Soot
Sulfur determination

Coal ash 524 (452.3) (521)
DT: Predates 1975
UF: Ash (coal)
BT: Coal byproducts
RT: Coal

Coal breakage 524
DT: January 1993
UF: Breakage (coal)
Coal preparation—Breakage*
BT: Coal preparation
RT: Crushing

*Coal briquets and briquetting**
USE: Briquets OR
Briquetting

Coal burning gas turbines
USE: Gas turbines

Coal byproducts 524 (522) (804)
DT: Predates 1975
BT: Byproducts
NT: Coal ash
RT: Coal

Coal carbonization 802.2 (522) (524)
DT: January 1993
BT: Carbonization
NT: Low temperature coal carbonization
RT: Coal
Coal tar
Coke
Coke ovens
Coking

*Coal carbonization—Low temperature**
USE: Low temperature coal carbonization

Coal cleaning
USE: Coal preparation

Coal combustion 521, 524
DT: January 1987
BT: Combustion

Coal combustion *(continued)*
RT: Coal
Coal dust
Coal fired boilers
Coal fueled furnaces
Coal fueled gas turbines

*Coal combustion—In situ**
USE: In situ coal combustion

Coal deposits 503 (482)
DT: Predates 1975
BT: Fossil fuel deposits
RT: Coal

Coal dust 451.1, 503, 524 (914.1)
DT: January 1993
BT: Dust
RT: Air pollution
Coal combustion

Coal fired boilers 524, 614.2 (521)
DT: January 1993
UF: Boiler firing—Coal*
BT: Boilers
RT: Coal combustion

Coal fueled furnaces
524 (532) (534) (535) (537) (642) (643)
DT: January 1993
UF: Furnaces—Coal fuel*
BT: Furnaces
RT: Coal
Coal combustion

Coal fueled gas turbines 524, 612.3
DT: January 1993
UF: Gas turbines—Coal burning*
BT: Gas turbines
RT: Coal
Coal combustion

Coal furnaces
USE: Furnaces

Coal gas 522, 524
DT: June 1990
BT: Gases
RT: Coal
Coal gasification
In situ coal combustion
Synthetic fuels

Coal gasification 522, 524
DT: May 1974
BT: Gasification
RT: Coal
Coal gas
Coal gasification plants
Coal hydrogenation
Fossil fuel power plants
Gas producers
Methanation
Pyrolysis
Synthetic fuels

Coal gasification plants 522, 524 (402.1)
DT: January 1983
BT: Industrial plants
RT: Coal gasification
Gas producers

Coal handling 524, 691.2
DT: Predates 1975
BT: Materials handling
RT: Coal
Coal transportation

*Coal handling—Gas plants**
USE: Gas plants

*Coal handling—Steam power plants**
USE: Steam power plants

Coal hydrogenation 522, 524, 802.1
DT: Predates 1975
BT: Hydrogenation
RT: Coal
Coal gasification

Coal industry 503, 524 (911) (913)
DT: Predates 1975
BT: Mineral industry
RT: Coal
Coal mines
Coal preparation

Coal liquefaction 523, 524
DT: May 1974
BT: Liquefaction
RT: Coal
Coal liquefaction plants
Synthetic fuels

Coal liquefaction plants 523, 524 (402.1)
DT: January 1983
BT: Industrial plants
RT: Coal liquefaction

Coal mines 503.1
DT: January 1993
UF: Coal mines and mining*
BT: Mines
NT: Lignite mines
RT: Coal
Coal industry
Coal tailings

*Coal mines and mining**
USE: Coal mines

Coal oil mixtures 523, 524
DT: January 1993
UF: Coal slurries—Coal oil mixtures*
BT: Fuels
Mixtures
RT: Coal
Coal slurries
Fuel oils

Coal preparation (503.2) (524)
DT: Predates 1975
UF: Coal cleaning
Coal washing
NT: Coal breakage
RT: Blending
Cleaning

* Former Ei Vocabulary term

Coal preparation *(continued)*
Coal
Coal industry
Coal pulverizers
Crushers
Heavy media separation

*Coal preparation—Blending**
USE: Blending

*Coal preparation—Breakage**
USE: Coal breakage

*Coal preparation—Classifiers**
USE: Classifiers

*Coal preparation—Heavy media separation**
USE: Heavy media separation

*Coal preparation—Screening**
USE: Screening

*Coal preparation—Tailings disposal**
USE: Tailings disposal

Coal pulverizers 524
DT: Predates 1975
BT: Grinding mills
RT: Coal preparation

Coal reclamation 452.3, 524
DT: January 1993
UF: Coal—Reclamation*
BT: Reclamation
RT: Coal

Coal research 524 (521)
DT: Predates 1975
BT: Research
RT: Coal

Coal slurries 524 (631.1)
DT: January 1986
BT: Slurries
RT: Coal
Coal oil mixtures
Coal water mixtures
Slurry pipelines

*Coal slurries—Coal oil mixtures**
USE: Coal oil mixtures

*Coal slurries—Coal water mixtures**
USE: Coal water mixtures

*Coal slurries—Flow**
USE: Flow of fluids

Coal storage 524, 694.4
DT: Predates 1975
BT: Fuel storage
RT: Coal

Coal tailings 524 (503.2) (452.3)
DT: January 1993
UF: Coal tailings disposal
BT: Tailings
RT: Coal
Coal mines

Coal tailings disposal
USE: Coal tailings AND
Tailings disposal

Coal tar 411.2
DT: Predates 1975
BT: Tar
RT: Coal
Coal carbonization

Coal transportation 503, 524 (432.3) (433.3) (434.3)
SN: Also use the mode of transportation, e.g., RAILROADS
DT: Predates 1975
BT: Freight transportation
RT: Coal handling

Coal washing
USE: Washing AND
Coal preparation

Coal water mixtures 524 (631.1)
DT: January 1993
UF: Coal slurries—Coal water mixtures*
BT: Mixtures
RT: Coal
Coal slurries

*Coal—Anthracite**
USE: Anthracite

*Coal—Coking properties**
USE: Coking properties

*Coal—Petrography**
USE: Petrography

*Coal—Reclamation**
USE: Coal reclamation

Coalescence 801.3
DT: January 1993
UF: Colloids—Coalescence*
RT: Colloids

Coastal engineering 407.3
DT: January 1977
BT: Civil engineering
RT: Bank protection
Beaches
Breakwaters
Coastal zones
Jetties
Port structures
Shore protection

*Coastal engineering—Bank protection**
USE: Bank protection

Coastal zones 471 (407.3)
DT: January 1986
UF: Coasts
Oceanography—Coastal zones*
Seashores
BT: Geographical regions
RT: Coastal engineering
Oceanography
Shore protection

* Former Ei Vocabulary term

Coasts
USE: Coastal zones

Coated fuel particles 621, 622.1
DT: January 1993
UF: Nuclear fuels—Coated particles*
BT: Fuels
RT: Coated materials
Nuclear fuels

Coated materials 813 (539)
SN: Also use the particular material or product, e.g., METALS
DT: January 1993
UF: Welding—Coated metals*
BT: Materials
NT: Aluminum coated steel
Clad metals
Galvanized metal
Precoated metals
RT: Abrasive coatings
Coated fuel particles
Coated wire electrodes
Coating techniques
Coatings

Coated wire electrodes 802.1 (704)
DT: January 1993
UF: Electrodes, Electrochemical—Coated wire*
BT: Electrochemical electrodes
RT: Coated materials
Welding electrodes
Welding rods

Coating techniques 813.1 (539)
DT: Predates 1975
UF: Covering techniques
Metal coating
Paper—Coating*
Paperboards—Coating*
Papermaking—Coating*
Sheet and strip metal—Coating*
BT: Deposition
NT: Cladding (coating)
Enameling
Encapsulation
Flame spraying
Hard facing
Lacquering
Metallizing
Painting
Paper coating
Plasma spraying
Plating
RT: Anodic oxidation
Coated materials
Coatings
Coloring
Electrodeposition
Epitaxial growth
Finishing
Laminating
Metal finishing
Precoated metals
Sealing (finishing)
Sputter deposition
Stripping (removal)
Surface treatment
Vapor deposition

Coatings 813.2 (539)
DT: January 1987
BT: Materials
NT: Abrasive coatings
Bituminous coatings
Black coatings
Boride coatings
Ceramic coatings
Chromate coatings
Diffusion coatings
Electrophoretic coatings
Electrostatic coatings
Enamels
Glazes
Inorganic coatings
Luminous coatings
Optical coatings
Organic coatings
Paint
Phosphate coatings
Plastic coatings
Powder coatings
Protective coatings
Reflective coatings
Rubber coatings
Silicone coatings
Sprayed coatings
Tape coatings
Temporary coatings
Vacuum deposited coatings
Water borne coatings
RT: Chemical vapor deposition
Coated materials
Coating techniques
Films
Solar absorbers
Solar control films
Solvents
Stripping (removal)
Substrates
Surface treatment
Thin films
Veneers

*Coatings—Antireflection coatings**
USE: Antireflection coatings

*Coatings—Black coatings**
USE: Black coatings

*Coatings—Diffusion coatings**
USE: Diffusion coatings

*Coatings—Luminous**
USE: Luminous coatings

*Coatings—Reflective**
USE: Reflective coatings

*Coatings—Solar control films**
USE: Solar control films

*Coatings—Sprayed**
USE: Sprayed coatings

*Coatings—Waterproofing**
USE: Waterproof coatings

Coax cables
USE: Coaxial cables

Coaxial cables (706.2) (718.1) (718.2)
DT: January 1993
UF: Co-ax cables
Coax cables
Concentric cables
Telecommunication cables, Coaxial*
BT: Telecommunication cables
RT: Cable television systems
Telegraph systems
Telephone lines
Television systems
Waveguides

Cobalt 549.3
DT: January 1993
UF: Co
Cobalt and alloys*
BT: Transition metals
RT: Cobalt alloys
Cobalt compounds
Cobalt deposits
Cobalt metallography
Cobalt metallurgy
Cobalt mines
Cobalt ore treatment
Cobalt plating
Strategic materials

Cobalt alloys 549.3
DT: January 1993
UF: Cobalt and alloys*
BT: Transition metal alloys
NT: Stellite
RT: Cobalt
Cobalt compounds

*Cobalt and alloys**
USE: Cobalt OR
Cobalt alloys

Cobalt compounds (804.1) (804.2)
DT: Predates 1975
BT: Transition metal compounds
RT: Cobalt
Cobalt alloys

Cobalt deposits 504.3, 549.3
DT: Predates 1975
BT: Ore deposits
RT: Cobalt
Cobalt mines

Cobalt metallography 531.2, 549.3
DT: Predates 1975
BT: Metallography
RT: Cobalt
Cobalt metallurgy

Cobalt metallurgy 531.1, 549.3
DT: Predates 1975
BT: Metallurgy
NT: Cobalt powder metallurgy
RT: Cobalt
Cobalt metallography
Cobalt ore treatment

Cobalt mines 504.3, 549.3
DT: January 1993
UF: Cobalt mines and mining*
BT: Mines
RT: Cobalt
Cobalt deposits

*Cobalt mines and mining**
USE: Cobalt mines

Cobalt ore treatment 533.1, 549.3
DT: January 1993
BT: Ore treatment
RT: Cobalt
Cobalt metallurgy

Cobalt plating 539.3, 549.3
DT: Predates 1975
BT: Plating
RT: Cobalt

Cobalt powder metallurgy 536.1, 549.3
DT: January 1993
UF: Powder metallurgy—Cobalt nickel*
Powder metallurgy—Cobalt*
BT: Cobalt metallurgy
Powder metallurgy

COBOL (programming language) 723.1.1
DT: January 1993
UF: Computer programming languages—COBOL*
BT: High level languages

Cockpits (aircraft) 652.1
DT: January 1993
UF: Aircraft cockpits
Aircraft—Cockpits*
BT: Aircraft parts and equipment

Code converters 723.2
DT: January 1993
UF: Converters (code)
Data processing—Code converters*
BT: Computer hardware
RT: Codes (symbols)
Data processing
Information theory
Telecommunication

Codes (safety)
USE: Codes (standards)

Codes (standards) 902.2
DT: January 1993
UF: Codes (safety)
Safety codes*
BT: Standards
NT: Boiler codes
Building codes
Electric codes

Codes (standards) *(continued)*
Elevator codes
Pipeline codes
Pressure vessel codes
Ventilation codes
Welding codes
RT: Accident prevention
Laws and legislation

Codes (symbols) 723.2
DT: January 1993
UF: Codes, Symbolic*
Electric machinery—Symbols*
Instruments—Symbols*
Servomechanisms—Symbols*
Steel—Symbols*
Symbolic codes
Symbols (codes)
NT: Bar codes
Binary codes
Block codes
Color codes
Convolutional codes
Map symbols
Telegraph codes
Trellis codes
RT: Code converters
Coding errors
Cryptography
Decoding
Encoding (symbols)
Error correction
Error detection
Information theory
Instruments

*Codes, Symbolic**
USE: Codes (symbols)

*Codes, Symbolic—Coding errors**
USE: Coding errors

*Codes, Symbolic—Decoding**
USE: Decoding

*Codes, Symbolic—Encoding**
USE: Encoding (symbols)

*Codes, Symbolic—Error correction**
USE: Error correction

*Codes, Symbolic—Error detection**
USE: Error detection

*Codes, Symbolic—Error statistics**
USE: Error statistics

Coding (images)
USE: Image coding

Coding (signals)
USE: Signal encoding

Coding (speech)
USE: Speech coding

Coding (symbols)
USE: Encoding (symbols)

Coding errors 723.2
DT: January 1993
UF: Codes, Symbolic—Coding errors*
BT: Errors
RT: Binary codes
Block codes
Codes (symbols)
Error analysis
Error compensation
Error correction
Error detection
Trellis codes

Coenzymes (461.2) (801.2) (804.1)
DT: January 1987
BT: Enzymes
NT: Adenosinetriphosphate

Coercive force 701.2
DT: January 1993
UF: Coercivity
Magnetic materials—Coercivity*
RT: Magnetic hysteresis
Magnetic materials
Magnetic properties
Magnetization
Remanence

Coercivity
USE: Coercive force

Coextrusion 816.1
DT: January 1993
BT: Extrusion
RT: Cladding (coating)
Extruders

Cofferdams (405.2)
DT: Predates 1975
BT: Construction equipment
Hydraulic structures
RT: Excavation

Cogeneration plants 614 (642) (643)
DT: January 1981
BT: Electric power plants
NT: Bottoming cycle systems
Topping cycle systems
RT: District heating

*Cogeneration plants—Bottoming systems**
USE: Bottoming cycle systems

*Cogeneration plants—Topping systems**
USE: Topping cycle systems

Cognitive systems (723.4) (731.1)
DT: January 1993
UF: Systems science and cybernetics—Cognitive systems*
BT: Cybernetics
RT: Adaptive systems
Artificial intelligence
Automata theory
Learning systems

Coherence
USE: Coherent light

* Former Ei Vocabulary term

Coherent light 741.1
DT: January 1993
UF: Coherence
Light coherence
Light—Coherent*
Optical coherence
BT: Light
NT: Laser beams
RT: Acoustooptical effects
Laser theory
Lasers
Optical properties
Quantum optics
Speckle

Coils (electric)
USE: Electric coils

Coin operated equipment (718.1)
SN: Including equipment which accepts paper currency or credit cards
DT: January 1993
UF: Coin operation
BT: Equipment
NT: Coin operated telephones
Vending machines

Coin operated telephones 718.1
SN: Including telephones which accept credit cards
DT: January 1993
UF: Pay telephones
Telephone apparatus—Coin operation*
BT: Coin operated equipment
Telephone apparatus
RT: Telephone sets

Coin operation
USE: Coin operated equipment

Coinage 535.2
DT: Predates 1975
BT: Metal stamping
RT: Mints

Coincidence circuits 713.5
DT: January 1993
UF: Circuits, Coincidence*
Electronic circuits, Coincidence*
BT: Pulse circuits
RT: Counting circuits
Synchronization

Coke 524
DT: Predates 1975
BT: Materials
NT: Metallurgical coke
Petroleum coke
RT: Carbon
Coal
Coal carbonization
Coke fired boilers
Coke fueled furnaces
Coke manufacture
Coking
Coking properties
Fuels
Furnace fuels
Graphitization

Coke fired boilers 524, 614.2
DT: January 1993
UF: Boiler firing—Coke*
BT: Boilers
RT: Coke

Coke fueled furnaces 524 (532) (534) (537.2) (642)
DT: January 1993
UF: Furnaces—Coke fuel*
BT: Furnaces
RT: Coke

Coke furnaces
USE: Furnaces

Coke manufacture 524 (522)
DT: Predates 1975
BT: Manufacture
RT: Coke
Coke plants

Coke ovens 524, 642.2 (522)
DT: Predates 1975
BT: Coke plants
Industrial ovens
RT: Coal carbonization

*Coke ovens—Metallurgical**
USE: Metallurgical furnaces

Coke plants 524 (402.1) (522)
DT: Predates 1975
UF: Coking plants
BT: Industrial plants
NT: Coke ovens
RT: Coke manufacture
Coke quenching towers
Coke quenching cars
Coking

*Coke plants—Quenching cars**
USE: Coke quenching cars

*Coke plants—Quenching towers**
USE: Coke quenching towers

Coke quenching cars 524 (522)
DT: January 1993
UF: Coke plants—Quenching cars*
RT: Coke plants

Coke quenching towers 524 (522)
DT: January 1993
UF: Coke plants—Quenching towers*
BT: Towers
RT: Coke plants

*Coke—Briquetting**
USE: Briquetting

*Coke—Graphitization**
USE: Graphitization

*Coke—Metallurgical**
USE: Metallurgical coke

Coking 524 (522)
DT: January 1993
BT: Carbonization
RT: Coal carbonization
Coke
Coke plants
Petroleum refining

Coking plants
USE: Coke plants

Coking properties 524
DT: January 1993
UF: Coal—Coking properties*
RT: Coal
Coke

Cold cathode tubes 714.1
DT: January 1993
UF: Electron tubes, Cold cathode*
BT: Electron tubes
NT: Counting tubes
Phototubes

Cold ductility
USE: Ductility

Cold effects (421) (443) (461.1)
DT: January 1993
BT: Low temperature effects
RT: Arctic buildings
Cold weather problems
Frost effects
Hypothermia
Low temperature phenomena
Low temperature properties
Temperature

Cold forming machines 535.2.1
DT: January 1993
UF: Plastics machinery—Cold forming machines*
BT: Plastics machinery
RT: Plastics forming

Cold fusion
USE: Fusion reactions

Cold heading 535.2.2 (605.2) (545.3)
DT: January 1993
UF: Bolts and nuts—Cold heading*
Metal forming—Cold heading*
Steel—Cold heading*
BT: Cold working
RT: Bolts
Steel
Upsetting (forming)

Cold region buildings
USE: Arctic buildings

Cold region engineering
USE: Arctic engineering

Cold region vehicles
USE: Arctic vehicles

Cold rolling 535.1.2
DT: January 1993
UF: Rolling mill practice—Cold rolling*
BT: Cold working
Rolling mill practice
RT: Cold rolling mills
Thread rolling

Cold rolling mills 535.1.1
DT: January 1993
BT: Rolling mills
RT: Cold rolling
Strip mills

Cold storage 644.3, 694.4
DT: Predates 1975
BT: Storage (materials)
NT: Refrigeration

Cold weather problems (409) (443) (483.1)
DT: January 1993
RT: Cold effects
Frost effects
Frozen soils
Muskeg
Snow
Snow and ice removal
Undersnow structures

Cold welding 538.2.1
DT: January 1993
UF: Welding—Cold method*
BT: Pressure welding

Cold working 535.2 (545.3)
DT: January 1993
UF: Metal forming—Cold working*
Steel—Cold working*
BT: Metal working
NT: Cold heading
Cold rolling
RT: Hot working
Metal spinning
Piercing
Punching
Sawing
Straightening
Strain hardening
Stretching
Swaging
Trimming
Upsetting (forming)

Coliform bacteria (461.2)
DT: January 1987
UF: Bacteriology—Coliforms*
BT: Bacteria

Collagen (804.1) (814) (461.2)
DT: January 1986
BT: Proteins

Collapsible containers (691.1) (694.2)
DT: January 1993
UF: Containers—Collapsible*
BT: Containers

Collapsible tubes 619.1 (632)
DT: January 1993
UF: Tubes—Collapsible*
BT: Tubes (components)

* Former Ei Vocabulary term

Collapsible warehouses 402.1 (694.4)
DT: January 1993
UF: Warehouses—Collapsible*
BT: Warehouses

Collection (refuse)
USE: Refuse collection

Collectors (dust)
USE: Dust collectors

Collectors (solar)
USE: Solar collectors

College buildings 402.2
DT: Predates 1975
UF: University buildings
BT: School buildings

Colliders
USE: Colliding beam accelerators

Colliding beam accelerators 932.1.1
DT: January 1993
UF: Accelerators—Colliding beam*
Colliders
BT: Particle accelerators

Collimators (optical)
USE: Optical collimators

Collision avoidance 914.1 (431.5)
DT: January 1995
UF: Obstacle avoidance
RT: Accident prevention
Air traffic control
Mobile robots
Motion planning
Navigation

Collision processes (plasmas)
USE: Plasma collision processes

Colloid chemistry 801.3
DT: January 1993
BT: Chemistry
RT: Colloids
Physical chemistry

Colloidal graphite 801.3
DT: January 1993
UF: Graphite—Colloidal*
BT: Colloids
Graphite

Colloids 801.3
DT: January 1986
BT: Dispersions
NT: Aerosols
Colloidal graphite
Gels
Hydrogels
Latexes
Micelles
Protective colloids
Sol-gels
Sols
RT: Brownian movement
Coalescence
Colloid chemistry
Critical micelle concentration
Dialysis
Electrophoresis
Flocculation
Foams
Gelation
Interfacial energy
Macromolecules
Non Newtonian liquids
Particles (particulate matter)
Phase interfaces
Precipitation (chemical)

*Colloids—Brownian movement**
USE: Brownian movement

*Colloids—Coalescence**
USE: Coalescence

*Colloids—Critical micelle concentration**
USE: Critical micelle concentration

*Colloids—Interfacial energy**
USE: Interfacial energy

*Colloids—Micelles**
USE: Micelles

*Colloids—Microemulsions**
USE: Microemulsions

*Colloids—Phase interfaces**
USE: Phase interfaces

*Colloids—Protective**
USE: Protective colloids

*Colloids—Thermodynamic stability**
USE: Thermodynamic stability

Color 741.1
DT: Predates 1975
UF: Colour
Enamel—Color*
BT: Optical properties
RT: Chromogenics
Color codes
Color computer graphics
Color fastness
Color image processing
Color matching
Color photocopying
Color photography
Color printing
Color removal (water treatment)
Color television
Color vision
Colorimeters
Electrochromism
Electromagnetic dispersion
Light
Photochromism
Photography
Spectrophotometers
Surface properties
Visibility

* Former Ei Vocabulary term

Color blindness
USE: Color vision

Color centers 933.1 (482) (801)
DT: January 1993
UF: Crystals—Color centers*
BT: Point defects
RT: Paramagnetic resonance

Color codes (706.2)
DT: January 1993
UF: Electric cables—Color codes*
BT: Codes (symbols)
RT: Color

Color comparison instruments
USE: Colorimeters

Color computer graphics (722.2) (723.5)
DT: January 1993
UF: Computer graphics—Color*
BT: Computer graphics
RT: Color

Color fastness (802.2) (803)
DT: January 1993
UF: Dyes and dyeing—Color fastness*
Paper—Color fastness*
Textiles—Color fastness*
RT: Color
Paper
Textiles

Color films 742.3
DT: January 1993
UF: Color plates (photography)
Photographic films and plates, Color*
BT: Photographic films
RT: Color photography
Photosensitivity

Color image processing 741.1
DT: January 1993
UF: Image processing—Color images*
BT: Image processing
RT: Color

Color matching (801) (803)
DT: January 1993
UF: Color—Matching*
Dyes and dyeing—Color matching*
RT: Color
Dyeing
Dyes

Color motion pictures 742.1
DT: January 1993
UF: Motion pictures, Color*
BT: Motion pictures
RT: Color photography

Color photocopying 745.2
DT: January 1993
UF: Photographic reproduction—Color*
BT: Photocopying
RT: Color

Color photography 742.1
DT: January 1993
UF: Photography, Color*
BT: Photography
RT: Color
Color films
Color motion pictures

Color plates (photography)
USE: Color films

Color printing 745.1
DT: January 1993
UF: Printing—Color*
BT: Printing
RT: Color

Color removal (water treatment) 445.1
DT: January 1993
UF: Water treatment—Color removal*
BT: Removal
Water treatment
RT: Chemicals removal (water treatment)
Color

Color television 716.4
DT: January 1993
UF: Color TV
Colour television
Television, Color*
BT: Television
RT: Color
Color television receivers
Color television transmitters
Television broadcasting

Color television receivers 716.4
DT: January 1993
UF: Television receivers, Color*
BT: Television receivers
RT: Color television
Color television transmitters

Color television transmitters 716.4
DT: January 1993
UF: Television transmitters, Color*
BT: Television transmitters
RT: Color television
Color television receivers

Color TV
USE: Color television

Color vision 741.2
DT: January 1993
UF: Color blindness
Vision, Color*
BT: Vision
RT: Color

*Color—Matching**
USE: Color matching

*Color—Terminology**
USE: Terminology

Colorimeters 941.3
DT: Predates 1975
UF: Chromometers
Color comparison instruments
BT: Comparators (optical)

* Former Ei Vocabulary term

Colorimeters *(continued)*
RT: Analytic equipment
Color
Colorimetric analysis
Colorimetry
Spectrophotometers

Colorimetric analysis 801
SN: Measurement of the concentration of known constituent in solution by comparison with the color of known concentrations of the constituent
DT: Predates 1975
BT: Chemical analysis
RT: Colorimeters
Spectrophotometers
Spectrophotometry

Colorimetry 941.4
SN: Measurement of colors by determining the intensities of the three primary colors that make up a given color
DT: Predates 1975
BT: Optical variables measurement
RT: Colorimeters
Spectrophotometers

Coloring 802.3 (812.3) (813.1)
DT: January 1993
BT: Finishing
NT: Dyeing
RT: Coating techniques
Decoration
Discoloration
Glazes
Painting

Colour
USE: Color

Colour television
USE: Color television

Columbium
USE: Niobium

Column chromatography 801
DT: January 1993
UF: Chromatographic analysis—Column*
BT: Chromatography

Columns (distillation)
USE: Distillation columns

Columns (structural) 408.2
DT: January 1993
UF: Columns*
BT: Structural members
RT: Beams and girders
Poles
Studs (structural members)
Supports

*Columns**
USE: Columns (structural)

Combinational circuits
USE: Combinatorial circuits

Combinational mathematics
USE: Combinatorial mathematics

Combinational switching
USE: Combinatorial switching

Combinatorial circuits 721.2 (713.5)
DT: January 1993
UF: Combinational circuits
Logic circuits, Combinatorial*
BT: Logic circuits
RT: Counting circuits
Sequential circuits

Combinatorial mathematics 921.4 (721.1)
DT: January 1993
UF: Combinational mathematics
Mathematical techniques—Combinatorial mathematics*
BT: Mathematical techniques
NT: Graph theory
RT: Information theory
Set theory
Topology

Combinatorial switching 721.1
DT: January 1993
UF: Combinational switching
Switching theory—Combinatorial switching*
BT: Switching
RT: Switching theory

Combined cycle power plants 614, 612
DT: January 1993
UF: Combined gas and steam cycle power plants
Power plants—Gas and steam turbine combined*
BT: Power plants
RT: Steam turbines

Combined gas and steam cycle power plants
USE: Combined cycle power plants

Combined sewers 452
SN: Systems that receive both runoff and sewage
DT: January 1993
UF: Sewers—Combined*
BT: Sewers
RT: Drainage
Rain
Runoff
Sanitary sewers
Storm sewers
Storms

Combines 821.1
DT: January 1993
UF: Agricultural machinery—Combines*
BT: Agricultural machinery
RT: Harvesting

Combustibility
USE: Flammability

Combustion 521.1
DT: Predates 1975
UF: Boiler control—Combustion*
Burning

* Former Ei Vocabulary term

Combustion *(continued)*
Incineration
BT: Chemical reactions
NT: Coal combustion
Fluidized bed combustion
Ignition
In situ combustion
Reverse combustion
Spontaneous combustion
Waste incineration
RT: Boiler firing
Calorific value
Combustion equipment
Combustion knock
Detonation
Fires
Flame research
Flammability
Furnaces
Heating
Oxidation
Oxygen blast enrichment
Reaction kinetics
Soot
Stratified charge engines
Temperature control
Water injection

Combustion chambers 654.2
DT: January 1993
UF: Rocket engines—Combustion chambers*
BT: Combustion equipment
Rocket engines
RT: Furnaces

Combustion equipment 521.2
DT: January 1993
BT: Equipment
NT: Combustion chambers
Combustors
Furnaces
RT: Combustion
Internal combustion engines

Combustion knock 521.1, 612.1
DT: January 1993
UF: Internal combustion engines—Combustion knock*
Knock (engines)
RT: Antiknock compounds
Antiknock rating
Combustion
Ignition systems
Internal combustion engines

*Combustion—Fluidized beds**
USE: Fluidized bed combustion

*Combustion—Noise**
USE: Acoustic noise

*Combustion—Reverse**
USE: Reverse combustion

*Combustion—Spontaneous**
USE: Spontaneous combustion

Combustors 521.2 (802.1)
DT: January 1987
UF: Gas turbines—Combustors*
BT: Combustion equipment
NT: Fluidized bed combustors
RT: Fuel burners
Gas turbines

*Combustors—Fluidized bed**
USE: Fluidized bed combustors

Command and control systems 731.1
DT: June 1990
UF: C3 systems
BT: Computer control systems
RT: Decision making
Military communications

Commerce
USE: International trade

Comminution (533.1) (802.3) (483)
DT: January 1993
UF: Milling (comminution)
Pulverization
Pulverization*
BT: Processing
NT: Crushing
Grinding (comminution)
RT: Disintegration
Granulation

Communicating robots
USE: Intelligent robots

Communication 716
SN: Communication in general, and personal and local communication. For communication at a distance use TELECOMMUNICATION. Very general term; prefer specific type of communication
DT: January 1993
NT: Biocommunications
Information dissemination
Signaling
Speech communication
Telecommunication
RT: Communication systems

Communication aids for nonvocal persons 461.5 (751.5)
DT: January 1993
UF: Human rehabilitation engineering—Communication aids for nonvocal
Nonvocal aids
Speech aids
BT: Human rehabilitation equipment
RT: Human rehabilitation engineering

Communication cables
USE: Telecommunication cables

Communication channels (information theory) 716.1
DT: January 1993
UF: Information theory—Communication channels*
BT: Information theory

* Former Ei Vocabulary term

Communication links
USE: Telecommunication links

Communication networks
USE: Telecommunication networks

Communication protocols
USE: Network protocols

Communication satellites 655.2 (716) (717)
SN: Extraterrestrial units of long-range communication relay systems
DT: January 1993
BT: Satellites
Telecommunication equipment
RT: Airborne television transmitters
Geostationary satellites
Satellite relay systems

Communication systems 716
DT: January 1993
UF: Civil defense—Communications*
NT: Signal systems
Telecommunication systems
RT: Communication
Military communications
Regional planning
Transportation

Communication theory
USE: Information theory

Communication traffic
USE: Telecommunication traffic

Communities (ecological)
USE: Ecosystems

Commutation
USE: Electric commutation

Commutators
USE: Electric commutators

Compact discs
USE: Compact disks

Compact disk players 752.3.1
DT: June 1990
UF: CD players
BT: Electronic equipment
RT: Audio equipment
CD-ROM
Computer peripheral equipment
Television equipment

Compact disks 722.1, 752.3.1
DT: January 1993
UF: Compact discs
Sound recording—Compact disks*
Sound reproduction—Compact disks*
NT: CD-ROM
RT: Phonograph records
ROM
Sound recording
Sound reproduction
Stereophonic recordings

Compaction (536.1) (802.3)
SN: The process of artificially increasing the density of material, as by compression or vibration
DT: Predates 1975
UF: Compression (materials)
Powder metallurgy—Compacting*
Waste disposal—Compaction*
BT: Processing
NT: Briquetting
Refuse compaction
RT: Compressibility
Consolidation
Densification
Forming
Hot pressing
Pelletizing
Preforming
Pressing (forming)
Vibrations (mechanical)
Waste disposal

Compandor circuits 713.5
DT: January 1993
UF: Compandors
Electronic circuits, Compandor*
BT: Networks (circuits)
Telecommunication equipment
RT: Interference suppression
Pulse code modulation
Telephone equipment
Telephone interference

Compandors
USE: Compandor circuits

Comparator circuits 713.5 (703.1)
DT: January 1993
UF: Electronic circuits, Comparator*
Voltage comparators
BT: Networks (circuits)
NT: Phase comparators
RT: Analog to digital conversion
Delay circuits
Discriminators
Multivibrators
Pulse circuits
Telephone lines

Comparators (optical) 741.3
DT: January 1993
UF: Comparators*
BT: Optical instruments
NT: Colorimeters
RT: Densitometers
Exposure meters
Photometers
Reflectometers

*Comparators**
USE: Comparators (optical)

Compasses (magnetic) 942.3
DT: January 1993
UF: Compasses*
BT: Instruments

* Former Ei Vocabulary term

Compasses (magnetic) *(continued)*
RT: Magnetic variables measurement
Surveying instruments

*Compasses**
USE: Compasses (magnetic)

Compensation (errors)
USE: Error compensation

Compensation (personnel) 912.4
DT: January 1993
NT: Pension plans
Wages
RT: Personnel

Competition 911.2
DT: January 1993
RT: Economics

Compilers (program)
USE: Program compilers

Complementary metal oxide semiconductor integrated circuits
USE: CMOS integrated circuits

Completion (oil well)
USE: Oil well completion

Completion (well)
USE: Well completion

Complex systems
USE: Large scale systems

Complexation 802.2
DT: January 1993
BT: Chemical reactions
RT: Chelation
Ions
Organometallics

Components (601.2)
DT: January 1993
UF: Parts
NT: Airfoils
Axles
Bellows
Bridge components
Bushings
Chains
Connectors (structural)
Couplings
Diaphragms
Electron tube components
Fasteners
Fiber optic components
Flanges
Hinges
Hydrofoils
Knobs
Lenses
Machine components
Mountings
Network components
Nozzles
Orifices
Pendulums
Pipe
Pistons
Plastic parts
Rings (components)
Seals
Shims
Springs (components)
Sprockets
Structural members
Supports
Suspensions (components)
Tubes (components)
Vents
Waveguide components
RT: Accessories
Equipment
Weibull distribution

Composite beams and girders 408.2, 412
DT: January 1993
UF: Beams and girders—Composite*
BT: Beams and girders
RT: Composite materials

Composite bridges 401.1 (415)
DT: January 1993
UF: Bridges, Composite*
BT: Bridges
Composite structures
NT: Masonry bridges

Composite materials (415) (811) (817.1)
SN: Also use the type of material or component, e.g., AIRCRAFT MATERIALS; FLOORS
DT: Predates 1975
UF: Aircraft materials—Composite materials*
Automobile materials—Composite materials*
Building materials—Composite materials*
Electric insulating materials—Composite materials*
Floors—Composite materials*
Poles—Composite materials*
Shipbuilding materials—Composite materials*
BT: Materials
NT: Carbon carbon composites
Ceramic matrix composites
Cermets
Concretes
Dental composites
Fiber reinforced materials
Laminated composites
Metallic matrix composites
Nonmetallic matrix composites
Reinforced plastics
Self lubricating composites
RT: Adhesion
Adhesives
Alloys
Composite beams and girders
Composite micromechanics
Composite structures
Intelligent materials
Mechanical alloying
Metals
Pultrusion

* Former Ei Vocabulary term

Composite materials *(continued)*
Reinforcement
Sandwich structures

*Composite materials—Fiber reinforced**
USE: Fiber reinforced materials

*Composite materials—Nonmetallic matrix composites**
USE: Nonmetallic matrix composites

*Composite materials—Self lubricating**
USE: Self lubricating composites

Composite micromechanics 931.1
DT: January 1995
UF: Micromechanics
RT: Composite materials
Mechanical properties

Composite structures 408.2 (415)
SN: Structures built of composite materials
DT: January 1977
BT: Structures (built objects)
NT: Composite bridges
RT: Composite materials

Composite systems
USE: Large scale systems

Composition (801)
DT: January 1993
UF: Chemical composition
Constitution of materials
Glass—Constitution*
NT: Phase composition
RT: Characterization
Chemical analysis
Composition effects
Electric properties
Glass
Mechanical properties
Physical properties
Stoichiometry

Composition effects (801)
SN: Property changes following changes in composition
DT: January 1993
BT: Effects
RT: Composition

Composting (821.5) (452)
DT: January 1993
UF: Waste disposal—Composting*
BT: Waste utilization
NT: Refuse composting
RT: Refuse disposal
Sewage
Waste disposal

Compound engines 612.1
DT: January 1993
UF: Internal combustion engines—Compounding*
BT: Engines
RT: Aircraft engines
Internal combustion engines
Steam engines

Compound helicopters 652.4
DT: January 1993
UF: Helicopters—Compound*
BT: Helicopters
RT: Helicopter rotors
VTOL/STOL aircraft

Compounding (chemical) 802.3 (816.1) (818.3)
DT: January 1993
UF: Plastics—Compounding*
BT: Chemical operations
NT: Rubber compounding
RT: Plastics

Compressed air (632.4) (804)
DT: Predates 1975
UF: Industrial plants—Compressed air*
Mines and mining—Compressed air*
BT: Air
RT: Compressed air motors
Compressors
Mining
Pneumatics
Pressure

Compressed air motors 632.4
DT: Predates 1975
UF: Air turbines
BT: Motors
RT: Compressed air
Compressors
Turbomachinery

*Compressed air—Receivers**
USE: Receivers (containers)

Compressibility
DT: January 1993
NT: Compressibility of gases
Compressibility of liquids
Compressibility of solids
RT: Compaction
Compressible flow
Compression testing
Compressors
Density (specific gravity)
Ductility
Elasticity
Mechanical properties

Compressibility of gases 931.2
DT: January 1993
UF: Gases—Compressibility*
BT: Compressibility
RT: Fluid mechanics
Gases

Compressibility of liquids 931.2
DT: January 1993
UF: Liquids—Compressibility*
BT: Compressibility
RT: Fluid mechanics
Liquids

Compressibility of solids 931.2 (421) (933)
DT: January 1995
BT: Compressibility

* Former Ei Vocabulary term

Compressibility of solids *(continued)*
Mechanical properties
RT: Compressive strength
Impact testing
Porosity
Solids

Compressible flow (631.1.2) (631.1.1)
DT: January 1993
UF: Flow of fluids—Compressible*
BT: Flow of fluids
RT: Aerodynamics
Compressibility

Compression (data)
USE: Data compression

Compression (image)
USE: Image compression

Compression (materials)
USE: Compaction

Compression molding (816.1)
DT: January 1993
UF: Plastics—Compression molding*
Pressure molding
BT: Molding
NT: Injection molding
RT: Plastics forming
Plastics molding

Compression testing (422.2) (943)
DT: January 1993
UF: Materials testing—Compression tests*
BT: Mechanical testing
RT: Compressibility
Compressive strength
High temperature testing
Impact testing
Tensile testing
Yield stress

Compressive strength (421)
DT: January 1995
UF: Crushing strength
BT: Strength of materials
RT: Compressibility of solids
Compression testing

Compressors 618.1
DT: Predates 1975
UF: Air compressors
BT: Machinery
NT: Free piston compressors
Superchargers
RT: Air
Air conditioning
Air engines
Blowers
Compressed air
Compressed air motors
Compressibility
Gas turbines
Pipeline compressor stations
Pumps
Refrigerating machinery
Refrigeration
Refrigerators
Turbomachinery
Vacuum pumps

*Compressors, Free piston**
USE: Free piston compressors

*Compressors—Portable**
USE: Portable equipment

Computability and decidability 721.1
DT: January 1993
UF: Automata theory—Computability and decidability*
BT: Formal logic
RT: Automata theory
Computational complexity

Computation
USE: Calculations

Computation theory 721.1 (921)
DT: January 1993
UF: Computer metatheory*
BT: Theory
NT: Automata theory
Binary sequences
Computational complexity
Computational linguistics
Formal logic
Programming theory
Theorem proving
RT: Algorithmic languages
Boolean algebra
Computer science
Switching theory

Computational complexity 721.1
DT: January 1993
UF: Computer metatheory—Computational complexity*
NP-complete problems
NP-hard problems
BT: Computation theory
RT: Computability and decidability

Computational fluid dynamics
723.5, 931.1 (631.1.1) (631.1.2) (921.6)
DT: January 1995
UF: CFD
BT: Fluid dynamics
RT: Computational methods
Computer applications
Differential equations
Equations of motion
Numerical analysis
Viscous flow

Computational geometry 723.5, 921.4
DT: January 1995
BT: Geometry
RT: Computer aided design
Computer graphics
Computer vision
Graph theory
Programming theory

* Former Ei Vocabulary term

Computational grammars 721.1
DT: January 1993
UF: Automata theory—Grammars*
BT: Computational linguistics
NT: Context free grammars
Context sensitive grammars
RT: Automata theory
Program compilers

Computational linguistics 721.1
DT: January 1993
UF: Automata theory—Computational linguistics*
Mathematical linguistics
Statistical linguistics
BT: Computation theory
Linguistics
NT: Computational grammars
RT: Formal languages

Computational methods (721.1) (921)
DT: January 1993
UF: Computers—Computational methods*
BT: Mathematical techniques
RT: Computational fluid dynamics
Computer science

Computer aided analysis 723.5
DT: January 1983
BT: Analysis
Computer applications
NT: Computer aided network analysis
RT: Computer aided design
Computer simulation

Computer aided design 723.5
DT: January 1977
UF: CAD
BT: Computer applications
Design
NT: Computer aided logic design
RT: Computational geometry
Computer aided analysis
Computer aided engineering
Computer aided manufacturing
Computer graphics
Computer hardware description languages
Computer simulation
Concurrent engineering
Design aids
Engineering
Interactive computer systems
Rapid prototyping
Simulation

Computer aided diagnosis 461.1, 723.5
DT: January 1993
UF: Biomedical engineering—Computer aided diagnosis*
BT: Computer applications
Diagnosis
RT: Biomedical engineering
Expert systems

Computer aided engineering 723.5
DT: January 1986
UF: CAE
BT: Computer applications
Engineering
RT: Computer aided design
Computer aided manufacturing
Expert systems

Computer aided instruction 723.5, 901.2
DT: June 1990
UF: CAI
Teaching machines*
BT: Computer applications
Education

Computer aided language translation 723.5 (721.1)
DT: January 1993
UF: Translating machines*
BT: Computer applications
Translation (languages)

Computer aided logic design 721.2, 723.5
DT: January 1993
UF: Logic CAD
Logic design—Computer aids*
BT: Computer aided design
Logic design

Computer aided manufacturing 723.5, 913.4.2
SN: Computer guidance of the automatic machinery used in production of goods
DT: January 1981
UF: CAM
BT: Computer applications
Manufacture
RT: Computer aided design
Computer aided engineering
Computer integrated manufacturing
Concurrent engineering
Factory automation
Rapid prototyping

Computer aided network analysis 703.1.1, 723.5
DT: January 1993
BT: Computer aided analysis
Electric network analysis

Computer aided software engineering 723.1, 723.5
DT: January 1993
UF: CASE
CASE tools
Software tools
BT: Software engineering
RT: Computer software
Computer software portability
Object oriented programming

Computer aided tomography
USE: Computerized tomography

Computer algorithms
USE: Algorithms

Computer applications 723.5
SN: For applications of computers to other equipment or technologies
DT: January 1993
UF: Computers—Special purpose application*

* Former Ei Vocabulary term

Computer applications *(continued)*
BT: Applications
NT: Computer aided analysis
Computer aided design
Computer aided diagnosis
Computer aided engineering
Computer aided instruction
Computer aided language translation
Computer aided manufacturing
Computer control systems
Computer generated holography
Computer graphics
Computer integrated manufacturing
Computer music
Computer simulation
Computerized tomography
Education computing
Medical computing
Natural sciences computing
Psychology computing
Social sciences computing
Software engineering
RT: Computational fluid dynamics
Computers
Data processing
Intelligent buildings
Intelligent vehicle highway systems
Programmable logic controllers
Special effects

Computer architecture (722) (723)
SN: Computer potential as revealed by its firmware, the combination of its hardware and software
DT: Predates 1975
UF: Architecture (of computers)
NT: Reduced instruction set computing
RT: Computer programming
Computers
Logic design
Microprogramming

*Computer architecture—Microprogramming**
USE: Microprogramming

*Computer architecture—Reduced instruction set computing**
USE: Reduced instruction set computing

Computer circuits 721.3 (703.1)
DT: January 1993
UF: Computers—Circuits*
BT: Computer hardware
Networks (circuits)
RT: Digital circuits
Logic circuits
Timing circuits

Computer control 723.5, 731.5 (922.2)
DT: January 1995
BT: Process control
RT: Statistical process control

Computer control systems 731.1, 723.5
DT: January 1993
UF: Computerized control systems
BT: Computer applications
Control systems
NT: Command and control systems
RT: Digital control systems
Numerical control systems

Computer crime (723) (902.3)
DT: June 1990
UF: Crime (computer)
NT: Computer viruses
Computer worms
RT: Computers
Security of data
Security systems

*Computer crime—Worms**
USE: Computer worms

Computer debugging 723.1 (722)
DT: January 1993
UF: Computers—Debugging*
Debugging (computers)
BT: Computer testing
RT: Program debugging

Computer generated holography 723.5, 743.1
DT: January 1993
BT: Computer applications
Holography
Image processing

Computer graphics 723.5 (921.4)
DT: Predates 1975
BT: Computer applications
NT: Color computer graphics
Interactive computer graphics
Three dimensional computer graphics
RT: Animation
Anti-aliasing
Computational geometry
Computer aided design
Computer graphics equipment
Computer software
Curve fitting
Fractals
Graphic methods
Graphical user interfaces
Object recognition
Simulation
Special effects
Virtual reality

Computer graphics equipment 722.2
DT: January 1993
UF: Computer peripheral equipment—Graphics*
BT: Computer peripheral equipment
RT: Computer graphics
Computer terminals
Computer workstations

*Computer graphics—Animation**
USE: Animation

*Computer graphics—Anti-aliasing**
USE: Anti-aliasing

*Computer graphics—Color**
USE: Color computer graphics

* Former Ei Vocabulary term

*Computer graphics—Interactive**
USE: Interactive computer graphics

*Computer graphics—Three dimensional graphics**
USE: Three dimensional computer graphics

Computer hardware 722
DT: January 1986
UF: Hardware (computer)
BT: Computers
NT: Code converters
Computer circuits
Computer peripheral equipment
Data storage equipment
Shift registers
RT: Computer systems
Data transfer

Computer hardware description languages 723.1.1 (722)
SN: Languages that facilitate development of digital systems
DT: January 1986
UF: Hardware description languages
Specification languages
BT: Computer programming languages
RT: Computer aided design
Expert systems
Formal languages
High level languages

Computer integrated manufacturing 723.5, 913.4.2
SN: Systems in which the functions of manufacturing enterprises are coordinated by computer-integrated operations
DT: January 1987
BT: Computer applications
Manufacture
RT: Computer aided manufacturing
Factory automation

*Computer interfaces**
USE: Interfaces (computer)

Computer keyboard substitutes (handicapped aid) 461.5, 722.2 (723.5)
DT: January 1993
UF: Braille keyboards
Human rehabilitation engineering—Computer keyboard substitutions*
Keyboard substitutes
BT: Human rehabilitation equipment
Interactive devices
RT: Computer keyboards
Human rehabilitation engineering
Voice activated input devices

Computer keyboards 722.2
DT: January 1993
UF: Computer peripheral equipment—Keyboards*
Keyboards (computer)
Keypads
BT: Interactive devices
RT: Computer keyboard substitutes (handicapped aid)
Typewriter keyboards

Computer languages
USE: Computer programming languages

Computer memory equipment
USE: Data storage equipment

*Computer metatheory**
USE: Computation theory

*Computer metatheory—Algorithmic languages**
USE: Algorithmic languages

*Computer metatheory—Binary sequences**
USE: Binary sequences

*Computer metatheory—Boolean algebra**
USE: Boolean algebra

*Computer metatheory—Boolean functions**
USE: Boolean functions

*Computer metatheory—Computational complexity**
USE: Computational complexity

*Computer metatheory—Equivalence classes**
USE: Equivalence classes

*Computer metatheory—Formal logic**
USE: Formal logic

*Computer metatheory—Majority logic**
USE: Majority logic

*Computer metatheory—Many valued logics**
USE: Many valued logics

*Computer metatheory—Probabilistic logics**
USE: Probabilistic logics

*Computer metatheory—Programming theory**
USE: Programming theory

*Computer metatheory—Threshold logic**
USE: Threshold logic

Computer monitors 722.2
DT: January 1993
UF: Monitors (computer)
BT: Display devices
Interactive devices

Computer music 723.5 (752)
DT: June 1990
UF: Music (computer)
Synthesized music
BT: Computer applications
RT: Computer software
Electronic musical instruments
Special effects

Computer networks (716) (717) (718) (723)
DT: January 1977
UF: Multiple processor systems
Networks (computer)
Queueing networks
NT: Gateways (computer networks)
Local area networks
Metropolitan area networks

* Former Ei Vocabulary term

Computer networks *(continued)*
Wide area networks
RT: Computer systems
Computers
Distributed database systems
Electric network analyzers
Network protocols
Packet networks
Packet switching

*Computer networks—Gateways**
USE: Gateways (computer networks)

*Computer networks—Local area networks**
USE: Local area networks

*Computer networks—Metropolitan area networks**
USE: Metropolitan area networks

*Computer networks—Protocols**
USE: Network protocols

*Computer networks—Wide area networks**
USE: Wide area networks

Computer operating procedures (722) (723)
DT: January 1993
UF: Computer operations
Computers—Operating procedures*
Operating procedures (computer)
RT: Computer operating systems
Computer system recovery
Computers
Management
Personnel

Computer operating systems (722) (723)
DT: Predates 1975
UF: Operating systems (computer)
BT: Computer software
NT: DOS
Input output programs
UNIX
RT: C (programming language)
Computer operating procedures
Computer systems programming
Computer system recovery
Distributed database systems
Report generators
Resource allocation
Storage allocation (computer)
Supervisory and executive programs
Virtual storage

*Computer operating systems—DOS**
USE: DOS

*Computer operating systems—Failure and recovery**
USE: Computer system recovery

*Computer operating systems—Program assemblers**
USE: Program assemblers

*Computer operating systems—Program compilers**
USE: Program compilers

*Computer operating systems—Program interpreters**
USE: Program interpreters

*Computer operating systems—Program processors**
USE: Program processors

*Computer operating systems—Program translators**
USE: Program translators

*Computer operating systems—Report generators**
USE: Report generators

*Computer operating systems—Storage allocation**
USE: Storage allocation (computer)

*Computer operating systems—UNIX**
USE: UNIX

Computer operations
USE: Computer operating procedures

Computer peripheral equipment 722.2
DT: Predates 1975
UF: Peripherals (computer)
BT: Computer hardware
NT: Character recognition equipment
Computer graphics equipment
Computer terminals
Interactive devices
Plotters (computer)
Printers (computer)
Punch card systems
Punch tape systems
Tape drives
RT: Compact disk players
Data storage equipment
Digital communication systems
Instrument displays
Interfaces (computer)
Optical data processing
Pattern recognition
Speech synthesis

*Computer peripheral equipment—Disk drives**
USE: Magnetic disk storage

*Computer peripheral equipment—Graphics**
USE: Computer graphics equipment

*Computer peripheral equipment—Keyboards**
USE: Computer keyboards

*Computer peripheral equipment—Light pens**
USE: Light pens

*Computer peripheral equipment—Mouse**
USE: Mice (computer peripherals)

*Computer peripheral equipment—Plotters**
USE: Plotters (computer)

*Computer peripheral equipment—Printers**
USE: Printers (computer)

*Computer peripheral equipment—Remote consoles**
USE: Remote consoles

*Computer peripheral equipment—Tape drives**
USE: Tape drives

*Computer peripheral equipment—Terminals**
USE: Computer terminals

Computer plotters
USE: Plotters (computer)

Computer printers
USE: Printers (computer)

Computer program listings 723.1
DT: January 1993
UF: Computer programs*
Listings (computer programs)
Program listings
BT: Computer software
RT: Computer programming
Subroutines

Computer programming 723.1
DT: Predates 1975
UF: Programming (computers)
NT: Computer systems programming
Heuristic programming
Logic programming
Microprogramming
Object oriented programming
Program debugging
Program diagnostics
Program documentation
Robot programming
Structured programming
RT: Computer architecture
Computer program listings
Computer software
Flowcharting
Macros
Programming theory
Subroutines
Systems science

Computer programming languages 723.1.1
DT: Predates 1975
UF: Computer languages
Programming languages
Robot programming languages
NT: Computer hardware description languages
EXAPT (programming language)
High level languages
Machine oriented languages
Report generators
RT: Algorithms
Flowcharting
Formal languages
Programming theory
Query languages

*Computer programming languages—Ada**
USE: Ada (programming language)

*Computer programming languages—ALGOL**
USE: ALGOL (programming language)

*Computer programming languages—APL**
USE: APL (programming language)

*Computer programming languages—BASIC**
USE: BASIC (programming language)

*Computer programming languages—C**
USE: C (programming language)

*Computer programming languages—COBOL**
USE: COBOL (programming language)

*Computer programming languages—EXAPT**
USE: EXAPT (programming language)

*Computer programming languages—Flowcharting**
USE: Flowcharting

*Computer programming languages—FORTH**
USE: FORTH (programming language)

*Computer programming languages—FORTRAN**
USE: FORTRAN (programming language)

*Computer programming languages—High level languages**
USE: High level languages

*Computer programming languages—LISP**
USE: LISP (programming language)

*Computer programming languages—List processing**
USE: List processing languages

*Computer programming languages—Machine orientation**
USE: Machine oriented languages

*Computer programming languages—Modula**
USE: Modula (programming language)

*Computer programming languages—Pascal**
USE: Pascal (programming language)

*Computer programming languages—PL/1**
USE: PL/1 (programming language)

*Computer programming languages—Problem orientation**
USE: Problem oriented languages

*Computer programming languages—Procedure orientation**
USE: Procedure oriented languages

*Computer programming languages—PROLOG**
USE: PROLOG (programming language)

*Computer programming—Algorithms**
USE: Algorithms

*Computer programming—Flowcharting**
USE: Flowcharting

* Former Ei Vocabulary term

*Computer programming—Logic programming**
USE: Logic programming

*Computer programming—Macros**
USE: Macros

*Computer programming—Object oriented programming**
USE: Object oriented programming

*Computer programming—Program debugging**
USE: Program debugging

*Computer programming—Program diagnostics**
USE: Program diagnostics

*Computer programming—Program documentation**
USE: Program documentation

*Computer programming—Spreadsheet**
USE: Spreadsheets

*Computer programming—Structured programming**
USE: Structured programming

*Computer programming—Subroutines**
USE: Subroutines

*Computer programs**
USE: Computer program listings

*Computer programs—Selection and evaluation**
USE: Computer software selection and evaluation

Computer reservation systems
USE: Reservation systems

Computer science (721) (722) (723)
DT: June 1990
RT: Computation theory
Computational methods
Computers
Heuristic programming
Information science

Computer security
USE: Security of data

Computer selection and evaluation 722 (912.2)
DT: January 1993
UF: Computers—Selection and evaluation*
RT: Computers
Evaluation
Management
Selection

Computer simulation 723.5
DT: October 1975
UF: Dynamic simulation
Modeling (computer)
Monte Carlo simulation
Numerical simulation
Simulation modeling
BT: Computer applications
Simulation
RT: Computer aided analysis
Computer aided design
Computer simulation languages
Models
Simulators

Computer simulation languages 723.1.1
DT: January 1986
UF: Simulation languages
BT: High level languages
RT: Computer simulation

Computer software 723
DT: January 1981
UF: Programs (computer)
Robot software
Robots, Industrial—Software*
Software (computer)
NT: Computer operating systems
Computer program listings
File editors
Macros
Parallel algorithms
Program processors
Subroutines
Supervisory and executive programs
Utility programs
RT: Computer aided software engineering
Computer graphics
Computer music
Computer programming
Computer software selection and evaluation
Computer systems
Computer systems programming
Computers
Industrial robots
Pipeline processing systems
Programming theory

Computer software portability 723
DT: January 1993
UF: Computer software—Portability*
Portability (software)
Software portability
RT: Computer aided software engineering
Standards

Computer software selection and evaluation 723 (912.2)
DT: January 1993
UF: Computer programs—Selection and evaluation*
RT: Computer software
Evaluation
Selection

*Computer software—Portability**
USE: Computer software portability

Computer storage equipment
USE: Data storage equipment

Computer subroutines
USE: Subroutines

Computer system recovery (722) (723)
DT: January 1993
UF: Computer operating systems—Failure and recovery*
Deadlocking (computers)
Error recovery (computers)
Failure (computers)
RT: Computer operating systems
Computer operating procedures
Computer systems
Fault tolerant computer systems

Computer systems (722) (723)
SN: The complete set of hardware and software of a computer or data processing system, including peripherals and interfaces to other computers, but excluding inter-computer networking connections
DT: January 1993
UF: Computer systems, Digital*
Digital computer systems
Systems (computer)
NT: Distributed computer systems
Fast response computer systems
Fault tolerant computer systems
Interactive computer systems
Online systems
Real time systems
Time sharing systems
RT: Computer hardware
Computer networks
Computer software
Computer system recovery
Computer systems programming
Computers
Digital computers
Response time (computer systems)

Computer systems programming 723.1
DT: Predates 1975
UF: System programming
Systems programming
BT: Computer programming
NT: Multiprogramming
RT: Computer operating systems
Computer software
Computer systems
Merging
Program processors
Program translators
Storage allocation (computer)
Supervisory and executive programs
System program documentation
Systems analysis

*Computer systems programming—Decision tables**
USE: Decision tables

*Computer systems programming—Documentation**
USE: System program documentation

*Computer systems programming—Input output programs**
USE: Input output programs

*Computer systems programming—Merging**
USE: Merging

*Computer systems programming—Multiprocessing programs**
USE: Multiprocessing programs

*Computer systems programming—Multiprogramming**
USE: Multiprogramming

*Computer systems programming—Sorting**
USE: Sorting

*Computer systems programming—Supervisory and executive programs.**
USE: Supervisory and executive programs

*Computer systems programming—Table lookup**
USE: Table lookup

*Computer systems programming—Time sharing programs**
USE: Time sharing programs

*Computer systems programming—Utility programs**
USE: Utility programs

*Computer systems, Digital**
USE: Computer systems

*Computer systems, Digital—Distributed**
USE: Distributed computer systems

*Computer systems, Digital—Fast response capability**
USE: Fast response computer systems

*Computer systems, Digital—Fault tolerant capability**
USE: Fault tolerant computer systems

*Computer systems, Digital—Interactive operation**
USE: Interactive computer systems

*Computer systems, Digital—Multiprocessing**
USE: Multiprocessing systems

*Computer systems, Digital—On line operation**
USE: Online systems

*Computer systems, Digital—Parallel processing**
USE: Parallel processing systems

*Computer systems, Digital—Pipeline processing**
USE: Pipeline processing systems

*Computer systems, Digital—Real time operation**
USE: Real time systems

*Computer systems, Digital—Time sharing**
USE: Time sharing systems

* Former Ei Vocabulary term

Computer terminals 722.2
DT: January 1993
UF: Computer peripheral equipment—Terminals*
Interactive terminals
Terminals (computer)
VDT
VDU
BT: Computer peripheral equipment
NT: Point of sale terminals
Remote consoles
RT: Computer graphics equipment
Computer workstations
Interactive devices
Mice (computer peripherals)

Computer testing (722)
DT: January 1993
BT: Electronic equipment testing
NT: Computer debugging
RT: Computers

Computer viruses (723)
DT: January 1993
UF: Viruses (computer crime)
BT: Computer crime
RT: Computer worms

Computer vision 741.2, 723.5 (731.6)
DT: June 1990
UF: Artificial vision
Machine vision
Robot vision
Robotics—Vision systems*
Robots, Industrial—Vision systems*
Vision (artificial)
BT: Image processing
Pattern recognition
RT: Computational geometry
Depth perception
Image sensors
Industrial robots
Object recognition
Pattern recognition systems
Robotics
Robots
Stereo vision
Vision

Computer workstations (722) (723)
DT: June 1990
UF: Workstations (computer)
BT: Computers
RT: Computer graphics equipment
Computer terminals
Microcomputers
Word processing

Computer worms (723)
DT: January 1993
UF: Computer crime—Worms*
Worms (computer crime)
BT: Computer crime
RT: Computer viruses

Computerized control systems
USE: Computer control systems

Computerized tomography 723.5 (461.1) (801) (531)
DT: June 1990
UF: CAT scans
Computer aided tomography
CT scans
BT: Computer applications
Radiography
RT: Biomedical engineering
Diagnostic radiography
Magnetic resonance imaging
Positron emission tomography

Computers (722) (723)
SN: For applications of computers use COMPUTER APPLICATIONS, plus the activity or equipment to which computers are applied
DT: Predates 1975
UF: Satellites—Computers*
Spacecraft—Computers*
BT: Electronic equipment
NT: Analog computers
Computer hardware
Computer workstations
Digital computers
Hybrid computers
RT: Automatic teller machines
Business machines
Computer applications
Computer architecture
Computer crime
Computer networks
Computer operating procedures
Computer science
Computer selection and evaluation
Computer software
Computer systems
Computer testing
Digital arithmetic
Information technology
Mathematical instruments
Numerical analysis
Office equipment
Pocket calculators
Systems engineering

*Computers, Analog**
USE: Analog computers

*Computers, Digital**
USE: Digital computers

*Computers, Hybrid**
USE: Hybrid computers

*Computers, Microcomputer**
USE: Microcomputers

*Computers, Microprocessor**
USE: Microcomputers

*Computers, Miniature**
USE: Minicomputers

*Computers, Minicomputer**
USE: Minicomputers

*Computers, Personal**
USE: Personal computers

*Computers, Supercomputer**
USE: Supercomputers

*Computers—Adders**
USE: Adders

*Computers—Carry logic**
USE: Carry logic

*Computers—Circuits**
USE: Computer circuits

*Computers—Computational methods**
USE: Computational methods

*Computers—Data communication equipment**
USE: Data communication equipment

*Computers—Data communication systems**
USE: Data communication systems

*Computers—Debugging**
USE: Computer debugging

*Computers—Differential analyzers**
USE: Analog differential analyzers

*Computers—Digital differential analyzers**
USE: Digital differential analyzers

*Computers—Direct analogs**
USE: Direct analogs

*Computers—Dividing circuits**
USE: Dividing circuits (arithmetic)

*Computers—Electrolytic tank**
USE: Electrolytic tanks

*Computers—Error compensation**
USE: Error compensation

*Computers—General purpose application**
USE: General purpose computers

*Computers—Multiplying circuits**
USE: Multiplying circuits

*Computers—Operating procedures**
USE: Computer operating procedures

*Computers—Selection and evaluation**
USE: Computer selection and evaluation

*Computers—Shift registers**
USE: Shift registers

*Computers—Special purpose application**
USE: Computer applications

*Computers—Summing circuits**
USE: Summing circuits

Concentration (process) (802.3)
DT: January 1993
BT: Separation
RT: Adsorption
Agglomeration
Coagulation
Crystallization
Dewatering
Extraction
Percolation (fluids)
Vaporization

Concentrators (solar)
USE: Solar concentrators

Concentrators (telecommunications)
USE: Multiplexing equipment

Concentric cables
USE: Coaxial cables

Concrete additives 412 (803)
DT: January 1993
UF: Concrete—Admixtures*
BT: Additives
RT: Concretes

Concrete aggregates 412
DT: Predates 1975
BT: Aggregates
Concretes
RT: Fly ash
Glass
Sawdust

*Concrete aggregates—Fly ash**
USE: Fly ash

*Concrete aggregates—Glass**
USE: Glass

*Concrete aggregates—Grading**
USE: Grading

*Concrete aggregates—Sawdust**
USE: Sawdust

Concrete beams and girders 408.2, 412
DT: January 1993
UF: Beams and girders—Concrete*
BT: Beams and girders
Concrete products
RT: Concrete construction

Concrete blocks 412, 408.2
DT: January 1993
UF: Concrete products—Blocks*
BT: Building materials
Concrete products
RT: Concrete construction

Concrete bridges 401.1, 412
DT: January 1993
UF: Bridges, Concrete*
BT: Masonry bridges
RT: Concrete construction

Concrete buildings 402, 412
DT: January 1993
UF: Buildings—Concrete*
BT: Buildings
RT: Concrete construction

Concrete construction 412 (405)
DT: Predates 1975
UF: Concrete structures
Floors—Concrete*
Footbridges—Concrete*

* Former Ei Vocabulary term

Concrete construction *(continued)*
Penstocks—Concrete*
Piles—Concrete*
Poles—Concrete*
Port structures—Concrete*
Pressure vessels—Concrete*
BT: Concrete industry
Masonry construction
NT: Concrete mixing
Concrete placing
Concrete vibrating
Shotcreting
Slip forming
RT: Anchorages (concrete construction)
Concrete beams and girders
Concrete blocks
Concrete bridges
Concrete buildings
Concrete dams
Concrete fences
Concrete mixers
Concrete pavements
Concrete shell roofs
Concrete slabs
Concrete tanks
Concretes
Forms (concrete)

*Concrete construction—Anchorages**
USE: Anchorages (concrete construction)

*Concrete construction—Blast resistance**
USE: Blast resistance

*Concrete construction—Forms**
USE: Forms (concrete)

*Concrete construction—Joints**
USE: Joints (structural components)

*Concrete construction—Precast**
USE: Precast concrete

*Concrete construction—Prestressing**
USE: Prestressed concrete

*Concrete construction—Pump placing**
USE: Concrete placing

*Concrete construction—Reinforced concrete**
USE: Reinforced concrete

*Concrete construction—Shotcreting**
USE: Shotcreting

*Concrete construction—Slip forming**
USE: Slip forming

*Concrete construction—Underwater**
USE: Underwater construction

*Concrete construction—Waterproofing**
USE: Waterproofing

Concrete dams 412, 441.1
DT: January 1993
UF: Dams, Concrete*
BT: Dams
RT: Concrete construction

Concrete fences 412 (402) (408.2)
DT: January 1993
UF: Fences—Concrete*
BT: Fences
RT: Concrete construction

Concrete forms
USE: Forms (concrete)

Concrete industry 412
DT: Predates 1975
BT: Industry
NT: Concrete construction
RT: Cement industry
Concretes
Construction industry

Concrete mixers 405.1, 412
DT: January 1993
UF: Concrete mixers and mixing*
BT: Mixers (machinery)
RT: Concrete construction
Concrete mixing
Roadbuilding machinery

*Concrete mixers and mixing**
USE: Concrete mixers OR
Concrete mixing

Concrete mixing 405.2, 412
DT: January 1993
UF: Concrete mixers and mixing*
BT: Concrete construction
Mixing
RT: Concrete mixers

Concrete pavements 406, 412
DT: January 1993
UF: Airport runways—Concrete*
Concrete runways
Pavements—Concrete*
Roads and streets—Concrete*
BT: Pavements
RT: Concrete construction
Concrete products
Road construction

Concrete pipe 412, 619.1
DT: January 1993
UF: Concrete pipelines
Pipe, Concrete*
Pipelines, Concrete*
BT: Concrete products
Pipe

Concrete pipelines
USE: Pipelines AND
Concrete pipe

Concrete placing 405.2, 412
DT: January 1993
UF: Concrete construction—Pump placing*
Placing (concrete)
BT: Concrete construction

Concrete products 412
DT: Predates 1975
NT: Concrete beams and girders

Concrete products *(continued)*
Concrete blocks
Concrete pipe
Concrete slabs
RT: Concrete pavements
Concrete reinforcements
Concrete shell roofs
Concretes

*Concrete products—Blocks**
USE: Concrete blocks

*Concrete products—Slabs**
USE: Concrete slabs

Concrete reinforcements 412.2
DT: Predates 1975
BT: Materials
Reinforcement
RT: Concrete products
Reinforced concrete

*Concrete reinforcements—Bond**
USE: Bond (masonry)

Concrete runways
USE: Concrete pavements AND
Airport runways

Concrete shell roofs 402, 408.2, 412
DT: January 1993
UF: Roofs—Concrete shell*
BT: Roofs
RT: Concrete construction
Concrete products

Concrete slabs 408.2, 412
DT: January 1993
UF: Concrete products—Slabs*
BT: Concrete products
RT: Concrete construction

Concrete structures
USE: Concrete construction

Concrete tanks 412, 619.2
DT: January 1993
UF: Oil tanks—Concrete*
Sewage tanks—Concrete*
Tanks—Concrete*
BT: Tanks (containers)
RT: Concrete construction

Concrete testing 412 (422.2)
DT: Predates 1975
BT: Materials testing
RT: Concretes

Concrete vibrating 405.2, 412
SN: The process, not the phenomenon
DT: January 1993
UF: Vibrating (concrete)
BT: Concrete construction
RT: Vibrations (mechanical)

*Concrete**
USE: Concretes

*Concrete—Admixtures**
USE: Concrete additives

*Concrete—Disintegration**
USE: Disintegration

*Concrete—Expansive**
USE: Expansive concrete

*Concrete—Light weight**
USE: Light weight concrete

*Concrete—Plaster adherence**
USE: Plaster AND
Adhesion

*Concrete—Seawater effects**
USE: Seawater effects

*Concrete—Termite proofing**
USE: Termite proofing

*Concrete—Waterproofing**
USE: Waterproofing

Concretes 412
DT: January 1993
UF: Concrete*
Mines and mining—Concrete lining*
Roadbuilding materials—Concrete*
BT: Composite materials
Masonry materials
NT: Concrete aggregates
Expansive concrete
Light weight concrete
Precast concrete
Prestressed concrete
Refractory concrete
Reinforced concrete
RT: Cements
Concrete additives
Concrete construction
Concrete industry
Concrete products
Concrete testing
Portland cement
Roadbuilding materials
Setting

Concurrency control (723.3) (731.3)
DT: January 1995
BT: Control
RT: Database systems
Distributed database systems
Multiprocessing programs
Multiprocessing systems
Parallel processing systems
Synchronization

Concurrent engineering 723.5, 913.6 (913.1)
DT: January 1995
UF: Simultaneous engineering
BT: Engineering
RT: Computer aided design
Computer aided manufacturing
Just in time production
Production engineering
Systems engineering

* Former Ei Vocabulary term

Condensate return lines 619.1 (642) (643.2)
DT: January 1993
UF: Return lines (condensate)
Steam pipelines—Condensate return*
BT: Steam pipelines
RT: Steam condensate
Steam condensers
Steam piping systems
Steam separators and traps

Condensate treatment 452.4
DT: January 1993
UF: Steam condensate treatment
Steam condensate—Treatment*
RT: Steam condensate
Water treatment

Condensation 802.3
DT: January 1993
BT: Phase transitions
NT: Polycondensation
RT: Dewatering
Distillation
Liquefaction of gases
Nucleation
Sublimation
Vapors

Condensation polymerization
USE: Polycondensation

Condensation reactions 802.2
DT: January 1993
BT: Chemical reactions

Condenser tubes (616.1) (644.3) (801.1) (619.1)
DT: January 1993
UF: Steam condensers—Tubes*
BT: Steam condensers
Tubes (components)

Condensers (electric)
USE: Capacitors

Condensers (liquefiers) (802.1) (644.3) (616.1)
DT: January 1993
UF: Chemical equipment—Condensers*
Heat exchangers—Condensers*
Liquefiers
Refrigerating machinery—Condensers*
BT: Chemical equipment
NT: Steam condensers
RT: Heat exchangers
Refrigerating machinery

Conditioners (soil)
USE: Soil conditioners

Conditioning (natural gas)
USE: Natural gas conditioning

Conductance (electric)
USE: Electric resistance

Conduction (heat)
USE: Heat conduction

Conductive films 708.2
DT: January 1993
UF: Films—Conducting*
BT: Conductive materials
Films

Conductive materials 708.2
DT: January 1993
BT: Materials
NT: Conductive films
Conductive plastics
Electrolytes
RT: Eddy currents
Electric conductivity
Electric conductors
Skin effect

Conductive plastics 708.2, 817.1
DT: January 1993
UF: Plastics—Conductive*
BT: Conductive materials
Plastics

Conductivity (electric)
USE: Electric conductivity

Conductor skin effect
USE: Skin effect

Conductors (electric)
USE: Electric conductors

Conduit flow
USE: Confined flow

Conduits (electric)
USE: Electric conduits

Cones (631.1)
DT: January 1993
UF: Flow of fluids—Cones*
NT: Nose cones
RT: Bodies of revolution

Conference calls 718.1
DT: January 1993
UF: Telephone—Conference calls*
BT: Telephone
RT: Teleconferencing

Confined flow 631.1
DT: January 1993
UF: Conduit flow
Flow confinement
Flow of fluids—Conduits*
BT: Flow of fluids
NT: Channel flow
Pipe flow
RT: Flow patterns

Confinement (plasmas)
USE: Plasma confinement

Conformal mapping (921)
DT: January 1993
UF: Mathematical techniques—Conformal mapping*
BT: Mapping
Mathematical transformations

* Former Ei Vocabulary term

Conformations 801.4 (931.3) (933.1)
DT: January 1995
RT: Atoms
Chemical bonds
Crystal orientation
Molecules

Congestion control (communication) (716) (717) (718)
DT: January 1995
BT: Control
RT: Data communication systems
Telecommunication networks

Conjugated polymers
USE: Organic polymers

Connecting rods (682.1.2) (601) (602)
DT: Predates 1975
UF: Locomotives—Connecting rods*
BT: Mechanisms
RT: Crankshafts
Internal combustion engines
Locomotives
Pistons

Connections (waveguide)
USE: Waveguide couplers

Connectors (electric)
USE: Electric connectors

Connectors (structural) (408.2)
DT: January 1993
UF: Wire rope—Connectors*
BT: Components
RT: Fasteners
Joints (structural components)

Conservation (454)
DT: January 1977
NT: Energy conservation
Soil conservation
Water conservation
RT: Natural resources

Consoles (remote)
USE: Remote consoles

Consolidation (483)
DT: January 1993
UF: Granular materials—Consolidation*
Soils—Consolidation*
RT: Compaction
Densification
Granular materials
Soils

Constitution of materials
USE: Composition

Constraint theory (721.1) (731.1)
DT: January 1993
UF: Systems science and cybernetics—Constraint theory*
BT: Systems science
RT: Dynamic programming
Linear programming
Nonlinear programming
Operations research
Optimization

Construction
DT: January 1993
UF: Helicopters—Electric line construction*
Kilns—Erection*
NT: Lift slab construction
Masonry construction
Modular construction
Pile driving
Prefabricated construction
Prestressing
Road construction
Shipbuilding
Steel construction
Underwater construction
Wooden construction
RT: Architecture
Blasting
Building moving
Buildings
Construction equipment
Construction industry
Excavation
Fabrication
Honeycomb structures
River basin projects
Surveying
Surveys
Tunneling (excavation)

Construction equipment 405.1
DT: Predates 1975
BT: Equipment
NT: Anchorages (concrete construction)
Caissons
Cofferdams
Military construction equipment
Pile drivers
Roadbuilding machinery
Scaffolds
Shovels
RT: Construction
Conveyors
Cranes
Dredges
Earthmoving machinery
Elevators
Forms (concrete)
Hand tools
Hoists

*Construction equipment—Military**
USE: Military construction equipment

*Construction equipment—Transmissions**
USE: Transmissions

Construction industry 405
DT: Predates 1975
UF: Helicopters—Construction industry*
BT: Industry
RT: Cement industry
Concrete industry
Construction

* Former Ei Vocabulary term

*Construction industry—Laws and regulations**
USE: Laws and legislation

Construction materials
USE: Building materials

Consumer electronics (715) (913)
DT: June 1990
BT: Consumer products
Electronics industry
RT: Electronic equipment

Consumer products (913)
DT: January 1977
NT: Consumer electronics

Contact angle 931.2 (631.1.1)
DT: January 1995
UF: Angle of contact
RT: Capillarity
Fluid mechanics
Surface tension
Wetting

Contact lenses 462.5, 741.3 (741.2)
DT: January 1993
UF: Vision—Contact lenses*
BT: Lenses
Optical devices
Vision aids

Contact sensors (732.2) (731.5)
DT: January 1993
UF: Robots, Industrial—Contact sensors*
BT: Sensors
RT: Industrial robots
Proximity sensors
Robots

Contactors (electric)
USE: Electric contactors

Contacts (electric)
USE: Electric contacts

Contacts (fluid mechanics) 631.1 (931.1)
DT: January 1993
BT: Interfaces (materials)
RT: Fluid mechanics

Container closures 694.2
DT: January 1993
UF: Caps (container)
Closures (container)
Containers—Closures*
Stoppers (container)
BT: Containers

Containers (694) (691)
DT: Predates 1975
NT: Aluminum containers
Autoclaves
Bins
Bottles
Collapsible containers
Container closures
Corrugated containers
Crucibles
Cylinders (containers)
Dispensers
Gas holders
Hoppers
Paper containers
Paperboard containers
Plastic containers
Receivers (containers)
Refrigerated containers
Rubber containers
Steel containers
Tanks (containers)
Transfer cases (vehicles)
Wooden containers
RT: Canning
Enclosures
Filling
Loading
Materials handling
Materials handling equipment
Packaging
Packaging materials
Pressure vessels
Pressurization

*Containers—Aluminum**
USE: Aluminum containers

*Containers—Closures**
USE: Container closures

*Containers—Collapsible**
USE: Collapsible containers

*Containers—Corrugated**
USE: Corrugated containers

*Containers—Dispensers**
USE: Dispensers

*Containers—Filling**
USE: Filling

*Containers—Fungus protection**
USE: Fungus protection

*Containers—Loading**
USE: Loading

*Containers—Paper and paperboard**
USE: Paper containers OR
Paperboard containers

*Containers—Plastics**
USE: Plastic containers

*Containers—Pressurization**
USE: Pressurization

*Containers—Refrigerated**
USE: Refrigerated containers

*Containers—Rubber**
USE: Rubber products

*Containers—Size determination**
USE: Size determination

*Containers—Skid resistant**
USE: Skid resistance

* Former Ei Vocabulary term

*Containers—Steel**
USE: Steel containers

*Containers—Unloading**
USE: Unloading

*Containers—Wood**
USE: Wooden containers

Containment vessels (621)
DT: January 1993
UF: Nuclear reactors—Containment vessels*
BT: Nuclear reactors

Contaminants
USE: Impurities

Contamination (454.2) (461.7) (714.2)
DT: January 1993
RT: Accidents
Clean rooms
Fallout
Fouling
Hazards
Health hazards
Impurities
Pollution
Toxicity

Content addressable memories
USE: Associative storage

Content addressable storage
USE: Associative storage

Context free grammars 721.1 (723.1.1)
DT: January 1993
UF: Automata theory—Context free grammars*
BT: Computational grammars
RT: Context free languages

Context free languages 723.1.1
DT: January 1993
UF: Automata theory—Context free languages*
BT: Formal languages
RT: Context free grammars

Context sensitive grammars 721.1
DT: January 1993
UF: Automata theory—Context sensitive grammars*
BT: Computational grammars
RT: Context sensitive languages

Context sensitive languages 723.1.1
DT: January 1993
UF: Automata theory—Context sensitive languages*
BT: Formal languages
RT: Context sensitive grammars

Continuous casting 534.2
DT: January 1993
UF: Foundry practice—Continuous casting*
Metals and alloys—Continuous casting*
BT: Metal casting
RT: Foundry practice
Permanent mold casting

Continuous cell culture (461.2) (801.2)
DT: January 1993
UF: Cell culture—Continuous*
BT: Cell culture
RT: Fermentation

Continuous miners 502.2 (503.3) (505.3)
DT: January 1993
UF: Mines and mining—Continuous miners*
BT: Cutter loaders

Continuous wave lasers 744.1
DT: January 1995
UF: CW lasers
BT: Lasers

Continuum mechanics 931.1
DT: January 1993
UF: Classical field theory
Classical mechanics of continuous media
Mechanics—Continuous media*
BT: Mechanics
RT: Elasticity
Fluid mechanics
Plasticity

Contour followers 603.1
DT: January 1993
UF: Machine tools—Contour followers*
BT: Machine tools

Contour measurement (943.3)
DT: January 1993
UF: Surfaces—Contour measurement*
BT: Surface measurement
RT: Surfaces

Contraction
USE: Shrinkage

Contracts 902.3 (912.2)
DT: Predates 1975
UF: Agreements
RT: Industrial plants
Management

Contrast media 803 (461.1) (461.6) (422.2)
DT: January 1993
UF: Radiography—Contrast media*
Radiopaque agents
BT: Agents
RT: Angiocardiography
Angiography
Barium compounds
Diagnostic radiography
Industrial radiography
Radiography

Control (731) (732)
DT: January 1993
UF: Automatic control
NT: Acoustic variables control
Automatic guidance (agricultural machinery)
Automatic train control
Boiler control
Budget control
Centralized signal control
Chemical variables control

* Former Ei Vocabulary term

Control *(continued)*
Climate control
Concurrency control
Congestion control (communication)
Decentralized control
Disease control
Dust control
Electric variables control
Feedback control
Flood control
Foam control
Fuzzy control
Ice control
Integrated control
Intelligent control
Inventory control
Magnetic variables control
Manual control
Mechanical variables control
Mine roof control
Moisture control
Odor control
Optical variables control
Papermaking slime control
Pest control
Pneumatic control
Pollution control
Press load control
Process control
Production control
Remote control
River control
Spatial variables control
Telecommunication control
Thermal variables control
Traffic control
Weight control
RT: Automation
Braking
Control equipment
Control facilities
Control systems
Control theory
Fluidics
Fog dispersal
Industrial plants
Industrial robots
Instruments
Redundancy
Remote consoles
Synchronization
Weather modification

Control algorithms
USE: Algorithms

Control equipment 732.1
DT: Predates 1975
UF: Controllers
BT: Equipment
NT: Boiler control instruments
Control rods
Cryostats
Electric control equipment
Final control devices
Mechanical control equipment
Pollution control equipment
Pressure regulators
Programmable logic controllers
Servomechanisms
Telecontrol equipment
RT: Control
Control rooms (power plants)
Control systems
Control towers
Indicators (instruments)
Industrial robots
Radio studios
Remote consoles
Remote control
Robots
Transducers

*Control equipment, Electric**
USE: Electric control equipment

*Control Equipment, Hydraulic**
USE: Hydraulic control equipment

*Control equipment, Hydraulic—Amplifiers**
USE: Hydraulic amplifiers

*Control equipment, Mechanical**
USE: Mechanical control equipment

*Control equipment, Pneumatic**
USE: Pneumatic control equipment

Control facilities (731) (732)
DT: Predates 1975
BT: Facilities
RT: Control
Control systems

Control nonlinearities 731.1
DT: January 1993
UF: Control systems—Nonlinearities*
Nonlinearities (control system)
BT: Control theory
RT: Bang bang control systems
Hysteresis
Nonlinear control systems
Nonlinear systems

Control rods 621 (732.1)
DT: January 1993
UF: Nuclear reactors—Control rods*
BT: Control equipment
Nuclear reactors

Control rooms (power plants) 732.1 (611) (613) (614) (615)
DT: January 1993
UF: Power plants—Control rooms*
BT: Power plants
RT: Control equipment

Control rooms (radio)
USE: Radio studios

Control system analysis 731.1
DT: January 1993
UF: Control systems—Analysis*
BT: Control theory
RT: Bode diagrams

* Former Ei Vocabulary term

Control system analysis *(continued)*
Control systems
Describing functions
Frequency domain analysis
Frequency response
Invariance
Linearization
Nyquist diagrams
Perturbation techniques
Phase space methods
Piecewise linear techniques
Root loci
Sensitivity analysis
State space methods
Step response
Time domain analysis
Transfer functions

Control system synthesis 731.1
DT: January 1993
UF: Control systems—Synthesis*
BT: Control theory
RT: Bode diagrams
Control systems
Describing functions
Frequency response
Invariance
Linearization
Nyquist diagrams
Perturbation techniques
Phase space methods
Piecewise linear techniques
Root loci
Sensitivity analysis
State space methods
Step response
Transfer functions

Control systems 731.1
DT: Predates 1975
NT: Adaptive control systems
Bang bang control systems
Cascade control systems
Closed loop control systems
Computer control systems
Delay control systems
Digital control systems
Discrete time control systems
Distributed parameter control systems
Electronic guidance systems
Linear control systems
Multivariable control systems
Nonlinear control systems
Numerical control systems
On-off control systems
Optimal control systems
Predictive control systems
Programmed control systems
Proportional control systems
Relay control systems
SCADA systems
Stochastic control systems
Three term control systems
Time varying control systems
Two term control systems
RT: Adaptive systems
Automation
Control
Control equipment
Control facilities
Control system analysis
Control system synthesis
Control theory
Controllability
Hierarchical systems
Identification (control systems)
Kalman filtering
Large scale systems
Observability
Optimal systems
Parameter estimation
Phase locked loops
Process control
Root loci
State estimation
Synchronization
System stability

*Control systems, Adaptive**
USE: Adaptive control systems

*Control systems, Bang bang**
USE: Bang bang control systems

*Control systems, Cascade**
USE: Cascade control systems

*Control systems, Delay**
USE: Delay control systems

*Control systems, Digital**
USE: Digital control systems

*Control systems, Direct digital**
USE: Direct digital control systems

*Control systems, Discrete time**
USE: Discrete time control systems

*Control systems, Distributed parameter**
USE: Distributed parameter control systems

*Control systems, Linear**
USE: Linear control systems

*Control systems, Multivariable**
USE: Multivariable control systems

*Control systems, Nonlinear**
USE: Nonlinear control systems

*Control systems, Numerical**
USE: Numerical control systems

*Control systems, On/off**
USE: On-off control systems

*Control systems, Optimal**
USE: Optimal control systems

*Control systems, Predictive**
USE: Predictive control systems

*Control systems, Programmed**
USE: Programmed control systems

* Former Ei Vocabulary term

*Control systems, Proportional**
USE: Proportional control systems

*Control systems, Relay**
USE: Relay control systems

*Control systems, Sampled data**
USE: Sampled data control systems

*Control systems, Self adjusting**
USE: Self adjusting control systems

*Control systems, Self tuning**
USE: Self tuning control systems

*Control systems, Stochastic**
USE: Stochastic control systems

*Control systems, Three term**
USE: Three term control systems

*Control systems, Time varying**
USE: Time varying control systems

*Control systems, Two term**
USE: Two term control systems

*Control systems—Analysis**
USE: Control system analysis

*Control systems—Bode diagrams**
USE: Bode diagrams

*Control systems—Controllability**
USE: Controllability

*Control systems—Describing functions**
USE: Describing functions

*Control systems—Frequency response**
USE: Frequency response

*Control systems—Identification**
USE: Identification (control systems)

*Control systems—Invariance**
USE: Invariance

*Control systems—Nonlinearities**
USE: Control nonlinearities

*Control systems—Nyquist diagrams**
USE: Nyquist diagrams

*Control systems—Observability**
USE: Observability

*Control systems—Robustness**
USE: Robustness (control systems)

*Control systems—Root loci**
USE: Root loci

*Control systems—Step response**
USE: Step response

*Control systems—Synthesis**
USE: Control system synthesis

*Control systems—Telecontrol**
USE: Remote control

*Control systems—Theory**
USE: Control theory

Control theory 731.1
DT: January 1993
UF: Control systems—Theory*
NT: Control nonlinearities
Control system analysis
Control system synthesis
Controllability
Invariance
Observability
Robustness (control systems)
Step response
RT: Control
Control systems
Cybernetics
Feedback
Models
Nonlinear systems
Petri nets
Riccati equations
Simulation
System stability

Control towers 431.4, 732.1, 402.4
DT: January 1993
UF: Airports—Control towers*
BT: Airport buildings
Towers
RT: Air traffic control
Control equipment

*Control, Acoustic variables**
USE: Acoustic variables control

*Control, Chemical variables**
USE: Chemical variables control

*Control, Electric variables**
USE: Electric variables control

*Control, Electric variables—Current**
USE: Electric current control

*Control, Electric variables—Frequency**
USE: Electric frequency control

*Control, Electric variables—Gain**
USE: Gain control

*Control, Electric variables—Phase**
USE: Phase control

*Control, Electric variables—Voltage**
USE: Voltage control

*Control, Magnetic variables**
USE: Magnetic variables control

Control, Mechanical variables (Density)
USE: Density control (specific gravity)

*Control, Mechanical variables**
USE: Mechanical variables control

*Control, Mechanical variables—Acceleration**
USE: Acceleration control

*Control, Mechanical variables—Flow**
USE: Flow control

*Control, Mechanical variables—Forces**
USE: Force control

* Former Ei Vocabulary term

*Control, Mechanical variables—Level**
USE: Level control

*Control, Mechanical variables—Motion**
USE: Motion control

*Control, Mechanical variables—Position**
USE: Position control

*Control, Mechanical variables—Strain**
USE: Strain control

*Control, Mechanical variables—Thickness**
USE: Thickness control

*Control, Mechanical variables—Torques**
USE: Torque control

*Control, Mechanical variables—Velocity**
USE: Velocity control

*Control, Mechanical variables—Volumes**
USE: Volume control (spatial)

*Control, Optical variables**
USE: Optical variables control

*Control, Power**
USE: Power control

*Control, Thermal variables**
USE: Thermal variables control

Controllability 731.1
DT: January 1993
UF: Control systems—Controllability*
BT: Control theory
RT: Control systems
Fuel sloshing
Stability
State space methods
System stability

Controlled atmospheres
USE: Protective atmospheres

Controlled drug delivery (461.6)
DT: January 1993
UF: Biomedical equipment—Microspheres*
Drug products—Controlled delivery*
BT: Drug therapy

Controlled vocabularies
USE: Vocabulary control

Controllers
USE: Control equipment

Controllers (computer peripherals)
USE: Interfaces (computer)

Convection (heat)
USE: Heat convection

Convergence of numerical methods 921.6
DT: January 1993
UF: Mathematical techniques—Convergence of numerical methods*
Numerical stability
Stability (numerical methods)
BT: Numerical methods

Conversion (biological)
USE: Bioconversion

Conversion (energy)
USE: Energy conversion

Conversion (ships)
USE: Ship conversion

Converters (analog to digital)
USE: Analog to digital conversion

Converters (Bessemer)
USE: Metallurgical furnaces

Converters (code)
USE: Code converters

Converters (digital to analog)
USE: Digital to analog conversion

Converters (electric)
USE: Electric converters

Converters (image)
USE: Image converters

Converters (magnetohydrodynamic)
USE: Magnetohydrodynamic converters

Converters (metal refining)
USE: Basic oxygen converters

Converters (torque)
USE: Torque converters

Convertible engines
USE: Dual fuel engines

Convertible helicopters 652.4
DT: January 1993
UF: Helicopters—Convertible*
BT: Helicopters

Convertors (electric)
USE: Electric converters

Conveying 692.1 (502.2) (503.3) (505.3)
DT: January 1993
UF: Mines and mining—Conveying*
BT: Materials handling

Conveyors 692.1
DT: Predates 1975
BT: Materials handling equipment
NT: Belt conveyors
Bridge conveyors
Chain conveyors
Gravity conveyors
Monorail conveyors
People movers
Pneumatic conveyors
Screw conveyors
Vibrating conveyors
RT: Construction equipment

*Conveyors—Belt**
USE: Belt conveyors

*Conveyors—Bridge**
USE: Bridge conveyors

*Conveyors—Chain**
USE: Chain conveyors

*Conveyors—Gravity**
USE: Gravity conveyors

*Conveyors—Monorail**
USE: Monorail conveyors

*Conveyors—Noise**
USE: Acoustic noise

*Conveyors—Passenger transportation**
USE: People movers

*Conveyors—Plastics parts**
USE: Plastic parts

*Conveyors—Pneumatic**
USE: Pneumatic conveyors

*Conveyors—Screw**
USE: Screw conveyors

*Conveyors—Shaking**
USE: Vibrating conveyors

*Conveyors—Vibrating**
USE: Vibrating conveyors

Convolutional codes 723.1
DT: January 1995
BT: Codes (symbols)
RT: Binary sequences
Encoding (symbols)
Error correction
Error detection

Cooking (pulp)
USE: Pulp cooking

Coolant loss accidents
USE: Loss of coolant accidents

Coolants 803 (607.1) (621) (631.1)
DT: January 1993
BT: Chemicals
NT: Cooling water
Refrigerants
RT: Cooling
Cooling systems
Cutting fluids

Cooling 641.2 (802.3)
DT: January 1993
UF: Aircraft carriers—Deck cooling*
NT: Freezing
Refrigeration
RT: Air conditioning
Climate control
Coolants
Cooling systems
Cryogenics
Environmental engineering
Heat pump systems
Heat transfer
Heating
Temperature
Temperature control
Thermal cycling
Thermal insulation
Ventilation
Wetting

Cooling systems (641.2) (616.1)
DT: January 1993
BT: Equipment
NT: Automobile cooling systems
Evaporative cooling systems
Refrigerating machinery
Water cooling systems
RT: Coolants
Cooling
Cryogenic equipment
Economizers
Heat exchangers
Heat sinks
Radiators
Refrigerants

Cooling towers (802.1) (616.1)
DT: October 1975
BT: Heat exchangers
Towers
NT: Water cooling towers
RT: Electric power plant equipment
Water supply

Cooling water (616)
DT: January 1987
BT: Coolants
Water

Coordination (electric insulation)
USE: Electric insulation coordination

Coordination reactions 802.2
DT: January 1993
UF: Polymerization—Coordination reactions*
BT: Chemical reactions

Coplanar waveguides
USE: Waveguides

Copolymerization 815.2
DT: January 1993
UF: Polymerization—Copolymerization*
BT: Polymerization
RT: Copolymers
Terpolymerization

Copolymers 815.1
DT: January 1977
BT: Organic polymers
NT: Block copolymers
Graft copolymers
Terpolymers
RT: Copolymerization

Copper 544.1
DT: January 1993
UF: Building materials—Copper*
Cast iron—Copper content*
Copper and alloys*
Cu
Springs—Copper*
BT: Transition metals
RT: Copper alloys

* Former Ei Vocabulary term

Copper *(continued)*
Copper clad steel
Copper compounds
Copper corrosion
Copper deposits
Copper foundry practice
Copper metallography
Copper metallurgy
Copper mines
Copper ore treatment
Copper pipe
Copper plating
Copper powder
Copper scrap

Copper alloys 544.2
DT: January 1993
UF: Copper and alloys*
BT: Transition metal alloys
NT: Aluminum copper alloys
Brass
Bronze
Monel metal
Nickel silver
Pewter
RT: Copper
Copper compounds
Soldering alloys

*Copper and alloys**
USE: Copper OR
Copper alloys

*Copper and alloys—Corrosion**
USE: Copper corrosion

Copper clad steel 545.3, 544.1 (535) (538)
DT: January 1993
UF: Steel—Copper clad*
BT: Clad metals
Steel
RT: Copper

Copper compounds (804.1) (804.2)
DT: Predates 1975
BT: Transition metal compounds
NT: Copper oxides
RT: Copper
Copper alloys

Copper corrosion 539.1, 544.1
DT: January 1993
UF: Copper and alloys—Corrosion*
BT: Corrosion
RT: Copper

Copper deposits 504.3, 544.1
DT: Predates 1975
BT: Ore deposits
NT: Copper lead deposits
Copper nickel deposits
Copper silver deposits
Copper zinc deposits
RT: Copper
Copper gold mines
Copper mines

Copper foundry practice 534.2, 544.1
DT: Predates 1975
UF: Brass foundry practice
BT: Copper metallurgy
Foundry practice
RT: Copper

Copper gold mines 504.3, 544.1, 547.1
DT: January 1993
UF: Copper gold mines and mining*
BT: Copper mines
Gold mines
RT: Copper deposits
Gold deposits

*Copper gold mines and mining**
USE: Copper gold mines

Copper lead deposits 504.3, 544.1, 546.1
DT: Predates 1975
BT: Copper deposits
Lead deposits
NT: Copper lead zinc deposits

Copper lead ore treatment 533.1, 544.1, 546.1
DT: Predates 1975
BT: Copper ore treatment
Lead ore treatment
NT: Copper lead zinc ore treatment

Copper lead zinc deposits 504.3, 544.1, 546.1, 546.3
DT: Predates 1975
BT: Copper lead deposits
Copper zinc deposits
RT: Copper lead zinc mines

Copper lead zinc mines 504.3, 544.1, 546.1, 546.3
DT: January 1993
UF: Copper lead zinc mines and mining*
BT: Copper zinc mines
Lead mines
RT: Copper lead zinc deposits

*Copper lead zinc mines and mining**
USE: Copper lead zinc mines

Copper lead zinc ore treatment 533.1, 544.1, 546.1, 546.3
DT: Predates 1975
BT: Copper lead ore treatment
Zinc ore treatment

Copper metallography 531.2, 544.1
DT: Predates 1975
BT: Metallography
RT: Copper
Copper metallurgy

Copper metallurgy 531.1, 544.1
DT: Predates 1975
BT: Metallurgy
NT: Copper foundry practice
Copper powder metallurgy
Copper refining
RT: Copper
Copper metallography

* Former Ei Vocabulary term

Copper metallurgy *(continued)*
Copper ore treatment
Copper smelting

Copper mines 504.3, 544.1
DT: January 1993
UF: Copper mines and mining*
BT: Mines
NT: Copper gold mines
Copper zinc mines
RT: Copper
Copper deposits
Copper silver deposits

*Copper mines and mining**
USE: Copper mines

Copper molybdenum ore treatment 533.1, 543.3, 544.1
DT: Predates 1975
BT: Copper ore treatment
Molybdenum ore treatment

Copper nickel deposits 504.3, 544.1, 548.1
DT: Predates 1975
BT: Copper deposits
Nickel deposits

Copper nickel ore treatment 533.1, 544.1, 548.1
DT: Predates 1975
BT: Copper ore treatment
Nickel ore treatment

Copper ore treatment 533.1, 544.1
DT: Predates 1975
BT: Ore treatment
NT: Copper lead ore treatment
Copper molybdenum ore treatment
Copper nickel ore treatment
Copper smelting
Copper zinc ore treatment
RT: Copper
Copper metallurgy

Copper oxides 804.2
DT: June 1990
BT: Copper compounds
Oxides

Copper pipe 619.1, 544.1
DT: January 1993
UF: Copper pipelines
Pipe, Copper*
Pipelines, Copper*
BT: Pipe
RT: Copper

Copper pipelines
USE: Pipelines AND
Copper pipe

Copper plating 539.3, 544.1
DT: Predates 1975
BT: Plating
RT: Copper

Copper powder 544.1 (536)
DT: Predates 1975
BT: Powder metals
RT: Copper
Powder metallurgy

Copper powder metallurgy 536.1, 544.1 (536.2)
DT: January 1993
UF: Powder metallurgy—Aluminum copper*
Powder metallurgy—Beryllium copper*
Powder metallurgy—Copper tin*
Powder metallurgy—Copper*
Powder metallurgy—Iron copper*
BT: Copper metallurgy
Powder metallurgy

Copper refining 533.2, 544.1
DT: Predates 1975
BT: Copper metallurgy

Copper scrap 544.1, 452.3
DT: Predates 1975
BT: Scrap metal
RT: Copper

Copper silver deposits 504.3, 544.1, 547.1
DT: Predates 1975
BT: Copper deposits
Silver deposits
RT: Copper mines
Silver mines

Copper smelting 533.2, 544.1
DT: Predates 1975
BT: Copper ore treatment
Smelting
RT: Copper metallurgy

Copper zinc deposits 504.3, 544.1, 546.3
DT: Predates 1975
BT: Copper deposits
Zinc deposits
NT: Copper lead zinc deposits
RT: Copper zinc mines

Copper zinc mines 504.3, 544.1, 546.3
DT: January 1993
UF: Copper zinc mines and mining*
BT: Copper mines
Zinc mines
NT: Copper lead zinc mines
RT: Copper zinc deposits

*Copper zinc mines and mining**
USE: Copper zinc mines

Copper zinc ore treatment 533.1, 544.1, 546.3
DT: Predates 1975
BT: Copper ore treatment
Zinc ore treatment

Copying 745.2 (903.2)
DT: January 1993
UF: Information dissemination—Reproduction*
NT: Photocopying
RT: Blueprints
Information technology
Office equipment

Copyrights 902.3 (903)
DT: June 1990
BT: Intellectual property
RT: Information dissemination

Cord
USE: Rope

Cordage
USE: Rope

Cordless telephones 718.1 (716.3)
DT: January 1993
UF: Telephone apparatus—Cordless*
BT: Telephone apparatus
RT: Radio telephone

Cords (tire)
USE: Tire cords

Core analysis 512.1.2
DT: January 1993
UF: Petroleum prospecting—Core analysis*
Petroleum reservoir engineering—Core analysis*
RT: Core samples
Petroleum prospecting
Petroleum reservoir engineering

Core disruptive accidents 621, 914.1
DT: January 1993
UF: Nuclear reactors—Core disruptive accident*
BT: Nuclear reactor accidents
NT: Core meltdown
RT: Nuclear reactors
Reactor cores

Core drilling 511.1
DT: January 1993
UF: Oil well drilling—Core*
BT: Drilling
RT: Core samples
Oil well drilling

Core meltdown 621, 914.1
DT: January 1993
UF: Meltdown (nuclear reactors)
Nuclear reactors—Core meltdown*
Reactor meltdown
BT: Core disruptive accidents
RT: Nuclear reactors
Reactor cores

Core samples (511.1)
DT: January 1993
RT: Core analysis
Core drilling
Geology
Sampling

Core sand
USE: Foundry sand

Core storage
USE: Magnetic core storage

Coremaking 534.2
DT: January 1993
UF: Foundry practice—Coremaking*
BT: Foundry practice

Cores (cable)
USE: Cable cores

Cores (reactor)
USE: Reactor cores

Corn sugar
USE: Dextrose

Corporate taxes
USE: Taxation

Correction (errors)
USE: Error correction

Correction (power factor)
USE: Electric power factor correction

Correction of vision
USE: Vision aids

Correlation (optical)
USE: Optical correlation

Correlation detectors (713.3) (716) (717) (718)
DT: January 1993
UF: Cross correlation detectors
Signal processing—Correlation detectors*
BT: Detectors
NT: Correlators
RT: Networks (circuits)
Signal processing

Correlation methods 922.2
DT: January 1993
UF: Mathematical techniques—Correlation methods*
BT: Statistical methods
NT: Optical correlation
RT: Correlation theory
Optimization
Quality control
Statistical tests

Correlation theory 716.1, 922.2
DT: January 1993
UF: Information theory—Correlation theory*
BT: Information theory
RT: Correlation methods
Statistical methods

Correlation-type receivers
USE: Correlators

Correlators (713.3) (716) (717) (718)
DT: Predates 1975
UF: Correlation-type receivers
BT: Correlation detectors

Corrosion (539.1) (802.2)
DT: Predates 1975
UF: Corrosion of alloys
Corrosion of metals
Metal corrosion
Metals and alloys—Corrosion*
NT: Aluminum corrosion
Atmospheric corrosion
Boiler corrosion
Cavitation corrosion
Copper corrosion
Electrochemical corrosion
Fretting corrosion

Corrosion *(continued)*
Pitting
Pollution induced corrosion
Seawater corrosion
Steel corrosion
Stress corrosion cracking
Sulfide corrosion cracking
Underground corrosion
RT: Alloys
Cathodic protection
Chemical attack
Corrosion fatigue
Corrosion prevention
Corrosion protection
Corrosion resistance
Corrosion resistant alloys
Corrosive effects
Degradation
Descaling
Deterioration
Erosion
Etching
Failure (mechanical)
Fouling
Hydrogen embrittlement
Internal oxidation
Metals
Oxidation
Scale (deposits)
Surface phenomena
Surface properties
Wear of materials
Weathering

Corrosion cracking
USE: Stress corrosion cracking

Corrosion fatigue 539.1 (421)
DT: January 1993
UF: Corrosion—Corrosion fatigue*
BT: Fatigue of materials
RT: Corrosion
Fretting corrosion
Hydrogen embrittlement

Corrosion inhibitors 539.2.1, 803
DT: January 1993
UF: Corrosion protection—Inhibitors*
Inhibitors (corrosion)
BT: Agents
RT: Corrosion protection

Corrosion of alloys
USE: Corrosion

Corrosion of metals
USE: Corrosion

Corrosion prevention 539.2
DT: January 1993
NT: Passivation
RT: Corrosion
Corrosion protection
Corrosion resistance
Oxidation resistance
Protective atmospheres

Corrosion protection 539.2
DT: Predates 1975
BT: Protection
NT: Anodic protection
Cathodic protection
RT: Corrosion
Corrosion inhibitors
Corrosion prevention
Corrosion resistance
Electric equipment protection
Protective coatings

*Corrosion protection, Anodic**
USE: Anodic protection

*Corrosion protection, Cathodic**
USE: Cathodic protection

*Corrosion protection—Inhibitors**
USE: Corrosion inhibitors

Corrosion resistance 539.1 (802.2)
DT: January 1993
NT: Oxidation resistance
RT: Acid resistance
Chemical resistance
Corrosion
Corrosion prevention
Corrosion protection
Corrosion resistant alloys
Corrosive effects
Ozone resistance

Corrosion resistant alloys 531, 539.1
DT: January 1993
UF: Cast iron—Corrosion resisting*
Metals and alloys—Corrosion resisting*
Welding—Corrosion resisting materials*
BT: Alloys
RT: Chemical resistant materials
Corrosion
Corrosion resistance
Metals
Oxidation resistance

*Corrosion—Atmospheric**
USE: Atmospheric corrosion

*Corrosion—Corrosion fatigue**
USE: Corrosion fatigue

*Corrosion—Electrochemical**
USE: Electrochemical corrosion

*Corrosion—Electrolytic**
USE: Electrochemical corrosion

*Corrosion—Fretting**
USE: Fretting corrosion

*Corrosion—Pitting**
USE: Pitting

*Corrosion—Pollution induced**
USE: Pollution induced corrosion

*Corrosion—Seawater**
USE: Seawater corrosion

* Former Ei Vocabulary term

*Corrosion—Stress corrosion cracking**
USE: Stress corrosion cracking

*Corrosion—Sulfide corrosion cracking**
USE: Sulfide corrosion cracking

*Corrosion—Tropics**
USE: Tropics

*Corrosion—Underground**
USE: Underground corrosion

Corrosive effects 539.1 (803) (804)
DT: January 1993
UF: Ammonia—Corrosive properties*
Antifreeze solutions—Corrosive properties*
Chemicals—Corrosive properties*
Corrosive properties
Fertilizers—Corrosive properties*
Food products—Corrosive properties*
Limestone—Corrosive properties*
Sulfuric acid—Corrosive properties*
BT: Effects
RT: Corrosion
Corrosion resistance

Corrosive properties
USE: Corrosive effects

Corrugated containers 694.1
DT: January 1993
UF: Containers—Corrugated*
BT: Containers
RT: Corrugated materials

Corrugated materials 694.2 (811.1)
DT: January 1993
BT: Materials
NT: Corrugated metal
Corrugated paperboard
RT: Corrugated containers

Corrugated metal 531 (694.2)
DT: January 1993
UF: Sewers—Corrugated metal*
BT: Corrugated materials
Sheet metal

Corrugated paperboard 811.1 (694.2)
DT: January 1993
UF: Paperboards—Corrugated*
BT: Corrugated materials
Paperboards

Corundum 482.2 (812.2) (606.1)
DT: Predates 1975
BT: Alumina
Oxide minerals
RT: Abrasives
Corundum deposits
Ruby
Sapphire

Corundum deposits 505.1
DT: Predates 1975
BT: Ore deposits
RT: Alumina
Corundum

Cosmic radiation
USE: Cosmic rays

Cosmic ray detectors 657, 944.7
DT: January 1993
BT: Particle detectors
RT: Cosmic ray measurement
Cosmic rays

Cosmic ray measurement 657, 944.8
DT: January 1993
BT: Measurements
RT: Cosmic ray detectors
Cosmic rays

Cosmic rays 657 (932)
DT: Predates 1975
UF: Cosmic radiation
Primary cosmic rays
BT: Elementary particles
Radiation
RT: Cosmic ray detectors
Cosmic ray measurement
Geophysics
Solar radiation

Cost accounting 911.1
DT: Predates 1975
UF: Electric utilities—Accounting*
Gas industry—Accounting*
Industrial plants—Accounting*
Petroleum industry—Accounting*
Sugar factories—Accounting*
Transportation—Accounting*
BT: Management
RT: Costs
Decision making
Depreciation
Industrial management

Cost effectiveness 911.2 (912.2) (912.3)
DT: January 1995
RT: Costs
Industrial management
Operations research

Costs 911
DT: January 1993
BT: Finance
RT: Cost accounting
Cost effectiveness
Industrial management

Cotton 821.4 (819.1)
DT: Predates 1975
UF: Cellulose—Cotton*
Pulp materials—Cotton*
BT: Agricultural products
RT: Cellulose
Cotton fabrics
Cotton fibers
Cotton yarn
Cottonseed oil
Pulp materials
Textiles

* Former Ei Vocabulary term

Cotton fabrics 819.5
DT: Predates 1975
BT: Textiles
RT: Cotton
Cotton fibers
Cotton yarn

Cotton fibers 819.1
DT: Predates 1975
BT: Natural fibers
RT: Cotton
Cotton fabrics
Cotton yarn

Cotton oil
USE: Cottonseed oil

Cotton yarn 819.1
DT: Predates 1975
BT: Yarn
RT: Cotton
Cotton fabrics
Cotton fibers

*Cotton—Growing**
USE: Cultivation

Cottonseed oil 822.3
DT: January 1993
UF: Cotton oil
Vegetable oils—Cottonseed*
BT: Vegetable oils
RT: Cotton

Coulometers 942.1
DT: Predates 1975
UF: Coulometry
BT: Electric measuring instruments
RT: Electrochemistry
Polarographs

Coulometry
USE: Coulometers

Countermeasures (crime)
USE: Electronic crime countermeasures

Countermeasures (military)
USE: Military electronic countermeasures

Countermeasures (radar)
USE: Radar countermeasures

Counters (Cerenkov)
USE: Cerenkov counters

Counters (crystal)
USE: Crystal counters

Counters (radiation)
USE: Radiation counters

Counting circuits 713.4
DT: January 1993
UF: Electronic circuits, Counting*
BT: Pulse circuits
RT: Adders
Analog to digital conversion
Coincidence circuits
Combinatorial circuits
Counting tubes
Digital arithmetic
Digital to analog conversion
Particle detectors

Counting tubes 714.1
DT: January 1993
UF: Electron tubes, Counting*
Scaling tubes
BT: Cold cathode tubes
RT: Counting circuits
Detectors
Geiger counters
Radiation counters

Coupled circuits (713.5)
DT: January 1993
UF: Circuit coupling
Coupling circuits
Electronic circuits, Coupler*
Electronic circuits—Coupling*
BT: Networks (circuits)
RT: Circuit theory
Waveguide couplers

Coupled fibers
USE: Optical fiber coupling

Coupled optical fibers
USE: Optical fiber coupling

Couplers (electric connectors)
USE: Electric connectors

Couplers (waveguide)
USE: Waveguide couplers

Coupling circuits
USE: Coupled circuits

Couplings (602)
DT: Predates 1975
UF: Locomotives—Couplings*
BT: Components
NT: Flexible couplings
Magnetic couplings
Marine couplings
Railroad car couplings
RT: Clutches
Splines

*Couplings—Flexible**
USE: Flexible couplings

*Couplings—Hydraulic**
USE: Hydraulic equipment

*Couplings—Magnetic**
USE: Magnetic couplings

*Couplings—Marine**
USE: Marine couplings

Covering techniques
USE: Coating techniques

Coverings (roof)
USE: Roof coverings

Cr
USE: Chromium

* Former Ei Vocabulary term

Crack growth
USE: Crack propagation

Crack initiation (421)
DT: January 1995
UF: Cracking (breaking)
Microcracking
RT: Crack propagation
Cracks
Microcracks

Crack propagation 421
DT: January 1993
UF: Crack growth
Propagation (crack)
RT: Crack initiation
Cracks
Crazing
Delamination
Failure (mechanical)
Fracture
Mechanical properties
Stress corrosion cracking

Cracking (breaking)
USE: Crack initiation

Cracking (chemical) 802.2
DT: January 1993
UF: Cracking*
Petroleum cracking
BT: Pyrolysis
NT: Catalytic cracking
Steam cracking
RT: Dehydrogenation
Hydrocarbons
Petroleum refining

*Cracking**
USE: Cracking (chemical)

Cracks (421)
DT: January 1993
UF: Fatigue cracks
BT: Defects
NT: Microcracks
RT: Crack initiation
Crack propagation
Failure (mechanical)
Fracture
Stress corrosion cracking
Sulfide corrosion cracking

Crane rails 693.1
DT: January 1993
UF: Cranes—Rails*
BT: Gantry cranes
Rails

Crane runways 693.1
DT: January 1993
UF: Cranes—Runways*
Runways (crane)
BT: Gantry cranes

Cranes 693.1
DT: Predates 1975
UF: Derricks
BT: Materials handling equipment
NT: Bridge cranes
Floating cranes
Gantry cranes
Helicopter cranes
Jib cranes
Locomotive cranes
Self erecting cranes
Ship cranes
Tower cranes
RT: Construction equipment
Hoists
Hooks
Winches

*Cranes—Braking**
USE: Braking

*Cranes—Bridge**
USE: Bridge cranes

*Cranes—Floating**
USE: Floating cranes

*Cranes—Gantry**
USE: Gantry cranes

*Cranes—Gears**
USE: Gears

*Cranes—Hooks**
USE: Hooks

*Cranes—Iron and steel plants**
USE: Iron and steel plants

*Cranes—Jib**
USE: Jib cranes

*Cranes—Locomotive**
USE: Locomotive cranes

*Cranes—Portable**
USE: Portable equipment

*Cranes—Rails**
USE: Crane rails

*Cranes—Runways**
USE: Crane runways

*Cranes—Self erecting**
USE: Self erecting cranes

*Cranes—Steam power plants**
USE: Steam power plants

*Cranes—Supports**
USE: Supports

*Cranes—Tower**
USE: Tower cranes

*Cranes—Traveling**
USE: Gantry cranes

Crankcases 612.1.1
DT: January 1993
UF: Internal combustion engines—Crankcases*
BT: Engines
RT: Internal combustion engines

* Former Ei Vocabulary term

Crankshafts 601.2
DT: Predates 1975
UF: Locomotives—Crankshafts*
BT: Shafts (machine components)
RT: Connecting rods
Engines
Internal combustion engines

Crashworthiness 914.1 (652.1) (622.1)
DT: January 1993
UF: Aircraft—Crashworthiness*
Automobiles—Crashworthiness*
RT: Accident prevention
Automobile driver simulators

Crating
USE: Packaging

Crazing (421)
DT: January 1993
BT: Surface phenomena
RT: Crack propagation

Creep (421)
DT: January 1993
BT: Deformation
RT: Anelastic relaxation
Creep testing
Flow of solids
High temperature testing
Mechanical properties
Plasticity
Residual stresses
Shear flow
Solids
Stress relaxation
Tensile properties
Viscoelasticity

Creep testing (421) (422.2)
DT: January 1993
UF: Materials testing—Creep*
BT: Mechanical testing
RT: Creep

Crew accommodations 671.1, 912.4
DT: January 1993
UF: Ship crew accommodations
Ships—Crew accommodations*
BT: Ships

Crime (computer)
USE: Computer crime

Crime countermeasures (electronic)
USE: Electronic crime countermeasures

Critical current density (superconductivity) 701.1
DT: January 1995
BT: Current density
RT: Critical currents
Superconducting materials

Critical currents 701.1 (708.3)
DT: January 1995
BT: Electric currents
RT: Critical current density (superconductivity)
Superconducting materials
Superconductivity

Critical materials
USE: Strategic materials

Critical micelle concentration 801.3
DT: January 1993
UF: Colloids—Critical micelle concentration*
RT: Colloids
Micelles

Critical path analysis (723.2) (912.2)
DT: January 1993
UF: Data processing—Critical path analysis*
BT: Management science
RT: Data processing
PERT
Scheduling
Systems analysis

Criticality (nuclear fission) 621, 932.2.1
DT: January 1993
UF: Nuclear reactors—Criticality*
RT: Nuclear physics
Nuclear reactors
Reactivity (nuclear)

CRO
USE: Cathode ray oscilloscopes

Crop growing
USE: Cultivation

Crops 821.4
DT: January 1993
UF: Farms—Crops*
RT: Agricultural products
Agriculture
Cultivation
Farms
Harvesting

Cross connections (water distribution) 446.2
DT: January 1993
UF: Water distribution systems—Cross connections*
BT: Water distribution systems

Cross correlation detectors
USE: Correlation detectors

Crossbar equipment 718.1
DT: January 1993
UF: Crossbar systems
Telephone switching equipment—Crossbar equipment*
BT: Telephone switching equipment

Crossbar systems
USE: Crossbar equipment

Crossing gates (railroad)
USE: Railroad crossings

Crossings (pipe and cable) (619.1) (706.2)
SN: Structures and equipment to carry pipelines and electric cables across obstacles such as rivers and railroad tracks, whether or not constructed for the purpose

Crossings (pipe and cable) *(continued)*
DT: January 1993
UF: Bridges—Pipeline crossings*
Electric cable crossings
Electric cables—Bridge crossings*
Electric cables—River crossings*
Pipelines—Bridge crossings*
Pipelines—River crossings*
Railroad plant and structures—Pipeline crossings*
BT: Structures (built objects)
RT: Pipelines
Railroad plant and structures

Crossings (railroad)
USE: Railroad crossings

Crosslinking 802.2
DT: January 1993
BT: Chemical reactions
RT: Curing
Polymerization

Crosstalk (711.1) (716.3) (718.1)
DT: January 1993
UF: Signal interference—Crosstalk*
BT: Signal interference
RT: Radio telephone
Telephone interference

Crown ethers 804.1
DT: January 1993
UF: Ethers—Crown ethers*
BT: Polyethers
RT: Ethers

CRT
USE: Cathode ray tubes

Crucible furnaces (532) (642.2)
DT: January 1993
UF: Furnaces—Crucible type*
BT: Melting furnaces
RT: Crucibles

Crucibles 532
DT: Predates 1975
BT: Containers
RT: Crucible furnaces
Heating

Crude oil
USE: Crude petroleum

Crude petroleum 512.1 (523)
DT: January 1993
UF: Crude oil
Heavy oils
Oil (crude)
Petroleum
Petroleum, Crude*
BT: Fossil fuels
RT: Enhanced recovery
Oil fields
Oil sands
Oil shale
Petroleum analysis
Petroleum chemistry
Petroleum deposits
Petroleum distillates
Petroleum engineering
Petroleum geology
Petroleum industry
Petroleum pipelines

Crushed stone plants 412 (402.1)
DT: Predates 1975
BT: Industrial plants
RT: Crushing

Crushers (533.1) (802.1)
DT: Predates 1975
UF: Crushing machines
BT: Machinery
RT: Coal preparation
Crushing
Granulators
Grinding mills
Materials handling equipment

Crushing (533.1) (802.3)
DT: January 1993
UF: Crushing and grinding*
BT: Comminution
RT: Coal breakage
Crushed stone plants
Crushers
Disintegration
Granulation
Ore treatment

*Crushing and grinding**
USE: Crushing OR
Grinding (comminution)

Crushing machines
USE: Crushers

Crushing strength
USE: Compressive strength

Crutches
USE: Walking aids

Cryoelectric data storage
USE: Cryoelectric storage

Cryoelectric storage 722.1 (644.4) (708.3)
DT: January 1993
UF: Cryoelectric data storage
Data storage, Cryoelectric*
BT: Digital storage
RT: Cryogenics
Cryotrons
Superconducting devices

Cryogenic cables
USE: Superconducting cables

Cryogenic equipment 644.5
DT: January 1993
UF: Bearings—Low temperature*
Heat exchangers—Cryogenic*
Low temperature bearings
Low temperature equipment
Rockets and missiles—Cryogenic equipment*
BT: Equipment

* Former Ei Vocabulary term

Cryogenic equipment *(continued)*
NT: Cryostats
Dewars
RT: Cooling systems
Cryogenics

Cryogenic liquids 644.4
DT: January 1993
UF: Liquids—Cryogenic*
BT: Liquids
RT: Cryogenics

Cryogenic phenomena
USE: Low temperature phenomena

*Cryogenic treatment**
USE: Cryogenics

Cryogenics 644.4
DT: Predates 1975
UF: Cryogenic treatment*
RT: Cooling
Cryoelectric storage
Cryogenic equipment
Cryogenic liquids
Cryotherapy
Cryotrons
Low temperature engineering
Low temperature operations
Low temperature phenomena
Superconductivity
Superfluid helium

*Cryogenics—Low temperature phenomena**
USE: Low temperature phenomena

Cryostats 644.5
DT: Predates 1975
BT: Control equipment
Cryogenic equipment
RT: Low temperature engineering
Superfluid helium
Temperature control
Thermostats

Cryosurgery 461.6, 644.4
DT: January 1993
UF: Biomedical engineering—Cryosurgery*
BT: Cryotherapy
Surgery
RT: Biomedical engineering

Cryotherapy 461.6, 644.4
DT: January 1993
UF: Biomedical engineering—Cryotherapy*
BT: Patient treatment
NT: Cryosurgery
RT: Biomedical engineering
Cryogenics

Cryotrons 704.1 (644.5) (708.3)
DT: Predates 1975
BT: Electric switches
Superconducting devices
RT: Cryoelectric storage
Cryogenics
Thin film circuits

Cryptography (723) (716) (717) (718)
DT: January 1983
UF: Cryptosystems
Encryption
RT: Codes (symbols)
Data communication systems
Data handling
Decoding
Encoding (symbols)
Security of data

Cryptosystems
USE: Cryptography

Crystal atomic structure 931.3, 933.1.1
DT: January 1993
UF: Atomic structure (crystals)
Crystals—Atomic structure*
BT: Crystal structure
RT: Atomic force microscopy
Crystallography
Crystals
Electronic structure
Lattice constants
Lattice vibrations
Solids
Superlattices

Crystal classes
USE: Crystal symmetry

Crystal counters 944.7 (933.1)
DT: Predates 1975
UF: Counters (crystal)
BT: Radiation counters

Crystal cutting (482.2.1)
DT: January 1993
UF: Gem cutting
Lapidary work
BT: Cutting
RT: Crystals

Crystal defects 933.1.1
DT: January 1993
UF: Crystals—Defects*
Crystals—Lattice defects*
Lattice defects*
Semiconductor defects
BT: Defects
NT: Dislocations (crystals)
Grain boundaries
Point defects
Stacking faults
RT: Crystal growth
Crystal impurities
Crystal microstructure
Crystal structure
Crystals
Inclusions
Lattice vibrations
Metallography
X ray crystallography

Crystal diodes
USE: Semiconductor diodes

* Former Ei Vocabulary term

Crystal dislocations
USE: Dislocations (crystals)

Crystal filters 703.2 (933.1)
DT: January 1993
UF: Electric filters, Crystal*
Piezoelectric filters
BT: Electromechanical filters
Piezoelectric devices
RT: Acoustic surface wave filters

Crystal growth 933.1.2
DT: January 1993
UF: Crystals—Growing*
BT: Growth (materials)
NT: Crystal growth from melt
Epitaxial growth
Twinning
RT: Crystal defects
Crystal structure
Crystallization
Crystals
Nucleation
Semiconductor growth

Crystal growth from melt 933.1.2
DT: January 1995
UF: Czochralski crystal growth
BT: Crystal growth
RT: Liquid phase epitaxy
Semiconductor materials

Crystal impurities 933.1.1
DT: January 1993
UF: Crystals—Impurities*
BT: Impurities
RT: Crystal defects
Crystals
Point defects

Crystal lattices 933.1.1
DT: January 1993
UF: Lattices (crystal)
BT: Crystal structure
NT: Superlattices
RT: Crystal whiskers
Lattice constants
Metallography
Molecular crystals
Percolation (solid state)
Point defects

Crystal microstructure 933.1.1
DT: January 1993
UF: Crystals—Microstructure*
Grain structure
BT: Crystal structure
Microstructure
RT: Crystal defects
Crystallography
Crystals
Fractography
Grain size and shape
Phase equilibria
Phase transitions
Precipitation (chemical)
Single crystals
Superlattices

Crystal orientation 933.1.1
DT: January 1993
UF: Crystals—Orientation*
Orientation (crystal)
BT: Crystal structure
RT: Conformations
Crystallography
Crystals
Molecular orientation

Crystal oscillators 713.2
DT: January 1993
UF: Oscillators, Crystal*
BT: Oscillators (electronic)

Crystal rectifiers
USE: Semiconductor diodes

Crystal resonators 713.5
DT: January 1993
UF: Resonators, Crystal*
BT: Piezoelectric devices
Resonators

Crystal structure 931.1.1 (801.4) (531.2)
DT: January 1993
UF: Crystals—Structure*
BT: Structure (composition)
NT: Crystal atomic structure
Crystal lattices
Crystal microstructure
Crystal orientation
Crystal symmetry
RT: Chemical bonds
Crystal defects
Crystal growth
Crystallography
Crystals
Polycrystals
Twinning
X ray crystallography

Crystal symmetry 933.1.1
DT: January 1993
UF: Crystal classes
Crystals—Symmetry*
Point groups
Symmetry (crystal)
BT: Crystal structure
RT: Crystallography
Crystals

Crystal twinning
USE: Twinning

Crystal whiskers 933.1.1 (801.4) (531.2)
DT: January 1993
UF: Crystals—Whiskers*
Whiskers (crystal)
BT: Single crystals
RT: Crystal lattices
Fibers
Metallography
Point contacts
Reinforcement

Crystalline materials 933.1 (482.2) (531.2)
DT: January 1993
UF: Crystalline*
BT: Materials
NT: Crystalline rocks
Polycrystalline materials
RT: Crystallization
Crystals

Crystalline rocks (481.1.2) (933.1) (482.2)
DT: January 1993
UF: Rock—Crystalline*
BT: Crystalline materials
Rocks
NT: Igneous rocks
Metamorphic rocks
RT: Petrology

*Crystalline**
USE: Crystalline materials

Crystalline-amorphous transformations
USE: Amorphization

Crystallization 802.3 (933.1)
DT: Predates 1975
UF: Recrystallization (separation)
RT: Chemical operations
Concentration (process)
Crystal growth
Crystalline materials
Crystallizers
Crystals
Nucleation
Precipitation (chemical)
Purification
Refining
Sedimentation
Separation
Supersaturation

Crystallizers 802.1
DT: January 1993
UF: Chemical equipment—Crystallizers*
BT: Chemical equipment
RT: Crystallization
Separators

Crystallography 933.1
DT: Predates 1975
BT: Natural sciences
NT: X ray crystallography
RT: Crystal atomic structure
Crystal microstructure
Crystal orientation
Crystal structure
Crystal symmetry
Crystals
Goniometers
Lattice constants
Metallography
Radiography
Textures

Crystals 933.1 (482.2) (801.4) (802.3)
DT: Predates 1975
BT: Materials
NT: Molecular crystals
Polycrystals
Single crystals
RT: Crystal atomic structure
Crystal cutting
Crystal defects
Crystal growth
Crystal impurities
Crystal microstructure
Crystal orientation
Crystal structure
Crystal symmetry
Crystalline materials
Crystallization
Crystallography
Electron energy levels
Eutectics
Quasicrystals
Semiconductor materials
Twinning

*Crystals, Liquid**
USE: Liquid crystals

*Crystals, Liquid—Cholesteric**
USE: Cholesteric liquid crystals

*Crystals, Liquid—Nematic**
USE: Nematic liquid crystals

*Crystals, Liquid—Smectic**
USE: Smectic liquid crystals

*Crystals, Liquid—Thermotropic**
USE: Thermotropic liquid crystals

*Crystals—Atomic structure**
USE: Crystal atomic structure

*Crystals—Color centers**
USE: Color centers

*Crystals—Defects**
USE: Crystal defects

*Crystals—Dislocations**
USE: Dislocations (crystals)

*Crystals—Electron states**
USE: Electron energy levels

*Crystals—Epitaxial growth**
USE: Epitaxial growth

*Crystals—Growing**
USE: Crystal growth

*Crystals—Impurities**
USE: Crystal impurities

*Crystals—Lattice defects**
USE: Crystal defects

*Crystals—Lattice vibrations**
USE: Lattice vibrations

*Crystals—Microstructure**
USE: Crystal microstructure

*Crystals—Orientation**
USE: Crystal orientation

*Crystals—Structure**
USE: Crystal structure

*Crystals—Symmetry**
USE: Crystal symmetry

*Crystals—Twinning**
USE: Twinning

*Crystals—Whiskers**
USE: Crystal whiskers

Cs
USE: Cesium

CT scans
USE: Computerized tomography

Cu
USE: Copper

Cubic boron nitride 812.1
DT: January 1993
UF: Cutting tools—Cubic boron nitride*
Grinding wheels—Cubic boron nitride*
BT: Boron compounds
Nitrides
RT: Abrasives
Ceramic materials
Cutting tools
Grinding wheels

Cultivation 821.3
DT: January 1993
UF: Cotton—Growing*
Crop growing
Growing (crops)
Sugar beets—Growing*
BT: Agriculture
RT: Crops

Culture (cells)
USE: Cell culture

Culverts (406) (632.1) (631.1)
DT: Predates 1975
BT: Hydraulic structures
RT: Bridges
Causeways
Drainage
Roads and streets

Cupola charging
USE: Charging (furnace) AND
Cupola practice

Cupola furnaces
USE: Cupolas

Cupola practice (534.2) (533.2) (532.5)
DT: Predates 1975
UF: Cupola charging
BT: Foundry practice
RT: Cupolas
Hot blast cupolas
Metal melting
Oxygen blast enrichment
Tapping (furnace)

*Cupola practice—Charging**
USE: Charging (furnace)

*Cupola practice—Hot blast**
USE: Hot blast cupolas

*Cupola practice—Oxygen blast enrichment**
USE: Oxygen blast enrichment

*Cupola practice—Tapping**
USE: Tapping (furnace)

Cupolas 532.5
DT: Predates 1975
UF: Cupola furnaces
Furnace cupolas
BT: Melting furnaces
NT: Hot blast cupolas
RT: Blast furnaces
Cupola practice

*Cupolas—Hot blast**
USE: Hot blast cupolas

Cuprate superconductors
USE: Oxide superconductors

Curbs 406.2
DT: January 1993
UF: Roads and streets—Curbs*
BT: Roads and streets
RT: Pavements

Curing 802.2 (815.1) (818.3)
DT: January 1993
BT: Chemical reactions
RT: Adhesion
Adhesives
Autoclaves
Bonding
Crosslinking
Dehydration
Drying
Polymerization
Vulcanization

Curium 622.1
DT: Predates 1975
UF: Cm
BT: Transuranium elements

Current (electric)
USE: Electric currents

Current carrying capacity (electric cables)
USE: Electric current carrying capacity (cables)

Current control
USE: Electric current control

Current density 701.1
DT: January 1995
NT: Critical current density (superconductivity)
RT: Electric currents
Electrolysis

Current limiters
USE: Limiters

Current limiting reactors 704
DT: January 1993
UF: Bus reactors
Electric reactors, Current limiting*
Feeder reactors
Synchronizing reactors
BT: Electric reactors
RT: Electric current control
Electric inductors

Current measurement
USE: Electric current measurement

Current transformers
USE: Electric instrument transformers

Current voltage characteristics (714) (701.1)
DT: January 1995
RT: Semiconductor materials

Currents (ocean)
USE: Ocean currents

Curve fitting 921.6 (921.4) (922.2)
DT: January 1993
UF: Mathematical techniques—Curve fitting*
BT: Numerical analysis
RT: Computer graphics
Least squares approximations

Curved beams and girders 408.2
DT: January 1993
UF: Beams and girders—Curved*
BT: Beams and girders

Curves (road) 406.2
DT: January 1993
UF: Roads and streets—Curves*
BT: Roads and streets
RT: Highway systems

Custom integrated circuits
USE: Application specific integrated circuits

Cutter loaders 502.2 (503.3) (504) (505.3)
DT: January 1993
UF: Mines and mining—Cutter loaders*
BT: Cutting machines (mining)
Loaders
NT: Continuous miners

Cutters (mining)
USE: Cutting machines (mining)

Cutting (604.1) (606.2)
DT: January 1993
NT: Abrasive cutting
Broaching
Crystal cutting
Deburring
Electrochemical cutting
Electron beam cutting
Gear cutting
Honing
Laser beam cutting
Metal cutting
Oxygen cutting
Plasma arc cutting
Sawing
Scarfing
Shearing
Slitting
Slotting
Spark cutting
Trimming
Ultrasonic cutting
Underwater cutting
RT: Boring
Cutting equipment
Cutting fluids
Drilling
Machining
Perforating

Cutting equipment (502.2) (604.1)
DT: January 1993
BT: Equipment
NT: Cutting machines (mining)
Cutting tools
Peat cutters
RT: Cutting
Cutting fluids

Cutting fluids (607.1)
DT: Predates 1975
BT: Fluids
RT: Coolants
Cutting
Cutting equipment
Drilling fluids

*Cutting fluids—Reclamation**
USE: Reclamation

Cutting machines (mining)
502.3 (503.5) (504) (505.3)
DT: January 1993
UF: Cutters (mining)
Mines and mining—Cutters*
BT: Cutting equipment
Mining machinery
NT: Cutter loaders
Mine augers

Cutting tools 603.2 (604.1) (605)
SN: Very general term; prefer specific tool
DT: Predates 1975
BT: Cutting equipment
Tools
NT: Bit holders
Bits
Boring tools
Broaches
Carbide cutting tools
Ceramic cutting tools
Dental cutting tools
Diamond cutting tools
Drills
Files and rasps
Gear cutters
Glass cutting machines
Grinding wheels
Jigs
Reamers
Saws

Cutting tools *(continued)*
Shearing machines
Slitting machines
Slotting machines
Taps
Thread cutting dies
RT: Cubic boron nitride
Dies
Hand tools
Metal working tools
Woodworking tools

*Cutting tools—Bits and holders**
USE: Bit holders OR
Bits

*Cutting tools—Carbide**
USE: Carbide cutting tools

*Cutting tools—Ceramic**
USE: Ceramic cutting tools

*Cutting tools—Cubic boron nitride**
USE: Cubic boron nitride

*Cutting tools—Diamond**
USE: Diamond cutting tools

*Cutting tools—Metal working**
USE: Metal working tools

*Cutting tools—Noise**
USE: Acoustic noise

*Cutting tools—Woodworking**
USE: Woodworking tools

CVD
USE: Chemical vapor deposition

CVD epitaxy
USE: Vapor phase epitaxy

CW lasers
USE: Continuous wave lasers

Cyanides 804.2
DT: January 1987
UF: Industrial wastes—Cyanides*
BT: Carbon inorganic compounds
Nitrogen compounds

Cyanite
USE: Kyanite

Cybernetics (461.9) (723.4) (731.1)
DT: January 1993
BT: Systems science
NT: Adaptive systems
Artificial intelligence
Biocontrol
Bionics
Brain models
Cognitive systems
Economic cybernetics
RT: Biocommunications
Control theory
Ergonomics
Feedback
Learning systems
Man machine systems
Neural networks
Pattern recognition
Process control
Robotics
Robots

Cyclic voltammetry 801.4.1 (942.2)
DT: January 1995
RT: Electrochemistry
Electrodes

Cyclical heating
USE: Thermal cycling

Cycloidal propulsion vehicles 434
DT: January 1993
UF: Vehicles—Cycloidal propulsion*
BT: Vehicles

Cyclone boilers 614.2 (802.1)
DT: January 1993
UF: Boilers—Cyclone*
BT: Boilers

Cyclone separators 802.1
DT: January 1993
UF: Chemical equipment—Cyclones*
BT: Separators

Cyclones
USE: Storms

Cyclotron radiation (932.1.1) (711)
DT: June 1990
BT: Electromagnetic waves

Cyclotron resonance (932.1.1) (701.1)
DT: January 1993
UF: Diamagnetic resonance
Electron tubes—Cyclotron resonance*
BT: Resonance
NT: Electron cyclotron resonance
RT: Diamagnetism
Electron tubes
Fermi surface
Plasma confinement
Plasma heating

Cyclotrons 932.1.1
DT: January 1993
UF: Accelerators, Cyclotron*
Phasotrons
BT: Particle accelerators
NT: Microtrons
DT: January 1993
UF: Cylinders*
BT: Containers
NT: Gas cylinders

Cylinders (engine)
USE: Engine cylinders

Cylinders (shapes) (408.2)
DT: January 1993
UF: Cylinders*
BT: Bodies of revolution
RT: Flow of fluids

*Cylinders**
USE: Cylinders (containers) OR
Cylinders (shapes)

Cylindrical antennas (716)
DT: January 1993
UF: Antennas, Cylindrical*
BT: Antennas

Cylindrical waveguides
USE: Circular waveguides

Cytology 461.2, 461.9
DT: January 1993
UF: Biomedical engineering—Cytology*
BT: Biology
RT: Biomedical engineering
Cells

Czochralski crystal growth
USE: Crystal growth from melt

D
USE: Deuterium

D layer
USE: D region

D region 443 (711.1)
DT: January 1993
UF: D layer
Ionosphere—D region*
BT: Ionosphere

Dairies 822.1 (402.1)
DT: Predates 1975
BT: Food products plants
RT: Dairy products
Farm buildings
Paint

Dairy products 822.3
DT: Predates 1975
BT: Agricultural products
Food products
NT: Casein
RT: Beverages
Dairies
Fermentation

*Dairy products—Casein**
USE: Casein

Damping 931.1
DT: January 1993
UF: Vibrations—Damping*
BT: Mechanics
RT: Dynamic response
Energy absorption
Hysteresis
Internal friction
Natural frequencies
Oscillations
Shock absorbers
Stability
System stability
Transfer functions
Vibration control
Vibrations (mechanical)

Dams 441.1
DT: Predates 1975
BT: Hydraulic structures
NT: Arch dams
Buttress dams
Concrete dams
Embankment dams
Fishways
Gravity dams
Movable dams
Weirs
RT: Flexible structures
Flood control
Hydroelectric power plants
Ice control
Reservoirs (water)
River basin projects
Seepage
Spillways
Uplift pressure
Water supply

*Dams, Arch**
USE: Arch dams

*Dams, Buttress**
USE: Buttress dams

*Dams, Concrete**
USE: Concrete dams

*Dams, Embankment**
USE: Embankment dams

*Dams, Gravity**
USE: Gravity dams

*Dams, Movable**
USE: Movable dams

*Dams—Energy dissipators**
USE: Energy dissipators

*Dams—Flexible structures**
USE: Flexible structures

*Dams—Ice control**
USE: Ice control

*Dams—Seepage**
USE: Seepage

*Dams—Slope protection**
USE: Slope protection

*Dams—Uplift pressure**
USE: Uplift pressure

Data acquisition 723.2
SN: Sensing of variables and recording of raw data on a computer-sensible medium for processing
DT: January 1993
UF: Data capture
Data entry
Data processing—Data acquisition*
BT: Data processing
NT: SCADA systems
RT: Data handling
Data recording
Indicators (instruments)

* Former Ei Vocabulary term

Data acquisition *(continued)*
Punch card systems
Punch tape systems
Surveying
Surveys

Data analysis
USE: Data reduction

Data and voice communications
USE: Voice/data communication systems

Data base systems
USE: Database systems

Data capture
USE: Data acquisition

Data communication equipment (722.3) (716) (717) (718)
DT: January 1993
UF: Computers—Data communication equipment*
Data transmission equipment*
Transmission equipment (for data)
BT: Telecommunication equipment
NT: Modems
RT: Data communication systems
Multiplexing equipment
Telegraph equipment
Telemetering equipment

Data communication systems (723) (716) (717) (718)
SN: Systems for communication of data between computers, typically at a distance. For specialized program control and hardware for transfer of data between a computer and its storage devices use DATA TRANSFER
DT: January 1993
UF: Computers—Data communication systems*
Data transmission systems
Data transmission*
BT: Telecommunication systems
NT: Image communication systems
RT: Congestion control (communication)
Cryptography
Data communication equipment
Data processing
Electronic mail
Intelligent networks
Intersymbol interference
Local area networks
Multiplexing
Network protocols
Optical data processing
Packet switching
Telegraph
Videotex

Data compression (716.1) (723.2) (722.3)
DT: January 1993
UF: Compression (data)
Information theory—Data compression*
BT: Data processing
Signal processing
NT: Image compression
RT: Information theory

*Data conversion, Analog to digital**
USE: Analog to digital conversion

*Data conversion, Digital to analog**
USE: Digital to analog conversion

Data description 723.2, 723.3
SN: Specification of the arrangement of data items within a database
DT: January 1993
UF: Data processing—Data description*
BT: Data processing
RT: Data structures
Database systems

Data dictionaries
USE: Database systems

Data entry
USE: Data acquisition

Data handling 723.2
DT: January 1993
UF: Data processing—Data handling*
BT: Data processing
NT: Data reduction
Merging
Sorting
Table lookup
RT: Cryptography
Data acquisition
List processing languages
Security of data

Data manipulation languages
USE: Query languages

Data models
USE: Data structures

Data processing 723.2
DT: Predates 1975
UF: Information processing
NT: Administrative data processing
Analog to digital conversion
Associative processing
Batch data processing
Data acquisition
Data compression
Data description
Data handling
Data recording
Data transfer
Digital to analog conversion
Hospital data processing
Humanities computing
Manufacturing data processing
Military data processing
Natural language processing systems
Optical data processing
Word processing
RT: Business machines
Character sets
Code converters
Computer applications
Critical path analysis
Data communication systems

* Former Ei Vocabulary term

Data processing *(continued)*
Data storage equipment
Education computing
Encoding (symbols)
File editors
File organization
Hybrid sensors
Mechanization
Medical computing
Natural sciences computing
Psychology computing
Security of data
Sensor data fusion
Social sciences computing

*Data processing, Business**
USE: Administrative data processing

*Data processing—Batch processing**
USE: Batch data processing

*Data processing—Character sets**
USE: Character sets

*Data processing—Code converters**
USE: Code converters

*Data processing—Critical path analysis**
USE: Critical path analysis

*Data processing—Data acquisition**
USE: Data acquisition

*Data processing—Data description**
USE: Data description

*Data processing—Data handling**
USE: Data handling

*Data processing—Data reduction and analysis**
USE: Data reduction

*Data processing—Data structures**
USE: Data structures

*Data processing—Data transfer**
USE: Data transfer

*Data processing—Distribution applications**
USE: Distribution of goods

*Data processing—Educational applications**
USE: Education computing

*Data processing—File editors**
USE: File editors

*Data processing—File organization**
USE: File organization

*Data processing—Financial applications**
USE: Financial data processing

*Data processing—Governmental applications**
USE: Government data processing

*Data processing—Hospital applications**
USE: Hospital data processing

*Data processing—Humanities applications**
USE: Humanities computing

*Data processing—Manufacturing applications**
USE: Manufacturing data processing

*Data processing—Medical information**
USE: Medical computing

*Data processing—Military purposes**
USE: Military data processing

*Data processing—Natural sciences applications**
USE: Natural sciences computing

*Data processing—PERT**
USE: PERT

*Data processing—Psychology applications**
USE: Psychology computing

*Data processing—Recorders**
USE: Recording instruments

*Data processing—Security of data**
USE: Security of data

*Data processing—Social and behavioral sciences applications**
USE: Social sciences computing

*Data processing—Word processing**
USE: Word processing

Data recording 723.2 (722.1)
DT: January 1993
UF: Magnetic data recording
Mass spectrometers—Data recording*
BT: Data processing
Recording
RT: Data acquisition
Digital storage
Magnetic recording
Mass spectrometers
Mass spectrometry
Recording instruments

Data reduction 723.2
SN: Includes data analysis
DT: January 1993
UF: Data analysis
Data processing—Data reduction and analysis*
Photogrammetry—Data reduction*
BT: Data handling
RT: Photogrammetry
Spectrum analysis

Data retrieval systems
USE: Nonbibliographic retrieval systems

Data security
USE: Security of data

Data storage equipment 722.1
DT: January 1993
UF: Computer memory equipment
Computer storage equipment
Data storage units*

* Former Ei Vocabulary term

Data storage equipment *(continued)*
Information storage
Information storage equipment
Memories
Memory devices
Storage (data)
Storage devices (data)
BT: Computer hardware
NT: Analog storage
Digital storage
Magnetic storage
Optical data storage
Semiconductor storage
RT: Computer peripheral equipment
Data processing
Data transfer
Percolation (computer storage)
Storage allocation (computer)

*Data storage units**
USE: Data storage equipment

*Data storage, Analog**
USE: Analog storage

*Data storage, Cryoelectric**
USE: Cryoelectric storage

*Data storage, Digital**
USE: Digital storage

*Data storage, Digital—Associative**
USE: Associative storage

*Data storage, Digital—Cellular arrays**
USE: Cellular arrays

*Data storage, Digital—Nondestructive**
USE: Nondestructive readout

*Data storage, Digital—Nonvolatile**
USE: Nonvolatile storage

*Data storage, Digital—Parallel search**
USE: Digital storage

*Data storage, Digital—PROM**
USE: PROM

*Data storage, Digital—Random access**
USE: Random access storage

*Data storage, Digital—ROM**
USE: ROM

*Data storage, Digital—Self organizing**
USE: Self organizing storage

*Data storage, Digital—Storage devices**
USE: Digital storage

*Data storage, Digital—Virtual**
USE: Virtual storage

*Data storage, Magnetic**
USE: Magnetic storage

*Data storage, Magnetic—Bubbles**
USE: Magnetic bubble memories

*Data storage, Magnetic—Core**
USE: Magnetic core storage

*Data storage, Magnetic—Disk**
USE: Magnetic disk storage

*Data storage, Magnetic—Film**
USE: Magnetic film storage

*Data storage, Magnetic—Floppy disk**
USE: Floppy disk storage

*Data storage, Magnetic—Hard disk**
USE: Hard disk storage

*Data storage, Magnetic—Plated wire**
USE: Plated wire storage

*Data storage, Magnetic—Storage devices**
USE: Magnetic storage

*Data storage, Magnetic—Tape**
USE: Magnetic tape storage

*Data storage, Optical**
USE: Optical data storage

*Data storage, Optical—Disk**
USE: Optical disk storage

*Data storage, Optical—Storage devices**
USE: Optical data storage

*Data storage, Semiconductor**
USE: Semiconductor storage

*Data storage, Semiconductor—Storage devices**
USE: Semiconductor storage

Data structures 723.2 (723.3)
DT: January 1993
UF: Data models
Data processing—Data structures*
BT: File organization
RT: Data description
Database systems
Relational database systems

Data transfer (723.2)
SN: Specialized program control and hardware for transfer of data between the computer and its storage devices. For transmission of data between computer systems use DATA COMMUNICATION SYSTEMS
DT: January 1993
UF: Data processing—Data transfer*
BT: Data processing
RT: Computer hardware
Data storage equipment

*Data transmission equipment**
USE: Data communication equipment

Data transmission systems
USE: Data communication systems

*Data transmission**
USE: Data communication systems

*Data transmission—Intersymbol interference**
USE: Intersymbol interference

*Data transmission—Packet switching**
USE: Packet switching

Database machines
USE: Database systems

Database management systems
USE: Database systems

Database systems 723.3
DT: January 1983
UF: Data base systems
Data dictionaries
Database machines
Database management systems
Databases
DBMS
NT: Distributed database systems
Query languages
Relational database systems
RT: Concurrency control
Data description
Data structures
Decision support systems
File organization
Geographic information systems
Management information systems

*Database systems—Distributed**
USE: Distributed database systems

*Database systems—Query languages**
USE: Query languages

*Database systems—Relational**
USE: Relational database systems

Databases
USE: Database systems

Dating (geology)
USE: Geochronology

Davits 671.2 (693.1)
DT: January 1993
UF: Ship equipment—Davits*
BT: Ship equipment

Daylight simulation 707.1
DT: January 1993
UF: Electric lamps—Daylight simulation*
BT: Electric lighting
RT: Daylighting

Daylighting (707) (402)
DT: January 1986
BT: Lighting
RT: Daylight simulation
Illuminating engineering
Solar radiation
Windows

DBMS
USE: Database systems

DC generator motors 705.2.2, 705.3.2
SN: Use for single direct-current machines capable of operating either as motors or as generators
DT: January 1993
UF: Electric generator motors, DC*
Motor generators (DC)
BT: DC generators
DC motors
RT: Pumped storage power plants

DC generators 705.2.2
DT: January 1993
UF: Dynamos
Electric generators, DC*
BT: DC machinery
Electric generators
NT: DC generator motors

DC machinery 705.1 (705.2.2) (705.3.2)
DT: January 1993
UF: Direct current machinery
Electric machinery, DC*
Turbogenerators, Direct current*
BT: Electric machinery
NT: DC generators
DC motors
Electric commutators

DC motors 705.3.2
DT: January 1993
UF: Electric motors, DC*
BT: DC machinery
Electric motors
NT: DC generator motors

DC power transmission 706.1.1
SN: Direct-current transmission
DT: January 1993
UF: Direct current power transmission
Electric power transmission, DC*
BT: Electric power transmission
NT: HVDC power transmission

DC to AC inverters
USE: Electric inverters

DC transformers (704.1) (706.2) (715)
DT: January 1993
UF: Direct current transformers
Electric transformers, DC*
BT: Electric transformers

DDA
USE: Digital differential analyzers

Deadlocking (computers)
USE: Computer system recovery

Debugging (computer programs)
USE: Program debugging

Debugging (computers)
USE: Computer debugging

Deburring 604.2, 606.2
DT: January 1993
UF: Metal finishing—Deburring*
BT: Cutting
Finishing
RT: Metal cleaning
Reaming
Scarfing
Trimming

* Former Ei Vocabulary term

Decarbonization 802.2 (537.1)
DT: January 1993
BT: Chemical reactions
RT: Carbon
Carbonization

Decarburization 537.1
DT: January 1993
UF: Heat treatment—Decarburization*
BT: Heat treatment
RT: Carbon

Decay (organic) 801.2, 802.2 (811.2)
DT: January 1993
UF: Wood—Decay*
BT: Decomposition
RT: Disintegration
Wood

Deceleration (654.1) (931.1)
DT: January 1993
UF: Rockets and missiles—Deceleration*
RT: Acceleration
Rockets
Thrust reversal
Velocity

Decentralized control 731.1
DT: January 1995
BT: Control
RT: Distributed parameter control systems
Multivariable control systems

Decision analysis
USE: Decision theory

Decision making 912.2
DT: January 1995
RT: Command and control systems
Cost accounting
Management
Systems engineering

Decision support systems 723, 912.2
DT: June 1990
UF: Group decision support systems
BT: Management information systems
RT: Database systems
Expert systems
Management
Management science

Decision tables 723.1
DT: January 1993
UF: Computer systems programming—Decision tables*
Tables (decision)
RT: Decision theory
Program documentation
Systems analysis

Decision theory (921) (922.2)
DT: January 1993
UF: Decision analysis
Decision theory and analysis*
BT: Systems science
RT: Decision tables
Dynamic programming
Game theory
Information theory
Mathematical models
Optimal control systems
Probability
Problem solving
Quality control
Random processes
Scheduling
Signal processing
Systems analysis

*Decision theory and analysis**
USE: Decision theory

Deck landing aircraft (652.1)
DT: January 1993
UF: Aircraft—Deck landing*
BT: Aircraft
RT: Aircraft carriers
Military aircraft

Decks (bridge)
USE: Bridge decks

Decks (ship) 671
DT: January 1993
UF: Aircraft carriers—Deck cooling*
BT: Ships

Decoding 723.2 (716.1)
DT: January 1993
UF: Codes, Symbolic—Decoding*
RT: Codes (symbols)
Cryptography
Demodulation

Decommissioning (nuclear reactors) 621
DT: January 1993
UF: Nuclear reactors—Decommissioning*
RT: Nuclear reactors

Decomposition 802.2
DT: January 1993
BT: Chemical reactions
NT: Decay (organic)
Hydrolysis
Oil shale retorting
Photolysis
Pyrolysis
RT: Degradation
Deterioration
Disintegration
Dissociation
Weathering

Decontamination (462) (802.3)
DT: January 1993
BT: Cleaning
RT: Protective coatings
Purging
Purification
Sterilization (cleaning)

Decoration
DT: January 1993
BT: Finishing
RT: Coloring

* Former Ei Vocabulary term

Decoration *(continued)*
Etching
Metallography
Painting

Deep drawing 535.2
DT: January 1993
UF: Metal drawing—Deep*
BT: Metal drawing
RT: Explosive forming
Hot working
Metal working
Pressing (forming)
Stretching

Deep level transient spectroscopy (801) (931.3) (932.2)
DT: January 1995
UF: DLTS
Transient capacitance spectroscopy
BT: Spectroscopy
RT: Capacitance
Electron energy levels

Deep oil well drilling 511.1
DT: January 1993
UF: Oil well drilling—Deep*
BT: Oil well drilling

Deep sea habitats
USE: Ocean habitats

Deep sea platforms
USE: Production platforms

Deep submergence vehicles
USE: Submersibles

Deep well disposal (452)
DT: January 1993
UF: Deepwell disposal
Waste disposal—Deepwell disposal*
BT: Waste disposal

Deep well pumps
USE: Well pumps

Deepwell disposal
USE: Deep well disposal

Defects (933.1) (531.2) (423)
DT: January 1993
UF: Flaws
Imperfections
NT: Cracks
Crystal defects
Discoloration
Leakage (fluid)
RT: Inclusions
Materials testing
Spalling

Defense rockets
USE: Military rockets

Defense satellites
USE: Military satellites

Defibrillators 462.2
DT: January 1993
UF: Biomedical equipment—Defibrillators*
BT: Biomedical equipment
RT: Cardiology

Deflected boreholes (501.1) (511.1)
DT: January 1993
UF: Boreholes—Deflected*
BT: Boreholes
RT: Deflected well drilling
Directional drilling

Deflected oil well drilling
USE: Oil well drilling

Deflected well drilling (501.1) (511.1)
DT: January 1993
UF: Oil well drilling—Deflected*
BT: Well drilling
RT: Deflected boreholes

Deflection (structures) 408.2 (421)
DT: January 1993
UF: Beams and girders—Deflection*
RT: Structural design

Deflection yokes 716.4 (714.1)
DT: January 1993
UF: Scanning yokes
Television receivers—Deflection yokes*
Yokes
BT: Electron tube components
RT: Television receivers

Deformability
USE: Formability

Deformation (421) (422)
SN: Unintended effect. For deformation as a forming process, use FORMING
DT: January 1993
NT: Bending (deformation)
Creep
Electrostriction
Plastic deformation
Shear deformation
RT: Buckling
Elasticity
Failure (mechanical)
Plasticity
Stiffness
Strain
Superplasticity
Torsional stress
Wear of materials

Defrosting (644.3)
DT: January 1993
UF: Refrigerators—Defrosting*
RT: Ice control
Refrigerators

Degassers
USE: Getters

Degassing (802.3)
DT: January 1993
UF: Outgassing
BT: Separation

* Former Ei Vocabulary term

Degaussing (672.1) (701.2)
DT: January 1993
UF: Warships—Degaussing*
BT: Demagnetization
RT: Warships

Degradation 802.2
DT: January 1993
BT: Chemical reactions
NT: Weathering
RT: Aging of materials
Corrosion
Decomposition
Deterioration
Discoloration
Durability
Failure (mechanical)
Fouling
Pyrolysis

Degrees of freedom (mechanics) 931.1
DT: January 1993
UF: Mechanisms—Degrees of freedom*
RT: Mechanics
Mechanisms

Degumming (822.2)
DT: January 1993
UF: Vegetable oils—Degumming*
BT: Processing
RT: Vegetable oils

Dehalogenation 802.2
DT: January 1993
BT: Chemical reactions
RT: Halogen compounds
Halogen elements
Halogenation

Dehydration 802.2
DT: January 1993
BT: Chemical reactions
RT: Curing
Dewatering
Drying
Efflorescence
Hydration

Dehydrogenation 802.2
DT: January 1993
BT: Chemical reactions
RT: Cracking (chemical)
Hydrogenation

Deinking (452.4) (811.1)
DT: January 1993
UF: Paper—Deinking*
BT: Recycling
RT: Paper

Delamination (421) (817.1)
DT: January 1995
RT: Crack propagation
Interfaces (materials)
Laminates
Peeling
Strength of materials

Delay circuits (703.1) (713.5)
DT: January 1993
UF: Electronic circuits, Delay type*
Time delay circuits
BT: Networks (circuits)
NT: Electric delay lines
RT: Comparator circuits

Delay control systems 731.1
DT: January 1993
UF: Control systems, Delay*
BT: Control systems

Delay lines
USE: Electric delay lines

Delta modulation (713.3) (716)
DT: Predates 1975
UF: Adaptive delta modulation
ADM
BT: Pulse modulation
RT: Pulse code modulation

Delta wing aircraft 652.1
DT: January 1993
UF: Aircraft—Delta wing*
BT: Aircraft

Demagnetization 701.2 (708.4)
DT: January 1993
UF: Magnetization—Demagnetization*
NT: Degaussing
RT: Magnetic fields
Magnetic properties
Magnetization

Demand management
USE: Electric load management

Demand meters
USE: Maximum demand indicators

Demodulation (713.3) (716)
DT: Predates 1975
UF: Detection (demodulation)
RT: Amplitude modulation
Decoding
Demodulators
Detector circuits
Detectors
Information theory
Intermodulation
Modulation
Phase modulation
Radio receivers
Signal detection
Signal receivers

Demodulators 713.3 (716)
DT: Predates 1975
UF: Radio detectors
BT: Electronic equipment
NT: Microwave demodulators
Modems
RT: Amplitude modulation
Demodulation
Detector circuits
Discriminators

* Former Ei Vocabulary term

Demodulators *(continued)*
Limiters
Mixer circuits
Modulators
Networks (circuits)
Phase locked loops
Phase modulation
Pulse amplitude modulation
Pulse code modulation
Pulse modulation
Pulse time modulation
Pulse width modulation
Radio equipment
Radio receivers
Signal receivers

*Demodulators, Microwave**
USE: Microwave demodulators

Demolition (402)
DT: January 1993
UF: Buildings—Demolition*
Chimneys—Demolition*
RT: Buildings

Demonstrations (901.2)
SN: Tutorials
DT: January 1993
UF: Education—Demonstrations*
RT: Education
Information dissemination
Technical presentations

Demulsification 802.3 (513.1)
DT: January 1993
UF: Liquids—Demulsification*
Petroleum refining—Demulsification*
BT: Separation
RT: Emulsification
Liquids

Denitration
USE: Denitrification

Denitrification 802.2
DT: January 1993
UF: Denitration
BT: Chemical reactions
RT: Nitrogen
Water treatment

Densification 802.3
DT: January 1993
RT: Agglomeration
Briquetting
Compaction
Consolidation
Pelletizing

Densitometers (941.3)
SN: Devices to measure optical density of material such as photographic film
DT: Predates 1975
UF: Opacity measurement
Optical density measuring instruments
BT: Density measuring instruments
Optical instruments
RT: Comparators (optical)
Nephelometers
Opacity
Photographic films
Photometers
Sensitometers

Density (optical) 741.1
DT: January 1993
UF: Density*
Optical density
BT: Optical properties
RT: Density measurement (optical)
Opacity

Density (specific gravity) 931.2
DT: January 1993
UF: Density*
Specific density
Specific gravity
Specific volume
Specific weight
BT: Mechanical properties
NT: Density of gases
Density of liquids
RT: Buoyancy
Compressibility
Density measurement (specific gravity)
Porosity

Density control (specific gravity) 731.3 (931.2)
DT: January 1993
UF: Control, Mechanical variables (Density)
BT: Mechanical variables control
RT: Density measurement (specific gravity)
Density of gases
Density of liquids
Volume control (spatial)

Density function
USE: Probability density function

Density measurement (electron)
USE: Electron density measurement

Density measurement (optical) (741.1) (941.4)
DT: January 1993
UF: Density measurement*
BT: Optical variables measurement
RT: Density (optical)

Density measurement (specific gravity) (931.2)
DT: January 1993
UF: Density measurement*
Mechanical variables measurement—Density*
BT: Mechanical variables measurement
RT: Density (specific gravity)
Density control (specific gravity)
Density of gases
Density of liquids
Gravitometers
Volume measurement

*Density measurement**
USE: Density measurement (specific gravity) OR
Density measurement (optical)

Density measuring instruments (943.1)
SN: Very general term; use more specific term when possible
DT: Predates 1975
BT: Instruments
NT: Densitometers
Gravitometers
Porosimeters
Turbidimeters

Density of gases 931.2
DT: January 1993
UF: Gases—Density*
Vapor density
BT: Density (specific gravity)
NT: Atmospheric density
Plasma density
RT: Density control (specific gravity)
Density measurement (specific gravity)
Gases

Density of liquids 931.2
DT: January 1993
UF: Liquids—Density*
BT: Density (specific gravity)
RT: Density control (specific gravity)
Density measurement (specific gravity)
Hydrometers
Liquids

*Density**
USE: Density (specific gravity) OR
Density (optical)

Dental alloys 462.3
DT: January 1993
UF: Dental equipment and supplies—Dental amalgams and alloys*
BT: Alloys
Dental materials

Dental amalgams 462.3
DT: January 1993
UF: Dental equipment and supplies—Dental amalgams and alloys*
BT: Dental materials
Mercury amalgams

Dental braces 462.3
DT: January 1993
UF: Braces (dental)
Orthotics—Braces*
BT: Dental orthoses

Dental cement 462.3
SN: Dental supplies, not the biological material
DT: January 1993
UF: Cement (dental)
Dental equipment and supplies—Dental cement*
BT: Dental materials

Dental composites 462.3
DT: January 1993
UF: Dental equipment and supplies—Dental composites*
BT: Composite materials
Dental materials

Dental cutting tools 462.3
DT: January 1993
UF: Dental equipment and supplies—Cutting tools*
BT: Cutting tools
Dental equipment

Dental equipment 462.3
DT: January 1993
UF: Dental equipment and supplies*
BT: Biomedical equipment
NT: Dental cutting tools
Dental orthoses
RT: Dentistry

*Dental equipment and supplies**
USE: Dental equipment OR
Dental materials

*Dental equipment and supplies—Bone cement**
USE: Bone cement

*Dental equipment and supplies—Cutting tools**
USE: Dental cutting tools

*Dental equipment and supplies—Dental amalgams and alloys**
USE: Dental alloys OR
Dental amalgams

*Dental equipment and supplies—Dental cement**
USE: Dental cement

*Dental equipment and supplies—Dental composites**
USE: Dental composites

*Dental equipment and supplies—Dental orthoses**
USE: Dental orthoses

*Dental equipment and supplies—Dental prostheses**
USE: Dental prostheses

*Dental equipment and supplies—Maxillofacial prostheses**
USE: Maxillofacial prostheses

Dental fillings
USE: Dental prostheses

Dental implants
USE: Dental prostheses

Dental materials 462.3
DT: January 1993
UF: Dental equipment and supplies*
BT: Biomaterials
NT: Dental alloys
Dental amalgams
Dental cement
Dental composites
Dental prostheses
RT: Bone cement
Dentistry

Dental orthoses 462.3
DT: January 1993
UF: Dental equipment and supplies—Dental orthoses*
BT: Dental equipment
Orthotics
NT: Dental braces

Dental prostheses 462.3
DT: January 1993
UF: Dental equipment and supplies—Dental prostheses*
Dental fillings
Dental implants
Dentures
False teeth
Fillings (dental)
Implants (dental)
BT: Dental materials
Prosthetics

Dentistry (462.3) (461.1) (461.6)
DT: January 1993
UF: Biomedical engineering—Dentistry*
BT: Medicine
RT: Biomedical engineering
Dental equipment
Dental materials
Tooth enamel

Dentures
USE: Dental prostheses

Deoxidants 803
DT: January 1993
UF: Deoxidizers
Steelmaking—Deoxidants*
BT: Agents
RT: Ferroalloys
Steelmaking

Deoxidizers
USE: Deoxidants

Deoxyribonucleic acid
USE: DNA

Department stores
USE: Retail stores

Deposition 802.3 (539.3) (813.1)
DT: January 1993
NT: Coating techniques
Electrodeposition
Sedimentation
Sputter deposition
Vapor deposition
RT: Optical coatings
Precipitation (chemical)

Deposits (532) (612) (617)
SN: Material deposited in or on equipment as a result of normal operations; not geological deposits
DT: January 1993
UF: Furnaces—Deposits*
Internal combustion engines—Deposits*
Turbomachinery—Deposits*
NT: Boiler deposits
Scale (deposits)
RT: Fouling

Depreciation (911.1)
DT: January 1993
UF: Buildings—Depreciation*
Industrial plants—Depreciation*
Machine tools—Depreciation*
Machinery—Depreciation*
Power plants—Depreciation*
Telephone equipment—Depreciation*
RT: Cost accounting
Management

Depth finders
USE: Depth indicators

Depth indicators (471) (671.2)
DT: January 1993
UF: Depth finders
Fishing vessels—Depth indicators*
BT: Indicators (instruments)
RT: Fishing vessels
Pressure gages
Pressure measurement

Depth measurement (ocean)
USE: Bathymetry

Depth perception (731.5) (741.2) (461.1)
DT: January 1993
UF: Robots, Industrial—Depth perception*
BT: Vision
RT: Computer vision
Industrial robots

Derailments 914.1, 682
DT: January 1993
UF: Railroad accidents—Derailment*
BT: Railroad accidents

Derivatives 804
DT: January 1993
UF: Chemical derivatives
BT: Chemical compounds
NT: Cellulose derivatives

Dermatitis 461.6, 461.2
DT: January 1993
UF: Occupational diseases—Dermatitis*
BT: Diseases
RT: Skin

Derricks
USE: Cranes

Desalination 445.1 (471.4) (802.3)
SN: Removal of salt from water
DT: October 1975
UF: Seawater—Salt removal*
Water treatment—Salt removal*
BT: Salt removal
RT: Distillation
Osmosis
Seawater
Sodium chloride
Vaporization
Water treatment

* Former Ei Vocabulary term

Desalting
USE: Salt removal

Descaling (802.3) (539.1) (606.2)
DT: January 1993
UF: Rolling mill practice—Scale removal*
Scale removal
Steel—Scale removal*
Wire—Scale removal*
RT: Cleaning
Corrosion
Finishing
Polishing
Rolling mill practice
Scale (deposits)
Steel
Ultrasonic cleaning

Describing functions (921) (731.1)
DT: January 1993
UF: Control systems—Describing functions*
BT: Functions
Mathematical techniques
RT: Control system analysis
Control system synthesis
Nonlinear systems
Numerical methods
Nyquist diagrams
On-off control systems

Design
DT: January 1993
UF: Design optimization
NT: Architectural design
Computer aided design
Logic design
Machine design
Optical design
Product design
Structural design
RT: Design aids
Planning

Design aids
DT: January 1993
UF: Engineering—Design aids*
BT: Equipment
RT: Computer aided design
Design

Design optimization
USE: Optimization AND
Design

Desktop computers
USE: Personal computers

Desktop publishing 723.5, 903.2
SN: Use of office computer equipment to produce published documents
DT: January 1993
UF: Information dissemination—Electronic publishing*
BT: Publishing
RT: Electronic publishing
Word processing

Desorption 802.3
DT: January 1983
BT: Sorption
RT: Absorption
Adsorption
Temperature programmed desorption

Desulfurization (802.2) (802.3)
DT: January 1993
UF: Desulfurizing
Sulfur removal
BT: Chemical operations
RT: Gas fuel purification
Refining
Smelting
Steelmaking
Sulfur
Sulfur determination

Desulfurizing
USE: Desulfurization

Detection (demodulation)
USE: Demodulation

Detection (errors)
USE: Error detection

Detector circuits 713.3
DT: January 1993
UF: Electronic circuits, Detector*
BT: Networks (circuits)
RT: Demodulation
Demodulators
Superregeneration

Detectors
DT: January 1993
BT: Instruments
NT: Correlation detectors
Fire detectors
Fish detectors
Gas detectors
Metal detectors
Obstacle detectors
Radar warning systems
Radiation detectors
Sensors
Sonar
RT: Alarm systems
Character recognition equipment
Counting tubes
Demodulation
Leak detection
Measurements
Monitoring
Pattern recognition
Pollution detection
Semiconductor devices
Signal receivers

Detergents 803, 804
DT: Predates 1975
UF: Sewage treatment—Detergents effects*
Water pollution—Detergents effects*
Water treatment—Detergents effects*
BT: Surface active agents

* Former Ei Vocabulary term

Detergents *(continued)*
NT: Soaps (detergents)
RT: Cleaning
Laundering

Deterioration (802.2)
DT: January 1993
NT: Disintegration
Embrittlement
RT: Corrosion
Decomposition
Degradation
Durability

Detonation (802.2) (502.2) (521)
DT: Predates 1975
UF: Explosives—Detonation*
RT: Combustion
Explosions
Explosives
Internal combustion engines
Shock waves

Deuterium (801.4.2) (804) (803)
SN: Hydrogen isotope, mass number 2
DT: Predates 1975
UF: D
BT: Isotopes
RT: Deuterium compounds
Hydrogen

Deuterium compounds 804.2
DT: Predates 1975
BT: Inorganic compounds
NT: Heavy water
RT: Deuterium
Hydrogen inorganic compounds
Radioisotopes

Deuterium oxide
USE: Heavy water

Developing countries
DT: January 1993
BT: Geographical regions

Dewar flasks
USE: Dewars

Dewars 633.1 (644.3)
DT: Predates 1975
UF: Dewar flasks
BT: Bottles
Cryogenic equipment

Dewatering 802.3
DT: January 1993
BT: Separation
RT: Beneficiation
Centrifugation
Concentration (process)
Condensation
Dehydration
Drainage
Drying
Evaporation
Extraction
Filtration

Dewaxing 802.3 (513.1)
DT: January 1993
UF: Petroleum refining—Dewaxing*
BT: Petroleum refining
Separation

Dextrose 804.1, 822.3
DT: January 1987
UF: Corn sugar
Grape sugar
BT: Sugars
RT: Food additives

DFB lasers
USE: Distributed feedback lasers

Diagnosis 461.6
DT: January 1993
UF: Biomedical engineering—Diagnosis*
NT: Computer aided diagnosis
Diagnostic radiography
Echocardiography
Electrocardiography
Electroencephalography
Electromyography
Endoscopy
Plethysmography
RT: Acoustic imaging
Biomedical engineering
Biomedical equipment
Diagnostic products
Magnetic resonance imaging
Medical imaging
Medicine
Noninvasive medical procedures
Radioactive tracers
Thermography (imaging)

Diagnostic products 461.7
DT: January 1993
UF: Health care—Diagnostic products*
RT: Diagnosis
Health care

Diagnostic radiography 461.7
DT: January 1993
UF: Radiography—Diagnostic applications*
BT: Diagnosis
Medical imaging
Radiography
NT: Angiography
RT: Computerized tomography
Contrast media
Positron emission tomography
Radiology

Diagnostics (computer programs)
USE: Program diagnostics

Diagnostics (plasma)
USE: Plasma diagnostics

Diakoptics
USE: Electric network topology

Dials (instrument)
USE: Instrument dials

* Former Ei Vocabulary term

Dialysis (802.2) (461.6)
DT: January 1987
BT: Separation
NT: Electrodialysis
RT: Colloids
Dialysis membranes
Diffusion
Hemodialyzers
Mass transfer
Mechanical permeability

Dialysis membranes (462.1) (802.2) (817.1)
DT: January 1993
UF: Membranes—Dialysis membranes*
BT: Membranes
RT: Dialysis

Diamagnetic resonance
USE: Cyclotron resonance

Diamagnetism 701.2
DT: January 1993
UF: Magnetic materials—Diamagnetism*
Magnetism—Diamagnetism*
BT: Magnetism
RT: Cyclotron resonance
Magnetic materials

Diamond cutting tools (605) (606.1) (482.2.1)
SN: Tools which use diamond as a cutting material
DT: January 1993
UF: Cutting tools—Diamond*
Grinding wheels—Diamond*
Saws—Diamond*
BT: Cutting tools
NT: Diamond drills
RT: Diamonds
Grinding wheels
Saws

Diamond deposits 505.1 (482.2.1)
DT: Predates 1975
BT: Mineral resources
RT: Diamonds

Diamond drilling (501.1) (482.2.1)
SN: Drilling using diamonds
DT: January 1993
UF: Boreholes—Diamond drilling*
BT: Drilling
RT: Diamond drills
Diamonds

Diamond drills (501.1) (603.2) (482.2.1)
SN: Drills using diamond as a cutting material
DT: January 1993
BT: Diamond cutting tools
Drills
RT: Diamond drilling

Diamond films (482.2) (804) (933.1)
DT: January 1995
BT: Films
Industrial diamonds

Diamond mines 505.1 (482.2.1)
DT: January 1993
UF: Diamond mines and mining*
BT: Mines
RT: Diamonds

*Diamond mines and mining**
USE: Diamond mines

Diamonds 482.2.1
DT: Predates 1975
BT: Carbon
Gems
NT: Industrial diamonds
Semiconducting diamonds
Synthetic diamonds
RT: Diamond cutting tools
Diamond deposits
Diamond drilling
Diamond mines
Single crystals

*Diamonds—Synthetic**
USE: Synthetic diamonds

Diaphragms 601.2
DT: Predates 1975
BT: Components
RT: Osmosis
Pressure transducers
Pumps
Seals
Valves (mechanical)

Diathermy
USE: Hyperthermia therapy

Diblock copolymers
USE: Block copolymers

Dictating machines 752.2.1, 752.3.1 (903.2)
DT: Predates 1975
BT: Business machines

Dictionaries
USE: Terminology

Die casting 534.2
DT: Predates 1975
BT: Permanent mold casting
NT: Vacuum die casting
RT: Die casting machines
Die castings
Dies

Die casting inserts 534.1
DT: January 1993
UF: Die casting—Inserts*
Die castings—Inserts*
Inserts (casting)
Screw threads—Inserts*
BT: Die castings

Die casting machines 534.1
DT: Predates 1975
BT: Metal forming machines
RT: Die casting

*Die casting—Dies**
USE: Dies

*Die casting—Inserts**
USE: Die casting inserts

* Former Ei Vocabulary term

*Die casting—Vacuum**
USE: Vacuum die casting

Die castings 534.1
DT: Predates 1975
BT: Metal castings
NT: Die casting inserts
RT: Die casting

*Die castings—Inserts**
USE: Die casting inserts

Dielectric amplifiers 713.1 (708.1)
DT: January 1993
UF: Amplifiers, Dielectric*
BT: Amplifiers (electronic)
Dielectric devices
RT: Ferroelectric devices

Dielectric constant
USE: Permittivity

Dielectric devices (704) (714) (708.1)
DT: Predates 1975
BT: Electric equipment
NT: Capacitors
Dielectric amplifiers
Dielectric waveguides
Electric insulators
Ferroelectric devices
Microstrip devices
Piezoelectric devices
RT: Ceramic products
Dielectric heating
Dielectric materials
Electrets

Dielectric dispersion
USE: Permittivity

Dielectric films 708.1
DT: January 1993
UF: Films—Dielectric*
BT: Dielectric materials
Films
RT: Optical films
Thin films

Dielectric heating (532.3) (642.1) (708.1)
DT: January 1993
UF: Electric heating, Dielectric*
BT: Electric heating
RT: Dielectric devices
Dielectric materials

Dielectric materials 708.1
DT: Predates 1975
UF: Dielectrics
BT: Materials
NT: Dielectric films
Electrets
Ferroelectric materials
Piezoelectric materials
RT: Ceramic materials
Dielectric devices
Dielectric heating
Electric insulating materials
Electric insulation
Glass
Microstrip lines
Plastics

Dielectric power factor
USE: Electric power factor

Dielectric properties 931.2 (708.1) (701)
DT: January 1977
BT: Electric properties
NT: Dielectric properties of gases
Dielectric properties of liquids
Dielectric properties of solids
Ferroelectricity
Permittivity
Piezoelectricity
Pyroelectricity
RT: Dielectric relaxation

Dielectric properties of gases
931.2 (708.1) (701)
DT: January 1993
UF: Gases—Dielectric properties*
BT: Dielectric properties
RT: Gases

Dielectric properties of liquids
931.2 (708.1) (701)
DT: January 1993
UF: Liquids—Dielectric properties*
BT: Dielectric properties
RT: Liquids
Solutions

Dielectric properties of solids
931.2 (708.1) (701)
DT: January 1993
UF: Solids—Dielectric properties*
BT: Dielectric properties
RT: Solids

Dielectric relaxation (931.2) (708.1) (701)
DT: January 1993
BT: Relaxation processes
RT: Dielectric properties

Dielectric waveguides 714.3 (701.1)
DT: January 1993
UF: Waveguides, Dielectric*
BT: Dielectric devices
Waveguides
NT: Optical waveguides

Dielectrics
USE: Dielectric materials

Dies 534.1, 603.2
DT: Predates 1975
UF: Die casting—Dies*
Forging dies
BT: Machine tools
NT: Bending dies
Carbide dies
Ceramic dies
Drawing dies
Extrusion dies
Plastics forming dies
Rubber forming dies

* Former Ei Vocabulary term

Dies *(continued)*
Thread cutting dies
Transfer dies
RT: Cutting tools
Die casting
Forging
Forging machines
Presses (machine tools)

*Dies—Bending**
USE: Bending dies

*Dies—Carbide**
USE: Carbide dies

*Dies—Ceramic**
USE: Ceramic dies

*Dies—Plastics**
USE: Plastics applications

*Dies—Rubber**
USE: Rubber products

*Dies—Transfer**
USE: Transfer dies

Diesel and gas turbine combined power plants 612.2, 612.3
DT: January 1993
UF: Power plants—Diesel and gas turbine combined*
BT: Power plants
RT: Diesel engines
Gas turbines

Diesel and hydroelectric combined power plants 611.1 (612.2)
DT: January 1993
UF: Electric power plants—Diesel and hydroelectric combined*
BT: Power plants
RT: Diesel engines

Diesel boat engines
USE: Diesel engines AND
Marine engines

Diesel dredges
USE: Dredges

Diesel electric locomotives
USE: Diesel locomotives

Diesel electric power plants 612.2
DT: Predates 1975
BT: Diesel power plants
Electric power plants

Diesel electric traction 612.2 (682.1.2)
DT: Predates 1975
BT: Electric traction
RT: Diesel engines
Diesel locomotives

Diesel engines 612.2
DT: Predates 1975
UF: Boats—Diesel*
Diesel boat engines
Dredges—Diesel*
Oil engines
BT: Internal combustion engines
RT: Diesel and gas turbine combined power plants
Diesel and hydroelectric combined power plants
Diesel electric traction
Diesel fuels
Diesel locomotives
Diesel propulsion

*Diesel engines—Starting**
USE: Starting

*Diesel engines—Traction applications**
USE: Traction motors

Diesel fuels 523 (612.2)
DT: Predates 1975
BT: Fuels
Petroleum products
RT: Alcohol fuels
Automotive fuels
Diesel engines
Diesel power plants
Ethanol fuels
Gas oils
Liquid fuels
Methanol fuels

Diesel locomotives 682.1.2 (612.2)
DT: January 1993
UF: Diesel electric locomotives
Locomotives, Diesel*
BT: Locomotives
RT: Diesel electric traction
Diesel engines

Diesel power plants 612.2
DT: January 1993
BT: Power plants
NT: Diesel electric power plants
RT: Diesel fuels

Diesel propulsion 612.2 (675.1) (682.1.2)
DT: January 1993
UF: Ship propulsion—Diesel*
BT: Propulsion
RT: Diesel engines
Ship propulsion

Diet
USE: Nutrition

Difference equations 921.6
DT: January 1993
UF: Mathematical techniques—Difference equations*
BT: Differential equations
Numerical analysis
RT: Boundary element method
Finite difference method
Finite element method
Optimal control systems
Piecewise linear techniques
Z transforms

* Former Ei Vocabulary term

Differential amplifiers 713.1
DT: January 1993
UF: Amplifiers, Differential*
BT: Amplifiers (electronic)

Differential analyzers (analog)
USE: Analog differential analyzers

Differential analyzers (digital)
USE: Digital differential analyzers

Differential calculus
USE: Differentiation (calculus)

Differential equations 921.2
DT: January 1993
UF: Mathematical techniques—Differential equations*
BT: Mathematical techniques
NT: Boundary value problems
Difference equations
Equations of motion
Green's function
Integrodifferential equations
Partial differential equations
Riccati equations
RT: Asymptotic stability
Boundary conditions
Boundary element method
Computational fluid dynamics
Differentiation (calculus)
Finite element method
Lagrange multipliers
Numerical analysis

Differential scanning calorimetry 944.6 (641.1) (801)
DT: January 1995
BT: Calorimetry
RT: Chemical reactions
Differential thermal analysis
Specific heat

Differential thermal analysis (801)
DT: Predates 1975
UF: DTA
BT: Thermoanalysis
RT: Differential scanning calorimetry

Differentiating circuits (713.5) (721.2) (721.3)
DT: January 1993
UF: Electronic circuits, Differentiating*
Peaking circuits
BT: Networks (circuits)

Differentiation (calculus) 921.2
DT: January 1993
UF: Differential calculus
Mathematical techniques—Differentiation*
BT: Mathematical techniques
RT: Differential equations
Numerical analysis

Diffraction (711.1) (741.1)
DT: January 1993
NT: Acoustic wave diffraction
Electromagnetic wave diffraction
Electron diffraction
Neutron diffraction
RT: Diffractometers
Holography

Diffraction gratings 741.3
DT: January 1993
UF: Gratings (diffraction)
Gratings (optical)
Optical gratings*
Optical instruments—Gratings*
BT: Optical devices
RT: Electromagnetic wave diffraction
Spectrometers

Diffractometers (741.3) (941.3)
DT: Predates 1975
BT: Optical instruments
RT: Diffraction

Diffusers (fluid) (631.1) (612.3)
DT: January 1993
UF: Flow of fluids—Diffusers*
Sugar factories—Diffusers*
Turbomachinery—Diffusers*
BT: Equipment
RT: Diffusion
Ducts
Flow of fluids

Diffusers (optical) 741.3
DT: January 1993
UF: Optical instruments—Diffusers*
BT: Optical devices
RT: Diffusion
Optical instruments

Diffusion (931.1)
DT: January 1977
UF: Diffusion coefficient
NT: Diffusion in gases
Diffusion in liquids
Diffusion in solids
Osmosis
Thermal diffusion
RT: Dialysis
Diffusers (fluid)
Diffusers (optical)
Drying
Electromigration
Evaporation
Ionic conduction
Mass transfer
Membranes
Mixing
Percolation (fluids)
Prandtl number
Reflection
Regain
Surface properties
Transport properties
Turbulence

Diffusion coatings 813.2 (813.1)
DT: January 1993
UF: Coatings—Diffusion coatings*
BT: Coatings

* Former Ei Vocabulary term

Diffusion coefficient
USE: Diffusion

Diffusion in gases (931.2)
DT: January 1993
UF: Gases—Diffusion*
BT: Diffusion
NT: Thermal diffusion in gases
RT: Gases

Diffusion in liquids (931.2)
DT: January 1993
UF: Liquids—Diffusion*
BT: Diffusion
NT: Thermal diffusion in liquids
RT: Liquids

Diffusion in solids (933) (931.2)
DT: January 1993
UF: Solids—Diffusion*
BT: Diffusion
NT: Interdiffusion (solids)
Thermal diffusion in solids
RT: Ionic conduction in solids
Solids

Digesters (pulp)
USE: Pulp digesters

Digestion (refuse)
USE: Refuse digestion

Digestion (sludge)
USE: Sludge digestion

Digital analog conversion
USE: Digital to analog conversion

Digital arithmetic 721.1, 921.6
DT: January 1993
UF: Arithmetic (digital)
Floating point arithmetic
Mathematical techniques—Digital arithmetic*
BT: Mathematical techniques
NT: Carry logic
RT: Adders
Computers
Counting circuits
Digital circuits
Digital filters
Dividing circuits (arithmetic)
Logic design
Multiplying circuits
Number theory

Digital circuits 721.3 (713.5) (722.4)
DT: January 1993
UF: Electronic circuits, Digital*
BT: Networks (circuits)
NT: Adders
Digital integrated circuits
Dividing circuits (arithmetic)
Multiplying circuits
RT: Computer circuits
Digital arithmetic
Digital devices
Logic circuits
Pulse circuits
Summing circuits
Switching circuits

Digital communication systems (722.3) (716) (717) (718)
DT: Predates 1975
BT: Telecommunication systems
NT: Voice/data communication systems
RT: Computer peripheral equipment
Intelligent networks
Intersymbol interference
Telegraph systems
Telephone systems

*Digital communication systems—Voice/data integrated services**
USE: Voice/data communication systems

Digital computer systems
USE: Computer systems

Digital computers 722.4
DT: January 1993
UF: Computers, Digital*
BT: Computers
NT: Digital differential analyzers
General purpose computers
Microcomputers
Minicomputers
Supercomputers
RT: Computer systems
Digital devices
Digital to analog conversion
Finite automata
Hybrid computers
Turing machines

Digital control systems 731.1 (722.4)
DT: January 1993
UF: Control systems, Digital*
BT: Control systems
NT: Direct digital control systems
RT: Computer control systems
Numerical control systems

Digital data storage
USE: Digital storage

Digital data storage devices
USE: Digital storage

Digital devices (713.5) (721)
DT: January 1981
BT: Electronic equipment
NT: Smart cards
RT: Digital circuits
Digital computers
Digital instruments
Digital voltmeters

Digital differential analyzers 722.4
DT: January 1993
UF: Computers—Digital differential analyzers*
DDA
Differential analyzers (digital)
BT: Digital computers

Digital filters 703.2
DT: January 1993
UF: Electric filters, Digital*
Finite impulse response filters
FIR filters
IIR filters
Infinite impulse response filters
Quadrature mirror filters
BT: Electric filters
RT: Adaptive filtering
Digital arithmetic
Signal processing

Digital image storage 722.1 (722.4) (741.3)
DT: January 1993
UF: Image archiving
Image storage (digital)
Image storage, Digital*
BT: Digital storage
RT: Image processing
Imaging techniques
Optical data storage

Digital instruments (941) (942) (943) (944)
DT: January 1993
UF: Instruments, Digital*
Instruments—Digital readout*
BT: Instruments
NT: Digital voltmeters
RT: Digital devices

Digital integrated circuits 714.2 (721.3)
DT: January 1993
UF: Integrated circuits, Digital*
BT: Digital circuits
Integrated circuits
NT: Cellular arrays
Microprocessor chips
RT: Application specific integrated circuits
Emitter coupled logic circuits
Integrated injection logic circuits
Transistor transistor logic circuits

Digital processing
USE: Digital signal processing

Digital signal processing (716.1) (722.4) (723.2)
DT: January 1993
UF: Digital processing
Information theory—Digital signals*
Signal processing—Digital techniques*
BT: Signal processing
RT: Information theory
Systolic arrays

Digital storage 722.1
DT: January 1993
UF: Data storage, Digital*
Data storage, Digital—Parallel search*
Data storage, Digital—Storage devices*
Digital data storage
Digital data storage devices
BT: Data storage equipment
NT: Associative storage
Cryoelectric storage
Digital image storage
Nonvolatile storage
Random access storage
ROM
Self organizing storage
Virtual storage
RT: Data recording
Flip flop circuits
Nondestructive readout
Recording

Digital storage oscilloscopes
USE: Cathode ray oscilloscopes

Digital to analog conversion 723.2
DT: January 1993
UF: Converters (digital to analog)
Data conversion, Digital to analog*
Digital analog conversion
Digital to analog converters
BT: Data processing
RT: Counting circuits
Digital computers
Hybrid computers
Interfaces (computer)
Plotters (computer)

Digital to analog converters
USE: Digital to analog conversion

Digital voltmeters 942.1
DT: January 1993
UF: DVM
Voltmeters, Digital*
BT: Digital instruments
Voltmeters
RT: Digital devices

Digitization
USE: Analog to digital conversion

Dikes
USE: Levees

Dilatometers 943.1
SN: For measuring changes in volume of solids or liquids due to thermal or chemical effects
DT: Predates 1975
UF: Extensometers
BT: Instruments
Transducers
RT: Gages
Size determination
Strain gages

Dimensional stability (421)
DT: January 1993
BT: Stability
RT: Shrinkage
Size determination
Spatial variables control
Swelling

Dimming (lamps) 707.2 (707.1)
DT: January 1993
UF: Electric lamps—Dimming*

* Former Ei Vocabulary term

Dining cars 682.1.1
DT: January 1993
UF: Cars—Dining*
BT: Railroad cars

Diode amplifiers 713.1 (714.2)
DT: January 1993
UF: Amplifiers, Diode*
BT: Microwave amplifiers
RT: Diodes

Diode lasers
USE: Semiconductor lasers

Diode logic circuits 721.2 (714.2) (721.3)
DT: January 1993
UF: Logic circuits, Diode*
BT: Logic circuits
NT: Diode transistor logic circuits

Diode transistor logic circuits 721.2 (714.2) (721.3)
DT: January 1993
UF: DTL circuits
Logic circuits, Diode transistor*
BT: Diode logic circuits
RT: Diodes
Switching circuits

Diodes (714.1) (714.2)
DT: January 1993
BT: Electronic equipment
NT: Electron tube diodes
Plasma diodes
Semiconductor diodes
RT: Diode amplifiers
Diode transistor logic circuits
Electric rectifiers
Electron tube rectifiers

Diplex transmission 718.2
DT: January 1993
UF: Telegraph—Diplex transmission*
Transmission (diplex)
BT: Carrier communication
RT: Telegraph

Dipole antennas (716)
DT: January 1993
UF: Antennas, Dipole*
BT: Antennas

Direct access storage
USE: Magnetic disk storage

Direct analogs (723)
DT: January 1993
UF: Computers—Direct analogs*
BT: Simulators
NT: Electrolytic tanks
RT: Analog computers

Direct coupled amplifiers 713.1
DT: January 1993
UF: Amplifiers, Direct coupled*
BT: Amplifiers (electronic)

Direct current machinery
USE: DC machinery

Direct current power transmission
USE: DC power transmission

Direct current transformers
USE: DC transformers

Direct digital control systems 731.1
DT: January 1993
UF: Control systems, Direct digital*
BT: Digital control systems

Direct energy conversion (702) (525.5)
DT: Predates 1975
BT: Electric power generation
Energy conversion
NT: Thermionic power generation
Thermoelectric energy conversion
RT: Direct energy converters
Electric energy storage
Electric generators
Electricity
Spacecraft power supplies
Transducers

Direct energy converters (702) (525.5)
DT: January 1993
BT: Electric equipment
NT: Electric batteries
Magnetohydrodynamic converters
Plasma diodes
Solar cells
RT: Direct energy conversion

Direct process refining 533.2 (545.3)
DT: January 1993
UF: Steelmaking—Direct process*
BT: Metal refining
RT: Steelmaking

Direct reduction process (545.3) (545.1)
DT: January 1993
UF: DR process
Flame reduction process
Steelmaking—Direct reduction process*
BT: Steelmaking

*Direction finding systems**
USE: Radio direction finding systems

*Direction finding systems—Adcock**
USE: Adcock direction finders

Directional drilling (511.1) (615.1)
DT: January 1993
UF: Geothermal wells—Directional drilling*
Oil well drilling—Directional*
BT: Drilling
RT: Deflected boreholes
Directional logging
Enhanced recovery
Fishing (oil wells)
Water wells

Directional logging 512.1.2
DT: January 1993
UF: Oil well logging—Directional*
BT: Well logging
RT: Directional drilling

* Former Ei Vocabulary term

Directional logging *(continued)*
Natural gas well logging
Oil well logging

Directional patterns (antenna) (716)
DT: January 1993
UF: Antenna lobe patterns
Antenna patterns
Antenna radiation patterns
Antennas—Directional patterns*
Field intensity patterns (antenna)
Lobes (distribution)
Radiation patterns (antenna)
BT: Antenna radiation
RT: Antenna reflectors
Antennas

Directional solidification
USE: Solidification

Directive antennas (716)
DT: January 1993
UF: Antennas, Directive*
BT: Antennas
NT: Loop antennas

Dirigibles
USE: Airships

Disabled persons
USE: Handicapped persons

Discharge (fluid mechanics) (631.1.1) (407) (444.1)
DT: January 1993
UF: Rivers—Discharge*
BT: Fluid mechanics
RT: Flow of water
Rivers
Stream flow

Discharge lamps 707.2
DT: January 1993
UF: Electric lamps, Discharge*
Gas discharge lamps
Glow discharge lamps
BT: Electric lamps
NT: Arc lamps
Ballasts (lamp)
Fluorescent lamps
Metal vapor lamps
RT: Electric discharges
Gas discharge tubes
Glow discharges

Discharge tubes
USE: Gas discharge tubes

Discharges (electric)
USE: Electric discharges

Discoloration (802.3)
DT: January 1993
BT: Defects
RT: Coloring
Degradation

Discrete control systems
USE: Discrete time control systems

Discrete time control systems 731.1
DT: January 1993
UF: Control systems, Discrete time*
Discrete control systems
BT: Control systems
NT: Sampled data control systems

Discriminators 713.3
DT: January 1993
UF: Electronic circuits, Discriminator*
BT: Networks (circuits)
RT: Comparator circuits
Demodulators

Disease control (461.7) (461.6)
DT: January 1993
BT: Control
NT: Malaria control
RT: Biomedical engineering
Diseases
Epidemiology
Mosquito control

Diseases (461.6) (461.7)
DT: January 1993
NT: Dermatitis
Occupational diseases
Pulmonary diseases
RT: Allergies
Disease control
Epidemiology
Medicine

Disinfectants 803 (801.2) (461.9)
DT: Predates 1975
BT: Biocides
RT: Biomedical equipment
Cleaning
Disinfection

Disinfection (802.3) (462) (822.2)
SN: Destruction of pathogens; distinguish from STERILIZATION, which also destroys all microbes
DT: January 1987
BT: Cleaning
RT: Disinfectants
Sterilization (cleaning)
Water treatment

Disintegration 802.3
DT: January 1993
UF: Concrete—Disintegration*
Refractory materials—Disintegration*
BT: Deterioration
RT: Comminution
Crushing
Decay (organic)
Decomposition
Grinding (comminution)
Pyrolysis
Weathering

Disk Operating System
USE: DOS

Disk storage (magnetic)
USE: Magnetic disk storage

Disks (machine components) 601.2
DT: January 1993
UF: Disks*
Heat transfer—Disks*
BT: Machine components
NT: Rotating disks
Rupture disks

Disks (structural components) 408.2
DT: January 1993
UF: Disks*
Heat transfer—Disks*
BT: Structural members
RT: Plates (structural components)

*Disks**
USE: Disks (machine components) OR
Disks (structural components)

*Disks—Rotating**
USE: Rotating disks

Dislocations (crystals) (531.2) (801.4) (933.1.1)
DT: January 1993
UF: Crystal dislocations
Crystals—Dislocations*
Misfit dislocations
BT: Crystal defects
RT: Stacking faults

Dispensers (607.2) (694.3)
SN: Containers capable of dispensing their contents
DT: January 1993
UF: Containers—Dispensers*
Lubricants—Dispensers*
BT: Containers
RT: Lubricators

Dispersion (waves) (711.1)
DT: January 1993
UF: Dispersion*
NT: Acoustic dispersion
Electromagnetic dispersion
RT: Refractive index
Wave propagation
Waves

Dispersion hardening 537.1 (536.1)
DT: January 1993
UF: Metals and alloys—Dispersion hardening*
Powder metallurgy—Dispersion hardening*
BT: Hardening
RT: Mechanical alloying
Metallurgy
Powder metallurgy
Rapid solidification

*Dispersion**
USE: Dispersion (waves)

Dispersions (801.3)
DT: January 1993
NT: Colloids
Emulsions
Smoke
Suspensions (fluids)
RT: Brownian movement
Dust
Mixtures

Display devices 722.2
DT: October 1975
UF: Display equipment
Displays
Monitors (displays)
BT: Equipment
NT: Computer monitors
Fluorescent screens
Instrument displays
Liquid crystal displays
Plasma display devices
RT: Cathode ray oscilloscopes
Cathode ray tubes
Gas discharge tubes
Light emitting diodes
Readout systems
Remote consoles
Television picture tubes
Three dimensional

*Display devices—Liquid crystal**
USE: Liquid crystal displays

*Display devices—Plasma**
USE: Plasma display devices

Display equipment
USE: Display devices

Display tubes (television)
USE: Television picture tubes

Displays
USE: Display devices

Disposability (452)
DT: January 1993
UF: Plastics—Disposability*
RT: Disposable biomedical equipment
Plastics
Waste disposal

Disposable biomedical equipment 462.1 (452)
DT: January 1993
UF: Biomedical equipment—Disposable*
BT: Biomedical equipment
RT: Disposability
Syringes
Waste disposal

Disposal (waste)
USE: Waste disposal

Dissimilar materials
DT: January 1993
UF: Bonding—Dissimilar materials*
BT: Materials
NT: Dissimilar metals

Dissimilar metals 531 (538)
DT: January 1993
UF: Brazing—Dissimilar base metals*
Welding—Dissimilar metals*

* Former Ei Vocabulary term

Dissimilar metals *(continued)*
BT: Dissimilar materials
RT: Metals

Dissipation (energy)
USE: Energy dissipation

Dissociation 802.2
DT: January 1993
BT: Chemical reactions
RT: Decomposition
Ionization
Pyrolysis

Dissolution 802.3
DT: January 1993
BT: Chemical operations

Dissolved oxygen sensors (801) (462.1)
DT: January 1993
UF: Clark sensors
Oxygen—Dissolved oxygen sensors*
BT: Oxygen sensors
RT: Biosensors
Oxygen

Distance measurement 943.2
DT: January 1993
UF: Mechanical variables measurement—Distance*
Proximity measurement
Range finding
Ranging
BT: Spatial variables measurement
NT: Radar measurement
RT: Interferometry
Micrometers
Navigation
Optical radar
Proximity indicators
Proximity sensors
Range finders
Sonar
Topology

Disthene
USE: Kyanite

Distillation 802.3
DT: Predates 1975
BT: Separation
RT: Condensation
Desalination
Distillation equipment
Distilleries
Fractionation
Phase transitions
Purification
Vaporization

Distillation columns 802.1, 802.3
DT: January 1993
UF: Chemical equipment—Distillation columns*
Columns (distillation)
BT: Distillation equipment

Distillation equipment 802.1, 802.3
DT: Predates 1975
BT: Chemical equipment
NT: Distillation columns
RT: Distillation
Distilleries

Distilleries 822.1 (402.1)
DT: January 1986
BT: Industrial plants
RT: Breweries
Distillation
Distillation equipment
Food products plants

Distortion (electric)
USE: Electric distortion

Distortion (waves) (716.1) (711.1) (751.1)
DT: January 1993
UF: Distortion*
NT: Acoustic distortion
Electric distortion
Signal distortion
RT: Wave propagation
Waves

*Distortion**
USE: Distortion (waves)

Distributed computer systems 722.4
DT: January 1993
UF: Computer systems, Digital—Distributed*
BT: Computer systems
NT: Multiprocessing systems
RT: Parallel algorithms

Distributed control systems
USE: Distributed parameter control systems

Distributed database systems 723.3
DT: January 1993
UF: Database systems—Distributed*
BT: Database systems
RT: Computer networks
Computer operating systems
Concurrency control
Relational database systems

Distributed feedback lasers 744.1
DT: January 1995
UF: DFB lasers
BT: Lasers
RT: Dye lasers
Integrated optics
Semiconductor lasers

Distributed parameter control systems 731.1
DT: January 1993
UF: Control systems, Distributed parameter*
Distributed control systems
BT: Control systems
RT: Decentralized control

Distributed parameter networks 703.1
DT: January 1993
UF: Electric networks, Distributed parameter*
BT: Networks (circuits)
RT: Circuit theory
Electric network analysis

* Former Ei Vocabulary term

Distributed parameter networks *(continued)*
Electric network synthesis
Microwave circuits
Transmission line theory

Distribution networks
USE: Electric power distribution

Distribution of goods (691)
DT: January 1993
UF: Data processing—Distribution applications*
BT: Materials handling

District heating 643.1
SN: Wide-area distribution of energy, from a central source via steam or hot-water pipelines, for space heating
DT: January 1993
UF: Heating—District*
BT: Space heating
RT: Cogeneration plants

Diversion (rivers)
USE: River diversion

Diversity reception 716.3
DT: January 1993
UF: Radio systems, Diversity*
BT: Radio
RT: Fading (radio)
Radio receivers
Radio systems
Radio transmission

Dividers (frequency)
USE: Frequency dividing circuits

Dividers (highway)
USE: Median dividers

Dividers (voltage)
USE: Voltage dividers

Dividing circuits (arithmetic) 721.3 (713.5)
DT: January 1993
UF: Computers—Dividing circuits*
BT: Digital circuits
RT: Digital arithmetic

Dividing circuits (frequency)
USE: Frequency dividing circuits

Diving apparatus 472
DT: Predates 1975
BT: Equipment
RT: Oceanographic equipment
Oceanography
Submersibles
Underwater construction

Diving bells
USE: Submersibles

DLTS
USE: Deep level transient spectroscopy

DNA 461.2 (801.2)
DT: January 1993
UF: Biological materials—DNA*
Deoxyribonucleic acid
BT: Nucleic acids
RT: DNA sequences
Genes
Proteins

DNA sequences 461.2 (801.2)
DT: January 1993
UF: Biochemistry—DNA sequences*
BT: Nucleic acid sequences
RT: Biochemistry
DNA

Docking (672) (674.1) (655.1)
DT: January 1993
UF: Boats—Docking*
Spacecraft—Docking*
BT: Joining
RT: Boats
Mooring
Mooring cables
Ships
Space rendezvous
Spacecraft

Docks 407.1
DT: Predates 1975
BT: Port structures
RT: Piers

Document retrieval systems
USE: Information retrieval systems

Documentation (computer programs)
USE: Program documentation

Documentation (systems programs)
USE: System program documentation

Dolphins (structures) 407.1
DT: January 1993
UF: Inland waterways—Dolphins*
Port structures—Dolphins*
BT: Port structures
RT: Inland waterways

Domains (magnetic)
USE: Magnetic domains

Domes 408.2
DT: January 1993
UF: Domes and shells*
Inflatable domes
BT: Structures (built objects)

*Domes and shells**
USE: Domes OR
Shells (structures)

*Domes and shells—Plastics**
USE: Plastics applications

Domestic appliances (522) (715) (704.2)
SN: Very general term; prefer specific appliance or type of appliance
DT: Predates 1975
UF: Appliances (domestic)
Household appliances
BT: Equipment
RT: Electric appliances

* Former Ei Vocabulary term

Domestic appliances *(continued)*
Gas appliances
Ventilation
Washing machines
Water heaters

Domestic furnaces
USE: Space heating furnaces

Door handles (402) (644.3)
DT: January 1993
UF: Handles (door)
Refrigerators—Door handles*
BT: Doors

*Door locks**
USE: Locks (fasteners)

Door to door service (freight) (691.2) (431) (432) (433) (434)
DT: January 1993
UF: Freight handling—Door to door*
BT: Cargo handling

Doors 402
DT: Predates 1975
UF: Buildings—Doors*
BT: Building components
NT: Door handles
RT: Air curtains
Hinges
Locks (fasteners)

*Doors—Air curtains**
USE: Air curtains

Doping (additives) (712.1) (714.2) (801)
DT: January 1993
UF: Doping*
BT: Additives
NT: Semiconductor doping
RT: Ion implantation

*Doping**
USE: Doping (additives)

Doppler broadening
USE: Doppler effect

Doppler effect (711.1) (741.1) (751.1)
DT: January 1977
UF: Doppler broadening
Doppler shift
RT: Acoustic radiators
Acoustic wave propagation
Acoustic waves
Doppler radar
Electromagnetic waves
Light
Navigation
Radar
Speed
Velocity
Waves

Doppler radar 716.2
DT: January 1995
BT: Radar
RT: Doppler effect
Radar tracking
Surveillance radar

Doppler shift
USE: Doppler effect

DOS 722.4
DT: January 1993
UF: Computer operating systems—DOS*
Disk Operating System
BT: Computer operating systems

Dosage (drug)
USE: Drug dosage

Dosemeters
USE: Dosimeters

Doses (radiation)
USE: Dosimetry

Dosimeters 944.7
DT: Predates 1975
UF: Dosemeters
Radiation dosemeters
BT: Nuclear instrumentation
RT: Dosimetry
Particle detectors
Radiation counters
Radiation detectors
Radiation effects
Radioactivity measurement

Dosimetry 944.8
DT: Predates 1975
UF: Doses (radiation)
Radiation doses
Radiation dosimetry
BT: Measurements
RT: Dosimeters
Particle detectors
Photodynamic therapy
Radiation effects
Radiation protection
Radiotherapy

Double refraction
USE: Birefringence

DR process
USE: Direct reduction process

Drafting practice 902.1
DT: Predates 1975
UF: Aircraft manufacture—Drafting practice*
Shipbuilding—Drafting*
BT: Graphic methods
RT: Blueprints

Drag (631.1) (651.1) (671.1) (931.1)
DT: January 1993
UF: Aerodynamics—Drag*
Flow of fluids—Drag*
Ships—Resistance*
BT: Fluid dynamics
RT: Aerodynamics
Hydrodynamics
Navier Stokes equations
Turbulence
Turbulent flow
Wakes

* Former Ei Vocabulary term

Drainage (401) (502) (406) (442)
DT: Predates 1975
RT: Combined sewers
Culverts
Dewatering
Flow of fluids
Outfalls
Ponding
Pumping plants
Roads and streets
Sewers
Storm sewers
Tunneling (excavation)

*Drainage—Pumping plants**
USE: Pumping plants

Drawbridges
USE: Movable bridges

Drawing (forming) (535.2) (816.1)
DT: January 1993
UF: Drawing*
BT: Forming
NT: Metal drawing
RT: Drawing dies
Extrusion
Metal pressing
Stamping
Stretching

Drawing (graphics) 902.1
DT: January 1993
UF: Drawing*
BT: Graphic methods

Drawing dies (535.2) (816.2)
DT: January 1993
UF: Metal drawing—Dies*
Wire drawing—Dies*
BT: Dies
RT: Drawing (forming)

*Drawing**
USE: Drawing (forming) OR
Drawing (graphics)

Dredges (405.1) (407) (502.2)
DT: Predates 1975
UF: Diesel dredges
BT: Excavators
NT: Pontoon dredges
RT: Construction equipment
Dredging

*Dredges—Diesel**
USE: Diesel engines

Dredging (405) (407) (502.1)
DT: January 1977
BT: Excavation
RT: Blasting
Dredges
Drilling
Inland waterways
River control

Dressing (mineral)
USE: Beneficiation

Dressing (wheels)
USE: Wheel dressing

Driers (equipment)
USE: Dryers (equipment)

Driers (materials) 803
DT: January 1993
UF: Dryers (materials)
Refrigerants—Dryers*
BT: Materials
RT: Drying oils
Refrigerants

Drill bits
USE: Bits

Drill pipe 511.2
DT: January 1993
UF: Oil well drilling—Drill pipe*
BT: Drilling equipment
Pipe
RT: Drills
Oil well drilling equipment

Drill steel
USE: Tool steel

Drilling (604.2)
DT: Predates 1975
NT: Core drilling
Diamond drilling
Directional drilling
Explosive drilling
Jet drilling
Rock drilling
Underwater drilling
Well drilling
RT: Cutting
Dredging
Drilling fluids
Drilling machines (machine tools)
Drills
Excavation
Metal cutting
Perforating
Piercing
Reaming
Shaft sinking
Tapping (threads)
Turbodrills
Vibrations (mechanical)

Drilling equipment (604.2)
SN: For borehole and well drilling. For machines for drilling holes in workpieces, use DRILLING MACHINES
DT: January 1993
UF: Drilling—Equipment*
BT: Equipment
NT: Drill pipe
Earth boring machines
Oil well drilling equipment
RT: Drilling fluids

Drilling equipment *(continued)*
Rock drills
Well drilling

Drilling fluids (511.2) (803)
DT: January 1993
UF: Oil well drilling—Drilling fluids*
Oil well drilling—Rotary mud*
Rotary mud
BT: Fluids
NT: Circulating media
RT: Cutting fluids
Drilling
Drilling equipment
Mud logging
Natural gas well drilling
Oil well drilling
Rock drilling
Well drilling

Drilling machines (machine tools) 603.1
SN: For drilling into workpieces. For equipment for drilling boreholes or wells, use DRILLING EQUIPMENT
DT: January 1993
BT: Machine tools
RT: Boring machines (machine tools)
Drilling
Lathes

Drilling rigs 511.2
DT: January 1993
UF: Oil well drilling—Rigs*
BT: Oil well drilling equipment
NT: Jack-up rigs

*Drilling—Equipment**
USE: Drilling equipment

Drills 603.2
DT: Predates 1975
BT: Cutting tools
NT: Diamond drills
Electric drills
Hydraulic drills
Mining drills
Oil well drills
Pneumatic drills
Rock drills
Turbodrills
RT: Bits
Drill pipe
Drilling
Well drilling

Drillships 674.2
DT: October 1975
BT: Ships
RT: Well drilling

Drinking water
USE: Potable water

Drive-in facilities 402.2
DT: January 1993
UF: Buildings—Drive-in facilities*
Drive-up facilities
BT: Facilities
NT: Outdoor motion picture theaters

Drive-in theaters
USE: Outdoor motion picture theaters

Drive-up facilities
USE: Drive-in facilities

Driver licensing 432, 902.3, 912.4
DT: January 1993
UF: Automobile drivers—Licensing*
Licensing (automobile drivers)
RT: Automobile drivers

Driver training 432, 912.4
DT: January 1993
UF: Electric railroads—Driver training*
Locomotives—Driver training*
BT: Personnel training
RT: Automobile drivers
Bus drivers
Electric railroads
Locomotives
Railroads
Truck drivers

Drivers (automobile)
USE: Automobile drivers

Drivers (bus)
USE: Bus drivers

Drivers (truck)
USE: Truck drivers

Drives (602.1) (632.2) (632.4) (705)
SN: Very general term; prefer specific type of drive. Also use the application. For drives that propel vehicles, use PROPULSION
DT: Predates 1975
BT: Machine components
NT: Electric drives
Hydraulic drives
Machine tool drives
Mechanical drives
Pneumatic drives
Torque converters
Variable speed drives
RT: Clutches
Electric power utilization
Gears
Propulsion

*Drives—Variable speed**
USE: Variable speed drives

Driveways 406.2
DT: January 1993
UF: Roads and streets—Driveways*
BT: Roads and streets
RT: Pavements

Drones (target)
USE: Target drones

Drop breakup 931.2
DT: January 1993
UF: Breakup (liquid drops)

* Former Ei Vocabulary term

Drop breakup *(continued)*
Liquids—Drop breakup*
BT: Fluid mechanics
RT: Liquids

Drop formation 931.2
DT: January 1993
UF: Liquids—Drop formation*
BT: Fluid mechanics
RT: Liquids

Drop presses
USE: Punch presses

Drop wires (718.1) (706.2)
DT: January 1993
UF: Telephone lines—Drop wires*
BT: Electric wire
RT: Electric lines
Telephone lines

Drought 443.3, 444
DT: January 1993
UF: Water resources—Drought*
RT: Climatology
Meteorology
Water resources

Drowned river mouths
USE: Estuaries

Drug dosage 461.6
DT: January 1993
UF: Dosage (drug)
Drug products—Dosage*
BT: Drug therapy

Drug infusion 461.6
DT: January 1993
UF: Drug products—Infusion*
Infusion (drugs)
BT: Drug therapy

Drug interactions 461.6, 802.2
DT: January 1993
UF: Drug products—Interactions*
Interactions (drug)
BT: Pharmacodynamics
RT: Drug therapy

Drug products (461.6) (804.1) (804.2)
DT: Predates 1975
UF: Pharmaceuticals
NT: Anesthetics
Antibiotics
Ionic drugs
Vaccines
Vitamins
RT: Drug products plants
Drug therapy

Drug products plants 802.1 (462.5) (402.1)
DT: Predates 1975
UF: Pharmaceutical plants
BT: Industrial plants
RT: Drug products

*Drug products—Anesthetics**
USE: Anesthetics

*Drug products—Antibiotics**
USE: Antibiotics

*Drug products—Bioassay**
USE: Bioassay

*Drug products—Controlled delivery**
USE: Controlled drug delivery

*Drug products—Dosage**
USE: Drug dosage

*Drug products—Infusion**
USE: Drug infusion

*Drug products—Interactions**
USE: Drug interactions

*Drug products—Ionic drugs**
USE: Ionic drugs

*Drug products—Pharmacodynamics**
USE: Pharmacodynamics

*Drug products—Pharmacokinetics**
USE: Pharmacokinetics

*Drug products—Vaccines**
USE: Vaccines

Drug therapy 461.6
DT: January 1993
UF: Biomedical engineering—Drug therapy*
Chemotherapy
Drug treatment
Pharmacotherapy
BT: Patient treatment
NT: Controlled drug delivery
Drug dosage
Drug infusion
RT: Biomedical engineering
Drug interactions
Drug products
Pharmacodynamics
Pharmacokinetics

Drug treatment
USE: Drug therapy

Dry cleaning 802, 819.5
DT: January 1993
UF: Textiles—Dry cleaning*
BT: Cleaning
RT: Textiles

Dry docks
USE: Drydocks

Dry etching 802.2 (714.2)
DT: January 1995
BT: Etching
RT: Semiconductor devices

Dry plasma etching
USE: Plasma etching

Drydocks 407.1, 673.3
DT: Predates 1975
UF: Dry docks
BT: Port structures
NT: Floating drydocks

* Former Ei Vocabulary term

Drydocks *(continued)*
RT: Maintenance
Outfitting (water craft)
Shipbuilding
Ships
Shipways
Shipyards

*Drydocks—Floating**
USE: Floating drydocks

Dryers (equipment) (642.2) (802.1)
DT: January 1993
UF: Driers (equipment)
Dryers*
Papermaking machinery—Dryers*
Textile machinery—Dryers*
BT: Separators
NT: Solar dryers
RT: Drying
Evaporators
Papermaking machinery
Textile machinery

Dryers (materials)
USE: Driers (materials)

*Dryers**
USE: Dryers (equipment)

*Dryers—Solar**
USE: Solar dryers

Drying (642.1) (802.3)
DT: Predates 1975
NT: Infrared drying
Low temperature drying
RT: Chemical operations
Chemical reactions
Curing
Dehydration
Dewatering
Diffusion
Dryers (equipment)
Evaporation
Heating
Industrial heating
Regain

Drying oils (813.2) (822.3)
DT: Predates 1975
BT: Vegetable oils
RT: Driers (materials)
Paint
Solvents
Varnish

*Drying—Infrared**
USE: Infrared drying

*Drying—Low temperature**
USE: Low temperature drying

DTA
USE: Differential thermal analysis

DTL circuits
USE: Diode transistor logic circuits

Dual fuel burners 521.3 (522) (523) (524)
DT: January 1993
UF: Fuel burners—Dual fuel*
Two fuel oil burners
BT: Fuel burners

Dual fuel engines 612.1 (522) (523)
DT: January 1993
UF: Convertible engines
Internal combustion engines—Dual fuel*
Two fuel engines
BT: Engines
RT: Internal combustion engines

Ducted fan engines 653.1
DT: January 1993
UF: Aircraft engines, Jet and turbine—Ducted fan*
BT: Jet engines
RT: Turbofan engines

Ductile iron
USE: Nodular iron

Ductility (421)
DT: January 1993
UF: Cold ductility
Hot ductility
Metals and alloys—Ductility*
Notch ductility
BT: Mechanical properties
RT: Brittle fracture
Brittleness
Compressibility
Embrittlement
Formability
Hardness
High temperature testing
Plastic deformation
Plasticity
Weldability

Ducts (641.2) (643.2) (643.4) (619.1)
DT: January 1993
UF: Heat transfer—Ducts*
BT: Equipment
NT: Air conditioning ducts
Ventilation ducts
RT: Diffusers (fluid)
Heat transfer
Orifices

Dummies (automobile crash testing)
USE: Automobile driver simulators

Dump cars 682.1.1
DT: January 1993
UF: Cars—Dumpers*
Dumpers (railroad cars)
BT: Freight cars

Dumpers (railroad cars)
USE: Dump cars

Durability (421)
DT: January 1993
UF: Endurance
RT: Degradation
Deterioration

* Former Ei Vocabulary term

Durability *(continued)*
Performance
Stability
Wear of materials

Durable press textiles 819.5
DT: January 1993
UF: Permanent press textiles
Textiles—Durable press*
BT: Textiles
RT: Wrinkle recovery

Duralumin
USE: Aluminum copper alloys

Dust 451.1
DT: Predates 1975
BT: Materials
NT: Coal dust
Mine dust
RT: Aerosols
Dispersions
Dust abatement
Dust collectors
Dust control
Particles (particulate matter)
Smoke

Dust abatement 451.2
DT: Predates 1975
UF: Abatement (dust)
BT: Air pollution control
Dust control
RT: Dust
Dust collectors

Dust collectors (451.2) (802.1)
DT: Predates 1975
UF: Collectors (dust)
BT: Air pollution control equipment
Separators
RT: Dust
Dust abatement
Dust control
Mine dust

*Dust collectors—Lining**
USE: Linings

Dust control 451.2
DT: January 1993
UF: Chemical laboratories—Dust control*
Dust problems*
Kilns—Dust control*
Mines and mining—Dust control*
Roads and streets—Dust control*
Sugar factories—Dust control*
BT: Control
NT: Dust abatement
RT: Dust
Dust collectors
Mine dust

Dust cores
USE: Powder magnetic cores

*Dust problems**
USE: Dust control

DVM
USE: Digital voltmeters

Dy
USE: Dysprosium

Dye lasers (744)
DT: January 1993
UF: Lasers, Dye*
BT: Lasers
RT: Distributed feedback lasers
Liquid lasers

Dye stripping
USE: Stripping (dyes)

Dyeing (802) (819.5) (819.4)
DT: January 1993
UF: Dyes and dyeing*
BT: Coloring
RT: Color matching
Dyes
Stripping (dyes)
Textile processing
Textile scouring
Textiles

Dyes 803
DT: January 1993
UF: Dyes and dyeing*
Industrial wastes—Dyes*
BT: Chemicals
NT: Azo dyes
Vat dyes
RT: Color matching
Dyeing
Industrial wastes
Stripping (dyes)

*Dyes and dyeing**
USE: Dyeing OR
Dyes

*Dyes and dyeing—Azo dyes**
USE: Azo dyes

*Dyes and dyeing—Color fastness**
USE: Color fastness

*Dyes and dyeing—Color matching**
USE: Color matching

*Dyes and dyeing—Mixed fibers**
USE: Textile blends

*Dyes and dyeing—Stripping**
USE: Stripping (dyes)

*Dyes and dyeing—Synthetic fibers**
USE: Synthetic fibers

*Dyes and dyeing—Vat dyes**
USE: Vat dyes

Dynamic loads 408.1
DT: January 1993
UF: Buildings—Dynamic loads*
Energy loads
BT: Loads (forces)
NT: Aerodynamic loads

* Former Ei Vocabulary term

RT: Bearing capacity
Soil structure interactions
Stresses

Dynamic positioning (671.2) (675.1) (723.5)
DT: January 1993
UF: Ships—Dynamic positioning*
BT: Ship propulsion

Dynamic programming 921.5
DT: January 1993
UF: Mathematical programming, Dynamic*
BT: Mathematical programming
NT: Integer programming
RT: Constraint theory
Decision theory
Maximum principle
Operations research
Optimal control systems

Dynamic response 408.1
DT: January 1993
UF: Structural analysis—Dynamic response*
RT: Damping
Sensitivity analysis
Structural analysis

Dynamic simulation
USE: Computer simulation

Dynamics (931.1)
DT: Predates 1975
UF: Nonlinear dynamics
BT: Mechanics
NT: Electrodynamics
Flow of solids
Fluid dynamics
Friction
Kinetic theory
Particle beam dynamics
Rotation
Vibrations (mechanical)
RT: Kinematics
Natural frequencies
Resonance

Dynamometers 943.1
DT: Predates 1975
UF: Electrodynamometers
BT: Electric measuring instruments
RT: Force measurement
Torque measurement
Torque meters

Dynamos
USE: DC generators

Dynamotors
USE: Rotary converters

Dysprosium 547.2
DT: Predates 1975
UF: Dy
BT: Rare earth elements
RT: Dysprosium alloys
Dysprosium compounds

Dysprosium alloys 547.2
DT: January 1993
BT: Rare earth alloys
RT: Dysprosium
Dysprosium compounds

Dysprosium compounds (804.1) (804.2)
DT: June 1990
BT: Rare earth compounds
RT: Dysprosium
Dysprosium alloys

E layer
USE: E region

E region 443.1 (711.1) (481.3)
DT: January 1993
UF: E layer
Heaviside layer
Ionosphere—E region*
Kennelly Heaviside layer
Sporadic E layer
BT: Ionosphere

E-mail
USE: Electronic mail

Ear protectors 462.1, 914.1
DT: Predates 1975
BT: Safety devices
RT: Audition
Protective clothing

Earphones 752.1 (462.1) (718.1)
SN: For insertion into ear. For devices which are held against the ear, use HEADPHONES
DT: January 1993
BT: Electroacoustic transducers
RT: Headphones
Hearing aids
Telephone apparatus
Telephone sets

Earth (planet) (657)
SN: Planet Earth from viewpoints of astronomy and astrophysics.
DT: January 1993
UF: Earth*
BT: Solar system
RT: Earth atmosphere
Earth sciences
Earthquakes
Geochemistry
Geology
Geomagnetism
Geophysics
Hydrology
Mineral industry
Mineral resources
Mineralogy
Natural resources
Natural resources exploration
Oceanography
Petrology
Seismology
Soils
Tectonics
Water resources

Earth atmosphere 443.1 (481.3)
DT: Predates 1975
UF: Atmosphere (earth)
Terrestrial atmosphere
NT: Clouds
Fog
Ozone layer
Troposphere
Upper atmosphere
RT: Air
Air pollution
Air quality
Atmospheric acoustics
Atmospheric aerosols
Atmospheric chemistry
Atmospheric composition
Atmospheric density
Atmospheric humidity
Atmospheric ionization
Atmospheric movements
Atmospheric optics
Atmospheric pressure
Atmospheric radiation
Atmospheric radioactivity
Atmospheric spectra
Atmospheric structure
Atmospheric temperature
Atmospheric thermodynamics
Climatology
Earth (planet)
Ionospheric electromagnetic wave propagation
Meteorology
Precipitation (meteorology)
Storms
Thermal stratification

*Earth atmosphere—Fallout**
USE: Fallout

*Earth atmosphere—Magnetosphere**
USE: Magnetosphere

*Earth atmosphere—Ozone layer**
USE: Ozone layer

*Earth atmosphere—Radiation belts**
USE: Radiation belts

*Earth atmosphere—Upper atmosphere**
USE: Upper atmosphere

Earth boring machines (405.1) (501.1)
DT: Predates 1975
UF: Boring machines (earth boring)
BT: Drilling equipment
RT: Rock drilling
Well drilling

Earth dams
USE: Embankment dams

Earth electrodes
USE: Grounding electrodes

Earth magnetism
USE: Geomagnetism

Earth sciences (443) (471) (481)
DT: January 1993
BT: Natural sciences
NT: Geology
Hydrology
Meteorology
Mineralogy
Oceanography
RT: Earth (planet)
Environmental engineering

*Earth**
USE: Earth (planet)

*Earth—Magnetism**
USE: Geomagnetism

Earthing
USE: Electric grounding

Earthmoving machinery (405.1) (502.2)
DT: Predates 1975
BT: Machinery
NT: Excavators
RT: Axles
Construction equipment
Excavation
Materials handling equipment
Roadbuilding machinery

*Earthmoving machinery—Axles**
USE: Axles

*Earthmoving machinery—Excavators**
USE: Excavators

*Earthmoving machinery—Fuel tanks**
USE: Fuel tanks

*Earthmoving machinery—Hydraulic equipment**
USE: Hydraulic equipment

*Earthmoving machinery—Transmissions**
USE: Vehicle transmissions

Earthquake effects (408.1) (484)
SN: For effects on structures use EARTHQUAKE RESISTANCE
DT: January 1993
BT: Effects
RT: Earthquake resistance
Earthquakes

Earthquake protection
USE: Earthquake resistance

Earthquake resistance 484.3 (405.2) (408) (483)
SN: Use for effects of earthquakes on structures, and design to improve resistance to earthquakes. For other effects, use EARTHQUAKE EFFECTS
DT: Predates 1975
UF: Earthquake protection
RT: Earthquake effects
Earthquakes
Protection
Shear walls
Structural design

* Former Ei Vocabulary term

Earthquakes 484
DT: Predates 1975
RT: Earth (planet)
Earthquake effects
Earthquake resistance
Landslides
Seismic waves
Seismographs
Seismology
Soil structure interactions
Tectonics
Tsunamis
Volcanoes

ECCM
USE: Military electronic countermeasures

ECG
USE: Electrocardiography

Echo cancellation
USE: Echo suppression

Echo suppression (716.1) (718.1)
SN: In telecommunication equipment
DT: Predates 1975
UF: Echo cancellation
RT: Signal interference
Telephone
Telephone equipment

Echocardiography 461.6
DT: January 1993
UF: Biomedical engineering—Echocardiography*
BT: Diagnosis
Ultrasonic imaging
RT: Biomedical engineering
Cardiology

ECL
USE: Emitter coupled logic circuits

ECM
USE: Military electronic countermeasures

Ecological communities
USE: Ecosystems

Ecology 454.3
DT: October 1975
RT: Biology
Ecosystems
Environmental engineering

Economic and social effects (901.4) (911.2)
DT: January 1993
UF: Economic effects
Robots, Industrial—Socioeconomic aspects*
Social effects
Socioeconomic effects
Sociological effects
Technology—Economic and sociological effects*
BT: Effects
RT: Behavioral research
Economics
Employment
Personnel
Social aspects
Social sciences
Technological forecasting
Technology
Technology transfer

Economic cybernetics (731.1) (461.9) (911.2)
DT: January 1993
UF: Systems science and cybernetics—Economic cybernetics*
BT: Cybernetics
RT: Economics

Economic effects
USE: Economic and social effects

Economics 911.2
DT: Predates 1975
BT: Social sciences
NT: Industrial economics
RT: Competition
Economic and social effects
Economic cybernetics
Efficiency
Employment
Energy policy
Finance
Management
Productivity
Public policy

Economizers 616.1
DT: Predates 1975
BT: Heat exchangers
RT: Cooling systems
Feedwater heaters
Power plants
Steam generators

Ecosystems 454.3
DT: January 1977
UF: Communities (ecological)
Ecological communities
RT: Ecology
Environmental engineering

Eddy current heating
USE: Induction heating

Eddy current testing (422.2) (423.2) (701.1)
DT: January 1993
UF: Nondestructive examination—Eddy current testing*
BT: Nondestructive examination

Eddy currents 701.1
DT: January 1981
UF: Foucault currents
BT: Electric currents
RT: Conductive materials
Electric conductors
Electromagnetism
Hysteresis
Induction heating

Edge detection (716) (723.5) (741.1) (751.1)
DT: January 1995
BT: Pattern recognition

* Former Ei Vocabulary term

Edge detection *(continued)*
RT: Feature extraction
Image processing

Edible fats
USE: Oils and fats

Edible oils
USE: Oils and fats

Editing machines (motion pictures)
USE: Motion picture editing machines

Editors (file)
USE: File editors

EDTV
USE: High definition television

Education (901.2)
DT: Predates 1975
UF: Television applications—Education*
NT: Computer aided instruction
Engineering education
Medical education
Personnel training
Teaching
RT: Demonstrations
Education computing
Educational motion pictures
Educational nuclear reactors
Information dissemination
Technical presentations

Education computing (723.2) (901.2)
DT: January 1993
UF: Data processing—Educational applications*
BT: Computer applications
RT: Data processing
Education

*Education—Demonstrations**
USE: Demonstrations

*Education—Teaching**
USE: Teaching

*Education—Technical presentation**
USE: Technical presentations

Educational motion pictures (742.1) (901.2)
DT: January 1993
UF: Motion pictures—Educational*
BT: Motion pictures
RT: Education
Information dissemination

Educational nuclear reactors 621, 901.2
DT: January 1993
UF: Nuclear reactors—Educational*
BT: Nuclear reactors
RT: Education

Eductors
USE: Ejectors (pumps)

EEG
USE: Electroencephalography

EELS
USE: Electron energy loss spectroscopy

Effects
DT: January 1993
NT: Composition effects
Corrosive effects
Earthquake effects
Economic and social effects
Electric field effects
Frost effects
Gravitational effects
High pressure effects
Hurricane effects
Low temperature effects
Magnetic field effects
Pesticide effects
pH effects
Pressure effects
Radiation effects
Seawater effects
Thermal effects
Wave effects
Wind effects

Efficiency
DT: January 1993
NT: Energy efficiency
RT: Economics
Industrial management
Management
Productivity

Efficiency (quantum)
USE: Quantum efficiency

Efflorescence (414.2) (812.1) (933.1)
SN: The tendency of hydrated crystals to lose water of hydration and crumble when exposed to air
DT: January 1993
UF: Brick—Efflorescence*
Ceramic materials—Efflorescence*
Clay products—Efflorescence*
RT: Dehydration

Effluent treatment 452.4
DT: January 1993
UF: Flue gases—Treatment*
Industrial plants—Effluent treatment*
BT: Industrial waste treatment
RT: Effluents
Flue gases

Effluents 452.3
DT: January 1993
UF: Industrial plants—Effluents*
BT: Wastes
NT: Gaseous effluents
Thermal effluents
RT: Effluent treatment
Filtration
Industrial plants
Industrial wastes
Sedimentation
Sewage

EHD
USE: Electrohydrodynamics

EHV power transmission 706.1.1
SN: Greater than 345 kv
DT: January 1993
UF: Electric power transmission, EHV*
Extra high voltage transmission
BT: Electric power transmission
NT: UHV power transmission

Eigenfunctions
USE: Eigenvalues and eigenfunctions

Eigenvalues and eigenfunctions (921.1)
DT: January 1993
UF: Eigenfunctions
Eigenvectors
Mathematical techniques—Eigenvalues and eigenfunctions*
BT: Linear algebra
RT: Numerical methods
Partial differential equations
Vectors

Eigenvectors
USE: Eigenvalues and eigenfunctions

Einsteinium 622.1
DT: January 1993
BT: Transuranium elements

Ejectors (air)
USE: Air ejectors

Ejectors (pumps) 618.2
DT: January 1993
UF: Eductors
Pumps—Ejectors*
Turbomachinery—Ejectors*
BT: Jet pumps
RT: Turbomachinery
Vacuum pumps

EKG
USE: Electrocardiography

Elastic moduli (421) (422.2)
DT: January 1995
UF: Bulk modulus
Elastic modulus
Modulus of elasticity
Tensile modulus
Young's modulus
BT: Mechanical properties
RT: Elasticity
Tensile testing
Torsion testing

Elastic modulus
USE: Elastic moduli

Elastic waves 931.1 (751.1)
DT: Predates 1975
UF: Flexural waves
Shear waves
Torsional waves
BT: Mechanical waves
NT: Acoustic waves
Seismic waves
Shock waves
RT: Elasticity
Elastohydrodynamics
Phonons
Vibrations (mechanical)

Elasticity (421) (422) (931.1)
DT: Predates 1975
UF: Acoustoelasticity
Elastostatics
Hydroelasticity
BT: Mechanical properties
RT: Anelastic relaxation
Compressibility
Continuum mechanics
Deformation
Elastic moduli
Elastic waves
Elasticity testing
Elastoplasticity
High temperature testing
Photoelasticity
Rheology
Strain
Thermoelasticity
Viscoelasticity

Elasticity testing (422.2)
DT: January 1993
BT: Materials testing
RT: Elasticity
Tensile testing

Elastohydrodynamic lubrication 607.2
DT: January 1993
UF: Lubrication—Elastohydrodynamic*
BT: Lubrication
RT: Elastohydrodynamics

Elastohydrodynamics (631.1.1) (931.1)
DT: January 1993
BT: Hydrodynamics
RT: Elastic waves
Elastohydrodynamic lubrication

Elastomers 818.2
DT: January 1986
UF: Chemical equipment—Elastomers*
BT: Organic polymers
NT: Synthetic rubber
Thermoplastic elastomers
RT: Adhesives
Latexes
Vulcanization

Elastoplasticity (421) (931.1)
DT: January 1977
BT: Mechanical properties
RT: Elasticity
Plasticity
Stresses

Elastostatics
USE: Elasticity

Electrets 708.1
DT: Predates 1975
BT: Dielectric materials

* Former Ei Vocabulary term

Electrets *(continued)*
RT: Capacitors
Ceramic materials
Dielectric devices
Electric fields
Electrostatics

Electric accidents 914.1 (704)
DT: Predates 1975
BT: Accidents
RT: Electricity

*Electric accidents—Resuscitation**
USE: Resuscitation

Electric admittance
USE: Electric impedance

Electric analogies
USE: Simulation

Electric appliances 704.2
DT: Predates 1975
UF: Appliances (electric)
BT: Electric equipment
NT: Electric stoves
Vacuum cleaners
RT: Domestic appliances
Electric codes
Electric ovens
Electric power utilization
Microwave ovens
Ovens
Stoves
Washing machines
Water heaters

*Electric appliances—Microwave ovens**
USE: Microwave ovens

*Electric appliances—Ovens**
USE: Electric ovens

*Electric appliances—Plastics parts**
USE: Plastic parts

Electric arc welding 538.2.1 (704.2)
DT: January 1993
UF: Arc welding
Welding, Electric arc*
BT: Electric welding
NT: Carbon dioxide arc welding
Flux core wire welding
Inert gas welding
Submerged arc welding
RT: Electric arcs
Pressure welding
Spot welding

Electric arcs 701.1
DT: Predates 1975
BT: Electric discharges
NT: Circuit breaking arcs
RT: Arc lamps
Electric arc welding
Electric breakdown
Electric corona
Electric fault currents
Electric insulating materials
Electric insulation
Electric sparks
Flashover
Ionization
Short circuit currents

*Electric arcs, Circuit breaking**
USE: Circuit breaking arcs

Electric attenuators (704.1)
DT: Predates 1975
UF: Attenuators (electric)
BT: Networks (circuits)
NT: Waveguide attenuators
RT: Attenuation
Electric shielding
Radio equipment

Electric automobiles 662.1 (702.1.2)
DT: January 1993
UF: Automobiles—Electric*
BT: Automobiles
Electric vehicles
RT: Linear motors

Electric autotransformers (704) (714)
DT: Predates 1975
BT: Electric transformers
Voltage dividers

Electric batteries 702.1
DT: Predates 1975
UF: Batteries (electric)
Electric cells
BT: Direct energy converters
NT: Fuel cells
Miniature batteries
Photoelectrochemical cells
Primary batteries
Secondary batteries
Silicon batteries
RT: Charging (batteries)
Electrochemical electrodes
Solar cells
Spacecraft power supplies

*Electric batteries, Primary**
USE: Primary batteries

*Electric batteries, Secondary**
USE: Secondary batteries

*Electric batteries—Charging**
USE: Charging (batteries)

*Electric batteries—Electrolytes**
USE: Electrolytes

*Electric batteries—Gas emission**
USE: Gas emissions

*Electric batteries—Lithium**
USE: Lithium batteries

*Electric batteries—Miniature**
USE: Miniature batteries

*Electric batteries—Nuclear**
USE: Nuclear batteries

* Former Ei Vocabulary term

*Electric batteries—Silicon**
USE: Silicon batteries

Electric brakes 602 (732.1) (704.2)
SN: Electrically actuated, including electromagnetic or electrostatic effects
DT: January 1993
UF: Brakes, Electric*
BT: Brakes
RT: Electric final control devices

Electric brazing 538.1.1 (704.2)
DT: January 1993
UF: Brazing—Electric*
BT: Brazing

Electric breakdown 701.1
SN: Prefer more specific terms
DT: Predates 1975
UF: Breakdown voltage
BT: Electric discharges
NT: Electric breakdown of gases
Electric breakdown of liquids
Electric breakdown of solids
RT: Avalanche diodes
Electric arcs
Electric corona
Electric fault currents
Electric insulation
Electric sparks
Flashover
Voltage distribution measurement

Electric breakdown of gases 701.1 (931.2)
DT: January 1993
UF: Gases—Electric breakdown*
BT: Electric breakdown
RT: Gases

Electric breakdown of liquids 701.1 (931.2)
DT: January 1993
UF: Liquids—Electric breakdown*
BT: Electric breakdown
RT: Liquids

Electric breakdown of solids 701.1 (931.2)
DT: January 1993
UF: Solids—Electric breakdown*
BT: Electric breakdown
NT: Zener effect
RT: Partial discharges
Solids

Electric busbars
USE: Busbars

Electric cable crossings
USE: Crossings (pipe and cable)

Electric cable laying 706.2
DT: January 1993
UF: Cable laying
Electric cables—Laying*
Laying (electric cables)
Ships—Cable laying application*
RT: Cable ships
Electric cables
Optical cables
Submarine cables
Telecommunication cables

Electric cables 706.2
SN: Very general term; prefer specific terms, e.g., TELECOMMUNICATION CABLES. For cables in use, use ELECTRIC LINES
DT: Predates 1975
BT: Cables
Electric conductors
NT: Firing cables
Flat cables
Gas filled cables
Oil filled cables
Submarine cables
Superconducting cables
Telecommunication cables
Underground cables
RT: Cable jointing
Cable sheathing
Cable shielding
Cable taping
Electric cable laying
Electric conduits
Electric connectors
Electric current carrying capacity (cables)
Electric fault currents
Electric fault location
Electric lines

*Electric cables—Bridge crossings**
USE: Crossings (pipe and cable)

*Electric cables—Color codes**
USE: Color codes

*Electric cables—Corona**
USE: Electric corona

*Electric cables—Current carrying capacity**
USE: Electric current carrying capacity (cables)

*Electric cables—Flat**
USE: Flat cables

*Electric cables—Gas filled**
USE: Gas filled cables

*Electric cables—Jointing**
USE: Cable jointing

*Electric cables—Laying**
USE: Electric cable laying

*Electric cables—Oil filled**
USE: Oil filled cables

*Electric cables—Reels**
USE: Reels

*Electric cables—River crossings**
USE: Crossings (pipe and cable)

*Electric cables—Sheathing**
USE: Cable sheathing

*Electric cables—Shielding**
USE: Cable shielding

*Electric cables—Submarine**
USE: Submarine cables

* Former Ei Vocabulary term

*Electric cables—Superconducting**
USE: Superconducting cables

*Electric cables—Taping**
USE: Cable taping

*Electric cables—Underground**
USE: Underground cables

Electric capacitors
USE: Capacitors

Electric capacity
USE: Capacitance

Electric cells
USE: Electric batteries

Electric charge 701.1
DT: January 1993
UF: Charge (electric)
Electric charge*
Electrostatics—Electric charge*
BT: Electricity
NT: Electric space charge
RT: Capacitance
Capacitors
Electric charge measurement
Electrostatics

Electric charge measurement 942.2
DT: January 1993
UF: Electric measurements—Charge*
BT: Electric variables measurement
RT: Electric charge
Electrometers

*Electric charge**
USE: Electric charge

Electric circuit breakers (706.2) (704.2)
DT: Predates 1975
UF: Circuit breakers
BT: Electric switches
NT: Air blast circuit breakers
Oil circuit breakers
Reclosing circuit breakers
Sulfur hexafluoride circuit breakers
RT: Circuit breaking arcs
Electric equipment protection
Electric relays
Electric switchgear
Networks (circuits)
Switching circuits

*Electric circuit breakers, Air blast**
USE: Air blast circuit breakers

*Electric circuit breakers, Oil**
USE: Oil circuit breakers

*Electric circuit breakers—Reclosing**
USE: Reclosing circuit breakers

*Electric circuit breakers—Sulfur hexafluoride**
USE: Sulfur hexafluoride circuit breakers

Electric circuits
USE: Networks (circuits)

Electric clocks 704.2 (943.3)
DT: January 1993
UF: Clocks, Electric*
BT: Clocks
Electric equipment
RT: Ancillary signal generators
Timing circuits

Electric clutches 602.1 (704.2)
DT: January 1993
UF: Clutches, Electric*
BT: Clutches
RT: Electric final control devices

Electric codes 902.2
DT: Predates 1975
BT: Codes (standards)
RT: Building codes
Building wiring
Electric appliances
Electric conductors
Electric wiring
Electricity
Laws and legislation

Electric coils (704.1) (715.1)
DT: Predates 1975
UF: Coils (electric)
Electric inductors—Coils*
BT: Electric equipment
NT: Solenoids
RT: Electric inductors
Electric transformers
Electric windings
Electromagnets
Magnetic circuits
Magnetic cores
Radio equipment
Saturable core reactors
Superconducting devices
Winding machines

*Electric coils—Windings**
USE: Electric windings

Electric communication
USE: Telecommunication

Electric commutation (704) (705)
DT: Predates 1975
UF: Commutation
RT: Electric commutators
Electric switches

Electric commutators (704) (705)
DT: Predates 1975
UF: Commutators
BT: DC machinery
RT: Brush conductors
Electric commutation
Electric contacts
Electric machinery components

Electric condensers
USE: Capacitors

Electric conductance
USE: Electric resistance

* Former Ei Vocabulary term

Electric conductivity 701.1
DT: Predates 1975
UF: Conductivity (electric)
Electric resistivity
Resistivity (electric)
BT: Electric properties
NT: Electric conductivity of gases
Electric conductivity of liquids
Electric conductivity of solids
Photoconductivity
Superconductivity
RT: Conductive materials
Electric conductivity measurement
Electric discharges
Electric field effects
Intercalation compounds
Ionic conduction
Limited space charge accumulation
Percolation (solid state)
Semiconductor materials

Electric conductivity measurement 942.2
DT: January 1993
UF: Electric measurements—Conductivity*
BT: Electric variables measurement
RT: Electric conductivity
Electric measuring bridges
Electric resistance measurement

Electric conductivity of gases 701.1 (931.2)
DT: January 1993
UF: Gases—Electric conductivity*
BT: Electric conductivity
RT: Gases

Electric conductivity of liquids 701.1 (931.2)
DT: January 1993
UF: Liquids—Electric conductivity*
BT: Electric conductivity
RT: Electrolytes
Liquids
Semiconducting liquids

Electric conductivity of solids 701.1 (931.2)
DT: January 1993
UF: Solids—Electric conductivity*
BT: Electric conductivity
RT: Ionic conduction in solids
Solids
Superconductivity

Electric conductors 704.1 (708.2)
DT: Predates 1975
UF: Conductors (electric)
BT: Electric equipment
NT: Bundled conductors
Busbars
Electric cables
Electric wire
Grounding electrodes
RT: Cable jointing
Conductive materials
Eddy currents
Electric codes
Electric conduits
Electric connectors
Electric contacts
Electric power transmission
Electric windings
Electric wiring
Electrodes
Resistors
Skin effect
Superconducting materials

*Electric conductors, Brush**
USE: Brush conductors

*Electric conductors, Busbar**
USE: Busbars

*Electric conductors, Insulated wire**
USE: Insulated wire

*Electric conductors, Wire**
USE: Electric wire

*Electric conductors—Joints**
USE: Joints (structural components)

*Electric conductors—Size determination**
USE: Size determination

*Electric conductors—Skin effect**
USE: Skin effect

Electric conduits 704.1 (706.2)
DT: Predates 1975
UF: Cable ducts
Conduits (electric)
BT: Electric equipment
RT: Cable sheathing
Electric cables
Electric conductors
Electric power distribution
Electric power transmission

Electric connectors 704.1 (706.2)
DT: Predates 1975
UF: Cable terminations
Connectors (electric)
Couplers (electric connectors)
Jointing (electric connectors)
Linkages (electric connectors)
Plugs (electric)
Sockets (electric)
BT: Electric equipment
RT: Cable jointing
Electric cables
Electric conductors
Electric contacts
Electric power transmission
Electric wire
Electric wiring
Optical interconnects
Radio equipment

Electric contactors (704) (715)
DT: Predates 1975
UF: Contactors (electric)
BT: Electric relays
RT: Electric switchgear

Electric contacts 704.1
DT: Predates 1975
UF: Contacts (electric)
Electric machinery—Contacts*
Electric relays—Contacts*
Radio equipment—Contacts*
Semiconductor devices—Contacts*
Telephone switching equipment—Contacts*
BT: Electric equipment
NT: Ohmic contacts
Point contacts
RT: Brush conductors
Electric commutators
Electric conductors
Electric connectors
Electric switches
Telephone switching equipment

*Electric contacts, Ohmic**
USE: Ohmic contacts

*Electric contacts, Point**
USE: Point contacts

Electric control equipment 732.1, 704.2
DT: January 1993
UF: Control equipment, Electric*
Ship equipment—Electric control*
BT: Control equipment
Electric equipment
NT: Electric final control devices
Electric fuses
Voltage regulators
RT: Electric generators

Electric converters (704.2) (705.3)
DT: Predates 1975
UF: Converters (electric)
Convertors (electric)
BT: Electric equipment
NT: Electric inverters
Frequency converters
Negative impedance converters
Power converters
Static converters

*Electric converters, Frequency**
USE: Frequency converters

*Electric converters, Power type**
USE: Power converters

*Electric converters, Rotary**
USE: Rotary converters

*Electric converters, Static**
USE: Static converters

Electric corona 701.1 (706.2)
DT: Predates 1975
UF: Electric cables—Corona*
Electric glow
Electric lines—Corona*
Electric transformers—Corona*
BT: Surface discharges
RT: Electric arcs
Electric breakdown
Electric fault currents
Electric insulating materials
Electric insulation
Electric sparks
Electric transformers
Flashover
Glow discharges
Ionization

Electric current carrying capacity (cables) 706.2
DT: January 1993
UF: Carrying capacity (electric cables)
Current carrying capacity (electric cables)
Electric cables—Current carrying capacity*
RT: Electric cables

Electric current collection (706.2)
DT: January 1993
UF: Electric traction—Current collection*
RT: Electric current collectors
Electric propulsion
Electric traction

Electric current collectors 704.2 (706.2)
DT: January 1993
UF: Electric railroads—Current collectors*
BT: Electric equipment
NT: Brush conductors
Pantographs
RT: Electric current collection
Electric railroads

Electric current control 731.3 (701.1)
DT: January 1993
UF: Control, Electric variables—Current*
Current control
Electric current regulation
BT: Electric variables control
RT: Current limiting reactors
Electric current regulators
Electric currents

Electric current distribution (701.1)
DT: January 1993
RT: Electric current distribution measurement
Electric currents

Electric current distribution measurement 942.2
DT: January 1993
UF: Electric measurements—Current distribution*
BT: Electric variables measurement
RT: Electric current measurement
Electric current distribution

Electric current measurement 942.2
DT: January 1993
UF: Current measurement
Electric measurements—Current*
BT: Electric variables measurement
RT: Ammeters
Amperometric sensors
Electric current distribution measurement
Galvanometers

Electric current regulation
USE: Electric current control

* Former Ei Vocabulary term

Electric current regulators 704.2, 732.1
DT: January 1993
UF: Electric lighting—Current regulators*
Regulators (electric current)
BT: Electric equipment
RT: Electric current control

Electric currents 701.1
DT: January 1993
UF: Current (electric)
BT: Electricity
NT: Critical currents
Eddy currents
Electric fault currents
Electric surges
Induced currents
RT: Current density
Electric current control
Electric current distribution
Electric fields
Skin effect

Electric delay lines 713.5
DT: Predates 1975
UF: Delay lines
Signal delay lines
BT: Delay circuits

Electric demand forecasting
USE: Electric load forecasting

Electric discharges 701.1
DT: Predates 1975
UF: Discharges (electric)
BT: Electric field effects
NT: Atmospherics
Electric arcs
Electric breakdown
Electric sparks
Flashover
Glow discharges
Lightning
Partial discharges
Surface discharges
RT: Discharge lamps
Electric conductivity
Electric fault currents
Electric insulation
Electric properties
Electric space charge
Ionization
Ionization of gases
Leakage currents
Short circuit currents

*Electric discharges—Partial discharges**
USE: Partial discharges

*Electric discharges—Surface phenomena**
USE: Surface discharges

Electric distortion 701.1 (711.1)
DT: January 1993
UF: Distortion (electric)
Electric waveforms—Distortion*
BT: Distortion (waves)
RT: Equalizers
Fading (radio)

Electric distribution
USE: Electric power distribution

Electric drills 704.2 (405.1) (603.2)
DT: January 1993
UF: Rock drills—Electric*
BT: Drills

*Electric drive**
USE: Electric drives

*Electric drive—Variable speed**
USE: Variable speed drives

Electric drives (705)
SN: For drives that propel vehicles, use PROPULSION
DT: January 1993
UF: Electric drive*
Machine tools—Electric drive*
BT: Drives
Electric equipment
RT: Electric motors
Electric propulsion
Machine tool drives
Variable speed drives

Electric energy storage (702)
DT: Predates 1975
BT: Energy storage
NT: Capacitor storage
RT: Direct energy conversion
Electricity
Storage tubes

Electric equipment 704.2
DT: October 1975
BT: Equipment
NT: Automobile electric equipment
Dielectric devices
Direct energy converters
Electric appliances
Electric clocks
Electric coils
Electric conductors
Electric conduits
Electric connectors
Electric contacts
Electric control equipment
Electric converters
Electric current collectors
Electric current regulators
Electric drives
Electric exciters
Electric fences
Electric furnaces
Electric heating elements
Electric lamps
Electric lighting
Electric machinery
Electric measuring instruments
Electric power supplies to apparatus
Electric rectifiers
Electric ship equipment
Electric switchboards
Electric switchgear

Electric equipment *(continued)*
Electric tools
Electric transformers
Electric windings
Electric wiring
Electromagnetic launchers
Electromagnets
Electromechanical devices
Electrostatic devices
Terminals (electric)
Thermoelectric equipment
Voltage dividers
RT: Electric equipment protection
Electric industry
Electric insulation
Electric manufacturing plants
Electric power systems
Electric power plant equipment
Electric tapes
Electronic equipment

Electric equipment protection 704.2
DT: Predates 1975
UF: Electric machinery protection
Electric motor protection
Generator protection
Machinery protection
BT: Protection
NT: Electric insulation coordination
Transformer protection
RT: Alarm systems
Corrosion protection
Electric circuit breakers
Electric equipment
Electric fault currents
Electric fuses
Electric generators
Electric grounding
Electric insulation
Electric insulating materials
Electric relays
Flashover
Grounding electrodes
Lightning protection
Machine windings
Radiation protection
Rotors (windings)
Stators
Voltage control

Electric excitation 701.1 (705.1)
SN: Excitation by electric energy
DT: Predates 1975
UF: Excitation (electric)
BT: Electricity
RT: Electric exciters
Electric machinery

Electric exciters 705.1
DT: Predates 1975
UF: Boosters (electric generators)
Electric machinery—Exciters*
Exciters (electric)
BT: Electric equipment
RT: Antennas
Electric excitation
Electric generators
Starters
Transmitters

Electric fault currents 701.1 (706.2)
DT: Predates 1975
UF: Electric faults
Fault currents
Faults (electric)
BT: Electric currents
NT: Leakage currents
Short circuit currents
RT: Electric arcs
Electric breakdown
Electric cables
Electric corona
Electric discharges
Electric equipment protection
Electric fault location
Electric grounding
Electric insulators
Electric insulation
Electric lines
Electric sparks
Transients

*Electric fault currents—Leakage currents**
USE: Leakage currents

*Electric fault currents—Short circuit currents**
USE: Short circuit currents

Electric fault location 706.2
DT: Predates 1975
UF: Fault location*
RT: Electric cables
Electric fault currents
Electric insulation testing
Electric lines

Electric faults
USE: Electric fault currents

Electric fences 402, 704.2
DT: January 1993
UF: Fences, Electrified*
BT: Electric equipment
Fences

Electric field effects 701.1
DT: Predates 1975
BT: Effects
NT: Acoustoelectric effects
Electric discharges
Electrohydrodynamics
Electroluminescence
Electromigration
Electrooptical effects
Electrostriction
Gunn effect
Limited space charge accumulation
Magnetoelectric effects
Piezoelectricity
RT: Electric conductivity
Electric fields
Electrochemistry

Electric field effects *(continued)*
Electromagnetic field effects
Electrophoresis
Particle optics

Electric field measurement 942.2
DT: Predates 1975
BT: Electric variables measurement
RT: Electric fields
Electromagnetic field measurement
Field plotting

Electric field meters
USE: Voltmeters

Electric fields 701.1
DT: Predates 1975
UF: Electrostatic fields
Fields (electric)
RT: Electrets
Electric currents
Electric field effects
Electric field measurement
Electric power distribution
Electric shielding
Electromagnetic waves
Electromagnetic fields
Electrostatic devices
Electrostatics
Field effect semiconductor devices
Maxwell equations

Electric filters 703.2
DT: Predates 1975
UF: Telephone filters
BT: Networks (circuits)
Wave filters
NT: Active filters
Bandpass filters
Digital filters
Electromechanical filters
High pass filters
Low pass filters
Microwave filters
Notch filters
Passive filters
Switched filters
RT: Circuit theory
Ladder networks
Radio equipment
Radio receivers
Signal processing

*Electric filters, Active**
USE: Active filters

*Electric filters, Bandpass**
USE: Bandpass filters

*Electric filters, Crystal**
USE: Crystal filters

*Electric filters, Digital**
USE: Digital filters

*Electric filters, Electromechanical**
USE: Electromechanical filters

*Electric filters, High pass**
USE: High pass filters

*Electric filters, Low pass**
USE: Low pass filters

*Electric filters, Microwave**
USE: Microwave filters

*Electric filters, Notch**
USE: Notch filters

*Electric filters, Passive**
USE: Passive filters

*Electric filters, Switched**
USE: Switched filters

Electric final control devices 732.1 (704.1)
DT: January 1993
UF: Final control devices, Electric*
BT: Electric control equipment
Final control devices
RT: Electric brakes
Electric clutches
Electric motors
Electric switches
Electric switchgear

Electric flashover
USE: Flashover

Electric forming
USE: Electroforming

Electric frequency control 731.3
DT: January 1993
UF: Control, Electric variables—Frequency*
Electric frequency regulation
Frequency control (electric)
BT: Electric variables control
RT: Electric frequency measurement
Frequency stability
Frequency standards

Electric frequency measurement 942.2
DT: January 1993
UF: Electric measurements—Frequency*
BT: Electric variables measurement
RT: Electric frequency control
Frequency meters
Frequency standards
Tachometers

Electric frequency meters
USE: Frequency meters

Electric frequency regulation
USE: Electric frequency control

Electric furnace process (532.3) (545.3)
DT: January 1993
UF: Steelmaking—Electric furnace process*
BT: Steelmaking

Electric furnaces (704.2) (532.3) (642.2)
DT: January 1993
UF: Furnaces, Electric*
BT: Electric equipment
Furnaces

* Former Ei Vocabulary term

Electric furnaces *(continued)*
NT: Vacuum furnaces
RT: Electric heating
Heating furnaces
Induction heating
Melting furnaces

Electric fuses 704.1
DT: Predates 1975
UF: Fuses (electric)
Fuses*
BT: Electric control equipment
RT: Electric equipment protection
Electric switches
Electric switchgear

*Electric generator motors, AC**
USE: AC generator motors

*Electric generator motors, DC**
USE: DC generator motors

Electric generators 705.2
DT: Predates 1975
UF: Generators (electric)
Power generators (electric)
BT: Electric machinery
NT: AC generators
DC generators
Electric motor generator sets
Electrostatic generators
Hydroelectric generators
Magnetos
Turbogenerators
RT: Direct energy conversion
Electric control equipment
Electric equipment protection
Electric exciters

*Electric generators, AC**
USE: AC generators

*Electric generators, Asynchronous**
USE: Asynchronous generators

*Electric generators, DC**
USE: DC generators

*Electric generators, Synchronous**
USE: Synchronous generators

*Electric generators—Hydroelectric**
USE: Hydroelectric generators

Electric glow
USE: Electric corona

Electric grounding (701.1) (704) (706)
DT: Predates 1975
UF: Earthing
Grounding (electric)
Grounding*
RT: Antenna grounds
Electric equipment protection
Electric fault currents
Electricity
Flashover
Grounding electrodes
Relay protection
Short circuit currents

Electric heat treatment 537.1 (704.2)
DT: January 1993
UF: Heat treatment—Electric*
BT: Heat treatment

Electric heating 642.1 (704.2)
SN: Very general term; prefer specific method or kind of heating
DT: Predates 1975
UF: Boilers—Electric heating*
Heating—Electric*
Plastics—Electric heating*
BT: Heating
NT: Dielectric heating
Electric space heating
Induction heating
Microwave heating
RT: Electric furnaces
Electric heating elements
Heat pump systems
Industrial heating
Ovens

Electric heating elements 704.1 (642.2)
DT: Predates 1975
UF: Heating elements
BT: Electric equipment
Heating equipment
RT: Electric heating

*Electric heating, Dielectric**
USE: Dielectric heating

*Electric heating, Induction**
USE: Induction heating

Electric impedance 701.1
DT: June 1990
UF: Admittance (electric)
Electric admittance
Impedance (electric)
BT: Electric properties
NT: Electric resistance
RT: Electric impedance measurement
Electric reactance measurement
Electric resistance measurement
Impedance matching (electric)

Electric impedance measurement 942.2
DT: January 1993
UF: Electric measurements—Impedance*
BT: Electric variables measurement
NT: Electric reactance measurement
Electric resistance measurement
RT: Electric impedance

*Electric impedance—Negative resistance**
USE: Negative resistance

Electric inductors (704.1) (714)
SN: Devices principally intended to introduce inductive reactance into circuits
DT: Predates 1975
UF: Chokes
Inductors (electric)
Variometers
BT: Electric reactors

* Former Ei Vocabulary term

Electric inductors *(continued)*
NT: Power inductors
RT: Current limiting reactors
Electric coils
Electric transformers
Resistors

*Electric inductors, Power type**
USE: Power inductors

*Electric inductors—Coils**
USE: Electric coils

Electric industry (704) (705) (706)
SN: Suppliers both of electric power and of the equipment used to generate and distribute it.
DT: Predates 1975
BT: Industry
RT: Electric equipment
Electric manufacturing plants
Electric power generation
Electric power systems
Electricity
Electronics industry

Electric instrument transformers 714 (704) (942.1)
DT: Predates 1975
UF: Current transformers
Instrument transformers
BT: Electric transformers
Instrument components

Electric insulating coatings (708.1) (413.1) (813.2)
DT: January 1993
UF: Electric insulating materials—Coatings*
BT: Electric insulating materials
Protective coatings

Electric insulating materials (413.1) (708.1)
DT: Predates 1975
BT: Insulating materials
NT: Electric insulating coatings
Liquid insulating materials
RT: Dielectric materials
Electric arcs
Electric corona
Electric equipment protection
Electric insulators
Electric insulation
Glass
Mica
Surface discharges
Varnish

*Electric insulating materials—Ceramic**
USE: Ceramic materials

*Electric insulating materials—Coatings**
USE: Electric insulating coatings

*Electric insulating materials—Composite materials**
USE: Composite materials

*Electric insulating materials—Gases**
USE: Gases

*Electric insulating materials—Glass**
USE: Glass

*Electric insulating materials—Liquids**
USE: Liquid insulating materials

*Electric insulating materials—Mica**
USE: Mica

*Electric insulating materials—Organic substances**
USE: Organic chemicals

*Electric insulating materials—Paper**
USE: Paper

*Electric insulating materials—Plastics**
USE: Plastics applications

*Electric insulating materials—Rubber**
USE: Rubber applications

*Electric insulating materials—Silicones**
USE: Silicones

*Electric insulating materials—Tape**
USE: Electric tapes

*Electric insulating materials—Varnish**
USE: Varnish

Electric insulation (704) (715)
SN: Processes or means used to insulate conductors. Also use the thing being insulated
DT: January 1993
BT: Insulation
NT: Electric machine insulation
Electric transformer insulation
RT: Dielectric materials
Electric arcs
Electric breakdown
Electric corona
Electric discharges
Electric equipment protection
Electric equipment
Electric fault currents
Electric insulators
Electric insulation coordination
Electric insulating materials
Electric insulation testing
Surface discharges

Electric insulation coordination (704) (715)
DT: January 1993
UF: Coordination (electric insulation)
Electric insulation—Coordination*
Insulation coordination
BT: Electric equipment protection
RT: Electric insulation

Electric insulation testing (708.1) (413.1)
DT: January 1993
UF: Electric insulation—Testing*
BT: Materials testing
NT: Impulse testing
RT: Electric fault location
Electric insulation

*Electric insulation—Coordination**
USE: Electric insulation coordination

*Electric insulation—Impulse testing**
USE: Impulse testing

*Electric insulation—Testing**
USE: Electric insulation testing

Electric insulators (704) (708.1)
DT: Predates 1985
UF: Insulators (electric)
Spacers (electric lines)
BT: Dielectric devices
RT: Bushings
Electric fault currents
Electric insulating materials
Electric insulation
Surface discharges

*Electric insulators—Bushings**
USE: Bushings

Electric inverters (704.2) (713.5)
DT: Predates 1975
UF: DC to AC inverters
Inverters (electric)
BT: Electric converters
RT: Choppers (circuits)
Electric machinery components
Power electronics

Electric lamps 707.2
DT: Predates 1975
UF: Lamps
BT: Electric equipment
Light sources
NT: Aircraft signal lights
Discharge lamps
Flashlights
Headlights
Incandescent lamps
Infrared lamps
Miners lamps
Searchlights
Ultraviolet lamps
Underwater lamps
RT: Electric lighting
Emergency lighting
Floodlighting
Light
Lighting
Lighting fixtures
Optics
Outdoor electric lighting

*Electric lamps, Arc**
USE: Arc lamps

*Electric lamps, Discharge**
USE: Discharge lamps

*Electric lamps, Fluorescent**
USE: Fluorescent lamps

*Electric lamps, Infrared**
USE: Infrared lamps

*Electric lamps, Mercury vapor**
USE: Mercury vapor lamps

*Electric lamps, Metal vapor**
USE: Metal vapor lamps

*Electric lamps, Ultraviolet**
USE: Ultraviolet lamps

*Electric lamps—Ballasts**
USE: Ballasts (lamp)

*Electric lamps—Daylight simulation**
USE: Daylight simulation

*Electric lamps—Dimming**
USE: Dimming (lamps)

*Electric lamps—Filaments**
USE: Filaments (lamp)

*Electric lamps—Flickering**
USE: Flickering

*Electric lamps—Underwater**
USE: Underwater lamps

Electric leakage currents
USE: Leakage currents

Electric lighting 707.1
DT: Predates 1975
UF: Industrial lighting
BT: Electric equipment
Lighting
NT: Daylight simulation
Emergency lighting
Floodlighting
Outdoor electric lighting
RT: Electric lamps
Glare
Illuminating engineering
Street lighting
Waste heat utilization

*Electric lighting—Current regulators**
USE: Electric current regulators

*Electric lighting—Emergency lighting**
USE: Emergency lighting

*Electric lighting—Glare**
USE: Glare

*Electric lighting—Heat utilization**
USE: Waste heat utilization

*Electric lighting—Outdoor**
USE: Outdoor electric lighting

*Electric lighting—Signs**
USE: Electric signs

Electric line supports 706.2
DT: January 1993
BT: Supports
NT: Electric towers
RT: Poles

Electric lines 706.2
DT: Predates 1975
UF: Helicopters—Electric line construction*
Spacers (electric lines)
Transmission lines

Electric lines *(continued)*
NT: Overhead lines
Superconducting electric lines
Telecommunication lines
RT: Bundled conductors
Carrier communication
Carrier transmission on power lines
Drop wires
Electric cables
Electric fault currents
Electric fault location
Electric surges
Poles
Transmission line theory

*Electric lines—Aerial conductors**
USE: Overhead lines

*Electric lines—Bundled conductors**
USE: Bundled conductors

*Electric lines—Carrier transmission**
USE: Carrier transmission on power lines

*Electric lines—Corona**
USE: Electric corona

*Electric lines—Losses**
USE: Electric losses

*Electric lines—Mechanical characteristics**
USE: Mechanical properties

*Electric lines—Superconducting**
USE: Superconducting electric lines

*Electric lines—Supports**
USE: Supports

*Electric lines—Surges**
USE: Electric surges

*Electric lines—Voltage regulation**
USE: Voltage control

Electric load dispatching 706.1.1
DT: January 1993
UF: Electric power systems—Load dispatching*
BT: Electric load management
RT: Electric load forecasting
Electric load shedding
Electric variables control

Electric load distribution 706.1.2
DT: January 1993
UF: Electric power systems—Load distribution*
BT: Electric load management
RT: Electric load forecasting
Electric load shedding

Electric load flow 706.1
DT: January 1993
UF: Electric power systems—Load flow analysis*
RT: Electric loads
Electric power transmission networks
Electric power distribution
Electric power transmission
Electric power systems
Transmission network calculations

Electric load forecasting 706.1
DT: January 1993
UF: Electric demand forecasting
Electric power systems—Load forecasting*
BT: Electric load management
Forecasting
RT: Electric load dispatching
Electric load distribution
Electric utilities

Electric load loss 706.1
DT: January 1993
UF: Electric power systems—Loss of load probability*
Load loss
Loss of load
BT: Electric transformer loads

Electric load management 706.1
DT: January 1993
UF: Demand management
Electric power systems—Load management*
Load management (electric)
BT: Management
NT: Electric load dispatching
Electric load distribution
Electric load forecasting
Electric load shedding
RT: Electric loads

Electric load shedding 706.1
DT: January 1993
UF: Electric power systems—Load shedding*
Shedding (electric loads)
BT: Electric load management
RT: Electric load dispatching
Electric load distribution
Electric power system protection

Electric loads 706.1
DT: Predates 1975
UF: Loads (electric)
NT: Electric power plant loads
Electric transformer loads
RT: Electric load flow
Electric load management
Electric power systems

Electric locomotives 682.1.2 (704.2)
DT: January 1993
UF: Locomotives, Electric*
BT: Electric vehicles
Locomotives
RT: Electric motors
Electric railroads

Electric logging 512.1.2
DT: January 1993
UF: Oil well logging—Electric*
BT: Well logging
NT: Induced polarization logging
Induction logging
Spontaneous potential logging
RT: Electric prospecting
Electromagnetic logging
Natural gas well logging
Oil well logging

* Former Ei Vocabulary term

Electric loss angle measurement 942.2
DT: January 1993
UF: Electric losses—Loss angle measurement*
BT: Electric variables measurement
RT: Electric losses
Electric reactance measurement

Electric losses (701.1)
DT: Predates 1975
UF: Electric lines—Losses*
Electric machinery—Losses*
Electric power transmission—Losses*
Electric transformers—Losses*
Losses (electric)
Waveguides—Losses*
BT: Energy dissipation
RT: Attenuation
Electric loss angle measurement
Electric transformers

*Electric losses—Loss angle measurement**
USE: Electric loss angle measurement

Electric machine insulation (705.1) (704)
DT: January 1993
UF: Electric machinery—Insulation*
BT: Electric insulation
Electric machinery components

Electric machine theory (705.1)
DT: January 1993
UF: Electric machinery—Theory*
Machine theory
BT: Theory
RT: Electric machinery

Electric machine tool drives
USE: Machine tool drives

Electric machinery (705.1)
DT: Predates 1975
UF: Electric machines
BT: Electric equipment
Machinery
NT: AC machinery
DC machinery
Electric generators
Electric machinery components
RT: Electric excitation
Electric machine theory
Electric machinery testing
Electric power utilization
Rotary converters

Electric machinery components (705.1)
DT: January 1993
UF: Electric machinery—Components*
BT: Electric machinery
NT: Electric machine insulation
Machine windings
RT: Brush conductors
Electric commutators
Electric inverters

Electric machinery protection
USE: Electric equipment protection

Electric machinery testing (705.1)
DT: January 1993
UF: Electric machinery—Testing*
BT: Equipment testing
RT: Electric machinery

*Electric machinery, AC**
USE: AC machinery

*Electric machinery, Asynchronous**
USE: Asynchronous machinery

*Electric machinery, DC**
USE: DC machinery

*Electric machinery, Synchronous**
USE: Synchronous machinery

*Electric machinery—Braking**
USE: Braking

*Electric machinery—Components**
USE: Electric machinery components

*Electric machinery—Contacts**
USE: Electric contacts

*Electric machinery—Exciters**
USE: Electric exciters

*Electric machinery—Explosionproof**
USE: Explosionproofing

*Electric machinery—Insulation**
USE: Electric machine insulation

*Electric machinery—Losses**
USE: Electric losses

*Electric machinery—Portable**
USE: Portable equipment

*Electric machinery—Rotors**
USE: Rotors (windings)

*Electric machinery—Shafts**
USE: Shafts (machine components)

*Electric machinery—Starters**
USE: Starters

*Electric machinery—Starting**
USE: Starting

*Electric machinery—Stators**
USE: Stators

*Electric machinery—Symbols**
USE: Codes (symbols)

*Electric machinery—Terminals**
USE: Terminals (electric)

*Electric machinery—Testing**
USE: Electric machinery testing

*Electric machinery—Theory**
USE: Electric machine theory

*Electric machinery—Windings**
USE: Machine windings

Electric machines
USE: Electric machinery

* Former Ei Vocabulary term

Electric manufacturing plants (913.1) (913.4) (402.1)
DT: Predates 1975
BT: Industrial plants
RT: Electric equipment
Electric industry

*Electric measurements**
USE: Electric variables measurement

*Electric measurements—Capacitance**
USE: Capacitance measurement

*Electric measurements—Charge**
USE: Electric charge measurement

*Electric measurements—Conductivity**
USE: Electric conductivity measurement

*Electric measurements—Current distribution**
USE: Electric current distribution measurement

*Electric measurements—Current**
USE: Electric current measurement

*Electric measurements—Frequency**
USE: Electric frequency measurement

*Electric measurements—Gain**
USE: Gain measurement

*Electric measurements—Impedance**
USE: Electric impedance measurement

*Electric measurements—Inductance**
USE: Inductance measurement

*Electric measurements—Permittivity**
USE: Permittivity measurement

*Electric measurements—Phase**
USE: Phase measurement

*Electric measurements—Power**
USE: Electric power measurement

*Electric measurements—Q factor**
USE: Q factor measurement

*Electric measurements—Reactance**
USE: Electric reactance measurement

*Electric measurements—Resistance**
USE: Electric resistance measurement

*Electric measurements—Voltage distribution**
USE: Voltage distribution measurement

*Electric measurements—Voltage**
USE: Voltage measurement

Electric measuring bridges 942.1
DT: Predates 1975
UF: Bridges (electric measurement)
BT: Electric measuring instruments
RT: Capacitance measurement
Electric conductivity measurement
Inductance measurement

Electric measuring instruments 942.1
SN: Instruments used to measure electrical quantities; prefer specific instruments
DT: January 1993
UF: Electric meters
BT: Electric equipment
Instruments
NT: Ammeters
Cathode ray oscilloscopes
Coulometers
Dynamometers
Electric measuring bridges
Electrometers
Frequency meters
Galvanometers
Maximum demand indicators
Ohmmeters
Oscillographs
Phase meters
Phase sequence indicators
Polarographs
Q meters
Standing wave meters
Voltmeters
Watt hour meters
Wattmeters
Wavemeters
RT: Electric variables measurement
Magnetometers
Permeameters (magnetic permeability)

Electric meters
USE: Electric measuring instruments

Electric motor generator sets 705.2, 705.3
SN: Two-unit electric machines, one unit driving the other
DT: Predates 1975
BT: Electric generators
Electric motors

Electric motor protection
USE: Electric equipment protection

Electric motors 705.3
DT: Predates 1975
BT: Motors
NT: AC motors
DC motors
Electric motor generator sets
Fractional horsepower motors
Linear motors
Printed circuit motors
Shaftless motors
Stepping motors
Submersible motors
Traction motors
RT: Electric drives
Electric final control devices
Electric locomotives

*Electric motors, AC**
USE: AC motors

*Electric motors, DC**
USE: DC motors

*Electric motors, Hysteresis type**
USE: Hysteresis motors

* Former Ei Vocabulary term

*Electric motors, Induction**
USE: Induction motors

*Electric motors, Linear**
USE: Linear motors

*Electric motors, Printed circuit**
USE: Printed circuit motors

*Electric motors, Reluctance type**
USE: Reluctance motors

*Electric motors, Stepping type**
USE: Stepping motors

*Electric motors, Synchronous**
USE: Synchronous motors

*Electric motors—Fractional horsepower**
USE: Fractional horsepower motors

*Electric motors—Shaftless**
USE: Shaftless motors

*Electric motors—Squirrel cage**
USE: Squirrel cage motors

*Electric motors—Submerged**
USE: Submersible motors

*Electric motors—Traction**
USE: Traction motors

Electric network analysis 703.1.1
DT: January 1993
UF: Circuit analysis
Network analysis
BT: Circuit theory
NT: Computer aided network analysis
Linear network analysis
Nonlinear network analysis
RT: Distributed parameter networks
Electric network synthesis
Electric network analyzers
Equivalent circuits
Frequency domain analysis
Graph theory
Gyrators
Operations research
State space methods
Time domain analysis
Topology
Transfer functions
Transmission network calculations
Vectors
Z transforms

Electric network analyzers (722.5) (706)
DT: Predates 1975
UF: AC network analyzers
Network analyzers
BT: Analog computers
RT: Computer networks
Electric network analysis
Local area networks
Networks (circuits)
Transmission network calculations

Electric network parameters 703.1
DT: January 1993
UF: Electric networks—Network parameters*
BT: Circuit theory
RT: Equivalent circuits
Matrix algebra

Electric network schematics
USE: Schematic diagrams

Electric network synthesis 703.1.2
DT: January 1993
BT: Circuit theory
NT: Linear network synthesis
Nonlinear network synthesis
RT: Distributed parameter networks
Electric network analysis
Equivalent circuits
Graph theory
State space methods
Switching theory
Topology
Vectors

Electric network topology 703.1
DT: January 1993
UF: Circuit topology
Diakoptics
Electric networks—Topology*
BT: Circuit theory
Topology

*Electric networks**
USE: Circuit theory

*Electric networks, Active**
USE: Active networks

*Electric networks, Communication**
USE: Telecommunication networks

*Electric networks, Distributed parameter**
USE: Distributed parameter networks

*Electric networks, Linear**
USE: Linear networks

*Electric networks, Lumped parameter**
USE: Lumped parameter networks

*Electric networks, Nonlinear**
USE: Nonlinear networks

*Electric networks, Passive**
USE: Passive networks

*Electric networks, Switching**
USE: Switching networks

*Electric networks, Time varying**
USE: Time varying networks

*Electric networks—Equivalent circuits**
USE: Equivalent circuits

*Electric networks—Ladder circuits**
USE: Ladder networks

*Electric networks—Negative impedance converters**
USE: Negative impedance converters

* Former Ei Vocabulary term

*Electric networks—Network parameters**
USE: Electric network parameters

*Electric networks—Reactive power**
USE: Reactive power

*Electric networks—Resonant circuits**
USE: Resonant circuits

*Electric networks—Schematic diagrams**
USE: Schematic diagrams

*Electric networks—Theory**
USE: Circuit theory

*Electric networks—Topology**
USE: Electric network topology

*Electric networks—Transmission line theory**
USE: Transmission line theory

Electric oil well pumping 511.1
DT: January 1993
UF: Oil well pumping—Electric*
BT: Oil well pumping

Electric ovens 704.2
DT: January 1993
UF: Electric appliances—Ovens*
BT: Ovens
NT: Microwave ovens
RT: Electric appliances
Electric stoves

Electric phase shift circuits
USE: Phase shifters

*Electric phase shifters**
USE: Phase shifters

Electric power distribution 706.1.2
DT: Predates 1975
UF: Distribution networks
Electric distribution
NT: Underground electric power distribution
RT: Electric conduits
Electric fields
Electric load flow
Electric power systems
Reactive power

*Electric power distribution—Underground installation**
USE: Underground electric power distribution

Electric power factor (703.1) (701.1) (706)
DT: Predates 1976
UF: Dielectric power factor
RT: Capacitance
Electric power factor correction
Electric power factor measurement
Networks (circuits)
Reactive power

Electric power factor correction (703.1) (706) (701.1)
DT: January 1993
UF: Correction (power factor)
Electric power factor—Correction*
RT: Electric power factor

Electric power factor measurement 942.2
DT: January 1993
UF: Electric power factor—Measurements*
BT: Electric variables measurement
RT: Electric power factor

*Electric power factor—Correction**
USE: Electric power factor correction

*Electric power factor—Measurements**
USE: Electric power factor measurement

Electric power generation (705.2) (615)
DT: Predates 1975
BT: Power generation
NT: Direct energy conversion
RT: Electric industry
Electric power plants
Electric power systems
Electric power utilization
Solar cells

*Electric power generation—Microwave energy**
USE: Microwave power transmission

Electric power measurement 942.2
DT: January 1993
UF: Electric measurements—Power*
Energy measurement
Power measurement (electric)
BT: Electric variables measurement
RT: Maximum demand indicators
Reactive power
Watt hour meters
Wattmeters

Electric power plant equipment (704.2)
SN: Use for general coverage of machines, regulating and control equipment, and electric circuits required to operate an electric power plant
DT: Predates 1975
BT: Electric power plants
Equipment
RT: Cooling towers
Electric equipment

Electric power plant loads (706.1)
DT: January 1993
UF: Electric power plants—Loads*
BT: Electric loads
RT: Electric power plants

Electric power plants 706.1 (402.1) (611) (612) (613) (614) (615)
SN: Plants for generating electric power
DT: Predates 1975
BT: Power plants
NT: Cogeneration plants
Diesel electric power plants
Electric power plant equipment
Hydroelectric power plants
Mobile electric power plants
Thermoelectric power plants
RT: Electric power generation
Electric power systems

Electric power plants *(continued)*
Electric power plant loads
Fossil fuel power plants
Gas turbine power plants
Nuclear power plants
Outages
Solar power plants
Standby power service
Steam power plants
Wind power

*Electric power plants, Mobile**
USE: Mobile electric power plants

*Electric power plants—Diesel and hydroelectric combined**
USE: Diesel and hydroelectric combined power plants

*Electric power plants—Hydroelectric and gas turbine combined**
USE: Hydroelectric and gas turbine combined power plants

*Electric power plants—Hydroelectric and nuclear combined**
USE: Hydroelectric and nuclear combined power plants

*Electric power plants—Hydroelectric and steam combined**
USE: Hydroelectric and steam combined power plants

*Electric power plants—Loads**
USE: Electric power plant loads

*Electric power plants—Thermoelectric**
USE: Thermoelectric power plants

Electric power supplies to apparatus (713.5) (715.2)
SN: Use for auxiliary or subsystem units, usually electronic, supplying power as required by apparatus. For utilization of electricity from private or public utility generating stations, use ELECTRIC POWER UTILIZATION.
DT: Predates 1975
UF: Electron tubes—Electric power supplies*
Electronic power supplies
Electroplating—Power supply*
Furnaces—Power supply*
Instruments—Power supply*
Machinery—Power supply*
Metal finishing—Power supply*
Power supplies to apparatus
Radar—Power supply*
Radio equipment—Power supply*
Radio telephone—Power supply*
Radio transmitters—Power supply*
Railroad signals and signaling—Power supply*
Telephone equipment—Power supply*
Television receivers—Power supply*
Uninterruptible power supplies
UPS
Welding—Power supply*
BT: Electric equipment
RT: Electric power utilization
Electric power systems
Power electronics
Power supply circuits
Spacecraft power supplies
Standby power systems

Electric power system interconnection 706.1
DT: January 1993
UF: Electric power systems—Interconnection*
Interconnected power systems
RT: Electric power transmission networks
Electric power systems

Electric power system protection 706.1 (914.1)
DT: January 1993
UF: Electric power systems—Protection*
BT: Protection
RT: Electric load shedding
Electric power systems
Security systems

Electric power systems 706.1
DT: Predates 1975
UF: Power supply systems
Supply systems (electric)
NT: Electric power transmission networks
Electric substations
Standby power systems
RT: Electric equipment
Electric industry
Electric load flow
Electric loads
Electric power utilization
Electric power plants
Electric power generation
Electric power distribution
Electric power system interconnection
Electric power system protection
Electric power transmission
Electric power supplies to apparatus
Electric utilities
Electricity
Reactive power

*Electric power systems—Interconnection**
USE: Electric power system interconnection

*Electric power systems—Load dispatching**
USE: Electric load dispatching

*Electric power systems—Load distribution**
USE: Electric load distribution

*Electric power systems—Load flow analysis**
USE: Electric load flow AND
Analysis

*Electric power systems—Load forecasting**
USE: Electric load forecasting

*Electric power systems—Load management**
USE: Electric load management

*Electric power systems—Load shedding**
USE: Electric load shedding

* Former Ei Vocabulary term

*Electric power systems—Loss of load probability**
USE: Probability AND
Electric load loss

*Electric power systems—Protection**
USE: Electric power system protection

*Electric power systems—Reactive power**
USE: Reactive power

Electric power transmission 706.1.1
DT: Predates 1975
UF: Electric transmission
Power transmission (electric)
NT: DC power transmission
EHV power transmission
Microwave power transmission
RT: Electric conductors
Electric conduits
Electric connectors
Electric load flow
Electric power transmission networks
Electric power systems
Electric rectifiers
Electric switchgear
Reactive power

Electric power transmission networks 706.1.1 (703.1)
DT: January 1993
UF: Electric power transmission—Networks*
Transmission networks
BT: Electric power systems
Networks (circuits)
RT: Electric load flow
Electric power system interconnection
Electric power transmission
Reactive power
Transmission network calculations

*Electric power transmission, DC**
USE: DC power transmission

*Electric power transmission, EHV**
USE: EHV power transmission

*Electric power transmission, HVDC**
USE: HVDC power transmission

*Electric power transmission, UHV**
USE: UHV power transmission

*Electric power transmission—Losses**
USE: Electric losses

*Electric power transmission—Microwave frequencies**
USE: Microwave power transmission

*Electric power transmission—Network calculations**
USE: Transmission network calculations

*Electric power transmission—Networks**
USE: Electric power transmission networks

Electric power utilisation
USE: Electric power utilization

Electric power utilization 706.1
SN: Utilization of electricity from private or public utility generating stations. For equipment power supplies use ELECTRIC POWER SUPPLIES TO APPARATUS
DT: Predates 1975
UF: Accelerators—Power supply*
Airports—Power supply*
Electric power utilisation
Electric railroads—Power supply*
Exhibitions—Power supply*
Industrial plants—Power supply*
Mines and mining—Power supply*
Power plants—Power supply*
Power utilization
Utilization (power)
RT: Air conditioning
Drives
Electric appliances
Electric machinery
Electric power supplies to apparatus
Electric power systems
Electric power generation
Electric traction
Electric utilities
Heating equipment
Lighting
Refrigeration

Electric properties (701.1) (931.2)
DT: January 1977
BT: Physical properties
NT: Capacitance
Dielectric properties
Electric conductivity
Electric impedance
Gunn effect
Ionic conduction
Photoelectricity
Reactive power
Zener effect
RT: Composition
Electric discharges
Electricity
Thermoelectricity
Transport properties

Electric propulsion (654.2) (675.1)
DT: January 1993
UF: Propulsion—Electric energy*
Rocket engines—Electric propulsion*
Ship propulsion—Electric*
BT: Propulsion
NT: Electromagnetic propulsion
Ion propulsion
RT: Electric current collection
Electric drives
Rocket engines
Ship propulsion
Storage battery vehicles

Electric prospecting 481.4 (701.1)
DT: January 1993
UF: Geophysical prospecting—Electrical*
BT: Geophysical prospecting

Electric prospecting *(continued)*
NT: Telluric prospecting
RT: Electric logging

Electric railroads 681 (706)
DT: Predates 1975
UF: Railroad electrification
BT: Railroads
RT: Driver training
Electric current collectors
Electric locomotives
Electric traction
Electric vehicles
Monorail cars
Monorails
Rapid transit
Subways
Traction motors

*Electric railroads—Current collectors**
USE: Electric current collectors

*Electric railroads—Driver training**
USE: Driver training

*Electric railroads—Monorail**
USE: Monorails

*Electric railroads—Power supply**
USE: Electric power utilization

Electric rates 706.1 (911.1.1)
DT: January 1993
UF: Electric utilities—Rates*
BT: Utility rates
RT: Electric utilities

Electric reactance measurement 942.2
DT: January 1993
UF: Electric measurements—Reactance*
Reactance measurement
BT: Electric impedance measurement
RT: Capacitance measurement
Electric impedance
Electric loss angle measurement
Inductance measurement

Electric reactors (704.1)
SN: Devices principally intended to introduce reactance into circuits
DT: Predates 1975
UF: Reactors (electric)
BT: Network components
NT: Capacitors
Current limiting reactors
Electric inductors
Saturable core reactors
RT: Reactive power

*Electric reactors, Current limiting**
USE: Current limiting reactors

*Electric reactors, Saturable core**
USE: Saturable core reactors

Electric rectifiers (704.2) (714.1) (714.2)
DT: January 1993
UF: Rectifiers (electric)
BT: Electric equipment
NT: Electron tube rectifiers
Mercury vapor rectifiers
Solid state rectifiers
RT: Diodes
Electric power transmission
Power supply circuits
Rectifier substations
Rectifying circuits
Rotary converters
Semiconductor devices
Thyratrons
Thyristors

*Electric rectifiers, Electron tube**
USE: Electron tube rectifiers

*Electric rectifiers, Mercury vapor**
USE: Mercury vapor rectifiers

*Electric rectifiers, Solid state**
USE: Solid state rectifiers

Electric relays 704.1 (704.2) (706.2) (714)
DT: Predates 1975
UF: Relays (electric)
Telephone relays
BT: Electric switches
NT: Electric contactors
Photoelectric relays
Reed relays
Semiconductor relays
RT: Electric circuit breakers
Electric equipment protection
Relay control systems
Relay protection
Solenoids
Switching circuits

*Electric relays, Reed**
USE: Reed relays

*Electric relays, Semiconductor**
USE: Semiconductor relays

*Electric relays—Contacts**
USE: Electric contacts

*Electric relays—Photoelectric**
USE: Photoelectric relays

Electric resistance 701.1
DT: January 1993
UF: Conductance (electric)
Electric conductance
Resistance (electric)
BT: Electric impedance
NT: Negative resistance
RT: Electric resistance measurement

Electric resistance measurement 942.2
DT: January 1993
UF: Electric measurements—Resistance*
BT: Electric impedance measurement
RT: Electric conductivity measurement
Electric impedance
Electric resistance
Ohmmeters

* Former Ei Vocabulary term

Electric resistance welding
USE: Resistance welding

Electric resistivity
USE: Electric conductivity

Electric resistors
USE: Resistors

Electric shielding (711) (701.1)
SN: Shielding against electrical fields
DT: Predates 1975
BT: Shielding
NT: Cable shielding
RT: Electric attenuators
Electric fields
Electromagnetic compatibility
Electromagnetic shielding
Electromagnetic waves
Signal interference

Electric ship equipment 671.2 (704.2)
DT: January 1993
UF: Ship equipment—Electric*
BT: Electric equipment
Ship equipment
RT: Ship instruments

Electric signal systems (704) (715)
SN: Device applications, not waveshaping or modulation
DT: Predates 1975
BT: Signal systems
NT: Traffic signals
RT: Electric signs
Marine signal systems
Railroad signal systems
Signaling
Subway signal systems
Touch tone telephone systems

Electric signs (707.2)
DT: Predates 1975
UF: Electric lighting—Signs*
BT: Signs
RT: Electric signal systems
Light sources

Electric slag process
USE: Electroslag process

Electric space charge 701.1
DT: Predates 1975
UF: Electron tubes—Space charge*
Space charge
BT: Electric charge
RT: Electric discharges
Electron tubes
Electrostatics
Plasmas

Electric space heating 643.1 (704.2)
DT: January 1993
BT: Electric heating
Space heating

Electric spark cutting
USE: Spark cutting

Electric spark gaps (704)
DT: Predates 1975
BT: Electric switches
RT: Electric sparks
Electrodes

*Electric spark gaps—Sparking**
USE: Electric sparks

Electric sparks 701.1
DT: Predates 1975
UF: Electric spark gaps—Sparking*
Sparking
Sparkover
Sparks
BT: Electric discharges
RT: Circuit breaking arcs
Electric arcs
Electric breakdown
Electric corona
Electric fault currents
Electric spark gaps
Ionization
Lightning

Electric stoves 704.2
DT: January 1993
UF: Stoves—Electric*
BT: Electric appliances
Stoves
RT: Electric ovens

Electric substations 706.1
DT: Predates 1975
UF: Substations (electric)
BT: Electric power systems
NT: Rectifier substations
Transformer substations

*Electric substations, Rectifier**
USE: Rectifier substations

*Electric substations, Transformer**
USE: Transformer substations

*Electric substations—Traction**
USE: Electric traction

Electric surges (706)
DT: January 1993
UF: Electric lines—Surges*
Electric transformers—Surges*
Surges (electric)
BT: Electric currents
RT: Electric lines
Electric transformers
Surge protection
Transients

Electric switchboards 704.2 (706.2)
DT: January 1993
UF: Switchboards (electric)
BT: Electric equipment
NT: Telephone switchboards
RT: Electric switchgear

Electric switches 704.2, 714
SN: For control of electric current flow. Very general term; prefer specific type of switch. Excludes electrically operated railroad track switches, for which use RAILROAD TRACK SWITCHES
DT: Predates 1975
BT: Switches
NT: Cryotrons
Electric circuit breakers
Electric relays
Electric spark gaps
Semiconductor switches
Telephone switching equipment
Thermostats
Time switches
RT: Electric commutation
Electric contacts
Electric final control devices
Electric fuses
Electric switchgear
Process control
Radio equipment
Switching circuits

*Electric switches, Semiconductor**
USE: Semiconductor switches

*Electric switches, Time**
USE: Time switches

Electric switchgear 704.2 (706.2)
DT: Predates 1975
UF: Locomotives—Switchgear*
Switchboxes
Switchgear
BT: Electric equipment
RT: Electric circuit breakers
Electric contactors
Electric final control devices
Electric fuses
Electric power transmission
Electric switchboards
Electric switches
Electric switchgear protection
Electric switchgear testing
Switching

Electric switchgear protection 704.2 (706.2)
DT: January 1993
UF: Electric switchgear—Protection*
BT: Protection
RT: Electric switchgear

Electric switchgear testing 704.2 (706.2)
DT: January 1993
UF: Electric switchgear—Testing*
BT: Equipment testing
RT: Electric switchgear

*Electric switchgear—Protection**
USE: Electric switchgear protection

*Electric switchgear—Testing**
USE: Electric switchgear testing

Electric tapes (708.1) (413.1)
DT: January 1993
UF: Electric insulating materials—Tape*
BT: Tapes
RT: Cable taping
Electric equipment

Electric terminals
USE: Terminals (electric)

Electric tools 704.2 (605.1) (603.1)
DT: January 1993
UF: Tools, jigs and fixtures—Electric*
BT: Electric equipment
Tools

Electric towers 706.2, 402.4
DT: January 1993
BT: Electric line supports
Towers

Electric traction (705)
DT: Predates 1975
UF: Electric substations—Traction*
Traction (electric)
NT: Diesel electric traction
RT: Electric current collection
Electric power utilization
Electric railroads
Electric vehicles
Flywheel propulsion
Linear motors
Pantographs
Traction motors

*Electric traction—Current collection**
USE: Electric current collection

*Electric traction—Flywheel propulsion**
USE: Flywheel propulsion

*Electric traction—Pantographs**
USE: Pantographs

Electric transformer insulation (715) (706.2) (704)
DT: January 1993
BT: Electric insulation
Electric transformers
RT: Insulating oil

Electric transformer loads (706.2) (715) (704)
DT: January 1993
UF: Electric transformers—Loads*
BT: Electric loads
NT: Electric load loss
RT: Electric transformers

Electric transformer testing (706.2) (704) (715)
DT: January 1993
UF: Electric transformers—Testing*
BT: Equipment testing
RT: Electric transformers

Electric transformers (704) (706.2) (715) (714)
DT: Predates 1975
UF: Electric transformers—Parallel operation*
Transformers (electric)
Welding machines—Transformers*

* Former Ei Vocabulary term

Electric transformers *(continued)*
BT: Electric equipment
NT: DC transformers
Electric autotransformers
Electric instrument transformers
Electric transformer insulation
High frequency transformers
Oil filled transformers
Pulse transformers
Transformer magnetic circuits
Transformer windings
RT: Amplifiers (electronic)
Electric coils
Electric corona
Electric inductors
Electric losses
Electric surges
Electric transformer loads
Electric transformer testing
Magnetic cores
Power supply circuits
Radio equipment
Transformer protection
Transformer substations
Voltage control

*Electric transformers, DC**
USE: DC transformers

*Electric transformers, High frequency**
USE: High frequency transformers

*Electric transformers, Pulse**
USE: Pulse transformers

*Electric transformers—Corona**
USE: Electric corona

*Electric transformers—Loads**
USE: Electric transformer loads

*Electric transformers—Losses**
USE: Electric losses

*Electric transformers—Magnetic circuit**
USE: Transformer magnetic circuits

*Electric transformers—Noise**
USE: Spurious signal noise

*Electric transformers—Oil filled**
USE: Oil filled transformers

*Electric transformers—Parallel operation**
USE: Electric transformers

*Electric transformers—Portable**
USE: Portable equipment

*Electric transformers—Surges**
USE: Electric surges

*Electric transformers—Temperature**
USE: Temperature

*Electric transformers—Testing**
USE: Electric transformer testing

Electric transmission
USE: Electric power transmission

Electric typewriters 704.2 (745.1.1)
DT: January 1993
UF: Typewriters—Electric*
BT: Typewriters
RT: Typewriter keyboards

Electric units
USE: Units of measurement

Electric utilities (706)
DT: Predates 1975
UF: Electricity supply industry
Power supply industry
BT: Public utilities
RT: Electric load forecasting
Electric power systems
Electric power utilization
Electric rates

*Electric utilities—Accounting**
USE: Cost accounting

*Electric utilities—Rates**
USE: Electric rates

*Electric utilities—Taxation**
USE: Taxation

*Electric utilities—Weather forecasting**
USE: Weather forecasting

*Electric utilities—Weather service**
USE: Meteorological problems

Electric variables control 701.1, 731.3
DT: January 1993
UF: Control, Electric variables*
BT: Control
NT: Electric current control
Electric frequency control
Gain control
Phase control
Power control
Voltage control
RT: Electric load dispatching
Electricity

Electric variables measurement 942.2
DT: January 1993
UF: Electric measurements*
BT: Measurements
NT: Capacitance measurement
Electric charge measurement
Electric conductivity measurement
Electric current distribution measurement
Electric current measurement
Electric field measurement
Electric frequency measurement
Electric impedance measurement
Electric loss angle measurement
Electric power factor measurement
Electric power measurement
Gain measurement
Inductance measurement
Microwave measurement
Permittivity measurement
Phase measurement
Q factor measurement

* Former Ei Vocabulary term

Electric variables measurement *(continued)*
Signal noise measurement
Voltage distribution measurement
Voltage measurement
RT: Electric measuring instruments

Electric vehicles
(702.1.2) (662.1) (662.2) (682.1.2) (432)
DT: January 1993
BT: Vehicles
NT: Electric automobiles
Electric locomotives
Storage battery vehicles
Trackless trolleys
RT: Electric railroads
Electric traction

Electric watches (715.1)
DT: January 1993
UF: Watches—Electric operation*
BT: Watches

Electric waveforms (701.1) (711.1)
SN: Use for time-functions of electrical energy
DT: Predates 1975
UF: Spectrum analysis—Electric waveforms*
RT: Electromagnetic waves
Spectrum analysis

*Electric waveforms—Distortion**
USE: Electric distortion

Electric waves
USE: Electromagnetic waves

Electric welding 538.2.1
DT: January 1993
UF: Welding, Electric*
BT: Welding
NT: Electric arc welding
Electroslag welding
Resistance welding
Seam welding
RT: Pressure welding
Welding electrodes
Welding rods

Electric windings (705) (704.1)
DT: Predates 1975
UF: Electric coils—Windings*
Windings (electric)
BT: Electric equipment
NT: Machine windings
Transformer windings
RT: AC generators
AC machinery
Asynchronous generators
Electric coils
Electric conductors
Electric wiring
Magnetic circuits

*Electric windings, Machine**
USE: Machine windings

*Electric windings, Transformer**
USE: Transformer windings

Electric wire 708.2
DT: January 1993
UF: Electric conductors, Wire*
BT: Electric conductors
Wire
NT: Drop wires
Insulated wire
Superconducting wire
RT: Electric connectors
Electric wiring

Electric wiring (708.2) (704.2)
DT: Predates 1975
UF: Radio equipment—Wiring*
Telephone equipment—Wiring*
Wiring
BT: Electric equipment
NT: Building wiring
RT: Electric codes
Electric conductors
Electric connectors
Electric windings
Electric wire
Printed circuits

*Electric wiring, Buildings**
USE: Building wiring

Electrical engineering 709
DT: Predates 1975
BT: Engineering

Electricity 701.1
DT: January 1993
NT: Atmospheric electricity
Bioelectric phenomena
Electric charge
Electric currents
Electric excitation
Static electricity
Thermoelectricity
Triboelectricity
RT: Direct energy conversion
Electric accidents
Electric codes
Electric energy storage
Electric grounding
Electric industry
Electric power systems
Electric properties
Electric variables control
Electrodynamics
Electromagnetism
Magnetism

Electricity supply industry
USE: Electric utilities

Electro-acoustic testing
USE: Electroacoustic testing

Electro-acoustic transducers
USE: Electroacoustic transducers

Electro-optical devices
USE: Electrooptical devices

* Former Ei Vocabulary term

Electro-optical effects
USE: Electrooptical effects

Electro-optical materials
USE: Electrooptical materials

Electroacoustic effects
USE: Acoustoelectric effects

Electroacoustic testing (422.2) (423.2) (701.1) (751.1)
DT: January 1993
UF: Acoustoelectric testing
Electro-acoustic testing
Materials testing—Electroacoustic*
BT: Materials testing

Electroacoustic transducers 752.1
DT: Predates 1975
UF: Acoustoelectric transducers
Electro-acoustic transducers
BT: Acoustic transducers
NT: Earphones
Headphones
Hydrophones
Loudspeakers
Microphones
RT: Telephone sets

Electroacupuncture 461.6, 701.1 (462.1)
DT: January 1993
UF: Electrotherapeutics—Electroacupuncture*
BT: Acupuncture
Electrotherapeutics

Electrocardiograms
USE: Electrocardiography

Electrocardiography 461.6, 701.1
DT: January 1993
UF: Biomedical engineering—Electrocardiography*
ECG
EKG
Electrocardiograms
BT: Diagnosis
RT: Bioelectric potentials
Biomedical engineering
Cardiology
Patient monitoring

Electrochemical corrosion 539.1, 801.4.1, 802.2
DT: January 1993
UF: Corrosion—Electrochemical*
Corrosion—Electrolytic*
Electrolytic corrosion
BT: Corrosion
RT: Electrochemistry
Electrolysis

Electrochemical cutting 801.4.1, 802.2, 604.1
DT: January 1993
UF: Electrolytic cutting
Metal cutting—Electrochemical*
BT: Cutting
RT: Electrochemistry
Electrolysis

Electrochemical electrodes (801.4.1) (802.1) (704.1)
DT: January 1993
UF: Electrodes, Electrochemical*
BT: Electrodes
NT: Coated wire electrodes
Gas sensing electrodes
Glass membrane electrodes
Ion selective electrodes
Liquid membrane electrodes
Polymer membrane electrodes
Urea electrodes
RT: Chemical modification
Electric batteries
Electrochemistry
Electropolymerization

*Electrochemical electrodes—Chemical modification**
USE: Chemical modification

Electrochemical impedance spectroscopy
USE: Spectroscopy

Electrochemical polishing
USE: Electrolytic polishing

Electrochemical polymerization
USE: Electropolymerization

Electrochemical sensors 801.4.1, 732.2, 801
DT: January 1993
UF: Sensors—Electrochemical*
BT: Chemical sensors
RT: Chemical modification
Electrolytic analysis

Electrochemistry 801.4.1
DT: Predates 1975
BT: Physical chemistry
RT: Coulometers
Cyclic voltammetry
Electric field effects
Electrochemical electrodes
Electrochemical corrosion
Electrochemical cutting
Electrodeposition
Electrolysis
Electrolytes
Electrolytic cleaning
Electrolytic polishing
Electrophoresis
Polarographic analysis

Electrochromism 701.1, 741.1
DT: January 1983
BT: Electrooptical effects
Optical properties
RT: Chromogenics
Color

Electrodeposition 539.3.1 (813.1)
DT: Predates 1975
UF: Electrolytic deposition
BT: Deposition
NT: Electroplating
RT: Coating techniques

Electrodeposition *(continued)*
Electrochemistry
Electroforming
Electrolysis
Plating

Electrodes (704.1) (714.1)
DT: Predates 1975
BT: Electronic equipment
NT: Anodes
Cathodes
Electrochemical electrodes
Graphite electrodes
Grounding electrodes
Microelectrodes
Welding electrodes
RT: Cyclic voltammetry
Electric conductors
Electric spark gaps
Electron emission
Electron tube components
Electron tubes
Enzyme sensors

*Electrodes, Electrochemical**
USE: Electrochemical electrodes

*Electrodes, Electrochemical—Coated wire**
USE: Coated wire electrodes

*Electrodes, Electrochemical—Gas sensing**
USE: Gas sensing electrodes

*Electrodes, Electrochemical—Glass membrane**
USE: Glass membrane electrodes

*Electrodes, Electrochemical—Ion selective**
USE: Ion selective electrodes

*Electrodes, Electrochemical—Liquid membrane**
USE: Liquid membrane electrodes

*Electrodes, Electrochemical—Polymer membrane**
USE: Polymer membrane electrodes

*Electrodes, Grounding**
USE: Grounding electrodes

Electrodialysis (801.4.1) (802.3)
DT: June 1990
BT: Dialysis
RT: Hydrometallurgy

Electrodynamics 701, 931.1
SN: The relations among electrical, magnetic, and mechanical phenomena
DT: Predates 1975
BT: Dynamics
RT: Electricity
Electromagnetism
Electron beams
Electron optics
Electron tubes
Magnetism
Particle optics
Skin effect

Electrodynamometers
USE: Dynamometers

Electroencephalograms
USE: Electroencephalography

Electroencephalography 461.6
DT: January 1993
UF: Biomedical engineering—Electroencephalography*
EEG
Electroencephalograms
BT: Diagnosis
RT: Bioelectric potentials
Biomedical engineering
Brain
Patient monitoring

Electrofluidynamics
USE: Electrohydrodynamics

Electrofluorescence
USE: Electroluminescence

Electroforming (535.2) (801.4.1) (539.3.1)
DT: Predates 1975
UF: Electric forming
BT: Forming
RT: Electrodeposition
Electrolysis
Electroplating

Electrogasdynamic power generation 615 (631.1.2)
DT: January 1993
UF: Power generation—Electrogasdynamic*
BT: Power generation
RT: Electrohydrodynamics

Electrohydraulic forming (631.1.1) (535.2)
DT: January 1993
UF: Metal forming—Electrohydraulic*
BT: Metal forming

Electrohydrodynamics (631.2) (701.1)
DT: Predates 1975
UF: EHD
Electrofluidynamics
BT: Electric field effects
Hydrodynamics
RT: Electrogasdynamic power generation

Electroless deposition
USE: Electroless plating

Electroless plating 539.3.2
DT: Predates 1975
UF: Electroless deposition
BT: Plating

Electroluminescence 701.1, 741
DT: Predates 1975
UF: Electrofluorescence
Electrophotoluminescence
BT: Electric field effects
Luminescence
RT: Electrooptical effects
Electrooptical devices
Glow discharges

* Former Ei Vocabulary term

Electroluminescent devices
USE: Luminescent devices

Electrolysis 801.4.1, 802.2 (702)
DT: Predates 1975
UF: Chemical reactions—Electrolytic*
Electrolytic reactions
BT: Chemical reactions
RT: Anodic polarization
Current density
Electrochemical cutting
Electrochemical corrosion
Electrochemistry
Electrodeposition
Electroforming
Electrolytes
Electrolytic analysis
Electrolytic capacitors
Electrolytic cells
Electrolytic cleaning
Electrolytic polishing
Electrolytic reduction
Electrolytic tanks
Electrometallurgy
Secondary batteries

Electrolytes 803, 804, 702
DT: Predates 1975
UF: Electric batteries—Electrolytes*
Electrolytic cells—Electrolytes*
Fuel cells—Electrolytes*
BT: Conductive materials
NT: Polyelectrolytes
Solid electrolytes
RT: Electric conductivity of liquids
Electrochemistry
Electrolysis
Electrolytic cells
Electrolytic tanks
Fused salts
Ionic strength
Ions

*Electrolytes, Solid**
USE: Solid electrolytes

Electrolytic analysis 801.4.1
DT: Predates 1975
BT: Chemical analysis
RT: Electrochemical sensors
Electrolysis
Gravimetric analysis

Electrolytic capacitors 704.1 (801.4.1)
DT: January 1993
UF: Capacitors, Electrolytic*
BT: Capacitors
RT: Electrolysis

Electrolytic cells (801.4.1) (802.1)
DT: Predates 1975
BT: Chemical equipment
RT: Electrolysis
Electrolytes

*Electrolytic cells—Electrolytes**
USE: Electrolytes

Electrolytic cleaning (539) (801.4.1) (802.3)
DT: January 1993
UF: Metal cleaning—Electrolytic*
BT: Chemical cleaning
RT: Electrochemistry
Electrolysis

Electrolytic corrosion
USE: Electrochemical corrosion

Electrolytic cutting
USE: Electrochemical cutting

Electrolytic deposition
USE: Electrodeposition

Electrolytic polishing (604.2) (801.4.1) (802.3)
DT: January 1993
UF: Electrochemical polishing
Electropolishing
Polishing—Electrolytic*
BT: Chemical polishing
RT: Electrochemistry
Electrolysis
Metal finishing

Electrolytic protection
USE: Cathodic protection

Electrolytic reactions
USE: Electrolysis

Electrolytic reduction 533.1, 802.2 (801.4.1)
DT: January 1993
UF: Ore treatment—Electrolytic reduction*
BT: Reduction
RT: Electrolysis
Ore treatment

Electrolytic tanks (801.4.1) (722.5)
DT: January 1993
UF: Computers—Electrolytic tank*
BT: Direct analogs
RT: Electrolysis
Electrolytes

Electromagnetic absorption
USE: Electromagnetic wave absorption

Electromagnetic compatibility 711.1 (706)
DT: Predates 1975
UF: EMC
RT: Electric shielding
Frequency allocation
Radio transmission
Signal interference
Telecommunication systems

Electromagnetic dispersion 711.1
DT: January 1993
UF: Electromagnetic wave dispersion
Electromagnetic waves—Dispersion*
BT: Dispersion (waves)
RT: Attenuation
Color
Refraction

Electromagnetic effects
USE: Magnetoelectric effects

Electromagnetic energy
USE: Electromagnetic waves

Electromagnetic field effects 701.1, 701.2
DT: January 1993
BT: Magnetic field effects
RT: Electric field effects

Electromagnetic field equations
USE: Maxwell equations

Electromagnetic field measurement 942.4
DT: Predates 1975
BT: Measurements
RT: Electric field measurement
Electromagnetic fields
Field plotting
Radio transmitters
Transmitters

Electromagnetic field theory 701
DT: Predates 1975
BT: Physics
Theory
RT: Electromagnetic waves
Electromagnetic fields
Electromagnetic wave propagation
Electromagnetism

Electromagnetic fields 701
DT: January 1981
RT: Electric fields
Electromagnetic field theory
Electromagnetic waves
Electromagnetic field measurement
Electromagnetism
Electron cyclotron resonance
Magnetic fields

Electromagnetic launchers 704.2 (404.1)
DT: June 1990
BT: Electric equipment
NT: Rail guns
RT: Ballistics
Electromagnetic propulsion
Military equipment
Projectiles

*Electromagnetic launchers—Rail guns**
USE: Rail guns

Electromagnetic logging 512.1.2, 701
DT: January 1993
UF: Oil well logging—Electromagnetic*
BT: Well logging
RT: Electric logging
Electromagnetic prospecting
Natural gas well logging
Oil well logging

Electromagnetic propulsion (654.2)
DT: January 1993
UF: Rocket engines—Electromagnetic propulsion*
BT: Electric propulsion
RT: Electromagnetic launchers
Ion propulsion
Rocket engines

Electromagnetic prospecting 481.4, 701 (501.1) (512)
DT: January 1993
UF: Geophysical prospecting—Electromagnetic*
BT: Geophysical prospecting
RT: Electromagnetic logging

Electromagnetic pulse 701
SN: The pulse of electromagnetic radiation generated by a large thermonuclear explosion
DT: January 1986
BT: Electromagnetic waves
RT: Nuclear explosions

Electromagnetic pumps 618.2, 701
DT: January 1993
UF: Pumps, Electromagnetic*
BT: Magnetic devices
Pumps

Electromagnetic radiation
USE: Electromagnetic waves

Electromagnetic shielding (701.1) (711)
DT: January 1983
UF: Radio equipment—Shielding*
Radio shielding
BT: Radiation shielding
NT: Radar shielding
RT: Electric shielding
Electromagnetic waves
Magnetic shielding

Electromagnetic shock tubes
USE: Shock tubes

Electromagnetic wave absorption 711
DT: January 1993
UF: Electromagnetic absorption
Electromagnetic waves—Absorption*
BT: Energy absorption
NT: Light absorption
RT: Electromagnetic wave attenuation
Electromagnetic wave propagation
Electromagnetic waves
Electron transitions
Fluorescence
Laser theory
Magnetic resonance
Phosphorescence
Transparency

Electromagnetic wave attenuation 711
DT: January 1993
UF: Electromagnetic waves—Attenuation*
BT: Attenuation
RT: Electromagnetic wave absorption
Electromagnetic wave scattering
Electromagnetic waves

Electromagnetic wave backscattering 711
DT: January 1993
UF: Electromagnetic waves—Backscattering*
Radio transmission—Backscattering*
BT: Backscattering
Electromagnetic wave scattering
RT: Electromagnetic waves

* Former Ei Vocabulary term

Electromagnetic wave diffraction 711
DT: January 1993
UF: Electromagnetic waves—Diffraction*
BT: Diffraction
NT: X ray diffraction
RT: Diffraction gratings
Electromagnetic wave propagation
Electromagnetic waves

Electromagnetic wave diffusion
USE: Electromagnetic wave scattering

Electromagnetic wave dispersion
USE: Electromagnetic dispersion

Electromagnetic wave emission 711
DT: January 1993
UF: Emission (electromagnetic waves)
NT: Light emission
RT: Electron transitions

Electromagnetic wave interference 711
DT: January 1993
BT: Wave interference
NT: Light interference
RT: Electromagnetic wave propagation
Interferometers
Interferometry
Moire fringes
Signal interference

Electromagnetic wave polarization 711
DT: January 1993
UF: Electromagnetic waves—Polarization*
BT: Polarization
NT: Light polarization
RT: Electromagnetic waves
Magnetooptical effects

Electromagnetic wave propagation 711
SN: In gases. For propagation in liquids and solids use ELECTROMAGNETIC WAVE TRANSMISSION
DT: January 1993
UF: Electromagnetic waves—Propagation*
BT: Wave propagation
NT: Electromagnetic wave propagation in plasma
Guided electromagnetic wave propagation
Light propagation
RT: Electromagnetic field theory
Electromagnetic wave absorption
Electromagnetic wave diffraction
Electromagnetic wave interference
Electromagnetic wave reflection
Electromagnetic wave refraction
Electromagnetic wave scattering
Light transmission
Radio transmission
Television transmission

Electromagnetic wave propagation in plasma 711.1 (932.3)
DT: January 1993
UF: Electromagnetic waves—Propagation in plasma*
Magnetic merging
BT: Electromagnetic wave propagation
NT: Ionospheric electromagnetic wave propagation
RT: Plasmas

Electromagnetic wave reflection 711
DT: January 1993
UF: Electromagnetic waves—Reflection*
BT: Reflection
NT: Light reflection
Radar reflection
RT: Electromagnetic wave propagation
Electromagnetic waves
Electromagnetic wave transmission

Electromagnetic wave refraction 711
DT: January 1993
UF: Electromagnetic waves—Refraction*
BT: Refraction
NT: Light refraction
RT: Electromagnetic wave propagation
Electromagnetic waves
Refractive index

Electromagnetic wave scattering 711
DT: January 1993
UF: Electromagnetic wave diffusion
Electromagnetic waves—Scattering*
BT: Scattering
NT: Electromagnetic wave backscattering
Light scattering
Rayleigh scattering
RT: Electromagnetic wave attenuation
Electromagnetic wave propagation
Electromagnetic waves
Radar cross section

Electromagnetic wave transmission 711
SN: In liquids and solids. For transmission in gases use ELECTROMAGNETIC WAVE PROPAGATION
DT: January 1993
BT: Wave transmission
NT: Light transmission
Television transmission
RT: Electromagnetic wave reflection
Radio transmission

Electromagnetic waves 711
DT: Predates 1975
UF: Electric waves
Electromagnetic energy
Electromagnetic radiation
EM waves
Hertzian waves
Radio waves
Radiowaves
BT: Radiation
Waves
NT: Antenna radiation
Cyclotron radiation
Electromagnetic pulse
Gamma rays
Heat radiation
Helicons
Light
Microwaves
Millimeter waves

* Former Ei Vocabulary term

Electromagnetic waves *(continued)*
Synchrotron radiation
X rays
RT: Atomic spectroscopy
Doppler effect
Electric fields
Electric shielding
Electric waveforms
Electromagnetic field theory
Electromagnetic fields
Electromagnetic wave reflection
Electromagnetic wave refraction
Electromagnetic wave absorption
Electromagnetic wave attenuation
Electromagnetic wave diffraction
Electromagnetic wave polarization
Electromagnetic wave scattering
Electromagnetic wave backscattering
Electromagnetic shielding
Electromagnetism
Magnetic shielding
Maxwell equations
Modulation
Molecular spectroscopy
Radio
Telecommunication
Wave filters
Wave plasma interactions

*Electromagnetic waves—Absorption**
USE: Electromagnetic wave absorption

*Electromagnetic waves—Attenuation**
USE: Electromagnetic wave attenuation

*Electromagnetic waves—Backscattering**
USE: Electromagnetic wave backscattering

*Electromagnetic waves—Diffraction**
USE: Electromagnetic wave diffraction

*Electromagnetic waves—Dispersion**
USE: Electromagnetic dispersion

*Electromagnetic waves—Polarization**
USE: Electromagnetic wave polarization

*Electromagnetic waves—Propagation in guides**
USE: Guided electromagnetic wave propagation

*Electromagnetic waves—Propagation in ionosphere**
USE: Ionospheric electromagnetic wave propagation

*Electromagnetic waves—Propagation in plasma**
USE: Electromagnetic wave propagation in plasma

*Electromagnetic waves—Propagation in troposphere**
USE: Troposphere

*Electromagnetic waves—Propagation**
USE: Electromagnetic wave propagation

*Electromagnetic waves—Radiation hazards**
USE: Radiation hazards

*Electromagnetic waves—Reflection**
USE: Electromagnetic wave reflection

*Electromagnetic waves—Refraction**
USE: Electromagnetic wave refraction

*Electromagnetic waves—Scattering**
USE: Electromagnetic wave scattering

Electromagnetism 701
DT: Predates 1975
BT: Magnetism
RT: Eddy currents
Electricity
Electrodynamics
Electromagnetic waves
Electromagnetic field theory
Electromagnetic fields
Electromagnets
Electrostatics
Magnetic leakage
Magnetic materials
Magnetoelectric effects
Magnetostatics
Maxwell equations
Photoelectromagnetic effects

Electromagnets 704.1
DT: Predates 1975
BT: Electric equipment
Magnets
NT: Lifting magnets
Superconducting magnets
RT: Actuators
Electric coils
Electromagnetism
Magnetic cores
Solenoids

Electromechanical devices (601) (732.1) (704.1)
DT: Predates 1975
BT: Electric equipment
NT: Electromechanical filters
Microelectromechanical devices

*Electromechanical devices—Microelectromechanical**
USE: Microelectromechanical devices

Electromechanical filters 703.2
DT: January 1993
UF: Electric filters, Electromechanical*
BT: Electric filters
Electromechanical devices
NT: Crystal filters

Electrometallurgy 531.1 (801.4.1)
DT: Predates 1975
UF: Metallurgy—Electrolytic*
BT: Extractive metallurgy
RT: Electrolysis

Electrometers 942.1
DT: Predates 1975
UF: Electrometry
BT: Electric measuring instruments
RT: Electric charge measurement
Voltage measurement

Electrometry
USE: Electrometers

* Former Ei Vocabulary term

Electromigration 701.1 (931.2)
DT: January 1995
BT: Electric field effects
RT: Diffusion
Integrated circuits
Interfaces (materials)
Surface properties

Electromyography (461.1) (461.6)
DT: January 1993
UF: Biomedical engineering—Electromyography*
BT: Diagnosis
RT: Bioelectric potentials
Biomedical engineering
Muscle

Electron absorption (931.3) (932)
DT: January 1993
UF: Electron beam absorption
Electrons—Absorption*
BT: Energy absorption
RT: Electrons

Electron accelerators
USE: Particle accelerators

Electron beam absorption
USE: Electron absorption

Electron beam brazing 538.1.1 (932)
DT: January 1993
UF: Brazing—Electron beam*
BT: Brazing
RT: Electron beams

Electron beam control
USE: Electron lenses

Electron beam cutting 604.1 (932)
DT: January 1993
UF: Metal cutting—Electron beam*
BT: Cutting
RT: Electron beams

Electron beam furnaces 642.2 (932)
DT: January 1993
UF: Furnaces—Electron beam*
BT: Melting furnaces
Vacuum furnaces
RT: Electron beams

Electron beam lithography (745.1) (932) (714.2)
DT: January 1993
UF: Electron lithography
Lithography—Electron beam*
BT: Lithography
RT: Electron beams

Electron beam melting (932) (531.1)
DT: January 1993
UF: Metal melting—Electron beam*
BT: Metal melting
RT: Electron beams

Electron beam pumping 744.1 (932)
DT: January 1993
BT: Pumping (laser)
RT: Electron beams
Lasers

Electron beam welding 538.2.1 (932)
DT: January 1993
UF: Welding—Electron beam*
BT: Welding
RT: Electron beams
Vacuum welding

Electron beams (931.3) (932)
DT: Predates 1975
UF: Electron streams
BT: Particle beams
RT: Betatrons
Cathode ray tubes
Cathode rays
Electrodynamics
Electron beam brazing
Electron beam cutting
Electron beam furnaces
Electron beam lithography
Electron beam melting
Electron beam pumping
Electron beam welding
Electron diffraction
Electron emission
Electron guns
Electron lenses
Electron optics
Electron tubes
Magnetic lenses

Electron compounds
USE: Intermetallics

Electron cyclotron resonance 931.3 (701)
DT: January 1995
BT: Cyclotron resonance
Electron resonance
RT: Electromagnetic fields
Electrons

Electron density
USE: Carrier concentration

Electron density measurement (932)
DT: January 1993
UF: Density measurement (electron)
Electrons—Density measurement*
BT: Measurements

Electron density of states
USE: Electronic density of states

Electron device manufacture (715) (913.4)
SN: For manufacturing of assembled equipment, use ELECTRONIC EQUIPMENT MANUFACTURE
DT: Predates 1975
BT: Electronic equipment manufacture
NT: Electron tube manufacture
Integrated circuit manufacture
Semiconductor device manufacture
RT: Electron devices
Electron device testing
Electronics industry

Electron device noise
USE: Spurious signal noise

* Former Ei Vocabulary term

Electron device testing (715)
DT: Predates 1975
BT: Electronic equipment testing
NT: Electron tube testing
Integrated circuit testing
Printed circuit testing
Semiconductor device testing
RT: Electron devices
Electron device manufacture

Electron devices (715)
DT: January 1993
BT: Electronic equipment
NT: Electron tubes
Solid state devices
RT: Electron device manufacture
Electron device testing

Electron diffraction (931.3) (932.2)
DT: January 1993
UF: Electrons—Diffraction*
BT: Diffraction
NT: High energy electron diffraction
Low energy electron diffraction
RT: Electron beams
Electron diffraction apparatus
Electron microscopes
Electron microscopy
Electron optics

Electron diffraction apparatus (715) (932.2)
DT: Predates 1975
BT: Electronic equipment
RT: Electron diffraction

Electron emission (931.3) (932.2) (714.1)
DT: January 1993
UF: Electron tubes—Electron emission*
Electrons—Emission*
Emission (electron)
RT: Cathodes
Electrodes
Electron beams
Electrons
Surface phenomena

Electron energy analyzers (932)
DT: January 1993
UF: Electrons—Energy analyzers*
BT: Instruments
RT: Electrons

Electron energy levels (931.3) (932) (933)
DT: January 1993
UF: Crystals—Electron states*
Electron energy states
Electron states
Energy levels
Energy states
NT: Band structure
Electronic density of states
Excitons
Fermi level
Fermi surface
RT: Binding energy
Crystals
Deep level transient spectroscopy
Electron energy loss spectroscopy
Electronic structure
Electrons
Low energy electron diffraction
Metals
Phonons
Semiconductor materials

Electron energy loss spectroscopy (801) (931.3) (932.2)
DT: January 1995
UF: EELS
BT: Electron spectroscopy
RT: Electron energy levels

Electron energy states
USE: Electron energy levels

Electron guns 714.1
DT: Predates 1975
BT: Electron tube components
Electrostatic devices
RT: Cathode ray tubes
Electron beams

Electron lenses 714.1
DT: Predates 1975
UF: Electron beam control
Quadrupole lenses
BT: Lenses
RT: Electron beams
Electrostatic lenses
Focusing
Magnetic lenses
Optics

Electron lithography
USE: Electron beam lithography

Electron microscopes (741.3) (931) (932)
DT: January 1993
UF: Microscopes, Electron*
BT: Microscopes
RT: Electron diffraction
Electron microscopy
Electron optics
Scanning electron microscopy
Transmission electron microscopy

Electron microscopy (931.2) (933)
DT: January 1995
UF: High resolution electron microscopy
BT: Microscopic examination
NT: Scanning electron microscopy
Transmission electron microscopy
RT: Atomic force microscopy
Electron diffraction
Electron microscopes
Electron optics
Metallography
Scanning tunneling microscopy
Specimen preparation

Electron multipliers 714.1
DT: January 1993
UF: Electron tubes, Electron multiplier*

* Former Ei Vocabulary term

Electron multipliers *(continued)*
Multipliers (electron)
BT: Electron tubes
NT: Photomultipliers

Electron optics (931.3) (701.1)
DT: Predates 1975
BT: Particle optics
RT: Beta ray spectrometers
Betatrons
Cathode ray tubes
Cathode rays
Electrodynamics
Electron beams
Electron diffraction
Electron microscopes
Electron microscopy
Electrooptical effects
Electrostatic lenses
Focusing
Magnetic lenses

Electron paramagnetic resonance
USE: Paramagnetic resonance

Electron reflection (931.3) (932.2)
DT: January 1993
UF: Electrons—Reflection*
BT: Reflection
RT: Electrons

Electron resonance (931.3) (932.2)
DT: January 1993
UF: Electrons—Resonance*
BT: Resonance
NT: Electron cyclotron resonance
RT: Electrons

Electron ring accelerators 932.1.1
DT: January 1993
UF: Accelerators, Electron ring*
BT: Particle accelerators

Electron scattering (931.3)
DT: January 1993
UF: Electrons—Scattering*
BT: Scattering
RT: Electrons

Electron sources (932.1)
DT: January 1993
UF: Electrons—Sources*
BT: Elementary particle sources
RT: Ion sources

Electron spectroscopy (801) (931.3) (932)
DT: January 1993
UF: Spectroscopy, Electron*
BT: Spectroscopy
NT: Auger electron spectroscopy
Electron energy loss spectroscopy
Electron spin resonance spectroscopy
Photoelectron spectroscopy

Electron spin resonance
USE: Paramagnetic resonance

Electron spin resonance spectroscopy 801 (931.3) (932.2)
DT: January 1995
BT: Electron spectroscopy
RT: Microwave spectroscopy
Paramagnetic resonance

Electron states
USE: Electron energy levels

Electron streams
USE: Electron beams

Electron transitions (711.1) (931.3)
DT: January 1993
UF: Electrons—Transitions*
Transitions (electron)
RT: Electromagnetic wave absorption
Electromagnetic wave emission
Electrons
Light absorption
Light emission

Electron transport properties (931.3) (932)
DT: January 1993
UF: Electrons—Transport properties*
Transport properties (electron)
RT: Electrons

Electron tube components 714.1
DT: Predates 1975
BT: Components
Electron tubes
NT: Deflection yokes
Electron guns
Getters
RT: Anodes
Cathodes
Electrodes
Photocathodes

Electron tube diodes 714.1
DT: January 1993
UF: Electron tubes, Diode*
BT: Diodes
Electron tubes

Electron tube manufacture 714.1 (913.4)
DT: January 1993
BT: Electron device manufacture
RT: Electron tubes

Electron tube rectifiers 714.1
DT: January 1993
UF: Electric rectifiers, Electron tube*
Rectifier tubes
Rectifier valves
BT: Electric rectifiers
RT: Diodes

Electron tube testing 714.1
DT: Predates 1975
BT: Electron device testing
RT: Electron tubes

Electron tubes 714.1
DT: Predates 1975
UF: Electron valves
Tubes (electron)

* Former Ei Vocabulary term

Electron tubes *(continued)*
Vacuum tubes
Valves (electron)
BT: Electron devices
NT: Cold cathode tubes
Electron multipliers
Electron tube components
Electron tube diodes
Gas discharge tubes
Microwave tubes
Nuvistors
Pentodes
Storage tubes
Tetrodes
Thermionic tubes
Triodes
RT: Cyclotron resonance
Electric space charge
Electrodes
Electrodynamics
Electron beams
Electron tube manufacture
Electron tube testing
Parametric devices
Spurious signal noise
Transconductance
Vacuum applications
Vacuum technology

*Electron tubes, Backward wave**
USE: Backward wave tubes

*Electron tubes, Cathode ray**
USE: Cathode ray tubes

*Electron tubes, Cold cathode**
USE: Cold cathode tubes

*Electron tubes, Counting**
USE: Counting tubes

*Electron tubes, Diode**
USE: Electron tube diodes

*Electron tubes, Electron multiplier**
USE: Electron multipliers

*Electron tubes, Electron wave**
USE: Electron wave tubes

*Electron tubes, Gas discharge**
USE: Gas discharge tubes

*Electron tubes, Image converter**
USE: Image converters

*Electron tubes, Image intensifier**
USE: Image intensifiers (electron tube)

*Electron tubes, Image storage**
USE: Image storage tubes

*Electron tubes, Klystron**
USE: Klystrons

*Electron tubes, Magnetron**
USE: Magnetrons

*Electron tubes, Microwave**
USE: Microwave tubes

*Electron tubes, Nuvistor**
USE: Nuvistors

*Electron tubes, Pentode**
USE: Pentodes

*Electron tubes, Photomultiplier**
USE: Photomultipliers

*Electron tubes, Phototube**
USE: Phototubes

*Electron tubes, Storage**
USE: Storage tubes

*Electron tubes, Television camera**
USE: Television camera tubes

*Electron tubes, Television picture**
USE: Television picture tubes

*Electron tubes, Tetrode**
USE: Tetrodes

*Electron tubes, Thermionic**
USE: Thermionic tubes

*Electron tubes, Thyratron**
USE: Thyratrons

*Electron tubes, Traveling wave**
USE: Traveling wave tubes

*Electron tubes, Triode**
USE: Triodes

*Electron tubes, Velocity modulation**
USE: Velocity modulation tubes

*Electron tubes—Cyclotron resonance**
USE: Cyclotron resonance

*Electron tubes—Distortion**
USE: Signal distortion

*Electron tubes—Electric power supplies**
USE: Electric power supplies to apparatus

*Electron tubes—Electron emission**
USE: Electron emission

*Electron tubes—Noise**
USE: Spurious signal noise

*Electron tubes—Space charge**
USE: Electric space charge

Electron tunneling (712.1) (714.2) (931.3) (932)
DT: January 1993
UF: Electrons—Tunneling*
Semiconductor devices—Tunneling*
Tunnel effect
Tunneling (electron)
Tunnelling (electron)
RT: Electrons
Scanning tunneling microscopy
Semiconductor materials
Tunnel diodes
Tunnel junctions
Zener effect

Electron valves
USE: Electron tubes

* Former Ei Vocabulary term

Electron velocity analyzers (942.1)
DT: January 1993
UF: Electrons—Velocity analyzers*
BT: Particle separators
RT: Electrons
Velocity
Velocity measurement

Electron wave tubes 714.1
DT: January 1993
UF: Electron tubes, Electron wave*
BT: Microwave tubes
Thermionic tubes
NT: Klystrons
Magnetrons
Traveling wave tubes
Velocity modulation tubes

Electronic circuit schematics
USE: Schematic diagrams

Electronic circuit tracking (713)
SN: The condition in which the tuned circuits in a receiver follow the frequency indicated by the dial
DT: Predates 1975
UF: Circuit tracking
Tracking (circuit)
RT: Networks (circuits)
Tuning

*Electronic circuits**
USE: Networks (circuits)

*Electronic circuits, Bridge**
USE: Bridge circuits

*Electronic circuits, Buffer**
USE: Buffer circuits

*Electronic circuits, Coincidence**
USE: Coincidence circuits

*Electronic circuits, Compandor**
USE: Compandor circuits

*Electronic circuits, Comparator**
USE: Comparator circuits

*Electronic circuits, Counting**
USE: Counting circuits

*Electronic circuits, Coupler**
USE: Coupled circuits

*Electronic circuits, Delay type**
USE: Delay circuits

*Electronic circuits, Detector**
USE: Detector circuits

*Electronic circuits, Differentiating**
USE: Differentiating circuits

*Electronic circuits, Digital**
USE: Digital circuits

*Electronic circuits, Discriminator**
USE: Discriminators

*Electronic circuits, Flip flop**
USE: Flip flop circuits

*Electronic circuits, Frequency converter**
USE: Frequency converter circuits

*Electronic circuits, Frequency dividing**
USE: Frequency dividing circuits

*Electronic circuits, Frequency multiplying**
USE: Frequency multiplying circuits

*Electronic circuits, Gyrator**
USE: Gyrators

*Electronic circuits, Integrating**
USE: Integrating circuits

*Electronic circuits, Limiter**
USE: Limiters

*Electronic circuits, Mixer**
USE: Mixer circuits

*Electronic circuits, Multistable**
USE: Multistable circuits

*Electronic circuits, Multivibrator**
USE: Multivibrators

*Electronic circuits, Phase changer**
USE: Phase changing circuits

*Electronic circuits, Phase comparator**
USE: Phase comparators

*Electronic circuits, Power supply**
USE: Power supply circuits

*Electronic circuits, Pulse analyzing**
USE: Pulse analyzing circuits

*Electronic circuits, Pulse shaping**
USE: Pulse shaping circuits

*Electronic circuits, Pulse signal**
USE: Pulse circuits

*Electronic circuits, Rectifying**
USE: Rectifying circuits

*Electronic circuits, Sweep**
USE: Sweep circuits

*Electronic circuits, Switching**
USE: Switching circuits

*Electronic circuits, Timing**
USE: Timing circuits

*Electronic circuits, Trigger**
USE: Trigger circuits

*Electronic circuits, Voltage stabilizing**
USE: Voltage stabilizing circuits

*Electronic circuits—Coupling**
USE: Coupled circuits

*Electronic circuits—Oscillations**
USE: Circuit oscillations

*Electronic circuits—Resonance**
USE: Circuit resonance

* Former Ei Vocabulary term

*Electronic circuits—Tuning**
USE: Tuning

Electronic counter-countermeasures
USE: Military electronic countermeasures

Electronic countermeasures (military)
USE: Military electronic countermeasures

Electronic crime countermeasures (715.1)
DT: October 1975
UF: Countermeasures (crime)
Crime countermeasures (electronic)
BT: Electronic equipment

Electronic density of states 931.3 (701.1)
DT: January 1995
UF: Electron density of states
BT: Electron energy levels
RT: Band structure
Fermi level
Fermi surface

Electronic equipment (715)
DT: October 1975
UF: Electronics
Surveying instruments—Electronics*
BT: Equipment
NT: Amplifiers (electronic)
Audio equipment
Automobile electronic equipment
Cascade connections
Compact disk players
Computers
Demodulators
Digital devices
Diodes
Electrodes
Electron devices
Electron diffraction apparatus
Electronic crime countermeasures
Electronic medical equipment
Electronic musical instruments
Electronic scales
Electronic ship equipment
Electronic timing devices
Electronic voltmeters
Electrooptical devices
Light amplifiers
Microwave devices
Modulators
Optoelectronic devices
Oscillators (electronic)
Parametric devices
Signal generators
Telecommunication equipment
Wave filters
Waveguides
RT: Atomic clocks
Consumer electronics
Electric equipment
Electronic equipment testing
Electronic equipment manufacture
Electronics engineering
Electronics industry
Industrial electronics
Printed circuits

Electronic equipment manufacture (715) (913.4)
DT: Predates 1975
BT: Manufacture
NT: Electron device manufacture
Electronics packaging
Printed circuit manufacture
Surface mount technology
RT: Assembly
Electronic equipment
Electronics industry

Electronic equipment testing (715)
DT: Predates 1975
BT: Equipment testing
NT: Computer testing
Electron device testing
RT: Electronic equipment

Electronic guidance systems (404.1) (654.1) (715.1)
DT: January 1993
UF: Guidance (electronic)
Missile guidance
Rocket guidance
Rockets and missiles—Guidance*
Spacecraft guidance
BT: Control systems
RT: Missiles
Rockets
Spacecraft

Electronic mail (723.5)
DT: January 1981
UF: E-mail
BT: Telecommunication services
RT: Administrative data processing
Data communication systems
Facsimile
Information dissemination
Office automation
Teleconferencing

Electronic medical equipment 462.1, 715.2
DT: January 1993
UF: Biomedical engineering—Electronics*
Medical electronics
BT: Biomedical equipment
Electronic equipment
RT: Biomedical engineering
Microelectrodes

Electronic musical instruments 715.1, 752.1, 752.4
DT: January 1993
UF: Musical instruments, Electronic*
BT: Electronic equipment
Frequency synthesizers
Musical instruments
RT: Computer music

Electronic office
USE: Office automation

Electronic power supplies
USE: Electric power supplies to apparatus

* Former Ei Vocabulary term

Electronic properties (701.1) (931.2)
DT: January 1977
BT: Physical properties

Electronic publishing 723.5, 903.2
SN: Publication of information in electronic form, e.g., as a database, whether or not the same information is also published in print
DT: January 1993
UF: Information dissemination—Electronic publishing*
BT: Publishing
RT: Desktop publishing

Electronic scales (943.3) (715)
DT: January 1993
UF: Scales and weighing—Electronic*
BT: Electronic equipment
Scales (weighing instruments)

Electronic ship equipment 671.2, 715.2
DT: January 1993
UF: Ship equipment—Electronic*
BT: Electronic equipment
Ship equipment
RT: Ship instruments

Electronic structure (931.1) (931.3) (933.1.1)
DT: January 1995
BT: Structure (composition)
RT: Crystal atomic structure
Electron energy levels
Electrons
Molecular structure

Electronic telephone exchanges
USE: Automatic telephone exchanges

Electronic timing devices 942.1
DT: January 1993
UF: Timing devices—Electronic*
BT: Electronic equipment
Timing devices
RT: Time measurement

Electronic voltmeters (715.1)
DT: January 1993
UF: Voltmeters—Electronic*
BT: Electronic equipment
Voltmeters

Electronic warfare 404.1 (715) (716)
DT: January 1993
UF: EW
Military engineering—Electronic warfare*
BT: Military operations
NT: Jamming
Military electronic countermeasures
RT: Military engineering
Radar

Electronics
USE: Electronic equipment

Electronics engineering
DT: January 1987
BT: Engineering
RT: Avionics
Electronic equipment
Telecommunication

Electronics industry (712) (713) (714) (715)
SN: Excludes ELECTRIC INDUSTRY, but includes all electron-device and electronic-equipment suppliers
DT: January 1993
UF: Radio industry
BT: Industry
NT: Consumer electronics
Industrial electronics
Microelectronics
RT: Electric industry
Electron device manufacture
Electronic equipment manufacture
Electronic equipment
Telecommunication

Electronics packaging (714) (715) (716)
SN: Grouping or combining electronic parts into a unit. Not PACKAGING.
DT: Predates 1975
BT: Electronic equipment manufacture
RT: Encapsulation
Integrated circuits
Multichip modules
Printed circuits
Printed circuit design
Printed circuit manufacture
Surface mount technology

*Electronics packaging—Surface mount technology**
USE: Surface mount technology

Electrons (701.1) (931.3)
DT: Predates 1975
BT: Charged particles
RT: Band structure
Cathode rays
Electron absorption
Electron cyclotron resonance
Electron emission
Electron energy analyzers
Electron energy levels
Electron reflection
Electron resonance
Electron scattering
Electron transitions
Electron transport properties
Electron tunneling
Electron velocity analyzers
Electronic structure
Quantum efficiency

*Electrons—Absorption**
USE: Electron absorption

*Electrons—Density measurement**
USE: Electron density measurement

*Electrons—Diffraction**
USE: Electron diffraction

*Electrons—Emission**
USE: Electron emission

* Former Ei Vocabulary term

*Electrons—Energy analyzers**
USE: Electron energy analyzers

*Electrons—Reflection**
USE: Electron reflection

*Electrons—Resonance**
USE: Electron resonance

*Electrons—Scattering**
USE: Electron scattering

*Electrons—Sources**
USE: Electron sources

*Electrons—Transitions**
USE: Electron transitions

*Electrons—Transport properties**
USE: Electron transport properties

*Electrons—Tunneling**
USE: Electron tunneling

*Electrons—Velocity analyzers**
USE: Electron velocity analyzers

Electrooptical devices 714, 741.3
DT: January 1977
UF: Electro-optical devices
BT: Electronic equipment
Optical devices
RT: Electroluminescence
Light modulators
Liquid crystal displays
Optical bistability
Q switched lasers
Q switching

Electrooptical effects 701.1, 741.1
DT: Predates 1975
UF: Electro-optical effects
Electrooptics
BT: Electric field effects
NT: Electrochromism
Kerr electrooptical effect
RT: Electroluminescence
Electron optics
Light modulation
Light modulators
Magnetooptical effects
Nonlinear optics
Optical properties
Polarization

Electrooptical Kerr effect
USE: Kerr electrooptical effect

Electrooptical materials (708.2) (741.3)
DT: January 1981
UF: Electro-optical materials
BT: Materials

Electrooptics
USE: Electrooptical effects

Electrophoresis 701.1, 801.3, 801.4.1
DT: Predates 1975
UF: Cataphoresis
RT: Colloids
Electric field effects
Electrochemistry
Electrophoretic coatings
Electroplating

Electrophoretic coatings 813.2, 801.4.1
DT: January 1993
UF: Protective coatings—Electrophoretic*
BT: Coatings
RT: Electrophoresis

Electrophotoluminescence
USE: Electroluminescence

Electrophysiology 461.1 (701.1)
DT: January 1993
UF: Biomedical engineering—Electrophysiology*
BT: Physiology
RT: Bioelectric phenomena
Biomedical engineering

Electroplated products 539.3.1
DT: Predates 1975
RT: Electroplating

Electroplating 539.3.1
SN: Also use type of material used for plating and/or material or thing upon which electroplating is done.
DT: Predates 1975
BT: Electrodeposition
Plating
RT: Electroforming
Electrophoresis
Electroplated products
Electroplating solutions
Electroplating shops

Electroplating shops 539.3.1, 402.1
DT: Predates 1975
BT: Industrial plants
RT: Electroplating

Electroplating solutions
539.3.1 (801) (802.3) (803) (804)
DT: January 1993
UF: Electroplating—Solutions*
BT: Solutions
RT: Electroplating

*Electroplating—Power supply**
USE: Electric power supplies to apparatus

*Electroplating—Solutions**
USE: Electroplating solutions

Electropolishing
USE: Electrolytic polishing

Electropolymerization 815.2 (801.4.1)
DT: January 1995
UF: Electrochemical polymerization
BT: Polymerization
RT: Catalysis
Electrochemical electrodes

Electroslag process 545.3
DT: January 1993
UF: Electric slag process
Steelmaking—Electroslag process*

* Former Ei Vocabulary term

Electroslag process *(continued)*
BT: Steelmaking
RT: Electroslag remelting

Electroslag remelting 531
DT: January 1993
UF: Metal melting—Electroslag remelting*
BT: Metal melting
RT: Electroslag process

Electroslag welding 538.2.1
DT: January 1993
UF: Welding—Electroslag*
BT: Electric welding

Electrostatic accelerators 932.1.1
DT: January 1993
UF: Accelerators, Electrostatic*
BT: Particle accelerators
NT: Van de Graaff accelerators

Electrostatic coatings (701.1) (813.2) (539.2.2)
DT: January 1993
UF: Protective coatings—Electrostatic*
BT: Coatings
RT: Paint spraying

Electrostatic devices (701.1) (704) (714)
DT: Predates 1975
BT: Electric equipment
NT: Capacitors
Electron guns
Electrostatic generators
Electrostatic lenses
Electrostatic loudspeakers
Electrostatic separators
RT: Electric fields
Electrostatics

Electrostatic fields
USE: Electric fields

Electrostatic generators (701.1) (704)
DT: Predates 1975
BT: Electric generators
Electrostatic devices
NT: Van de Graaff generators
RT: Electrostatics

*Electrostatic generators, Van de Graaff**
USE: Van de Graaff generators

Electrostatic lenses (701.1) (932.1)
DT: Predates 1975
BT: Electrostatic devices
Lenses
RT: Electron lenses
Electron optics
Particle beams
Particle optics

Electrostatic loudspeakers 752.1 (701.1)
DT: January 1993
UF: Loudspeakers—Electrostatic actuation*
BT: Electrostatic devices
Loudspeakers

Electrostatic ore treatment 533.1, 701.1
DT: January 1993
UF: Ore treatment—Electrostatic*
BT: Ore treatment
RT: Electrostatics

Electrostatic printing 745.1, 701.1
DT: January 1993
UF: Printing—Electrostatic*
BT: Printing
RT: Electrostatics

Electrostatic reproduction 745.2, 701.1
DT: January 1993
UF: Photographic reproduction—Electrostatic*
BT: Photocopying
NT: Xerography
RT: Electrostatics

Electrostatic separators 802.1 (701.1)
DT: January 1993
UF: Separators—Electrostatic*
BT: Electrostatic devices
Separators
RT: Smoke abatement

Electrostatics 701.1
DT: Predates 1975
BT: Physics
RT: Atmospheric electricity
Capacitance
Electrets
Electric charge
Electric fields
Electric space charge
Electromagnetism
Electrostatic devices
Electrostatic generators
Electrostatic ore treatment
Electrostatic printing
Electrostatic reproduction
Ionic strength
Static electricity
Triboelectricity

*Electrostatics—Electric charge**
USE: Electric charge

Electrostriction 701.1
DT: January 1981
BT: Deformation
Electric field effects
RT: Piezoelectric materials

Electrosurgery 461.6, 701.1
DT: January 1993
UF: Biomedical engineering—Electrosurgery*
BT: Surgery
RT: Biomedical engineering

Electrotherapeutics 461.6, 701.1 (462.1)
DT: Predates 1975
BT: Patient treatment
NT: Electroacupuncture
RT: Biomedical engineering

*Electrotherapeutics—Electroacupuncture**
USE: Electroacupuncture

Elementary particle sources (931.3) (932.1.1)
DT: January 1993
UF: Particle sources (elementary particles)
Sources (particle)
NT: Electron sources
Ion sources
Neutron sources
RT: Atomic physics
Elementary particles

Elementary particles 931.3
DT: January 1993
UF: Particles (elementary)
NT: Charged particles
Cosmic rays
Neutrons
Particle beams
Photons
RT: Atomic physics
Elementary particle sources
Green's function
High energy physics
Particle accelerators
Particle optics
Particle separators
Particle spectrometers
Phonons
Quantum theory

Elevator cables 692.2
DT: January 1993
UF: Elevators—Cables*
BT: Cables
Elevators

Elevator codes 692.2, 902.2
DT: January 1993
UF: Elevators—Codes*
BT: Codes (standards)
RT: Building codes
Elevators
Laws and legislation

Elevators 692.2
SN: For vertical transport of materials or people. Excludes GRAIN ELEVATORS
DT: Predates 1975
UF: Ships—Elevators*
BT: Machinery
NT: Elevator cables
RT: Construction equipment
Elevator codes
Escalators
Materials handling equipment
Winches

*Elevators—Cables**
USE: Elevator cables

*Elevators—Codes**
USE: Elevator codes

Ellipsometry 941.4 (741.1)
DT: January 1993
UF: Optical ellipsometry
BT: Optical variables measurement
RT: Polarimeters

EM waves
USE: Electromagnetic waves

Embankment dams 441.1
DT: January 1993
UF: Dams, Embankment*
Earth dams
Rock fill dams
BT: Dams
Embankments

Embankments 405, 483 (441)
DT: Predates 1975
UF: Railroad plant and structures—Embankments*
Roads and streets—Embankments*
BT: Structures (built objects)
NT: Embankment dams
Levees
RT: Geotextiles
Hydraulic structures
Railroad plant and structures
Retaining walls
Revetments
Road construction
Roads and streets
Roadsides

Embrittlement (421)
DT: January 1993
BT: Deterioration
NT: Hydrogen embrittlement
RT: Brittleness
Ductility
High temperature testing

EMC
USE: Electromagnetic compatibility

Emergency exits (aircraft)
USE: Aircraft emergency exits

Emergency lighting 707.1, 914.1
DT: January 1993
UF: Electric lighting—Emergency lighting*
BT: Electric lighting
RT: Electric lamps

Emergency rooms 462.2 (402.2)
DT: January 1993
UF: Hospitals—Emergency rooms*
BT: Hospitals

Emergency runways 431.4, 914.1
DT: January 1993
UF: Airport runways—Emergency*
BT: Airport runways
RT: Landing mats

Emergency traffic control 432.4, 914.1 (731.2)
DT: January 1993
UF: Street traffic control—Emergency measures*
BT: Street traffic control

Emergency vehicles 662, 914.1 (462)
DT: January 1993
BT: Vehicles
NT: Ambulances

* Former Ei Vocabulary term

Emission (electromagnetic waves)
USE: Electromagnetic wave emission

Emission (electron)
USE: Electron emission

Emission (light)
USE: Light emission

Emission (neutron)
USE: Neutron emission

Emission spectroscopy (741.1) (801) (931.3)
DT: January 1993
UF: Spectroscopy, Emission*
BT: Spectroscopy

Emissions (acoustic)
USE: Acoustic emissions

Emissions (industrial)
USE: Industrial emissions

Emissions (particulate)
USE: Particulate emissions

Emitter coupled logic circuits 721.2 (714.2)
DT: January 1993
UF: CML
ECL
Logic circuits, Emitter coupled*
Logic devices—Current mode*
BT: Logic circuits
RT: Digital integrated circuits
Switching circuits

Empennages 652.1
DT: January 1993
UF: Aircraft—Empennage*
Tail assemblies (aircraft)
BT: Aircraft parts and equipment

Employees
USE: Personnel

Employment (912) (913)
DT: Predates 1975
RT: Economic and social effects
Economics
Industrial management
Management
Personnel
Technological forecasting

Emulsification 802.3
DT: January 1993
BT: Chemical operations
RT: Demulsification
Emulsions
Flocculation
Mixing

Emulsion polymerization 815.2 (801.3)
DT: January 1995
BT: Polymerization
RT: Emulsions

Emulsions 804 (801.3)
DT: Predates 1975
BT: Dispersions
NT: Microemulsions
Photographic emulsions
RT: Coagulation
Emulsification
Emulsion polymerization

*Enamel**
USE: Enamels

*Enamel—Color**
USE: Color

Enameling 813.1 (812) (817.2)
SN: Protective coatings
DT: Predates 1975
UF: Wire—Enameling*
BT: Coating techniques
RT: Painting

Enamels 813.2 (812) (817.2)
SN: Vitreous, baked-on coatings; for glossy paints use PAINT
DT: January 1993
UF: Enamel*
BT: Coatings
RT: Ceramic products
Paint
Pigments
Varnish

Encapsulation (714.2) (813.2) (817.2) (461.6)
DT: Predates 1975
UF: Wood products—Plastics encasing*
BT: Coating techniques
NT: Radioactive waste encapsulation
RT: Electronics packaging
Sealing (closing)

Enclosures
SN: Means for preventing access to devices or machines
DT: Predates 1975
BT: Equipment
NT: Fences
RT: Accident prevention
Containers
Security systems
Walls (structural partitions)

Encoding (images)
USE: Image coding

Encoding (signal)
USE: Signal encoding

Encoding (symbols) 723.2
DT: January 1993
UF: Codes, Symbolic—Encoding*
Coding (symbols)
RT: Binary codes
Block codes
Codes (symbols)
Convolutional codes
Cryptography
Data processing
Pulse code modulation
Random processes
Signal processing

* Former Ei Vocabulary term

Encoding (symbols) *(continued)*
Speech coding
Trellis codes

Encryption
USE: Cryptography

End effectors 731.5 (731.6)
DT: January 1995
UF: Hands (robotic)
Robotic hands
BT: Robots
RT: Grippers

Endocrinology 461.6
DT: January 1993
UF: Biomedical engineering—Endocrinology*
BT: Medicine
RT: Biomedical engineering

Endoscopes
USE: Endoscopy

Endoscopy 461.6 (462.1)
DT: January 1993
UF: Biomedical engineering—Endoscopy*
Endoscopes
BT: Diagnosis
RT: Biomedical engineering
Optical instruments
Photodynamic therapy

Endurance
USE: Durability

Energetic materials
USE: Explosives

Energy absorption (931.3)
DT: January 1993
UF: Absorption (of energy)
NT: Acoustic wave absorption
Electromagnetic wave absorption
Electron absorption
Neutron absorption
RT: Attenuation
Damping
Heat sinks
Reflection
Solar absorbers

Energy bands
USE: Band structure

Energy conservation 525.2
DT: January 1977
BT: Conservation
RT: Energy efficiency
Energy management
Energy policy
Energy utilization
Recycling

Energy consumption
USE: Energy utilization

Energy control
USE: Power control

Energy conversion 525.5
DT: June 1990
UF: Conversion (energy)
NT: Direct energy conversion
Ocean thermal energy conversion
Power generation
Wave energy conversion

Energy dissipation 525.4
DT: January 1986
UF: Dissipation (energy)
Energy losses
Loss (energy)
NT: Electric losses
Heat losses
RT: Energy dissipators
Energy transfer

Energy dissipators (616.1) (641.2) (714.2)
DT: January 1993
UF: Dams—Energy dissipators*
Hydraulic structures—Energy dissipators*
BT: Structures (built objects)
NT: Stilling basins
RT: Energy dissipation
Hydraulic structures

Energy efficiency 525.2
DT: January 1995
BT: Efficiency
RT: Energy conservation
Energy utilization

Energy gap (712.1) (931.3)
DT: January 1993
UF: Semiconductor materials—Energy gap*
BT: Band structure

Energy levels
USE: Electron energy levels

Energy loads
USE: Dynamic loads

Energy losses
USE: Energy dissipation

Energy management 525
DT: January 1981
BT: Management
RT: Energy conservation
Energy resources
Intelligent buildings

Energy measurement
USE: Electric power measurement

Energy policy 525.6
DT: January 1977
BT: Public policy
RT: Economics
Energy conservation
Energy resources
Energy utilization

Energy resources 525.1
DT: October 1975
NT: Fossil fuels
Nuclear fuels

* Former Ei Vocabulary term

Energy resources *(continued)*
Renewable energy resources
Wood fuels
RT: Energy management
Energy policy
Energy utilization
Fossil fuel deposits
Natural resources

*Energy resources—Renewable**
USE: Renewable energy resources

Energy states
USE: Electron energy levels

Energy storage 525.7 (702)
DT: January 1977
UF: Storage (energy)
NT: Electric energy storage
Heat storage
RT: Flywheels

Energy transfer (525.4) (641.2)
DT: January 1995
RT: Energy dissipation
Heat transfer

Energy utilization 525.3
DT: January 1977
UF: Energy consumption
RT: Energy conservation
Energy efficiency
Energy policy
Energy resources

Engine afterburners
USE: Afterburners (engine)

Engine cylinders 612.1.1
DT: January 1993
UF: Cylinders (engine)
Engines—Cylinders*
BT: Internal combustion engines
RT: Scavenging

Engine exhausts
USE: Exhaust systems (engine)

Engine indicators
USE: Working fluid pressure indicators

Engine mountings (653.2) (661.2)
DT: January 1993
UF: Aircraft—Engine mounting*
Automobiles—Engine mounting*
BT: Mountings
RT: Antivibration mountings
Engines

Engine pistons 612.1.1
DT: January 1993
BT: Internal combustion engines
Machine components
Pistons

Engineering (901)
SN: Very general term. Prefer specific type of engineering
DT: Predates 1975
NT: Aerospace engineering
Agricultural engineering
Arctic engineering
Automotive engineering
Biomedical engineering
Chemical engineering
Civil engineering
Computer aided engineering
Concurrent engineering
Electrical engineering
Electronics engineering
Engineering geology
Engineering research
Environmental engineering
Gas engineering
High pressure engineering
High temperature engineering
Human engineering
Human rehabilitation engineering
Illuminating engineering
Industrial engineering
Low temperature engineering
Marine engineering
Mechanical engineering
Military engineering
Mining engineering
Municipal engineering
Nuclear engineering
Ocean engineering
Optical engineering
Petroleum engineering
Precision engineering
Process engineering
Production engineering
Steam engineering
Systems engineering
Tropical engineering
Value engineering
RT: Computer aided design
Engineering education
Engineering exhibitions
Engineering facilities
Engineers
Handbooks
History
Machine design
Maintainability
Patents and inventions
Philosophical aspects
Project management
Quality assurance
Research
Structural design
Technical writing
Technology
Theory

Engineering education 901.2
DT: Predates 1975
BT: Education
RT: Engineering
Personnel training

Engineering exhibitions (901.2)
DT: January 1993
UF: Engineering—Exhibitions*
BT: Exhibitions
RT: Engineering

Engineering facilities (901.2) (901.3) (402.2)
DT: Predates 1975
BT: Facilities
RT: Engineering
Laboratories

Engineering geology (481)
DT: January 1993
UF: Geology—Engineering*
BT: Engineering
Geology
RT: Soil structure interactions

Engineering handbooks
USE: Handbooks

Engineering research 901.3
DT: Predates 1975
BT: Engineering
Research

*Engineering research—Experiments**
USE: Experiments

*Engineering writing**
USE: Technical writing

*Engineering—Design aids**
USE: Design aids

*Engineering—Exhibitions**
USE: Engineering exhibitions

*Engineering—Handbooks**
USE: Handbooks

*Engineering—Professional aspects**
USE: Professional aspects

*Engineering—Project management**
USE: Project management

*Engineering—Public policy**
USE: Public policy

*Engineering—Purchasing**
USE: Purchasing

*Engineering—Record preservation**
USE: Records management

*Engineering—Social aspects**
USE: Social aspects

*Engineering—Textbooks**
USE: Textbooks

Engineers (901.1) (912.4)
DT: January 1993
BT: Personnel
RT: Biographies
Engineering
Registration of engineers

*Engineers—Biographies**
USE: Biographies

*Engineers—Registration**
USE: Registration of engineers

Engines (612) (617.3) (654.2)
SN: Machines with self-contained energy sources; for machines operated by external sources energy, use MOTORS
DT: January 1993
BT: Machinery
NT: Adiabatic engines
Compound engines
Crankcases
Dual fuel engines
Exhaust systems (engine)
Flywheels
Heat engines
Jet engines
Marine engines
Propellers
Rocket engines
Rotary engines
Rotors
RT: Crankshafts
Engine mountings
Fuel pumps
Motors
Thermodynamics

*Engines—Cylinders**
USE: Engine cylinders

Enhanced recovery 511.1
DT: January 1993
UF: Oil well production—Enhanced recovery*
Oil well production—Secondary*
Oil well production—Tertiary*
Secondary recovery
Tertiary recovery
BT: Production
NT: Thermal oil recovery
Well flooding
RT: Acidization
Crude petroleum
Directional drilling
Natural gas
Natural gas well production
Oil well production
Well workover

Enhancement (images)
USE: Image enhancement

Enrichment (blast)
USE: Blast enrichment

Enthalpy 641.1
DT: January 1995
UF: Heat content
Total heat
BT: Thermodynamic properties
RT: Gibbs free energy
Thermodynamics

Entrainment (air)
USE: Air entrainment

* Former Ei Vocabulary term

Entropy 641.1
DT: January 1993
UF: Thermodynamics—Entropy*
BT: Thermodynamic properties
RT: Free energy
Gibbs free energy
Heating

Environmental chambers (454) (656.2)
DT: Predates 1975
BT: Simulators
NT: Space simulators
RT: Environmental engineering
Satellite simulators

Environmental degradation
USE: Weathering

Environmental effects
USE: Environmental impact

Environmental engineering (454)
DT: Predates 1975
UF: Remote sensing—Environmental applications*
BT: Engineering
RT: Accident prevention
Air conditioning
Air pollution
Air quality
Bank protection
Biomedical engineering
Clean rooms
Cooling
Earth sciences
Ecology
Ecosystems
Environmental chambers
Environmental impact
Environmental protection
Environmental testing
Ergonomics
Flood control
Forestry
Heating
Human engineering
Illuminating engineering
Land reclamation
Lighting
Meteorology
Noise abatement
Pollution control
Refuse disposal
Regional planning
Sanitary engineering
Shore protection
Smoke abatement
Space heating
Temperature control
Temperature distribution
Ventilation
Water pollution
Water quality

Environmental impact 454.2
DT: January 1977
UF: Environmental effects
RT: Environmental engineering
Environmental protection

Environmental protection 454.2
DT: October 1975
BT: Protection
RT: Environmental engineering
Environmental impact

Environmental testing (423.2) (454)
SN: Use for testing materials or equipment under specific environment conditions, and for investigations of specific environments, such as those of buildings.
DT: January 1987
BT: Testing
RT: Environmental engineering
Temperature control
Temperature distribution
Ventilation
Weathering

Enzyme immobilization (461.8) (801.2)
DT: January 1993
UF: Enzymes—Immobilization*
Immobilization (enzymes)
RT: Biotechnology
Enzymes

Enzyme inhibition (461.8) (801)
DT: January 1993
UF: Enzymes—Inhibition*
Inhibition (enzyme)
RT: Enzymes

Enzyme kinetics (461.8) (801.2) (802.2)
DT: January 1993
UF: Enzymes—Enzyme kinetics*
Kinetics (enzyme)
RT: Enzymes

Enzyme sensors (461.2) (801.2)
DT: January 1993
BT: Biosensors
RT: Electrodes
Enzymes

Enzymes (461.2) (801.2) (804.1)
DT: January 1986
BT: Biocatalysts
Biological materials
Proteins
NT: Coenzymes
RT: Enzyme immobilization
Enzyme inhibition
Enzyme kinetics
Enzyme sensors
Fermentation
Metabolism
Saccharification

*Enzymes—Enzyme kinetics**
USE: Enzyme kinetics

*Enzymes—Immobilization**
USE: Enzyme immobilization

* Former Ei Vocabulary term

*Enzymes—Inhibition**
USE: Enzyme inhibition

Epicyclic gears 601.2
DT: January 1993
UF: Gears—Epicyclic*
BT: Gears

Epidemiology 461.7
DT: January 1993
UF: Health care—Epidemiology*
BT: Medicine
RT: Disease control
Diseases
Health care

Epitaxial growth 802.3, 933.1.2 (531.2) (712.1)
DT: January 1993
UF: Crystals—Epitaxial growth*
Epitaxy
BT: Crystal growth
NT: Liquid phase epitaxy
Vapor phase epitaxy
RT: Coating techniques
Recrystallization (metallurgy)
Substrates
Thin films
Vapor deposition

Epitaxy
USE: Epitaxial growth

Epoxy resins 815.1.1
DT: Predates 1975
UF: Roadbuilding materials—Epoxy resins*
BT: Polyethers
RT: Roadbuilding materials

EPR
USE: Paramagnetic resonance

Equalisers
USE: Equalizers

Equalizers 713.5
DT: Predates 1975
UF: Equalisers
BT: Networks (circuits)
NT: Attenuation equalizers
RT: Electric distortion
Telephone equipment

*Equalizers, Attenuation**
USE: Attenuation equalizers

Equations of motion 921.2 (931.1)
DT: January 1986
BT: Differential equations
NT: Navier Stokes equations
RT: Computational fluid dynamics
Stability

Equations of state (931.2) (921)
DT: Predates 1975
BT: Mathematical techniques
NT: Equations of state of gases
Equations of state of liquids
Equations of state of solids
RT: Phase transitions
Thermodynamics

Equations of state of gases (931.2) (921)
DT: January 1993
UF: Equations of state—Gases*
BT: Equations of state
RT: Gases
Kinetic theory of gases

Equations of state of liquids (931.2) (921)
DT: January 1993
UF: Equations of state—Liquids*
BT: Equations of state
RT: Liquids

Equations of state of solids (931.2) (933) (921)
DT: January 1993
UF: Equations of state—Solids*
BT: Equations of state
RT: Solids

*Equations of state—Gases**
USE: Equations of state of gases

*Equations of state—Liquids**
USE: Equations of state of liquids

*Equations of state—Solids**
USE: Equations of state of solids

Equilibrium (phase)
USE: Phase equilibria

Equipment
DT: January 1993
NT: Acoustic equipment
Agricultural implements
Air cleaners
Air ejectors
Airport ground equipment
Applicators
Auxiliary equipment
Biomedical equipment
Boat lifts
Brakes
Chemical equipment
Clutches
Coin operated equipment
Combustion equipment
Construction equipment
Control equipment
Cooling systems
Cryogenic equipment
Cutting equipment
Design aids
Diffusers (fluid)
Display devices
Diving apparatus
Domestic appliances
Drilling equipment
Ducts
Electric equipment
Electric power plant equipment
Electronic equipment
Enclosures
Explosive actuated devices
Fire fighting equipment

Equipment *(continued)*
Fixtures (tooling)
Fluidic devices
Fuel systems
Gas appliances
Gas fired boilers
Gas generators
Gas producers
Guides (mechanical)
Guideways
Guns (armament)
Hardware
Heat exchangers
Heating equipment
Hose
Hydraulic equipment
Inflatable equipment
Inspection equipment
Instrument panels
Instruments
Intake systems
Jacks
Jets
Keys (for locks)
Ladders
Ladles
Landing mats
Light weight equipment
Lighting fixtures
Lightning arresters
Logic devices
Lubricators
Magnetic devices
Materials handling equipment
Materials testing apparatus
Military equipment
Mining equipment
Mist eliminators
Molds
Motion picture screens
Navigation systems
Needles
Neutron irradiation apparatus
Oceanographic equipment
Office equipment
Oil field equipment
Optical devices
Oscillators (mechanical)
Parachutes
Particle accelerators
Photoemissive devices
Photographic equipment
Plasma devices
Plumbing fixtures
Pneumatic equipment
Portable equipment
Printing plates
Probes
Projection screens
Protective clothing
Pumps
Rails
Resonators
Robots
Rockets
Safety devices
Sanding equipment
Schlieren systems
Seats
Security systems
Separators
Shielding
Ship equipment
Shipways
Shock absorbers
Signaling equipment
Snow melting systems
Snow plows
Solar equipment
Spacecraft equipment
Spectrum analyzers
Superconducting devices
Switches
Tools
Transducers
Underground equipment
Underwater equipment
Vehicle parts and equipment
Ventilation exhausts
Well equipment
Wheels
X ray apparatus
RT: Components
Industrial plants
Obsolescence

Equipment mountings
USE: Mountings

Equipment testing
DT: January 1993
BT: Testing
NT: Automobile testing
Electric machinery testing
Electric switchgear testing
Electric transformer testing
Electronic equipment testing
Instrument testing
Ship testing
RT: Automatic testing
Flammability testing
High temperature testing
Impact testing
Laser diagnostics
Low temperature testing
Measurements
Nondestructive examination
Safety testing
Shock testing
Track test cars

Equivalence classes (921) (721.1)
DT: January 1993
UF: Computer metatheory—Equivalence classes*
BT: Mathematical techniques
RT: Automata theory

Equivalent circuits (703.1)
DT: January 1993
UF: Electric networks—Equivalent circuits*

Equivalent circuits *(continued)*
BT: Networks (circuits)
RT: Electric network analysis
Electric network synthesis
Electric network parameters
Semiconductor device models

Erbium 547.2
DT: January 1993
UF: Erbium and alloys*
BT: Rare earth elements
RT: Erbium alloys
Erbium compounds

Erbium alloys 547.2
DT: January 1993
UF: Erbium and alloys*
BT: Rare earth alloys
RT: Erbium
Erbium compounds

*Erbium and alloys**
USE: Erbium OR
Erbium alloys

Erbium compounds (804.1) (804.2)
DT: January 1993
BT: Rare earth compounds
RT: Erbium
Erbium alloys

Ergometers
USE: Exercise equipment

Ergonomics (461.4)
DT: January 1987
BT: Human engineering
RT: Cybernetics
Environmental engineering
Job analysis
Man machine systems
Physiology
Systems engineering

Erosion (407) (483) (539.1)
DT: Predates 1975
UF: Aircraft—Rain erosion*
Inland waterways—Bank erosion*
Soils—Erosion*
NT: Cavitation corrosion
RT: Ablation
Bank protection
Beaches
Cavitation
Corrosion
Floods
Inland waterways
Scour
Shore protection
Soil conservation
Soils
Subsidence
Water
Wear of materials
Weathering
Wind effects

Error analysis (921.6)
DT: January 1993
UF: Mathematical techniques—Error analysis*
BT: Numerical analysis
RT: Bit error rate
Coding errors
Error correction
Error detection
Errors
Instrument errors

Error compensation (721.1)
DT: January 1993
UF: Compensation (errors)
Computers—Error compensation*
RT: Analog computers
Coding errors
Error correction
Error detection
Errors
Instrument errors

Error correction (721.1)
DT: January 1993
UF: Codes, Symbolic—Error correction*
Correction (errors)
RT: Binary codes
Bit error rate
Block codes
Codes (symbols)
Coding errors
Convolutional codes
Error analysis
Error compensation
Errors
Fault tolerant computer systems
Information theory
Instrument errors
Trellis codes

Error detection (721.1)
DT: January 1993
UF: Codes, Symbolic—Error detection*
Detection (errors)
RT: Binary codes
Bit error rate
Block codes
Codes (symbols)
Coding errors
Convolutional codes
Error analysis
Error compensation
Errors
Instrument errors
Program debugging
Trellis codes

Error probability
USE: Errors AND
Probability

Error recovery (computers)
USE: Computer system recovery

Error statistics 922.2
DT: January 1993
UF: Codes, Symbolic—Error statistics*
BT: Statistics
RT: Errors

Errors
DT: January 1993
UF: Error probability
NT: Coding errors
Instrument errors
Measurement errors
RT: Error analysis
Error compensation
Error correction
Error detection
Error statistics
Quality control
Testing

Escalators (692)
DT: Predates 1975
BT: Machinery
RT: Elevators
Stairs

Escape devices (aircraft)
USE: Aircraft escape devices

Escape ramps (406) (914.1)
DT: January 1993
UF: Roads and streets—Escape ramps*
BT: Roads and streets
RT: Accident prevention

Escape systems (spacecraft)
USE: Spacecraft escape systems

ESR
USE: Paramagnetic resonance

Essential oils 804.1 (822.3)
DT: Predates 1975
BT: Vegetable oils
RT: Fragrances

Esterification 802.2
DT: January 1993
BT: Chemical reactions
RT: Esters

Esters 804.1
DT: January 1986
BT: Organic compounds
NT: Adenosinetriphosphate
Nitrocellulose
RT: Esterification
Lipids
Salts

Estimation 921
DT: January 1993
UF: Method of moments
BT: Mathematical techniques
RT: Riccati equations

Estuaries 407.2
DT: January 1993
UF: Drowned river mouths
Estuarine water resources
Rivers—Estuaries*
Water resources—Estuarine*
BT: Surface waters
RT: Rivers
Surface water resources

Estuarine water resources
USE: Surface water resources AND Estuaries

Etalons 941.3
DT: January 1993
BT: Interferometers
RT: Fabry-Perot interferometers

Etching 802.2 (531) (714.2)
DT: Predates 1975
BT: Surface treatment
NT: Dry etching
Plasma etching
Reactive ion etching
RT: Chemical finishing
Corrosion
Decoration
Metallography
Printed circuits
Semiconductor device manufacture

Ethane 804.1
DT: Predates 1975
BT: Paraffins

Ethanoic acid
USE: Acetic acid

Ethanol 804.1 (523)
DT: January 1981
UF: Ethyl alcohol
Fermentation alcohol
Grain alcohol
BT: Alcohols
RT: Ethanol fuels
Gasohol
Solvents

Ethanol fuels 523, 804.1
DT: January 1993
BT: Alcohol fuels
RT: Automotive fuels
Diesel fuels
Ethanol
Gasohol

Ethers 804.1
DT: January 1986
BT: Hydrocarbons
RT: Crown ethers
Salts
Solvents

*Ethers—Crown ethers**
USE: Crown ethers

Ethyl alcohol
USE: Ethanol

Ethylene 804.1
DT: Predates 1975
BT: Olefins
RT: Polyethylenes

Europium 547.2
DT: January 1981
BT: Rare earth elements
RT: Europium alloys
Europium compounds

Europium alloys 547.2
DT: January 1993
BT: Rare earth alloys
RT: Europium
Europium compounds

Europium compounds (804.1) (804.2)
DT: Predates 1975
BT: Rare earth compounds
RT: Europium
Europium alloys

Eutectics 531.2 (801.4)
DT: January 1993
BT: Metallography
RT: Binary alloys
Crystals
Graphitization
Metallographic microstructure
Solid solutions
Solutions
Ternary systems

Eutectoid steel
USE: Carbon steel

Evaluation
DT: January 1993
NT: Personnel rating
Petroleum reservoir evaluation
RT: Computer selection and evaluation
Computer software selection and evaluation
Inspection
Performance
Queueing theory
Specifications
Testing

Evaporation 802.3
DT: Predates 1975
BT: Vaporization
RT: Dewatering
Diffusion
Drying
Evaporators
Evapotranspiration
Sublimation

Evaporative cooling systems (643.4) (644.3)
DT: January 1993
UF: Air conditioning—Evaporative cooling*
BT: Cooling systems
RT: Air conditioning
Evaporators

Evaporators 802.1
DT: Predates 1975
UF: Refrigerating machinery—Evaporators*
Ship equipment—Evaporators*
Sugar factories—Evaporators*
BT: Chemical equipment
RT: Dryers (equipment)
Evaporation
Evaporative cooling systems
Refrigerating machinery

Evapotranspiration (802.3) (444.1)
DT: January 1993
UF: Water resources—Evapotranspiration*
RT: Evaporation
Surface waters
Water resources

Evoked potentials
USE: Bioelectric potentials

EW
USE: Electronic warfare

EXAPT (programming language) 723.1.1
DT: January 1993
UF: Computer programming languages—EXAPT*
BT: Computer programming languages

Excavation (405.2) (502.1)
DT: Predates 1975
NT: Dredging
Shaft sinking
Trenching
Tunneling (excavation)
RT: Boreholes
Boring
Cofferdams
Construction
Drilling
Earthmoving machinery
Excavators
Explosions
Foundations
Mining
Nuclear explosions

Excavators (405.1) (502.2)
DT: January 1993
UF: Earthmoving machinery—Excavators*
BT: Earthmoving machinery
NT: Dredges
RT: Excavation

Exchange (ions)
USE: Ion exchange

Exchangers (ion)
USE: Ion exchangers

Exchanges (telephone)
USE: Telephone exchanges

Excimer lasers (744.2)
DT: January 1993
UF: Lasers, Excimer*
BT: Lasers
RT: Gas lasers

Excitation (electric)
USE: Electric excitation

* Former Ei Vocabulary term

Exciters (electric)
USE: Electric exciters

Excitons (712.1) (931.3)
DT: January 1993
BT: Electron energy levels
RT: Semiconductor materials

Executive programs
USE: Supervisory and executive programs

Exercise bicycles
USE: Exercise equipment

Exercise equipment (462.1) (461.7)
SN: Includes exercise bicycles, ergometers, rowing machines, etc.
DT: January 1993
UF: Biomedical equipment—Exercisers*
Ergometers
Exercise bicycles
Rowing machines
Treadmills
Weight machines
BT: Biomedical equipment
RT: Bicycles
Human rehabilitation equipment

Exhaust gases 451.1 (612.1) (642.2)
DT: January 1993
UF: Internal combustion engines—Exhaust gases*
Ovens, Industrial—Exhaust gases*
BT: Gases
Wastes
NT: Flue gases
RT: Industrial ovens
Industrial wastes
Internal combustion engines
Smoke
Vapors

Exhaust systems (engine) (612.1.1) (612.3) (654.2)
DT: January 1993
UF: Engine exhausts
Internal combustion engines—Exhaust systems*
Rocket engines—Exhausts*
Turbomachinery—Exhausts*
BT: Engines
RT: Afterburners (engine)
Internal combustion engines
Mufflers
Rocket engines

Exhausts (ventilation)
USE: Ventilation exhausts

Exhibition buildings 402.2 (902.2)
SN: Buildings for holding exhibitions
DT: Predates 1975
BT: Buildings
RT: Exhibitions

Exhibitions (902.2)
DT: Predates 1975
NT: Aircraft exhibitions
Automobile exhibitions
Chemical exhibitions
Engineering exhibitions
Machinery exhibitions
RT: Exhibition buildings

*Exhibitions—Power supply**
USE: Electric power utilization

Expandable tubes (619.1) (632)
DT: January 1993
UF: Tubes—Expandable*
BT: Tubes (components)

Expanded plastics
USE: Foamed plastics

Expansion
SN: Of materials
DT: January 1993
NT: Thermal expansion
RT: Expansion joints
Swelling

Expansion (industrial plants)
USE: Plant expansion

Expansion joints 408.2 (401.1) (619.1)
DT: January 1993
UF: Bridges—Expansion joints*
Pipelines—Expansion joints*
BT: Joints (structural components)
RT: Expansion
Pipe joints

Expansive cellulose 811.3
DT: January 1993
UF: Cellulose—Expansive*
BT: Cellulose

Expansive concrete 412.1
DT: January 1993
UF: Concrete—Expansive*
BT: Concretes

Experimental mineralogy 482.1
DT: January 1993
UF: Mineralogy—Experimental*
BT: Mineralogy

Experimental reactors 621
DT: January 1993
UF: Nuclear reactors, Experimental*
BT: Nuclear reactors

Experiments
SN: Very general. Whenever possible use a more specific term. Also use Treatment Code X
DT: January 1993
UF: Engineering research—Experiments*
BT: Research
NT: Cadaveric experiments
Living systems studies

Expert systems 723.4.1
DT: June 1990
UF: Blackboards (artificial intelligence)
BT: Knowledge based systems
NT: Inference engines
RT: Artificial intelligence

* Former Ei Vocabulary term

Expert systems *(continued)*
Computer aided diagnosis
Computer aided engineering
Computer hardware description languages
Decision support systems
Industrial robots
Intelligent robots

*Expert systems—Inference engines**
USE: Inference engines

*Expert systems—Knowledge bases**
USE: Knowledge based systems

Exploding wires 704, 708.2 (502.2) (932.1)
DT: January 1987
BT: Wire
RT: Explosions

Exploration (natural resources)
USE: Natural resources exploration

*Exploration**
USE: Natural resources exploration

Exploratory boreholes (501.1) (512.1.2) (512.2.2)
DT: January 1993
UF: Boreholes—Exploratory*
BT: Boreholes
RT: Exploratory oil well drilling
Petroleum prospecting

Exploratory geochemistry 481.2 (501.1)
DT: January 1993
UF: Geochemistry—Exploratory*
BT: Geochemistry
RT: Natural resources exploration

Exploratory oil well drilling 511.1
DT: January 1993
UF: Oil well drilling—Exploratory*
BT: Oil well drilling
Petroleum prospecting
RT: Exploratory boreholes

Explosion testing (422.2) (423.2) (421)
DT: January 1993
UF: Materials testing—Explosion*
BT: Materials testing
RT: Explosions

Explosion welding
USE: Explosive welding

Explosionproofing 914.1
DT: January 1993
UF: Electric machinery—Explosionproof*
Instruments—Explosionproof*
BT: Protection

Explosions (502.1) (802.2) (914.1)
DT: Predates 1975
NT: Nuclear explosions
Underground explosions
Underwater explosions
RT: Accidents
Blast resistance
Blasting
Detonation
Excavation
Exploding wires
Explosion testing
Gas hazards
Gases
Reaction kinetics
Shock waves
Spontaneous combustion

*Explosions—Underground**
USE: Underground explosions

*Explosions—Underwater**
USE: Underwater explosions

Explosive actuated devices 605.1
DT: January 1993
BT: Equipment
NT: Explosive actuated fasteners
Explosive actuated tools

Explosive actuated fasteners 605
DT: January 1993
UF: Fasteners—Explosive*
BT: Explosive actuated devices
Fasteners

Explosive actuated tools 605.1
DT: January 1993
UF: Tools, jigs and fixtures—Explosive*
BT: Explosive actuated devices
Tools

Explosive bonding (538.1)
DT: January 1993
UF: Metals and alloys—Explosive bonding*
BT: Bonding
RT: Explosive welding

Explosive drilling (502.1) (511.1)
DT: January 1993
UF: Rock drilling—Explosive*
BT: Drilling
NT: Explosive oil well drilling
RT: Rock drilling

Explosive forging
USE: Explosive forming

Explosive forming 535.2.2
DT: January 1993
UF: Explosive forging
Metal forming—Explosive*
BT: High energy forming
RT: Deep drawing

Explosive oil well drilling 511.1
DT: January 1993
UF: Oil well drilling—Explosive*
BT: Explosive drilling
Oil well drilling

Explosive welding 538.2.1
DT: January 1993
UF: Explosion welding
Welding—Explosive*
BT: Pressure welding
RT: Explosive bonding

Explosive well stimulation (512.1.2) (512.2.2)
DT: January 1993
UF: Natural gas wells—Nuclear stimulation*
Nuclear stimulation
Oil well shooting*
Shooting (well)
Well shooting
BT: Well stimulation
RT: Oil well production

Explosives (404.1) (405.2) (502.2) (804)
DT: Predates 1975
UF: Energetic materials
Mines and mining—Explosives*
Powder metallurgy—Explosives applications*
BT: Materials
NT: Shaped charges
RT: Detonation
Mining
Powder metallurgy
Propellants

*Explosives—Detonation**
USE: Detonation

*Explosives—Shaped charges**
USE: Shaped charges

Exposure controls 742.1
DT: January 1993
UF: Cameras—Exposure control*
BT: Cameras
RT: Photometry

Exposure meters 742.2
DT: January 1993
UF: Light meters (exposure meters)
Photography—Exposure meters*
BT: Photometers
RT: Cameras
Comparators (optical)
Photographic accessories
Photography

Extendable tubes 619.1
DT: January 1993
UF: Tubes—Extendable*
BT: Tubes (components)

Extended definition television
USE: High definition television

Extension pipelines 619.1
DT: January 1993
UF: Pipelines—Extension*
BT: Pipelines

Extensometers
USE: Dilatometers

Extinction (light)
USE: Light extinction

Extra high voltage transmission
USE: EHV power transmission

Extraction 802.3
SN: Limit to such disengagement of substances as is accomplished by chemical engineering processes. Exclude metallurgical processes, for which use EXTRACTIVE METALLURGY
DT: Predates 1975
BT: Separation
NT: Solvent extraction
RT: Centrifugation
Concentration (process)
Dewatering
Leaching
Solutions
Solvents

Extraction (beam)
USE: Particle beam extraction

*Extraction—Supercritical fluid**
USE: Supercritical fluid extraction

Extractive metallurgy 533.1, 531.1
DT: January 1977
BT: Metallurgy
NT: Electrometallurgy
Hydrometallurgy
RT: Metal refining
Ore treatment
Reduction
Refining

Extraordinary ray
USE: Birefringence

Extraterrestrial atmospheres 657.2
DT: Predates 1975
UF: Atmospheres (extraterrestrial)
RT: Atmospheric acoustics
Atmospheric chemistry
Atmospheric composition
Atmospheric density
Atmospheric electricity
Atmospheric humidity
Atmospheric movements
Atmospheric optics
Atmospheric pressure
Atmospheric radiation
Atmospheric radioactivity
Atmospheric spectra
Atmospheric structure
Atmospheric temperature
Atmospheric thermodynamics
Meteorology
Planets
Precipitation (meteorology)
Storms

Extraterrestrial communication links 657.2 (716)
DT: January 1993
UF: Interplanetary communication links
Telecommunication links, Extraterrestrial*
BT: Telecommunication links
NT: Satellite links
RT: Interplanetary flight
Moon bases
Space flight
Space research

* Former Ei Vocabulary term

Extruders (535.2.1) (816.2)
DT: Predates 1975
UF: Extrusion machinery
Extrusion presses
BT: Machinery
NT: Extrusion dies
Plastics extruders
Wire insulating extruders
RT: Coextrusion
Extrusion
Extrusion molding
Metal extrusion
Plastics machinery
Presses (machine tools)

Extrusion (535.2.2) (816.1)
DT: Predates 1975
BT: Forming
NT: Coextrusion
Extrusion molding
Metal extrusion
RT: Drawing (forming)
Extruders
High energy forming
Piercing
Pressing (forming)
Pultrusion
Rolling

Extrusion dies (535.2.1) (816.2)
DT: January 1993
UF: Wire insulating extruders—Dies*
BT: Dies
Extruders

Extrusion machinery
USE: Extruders

Extrusion molding 816, 535.2.2
DT: January 1993
UF: Plastics—Extrusion molding*
BT: Extrusion
Molding
RT: Extruders
Plastics extruders
Plastics forming
Plastics molding

Extrusion presses
USE: Extruders

Eye controlled devices 461.5
DT: January 1993
UF: Human rehabilitation engineering—Eye-controlled devices*
BT: Human rehabilitation equipment
RT: Human rehabilitation engineering

Eye movements (461.1)
DT: January 1993
UF: Nystagmus
Saccades
Vision—Eye movements*
RT: Vision

Eye protection 914.1 (461.4) (741.2)
DT: Predates 1975
UF: Personnel—Eye protection*
BT: Protection
RT: Accident prevention
Goggles
Personnel
Vision

Eyeglasses 741.3 (461.5)
DT: January 1993
UF: Vision—Eyeglasses*
BT: Vision aids
RT: Goggles

F layer
USE: F region

F region 443.1 (711.1)
DT: January 1993
UF: F layer
Ionosphere—F region*
BT: Ionosphere

Fabrication
DT: January 1993
BT: Manufacture
NT: Optical fiber fabrication
RT: Assembly
Construction

Fabrics 819.4
DT: January 1993
UF: Cloth
BT: Textiles
NT: Aircraft fabrics
Automobile fabrics
Knit fabrics
Linen
Nonwoven fabrics
Rayon fabrics
Woolen and worsted fabrics
RT: Carpet manufacture
Fibers
Paper
Yarn

Fabry-Perot interferometers 941.3
DT: January 1993
BT: Interferometers
RT: Etalons

Facades 402, 408.2
DT: January 1993
UF: Architectural design—Facades*
BT: Building components
RT: Facings

Facilities (402)
SN: Very general term; prefer specific type of facility, structure, or building
DT: January 1993
NT: Airports
Auditoriums
Automobile proving grounds
Bus terminals
Clean rooms
Control facilities
Drive-in facilities

Facilities *(continued)*
Engineering facilities
Ferry terminals
Fusion reactors
Garages (parking)
Geological repositories
Greenhouses
Gymnasiums
Kitchens
Laboratories
Libraries
Marinas
Military bases
Mines
Mints
Moon bases
Motion picture theaters
Museums
Nuclear reactors
Observatories
Oil terminals
Opera houses
Pipeline terminals
Planetariums
Port terminals
Ports and harbors
Post offices
Prisons
Projection rooms
Pumping plants
Railroad plant and structures
Recreation centers
Recreational facilities
Sewers
Shelters (civil defense)
Skating rinks
Ski jumps
Stadiums
Stages
Studios
Swimming pools
Tennis courts
Test facilities
Theaters (legitimate)
Transfer stations
Truck terminals
Warehouses
Water treatment plants
Waterworks
Weigh stations
RT: Buildings
Public works
Site selection
Structures (built objects)

Facings 402, 408.2
DT: January 1993
UF: Buildings—Facings*
BT: Building components
RT: Facades
Revetments
Veneers

Facsimile 718.3
DT: Predates 1975
UF: Facsimile transmission
Fax
BT: Telecommunication services
RT: Electronic mail
Facsimile equipment
Office automation
Scanning

Facsimile equipment 718.3
DT: Predates 1975
BT: Telecommunication equipment
RT: Facsimile
Telephone apparatus

Facsimile transmission
USE: Facsimile

Factories
USE: Industrial plants

Factory automation 913.4.2
DT: January 1993
BT: Automation
RT: Computer aided manufacturing
Computer integrated manufacturing
Industrial plants
Industrial robots
Information technology

Factory management
USE: Plant management

Fading (radio) 716.3
DT: January 1993
UF: Radio transmission—Fading*
Rayleigh fading
BT: Radio transmission
RT: Diversity reception
Electric distortion
Ionospheric electromagnetic wave propagation
Signal distortion

Failure (computers)
USE: Computer system recovery

Failure (mechanical) (421)
DT: January 1993
UF: Failure*
NT: Buckling
Fracture
RT: Accidents
Corrosion
Crack propagation
Cracks
Deformation
Degradation
Failure analysis
Reliability
Strength of materials
Structural design
Wear of materials

Failure analysis (421) (921)
DT: Predates 1975
UF: Fault tree analysis
BT: Analysis
RT: Availability
Failure (mechanical)

Failure analysis *(continued)*
Maintainability
Maintenance
Reliability
Reliability theory

*Failure**
USE: Failure (mechanical)

Fallout 622.1 (451.1) (453.1) (914.1)
DT: January 1993
UF: Earth atmosphere—Fallout*
Radioactive fallout
RT: Accidents
Aerosols
Air pollution
Contamination
Nuclear explosions

False teeth
USE: Dental prostheses

Fans 618.3
DT: Predates 1975
BT: Machinery
NT: Blowers
RT: Radiators
Refrigerating machinery
Ventilation

Faraday effect 701.1
DT: January 1993
BT: Magnetooptical effects
RT: Light polarization
Optical rotation

Fares (transportation)
USE: Transportation charges

Farm buildings 402.1, 821.6
DT: Predates 1975
BT: Buildings
Farms
NT: Silos (agricultural)
RT: Dairies
Grain elevators
Greenhouses

Farm equipment
USE: Agricultural machinery

Farms 821
DT: Predates 1975
UF: Materials handling—Farms*
NT: Farm buildings
Orchards
Rubber plantations
RT: Agricultural machinery
Agriculture
Crops
Rural areas

*Farms—Crops**
USE: Crops

Fast Fourier transforms 921.3
DT: January 1993
UF: FFT
Mathematical transformations—Fast Fourier transforms*
BT: Fourier transforms

Fast reactors 621.1
DT: January 1993
UF: Nuclear reactors, Fast*
BT: Nuclear reactors

Fast response computer systems 722.4
DT: January 1993
UF: Computer systems, Digital—Fast response capability*
BT: Computer systems

Fasteners 605
DT: Predates 1975
UF: Zippers
BT: Components
NT: Bolts
Explosive actuated fasteners
Hooks
Locks (fasteners)
Nails
Nuts (fasteners)
Rivets
Screws
Studs (fasteners)
Washers
RT: Chains
Connectors (structural)
Hinges
Joining
Joints (structural components)
Splines
Strapping
Welding

*Fasteners—Explosive**
USE: Explosive actuated fasteners

Fatigue cracks
USE: Cracks AND
Fatigue of materials

Fatigue life
USE: Fatigue of materials

Fatigue of materials (421)
DT: Predates 1975
UF: Fatigue cracks
Fatigue life
Fatigue strength
Fatigue*
BT: Wear of materials
NT: Corrosion fatigue
RT: Fatigue testing
Fretting corrosion
High temperature testing
Mechanical properties
Prestressing
Shear strength
Shear stress
Spalling
Stress corrosion cracking
Tensile properties

Fatigue strength
USE: Strength of materials AND
Fatigue of materials

Fatigue testing (422.2)
DT: January 1993
UF: Materials testing—Fatigue*
BT: Mechanical testing
RT: Fatigue of materials
Hardness testing
Impact testing
Stress concentration
Tensile testing

*Fatigue**
USE: Fatigue of materials

Fats
USE: Oils and fats

Fatty acids 804.1
DT: Predates 1975
UF: Oleic acid
BT: Carboxylic acids
NT: Acetic acid
RT: Lipids

Fault currents
USE: Electric fault currents

*Fault location**
USE: Electric fault location

Fault tolerance
USE: Fault tolerant computer systems

Fault tolerant computer systems 722.4
DT: January 1993
UF: Computer systems, Digital—Fault tolerant capability*
Fault tolerance
Fault tolerant computing
BT: Computer systems
RT: Computer system recovery
Error correction
Reliability

Fault tolerant computing
USE: Fault tolerant computer systems

Fault tree analysis
USE: Failure analysis

Faults (electric)
USE: Electric fault currents

Fax
USE: Facsimile

FDM
USE: Frequency division multiplexing

Fe
USE: Iron

FEA
USE: Finite element method

Feature extraction (716) (723.5) (741.1) (751.1)
DT: January 1995
BT: Pattern recognition
RT: Edge detection
Image processing

Federal taxes
USE: Taxation

Feedback 731.1
DT: January 1995
UF: Negative feedback
Positive feedback
State feedback
RT: Cascade control systems
Closed loop control systems
Control theory
Cybernetics
Feedback amplifiers
Feedback control

Feedback amplifiers 713.1
DT: January 1993
UF: Amplifiers, Feedback*
BT: Amplifiers (electronic)
RT: Feedback

Feedback control 731.1
DT: January 1995
BT: Control
RT: Feedback

Feeder reactors
USE: Current limiting reactors

Feedforward neural networks (461.1) (723.4)
DT: January 1995
BT: Neural networks
RT: Learning systems

Feeding 691.2
SN: Of materials to processes
DT: January 1993
UF: Machine tools—Feeding*
BT: Materials handling
RT: Machine tools

Feeds (antenna)
USE: Antenna feeders

Feedstocks (513.3) (803) (804)
DT: January 1983
BT: Raw materials

Feedwater analysis 445.2 (614.2) (802.1)
DT: Predates 1975
BT: Water analysis
RT: Boilers

Feedwater heaters 614.2, 616.1, 642.2
DT: Predates 1975
UF: Preheaters (steam)
Steam preheaters
BT: Water heaters
RT: Boilers
Economizers
Heat exchangers
Power plants
Steam

* Former Ei Vocabulary term

Feedwater pumps 614.2, 618.2
DT: Predates 1975
BT: Pumps
RT: Boilers
Power plants

Feedwater treatment 445.1.2 (614.2)
DT: Predates 1975
BT: Industrial water treatment
RT: Boilers
Steam power plants

Fees and charges (911.1)
DT: January 1993
UF: Charges (fees)
BT: Finance
NT: Transportation charges
Utility rates
RT: Taxation
Toll bridges
Toll highways

FEL
USE: Free electron lasers

Feldspar 482.2 (812)
DT: Predates 1975
BT: Aluminum compounds
Silicate minerals
RT: Feldspar deposits
Granite

Feldspar deposits 505.1
DT: Predates 1975
BT: Mineral resources
RT: Feldspar

Felt 819.4, 811.1
DT: Predates 1975
BT: Nonwoven fabrics
RT: Wool

Felts 811.1.2
DT: January 1993
UF: Papermaking machinery—Felts*
BT: Machine components

FEM
USE: Finite element method

Fences (402) (408.2)
DT: Predates 1975
BT: Enclosures
NT: Concrete fences
Electric fences
Wooden fences

*Fences, Electrified**
USE: Electric fences

*Fences—Concrete**
USE: Concrete fences

*Fences—Wood**
USE: Wooden fences

Fenders (boat) (671.2)
DT: January 1993
UF: Boat fenders
Boats—Fenders*
Ship fenders
BT: Boat equipment

Fenders (port structures) 407.1
DT: January 1993
UF: Port structures—Fenders*
BT: Port structures

Fermentation (461.8) (801.2) (802.2) (805.1.1)
DT: June 1990
BT: Chemical reactions
RT: Antibiotics
Biogas
Biotechnology
Continuous cell culture
Dairy products
Enzymes
Fermenters
Metabolism
Methanogens
Wine
Yeast

Fermentation alcohol
USE: Ethanol

Fermenters 802.1 (461.8) (462.1)
DT: January 1993
UF: Bioreactors—Fermenters*
BT: Bioreactors
RT: Fermentation

Fermi energy
USE: Fermi level

Fermi level 931.3 (701.1)
DT: January 1995
UF: Fermi energy
BT: Electron energy levels
RT: Electronic density of states

Fermi surface 931.3
DT: January 1995
BT: Electron energy levels
RT: Band structure
Cyclotron resonance
Electronic density of states

Fermium 622.1
DT: January 1993
BT: Transuranium elements

Ferrimagnetic materials 708.4
DT: January 1977
UF: Magnetic materials—Ferrimagnetism*
BT: Magnetic materials
NT: Ferrites
RT: Ferrimagnetism
Magnetic semiconductors

Ferrimagnetic resonance 701.2, 708.4
DT: January 1993
BT: Magnetic resonance
RT: Ferrimagnetism

Ferrimagnetism 701.2, 708.4
DT: January 1993
UF: Magnetism—Ferrimagnetism*
BT: Magnetism

Ferrimagnetism *(continued)*
RT: Ferrimagnetic materials
Ferrimagnetic resonance
Ferrites
Magnetic materials

Ferrite 531.2 (545.3)
DT: January 1993
BT: Metallographic phases
RT: Steel metallography

Ferrite applications 708.4
DT: Predates 1975
BT: Applications
RT: Ferrite devices

Ferrite devices (704) (714) (708.4)
SN: Devices based on ferrites
DT: Predates 1975
BT: Magnetic devices
RT: Ferrite applications
Ferrites
Gyrators
Magnetic cores
Phase shifters
Powder magnetic cores
Waveguide components

Ferrites 708.4
SN: Ferrimagnetic materials having high electrical resistivity, a spinel structure, and a chemical formula of XFe2O4
DT: Predates 1975
BT: Ferrimagnetic materials
RT: Ferrimagnetism
Ferrite devices
Powder magnetic cores
Zinc compounds

Ferroalloys 545.2
SN: Use only for alloying additives or deoxidizers, as used in steelmaking
DT: Predates 1975
BT: Additives
Alloys
RT: Alloying
Alloying elements
Deoxidants
Iron alloys

Ferrocement
USE: Reinforced concrete

Ferroelectric devices (708.1) (714)
DT: Predates 1975
BT: Dielectric devices
RT: Dielectric amplifiers
Ferroelectric materials
Ferroelectricity

Ferroelectric materials 708.1
DT: Predates 1975
UF: Ceramic materials—Ferroelectric*
Ferroelectrics
BT: Dielectric materials
RT: Barium titanate
Ceramic materials
Ferroelectric devices
Ferroelectricity
Piezoelectricity

Ferroelectric phenomena
USE: Ferroelectricity

Ferroelectricity 701.1
DT: Predates 1975
UF: Ferroelectric phenomena
BT: Dielectric properties
RT: Ferroelectric devices
Ferroelectric materials
Polarization

Ferroelectrics
USE: Ferroelectric materials

Ferromagnetic materials 708.4
DT: June 1990
UF: Ferromagnetic metals
Magnetic materials—Ferromagnetism*
BT: Magnetic materials
NT: Magnetite
RT: Ferromagnetism
Magnetic cores
Magnetic semiconductors
Magnets

Ferromagnetic metals
USE: Ferromagnetic materials

Ferromagnetic resonance 701.2, 708.4
DT: January 1993
UF: FMR
BT: Magnetic resonance
RT: Ferromagnetism

Ferromagnetism 701.2, 708.4
DT: January 1993
UF: Magnetism—Ferromagnetism*
BT: Magnetism
RT: Ferromagnetic materials
Ferromagnetic resonance
Magnetic materials
Metamagnetism

Ferrous alloys
USE: Iron alloys

Ferry boats 434.2, 674.1
DT: Predates 1975
UF: Automobile ferries
Car ferries
BT: Boats
NT: Train ferries

*Ferry boats—Constructing and outfitting**
USE: Outfitting (water craft) OR
Shipbuilding

Ferry terminals 407.1
DT: Predates 1975
UF: Terminals (ferry)
BT: Facilities
RT: Port terminals

* Former Ei Vocabulary term

Fertilizers 804, 821.2
DT: Predates 1975
BT: Materials
NT: Nitrogen fertilizers
Phosphate fertilizers
Potassium fertilizers
RT: Agriculture
Water pollution

*Fertilizers—Corrosive properties**
USE: Corrosive effects

*Fertilizers—Nitrogen**
USE: Nitrogen fertilizers

*Fertilizers—Phosphates**
USE: Phosphate fertilizers

*Fertilizers—Potassium**
USE: Potassium fertilizers

*Fertilizers—Spreaders**
USE: Spreaders

*Fertilizers—Urea**
USE: Urea fertilizers

Fetal monitoring 461.6
DT: January 1993
UF: Biomedical engineering—Fetal monitoring*
BT: Patient monitoring
RT: Biomedical engineering

FFT
USE: Fast Fourier transforms

FHP motors
USE: Fractional horsepower motors

Fiber amplifiers
USE: Fiber lasers

Fiber fabrication
USE: Optical fiber fabrication

Fiber lasers 744.4 (741.1.2)
DT: January 1993
UF: Fiber amplifiers
Fibre lasers
Optical fiber lasers
BT: Solid state lasers
RT: Fiber optics
Optical fibers

Fiber optic cables
USE: Optical cables

Fiber optic chemical sensors 741.1.2 (732.2) (801)
DT: January 1993
UF: Optical fiber chemical sensors
Sensors—Fiber optic chemical sensors*
BT: Fiber optic sensors
Optical instruments
RT: Fiber optics

Fiber optic components 741.1.2, 741.3
DT: January 1993
BT: Components
RT: Fiber optics
Optical fibers

Fiber optic coupling
USE: Optical fiber coupling

Fiber optic networks 741.1.2 (717)
DT: January 1993
BT: Networks (circuits)
RT: Optical communication

Fiber optic sensors 741.1.2 (732.2)
DT: January 1993
BT: Optical sensors
NT: Fiber optic chemical sensors
RT: Fiber optics
Pollution

Fiber optics 741.1.2
SN: Use for light transmission by filaments. For properties of fibers, use OPTICAL FIBERS
DT: Predates 1975
UF: Light pipes
BT: Optics
RT: Fiber lasers
Fiber optic chemical sensors
Fiber optic components
Fiber optic sensors
Glass fibers
Light
Light transmission
Optical communication
Optical communication equipment
Optical fibers
Optical glass
Photodynamic therapy

Fiber optics in medicine
USE: Medical applications

Fiber orientation
USE: Fiber reinforced materials

Fiber reinforced materials (415) (531) (811) (817.1)
DT: January 1993
UF: Composite materials—Fiber reinforced*
Fiber orientation
Fiber reinforcement*
BT: Composite materials
NT: Fiber reinforced metals
Fiber reinforced plastics
RT: Fibers

Fiber reinforced metals 415.1
DT: January 1993
BT: Fiber reinforced materials
RT: Metallic matrix composites
Metals

Fiber reinforced plastics 415.2
DT: January 1993
BT: Fiber reinforced materials
Reinforced plastics
NT: Carbon fiber reinforced plastics
Glass fiber reinforced plastics

*Fiber reinforcement**
USE: Fiber reinforced materials

* Former Ei Vocabulary term

Fibers (812) (817) (819.4)
DT: January 1993
UF: Fibers, Nontextile*
Fibres
Filaments (fibers)
Nontextile fibers
BT: Materials
NT: Natural fibers
Synthetic fibers
Textile fibers
RT: Crystal whiskers
Fabrics
Fiber reinforced materials
Metallic matrix composites
Spinning (fibers)
Zein

*Fibers, Nontextile**
USE: Fibers

Fibre lasers
USE: Fiber lasers

Fibres
USE: Fibers

Fibrous membranes 819.4 (631.1) (802.1)
DT: January 1993
UF: Membranes—Fibrous*
BT: Membranes

Field effect semiconductor devices 714.2
DT: January 1993
UF: Semiconductor devices, Field effect*
BT: Semiconductor devices
NT: Field effect transistors
MIS devices
RT: Electric fields

Field effect transistors 714.2
DT: January 1993
UF: Transistors, Field effect*
BT: Field effect semiconductor devices
Transistors
NT: Chemically sensitive field effect transistors
Gates (transistor)
High electron mobility transistors
Ion sensitive field effect transistors
Junction gate field effect transistors
MESFET devices
MISFET devices
MOSFET devices
RT: Bipolar transistors
MIS devices

Field emission cathodes 714.1
DT: January 1993
UF: Cathodes, Field emission*
BT: Cathodes

Field emission microscopes 741.3
DT: January 1993
UF: Microscopes—Field emission*
BT: Microscopes

Field intensity patterns (antenna)
USE: Directional patterns (antenna)

Field plotting (701.1) (701.2) (942)
SN: Of static or dynamic fields
DT: Predates 1975
BT: Graphic methods
RT: Electric field measurement
Electromagnetic field measurement
Magnetic field measurement

Fields (electric)
USE: Electric fields

Fighter aircraft 652.1.2
DT: January 1993
UF: Aircraft—Fighter*
Fighters
BT: Military aircraft
RT: Supersonic aircraft

Fighters
USE: Fighter aircraft

Filament lamps
USE: Incandescent lamps

Filament winding 816.1
DT: January 1993
UF: Plastics—Filament winding*
BT: Winding
RT: Plastic filaments
Weaving

Filaments (fibers)
USE: Fibers

Filaments (lamp) 707.2
DT: January 1993
UF: Electric lamps—Filaments*
Lamp filaments
BT: Incandescent lamps

Filaments (plastic)
USE: Plastic filaments

File editors 723.2
DT: January 1993
UF: Data processing—File editors*
Editors (file)
BT: Computer software
RT: Data processing

File organization 723.2 (723.3)
DT: January 1993
UF: Data processing—File organization*
NT: Data structures
RT: Data processing
Database systems
Virtual storage

Files and rasps 605.2
DT: Predates 1975
UF: Rasps
BT: Cutting tools
RT: Hand tools

Filler metals 538.2
DT: January 1993
UF: Welding—Filler metals*
BT: Materials
NT: Brazing filler metals

* Former Ei Vocabulary term

Filler metals *(continued)*
Soldering alloys
RT: Metals
Welding
Welding electrodes
Welding rods

Fillers (803) (804.1) (804.2)
DT: January 1986
BT: Chemicals
NT: Plastics fillers
RT: Additives
Grouting
Sizing (finishing operation)
Talc

Filling 691.2
DT: January 1993
UF: Containers—Filling*
BT: Materials handling
RT: Charging (furnace)
Containers
Loading

Filling stations (402.1) (432)
DT: Predates 1975
UF: Garages (filling stations)
Service stations
BT: Buildings

Fillings (dental)
USE: Dental prostheses

Film growth (712.1) (813.1)
DT: January 1993
UF: Films—Growing*
BT: Growth (materials)
RT: Films

Film preparation (712.1) (813.1)
DT: January 1993
UF: Films—Preparation*
BT: Manufacture
RT: Films

Films
DT: Predates 1975
BT: Materials
NT: Amorphous films
Biofilms
Cellulose films
Conductive films
Diamond films
Dielectric films
Langmuir Blodgett films
Magnetic films
Metallic films
Optical films
Photographic films
Plastic films
Rubber films
Semiconducting films
Solar control films
Superconducting films
Thick films
Thin films
RT: Chemical vapor deposition
Coatings
Film growth
Film preparation
Multilayers
Protective coatings
Substrates
Surface phenomena

*Films—Amorphous**
USE: Amorphous films

*Films—Biofilms**
USE: Biofilms

*Films—Conducting**
USE: Conductive films

*Films—Dielectric**
USE: Dielectric films

*Films—Growing**
USE: Film growth

*Films—Langmuir-Blodgett**
USE: Langmuir Blodgett films

*Films—Preparation**
USE: Film preparation

Filter beds
USE: Biological filter beds

Filtering
USE: Filtration

Filtering and prediction theory
USE: Signal filtering and prediction

Filters (for fluids) (445.1) (451.2) (802.1)
SN: Limited to equipment for mechanical or chemical filtration of fluids
DT: January 1993
BT: Separators
NT: Air filters
Biological filter beds
Cigarette filters
Fuel filters
Magnetic filters
RT: Filtration
Ultrafiltration

Filters (wave)
USE: Wave filters

*Filters—Magnetic**
USE: Magnetic filters

Filtration 802.3
SN: Limited to separation of substances, as of suspended solids from liquids
DT: Predates 1975
UF: Filtering
BT: Separation
NT: Trickling filtration
Ultrafiltration
Water filtration
RT: Air filters
Centrifugation
Dewatering
Effluents

Filtration *(continued)*
Filters (for fluids)
Percolation (fluids)
Precipitation (chemical)
Size separation

Final control devices 732.1
DT: Predates 1975
BT: Control equipment
NT: Electric final control devices
Nonelectric final control devices
RT: Actuators
Brakes
Clutches
Speed regulators
Switches

*Final control devices, Electric**
USE: Electric final control devices

*Final control devices, Nonelectric**
USE: Nonelectric final control devices

Finance (911.1) (911.2)
DT: January 1993
UF: Highway administration—Financing*
NT: Costs
Fees and charges
Insurance
Taxation
RT: Economics
Financial data processing
Management

Financial data processing (911.1)
DT: January 1993
UF: Data processing—Financial applications*
BT: Administrative data processing
RT: Finance
Spreadsheets

Finishing
DT: Predates 1975
UF: Aircraft manufacture—Finishing*
Automobile manufacture—Finishing*
Building materials—Finishing*
Furniture manufacture—Finishing*
Machine tools—Finishing*
Office equipment—Finishing*
Papermaking—Finishing*
Plastics—Finishing*
Wire—Finishing*
Wood products—Finishing*
Wood—Finishing*
Woodworking—Finishing*
BT: Processing
NT: Barreling
Buffing
Burnishing
Calendering
Chemical finishing
Coloring
Deburring
Decoration
Honing
Metal finishing
Polishing
Sealing (finishing)
Shot peening
Sizing (finishing operation)
Textile finishing
Vibratory finishing
RT: Cleaning
Coating techniques
Descaling
Forming
Machine tools
Surface cleaning
Surface treatment

Finite automata 721.1
DT: January 1993
UF: Automata theory—Finite automata*
Finite state automata
BT: Automata theory
RT: Digital computers
Petri nets
Sequential switching

Finite difference method 921.6
DT: January 1993
UF: Mathematical techniques—Finite difference method*
BT: Numerical analysis
RT: Difference equations

Finite element analysis
USE: Finite element method

Finite element method 921.6
DT: January 1993
UF: FEA
FEM
Finite element analysis
Mathematical techniques—Finite element method*
BT: Numerical analysis
RT: Approximation theory
Boundary element method
Boundary value problems
Difference equations
Differential equations
Function evaluation

Finite impulse response filters
USE: Digital filters

Finite state automata
USE: Finite automata

Finite volume method 921.6 (631.1)
DT: January 1995
BT: Mathematical techniques
RT: Flow of fluids

Fins (heat exchange) 616.1
DT: January 1993
UF: Heat exchangers—Finned tubes*
Heat transfer—Fins*
BT: Heat sinks
RT: Heat transfer

FIR filters
USE: Digital filters

* Former Ei Vocabulary term

Fire alarm systems 914.2
DT: Predates 1975
BT: Alarm systems
RT: Fire protection
Safety devices

Fire boats 674.1, 914.2
DT: Predates 1975
BT: Boats
Fire fighting equipment

*Fire boats—Constructing and outfitting**
USE: Outfitting (water craft) OR
Shipbuilding

Fire control systems 731.2, 914.2
DT: January 1993
UF: Aviation, Military—Fire control systems*
Guns—Fire control*
BT: Military equipment
RT: Guns (armament)
Missiles

Fire detectors 914.2
DT: January 1993
UF: Fire protection—Detectors*
BT: Detectors
Safety devices
NT: Smoke detectors
RT: Fire protection
Infrared detectors

Fire extinguishers 914.2
DT: Predates 1975
BT: Fire fighting equipment
RT: Foams
Safety devices

*Fire extinguishers—Foam**
USE: Foams

Fire fighting equipment 914.2
DT: Predates 1975
BT: Equipment
NT: Fire boats
Fire extinguishers
Hydrants
Sprinkler systems (fire fighting)
RT: Fire houses
Fire protection
Safety devices

*Fire fighting equipment—Hydrants**
USE: Hydrants

*Fire fighting equipment—Sprinkler systems**
USE: Sprinkler systems (fire fighting)

Fire hazards 914.2
DT: January 1993
UF: Industrial plants—Fire hazards*
Magnesium and alloys—Fire hazards*
Oxygen cutting—Fire hazards*
Photographic films and plates—Fire hazards*
Powder metallurgy—Fire hazards*
BT: Hazards
RT: Accidents
Fire protection
Fires
Magnesium
Oil well fires
Powder metallurgy
Spontaneous combustion

Fire houses 402.2, 914.2
DT: Predates 1975
UF: Firehouses
BT: Buildings
RT: Fire fighting equipment

Fire hydrants
USE: Hydrants

Fire insurance 914.2 (911.1)
DT: January 1993
UF: Fires—Insurance*
BT: Insurance
RT: Fire protection

Fire protection 914.2
DT: Predates 1975
UF: Ships—Fire prevention*
BT: Protection
NT: Fireproofing
Runway foaming
RT: Fire alarm systems
Fire detectors
Fire fighting equipment
Fire hazards
Fire insurance
Fire resistance
Fires
Flame retardants
Smoke detectors
Spontaneous combustion
Sprinkler systems (fire fighting)

*Fire protection—Detectors**
USE: Fire detectors

*Fire protection—Sprinkler systems**
USE: Sprinkler systems (fire fighting)

Fire resistance 914.2
DT: January 1993
NT: Flame resistance
RT: Fire protection
Fireproofing
Fires
Flame retardants
Flammability
Structural design

Fire retardants
USE: Flame retardants

Fire testing
USE: Flammability testing

Fire tube boilers 614.2
DT: January 1993
UF: Boilers—Fire tube*
BT: Boilers

Fireclay 812.2
DT: January 1993
UF: Refractory materials—Fireclay*
BT: Refractory materials

* Former Ei Vocabulary term

Firedamp (503.1) (522) (804.1) (914.2)
DT: January 1993
UF: Mines and mining—Firedamp*
BT: Gases
RT: Gas hazards
Methane
Mines
Mining

Firehouses
USE: Fire houses

Fireproofing 914.2
DT: January 1993
BT: Fire protection
RT: Fire resistance

Fires 914.2
DT: Predates 1975
NT: Mine fires
Oil well fires
RT: Accidents
Combustion
Fire hazards
Fire protection
Fire resistance
Flame research
Spontaneous combustion

*Fires—Insurance**
USE: Fire insurance

Firing (boilers)
USE: Boiler firing

Firing (of materials) (414.1) (812.1) (812.2)
DT: January 1993
UF: Refractory materials—Firing*
RT: Ceramic materials
Refractory materials

Firing cables 706.2 (654.1)
DT: January 1993
UF: Rockets and missiles—Firing cables*
Shot firing cables
BT: Electric cables
RT: Rockets

Fischer-Tropsch synthesis 802.2
DT: January 1993
BT: Catalysis
Synthesis (chemical)
RT: Hydrocarbons

Fish detectors (471.5) (674.1)
DT: January 1993
UF: Fishing vessels—Fish detection*
Fishing vessels—Fish detectors*
BT: Detectors
RT: Fisheries
Fishing vessels

Fish ladders
USE: Fishways

Fish ponds 821.3 (471.5)
DT: January 1993
UF: Aquaculture—Fish ponds*
RT: Aquaculture
Surface waters

Fisheries (471.5)
DT: Predates 1975
RT: Aquaculture
Fish detectors
Fishing vessels
Marine biology

Fishing (oil wells) (501.1) (512.1.2)
DT: January 1993
UF: Boreholes—Fishing*
Oil well drilling—Fishing*
RT: Boreholes
Directional drilling
Oil well casings
Oil well drilling equipment

Fishing vessels 674.1 (471.5)
DT: Predates 1975
BT: Boats
RT: Depth indicators
Fish detectors
Fisheries

*Fishing vessels—Constructing and outfitting**
USE: Outfitting (water craft) OR
Shipbuilding

*Fishing vessels—Depth indicators**
USE: Depth indicators

*Fishing vessels—Fish detection**
USE: Fish detectors

*Fishing vessels—Fish detectors**
USE: Fish detectors

Fishways (407.2) (441.3) (821.3)
DT: Predates 1975
UF: Fish ladders
BT: Dams

Fission products 622.1 (932.2.1)
DT: January 1993
UF: Nuclear reactors—Fission products*
BT: Radioactive materials
RT: Fission reactions
Nuclear reactors
Radioactive wastes
Radioactivity
Spent fuels

Fission reactions 932.2.1 (621.1)
DT: January 1993
UF: Nuclear energy—Fission reactions*
Nuclear fission
Nuclear fuels—Fission*
Uranium and alloys—Fission*
RT: Fission products
Nuclear energy
Nuclear explosions
Nuclear physics
Nuclear reactors

Fission reactors
USE: Nuclear reactors

* Former Ei Vocabulary term

Fits and tolerances (601) (902.2)
DT: Predates 1975
UF: Tolerances
RT: Inspection
Materials testing
Quality control
Reliability

Fittings (hose)
USE: Hose fittings

Fittings (pipe)
USE: Pipe fittings

Fixation (nitrogen)
USE: Nitrogen fixation

Fixtures (plumbing)
USE: Plumbing fixtures

Fixtures (tooling) 603.2 (604.2)
DT: January 1993
UF: Fixtures—Brazing*
Industrial plants—Tools, jigs and fixtures*
Tools, jigs and fixtures*
BT: Equipment
RT: Clamping devices
Indexing (materials working)
Jigs
Tools

*Fixtures—Brazing**
USE: Fixtures (tooling)

Flame cleaning (539)
DT: January 1993
UF: Metal cleaning—Flame method*
BT: Metal cleaning

Flame deflectors (654.2) (914.1) (914.2)
DT: January 1993
UF: Rocket engines—Flame deflectors*
BT: Rocket engines

Flame hardening 537.1 (545.3)
DT: January 1993
UF: Steel heat treatment—Flame hardening*
BT: Hardening
Heat treatment
RT: Case hardening
Spark hardening

Flame reduction process
USE: Direct reduction process

Flame research 521.4 (914.2)
DT: Predates 1975
BT: Research
RT: Combustion
Fires
Fuels

Flame resistance 914.2
DT: January 1993
UF: Paper—Flame resistance*
Plastics—Flame resistance*
Textiles—Flame resistance*
BT: Fire resistance
RT: Flame retardants

Flame retardants 803, 914.2
DT: January 1981
UF: Fire retardants
BT: Agents
RT: Additives
Fire protection
Fire resistance
Flame resistance

Flame spraying 813.1, 817.2
DT: January 1993
UF: Protective coatings—Flame spraying*
BT: Coating techniques

Flammability 521.4, 914.2
DT: January 1993
UF: Combustibility
Inflammability
BT: Physical properties
RT: Combustion
Fire resistance
Flammability testing
Flammable materials
Spontaneous combustion

Flammability testing 914.2, 521.4
DT: January 1993
UF: Building materials—Flammability tests*
Fire testing
BT: Materials testing
Safety testing
RT: Equipment testing
Flammability

Flammable materials 521.4, 914.2
DT: Predates 1975
UF: Inflammable materials
BT: Materials
RT: Flammability

Flanges 619.1.1
DT: January 1993
UF: Pipe joints—Flanges*
BT: Components
RT: Pipe joints

Flare stacks 513.2, 402.1
DT: January 1993
UF: Petroleum refineries—Flare stacks*
Smokestacks (petroleum refinery)
BT: Petroleum refineries

Flash welding 538.2.1
DT: January 1993
UF: Welding—Flash*
BT: Pressure welding
Resistance welding

Flashlights 707
DT: Predates 1975
BT: Electric lamps

Flashover 701.1
DT: Predates 1975
UF: Electric flashover
BT: Electric discharges
RT: Electric arcs
Electric breakdown

* Former Ei Vocabulary term

Flashover *(continued)*
Electric corona
Electric equipment protection
Electric grounding
Surface discharges

Flat cables 706.2
DT: January 1993
UF: Electric cables—Flat*
Ribbon cables
BT: Electric cables

Flavors 822.3
DT: January 1993
UF: Flavors and fragrances*
BT: Food additives

*Flavors and fragrances**
USE: Flavors OR
Fragrances

Flaws
USE: Defects

Flax 819.1
DT: Predates 1975
BT: Natural fibers
RT: Linen

Fleet operations (432)
SN: Motor transportation only
DT: January 1993
BT: Motor transportation
RT: Truck drivers

Flexible couplings 602.2
DT: January 1993
UF: Couplings—Flexible*
BT: Couplings

Flexible foamed plastics 817.1
DT: January 1993
UF: Plastics, Foamed—Flexible*
BT: Foamed plastics

Flexible manufacturing systems 913.4.1
DT: June 1990
BT: Manufacture
RT: Industrial plants

Flexible structures 408.2 (441.1) (655.1)
DT: January 1993
UF: Dams—Flexible structures*
Spacecraft—Flexible structures*
BT: Structures (built objects)
RT: Dams
Inflatable structures

Flexural strength
USE: Bending strength

Flexural waves
USE: Elastic waves

Flickering 741.1
DT: January 1993
UF: Electric lamps—Flickering*
Visibility—Flicker effects*
RT: Visibility

Flight dynamics (652.1) (654.1) (931.1)
DT: January 1993
UF: Aircraft—Flight dynamics*
BT: Aerodynamics
RT: Aircraft

Flight simulators (652.1) (652.4) (656.1) (912.4)
DT: January 1993
UF: Aircraft simulators
Aircraft—Flight simulators*
Helicopters—Flight simulators*
BT: Simulators
RT: Aircraft
Helicopters
Space simulators
Virtual reality

Flip chip devices 714.2
DT: January 1993
UF: Integrated circuits—Flip chip construction*
BT: Hybrid integrated circuits
RT: Thin film circuits

Flip flop circuits 713.4
DT: January 1993
UF: Bistable multivibrators
Electronic circuits, Flip flop*
Flip flops
Latches (circuits)
BT: Multivibrators
RT: Digital storage
Logic circuits
Oscillators (electronic)
Trigger circuits

Flip flops
USE: Flip flop circuits

Floating breakwaters 407.1
DT: January 1993
UF: Breakwaters—Floating*
BT: Breakwaters

Floating cranes 693.1
DT: January 1993
UF: Cranes—Floating*
BT: Cranes

Floating drydocks 673.3
DT: January 1993
UF: Drydocks—Floating*
BT: Drydocks

Floating point arithmetic
USE: Digital arithmetic

Floating power plants 614.2
DT: January 1993
UF: Power plants—Floating*
BT: Power plants

Floating pumping plants 446 (402.1)
DT: January 1993
UF: Pumping plants—Floating*
BT: Pumping plants

Flocculation 802.3
DT: January 1993
BT: Agglomeration

* Former Ei Vocabulary term

Flocculation *(continued)*
RT: Colloids
Emulsification
Flotation
Sedimentation
Sewage treatment
Water treatment

Flood control 442.1, 914.1 (454.1)
DT: Predates 1975
BT: Control
RT: Dams
Environmental engineering
Floods
Hydrology
Levees
Regional planning
River control
Rivers
Runoff
Water supply

Flood damage (406) (422.1) (914.1)
DT: January 1993
UF: Roads and streets—Flood damage*
RT: Flood insurance
Floods

Flood insurance (442.1) (914.1) (911.1)
DT: January 1993
UF: Floods—Insurance*
BT: Insurance
RT: Flood damage
Floods

Flooding (mines)
USE: Mine flooding

Flooding (oil wells)
USE: Oil well flooding

Floodlighting 707.1
DT: Predates 1975
BT: Electric lighting
RT: Electric lamps

Floods (442.1) (914.1)
DT: Predates 1975
RT: Erosion
Flood control
Flood damage
Flood insurance
Hydrology
Mine flooding
Precipitation (meteorology)
Scour
Surface waters
Tides

*Floods—Insurance**
USE: Flood insurance

Floors 402
DT: Predates 1975
UF: Buildings—Floors*
Cars—Floors*
BT: Structural members
NT: Wooden floors
RT: Uplift pressure
Walls (structural partitions)

Floors (bridge)
USE: Bridge decks

*Floors—Composite materials**
USE: Composite materials

*Floors—Concrete**
USE: Concrete construction

*Floors—Wood**
USE: Wooden floors

Floppy disk drives
USE: Floppy disk storage

Floppy disk storage 722.1
DT: January 1993
UF: Data storage, Magnetic—Floppy disk*
Floppy disk drives
Floppy disks
BT: Magnetic disk storage

Floppy disks
USE: Floppy disk storage

Flotation 802.3 (533.1)
DT: Predates 1975
BT: Separation
NT: Froth flotation
RT: Beneficiation
Flocculation
Flotation agents
Ore treatment
Sedimentation
Sewage treatment
Size separation
Water treatment

Flotation agents 803 (533.1) (804)
DT: January 1993
UF: Flotation—Agents*
BT: Agents
RT: Flotation

*Flotation—Agents**
USE: Flotation agents

Flow confinement
USE: Confined flow

Flow control 631.1, 731.3
DT: January 1993
UF: Control, Mechanical variables—Flow*
Flow of fluids—Control*
BT: Mechanical variables control
RT: Flow measurement
Flow of fluids

Flow instrumentation
USE: Flow measuring instruments

Flow interactions 631.1
DT: January 1993
UF: Flow of fluids—Flow interactions*
Interactions (flow)
BT: Flow of fluids
NT: Fluid structure interaction
RT: Pressure drop

Flow measurement 631.1, 943.2
DT: January 1993
UF: Flow of fluids—Measurement*
Mechanical variables measurement—Flow*
BT: Mechanical variables measurement
RT: Anemometers
Flow control
Flow measuring instruments
Flow of fluids
Flowmeters
Lysimeters
Orifices
Permeameters (mechanical permeability)
Pressure measurement
Purging
Rheology

Flow measuring instruments 943.1
DT: January 1993
UF: Flow instrumentation
BT: Instruments
NT: Anemometers
Flowmeters
Lysimeters
Permeameters (mechanical permeability)
RT: Flow measurement
Flow of fluids

Flow of fluids 631.1 (931.1)
DT: Predates 1975
UF: Coal slurries—Flow*
Flow of gases
Flow of liquids
Fluid flow
Furnaces—Flow*
Gaseous flow
Internal combustion engines—Flow problems*
Natural gas wells—Flow*
Oil well production—Flow*
Pulp manufacture—Flow*
Refrigerants—Flow*
Sewers—Flow*
BT: Fluid dynamics
NT: Axial flow
Boundary layer flow
Capillary flow
Compressible flow
Confined flow
Flow interactions
Flow of water
Hypersonic flow
Laminar flow
Multiphase flow
Newtonian flow
Non Newtonian flow
Percolation (fluids)
Plasma flow
Potential flow
Pulsatile flow
Radial flow
Rotational flow
Subsonic flow
Supersonic flow
Transition flow
Transonic flow
Turbulent flow
Unsteady flow
Viscous flow
Vortex flow
Wall flow
RT: Aerodynamics
Bodies of revolution
Bubble formation
Cylinders (shapes)
Diffusers (fluid)
Drainage
Finite volume method
Flow control
Flow measurement
Flow measuring instruments
Flow of solids
Flow visualization
Gases
Granular materials
Headboxes
Heat convection
Hydraulics
Hydrodynamics
Intake systems
Jets
Liquids
Nozzles
Packed beds
Pipe
Reynolds number
Sediment transport
Seepage
Shear flow
Spheres
Turbulence

*Flow of fluids—Axial**
USE: Axial flow

*Flow of fluids—Bodies of revolution**
USE: Bodies of revolution

*Flow of fluids—Boiling fluids**
USE: Boiling liquids

*Flow of fluids—Boundary layer**
USE: Boundary layer flow

*Flow of fluids—Bubble formation**
USE: Bubble formation

*Flow of fluids—Capillaries**
USE: Capillary flow

*Flow of fluids—Cascades**
USE: Cascades (fluid mechanics)

*Flow of fluids—Channel flow**
USE: Channel flow

*Flow of fluids—Compressible**
USE: Compressible flow

*Flow of fluids—Conduits**
USE: Confined flow

*Flow of fluids—Cones**
USE: Cones

* Former Ei Vocabulary term

*Flow of fluids—Control**
USE: Flow control

*Flow of fluids—Diffusers**
USE: Diffusers (fluid)

*Flow of fluids—Drag**
USE: Drag

*Flow of fluids—Flow interactions**
USE: Flow interactions

*Flow of fluids—Granular materials**
USE: Granular materials

*Flow of fluids—Hypersonic**
USE: Hypersonic flow

*Flow of fluids—Inlets**
USE: Intake systems

*Flow of fluids—Jets**
USE: Jets

*Flow of fluids—Laminar**
USE: Laminar flow

*Flow of fluids—Measurement**
USE: Flow measurement

*Flow of fluids—Mixing**
USE: Mixing

*Flow of fluids—Multiphase**
USE: Multiphase flow

*Flow of fluids—Newtonian**
USE: Newtonian flow

*Flow of fluids—Non Newtonian**
USE: Non Newtonian flow

*Flow of fluids—Open channels**
USE: Open channel flow

*Flow of fluids—Orifices**
USE: Orifices

*Flow of fluids—Packed beds**
USE: Packed beds

*Flow of fluids—Pipes**
USE: Pipe flow

*Flow of fluids—Plasmas**
USE: Plasma flow

*Flow of fluids—Porous materials**
USE: Porous materials

*Flow of fluids—Potential flow**
USE: Potential flow

*Flow of fluids—Pulsatile flow**
USE: Pulsatile flow

*Flow of fluids—Rotating**
USE: Rotational flow

*Flow of fluids—Screens**
USE: Screens (sizing)

*Flow of fluids—Sediment transport**
USE: Sediment transport

*Flow of fluids—Spheres**
USE: Spheres

*Flow of fluids—Subsonic**
USE: Subsonic flow

*Flow of fluids—Supersonic**
USE: Supersonic flow

*Flow of fluids—Transition flow**
USE: Transition flow

*Flow of fluids—Transonic**
USE: Transonic flow

*Flow of fluids—Tubes**
USE: Pipe flow

*Flow of fluids—Turbulent**
USE: Turbulent flow

*Flow of fluids—Two phase flow**
USE: Two phase flow

*Flow of fluids—Unsteady flow**
USE: Unsteady flow

*Flow of fluids—Viscous**
USE: Viscous flow

*Flow of fluids—Visualization**
USE: Flow visualization

*Flow of fluids—Vortex flow**
USE: Vortex flow

*Flow of fluids—Wakes**
USE: Wakes

Flow of gases
USE: Flow of fluids

Flow of liquids
USE: Flow of fluids

Flow of solids 931.1
SN: Excludes transport of solids in fluid media
DT: Predates 1975
UF: Solid flow
BT: Dynamics
NT: Plastic flow
RT: Creep
Flow of fluids
Granular materials
Materials handling
Plasticity
Rheology
Sediment transport
Shear flow
Solids
Two phase flow
Unsteady flow

*Flow of solids—Granular materials**
USE: Granular materials

Flow of water 631.1.1
DT: Predates 1975
BT: Flow of fluids
NT: Groundwater flow
Stream flow

* Former Ei Vocabulary term

Flow of water *(continued)*
RT: Discharge (fluid mechanics)
Hydrodynamics
Ocean currents
Water

*Flow of water—Underground**
USE: Groundwater flow

Flow patterns 631.1
DT: January 1995
RT: Channel flow
Confined flow
Pipe flow
Two phase flow

Flow stress
USE: Plastic flow

Flow visualization 631.1
DT: January 1993
UF: Flow of fluids—Visualization*
Visualization (flow)
Wind tunnels—Visualization*
BT: Visualization
RT: Aerosols
Bubbles (in fluids)
Flow of fluids
Wind tunnels

Flowcharting 723.1
DT: January 1993
UF: Computer programming languages—Flowcharting*
Computer programming—Flowcharting*
Flowsheets
Industrial plants—Flow sheets*
BT: Systems analysis
RT: Computer programming
Computer programming languages
Industrial plants
Mathematical models

Flowlines 511.2, 619.1
DT: January 1993
UF: Oil well production—Flowlines*
BT: Petroleum pipelines

Flowmeters 943.1
SN: For fluids flowing in pipes
DT: Predates 1975
UF: Fluid meters
BT: Flow measuring instruments
Gages
NT: Gas meters
Magnetic flowmeters
Rheometers
Ultrasonic flowmeters
Water meters
RT: Flow measurement
Orifices
Permeameters (mechanical permeability)
Porosimeters
Velocimeters

*Flowmeters—Magnetic**
USE: Magnetic flowmeters

*Flowmeters—Ultrasonic**
USE: Ultrasonic flowmeters

Flowsheets
USE: Flowcharting

Flue gases 451.1 (521)
DT: Predates 1975
BT: Exhaust gases
RT: Effluent treatment

*Flue gases—Treatment**
USE: Effluent treatment

Flues (521) (402.1) (614.1)
DT: January 1993
UF: Gas appliances—Flues*
BT: Chimneys
RT: Gas appliances

Fluid catalytic cracking 802.2 (513.1)
DT: January 1993
BT: Catalytic cracking
RT: Petroleum refining

Fluid dynamics 931.1 (631.1)
DT: January 1977
BT: Dynamics
Fluid mechanics
NT: Computational fluid dynamics
Drag
Flow of fluids
Gas dynamics
Hydrodynamics
Surges (fluid)
Turbulence
RT: Cascades (fluid mechanics)
Fluids
Jets
Liquid waves
Navier Stokes equations
Shock waves
Viscosity of gases

Fluid flow
USE: Flow of fluids

Fluid mechanics 931.1 (631.1)
DT: October 1975
BT: Mechanics
NT: Bubble formation
Cascades (fluid mechanics)
Cavitation
Discharge (fluid mechanics)
Drop breakup
Drop formation
Fluid dynamics
Hydraulics
Pneumatics
RT: Bubbles (in fluids)
Buoyancy
Capillarity
Compressibility of gases
Compressibility of liquids
Contact angle
Contacts (fluid mechanics)
Continuum mechanics

Fluid mechanics *(continued)*
Fluids
Gases
Hydrostatic pressure
Liquids

Fluid meters
USE: Flowmeters

Fluid structure interaction 631.1 (931.1)
DT: January 1993
BT: Flow interactions
RT: Hydrodynamics
Vortex shedding

Fluidic amplifiers 632.2, 732.1
DT: Predates 1975
UF: Amplifiers (fluidic)
BT: Fluidic devices
Hydraulic control equipment
NT: Hydraulic amplifiers
RT: Nonelectric final control devices

Fluidic devices 632.2, 732.1
DT: Predates 1975
BT: Equipment
NT: Fluidic amplifiers
Fluidic logic devices
RT: Fluidics

Fluidic logic devices 632.2, 732.1 (721.2)
DT: January 1993
UF: Logic devices—Fluidic elements*
BT: Fluidic devices
Logic devices
RT: Fluidics

Fluidics 632.1, 732.1
DT: Predates 1975
RT: Control
Fluidic devices
Fluidic logic devices
Hydraulics

Fluidity 931.2
DT: January 1993
UF: Fluidity testing
Foundry practice—Fluidity testing*
Liquid metals—Fluidity testing*
BT: Physical properties
RT: Viscosity

Fluidity testing
USE: Metal testing AND
Fluidity

Fluidization 802.3
DT: January 1993
UF: Fluidizing
BT: Chemical operations
RT: Fluidized beds

Fluidized bed combustion 521.1
DT: January 1993
UF: Combustion—Fluidized beds*
BT: Combustion
RT: Fluidized bed combustors
Fluidized bed furnaces
Fluidized beds

Fluidized bed combustors 521.2
DT: January 1993
UF: Combustors—Fluidized bed*
BT: Combustors
RT: Fluidized bed combustion
Fluidized bed furnaces
Fluidized beds

Fluidized bed furnaces 642.2
DT: January 1993
UF: Furnaces—Fluidized bed*
BT: Furnaces
RT: Fluidized bed combustion
Fluidized bed combustors
Fluidized bed melting
Fluidized bed process
Fluidized beds

Fluidized bed melting (532.5) (534.1)
DT: January 1993
UF: Metal melting—Fluidized bed*
BT: Metal melting
RT: Fluidized bed furnaces
Fluidized bed process
Fluidized beds

Fluidized bed process (533.1)
DT: January 1993
UF: Ore treatment—Fluidized bed process*
BT: Ore treatment
RT: Fluidized bed furnaces
Fluidized bed melting
Fluidized beds

Fluidized beds (802.1) (521.2) (642.2)
DT: January 1993
UF: Chemical equipment—Fluidized beds*
Heat transfer—Fluidized beds*
BT: Chemical equipment
RT: Fluidization
Fluidized bed combustion
Fluidized bed combustors
Fluidized bed furnaces
Fluidized bed melting
Fluidized bed process
Heat transfer
Packed beds

Fluidizing
USE: Fluidization

Fluids (931.1) (932.2) (631)
DT: January 1977
UF: Heat transfer—Fluids*
BT: Materials
NT: Air
Cutting fluids
Drilling fluids
Fracturing fluids
Gases
Hydraulic fluids
Liquids
Magnetic fluids
RT: Fluid dynamics
Fluid mechanics

* Former Ei Vocabulary term

Fluorescence 741.1
DT: Predates 1975
BT: Luminescence
RT: Electromagnetic wave absorption
Fluorescent lamps
Fluorescent screens
Fluorometers
Phosphorescence

Fluorescent lamps 707.2 (741.1)
DT: January 1993
UF: Electric lamps, Fluorescent*
Street lighting—Fluorescent*
BT: Discharge lamps
Luminescent devices
RT: Fluorescence
Street lighting

Fluorescent screens 741.3
DT: Predates 1975
UF: Cathode ray tube screens
Screens (display)
Video display screens
BT: Display devices
Luminescent devices
RT: Cathode ray tubes
Cathode rays
Fluorescence
Interactive devices
Radar displays
Television receivers

Fluoridation 445.1
SN: Of water
DT: January 1993
UF: Water treatment—Fluoridation*
BT: Water treatment

Fluoride minerals 482.2
DT: January 1993
UF: Mineralogy—Fluorides*
BT: Fluorine compounds
Halide minerals
RT: Fluorine

Fluorimeters
USE: Fluorometers

Fluorine 804
DT: Predates 1975
BT: Halogen elements
RT: Fluoride minerals

Fluorine compounds (804.1) (804.2)
DT: Predates 1975
BT: Halogen compounds
NT: Fluoride minerals
Fluorine containing polymers
Fluorocarbons
Freons
Hydrofluoric acid

Fluorine containing polymers 815.1
SN: Includes copolymers
DT: Predates 1975
UF: Fluoropolymers
BT: Fluorine compounds
Organic polymers
NT: Polytetrafluoroethylenes

Fluorite
USE: Fluorspar

Fluorocarbon based blood substitutes
USE: Fluorocarbons AND
Blood substitutes

Fluorocarbons 804.1
DT: January 1993
UF: Fluorocarbon based blood substitutes
Fluorohydrocarbons*
BT: Fluorine compounds
RT: Freons
Hydrocarbons

*Fluorohydrocarbons**
USE: Fluorocarbons

Fluorometers 941.3
DT: Predates 1975
UF: Fluorimeters
BT: Optical instruments
RT: Analytic equipment
Fluorescence

Fluoropolymers
USE: Fluorine containing polymers

Fluorspar 482.2
DT: Predates 1975
UF: Fluorite
BT: Calcium compounds
Ores
RT: Fluorspar deposits

Fluorspar deposits 505.1
DT: Predates 1975
BT: Mineral resources
RT: Fluorspar

Flutter (aerodynamics) 651.1 (652.4)
DT: January 1993
UF: Aerodynamics—Flutter*
Aeronautical flutter
Helicopters—Flutter*
BT: Aerodynamics
RT: Helicopters

Flux core wire welding 538.2.1
DT: January 1993
UF: Fluxed core welding
Welding—Fluxed core*
BT: Electric arc welding

Fluxed core welding
USE: Flux core wire welding

Fluxes (804) (538)
DT: January 1993
UF: Welding—Fluxes*
BT: Materials
RT: Agents
Brazing
Oxygen cutting
Soldering
Welding

* Former Ei Vocabulary term

Fluxmeters 942.3
SN: Instruments for measuring magnetic flux
DT: Predates 1975
UF: Magnetic flux measuring instruments
BT: Magnetic measuring instruments
RT: Magnetic field measurement

Fly ash (451.1) (521.1) (804)
DT: Predates 1975
UF: Brickmaking—Fly ash*
Concrete aggregates—Fly ash*
Roadbuilding materials—Fly ash*
BT: Industrial wastes
RT: Aerosols
Brickmaking
Concrete aggregates
Roadbuilding materials

Flywheel propulsion (601) (705)
DT: January 1993
UF: Electric traction—Flywheel propulsion*
BT: Propulsion
RT: Electric traction

Flywheels 601.1
DT: Predates 1975
BT: Engines
NT: Hydraulic accumulators
RT: Energy storage

FMR
USE: Ferromagnetic resonance

Foam control (802) (931.2)
DT: January 1993
UF: Liquids—Foam control*
BT: Control
RT: Foams
Liquids

Foam rubber
USE: Foamed rubber

Foamed plastics 817.1
DT: January 1993
UF: Cellular plastics
Expanded plastics
Plastic foams
Plastics, Foamed*
BT: Foams
Plastics
NT: Flexible foamed plastics
Rigid foamed plastics
RT: Blowing agents
Foamed products
Insulating materials

Foamed products (413.2) (804)
DT: Predates 1975
UF: Heat insulating materials—Foamed products*
RT: Foamed plastics
Foamed rubber
Thermal insulating materials

Foamed rubber 818.5
DT: January 1993
UF: Foam rubber
Rubber, Foamed*
BT: Foams
RT: Blowing agents
Foamed products
Rubber
Synthetic rubber

Foaming (runways)
USE: Runway foaming

Foaming agents
USE: Blowing agents

Foams (804)
DT: Predates 1975
UF: Fire extinguishers—Foam*
BT: Porous materials
NT: Ceramic foams
Foamed plastics
Foamed rubber
RT: Aerosols
Colloids
Fire extinguishers
Foam control

Focus
USE: Focusing

Focusing (711.1) (741.1) (751.1)
DT: Predates 1975
UF: Focus
RT: Cameras
Cathode rays
Electron lenses
Electron optics
Lenses
Magnetic lenses
Optics
Particle accelerators
Particle optics
Thermal blooming

Fog 443.1
DT: January 1993
UF: Meteorology—Fog*
Mist
BT: Earth atmosphere
RT: Fog dispersal
Meteorology
Steam

Fog dispersal 443.1 (431.4)
DT: January 1993
UF: Airports—Fog dispersal*
RT: Airports
Control
Fog
Visibility

Foil
USE: Metal foil

Foil bearings 601.2
DT: January 1993
UF: Bearings—Foil*
BT: Bearings (machine parts)
RT: Journal bearings
Metal foil

* Former Ei Vocabulary term

Food additives 822.3, 803 (804)
DT: January 1993
UF: Food products—Chemical additives*
BT: Additives
Food products
NT: Flavors
Food preservatives
RT: Dextrose
Fructose
Sugar substitutes

Food preservation 822.2
DT: January 1993
UF: Food products—Preservation*
Preservation (food)
BT: Food processing
RT: Food preservatives

Food preservatives 822.3, 803 (804)
DT: January 1993
UF: Food products—Preservatives*
Preservatives (food)
BT: Food additives
RT: Food preservation

Food processing 822.2
DT: January 1993
UF: Food products—Processing*
NT: Canning
Food preservation
RT: Food products
Food products plants
Food storage
Freezing
Heating
Kitchens
Sterilization (cleaning)
Sugar manufacture

Food products 822.3
DT: Predates 1975
NT: Beverages
Dairy products
Food additives
Fruits
Meats
Molasses
Sugar (sucrose)
Synthetic foods
RT: Agricultural products
Carbohydrates
Food processing
Food products plants
Food storage
Food supply
Nutrition
Proteins
Starch
Vegetable oils
Vitamins

Food products plants 822.1 (402.1)
DT: Predates 1975
BT: Industrial plants
NT: Bakeries
Bottling plants
Breweries
Dairies
Packing plants
Sugar factories
RT: Distilleries
Food processing
Food products

*Food products—Algae**
USE: Algae

*Food products—Canning**
USE: Canning

*Food products—Chemical additives**
USE: Food additives

*Food products—Corrosive properties**
USE: Corrosive effects

*Food products—Fats**
USE: Oils and fats

*Food products—Fruit juices**
USE: Fruit juices

*Food products—Fruits**
USE: Fruits

*Food products—Meats**
USE: Meats

*Food products—Preservation**
USE: Food preservation

*Food products—Preservatives**
USE: Food preservatives

*Food products—Processing**
USE: Food processing

*Food products—Storage**
USE: Food storage

*Food products—Synthetics**
USE: Synthetic foods

*Food products—Thawing**
USE: Thawing

Food storage 694.4, 822.1
DT: January 1993
UF: Food products—Storage*
BT: Storage (materials)
RT: Food processing
Food products
Grain elevators
Refrigeration

Food supply 822.3 (404.2)
DT: January 1993
UF: Civil defense—Food supply*
RT: Food products

Footage counters 742.2 (943.3)
DT: January 1993
UF: Motion pictures—Footage counters*
BT: Instruments
RT: Motion picture cameras
Motion picture editing machines

Footbridges 401.1 (406.1)
DT: Predates 1975
UF: Pedestrian bridges
Pedestrian overpasses
BT: Bridges
RT: Highway systems
Pedestrian safety

*Footbridges—Aluminum**
USE: Aluminum

*Footbridges—Concrete**
USE: Concrete construction

*Footbridges—Welded steel**
USE: Welded steel structures

Force control 731.3
DT: January 1993
UF: Control, Mechanical variables—Forces*
BT: Mechanical variables control
RT: Force measurement

Force measurement 943.2
DT: January 1993
UF: Mechanical variables measurement—Forces*
Metal cutting—Force measurement*
BT: Mechanical variables measurement
RT: Dynamometers
Force control
Metal cutting
Transducers

Forced convection
USE: Heat convection

Forecasting (912.2) (921) (922.2)
DT: January 1993
UF: Prediction*
NT: Electric load forecasting
Technological forecasting
Weather forecasting
RT: Management
Management science
Mathematical models
Planning
Research and development management
Risk assessment
Risks
Scheduling
Statistical methods

Foreign trade
USE: International trade

Foremen
USE: Supervisory personnel

Forestry 821
DT: Predates 1975
UF: Aircraft—Forestry applications*
Photogrammetry—Forestry applications*
NT: Logging (forestry)
RT: Agriculture
Environmental engineering

*Forestry—Logging**
USE: Logging (forestry)

Forge welding
USE: Pressure welding

Forging 535.2.2
SN: Use for the process; for the product, use FORGINGS
DT: Predates 1975
BT: Metal forming
NT: Upsetting (forming)
RT: Dies
Forging machines
Forgings
Hot working
Metal stamping
Rolling
Swaging

Forging dies
USE: Dies

Forging machines 535.2.1
DT: Predates 1975
BT: Metal forming machines
RT: Dies
Forging

Forgings 535.2
SN: Use for the product; for the process, use FORGING
DT: Predates 1975
RT: Forging

Formability (535) (931.2)
SN: Includes drawability, forgeability, etc.
DT: January 1993
UF: Bendability
Deformability
Metals and alloys—Formability*
BT: Mechanical properties
NT: Machinability
Weldability
RT: Anisotropy
Bending (forming)
Ductility
Forming
Plasticity

Formal languages 721.1, 723.1.1
DT: January 1993
UF: Automata theory—Formal languages*
Languages (formal)
Regular languages
NT: Algorithmic languages
Context free languages
Context sensitive languages
RT: Computational linguistics
Computer hardware description languages
Computer programming languages
Formal logic
Query languages

Formal logic 721.1
DT: January 1993
UF: Computer metatheory—Formal logic*
Logic (formal)
Mathematical logic
BT: Computation theory

* Former Ei Vocabulary term

Formal logic *(continued)*
NT: Boolean functions
Computability and decidability
Majority logic
Many valued logics
Probabilistic logics
Recursive functions
Threshold logic
RT: Artificial intelligence
Automata theory
Boolean algebra
Formal languages
Fuzzy sets
Switching theory
Theorem proving
Turing machines

Formaldehyde 804.1
DT: Predates 1975
UF: Formalin
BT: Aldehydes
RT: Acetal resins
Melamine formaldehyde resins
Urea formaldehyde resins

Formalin
USE: Formaldehyde

Formations (oil bearing)
USE: Oil bearing formations

Forming (535.2) (816.1)
DT: January 1993
UF: Materials working
Pressure forming
BT: Manufacture
NT: Bending (forming)
Casting
Drawing (forming)
Electroforming
Extrusion
Metal forming
Molding
Perforating
Photochemical forming
Piercing
Plastics forming
Preforming
Pressing (forming)
Pultrusion
Punching
Stamping
Straightening
Stretching
Thermomechanical treatment
RT: Compaction
Finishing
Formability

Forms (concrete) (412)
DT: January 1993
UF: Concrete construction—Forms*
Concrete forms
BT: Molds
RT: Concrete construction
Construction equipment
Slip forming

FORTH (programming language) 723.1.1
DT: January 1993
UF: Computer programming languages—FORTH*
BT: High level languages

FORTRAN (programming language) 723.1.1
DT: January 1993
UF: Computer programming languages—FORTRAN*
BT: Procedure oriented languages

Fossil fuel deposits (503) (512)
DT: January 1993
BT: Mineral resources
NT: Coal deposits
Natural gas deposits
Oil sands
Oil shale
Petroleum deposits
RT: Energy resources
Fossil fuels

Fossil fuel power plants 614 (402.1) (706.1)
DT: January 1983
BT: Power plants
RT: Coal gasification
Electric power plants
Fossil fuels

Fossil fuels (522) (523) (524)
DT: January 1986
BT: Energy resources
Fuels
NT: Coal
Crude petroleum
Peat
Shale oil
RT: Fossil fuel deposits
Fossil fuel power plants
Hydrocarbons
Oil sands
Oil shale

Foucault currents
USE: Eddy currents

Fouling (539.1)
DT: January 1995
RT: Contamination
Corrosion
Degradation
Deposits

Foundations 483.2 (405)
DT: Predates 1975
BT: Structures (built objects)
NT: Anchorages (foundations)
Pile foundations
Underwater foundations
RT: Excavation
Grouting
Settlement of structures
Soil structure interactions
Subsidence
Underpinnings

* Former Ei Vocabulary term

*Foundations—Anchorages**
USE: Anchorages (foundations)

*Foundations—Frost effect**
USE: Frost effects

*Foundations—Grouting**
USE: Grouting

*Foundations—Permafrost**
USE: Permafrost

*Foundations—Piles**
USE: Pile foundations

*Foundations—Settlement**
USE: Settlement of structures

*Foundations—Soil structure interaction**
USE: Soil structure interactions

*Foundations—Subsidence**
USE: Subsidence

*Foundations—Underpinning**
USE: Underpinnings

*Foundations—Underwater**
USE: Underwater foundations

Foundries 534.1 (402.1)
DT: Predates 1975
BT: Industrial plants
RT: Foundry practice
Furnaces
Metal casting
Molding

*Foundries—Scrap reclamation**
USE: Scrap metal reprocessing

Foundry practice 534.2
DT: Predates 1975
NT: Aluminum foundry practice
Bronze foundry practice
Copper foundry practice
Coremaking
Cupola practice
Iron foundry practice
Lead foundry practice
Magnesium foundry practice
Steel foundry practice
Titanium foundry practice
Zinc foundry practice
RT: Carbon dioxide process
Casting
Centrifugal casting
Continuous casting
Foundries
Foundry sand
Full mold process
Gating and feeding
Investment casting
Light metals
Metal casting
Metal cleaning
Metal melting
Metallurgy
Molding
Patternmaking
Precision casting
Pressure pouring
Protective atmospheres
Pyrometry
Sealing (finishing)
Shaw process
Shell mold casting
Sodium silicate process
Zero gravity casting

*Foundry practice—Carbon dioxide process**
USE: Carbon dioxide process

*Foundry practice—Casting**
USE: Casting

*Foundry practice—Centrifugal casting**
USE: Centrifugal casting

*Foundry practice—Continuous casting**
USE: Continuous casting

*Foundry practice—Coremaking**
USE: Coremaking

*Foundry practice—Fluidity testing**
USE: Fluidity

*Foundry practice—Full mold process**
USE: Full mold process

*Foundry practice—Gating and feeding**
USE: Gating and feeding

*Foundry practice—Investment casting**
USE: Investment casting

*Foundry practice—Molding**
USE: Metal molding

*Foundry practice—Patternmaking**
USE: Patternmaking

*Foundry practice—Permanent mold**
USE: Permanent mold casting

*Foundry practice—Precision casting**
USE: Precision casting

*Foundry practice—Pressure pouring**
USE: Pressure pouring

*Foundry practice—Pyrometry**
USE: Pyrometry

*Foundry practice—Sealing**
USE: Sealing (finishing)

*Foundry practice—Shaw process**
USE: Shaw process

*Foundry practice—Shell process**
USE: Shell mold casting

*Foundry practice—Sodium silicate process**
USE: Sodium silicate process

*Foundry practice—Zero gravity**
USE: Zero gravity casting

* Former Ei Vocabulary term

Foundry sand 534.2
DT: January 1993
UF: Core sand
Molding sand
Sand, Foundry*
BT: Sand
RT: Bentonite
Binders
Foundry practice
Molding
Molds
Olivine

Fountains (403.1)
SN: Ornamental structures, not drinking fountains
DT: Predates 1975
BT: Structures (built objects)

Four wave mixing (711.1) (741.1)
DT: January 1993
UF: Multiwave mixing
BT: Nonlinear optics

Fourdrinier machines 811.1.2
DT: January 1993
UF: Papermaking machinery—Fourdrinier machines*
BT: Papermaking machinery

Fourier optics 741.1
DT: January 1993
BT: Optics
RT: Fourier transforms
Image processing
Optical data processing

Fourier transform infrared spectroscopy 801 (921.4) (931.1)
DT: January 1995
BT: Infrared spectroscopy
RT: Fourier transforms

Fourier transforms 921.3
DT: January 1993
UF: Mathematical transformations—Fourier transforms*
BT: Mathematical transformations
NT: Fast Fourier transforms
RT: Fourier optics
Fourier transform infrared spectroscopy

Fourth generation languages
USE: Query languages

Fractals 921
DT: January 1993
RT: Computer graphics
Geometry
Mathematical techniques
Topology

Fractional horsepower motors 705.3
DT: January 1993
UF: Electric motors—Fractional horsepower*
FHP motors
BT: Electric motors

Fractionating units 802.1
DT: January 1993
UF: Petroleum refineries—Fractionating units*
BT: Separators
RT: Fractionation
Petroleum refineries

Fractionation 802.3
DT: January 1993
BT: Separation
RT: Distillation
Fractionating units
Petroleum refining
Volume fraction

Fractography (421) (422.2)
DT: January 1995
BT: Microscopic examination
RT: Crystal microstructure
Fracture
Metallography
Radiography
Specimen preparation

Fracture (421)
DT: January 1993
BT: Failure (mechanical)
NT: Brittle fracture
RT: Crack propagation
Cracks
Fractography
Fracture mechanics
Fracture testing
Fracture toughness
High temperature testing
Mechanical properties
Stress corrosion cracking

Fracture fixation 461.1
DT: January 1993
UF: Biomedical engineering—Fracture fixation*
BT: Patient treatment
RT: Biomedical engineering

Fracture mechanics 931.1 (421)
DT: October 1975
BT: Mechanics
RT: Fracture

Fracture strength
USE: Fracture toughness

Fracture testing (421) (422.2)
DT: January 1993
UF: Materials testing—Fracture*
BT: Mechanical testing
RT: Fracture
Stress concentration
Tensile testing

Fracture toughness (421) (422.2)
DT: January 1995
UF: Fracture strength
Impact toughness
BT: Toughness
RT: Fracture
Impact resistance

Fracturing (oil wells) 511.1
DT: January 1993
UF: Oil wells—Fracturing*
BT: Production
NT: Hydraulic fracturing
RT: Fracturing fluids
Oil wells

Fracturing fluids (511.1)
DT: January 1993
UF: Oil wells—Fracturing fluids*
BT: Fluids
RT: Fracturing (oil wells)
Hydraulic fracturing
Oil wells

Fragrances (822.3) (804.1)
DT: January 1993
UF: Flavors and fragrances*
Perfumes
BT: Additives
RT: Essential oils

Frames (automobile)
USE: Automobile frames

Frames (structural)
USE: Structural frames

Francis turbines 617.1
DT: January 1993
UF: Hydraulic turbines—Francis*
BT: Hydraulic turbines

Francium 549.1, 622.1
DT: January 1993
BT: Alkali metals

Free convection
USE: Natural convection

Free electron lasers 744.5
DT: January 1993
UF: FEL
Lasers, Free electron*
BT: Lasers

Free energy 641.1
DT: January 1995
UF: Helmholtz free energy
BT: Thermodynamic properties
RT: Entropy
Gibbs free energy

Free piston compressors 618.1 (682.1.2)
DT: January 1993
UF: Compressors, Free piston*
Locomotives—Free piston compressors*
BT: Compressors
RT: Free piston engines
Locomotives

Free piston engines 612.1
DT: January 1993
UF: Internal combustion engines—Free piston engines*
BT: Internal combustion engines
RT: Free piston compressors
Gas turbines

Free radical polymerization 815.2
DT: January 1993
UF: Polymerization—Free radical polymerization*
BT: Free radical reactions
Polymerization
RT: Free radicals

Free radical reactions 802.2
DT: January 1993
BT: Chemical reactions
NT: Free radical polymerization
RT: Free radicals

Free radicals (804)
DT: January 1993
RT: Atoms
Free radical polymerization
Free radical reactions
Molecules

Free wing aircraft
USE: Freewing aircraft

Freeways
USE: Highway systems

Freewing aircraft 652.1
DT: January 1993
UF: Aircraft—Freewing*
Free wing aircraft
BT: Aircraft

Freezing (644.1) (802.3) (822.2)
DT: January 1993
UF: Shaft sinking—Freezing*
BT: Cooling
Phase transitions
RT: Food processing
Ice
Liquids
Nucleation
Sea ice
Solidification
Zone melting

Freight aircraft
USE: Cargo aircraft

Freight cars 682.1.1 (433)
DT: January 1993
UF: Cars—Freight*
BT: Railroad cars
NT: Baggage cars
Dump cars
Hopper cars
Postal cars
Refrigerator cars
Tank cars
RT: Freight transportation
Mine cars

*Freight handling**
USE: Cargo handling

*Freight handling—Door to door**
USE: Door to door service (freight)

*Freight handling—Piggyback**
USE: Piggyback freight handling

* Former Ei Vocabulary term

*Freight handling—Refrigerated**
USE: Refrigerated freight handling

*Freight handling—Shock problems**
USE: Shock problems

Freight transportation (431) (432) (433) (434)
DT: January 1993
UF: Shipping
Transportation—Freight*
BT: Transportation
NT: Coal transportation
RT: Cargo aircraft
Cargo handling
Freight cars
Loading

Freons 804 (644.2)
DT: Predates 1975
UF: Refrigerants—Freons*
BT: Fluorine compounds
Hydrocarbons
RT: Aerosols
Chlorofluorocarbons
Fluorocarbons
Propellants
Refrigerants

Frequencies (711.1) (741.1) (751.1)
DT: January 1995
NT: Natural frequencies

Frequency agility 716.2 (716.1)
DT: January 1993
UF: Radar—Frequency agility*
RT: Frequency hopping
Information theory
Radar

Frequency allocation 716.3, 716.4
DT: January 1993
UF: Radio—Frequency allocation*
RT: Electromagnetic compatibility
Radio
Radio broadcasting
Radio communication
Signal interference

Frequency changers
USE: Frequency converters

Frequency control (electric)
USE: Electric frequency control

Frequency conversion (optical)
USE: Optical frequency conversion

Frequency converter circuits 713.5
DT: January 1993
UF: Electronic circuits, Frequency converter*
BT: Frequency converters
Networks (circuits)
NT: Frequency dividing circuits
Frequency multiplying circuits
RT: Mixer circuits

Frequency converters (705.1) (713.5)
DT: January 1993
UF: Electric converters, Frequency*
Frequency changers
BT: Electric converters
NT: Frequency converter circuits
RT: Optical frequency conversion

Frequency dividing circuits 713.5
DT: January 1993
UF: Dividers (frequency)
Dividing circuits (frequency)
Electronic circuits, Frequency dividing*
BT: Frequency converter circuits
RT: Multivibrators

Frequency division multiplexing (716) (717) (718)
DT: January 1993
UF: FDM
Multiplexing, Frequency division*
Wavelength division multiplexing
WDM
BT: Multiplexing

Frequency domain analysis
921.3 (703.1.1) (731.1)
DT: January 1993
UF: Mathematical techniques—Frequency domain analysis*
BT: Mathematical techniques
RT: Control system analysis
Electric network analysis
Nyquist diagrams
Root loci

Frequency doublers
USE: Frequency multiplying circuits

Frequency function
USE: Probability density function

Frequency hopping 716.1
DT: January 1995
RT: Carrier communication
Frequency agility
Spread spectrum communication

Frequency meters 942.1
DT: Predates 1975
UF: Electric frequency meters
BT: Electric measuring instruments
RT: Atomic clocks
Electric frequency measurement
Time measurement
Wavemeters

Frequency modulation (716) (717) (718)
DT: Predates 1975
BT: Modulation
NT: Frequency shift keying
RT: Modulators
Vocoders

*Frequency modulation—Frequency shift keying**
USE: Frequency shift keying

Frequency multiplying circuits 713.5
DT: January 1993
UF: Electronic circuits, Frequency multiplying*
Frequency doublers
Frequency triplers

Frequency multiplying circuits *(continued)*
Multipliers (frequency)
BT: Frequency converter circuits
RT: Multivibrators
Signal processing
Varactors

Frequency response (703.1) (731.1)
DT: January 1993
UF: Control systems—Frequency response*
RT: Bode diagrams
Control system analysis
Control system synthesis
Networks (circuits)
Phase space methods
Sensitivity analysis
Transfer functions

Frequency shift keying (716) (717) (718)
DT: January 1993
UF: Frequency modulation—Frequency shift keying*
FSK
BT: Frequency modulation

Frequency stability (703.1) (731.1)
DT: January 1993
UF: System stability—Frequency stability*
BT: System stability
RT: Electric frequency control
Oscillators (electronic)

Frequency standards (902.2) (713.2)
SN: Stable oscillators, used for calibration of frequencies
DT: October 1975
BT: Standards
RT: Calibration
Electric frequency control
Electric frequency measurement
Oscillators (electronic)
Oscillators (mechanical)

Frequency synthesizers 713.2
DT: Predates 1975
BT: Signal generators
NT: Electronic musical instruments
Vocoders
RT: Oscillators (electronic)

Frequency triplers
USE: Frequency multiplying circuits

Fretting corrosion (539.1) (802.2)
DT: January 1993
UF: Corrosion—Fretting*
BT: Corrosion
RT: Corrosion fatigue
Fatigue of materials

Friction (931.1)
DT: Predates 1975
BT: Dynamics
NT: Internal friction
Rolling resistance
Stiction
Traction (friction)
RT: Friction materials
Friction meters
Mechanical properties
Pressure drop
Seizing
Surface phenomena
Surface properties
Triboelectricity
Tribology
Triboluminescence
Wear of materials

Friction materials (601.1) (931.2)
DT: January 1983
BT: Materials
RT: Brakes
Friction
Tribology

Friction meters 943.1
DT: Predates 1975
BT: Instruments
RT: Friction
Tribology

Friction theory
USE: Tribology

Friction welding 538.2.1
DT: January 1993
UF: Welding—Friction*
BT: Pressure welding

Frictional electricity
USE: Triboelectricity

Frictional light
USE: Triboluminescence

Friedel-Crafts reaction 802.2
DT: January 1993
BT: Substitution reactions
RT: Catalysis
Organic compounds

Front wheel drive automobiles 662.1
DT: January 1993
UF: Automobiles—Front wheel drive*
BT: Automobiles

Frost effects (421) (483) (405.1) (405.2)
DT: January 1993
UF: Foundations—Frost effect*
Roads and streets—Frost effect*
Soils—Frost effect*
BT: Effects
RT: Cold effects
Cold weather problems
Frost protection
Frost resistance
Frozen soils
Roads and streets
Soils

Frost protection (821.3)
DT: January 1993
UF: Orchards—Frost protection*
BT: Protection

Frost protection *(continued)*
RT: Frost effects
Frost resistance
Orchards

Frost resistance (421) (405.1) (405.2)
DT: January 1993
RT: Frost effects
Frost protection

Froth flotation (533.1) (802.3)
DT: January 1993
BT: Flotation

Frozen soils 483.1
DT: January 1993
UF: Sewers—Frozen ground*
Soils—Frozen*
BT: Soils
NT: Permafrost
RT: Cold weather problems
Frost effects

Fructose 804.1 (822.3)
DT: Predates 1975
BT: Sugars
RT: Food additives

Fruit juices 822.3
DT: January 1993
UF: Food products—Fruit juices*
BT: Beverages
RT: Fruits

Fruits 821.4
DT: January 1993
UF: Food products—Fruits*
BT: Agricultural products
Food products
RT: Fruit juices
Orchards

FSK
USE: Frequency shift keying

Fuel additives 803 (522) (523) (524)
DT: January 1993
UF: Fuels—Additive compounds*
BT: Additives
Fuels
NT: Antiknock compounds

Fuel burners 521.3
DT: Predates 1975
UF: Burners
BT: Furnaces
NT: Dual fuel burners
Gas burners
Oil burners
RT: Combustors
Fuel systems

*Fuel burners—Dual fuel**
USE: Dual fuel burners

*Fuel burners—Noise**
USE: Acoustic noise

*Fuel burners—Pulverized coal**
USE: Pulverized fuel AND
Coal

Fuel cells 702.2
DT: Predates 1975
UF: Batteries (fuel cells)
Spacecraft—Fuel cells*
BT: Electric batteries
RT: Spacecraft power supplies

*Fuel cells—Electrolytes**
USE: Electrolytes

Fuel consumption 525.3 (521)
DT: January 1993
RT: Fuel economy
Fuels
Performance

Fuel economy 525.2 (521)
DT: Predates 1975
RT: Fuel consumption
Fuels
Gasohol
Performance

Fuel elements (nuclear fuel)
USE: Nuclear fuel elements

Fuel filters (521.3) (522) (523)
DT: January 1993
UF: Fuels—Filters*
BT: Filters (for fluids)
RT: Fuel systems

Fuel gages (521.3) (943.1)
DT: January 1993
UF: Aircraft instruments—Fuel gages*
BT: Liquid level indicators
RT: Aircraft instruments
Fuel systems

Fuel gases
USE: Gas fuels

Fuel injection (521)
DT: January 1993
UF: Blast furnace practice—Fuel injection*
Internal combustion engines—Fuel injection*
Rocket engines—Fuel injection*
RT: Blast furnace practice
Fuel systems
Internal combustion engines
Oxygen blast enrichment
Rocket engines
Stratified charge engines

Fuel oil
USE: Fuel oils

Fuel oils 523
DT: January 1993
UF: Fuel oil
Oil fuel*
BT: Liquid fuels
NT: Residual fuels
RT: Coal oil mixtures
Furnace fuels

* Former Ei Vocabulary term

Fuel oils *(continued)*
Oil burners
Oil fired boilers
Oil furnaces
Petroleum industry

Fuel pellets (nuclear fuel)
USE: Nuclear fuel pellets

Fuel pumps 618.2 (522) (523)
DT: Predates 1975
BT: Pumps
RT: Engines
Fuel systems
Furnaces
Internal combustion engines

Fuel purification (522) (523) (524)
DT: January 1993
UF: Fuels—Purification*
BT: Purification
NT: Gas fuel purification
RT: Fuels

Fuel reprocessing
USE: Nuclear fuel reprocessing

Fuel sloshing (523) (654.1)
DT: January 1993
UF: Liquid sloshing
Rockets and missiles—Fuel sloshing*
Sloshing (fuel)
RT: Controllability
Fuel tanks
Missiles
Rockets
Stability

Fuel storage 694.4 (522) (523) (524)
DT: January 1993
UF: Fuels—Storage*
BT: Storage (materials)
NT: Coal storage
Gas fuel storage
RT: Fuels

Fuel systems (522) (523) (524)
DT: January 1993
BT: Equipment
RT: Fuel burners
Fuel filters
Fuel gages
Fuel injection
Fuel pumps
Fuel tanks
Fuels
Internal combustion engines
Vapor lock

Fuel tanks 619.2 (522) (523)
DT: January 1993
UF: Earthmoving machinery—Fuel tanks*
Rockets and missiles—Fuel tanks*
BT: Tanks (containers)
NT: Aircraft fuel tanks
Automobile fuel tanks
RT: Fuel sloshing
Fuel systems

Fueling (522) (523) (524) (691.2)
DT: January 1993
BT: Materials handling
NT: Aircraft fueling
Ship fueling
RT: Automobiles
Boiler firing
Ships
Vehicles

Fueling equipment (aircraft)
USE: Aircraft fueling equipment

Fuels (522) (523) (524)
DT: Predates 1975
BT: Materials
NT: Aircraft fuels
Automotive fuels
Coal oil mixtures
Coated fuel particles
Diesel fuels
Fossil fuels
Fuel additives
Furnace fuels
Gas fuels
Liquid fuels
Natural gas substitutes
Nuclear fuels
Pulverized fuel
Refuse derived fuels
Synthetic fuels
Wood fuels
RT: Antiknock rating
Atomization
Briquets
Butane
Calorific value
Charcoal
Coke
Flame research
Fuel consumption
Fuel economy
Fuel purification
Fuel storage
Fuel systems
Gas burners
Gas fuel storage
Petroleum distillates
Power generation
Sulfur determination
Wood

*Fuels—Additive compounds**
USE: Fuel additives

*Fuels—Antiknock rating**
USE: Antiknock rating

*Fuels—Atomization**
USE: Atomization

*Fuels—Calorific value**
USE: Calorific value

* Former Ei Vocabulary term

*Fuels—Distillates**
USE: Petroleum distillates

*Fuels—Filters**
USE: Fuel filters

*Fuels—Purification**
USE: Fuel purification

*Fuels—Refuse derived fuels**
USE: Refuse derived fuels

*Fuels—Storage**
USE: Fuel storage

*Fuels—Sulfur determination**
USE: Sulfur determination

*Fuels—Vapor lock**
USE: Vapor lock

Full mold process 534.2
DT: January 1993
UF: Cavityless casting
Foundry practice—Full mold process*
BT: Casting
RT: Foundry practice
Metal casting

Fullerenes 804 (708.3.1)
DT: January 1995
UF: Buckminsterfullerene
Buckyballs
BT: Carbon
RT: Intercalation compounds
Superconducting materials

Fullers earth 803 (483.1)
DT: Predates 1975
BT: Adsorbents
RT: Clay
Soils

Fume control 451.2
DT: January 1993
UF: Furnaces—Fume control*
Industrial plants—Fume control*
Metal finishing—Fume control*
Ore treatment—Fume control*
Processing fumes control
BT: Air pollution control
RT: Aerosols
Industrial poisons
Ore treatment
Smoke
Smoke abatement

Fumigation (451.1) (821.2) (821.3)
DT: January 1993
UF: Air pollution—Fumigation*
BT: Pest control
RT: Air pollution
Biocides
Pesticides
Sterilization (cleaning)

Function evaluation 921.6
DT: January 1993
UF: Mathematical techniques—Function evaluation*
BT: Numerical analysis
RT: Boolean functions
Finite element method

Function generators 713.5
DT: Predates 1975
BT: Signal generators
NT: Pulse generators
Square wave generators
RT: Analog computers

Functional assessment 461.5
DT: January 1993
UF: Human rehabilitation engineering—Functional assessment*
BT: Patient rehabilitation
RT: Gait analysis
Human rehabilitation engineering

Functional electric stimulation 461.5
DT: January 1993
UF: Human rehabilitation engineering—Functional electric stimulation*
BT: Patient rehabilitation
RT: Functional neural stimulation

Functional neural stimulation 461.5
DT: January 1993
UF: Human rehabilitation engineering—Functional neural stimulation*
BT: Neuromuscular rehabilitation
RT: Functional electric stimulation
Human rehabilitation engineering
Orthotics

Functions 921
DT: January 1995
BT: Mathematical techniques
NT: Boolean functions
Describing functions
Green's function
Mathematical transformations
Membership functions
Probability density function
Recursive functions
Switching functions
Transfer functions

Fungi (461.9) (801.2)
DT: January 1987
UF: Mold
Mycology
BT: Plants (botany)
NT: Yeast
RT: Aflatoxins
Fungicides
Fungus attack
Fungus protection
Fungus resistance
Microbiology
Microorganisms

Fungicides 804.1 (821.2)
DT: Predates 1975
BT: Pesticides

* Former Ei Vocabulary term

Fungicides *(continued)*
RT: Fungi
Fungus attack
Fungus protection

*Fungicides—Applicators**
USE: Applicators

Fungus attack (801.2) (811.2) (461.9)
DT: January 1993
UF: Wood—Fungus attack*
RT: Fungi
Fungicides
Fungus protection
Fungus resistance
Wood

Fungus protection (461.9) (801.2)
DT: January 1993
UF: Containers—Fungus protection*
Leather—Fungus protection*
Optical instruments—Fungus protection*
BT: Protection
RT: Fungi
Fungicides
Fungus attack
Fungus resistance
Leather

Fungus resistance (461.9) (801.2)
DT: January 1993
UF: Plastics—Fungus resistance*
RT: Fungi
Fungus attack
Fungus protection

Furan resins 815.1.1
DT: January 1987
BT: Thermosets

Furfural 804.1
DT: Predates 1975
BT: Organic compounds

Furnace cupolas
USE: Cupolas

Furnace fuels (522) (523) (524)
DT: January 1993
BT: Fuels
RT: Coal
Coke
Fuel oils
Furnaces

Furnace linings (532) (642.2) (812.2)
DT: January 1993
UF: Furnaces—Lining*
BT: Furnaces
Linings

Furnaces (532) (642.2) (643.2)
DT: Predates 1975
UF: Coal furnaces
Coke furnaces
BT: Combustion equipment
Heating equipment
NT: Charging equipment (furnaces)
Coal fueled furnaces
Coke fueled furnaces
Electric furnaces
Fluidized bed furnaces
Fuel burners
Furnace linings
Gas furnaces
Glass furnaces
Heat treating furnaces
Heating furnaces
Industrial furnaces
Kilns
Laboratory furnaces
Melting furnaces
Metallurgical furnaces
Oil furnaces
Recuperators
Refuse incinerators
Soaking pits
Space heating furnaces
RT: Boilers
Charging (furnace)
Chimneys
Combustion
Combustion chambers
Foundries
Fuel pumps
Furnace fuels
Industrial ovens
Modernization
Protective atmospheres
Stokers

*Furnaces, Electric**
USE: Electric furnaces

*Furnaces, Heat treating**
USE: Heat treating furnaces

*Furnaces, Heating**
USE: Heating furnaces

*Furnaces, Industrial**
USE: Industrial furnaces

*Furnaces, Laboratory**
USE: Laboratory furnaces

*Furnaces, Melting**
USE: Melting furnaces

*Furnaces, Metallurgical**
USE: Metallurgical furnaces

*Furnaces, Space heating**
USE: Space heating furnaces

*Furnaces—Blowers**
USE: Blowers

*Furnaces—Charging**
USE: Charging (furnace)

*Furnaces—Coal fuel**
USE: Coal fueled furnaces

*Furnaces—Coke fuel**
USE: Coke fueled furnaces

*Furnaces—Crucible type**
USE: Crucible furnaces

*Furnaces—Deposits**
USE: Deposits

*Furnaces—Electron beam**
USE: Electron beam furnaces

*Furnaces—Flow**
USE: Flow of fluids

*Furnaces—Fluidized bed**
USE: Fluidized bed furnaces

*Furnaces—Fume control**
USE: Fume control

*Furnaces—Gas fuel**
USE: Gas furnaces

*Furnaces—Glass bath**
USE: Industrial furnaces

*Furnaces—Infrared energy**
USE: Infrared furnaces

*Furnaces—Lining**
USE: Furnace linings

*Furnaces—Modernization**
USE: Modernization

*Furnaces—Oil fuel**
USE: Oil furnaces

*Furnaces—Optical image**
USE: Industrial furnaces

*Furnaces—Power supply**
USE: Electric power supplies to apparatus

*Furnaces—Radiation**
USE: Heat radiation

*Furnaces—Recuperators**
USE: Recuperators

*Furnaces—Regenerators**
USE: Regenerators

*Furnaces—Salt bath**
USE: Salt bath furnaces

*Furnaces—Solar**
USE: Solar equipment

*Furnaces—Steelmaking**
USE: Steelmaking furnaces

*Furnaces—Vacuum**
USE: Vacuum furnaces

Furniture manufacture (811.2) (814.1) (817.1) (913.4)
DT: Predates 1975
BT: Manufacture
RT: Veneers
Wood products
Woodworking

*Furniture manufacture—Finishing**
USE: Finishing

Fused salts 804.2
DT: January 1993
UF: Salts—Fused*
BT: Salts
RT: Electrolytes

Fused silica 812.3
DT: January 1993
UF: Silica glass
Silica—Fused*
Vitreous silica
BT: Silica

Fuselages 652.1
DT: January 1993
UF: Aircraft—Fuselage*
BT: Aircraft parts and equipment

Fuses (electric)
USE: Electric fuses

*Fuses**
USE: Electric fuses

Fusion
USE: Fusion reactions

Fusion reactions 932.2.1 (621.2)
DT: January 1993
UF: Cold fusion
Fusion
Nuclear energy—Fusion reactions*
Nuclear fuels—Fusion*
Nuclear fusion reactions
NT: Inertial confinement fusion
Laser fusion
Thermonuclear reactions
RT: Fusion reactors
Nuclear energy
Nuclear explosions
Nuclear physics
Nuclear reactors

Fusion reactors 621.2 (932.2.1)
DT: January 1993
UF: Nuclear reactors, Fusion*
Thermonuclear reactors
BT: Facilities
NT: Tokamak devices
RT: Fusion reactions
Inertial confinement fusion
Plasma confinement
Plasma heating

Fuzzy control 731.1 (921)
DT: January 1995
BT: Control
RT: Fuzzy sets
Intelligent control

Fuzzy logic
USE: Fuzzy sets

Fuzzy sets (921)
DT: January 1993
UF: Fuzzy logic
Mathematical techniques—Fuzzy sets*
BT: Set theory

* Former Ei Vocabulary term

Fuzzy sets *(continued)*
RT: Formal logic
Fuzzy control
Membership functions
Probabilistic logics

Gadolinium 547.2
DT: January 1993
UF: Gadolinium and alloys*
BT: Rare earth elements
RT: Gadolinium alloys
Gadolinium compounds

Gadolinium alloys 547.2
DT: January 1993
UF: Gadolinium and alloys*
BT: Rare earth alloys
RT: Gadolinium
Gadolinium compounds

*Gadolinium and alloys**
USE: Gadolinium OR
Gadolinium alloys

Gadolinium compounds (804.1) (804.2)
DT: Predates 1975
BT: Rare earth compounds
RT: Gadolinium
Gadolinium alloys

Gage blocks 943.3 (913.4)
DT: January 1993
UF: Block gages
Gages—Block*
BT: Gages

Gages 943.3
DT: Predates 1975
UF: Gauges
BT: Instruments
NT: Flowmeters
Gage blocks
Liquid level indicators
Manometers
Pneumatic gages
Pressure gages
Radioisotope gages
Ring gages
Screw thread gages
Snow gages
Strain gages
Temperature measuring instruments
Thickness gages
Tide gages
RT: Dilatometers
Gaging

*Gages—Block**
USE: Gage blocks

*Gages—Pneumatic**
USE: Pneumatic gages

*Gages—Pressure measurement**
USE: Pressure gages

*Gages—Radioisotope**
USE: Radioisotope gages

*Gages—Ring**
USE: Ring gages

*Gages—Screw thread**
USE: Screw thread gages

*Gages—Thickness measurement**
USE: Thickness gages

Gaging (523) (943)
DT: January 1993
UF: Oil tanks—Gaging*
Wire—Gaging*
BT: Measurements
RT: Gages

Gain control 731.3
DT: January 1993
UF: Control, Electric variables—Gain*
Gain regulation
Volume control (gain)
BT: Electric variables control
RT: Amplification
Amplifiers (electronic)
Gain measurement
Signal receivers
Transmitters
Tuning

Gain measurement 942.2
DT: January 1993
UF: Amplification measurement
Electric measurements—Gain*
BT: Electric variables measurement
RT: Amplification
Amplifiers (electronic)
Gain control

Gain regulation
USE: Gain control

Gait analysis 461.3
DT: January 1993
UF: Biomechanics—Gait analysis*
BT: Biped locomotion
RT: Functional assessment

Galaxies (657.2)
DT: January 1986
RT: Astronomy
Astrophysics

Gallium 549.3
DT: January 1993
UF: Gallium and alloys*
BT: Nonferrous metals
NT: Semiconducting gallium
RT: Gallium alloys
Gallium compounds

Gallium alloys 549.3
DT: January 1993
UF: Gallium and alloys*
BT: Alloys
RT: Gallium
Gallium compounds

* Former Ei Vocabulary term

*Gallium and alloys**
USE: Gallium OR
Gallium alloys

Gallium arsenide materials
USE: Semiconducting gallium arsenide

Gallium compounds (804.1) (804.2)
DT: Predates 1975
BT: Metallic compounds
NT: Semiconducting gallium compounds
RT: Gallium
Gallium alloys

Gallows frames
USE: Head frames

Galvanized metal (539.3) (545) (546.3)
DT: Predates 1975
UF: Welding—Galvanized metal*
BT: Coated materials
Materials
RT: Galvanizing
Metals

Galvanizing 539.3, 546.3
DT: Predates 1975
BT: Surface treatment
Zinc plating
RT: Galvanized metal
Protective coatings
Zinc

Galvanomagnetic effects 701.2
DT: January 1995
BT: Magnetic field effects
NT: Hall effect
Magnetoresistance

Galvanomagnetic transducers
USE: Hall effect transducers

Galvanometers 942.1
DT: Predates 1975
BT: Electric measuring instruments
RT: Electric current measurement

Game theory 922.1
DT: January 1993
UF: Probability—Game theory*
BT: Statistical methods
RT: Decision theory
Information theory
Mathematical programming
Optimal control systems
Probability
Scheduling

Gamma radiation
USE: Gamma rays

Gamma radiography (931.3) (461.6) (932.1)
DT: January 1993
UF: Gamma ray radiography
Radiography—Gamma ray*
BT: Radiography
RT: Gamma rays
Nondestructive examination

Gamma ray production 931.3, 932.1
DT: January 1993
UF: X-ray and gamma ray production*
RT: Gamma rays

Gamma ray radiography
USE: Gamma radiography

Gamma ray spectrometers 944.7 (931.3) (932.1)
DT: January 1993
UF: Spectrometers, Gamma ray*
BT: Particle spectrometers
RT: Gamma rays

Gamma rays 931.3, 932.1
DT: Predates 1975
UF: Gamma radiation
BT: Electromagnetic waves
RT: Gamma radiography
Gamma ray production
Gamma ray spectrometers
Photons
Radioactivity
Radiography

Gantry cranes 693.1
DT: January 1993
UF: Cranes—Gantry*
Cranes—Traveling*
Traveling cranes
BT: Cranes
NT: Crane rails
Crane runways

Garages (filling stations)
USE: Filling stations

Garages (parking) 402.2
DT: January 1993
UF: Garages*
Garages—Mechanical parking*
Parking garages
BT: Facilities
NT: Bus garages
RT: Parking

*Garages**
USE: Garages (parking)

*Garages—Mechanical parking**
USE: Garages (parking)

*Garages—Motor bus**
USE: Bus garages

Garbage disposal
USE: Refuse disposal

Garbage trucks 452, 663.1
DT: January 1993
UF: Motor trucks—Refuse collecting*
BT: Trucks
RT: Refuse collection

Garment manufacture 819.5, 913.4
DT: Predates 1975
UF: Clothing manufacture
BT: Manufacture
NT: Hosiery manufacture

* Former Ei Vocabulary term

Garment manufacture *(continued)*
RT: Shoe manufacture
Textile industry
Textile mills
Textile processing

Garnet
USE: Garnets

Garnets (482.2) (606.1)
DT: Predates 1975
UF: Garnet
BT: Gems
Silicates
RT: Industrial gems
Magnetic materials

Gas
USE: Gases

Gas absorption 802.3 (931.2)
DT: January 1993
UF: Gases—Absorption*
BT: Absorption
RT: Gases

Gas adsorption 802.3 (931.2)
DT: January 1993
UF: Gases—Adsorption*
BT: Adsorption
RT: Gases

Gas alloying 531.1
DT: January 1993
UF: Metals and alloys—Gas alloying*
BT: Alloying
RT: Vapor deposition

*Gas analysis**
USE: Gas fuel analysis

Gas appliances (522) (643.2) (642.2)
DT: Predates 1975
UF: Refrigerators—Gas fuel*
Water heaters—Gas*
BT: Equipment
NT: Gas stoves
Pilot lights
RT: Domestic appliances
Flues
Gas fuels
Pressure regulators
Vents

*Gas appliances—Flues**
USE: Flues

*Gas appliances—Pilot lights**
USE: Pilot lights

*Gas appliances—Pressure regulators**
USE: Pressure regulators

*Gas appliances—Safety devices**
USE: Safety devices

*Gas appliances—Vents**
USE: Vents

Gas burners 521.3, 522 (642.2) (643.2)
DT: Predates 1975
BT: Fuel burners
RT: Fuels
Gas fuels
Gas furnaces

Gas bursts 502.1, 914.1
DT: January 1993
UF: Mines and mining—Gas bursts*
RT: Hazards
Mining
Rock bursts

Gas chromatography 802.3 (801)
DT: January 1993
UF: Chromatographic analysis—Gas*
BT: Chromatography
RT: Gas fuel analysis

Gas condensate liquids
USE: Gas condensates

*Gas condensate**
USE: Gas condensates

Gas condensates 522 (512.2)
DT: January 1993
UF: Gas condensate liquids
Gas condensate*
BT: Hydrocarbons
NT: Propane
RT: Butane
Natural gas

Gas cooled reactors 621.1
DT: January 1993
UF: Nuclear reactors, Gas cooled*
BT: Nuclear reactors

Gas cutting
USE: Oxygen cutting

Gas cyaniding
USE: Carbonitriding

Gas cylinders 619.2, 691.1, 694.4
SN: Containers for handling, shipping, and temporarily storing gases, usually under pressure
DT: Predates 1975
BT: Cylinders (containers)
RT: Gas fuel storage

Gas detectors 943.3, 914.1 (503.2) (522) (801)
SN: Detectors of noxious gases
DT: Predates 1975
BT: Detectors
RT: Hazardous materials
Safety devices

Gas discharge lamps
USE: Discharge lamps

Gas discharge tubes 714.1
DT: January 1993
UF: Discharge tubes
Electron tubes, Gas discharge*
BT: Electron tubes

Gas discharge tubes *(continued)*
NT: Mercury vapor rectifiers
Thyratrons
RT: Discharge lamps
Display devices

Gas dynamic lasers 744.2
DT: January 1993
UF: Gasdynamic lasers
Lasers, Gas dynamic*
BT: Gas lasers

Gas dynamics 631.1.2
DT: January 1977
BT: Fluid dynamics
NT: Aerodynamics
RT: Gases

Gas emissions 451.1
DT: January 1993
UF: Electric batteries—Gas emission*
Gaseous emissions
RT: Gaseous effluents
Gases

Gas engineering 522 (901)
DT: Predates 1975
BT: Engineering

Gas engines (612.3)
DT: Predates 1975
BT: Internal combustion engines
NT: Air engines
Gas turbines

Gas filled cables 706.2
DT: January 1993
UF: Electric cables—Gas filled*
Gas insulated cables
BT: Electric cables

Gas fired boilers 522 (521.1) (614.2)
DT: January 1993
UF: Boilers—Gas firing*
BT: Boilers
Equipment

Gas flow integrators
USE: Gas meters

Gas fuel analysis 522, 801
DT: January 1993
UF: Gas analysis*
Gases—Analysis*
BT: Chemical analysis
RT: Gas chromatography
Gas fuels
Gases
Volumetric analysis

Gas fuel manufacture 522 (913.4)
DT: January 1993
UF: Gas manufacture*
BT: Manufacture
NT: Odorization
Synthesis gas manufacture
Underground gas manufacture
RT: Biogas
Catalysts
Catalytic cracking
Gas fuel purification
Gas fuels
Gas generators
Gas plants
Gases
Mixed gas fuels
Waste liquor utilization

Gas fuel measurement 522 (943.2)
DT: January 1993
UF: Gas measurement*
BT: Measurements
RT: Gas fuels
Gas meters

Gas fuel purification 522 (802.3)
SN: Includes cleaning.
DT: January 1993
UF: Gas purification*
BT: Fuel purification
RT: Desulfurization
Gas fuel manufacture
Gas fuels
Gases
Scrubbers

Gas fuel storage 522, 694.4
DT: January 1993
UF: Gas storage*
BT: Fuel storage
NT: Underground gas storage
RT: Fuels
Gas cylinders
Gas fuels
Gas holders
Gases
Natural gas
Tank cars
Tank trucks

Gas fueled air conditioning 522, 643.4
DT: January 1993
UF: Air conditioning—Gas fuel*
BT: Air conditioning
RT: Gas fuels

Gas fuels 522
DT: January 1993
UF: Fuel gases
BT: Fuels
NT: Mixed gas fuels
Natural gas
RT: Gas appliances
Gas burners
Gas fuel analysis
Gas fuel manufacture
Gas fuel measurement
Gas fuel purification
Gas fuel storage
Gas fueled air conditioning
Gas furnaces

* Former Ei Vocabulary term

Gas furnaces
522 (532) (534.4) (537.2) (642.2) (643.2)
DT: January 1993
UF: Furnaces—Gas fuel*
BT: Furnaces
RT: Gas burners
Gas fuels

Gas generators 522 (802.1)
SN: Generators of gases
DT: January 1993
UF: Gases—Generators*
Generators (gas)
BT: Equipment
RT: Gas fuel manufacture
Gas producers
Gases

Gas hazards 522, 914.1 (511.1) (512.2)
DT: January 1993
UF: Petroleum pipelines—Gas hazards*
BT: Hazards
RT: Explosions
Firedamp
Petroleum pipelines

Gas heating 522 (642.1) (643.1)
DT: January 1993
UF: Heating—Gas fuel*
BT: Heating
RT: Industrial heating
Space heating

Gas holders 694.4, 522
SN: Tanklike structures for collecting and storing gas for distribution
DT: Predates 1975
UF: Gasometers
BT: Containers
RT: Gas fuel storage

*Gas holders—Welded steel**
USE: Welded steel structures

Gas hydrates 512.2, 522
DT: January 1993
BT: Hydrates
RT: Natural gas
Natural gas wells

Gas industry 522 (911.2)
DT: Predates 1975
BT: Industry
RT: Gas loads

*Gas industry—Accounting**
USE: Cost accounting

*Gas industry—Load**
USE: Gas loads

Gas insulated cables
USE: Gas filled cables

Gas ionization
USE: Ionization of gases

Gas kinetics
USE: Kinetic theory of gases

Gas lasers 744.2
DT: January 1993
UF: Lasers, Gas*
BT: Lasers
NT: Carbon dioxide lasers
Gas dynamic lasers
Helium neon lasers
RT: Chemical lasers
Excimer lasers

Gas lifts 511.1
DT: January 1993
UF: Oil well production—Gas lift*
BT: Oil well production

Gas loads 522
DT: January 1993
UF: Gas industry—Load*
Loads (gas)
RT: Gas industry
Gas supply

Gas lubricated bearings 601.2, 607.2
DT: January 1993
UF: Bearings—Gas lubrication*
BT: Antifriction bearings
RT: Lubrication

*Gas manufacture**
USE: Gas fuel manufacture

*Gas manufacture—Agricultural wastes**
USE: Agricultural wastes

*Gas manufacture—Catalytic cracking process**
USE: Catalytic cracking

*Gas manufacture—Mixed gas**
USE: Mixed gas fuels

*Gas manufacture—Odorizing**
USE: Odorization

*Gas manufacture—Synthesis gas**
USE: Synthesis gas manufacture

*Gas manufacture—Underground**
USE: Underground gas manufacture

*Gas manufacture—Waste liquor utilization**
USE: Waste liquor utilization

Gas masks 914.1
DT: Predates 1976
BT: Safety devices
RT: Military equipment
Protective clothing

*Gas measurement**
USE: Gas fuel measurement

Gas metal arc welding
USE: Carbon dioxide arc welding

Gas meters 522, 943.3
DT: Predates 1975
UF: Gas flow integrators
BT: Flowmeters
RT: Gas fuel measurement

* Former Ei Vocabulary term

Gas oil
USE: Gas oils

Gas oils 523
DT: January 1993
UF: Gas oil
BT: Petroleum products
RT: Diesel fuels
Kerosene
Steam cracking

Gas permeable membranes (462.1) (802.1) (631.1)
DT: January 1993
UF: Membranes—Gas permeable*
BT: Membranes
NT: Oxygen permeable membranes

Gas pipelines 522, 619.1
SN: Local distribution mains for manufactured or natural gas
DT: Predates 1975
BT: Pipelines
RT: Gas supply
Gases
Pipeline changeover

*Gas pipelines—Changeover**
USE: Pipeline changeover

Gas piping systems 522, 619.1
DT: January 1993
UF: Gases—Piping systems*
BT: Piping systems
RT: Gases

Gas plants 522 (402.1)
DT: Predates 1975
UF: Coal handling—Gas plants*
BT: Industrial plants
RT: Gas fuel manufacture

Gas producers 522
SN: Devices for coal gasification which use the air and water-gas reactions simultaneously
DT: Predates 1975
BT: Equipment
RT: Coal gasification
Coal gasification plants
Gas generators

*Gas purification**
USE: Gas fuel purification

*Gas purification—Scrubbers**
USE: Scrubbers

Gas sensing electrodes (704.1) (732.2) (801.4.1)
DT: January 1993
UF: Electrodes, Electrochemical—Gas sensing*
BT: Electrochemical electrodes

Gas sensors
USE: Chemical sensors

*Gas storage**
USE: Gas fuel storage OR
Gases AND
Storage (materials)

*Gas storage—Underground**
USE: Underground gas storage

Gas stoves 522 (642.1) (643.2)
DT: January 1993
UF: Stoves—Gas*
BT: Gas appliances
Stoves
RT: Pilot lights

Gas supply 522 (619.1)
DT: January 1993
UF: Buildings—Gas supply*
RT: Gas loads
Gas pipelines
Natural gas

Gas turbine locomotives 522, 612.3, 682.1.2
DT: January 1993
UF: Locomotives, Gas turbine*
BT: Locomotives
RT: Gas turbines

Gas turbine power plants 522, 612.3 (402)
DT: Predates 1975
BT: Power plants
RT: Electric power plants
Gas turbines
Power generation

Gas turbines 612.3
DT: Predates 1975
UF: Coal burning gas turbines
Ship propulsion—Gas and steam turbine combined*
Ship propulsion—Gas turbine*
BT: Gas engines
Turbines
NT: Coal fueled gas turbines
RT: Brayton cycle
Closed cycle machinery
Combustors
Compressors
Diesel and gas turbine combined power plants
Free piston engines
Gas turbine locomotives
Gas turbine power plants
Hydroelectric and gas turbine combined power plants
Steam turbines
Turbogenerators
Turbojet engines

*Gas turbines—Automotive applications**
USE: Automobile engines

*Gas turbines—Closed cycle**
USE: Closed cycle machinery

*Gas turbines—Coal burning**
USE: Coal fueled gas turbines

*Gas turbines—Combustors**
USE: Combustors

Gas warfare
USE: Chemical warfare

* Former Ei Vocabulary term

Gas welding 538.2.1
DT: January 1993
UF: Welding, Gas*
BT: Welding
RT: Pressure welding

Gasdynamic lasers
USE: Gas dynamic lasers

Gaseous effluents 451.1
DT: January 1993
UF: Air pollution—Gaseous effluents*
Gaseous wastes
BT: Effluents
RT: Air pollution
Gas emissions

Gaseous emissions
USE: Gas emissions

Gaseous flow
USE: Flow of fluids

Gaseous wastes
USE: Gaseous effluents

Gases (931.2)
SN: Very general term; prefer specific gas if applicable
DT: Predates 1975
UF: Electric insulating materials—Gases*
Gas
Gas storage*
Heat transfer—Gases*
Metals—Gases*
BT: Fluids
NT: Coal gas
Exhaust gases
Firedamp
Inert gases
Radiogenic gases
Synthesis gas
Vapors
RT: Aerodynamics
Air
Compressibility of gases
Density of gases
Dielectric properties of gases
Diffusion in gases
Electric breakdown of gases
Electric conductivity of gases
Equations of state of gases
Explosions
Flow of fluids
Fluid mechanics
Gas absorption
Gas adsorption
Gas dynamics
Gas emissions
Gas fuel analysis
Gas fuel manufacture
Gas fuel purification
Gas fuel storage
Gas generators
Gas pipelines
Gas piping systems
Ignition
Ionization of gases
Kinetic theory of gases
Liquefaction of gases
Liquefied gases
Luminescence of gases
Natural gas
Photoionization
Refractive index
Specific heat of gases
Thermal conductivity of gases
Thermal diffusion in gases
Thermodynamics
Viscosity of gases
Volume fraction

*Gases, Inert**
USE: Inert gases

*Gases, Liquefied**
USE: Liquefied gases

*Gases—Absorption**
USE: Gas absorption

*Gases—Adsorption**
USE: Gas adsorption

*Gases—Analysis**
USE: Gas fuel analysis

*Gases—Compressibility**
USE: Compressibility of gases

*Gases—Density**
USE: Density of gases

*Gases—Dielectric properties**
USE: Dielectric properties of gases

*Gases—Diffusion**
USE: Diffusion in gases

*Gases—Electric breakdown**
USE: Electric breakdown of gases

*Gases—Electric conductivity**
USE: Electric conductivity of gases

*Gases—Generators**
USE: Gas generators

*Gases—Ionization**
USE: Ionization of gases

*Gases—Kinetic theory**
USE: Kinetic theory of gases

*Gases—Liquefaction**
USE: Liquefaction of gases

*Gases—Manufacture**
USE: Manufacture

*Gases—Measurements**
USE: Measurements

*Gases—Piping systems**
USE: Gas piping systems

*Gases—Purification**
USE: Purification

* Former Ei Vocabulary term

*Gases—Refractive index**
USE: Refractive index

*Gases—Thermal conductivity**
USE: Thermal conductivity of gases

*Gases—Viscosity**
USE: Viscosity of gases

Gasification 802.3
DT: January 1993
BT: Phase transitions
NT: Coal gasification
RT: Vaporization

Gaskets 619.1.1 (601.3) (612.2.1)
DT: Predates 1975
UF: Internal combustion engines—Gaskets*
O rings
BT: Seals
RT: Internal combustion engines
Pipe joints

Gasohol 523
DT: June 1990
UF: Automotive fuels—Gasohol*
BT: Liquid fuels
RT: Alcohol fuels
Alcohols
Automotive fuels
Ethanol
Ethanol fuels
Fuel economy
Methanol
Methanol fuels

Gasoline 523
DT: Predates 1975
UF: Benzines
BT: Liquid fuels
Petroleum products
RT: Antiknock rating
Automotive fuels
Gasoline refining

Gasoline refining 513.1, 523
DT: Predates 1975
BT: Petroleum refining
RT: Gasoline

*Gasoline refining—Sulfur compounds**
USE: Sulfur compounds

Gasometers
USE: Gas holders

Gastroenterology 461.6
DT: January 1993
UF: Biomedical engineering—Gastroenterology*
BT: Medicine
RT: Biomedical engineering

Gate turn-off devices
USE: Thyristors

Gates (canal)
USE: Canal gates

Gates (hydraulic)
USE: Hydraulic gates

Gates (railroad crossing)
USE: Railroad crossings

Gates (spillway)
USE: Spillway gates

Gates (transistor) 714.2
DT: January 1993
UF: Transistors, Field effect—Gates*
BT: Field effect transistors

Gateways (computer networks) (723) (716) (717) (718)
DT: January 1993
UF: Computer networks—Gateways*
BT: Computer networks
RT: Interfaces (computer)

Gating and feeding 534.2
DT: January 1993
UF: Foundry practice—Gating and feeding*
BT: Molding
RT: Foundry practice

Gauges
USE: Gages

Gaussian noise (electronic)
USE: Spurious signal noise

Ge
USE: Germanium

Gear cutters 603.1 (604.1)
SN: For machining of gears
DT: Predates 1975
BT: Cutting tools
RT: Gear cutting
Gear cutting machines
Gear manufacture

Gear cutting 604.1
DT: Predates 1975
BT: Cutting
RT: Gear cutters
Gear cutting machines
Gear manufacture
Gears
Metal cutting
Milling (machining)

Gear cutting machines 603.1 (604.1)
DT: Predates 1975
BT: Machine tools
RT: Gear cutters
Gear cutting
Gear manufacture
Gears

Gear manufacture 601.2 (913.4)
DT: Predates 1975
BT: Manufacture
RT: Gear cutters
Gear cutting
Gear cutting machines
Gears

*Gear manufacture—Powder metals**
USE: Powder metal products

Gear pumps 618.2
DT: January 1993
UF: Pumps, Gear*
BT: Rotary pumps
RT: Centrifugal pumps
Reciprocating pumps
Vane pumps

Gear teeth 601.2
DT: January 1993
UF: Gears—Teeth*
Teeth (gears)
BT: Gears

Gearing
USE: Gears

Gears 601.2
DT: Predates 1975
UF: Cranes—Gears*
Gearing
Instruments—Gears*
Machine tools—Gears*
Machinery—Gears*
Rolling mills—Gears*
Tractors—Gears*
BT: Machine components
NT: Bevel gears
Epicyclic gears
Gear teeth
Helical gears
Nonmetallic gears
Speed reducers
Spur gears
Worm gears
RT: Drives
Gear cutting
Gear cutting machines
Gear manufacture
Instruments
Machine tools
Mechanical drives
Sprockets
Transmissions
Wheels

*Gears—Bevel**
USE: Bevel gears

*Gears—Epicyclic**
USE: Epicyclic gears

*Gears—Helical**
USE: Helical gears

*Gears—Noise**
USE: Acoustic noise

*Gears—Nonmetallic**
USE: Nonmetallic gears

*Gears—Teeth**
USE: Gear teeth

*Gears—Worm**
USE: Worm gears

Geiger counters 944.7
DT: Predates 1975
UF: Geiger Mueller counters
Geiger Muller counters
BT: Radiation counters
RT: Counting tubes
Ionization chambers

Geiger Mueller counters
USE: Geiger counters

Geiger Muller counters
USE: Geiger counters

Gel permeation chromatography 802.3 (801)
DT: January 1993
UF: Chromatographic analysis—Gel permeation*
BT: Liquid chromatography

Gelation 802.3
DT: January 1993
BT: Chemical operations
RT: Colloids
Gels

Gels 804, 801.3
DT: Predates 1975
BT: Colloids
NT: Silica gel
RT: Gelation
Macromolecules
Sol-gels

GEM (ground effect machines)
USE: Air cushion vehicles

Gem cutting
USE: Crystal cutting

Gems 482.2.1
DT: Predates 1975
UF: Jewels (precious)
Precious stones
BT: Materials
NT: Diamonds
Garnets
Industrial gems
Natural gems
Ruby
Sapphire
Synthetic gems
Topaz
RT: Amber
Jewelry manufacture
Quartz
Zircon

*Gems—Natural origin**
USE: Natural gems

*Gems—Synthetic origin**
USE: Synthetic gems

Gene transfer
USE: Genetic engineering

General purpose computers 722
DT: January 1993
UF: Computers—General purpose application*
BT: Digital computers

* Former Ei Vocabulary term

Generator protection
USE: Electric equipment protection

Generators (acoustic)
USE: Acoustic generators

Generators (electric)
USE: Electric generators

Generators (gas)
USE: Gas generators

Genes 461.2
DT: January 1993
UF: Biological materials—Genes*
BT: Chromosomes
RT: DNA

Genetic algorithms (723) (921)
DT: January 1995
BT: Algorithms
RT: Optimization
Simulated annealing

Genetic engineering 461.8.1
DT: January 1981
UF: Gene transfer
Recombinant DNA technology
BT: Biotechnology
NT: Cloning
RT: Mutagenesis

Geochemistry 481.2
DT: Predates 1975
BT: Chemistry
NT: Analytical geochemistry
Exploratory geochemistry
Groundwater geochemistry
Natural water geochemistry
RT: Earth (planet)
Geophysics
Hydrology
Mineralogy
Petrology
Physical chemistry
Volcanic rocks

*Geochemistry—Analytical**
USE: Analytical geochemistry

*Geochemistry—Exploratory**
USE: Exploratory geochemistry

*Geochemistry—Groundwater**
USE: Groundwater geochemistry

*Geochemistry—Natural waters**
USE: Natural water geochemistry

*Geochemistry—Organic compounds**
USE: Organic compounds

*Geochemistry—Radioactive elements**
USE: Radioactive elements

*Geochemistry—Rare earths**
USE: Rare earths

*Geochemistry—Trace elements**
USE: Trace elements

*Geochemistry—Volcanic rocks**
USE: Volcanic rocks

Geochronology 481.1, 481.3
DT: January 1993
UF: Dating (geology)
Geology—Dating*
BT: Geology
RT: Geophysics

Geodesy 405.3, 481.3
DT: January 1993
UF: Geophysics—Geodesy*
BT: Geophysics
RT: Geodetic satellites
Photogrammetry
Photomapping
Surveying
Surveys

Geodetic satellites 405.3, 655.2
DT: January 1993
UF: Satellites—Geodetic*
BT: Satellites
RT: Geodesy

Geofabrics
USE: Geotextiles

Geographic information systems 903.3 (723.3)
DT: January 1995
UF: GIS
BT: Nonbibliographic retrieval systems
RT: Database systems
Mapping

Geographical regions
DT: January 1993
NT: Arid regions
Coastal zones
Developing Countries
Tropics

Geologic models 481.1
DT: January 1993
BT: Models
RT: Geology

Geologic surveys
USE: Geological surveys

Geological repositories 481.1, 622.5 (452.4)
SN: For radioactive wastes
DT: January 1993
UF: Radioactive wastes—Geological repositories*
Repository disposal
BT: Facilities
RT: Radioactive waste disposal

Geological surveys 481.1
DT: Predates 1975
UF: Geologic surveys
BT: Surveys
RT: Geology
Mapping
Natural resources exploration
Soil surveys

* Former Ei Vocabulary term

Geology 481.1
DT: Predates 1975
UF: Geology—Subcrustal*
Subcrustal geology
BT: Earth sciences
NT: Engineering geology
Geochronology
Geomorphology
Glacial geology
Lithology
Military geology
Ore deposit geology
Petrography
Petroleum geology
Petrology
Sedimentology
Stratigraphy
Submarine geology
Tectonics
RT: Aquifers
Biomarkers
Coal
Core samples
Earth (planet)
Geologic models
Geological surveys
Geophysics
Geothermal energy
Landforms
Natural resources
Oil bearing formations
Photomapping
Rock mechanics
Rocks
Sand consolidation
Soils
Volcanoes
Weathering

*Geology—Caves**
USE: Caves

*Geology—Coal**
USE: Coal

*Geology—Dating**
USE: Geochronology

*Geology—Engineering**
USE: Engineering geology

*Geology—Geomorphology**
USE: Geomorphology

*Geology—Glacial**
USE: Glacial geology

*Geology—Lithology**
USE: Lithology

*Geology—Military**
USE: Military geology

*Geology—Reefs**
USE: Reefs

*Geology—Sedimentology**
USE: Sedimentology

*Geology—Stratigraphy**
USE: Stratigraphy

*Geology—Subaqueous**
USE: Submarine geology

*Geology—Subcrustal**
USE: Geology

*Geology—Tectonics**
USE: Tectonics

Geomagnetism 481.3.2
DT: January 1993
UF: Earth magnetism
Earth—Magnetism*
Terrestrial magnetism
BT: Magnetism
RT: Earth (planet)
Magnetosphere

Geometrical optics 741.1
DT: January 1993
UF: Optics—Geometrical*
BT: Optics
RT: Gradient index optics

Geometry 921
DT: January 1993
UF: Mathematical techniques—Geometry*
BT: Mathematical techniques
NT: Bodies of revolution
Computational geometry
RT: Fractals
Surfaces
Surveying
Surveys
Three dimensional

Geomorphology 481.1.1
DT: January 1993
UF: Geology—Geomorphology*
BT: Geology
RT: Landforms
Stratigraphy

Geophysical prospecting 481.4 (501) (512)
DT: June 1990
BT: Natural resources exploration
NT: Electric prospecting
Electromagnetic prospecting
Geothermal prospecting
Gravitational prospecting
Magnetic prospecting
Radioactive prospecting
Seismic prospecting
RT: Geophysics
Petroleum prospecting
Remote sensing
Well logging

*Geophysical prospecting—Electrical**
USE: Electric prospecting

*Geophysical prospecting—Electromagnetic**
USE: Electromagnetic prospecting

*Geophysical prospecting—Gravitational**
USE: Gravitational prospecting

* Former Ei Vocabulary term

*Geophysical prospecting—Magnetic**
USE: Magnetic prospecting

*Geophysical prospecting—Radioactive methods**
USE: Radioactive prospecting

*Geophysical prospecting—Seismic**
USE: Seismic prospecting

*Geophysical prospecting—Telluric**
USE: Telluric prospecting

Geophysics 481.3
DT: Predates 1975
UF: Geophysics—Subcrustal*
Subcrustal geophysics
BT: Physics
NT: Geodesy
Seismology
Submarine geophysics
RT: Cosmic rays
Earth (planet)
Geochemistry
Geochronology
Geology
Geophysical prospecting
Gravimeters
Petrology
Photogrammetry
Surveying
Surveys
Tectonics

*Geophysics—Geodesy**
USE: Geodesy

*Geophysics—Geothermal**
USE: Geothermal prospecting

*Geophysics—Radioactivity investigations**
USE: Radioactive prospecting

*Geophysics—Rock properties**
USE: Petrology

*Geophysics—Subaqueous**
USE: Submarine geophysics

*Geophysics—Subcrustal**
USE: Geophysics

Geostationary satellites 655.2
DT: January 1993
UF: Satellites—Geostationary*
BT: Satellites
RT: Communication satellites

Geotextiles 819.5 (405) (483.1)
DT: January 1986
UF: Geofabrics
Mesh (geotextile)
BT: Textiles
RT: Embankments

Geothermal energy 481.3.1, 615.1
DT: July 1974
BT: Natural resources
Renewable energy resources
NT: Geothermal fields
RT: Geology
Geothermal heating
Geothermal power plants
Geothermal springs
Geothermal water resources
Volcanoes

Geothermal fields 481.3.1, 615.1
DT: June 1990
BT: Geothermal energy
NT: Hot dry rock systems
RT: Geothermal springs
Geothermal water resources
Geothermal wells

*Geothermal fields—Hot dry rock systems**
USE: Hot dry rock systems

Geothermal heating 615.1 (643.2) (642.1)
DT: January 1993
UF: Heating—Geothermal*
BT: Heating
RT: Geothermal energy

Geothermal logging 481.4 (481.3.1) (512.1.2)
DT: January 1993
UF: Oil well logging—Geothermal*
BT: Well logging
RT: Geothermal prospecting

Geothermal power plants
481.3.1, 615.1 (402.1) (614.2)
DT: January 1993
UF: Power plants—Geothermal energy*
BT: Power plants
RT: Geothermal energy

Geothermal prospecting 481.3.1, 481.4
DT: January 1993
UF: Geophysics—Geothermal*
BT: Geophysical prospecting
RT: Geothermal logging
Geothermal water resources

Geothermal springs 481.3.1, 615.1
DT: June 1990
UF: Thermal springs
Warm springs
BT: Springs (water)
RT: Geothermal energy
Geothermal fields
Geothermal water resources
Mineral springs

*Geothermal springs—Geysers**
USE: Geysers

*Geothermal springs—Hot springs**
USE: Hot springs

Geothermal water resources 444.2, 481.3.1
DT: January 1993
UF: Water resources—Thermal*
BT: Water resources
NT: Geothermal wells
RT: Geothermal energy
Geothermal fields
Geothermal prospecting

Geothermal water resources *(continued)*
Geothermal springs
Geysers
Hot springs

Geothermal wells 481.3.1, 615.1 (444.2)
DT: January 1983
BT: Geothermal water resources
Water wells
RT: Brines
Geothermal fields
Recharging (underground waters)
Scale (deposits)
Well drilling
Well pressure

*Geothermal wells—Brines**
USE: Brines

*Geothermal wells—Directional drilling**
USE: Directional drilling

*Geothermal wells—Recharging**
USE: Recharging (underground waters)

*Geothermal wells—Scale problems**
USE: Scale (deposits)

*Geothermal wells—Well pressure**
USE: Well pressure

Geriatrics 461.6
DT: January 1993
UF: Human engineering—Geriatrics*
BT: Medicine
RT: Human engineering

Germanate minerals 482.2
DT: January 1993
UF: Mineralogy—Germanates*
BT: Germanium compounds
Minerals

Germanium 549.3 (804)
DT: January 1993
UF: Ge
Germanium and alloys*
BT: Metalloids
NT: Semiconducting germanium
RT: Germanium alloys
Germanium compounds
Germanium deposits
Germanium metallurgy
Germanium metallography
Metals

Germanium alloys 549.3 (804)
DT: January 1993
UF: Germanium and alloys*
BT: Alloys
RT: Germanium
Germanium compounds

*Germanium and alloys**
USE: Germanium OR
Germanium alloys

Germanium compounds (804.1) (804.2)
DT: Predates 1975
BT: Metallic compounds
NT: Germanate minerals
Semiconducting germanium compounds
RT: Germanium
Germanium alloys

Germanium deposits 504.1
DT: Predates 1975
BT: Ore deposits
RT: Germanium

Germanium metallography (531.2) (549.3)
DT: Predates 1975
BT: Metallography
RT: Germanium
Germanium metallurgy

Germanium metallurgy 531.1 (549.3)
DT: Predates 1975
BT: Metallurgy
RT: Germanium
Germanium metallography

Getters 714.1
DT: January 1993
UF: Degassers
Vacuum technology—Getters*
BT: Electron tube components
RT: Thallium
Vacuum applications

Geysers 481.3.1 (615.1)
DT: January 1993
UF: Geothermal springs—Geysers*
BT: Hot springs
RT: Geothermal water resources

Giant pulsed lasers
USE: Q switched lasers

Gibbs energy
USE: Gibbs free energy

Gibbs free energy 641.1
DT: January 1995
UF: Gibbs energy
BT: Thermodynamic properties
RT: Enthalpy
Entropy
Free energy

Girders
USE: Beams and girders

GIS
USE: Geographic information systems

Glacial geology 481.1
DT: January 1993
UF: Geology—Glacial*
BT: Geology
RT: Glaciers

Glaciers (443) (444) (481.1)
DT: Predates 1975
RT: Arctic engineering
Glacial geology
Ice
Water

* Former Ei Vocabulary term

*Glaciers—Mapping**
USE: Mapping

Glare (707) (741.1)
DT: January 1993
UF: Electric lighting—Glare*
Street lighting—Glare*
BT: Light
RT: Electric lighting
Glare effects
Street lighting

Glare effects (707) (741.1)
DT: January 1993
UF: Visibility—Glare effects*
BT: Visibility
RT: Glare

Glass 812.3
DT: Predates 1975
UF: Concrete aggregates—Glass*
Electric insulating materials—Glass*
Heat exchangers—Glass*
Heat transfer—Glass*
BT: Amorphous materials
NT: Borophosphate glass
Borosilicate glass
Glass fibers
Light control glass
Meteoritic glass
Optical glass
Photosensitive glass
Safety glass
Semiconducting glass
Stained glass
RT: Bending (deformation)
Ceramic products
Composition
Concrete aggregates
Dielectric materials
Electric insulating materials
Glass bonding
Glass bottles
Glass industry
Glass manufacture
Glass membrane electrodes
Glass transition
Infrared transmission
Light transmission
Metallic glass
Sol-gels
Surface tension
Vitrification

Glass bonding 812.3
DT: January 1993
UF: Glass—Bonding*
Wire—Glass bonding*
BT: Bonding
RT: Glass
Wire

Glass bottles 812.3, 694.2
DT: January 1993
UF: Bottles—Glass*
BT: Bottles
RT: Glass

Glass cutting machines 812.3 (603.1)
DT: January 1993
UF: Machine tools—Glass cutting*
BT: Cutting tools
Machine tools

Glass fiber reinforced plastics 812.3, 817.1
DT: January 1993
UF: Plastics, Reinforced—Glass fiber*
BT: Fiber reinforced plastics
RT: Glass fibers

*Glass fiber**
USE: Glass fibers

Glass fibers 812.3 (413) (819.2) (741.1.2)
DT: January 1993
UF: Glass fiber*
Glass filaments
BT: Glass
Synthetic fibers
RT: Ceramic fibers
Fiber optics
Glass fiber reinforced plastics
Mineral wool
Optical fibers
Optical glass

Glass filaments
USE: Glass fibers

Glass forming machines 812.3
DT: Predates 1975
BT: Machinery
RT: Glass manufacture

Glass furnaces 812.3, 642.2
DT: Predates 1975
BT: Furnaces
RT: Glass manufacture

Glass industry 812.3 (912) (913)
DT: Predates 1975
BT: Industry
RT: Glass
Glass manufacture
Glass plants

Glass manufacture 812.3, 913.4
DT: Predates 1975
UF: Glass—Pressing*
BT: Manufacture
RT: Annealing
Glass
Glass forming machines
Glass furnaces
Glass industry
Glass plants
Molds
Opacifiers

*Glass manufacture—Annealing**
USE: Annealing

*Glass manufacture—Molds**
USE: Molds

* Former Ei Vocabulary term

*Glass manufacture—Opacifiers**
USE: Opacifiers

Glass membrane electrodes 812.3, 704.1
DT: January 1993
UF: Electrodes, Electrochemical—Glass membrane*
BT: Electrochemical electrodes
RT: Glass

Glass plants 812.3 (402.1)
DT: Predates 1975
BT: Industrial plants
RT: Glass industry
Glass manufacture

Glass transition 802.3 (815.1)
DT: January 1993
UF: Glass transition temperature
Plastics—Glass transition*
BT: Phase transitions
RT: Glass
Plastics
Vitrification

Glass transition temperature
USE: Glass transition AND Temperature

*Glass, Metallic**
USE: Metallic glass

*Glass, Polymeric**
USE: Polymeric glass

*Glass—Bending**
USE: Bending (forming)

*Glass—Bonding**
USE: Glass bonding

*Glass—Borophosphate**
USE: Borophosphate glass

*Glass—Borosilicate**
USE: Borosilicate glass

*Glass—Constitution**
USE: Composition

*Glass—Infrared transmission**
USE: Infrared transmission

*Glass—Light control**
USE: Light control glass

*Glass—Meteoric origin**
USE: Meteoritic glass

*Glass—Optical quality**
USE: Optical glass

*Glass—Organic compounds**
USE: Organic compounds

*Glass—Photosensitive**
USE: Photosensitive glass

*Glass—Pressing**
USE: Glass manufacture AND Pressing (forming)

*Glass—Safety**
USE: Safety glass

*Glass—Spectral properties**
USE: Spectrum analysis AND Optical properties

*Glass—Surface tension**
USE: Surface tension

Glazes 813.2 (812)
DT: Predates 1975
BT: Coatings
RT: Coloring
Paint
Varnish

Gliders 652.5
DT: Predates 1975
BT: Aircraft

Global positioning system (716.3) (655.2.1) (943.3)
DT: January 1993
UF: GPS
Radio navigation—Global positioning system*
BT: Satellite navigation aids
RT: Radio navigation
Satellite relay systems

Global warming 443.1 (451)
DT: January 1995
BT: Climate change

Gloss measurement (741.1) (811.1) (941.4)
DT: Predates 1975
UF: Paper—Gloss measurement*
BT: Measurements
RT: Paper

*Glossaries**
USE: Terminology

Glow discharge lamps
USE: Discharge lamps

Glow discharges 701.1
DT: Predates 1975
BT: Electric discharges
RT: Discharge lamps
Electric corona
Electroluminescence

Glucose 804.1 (822.3)
DT: January 1987
UF: Biochemistry—Glucose in body fluids*
Carbohydrates—Glucose*
BT: Sugars
RT: Glucose sensors
Saccharification

Glucose sensors 462.1 (801.2)
DT: January 1993
UF: Sensors—Glucose sensors*
BT: Biosensors
RT: Amperometric sensors
Glucose

Glues
USE: Adhesives

* Former Ei Vocabulary term

Gluing (802.3) (811.2)
DT: January 1993
UF: Wood products—Gluing*
BT: Bonding
RT: Adhesive joints
Adhesives
Laminating
Sealing (closing)
Wood products

Glycerin
USE: Glycerol

Glycerol 804.1
DT: Predates 1975
UF: Glycerin
BT: Alcohols
RT: Textile auxiliary materials

Glycols 804.1
DT: Predates 1975
BT: Alcohols

Goggles (914.1)
DT: Predates 1975
BT: Safety devices
RT: Eye protection
Eyeglasses
Safety glass

Gold 547.1
DT: January 1993
UF: Gold and alloys*
BT: Precious metals
RT: Gold alloys
Gold compounds
Gold deposits
Gold metallography
Gold metallurgy
Gold mines
Gold ore treatment
Gold plating

Gold alloys 547.1
DT: January 1993
UF: Gold and alloys*
BT: Precious metal alloys
RT: Gold
Gold compounds

*Gold and alloys**
USE: Gold OR
Gold alloys

Gold compounds (804.1) (804.2)
DT: January 1981
BT: Precious metal compounds
RT: Gold
Gold alloys

Gold deposits 504.3, 547.1
DT: Predates 1975
BT: Ore deposits
RT: Copper gold mines
Gold
Gold mines
Silver gold mines

Gold metallography 531.2, 547.1
DT: Predates 1975
BT: Metallography
RT: Gold
Gold metallurgy

Gold metallurgy 531.1, 547.1
DT: January 1993
BT: Metallurgy
NT: Gold refining
RT: Gold
Gold metallography
Gold ore treatment

Gold mines 504.3, 547.1
DT: January 1993
UF: Gold mines and mining*
BT: Mines
NT: Copper gold mines
Silver gold mines
RT: Gold
Gold deposits

*Gold mines and mining**
USE: Gold mines

Gold ore treatment 533.1, 547.1
DT: Predates 1975
BT: Ore treatment
RT: Gold
Gold metallurgy

Gold plating 539.3, 547.1
DT: Predates 1975
BT: Plating
RT: Gold

Gold refining 533.2, 547.1
DT: Predates 1975
BT: Gold metallurgy
Metal refining

Goniometers 943.3
DT: Predates 1975
UF: Radiogoniometers
BT: Instruments
RT: Crystallography
Radio direction finding systems

Government data processing 723.2 (902.3)
DT: January 1993
UF: Data processing—Governmental applications*
BT: Administrative data processing

Government policy
USE: Public policy

Governors 602.1 (612.1.1)
DT: Predates 1975
BT: Speed regulators
RT: Speed indicators

GPS
USE: Global positioning system

Graded index optics
USE: Gradient index optics

* Former Ei Vocabulary term

Gradient index optics 741.1
DT: January 1993
UF: Graded index optics
GRIN optics
BT: Optics
RT: Geometrical optics
Integrated optics
Lenses
Optical fibers
Optical materials
Refractive index

Grading (913.3)
DT: January 1993
UF: Concrete aggregates—Grading*
Lumber—Grading*
Tobacco—Grading*
Wood—Grading*
RT: Quality control
Size determination
Standards

Gradiometers (magnetic)
USE: Magnetometers

Graft copolymers 815.1
DT: January 1986
UF: Graft polymers
BT: Copolymers
RT: Grafting (chemical)
Ion exchange
Polymer blends
Terpolymers

Graft polymers
USE: Graft copolymers

Graft vs. host reactions 461.9.1
DT: January 1993
UF: Graft-host reactions
Immunology—Graft vs. host reaction*
RT: Antigen-antibody reactions
Grafts
Immunology
Transplantation (surgical)
Transplants

Graft-host reactions
USE: Graft vs. host reactions

Grafting (chemical) 802.2 (815.1)
DT: January 1993
UF: Grafting*
BT: Chemical reactions
RT: Graft copolymers

*Grafting**
USE: Grafting (chemical)

Grafts 462.4
DT: January 1993
UF: Prosthetics—Grafts*
BT: Transplants
RT: Biomaterials
Graft vs. host reactions
Implants (surgical)
Prosthetics

Grain (agricultural product) 821.4
DT: January 1993
UF: Grain*
BT: Agricultural products
RT: Grain elevators

Grain alcohol
USE: Ethanol

Grain boundaries (531.2) (933.1)
DT: January 1993
BT: Crystal defects
Interfaces (materials)
RT: Grain growth
Grain size and shape
Metallography
Nanostructured materials

Grain elevators 694.4, 821.4 (402.1)
DT: Predates 1975
BT: Buildings
RT: Farm buildings
Food storage
Grain (agricultural product)

Grain growth 933.1.2 (531.2) (802.3)
DT: January 1993
BT: Growth (materials)
RT: Grain boundaries
Grain size and shape
Metallography
Recrystallization (metallurgy)

Grain size and shape (531.2) (801.4) (933.1.1)
DT: January 1993
RT: Crystal microstructure
Grain boundaries
Grain growth
Metallography
Nanostructured materials

Grain structure
USE: Crystal microstructure

*Grain**
USE: Grain (agricultural product)

*Grain—Handling**
USE: Materials handling

Gramophones
USE: Phonographs

Grandstands
USE: Stadiums

Granite (481.1.2) (481.1)
DT: Predates 1975
BT: Igneous rocks
RT: Feldspar
Mica
Quartz

Granular materials
DT: Predates 1975
UF: Flow of fluids—Granular materials*
Flow of solids—Granular materials*
Heat transfer—Granular materials*
Materials handling—Granular materials*

* Former Ei Vocabulary term

Granular materials *(continued)*
BT: Materials
RT: Consolidation
Flow of fluids
Flow of solids
Materials handling
Mechanical permeability
Particles (particulate matter)
Pelletizing
Powder metals
Powders
Sand
Soils
Surface measurement

*Granular materials—Consolidation**
USE: Consolidation

*Granular materials—Permeability**
USE: Mechanical permeability

*Granular materials—Size determination**
USE: Size determination

*Granular materials—Surface measurement**
USE: Surface measurement

*Granular materials—Washing**
USE: Washing

Granulation 802.3
DT: January 1993
RT: Agglomeration
Atomization
Comminution
Crushing
Granulators

Granulators (816.2) (802.1) (822.1)
DT: January 1993
UF: Plastics machinery—Granulators*
BT: Machinery
RT: Ball mills
Crushers
Granulation
Grinding mills
Plastics machinery

Grape sugar
USE: Dextrose

Graph theory 921.4
DT: January 1993
UF: Mathematical techniques—Graph theory*
BT: Combinatorial mathematics
Topology
NT: Petri nets
Trees (mathematics)
RT: Computational geometry
Electric network analysis
Electric network synthesis
Information theory
Mathematical models

Graphic methods (723.5) (902.1)
DT: Predates 1975
NT: Drafting practice
Drawing (graphics)
Field plotting
Mapping
RT: Bode diagrams
Computer graphics
Information dissemination
Maps
Navigation charts
Nomograms
Nyquist diagrams
Phase diagrams
Photography
Printing
Schematic diagrams
Spreadsheets

Graphical user interfaces 722.2
DT: January 1995
UF: GUI
Icon based interfaces
BT: User interfaces
RT: Computer graphics

Graphite (482.2) (812.2) (813.2) (804)
DT: Predates 1975
UF: Aircraft materials—Graphite*
Powder metallurgy—Iron graphite*
BT: Carbon
Materials
NT: Artificial graphite
Colloidal graphite
Graphite fibers
RT: Carbonaceous refractories
Graphite electrodes
Graphitization
Moderators

Graphite electrodes (704.1)
DT: January 1993
BT: Electrodes
RT: Graphite

Graphite fiber reinforced plastics 817.1
DT: January 1993
UF: Plastics, Reinforced—Graphite fiber*
BT: Carbon fiber reinforced plastics
RT: Graphite fibers

Graphite fibers (804)
DT: January 1993
UF: Graphite—Fibers*
BT: Carbon fibers
Graphite
RT: Graphite fiber reinforced plastics

Graphite intercalation compounds
USE: Intercalation compounds

*Graphite—Artificial**
USE: Artificial graphite

*Graphite—Colloidal**
USE: Colloidal graphite

*Graphite—Fibers**
USE: Graphite fibers

Graphitic steel
USE: Carbon steel

* Former Ei Vocabulary term

Graphitization 802.2
DT: January 1993
UF: Cast iron—Graphitization*
Coke—Graphitization*
Iron and steel metallography—Graphitization*
Metallography—Graphitization*
RT: Cast iron
Coke
Eutectics
Graphite
Iron metallography
Metallography
Phase transitions
Steel metallography

Gratings (diffraction)
USE: Diffraction gratings

Gratings (optical)
USE: Diffraction gratings

Gravel (483.2) (481.1)
DT: January 1993
UF: Roadbuilding materials—Gravel*
Sand and gravel*
BT: Materials
RT: Building materials
Quarries
Quarrying
Roadbuilding materials
Rocks
Sand
Sand and gravel plants
Soils

Gravel roads 406.2
DT: January 1993
UF: Roads and streets—Gravel*
BT: Roads and streets

Gravimeters (931.5) (943.3)
SN: Devices intended to measure gravitation. For measuring specific gravity use GRAVITOMETERS
DT: Predates 1975
UF: Gravity meters
BT: Instruments
RT: Accelerometers
Geophysics
Gravitation

Gravimetric analysis (801)
DT: Predates 1975
UF: Gravimetry
BT: Chemical analysis
NT: Thermogravimetric analysis
RT: Electrolytic analysis
Microanalysis
Volumetric analysis

Gravimetry
USE: Gravimetric analysis

Gravitation 931.5
DT: Predates 1975
UF: Gravity
RT: Gravimeters
Gravitational effects
Gravity waves
Microgravity processing
Pendulums
Weightlessness

Gravitational effects 931.5
DT: June 1990
BT: Effects
RT: Gravitation

Gravitational pressure
USE: Hydrostatic pressure

Gravitational prospecting 481.4, 931.5
DT: January 1993
UF: Geophysical prospecting—Gravitational*
BT: Geophysical prospecting

Gravitational waves
USE: Gravity waves

Gravitometers 943.1
SN: Direct-reading instruments intended to measure specific gravity of liquids, solids, or gases. For instruments to measure gravitation, see GRAVIMETERS
DT: Predates 1975
UF: Specific gravity measuring instruments
BT: Density measuring instruments
NT: Hydrometers
RT: Density measurement (specific gravity)

Gravity
USE: Gravitation

Gravity conveyors 691.1
DT: January 1993
UF: Conveyors—Gravity*
BT: Conveyors

Gravity dams 441.1
DT: January 1993
UF: Dams, Gravity*
BT: Dams

Gravity meters
USE: Gravimeters

Gravity waves 931.5
DT: Predates 1975
UF: Gravitational waves
BT: Mechanical waves
RT: Gravitation

Greases
USE: Lubricating greases

Green's function 921 (931.4)
DT: January 1995
UF: Green's function methods
BT: Differential equations
Functions
RT: Boundary value problems
Elementary particles
Quantum theory

Green's function methods
USE: Green's function

Greenhouse effect 443.1, 451
DT: June 1990
BT: Climate change
RT: Air pollution
Atmospheric temperature
Carbon dioxide
Climatology
Heat radiation
Heating

Greenhouses 821.6 (402.1)
DT: Predates 1975
BT: Facilities
RT: Farm buildings

GRIN optics
USE: Gradient index optics

Grinders
USE: Grinding mills

Grinding (comminution) 802.3
DT: January 1993
UF: Crushing and grinding*
BT: Comminution
RT: Disintegration
Grinding mills

Grinding (machining) 604.2 (606.2)
SN: Use for removal of material. For comminution, use CRUSHING or PULVERIZATION
DT: January 1993
UF: Abrasive grinding
Grinding*
BT: Abrasive cutting
Machining
NT: Ball milling
Centerless grinding
RT: Buffing
Grinding machines
Grinding wheels
Metal cutting
Milling (machining)
Stress relief
Trimming
Wear of materials

Grinding machines (603.1) (606.2)
SN: For removing material. For comminuting machines, use GRINDING MILLS
DT: Predates 1975
BT: Machine tools
NT: Thread grinders
RT: Grinding (machining)
Lathes
Sanders

*Grinding machines—Thread grinding**
USE: Thread grinders

Grinding mills (802.1) (603.1)
SN: For comminuting material. For machines to remove material by abrasion, use GRINDING MACHINES
DT: Predates 1975
UF: Grinders
Pulverizers
Refuse incinerators—Grinding mills*
BT: Machinery
NT: Ball mills
Coal pulverizers
RT: Crushers
Granulators
Grinding (comminution)
Refuse incinerators

*Grinding mills—Ball milling**
USE: Ball milling

Grinding tools
USE: Grinding wheels

Grinding wheels (606.2) (603.1)
SN: For removal of material by abrasion. For comminution of material, use GRINDING MILLS
DT: Predates 1975
UF: Abrasive wheels
Grinding tools
BT: Cutting tools
Wheels
RT: Abrasives
Cubic boron nitride
Diamond cutting tools
Grinding (machining)
Guards (shields)
Sanders
Wheel dressing

*Grinding wheels—Cubic boron nitride**
USE: Cubic boron nitride

*Grinding wheels—Diamond**
USE: Diamond cutting tools

*Grinding wheels—Dressing**
USE: Wheel dressing

*Grinding wheels—Guards**
USE: Guards (shields)

*Grinding wheels—Mounting**
USE: Mountings

*Grinding**
USE: Grinding (machining)

*Grinding—Centerless**
USE: Centerless grinding

*Grinding—Stress relief**
USE: Stress relief

Grippers 731.5
SN: Robot components
DT: January 1993
UF: Robots, Industrial—Grippers*
BT: Robots
RT: Industrial robots
End effectors

Grit chambers (452.2) (452.4)
DT: January 1993
UF: Sewage treatment plants—Grit chambers*
BT: Sewage treatment plants

Groins (hydraulic structures)
USE: Jetties

* Former Ei Vocabulary term

*Groins**
USE: Jetties

Ground based space surveillance
USE: Space surveillance

Ground effect (651.1)
DT: January 1993
UF: Aerodynamics—Ground effect*
BT: Aerodynamics
RT: Air cushion vehicles
Lift
Wakes

Ground effect vehicles
USE: Air cushion vehicles

Ground electrodes
USE: Grounding electrodes

Ground operations 404.1 (652.1.2)
DT: January 1993
UF: Aviation, Military—Ground operations*
BT: Military operations
RT: Military aviation

Ground subsidence
USE: Subsidence

Ground support (aerospace)
USE: Aerospace ground support

Ground supports (401.2) (483) (502.1)
DT: January 1993
UF: Tunnels and tunneling—Ground supports*
BT: Supports
Tunnels

Ground traffic (airports)
USE: Airport vehicular traffic

Ground vehicle parts and equipment (662.4) (663.2)
DT: January 1993
BT: Vehicle parts and equipment
NT: Automobile parts and equipment

Ground vehicles (662) (663)
DT: January 1993
BT: Vehicles
NT: All wheel drive vehicles
Automobiles
Bicycles
Buses
Magnetic levitation vehicles
Motorcycles
Off road vehicles
Railroad rolling stock
Tanks (military)
Tractors (agricultural)
Trucks
RT: Land vehicle propulsion

Ground water
USE: Groundwater

Grounding (electric)
USE: Electric grounding

Grounding electrodes (704.1) (706) (714) (914.1)
SN: Electrodes for electrical grounding
DT: January 1993
UF: Earth electrodes
Electrodes, Grounding*
Ground electrodes
BT: Electric conductors
Electrodes
RT: Electric equipment protection
Electric grounding

*Grounding**
USE: Electric grounding

Grounds (antenna)
USE: Antenna grounds

Groundwater 444.2
DT: January 1993
UF: Ground water
Underground water
Water, Underground*
BT: Water
RT: Aquifers
Groundwater flow
Groundwater geochemistry
Groundwater pollution
Groundwater resources
Hydraulic conductivity
Recharging (underground waters)
Springs (water)
Surface waters

Groundwater flow 444.2, 631.1
DT: January 1993
UF: Flow of water—Underground*
BT: Flow of water
RT: Groundwater
Groundwater resources

Groundwater geochemistry 444.2, 481.2
DT: January 1993
UF: Geochemistry—Groundwater*
BT: Geochemistry
RT: Groundwater
Natural water geochemistry

Groundwater pollution 444.2, 453.1
DT: January 1993
UF: Water pollution—Underground*
BT: Water pollution
RT: Groundwater
Groundwater resources

Groundwater resources 444.2
DT: January 1993
UF: Water resources—Groundwater*
Water resources—Underground*
BT: Water resources
RT: Aquifers
Groundwater
Groundwater flow
Groundwater pollution
Recharging (underground waters)
Water wells

* Former Ei Vocabulary term

Groundwood pulp
USE: Mechanical pulp

Group decision support systems
USE: Decision support systems

Grouting 405.2 (412) (502.1)
DT: Predates 1975
UF: Foundations—Grouting*
Mines and mining—Grouting*
Shaft sinking—Grouting*
BT: Sealing (closing)
RT: Bonding
Cementing (shafts)
Cements
Fillers
Foundations
Mining
Seepage
Shaft sinking
Well cementing

Growing (crops)
USE: Cultivation

Growth (materials) (802.3) (933.1.2) (712.1)
DT: January 1993
UF: Growth*
NT: Crystal growth
Film growth
Grain growth
Semiconductor growth
RT: Materials
Shrinkage
Swelling

Growth kinetics (461.2) (461.8) (801.2)
DT: January 1993
UF: Cell culture—Growth kinetics*
RT: Biology
Cell culture

*Growth**
USE: Growth (materials)

GTO devices
USE: Thyristors

Guard rails (406)
DT: January 1993
UF: Roads and streets—Guard rails*
BT: Roads and streets
RT: Highway systems
Safety devices

Guards (shields) 914.1 (601.2)
DT: January 1993
UF: Grinding wheels—Guards*
Machine tools—Guards*
Machinery—Guards*
BT: Safety devices
RT: Grinding wheels
Machine tools
Machinery

GUI
USE: Graphical user interfaces

Guidance (electronic)
USE: Electronic guidance systems

Guided electromagnetic wave propagation 711.1 (714.3)
DT: January 1993
UF: Electromagnetic waves—Propagation in guides*
Guided waves
BT: Electromagnetic wave propagation
RT: Waveguides

Guided waves
USE: Guided electromagnetic wave propagation

Guides (mechanical) (502.2) (535.1.1) (601.2)
DT: January 1993
UF: Mine hoists—Guides*
Rolling mills—Guides*
BT: Equipment
RT: Chucks
Hoists
Jigs

Guideways (433) (681.1)
DT: January 1993
UF: Monorails—Guideways*
Transportation—Guideways*
Vehicles—Guideways*
BT: Equipment
RT: Monorails
Railroad transportation
Vehicles

Guideways (machine tools)
USE: Slideways

Gun perforators 511.2
DT: January 1993
UF: Oil well casing—Gun perforators*
BT: Perforators
RT: Oil well casings
Well perforation

Gunn devices 714.2
DT: January 1993
UF: Gunn effect devices
Semiconductor devices, Gunn effect*
BT: Semiconductor devices
NT: Gunn diodes
RT: Gunn effect
Gunn oscillators

Gunn diodes 714.2
DT: January 1993
UF: Semiconductor diodes, Gunn*
BT: Gunn devices
Semiconductor diodes
RT: Gunn effect
Gunn oscillators

Gunn effect 701.1 (712.1) (714.2)
DT: January 1993
UF: Semiconductor materials—Gunn effect*
BT: Electric field effects
Electric properties
RT: Gunn devices
Gunn diodes

* Former Ei Vocabulary term

Gunn effect *(continued)*
Gunn oscillators
Semiconductor materials

Gunn effect devices
USE: Gunn devices

Gunn oscillators 713.2, 714.2
DT: January 1993
UF: Oscillators, Gunn*
BT: Microwave oscillators
RT: Gunn devices
Gunn diodes
Gunn effect

Guns (armament) (404)
SN: Includes artillery and small arms, both military and nonmilitary
DT: January 1993
UF: Guns*
Pistols
Rifles
Weapons (guns)
BT: Equipment
NT: Gunsights
RT: Fire control systems
Ordnance

Guns (plasma)
USE: Plasma guns

*Guns**
USE: Guns (armament)

*Guns—Fire control**
USE: Fire control systems

*Guns—Sights**
USE: Gunsights

Gunsights 741.3 (404)
DT: January 1993
UF: Guns—Sights*
BT: Guns (armament)
RT: Range finders

Gust loads
USE: Wind effects

Gymnasiums 402.2
DT: Predates 1975
UF: Athletic facilities
BT: Facilities
RT: Recreational facilities

Gynecology 461.6
DT: January 1993
UF: Biomedical engineering—Gynecology*
BT: Medicine
RT: Biomedical engineering

Gypsum 482.2 (412) (414) (813) (821.2) (415.4)
DT: Predates 1975
BT: Sulfate minerals
RT: Gypsum plants
Plaster

Gypsum plants (402.1) (415.4)
DT: Predates 1975
BT: Industrial plants
RT: Gypsum

Gypsum plaster
USE: Plaster

Gyrators (714.3) (708.4)
DT: Predates 1975
UF: Electronic circuits, Gyrator*
Positive impedance inverters
BT: Networks (circuits)
RT: Electric network analysis
Ferrite devices
Phase shifters
Waveguides

Gyroscopes (943.1)
DT: Predates 1975
BT: Instruments
NT: Nuclear gyroscopes
RT: Aircraft instruments
Navigation
Ring lasers

*Gyroscopes—Nuclear**
USE: Nuclear gyroscopes

Gyrotrons 714.1 (711.1)
DT: January 1983
BT: Microwave tubes
RT: Microwave amplifiers
Microwave generation

H
USE: Hydrogen

Habitats (ocean)
USE: Ocean habitats

Hafnium 549.3
DT: January 1993
UF: Hafnium and alloys*
BT: Nonferrous metals
Transition metals
RT: Hafnium alloys
Hafnium compounds

Hafnium alloys 549.3
DT: January 1993
UF: Hafnium and alloys*
BT: Transition metal alloys
RT: Hafnium
Hafnium compounds

*Hafnium and alloys**
USE: Hafnium OR
Hafnium alloys

Hafnium compounds (804.1) (804.2)
DT: Predates 1975
BT: Transition metal compounds
RT: Hafnium
Hafnium alloys

Half adders
USE: Adders

Halide minerals 482.2
DT: January 1993
UF: Mineralogy—Halides*
BT: Halogen compounds

* Former Ei Vocabulary term

Halide minerals *(continued)*
Minerals
NT: Chloride minerals
Fluoride minerals
RT: Halogen elements

Halite
USE: Sodium chloride

Halite deposits
USE: Salt deposits

Hall constant
USE: Hall effect

Hall effect 701.1
DT: Predates 1975
UF: Hall constant
BT: Galvanomagnetic effects
RT: Hall effect devices
Magnetoelectric effects

Hall effect devices (708.4) (714)
DT: Predates 1975
UF: Hall generators
BT: Magnetic devices
NT: Hall effect transducers
RT: Hall effect
Semiconductor devices

Hall effect transducers (708.4) (714)
DT: Predates 1975
UF: Galvanomagnetic transducers
BT: Hall effect devices
Transducers

Hall generators
USE: Hall effect devices

Halogen compounds (804.1) (804.2)
DT: January 1986
BT: Chemical compounds
NT: Bromine compounds
Chlorine compounds
Chlorofluorocarbons
Fluorine compounds
Halide minerals
Iodine compounds
RT: Dehalogenation
Halogen elements
Halogenation

Halogen elements 804
DT: January 1993
BT: Nonmetals
NT: Astatine
Bromine
Chlorine
Fluorine
Iodine
RT: Dehalogenation
Halide minerals
Halogen compounds
Halogenation

Halogenation 802.2
DT: January 1993
BT: Chemical reactions
NT: Chlorination
RT: Dehalogenation
Halogen compounds
Halogen elements

Hammers 605.2
SN: Hand-held hammers. For power hammers not hand-held, use FORGING MACHINES, or PILE DRIVERS
DT: Predates 1975
BT: Hand tools
RT: Presses (machine tools)

Hand held computers
USE: Personal computers

Hand tools 605.2
DT: January 1993
UF: Tools, Hand*
Wrenches
BT: Tools
NT: Hammers
RT: Buffing machines
Construction equipment
Cutting tools
Files and rasps
Jacks
Reamers
Riveting machines
Sanders
Saws
Shovels
Spray guns
Taps

Handbooks 903.2 (901.2)
DT: January 1993
UF: Engineering handbooks
Engineering—Handbooks*
RT: Engineering
Information dissemination
Steam tables and charts
Textbooks

Handicapped aids
USE: Human rehabilitation equipment

Handicapped persons 461.5 (912.4)
DT: January 1993
UF: Disabled persons
Personnel—Handicapped persons*
RT: Human rehabilitation engineering
Patient rehabilitation
Personnel

Handles (door)
USE: Door handles

Handrails
USE: Railings

Hands (robotic)
USE: End effectors

Handsets (telephone)
USE: Telephone sets

* Former Ei Vocabulary term

Hangars 431.4 (402.1)
DT: January 1993
UF: Aircraft—Shelters*
BT: Airport buildings

Harbors
USE: Ports and harbors

Hard coal
USE: Anthracite

Hard disk storage 722.1
DT: January 1993
UF: Data storage, Magnetic—Hard disk*
Hard disks
BT: Magnetic disk storage

Hard disks
USE: Hard disk storage

Hard facing 531.1, 813.1
DT: January 1993
UF: Hard surfacing
Metals and alloys—Hard facing*
Surfacing
BT: Coating techniques
RT: Cladding (coating)
Metals
Plating
Welding

Hard surfacing
USE: Hard facing

Hardening 537.1
DT: January 1993
NT: Age hardening
Boriding
Case hardening
Dispersion hardening
Flame hardening
Spark hardening
Strain hardening
RT: Heat treatment
Martensite
Protective atmospheres
Quenching
Strengthening (metal)
Tempering

Hardness (421)
DT: January 1993
UF: Microhardness
BT: Mechanical properties
RT: Brittleness
Ductility
Machinability
Plasticity
Wear of materials
Wear resistance

Hardness testing (422.2)
DT: January 1993
UF: Materials testing—Hardness*
BT: Mechanical testing
RT: Fatigue testing
High temperature testing
Impact testing
Metal testing
Metals
Tensile testing

Hardware 605
DT: Predates 1975
BT: Equipment
NT: Automobile hardware
RT: Tools

Hardware (computer)
USE: Computer hardware

Hardware description languages
USE: Computer hardware description languages

Harmonic analysis 921.6
DT: January 1993
UF: Mathematical techniques—Harmonic analysis*
BT: Mathematical techniques
RT: Waveform analysis

Harmonic generation (713.2) (751.1)
DT: January 1977
NT: Second harmonic generation
RT: Natural frequencies
Oscillators (electronic)
Varactors

Harvesters 821.1
DT: January 1993
UF: Agricultural machinery—Harvesters*
BT: Agricultural machinery

Harvesting 821.3
DT: January 1993
UF: Sugar cane—Harvesting*
BT: Agriculture
RT: Combines
Crops

Hatch covers (671.2)
DT: January 1993
UF: Ships—Hatch covers*
BT: Ship equipment
RT: Hatches

Hatches (671.2)
DT: January 1993
UF: Ships—Hatches*
BT: Ship equipment
RT: Hatch covers

Hazardous materials (804) (914.1)
DT: January 1977
UF: Industrial wastes—Hazardous materials*
Toxic materials
Toxic wastes
BT: Materials
NT: Aflatoxins
Industrial poisons
RT: Asbestos
Carcinogens
Chemical wastes
Chemicals
Gas detectors
Hazardous materials spills
Hazards

* Former Ei Vocabulary term

Hazardous materials *(continued)*
Health hazards
Industrial wastes
Radioactive materials

Hazardous materials spills (451.1) (453.1) (804) (914.1)
DT: January 1993
UF: Hazardous materials—Spills*
Spills
NT: Oil spills
RT: Accidents
Hazardous materials
Hazards
Leakage (fluid)
Oil booms
Soil pollution
Water pollution

*Hazardous materials—Spills**
USE: Hazardous materials spills

Hazards 914.1
DT: January 1993
UF: Protective coatings—Hazards*
NT: Fire hazards
Gas hazards
Health hazards
Radiation hazards
RT: Accidents
Contamination
Gas bursts
Hazardous materials
Hazardous materials spills
Industrial plants
Mine dust
Mine explosions
Mine fires
Mine flooding
Mine rescue
Nuclear materials safeguards
Prevention
Product liability
Protection
Protective coatings
Rock bursts
Safety devices

Hazards and race conditions 721.1
DT: January 1993
UF: Race conditions
Switching theory—Hazards and race conditions*
BT: Switching theory

HBT
USE: Heterojunction bipolar transistors

HCI
USE: Human computer interaction

HDPE
USE: High density polyethylenes

HDTV
USE: High definition television

He
USE: Helium

Head frames 502.2
DT: January 1993
UF: Gallows frames
Headframes
Headstocks
Hoist frames
Mine hoists—Head frames*
BT: Mine hoists

Headboxes 601.2 (631.1) (811.1.2)
SN: Devices which control the flow of suspensions of solids into machines
DT: January 1993
UF: Papermaking machinery—Headbox*
BT: Machine components
RT: Flow of fluids
Papermaking machinery

Headframes
USE: Head frames

Headlights 662.4, 707.2
DT: January 1993
UF: Automobile headlights
Automobiles—Headlights*
BT: Electric lamps
RT: Automobile parts and equipment

Headphones 752.1 (718.1)
SN: Devices held against the ear; for devices which are to be inserted into the ear, use EARPHONES
DT: Predates 1975
BT: Audio equipment
Electroacoustic transducers
RT: Earphones
Telephone apparatus

Heads (magnetic)
USE: Magnetic heads

Headstocks
USE: Head frames

Health 461.6 (914.3)
DT: January 1993
UF: Personnel—Health*
RT: Health care
Health risks
Industrial hygiene
Medical problems
Medicine
Nutrition
Personnel

Health care 461.7 (912.4)
DT: January 1977
UF: Medical care
NT: Nutrition
RT: Ambulance cars
Ambulances
Diagnostic products
Epidemiology
Health
Health hazards

* Former Ei Vocabulary term

Health care *(continued)*
Health insurance
Health risks
Hospitals
Industrial hygiene
Medical problems
Medicine
Nursing

*Health care—Diagnostic products**
USE: Diagnostic products

*Health care—Epidemiology**
USE: Epidemiology

*Health care—Insurance**
USE: Health insurance

*Health care—Nursing**
USE: Nursing

*Health care—Nutrition**
USE: Nutrition

Health hazards 461.7 (914.1) (914.3.1)
DT: January 1993
BT: Hazards
RT: Accidents
Cholesterol
Contamination
Hazardous materials
Health care
Health risks
Heavy metals
Industrial hygiene
Mine dust
Occupational diseases
Pollution
Radioactive materials
Toxicity

Health insurance 461.7 (912.4) (911.1)
DT: January 1993
UF: Health care—Insurance*
BT: Insurance
RT: Health care

Health problems
USE: Medical problems

Health risks 461.7 (912.4) (914.3.1)
DT: January 1993
UF: Risk studies—Health risks*
BT: Risks
RT: Health
Health care
Health hazards
Industrial hygiene
Occupational risks

Hearing
USE: Audition

Hearing aids 752.1, 461.5 (718.1) (462.1)
DT: Predates 1975
BT: Sensory aids
NT: Telephone hearing aids
RT: Audition
Earphones
Prosthetics
Sound reproduction

Heart valve prostheses 462.4
DT: January 1993
UF: Prosthetics—Heart valves*
BT: Biomaterials
Prosthetics

Heat affected zone 538.2
SN: In welding
DT: January 1993
UF: Welds—Heat affected zone*
RT: Welding
Welds

Heat capacity
USE: Specific heat

Heat conduction 641.2
DT: January 1993
UF: Conduction (heat)
Heat transfer—Conduction*
Heat transfer—Contacts*
BT: Heat transfer
RT: Nusselt number
Thermal conductivity

Heat conductivity
USE: Thermal conductivity

Heat content
USE: Enthalpy

Heat convection 641.2
DT: January 1993
UF: Convection (heat)
Forced convection
Heat transfer—Convection*
Mixed convection
BT: Heat transfer
NT: Natural convection
RT: Flow of fluids
Nusselt number
Prandtl number

Heat dissipation
USE: Heat losses

Heat effects
USE: Thermal effects

Heat engines (612.1)
DT: Predates 1975
UF: Stirling cycle engines
BT: Engines
NT: Internal combustion engines
Steam engines
RT: Stirling cycle

Heat exchangers 616.1
SN: Includes unfired steam-generating units
DT: Predates 1975
BT: Equipment
NT: Air preheaters
Cooling towers
Economizers
Heat pipes
Heat pump systems

Heat exchangers *(continued)*
Heat sinks
Radiators
Regenerators
RT: Condensers (liquefiers)
Cooling systems
Feedwater heaters
Heat transfer
Heating equipment
Steam condensers
Steam generators
Superheaters
Tubes (components)
Waste heat utilization
Water heaters

*Heat exchangers—Condensers**
USE: Condensers (liquefiers)

*Heat exchangers—Cryogenic**
USE: Cryogenic equipment

*Heat exchangers—Finned tubes**
USE: Tubes (components) AND
Fins (heat exchange)

*Heat exchangers—Glass**
USE: Glass

*Heat exchangers—Regenerators**
USE: Regenerators

*Heat exchangers—Scale formation**
USE: Scale (deposits)

*Heat exchangers—Tubes**
USE: Tubes (components)

Heat flux 641.2
DT: January 1995
RT: Heat transfer
Heat transfer coefficients

*Heat insulating materials**
USE: Thermal insulating materials

*Heat insulating materials—Asbestos**
USE: Asbestos

*Heat insulating materials—Foamed products**
USE: Foamed products

*Heat insulating materials—Metal foil**
USE: Metal foil

*Heat insulating materials—Plastics**
USE: Plastics applications

Heat insulation
USE: Thermal insulation

Heat load
USE: Thermal load

Heat losses 641.2
DT: January 1993
UF: Heat dissipation
Roofs—Heat losses*
Tanks—Heat losses*
BT: Energy dissipation
RT: Heating
Roofs
Tanks (containers)
Thermal insulation

Heat measurement
USE: Thermal variables measurement

Heat pipes 619.1 (616.1) (641.2)
SN: Sealed pipes within which liquid metal or other fluid is evaporated to transport heat
DT: Predates 1975
BT: Heat exchangers
Pipe
RT: Heat transfer

Heat problems (654.2) (655.2) (651.1)
SN: Problems arising from aerodynamic heating. For problems caused by hot weather, use HOT WEATHER PROBLEMS
DT: January 1993
UF: Heating problems
Rockets and missiles—Heat problems*
Satellites—Heat problems*
Thermal problems
RT: Aerodynamic heating
Missiles
Rockets
Satellites
Thermal effects

Heat pump systems 616.1 (641.2)
DT: Predates 1975
BT: Heat exchangers
RT: Cooling
Electric heating
Heat transfer
Pumps

Heat radiation 641.2
DT: January 1993
UF: Furnaces—Radiation*
Heat transfer—Radiation*
Radiation (heat)
Thermal radiation
BT: Electromagnetic waves
Heat transfer
NT: Atmospheric radiation
RT: Bolometers
Greenhouse effect
Infrared radiation

Heat recovery
USE: Waste heat utilization

Heat resistance (931.2)
DT: January 1993
UF: Heat resisting*
Materials—Heat resistance*
Textiles—Heat resistance*
Thermal resistance
RT: Refractory materials
Superalloys

Heat resistant alloys
USE: Superalloys

*Heat resisting**
USE: Heat resistance

* Former Ei Vocabulary term

Heat shielding (413.2) (641.2) (654.1) (655.1)
DT: January 1993
UF: Rockets and missiles—Heat shields*
Thermal shielding
BT: Shielding
RT: Ablation
Heat transfer
Missiles
Rockets
Thermal insulation

Heat sinks 616.1 (714.2)
DT: Predates 1975
BT: Heat exchangers
NT: Fins (heat exchange)
RT: Cooling systems
Energy absorption
Thermal insulation

Heat stabilizers 803 (817.1)
DT: January 1993
UF: Plastics—Heat stabilizers*
BT: Stabilizers (agents)
RT: Plastics

Heat storage (641.2) (525.7)
DT: June 1990
UF: Heating—Heat storage*
BT: Energy storage
RT: Heating
Steam accumulators

Heat transfer 641.2
DT: Predates 1975
UF: Heat transfer—Low density*
NT: Heat conduction
Heat convection
Heat radiation
Underground heat transfer
Underwater heat transfer
RT: Aerodynamics
Boiling liquids
Boundary layers
Channel flow
Cooling
Ducts
Energy transfer
Fins (heat exchange)
Fluidized beds
Heat exchangers
Heat flux
Heat pipes
Heat pump systems
Heat shielding
Heat transfer coefficients
Heating
Mass transfer
Nucleate boiling
Nusselt number
Packed beds
Radiators
Steam condensers
Steam generators
Thermal conductivity
Thermal diffusion
Thermal insulation
Thermal stratification
Two phase flow
Unsteady flow
Vapors
Wall flow

Heat transfer coefficients 641.2
DT: January 1995
RT: Heat flux
Heat transfer

*Heat transfer—Boiling liquids**
USE: Boiling liquids

*Heat transfer—Boundary layer**
USE: Boundary layers

*Heat transfer—Channel flow**
USE: Channel flow

*Heat transfer—Conduction**
USE: Heat conduction

*Heat transfer—Contacts**
USE: Heat conduction

*Heat transfer—Convection**
USE: Heat convection

*Heat transfer—Disks**
USE: Disks (machine components) OR
Disks (structural components)

*Heat transfer—Ducts**
USE: Ducts

*Heat transfer—Fins**
USE: Fins (heat exchange)

*Heat transfer—Fluidized beds**
USE: Fluidized beds

*Heat transfer—Fluids**
USE: Fluids

*Heat transfer—Gases**
USE: Gases

*Heat transfer—Glass**
USE: Glass

*Heat transfer—Granular materials**
USE: Granular materials

*Heat transfer—Joints**
USE: Joints (structural components)

*Heat transfer—Liquid metals**
USE: Liquid metals

*Heat transfer—Liquids**
USE: Liquids

*Heat transfer—Low density**
USE: Heat transfer

*Heat transfer—Medical problems**
USE: Medical problems

*Heat transfer—Packed beds**
USE: Packed beds

*Heat transfer—Pipes**
USE: Pipe

* Former Ei Vocabulary term

*Heat transfer—Plates**
USE: Plates (structural components)

*Heat transfer—Porous materials**
USE: Porous materials

*Heat transfer—Radiation**
USE: Heat radiation

*Heat transfer—Rubber**
USE: Rubber

*Heat transfer—Textiles**
USE: Textiles

*Heat transfer—Tubes**
USE: Tubes (components)

*Heat transfer—Underground**
USE: Underground heat transfer

*Heat transfer—Underwater**
USE: Underwater heat transfer

*Heat transfer—Vapors**
USE: Vapors

*Heat transfer—Walls**
USE: Walls (structural partitions)

Heat treating furnaces 532.4
DT: January 1993
UF: Furnaces, Heat treating*
BT: Furnaces
NT: Salt bath furnaces
RT: Heat treatment
Heating furnaces
Ovens
Vacuum furnaces

Heat treatment 537.1
DT: Predates 1975
UF: Heat treatment—Lead bath*
Thermal treatment
NT: Age hardening
Annealing
Boriding
Carburizing
Case hardening
Decarburization
Electric heat treatment
Flame hardening
Nitriding
Quenching
Recrystallization (metallurgy)
Sintering
Spark hardening
Steel heat treatment
Tempering
Thermomechanical treatment
RT: Aging of materials
Hardening
Heat treating furnaces
Metallurgy
Nucleation
Protective atmospheres
Strengthening (metal)
Stress relaxation
Stress relief
Supersaturation
Welds

*Heat treatment—Annealing**
USE: Annealing

*Heat treatment—Boriding**
USE: Boriding

*Heat treatment—Carbonitriding**
USE: Carbonitriding

*Heat treatment—Carburizing**
USE: Carburizing

*Heat treatment—Case hardening**
USE: Case hardening

*Heat treatment—Decarburization**
USE: Decarburization

*Heat treatment—Electric**
USE: Electric heat treatment

*Heat treatment—Lead bath**
USE: Heat treatment

*Heat treatment—Low temperature**
USE: Low temperature operations

*Heat treatment—Nitriding**
USE: Nitriding

*Heat treatment—Quenching**
USE: Quenching

*Heat treatment—Shrinking**
USE: Shrinkage

*Heat treatment—Stress relief**
USE: Stress relief

*Heat treatment—Tempering**
USE: Tempering

Heating (643.1) (642.1)
SN: Scope formerly limited to space heating.
DT: Predates 1975
NT: Aerodynamic heating
Electric heating
Gas heating
Geothermal heating
Hot air heating
Hot water heating
Industrial heating
Plasma heating
Radiant heating
Space heating
RT: Air conditioning
Boilers
Climate control
Combustion
Cooling
Crucibles
Drying
Entropy
Environmental engineering
Food processing
Greenhouse effect
Heat losses
Heat storage

* Former Ei Vocabulary term

Heating *(continued)*
Heat transfer
Heating equipment
Ocean thermal energy conversion
Temperature
Temperature control
Thermal cycling
Thermal insulation
Thermal load
Thermal variables measurement
Thermal variables control
Thermodynamic properties
Thermodynamics
Vaporization
Waste heat
Waste heat utilization

Heating elements
USE: Electric heating elements

Heating equipment (642.2) (643.2)
DT: January 1993
UF: Automobiles—Heaters*
Chemical operations—Heaters*
Oil field equipment—Heaters*
Oil tanks—Heaters*
Petroleum refineries—Heaters*
Wind tunnels—Heaters*
BT: Equipment
NT: Automobile heaters
Electric heating elements
Furnaces
Ovens
Steam generators
Stokers
Stoves
Water heaters
RT: Chemical operations
Electric power utilization
Heat exchangers
Heating
Oil tanks
Petroleum refineries
Steam condensers
Wind tunnels

Heating furnaces (532)
SN: Excludes furnaces for space heating, for which use SPACE HEATING FURNACES, and for process heating, for which use INDUSTRIAL FURNACES
DT: January 1993
UF: Furnaces, Heating*
BT: Furnaces
RT: Electric furnaces
Heat treating furnaces
Ovens

Heating problems
USE: Heat problems

*Heating—Cold vapor systems**
USE: Humidity control

*Heating—District**
USE: District heating

*Heating—Electric**
USE: Electric heating

*Heating—Gas fuel**
USE: Gas heating

*Heating—Geothermal**
USE: Geothermal heating

*Heating—Heat storage**
USE: Heat storage

*Heating—Hot air systems**
USE: Hot air heating

*Heating—Hot water systems**
USE: Hot water heating

*Heating—Infrared**
USE: Infrared heating

*Heating—Pipelines**
USE: Pipelines

Heating—Piping systems
USE: Pipelines

*Heating—Radiant**
USE: Radiant heating

*Heating—Solar**
USE: Solar heating

Heaviside layer
USE: E region

Heavy fuels
USE: Residual fuels

Heavy media separation 802.3 (524) (533.1)
DT: January 1993
UF: Coal preparation—Heavy media separation*
Ore treatment—Heavy media separation*
BT: Separation
RT: Coal preparation
Ore treatment

Heavy metal alloys 531
DT: January 1993
BT: Alloys
NT: Antimony alloys
Bismuth alloys
Cadmium alloys
Lead alloys
Mercury amalgams
Thallium alloys
Tin alloys
RT: Heavy metal compounds
Heavy metals

Heavy metal compounds (804.1) (804.2)
DT: January 1983
BT: Metallic compounds
NT: Antimony compounds
Bismuth compounds
Cadmium compounds
Lead compounds
Mercury compounds
Metallic soaps
Thallium compounds
Tin compounds

* Former Ei Vocabulary term

Heavy metal compounds *(continued)*
RT: Heavy metal alloys
Heavy metals

Heavy metals 531
DT: January 1977
BT: Nonferrous metals
NT: Antimony
Bismuth
Cadmium
Lead
Mercury (metal)
Thallium
Tin
RT: Health hazards
Heavy metal alloys
Heavy metal compounds

Heavy oil production 511.1 (512.1.2)
DT: January 1993
UF: Oil well production—Heavy oil*
BT: Oil well production

Heavy oils
USE: Crude petroleum

Heavy water 804
DT: Predates 1975
UF: Deuterium oxide
BT: Deuterium compounds
Water
RT: Moderators
Radioisotopes

Heavy water reactors 621.1
DT: January 1993
UF: Nuclear reactors, Heavy water*
BT: Water cooled reactors

HEED
USE: High energy electron diffraction

Helical antennas (716)
DT: January 1993
UF: Antennas, Helical*
BT: Antennas

Helical gears 601.2
DT: January 1993
UF: Gears—Helical*
BT: Gears
RT: Spur gears
Worm gears

Helical springs 601.2
DT: January 1993
UF: Springs—Helical*
BT: Springs (components)

Helical waveguides 714.3
DT: January 1993
UF: Waveguides, Helical*
BT: Waveguides
RT: Circular waveguides

Helicons 711.1
SN: Low-frequency electromagnetic waves
DT: Predates 1975
BT: Electromagnetic waves

Helicopter cranes 652.4, 693.1
DT: January 1993
UF: Helicopters—Cranes*
BT: Cranes
Helicopters

Helicopter rescue services 652.4, 914.1
DT: January 1993
UF: Helicopters—Rescue service*
Rescue services (helicopter)
BT: Helicopter services
RT: Helicopters

Helicopter rotors 652.4, 601.1
DT: January 1993
UF: Helicopters—Rotors*
Rotors (helicopter)
BT: Helicopters
Rotors
RT: Compound helicopters

Helicopter services 652.4 (914.1)
DT: January 1993
UF: Airports—Helicopter services*
BT: Air transportation
NT: Helicopter rescue services
RT: Helicopters
Heliports

Helicopters 652.4
DT: Predates 1975
BT: Aircraft
NT: Compound helicopters
Convertible helicopters
Helicopter cranes
Helicopter rotors
Jet propelled helicopters
Military helicopters
RT: Aircraft communication
Flight simulators
Flutter (aerodynamics)
Helicopter rescue services
Helicopter services

*Helicopters, Military**
USE: Military helicopters

*Helicopters—Compound**
USE: Compound helicopters

*Helicopters—Construction industry**
USE: Construction industry

*Helicopters—Convertible**
USE: Convertible helicopters

*Helicopters—Cranes**
USE: Helicopter cranes

*Helicopters—Electric line construction**
USE: Construction AND
Electric lines

*Helicopters—Flight simulators**
USE: Flight simulators

*Helicopters—Flutter**
USE: Flutter (aerodynamics)

*Helicopters—Jet propelled**
USE: Jet propelled helicopters

*Helicopters—Logging applications**
USE: Logging (forestry)

*Helicopters—Petroleum industry**
USE: Petroleum industry

*Helicopters—Rescue service**
USE: Helicopter rescue services

*Helicopters—Rotors**
USE: Helicopter rotors

*Helicopters—Training applications**
USE: Personnel training

*Helicopters—Transmissions**
USE: Vehicle transmissions

Heliographs (941.3) (941.4) (657.1)
SN: Instruments which record the amount and duration of sunshine on blueprint paper
DT: Predates 1975
BT: Recording instruments

Heliostats (instruments) (657.1) (941.3)
SN: Instruments which automatically point in the direction of the sun
DT: January 1993
UF: Solar radiation—Heliostats*
BT: Instruments
RT: Solar radiation

Heliostats (solar concentrators)
USE: Solar concentrators

Heliports (431.4) (652.4)
DT: January 1987
BT: Airports
RT: Helicopter services

Helium 804
DT: Predates 1975
UF: He
BT: Inert gases
NT: Superfluid helium

Helium neon lasers 744.2
DT: January 1993
BT: Gas lasers

Helmholtz free energy
USE: Free energy

Hemodialyzers 462.1
DT: January 1993
UF: Biomedical equipment—Hemodialyzers*
Blood dialysis equipment
Kidney dialysis equipment
BT: Biomedical equipment
RT: Dialysis

Hemodynamics 461.1 (631.1)
DT: January 1993
UF: Biomedical engineering—Hemodynamics*
Circulation (blood)
NT: Microcirculation
RT: Biomedical engineering
Capillary flow
Cardiovascular system
Physiology

Hemoglobin 461.2, 804
DT: January 1995
UF: Hemoglobin based blood substitutes
BT: Pigments
Proteins
RT: Blood

Hemoglobin based blood substitutes
USE: Blood substitutes AND Hemoglobin

Hemoglobin oxygen saturation 801.2, 461.2
DT: January 1993
UF: Biochemistry—Hemoglobin oxygen saturation*
RT: Blood gas analysis
Oximeters
Oxygen

Hemp 821.4 (819.1)
DT: Predates 1975
BT: Agricultural products
RT: Hemp fibers
Textile fibers

Hemp fibers 819.1 (821.4)
DT: January 1993
BT: Bast fibers
RT: Hemp
Textile fibers

HEMT
USE: High electron mobility transistors

Herbicides 804.1 (804.2) (803) (821.2)
DT: Predates 1975
BT: Pesticides
RT: Algae control
Weed control

*Herbicides—Applicators**
USE: Applicators

Hermetic motors
USE: Induction motors

Hertzian waves
USE: Electromagnetic waves

Heterodyning (713.3) (716)
DT: January 1993
BT: Signal processing
RT: Mixer circuits

Heterogeneous catalysis
USE: Catalysis

Heterojunction bipolar transistors 714.2
DT: January 1995
UF: HBT
HJBT
BT: Bipolar transistors
RT: Bipolar integrated circuits
Semiconductor superlattices

Heterojunctions 714.2 (712.1)
SN: Boundaries between different types of semiconductor materials

* Former Ei Vocabulary term

Heterojunctions *(continued)*
DT: January 1993
UF: Heterostructures
Semiconductor devices—Heterojunctions*
BT: Semiconductor junctions
RT: Semiconductor devices

Heterostructures
USE: Heterojunctions

Heuristic methods (921)
DT: January 1993
UF: Mathematical techniques—Heuristic*
BT: Mathematical techniques
RT: Heuristic programming

Heuristic programming 723.1 (921)
DT: January 1993
UF: Heuristic routines
Systems science and cybernetics—Heuristic programming*
BT: Computer programming
RT: Artificial intelligence
Automata theory
Computer science
Heuristic methods
Operations research
Probability
Systems science

Heuristic routines
USE: Heuristic programming

Hg
USE: Mercury (metal)

Hierarchical systems (731.1)
DT: January 1993
UF: Systems science and cybernetics—Hierarchical systems*
BT: Systems science
RT: Control systems
Multivariable control systems
Multivariable systems

High definition television 716.4
DT: January 1993
UF: EDTV
Extended definition television
HDTV
Television—High definition*
BT: Television
RT: Television broadcasting
Television equipment
Television standards
Television systems

High density polyethylenes 815.1.1
DT: January 1993
UF: HDPE
Polyethylenes—High density*
BT: Polyethylenes

High electron mobility transistors 714.2
DT: January 1993
UF: HEMT
Transistors, High electron mobility*
BT: Field effect transistors
RT: Quantum interference devices
Semiconductor superlattices

High energy electron diffraction 931.3, 932.1
DT: January 1995
UF: HEED
BT: Electron diffraction
NT: Reflection high energy electron diffraction

High energy forming 535.2
DT: January 1993
UF: High energy rate forming
Metal forming—High energy*
BT: Metal forming
NT: Explosive forming
RT: Extrusion

High energy lasers 744.1
DT: January 1993
BT: Lasers

High energy milling
USE: Mechanical alloying

High energy physics 932.1
DT: January 1993
UF: Particle physics
Physics—High energy*
BT: Physics
RT: Elementary particles
Particle accelerators
Particle optics

High energy rate forming
USE: High energy forming

High frequency amplifiers 713.1
DT: January 1993
UF: Amplifiers, High frequency*
BT: Amplifiers (electronic)

High frequency telecommunication lines (716) (718)
DT: January 1993
UF: Radio lines
Telecommunication lines, High frequency*
BT: Telecommunication lines
RT: Radio
Radio systems

High frequency transformers (714)
DT: January 1993
UF: Electric transformers, High frequency*
BT: Electric transformers

High intensity light 741.1
DT: January 1993
UF: Light—High intensity*
BT: Light

High level languages 723.1.1
DT: January 1993
UF: Computer programming languages—High level languages*
BT: Computer programming languages
NT: Algorithmic languages
APL (programming language)
C (programming language)
COBOL (programming language)
Computer simulation languages

* Former Ei Vocabulary term

High level languages *(continued)*
FORTH (programming language)
List processing languages
Modula (programming language)
PL/1 (programming language)
Problem oriented languages
Procedure oriented languages
PROLOG (programming language)
RT: Computer hardware description languages

High modulus textile fibers (819) (819.2)
DT: January 1993
UF: Textile fibers—High modulus*
BT: Textile fibers

High pass filters 703.2
DT: January 1993
UF: Electric filters, High pass*
BT: Electric filters

High performance liquid chromatography (801) (802.3)
DT: January 1993
UF: Chromatographic analysis—High performance liquid*
BT: Liquid chromatography

High power lasers 744.1
DT: January 1993
BT: Lasers

High pressure boilers 614.2
DT: January 1993
UF: Boilers—High pressure*
BT: Boilers

High pressure effects (931.2)
DT: January 1993
UF: High pressure phenomena
BT: Effects
NT: High pressure effects in solids
RT: High pressure engineering
Pressure

High pressure effects in solids (931.2)
DT: January 1993
UF: Solids—High pressure effects*
BT: High pressure effects

High pressure engineering (901) (931.2)
DT: Predates 1975
BT: Engineering
RT: High pressure effects

High pressure liquid chromatography (801) (802.3)
DT: January 1993
UF: Chromatographic analysis—High pressure liquid*
BT: Liquid chromatography

High pressure phenomena
USE: High pressure effects

High pressure pipelines 619.1
DT: January 1993
UF: Steam pipelines—High pressure*
BT: Pipelines
RT: Steam pipelines

High pressure piping 619.1
DT: October 1975
BT: Piping systems

High pressure turbomachinery (612.3) (617)
DT: January 1993
UF: Turbomachinery—High pressure*
BT: Turbomachinery

High resolution electron microscopy
USE: Electron microscopy

High speed cameras 742.2
DT: January 1993
UF: Cameras—High speed*
BT: Cameras
RT: High speed photography

High speed photography 742.1
DT: January 1993
UF: Photography, High speed*
BT: Photography
RT: High speed cameras

High temperature applications
DT: January 1993
UF: Bearings—High temperature*
High temperature applications*
Metal cutting—High temperature*
Radio equipment—High temperature*
BT: Applications
RT: High temperature effects
High temperature engineering
High temperature operations
High temperature properties
High temperature reactors
High temperature superconductors
High temperature testing

*High temperature applications**
USE: High temperature applications

High temperature effects 931.2
DT: January 1993
BT: Thermal effects
RT: High temperature applications
High temperature engineering
High temperature operations
Temperature

High temperature engineering (901) (931.2)
DT: January 1993
BT: Engineering
RT: High temperature effects
High temperature properties
High temperature applications
High temperature operations

High temperature operations (931.2)
DT: January 1993
RT: High temperature applications
High temperature engineering
High temperature effects
High temperature reactors
High temperature superconductors
High temperature testing
Temperature

* Former Ei Vocabulary term

High temperature properties (931.2)
DT: January 1993
BT: Physical properties
RT: High temperature engineering
High temperature applications

High temperature reactors 621.1
DT: January 1993
UF: Nuclear reactors, High temperature*
BT: Nuclear reactors
RT: High temperature applications
High temperature operations

High temperature superconductors 708.3.1
DT: June 1990
BT: Superconducting materials
RT: High temperature applications
High temperature operations
Superconducting transition temperature

*High temperature superconductors—Transition temperature**
USE: Superconducting transition temperature

High temperature testing
DT: January 1993
UF: Materials testing—High temperature*
BT: Testing
RT: Buckling
Calorific value
Compression testing
Creep
Ductility
Elasticity
Embrittlement
Equipment testing
Fatigue of materials
Fracture
Hardness testing
High temperature operations
High temperature applications
Materials testing
Nondestructive examination
Plasticity
Quality control
Tensile testing
Thermodynamic properties
Viscoelasticity

High voltage DC transmission
USE: HVDC power transmission

Highway accidents 432.1, 914.1
DT: Predates 1975
UF: Road accidents
Traffic accidents
BT: Accidents
RT: Accident prevention
Skid resistance
Skidding

Highway administration 432.1, 912.2 (406.1)
DT: Predates 1975
BT: Management
RT: Highway engineering
Highway planning
Highway systems
Rights of way
Roads and streets

*Highway administration—Financing**
USE: Finance

Highway bridges 401.1, 406.1
SN: Also use material or type of bridge
DT: January 1993
UF: Bridges, Highway*
BT: Bridges
Highway systems

Highway design
USE: Highway engineering

Highway engineering (406)
DT: Predates 1975
UF: Highway design
BT: Civil engineering
RT: Highway administration
Highway systems
Roads and streets

Highway interchanges
USE: Interchanges

Highway markings 406.1, 432.4
DT: January 1993
UF: Highway signs, signals and markings*
Markings (highway)
Traffic signs, signals and markings*
RT: Highway systems
Road and street markings
Roads and streets
Runway markings
Street traffic control
Striping machines
Traffic signals

Highway overpasses
USE: Overpasses

Highway planning 432.1, 912.2 (406.1)
DT: January 1993
BT: Planning
RT: Highway administration
Regional planning

Highway service areas (406.1)
DT: January 1993
UF: Highway systems—Service areas*
Rest areas (highway)
Service areas (highway)
BT: Highway systems

Highway signals
USE: Traffic signals

Highway signs
USE: Traffic signs

*Highway signs, signals and markings**
USE: Traffic signals OR
Highway markings OR
Traffic signs

Highway systems 406.1
DT: Predates 1975
UF: Freeways

* Former Ei Vocabulary term

Highway systems *(continued)*
Interstate highways
Parkways
Thruways
BT: Public works
NT: Highway bridges
Highway service areas
Intelligent vehicle highway systems
Interchanges
Toll highways
Vehicular tunnels
RT: Airport vehicular traffic
Curves (road)
Footbridges
Guard rails
Highway administration
Highway engineering
Highway markings
Highway traffic control
Ice control
Intersections
Median dividers
Overpasses
Pavements
Reversible lanes
Rights of way
Road construction
Roads and streets
Roadsides
Snow and ice removal
Street lighting
Traffic signals
Traffic signs
Traffic surveys
Transportation routes
Underpasses
Urban planning
Weed control
Weigh stations

*Highway systems—Adjacent land areas**
USE: Roadsides

*Highway systems—Interchanges**
USE: Interchanges

*Highway systems—Median dividers**
USE: Median dividers

*Highway systems—Reversible lanes**
USE: Reversible lanes

*Highway systems—Right of way**
USE: Rights of way

*Highway systems—Roadside improvement**
USE: Improvement AND
Roadsides

*Highway systems—Service areas**
USE: Highway service areas

*Highway systems—Toll systems**
USE: Toll highways

Highway traffic control 432.4, 406.1
DT: Predates 1975
BT: Traffic control
RT: Highway systems
Intelligent vehicle highway systems
Street traffic control
Traffic signals
Traffic signs
Traffic surveys

Highway transportation
USE: Motor transportation

Highway underpasses
USE: Underpasses

Hinges (605)
DT: Predates 1975
BT: Components
RT: Doors
Fasteners

HIP
USE: Hot isostatic pressing

Hip prostheses 462.4
DT: January 1993
UF: Artificial hip joints
Prosthetics—Hip prostheses*
BT: Joint prostheses
RT: Biomaterials

History
DT: January 1993
RT: Biographies
Engineering
Technology

HJBT
USE: Heterojunction bipolar transistors

HOE
USE: Holographic optical elements

Hoist frames
USE: Head frames

Hoists 691.1
DT: Predates 1975
BT: Materials handling equipment
NT: Mine hoists
RT: Construction equipment
Cranes
Guides (mechanical)
Winches

Holding magnets
USE: Lifting magnets

Holmium 547.2
DT: January 1981
BT: Rare earth elements
RT: Holmium alloys
Holmium compounds

Holmium alloys 547.2
DT: January 1993
BT: Rare earth alloys
RT: Holmium
Holmium compounds

Holmium compounds (804.1) (804.2)
DT: June 1990
BT: Rare earth compounds
RT: Holmium
Holmium alloys

Holograms 743 (742.3)
DT: January 1986
RT: Holography
Optical image storage
Three dimensional

Holographic interferometry (743.2) (941.4)
DT: January 1993
UF: Interferometry, Holographic*
BT: Holography
Interferometry
RT: Speckle

Holographic optical elements (743.1.1)
DT: January 1993
UF: HOE
BT: Optical devices
RT: Holography
Optical interconnects

Holography 743
DT: Predates 1975
BT: Imaging techniques
NT: Acoustic holography
Computer generated holography
Holographic interferometry
Microwave holography
RT: Diffraction
Holograms
Holographic optical elements
Lasers
Optical correlation
Optical data processing
Optical image storage
Optics
Photography
Photomapping
Synthetic apertures

Honeycomb construction
USE: Honeycomb structures

Honeycomb structures 408.2
DT: June 1990
UF: Honeycomb construction
BT: Structures (built objects)
RT: Construction
Insulation
Laminates
Sandwich structures

Honing (606.2) (604.2)
DT: Predates 1975
BT: Cutting
Finishing
RT: Honing machines

Honing machines (606.2) (604.2) (603.1)
DT: Predates 1975
BT: Machine tools
RT: Honing

Hooks (691.1) (693.1)
DT: January 1993
UF: Cranes—Hooks*
Materials handling—Hooks*
BT: Fasteners
RT: Cranes
Materials handling equipment

Hopper cars 682.1.1
DT: January 1993
BT: Freight cars
RT: Hoppers
Mine cars

Hoppers 691.1
DT: January 1993
UF: Materials handling—Hoppers*
BT: Containers
RT: Hopper cars

Horizontal wells 512.1.1
DT: January 1995
BT: Oil wells
RT: Oil well drilling

Hormones 461.2 (801.2)
DT: January 1987
BT: Biopolymers
NT: Insulin

Horn antennas (716)
DT: January 1993
UF: Antennas, Horn*
BT: Antennas

Hose 619.1
SN: Flexible tubing for conducting fluids
DT: Predates 1975
BT: Equipment
NT: Brake hoses
Hose fittings
RT: Tubes (components)
Tubing

Hose fittings 619.1.1
DT: January 1993
UF: Brakes—Hose couplings*
Fittings (hose)
Hose—Fittings*
BT: Hose

*Hose—Fittings**
USE: Hose fittings

*Hose—Plastics**
USE: Plastics applications

*Hose—Rubber**
USE: Rubber products

Hosiery manufacture 819.5, 913.4
DT: Predates 1975
BT: Garment manufacture
RT: Hosiery mills

Hosiery mills 819.6 (402.1)
DT: Predates 1975
BT: Textile mills
RT: Hosiery manufacture
Textile industry

Hospital beds 462.2
DT: January 1993
UF: Beds (hospital)
Hospitals—Beds*
BT: Biomedical equipment
RT: Hospitals

Hospital data processing 462.2, 723.2
DT: January 1993
UF: Data processing—Hospital applications*
BT: Data processing
RT: Hospitals

Hospitals 462.2 (402.2)
DT: Predates 1975
BT: Buildings
NT: Emergency rooms
Intensive care units
Operating rooms
RT: Autoclaves
Clinical laboratories
Health care
Hospital beds
Hospital data processing
Medicine
Public works
Sterilizers

*Hospitals—Beds**
USE: Hospital beds

*Hospitals—Clinical laboratories**
USE: Clinical laboratories

*Hospitals—Emergency rooms**
USE: Emergency rooms

*Hospitals—Intensive care units**
USE: Intensive care units

*Hospitals—Operating rooms**
USE: Operating rooms

Hot air heating 643.1
DT: January 1993
UF: Heating—Hot air systems*
BT: Heating
RT: Space heating

Hot blast cupolas 532.5
DT: January 1993
UF: Blast cupolas
Cupola practice—Hot blast*
Cupolas—Hot blast*
Recuperative cupolas
BT: Cupolas
RT: Cupola practice

Hot carriers 701.1 (712.1)
DT: January 1993
UF: Semiconductor materials—Hot carriers*
BT: Charge carriers

Hot compression
USE: Hot pressing

Hot dry rock systems 481.3.1
DT: January 1993
UF: Geothermal fields—Hot dry rock systems*
BT: Geothermal fields

Hot ductility
USE: Ductility

Hot isostatic pressing (536.1) (812.1) (812.2)
DT: January 1995
UF: HIP
BT: Hot pressing
RT: Powder metallurgy
Sintering

Hot melt adhesives (804.1) (817.1) (818.5)
DT: January 1993
UF: Adhesives—Hot melt*
BT: Adhesives

Hot pressing (536.1) (812.1) (812.2)
DT: January 1993
UF: Ceramic materials—Hot pressing*
Hot compression
Powder metallurgy—Hot pressing*
Refractory materials—Hot pressing*
BT: Pressing (forming)
NT: Hot isostatic pressing
RT: Ceramic materials
Compaction
Hot working
Metal pressing
Powder metallurgy
Refractory materials

Hot reduction
USE: Hot working

Hot rolling 535.1.2
DT: January 1993
UF: Rolling mill practice—Hot rolling*
BT: Hot working
Rolling mill practice
RT: Hot rolling mills

Hot rolling mills 535.1.1
DT: January 1993
BT: Rolling mills
NT: Billet mills
Blooming mills
Slab mills
RT: Hot rolling
Strip mills

Hot springs 481.3.1
DT: January 1993
UF: Geothermal springs—Hot springs*
BT: Springs (water)
NT: Geysers
RT: Geothermal water resources
Mineral springs

Hot topping 534.2 (545.3)
DT: January 1993
UF: Steel ingots—Hot topping*
Topping (hot)
BT: Metal casting
RT: Ingots
Steel ingots

* Former Ei Vocabulary term

Hot water distribution systems (402) (643.2) (446.1)
DT: January 1993
UF: Hot water supply systems*
BT: Water distribution systems
RT: Hot water heating

Hot water heating 643.1
SN: Space heating only
DT: January 1993
UF: Heating—Hot water systems*
Hydronic heating
BT: Heating
RT: Hot water distribution systems
Space heating
Water heaters

*Hot water supply systems**
USE: Hot water distribution systems

Hot weather problems (443.1) (409)
DT: January 1993
RT: Thermal effects
Tropical buildings
Tropical engineering
Tropics

Hot working 535.2
DT: January 1993
UF: Hot reduction
BT: Metal working
NT: Hot rolling
RT: Cold working
Deep drawing
Forging
Hot pressing
Metal spinning
Metal stamping
Straightening
Stretching
Swaging
Upsetting (forming)

Hotels 402.2
DT: Predates 1975
BT: Housing
RT: Motels
Reservation systems

Household appliances
USE: Domestic appliances

Houses 402.3
DT: Predates 1975
BT: Housing
NT: Mobile homes
RT: Apartment houses

*Houses—Mobile**
USE: Mobile homes

Housing 403.1
SN: Very general term; prefer specific type of housing
DT: Predates 1975
BT: Buildings
NT: Apartment houses
Hotels
Houses
Motels
RT: Intelligent buildings
Urban planning

Hovercraft
USE: Air cushion vehicles

Hulls (seed coverings) 821.5
DT: January 1993
UF: Agricultural wastes—Hulls*
BT: Agricultural wastes

Hulls (ship) 671.1
DT: January 1993
UF: Ships—Hulls*
BT: Ships
RT: Air cushion vehicles
Boats

Human computer interaction (461.4) (722.2)
DT: January 1995
UF: HCI
Man machine interaction
RT: Man machine systems
User interfaces

Human engineering 461.4
DT: Predates 1975
UF: Bioengineering (human)
Human engineering—Sensory verbal descriptor scales*
Human factors engineering
Human factors*
BT: Engineering
NT: Ergonomics
RT: Ability testing
Acoustic noise
Acoustic wave effects
Arousal
Behavioral research
Biofeedback
Biomedical engineering
Environmental engineering
Geriatrics
Human form models
Human reaction time
Job analysis
Man machine systems
Manual control
Noise abatement
Noise pollution
Personnel testing
Psychophysiology
Sensory perception
Sleep research
Space flight
Speech intelligibility
Subjective testing
Time and motion study
Weightlessness
Work-rest schedules

*Human engineering—Arousal**
USE: Arousal

* Former Ei Vocabulary term

*Human engineering—Behavioral research**
USE: Behavioral research

*Human engineering—Biofeedback**
USE: Biofeedback

*Human engineering—Geriatrics**
USE: Geriatrics

*Human engineering—Manual control**
USE: Manual control

*Human engineering—Psychophysiology**
USE: Psychophysiology

*Human engineering—Reaction time**
USE: Human reaction time

*Human engineering—Sensory perception**
USE: Sensory perception

*Human engineering—Sensory verbal descriptor scales**
USE: Human engineering

*Human engineering—Sleep studies**
USE: Sleep research

*Human engineering—Subjective tests**
USE: Subjective testing

*Human engineering—Work-rest schedules**
USE: Work-rest schedules

Human factors engineering
USE: Human engineering

*Human factors**
USE: Human engineering

Human form models (914.1) (652.1) (662.1)
DT: January 1983
BT: Models
RT: Automobile driver simulators
Human engineering
Physiological models

Human machine systems
USE: Man machine systems

Human powered aircraft
USE: Man powered aircraft

Human reaction time 461.4
DT: January 1993
UF: Human engineering—Reaction time*
Reaction time
RT: Human engineering

Human rehabilitation engineering 461.5
DT: January 1987
BT: Engineering
RT: Artificial limbs
Breath controlled devices
Communication aids for nonvocal persons
Computer keyboard substitutes (handicapped aid)
Eye controlled devices
Functional assessment
Functional neural stimulation
Handicapped persons
Human rehabilitation equipment
Independent living systems
Learning aids for handicapped persons
Mobility aids for blind persons
Neuromuscular rehabilitation
Nutrition
Occupational therapy
Patient rehabilitation
Physical therapy
Seating for disabled persons
Sensory aids
Sensory feedback
Speech production aids
Tactile reading aids
Vibrotactile aids
Walking aids
Wheelchairs

*Human rehabilitation engineering—Breath-controlled devices**
USE: Breath controlled devices

*Human rehabilitation engineering—Communication aids for nonvocal**
USE: Communication aids for nonvocal persons

*Human rehabilitation engineering—Computer keyboard substitutions**
USE: Computer keyboard substitutes (handicapped aid)

*Human rehabilitation engineering—Eye-controlled devices**
USE: Eye controlled devices

*Human rehabilitation engineering—Functional assessment**
USE: Functional assessment

*Human rehabilitation engineering—Functional electric stimulation**
USE: Functional electric stimulation

*Human rehabilitation engineering—Functional neural stimulation**
USE: Functional neural stimulation

*Human rehabilitation engineering—Independent living systems**
USE: Independent living systems

*Human rehabilitation engineering—Learning aids for handicapped**
USE: Learning aids for handicapped persons

*Human rehabilitation engineering—Mobility aids for blind**
USE: Mobility aids for blind persons

*Human rehabilitation engineering—Neuromuscular rehabilitation**
USE: Neuromuscular rehabilitation

*Human rehabilitation engineering—Occupational therapy**
USE: Occupational therapy

*Human rehabilitation engineering—Physical therapy**
USE: Physical therapy

*Human rehabilitation engineering—Seating for disabled**
USE: Seating for disabled persons

*Human rehabilitation engineering—Sensory feedback techniques**
USE: Sensory feedback

*Human rehabilitation engineering—Speech production aids**
USE: Speech production aids

*Human rehabilitation engineering—Tactile reading aids**
USE: Tactile reading aids

*Human rehabilitation engineering—Vibrotactile aids**
USE: Vibrotactile aids

*Human rehabilitation engineering—Walking aids**
USE: Walking aids

Human rehabilitation equipment 461.5, 462.2
DT: January 1993
UF: Aids for handicapped persons
Handicapped aids
Rehabilitation equipment
BT: Biomedical equipment
NT: Breath controlled devices
Communication aids for nonvocal persons
Computer keyboard substitutes (handicapped aid)
Eye controlled devices
Learning aids for handicapped persons
Mobility aids for blind persons
Speech production aids
Tactile reading aids
Vibrotactile aids
Walking aids
Wheelchairs
RT: Artificial limbs
Exercise equipment
Human rehabilitation engineering
Independent living systems
Orthotics
Patient rehabilitation
Prosthetics
Seating for disabled persons

Humanities computing 723.2
DT: January 1993
UF: Data processing—Humanities applications*
Literary computing
BT: Data processing

Humidity
USE: Atmospheric humidity

Humidity control (643.3) (402)
SN: Use for control of moisture content of an atmosphere, as in a room or other confined space. For control of moisture in substances use MOISTURE CONTROL
DT: Predates 1975
UF: Heating—Cold vapor systems*
BT: Moisture control
RT: Air conditioning
Atmospheric humidity
Regain
Tropical engineering

Hurricane effects (914.1) (443)
DT: January 1993
UF: Buildings—Hurricane effects*
BT: Effects
RT: Buildings
Hurricane resistance
Hurricanes
Meteorology
Structural design
Structures (built objects)
Water wave effects
Wind effects
Wind stress

Hurricane resistance (914.1) (402) (408.1) (443)
DT: January 1993
RT: Buildings
Hurricane effects
Hurricanes
Structural design
Structures (built objects)

Hurricanes (443)
DT: Predates 1975
BT: Storms
RT: Hurricane effects
Hurricane resistance
Meteorology
Rain
Thunderstorms
Water waves
Wind effects

HVDC power transmission 706.1.1
DT: January 1993
UF: Electric power transmission, HVDC*
High voltage DC transmission
BT: DC power transmission

Hybrid computers 722.5
DT: January 1993
UF: Analog digital computers
Computers, Hybrid*
BT: Computers
RT: Analog computers
Analog to digital conversion
Digital computers
Digital to analog conversion

Hybrid integrated circuits 714.2
DT: January 1993
UF: Integrated circuits, Hybrid*
BT: Integrated circuits
NT: Flip chip devices
RT: Beam lead devices

Hybrid sensors (723.2) (732.2)
SN: Devices which include data processing capability
DT: January 1993
UF: Sensors—Hybrid construction*

Hybrid sensors *(continued)*
BT: Sensors
RT: Data processing

Hydrants 914.2 (446.1)
DT: January 1993
UF: Fire fighting equipment—Hydrants*
Fire hydrants
BT: Fire fighting equipment

Hydrated alumina 804.2 (812.1)
DT: January 1993
UF: Alumina trihydrate
Alumina—Hydrated*
Aluminum hydrate
Aluminum hydroxide
BT: Aluminum compounds
RT: Alumina

Hydrated lime 804.2 (412)
DT: January 1993
UF: Calcium hydroxide
Lime—Hydrated*
Slaked lime
BT: Calcium compounds
Hydrogen inorganic compounds
RT: Lime

Hydrates (512.2.2) (804)
DT: January 1993
UF: Natural gas wells—Hydrates*
BT: Chemical compounds
NT: Gas hydrates
RT: Hydration
Salts

Hydration (802.2)
DT: January 1993
BT: Chemical reactions
RT: Dehydration
Hydrates

Hydraulic accumulators 632.2 (601.1)
DT: Predates 1975
BT: Flywheels
Hydraulic machinery
RT: Surge tanks

Hydraulic amplifiers 632.2, 732.1
DT: January 1993
UF: Control equipment, Hydraulic—Amplifiers*
BT: Fluidic amplifiers

Hydraulic brakes 602, 632.2
SN: Hydraulically actuated mechanical brakes
DT: January 1993
UF: Brakes, Hydraulic*
BT: Brakes
Hydraulic control equipment
RT: Nonelectric final control devices

Hydraulic conductivity 632.1
DT: January 1995
UF: Permeability coefficient
RT: Groundwater
Mechanical permeability
Percolation (fluids)

Hydraulic control equipment 632.2, 732.1 (671.2)
DT: January 1993
UF: Control Equipment, Hydraulic*
Ship equipment—Hydraulic control*
BT: Hydraulic equipment
Mechanical control equipment
NT: Fluidic amplifiers
Hydraulic brakes
Hydraulic motors
Hydraulic rams
Hydraulic servomechanisms
RT: Hydraulic machinery
Hydraulics

Hydraulic drills 632.2 (405.1) (502.2)
DT: January 1993
UF: Rock drills—Hydraulic*
BT: Drills
Hydraulic equipment

*Hydraulic drive**
USE: Hydraulic drives

*Hydraulic drive—Variable speed**
USE: Variable speed drives

Hydraulic drives 602.1, 632.2
SN: For drives that propel vehicles, use HYDRAULIC PROPULSION
DT: Predates 1975
UF: Hydraulic drive*
Hydraulic transmission
BT: Drives
Hydraulic machinery
NT: Hydraulic torque converters
RT: Hydraulic fluids
Hydraulic jet propulsion
Variable speed drives

Hydraulic equipment 632.2
DT: January 1993
UF: Agricultural machinery—Hydraulic equipment*
Aircraft—Hydraulic equipment*
Automobile hydraulic equipment
Automobiles—Hydraulic equipment*
Couplings—Hydraulic*
Earthmoving machinery—Hydraulic equipment*
Industrial plants—Hydraulic equipment*
Materials handling—Hydraulic equipment*
Mines and mining—Hydraulic equipment*
Rockets and missiles—Hydraulic equipment*
Separators—Hydraulic*
Ship equipment—Hydraulic*
BT: Equipment
NT: Hydraulic control equipment
Hydraulic drills
Hydraulic tools
RT: Hydraulic fluids
Hydraulic machinery
Hydraulic mining
Hydraulic structures
Valves (mechanical)
Water hammer

Hydraulic fluids (632.1) (632.2)
SN: Working fluids for hydraulically actuated devices
DT: Predates 1975
BT: Fluids
RT: Hydraulic drives
Hydraulic equipment
Hydraulic machinery

Hydraulic fracturing 512.1.2
DT: January 1993
UF: Oil wells—Hydraulic fracturing*
Water wells—Hydraulic fracturing*
BT: Fracturing (oil wells)
RT: Fracturing fluids
Oil wells
Water wells

Hydraulic gates (434.1) (407.2)
DT: Predates 1975
UF: Gates (hydraulic)
BT: Hydraulic structures
RT: Canal gates
Spillway gates

Hydraulic jet propulsion (632.2) (675.1)
DT: January 1993
UF: Ship propulsion—Hydraulic jet*
BT: Ship propulsion
RT: Hydraulic drives

Hydraulic jump (632.1) (631.1)
DT: January 1993
RT: Hydraulics

Hydraulic laboratories 632.1 (402.1)
DT: January 1993
BT: Laboratories

Hydraulic machinery 632.2
DT: Predates 1975
UF: Machinery—Hydraulic*
Presses—Hydraulic*
BT: Machinery
NT: Hydraulic accumulators
Hydraulic drives
RT: Hydraulic control equipment
Hydraulic equipment
Hydraulic fluids
Hydraulic mining
Hydraulic models

Hydraulic mining 502.1 (632.1)
DT: January 1993
UF: Mines and mining—Hydraulic process*
BT: Mining
RT: Hydraulic equipment
Hydraulic machinery

Hydraulic models 632.1
DT: Predates 1975
BT: Models
RT: Hydraulic machinery
Hydraulics

Hydraulic motors 632.2
SN: Excludes HYDRAULIC TURBINES
DT: Predates 1975
BT: Hydraulic control equipment
Motors

Hydraulic rams 632.2, 732.1
DT: Predates 1975
BT: Hydraulic control equipment
RT: Actuators

Hydraulic servomechanisms 632.2, 732.1
DT: January 1993
UF: Servomechanisms—Hydraulic*
BT: Hydraulic control equipment
Servomechanisms

Hydraulic structures (407.2) (441) (446.2) (611)
SN: Very broad term; prefer term for specific type of structure
DT: Predates 1975
BT: Structures (built objects)
NT: Breakwaters
Caissons
Canal gates
Cofferdams
Culverts
Dams
Hydraulic gates
Jetties
Levees
Locks (on waterways)
Pontoons
Reservoirs (water)
Spillways
Stilling basins
Surge tanks
RT: Bridges
Causeways
Embankments
Energy dissipators
Hydraulic equipment
Port structures
Tanks (containers)
Water supply

*Hydraulic structures—Energy dissipators**
USE: Energy dissipators

Hydraulic tools 605.1, 632.2
DT: January 1993
UF: Jacks—Hydraulic*
Tools, jigs and fixtures—Hydraulic*
BT: Hydraulic equipment
Tools

Hydraulic torque converters 602, 632.2
DT: January 1993
UF: Hydrodynamic torque converters
Torque converters, Hydraulic*
Torque converting turbomachinery
BT: Hydraulic drives
Torque converters
RT: Turbomachinery

Hydraulic transmission
USE: Hydraulic drives

Hydraulic turbine generators
USE: Hydroelectric generators

Hydraulic turbines 617.1
DT: Predates 1975
BT: Turbines
NT: Bulb turbines
Francis turbines
Kaplan turbines
Tubular turbines
RT: Hydroelectric generators
Pumped storage power plants

*Hydraulic turbines—Bulb**
USE: Bulb turbines

*Hydraulic turbines—Francis**
USE: Francis turbines

*Hydraulic turbines—Kaplan**
USE: Kaplan turbines

*Hydraulic turbines—Tubular**
USE: Tubular turbines

Hydraulics 632.1
DT: Predates 1975
UF: Photogrammetry—Hydraulics applications*
Propulsion—Hydraulic energy*
BT: Fluid mechanics
RT: Flow of fluids
Fluidics
Hydraulic control equipment
Hydraulic jump
Hydraulic models
Hydrodynamics

Hydrazine 804.2
DT: Predates 1975
BT: Hydrogen inorganic compounds
Nitrogen compounds

Hydrides 804.2
DT: June 1990
BT: Hydrogen inorganic compounds
NT: Silanes

Hydrocarbon refining 802.3, 513.1 (804.1)
DT: January 1993
UF: Hydrocarbons—Refining*
BT: Refining
NT: Benzene refining
Oil shale refining
Petroleum refining
RT: Hydrocarbons

Hydrocarbons 804.1 (803)
DT: Predates 1975
BT: Organic compounds
NT: Acetylene
Aromatic hydrocarbons
Biogas
Butadiene
Ethers
Freons
Gas condensates
Insulating oil
Methane
Natural gasoline
Olefins
Paraffin waxes
Paraffins
RT: Cracking (chemical)
Fischer-Tropsch synthesis
Fluorocarbons
Fossil fuels
Hydrocarbon refining
Hydrocracking

*Hydrocarbons—Aromatic**
USE: Aromatic hydrocarbons

*Hydrocarbons—Refining**
USE: Hydrocarbon refining

*Hydrocarbons—Vapor pressure**
USE: Vapor pressure

Hydrochloric acid 804.2 (803)
DT: Predates 1975
UF: Muriatic acid
BT: Inorganic acids

Hydrocracking 802.2 (513.1)
DT: January 1993
BT: Catalytic cracking
RT: Hydrocarbons
Petroleum refining

Hydrodynamic torque converters
USE: Hydraulic torque converters

Hydrodynamic welding 538.2.1
SN: Pulsed-pressure welding
DT: January 1993
UF: Pulsed pressure welding
Welding—Hydrodynamic*
BT: Welding
RT: Hydrodynamics

Hydrodynamics 631.1.1
SN: Very general term; prefer more specific term
DT: Predates 1975
BT: Fluid dynamics
NT: Elastohydrodynamics
Electrohydrodynamics
Magnetohydrodynamics
RT: Aerodynamics
Atmospheric movements
Drag
Flow of fluids
Flow of water
Fluid structure interaction
Hydraulics
Hydrodynamic welding
Hydrofoil boats
Hydrofoils
Jets
Liquid waves
Liquids
Slamming (ships)
Viscosity of liquids
Water hammer
Water waves

Hydroelasticity
USE: Elasticity

Hydroelectric and gas turbine combined power plants 611.1, 612.3 (402.1)
DT: January 1993
UF: Electric power plants—Hydroelectric and gas turbine combined*
BT: Power plants
RT: Gas turbines

Hydroelectric and nuclear combined power plants 611.1, 613 (402.1)
DT: January 1993
UF: Electric power plants—Hydroelectric and nuclear combined*
BT: Power plants

Hydroelectric and steam combined power plants 611.1, 614 (402.1)
DT: January 1993
UF: Electric power plants—Hydroelectric and steam combined*
BT: Power plants
RT: Steam turbines

Hydroelectric generators 611.1, 705.2
DT: January 1993
UF: Electric generators—Hydroelectric*
Hydraulic turbine generators
BT: Electric generators
RT: Hydraulic turbines
Turbogenerators

Hydroelectric power 611.1
DT: January 1993
UF: Hydroelectricity
BT: Water power
RT: Hydroelectric power plants
Pumped storage power plants

Hydroelectric power plants 611.1
DT: Predates 1975
BT: Electric power plants
NT: Pumped storage power plants
Tidal power plants
RT: Dams
Hydroelectric power
Penstocks
Power plant intakes
River basin projects
Spillways
Turbines

*Hydroelectric power plants—Ice control**
USE: Ice control

*Hydroelectric power plants—Intakes**
USE: Power plant intakes

*Hydroelectric power plants—Pumped storage**
USE: Pumped storage power plants

Hydroelectricity
USE: Hydroelectric power

Hydrofluoric acid 804.2
DT: Predates 1975
BT: Fluorine compounds
Inorganic acids

Hydrofoil boats 674.1
DT: January 1993
UF: Hydrofoils*
BT: Motor boats
RT: Air cushion vehicles
Airfoils
Hydrodynamics
Hydrofoil propulsion
Hydrofoils

Hydrofoil propulsion 675.1
DT: January 1993
UF: Hydrofoils—Propulsion systems*
BT: Ship propulsion
RT: Hydrofoil boats
Hydrofoils

Hydrofoils (631.1.1) (674.1)
SN: Use for hydroplane surfaces; for boats or ships use HYDROFOIL BOATS
DT: Predates 1975
UF: Hydrofoils*
BT: Components
RT: Air cushion vehicles
Airfoils
Hydrodynamics
Hydrofoil boats
Hydrofoil propulsion
Seaplanes

*Hydrofoils**
USE: Hydrofoil boats OR
Hydrofoils

*Hydrofoils—Propulsion systems**
USE: Hydrofoil propulsion

Hydrogels 801.3, 804
DT: January 1995
BT: Colloids
RT: Viscosity

Hydrogen 804
DT: Predates 1975
UF: Cast iron—Hydrogen content*
H
BT: Nonmetals
RT: Deuterium
Hydrogen fuels
Hydrogen inorganic compounds
Protons
Tritium

Hydrogen bonds 801.4
DT: January 1995
BT: Chemical bonds
RT: Molecular structure

Hydrogen compounds (inorganic)
USE: Hydrogen inorganic compounds

Hydrogen compounds (organic)
USE: Organic compounds

Hydrogen embrittlement 531.1
DT: January 1993
UF: Metals and alloys—Hydrogen embrittlement*
BT: Embrittlement
RT: Corrosion
Corrosion fatigue

Hydrogen fuels 522, 804
DT: January 1977
BT: Synthetic fuels
RT: Automotive fuels
Hydrogen
Propellants

Hydrogen inorganic compounds 804.2
DT: Predates 1975
UF: Hydrogen compounds (inorganic)
BT: Inorganic compounds
NT: Caustic soda
Hydrated lime
Hydrazine
Hydrides
Hydrogen peroxide
Hydrogen sulfide
Inorganic acids
Water
RT: Deuterium compounds
Hydrogen

*Hydrogen ion concentration**
USE: pH

*Hydrogen ion concentration—pH meters**
USE: pH meters

*Hydrogen ion concentration—Sensors**
USE: pH sensors

Hydrogen ion sensors
USE: pH sensors

Hydrogen organic compounds
USE: Organic compounds

Hydrogen peroxide 804.2
DT: Predates 1975
BT: Hydrogen inorganic compounds
Peroxides

Hydrogen sulfide 804.2
DT: January 1986
BT: Hydrogen inorganic compounds
Sulfur compounds
RT: Hydrogen sulfide removal (water treatment)
Sour gas

Hydrogen sulfide removal (water treatment) 445.1, 804.2
DT: January 1993
UF: Water treatment—Hydrogen sulfide removal*
BT: Chemicals removal (water treatment)
RT: Hydrogen sulfide

Hydrogenation 802.2
DT: January 1993
BT: Chemical reactions
NT: Coal hydrogenation
RT: Dehydrogenation
Oils and fats
Petroleum refining

*Hydrographic surveying**
USE: Hydrographic surveys

Hydrographic surveys (471.1) (471.4)
DT: January 1993
UF: Hydrographic surveying*
BT: Surveys
RT: Ocean engineering
Oceanography

Hydrologic instruments (444) (471.2) (943.3)
DT: January 1993
BT: Instruments
RT: Hydrology

Hydrology (444) (471)
DT: January 1993
BT: Earth sciences
RT: Earth (planet)
Flood control
Floods
Geochemistry
Hydrologic instruments
Oceanography
Recharging (underground waters)
Remote sensing
Surface waters
Water resources
Water supply

Hydrolysis 802.2
DT: January 1993
BT: Decomposition
NT: Saccharification
RT: Saponification

Hydrometallurgy 531.1, 533.1
DT: January 1977
BT: Extractive metallurgy
RT: Electrodialysis
Ion exchange
Metal refining
Pyrometallurgy
Refining

Hydrometers 943.3
SN: Instruments used to measure density or specific gravity of liquids
DT: Predates 1975
BT: Gravitometers
RT: Density of liquids

Hydronic air conditioning systems
USE: Air conditioning

Hydronic heating
USE: Hot water heating

Hydrophobic chromatography (802.3) (801)
DT: January 1993
UF: Chromatographic analysis—Hydrophobic*
BT: Chromatography

Hydrophones 752.1
DT: Predates 1975
BT: Electroacoustic transducers
RT: Microphones
Oceanography
Sonar

* Former Ei Vocabulary term

Hydrostatic bearings 601.2 (631.1.1)
DT: January 1993
UF: Bearings—Hydrostatic*
BT: Bearings (machine parts)

Hydrostatic pressing (536.1) (812.1)
DT: January 1993
UF: Ceramic materials—Hydrostatic pressing*
Powder metallurgy—Hydrostatic pressing*
Refractory materials—Hydrostatic pressing*
BT: Pressing (forming)
RT: Ceramic materials
Powder metallurgy
Refractory materials

Hydrostatic pressure 631.1.1
DT: January 1995
UF: Gravitational pressure
BT: Pressure
RT: Fluid mechanics

Hydrotherapy
USE: Physical therapy

Hydroxylation 802.2
DT: January 1993
BT: Chemical reactions

Hygrometers 443.2, 944.1
SN: Instruments used to measure the relative humidity of the atmosphere
DT: Predates 1975
BT: Meteorological instruments
NT: Psychrometers
RT: Moisture meters

Hyperbaric chambers 462.1
DT: January 1977
BT: Biomedical equipment

Hypersonic aerodynamics 651.1
DT: January 1993
UF: Aerodynamics—Hypersonic*
BT: Aerodynamics
RT: Hypersonic flow
Supersonic aerodynamics

Hypersonic flow 631.1
DT: January 1993
UF: Flow of fluids—Hypersonic*
Wind tunnels—Hypersonic flow*
BT: Flow of fluids
RT: Hypersonic aerodynamics
Shock tubes
Supersonic flow
Wind tunnels

Hyperthermia therapy 461.6
DT: January 1993
UF: Biomedical engineering—Hyperthermia therapy*
Diathermy
BT: Patient treatment
RT: Biomedical engineering

Hypothermia 461.6
DT: January 1993
UF: Biomedical engineering—Hypothermia*
RT: Biomedical engineering
Cold effects

Hysteresigraphs
USE: Magnetic measuring instruments

Hysteresis (701.2) (931.2)
DT: Predates 1975
NT: Magnetic hysteresis
RT: Control nonlinearities
Damping
Eddy currents
Hysteresis motors
Internal friction
Tensile properties

Hysteresis motors 705.3.1
DT: January 1993
UF: Electric motors, Hysteresis type*
BT: Synchronous motors
RT: Hysteresis

I/O programs
USE: Input output programs

IC
USE: Integrated circuits

Ice (443)
DT: Predates 1975
UF: Inland waterways—Ice formation*
BT: Materials
NT: Sea ice
RT: Freezing
Glaciers
Ice control
Ice problems
Meteorology
Refrigerants
Snow and ice removal
Thermal expansion
Water

Ice control
DT: January 1993
UF: Airport runways—Ice control*
Dams—Ice control*
Hydroelectric power plants—Ice control*
Inland waterways—Ice control*
Port structures—Ice control*
Ports and harbors—Ice control*
BT: Control
RT: Airport runways
Dams
Defrosting
Highway systems
Ice
Ice problems
Inland waterways
Port structures
Ports and harbors
River control
Roads and streets
Sanding equipment
Snow and ice removal

* Former Ei Vocabulary term

Ice problems (443)
DT: January 1993
UF: Ships—Icing*
RT: Ice
Ice control
Ships
Snow and ice removal

Ice removal
USE: Snow and ice removal

Icebreakers (672)
DT: January 1993
UF: Ships—Icebreakers*
BT: Ships

Icon based interfaces
USE: Graphical user interfaces

ICUs
USE: Intensive care units

Identification (control systems) 731.1
DT: January 1993
UF: Control systems—Identification*
System identification
Systems science and cybernetics—Identification*
NT: Parameter estimation
State estimation
RT: Control systems
Models
Optimal control systems
Simulation
Step response
System theory
Systems science

Igneous rocks (481.1)
DT: January 1993
UF: Rock—Igneous*
BT: Crystalline rocks
NT: Granite
Nepheline syenite
Volcanic rocks
RT: Petrology

Ignition 521.1
DT: January 1993
BT: Combustion
NT: Preignition
RT: Gases
Ignition systems
Spontaneous combustion
Starting

Ignition systems 612.1, 521.1
DT: January 1993
UF: Internal combustion engines—Ignition systems*
BT: Internal combustion engines
RT: Combustion knock
Ignition
Starters
Starting

Ignitrons
USE: Mercury vapor rectifiers

IIR filters
USE: Digital filters

Illuminating engineering 707
DT: Predates 1975
BT: Engineering
RT: Daylighting
Electric lighting
Environmental engineering
Light sources
Photometry
Visibility
Vision

Illumination
USE: Lighting

Illumination meters 941.3
DT: January 1993
UF: Illuminometers
Photometers—Illumination meters*
BT: Photometers

Illuminometers
USE: Illumination meters

Ilmenite 482.2
DT: Predates 1975
BT: Ores
Oxide minerals
RT: Ilmenite ore treatment

Ilmenite ore treatment 533.1
DT: January 1981
BT: Titanium ore treatment
RT: Ilmenite

Image amplifiers (electron tube)
USE: Image intensifiers (electron tube)

Image analysis (741) (723.2)
DT: January 1993
UF: Image processing—Image analysis*
BT: Analysis
Image processing
RT: Optical flows

Image archiving
USE: Digital image storage OR
Optical image storage

Image coding (741) (723.2)
DT: January 1993
UF: Coding (images)
Encoding (images)
Image processing—Image coding*
Video coding
BT: Image processing
RT: Vector quantization

Image communication systems
(716) (717) (718) (722.3) (741.1)
DT: January 1995
UF: Image transmission
BT: Data communication systems
RT: Image processing

Image compression (741) (723.2)
DT: January 1993
UF: Compression (image)
Video compression
BT: Data compression
Image processing
NT: Vector quantization
RT: Image quality

Image converters 714.1
DT: January 1993
UF: Converters (image)
Electron tubes, Image converter*
BT: Cathode ray tubes
NT: Image intensifiers (electron tube)
RT: Image sensors

Image enhancement (741) (723.2)
DT: January 1993
UF: Enhancement (images)
Image processing—Enhancement*
BT: Image processing
RT: Image quality

Image formation
USE: Image processing

Image intensifiers (electron tube) 714.1
DT: January 1993
UF: Electron tubes, Image intensifier*
Image amplifiers (electron tube)
Intensifiers (electron tube)
Microscopes—Image intensifiers*
BT: Image converters
RT: Image sensors
Light amplifiers
Microscopes
Phosphors
Photocathodes
Photomultipliers
Telescopes
X ray apparatus

Image intensifiers (solid state) 714.2
DT: January 1993
UF: Intensifiers (solid state)
Microscopes—Image intensifiers*
BT: Optoelectronic devices
RT: Image sensors
Light amplifiers
Microscopes
Photomultipliers
Telescopes

Image processing (741) (723.2)
DT: October 1975
UF: Image formation
Picture processing
BT: Signal processing
NT: Anti-aliasing
Color image processing
Computer generated holography
Computer vision
Image analysis
Image coding
Image compression
Image enhancement
Image reconstruction
Image segmentation
Image understanding
Object recognition
RT: Adaptive algorithms
Analog to digital conversion
Digital image storage
Edge detection
Feature extraction
Fourier optics
Image communication systems
Image quality
Image recording
Imaging systems
Imaging techniques
Information technology
Optical character recognition
Optical image storage
Pattern recognition
Radiography
Video signal processing

*Image processing—Color images**
USE: Color image processing

*Image processing—Enhancement**
USE: Image enhancement

*Image processing—Image analysis**
USE: Image analysis

*Image processing—Image coding**
USE: Image coding

*Image processing—Reconstruction**
USE: Image reconstruction

Image quality (741)
DT: January 1993
UF: Quality (images)
RT: Image compression
Image enhancement
Image processing
Image reconstruction

Image reconstruction (741) (723.2)
DT: January 1993
UF: Image processing—Reconstruction*
Image restoration
Reconstruction (images)
Restoration (images)
BT: Image processing
RT: Image quality

Image recording (741) (716.4)
DT: January 1993
BT: Recording
RT: Image processing
Imaging systems
Imaging techniques

Image restoration
USE: Image reconstruction

* Former Ei Vocabulary term

Image segmentation (723.2) (741.1)
DT: January 1995
UF: Segmentation (image)
BT: Image processing

Image sensors (741.3)
DT: January 1977
UF: Imagers
BT: Optoelectronic devices
Sensors
RT: Computer vision
Image converters
Image intensifiers (electron tube)
Image intensifiers (solid state)
Infrared imaging
Optical sensors

Image storage (digital)
USE: Digital image storage

Image storage (optical)
USE: Optical image storage

Image storage tubes 714.1
DT: January 1993
UF: Electron tubes, Image storage*
Image tubes
BT: Cathode ray tubes
Storage tubes
RT: Television camera tubes

*Image storage, Digital**
USE: Digital image storage

*Image storage, Optical**
USE: Optical image storage

Image transmission
USE: Image communication systems

Image tubes
USE: Image storage tubes

Image understanding (741) (723.2)
DT: January 1993
BT: Image processing

Imagers
USE: Image sensors

Imaging
USE: Imaging techniques

Imaging systems (741) (723.2)
DT: January 1993
RT: Image processing
Image recording
Imaging techniques

Imaging techniques (741) (723.2)
DT: January 1977
UF: Imaging
NT: Acoustic imaging
Holography
Infrared imaging
Magnetic resonance imaging
Medical imaging
Photography
Radar imaging
Radiography
RT: Acoustic microscopes
Digital image storage
Image processing
Image recording
Imaging systems
Optical image storage
Synthetic apertures

Immiscibility
USE: Solubility

Immobilization (cell cultures)
USE: Cell immobilization

Immobilization (enzymes)
USE: Enzyme immobilization

Immobilization (radioactive wastes)
USE: Radioactive waste vitrification

Immune reactions
USE: Antigen-antibody reactions

Immunization 461.7 (461.9.1)
DT: January 1987
BT: Patient treatment
RT: Antibodies
Immunology
Vaccines

Immunology 461.9.1
DT: October 1975
BT: Biology
RT: Allergies
Antibodies
Antigen-antibody reactions
Antigens
Biocompatibility
Graft vs. host reactions
Immunization
Immunosensors
Interferons
Monoclonal antibodies
Vaccines

*Immunology—Antigen-antibody reactions**
USE: Antigen-antibody reactions

*Immunology—Graft vs. host reaction**
USE: Graft vs. host reactions

*Immunology—Monoclonal antibody**
USE: Monoclonal antibodies

*Immunology—Sensors**
USE: Immunosensors

Immunosensors (461.9.1) (801) (462.1)
DT: January 1993
UF: Immunology—Sensors*
BT: Biosensors
RT: Immunology

Impact avalanche transit time diodes
USE: IMPATT diodes

Impact resistance (408.1) (421)
DT: January 1993
UF: Structural design—Impact resistance*
RT: Fracture toughness

* Former Ei Vocabulary term

Impact resistance *(continued)*
Impact testing
Structural design

Impact testing (422.2)
DT: January 1993
UF: Materials testing—Impact*
BT: Testing
RT: Compressibility of solids
Compression testing
Equipment testing
Fatigue testing
Hardness testing
Impact resistance
Materials testing
Shock testing
Strain rate
Stress concentration
Tensile strength

Impact toughness
USE: Fracture toughness

IMPATT diodes 714.2
SN: Impact ionization avalanche transit time diodes
DT: January 1993
UF: Impact avalanche transit time diodes
Semiconductor diodes, IMPATT*
BT: Avalanche diodes
Transit time devices

Impedance (acoustic)
USE: Acoustic impedance

Impedance (electric)
USE: Electric impedance

Impedance matching (acoustic) (751.1)
DT: January 1993
UF: Impedance matching*
RT: Acoustic impedance

Impedance matching (electric) (703.1) (701.1)
DT: January 1993
UF: Impedance matching*
RT: Electric impedance
Standing wave meters

*Impedance matching**
USE: Impedance matching (acoustic) OR
Impedance matching (electric)

Imperfections
USE: Defects

Implantation (ions)
USE: Ion implantation

Implants (dental)
USE: Dental prostheses

Implants (surgical) 462.4
DT: January 1993
UF: Biomedical engineering—Surgical implants*
Prosthetic implants
Surgical implants
BT: Biomaterials
RT: Biomedical engineering
Grafts
Intraocular lenses
Prosthetics
Surgery
Transplantation (surgical)

Impregnation (811.2) (802.3)
DT: January 1993
RT: Infiltration
Lubrication
Porosity
Porous materials
Wood

Improvement
DT: January 1993
UF: Highway systems—Roadside improvement*
Inland waterways—Improvement*
NT: Widening (transportation arteries)
RT: Inland waterways
Maintenance

Impulse testing (413.1)
DT: January 1993
UF: Electric insulation—Impulse testing*
BT: Electric insulation testing

Impurities
DT: January 1993
UF: Contaminants
BT: Materials
NT: Crystal impurities
RT: Chemical analysis
Contamination
Inclusions

In situ coal combustion 521.1, 524
DT: January 1993
UF: Coal combustion—In situ*
BT: In situ combustion
RT: Coal gas

In situ combustion 521.1 (511.1)
DT: January 1993
UF: Oil well production—In situ combustion*
BT: Combustion
In situ processing
NT: In situ coal combustion
RT: Oil well production
Reverse combustion
Thermal oil recovery

In situ processing
DT: January 1993
BT: Processing
NT: In situ combustion

Incandescent lamps 707.2
DT: January 1993
UF: Filament lamps
BT: Electric lamps
NT: Filaments (lamp)

Incineration
USE: Combustion

Incineration (refuse)
USE: Refuse incineration

* Former Ei Vocabulary term

Incineration (waste)
USE: Waste incineration

Incinerators
USE: Refuse incinerators

Inclusions
DT: January 1993
UF: Metals—Gases*
RT: Crystal defects
Defects
Impurities
Metallographic microstructure
Segregation (metallography)
Stress analysis

Independent living systems (461.5)
DT: January 1993
UF: Human rehabilitation engineering—Independent living systems*
RT: Human rehabilitation engineering
Human rehabilitation equipment

Indexing (materials working) (603.2) (604.2)
DT: January 1993
UF: Machine tools—Indexing*
Tools, jigs and fixtures—Indexing*
BT: Processing
RT: Fixtures (tooling)
Machine tools
Tools

Indexing (of information) 903.1
DT: January 1993
UF: Information science—Indexing*
BT: Information analysis
RT: Classification (of information)
Vocabulary control

Indicators (chemical) 801, 804
DT: January 1993
UF: Chemical analysis—Indicators*
Chemical indicators
BT: Chemicals
RT: Chemical analysis

Indicators (instruments) (944.3)
DT: January 1993
UF: Indicators*
BT: Instruments
NT: Depth indicators
Liquid level indicators
Proximity indicators
Speed indicators
Standing wave meters
Working fluid pressure indicators
RT: Alarm systems
Control equipment
Data acquisition
Readout systems

*Indicators**
USE: Indicators (instruments)

*Indicators—Working fluid pressure**
USE: Working fluid pressure indicators

Indium 549.3
DT: January 1993
UF: Indium and alloys*
BT: Nonferrous metals
NT: Semiconducting indium
RT: Indium alloys
Indium compounds
Indium metallography
Indium metallurgy
Indium plating

Indium alloys 549.3
DT: January 1993
UF: Indium and alloys*
BT: Alloys
RT: Indium
Indium compounds

*Indium and alloys**
USE: Indium OR
Indium alloys

Indium compounds (804.1) (804.2)
DT: Predates 1975
BT: Metallic compounds
NT: Semiconducting indium compounds
RT: Indium
Indium alloys

Indium metallography 531.2, 549.3
DT: Predates 1975
BT: Metallography
RT: Indium
Indium metallurgy

Indium metallurgy 531.1, 549.3
DT: January 1993
BT: Metallurgy
RT: Indium
Indium metallography

Indium plating 539.3, 549.3
DT: Predates 1975
BT: Plating
RT: Indium

Indoor air pollution 451
DT: January 1993
UF: Air pollution—Indoor*
BT: Air pollution
RT: Air cleaners

Induced currents 701.1
DT: January 1993
UF: Pipelines—Induced currents*
BT: Electric currents

Induced polarization logging 512.1.2
DT: January 1993
UF: Oil well logging—Induced polarization*
BT: Electric logging
RT: Natural gas well logging
Oil well logging

Inductance measurement 942.2
DT: January 1993
UF: Electric measurements—Inductance*
BT: Electric variables measurement

* Former Ei Vocabulary term

Inductance measurement *(continued)*
RT: Electric measuring bridges
Electric reactance measurement

Induction generators
USE: Asynchronous generators

Induction heating 642.1 (701.1) (704.2)
DT: January 1993
UF: Eddy current heating
Electric heating, Induction*
BT: Electric heating
RT: Eddy currents
Electric furnaces

Induction logging 512.1.2
DT: January 1993
UF: Oil well logging—Induction*
BT: Electric logging
RT: Natural gas well logging
Oil well logging

Induction machinery
USE: Asynchronous machinery

Induction motors 705.3.1
DT: January 1993
UF: Asynchronous motors
Electric motors, Induction*
Hermetic motors
BT: AC motors
Asynchronous machinery
NT: Squirrel cage motors

Inductors (electric)
USE: Electric inductors

Industrial applications
DT: January 1993
UF: Industrial applications*
Motion pictures—Industrial*
Television applications—Industry*
BT: Applications
RT: Industrial optics
Industrial plants
Programmable logic controllers

*Industrial applications**
USE: Industrial applications

Industrial design
USE: Product design

Industrial diamonds (482.2) (604) (606)
DT: January 1993
BT: Diamonds
Industrial gems
NT: Diamond films
RT: Abrasives

Industrial economics 911.2
DT: Predates 1975
BT: Economics
RT: Industry
Manufacturing data processing

*Industrial economics—Strategic materials**
USE: Strategic materials

Industrial electronics (714) (715) (716)
DT: Predates 1975
BT: Electronics industry
RT: Electronic equipment

Industrial emissions 451.1
DT: January 1993
UF: Emissions (industrial)
Industrial plants—Emissions*
BT: Industrial wastes
RT: Industrial plants
Particulate emissions
Volatile organic compounds

Industrial engineering 912.1
DT: Predates 1975
BT: Engineering
RT: Industry
Time and motion study
Work simplification

Industrial furnaces 642.2
SN: For process heating; for space heating use SPACE HEATING FURNACES
DT: January 1993
UF: Furnaces, Industrial*
Furnaces—Glass bath*
Furnaces—Optical image*
BT: Furnaces
NT: Infrared furnaces
RT: Industrial heating

Industrial gems 482.2.1
DT: January 1993
BT: Gems
NT: Industrial diamonds
RT: Garnets
Topaz

Industrial heating 642.1
SN: Heat used for processing
DT: Predates 1975
UF: Process heating*
BT: Heating
NT: Preheating
RT: Drying
Electric heating
Gas heating
Industrial furnaces
Infrared drying
Infrared heating
Microwave heating
Ovens

Industrial hygiene 914.3, 461.7
DT: Predates 1975
RT: Health
Health care
Health hazards
Health risks
Industry
Laws and legislation
Medical problems
Medicine
Occupational diseases
Personnel

* Former Ei Vocabulary term

Industrial insurance (911.1) (911.2)
DT: Predates 1975
BT: Insurance
RT: Industrial management
Industrial plants
Industry

Industrial laboratories (402.1) (912.1) (901.3)
DT: January 1993
UF: Industrial plants—Laboratories*
BT: Laboratories
RT: Industrial plants
Test facilities

Industrial lighting
USE: Electric lighting

Industrial locomotives 682.1.2
DT: January 1993
UF: Locomotives—Industrial*
BT: Locomotives
NT: Mine locomotives
RT: Industrial railroads
Industry

Industrial management 912.2
DT: Predates 1975
BT: Management
NT: Industrial relations
Plant management
Production control
RT: Administrative data processing
Budget control
Cost accounting
Cost effectiveness
Costs
Efficiency
Employment
Industrial insurance
Industrial plants
Inventory control
Operations research
Process control
Public relations

Industrial optics 741.1 (912.1)
DT: January 1993
BT: Optics
RT: Industrial applications

Industrial ovens 642.2
DT: January 1993
UF: Bakeries—Ovens*
Ovens, Industrial*
BT: Ovens
NT: Afterburners (oven)
Coke ovens
RT: Bakeries
Exhaust gases
Furnaces
Industrial stoves
Kilns

Industrial plants (912.1) (402.1)
DT: Predates 1975
UF: Factories
NT: Aircraft plants
Aluminum plants
Asphalt plants
Automobile plants
Cement plants
Ceramic plants
Chemical plants
Coal gasification plants
Coal liquefaction plants
Coke plants
Crushed stone plants
Distilleries
Drug products plants
Electric manufacturing plants
Electroplating shops
Food products plants
Foundries
Gas plants
Glass plants
Gypsum plants
Iron and steel plants
Laundries
Machine shops
Medical departments (industrial plants)
Metal refineries
Natural gasoline plants
Paper and pulp mills
Petroleum refineries
Pilot plants
Plastics plants
Portable industrial plants
Power plants
Rock products plants
Rolling mills
Rubber factories
Sand and gravel plants
Sawmills
Sewage treatment plants
Shipyards
Textile mills
Welding shops
Wire mills
RT: Accident prevention
Assembly machines
Buildings
Byproducts
Contracts
Control
Effluents
Equipment
Factory automation
Flexible manufacturing systems
Flowcharting
Hazards
Industrial applications
Industrial emissions
Industrial insurance
Industrial laboratories
Industrial management
Industrial poisons
Industrial relations
Industrial research
Industrial trucks
Industrial wastes
Industry

* Former Ei Vocabulary term

Industrial plants *(continued)*
Integrated control
Inventory control
Location
Machinery
Manufacturing data processing
Modernization
Personnel
Personnel training
Plant expansion
Plant layout
Plant life extension
Plant management
Plant shutdowns
Plant startup
Production control
Public relations
Quality control
Site selection
Structures (built objects)
Time and motion study

*Industrial plants—Accounting**
USE: Cost accounting

*Industrial plants—Assembly machines**
USE: Assembly machines

*Industrial plants—Auxiliary equipment**
USE: Auxiliary equipment

*Industrial plants—Compressed air**
USE: Compressed air

*Industrial plants—Depreciation**
USE: Depreciation

*Industrial plants—Effluent treatment**
USE: Effluent treatment

*Industrial plants—Effluents**
USE: Effluents

*Industrial plants—Emissions**
USE: Industrial emissions

*Industrial plants—Expansion**
USE: Plant expansion

*Industrial plants—Fire hazards**
USE: Fire hazards

*Industrial plants—Flow sheets**
USE: Flowcharting

*Industrial plants—Fume control**
USE: Fume control

*Industrial plants—Hydraulic equipment**
USE: Hydraulic equipment

*Industrial plants—Integrated control**
USE: Integrated control

*Industrial plants—Laboratories**
USE: Industrial laboratories

*Industrial plants—Layout**
USE: Plant layout

*Industrial plants—Life extension**
USE: Plant life extension

*Industrial plants—Lighting**
USE: Lighting

*Industrial plants—Location**
USE: Location

*Industrial plants—Machinery**
USE: Machinery

*Industrial plants—Medical departments**
USE: Medical departments (industrial plants)

*Industrial plants—Mist eliminators**
USE: Mist eliminators

*Industrial plants—Modernization**
USE: Modernization

*Industrial plants—Noise**
USE: Acoustic noise

*Industrial plants—Pilot plants**
USE: Pilot plants

*Industrial plants—Pneumatic control**
USE: Pneumatic control

*Industrial plants—Portable**
USE: Portable industrial plants

*Industrial plants—Power supply**
USE: Electric power utilization

*Industrial plants—Presses**
USE: Presses (machine tools)

*Industrial plants—Public relations**
USE: Public relations

*Industrial plants—Safety devices**
USE: Safety devices

*Industrial plants—Sanitation**
USE: Sanitation

*Industrial plants—Shutdown**
USE: Plant shutdowns

*Industrial plants—Site selection**
USE: Site selection

*Industrial plants—Startup problems**
USE: Plant startup

*Industrial plants—Time and motion study**
USE: Time and motion study

*Industrial plants—Tools, jigs and fixtures**
USE: Tools OR
Jigs OR
Fixtures (tooling)

*Industrial plants—Water cooling systems**
USE: Water cooling systems

*Industrial plants—Water recycling**
USE: Water recycling

Industrial poisons (914.1) (914.3.1) (804.1) (804.2)
DT: Predates 1975
UF: Lead poisoning

Industrial poisons *(continued)*
Mercury poisoning
Poisons (industrial)
BT: Hazardous materials
RT: Fume control
Industrial plants
Occupational diseases
Toxicity

Industrial property
USE: Intellectual property

Industrial radioactive wastes
USE: Industrial wastes AND Radioactive wastes

Industrial radiography (944.8) (912.1) (901.3)
DT: January 1993
BT: Radiography
RT: Contrast media
Inspection
Nondestructive examination

Industrial railroads 681.1 (433.3)
DT: January 1993
UF: Railroads—Industrial*
BT: Railroads
RT: Industrial locomotives
Industry

Industrial refrigeration (644.1) (912.1)
DT: January 1993
UF: Refrigeration—Industrial*
BT: Refrigeration

Industrial relations 912.2 (912.4)
DT: Predates 1975
UF: Personnel relations
BT: Industrial management
RT: Industrial plants
Personnel

Industrial research 901.3, 912.1
DT: January 1993
BT: Research
RT: Industrial plants
Operations research
Research and development management

Industrial research management
USE: Research and development management

Industrial robots 731.6 (912.1)
DT: January 1993
UF: Robots, Industrial*
BT: Robots
RT: Anthropomorphic robots
Computer software
Computer vision
Contact sensors
Control
Control equipment
Depth perception
Expert systems
Factory automation
Grippers
Manipulators
Materials handling
Mobile robots
Modular robots
Multipurpose robots
Processing
Programmable robots
Proximity sensors
Robot applications
Robot learning
Robot programming
Robotic arms
Robotic assembly
Robotics

Industrial stoves 642.2
DT: January 1993
BT: Stoves
RT: Industrial ovens

Industrial trucks (691.1) (663.1)
SN: Materials handling equipment
DT: Predates 1975
BT: Materials handling equipment
Trucks
NT: Mine trucks
RT: Industrial plants

*Industrial trucks—Attachments**
USE: Accessories

Industrial waste disposal 452.4
SN: Disposal of industrial-process wastes
DT: January 1993
UF: Metal cutting—Chip disposal*
BT: Waste disposal
NT: Radioactive waste disposal
Tailings disposal
RT: Industrial wastes

Industrial waste treatment 452.4
DT: January 1993
UF: Industrial wastes—Treatment*
BT: Waste treatment
NT: Effluent treatment
RT: Industrial wastes
Waste disposal
Wastewater treatment

Industrial wastes 452.3
SN: Also use the particular material, e.g., DYES, if appropriate
DT: Predates 1975
UF: Industrial radioactive wastes
Metal cutting—Chip formation*
Radioactive industrial wastes
BT: Wastes
NT: Chemical wastes
Fly ash
Industrial emissions
Slags
Tailings
RT: Air pollution
Dyes
Effluents
Exhaust gases
Hazardous materials
Industrial plants

* Former Ei Vocabulary term

Industrial wastes *(continued)*
Industrial waste treatment
Industrial waste disposal
Phenols
Plastics
Radioactive wastes
Scrap metal
Sewage
Waste disposal
Waste utilization
Wastewater reclamation
Water pollution

*Industrial wastes—Chemicals**
USE: Chemical wastes

*Industrial wastes—Cyanides**
USE: Cyanides

*Industrial wastes—Dyes**
USE: Dyes

*Industrial wastes—Hazardous materials**
USE: Hazardous materials

*Industrial wastes—Phenols**
USE: Phenols

*Industrial wastes—Plastics**
USE: Plastics

*Industrial wastes—Radioactive materials**
USE: Radioactive wastes

*Industrial wastes—Treatment**
USE: Industrial waste treatment

*Industrial wastes—Water reclamation**
USE: Wastewater reclamation

Industrial water treatment 445.1.2
SN: Treatment of water for industrial rather than potable use
DT: January 1993
UF: Water treatment, Industrial*
BT: Water treatment
NT: Feedwater treatment

Industry
DT: January 1993
NT: Agriculture
Aquaculture
Cement industry
Chemical industry
Concrete industry
Construction industry
Electric industry
Electronics industry
Gas industry
Glass industry
Iron and steel industry
Mineral industry
Nuclear industry
Paper and pulp industry
Plastics industry
Public utilities
Rubber industry
Sugar industry
Textile industry
Tire industry
RT: Industrial economics
Industrial engineering
Industrial hygiene
Industrial insurance
Industrial locomotives
Industrial plants
Industrial railroads
Occupational diseases
Occupational risks

Inert gas welding 538.2.1
DT: January 1993
UF: Welding—Inert gas*
BT: Electric arc welding
RT: Inert gases
Pressure welding
Spot welding
Stud welding

Inert gases 804.2 (931.2)
DT: January 1993
UF: Gases, Inert*
Noble gases
Rare gases
BT: Gases
Nonmetals
NT: Argon
Helium
Krypton
Neon
Radon
Xenon
RT: Inert gas welding
Protective atmospheres

Inertial confinement fusion 932.2.1 (621.2)
DT: June 1990
BT: Fusion reactions
RT: Fusion reactors

Inertial navigation systems (431.5)
DT: January 1993
UF: Air navigation—Inertial systems*
Inertial navigators
Navigation—Inertial systems*
BT: Navigation systems
RT: Air navigation
Navigation

Inertial navigators
USE: Inertial navigation systems

Inference engines 723.4.1
DT: January 1993
UF: Expert systems—Inference engines*
BT: Expert systems

Infiltration (452.1) (483.1) (802.3)
DT: January 1993
UF: Sewers—Infiltration*
Soils—Infiltration*
RT: Impregnation
Leaching
Mechanical permeability
Percolation (fluids)
Porosity

Infiltration *(continued)*
Porous materials
Seepage
Soils

Infinite impulse response filters
USE: Digital filters

Inflammability
USE: Flammability

Inflammable materials
USE: Flammable materials

Inflatable boats
USE: Inflatable equipment AND
Boats

Inflatable domes
USE: Inflatable structures AND
Domes

Inflatable equipment
DT: January 1993
UF: Boats—Inflatable*
Inflatable boats
Inflatable satellites
Satellites—Inflatable*
Spacecraft—Inflatable equipment*
BT: Equipment
RT: Inflatable structures

Inflatable roofs
USE: Inflatable structures AND
Roofs

Inflatable satellites
USE: Inflatable equipment AND
Satellites

Inflatable shells
USE: Shells (structures) AND
Inflatable structures

Inflatable structures (408.2)
DT: Predates 1975
UF: Air supported structures
Inflatable domes
Inflatable roofs
Inflatable shells
BT: Structures (built objects)
RT: Balloons
Flexible structures
Inflatable equipment
Space stations

Information analysis 903.1
DT: January 1993
BT: Information science
NT: Abstracting
Classification (of information)
Indexing (of information)
Information retrieval
Translation (languages)
Vocabulary control
RT: Information retrieval systems
Management information systems

Information capacity
USE: Channel capacity

Information centers
USE: Information services

Information dissemination 903.2
DT: Predates 1975
BT: Communication
NT: Publishing
Technical writing
RT: Bibliographies
Copyrights
Demonstrations
Education
Educational motion pictures
Electronic mail
Graphic methods
Handbooks
Information retrieval systems
Information services
Libraries
Microforms
Printing
Speech communication
Speech transmission
Teaching
Technical presentations
Textbooks
Videotex

*Information dissemination—Electronic publishing**
USE: Electronic publishing OR
Desktop publishing

*Information dissemination—Microforms**
USE: Microforms

*Information dissemination—Publishing**
USE: Publishing

*Information dissemination—Reproduction**
USE: Copying

*Information dissemination—Speech communication**
USE: Speech communication

*Information dissemination—Technical writing**
USE: Technical writing

Information management (912.2) (903.2)
DT: January 1993
BT: Management
RT: Information retrieval systems
Information science
Management information systems

Information processing
USE: Data processing

Information retrieval 903.3
DT: January 1993
UF: Information science—Information retrieval*
BT: Information analysis
NT: Online searching

* Former Ei Vocabulary term

Information retrieval *(continued)*
RT: Information retrieval systems
Query languages

Information retrieval languages
USE: Query languages

Information retrieval systems 903.3
DT: Predates 1975
UF: Document retrieval systems
Information storage and retrieval systems
NT: Bibliographic retrieval systems
Nonbibliographic retrieval systems
RT: Information analysis
Information dissemination
Information management
Information retrieval
Management
Online searching
Query languages
Records management

*Information retrieval systems—Bibliographic**
USE: Bibliographic retrieval systems

*Information retrieval systems—Nonbibliographic**
USE: Nonbibliographic retrieval systems

*Information retrieval systems—Online searching**
USE: Online searching

*Information retrieval systems—Teletext and videotex**
USE: Videotex

Information science (903)
DT: Predates 1975
NT: Information analysis
Information use
Linguistics
RT: Computer science
Information management
Information technology
Information theory
Management

*Information science—Abstracting**
USE: Abstracting

*Information science—Indexing**
USE: Indexing (of information)

*Information science—Information retrieval**
USE: Information retrieval

*Information science—Information use**
USE: Information use

*Information science—Language translation and linguistics**
USE: Linguistics OR
Translation (languages)

*Information science—Vocabulary control**
USE: Vocabulary control

Information security
USE: Security of data

Information services 903.4
DT: Predates 1975
UF: Information centers
NT: Weather information services
RT: Information dissemination
Libraries
Telecommunication services
Videotex

Information storage
USE: Data storage equipment

Information storage and retrieval systems
USE: Information retrieval systems

Information storage equipment
USE: Data storage equipment

Information technology 723.5, 903
DT: January 1995
RT: Computers
Copying
Factory automation
Image processing
Information science
Microforms
Office automation
Telecommunication
Telecommunication equipment

Information theory 716.1
DT: Predates 1975
UF: Communication theory
NT: Clutter (information theory)
Communication channels (information theory)
Correlation theory
RT: Bandwidth compression
Channel capacity
Code converters
Codes (symbols)
Combinatorial mathematics
Data compression
Decision theory
Demodulation
Digital signal processing
Error correction
Frequency agility
Game theory
Graph theory
Information science
Modulation
Operations research
Probability
Random processes
Redundancy
Signal detection
Signal filtering and prediction
Signal processing
Signal theory
Speech intelligibility
System theory
Systems analysis

*Information theory—Bandwidth compression**
USE: Bandwidth compression

* Former Ei Vocabulary term

*Information theory—Channel capacity**
USE: Channel capacity

*Information theory—Clutter**
USE: Clutter (information theory)

*Information theory—Communication channels**
USE: Communication channels (information theory)

*Information theory—Correlation theory**
USE: Correlation theory

*Information theory—Data compression**
USE: Data compression

*Information theory—Digital signals**
USE: Digital signal processing

*Information theory—Signal acquisition**
USE: Signal detection

Information use 903.3
DT: January 1993
UF: Information science—Information use*
BT: Information science

Infrared
USE: Infrared radiation

Infrared detectors 944.7 (741.3) (914.1)
DT: Predates 1975
BT: Infrared devices
Photodetectors
RT: Bolometers
Fire detectors
Radiometers
Radiometry
Thermal variables measurement

Infrared devices (741.3)
DT: January 1983
BT: Optical devices
NT: Infrared detectors
Infrared spectrographs
Infrared spectrometers
RT: Infrared instruments
Infrared lamps
Infrared radiation
Infrared television

Infrared drying (642.1) (802.3)
DT: January 1993
UF: Drying—Infrared*
BT: Drying
RT: Industrial heating
Infrared furnaces
Infrared heating

Infrared furnaces 642.2
DT: January 1993
UF: Furnaces—Infrared energy*
BT: Industrial furnaces
RT: Infrared drying
Infrared heating
Infrared radiation

Infrared heating 642.1 (741.3)
DT: Predates 1975
UF: Heating—Infrared*
BT: Radiant heating
RT: Industrial heating
Infrared drying
Infrared furnaces
Infrared lamps
Infrared radiation

Infrared imaging (741.3)
DT: Predates 1975
UF: Infrared photography
Photogrammetry—Infrared radiation*
Photography—Infrared radiation*
Thermal imaging
BT: Imaging techniques
NT: Thermography (imaging)
RT: Image sensors
Infrared radiation
Photogrammetry
Remote sensing

Infrared instruments (941.3) (741.3)
DT: January 1993
UF: Optical instruments—Infrared*
BT: Optical instruments
RT: Infrared devices

Infrared lamps 707.2 (741.1)
DT: January 1993
UF: Electric lamps, Infrared*
BT: Electric lamps
RT: Infrared devices
Infrared heating

Infrared photography
USE: Infrared imaging

Infrared radiation 741.1
DT: Predates 1975
UF: Infrared
Infrared waves
BT: Light
RT: Atmospheric radiation
Heat radiation
Infrared devices
Infrared furnaces
Infrared heating
Infrared imaging
Infrared television
Infrared transmission
Optical radar
Solar radiation

Infrared spectrographs (741.3) (742.1)
SN: Infrared spectroscopes that produce photographic or fluorescent-screen spectrograms
DT: January 1993
UF: Spectrographs, Infrared*
BT: Infrared devices
Spectrographs
RT: Infrared spectroscopy

Infrared spectrometers (741.3)
SN: Spectroscopes capable of measuring angular deviation of infrared radiation
DT: January 1993
UF: Spectrometers, Infrared*

Infrared spectrometers *(continued)*
BT: Infrared devices
Spectrometers
NT: Infrared spectrophotometers
RT: Infrared spectroscopy

Infrared spectrophotometers (741.3)
SN: Spectrometers for measuring infrared spectral energy distributions
DT: January 1993
UF: Spectrophotometers, Infrared*
BT: Infrared spectrometers
Spectrophotometers
RT: Infrared spectroscopy
Spectrophotometry

Infrared spectroscopy (941.4) (741.1) (931.2)
DT: January 1993
UF: IR spectroscopy
Spectroscopy, Infrared*
BT: Spectroscopy
NT: Fourier transform infrared spectroscopy
RT: Infrared spectrographs
Infrared spectrometers
Infrared spectrophotometers
Raman spectroscopy

Infrared television 716.4
DT: January 1993
UF: Television—Infrared*
BT: Television
RT: Infrared devices
Infrared radiation

Infrared transmission (741.1)
DT: January 1993
UF: Glass—Infrared transmission*
BT: Light transmission
RT: Glass
Infrared radiation

Infrared waves
USE: Infrared radiation

Infusion (drugs)
USE: Drug infusion

Ingestion (engines) 653.1
DT: January 1993
UF: Aircraft engines, Jet and Turbine—Ingestion*
RT: Aircraft engines
Intake systems
Jet engines

Ingot molds 534.2
DT: Predates 1975
BT: Molds
RT: Ingots
Metal castings

*Ingot molds—Iron construction**
USE: Iron

Ingots 534.2
DT: June 1990
BT: Metal castings
NT: Steel ingots
RT: Billets (metal bars)
Blooms (metal)
Hot topping
Ingot molds

Inhibition (enzyme)
USE: Enzyme inhibition

Inhibitors (corrosion)
USE: Corrosion inhibitors

Initiators (chemical) 803
DT: January 1995
BT: Agents
RT: Catalysts

Injection (beam)
USE: Particle beam injection

Injection (oil wells) 511.1
DT: January 1993
BT: Oil well production

Injection lasers 744.4.1
DT: January 1993
UF: Lasers, Injection*
BT: Semiconductor lasers

Injection molding 816.1
DT: January 1993
BT: Compression molding
NT: Reaction injection molding
RT: Plastics forming
Plastics molding

Ink 804
SN: Includes printing ink
DT: Predates 1975
UF: Printing ink
BT: Materials
RT: Printing

Inland waterways 407.2 (434)
DT: Predates 1975
NT: Canals
RT: Bank protection
Dolphins (structures)
Dredging
Erosion
Ice control
Improvement
Lakes
Locks (on waterways)
Marinas
Revetments
Rivers
Salt water barriers
Seepage
Surface waters
Transportation routes
Water levels
Water pollution
Waterway transportation
Weed control
Widening (transportation arteries)

*Inland waterways—Bank erosion**
USE: Erosion

* Former Ei Vocabulary term

*Inland waterways—Bank protection**
USE: Bank protection

*Inland waterways—Dolphins**
USE: Dolphins (structures)

*Inland waterways—Ice control**
USE: Ice control

*Inland waterways—Ice formation**
USE: Ice

*Inland waterways—Improvement**
USE: Improvement

*Inland waterways—Locks**
USE: Locks (on waterways)

*Inland waterways—Revetments**
USE: Revetments

*Inland waterways—Salt water barriers**
USE: Salt water barriers

*Inland waterways—Seepage**
USE: Seepage

*Inland waterways—Water level**
USE: Water levels

*Inland waterways—Weed control**
USE: Weed control

*Inland waterways—Widening**
USE: Widening (transportation arteries)

Inlets (devices)
USE: Intake systems

Inlets (reservoir)
USE: Reservoir inlets

Inorganic acids 804.2
DT: January 1993
UF: Acids—Inorganic*
BT: Acids
Hydrogen inorganic compounds
NT: Hydrochloric acid
Hydrofluoric acid
Nitric acid
Phosphoric acid
Sulfamic acid
Sulfuric acid

Inorganic chemicals 804.2
DT: January 1986
BT: Chemicals

Inorganic coatings 813.2 (804.2)
DT: January 1993
UF: Protective coatings—Inorganic*
BT: Coatings

Inorganic compounds 804.2
DT: January 1977
BT: Chemical compounds
NT: Carbon inorganic compounds
Deuterium compounds
Hydrogen inorganic compounds
Inorganic polymers
RT: Luminescence of inorganic solids
Metallic compounds

Inorganic polymers 815.1.2 (804.2)
SN: Polymers with chains based on elements other than carbon
DT: Predates 1975
BT: Inorganic compounds
Polymers
NT: Polysilanes
Silicones

Input output programs 723.1
DT: January 1993
UF: Computer systems programming—Input output programs*
I/O programs
BT: Computer operating systems

Insect control (461.7) (821.2) (804.1) (804.2)
DT: Predates 1975
UF: Sewage treatment plants—Fly control*
Water treatment plants—Fly control*
BT: Pest control
NT: Mosquito control
Mothproofing
Termite proofing
RT: Insecticides
Protection
Termite resistance

Insecticides 803 (821.2) (804.2) (804.1)
DT: Predates 1975
BT: Pesticides
RT: Insect control
Malaria control

*Insecticides—Applicators**
USE: Applicators

*Insecticides—Spraying**
USE: Spraying

Inserts (casting)
USE: Die casting inserts

Insolubility
USE: Solubility

Inspection 913.3.1
DT: Predates 1975
UF: Railroad plant and structures—Track inspection*
RT: Analysis
Calibration
Evaluation
Fits and tolerances
Industrial radiography
Inspection equipment
Measurements
Mechanical properties
Monitoring
Nondestructive examination
Nuclear materials safeguards
Physical properties
Quality control
Reliability
Sampling
Specifications

* Former Ei Vocabulary term

Inspection *(continued)*
Standards
Testing

Inspection equipment 913.3.1
DT: January 1993
UF: Bridges—Inspection equipment*
BT: Equipment
RT: Inspection

Instability
USE: Stability

Instability (plasma)
USE: Plasma stability

Installation
DT: January 1993
UF: Machinery—Installations*
RT: Assembly
Maintenance

Institutions
USE: Societies and institutions

Instrument circuits
(713) (941) (942) (943) (944) (632)
DT: January 1993
UF: Instruments—Circuits*
BT: Instrument components
Networks (circuits)

Instrument components (941) (942) (943) (944)
DT: January 1993
UF: Instruments—Components*
BT: Instruments
NT: Electric instrument transformers
Instrument circuits
Instrument dials
Instrument displays
Readout systems
RT: Knobs

Instrument dials (941) (942) (943) (944)
DT: January 1993
UF: Dials (instrument)
Instruments—Dials*
BT: Instrument components
RT: Readout systems

Instrument displays (941) (942) (943) (944)
DT: January 1993
UF: Instruments—Display systems*
Optical instruments—Display systems*
BT: Display devices
Instrument components
NT: Radar displays
RT: Computer peripheral equipment
Optical instruments
Readout systems

Instrument errors (941) (942) (943) (944)
DT: January 1993
UF: Instruments—Errors*
BT: Errors
RT: Calibration
Error analysis
Error compensation
Error correction
Error detection
Instruments
Reliability

Instrument panels (941) (942) (943) (944)
DT: January 1993
UF: Instruments—Panels*
BT: Equipment
NT: Automobile instrument panels
RT: Instruments

Instrument readouts
USE: Readout systems

Instrument scales (941) (942) (943) (944)
DT: January 1993
UF: Instruments—Scales*
Scales (readouts)
BT: Readout systems

Instrument testing (941) (942) (943) (944)
DT: January 1993
UF: Instruments—Testing*
BT: Equipment testing
RT: Instruments

Instrument transformers
USE: Electric instrument transformers

Instruments
SN: Very general term; prefer specific type of instrument
DT: Predates 1975
UF: Meters
Textile measuring instruments*
BT: Equipment
NT: Accelerometers
Acoustic measuring instruments
Aircraft instruments
Audiometers
Boat instruments
Calorimeters
Compasses (magnetic)
Density measuring instruments
Detectors
Digital instruments
Dilatometers
Electric measuring instruments
Electron energy analyzers
Flow measuring instruments
Footage counters
Friction meters
Gages
Goniometers
Gravimeters
Gyroscopes
Heliostats (instruments)
Hydrologic instruments
Indicators (instruments)
Instrument components
Magnetic measuring instruments
Mathematical instruments
Meteorological instruments
Micrometers
Microscopes
Miniature instruments

* Former Ei Vocabulary term

Instruments *(continued)*
Moisture meters
Neutron diffraction apparatus
Nuclear instrumentation
Oceanographic instruments
Optical instruments
Oxygen regulators
pH meters
Planimeters
Radio direction finding systems
Range finders
Recording instruments
Reflectometers
Refractometers
Scales (weighing instruments)
Sediment traps
Seismographs
Sextants
Spacecraft instruments
Stroboscopes
Surveying instruments
Taxi meters
Telescopes
Timing devices
Torque meters
Velocimeters
Viscometers
RT: Acoustic devices
Calibration
Codes (symbols)
Control
Gears
Instrument errors
Instrument panels
Instrument testing
Light sources
Probes

*Instruments, Digital**
USE: Digital instruments

*Instruments—Circuits**
USE: Instrument circuits

*Instruments—Components**
USE: Instrument components

*Instruments—Dials**
USE: Instrument dials

*Instruments—Digital readout**
USE: Digital instruments

*Instruments—Display systems**
USE: Instrument displays

*Instruments—Errors**
USE: Instrument errors

*Instruments—Explosionproof**
USE: Explosionproofing

*Instruments—Gears**
USE: Gears

*Instruments—Knobs**
USE: Knobs

*Instruments—Light sources**
USE: Light sources

*Instruments—Miniature**
USE: Miniature instruments

*Instruments—Noise**
USE: Acoustic noise

*Instruments—Panels**
USE: Instrument panels

*Instruments—Power supply**
USE: Electric power supplies to apparatus

*Instruments—Probes**
USE: Probes

*Instruments—Readout systems**
USE: Readout systems

*Instruments—Recording**
USE: Recording instruments

*Instruments—Remote reading**
USE: Remote readouts

*Instruments—Scales**
USE: Instrument scales

*Instruments—Symbols**
USE: Codes (symbols)

*Instruments—Terminology**
USE: Terminology

*Instruments—Testing**
USE: Instrument testing

*Instruments—Waterproofing**
USE: Waterproofing

Insulated gate field effect transistors
USE: MOSFET devices

Insulated wire (413.1) (704.2) (708.2)
DT: January 1993
UF: Electric conductors, Insulated wire*
BT: Electric wire
RT: Wire insulating extruders

Insulating materials 413
DT: January 1993
BT: Materials
NT: Electric insulating materials
Sound insulating materials
Thermal insulating materials
RT: Foamed plastics
Insulation
Permittivity
Semiconductor insulator boundaries

Insulating oil 607.1 (413.1)
DT: Predates 1975
UF: Transformer oil
BT: Hydrocarbons
Liquid insulating materials
RT: Electric transformer insulation
Oil circuit breakers
Oil filled cables

* Former Ei Vocabulary term

Insulating oil *(continued)*
Oil filled transformers
Petroleum products

*Insulating oil—Reclamation**
USE: Reclamation

Insulation 413
DT: Predates 1975
BT: Materials
NT: Electric insulation
Sound insulation
Thermal insulation
RT: Building components
Honeycomb structures
Insulating materials
Permittivity
Protection
Wire insulating extruders

Insulation coordination
USE: Electric insulation coordination

Insulator semiconductor boundaries
USE: Semiconductor insulator boundaries

Insulators (electric)
USE: Electric insulators

Insulin 461.2 (804.1) (801.2)
DT: January 1993
BT: Hormones

Insurance (911.1)
DT: January 1993
BT: Finance
NT: Fire insurance
Flood insurance
Health insurance
Industrial insurance
Marine insurance

Intake silencers 612.1.1 (661.2)
DT: January 1993
UF: Automobile engines—Intake silencers*
Silencers (intake)
BT: Intake systems
Internal combustion engines

Intake systems 631.1
DT: January 1993
UF: Flow of fluids—Inlets*
Inlets (devices)
Turbomachinery—Inlets*
BT: Equipment
NT: Intake silencers
RT: Flow of fluids
Ingestion (engines)
Superchargers
Turbomachinery

Integer programming 921.5
DT: January 1995
BT: Dynamic programming
RT: Operations research

Integral equations 921.2
DT: January 1993
UF: Mathematical techniques—Integral equations*
BT: Mathematical techniques
NT: Integrodifferential equations
RT: Numerical analysis

Integrated circuit layout 714.2
DT: January 1993
UF: Circuit layout
Integrated circuits—Layout*
Layout (integrated circuits)
BT: Integrated circuit manufacture
Product design
RT: Integrated circuits
Printed circuit design

Integrated circuit manufacture 714.2
DT: Predates 1975
BT: Electron device manufacture
NT: Integrated circuit layout
RT: Clean rooms
Integrated circuits
Integrated optoelectronics
Lithography
Microelectronic processing
Nanotechnology
Optical interconnects
Photolithography
Photoresists
Plasma etching
Printed circuit manufacture
Reactive ion etching
Substrates
Surface mount technology
X ray lithography

Integrated circuit testing 714.2
DT: Predates 1975
BT: Electron device testing
RT: Integrated circuits
Printed circuit testing

Integrated circuits 714.2
SN: Very general term; prefer specific type
DT: Predates 1975
UF: IC
BT: Networks (circuits)
Semiconductor devices
NT: Bipolar integrated circuits
Digital integrated circuits
Hybrid integrated circuits
Integrated optoelectronics
Linear integrated circuits
Microwave integrated circuits
Monolithic integrated circuits
Thick film circuits
Thin film circuits
RT: Beam lead devices
Electromigration
Electronics packaging
Integrated circuit manufacture
Integrated circuit testing
Integrated circuit layout
Masks
Microelectronics
Multichip modules
Radio receivers
Substrates
Television receivers

* Former Ei Vocabulary term

*Integrated circuits, CMOS**
USE: CMOS integrated circuits

*Integrated circuits, Digital**
USE: Digital integrated circuits

*Integrated circuits, Hybrid**
USE: Hybrid integrated circuits

*Integrated circuits, Linear**
USE: Linear integrated circuits

*Integrated circuits, LSI**
USE: LSI circuits

*Integrated circuits, Monolithic**
USE: Monolithic integrated circuits

*Integrated circuits, Thick film**
USE: Thick film circuits

*Integrated circuits, Thin film**
USE: Thin film circuits

*Integrated circuits, ULSI**
USE: ULSI circuits

*Integrated circuits, VLSI**
USE: VLSI circuits

*Integrated circuits, WSI**
USE: WSI circuits

*Integrated circuits—Beam lead construction**
USE: Beam lead devices

*Integrated circuits—Flip chip construction**
USE: Flip chip devices

*Integrated circuits—Layout**
USE: Integrated circuit layout

*Integrated circuits—Masks**
USE: Masks

Integrated control 731.1
DT: January 1993
UF: Centralized plant control
Industrial plants—Integrated control*
BT: Control
RT: Industrial plants
Process control

Integrated injection logic circuits 714.2, 721.2 (721.3)
DT: January 1993
UF: Logic circuits, Integrated injection*
BT: Logic circuits
RT: Digital integrated circuits
LSI circuits

Integrated optics (741.3)
DT: October 1975
BT: Optics
RT: Distributed feedback lasers
Gradient index optics
Integrated optoelectronics
Optical films
Optical interconnects
Optical waveguides

Integrated optoelectronic circuits
USE: Integrated optoelectronics

Integrated optoelectronics (714.2) (741.3)
DT: January 1993
UF: Integrated optoelectronic circuits
OEIC
Optoelectronic integrated circuits
BT: Integrated circuits
Optoelectronic devices
RT: Integrated circuit manufacture
Integrated optics
Light emitting diodes
Monolithic integrated circuits
Optical communication equipment
Optical interconnects
Photodetectors
Photodiodes
Phototransistors

Integrated services digital network
USE: Voice/data communication systems

Integrated voice/data services
USE: Voice/data communication systems

Integrating circuits 713.5
DT: January 1993
UF: Electronic circuits, Integrating*
BT: Networks (circuits)
RT: Signal processing

Integration 921.2
DT: January 1993
UF: Mathematical techniques—Integration*
BT: Mathematical techniques
RT: Numerical analysis

Integro-differential equations
USE: Integrodifferential equations

Integrodifferential equations 921.2
DT: January 1993
UF: Integro-differential equations
Mathematical techniques—Integrodifferential equations*
BT: Differential equations
Integral equations

Intellectual property 902.3
DT: January 1993
UF: Industrial property
NT: Copyrights
Patents and inventions
Trademarks
RT: Laws and legislation

Intelligent buildings 402, 723.5, 731.1
SN: Buildings in which major functions of management, such as climate control, lighting, and elevator operation, are controlled by a central computerized facility
DT: January 1993
UF: Smart buildings
Smart houses
BT: Buildings
RT: Apartment houses
Automation

* Former Ei Vocabulary term

Intelligent buildings *(continued)*
Computer applications
Energy management
Housing
Office buildings

Intelligent control 723.4.1, 731.1
DT: January 1995
BT: Control
RT: Adaptive control systems
Fuzzy control
Intelligent structures
Knowledge based systems

Intelligent highway systems
USE: Intelligent vehicle highway systems

Intelligent materials (415) (931.2)
DT: January 1995
UF: Smart materials
BT: Materials
RT: Composite materials
Intelligent structures
Piezoelectric materials
Shape memory effect

Intelligent networks (716) (717) (718) (723.4)
DT: January 1995
BT: Telecommunication networks
RT: Data communication systems
Digital communication systems
Switching systems
Telecommunication services

Intelligent robots 731.6
DT: January 1993
UF: Adaptive robots
Communicating robots
Robots, Industrial—Intelligent*
BT: Robots
RT: Expert systems
Programmable robots

Intelligent structures 409, 723.4
DT: January 1995
UF: Smart structures
BT: Structures (built objects)
RT: Intelligent control
Intelligent materials

Intelligent vehicle highway systems 406.1, 723.5
DT: January 1995
UF: Intelligent highway systems
Intelligent vehicles
IVHS
BT: Highway systems
RT: Automobiles
Computer applications
Highway traffic control
Radio navigation
Vehicles

Intelligent vehicles
USE: Intelligent vehicle highway systems

Intelligibility (speech)
USE: Speech intelligibility

Intensifiers (electron tube)
USE: Image intensifiers (electron tube)

Intensifiers (solid state)
USE: Image intensifiers (solid state)

Intensive care units 462.2
DT: January 1993
UF: Hospitals—Intensive care units*
ICUs
BT: Hospitals

Interactions (drug)
USE: Drug interactions

Interactions (flow)
USE: Flow interactions

Interactions (plasmas)
USE: Beam plasma interactions

Interactive computer graphics 723.5 (921.4)
DT: January 1993
UF: Computer graphics—Interactive*
BT: Computer graphics
RT: Interactive computer systems

Interactive computer systems 722.4
DT: January 1993
UF: Computer systems, Digital—Interactive operation*
BT: Computer systems
RT: Computer aided design
Interactive computer graphics
Interactive devices
Online systems
User interfaces

Interactive devices 722.2
DT: January 1993
BT: Computer peripheral equipment
NT: Computer keyboard substitutes (handicapped aid)
Computer keyboards
Computer monitors
Light pens
Mice (computer peripherals)
Voice activated input devices
RT: Computer terminals
Fluorescent screens
Interactive computer systems
User interfaces

Interactive terminals
USE: Computer terminals

Intercalation compounds (701.1) (804.1) (804.2)
DT: January 1995
UF: Graphite intercalation compounds
BT: Chemical compounds
RT: Electric conductivity
Fullerenes

Interception (satellite)
USE: Satellite interception

Interchanges 406.1
DT: January 1993
UF: Highway interchanges

Interchanges *(continued)*
Highway systems—Interchanges*
BT: Highway systems
RT: Intersections

Intercom systems 718.1
DT: January 1993
UF: Intercommunication systems
Telephone—Intercommunication*
BT: Telephone systems
RT: Telephone

Intercommunication systems
USE: Intercom systems

Interconnected power systems
USE: Electric power system interconnection

Interconnected systems
USE: Large scale systems

Interconnection networks (703.1) (721)
DT: January 1995
UF: Multiprocessor interconnection networks
BT: Switching networks
RT: Multiprocessing systems
Parallel processing systems

Interconnects (optical)
USE: Optical interconnects

Interdiffusion (solids) 933 (931.2)
DT: January 1995
BT: Diffusion in solids
RT: Interfaces (materials)
Semiconductor materials

Interface tension
USE: Surface tension

Interfaces (computer) 722.2
DT: January 1993
UF: Computer interfaces*
Controllers (computer peripherals)
Network interfaces
Peripheral controllers
NT: User interfaces
RT: Computer peripheral equipment
Digital to analog conversion
Gateways (computer networks)
Network protocols
Optical data processing
Program processors
Telecommunication links

Interfaces (materials) (931.2)
DT: January 1993
UF: Interfaces*
NT: Contacts (fluid mechanics)
Grain boundaries
Phase interfaces
RT: Delamination
Electromigration
Interdiffusion (solids)
Interfacial energy
Surface phenomena
Surface properties
Surface tension
Surfaces

*Interfaces**
USE: Interfaces (materials)

Interfacial energy (931.2)
DT: January 1993
UF: Colloids—Interfacial energy*
Surface energy
BT: Surface phenomena
Thermodynamic properties
RT: Colloids
Interfaces (materials)
Surface tension

Interfacial tension
USE: Surface tension

Interference (radio)
USE: Radio interference

Interference (signal)
USE: Signal interference

Interference (wave)
USE: Wave interference

Interference nulling
USE: Interference suppression

Interference suppression (711) (716.1)
DT: January 1993
UF: Interference nulling
Signal interference—Suppression*
BT: Signal processing
RT: Compandor circuits
Signal interference
Spurious signal noise
Telecommunication systems

Interferometers 941.3
DT: Predates 1975
BT: Optical instruments
NT: Etalons
Fabry-Perot interferometers
RT: Electromagnetic wave interference
Interferometry

Interferometry 941.4
DT: October 1975
BT: Measurements
NT: Holographic interferometry
RT: Distance measurement
Electromagnetic wave interference
Interferometers
Micrometers
Schlieren systems

*Interferometry, Holographic**
USE: Holographic interferometry

Interferons 461.2 (461.9.1) (801.2)
DT: January 1987
BT: Proteins
RT: Biomedical engineering
Immunology
Viruses

* Former Ei Vocabulary term

Interiors (building) 402
DT: January 1993
UF: Architectural design—Interiors*
BT: Building components

Interlocking signals 681.3 (433.4)
DT: January 1993
UF: Railroad signals and signaling—Interlocking*
BT: Railroad signal systems

Intermediate frequency amplifiers 713.1
DT: January 1993
UF: Amplifiers, Intermediate frequency*
BT: Amplifiers (electronic)

Intermetallic compounds
USE: Intermetallics

Intermetallics 531.1
DT: Predates 1975
UF: Electron compounds
Intermetallic compounds
BT: Alloys
NT: Semiconducting intermetallics
RT: Metallic compounds
Metalloids
Metals
Platinum compounds
Zinc compounds

Intermodulation (716.1)
SN: Modulation of the components of a complex wave by each other
DT: Predates 1975
BT: Modulation
RT: Demodulation
Intermodulation measurement
Signal distortion

Intermodulation measurement 942.2
DT: January 1993
UF: Intermodulation—Measurements*
BT: Measurements
RT: Intermodulation

*Intermodulation—Measurements**
USE: Intermodulation measurement

Internal combustion engines 612.1
DT: Predates 1975
BT: Heat engines
NT: Afterburners (engine)
Automobile engines
Carburetors
Catalytic converters
Diesel engines
Engine cylinders
Engine pistons
Free piston engines
Gas engines
Ignition systems
Intake silencers
Mufflers
Spark plugs
Stratified charge engines
Wankel engines
RT: Adiabatic engines
Aircraft engines
Antivibration mountings
Combustion equipment
Combustion knock
Compound engines
Connecting rods
Crankcases
Crankshafts
Detonation
Dual fuel engines
Exhaust gases
Exhaust systems (engine)
Fuel injection
Fuel pumps
Fuel systems
Gaskets
Marine engines
Piston rings
Preignition
Scavenging
Starters
Starting
Superchargers
Vapor lock
Water injection

*Internal combustion engines—Adiabatic**
USE: Adiabatic engines

*Internal combustion engines—Afterburners**
USE: Afterburners (engine)

*Internal combustion engines—Antivibration mountings**
USE: Antivibration mountings

*Internal combustion engines—Camshafts**
USE: Camshafts

*Internal combustion engines—Carburetors**
USE: Carburetors

*Internal combustion engines—Catalytic converters**
USE: Catalytic converters

*Internal combustion engines—Combustion knock**
USE: Combustion knock

*Internal combustion engines—Compounding**
USE: Compound engines

*Internal combustion engines—Crankcases**
USE: Crankcases

*Internal combustion engines—Deposits**
USE: Deposits

*Internal combustion engines—Dual fuel**
USE: Dual fuel engines

*Internal combustion engines—Exhaust gases**
USE: Exhaust gases

*Internal combustion engines—Exhaust systems**
USE: Exhaust systems (engine)

* Former Ei Vocabulary term

*Internal combustion engines—Flow problems**
USE: Flow of fluids

*Internal combustion engines—Free piston engines**
USE: Free piston engines

*Internal combustion engines—Fuel injection**
USE: Fuel injection

*Internal combustion engines—Gaskets**
USE: Gaskets

*Internal combustion engines—Ignition systems**
USE: Ignition systems

*Internal combustion engines—Mufflers**
USE: Mufflers

*Internal combustion engines—Noise**
USE: Acoustic noise

*Internal combustion engines—Piston rings**
USE: Piston rings

*Internal combustion engines—Portable types**
USE: Portable equipment

*Internal combustion engines—Rotary**
USE: Rotary engines

*Internal combustion engines—Scavenging**
USE: Scavenging

*Internal combustion engines—Stratified charge**
USE: Stratified charge engines

*Internal combustion engines—Wankel**
USE: Wankel engines

*Internal combustion engines—Water injection**
USE: Water injection

Internal friction (931.2) (931.1)
DT: January 1993
BT: Friction
RT: Anelastic relaxation
Attenuation
Damping
Hysteresis
Plastic flow
Viscoelasticity
Viscosity

Internal oxidation 802.2 (539.1)
DT: January 1993
UF: Metals and alloys—Internal oxidation*
BT: Oxidation
RT: Corrosion
Metals

Internal stresses
USE: Residual stresses

*International agreements**
USE: International cooperation

International cooperation (902.3) (912)
DT: January 1993
UF: International agreements*
Nuclear energy—International cooperation*
Treaties
RT: International law
International trade
Nuclear energy

International law 902.3
DT: January 1993
UF: Mining laws and regulations—International law*
BT: Laws and legislation
RT: International cooperation

International system of units
USE: Metric system

International trade (902.3) (911.4)
DT: January 1993
UF: Commerce
Foreign trade
Marketing—International trade*
RT: International cooperation
Marketing

Interpenetrating networks
USE: Interpenetrating polymer networks

Interpenetrating polymer networks 815.1
DT: January 1986
UF: Interpenetrating networks
BT: Polymer blends
RT: Molecular structure

Interplanetary communication links
USE: Extraterrestrial communication links

Interplanetary flight 656.1
DT: January 1993
UF: Planetary space flight
Space flight—Interplanetary flight*
BT: Space flight
RT: Extraterrestrial communication links
Interplanetary spacecraft
Manned space flight
Orbits
Planetary landers
Planets
Solar system
Space platforms
Space probes
Space stations
Trajectories

Interplanetary propulsion
USE: Interplanetary spacecraft

Interplanetary spacecraft 655.1
DT: January 1993
UF: Interplanetary propulsion
Planetary spacecraft
Spacecraft—Interplanetary*
BT: Spacecraft
RT: Interplanetary flight
Planetary landers
Satellites

* Former Ei Vocabulary term

Interpolation 921.6
DT: January 1993
UF: Mathematical techniques—Interpolation*
BT: Numerical analysis
RT: Approximation theory

Interpreters (computer program)
USE: Program interpreters

Intersections (406)
DT: January 1993
UF: Roads and streets—Intersections*
BT: Roads and streets
RT: Highway systems
Interchanges
Overpasses

Interstate highways
USE: Highway systems

Intersymbol interference (716) (717) (718)
DT: January 1993
UF: Data transmission—Intersymbol interference*
BT: Signal interference
RT: Data communication systems
Digital communication systems
Pulse modulation

Intraocular lenses 462.4, 741.3, 462.5 (461.5)
SN: Surgically implanted
DT: January 1993
UF: Vision—Intraocular lenses*
BT: Biomaterials
Lenses
Optical devices
Prosthetics
Vision aids
RT: Implants (surgical)

Invariance (731.1)
DT: January 1993
UF: Control systems—Invariance*
BT: Control theory
RT: Control system analysis
Control system synthesis

Inventions
USE: Patents and inventions

Inventory control 911.3
SN: Includes stock control and stores control
DT: Predates 1975
UF: Stock control
Stores control
BT: Control
Management
RT: Industrial management
Industrial plants
Just in time production
Nuclear fuel accounting
Production control

Inverse kinematics 931.1
DT: January 1995
BT: Kinematics
RT: Inverse problems

Inverse problems (921)
DT: January 1993
BT: Mathematical techniques
RT: Inverse kinematics

Inverters (electric)
USE: Electric inverters

Investment casting 534.2
DT: January 1993
UF: Foundry practice—Investment casting*
BT: Metal casting
RT: Centrifugal casting
Foundry practice
Precision casting

Iodine 804
DT: Predates 1975
BT: Halogen elements
RT: Iodine compounds

Iodine compounds (804.1) (804.2)
DT: Predates 1975
BT: Halogen compounds
RT: Iodine

Ion acoustic waves 751.1
DT: January 1977
BT: Acoustic waves
RT: Wave propagation

Ion beam lithography (745.1) (932.1)
DT: January 1993
UF: Ion lithography
Lithography—Ion beam*
BT: Lithography
RT: Ion beams
Semiconductor device manufacture

Ion beams 932.1 (931.3)
DT: Predates 1975
BT: Particle beams
RT: Ion beam lithography
Ion bombardment
Ion microscopes
Magnetic lenses
Sputtering

Ion bombardment 932.1
DT: January 1995
BT: Irradiation
RT: Ion beams
Ion sources
Ions
Neutron irradiation

Ion chambers
USE: Ionization chambers

Ion counters
USE: Radiation counters

Ion exchange 802.2
DT: Predates 1975
UF: Exchange (ions)
Ion exchanging
BT: Chemical reactions
RT: Graft copolymers
Hydrometallurgy

* Former Ei Vocabulary term

Ion exchange *(continued)*
Ion exchange membranes
Ion exchange resins
Ion exchangers
Ions
Purification
Separation
Solutions
Water treatment

Ion exchange membranes 802.1
DT: January 1993
UF: Membranes—Ion exchange membranes*
BT: Ion exchangers
Membranes
RT: Ion exchange

Ion exchange resins 815.1.1 (802.1)
DT: January 1981
BT: Ion exchangers
Organic polymers
RT: Ion exchange

Ion exchangers 802.1
DT: Predates 1975
UF: Exchangers (ion)
BT: Separators
NT: Ion exchange membranes
Ion exchange resins
Organic ion exchangers
RT: Clay
Clay minerals
Ion exchange
Polyelectrolytes
Zeolites

*Ion exchangers—Organic**
USE: Organic ion exchangers

Ion exchanging
USE: Ion exchange

Ion guns
USE: Ion sources

Ion implantation (712.1) (932.1)
DT: January 1993
UF: Implantation (ions)
RT: Doping (additives)
Ions
Semiconductor doping
Semiconductor materials

Ion lithography
USE: Ion beam lithography

Ion microscopes 741.3
DT: January 1993
UF: Microscopes, Ion*
BT: Microscopes
RT: Ion beams

Ion propulsion (932.1) (654.2)
DT: January 1993
UF: Propulsion—Ion energy*
Rocket engines—Ion propulsion*
BT: Electric propulsion
RT: Electromagnetic propulsion
Ions
Rocket engines
Spacecraft propulsion

Ion selective electrodes 802.1 (801.4.1)
DT: January 1993
UF: Electrodes, Electrochemical—Ion selective*
BT: Electrochemical electrodes
RT: Chemical analysis
Ions
Sensors

Ion selective membranes 802.1 (801.4.1)
DT: January 1993
UF: Membranes—Ion selective*
BT: Membranes
RT: Ions

Ion sensitive field effect transistors 714.2
DT: January 1993
UF: ISFET
Transistors, Field effect—Ion sensitive*
BT: Field effect transistors
RT: Ions
Sensors

Ion sources 932.1
DT: Predates 1975
UF: Ion guns
Ionization source
BT: Elementary particle sources
RT: Electron sources
Ion bombardment
Ions
Mass spectrometers
Mass spectrometry
Particle accelerators
Plasma sources

Ionic conduction 701.1
DT: January 1993
BT: Electric properties
NT: Ionic conduction in solids
RT: Diffusion
Electric conductivity
Ions

Ionic conduction in solids 701.1 (933)
DT: January 1993
UF: Solids—Ionic conduction*
BT: Ionic conduction
RT: Diffusion in solids
Electric conductivity of solids
Solids

Ionic drug sensors 462.1 (801.2)
DT: January 1993
UF: Biosensors—Ionic drug sensors*
BT: Biosensors
RT: Ionic drugs

Ionic drugs (461.6) (804.1) (804.2)
DT: January 1993
UF: Drug products—Ionic drugs*
BT: Drug products
RT: Ionic drug sensors
Ions

* Former Ei Vocabulary term

Ionic strength 801.4
DT: January 1995
RT: Electrolytes
Electrostatics
Ions

Ionisation
USE: Ionization

Ionization 802.2
DT: Predates 1975
UF: Ionisation
BT: Chemical reactions
NT: Ionization of gases
Ionization of liquids
Ionization of solids
Photoionization
RT: Charge transfer
Dissociation
Electric arcs
Electric corona
Electric discharges
Electric sparks
Ionization chambers
Ionomers
Ions
Magnetohydrodynamics
Plasma heating

Ionization chambers (932.1) (944.7)
SN: Devices for measuring ionizing radiation
DT: Predates 1975
UF: Ion chambers
BT: Particle detectors
RT: Bubble chambers
Cloud chambers
Geiger counters
Ionization
Ions
Proportional counters
Spark chambers

Ionization of gases 802.2 (931.2)
DT: January 1993
UF: Gas ionization
Gases—Ionization*
BT: Ionization
NT: Atmospheric ionization
RT: Electric discharges
Gases

Ionization of liquids 802.2 (931.2)
DT: January 1993
UF: Liquids—Ionization*
BT: Ionization
RT: Liquids

Ionization of solids 802.2 (931.2)
DT: January 1993
UF: Solids—Ionization*
BT: Ionization
RT: Solids

Ionization source
USE: Ion sources

Ionizing radiation
USE: Radiation

Ionomers 815.1.1
SN: Ionic, synthetic polymers
DT: Predates 1975
BT: Organic polymers
RT: Ionization
Ions

Ionosphere 443.1 (481.3) (711.1)
DT: Predates 1975
BT: Upper atmosphere
NT: D region
E region
F region
RT: Atmospheric electricity
Atmospheric ionization
Ionospheric electromagnetic wave propagation
Ionospheric measurement
Magnetosphere

*Ionosphere—D region**
USE: D region

*Ionosphere—E region**
USE: E region

*Ionosphere—F region**
USE: F region

Ionospheric electromagnetic wave propagation 711.1 (443.1)
DT: January 1993
UF: Electromagnetic waves—Propagation in ionosphere*
Ionospheric propagation
BT: Electromagnetic wave propagation in plasma
RT: Earth atmosphere
Fading (radio)
Ionosphere
Radio transmission

Ionospheric measurement 443.1 (711.1) (944.8)
DT: January 1993
BT: Measurements
RT: Ionosphere

Ionospheric propagation
USE: Ionospheric electromagnetic wave propagation

Ions (801) (701.1) (931.3)
DT: Predates 1975
BT: Charged particles
RT: Alpha particles
Atoms
Chelation
Chemical reactions
Complexation
Electrolytes
Ion bombardment
Ion exchange
Ion implantation
Ion propulsion
Ion selective electrodes
Ion selective membranes
Ion sensitive field effect transistors
Ion sources

* Former Ei Vocabulary term

Ions *(continued)*
Ionic conduction
Ionic drugs
Ionic strength
Ionization
Ionization chambers
Ionomers
Protons
Reactive ion etching

IR spectroscopy
USE: Infrared spectroscopy

Iridium 547.1
DT: January 1993
UF: Iridium and alloys*
BT: Platinum metals
RT: Iridium alloys
Iridium compounds
Iridium plating

Iridium alloys 547.1
DT: January 1993
UF: Iridium and alloys*
BT: Platinum metal alloys
RT: Iridium
Iridium compounds

*Iridium and alloys**
USE: Iridium OR
Iridium alloys

Iridium compounds (804.1) (804.2)
DT: Predates 1975
BT: Platinum metal compounds
RT: Iridium
Iridium alloys

Iridium plating 539.3, 547.1
DT: Predates 1975
BT: Plating
RT: Iridium

Iron 545.1
DT: January 1993
UF: Fe
Ingot molds—Iron construction*
Iron and alloys*
Slags—Iron recovery*
BT: Transition metals
RT: Cast iron
Iron analysis
Iron and steel plants
Iron and steel industry
Iron compounds
Iron foundry practice
Iron metallography
Iron metallurgy
Iron mines
Iron ore treatment
Iron plating
Iron powder
Iron removal (water treatment)
Iron research
Iron scrap
Steel

Iron alloys 545.2 (545.3)
DT: January 1993
UF: Ferrous alloys
Iron and alloys*
BT: Transition metal alloys
NT: Cast iron
Pig iron
Sponge iron
Steel
Wrought iron
RT: Ferroalloys
Iron compounds

Iron analysis 545.1 (801)
DT: January 1993
UF: Iron and steel analysis*
BT: Metal analysis
RT: Iron
Steel analysis

*Iron and alloys**
USE: Iron OR
Iron alloys

*Iron and steel analysis**
USE: Iron analysis OR
Steel analysis

Iron and steel industry 545
DT: Predates 1975
UF: Steel industry
BT: Industry
RT: Iron
Iron and steel plants
Iron mines
Iron research
Steel
Steel research
Steelmaking

*Iron and steel metallography**
USE: Iron metallography OR
Steel metallography

*Iron and steel metallography—Austenite**
USE: Austenite

*Iron and steel metallography—Bainite**
USE: Bainite

*Iron and steel metallography—Graphitization**
USE: Graphitization

*Iron and steel metallography—Martensite**
USE: Martensite

*Iron and steel metallography—Pearlite**
USE: Pearlite

*Iron and steel metallurgy**
USE: Iron metallurgy OR
Steel metallurgy

Iron and steel plants 545 (402.1)
DT: Predates 1975
UF: Cranes—Iron and steel plants*
Steel mills
Steel plants
BT: Industrial plants

Iron and steel plants *(continued)*
RT: Iron
Iron and steel industry
Iron metallurgy
Rolling mills
Steel
Steel metallurgy
Steelmaking

*Iron and steel plants—Oxygen supply**
USE: Oxygen supply

*Iron and steel plants—Pyrometry**
USE: Pyrometry

*Iron and steel research**
USE: Iron research OR
Steel research

*Iron and steel scrap**
USE: Iron scrap OR
Steel scrap

Iron castings
USE: Cast iron

Iron compounds (804.1) (804.2)
DT: Predates 1975
BT: Transition metal compounds
NT: Chlorite minerals
Chromite
Iron oxides
Pyrites
RT: Iron
Iron alloys

Iron deposits 504.3, 545.1
DT: Predates 1975
BT: Ore deposits
RT: Iron mines
Magnetite

Iron foundry practice 534.2, 545.1
DT: Predates 1975
BT: Foundry practice
Iron metallurgy
NT: Malleable iron foundry practice
RT: Iron

Iron metallography 531.2, 545.1
DT: January 1993
UF: Iron and steel metallography*
BT: Metallography
RT: Graphitization
Iron
Iron metallurgy
Steel metallography

Iron metallurgy 531.1, 545.1
DT: January 1993
UF: Iron and steel metallurgy*
BT: Metallurgy
NT: Iron foundry practice
Iron powder metallurgy
RT: Iron
Iron and steel plants
Iron metallography
Iron ore treatment
Steel metallurgy

Iron mines 504.3, 545.1
DT: January 1993
UF: Iron mines and mining*
BT: Mines
RT: Iron
Iron and steel industry
Iron deposits

*Iron mines and mining**
USE: Iron mines

Iron ore pellets 533.1, 545.1
DT: January 1977
BT: Ore pellets
RT: Iron ores

Iron ore reduction 533.1, 545.1
DT: Predates 1975
BT: Ore reduction
RT: Iron ores

Iron ore sinter 533.1, 545.1
DT: January 1977
BT: Ore sinter
RT: Iron ores

Iron ore treatment 533.1, 545.1
DT: Predates 1975
BT: Ore treatment
NT: Blast furnace practice
RT: Iron
Iron metallurgy
Iron ores

Iron ores 504.3, 545.1
DT: January 1993
BT: Ores
NT: Magnetite
Pyrites
RT: Iron ore pellets
Iron ore reduction
Iron ore sinter
Iron ore treatment

Iron oxides 804.1
DT: January 1986
BT: Iron compounds
Oxides
NT: Magnetite

Iron plating 539.3, 545.1
DT: Predates 1975
BT: Plating
RT: Iron

Iron powder 545.1 (536.1)
DT: Predates 1975
BT: Powder metals
RT: Iron
Powder metallurgy

Iron powder metallurgy 536.1, 545.1
DT: January 1993
UF: Powder metallurgy—Iron copper*
Powder metallurgy—Iron graphite*
Powder metallurgy—Iron*
BT: Iron metallurgy

* Former Ei Vocabulary term

Iron powder metallurgy *(continued)*
Powder metallurgy
RT: Steel powder metallurgy

Iron pyrites
USE: Pyrites

Iron removal (water treatment) 445.1 (802.2) (802.3)
DT: January 1993
UF: Water treatment—Iron removal*
BT: Chemicals removal (water treatment)
RT: Iron

Iron research 545.1
DT: January 1993
UF: Iron and steel research*
BT: Research
RT: Iron
Iron and steel industry
Steel research

Iron scrap 545.1 (452.3)
DT: January 1993
UF: Iron and steel scrap*
BT: Scrap metal
RT: Iron
Steel scrap

Irradiation (622.2) (711.1) (741.1) (932.1)
DT: January 1995
NT: Ion bombardment
Neutron irradiation
RT: Neutron irradiation apparatus
Radiation

Irradiation apparatus (neutron)
USE: Neutron irradiation apparatus

Irrigation 821.3
DT: Predates 1975
BT: Agriculture
RT: Irrigation canals
Pumping plants
Sprinkler systems (irrigation)
Surface waters
Water pipelines
Water piping systems
Water supply

Irrigation canals (446.1) (821.3)
DT: Predates 1975
BT: Canals
RT: Agriculture
Irrigation

*Irrigation—Pumping plants**
USE: Pumping plants

*Irrigation—Sprinkler systems**
USE: Sprinkler systems (irrigation)

ISDN
USE: Voice/data communication systems

ISFET
USE: Ion sensitive field effect transistors

Isolators (microwave)
USE: Microwave isolators

Isomerization 802.2
DT: January 1993
BT: Chemical reactions
RT: Molecular structure
Refining

Isotherms (641.1) (801.4) (931.2)
DT: January 1995
NT: Adsorption isotherms
RT: Temperature

Isotopes (803) (804)
DT: Predates 1975
UF: Nuclides*
NT: Deuterium
Radioisotopes
RT: Atoms
Radioactive tracers

*Isotopes—Radioactivity**
USE: Radioisotopes

Iterated switching networks (721) (703.1)
DT: January 1993
UF: Switching theory—Iterated switching networks*
BT: Switching networks
RT: Switching theory

Iterative methods 921.6
DT: January 1993
UF: Mathematical techniques—Iterative methods*
BT: Numerical analysis

IVHS
USE: Intelligent vehicle highway systems

Jack-up rigs 511.2
DT: January 1993
UF: Oil rigs, Jack-up*
BT: Drilling rigs

Jacks (605.2) (405.1)
SN: Lifting devices. For electrical jacks use ELECTRIC CONNECTORS
DT: Predates 1975
BT: Equipment
RT: Actuators
Hand tools

*Jacks—Hydraulic**
USE: Hydraulic tools

Jails
USE: Prisons

Jamming 711 (404)
DT: January 1993
UF: Signal interference—Jamming*
BT: Electronic warfare
Signal interference
RT: Radar
Radar clutter
Radar countermeasures
Radar interference
Radar warning systems
Radio interference
White noise

* Former Ei Vocabulary term

Jet aircraft 652.1
DT: January 1993
UF: Aircraft—Jet propelled*
BT: Aircraft
NT: Jet engines
Jet propelled helicopters
RT: Reconnaissance aircraft
Supersonic aircraft

Jet augmented wing flaps
USE: Jet flaps

Jet drilling (405.2) (502.1)
DT: January 1993
UF: Rock drilling—Jet*
Wash boring
BT: Drilling
RT: Jets
Rock drilling
Well drilling

Jet engines 653.1
DT: January 1993
UF: Aircraft engines, Jet and turbine*
BT: Engines
Jet aircraft
NT: Ducted fan engines
Ramjet engines
Turbojet engines
RT: Aircraft engines
Ingestion (engines)
Jets
Nuclear propulsion

Jet flaps 651.1
DT: January 1993
UF: Aircraft—Jet flap*
Jet augmented wing flaps
BT: Airfoils

Jet propelled helicopters 652.4
DT: January 1993
UF: Helicopters—Jet propelled*
BT: Helicopters
Jet aircraft

Jet pumps 618.2
DT: January 1993
UF: Pumps, Jet*
BT: Pumps
NT: Ejectors (pumps)
RT: Jets

Jets 631.1
DT: January 1993
UF: Flow of fluids—Jets*
BT: Equipment
NT: Plasma jets
RT: Aerodynamics
Air ejectors
Flow of fluids
Fluid dynamics
Hydrodynamics
Jet drilling
Jet engines
Jet pumps
Nozzles

Jetties 407.1
DT: Predates 1975
UF: Groins (hydraulic structures)
Groins*
BT: Hydraulic structures
RT: Breakwaters
Coastal engineering
Shore protection

Jewel bearings 601.2 (482.2)
DT: January 1993
UF: Bearings—Jewels*
Jewels (bearing)
BT: Bearings (machine parts)

Jewelry manufacture (547.1) (482.2.1)
DT: Predates 1975
BT: Manufacture
RT: Gems

Jewels (bearing)
USE: Jewel bearings

Jewels (precious)
USE: Gems

JFET
USE: Junction gate field effect transistors

Jib cranes 693.1
DT: January 1993
UF: Cranes—Jib*
BT: Cranes

Jigs (605)
DT: January 1993
UF: Industrial plants—Tools, jigs and fixtures*
Tools, jigs and fixtures*
BT: Cutting tools
RT: Fixtures (tooling)
Guides (mechanical)

Job analysis (912) (913.1)
DT: Predates 1975
BT: Management
RT: Ergonomics
Human engineering
Operations research
Personnel
Production control
Time and motion study
Work simplification

Job satisfaction 912.4
DT: January 1993
UF: Personnel—Job satisfaction*
RT: Management
Motivation
Personnel

Joining
DT: January 1993
UF: Rails—Fastening*
NT: Cable jointing
Docking
Mooring
Riveting
Rock bolting

* Former Ei Vocabulary term

Joining *(continued)*
RT: Fasteners
Joints (structural components)
Shrinkfitting

Joint prostheses 462.4
DT: January 1993
UF: Prosthetics—Joint prostheses*
BT: Prosthetics
NT: Hip prostheses
RT: Biomaterials

Jointing (cables)
USE: Cable jointing

Jointing (electric connectors)
USE: Electric connectors

Joints (anatomy) 461.3
DT: January 1993
UF: Biomechanics—Joints*
BT: Musculoskeletal system
RT: Biomechanics

Joints (structural components) 408.2
DT: January 1993
UF: Bridges—Joints*
Concrete construction—Joints*
Electric conductors—Joints*
Heat transfer—Joints*
Joints*
Machine tools—Joints*
Pressure transducers—Joints*
Pressure vessels—Joints*
Rails—Joints*
Roads and streets—Joints*
Sewers—Joints*
Steel structures—Connections*
Wooden construction—Connections*
BT: Structural members
NT: Adhesive joints
Bolted joints
Expansion joints
Pipe joints
Soldered joints
Welds
RT: Bonding
Connectors (structural)
Fasteners
Joining
Sealants
Seals
Steel structures

Joints (universal)
USE: Universal joints

*Joints**
USE: Joints (structural components)

*Joints, Adhesive**
USE: Adhesive joints

*Joints, Adhesive—Plastics adhesives**
USE: Plastic adhesives

*Joints, Adhesive—Wood products**
USE: Wood products

*Joints—Bolted**
USE: Bolted joints

*Joints—Sealants**
USE: Sealants

*Joints—Soldered**
USE: Soldered joints

*Joints—Welded**
USE: Welds

Josephson junction devices (708.3) (704.2) (715.1)
DT: January 1993
UF: Superconducting devices—Josephson junctions*
Superconducting junction devices
BT: Superconducting devices
RT: Tunnel junctions

Journal bearings 601.2
DT: January 1993
UF: Bearings—Journal*
BT: Bearings (machine parts)
RT: Antifriction bearings
Foil bearings

Junction gate field effect transistors 714.2
DT: January 1993
UF: JFET
Transistors, Field effect—Junction gate*
BT: Field effect transistors

Junction lasers
USE: Semiconductor lasers

Junctions (semiconductor)
USE: Semiconductor junctions

Junctions (waveguide)
USE: Waveguide junctions

Just in time production (913.1) (913.2)
DT: January 1993
UF: Just-in-time production*
BT: Production control
RT: Concurrent engineering
Inventory control

*Just-in-time production**
USE: Just in time production

Jute fibers 819.1
SN: Vegetable bast fiber
DT: January 1993
UF: Jute*
BT: Bast fibers
RT: Textile fibers

*Jute**
USE: Jute fibers

K
USE: Potassium

Kalman filtering (703.2) (731.1) (716.1)
DT: January 1993
UF: Kalman filters
Signal filtering and prediction—Kalman filtering*

* Former Ei Vocabulary term

Kalman filtering *(continued)*
BT: Signal filtering and prediction
RT: Control systems

Kalman filters
USE: Kalman filtering

Kanaf
USE: Kenaf fibers

Kaolin 482.2
DT: Predates 1975
UF: Bolus alba
White clay
BT: Clay minerals

Kaplan turbines 617.1
DT: January 1993
UF: Hydraulic turbines—Kaplan*
BT: Hydraulic turbines

Kelvin skin effect
USE: Skin effect

Kenaf fibers 819.1 (811.1)
DT: January 1993
UF: Kanaf
Pulp materials—Kenaf*
BT: Bast fibers
RT: Pulp materials
Textile fibers

Kennelly Heaviside layer
USE: E region

Kerogen 804.1 (512.1) (513.3)
DT: June 1990
BT: Bituminous materials
RT: Lubricants
Oil shale

Kerosene 523, 513.3
DT: Predates 1975
UF: Paraffin oils
BT: Liquid fuels
RT: Gas oils
Paraffins

Kerr effect (optical)
USE: Optical Kerr effect

Kerr electrooptical effect 701.1, 741.1
DT: January 1995
UF: Electrooptical Kerr effect
BT: Electrooptical effects
RT: Birefringence
Optical Kerr effect

Kerr magnetooptical effect 701.2, 741.1
DT: January 1995
UF: Magnetooptical Kerr effect
BT: Magnetooptical effects
RT: Optical recording
Optical rotation

Ketones 804.1
DT: January 1986
BT: Organic compounds
NT: Acetone
RT: Salts

Keyboard substitutes
USE: Computer keyboard substitutes (handicapped aid)

Keyboards (computer)
USE: Computer keyboards

Keyboards (typewriter)
USE: Typewriter keyboards

Keypads
USE: Computer keyboards

Keys (for locks) (601.2)
DT: January 1993
UF: Keys and keyways*
BT: Equipment
RT: Locks (fasteners)

*Keys and keyways**
USE: Keys (for locks) OR
Keyways

Keyways (601.2)
DT: January 1993
UF: Keys and keyways*
BT: Locks (fasteners)

Kidney dialysis equipment
USE: Hemodialyzers

Kilns 642.2
DT: Predates 1975
BT: Furnaces
NT: Rotary kilns
RT: Brickmaking
Industrial ovens

*Kilns—Dust control**
USE: Dust control

*Kilns—Erection**
USE: Construction

*Kilns—Rotary**
USE: Rotary kilns

Kinematics 931.1
DT: Predates 1975
BT: Mechanics
NT: Ballistics
Inverse kinematics
RT: Dynamics
Kinetic theory

Kinetic theory (931.1)
DT: January 1993
UF: Kinetics*
BT: Dynamics
NT: Kinetic theory of gases
RT: Kinematics
Molecules
Statistical mechanics
Transport properties

Kinetic theory of gases (931.1) (931.2)
DT: January 1993
UF: Gas kinetics
Gases—Kinetic theory*
BT: Kinetic theory

* Former Ei Vocabulary term

Kinetic theory of gases *(continued)*
RT: Brownian movement
Equations of state of gases
Gases
Navier Stokes equations

Kinetics (enzyme)
USE: Enzyme kinetics

Kinetics of chemical reactions
USE: Reaction kinetics

*Kinetics**
USE: Kinetic theory

Kitchens (402)
DT: Predates 1975
BT: Facilities
RT: Food processing

Klystrons 714.1
DT: January 1993
UF: Electron tubes, Klystron*
BT: Electron wave tubes
RT: Cavity resonators
Oscillators (electronic)

Knit fabrics 819.5
DT: Predates 1975
BT: Fabrics
RT: Knitting machinery
Textile pilling

Knitting machinery 819.6
DT: Predates 1975
BT: Textile machinery
RT: Knit fabrics

Knobs (601.2)
DT: January 1993
UF: Instruments—Knobs*
BT: Components
RT: Instrument components

Knock (engines)
USE: Combustion knock

Knowledge acquisition 723.4
DT: January 1995
UF: Knowledge elicitation
BT: Knowledge engineering
RT: Knowledge based systems

Knowledge based systems 723.4.1
DT: January 1993
UF: Expert systems—Knowledge bases*
Rule based systems
NT: Expert systems
RT: Intelligent control
Knowledge acquisition
Knowledge engineering
Knowledge representation
Motion planning

Knowledge elicitation
USE: Knowledge acquisition

Knowledge engineering 723.4
DT: January 1995
BT: Artificial intelligence
NT: Knowledge acquisition
Knowledge representation
RT: Knowledge based systems

Knowledge representation 723.4
DT: January 1995
BT: Knowledge engineering
RT: Knowledge based systems

Koepe hoists
USE: Koepe winders

Koepe winders 502.2
DT: January 1993
UF: Koepe hoists
Mine hoists—Koepe system*
BT: Mine hoists

Kr
USE: Krypton

Kraft paper 811.1
DT: January 1993
BT: Paper
RT: Kraft paperboards
Kraft process
Kraft pulp

Kraft paperboards 811.1
DT: January 1993
UF: Paperboards—Kraft*
BT: Paperboards
RT: Kraft paper

Kraft process 811.1.1
DT: January 1993
UF: Kraft pulping
Pulp manufacture—Kraft process*
Sulfate pulping
BT: Pulp manufacture
RT: Kraft paper
Kraft pulp
Sulfate pulp

Kraft pulp 811.1
DT: January 1993
UF: Pulp—Kraft*
BT: Pulp
RT: Kraft paper
Kraft process

Kraft pulping
USE: Kraft process

Krypton 804
DT: Predates 1975
UF: Kr
BT: Inert gases

Kurchatovium 622.1
DT: January 1993
BT: Transuranium elements

Kyanite 482.2 (812)
DT: Predates 1975
UF: Cyanite
Disthene
Sappare
BT: Silicate minerals

* Former Ei Vocabulary term

Kyanite *(continued)*
RT: Kyanite deposits
Refractory materials

Kyanite deposits 505.1
DT: Predates 1975
BT: Mineral resources
RT: Kyanite

Labeling 694.1
DT: Predates 1975
UF: Labelling
RT: Labels
Laws and legislation
Marking machines
Materials handling
Numbering systems
Packaging
Printing
Textile processing

Labelling
USE: Labeling

Labels 694.2
DT: Predates 1975
NT: Nameplates
RT: Labeling
Signs

Labor
USE: Personnel

Laboratories (402.1) (801) (802.1) (901.3)
DT: January 1993
BT: Facilities
NT: Acoustics laboratories
Chemical laboratories
Clinical laboratories
Hydraulic laboratories
Industrial laboratories
Materials testing laboratories
Orbital laboratories
Power plant laboratories
Research laboratories
X ray laboratories
RT: Engineering facilities
Laboratory furnaces
Ship model tanks
Sterilizers
Test facilities
Wind tunnels

Laboratory furnaces 642.2 (802.1) (532)
DT: January 1993
UF: Furnaces, Laboratory*
BT: Furnaces
RT: Laboratories

*Lacquer and lacquering**
USE: Lacquers OR
Lacquering

Lacquering 813.1
DT: January 1993
UF: Lacquer and lacquering*
BT: Coating techniques
RT: Painting

Lacquers 813.2
DT: January 1993
UF: Lacquer and lacquering*
BT: Organic coatings
RT: Paint
Pigments
Solvents
Varnish

Ladder filters
USE: Ladder networks

Ladder networks 703.1 (703.2)
DT: January 1993
UF: Electric networks—Ladder circuits*
Ladder filters
BT: Networks (circuits)
RT: Electric filters

Ladders 405.1
DT: Predates 1975
BT: Equipment
RT: Stairs

Ladle metallurgy 531.1 (545.3)
DT: January 1993
UF: Steelmaking—Ladle process*
BT: Metallurgy
RT: Steelmaking

Ladles (545.3) (534.1) (531.1)
DT: Predates 1975
BT: Equipment

Lagoons (sewage)
USE: Sewage lagoons

Lagrange multipliers (921)
DT: January 1995
BT: Mathematical techniques
RT: Differential equations
Operations research
Optimization

Lakes (407) (444.1)
DT: Predates 1975
BT: Surface waters
RT: Banks (bodies of water)
Beaches
Inland waterways
Reservoirs (water)
River basin projects
Surface water resources
Thermal stratification
Water supply
Watersheds

Laminar flow 631.1
DT: January 1993
UF: Flow of fluids—Laminar*
BT: Flow of fluids
RT: Transition flow
Turbulent flow
Two phase flow
Unsteady flow
Viscous flow
Wall flow

* Former Ei Vocabulary term

Laminated composites (415)
DT: January 1993
UF: Laminated products—Composites*
BT: Composite materials
Laminates
RT: Laminating

*Laminated products**
USE: Laminates

*Laminated products—Composites**
USE: Laminated composites

Laminates (415)
DT: January 1993
UF: Laminated products*
BT: Materials
NT: Laminated composites
Magnetic laminates
Paper laminates
Plastic laminates
Wood laminates
RT: Bilaminate membranes
Delamination
Honeycomb structures
Laminating
Plating
Substrates

Laminating 816.1 (813.1)
DT: January 1993
UF: Plastics—Laminating*
BT: Bonding
RT: Adhesion
Adhesives
Cladding (coating)
Coating techniques
Gluing
Laminated composites
Laminates
Laminating machinery
Roll bonding

Laminating machinery (816.2)
DT: January 1993
UF: Plastics machinery—Laminating machines*
BT: Machinery
RT: Laminating
Plastics machinery

Lamp filaments
USE: Filaments (lamp)

Lamps
USE: Electric lamps

LAN
USE: Local area networks

Land fill 452
DT: June 1990
UF: Landfills
Refuse disposal—Land fill*
Sanitary landfills
BT: Refuse disposal
NT: Leachate treatment
RT: Land reclamation
Landfill linings
Transfer stations
Volatile organic compounds

*Land fill—Leachate treatment**
USE: Leachate treatment

Land reclamation 442.2
DT: Predates 1975
UF: Reclamation (land)
BT: Reclamation
NT: Revegetation
RT: Environmental engineering
Land fill
Land use
Mines
Mining
Regional planning
Soil conservation
Urban planning

*Land reclamation—Revegetation**
USE: Revegetation

Land transportation (motor)
USE: Motor transportation

Land transportation (rail)
USE: Railroad transportation

Land use 403 (442.2)
DT: January 1993
UF: Regional planning—Land use*
Urban planning—Land use*
RT: Land reclamation
Parks
Regional planning
Roadsides
Urban planning
Watersheds
Zoning

Land vehicle propulsion (661.1) (682.1.2)
DT: January 1993
UF: Propulsion—Land vehicle applications*
BT: Propulsion
RT: Ground vehicles

Landers (planetary)
USE: Planetary landers

Landfill linings 452 (453.2) (454.2)
DT: January 1993
BT: Linings
RT: Land fill
Pollution control
Water pollution

Landfills
USE: Land fill

Landforms 481.1
DT: January 1993
NT: Beaches
Caves
Muskeg
Reefs
Volcanoes
RT: Aquifers
Geology

* Former Ei Vocabulary term

Landforms *(continued)*
Geomorphology
Oil bearing formations
Stratigraphy
Tectonics

Landing (652.1) (656.1)
DT: January 1993
NT: Aircraft landing
Spacecraft landing
RT: Aircraft landing systems
Airport runways
Aviation
Landing gear (aircraft)

Landing gear (aircraft) 652.3
DT: January 1993
UF: Aircraft landing gear*
BT: Aircraft parts and equipment
NT: Arresting devices (aircraft)
Landing gear retracting devices
RT: Landing

Landing gear retracting devices 652.3
DT: January 1993
UF: Aircraft landing gear—Retracting devices*
BT: Landing gear (aircraft)

Landing mats 431.4
DT: January 1993
UF: Airport runways—Landing mats*
BT: Equipment
RT: Aircraft landing
Airport runways
Emergency runways
Temporary runways

Landing systems (aircraft)
USE: Aircraft landing systems

Landslides (483.1) (484.1)
DT: Predates 1975
RT: Avalanches (snowslides)
Earthquakes
Mining
Rock bolting
Soil liquefaction
Soils

*Landslides—Rock bolting**
USE: Rock bolting

Langmuir Blodgett films (801.4)
DT: January 1993
UF: Films—Langmuir-Blodgett*
Langmuir films
LB films
BT: Films
RT: Monolayers

Langmuir films
USE: Langmuir Blodgett films

Languages (formal)
USE: Formal languages

Languages (query)
USE: Query languages

Lanthanides
USE: Rare earth elements

Lanthanum 547.2
DT: January 1993
UF: Lanthanum and alloys*
BT: Rare earth elements
RT: Lanthanum alloys
Lanthanum compounds

Lanthanum alloys 547.2
DT: January 1993
UF: Lanthanum and alloys*
BT: Rare earth alloys
RT: Lanthanum
Lanthanum compounds

*Lanthanum and alloys**
USE: Lanthanum OR
Lanthanum alloys

Lanthanum compounds (804.1) (804.2)
DT: Predates 1975
BT: Rare earth compounds
RT: Lanthanum
Lanthanum alloys

Lapidary work
USE: Crystal cutting

Laplace transforms 921.3
DT: January 1993
UF: Mathematical transformations—Laplace transforms*
BT: Mathematical transformations
RT: Circuit theory

Lapping 604.2, 606.2
DT: Predates 1975
BT: Machining
Metal finishing
Polishing
RT: Lapping machines

Lapping machines 603.1
DT: Predates 1975
BT: Machine tools
RT: Lapping
Polishing machines

Laptop computers
USE: Personal computers

Large scale integration
USE: LSI circuits

Large scale systems (461.1) (731.1) (912.3)
DT: January 1993
UF: Complex systems
Composite systems
Interconnected systems
Systems science and cybernetics—Large scale systems*
BT: Systems science
RT: Control systems

* Former Ei Vocabulary term

Large screen projection television 716.4
DT: January 1993
UF: Television—Large screen projection*
BT: Television receivers

Laser ablation 641.2, 744.8
DT: January 1995
BT: Ablation
Laser beam effects
RT: Laser beams
Lasers

Laser accessories 744.7
DT: January 1993
UF: Lasers—Accessories*
BT: Accessories
NT: Laser windows
RT: Lasers
Light modulation
Light modulators
Mirrors
Optical waveguides

Laser applications 744.9
SN: For applications of lasers to other equipment or technologies
DT: January 1993
BT: Applications
NT: Laser beam cutting
Laser beam welding
Laser recording
Laser surgery
Pulsed laser applications
RT: Laser beam effects
Laser chemistry
Laser diagnostics
Laser Doppler velocimeters
Laser produced plasmas
Laser pulses
Laser resonators
Laser welding machines
Lasers
Optical communication
Optical links
Photodynamic therapy
Raman spectroscopy
Videodisks

Laser beam cutting 604.1, 744.9
DT: January 1993
UF: Metal cutting—Laser beam*
BT: Cutting
Laser applications
RT: Laser beams
Lasers

Laser beam effects 744.8
SN: Produced by laser beams
DT: January 1993
UF: Laser beams—Effects*
Laser damage
Laser induced damage
BT: Radiation effects
NT: Laser ablation
Laser tissue interaction
RT: Laser applications
Laser beams

Laser beam welding 538.2.1, 744.9
DT: January 1993
UF: Laser welding
Welding—Laser*
BT: Laser applications
Welding
RT: Laser beams
Laser welding machines
Lasers
Vacuum welding

Laser beams 744.8
DT: Predates 1975
BT: Coherent light
NT: Laser pulses
RT: Laser ablation
Laser beam cutting
Laser beam effects
Laser beam welding
Laser optics
Optical communication
Optical radar
Thermal blooming

*Laser beams—Effects**
USE: Laser beam effects

Laser chemistry 744.1 (801)
DT: January 1993
BT: Physical chemistry
RT: Laser applications
Lasers

Laser communication
USE: Optical links

Laser damage
USE: Laser beam effects

Laser diagnostics 744.9
DT: January 1993
BT: Testing
RT: Equipment testing
Laser applications
Lasers
Materials testing

Laser diodes
USE: Semiconductor lasers

Laser disks
USE: Videodisks

Laser Doppler velocimeters 744.9, 943.1
DT: January 1993
UF: Velocimeters—Laser Doppler*
BT: Velocimeters
RT: Anemometers
Laser applications

Laser fusion 744.9, 932.2.1 (621.2)
DT: January 1993
UF: Laser induced fusion
Lasers—Fusion applications*
BT: Fusion reactions
RT: Nuclear physics

Laser gyros
USE: Ring lasers

Laser induced damage
USE: Laser beam effects

Laser induced fusion
USE: Laser fusion

Laser mode locking 744.1
DT: January 1993
UF: Lasers—Mode locking*
Mode locking
RT: Laser modes
Lasers
Light modulation
Light modulators

Laser modes 744.1
DT: January 1993
UF: Lasers—Modes*
Modes (laser)
RT: Laser mode locking
Lasers

Laser optics 741.1, 744.1
DT: January 1986
BT: Optics
RT: Laser beams
Lasers

Laser produced plasmas 744.9, 932.3
DT: January 1993
UF: Plasmas—Laser-produced*
BT: Plasmas
RT: Laser applications

Laser pulses 744.1
DT: January 1986
UF: Pulses (laser)
BT: Laser beams
RT: Laser applications
Lasers
Pulsed laser applications
Ultrafast phenomena

Laser recording 744.9 (741.1) (716.3) (716.4)
DT: January 1993
BT: Laser applications
Optical recording
RT: Lasers

Laser resonators 744.7
DT: January 1993
UF: Lasers—Resonators*
BT: Lasers
Optical resonators
RT: Laser applications

Laser safety 744.1, 914.1
DT: January 1993
BT: Accident prevention

Laser surgery 461.6, 744.9
DT: January 1993
BT: Laser applications
Surgery
RT: Laser tissue interaction
Lasers

Laser theory 744.1
DT: January 1993
UF: Lasers—Theory*
BT: Physics
Theory
RT: Coherent light
Electromagnetic wave absorption
Lasers

Laser tissue interaction 461.2, 744.8
DT: January 1993
BT: Laser beam effects
RT: Laser surgery

Laser tuning 744.1
DT: January 1993
UF: Lasers—Tuning*
BT: Tuning
RT: Lasers

Laser welding
USE: Laser beam welding

Laser welding machines 538.2.2, 744.9
DT: January 1993
UF: Welding machines—Lasers*
BT: Welding machines
RT: Laser applications
Laser beam welding

Laser windows 744.7
DT: January 1993
UF: Optical devices—Laser windows*
BT: Laser accessories
Optical devices
RT: Lenses

Lasers 744.1
DT: Predates 1975
UF: Optical masers
BT: Light amplifiers
Light sources
NT: Chemical lasers
Continuous wave lasers
Distributed feedback lasers
Dye lasers
Excimer lasers
Free electron lasers
Gas lasers
High energy lasers
High power lasers
Laser resonators
Liquid lasers
Optically pumped lasers
Q switched lasers
Ring lasers
Solid state lasers
X ray lasers
RT: Coherent light
Electron beam pumping
Holography
Laser ablation
Laser accessories
Laser applications
Laser beam cutting
Laser beam welding

* Former Ei Vocabulary term

Lasers *(continued)*
Laser chemistry
Laser diagnostics
Laser mode locking
Laser modes
Laser optics
Laser pulses
Laser recording
Laser surgery
Laser theory
Laser tuning
Light modulation
Light modulators
Light pulse generators
Masers
Optical pumping
Pollution
Pumping (laser)
Quantum electronics
Second harmonic generation

*Lasers, Carbon dioxide**
USE: Carbon dioxide lasers

*Lasers, Chemical**
USE: Chemical lasers

*Lasers, Dye**
USE: Dye lasers

*Lasers, Excimer**
USE: Excimer lasers

*Lasers, Free electron**
USE: Free electron lasers

*Lasers, Gas dynamic**
USE: Gas dynamic lasers

*Lasers, Gas**
USE: Gas lasers

*Lasers, Injection**
USE: Injection lasers

*Lasers, Liquid**
USE: Liquid lasers

*Lasers, Ring**
USE: Ring lasers

*Lasers, Semiconductor**
USE: Semiconductor lasers

*Lasers, Solid state**
USE: Solid state lasers

*Lasers, X-ray**
USE: X ray lasers

*Lasers—Accessories**
USE: Laser accessories

*Lasers—Fusion applications**
USE: Laser fusion

*Lasers—Mode locking**
USE: Laser mode locking

*Lasers—Modes**
USE: Laser modes

*Lasers—Optical pumping**
USE: Optically pumped lasers

*Lasers—Q switching**
USE: Q switched lasers AND
Q switching

*Lasers—Resonators**
USE: Laser resonators

*Lasers—Theory**
USE: Laser theory

*Lasers—Tuning**
USE: Laser tuning

Latches (circuits)
USE: Flip flop circuits

Latexes 801.3 (817) (818)
SN: Dispersions of polymeric substances in essentially aqueous media
DT: Predates 1975
UF: Papermaking—Latex applications*
Plastics—Latex*
BT: Colloids
RT: Elastomers
Plastics
Rubber
Synthetic rubber

Lathes 603.1
DT: Predates 1975
BT: Machine tools
NT: Chucking machines
Turret lathes
RT: Boring machines (machine tools)
Drilling machines (machine tools)
Grinding machines
Turning

*Lathes, Turret**
USE: Turret lathes

Lattice constants 933.1.1
DT: January 1995
UF: Lattice parameters
RT: Crystal atomic structure
Crystal lattices
Crystallography

*Lattice defects**
USE: Crystal defects

Lattice dynamics
USE: Lattice vibrations

Lattice mechanics
USE: Lattice vibrations

Lattice parameters
USE: Lattice constants

Lattice vibrations 933.1.1
DT: January 1993
UF: Acoustic mode (crystal)
Atomic scattering factors
Crystals—Lattice vibrations*
Lattice dynamics
Lattice mechanics

* Former Ei Vocabulary term

Lattice vibrations *(continued)*
Vibrations (lattice)
RT: Acoustic waves
Crystal atomic structure
Crystal defects
Phonons

Lattices (crystal)
USE: Crystal lattices

Launchers (aircraft)
USE: Catapults (aircraft launchers)

Launching (655.1) (672) (674.1)
DT: January 1993
UF: Rockets and missiles—Launching*
Satellites—Launching*
Ships—Launching*
Torpedoes—Launching*
RT: Missile launching systems
Missile silos
Missiles
Rockets
Satellites
Ships
Spacecraft
Starting
Takeoff
Torpedoes

Laundering 819.5 (802.3)
DT: January 1993
UF: Textiles—Laundering*
BT: Washing
RT: Detergents
Laundries
Soaps (detergents)
Textile pilling
Textiles

Laundries 819.6 (402.1)
DT: Predates 1975
BT: Industrial plants
RT: Laundering

Lawn mowers (605) (821.1)
DT: Predates 1975
UF: Mowers
BT: Machinery
RT: Agricultural machinery

Lawrencium 622.1
DT: January 1993
BT: Transuranium elements

Laws and legislation 902.3
DT: January 1993
UF: Construction industry—Laws and regulations*
Legislation*
Regulations
NT: International law
Mining laws and regulations
RT: Boiler codes
Building codes
Codes (standards)
Electric codes
Elevator codes
Industrial hygiene
Intellectual property
Labeling
License plates (automobile)
Nuclear reactor licensing
Pipeline codes
Pressure vessel codes
Professional aspects
Public policy
Regional planning
Rights of way
Standards
Ventilation codes
Weigh stations
Welding codes
Zoning

Laying (electric cables)
USE: Electric cable laying

Laying (pipelines)
USE: Pipeline laying

Laying (rails)
USE: Rail laying

Laying (track)
USE: Rail laying

Layout (industrial plants)
USE: Plant layout

Layout (integrated circuits)
USE: Integrated circuit layout

LB films
USE: Langmuir Blodgett films

LCD
USE: Liquid crystal displays

LDPE
USE: Low density polyethylenes

Leachate treatment
(452.2) (452.4) (453.2) (483.1) (802.2) (802.3)
DT: January 1993
UF: Land fill—Leachate treatment*
Refuse disposal—Leachate treatment*
BT: Land fill

Leaching 802.3 (533.1)
DT: January 1993
BT: Chemical operations
RT: Extraction
Infiltration
Mechanical permeability
Ore treatment
Percolation (fluids)
Purification
Separation
Solution mining
Solutions
Solvent extraction
Solvents
Washing
Weathering

* Former Ei Vocabulary term

Lead 546.1
DT: January 1993
UF: Cast iron—Lead content*
Lead and alloys*
Pb
BT: Heavy metals
RT: Lead alloys
Lead compounds
Lead deposits
Lead foundry practice
Lead metallography
Lead metallurgy
Lead mines
Lead ore treatment
Lead plating
Lead refining
Lead removal (water treatment)

Lead alloys 546.1
DT: January 1993
UF: Lead and alloys*
BT: Heavy metal alloys
RT: Lead
Lead compounds
Soldering alloys

*Lead and alloys**
USE: Lead OR
Lead alloys

Lead compounds (804.1) (804.2)
DT: Predates 1975
BT: Heavy metal compounds
NT: Semiconducting lead compounds
RT: Lead
Lead alloys
Oxide superconductors

Lead deposits 504.3, 546.1
DT: Predates 1975
BT: Ore deposits
NT: Copper lead deposits
Lead zinc deposits
RT: Lead

Lead foundry practice 534.2, 546.1
DT: Predates 1975
BT: Foundry practice
Lead metallurgy
RT: Lead

Lead metallography 531.2, 546.1
DT: Predates 1975
BT: Metallography
RT: Lead
Lead metallurgy

Lead metallurgy 531.1, 546.1
DT: Predates 1975
BT: Metallurgy
NT: Lead foundry practice
Lead powder metallurgy
RT: Lead
Lead metallography
Lead ore treatment
Lead smelting

Lead mines 504.3, 546.1
DT: January 1993
UF: Lead mines and mining*
BT: Mines
NT: Copper lead zinc mines
RT: Lead
Lead zinc deposits

*Lead mines and mining**
USE: Lead mines

Lead ore treatment 533.1, 546.1
DT: Predates 1975
BT: Ore treatment
NT: Copper lead ore treatment
Lead smelting
RT: Lead
Lead metallurgy

Lead plating 539.3, 546.1
DT: Predates 1975
BT: Plating
RT: Lead

Lead poisoning
USE: Industrial poisons

Lead powder metallurgy 536.1, 546.1
DT: January 1993
UF: Powder metallurgy—Lead*
BT: Lead metallurgy
Powder metallurgy

Lead refining 533.2, 546.1
DT: Predates 1975
BT: Metal refining
RT: Lead

Lead removal (water treatment) 445.1
DT: January 1993
UF: Water treatment—Lead removal*
BT: Chemicals removal (water treatment)
RT: Lead

Lead screws 546.1, 605
DT: January 1993
UF: Machine tools—Lead screws*
BT: Machine tools
Shafts (machine components)

Lead smelting 533.2, 546.1
DT: Predates 1975
BT: Lead ore treatment
Smelting
RT: Lead metallurgy

Lead zinc deposits 504.3, 546.1, 546.3
DT: January 1981
BT: Lead deposits
Zinc deposits
RT: Lead mines
Zinc mines

Leak detection (631.1) (914.1) (943)
SN: Detection of fluid leakage
DT: Predates 1975
RT: Accident prevention
Accidents
Detectors
Leakage (fluid)

Leakage (fluid) (452.1) (452.3) (483.1) (619.1)
DT: January 1993
UF: Pipelines—Leakage*
Sewers—Leakage*
BT: Defects
RT: Hazardous materials spills
Leak detection
Linings
Pipelines
Ponding
Porosity
Seepage

Leakage (magnetic)
USE: Magnetic leakage

Leakage currents 701.1
DT: January 1993
UF: Electric fault currents—Leakage currents*
Electric leakage currents
BT: Electric fault currents
RT: Electric discharges

Learning aids for handicapped persons 461.5
DT: January 1993
UF: Human rehabilitation engineering—Learning aids for handicapped*
BT: Human rehabilitation equipment
RT: Human rehabilitation engineering

Learning algorithms (723)
DT: January 1995
BT: Algorithms
RT: Learning systems

Learning systems (461.4) (723.4) (723.5) (731.5)
DT: June 1990
UF: Machine learning
BT: Systems science
NT: Robot learning
RT: Adaptive systems
Artificial intelligence
Automata theory
Backpropagation
Biocommunications
Brain models
Cognitive systems
Cybernetics
Feedforward neural networks
Learning algorithms
Man machine systems
Neural networks

Least squares approximations 921.6
DT: January 1993
UF: Mathematical techniques—Least squares approximations*
BT: Approximation theory
Numerical analysis
RT: Curve fitting
Optimization

Leather 814.1
DT: Predates 1975
BT: Materials
RT: Fungus protection
Synthetic leather
Tanning
Textiles

*Leather—Fungus protection**
USE: Fungus protection

*Leather—Synthetic**
USE: Synthetic leather

*Leather—Tanning**
USE: Tanning

*Leather—Water absorption**
USE: Water absorption

*Leather—Waterproofing**
USE: Waterproofing

LED
USE: Light emitting diodes

LEED
USE: Low energy electron diffraction

*Legislation**
USE: Laws and legislation

Lenses 741.3
DT: Predates 1975
BT: Components
NT: Contact lenses
Electron lenses
Electrostatic lenses
Intraocular lenses
Magnetic lenses
Optical instrument lenses
Plastic lenses
RT: Aberrations
Acoustic imaging
Aspherics
Focusing
Gradient index optics
Laser windows
Optical devices
Optical materials
Optical shutters
Polishing
Telescopes

*Lenses—Plastics**
USE: Plastic lenses

Levees 442.1
DT: Predates 1975
UF: Dikes
BT: Embankments
Hydraulic structures
RT: Flood control
Retaining walls
Shore protection

Level control 731.3
DT: January 1993
UF: Boiler control—Water level*
Control, Mechanical variables—Level*
BT: Spatial variables control

* Former Ei Vocabulary term

Level measurement 943.2
DT: January 1993
UF: Mechanical variables measurement—Level*
BT: Spatial variables measurement

Leveling (machinery) 603.1
SN: Positioning
DT: January 1993
UF: Machinery—Leveling*
RT: Alignment
Machinery

Levitation melting 531.1, 701.2
DT: January 1993
UF: Metal melting—Levitation*
BT: Metal melting
RT: Zone melting

Li
USE: Lithium

Liability (product)
USE: Product liability

Libraries 903.4.1 (402.2)
DT: Predates 1975
BT: Facilities
RT: Information dissemination
Information services

License plates (automobile) 662.1 (902.3)
DT: January 1993
UF: Automobiles—License plates*
RT: Automobiles
Laws and legislation

Licensing (automobile drivers)
USE: Driver licensing

Licensing (reactors)
USE: Nuclear reactor licensing

Lidar
USE: Optical radar

Life extension (industrial plants)
USE: Plant life extension

Life in service
USE: Service life

Life support systems (spacecraft) 655.1
DT: January 1993
UF: Spacecraft—Life support systems*
BT: Spacecraft equipment
RT: Ventilation
Weightlessness

Lifeboats 674.1
DT: Predates 1975
BT: Boats
Lifesaving equipment

Lifesaving equipment 671.2 (914.1)
SN: Ship equipment
DT: January 1993
UF: Lifesaving equipment*
Ship equipment—Lifesaving*
BT: Ship equipment
NT: Lifeboats
RT: Safety devices

*Lifesaving equipment**
USE: Lifesaving equipment

Lift 651.1
DT: January 1993
UF: Aerodynamics—Lift*
BT: Aerodynamics
RT: Airfoils
Ground effect

Lift bridges 401.1
DT: January 1993
UF: Bridges, Lift*
BT: Movable bridges

Lift slab construction 405.2
DT: January 1993
UF: Buildings—Lift slab construction*
BT: Construction

Lifting magnets 691.1, 701.2
DT: January 1993
UF: Holding magnets
Magnets—Lifting*
BT: Electromagnets
Materials handling equipment

Ligaments 461.2
DT: January 1993
UF: Biological materials—Ligaments*
BT: Musculoskeletal system
Tissue

Light 741.1
SN: The visible spectrum, as well as very general treatments of the range from infrared through ultraviolet
DT: Predates 1975
UF: Beams (light)
BT: Electromagnetic waves
NT: Coherent light
Glare
High intensity light
Infrared radiation
Stray light
Ultraviolet radiation
RT: Acoustooptical effects
Birefringence
Color
Doppler effect
Electric lamps
Fiber optics
Light absorption
Light emission
Light polarization
Light reflection
Light refraction
Light scattering
Light transmission
Light velocity
Lighting
Opacity
Optical communication
Optical instruments
Optical links

* Former Ei Vocabulary term

Light *(continued)*
Optical resonators
Optical rotation
Optical variables measurement
Optical waveguides
Optics
Photochemical reactions
Photodetectors
Photometry
Photons
Spectrum analysis
Visibility
Wavefronts

Light absorption 741.1
DT: January 1993
UF: Light—Absorption*
Optical absorption
BT: Electromagnetic wave absorption
NT: Light extinction
RT: Atmospheric optics
Electron transitions
Light
Light propagation
Light transmission
Optical films
Optical filters
Photoluminescence

Light amplifiers (714.2) (741.3) (744.1)
DT: January 1993
UF: Amplifiers (light)
Light intensifiers
Light—Amplifiers*
Optical amplifiers
BT: Electronic equipment
NT: Lasers
RT: Amplification
Image intensifiers (electron tube)
Image intensifiers (solid state)
Telecommunication repeaters

Light coherence
USE: Coherent light

Light control films
USE: Solar control films

Light control glass 812.3, 741.1
DT: January 1993
UF: Glass—Light control*
BT: Glass
RT: Optical properties

Light detectors
USE: Photodetectors

Light emission 741.1
DT: January 1993
UF: Emission (light)
Light—Emission*
BT: Electromagnetic wave emission
NT: Luminescence
RT: Electron transitions
Light

Light emitting diodes 714.2 (741.1)
DT: January 1993
UF: LED
Semiconductor diode light emitters
Semiconductor diodes, Light emitting*
BT: Light sources
Optoelectronic devices
Semiconductor diodes
RT: Display devices
Integrated optoelectronics

Light extinction 741.1
DT: January 1993
UF: Extinction (light)
BT: Light absorption
RT: Light polarization

Light filters
USE: Optical filters

Light intensifiers
USE: Light amplifiers

Light interference 741.1
DT: January 1993
UF: Light—Interference*
Optical interference
BT: Electromagnetic wave interference
RT: Moire fringes
Optical films
Optical filters
Speckle

Light measurement 941.4
DT: January 1993
UF: Light—Measurements*
BT: Optical variables measurement

Light metals (541.1) (542.1) (542.2) (542.3)
DT: Predates 1975
UF: Aircraft materials—Light metals*
Automobile engines—Light metals*
Automobile materials—Light metals*
Rolling mill practice—Light metals*
Welding—Light metals*
BT: Nonferrous metals
NT: Aluminum
Beryllium
Magnesium
Titanium
RT: Aircraft materials
Automobile materials
Foundry practice

Light meters (exposure meters)
USE: Exposure meters

Light meters (photometers)
USE: Photometers

Light microscopy
USE: Optical microscopy

Light modulation 741.1
DT: January 1993
UF: Light—Modulation*
Optical modulation
Spatial light modulators

* Former Ei Vocabulary term

Light modulation *(continued)*
BT: Modulation
RT: Acoustooptical effects
Electrooptical effects
Laser accessories
Laser mode locking
Lasers
Light modulators
Optics

Light modulators (717.2) (741.3)
DT: January 1993
UF: Light—Modulators*
Optical modulators
BT: Modulators
RT: Acoustooptical effects
Electrooptical devices
Electrooptical effects
Laser accessories
Laser mode locking
Lasers
Light modulation
Optical communication equipment
Optics

Light pens 722.2 (714.1) (714.2)
DT: January 1993
UF: Beam pens
Computer peripheral equipment—Light pens*
Pens (light)
BT: Interactive devices
RT: Mice (computer peripherals)

Light pipes
USE: Fiber optics

Light polarization 741.1
DT: January 1993
UF: Light—Polarization*
Optical polarization
Polarized light
BT: Electromagnetic wave polarization
RT: Birefringence
Faraday effect
Light
Light extinction
Magnetooptical effects
Optical Kerr effect
Optical rotation
Photoelasticity
Polarimeters
Prisms

Light propagation 741.1
SN: In gases. For propagation in liquids and solids use LIGHT TRANSMISSION
DT: January 1993
UF: Light—Propagation*
Optical propagation
BT: Electromagnetic wave propagation
RT: Light absorption
Light reflection
Light refraction
Light transmission

Light pulse generators (741.1) (741.3)
DT: January 1993
UF: Light—Pulse generators*
BT: Pulse generators
RT: Lasers

Light rail transit 433.1 (682)
DT: January 1986
BT: Railroads
NT: Monorails
RT: Rapid transit
Subways
Trackless trolleys
Trolley cars
Urban planning

Light reflection 741.1
DT: January 1993
BT: Electromagnetic wave reflection
RT: Atmospheric optics
Light
Light propagation
Light transmission

Light reflectometers
USE: Reflectometers

Light refraction 741.1
DT: January 1993
BT: Electromagnetic wave refraction
RT: Light
Light propagation
Refractive index

Light refractometers
USE: Refractometers

Light resonators
USE: Optical resonators

Light scattering 741.1
DT: January 1993
BT: Electromagnetic wave scattering
NT: Brillouin scattering
Raman scattering
Speckle
RT: Atmospheric optics
Light
Turbidimeters

Light sensitive materials (741.3)
DT: January 1993
UF: Optical materials—Light sensitive materials*
BT: Optical materials
NT: Photographic emulsions
Photographic films
RT: Photochromism
Photosensitivity

Light sensitive resistors
USE: Photoresistors

Light sensitivity
USE: Photosensitivity

Light sources
(707.2) (714.2) (741.1) (741.3) (744)
DT: Predates 1975
UF: Chemical equipment—Light sources*

* Former Ei Vocabulary term

Light sources *(continued)*
Instruments—Light sources*
Light—Sources*
Motion pictures—Light sources*
Optical instruments—Light sources*
Optical sources
Sources (light)
NT: Electric lamps
Lasers
Light emitting diodes
Luminescent devices
RT: Electric signs
Illuminating engineering
Instruments
Lighting
Lighting fixtures
Monochromators
Motion pictures
Optical instruments
Phosphorescence
Photography

Light speed
USE: Light velocity

Light trailers 662.2 (402.3)
DT: January 1993
UF: Camping trailers
Trailers*
BT: Light weight vehicles
RT: Mobile homes

Light transmission 741.1
SN: In liquids and solids. For transmission in gases use LIGHT PROPAGATION
DT: January 1993
UF: Light—Transmission*
Optical transmission
BT: Electromagnetic wave transmission
NT: Infrared transmission
RT: Electromagnetic wave propagation
Fiber optics
Glass
Light
Light absorption
Light propagation
Light reflection
Optical filters
Optical waveguides
Visibility

Light velocity 741.1
DT: January 1993
UF: Light speed
Light—Velocity*
Speed of light
Velocity of light
BT: Velocity
RT: Light

Light water reactors (621.1) (621.2)
DT: January 1993
UF: Nuclear reactors, Light water*
BT: Water cooled reactors

Light waveguides
USE: Optical waveguides

Light weight building materials (415)
DT: January 1993
UF: Building materials—Light weight*
BT: Building materials
NT: Light weight concrete
RT: Light weight structures

Light weight concrete (412)
DT: January 1993
UF: Concrete—Light weight*
BT: Concretes
Light weight building materials
RT: Light weight structures

Light weight equipment
DT: January 1993
BT: Equipment
NT: Light weight rolling stock

Light weight rolling stock 682.1.1
DT: January 1993
UF: Cars—Light weight*
Railroad rolling stock—Light weight*
BT: Light weight equipment
Railroad rolling stock

Light weight structures 408.2 (415)
DT: January 1993
UF: Structural design—Light weight*
BT: Structures (built objects)
RT: Light weight building materials
Light weight concrete
Structural design

Light weight vehicles (415) (662.2)
DT: January 1993
UF: Vehicles—Light weight*
BT: Vehicles
NT: Light trailers

*Light—Absorption**
USE: Light absorption

*Light—Acoustooptical effects**
USE: Acoustooptical effects

*Light—Amplifiers**
USE: Light amplifiers

*Light—Birefringence**
USE: Birefringence

*Light—Brillouin scattering**
USE: Brillouin scattering

*Light—Coherent**
USE: Coherent light

*Light—Emission**
USE: Light emission

*Light—High intensity**
USE: High intensity light

*Light—Interference**
USE: Light interference

* Former Ei Vocabulary term

*Light—Measurements**
USE: Light measurement

*Light—Modulation**
USE: Light modulation

*Light—Modulators**
USE: Light modulators

*Light—Nonlinear optical effects**
USE: Nonlinear optics

*Light—Optical resonators**
USE: Optical resonators

*Light—Optical rotation**
USE: Optical rotation

*Light—Polarization**
USE: Light polarization

*Light—Propagation**
USE: Light propagation

*Light—Pulse generators**
USE: Light pulse generators

*Light—Sources**
USE: Light sources

*Light—Speckle**
USE: Speckle

*Light—Transmission**
USE: Light transmission

*Light—Velocity**
USE: Light velocity

Lighthouses 402.4 (407.1) (434.1)
DT: Predates 1975
BT: Buildings

Lighting (707)
DT: January 1993
UF: Automobiles—Lighting*
Buildings—Lighting*
Illumination
Industrial plants—Lighting*
Machinery—Lighting*
Mines and mining—Lighting*
Ships—Lighting*
NT: Daylighting
Electric lighting
Street lighting
RT: Building components
Electric lamps
Electric power utilization
Environmental engineering
Light
Light sources
Lighting fixtures
Machinery
Photometry

Lighting fixtures (707.2)
DT: Predates 1975
UF: Luminaires
BT: Equipment
NT: Troffers
RT: Electric lamps
Light sources
Lighting

*Lighting fixtures—Plastics**
USE: Plastics applications

*Lighting fixtures—Troffers**
USE: Troffers

Lightning 443.1 (701.1)
DT: Predates 1975
BT: Atmospheric electricity
Electric discharges
RT: Atmospherics
Electric sparks
Lightning arresters
Lightning protection
Thunderstorms

Lightning arresters 704.2, 914.1 (706)
DT: Predates 1975
BT: Equipment
RT: Lightning
Lightning protection
Safety devices

Lightning protection 914.1 (443.1)
DT: Predates 1975
BT: Protection
RT: Electric equipment protection
Lightning
Lightning arresters
Surge protection

Lignin 811.3 (815.1.1)
DT: Predates 1975
BT: Natural polymers
RT: Biomass
Cellulose

Lignite 524
DT: Predates 1975
UF: Brown coal
BT: Coal
RT: Briquets
Lignite liquefaction
Lignite mines
Liquefaction

Lignite liquefaction (523)
DT: January 1993
UF: Lignite—Liquefaction*
BT: Liquefaction
RT: Lignite

Lignite mines 503.1
DT: January 1993
UF: Lignite mines and mining*
BT: Coal mines
RT: Lignite

*Lignite mines and mining**
USE: Lignite mines

*Lignite—Briquets**
USE: Briquets

*Lignite—Liquefaction**
USE: Lignite liquefaction

* Former Ei Vocabulary term

Lime 804.2 (412)
DT: Predates 1975
UF: Calcium oxide
Quicklime
Roadbuilding materials—Lime*
Water treatment plants—Lime handling*
Water treatment—Lime*
BT: Calcium compounds
RT: Hydrated lime
Limestone
Roadbuilding materials
Slag cement
Water treatment

Lime brick 812.2
DT: January 1993
UF: Refractory materials—Lime brick*
BT: Brick
Refractory materials
RT: Silica brick

*Lime—Hydrated**
USE: Hydrated lime

*Lime—Slag admixture**
USE: Slag cement

Limestone (414) (482.2)
DT: Predates 1975
UF: Roadbuilding materials—Limestone*
BT: Sedimentary rocks
RT: Lime
Marble
Roadbuilding materials

*Limestone—Agricultural applications**
USE: Agriculture

*Limestone—Corrosive properties**
USE: Corrosive effects

Limited space charge accumulation (701.1) (714.2)
DT: January 1993
UF: Semiconductor diodes—Limited space charge accumulation*
BT: Electric field effects
RT: Electric conductivity
Semiconductor diodes

Limiters 713.3
DT: January 1993
UF: Amplitude limiting circuits
Amplitude selectors
Clipping circuits
Current limiters
Electronic circuits, Limiter*
Peak limiters
BT: Networks (circuits)
NT: Microwave limiters
RT: Demodulators
Signal receivers

Line concentrators (718.1) (718.2)
SN: Telephone or telegraph switching equipment
DT: Predates 1975
BT: Multiplexing equipment
Switching systems
RT: Telegraph equipment
Telephone lines
Telephone switching equipment

Linear accelerators 932.1.1
DT: January 1993
UF: Accelerators, Linear*
BT: Particle accelerators

Linear algebra 921.1
DT: January 1993
UF: Mathematical techniques—Linear algebra*
BT: Algebra
NT: Eigenvalues and eigenfunctions
Matrix algebra
Tensors
Vectors
RT: Numerical analysis

Linear control systems 731.1
DT: January 1993
UF: Control systems, Linear*
BT: Control systems
RT: Nonlinear systems

Linear integrated circuits 714.2
DT: January 1993
UF: Integrated circuits, Linear*
BT: Integrated circuits

Linear low density polyethylenes 815.1.1
DT: January 1993
UF: LLDPE
Polyethylenes—Linear low density*
BT: Low density polyethylenes

Linear motors 705.3
DT: January 1993
UF: Electric motors, Linear*
BT: Electric motors
RT: Electric automobiles
Electric traction

Linear network analysis 703.1.1
DT: January 1993
BT: Electric network analysis
RT: Linear network synthesis
Linear networks

Linear network synthesis 703.1.2
DT: January 1993
BT: Electric network synthesis
RT: Linear network analysis
Linear networks

Linear networks 703.1
DT: January 1993
UF: Electric networks, Linear*
BT: Networks (circuits)
RT: Linear network analysis
Linear network synthesis

Linear programming (723.1) (921.5) (922)
DT: January 1993
UF: Mathematical programming, Linear*
BT: Mathematical programming
RT: Constraint theory
Operations research

* Former Ei Vocabulary term

Linearization (731.1) (921)
DT: January 1993
UF: Mathematical techniques—Linearization*
BT: Approximation theory
RT: Control system analysis
Control system synthesis

Linen 819.5
DT: Predates 1975
BT: Fabrics
RT: Flax

Lines (telephone)
USE: Telephone lines

Linguistics (903.2)
DT: January 1993
UF: Information science—Language translation and linguistics*
BT: Information science
NT: Computational linguistics
RT: Speech

Linings
DT: January 1993
UF: Chemical equipment—Lining*
Dust collectors—Lining*
Mines and mining—Lining*
Pulp digesters—Lining*
Shaft sinking—Lining*
NT: Brake linings
Canal linings
Chimney linings
Furnace linings
Landfill linings
Mine shaft linings
Pipe linings
Pump linings
Reservoir linings
Sewer linings
Tank linings
Tunnel linings
RT: Leakage (fluid)
Sealing (closing)
Shaft sinking

Linkages (electric connectors)
USE: Electric connectors

Links (telecommunication)
USE: Telecommunication links

Linoleum (819)
DT: Predates 1975
RT: Carpet manufacture

Lipids 804.1
DT: January 1987
BT: Organic compounds
NT: Waxes
RT: Cholesterol
Esters
Fatty acids
Oils and fats

Liquefaction 802.3
DT: January 1993
NT: Coal liquefaction
Lignite liquefaction
Liquefaction of gases
Soil liquefaction
RT: Lignite
Liquid fuels

Liquefaction of gases 802.3 (931.2)
DT: January 1993
UF: Gases—Liquefaction*
BT: Liquefaction
Phase transitions
RT: Condensation
Gases
Liquefied gases

Liquefied gases (931.2)
DT: January 1993
UF: Gases, Liquefied*
BT: Liquids
NT: Liquefied natural gas
Liquefied petroleum gas
Liquid methane
Superfluid helium
RT: Gases
Liquefaction of gases

Liquefied natural gas 523
DT: January 1993
UF: LNG
Natural gas, Liquefied*
BT: Liquefied gases
Natural gas
RT: Liquid fuels
Liquid methane
Natural gasoline
Petroleum industry

Liquefied petroleum gas 523
DT: January 1993
UF: LPG
Petroleum gas, Liquefied*
BT: Liquefied gases
Natural gasoline
Petroleum products
RT: Butane
Liquid fuels
Petroleum industry
Propane

Liquefiers
USE: Condensers (liquefiers)

Liquid chromatography (801) (802.3)
DT: January 1993
UF: Chromatographic analysis—Liquid*
BT: Chromatography
NT: Gel permeation chromatography
High performance liquid chromatography
High pressure liquid chromatography

Liquid crystal displays (722.2) (741.3)
DT: January 1993
UF: Display devices—Liquid crystal*
LCD
BT: Display devices
RT: Electrooptical devices
Liquid crystals

Liquid crystal polymers 815.1.1 (931.2)
DT: January 1995
UF: Liquid crystalline polymers
BT: Liquid crystals
Polymers

Liquid crystalline polymers
USE: Liquid crystal polymers

Liquid crystals (804) (931.2)
DT: January 1993
UF: Crystals, Liquid*
NT: Cholesteric liquid crystals
Liquid crystal polymers
Nematic liquid crystals
Smectic liquid crystals
Thermotropic liquid crystals
RT: Liquid crystal displays
Molecular orientation

Liquid fuels 523 (513.3)
DT: Predates 1975
BT: Fuels
NT: Fuel oils
Gasohol
Gasoline
Kerosene
Liquid propellants
RT: Aircraft fuels
Atomization
Automotive fuels
Diesel fuels
Liquefaction
Liquefied natural gas
Liquefied petroleum gas

Liquid insulating materials 413.1
DT: January 1993
UF: Electric insulating materials—Liquids*
BT: Electric insulating materials
NT: Insulating oil
RT: Polychlorinated biphenyls

Liquid lasers 744.3
DT: January 1993
UF: Lasers, Liquid*
BT: Lasers
RT: Dye lasers
Liquids

Liquid level indicators 943.3
DT: Predates 1975
BT: Gages
Indicators (instruments)
NT: Fuel gages
Rain gages

Liquid membrane electrodes (704.1) (801.4.1) (802.1)
DT: January 1993
UF: Electrodes, Electrochemical—Liquid membrane*
BT: Electrochemical electrodes
RT: Liquid membranes
Sensors

Liquid membranes (631.1) (802.1)
DT: January 1993
UF: Membranes—Liquid*
BT: Membranes
RT: Liquid membrane electrodes

Liquid metal cooled reactors 621.1
DT: January 1993
UF: Nuclear reactors, Liquid metal cooled*
BT: Nuclear reactors

Liquid metals 531.1
SN: Includes molten metals
DT: Predates 1975
UF: Heat transfer—Liquid metals*
Metals—Molten*
Molten metals
Pumps—Liquid metals*
BT: Liquids
RT: Mercury (metal)
Metals

*Liquid metals—Fluidity testing**
USE: Fluidity

Liquid methane 523, 804.1
DT: January 1993
UF: Methane—Liquid*
BT: Liquefied gases
Methane
RT: Liquefied natural gas

Liquid phase epitaxial growth
USE: Liquid phase epitaxy

Liquid phase epitaxy 933.1.2
DT: January 1995
UF: Liquid phase epitaxial growth
BT: Epitaxial growth
RT: Crystal growth from melt

Liquid propellants 523 (804)
DT: January 1993
UF: Propellants—Liquid*
BT: Liquid fuels
Propellants

Liquid rosin
USE: Tall oil

Liquid sloshing
USE: Fuel sloshing

Liquid sugar 822.3
DT: January 1993
UF: Sugar—Liquid*
BT: Liquids
Sugar (sucrose)

Liquid waves 931.1
DT: January 1993
UF: Liquids—Waves*
BT: Mechanical waves
NT: Water waves
RT: Fluid dynamics
Hydrodynamics
Liquids

* Former Ei Vocabulary term

Liquids (931.2)
SN: Very general term; prefer specific substance
DT: Predates 1975
UF: Heat transfer—Liquids*
BT: Fluids
NT: Boiling liquids
Cryogenic liquids
Liquefied gases
Liquid metals
Liquid sugar
Molten materials
Newtonian liquids
Non Newtonian liquids
Semiconducting liquids
Solutions
RT: Atomization
Bubble formation
Bubbles (in fluids)
Compressibility of liquids
Demulsification
Density of liquids
Dielectric properties of liquids
Diffusion in liquids
Drop breakup
Drop formation
Electric breakdown of liquids
Electric conductivity of liquids
Equations of state of liquids
Flow of fluids
Fluid mechanics
Foam control
Freezing
Hydrodynamics
Ionization of liquids
Liquid lasers
Liquid waves
Luminescence of liquids and solutions
Percolation (fluids)
Specific heat of liquids
Surface tension
Thermal conductivity of liquids
Thermal diffusion in liquids
Viscosity of liquids
Volume fraction
Water

*Liquids—Atomization**
USE: Atomization

*Liquids—Bubble formation**
USE: Bubble formation

*Liquids—Compressibility**
USE: Compressibility of liquids

*Liquids—Cryogenic**
USE: Cryogenic liquids

*Liquids—Demulsification**
USE: Demulsification

*Liquids—Density**
USE: Density of liquids

*Liquids—Dielectric properties**
USE: Dielectric properties of liquids

*Liquids—Diffusion**
USE: Diffusion in liquids

*Liquids—Drop breakup**
USE: Drop breakup

*Liquids—Drop formation**
USE: Drop formation

*Liquids—Electric breakdown**
USE: Electric breakdown of liquids

*Liquids—Electric conductivity**
USE: Electric conductivity of liquids

*Liquids—Foam control**
USE: Foam control

*Liquids—Ionization**
USE: Ionization of liquids

*Liquids—Newtonian**
USE: Newtonian liquids

*Liquids—Non Newtonian**
USE: Non Newtonian liquids

*Liquids—Percolation**
USE: Percolation (fluids)

*Liquids—Surface tension**
USE: Surface tension

*Liquids—Thermal conductivity**
USE: Thermal conductivity of liquids

*Liquids—Viscosity**
USE: Viscosity of liquids

*Liquids—Waves**
USE: Liquid waves

LISP (programming language) 723.1.1
DT: January 1993
UF: Computer programming languages—LISP*
BT: List processing languages

List processing languages 723.1.1
DT: January 1993
UF: Computer programming languages—List processing*
BT: High level languages
NT: LISP (programming language)
RT: Data handling

Listings (computer programs)
USE: Computer program listings

Literary computing
USE: Humanities computing

Lithium 542.4, 549.1
DT: January 1993
UF: Li
Lithium and alloys*
BT: Alkali metals
RT: Lithium alloys
Lithium batteries
Lithium compounds
Lithium deposits
Lithium metallography

Lithium alloys 542.4, 549.1
DT: January 1993
UF: Lithium and alloys*
BT: Alkali metal alloys
RT: Lithium
Lithium compounds

*Lithium and alloys**
USE: Lithium OR
Lithium alloys

Lithium batteries 702.1.1 (542.4) (549.1)
DT: January 1993
UF: Electric batteries—Lithium*
BT: Primary batteries
RT: Lithium

Lithium compounds (804.1) (804.2)
DT: Predates 1975
BT: Alkali metal compounds
RT: Lithium
Lithium alloys

Lithium deposits 504.1, 542.4, 549.1
DT: Predates 1975
BT: Ore deposits
RT: Lithium

Lithium metallography 531.2, 542.4, 549.1
DT: Predates 1975
BT: Metallography
RT: Lithium

Lithography (714.2) (745.1)
DT: January 1977
UF: Microlithography
NT: Electron beam lithography
Ion beam lithography
Photolithography
RT: Integrated circuit manufacture
Masks
Offset printing
Printed circuits
Semiconductor devices

*Lithography—Electron beam**
USE: Electron beam lithography

*Lithography—Ion beam**
USE: Ion beam lithography

*Lithography—Photolithography**
USE: Photolithography

*Lithography—X-ray**
USE: X ray lithography

Lithology 481.1
DT: January 1993
UF: Geology—Lithology*
BT: Geology
RT: Rocks

Living systems studies 461.1
DT: January 1993
UF: Biomedical engineering—Living systems studies*
BT: Experiments
RT: Biomedical engineering

LLDPE
USE: Linear low density polyethylenes

LNG
USE: Liquefied natural gas

Load control (presses)
USE: Press load control

Load limits 408.1 (401.1) (914.1)
DT: January 1993
UF: Bridges—Load limits*
BT: Safety factor
RT: Bearing capacity
Bridges
Structural loads
Wind stress

Load loss
USE: Electric load loss

Load management (electric)
USE: Electric load management

Load testing (422.2) (483.1)
DT: January 1993
UF: Soil load testing
Soils—Load testing*
BT: Materials testing
RT: Soil mechanics

Loaders 691.1 (502.2)
DT: January 1993
UF: Mines and mining—Loaders*
BT: Materials handling equipment
NT: Cutter loaders

Loading 691.2 (663.1) (672)
DT: January 1993
UF: Containers—Loading*
Materials handling—Loading*
Motor trucks—Loading*
Ships—Loading*
BT: Materials handling
RT: Charging (furnace)
Containers
Filling
Freight transportation
Ships
Transportation
Trucks
Turnaround time
Unloading

Loads (electric)
USE: Electric loads

Loads (forces) 408 (421)
DT: January 1993
NT: Dynamic loads
Structural loads
RT: Bearing capacity
Buildings
Mechanical properties
Stresses
Tensile strength

Loads (gas)
USE: Gas loads

* Former Ei Vocabulary term

Loads (structural)
USE: Structural loads

Lobes (distribution)
USE: Directional patterns (antenna)

LOCA
USE: Loss of coolant accidents

Local area networks (716) (717) (718) (723)
DT: January 1993
UF: Computer networks—Local area networks*
LAN
BT: Computer networks
Networks (circuits)
RT: Broadband networks
Data communication systems
Electric network analyzers

Location
SN: Of facilities
DT: January 1993
UF: Industrial plants—Location*
Pipelines—Location*
Plant location
NT: Relocation
RT: Industrial plants
Pipelines
Site selection

Locks (fasteners) (601.3)
DT: January 1993
UF: Door locks*
Locks*
BT: Fasteners
NT: Automobile door locks
Keyways
RT: Doors
Keys (for locks)

Locks (on waterways) 407.2 (434.1)
DT: January 1993
UF: Inland waterways—Locks*
BT: Hydraulic structures
RT: Canals
Inland waterways

*Locks**
USE: Locks (fasteners)

Locomotion (biped)
USE: Biped locomotion

Locomotive cranes (682.1.2) (693.1)
DT: January 1993
UF: Cranes—Locomotive*
Rail cranes
BT: Cranes
Railroad rolling stock

Locomotives 682.1.2
DT: Predates 1975
BT: Railroad rolling stock
NT: Diesel locomotives
Electric locomotives
Gas turbine locomotives
Industrial locomotives
Steam locomotives
RT: Axles
Bogies (railroad rolling stock)
Connecting rods
Driver training
Free piston compressors
Nuclear powered vehicles
Switching (rolling stock)
Traction (friction)

*Locomotives, Diesel**
USE: Diesel locomotives

*Locomotives, Electric**
USE: Electric locomotives

*Locomotives, Gas turbine**
USE: Gas turbine locomotives

*Locomotives—Axles**
USE: Axles

*Locomotives—Bogies**
USE: Bogies (railroad rolling stock)

*Locomotives—Connecting rods**
USE: Connecting rods

*Locomotives—Couplings**
USE: Couplings

*Locomotives—Crankshafts**
USE: Crankshafts

*Locomotives—Driver training**
USE: Driver training

*Locomotives—Free piston compressors**
USE: Free piston compressors

*Locomotives—Industrial**
USE: Industrial locomotives

*Locomotives—Mine**
USE: Mine locomotives

*Locomotives—Noise**
USE: Acoustic noise

*Locomotives—Nuclear power**
USE: Nuclear powered vehicles

*Locomotives—Sanding equipment**
USE: Sanding equipment

*Locomotives—Springs and suspensions**
USE: Vehicle springs OR
Vehicle suspensions

*Locomotives—Steam power**
USE: Steam locomotives

*Locomotives—Switchgear**
USE: Electric switchgear

*Locomotives—Traction**
USE: Traction (friction)

*Locomotives—Transmissions**
USE: Vehicle transmissions

*Locomotives—Valve gear**
USE: Valves (mechanical)

* Former Ei Vocabulary term

Logarithmic amplifiers 713.1
DT: January 1993
UF: Amplifiers, Logarithmic*
BT: Amplifiers (electronic)

Logging (forestry) (811.2) (821.3)
SN: Includes felling trees, cutting them into logs, and transporting logs
DT: January 1993
UF: Forestry—Logging*
Helicopters—Logging applications*
Logging*
Lumbering
BT: Forestry
RT: Lumber
Sawmills

Logging (wells)
USE: Well logging

*Logging**
USE: Logging (forestry)

Logic (formal)
USE: Formal logic

Logic CAD
USE: Computer aided logic design

Logic circuit elements
USE: Logic devices

Logic circuits 721.2 (721.3)
DT: Predates 1975
BT: Networks (circuits)
NT: Combinatorial circuits
Diode logic circuits
Emitter coupled logic circuits
Integrated injection logic circuits
NAND circuits
Sequential circuits
Transistor transistor logic circuits
RT: Adders
Carry logic
Computer circuits
Digital circuits
Flip flop circuits
Logic design
Logic devices
Logic gates
Majority logic
Many valued logics
Probabilistic logics
Pulse circuits
Shift registers
Switching circuits
Switching theory
Threshold logic
Trigger circuits

*Logic circuits, Combinatorial**
USE: Combinatorial circuits

*Logic circuits, Diode transistor**
USE: Diode transistor logic circuits

*Logic circuits, Diode**
USE: Diode logic circuits

*Logic circuits, Emitter coupled**
USE: Emitter coupled logic circuits

*Logic circuits, Integrated injection**
USE: Integrated injection logic circuits

*Logic circuits, NAND**
USE: NAND circuits

*Logic circuits, Sequential**
USE: Sequential circuits

*Logic circuits, Transistor transistor**
USE: Transistor transistor logic circuits

*Logic circuits—Design**
USE: Logic design

Logic design 721.2
DT: Predates 1975
UF: Logic circuits—Design*
BT: Design
NT: Computer aided logic design
RT: Application specific integrated circuits
Carry logic
Computer architecture
Digital arithmetic
Logic circuits
Logic devices
Minimization of switching nets
Switching theory

*Logic design—Computer aids**
USE: Computer aided logic design

Logic devices 721.2
DT: Predates 1975
UF: Logic circuit elements
BT: Equipment
NT: Fluidic logic devices
Logic gates
Magnetic logic devices
Threshold elements
RT: Logic circuits
Logic design

*Logic devices—Current mode**
USE: Emitter coupled logic circuits

*Logic devices—Fluidic elements**
USE: Fluidic logic devices

*Logic devices—Gates**
USE: Logic gates

*Logic devices—Magnetic elements**
USE: Magnetic logic devices

*Logic devices—Threshold elements**
USE: Threshold elements

Logic gates 721.2 (721.3)
DT: January 1993
UF: Logic devices—Gates*
BT: Logic devices
RT: Logic circuits

Logic programming 723.1 (721.2)
DT: January 1993
UF: Computer programming—Logic programming*
Rule based programming

* Former Ei Vocabulary term

Logic programming *(continued)*
BT: Computer programming
RT: Artificial intelligence
PROLOG (programming language)

Logistics 404.1
DT: January 1993
UF: Military engineering—Logistics*
BT: Military operations
RT: Military engineering

Long distance telephone systems 718.1
DT: January 1993
UF: Telephone systems—Long distance*
BT: Telephone systems

Longwall mining 502.2
DT: January 1993
UF: Mines and mining—Longwall*
BT: Mining

Lookup tables
USE: Table lookup

Looms 819.6
DT: January 1993
UF: Textile machinery—Looms*
BT: Textile machinery
RT: Weaving

Loop antennas (716)
DT: January 1993
UF: Antennas, Loop*
BT: Directive antennas

Loss (energy)
USE: Energy dissipation

Loss of coolant accidents 914.1 (621.1) (621.2)
DT: January 1993
UF: Coolant loss accidents
LOCA
Nuclear reactors—Loss of coolant accident*
BT: Nuclear reactor accidents
RT: Nuclear reactors

Loss of load
USE: Electric load loss

Losses
SN: Of material and equipment. For electric current losses, use ELECTRIC LOSSES; for heat loss use HEAT LOSSES; for magnetic losses use MAGNETIC LEAKAGE
DT: January 1993
UF: Oil tanks—Losses*
Petroleum industry—Losses*
Ships—Losses*

Losses (electric)
USE: Electric losses

Losses (magnetic)
USE: Magnetic leakage

Loudspeakers 752.1, 752.3.1
DT: Predates 1975
BT: Acoustic generators
Audio equipment
Electroacoustic transducers
NT: Electrostatic loudspeakers
RT: Radio equipment
Radio receivers
Sound reproduction

*Loudspeakers—Electrostatic actuation**
USE: Electrostatic loudspeakers

Low density polyethylenes 815.1.1
DT: January 1993
UF: LDPE
Polyethylenes—Low density*
BT: Polyethylenes
NT: Linear low density polyethylenes

Low energy electron diffraction 931.3, 932.1
DT: January 1995
UF: LEED
BT: Electron diffraction
RT: Electron energy levels
Surface structure
Surfaces

Low grade fuel firing 521.1 (821.5)
SN: Includes bagasse, sulfite pulp, bark, cane, corn cobs, and wastes
DT: January 1993
UF: Boiler firing—Low grade fuel*
BT: Boiler firing
RT: Bagasse
Renewable energy resources
Wastes

Low level languages
USE: Machine oriented languages

Low pass filters 703.2
DT: January 1993
UF: Electric filters, Low pass*
BT: Electric filters

Low permeability reservoirs 512.1 (481.1)
DT: January 1995
RT: Mechanical permeability
Petroleum reservoirs
Underground reservoirs

Low shaft blast furnaces 532.2
DT: January 1993
UF: Blast furnaces—Low shaft*
BT: Blast furnaces

Low temperature bearings
USE: Bearings (machine parts) AND Cryogenic equipment

Low temperature coal carbonization 802.2 (522) (524) (644.5)
DT: January 1993
UF: Coal carbonization—Low temperature*
BT: Coal carbonization
Low temperature operations

Low temperature drying 802.3
DT: January 1993
UF: Drying—Low temperature*
BT: Drying
Low temperature operations

* Former Ei Vocabulary term

Low temperature effects (644.4)
DT: January 1993
BT: Effects
NT: Cold effects
RT: Low temperature engineering
Low temperature operations
Low temperature production

Low temperature engineering (644.4)
DT: Predates 1975
BT: Engineering
RT: Arctic engineering
Cryogenics
Cryostats
Low temperature operations
Low temperature production
Low temperature effects
Low temperature phenomena
Low temperature properties

*Low temperature engineering—Low temperature production**
USE: Low temperature production

Low temperature equipment
USE: Cryogenic equipment

Low temperature operations (802.2) (802.3)
DT: January 1993
UF: Heat treatment—Low temperature*
NT: Low temperature coal carbonization
Low temperature drying
RT: Cryogenics
Low temperature engineering
Low temperature effects
Low temperature phenomena
Low temperature testing

Low temperature phenomena (644.4)
DT: January 1993
UF: Cryogenic phenomena
Cryogenics—Low temperature phenomena*
RT: Cold effects
Cryogenics
Low temperature engineering
Low temperature operations
Low temperature production

Low temperature production (644.4)
DT: January 1993
UF: Low temperature engineering—Low temperature production*
BT: Production
RT: Low temperature engineering
Low temperature effects
Low temperature phenomena

Low temperature properties (931.2)
DT: January 1993
BT: Physical properties
RT: Cold effects
Low temperature engineering

Low temperature testing
DT: January 1993
UF: Materials testing—Low temperature*
BT: Testing
RT: Equipment testing
Low temperature operations
Materials testing
Plasticity testing
Tensile testing
Thermal conductivity
Thermodynamic properties
Viscoelasticity

LPG
USE: Liquefied petroleum gas

LSI circuits 714.2
DT: January 1993
UF: Integrated circuits, LSI*
Large scale integration
BT: Monolithic integrated circuits
NT: VLSI circuits
RT: Integrated injection logic circuits
Microprocessor chips

Lubricants 607.1
DT: Predates 1975
BT: Materials
NT: Lubricating greases
Lubricating oils
Solid lubricants
Synthetic lubricants
RT: Additives
Kerogen
Lubrication
Lubricators
Maintenance
Metal cutting
Metal drawing
Metal forming
Petroleum products

*Lubricants—Dispensers**
USE: Dispensers

*Lubricants—Metal cutting**
USE: Metal cutting

*Lubricants—Metal drawing**
USE: Metal drawing

*Lubricants—Metal forming**
USE: Metal forming

*Lubricants—Reclamation**
USE: Reclamation

*Lubricants—Solid films**
USE: Solid lubricants

*Lubricants—Synthetic products**
USE: Synthetic lubricants

Lubricating greases 607.1
DT: Predates 1975
UF: Greases
BT: Lubricants
RT: Lubricating oils

Lubricating oils 607.1
DT: Predates 1975
BT: Lubricants
RT: Lubricating greases

* Former Ei Vocabulary term

Lubricating oils *(continued)*
Oils and fats
Petroleum products

Lubrication 607.2
DT: Predates 1975
NT: Air lubrication
Elastohydrodynamic lubrication
RT: Bearings (machine parts)
Gas lubricated bearings
Impregnation
Lubricants
Lubricators
Self lubricating composites
Tribology
Viscosity of liquids
Wear of materials

*Lubrication—Elastohydrodynamic**
USE: Elastohydrodynamic lubrication

Lubricators 601.2 (607.2)
DT: Predates 1975
BT: Equipment
RT: Dispensers
Lubricants
Lubrication

Lumber 811.2 (821.4)
DT: Predates 1975
BT: Wood products
RT: Bark stripping
Logging (forestry)
Sawmills
Wood

Lumber mills
USE: Sawmills

*Lumber—Grading**
USE: Grading

*Lumber—Handling**
USE: Materials handling

Lumbering
USE: Logging (forestry)

Luminaires
USE: Lighting fixtures

Luminescence 741.1
DT: Predates 1975
UF: Pigments—Luminescence*
BT: Light emission
NT: Cathodoluminescence
Chemiluminescence
Electroluminescence
Fluorescence
Luminescence of gases
Luminescence of liquids and solutions
Luminescence of solids
Phosphorescence
Photoluminescence
Sonoluminescence
Thermoluminescence
Triboluminescence
RT: Luminescent devices
Phosphors
Pigments

Luminescence of gases 741.1, 931.2
DT: January 1993
UF: Luminescence—Gases*
BT: Luminescence
RT: Gases

Luminescence of inorganic solids 741.1, 804.2 (931.2)
DT: January 1993
UF: Luminescence—Inorganic solids*
BT: Luminescence of solids
RT: Inorganic compounds

Luminescence of liquids and solutions 741.1, 931.2
DT: January 1993
UF: Luminescence—Liquids and solutions*
BT: Luminescence
RT: Liquids
Solutions

Luminescence of organic solids 741.1, 804.1, 931.2
DT: January 1993
UF: Luminescence—Organic solids*
BT: Luminescence of solids
RT: Organic compounds

Luminescence of solids 741.1, 931.2
DT: January 1993
UF: Luminescence—Solids*
BT: Luminescence
NT: Luminescence of inorganic solids
Luminescence of organic solids
RT: Solids

*Luminescence—Gases**
USE: Luminescence of gases

*Luminescence—Inorganic solids**
USE: Luminescence of inorganic solids

*Luminescence—Liquids and solutions**
USE: Luminescence of liquids and solutions

*Luminescence—Organic solids**
USE: Luminescence of organic solids

*Luminescence—Solids**
USE: Luminescence of solids

Luminescent devices 741.3
DT: Predates 1975
UF: Electroluminescent devices
BT: Light sources
NT: Fluorescent lamps
Fluorescent screens
Scintillation counters
RT: Luminescence
Phosphors

Luminophors
USE: Phosphors

Luminors
USE: Phosphors

* Former Ei Vocabulary term

Luminous coatings 813.2 (741.3)
DT: January 1993
UF: Coatings—Luminous*
BT: Coatings
Luminous materials
NT: Luminous paint
RT: Luminous railings

Luminous materials (741.3) (813.2)
DT: January 1993
BT: Materials
NT: Luminous coatings

Luminous paint 813.2 (741.3)
DT: January 1993
UF: Paint—Luminous*
BT: Luminous coatings
Paint

Luminous railings (401.1) (741.3) (813.2)
DT: January 1993
UF: Bridges—Luminous railings*
BT: Railings
RT: Luminous coatings

Lumped parameter networks 703.1
DT: January 1993
UF: Electric networks, Lumped parameter*
BT: Networks (circuits)
RT: Circuit theory

Lunar bases
USE: Moon bases

Lunar exploration
USE: Lunar missions

Lunar landing (655.1) (656.1) (657.2)
DT: January 1993
UF: Spacecraft—Lunar landing*
BT: Spacecraft landing
RT: Lunar missions
Moon
Moon bases
Planetary landers
Spacecraft

Lunar missions (655.1) (656.1) (657.2)
DT: January 1993
UF: Lunar exploration
Spacecraft—Lunar missions*
RT: Lunar landing
Lunar surface analysis
Moon
Moon bases
Orbits
Space flight
Spacecraft
Trajectories

Lunar surface analysis (657.2) (801)
DT: January 1993
UF: Moon surface analysis
Moon—Surface analysis*
Surface analysis (moon)
BT: Analysis
RT: Chemical analysis
Lunar missions
Moon

Lung diseases
USE: Pulmonary diseases

Lutetium 547.2
DT: January 1981
BT: Rare earth elements
RT: Lutetium alloys
Lutetium compounds

Lutetium alloys 547.2
DT: January 1993
BT: Rare earth alloys
RT: Lutetium

Lutetium compounds (804.1) (804.2)
DT: June 1990
BT: Rare earth compounds
RT: Lutetium

Lyapunov methods (731.1) (921)
DT: January 1993
UF: System stability—Lyapunov methods*
RT: System stability

Lysimeters 943.3 (483.1) (631.1)
SN: For measuring percolation of water through soils
DT: Predates 1975
BT: Flow measuring instruments
RT: Flow measurement
Mechanical permeability
Percolation (fluids)
Permeameters (mechanical permeability)

Machinability (604.2)
DT: January 1993
UF: Metals and alloys—Machinability*
BT: Formability
RT: Alloys
Hardness
Machining
Metals

Machine components 601.2
DT: October 1975
UF: Machine parts
BT: Components
Machinery
NT: Bearings (machine parts)
Belts
Cams
Camshafts
Disks (machine components)
Drives
Engine pistons
Felts
Gears
Headboxes
Rollers (machine components)
Rolls (machine components)
Shafts (machine components)
Splines
Starters
Universal joints
RT: Clutches

* Former Ei Vocabulary term

Machine design 601
DT: Predates 1975
BT: Design
RT: Engineering
Machinery

Machine learning
USE: Learning systems

Machine oriented languages 723.1.1
DT: January 1993
UF: Computer programming languages—Machine orientation*
Low level languages
BT: Computer programming languages

Machine parts
USE: Machine components

Machine shop practice 604.2
DT: January 1987
RT: Machine shops
Machining
Metal cutting
Metal forming

Machine shops 604.2 (402.1)
DT: Predates 1975
BT: Industrial plants
RT: Machine shop practice
Machine tools
Machining

Machine theory
USE: Electric machine theory

Machine tool attachments 603.2
DT: January 1993
UF: Machine tools—Attachments*
BT: Accessories
Machine tools

Machine tool drives 602.1
DT: January 1993
UF: Electric machine tool drives
Mechanical machine tool drives
BT: Drives
Machine tools
RT: Electric drives
Mechanical drives

Machine tools 603.1
DT: Predates 1975
UF: Machine tools—Columns*
BT: Machinery
Tools
NT: Boring machines (machine tools)
Broaching machines
Buffing machines
Contour followers
Dies
Drilling machines (machine tools)
Gear cutting machines
Glass cutting machines
Grinding machines
Honing machines
Lapping machines
Lathes
Lead screws
Machine tool attachments
Machine tool drives
Machining centers
Metal working tools
Milling machines
Oxygen cutting machines
Planers
Polishing machines
Punch presses
Riveting machines
Shearing machines
Slideways
Slitting machines
Slotting machines
Special purpose machine tools
Ultrasonic machine tools
Welding machines
Woodworking machinery
RT: Alignment
Antivibration mountings
Boring tools
Feeding
Finishing
Gears
Guards (shields)
Indexing (materials working)
Machine shops
Machining
Obsolescence
Saws
Stiction

*Machine tools—Alignment**
USE: Alignment

*Machine tools—Antivibration mountings**
USE: Antivibration mountings

*Machine tools—Attachments**
USE: Machine tool attachments

*Machine tools—Columns**
USE: Machine tools

*Machine tools—Contour followers**
USE: Contour followers

*Machine tools—Depreciation**
USE: Depreciation

*Machine tools—Electric drive**
USE: Electric drives

*Machine tools—Feeding**
USE: Feeding

*Machine tools—Finishing**
USE: Finishing

*Machine tools—Gears**
USE: Gears

*Machine tools—Glass cutting**
USE: Glass cutting machines

*Machine tools—Guards**
USE: Guards (shields)

* Former Ei Vocabulary term

*Machine tools—Indexing**
USE: Indexing (materials working)

*Machine tools—Joints**
USE: Joints (structural components)

*Machine tools—Lead screws**
USE: Lead screws

*Machine tools—Machining centers**
USE: Machining centers

*Machine tools—Mechanical drive**
USE: Mechanical drives

*Machine tools—Obsolescence**
USE: Obsolescence

*Machine tools—Slideways**
USE: Slideways

*Machine tools—Special purpose**
USE: Special purpose machine tools

*Machine tools—Stiction**
USE: Stiction

*Machine tools—Ultrasonic**
USE: Ultrasonic machine tools

Machine vibrations 601.3 (931.1)
DT: January 1977
UF: Machinery—Vibrations*
BT: Vibrations (mechanical)
RT: Antivibration mountings
Machinery

Machine vision
USE: Computer vision

Machine windings (704.1) (705)
DT: January 1993
UF: Electric machinery—Windings*
Electric windings, Machine*
BT: Electric machinery components
Electric windings
NT: Rotors (windings)
Stators
RT: AC generators
AC machinery
AC motors
Asynchronous generators
Asynchronous machinery
Electric equipment protection

Machinery
DT: Predates 1975
UF: Industrial plants—Machinery*
NT: Agricultural machinery
Assembly machines
Balancing machines
Business machines
Calenders
Closed cycle machinery
Compressors
Crushers
Earthmoving machinery
Electric machinery
Elevators
Engines
Escalators
Extruders
Fans
Glass forming machines
Granulators
Grinding mills
Hydraulic machinery
Laminating machinery
Lawn mowers
Machine components
Machine tools
Marking machines
Mechanisms
Metal forming machines
Mining machinery
Mixers (machinery)
Motion picture editing machines
Motors
Packaging machines
Papermaking machinery
Plastics machinery
Printing machinery
Refrigerating machinery
Roadbuilding machinery
Rotating machinery
Rubber machinery
Spinning machines
Striping machines
Textile machinery
Tunneling machines
Turbomachinery
Vending machines
Voting machines
Washing machines
Winding machines
RT: Alignment
Antivibration mountings
Guards (shields)
Industrial plants
Leveling (machinery)
Lighting
Machine design
Machine vibrations
Machinery exhibitions
Materials handling equipment
Mechanization
Modernization
Obsolescence
Plastic parts

Machinery exhibitions (911.4)
DT: Predates 1975
BT: Exhibitions
RT: Aircraft exhibitions
Automobile exhibitions
Machinery

Machinery protection
USE: Electric equipment protection

*Machinery—Alignment**
USE: Alignment

*Machinery—Antivibration mountings**
USE: Antivibration mountings

*Machinery—Depreciation**
USE: Depreciation

*Machinery—Gears**
USE: Gears

*Machinery—Guards**
USE: Guards (shields)

*Machinery—Hydraulic**
USE: Hydraulic machinery

*Machinery—Installations**
USE: Installation

*Machinery—Leveling**
USE: Leveling (machinery)

*Machinery—Lighting**
USE: Lighting

*Machinery—Modernization**
USE: Modernization

*Machinery—Noise**
USE: Acoustic noise

*Machinery—Plastics parts**
USE: Plastic parts

*Machinery—Power supply**
USE: Electric power supplies to apparatus

*Machinery—Rotating**
USE: Rotating machinery

*Machinery—Vibrations**
USE: Machine vibrations

Machining 604.2
DT: January 1993
NT: Grinding (machining)
Lapping
Micromachining
Milling (machining)
Reaming
Thread cutting
Turning
RT: Boring
Broaching
Cutting
Machinability
Machine shop practice
Machine shops
Machine tools
Metallurgy
Thread rolling
Trimming

Machining centers 603.1
DT: January 1993
UF: Machine tools—Machining centers*
BT: Machine tools

Macromolecules 815.1, 931.3 (801.3) (804.1)
DT: January 1995
BT: Molecules
NT: Nucleic acids
RT: Colloids
Gels
Organic compounds
Polymers
Proteins

Macros 723.1
DT: January 1993
UF: Computer programming—Macros*
BT: Computer software
RT: Computer programming
Subroutines

Made for television motion pictures
USE: Motion pictures

Mag-lev vehicles
USE: Magnetic levitation vehicles

Maglev vehicles
USE: Magnetic levitation vehicles

Magnesia 804.2 (812.2)
DT: Predates 1975
UF: Magnesium oxide
BT: Magnesium compounds
Oxides
RT: Magnesia refractories

Magnesia refractories 812.2 (804.2)
DT: January 1993
UF: Refractory materials—Magnesia*
BT: Refractory materials
RT: Magnesia
Magnesium compounds

Magnesite 482.2 (812.2)
DT: Predates 1975
BT: Carbonate minerals
Magnesium compounds
RT: Magnesite mines
Magnesite ore treatment
Magnesite refractories

Magnesite mines 504.1 (482.2)
DT: January 1993
UF: Magnesite mines and mining*
BT: Mines
RT: Magnesite
Magnesium

*Magnesite mines and mining**
USE: Magnesite mines

Magnesite ore treatment 533.1
DT: Predates 1975
BT: Ore treatment
RT: Magnesite
Magnesium
Magnesium metallurgy

Magnesite refractories 812.2
DT: January 1993
UF: Refractory materials—Magnesite*
BT: Refractory materials
RT: Magnesite
Magnesium compounds

Magnesium 542.2, 549.2
DT: January 1993
UF: Magnesium and alloys*
Mg

Magnesium *(continued)*
Tools, jigs and fixtures—Magnesium*
BT: Alkaline earth metals
Light metals
RT: Fire hazards
Magnesite mines
Magnesite ore treatment
Magnesium alloys
Magnesium castings
Magnesium compounds
Magnesium deposits
Magnesium foundry practice
Magnesium metallurgy
Magnesium metallography
Magnesium powder
Magnesium printing plates

Magnesium alloys 542.2, 549.2
DT: January 1993
UF: Magnesium and alloys*
BT: Alkaline earth metal alloys
RT: Magnesium
Magnesium compounds

*Magnesium and alloys**
USE: Magnesium OR
Magnesium alloys

*Magnesium and alloys—Fire hazards**
USE: Fire hazards

Magnesium castings 534.2 (542.2) (549.2)
DT: Predates 1975
BT: Metal castings
RT: Magnesium
Magnesium metallurgy

Magnesium compounds (804.1) (804.2)
DT: Predates 1975
BT: Alkaline earth metal compounds
NT: Chlorite minerals
Magnesia
Magnesite
RT: Magnesia refractories
Magnesite refractories
Magnesium
Magnesium alloys

Magnesium deposits 504.1, 542.2, 549.2
DT: Predates 1975
BT: Ore deposits
RT: Magnesium

Magnesium foundry practice
534.2 (542.2) (549.2)
DT: Predates 1975
BT: Foundry practice
Magnesium metallurgy
RT: Magnesium

Magnesium metallography 531.2 (542.2) (549.2)
DT: Predates 1975
BT: Metallography
RT: Magnesium
Magnesium metallurgy

Magnesium metallurgy 531.1 (542.2) (549.2)
DT: Predates 1975
BT: Metallurgy
NT: Magnesium foundry practice
Magnesium powder metallurgy
RT: Magnesite ore treatment
Magnesium
Magnesium castings
Magnesium metallography

Magnesium oxide
USE: Magnesia

Magnesium powder 542.2, 549.2 (536.1)
DT: Predates 1975
BT: Powder metals
RT: Magnesium
Powder metallurgy

Magnesium powder metallurgy
(536) (542.2) (549.2) (804)
DT: January 1993
UF: Powder metallurgy—Magnesium*
BT: Magnesium metallurgy
Powder metallurgy

Magnesium printing plates
745.1.1 (542.2) (549.2)
DT: January 1993
UF: Magnesium*
Printing plates—Magnesium*
BT: Printing plates
RT: Magnesium

*Magnesium**
USE: Magnesium printing plates

Magnetic amplifiers 704.1 (708.4)
DT: January 1993
UF: Amplifiers, Magnetic*
BT: Magnetic devices
Power amplifiers
NT: Rotating magnetic amplifiers
RT: Saturable core reactors

Magnetic anisotropy 701.2, 931.2
DT: January 1995
BT: Anisotropy
Magnetic properties

Magnetic bearings 601.2 (708.4)
DT: January 1993
UF: Bearings—Magnetic*
BT: Bearings (machine parts)
Magnetic devices

Magnetic bubble devices (708.4) (722.1)
DT: January 1993
UF: Magnetic devices—Bubbles*
BT: Magnetic devices
NT: Magnetic bubble memories
RT: Magnetic bubbles
Magnetic thin film devices

Magnetic bubble memories 722.1 (708.4)
DT: January 1993
UF: Bubble memories
Data storage, Magnetic—Bubbles*
BT: Magnetic bubble devices
Magnetic storage

Magnetic bubble memories *(continued)*
Nonvolatile storage
RT: Magnetic bubbles
Magnetic film storage
Magnetic thin film devices

Magnetic bubbles 701.2 (708.4)
DT: January 1993
UF: Bubbles (magnetic)
Magnetic materials—Bubbles*
RT: Magnetic bubble memories
Magnetic bubble devices
Magnetic devices
Magnetic domains
Magnetic materials
Magnetization

Magnetic circuits 701.2
DT: Predates 1975
BT: Networks (circuits)
NT: Transformer magnetic circuits
RT: Electric coils
Electric windings
Magnetic cores
Magnetic fields
Magnets

Magnetic core storage 722.1 (708.4)
DT: January 1993
UF: Core storage
Data storage, Magnetic—Core*
BT: Magnetic storage
RT: Magnetic cores

Magnetic cores (708.4) (722.1)
SN: Ferrous material used in coils and transformers to provide a better path than air for magnetic flux
DT: Predates 1975
UF: Transfluxors
BT: Magnetic materials
NT: Powder magnetic cores
RT: Electric coils
Electric transformers
Electromagnets
Ferrite devices
Ferromagnetic materials
Magnetic circuits
Magnetic core storage
Magnetic devices
Magnets
Saturable core reactors

*Magnetic cores, Powder**
USE: Powder magnetic cores

Magnetic couplings 602 (701.2) (708.4)
DT: January 1993
UF: Couplings—Magnetic*
BT: Couplings

Magnetic data recording
USE: Data recording

Magnetic data storage
USE: Magnetic storage

Magnetic devices (704) (708.4) (714)
DT: Predates 1975
BT: Equipment
NT: Electromagnetic pumps
Ferrite devices
Hall effect devices
Magnetic amplifiers
Magnetic bearings
Magnetic bubble devices
Magnetic filters
Magnetic flowmeters
Magnetic lenses
Magnetic logic devices
Magnetic scales
Magnetic separators
Magnetic storage
Magnetic thin film devices
Magnetooptical devices
Magnetos
Magnetostrictive devices
Magnetrons
Magnets
RT: Magnetic bubbles
Magnetic cores
Magnetic materials
Magnetic rubber products
Solid state devices

*Magnetic devices, Thin film**
USE: Magnetic thin film devices

*Magnetic devices—Bubbles**
USE: Magnetic bubble devices

Magnetic dipole moments
USE: Magnetic moments

Magnetic disk storage 722.1 (708.4)
DT: January 1993
UF: Computer peripheral equipment—Disk drives*
Data storage, Magnetic—Disk*
Direct access storage
Disk storage (magnetic)
BT: Magnetic storage
Nonvolatile storage
NT: Floppy disk storage
Hard disk storage
RT: Magnetic recording

Magnetic domains 701.2
DT: January 1993
UF: Domains (magnetic)
Magnetic materials—Domains*
RT: Magnetic bubbles
Magnetic materials
Magnetization

Magnetic field effects 701.2
DT: Predates 1975
UF: Spacecraft—Magnetic effects*
BT: Effects
NT: Electromagnetic field effects
Galvanomagnetic effects
Magnetic levitation
Magnetic separation
Magnetoacoustic effects

* Former Ei Vocabulary term

Magnetic field effects *(continued)*
Magnetoelectric effects
Magnetohydrodynamics
Magnetooptical effects
Magnetostriction
Photoelectromagnetic effects
RT: Magnetic fields
Magnetic properties
Magnetic variables measurement
Particle optics
Superconductivity

Magnetic field measurement 942.4
DT: Predates 1975
BT: Magnetic variables measurement
RT: Field plotting
Fluxmeters
Magnetic fields
Magnetometers

Magnetic fields 701.2
DT: Predates 1975
RT: Demagnetization
Electromagnetic fields
Magnetic circuits
Magnetic field effects
Magnetic field measurement
Magnetic lenses
Magnetic levitation
Magnetic shielding
Magnetism
Magnetization
Magnetostatics
Magnets
Maxwell equations
Pinch effect
Plasma confinement

Magnetic film storage 722.1 (708.4)
DT: January 1993
UF: Data storage, Magnetic—Film*
Magnetic thin film storage
Thin film storage
BT: Magnetic storage
Magnetic thin film devices
Nonvolatile storage
NT: Plated wire storage
RT: Magnetic bubble memories

Magnetic films 708.4
DT: January 1993
BT: Films
Magnetic materials
NT: Magnetic thick films
Magnetic thin films

Magnetic filters 802.1 (701.2) (708.4)
DT: January 1993
UF: Filters—Magnetic*
BT: Filters (for fluids)
Magnetic devices

Magnetic flowmeters 943.1 (708.4)
DT: January 1993
UF: Flowmeters—Magnetic*
BT: Flowmeters
Magnetic devices

Magnetic fluids 708.4
DT: June 1990
BT: Fluids
Magnetic materials

Magnetic flux measuring instruments
USE: Fluxmeters

Magnetic forming 535.2, 701.2
DT: January 1993
UF: Metal forming—Magnetic*
BT: Metal forming
RT: Magnetism

Magnetic heads (708.4) (722.1) (752)
DT: June 1990
UF: Heads (magnetic)
Magnetic read/write heads
BT: Magnetic storage
RT: Magnetic recording

Magnetic hysteresis 701.2
DT: January 1993
BT: Hysteresis
RT: Coercive force
Magnetic measuring instruments
Magnetic permeability
Magnetization
Remanence

Magnetic laminates 708.4
DT: January 1993
UF: Magnetic materials—Laminations*
BT: Laminates
Magnetic materials

Magnetic leakage 701.2 (708.4)
DT: Predates 1975
UF: Leakage (magnetic)
Losses (magnetic)
Magnetic losses
RT: Electromagnetism
Magnetic shielding
Magnetism

Magnetic lenses (701.2) (932.1)
DT: Predates 1975
BT: Lenses
Magnetic devices
RT: Cathode ray tubes
Electron beams
Electron lenses
Electron optics
Focusing
Ion beams
Magnetic fields
Particle optics
Plasma guns

Magnetic levitation 701.2 (708.4)
DT: January 1977
BT: Magnetic field effects
RT: Magnetic fields
Magnetic levitation vehicles

* Former Ei Vocabulary term

Magnetic levitation vehicles (682) (701.2) (708.4)
DT: January 1993
UF: Mag-lev vehicles
Maglev vehicles
Magnetic suspension vehicles
Vehicles—Magnetic levitation*
BT: Ground vehicles
RT: Magnetic levitation
Railroads
Transportation

Magnetic logic devices 721.2 (708.4)
DT: January 1993
UF: Logic devices—Magnetic elements*
BT: Logic devices
Magnetic devices

Magnetic losses
USE: Magnetic leakage

Magnetic materials 708.4
SN: Use more general terms, such as MAGNETISM, for theoretical papers.
DT: Predates 1975
BT: Materials
NT: Antiferromagnetic materials
Ferrimagnetic materials
Ferromagnetic materials
Magnetic cores
Magnetic films
Magnetic fluids
Magnetic laminates
Magnetic semiconductors
Magnetic tape
Magnetic thin films
RT: Coercive force
Diamagnetism
Electromagnetism
Ferrimagnetism
Ferromagnetism
Garnets
Magnetic bubbles
Magnetic devices
Magnetic domains
Magnetic permeability
Magnetic properties
Magnetic rubber products
Magnetic variables measurement
Magnetism
Magnetostriction
Magnets
Metamagnetism
Paramagnetism
Remanence

*Magnetic materials—Antiferromagnetism**
USE: Antiferromagnetic materials

*Magnetic materials—Bubbles**
USE: Magnetic bubbles

*Magnetic materials—Coercivity**
USE: Coercive force

*Magnetic materials—Diamagnetism**
USE: Diamagnetism

*Magnetic materials—Domains**
USE: Magnetic domains

*Magnetic materials—Ferrimagnetism**
USE: Ferrimagnetic materials

*Magnetic materials—Ferromagnetism**
USE: Ferromagnetic materials

*Magnetic materials—Laminations**
USE: Magnetic laminates

*Magnetic materials—Magnetostriction**
USE: Magnetostriction

*Magnetic materials—Measurements**
USE: Magnetic variables measurement

*Magnetic materials—Metamagnetism**
USE: Metamagnetism

*Magnetic materials—Paramagnetism**
USE: Paramagnetism

*Magnetic materials—Remanence**
USE: Remanence

*Magnetic materials—Thin films**
USE: Magnetic thin films

*Magnetic measurements**
USE: Magnetic variables measurement

*Magnetic measurements—Permeability**
USE: Magnetic permeability measurement

*Magnetic measurements—Relaxation**
USE: Magnetic relaxation measurement

*Magnetic measurements—Resonance**
USE: Magnetic resonance measurement

Magnetic measuring instruments 942.3
SN: For measuring magnetic quantities
DT: Predates 1975
UF: Hysteresigraphs
BT: Instruments
NT: Fluxmeters
Magnetometers
Permeameters (magnetic permeability)
RT: Magnetic hysteresis
Magnetic variables measurement

Magnetic merging
USE: Electromagnetic wave propagation in plasma

Magnetic moments 701.2
DT: January 1995
UF: Magnetic dipole moments
RT: Magnetic properties
Magnetic resonance
Magnetism
Magnetization

Magnetic ordering
USE: Magnetization

Magnetic permeability 701.2
DT: January 1993
UF: Magnetic susceptibility
Permeability (magnetic)
Permeability, Magnetic*

* Former Ei Vocabulary term

Magnetic permeability *(continued)*
BT: Magnetic properties
RT: Magnetic hysteresis
Magnetic materials
Magnetic permeability measurement

Magnetic permeability measurement 942.4
DT: January 1993
UF: Magnetic measurements—Permeability*
BT: Magnetic variables measurement
RT: Magnetic permeability
Permeameters (magnetic permeability)

Magnetic permeability permeameters
USE: Permeameters (magnetic permeability)

Magnetic properties 701.2
DT: Predates 1975
BT: Physical properties
NT: Magnetic anisotropy
Magnetic permeability
RT: Coercive force
Demagnetization
Magnetic field effects
Magnetic materials
Magnetic moments
Magnetic relaxation
Magnetic resonance
Magnetic variables measurement
Magnetic variables control
Magnetism
Magnetization
Magnetoelectric effects
Magnetooptical effects

Magnetic prospecting 481.4, 701.2, 708.4 (942.3)
DT: January 1993
UF: Geophysical prospecting—Magnetic*
BT: Geophysical prospecting
RT: Magnetic variables measurement

Magnetic read/write heads
USE: Magnetic heads

Magnetic recording (708.4) (722.1) (752.2)
DT: January 1995
BT: Recording
RT: Data recording
Magnetic disk storage
Magnetic heads
Magnetic tape
Tape recorders

Magnetic relaxation 701.2
DT: January 1993
BT: Relaxation processes
RT: Magnetic properties
Magnetic relaxation measurement
Magnetic resonance
Magnetism

Magnetic relaxation measurement 942.4
DT: January 1993
UF: Magnetic measurements—Relaxation*
BT: Magnetic variables measurement
RT: Magnetic relaxation

Magnetic resonance 701.2
DT: Predates 1975
BT: Resonance
NT: Ferrimagnetic resonance
Ferromagnetic resonance
Nuclear magnetic resonance
Paramagnetic resonance
RT: Electromagnetic wave absorption
Magnetic moments
Magnetic properties
Magnetic relaxation
Magnetic resonance imaging
Magnetic resonance spectrometers
Magnetic resonance measurement
Magnetic resonance spectroscopy
Radiofrequency spectroscopy

Magnetic resonance imaging
701.2 (461.1) (711.1) (932.2)
DT: June 1990
UF: MRI
BT: Imaging techniques
RT: Biomedical engineering
Computerized tomography
Diagnosis
Magnetic resonance
Nuclear magnetic resonance
Nuclear magnetic resonance spectroscopy

Magnetic resonance measurement 942.4
DT: January 1993
UF: Magnetic measurements—Resonance*
BT: Magnetic variables measurement
RT: Magnetic resonance

Magnetic resonance spectrometers 942.3
SN: Spectrometers for investigating magnetic-resonance phenomena. The frequency of radio excitation may extend into the microwave region
DT: January 1993
UF: Spectrometers, Magnetic resonance*
BT: Spectrometers
RT: Magnetic resonance
Magnetic resonance spectroscopy
Microwave spectrometers
Nuclear magnetic resonance spectroscopy

Magnetic resonance spectroscopy
(701.2) (801) (931.3) (932.2)
DT: January 1995
BT: Spectroscopy
NT: Nuclear magnetic resonance spectroscopy
RT: Magnetic resonance spectrometers
Magnetic resonance
Microwave spectroscopy

Magnetic rubber products 708.4, 818.5
DT: January 1993
UF: Rubber products—Magnetic*
BT: Rubber products
RT: Magnetic devices
Magnetic materials

Magnetic scales 701.2, 943.3
DT: January 1993
UF: Scales and weighing—Magnetic*

* Former Ei Vocabulary term

Magnetic scales *(continued)*
BT: Magnetic devices
Scales (weighing instruments)

Magnetic semiconductors 708.4, 712.1
DT: January 1981
BT: Magnetic materials
Semiconductor materials
RT: Antiferromagnetic materials
Ferrimagnetic materials
Ferromagnetic materials

Magnetic separation 701.2, 708.4 (802.3)
DT: January 1993
BT: Magnetic field effects
Separation
RT: Magnetic separators

Magnetic separators 708.4, 701.2 (802.1)
DT: January 1993
UF: Separators—Magnetic*
BT: Magnetic devices
Separators
RT: Magnetic separation

Magnetic shielding 701.2 (711.1)
SN: Shielding against magnetic fields
DT: January 1981
BT: Shielding
RT: Electromagnetic shielding
Electromagnetic waves
Magnetic fields
Magnetic leakage
Magnetism

Magnetic storage 722.1 (708.4)
DT: January 1993
UF: Data storage, Magnetic*
Data storage, Magnetic—Storage devices*
Magnetic data storage
BT: Data storage equipment
Magnetic devices
NT: Magnetic bubble memories
Magnetic core storage
Magnetic disk storage
Magnetic film storage
Magnetic heads
Magnetic tape storage
RT: Recording

Magnetic susceptibility
USE: Magnetic permeability

Magnetic suspension vehicles
USE: Magnetic levitation vehicles

Magnetic tape 708.4, 722.1
DT: Predates 1975
BT: Magnetic materials
Tapes
RT: Magnetic recording
Magnetic tape storage

Magnetic tape storage 722.1 (708.4)
DT: January 1993
UF: Data storage, Magnetic—Tape*
Tape storage (of data)
BT: Magnetic storage
Nonvolatile storage
RT: Magnetic tape
Tape drives
Tape recorders

*Magnetic tape—Spools**
USE: Reels

Magnetic thick films 708.4
DT: January 1993
BT: Magnetic films
Thick films
RT: Magnetic thin films

Magnetic thin film devices 708.4 (722.1)
DT: January 1993
UF: Magnetic devices, Thin film*
BT: Magnetic devices
Thin film devices
NT: Magnetic film storage
RT: Magnetic bubble devices
Magnetic bubble memories
Magnetic thin films

Magnetic thin film storage
USE: Magnetic film storage

Magnetic thin films 708.4 (722.1)
DT: January 1993
UF: Magnetic materials—Thin films*
BT: Magnetic films
Magnetic materials
Thin films
RT: Magnetic thick films
Magnetic thin film devices

Magnetic variables control 701.2 (731.3)
DT: January 1993
UF: Control, Magnetic variables*
BT: Control
RT: Magnetic properties
Magnetism

Magnetic variables measurement 942.4
DT: January 1993
UF: Magnetic materials—Measurements*
Magnetic measurements*
BT: Measurements
NT: Magnetic field measurement
Magnetic permeability measurement
Magnetic relaxation measurement
Magnetic resonance measurement
RT: Compasses (magnetic)
Magnetic field effects
Magnetic materials
Magnetic measuring instruments
Magnetic properties
Magnetic prospecting
Magnetism

Magnetisation
USE: Magnetization

Magnetism 701.2
SN: The phenomenon. For the process or state, use MAGNETIZATION
DT: Predates 1975
NT: Antiferromagnetism

* Former Ei Vocabulary term

Magnetism *(continued)*
Diamagnetism
Electromagnetism
Ferrimagnetism
Ferromagnetism
Geomagnetism
Metamagnetism
Paramagnetism
RT: Electricity
Electrodynamics
Magnetic fields
Magnetic forming
Magnetic leakage
Magnetic materials
Magnetic moments
Magnetic properties
Magnetic relaxation
Magnetic shielding
Magnetic variables control
Magnetic variables measurement
Magnetization
Magnetostatics
Magnets

*Magnetism—Antiferromagnetism**
USE: Antiferromagnetism

*Magnetism—Diamagnetism**
USE: Diamagnetism

*Magnetism—Ferrimagnetism**
USE: Ferrimagnetism

*Magnetism—Ferromagnetism**
USE: Ferromagnetism

*Magnetism—Metamagnetism**
USE: Metamagnetism

*Magnetism—Paramagnetism**
USE: Paramagnetism

Magnetite (482.2) (545.1) (708.4)
DT: June 1990
BT: Ferromagnetic materials
Iron ores
Iron oxides
Oxide minerals
RT: Iron deposits

Magnetization 701.2
DT: Predates 1975
UF: Magnetic ordering
Magnetisation
Magnetization—Process*
Magnetization—State*
RT: Coercive force
Demagnetization
Magnetic bubbles
Magnetic domains
Magnetic fields
Magnetic hysteresis
Magnetic moments
Magnetic properties
Magnetism
Magnets
Remanence

*Magnetization—Demagnetization**
USE: Demagnetization

*Magnetization—Process**
USE: Magnetization

*Magnetization—Remanence**
USE: Remanence

*Magnetization—State**
USE: Magnetization

Magneto-electric effects
USE: Magnetoelectric effects

Magneto-optical devices
USE: Magnetooptical devices

Magneto-optical effects
USE: Magnetooptical effects

Magneto-optics
USE: Magnetooptical effects

Magnetoacoustic effects 701.2, 751.1
DT: Predates 1975
UF: Acoustomagnetic effects
BT: Magnetic field effects
RT: Acoustic wave effects

Magnetoelectric effects 701.1, 701.2
DT: Predates 1975
UF: Electromagnetic effects
Magneto-electric effects
BT: Electric field effects
Magnetic field effects
RT: Electromagnetism
Hall effect
Magnetic properties

Magnetoelectric generators
USE: Magnetos

Magnetographs
USE: Magnetometers

Magnetohydrodynamic converters 615.3
DT: Predates 1975
UF: Converters (magnetohydrodynamic)
Magnetohydrodynamic generators
MHD converters
BT: Direct energy converters
RT: Magnetohydrodynamics
Magnetohydrodynamic power plants
Magnetohydrodynamic power generation
Plasma devices

Magnetohydrodynamic generators
USE: Magnetohydrodynamic converters

Magnetohydrodynamic power generation 615.3
DT: January 1993
UF: MHD power generation
Power generation—Magnetohydrodynamic*
BT: Power generation
RT: Magnetohydrodynamic power plants
Magnetohydrodynamics
Magnetohydrodynamic converters

* Former Ei Vocabulary term

Magnetohydrodynamic power plants 615.3
DT: January 1993
UF: MHD power plants
Power plants—Magnetohydrodynamic*
BT: Power plants
RT: Magnetohydrodynamic converters
Magnetohydrodynamics
Magnetohydrodynamic power generation

Magnetohydrodynamics 615.3
DT: Predates 1975
UF: MHD
BT: Hydrodynamics
Magnetic field effects
RT: Ionization
Magnetohydrodynamic converters
Magnetohydrodynamic power plants
Magnetohydrodynamic power generation

Magnetometers 942.3
DT: Predates 1975
UF: Gradiometers (magnetic)
Magnetographs
BT: Magnetic measuring instruments
RT: Electric measuring instruments
Magnetic field measurement
SQUIDs

Magnetooptical devices 741.3, 701.2
DT: January 1981
UF: Magneto-optical devices
BT: Magnetic devices
Optical devices
RT: Magnetooptical effects

Magnetooptical effects 701.2, 741.1
DT: Predates 1975
UF: Magneto-optical effects
Magneto-optics
Magnetooptics
BT: Magnetic field effects
NT: Faraday effect
Kerr magnetooptical effect
RT: Birefringence
Electromagnetic wave polarization
Electrooptical effects
Light polarization
Magnetic properties
Magnetooptical devices
Optical properties
Polarization

Magnetooptical Kerr effect
USE: Kerr magnetooptical effect

Magnetooptics
USE: Magnetooptical effects

Magnetoplasma 701.2, 932.3
DT: January 1993
UF: Plasmas—Magnetoplasma*
BT: Plasmas

Magnetoresistance 701.2
DT: January 1995
UF: Magnetoresistivity
Magnetoresistors
Shubnikov-de Haas effect
BT: Galvanomagnetic effects

Magnetoresistivity
USE: Magnetoresistance

Magnetoresistors
USE: Magnetoresistance

Magnetos 705.2, 708.4
DT: Predates 1975
UF: Magnetoelectric generators
BT: Electric generators
Magnetic devices

Magnetosphere 443.1, 701.2
DT: January 1993
UF: Earth atmosphere—Magnetosphere*
BT: Upper atmosphere
NT: Radiation belts
RT: Geomagnetism
Ionosphere

Magnetostatics 701.2
DT: January 1981
BT: Physics
RT: Electromagnetism
Magnetic fields
Magnetism

Magnetostriction 701.2
DT: Predates 1975
UF: Magnetic materials—Magnetostriction*
BT: Magnetic field effects
RT: Magnetic materials
Magnetostrictive devices

Magnetostrictive devices (704) (708.4) (714)
DT: Predates 1975
BT: Magnetic devices
RT: Acoustic variables measurement
Magnetostriction
Mechanical variables measurement
Sonar
Transducers

Magnetron sputtering
(539.3) (715.1) (813.1) (932.1)
DT: January 1995
BT: Sputtering
RT: Magnetrons

Magnetrons 714.1
DT: January 1993
UF: Electron tubes, Magnetron*
BT: Electron wave tubes
Magnetic devices
RT: Cavity resonators
Magnetron sputtering
Oscillators (electronic)

Magnets (704) (708.4)
DT: Predates 1975
BT: Magnetic devices
NT: Accelerator magnets
Electromagnets
Permanent magnets
RT: Ferromagnetic materials

* Former Ei Vocabulary term

Magnets *(continued)*
Magnetic circuits
Magnetic cores
Magnetic fields
Magnetic materials
Magnetism
Magnetization

*Magnets—Lifting**
USE: Lifting magnets

*Magnets—Powder metal**
USE: Powder metal products

Mail cars
USE: Postal cars

Mail handling (691)
DT: Predates 1975
BT: Materials handling
RT: Post offices
Postal cars
Postal services

Maintainability 913.5
SN: Use for qualities of equipment which will affect its capability of being maintained. For the process of maintenance, use MAINTENANCE
DT: Predates 1975
UF: Maintenance engineering
RT: Availability
Engineering
Failure analysis
Maintenance
Performance
Reliability

Maintenance 913.5
SN: Use for the process of maintenance. For qualities of equipment which affect its capability of being maintained, use MAINTAINABILITY
DT: Predates 1975
NT: Maintenance of way
Wheel dressing
RT: Alignment
Cleaning
Drydocks
Failure analysis
Improvement
Installation
Lubricants
Maintainability
Modernization
Outages
Pavement overlays
Performance
Prevention
Reconstruction (structural)
Reliability
Repair
Restoration
Retrofitting
Specifications

Maintenance engineering
USE: Maintainability

Maintenance of way (681)
SN: Railroads
DT: January 1993
UF: Railroad plant and structures—Maintenance of way*
BT: Maintenance
RT: Railroad plant and structures
Snow and ice removal
Weed control

Majority logic 721.1
DT: January 1993
UF: Computer metatheory—Majority logic*
BT: Formal logic
RT: Logic circuits
Redundancy
Switching theory

Malaria control 461.7
DT: Predates 1975
BT: Disease control
RT: Biomedical engineering
Insecticides
Mosquito control
Tropics

Malcolmizing
USE: Nitriding

Malleable iron castings 545.1 (534.2)
DT: Predates 1975
BT: Cast iron
RT: Malleable iron foundry practice

Malleable iron foundry practice 534.2, 545.1
DT: Predates 1975
BT: Iron foundry practice
RT: Malleable iron castings

Malls
USE: Shopping centers

Malt sugar
USE: Maltose

Maltose (804.1) (822.3)
DT: January 1987
UF: Malt sugar
BT: Sugars

Man machine interaction
USE: Human computer interaction

Man machine systems (731.1)
DT: January 1993
UF: Human machine systems
Systems science and cybernetics—Man machine systems*
BT: Systems science
RT: Cybernetics
Ergonomics
Human computer interaction
Human engineering
Learning systems
Mechanization
Systems engineering
User interfaces

* Former Ei Vocabulary term

Man powered aircraft 652.1
DT: January 1993
UF: Aircraft—Man powered*
Human powered aircraft
BT: Aircraft

Management 912.2
SN: Very general term. Prefer specific kind of management or specific management activity. For management aspects of particular topics, use Treatment Code M
DT: Predates 1975
UF: Administration
NT: Budget control
Cost accounting
Electric load management
Energy management
Highway administration
Industrial management
Information management
Inventory control
Job analysis
Project management
Records management
Research and development management
Time and motion study
Total quality management
RT: Administrative data processing
Computer operating procedures
Computer selection and evaluation
Contracts
Decision making
Decision support systems
Depreciation
Economics
Efficiency
Employment
Finance
Forecasting
Information retrieval systems
Information science
Job satisfaction
Management information systems
Management science
Office automation
Pension plans
Personnel
Planning
Purchasing
Quality assurance
Resource allocation
Strategic planning
Supervisory personnel
Systems engineering
Turnaround time

Management information systems 723.2 (912.2)
DT: January 1993
UF: Management—Information systems*
MIS
BT: Administrative data processing
NT: Decision support systems
RT: Database systems
Information analysis
Information management
Management
Management science

Management science 912.2
SN: Management techniques and theory
DT: Predates 1975
BT: Social sciences
NT: Critical path analysis
PERT
RT: Administrative data processing
Decision support systems
Forecasting
Management
Management information systems
Operations research

*Management—Information systems**
USE: Management information systems

*Management—Research and development application**
USE: Research and development management

Managers
USE: Supervisory personnel

Mandelstam-Brillouin scattering
USE: Brillouin scattering

Maneuverability (652.1) (655.1) (671.1)
DT: January 1993
UF: Ships—Maneuverability*
RT: Aircraft
Ships
Spacecraft
Steering

Manganese 543.2
DT: January 1993
UF: Manganese and alloys*
Slags—Manganese recovery*
BT: Nonferrous metals
Transition metals
RT: Manganese alloys
Manganese compounds
Manganese deposits
Manganese metallurgy
Manganese metallography
Manganese mines
Manganese nodules
Manganese ore treatment
Manganese removal (water treatment)
Strategic materials

Manganese alloys 543.2
DT: January 1993
UF: Manganese and alloys*
BT: Transition metal alloys
RT: Manganese
Manganese compounds

*Manganese and alloys**
USE: Manganese OR
Manganese alloys

Manganese compounds (804.1) (804.2)
DT: Predates 1975
BT: Transition metal compounds

* Former Ei Vocabulary term

Manganese compounds *(continued)*
NT: Semiconducting manganese compounds
RT: Manganese
Manganese alloys

Manganese deposits 504.1, 543.2
DT: Predates 1975
BT: Ore deposits
NT: Manganese nodules
RT: Manganese
Manganese mines

*Manganese deposits—Undersea nodules**
USE: Manganese nodules

Manganese metallography 531.2, 543.2
DT: Predates 1975
BT: Metallography
RT: Manganese
Manganese metallurgy

Manganese metallurgy 531.1, 543.2
DT: Predates 1975
BT: Metallurgy
RT: Manganese
Manganese metallography
Manganese ore treatment

Manganese mines 504.1, 543.2
DT: January 1993
UF: Manganese mines and mining*
BT: Mines
RT: Manganese
Manganese deposits

*Manganese mines and mining**
USE: Manganese mines

Manganese nodules 543.2, 471.5
DT: January 1993
UF: Manganese deposits—Undersea nodules*
Ocean floor nodules
Sea floor nodules
Undersea manganese nodules
BT: Manganese deposits
Underwater mineral resources
RT: Manganese

Manganese ore treatment 533.1, 543.2
DT: Predates 1975
BT: Ore treatment
RT: Manganese
Manganese metallurgy

Manganese removal (water treatment) 445.1
DT: January 1993
UF: Water treatment—Manganese removal*
BT: Chemicals removal (water treatment)
RT: Manganese

Manifolds (automobile engines)
USE: Automobile engine manifolds

Manipulators 691.1, 731.5
DT: January 1993
UF: Materials handling—Manipulators*
Robots, Industrial—Manipulators*
BT: Materials handling equipment
RT: Industrial robots
Motion planning

Manned space flight 656.1
DT: January 1993
UF: Space flight—Manned flight*
BT: Space flight
RT: Interplanetary flight
Space shuttles
Space stations

Manometers 944.3
DT: Predates 1975
BT: Gages
Pressure transducers

Manual control (461.4) (731.1)
DT: January 1993
UF: Human engineering—Manual control*
BT: Control
RT: Human engineering

Manufacture 913.4
DT: January 1993
UF: Gases—Manufacture*
Manufacturing
BT: Production
NT: Aircraft manufacture
Automobile manufacture
Brickmaking
Carpet manufacture
Cement manufacture
Cigarette manufacture
Coke manufacture
Computer aided manufacturing
Computer integrated manufacturing
Electronic equipment manufacture
Fabrication
Film preparation
Flexible manufacturing systems
Forming
Furniture manufacture
Garment manufacture
Gas fuel manufacture
Gear manufacture
Glass manufacture
Jewelry manufacture
Papermaking
Processing
Pulp manufacture
Shoe manufacture
Sugar manufacture
Toy manufacture
Woodworking
RT: Manufacturing data processing
Quality assurance
Raw materials
Synthesis (chemical)

Manufacturing
USE: Manufacture

Manufacturing data processing 723.2, 913.4
DT: January 1993
UF: Data processing—Manufacturing applications*
BT: Data processing

* Former Ei Vocabulary term

Manufacturing data processing *(continued)*
RT: Industrial economics
Industrial plants
Manufacture
Statistical process control

Manures 821.5 (452.3)
DT: January 1993
UF: Agricultural wastes—Manures*
BT: Agricultural wastes

Many valued logics 721.1
DT: January 1993
UF: Computer metatheory—Many valued logics*
BT: Formal logic
RT: Logic circuits

Map symbols 405.3 (902.1)
DT: January 1993
UF: Maps and mapping—Symbols*
BT: Codes (symbols)
RT: Mapping

Mapping 405.3 (902.1)
SN: Cartographic mapping. For the mathematical sense, use MATHEMATICAL TECHNIQUES
DT: January 1993
UF: Cartography
Glaciers—Mapping*
Maps and mapping*
Watersheds—Mapping*
BT: Graphic methods
NT: Conformal mapping
Military mapping
Photomapping
RT: Geographic information systems
Geological surveys
Map symbols
Maps
Photogrammetry
Surveying
Surveys

Maps 405.3 (902.1)
SN: Cartographic maps
DT: January 1993
UF: Maps and mapping*
Pipelines—Maps*
Sewers—Maps*
RT: Graphic methods
Mapping
Navigation charts
Surveying
Surveys

*Maps and mapping**
USE: Maps OR
Mapping

*Maps and mapping—Military**
USE: Military mapping

*Maps and mapping—Photomaps**
USE: Photomapping

*Maps and mapping—Symbols**
USE: Map symbols

Maraging steel 545.3
DT: Predates 1975
UF: Aircraft materials—Maraging steel*
Welding—Maraging steel*
BT: Steel
RT: Stainless steel

Marble (414) (481.1.2)
DT: Predates 1975
BT: Metamorphic rocks
RT: Building materials
Limestone
Synthetic marble

*Marble—Synthetic**
USE: Synthetic marble

Marinas 407.1
DT: Predates 1975
BT: Facilities
RT: Inland waterways
Port structures
Ports and harbors

Marine applications (471.1) (472)
DT: January 1993
UF: Oceanographic applications
BT: Applications
RT: Marine engineering
Ocean engineering
Oceanography

Marine biology 461.9 (471.1)
DT: January 1977
BT: Biology
RT: Algae
Fisheries
Marine borers
Marine pollution
Oceanography
Thermal pollution

Marine boilers (614.2) (671.2)
DT: January 1993
UF: Boilers—Marine*
BT: Boilers
Water craft parts and equipment

Marine borers 461.9 (471.1)
SN: Aquatic animals
DT: Predates 1975
RT: Antifouling paint
Marine biology
Pest control
Port structures
Ships

Marine couplings (602) (671.2)
DT: January 1993
UF: Couplings—Marine*
BT: Couplings
RT: Water craft parts and equipment

Marine engineering 675
SN: Primarily concerned with propulsive and auxiliary equipment and other fittings of waterborne craft. Distinguish from NAVAL ARCHITECTURE
DT: Predates 1975
BT: Engineering
RT: Boats
Marine applications
Marine engines
Marine insurance
Merchant marine
Naval architecture
Ships

Marine engines 671.2 (612)
DT: Predates 1975
UF: Boats—Engines*
Diesel boat engines
BT: Engines
Water craft parts and equipment
RT: Internal combustion engines
Marine engineering
Ship propulsion

Marine insurance 434.1 (911.1)
DT: Predates 1975
BT: Insurance
RT: Boats
Marine engineering
Offshore structures
Production platforms
Ships

Marine missiles 654.1, 404.1 (672.1)
DT: January 1993
UF: Rockets and missiles—Marine*
Ship launched missiles
BT: Missiles
NT: Underwater launched missiles
RT: Naval warfare

*Marine platforms**
USE: Production platforms

Marine pollution 453
DT: January 1993
UF: Ocean pollution
Water pollution—Marine pollution*
BT: Water pollution
RT: Marine biology
Ocean dumping
Oil spills
Seawater
Thermal pollution

Marine power plants (612) (613) (614)
DT: January 1993
UF: Power plants—Marine*
BT: Power plants

Marine radar 716.2
DT: January 1993
UF: Radar—Marine*
BT: Radar

Marine risers 674.2, 511.2, 619.1
DT: January 1977
UF: Risers (marine)
BT: Pipe
RT: Offshore drilling
Production platforms

Marine signal systems 434.4 (914.1)
DT: January 1993
UF: Marine signals and signaling*
BT: Signal systems
NT: Buoys
RT: Electric signal systems
Signaling
Water craft parts and equipment

*Marine signals and signaling**
USE: Marine signal systems

Marine terminals
USE: Port terminals

Marketing 911.4
DT: Predates 1975
UF: Advertising
Sales
Selling
RT: International trade
Point of sale terminals

*Marketing—International trade**
USE: International trade

Marking machines 601.1, 694.1
DT: Predates 1975
BT: Machinery
RT: Bar codes
Labeling

Markings (highway)
USE: Highway markings

Markings (runway)
USE: Runway markings

Markov chains
USE: Markov processes

Markov processes 922.1
DT: January 1995
UF: Markov chains
BT: Random processes
RT: Probability

Martensite 531.2 (545.3)
DT: January 1993
UF: Iron and steel metallography—Martensite*
Metallography—Martensite*
BT: Metallographic phases
RT: Hardening
Martensitic transformations
Steel metallography

Martensitic transformations 531.2 (545.3)
DT: January 1993
BT: Phase transitions
RT: Martensite

Masers 714 (711) (931.3)
DT: Predates 1975
BT: Microwave amplifiers
RT: Atomic clocks
Cavity resonators
Lasers
Oscillators (electronic)
Parametric devices
Quantum electronics
Time measurement

*Masers—Noise**
USE: Spurious signal noise

Masking (acoustic)
USE: Speech intelligibility

Masks 714.2
DT: January 1993
UF: Integrated circuits—Masks*
RT: Integrated circuits
Lithography
Photolithography
Schematic diagrams
Semiconductor devices
X ray lithography

Masonry bridges 401.1, 414
DT: January 1993
UF: Bridges, Masonry*
BT: Composite bridges
NT: Concrete bridges
RT: Masonry construction

Masonry construction 405.2, 414
DT: June 1990
BT: Construction
NT: Brick construction
Concrete construction
RT: Masonry bridges
Masonry materials

Masonry materials 414
DT: January 1977
BT: Building materials
NT: Brick
Concretes
Mortar
RT: Masonry construction

Mass spectrometers (943.3)
SN: Mass spectroscopes in which a slit moves across the paths of particles, and an electrical detector behind the slit records the intensity distribution of the masses of the particles.
DT: Predates 1975
BT: Spectrometers
RT: Data recording
Ion sources
Mass spectrometry
Neutron activation analysis
Secondary ion mass spectrometry

*Mass spectrometers—Data recording**
USE: Data recording

Mass spectrometry 801 (943.3)
DT: June 1990
BT: Spectrometry
NT: Secondary ion mass spectrometry
RT: Data recording
Ion sources
Mass spectrometers
Neutron activation analysis

Mass transfer 641.3
DT: Predates 1975
NT: Runoff
RT: Dialysis
Diffusion
Heat transfer
Transport properties

*Mass transfer—Packed beds**
USE: Packed beds

Mass transportation
(431.2) (432.2) (433.2) (434.2)
DT: January 1986
BT: Transportation
RT: Rapid transit
Subways

Mastication 461.3
DT: January 1993
UF: Biomechanics—Mastication*
Chewing
BT: Biomechanics

Masts (boat) 674.1
DT: January 1993
UF: Boats—Masts*
BT: Boat equipment

Materials
SN: Very general term; prefer specific term if applicable
DT: Predates 1975
UF: Materials with memory*
Materials*
NT: Ablative materials
Abrasives
Adhesives
Adsorbents
Aggregates
Aircraft materials
Alloys
Amorphous materials
Asbestos
Automobile materials
Bagasse
Ballast (railroad track)
Bimetals
Binders
Biological materials
Biomass
Biomaterials
Bituminous materials
Briquets
Building materials
Byproducts
Ceramic materials
Chemical resistant materials
Chemicals

Materials *(continued)*
Coated materials
Coatings
Coke
Composite materials
Concrete reinforcements
Conductive materials
Corrugated materials
Crystalline materials
Crystals
Dielectric materials
Dissimilar materials
Driers (materials)
Dust
Electrooptical materials
Explosives
Fertilizers
Fibers
Filler metals
Films
Flammable materials
Fluids
Fluxes
Friction materials
Fuels
Galvanized metal
Gems
Granular materials
Graphite
Gravel
Hazardous materials
Ice
Impurities
Ink
Insulating materials
Insulation
Intelligent materials
Laminates
Leather
Lubricants
Luminous materials
Magnetic materials
Membranes
Minerals
Mixtures
Moderators
Nanostructured materials
Nonfibrous paper materials
Nonmetallic materials
Optical materials
Organic conductors
Packaging materials
Paper
Paperboards
Petroleum distillates
Petroleum products
Phosphors
Photoconducting materials
Pigments
Plaster
Plastics
Plate metal
Polymers
Porous materials
Powders
Propellants
Pulp
Radioactive materials
Raw materials
Refractory materials
Rocks
Rope
Rubber
Sand
Scrap metal
Sealants
Sediments
Semiconductor materials
Sheet metal
Sheet molding compounds
Soil conditioners
Solar absorbers
Solids
Steam condensate
Strategic materials
Strip metal
Structural metals
Superconducting materials
Synthetic marble
Talc
Tall oil
Tapes
Tar
Textile auxiliary materials
Textiles
Tubing
Twine
Wire
Wood
Yarn
RT: Casting
Growth (materials)
Materials handling
Materials science
Materials testing
Substrates

Materials casting
USE: Casting

Materials handling (691)
SN: Very general term; prefer term for specific type of material if available, e.g., COAL HANDLING
DT: Predates 1975
UF: Cement handling*
Grain—Handling*
Lumber—Handling*
Ore handling*
Pipe—Handling*
Sugar handling*
Warships—Ammunition handling*
NT: Ash handling
Baggage handling
Cargo handling
Coal handling
Conveying
Distribution of goods
Feeding

* Former Ei Vocabulary term

Materials handling *(continued)*
Filling
Fueling
Loading
Mail handling
Safe handling
Unloading
Winding
RT: Assembly
Containers
Flow of solids
Granular materials
Industrial robots
Labeling
Materials
Materials handling equipment
Monorails
Packaging
Petroleum transportation
Pipelines
Shock problems
Slurry pipelines
Storage (materials)
Tanks (containers)
Transportation
Warehouses

Materials handling equipment 691.1
DT: January 1993
UF: Robots, Industrial—Materials handling applications*
BT: Equipment
NT: Cableways
Conveyors
Cranes
Hoists
Industrial trucks
Lifting magnets
Loaders
Manipulators
Pallets
Pneumatic materials handling equipment
Reels
Skids
Slings
Winches
RT: Cages (mine hoists)
Containers
Crushers
Earthmoving machinery
Elevators
Hooks
Machinery
Materials handling
Mine cars
Packaging machines
Pipe
Trucks
Vibrators
Warehouses

*Materials handling—Department stores**
USE: Retail stores

*Materials handling—Farms**
USE: Farms

*Materials handling—Granular materials**
USE: Granular materials

*Materials handling—Hooks**
USE: Hooks

*Materials handling—Hoppers**
USE: Hoppers

*Materials handling—Hydraulic equipment**
USE: Hydraulic equipment

*Materials handling—Loading**
USE: Loading

*Materials handling—Manipulators**
USE: Manipulators

*Materials handling—Pallets**
USE: Pallets

*Materials handling—Pipe**
USE: Pipe

*Materials handling—Pneumatic**
USE: Pneumatic materials handling equipment

*Materials handling—Reels**
USE: Reels

*Materials handling—Scrap metal**
USE: Scrap metal

*Materials handling—Sheet metal**
USE: Sheet metal

Materials handling—Shock problems
USE: Shock problems

*Materials handling—Skids**
USE: Skids

*Materials handling—Slings**
USE: Slings

*Materials handling—Unloading**
USE: Unloading

*Materials handling—Vacuum**
USE: Vacuum applications

*Materials handling—Vibrators**
USE: Vibrators

*Materials handling—Warehouses**
USE: Warehouses

Materials science (931.2)
DT: October 1975
NT: Morphology
RT: Materials
Physics
Steel research

Materials testing (422) (423)
DT: Predates 1975
BT: Testing
NT: Concrete testing
Elasticity testing
Electric insulation testing

* Former Ei Vocabulary term

Materials testing *(continued)*
Electroacoustic testing
Explosion testing
Flammability testing
Load testing
Metal testing
Moisture determination
Rubber testing
Surface testing
RT: Automatic testing
Defects
Fits and tolerances
High temperature testing
Impact testing
Laser diagnostics
Low temperature testing
Materials
Materials testing laboratories
Materials testing apparatus
Measurements
Nondestructive examination
Safety testing
Shock testing
Temperature programmed desorption

Materials testing apparatus (422.1) (423.1)
DT: Predates 1975
BT: Equipment
NT: Track test cars
RT: Materials testing
Materials testing laboratories

Materials testing laboratories
(402.1) (422.1) (423.1) (901.3)
DT: Predates 1975
BT: Laboratories
RT: Materials testing
Materials testing apparatus

*Materials testing—Compression tests**
USE: Compression testing

*Materials testing—Creep**
USE: Creep testing

*Materials testing—Electroacoustic**
USE: Electroacoustic testing

*Materials testing—Explosion**
USE: Explosion testing

*Materials testing—Fatigue**
USE: Fatigue testing

*Materials testing—Fracture**
USE: Fracture testing

*Materials testing—Hardness**
USE: Hardness testing

*Materials testing—High temperature**
USE: High temperature testing

*Materials testing—Impact**
USE: Impact testing

*Materials testing—Low temperature**
USE: Low temperature testing

*Materials testing—Notched bar**
USE: Notched bar tensile testing

*Materials testing—Surface**
USE: Surface testing

*Materials testing—Tensile tests**
USE: Tensile testing

*Materials testing—Torsion tests**
USE: Torsion testing

*Materials with memory**
USE: Shape memory effect AND
Materials

Materials working
USE: Forming

*Materials**
USE: Materials

*Materials—Amorphous**
USE: Amorphous materials

*Materials—Casting**
USE: Casting

*Materials—Heat resistance**
USE: Heat resistance

*Materials—Residual stresses**
USE: Residual stresses

Mathematical instruments (722.4) (921)
DT: Predates 1975
UF: Slide rules*
BT: Instruments
NT: Pocket calculators
RT: Computers

*Mathematical instruments—Pocket calculators**
USE: Pocket calculators

Mathematical linguistics
USE: Computational linguistics

Mathematical logic
USE: Formal logic

Mathematical models (921)
DT: Predates 1975
UF: Models (mathematical)
NT: Semiconductor device models
RT: Chaos theory
Decision theory
Flowcharting
Forecasting
Graph theory
Mathematical programming
Nonlinear programming
Numerical analysis
Operations research
Risks
Simulation
Simulators
Systems analysis
Systems engineering

Mathematical morphology (921)
DT: January 1993
RT: Mathematical techniques

Mathematical operators (921)
DT: January 1993
UF: Mathematical techniques—Operators*
Operators (mathematics)
BT: Mathematical techniques

Mathematical programming (723.1) (921.5) (922)
DT: Predates 1975
UF: Programming (mathematical)
BT: Optimization
NT: Dynamic programming
Linear programming
Nonlinear programming
RT: Game theory
Mathematical models
Models

*Mathematical programming, Dynamic**
USE: Dynamic programming

*Mathematical programming, Linear**
USE: Linear programming

*Mathematical programming, Nonlinear**
USE: Nonlinear programming

*Mathematical statistics**
USE: Statistical methods

*Mathematical statistics—Monte Carlo methods**
USE: Monte Carlo methods

*Mathematical statistics—Random number generation**
USE: Random number generation

Mathematical techniques (921)
DT: Predates 1975
UF: Mathematics
NT: Algebra
Algorithms
Approximation theory
Bifurcation (mathematics)
Combinatorial mathematics
Computational methods
Describing functions
Differential equations
Differentiation (calculus)
Digital arithmetic
Equations of state
Equivalence classes
Estimation
Finite volume method
Frequency domain analysis
Functions
Geometry
Harmonic analysis
Heuristic methods
Integral equations
Integration
Inverse problems
Lagrange multipliers
Mathematical operators
Mathematical transformations
Modal analysis
Nonlinear equations
Number theory
Numerical analysis
Numerical methods
Perturbation techniques
Phase space methods
Piecewise linear techniques
Poles and zeros
Probability
Random number generation
Sensitivity analysis
Set theory
State space methods
Statistical methods
Time domain analysis
Topology
Transfer functions
Variational techniques
RT: Calculations
Fractals
Mathematical morphology
Maximum principle
Structural analysis

*Mathematical techniques—Algebra**
USE: Algebra

*Mathematical techniques—Algorithms**
USE: Algorithms

*Mathematical techniques—Approximation theory**
USE: Approximation theory

*Mathematical techniques—Boundary element method**
USE: Boundary element method

*Mathematical techniques—Boundary value problems**
USE: Boundary value problems

*Mathematical techniques—Chebyshev approximation**
USE: Chebyshev approximation

*Mathematical techniques—Combinatorial mathematics**
USE: Combinatorial mathematics

*Mathematical techniques—Conformal mapping**
USE: Conformal mapping

*Mathematical techniques—Convergence of numerical methods**
USE: Convergence of numerical methods

*Mathematical techniques—Correlation methods**
USE: Correlation methods

*Mathematical techniques—Curve fitting**
USE: Curve fitting

* Former Ei Vocabulary term

*Mathematical techniques—Difference equations**
USE: Difference equations

*Mathematical techniques—Differential equations**
USE: Differential equations

*Mathematical techniques—Differentiation**
USE: Differentiation (calculus)

*Mathematical techniques—Digital arithmetic**
USE: Digital arithmetic

*Mathematical techniques—Eigenvalues and eigenfunctions**
USE: Eigenvalues and eigenfunctions

*Mathematical techniques—Error analysis**
USE: Error analysis

*Mathematical techniques—Finite difference method**
USE: Finite difference method

*Mathematical techniques—Finite element method**
USE: Finite element method

*Mathematical techniques—Frequency domain analysis**
USE: Frequency domain analysis

*Mathematical techniques—Function evaluation**
USE: Function evaluation

*Mathematical techniques—Fuzzy sets**
USE: Fuzzy sets

*Mathematical techniques—Geometry**
USE: Geometry

*Mathematical techniques—Graph theory**
USE: Graph theory

*Mathematical techniques—Harmonic analysis**
USE: Harmonic analysis

*Mathematical techniques—Heuristic**
USE: Heuristic methods

*Mathematical techniques—Integral equations**
USE: Integral equations

*Mathematical techniques—Integration**
USE: Integration

*Mathematical techniques—Integrodifferential equations**
USE: Integrodifferential equations

*Mathematical techniques—Interpolation**

USE: Interpolation

*Mathematical techniques—Iterative methods**
USE: Iterative methods

*Mathematical techniques—Least squares approximations**
USE: Least squares approximations

*Mathematical techniques—Linear algebra**
USE: Linear algebra

*Mathematical techniques—Linearization**
USE: Linearization

*Mathematical techniques—Matrix algebra**
USE: Matrix algebra

*Mathematical techniques—Maximum principle**
USE: Maximum principle

*Mathematical techniques—Nonlinear equations**
USE: Nonlinear equations

*Mathematical techniques—Number theory**
USE: Number theory

*Mathematical techniques—Numerical analysis**
USE: Numerical analysis

*Mathematical techniques—Numerical methods**
USE: Numerical methods

*Mathematical techniques—Operators**
USE: Mathematical operators

*Mathematical techniques—Perturbation techniques**
USE: Perturbation techniques

*Mathematical techniques—Petri nets**
USE: Petri nets

*Mathematical techniques—Phase space methods**
USE: Phase space methods

*Mathematical techniques—Piecewise linear techniques**
USE: Piecewise linear techniques

*Mathematical techniques—Poles and zeros**
USE: Poles and zeros

*Mathematical techniques—Polynomials**
USE: Polynomials

*Mathematical techniques—Sensitivity analysis**
USE: Sensitivity analysis

*Mathematical techniques—Set theory**
USE: Set theory

*Mathematical techniques—State space methods**
USE: State space methods

*Mathematical techniques—Tensors**
USE: Tensors

*Mathematical techniques—Time domain analysis**
USE: Time domain analysis

*Mathematical techniques—Topology**
USE: Topology

*Mathematical techniques—Transfer functions**
USE: Transfer functions

* Former Ei Vocabulary term

*Mathematical techniques—Trees**
USE: Trees (mathematics)

*Mathematical techniques—Variational techniques**
USE: Variational techniques

*Mathematical techniques—Vectors**
USE: Vectors

Mathematical transformations 921.3
DT: Predates 1975
UF: Transformations (mathematical)
Transforms
BT: Functions
Mathematical techniques
NT: Conformal mapping
Fourier transforms
Laplace transforms
Walsh transforms
Wavelet transforms
Z transforms

*Mathematical transformations—Fast Fourier transforms**
USE: Fast Fourier transforms

*Mathematical transformations—Fourier transforms**
USE: Fourier transforms

*Mathematical transformations—Laplace transforms**
USE: Laplace transforms

*Mathematical transformations—Walsh transforms**
USE: Walsh transforms

*Mathematical transformations—Z transforms**
USE: Z transforms

Mathematics
USE: Mathematical techniques

Matrices
USE: Matrix algebra

Matrix algebra 921.1
DT: January 1993
UF: Mathematical techniques—Matrix algebra*
Matrices
Matroids
BT: Linear algebra
NT: Stiffness matrix
RT: Electric network parameters
Tensors

Matroids
USE: Matrix algebra

Maxillofacial prostheses 462.4
DT: January 1993
UF: Dental equipment and supplies—Maxillofacial prostheses*
BT: Prosthetics

Maximum demand indicators 942.1
DT: Predates 1975
UF: Demand meters
BT: Electric measuring instruments
RT: Electric power measurement

Maximum principle (731.1) (921.5)
SN: Of Pontryagin
DT: January 1993
UF: Mathematical techniques—Maximum principle*
Pontryagin maximum principle
RT: Dynamic programming
Mathematical techniques
Optimal control systems

Maxwell equations 701.1, 701.2, 921.2
DT: January 1995
UF: Electromagnetic field equations
Maxwell field equations
Maxwell's equations
BT: Partial differential equations
RT: Electric fields
Electromagnetic waves
Electromagnetism
Magnetic fields

Maxwell field equations
USE: Maxwell equations

Maxwell's equations
USE: Maxwell equations

MBE
USE: Molecular beam epitaxy

Measurement
USE: Measurements

Measurement errors (922)
DT: Predates 1975
UF: Systematic errors
BT: Errors
RT: Measurement theory
Measurements
Statistical methods
Statistical tests
Statistics

Measurement theory (922)
DT: Predates 1975
BT: Theory
RT: Measurement errors
Measurements

Measurement units
USE: Units of measurement

Measurements
SN: Very general term; prefer terms for specific measurements or classes of measurement
DT: Predates 1975
UF: Gases—Measurements*
Measurement
Mensuration
Metrology
NT: Acoustic variables measurement
Anthropometry
Bathymetry
Calorimetry
Cosmic ray measurement
Dosimetry

* Former Ei Vocabulary term

Measurements *(continued)*
Electric variables measurement
Electromagnetic field measurement
Electron density measurement
Gaging
Gas fuel measurement
Gloss measurement
Interferometry
Intermodulation measurement
Ionospheric measurement
Magnetic variables measurement
Mechanical variables measurement
Optical variables measurement
Plasma diagnostics
Radioactivity measurement
Radiometry
Salinity measurement
Snowfall measurement
Spatial variables measurement
Spectrometry
Thermal variables measurement
Time measurement
Viscosity measurement
Weighing
RT: Calibration
Chemical analysis
Detectors
Equipment testing
Inspection
Materials testing
Measurement errors
Measurement theory
Metric system
Monitoring
Surveying
Surveys
Testing
Units of measurement
Volumetric analysis

Meat packing plants
USE: Packing plants

Meats 822.3
DT: January 1993
UF: Food products—Meats*
BT: Food products
RT: Packing plants

Mechanical alloying 531 (536.1)
DT: January 1995
UF: High energy milling
BT: Alloying
Mixing
RT: Ball milling
Composite materials
Dispersion hardening
Powder metallurgy

Mechanical clocks 943.3
DT: January 1993
UF: Clocks, Mechanical*
BT: Clocks
NT: Chronometers

Mechanical control equipment (732.1)
DT: January 1993
UF: Control equipment, Mechanical*
BT: Control equipment
NT: Hydraulic control equipment
Pneumatic control equipment
Valves (mechanical)
RT: Nonelectric final control devices

*Mechanical drive**
USE: Mechanical drives

*Mechanical drive—Variable speed**
USE: Variable speed drives

Mechanical drives 602.1
DT: January 1993
UF: Belt drives
Chain drives
Machine tools—Mechanical drive*
Mechanical drive*
Mechanical machine tool drives
BT: Drives
NT: Transmissions
RT: Belts
Chains
Gears
Machine tool drives
Variable speed drives
Wheels

Mechanical engineering 608
DT: Predates 1975
BT: Engineering
RT: Mechanical properties
Mechanical testing
Mechanics
Stress analysis
Structural design

Mechanical machine tool drives
USE: Mechanical drives AND
Machine tool drives

Mechanical permeability (931.2)
DT: January 1993
UF: Brick—Permeability*
Granular materials—Permeability*
Packaging materials—Permeability*
Paper—Permeability*
Permeability (mechanical)
Permeability, Mechanical*
Plastics—Permeability*
Protective coatings—Permeability*
Wood—Permeability*
BT: Mechanical properties
NT: Air permeability
RT: Dialysis
Granular materials
Hydraulic conductivity
Infiltration
Leaching
Low permeability reservoirs
Lysimeters

* Former Ei Vocabulary term

Mechanical permeability *(continued)*
Percolation (fluids)
Permeameters (mechanical permeability)
Porosity
Porous materials

Mechanical permeability permeameters
USE: Permeameters (mechanical permeability)

Mechanical properties
DT: January 1993
UF: Electric lines—Mechanical characteristics*
Mechanical variables
NT: Bearing capacity
Blast resistance
Brittleness
Compressibility of solids
Density (specific gravity)
Ductility
Elastic moduli
Elasticity
Elastoplasticity
Formability
Hardness
Mechanical permeability
Photoelasticity
Plasticity
Shape memory effect
Stiffness
Strain
Strength of materials
Tensile properties
Toughness
Viscoelasticity
Viscoplasticity
Wear resistance
Yield stress
RT: Composite micromechanics
Composition
Compressibility
Crack propagation
Creep
Fatigue of materials
Fracture
Friction
Inspection
Loads (forces)
Mechanical engineering
Mechanical variables control
Mechanical variables measurement
Mechanics
Stress relaxation
Surface properties
Textures
Torque
Velocity
Wear of materials

Mechanical pulp 811.1
DT: January 1993
UF: Groundwood pulp
Pulp—Mechanical*
BT: Pulp
NT: Thermomechanical pulp

Mechanical relaxation
USE: Anelastic relaxation

Mechanical strength
USE: Strength of materials

Mechanical testing
DT: January 1993
BT: Testing
NT: Compression testing
Creep testing
Fatigue testing
Fracture testing
Hardness testing
Plasticity testing
Shock testing
Soil testing
Tensile testing
Torsion testing
RT: Mechanical engineering
Piezoelectric devices
Strain gages
Strain measurement
Stresses
Transducers

Mechanical torque converters 602.2
DT: January 1993
UF: Torque converters, Mechanical*
BT: Torque converters

Mechanical variables
USE: Mechanical properties

Mechanical variables control 731.3 (943.2)
DT: January 1993
UF: Control, Mechanical variables*
BT: Control
NT: Acceleration control
Density control (specific gravity)
Flow control
Force control
Motion control
Pressure control
Speed control
Strain control
Torque control
Velocity control
Vibration control
RT: Mechanical properties
Mechanical variables measurement
Mechanics
Moisture control
Spatial variables control

Mechanical variables measurement 943.2
DT: Predates 1975
BT: Measurements
NT: Acceleration measurement
Density measurement (specific gravity)
Flow measurement
Force measurement
Pressure measurement
Strain measurement
Torque measurement

* Former Ei Vocabulary term

Mechanical variables measurement *(continued)*
Velocity measurement
Vibration measurement
RT: Magnetostrictive devices
Mechanical properties
Mechanical variables control
Mechanics
Spatial variables measurement

*Mechanical variables measurement—Acceleration**
USE: Acceleration measurement

*Mechanical variables measurement—Angles**
USE: Angle measurement

*Mechanical variables measurement—Density**
USE: Density measurement (specific gravity)

*Mechanical variables measurement—Distance**
USE: Distance measurement

*Mechanical variables measurement—Flow**
USE: Flow measurement

*Mechanical variables measurement—Forces**
USE: Force measurement

*Mechanical variables measurement—Level**
USE: Level measurement

*Mechanical variables measurement—Position**
USE: Position measurement

*Mechanical variables measurement—Strain**
USE: Strain measurement

*Mechanical variables measurement—Torques**
USE: Torque measurement

*Mechanical variables measurement—Velocity**
USE: Velocity measurement

*Mechanical variables measurement—Volumes**
USE: Volume measurement

Mechanical waves (931.1)
DT: Predates 1975
BT: Waves
NT: Elastic waves
Gravity waves
Liquid waves
RT: Surface waves

Mechanical wear
USE: Wear of materials

Mechanics 931.1
DT: Predates 1975
UF: Applied mechanics
NT: Biomechanics
Continuum mechanics
Damping
Dynamics
Fluid mechanics
Fracture mechanics
Kinematics
Rheology
Rock mechanics
Soil mechanics
Statistical mechanics
Van der Waals forces
RT: Degrees of freedom (mechanics)
Mechanical engineering
Mechanical properties
Mechanical variables control
Mechanical variables measurement
Photomechanics
Quantum theory
Shock waves
Structural design

*Mechanics—Continuous media**
USE: Continuum mechanics

Mechanisms 601.3
DT: Predates 1975
BT: Machinery
NT: Connecting rods
RT: Degrees of freedom (mechanics)
Models

*Mechanisms—Degrees of freedom**
USE: Degrees of freedom (mechanics)

Mechanization
SN: Use for processes or activities in which manual operation is to be replaced by non-manual actuation, and automatic control is not included. For automatically controlled mechanization, use AUTOMATION
DT: January 1993
RT: Automation
Data processing
Machinery
Man machine systems
Systems engineering
Tools

Mechanoluminescence
USE: Triboluminescence

Median dividers 406.1
DT: January 1993
UF: Dividers (highway)
Highway systems—Median dividers*
BT: Roads and streets
RT: Highway systems

Medical applications (461.1)
DT: January 1993
UF: Fiber optics in medicine
BT: Applications
RT: Biomedical engineering
Medicine

Medical care
USE: Health care

Medical computing 723.5, 461.1
DT: January 1993
UF: Data processing—Medical information*
BT: Computer applications
RT: Data processing
Medicine

Medical departments (industrial plants) (402.2) (461.7)
DT: January 1993
UF: Industrial plants—Medical departments*
BT: Industrial plants
RT: Medicine
Occupational diseases
Patient treatment

Medical education 461.6, 901.2
DT: June 1990
BT: Education
RT: Medicine

Medical electronics
USE: Electronic medical equipment

Medical equipment
USE: Biomedical equipment

Medical imaging 461.1 (723.2) (741)
DT: January 1993
UF: Clinical imaging
BT: Imaging techniques
NT: Diagnostic radiography
RT: Diagnosis
Medicine
Patient treatment

Medical problems (461.6) (461.7)
DT: January 1993
UF: Aviation—Medical problems*
Aviators—Medical problems*
Health problems
Heat transfer—Medical problems*
RT: Health
Health care
Industrial hygiene
Medicine

Medical supplies
USE: Biomedical equipment

Medicine 461.6
DT: January 1993
UF: Television applications—Medicine*
NT: Anesthesiology
Aviation medicine
Cardiology
Dentistry
Endocrinology
Epidemiology
Gastroenterology
Geriatrics
Gynecology
Neurology
Nuclear medicine
Obstetrics
Oncology
Ophthalmology
Orthopedics
Pediatrics
Radiology
Sports medicine
Urology
RT: Allergies
Biomedical engineering
Biomedical equipment
Diagnosis
Diseases
Health
Health care
Hospitals
Industrial hygiene
Medical applications
Medical computing
Medical departments (industrial plants)
Medical education
Medical imaging
Medical problems
Natural sciences
Nursing
Patient rehabilitation
Patient treatment
Physiology
Radiation protection
Surgery

Medium density polyethylenes 815.1.1
DT: January 1993
UF: Polyethylenes—Medium density*
BT: Polyethylenes

Meehanite
USE: Cast iron

Melamine 804.1
DT: January 1986
BT: Organic compounds
RT: Melamine formaldehyde resins

Melamine formaldehyde resins 815.1.1
DT: Predates 1975
UF: Amino resins (melamine)
BT: Thermosets
RT: Formaldehyde
Melamine
Urea formaldehyde resins

Melt spinning (535.2.2)
DT: January 1995
BT: Casting
RT: Metallic glass
Quenching
Rapid solidification

Meltdown (nuclear reactors)
USE: Core meltdown

Melting 802.3 (531.1)
DT: January 1993
BT: Phase transitions
NT: Metal melting
Thawing
RT: Ablation
Molten materials
Tapping (furnace)

Melting furnaces 532.5 (534.2) (642.2)
DT: January 1993
UF: Furnaces, Melting*

* Former Ei Vocabulary term

Melting furnaces *(continued)*
BT: Furnaces
NT: Crucible furnaces
Cupolas
Electron beam furnaces
Open hearth furnaces
RT: Electric furnaces
Metal melting

Membership functions 921
DT: January 1995
BT: Functions
RT: Fuzzy sets

Membranes
DT: Predates 1975
BT: Materials
NT: Bilaminate membranes
Biological membranes
Dialysis membranes
Fibrous membranes
Gas permeable membranes
Ion exchange membranes
Ion selective membranes
Liquid membranes
Osmosis membranes
Permselective membranes
Photosynthetic membranes
Polymeric membranes
RT: Diffusion
Pervaporation

*Membranes—Bilaminate**
USE: Bilaminate membranes

*Membranes—Biological**
USE: Biological membranes

*Membranes—Cell membrane**
USE: Cell membranes

*Membranes—Dialysis membranes**
USE: Dialysis membranes

*Membranes—Fibrous**
USE: Fibrous membranes

*Membranes—Gas permeable**
USE: Gas permeable membranes

*Membranes—Ion exchange membranes**
USE: Ion exchange membranes

*Membranes—Ion selective**
USE: Ion selective membranes

*Membranes—Liquid**
USE: Liquid membranes

*Membranes—Osmosis membranes**
USE: Osmosis membranes

*Membranes—Oxygen permeable**
USE: Oxygen permeable membranes

*Membranes—Permselective**
USE: Permselective membranes

*Membranes—Photosynthetic membranes**
USE: Photosynthetic membranes

*Membranes—Polymeric**
USE: Polymeric membranes

Memories
USE: Data storage equipment

Memory (shape)
USE: Shape memory effect

Memory allocation
USE: Storage allocation (computer)

Memory devices
USE: Data storage equipment

Memory expansion boards
USE: Semiconductor storage

Mendelevium 622.1
DT: January 1993
BT: Transuranium elements

Mensuration
USE: Measurements

Mercerization 819.5
DT: January 1993
UF: Textile finishing—Mercerization*
BT: Textile finishing

Merchant marine (434.1)
DT: Predates 1975
RT: Marine engineering
Ships
Waterway transportation

Mercury (metal) 549.3
DT: January 1993
UF: Hg
Mercury and amalgams*
Mercury poisoning
BT: Heavy metals
Transition metals
RT: Liquid metals
Mercury amalgams
Mercury compounds
Mercury deposits
Mercury mines
Mercury vapor lamps
Mercury vapor rectifiers

Mercury amalgams 549.3
DT: January 1993
UF: Mercury and amalgams*
BT: Heavy metal alloys
Transition metal alloys
NT: Dental amalgams
RT: Mercury (metal)
Mercury compounds

*Mercury and amalgams**
USE: Mercury (metal) OR
Mercury amalgams

Mercury arc rectifiers
USE: Mercury vapor rectifiers

Mercury compounds (804.1) (804.2)
DT: January 1986
BT: Heavy metal compounds

* Former Ei Vocabulary term

Mercury compounds *(continued)*
Transition metal compounds
RT: Mercury (metal)
Mercury amalgams

Mercury deposits 504.3, 549.3
DT: Predates 1975
BT: Ore deposits
RT: Mercury (metal)
Mercury mines

Mercury lamps
USE: Mercury vapor lamps

Mercury mines 504.3, 549.3
DT: January 1993
UF: Mercury mines and mining*
BT: Mines
RT: Mercury (metal)
Mercury deposits

*Mercury mines and mining**
USE: Mercury mines

Mercury poisoning
USE: Mercury (metal) AND
Industrial poisons

Mercury vapor lamps 707.2
DT: January 1993
UF: Electric lamps, Mercury vapor*
Mercury lamps
BT: Metal vapor lamps
RT: Mercury (metal)

Mercury vapor rectifiers 714.1
DT: January 1993
UF: Electric rectifiers, Mercury vapor*
Ignitrons
Mercury arc rectifiers
BT: Electric rectifiers
Gas discharge tubes
Power converters
RT: Mercury (metal)

Merging (723.1) (723.2)
DT: January 1993
UF: Computer systems programming—Merging*
BT: Data handling
RT: Computer systems programming
Sorting

MESFET devices 714.2
SN: Metal semiconductor field-effect transistors
DT: January 1993
UF: Metal semiconductor field effect transistors
Schottky gate field effect transistors
Semiconductor devices, MESFET*
BT: Field effect transistors

Mesh (geotextile)
USE: Geotextiles

Metabolism (461.9) (801.2)
DT: January 1987
UF: Biochemistry—Metabolism*
RT: Biochemistry
Enzymes
Fermentation
Metabolites

Metabolites (461.2) (801.2)
DT: June 1990
BT: Biological materials
RT: Metabolism

Metal analysis (531) (801)
DT: Predates 1975
BT: Chemical analysis
NT: Iron analysis
Steel analysis
RT: Alloys
Metallography
Metallurgy
Metals

Metal casting 534.2
SN: Use for the process; for the product, use
METAL CASTINGS
DT: January 1993
BT: Casting
Metallurgy
NT: Carbon dioxide process
Continuous casting
Hot topping
Investment casting
Permanent mold casting
Precision casting
Pressure pouring
Shaw process
Shell mold casting
Sodium silicate process
Zero gravity casting
RT: Aluminum castings
Cast iron
Foundries
Foundry practice
Full mold process
Metal castings
Metal molding

Metal castings 534.2
SN: Use for the product; for the process, use
METAL CASTING
DT: January 1993
UF: Castings*
BT: Castings
NT: Aluminum castings
Cast iron
Die castings
Ingots
Magnesium castings
Steel castings
Titanium castings
Zinc castings
RT: Casting
Ingot molds
Metal casting

Metal ceramics
USE: Cermets

Metal cladding (535.1) (538.2)
SN: Limited to covering one metal with another metal by a metalworking process such as pressure welding, pressure rolling, or casting. Excludes metal coating by electrodeposition, chemical replacement, and dipping processes
DT: Predates 1975
UF: Metals and alloys—Cladding*
Wire—Cladding*
BT: Cladding (coating)
NT: Aluminum cladding
RT: Alloys
Clad metals
Metal finishing
Metallizing
Rolling

Metal cleaning (539) (802.3)
DT: Predates 1975
BT: Metal finishing
Surface cleaning
NT: Blast cleaning
Flame cleaning
RT: Deburring
Foundry practice
Pickling
Shot peening

*Metal cleaning—Blast**
USE: Blast cleaning

*Metal cleaning—Chemical**
USE: Chemical cleaning

*Metal cleaning—Electrolytic**
USE: Electrolytic cleaning

*Metal cleaning—Flame method**
USE: Flame cleaning

*Metal cleaning—Shot peening**
USE: Shot peening

*Metal cleaning—Ultrasonic**
USE: Ultrasonic cleaning

Metal coating
USE: Metals AND
Coating techniques

Metal corrosion
USE: Corrosion

Metal cutting 604.1
DT: Predates 1975
UF: Lubricants—Metal cutting*
BT: Cutting
RT: Broaching
Drilling
Force measurement
Gear cutting
Grinding (machining)
Lubricants
Machine shop practice
Metal finishing
Metal forming
Metallurgy
Milling (machining)
Oxygen cutting
Plasma arc cutting
Reaming
Shearing
Slotting
Tapping (threads)
Thread cutting
Turning

*Metal cutting—Abrasive**
USE: Abrasive cutting

*Metal cutting—Chip disposal**
USE: Industrial waste disposal

*Metal cutting—Chip formation**
USE: Industrial wastes

*Metal cutting—Electric spark**
USE: Spark cutting

*Metal cutting—Electrochemical**
USE: Electrochemical cutting

*Metal cutting—Electron beam**
USE: Electron beam cutting

*Metal cutting—Force measurement**
USE: Force measurement

*Metal cutting—High temperature**
USE: High temperature applications

*Metal cutting—Laser beam**
USE: Laser beam cutting

*Metal cutting—Micromachining**
USE: Micromachining

*Metal cutting—Ultrasonic**
USE: Ultrasonic cutting

*Metal cutting—Underwater**
USE: Underwater cutting

Metal detectors (943.3)
SN: To locate metals
DT: Predates 1975
BT: Detectors
RT: Alloys
Metals

Metal drawing 535.2
DT: Predates 1975
UF: Lubricants—Metal drawing*
BT: Drawing (forming)
Metal forming
NT: Deep drawing
Wire drawing
RT: Lubricants
Metal extrusion
Metal stamping

*Metal drawing—Deep**
USE: Deep drawing

*Metal drawing—Dies**
USE: Drawing dies

* Former Ei Vocabulary term

Metal extrusion 535.2
DT: Predates 1975
UF: Powder metal extrusion
Powder metal products—Extruding*
BT: Extrusion
Metal forming
RT: Alloys
Extruders
Metal drawing
Metals

Metal fatigue
USE: Metals

Metal finishing 604.2
SN: Includes chemical and mechanical surface finishing, but excludes surface coating
DT: Predates 1975
UF: Aluminum castings—Finishing*
Powder metal finishing
Powder metal products—Finishing*
Screw threads—Finishing*
Sheet and strip metal—Finishing*
Welds—Finishing*
BT: Finishing
NT: Blast finishing
Lapping
Metal cleaning
Wire pointing
RT: Alloys
Billets (metal bars)
Buffing
Burnishing
Coating techniques
Electrolytic polishing
Metal cladding
Metal cutting
Metals
Protective coatings
Rollers (machine components)
Surface treatment

*Metal finishing—Barrel**
USE: Barreling

*Metal finishing—Blast**
USE: Blast finishing

*Metal finishing—Burnishing**
USE: Burnishing

*Metal finishing—Deburring**
USE: Deburring

*Metal finishing—Fume control**
USE: Fume control

*Metal finishing—Power supply**
USE: Electric power supplies to apparatus

*Metal finishing—Rollers**
USE: Rollers (machine components)

*Metal finishing—Tumbling**
USE: Barreling

*Metal finishing—Vibration**
USE: Vibratory finishing

Metal foil (535.1)
DT: Predates 1975
UF: Foil
Heat insulating materials—Metal foil*
BT: Sheet metal
NT: Aluminum foil
Tin foil
RT: Foil bearings

Metal forming 535.2
DT: Predates 1975
UF: Lubricants—Metal forming*
BT: Forming
NT: Electrohydraulic forming
Forging
High energy forming
Magnetic forming
Metal drawing
Metal extrusion
Metal molding
Metal pressing
Metal spinning
Metal stamping
Rolling
Swaging
RT: Alloys
Bending (forming)
Lubricants
Machine shop practice
Metal cutting
Metal forming machines
Metal working
Metals

Metal forming machines 535.2.1
DT: October 1975
BT: Machinery
NT: Bending machines
Die casting machines
Forging machines
Presses (machine tools)
Straightening machines
Swaging machines
Wire forming machines
RT: Metal forming

*Metal forming—Bending**
USE: Bending (forming)

*Metal forming—Cold heading**
USE: Cold heading

*Metal forming—Cold working**
USE: Cold working

*Metal forming—Electrohydraulic**
USE: Electrohydraulic forming

*Metal forming—Explosive**
USE: Explosive forming

*Metal forming—High energy**
USE: High energy forming

*Metal forming—Magnetic**
USE: Magnetic forming

*Metal forming—Photochemical**
USE: Photochemical forming

*Metal forming—Pressing**
USE: Metal pressing

*Metal forming—Punching**
USE: Punching

*Metal forming—Spinning**
USE: Metal spinning

*Metal forming—Stretching**
USE: Stretching

*Metal forming—Upsetting**
USE: Upsetting (forming)

Metal insulator boundaries (714.2)
DT: January 1993
UF: Boundaries (metal insulator)
Solid state devices—Metal/insulator boundaries*
RT: Semiconductor devices
Solid state devices

Metal insulator metal devices
USE: MIM devices

Metal insulator semiconductor field effect transistors
USE: MISFET devices

Metal melting (531.1) (534.2)
DT: Predates 1975
BT: Melting
Metallurgy
NT: Electron beam melting
Electroslag remelting
Fluidized bed melting
Levitation melting
Plasma arc melting
Remelting
Zone melting
RT: Cupola practice
Foundry practice
Melting furnaces
Pyrometallurgy

*Metal melting—Electron beam**
USE: Electron beam melting

*Metal melting—Electroslag remelting**
USE: Electroslag remelting

*Metal melting—Fluidized bed**
USE: Fluidized bed melting

*Metal melting—Levitation**
USE: Levitation melting

*Metal melting—Plasma arc**
USE: Plasma arc melting

Metal molding 534.2
DT: January 1993
UF: Foundry practice—Molding*
BT: Metal forming
Molding
RT: Metal casting

Metal oxide semiconductor devices
USE: MOS devices

Metal oxide semiconductor field effect transistors
USE: MOSFET devices

Metal plating
USE: Plating

Metal powders
USE: Powder metals

Metal pressing 535.2
DT: January 1993
UF: Metal forming—Pressing*
BT: Metal forming
Pressing (forming)
RT: Drawing (forming)
Hot pressing
Metal stamping
Perforating
Piercing
Preforming
Punching

Metal recovery (531)
DT: January 1993
UF: Metals and alloys—Recovery*
Slags—Aluminum recovery*
Slags—Iron recovery*
Slags—Manganese recovery*
Slags—Vanadium recovery*
Slags—Zinc recovery*
BT: Recovery
NT: Scrap metal reprocessing
RT: Slags
Waste utilization

Metal refineries 533.2 (402.1)
DT: January 1993
UF: Refineries (metal)
BT: Industrial plants
NT: Metallurgical furnaces
RT: Metal refining

Metal refining 533.2
DT: Predates 1975
BT: Metallurgy
Refining
NT: Aluminum refining
Direct process refining
Gold refining
Lead refining
Nickel refining
Niobium refining
Silver refining
Tin refining
Zinc refining
RT: Alloys
Beneficiation
Extractive metallurgy
Hydrometallurgy
Metal refineries
Metals

Metal refining *(continued)*
Ore treatment
Purification
Pyrometallurgy

Metal semiconductor boundaries
USE: Semiconductor metal boundaries

Metal semiconductor field effect transistors
USE: MESFET devices

Metal sheet
USE: Sheet metal

Metal spinning 535.2
DT: January 1993
UF: Metal forming—Spinning*
Spinning (metal)
BT: Metal forming
RT: Cold working
Hot working

Metal stamping 535.2
DT: June 1990
BT: Metal forming
Stamping
NT: Coinage
RT: Forging
Hot working
Metal drawing
Metal pressing
Punching

Metal strip
USE: Strip metal

Metal testing (531)
DT: January 1993
UF: Fluidity testing
Metals testing
BT: Materials testing
NT: Notched bar tensile testing
Steel testing
RT: Hardness testing
Metals

Metal vapor lamps 707.2
DT: January 1993
UF: Electric lamps, Metal vapor*
BT: Discharge lamps
NT: Mercury vapor lamps

Metal working 535.2.2
DT: January 1993
BT: Processing
NT: Cold working
Hot working
Wire flattening
RT: Deep drawing
Metal forming
Metal working tools
Metallurgy
Soldering
Straightening
Strip mills
Welding

Metal working planers 603.1
DT: January 1993
UF: Planers, Metal working*
BT: Metal working tools
Planers

Metal working saws (605)
DT: January 1993
UF: Power saws
Saws—Metal working*
BT: Metal working tools
Saws

Metal working shapers 603.1
DT: January 1993
UF: Shapers (metal working)
Shapers, Metal working*
BT: Metal working tools

Metal working tools 603.1
DT: January 1993
UF: Cutting tools—Metal working*
BT: Machine tools
NT: Metal working planers
Metal working saws
Metal working shapers
Milling cutters
RT: Cutting tools
Metal working

Metallic compounds (804.1) (804.2)
SN: Excludes metal organic compounds, for which use ORGANOMETALLICS
DT: Predates 1975
BT: Chemical compounds
NT: Alkali metal compounds
Alkaline earth metal compounds
Aluminum compounds
Gallium compounds
Germanium compounds
Heavy metal compounds
Indium compounds
Plutonium compounds
Rare earth compounds
Technetium compounds
Thorium compounds
Transition metal compounds
Uranium compounds
RT: Alloys
Inorganic compounds
Intermetallics
Metals
Sulfide minerals

Metallic films (531) (539)
DT: June 1990
BT: Films
RT: Metallic superlattices
Metals
Thick films
Thin films

Metallic glass 531 (933.2)
DT: January 1993
UF: Glass, Metallic*
BT: Amorphous materials

* Former Ei Vocabulary term

Metallic glass *(continued)*
RT: Alloys
Glass
Melt spinning
Metals
Rapid solidification

Metallic matrix composites 531 (415.1)
DT: January 1993
UF: Metals and alloys—Metallic matrix composites*
BT: Composite materials
RT: Alloys
Fiber reinforced metals
Fibers
Metals

Metallic refractories
USE: Refractory metals

Metallic soaps 804.1
DT: January 1993
UF: Soap—Metallic*
Soaps (metallic)
BT: Heavy metal compounds
RT: Saponification

Metallic superlattices 531.2 (933.1.1)
DT: January 1993
UF: Superlattices—Metallic*
BT: Superlattices
RT: Metallic films

Metallic textiles 531, 819.5
DT: January 1993
UF: Textiles—Metallic materials*
BT: Textiles
RT: Metals

Metallizing 539.3, 813.1
DT: Predates 1975
BT: Coating techniques
RT: Cladding (coating)
Metal cladding
Plating
Sputter deposition
Substrates
Vapor deposition

Metallographic microstructure 531.2
DT: January 1993
BT: Microstructure
RT: Eutectics
Inclusions
Metallography
Recrystallization (metallurgy)

Metallographic phases 531.2
DT: January 1993
NT: Austenite
Bainite
Ferrite
Martensite
Pearlite

Metallography 531.2
DT: Predates 1975
UF: Nuclear fuels—Metallography*
NT: Aluminum metallography
Barium metallography
Beryllium metallography
Bismuth metallography
Cadmium metallography
Chromium metallography
Cobalt metallography
Copper metallography
Eutectics
Germanium metallography
Gold metallography
Indium metallography
Iron metallography
Lead metallography
Lithium metallography
Magnesium metallography
Manganese metallography
Molybdenum metallography
Nickel metallography
Niobium metallography
Rhodium metallography
Segregation (metallography)
Silver metallography
Sodium metallography
Steel metallography
Tantalum metallography
Thallium metallography
Thorium metallography
Tin metallography
Titanium metallography
Tungsten metallography
Uranium metallography
Vanadium metallography
Yttrium metallography
Zinc metallography
Zirconium metallography
RT: Alloys
Anisotropy
Crystal defects
Crystal lattices
Crystal whiskers
Crystallography
Decoration
Electron microscopy
Etching
Fractography
Grain boundaries
Grain growth
Grain size and shape
Graphitization
Metal analysis
Metallographic microstructure
Metals
Optical microscopy
Phase transitions
Radiography
Specimen preparation
X ray diffraction

*Metallography—Austenite**
USE: Austenite

*Metallography—Bainite**
USE: Bainite

* Former Ei Vocabulary term

*Metallography—Graphitization**
USE: Graphitization

*Metallography—Martensite**
USE: Martensite

*Metallography—Pearlite**
USE: Pearlite

*Metallography—Segregation**
USE: Segregation (metallography)

*Metallography—Specimen preparation**
USE: Specimen preparation

*Metallography—Transformations**
USE: Phase transitions

Metalloids (804)
DT: January 1993
UF: Metals and alloys—Metalloids*
Semimetals
BT: Nonmetals
NT: Arsenic
Boron
Germanium
Polonium
Silicon
Tellurium
RT: Alloys
Intermetallics
Metals

Metallorganic chemical vapor deposition 802.2
DT: January 1995
UF: Metalorganic chemical vapor deposition
MOCVD
BT: Chemical vapor deposition
RT: Organometallics
Semiconducting films
Semiconductor growth

Metallorganic compounds
USE: Organometallics

Metallorganic polymers 815.1.1
SN: Polymers with organic groups coordinated or complexed to metal atoms
DT: January 1983
UF: Metalorganic polymers
BT: Organic polymers
Organometallics

Metallorganic vapor phase epitaxy 933.1.2
DT: January 1995
UF: Metalorganic vapor phase epitaxy
MOVPE
OMVPE
BT: Vapor phase epitaxy
RT: Organometallics
Semiconducting films
Semiconductor growth

Metallurgical coke 524
DT: January 1993
UF: Coke—Metallurgical*
BT: Coke
RT: Metallurgy

Metallurgical furnaces (532)
DT: January 1993
UF: Bessemer converters
Coke ovens—Metallurgical*
Converters (Bessemer)
Furnaces, Metallurgical*
BT: Furnaces
Metal refineries
NT: Basic oxygen converters
Blast furnaces
Open hearth furnaces
Steelmaking furnaces
RT: Ore treatment

Metallurgy 531.1
DT: Predates 1975
UF: Nuclear fuels—Metallurgy*
Welds—Metallurgy*
NT: Aluminum metallurgy
Antimony metallurgy
Beryllium metallurgy
Bismuth metallurgy
Cadmium metallurgy
Chromium metallurgy
Cobalt metallurgy
Copper metallurgy
Extractive metallurgy
Germanium metallurgy
Gold metallurgy
Indium metallurgy
Iron metallurgy
Ladle metallurgy
Lead metallurgy
Magnesium metallurgy
Manganese metallurgy
Metal casting
Metal melting
Metal refining
Molybdenum metallurgy
Nickel metallurgy
Niobium metallurgy
Powder metallurgy
Pyrometallurgy
Silver metallurgy
Steel metallurgy
Strengthening (metal)
Tapping (furnace)
Tin metallurgy
Titanium metallurgy
Tungsten metallurgy
Uranium metallurgy
Vanadium metallurgy
Zinc metallurgy
Zirconium metallurgy
RT: Aging of materials
Alloying
Dispersion hardening
Foundry practice
Heat treatment
Machining
Metal analysis
Metal cutting
Metal working
Metallurgical coke

* Former Ei Vocabulary term

Metallurgy *(continued)*
Phase diagrams
Physical chemistry
Rare earth additions
Smelting
Surface treatment
Welds

*Metallurgy—Electrolytic**
USE: Electrometallurgy

Metalorganic chemical vapor deposition
USE: Metallorganic chemical vapor deposition

Metalorganic compounds
USE: Organometallics

Metalorganic polymers
USE: Metallorganic polymers

Metalorganic vapor phase epitaxy
USE: Metallorganic vapor phase epitaxy

Metals (531)
DT: January 1993
UF: Metal coating
Metal fatigue
Metals and alloys*
BT: Chemical elements
NT: Nonferrous metals
Transition metals
RT: Alloys
Clad metals
Composite materials
Corrosion
Corrosion resistant alloys
Dissimilar metals
Electron energy levels
Fiber reinforced metals
Filler metals
Galvanized metal
Germanium
Hard facing
Hardness testing
Intermetallics
Internal oxidation
Liquid metals
Machinability
Metal analysis
Metal detectors
Metal extrusion
Metal finishing
Metal forming
Metal refining
Metal testing
Metallic compounds
Metallic films
Metallic glass
Metallic matrix composites
Metallic textiles
Metallography
Metalloids
Plate metal
Plating
Powder metals
Scrap metal
Semiconductor metal boundaries
Sheet metal
Strip metal
Structural metals
Welding

*Metals and alloys**
USE: Metals OR
Alloys

*Metals and alloys—Age hardening**
USE: Age hardening

*Metals and alloys—Anodic oxidation**
USE: Anodic oxidation

*Metals and alloys—Bimetals**
USE: Bimetals

*Metals and alloys—Cladding**
USE: Metal cladding

*Metals and alloys—Continuous casting**
USE: Continuous casting

*Metals and alloys—Corrosion resisting**
USE: Corrosion resistant alloys

*Metals and alloys—Corrosion**
USE: Corrosion

*Metals and alloys—Dispersion hardening**
USE: Dispersion hardening

*Metals and alloys—Ductility**
USE: Ductility

*Metals and alloys—Explosive bonding**
USE: Explosive bonding

*Metals and alloys—Formability**
USE: Formability

*Metals and alloys—Gas alloying**
USE: Gas alloying

*Metals and alloys—Hard facing**
USE: Hard facing

*Metals and alloys—Hydrogen embrittlement**
USE: Hydrogen embrittlement

*Metals and alloys—Internal oxidation**
USE: Internal oxidation

*Metals and alloys—Machinability**
USE: Machinability

*Metals and alloys—Metallic matrix composites**
USE: Metallic matrix composites

*Metals and alloys—Metalloids**
USE: Metalloids

*Metals and alloys—Plating**
USE: Plating

*Metals and alloys—Rare earth additions**
USE: Rare earth additions

*Metals and alloys—Recovery**
USE: Metal recovery

* Former Ei Vocabulary term

*Metals and alloys—Sealing**
USE: Sealing (finishing)

*Metals and alloys—Sputtering**
USE: Sputter deposition

*Metals and alloys—Strengthening**
USE: Strengthening (metal)

*Metals and alloys—Stress corrosion cracking**
USE: Stress corrosion cracking

*Metals and alloys—Structural**
USE: Structural metals

*Metals and alloys—Weldability**
USE: Weldability

*Metals and alloys—Zone melting**
USE: Zone melting

Metals testing
USE: Metal testing

*Metals—Gases**
USE: Inclusions AND
Gases

*Metals—Molten**
USE: Liquid metals

Metamagnetism 701.2 (708.4)
DT: January 1993
UF: Magnetic materials—Metamagnetism*
Magnetism—Metamagnetism*
BT: Magnetism
RT: Antiferromagnetism
Ferromagnetism
Magnetic materials

Metamorphic rocks (481.1.2)
DT: January 1993
UF: Rock—Metamorphic*
BT: Crystalline rocks
NT: Marble
Slate
Steatite
RT: Petrology

Meteor burst communication 716.3
DT: January 1993
UF: Meteor scatter communication
Radio communication—Meteor bursts*
BT: Radio communication
RT: Radio links

Meteor impacts (655) (657.2) (914.1)
DT: January 1993
UF: Satellites—Meteor impact*
Spacecraft—Meteor impact*
RT: Accidents
Meteorites
Satellites
Spacecraft

Meteor scatter communication
USE: Meteor burst communication

Meteorites 657.2
DT: Predates 1975
RT: Meteor impacts
Meteoritic glass

Meteoritic glass 812.3 (657.2)
DT: January 1993
UF: Glass—Meteoric origin*
BT: Glass
RT: Meteorites

Meteorological balloons 443.2, 652.5
DT: January 1993
UF: Balloons—Meteorology*
Weather balloons
BT: Balloons
RT: Meteorological instruments
Weather forecasting

Meteorological instruments 443.2
DT: Predates 1975
BT: Instruments
NT: Anemometers
Barometers
Hygrometers
RT: Meteorological balloons
Meteorological radar
Meteorology
Radiosondes
Sounding rockets
Weather satellites

Meteorological problems 443.1
DT: January 1993
UF: Aircraft—Meteorological problems*
Electric utilities—Weather service*
Spacecraft—Meteorological problems*
Weather problems
RT: Aircraft
Meteorology
Spacecraft
Weather forecasting
Weather modification

Meteorological radar 443.2, 716.2
DT: January 1993
UF: Radar—Meteorological*
Weather radar
BT: Radar
RT: Meteorological instruments
Meteorology

Meteorological rockets
USE: Sounding rockets

Meteorological satellites
USE: Weather satellites

Meteorology (443)
DT: Predates 1975
UF: Aviation—Meteorology*
BT: Earth sciences
NT: Climatology
RT: Atmospheric chemistry
Atmospheric composition
Atmospheric corrosion
Atmospheric density
Atmospheric electricity
Atmospheric humidity

Meteorology *(continued)*
Atmospheric ionization
Atmospheric movements
Atmospheric optics
Atmospheric pressure
Atmospheric radiation
Atmospheric structure
Atmospheric temperature
Atmospheric thermodynamics
Atmospheric turbulence
Cloud seeding
Clouds
Drought
Earth atmosphere
Environmental engineering
Extraterrestrial atmospheres
Fog
Hurricane effects
Hurricanes
Ice
Meteorological instruments
Meteorological problems
Meteorological radar
Precipitation (meteorology)
Rain
Snow
Snowfall measurement
Storms
Thunderstorms
Tornadoes
Weather forecasting
Weather modification

*Meteorology—Atmospheric precipitation**
USE: Precipitation (meteorology)

*Meteorology—Climatology**
USE: Climatology

*Meteorology—Clouds**
USE: Clouds

*Meteorology—Fog**
USE: Fog

*Meteorology—Storms**
USE: Storms

*Meteorology—Thunderstorms**
USE: Thunderstorms

*Meteorology—Weather forecasting**
USE: Weather forecasting

*Meteorology—Weather modification**
USE: Weather modification

Meters
USE: Instruments

Methanation 802.2
DT: January 1993
BT: Chemical reactions
RT: Coal gasification
Methane
Synthesis gas
Synthesis gas manufacture

Methane 804.1 (522)
DT: Predates 1975
BT: Hydrocarbons
NT: Liquid methane
RT: Biogas
Firedamp
Methanation
Methanogens
Natural gas
Paraffins

*Methane—Liquid**
USE: Liquid methane

Methanogenic bacteria
USE: Methanogens

Methanogens 461.9 (801.2)
DT: January 1987
UF: Methanogenic bacteria
BT: Microorganisms
RT: Bacteria
Fermentation
Methane

Methanol 804.1
DT: January 1981
UF: Methyl alcohol
Wood alcohol
BT: Alcohols
RT: Gasohol
Methanol fuels
Organic solvents
Solvents

Methanol fuels 523 (804.1)
DT: January 1993
BT: Alcohol fuels
RT: Automotive fuels
Diesel fuels
Gasohol
Methanol

Method of moments
USE: Estimation

Methyl alcohol
USE: Methanol

Methyl ethylene
USE: Propylene

Metric system (902.2)
DT: Predates 1975
UF: International system of units
BT: Units of measurement
RT: Measurements

Metrology
USE: Measurements

Metropolitan area networks (716) (717) (718) (723)
DT: January 1993
UF: Computer networks—Metropolitan area networks*
BT: Computer networks

* Former Ei Vocabulary term

Mg
USE: Magnesium

MHD
USE: Magnetohydrodynamics

MHD converters
USE: Magnetohydrodynamic converters

MHD power generation
USE: Magnetohydrodynamic power generation

MHD power plants
USE: Magnetohydrodynamic power plants

Mica 482.2 (413.1)
DT: Predates 1975
UF: Electric insulating materials—Mica*
BT: Silicate minerals
RT: Electric insulating materials
Granite
Mica deposits

Mica deposits 505.1
DT: Predates 1975
BT: Mineral resources
RT: Mica

*Mica—Synthesizing**
USE: Synthesis (chemical)

Mice (computer peripherals) 722.2
DT: January 1993
UF: Computer peripheral equipment—Mouse*
Mouse (computer peripheral)
BT: Interactive devices
RT: Computer terminals
Light pens

Micelles 801.3
DT: January 1993
UF: Colloids—Micelles*
BT: Colloids
RT: Critical micelle concentration

Micro-optics
USE: Optics

Microanalysis (801)
DT: Predates 1975
BT: Chemical analysis
RT: Gravimetric analysis
Neutron activation analysis
Volumetric analysis

Microbial electrodes (461.8) (462.1) (801.2)
SN: Microbes used as electrodes for purposes of detection
DT: January 1993
UF: Biosensors—Microbial electrodes*
BT: Biosensors
Microorganisms

Microbiology 461.9 (801.2)
DT: January 1987
BT: Biology
NT: Bacteriology
RT: Fungi
Microorganisms
Viruses

Microcirculation 461.1
DT: January 1993
UF: Biomedical engineering—Microcirculation*
BT: Hemodynamics
RT: Biomedical engineering
Capillary flow

Microcomputers 722.4
DT: January 1993
UF: Computers, Microcomputer*
Computers, Microprocessor*
BT: Digital computers
NT: Personal computers
RT: Computer workstations
Microprocessor chips
Microprogramming
Minicomputers
Pocket calculators
Transputers

Microcracking
USE: Crack initiation

Microcracks (421)
DT: January 1995
BT: Cracks
RT: Crack initiation

Microelectrodes (462.1) (801.2)
DT: January 1993
UF: Biomedical equipment—Microelectrodes*
BT: Electrodes
RT: Biomedical equipment
Biosensors
Electronic medical equipment
Probes

Microelectromechanical devices (601.1) (704.1) (732.1)
DT: January 1993
UF: Electromechanical devices—Microelectromechanical*
BT: Electromechanical devices

Microelectronic processing (714)
DT: January 1993
BT: Processing
RT: Integrated circuit manufacture
Microelectronics
Semiconductor device manufacture

Microelectronics (713) (714)
DT: January 1987
BT: Electronics industry
RT: Integrated circuits
Microelectronic processing

Microemulsions 804 (801.3)
DT: January 1993
UF: Colloids—Microemulsions*
BT: Emulsions

Microfilm 742.3
DT: Predates 1975
BT: Microforms

Microforms 742.3 (903.2)
DT: January 1993
UF: Information dissemination—Microforms*

* Former Ei Vocabulary term

Microforms *(continued)*
BT: Photographic films
NT: Microfilm
RT: Information dissemination
Information technology
Publishing

Microgravity processing 656.2, 931.5
DT: January 1993
UF: Satellites—Microgravity materials processing*
BT: Processing
RT: Gravitation
Satellites
Weightlessness

Microhardness
USE: Hardness

Microlithography
USE: Lithography

Micromachining 604.2
DT: January 1993
UF: Metal cutting—Micromachining*
BT: Machining

Micromechanics
USE: Composite micromechanics

Micrometers 943.1
DT: Predates 1975
UF: Micrometry
BT: Instruments
RT: Angle measurement
Distance measurement
Interferometry
Spatial variables measurement
Thickness measurement

Micrometry
USE: Micrometers

Microoptics
USE: Optics

Microorganisms 461.9 (801.2)
DT: January 1987
NT: Bacteria
Methanogens
Microbial electrodes
Protozoa
Viruses
RT: Algae
Antibiotics
Bacteriology
Biological materials
Fungi
Microbiology
Yeast

*Microorganisms—Protozoa**
USE: Protozoa

Microphones 752.1
DT: Predates 1975
BT: Audio equipment
Electroacoustic transducers
RT: Acoustic receivers
Hydrophones
Radio studios
Radio transmitters
Sound recording
Sound reproduction
Television receivers
Transmitters

Microphonics
USE: Microphonism

Microphonism 714.1 (701.1) (716.1)
DT: January 1993
UF: Microphonics
Noise, Spurious signal—Microphonism*
BT: Spurious signal noise

Microprocessor chips (714.2) (721) (722)
DT: June 1990
UF: Chips (microprocessor)
Microprocessors
BT: Digital integrated circuits
NT: Transputers
RT: LSI circuits
Microcomputers
Smart cards

*Microprocessor chips—Smart cards**
USE: Smart cards

Microprocessors
USE: Microprocessor chips

Microprogramming 723.1
DT: January 1993
UF: Computer architecture—Microprogramming*
BT: Computer programming
RT: Computer architecture
Microcomputers

Microscopes 741.3
DT: Predates 1975
BT: Instruments
NT: Acoustic microscopes
Electron microscopes
Field emission microscopes
Ion microscopes
Photoelectric microscopes
X ray microscopes
RT: Image intensifiers (solid state)
Image intensifiers (electron tube)
Microscopic examination

*Microscopes, Acoustic**
USE: Acoustic microscopes

*Microscopes, Electron**
USE: Electron microscopes

*Microscopes, Ion**
USE: Ion microscopes

*Microscopes—Field emission**
USE: Field emission microscopes

*Microscopes—Image intensifiers**
USE: Image intensifiers (solid state) OR
Image intensifiers (electron tube)

*Microscopes—Photoelectric**
USE: Photoelectric microscopes

* Former Ei Vocabulary term

*Microscopes—X-ray**
USE: X ray microscopes

Microscopic examination (741.3)
DT: Predates 1975
UF: Microscopy
NT: Atomic force microscopy
Electron microscopy
Fractography
Optical microscopy
Scanning tunneling microscopy
RT: Microscopes
Particle size analysis
Specimen preparation

*Microscopic examination—Scanning electron microscopy**
USE: Scanning electron microscopy

*Microscopic examination—Specimen preparation**
USE: Specimen preparation

*Microscopic examination—Transmission electron microscopy**
USE: Transmission electron microscopy

Microscopy
USE: Microscopic examination

Microstrip antennas (716) (708.1)
DT: January 1993
UF: Antennas, Microstrip*
BT: Antennas
Microstrip devices
RT: Microwave antennas

Microstrip devices (708.1) (716)
DT: January 1977
BT: Dielectric devices
Thin film devices
NT: Microstrip antennas
RT: Microstrip lines

Microstrip lines (716)
DT: January 1995
BT: Strip telecommunication lines
RT: Dielectric materials
Microstrip devices
Microwave circuits
Thin films

Microstructure (531.2) (801.4) (933)
DT: January 1993
NT: Crystal microstructure
Metallographic microstructure
Textures
RT: Molecular structure
Porosity

Microtrons 932.1.1
DT: January 1993
UF: Accelerators, Microtron*
BT: Cyclotrons

Microwave acoustic devices 752.1
DT: January 1993
UF: Acoustic devices—Microwave frequencies*
BT: Acoustic devices
Microwave devices
RT: Microwave acoustics
Ultrasonic transducers

Microwave acoustic frequencies
USE: Microwave acoustics

Microwave acoustics (751.1) (931.1)
DT: January 1993
UF: Microwave acoustic frequencies
BT: Acoustics
RT: Microwave acoustic devices

Microwave amplifiers 713.1
DT: January 1993
UF: Amplifiers, Microwave*
BT: Amplifiers (electronic)
Microwave devices
NT: Diode amplifiers
Masers
RT: Gyrotrons
Microwave circuits

Microwave antennas (716)
DT: January 1993
UF: Antennas—Microwave*
BT: Antennas
Microwave devices
RT: Microstrip antennas

Microwave circuits (714.2)
DT: January 1995
BT: Networks (circuits)
NT: Microwave integrated circuits
RT: Distributed parameter networks
Microstrip lines
Microwave amplifiers
Microwave filters
Microwave limiters
Microwave oscillators
Passive networks

Microwave circulators 714.3
DT: Predates 1975
UF: Circulators (microwave)
BT: Microwave devices
RT: Waveguide circulators

Microwave demodulators 713.3
DT: January 1993
UF: Demodulators, Microwave*
BT: Demodulators
Microwave devices

Microwave devices (714) (715)
DT: Predates 1975
UF: Microwave radio equipment
Radio receivers—Microwave frequencies*
Radio transmitters—Microwave frequencies*
BT: Electronic equipment
NT: Microwave acoustic devices
Microwave amplifiers
Microwave antennas
Microwave circulators
Microwave demodulators
Microwave filters
Microwave isolators

Microwave devices *(continued)*
Microwave limiters
Microwave oscillators
Microwave ovens
Microwave tubes
Millimeter wave devices
RT: Microwaves
Radio equipment
Reflectometers
Waveguides

Microwave filters 703.2
DT: January 1993
UF: Electric filters, Microwave*
BT: Electric filters
Microwave devices
RT: Cavity resonators
Microwave circuits
Radar equipment
Waveguide components

Microwave generation 713.2 (714.1) (714.2)
DT: Predates 1975
RT: Gyrotrons
Microwave oscillators
Microwaves

Microwave heaters
USE: Microwave heating

Microwave heating 642.1, 711.1
DT: January 1995
UF: Microwave heaters
BT: Electric heating
RT: Industrial heating

Microwave holography 743.1.3
DT: January 1977
BT: Holography
RT: Microwaves

Microwave integrated circuits 714.2
DT: January 1995
BT: Integrated circuits
Microwave circuits
NT: Monolithic microwave integrated circuits
RT: Strip telecommunication lines

Microwave isolators 714.3
DT: Predates 1975
UF: Isolators (microwave)
BT: Microwave devices
RT: Waveguide components
Waveguide isolators

Microwave limiters 713.3
DT: Predates 1975
BT: Limiters
Microwave devices
RT: Microwave circuits
Waveguide components

Microwave links (716.3)
DT: January 1993
UF: Telecommunication links, Microwave*
BT: Radio links

Microwave measurement 942.2
DT: January 1993
UF: Microwave measurements*
BT: Electric variables measurement
RT: Microwaves
Radiometry
Wavemeters

*Microwave measurements**
USE: Microwave measurement

Microwave oscillators 713.2
DT: January 1993
UF: Oscillators, Microwave*
BT: Microwave devices
Oscillators (electronic)
NT: Gunn oscillators
RT: Microwave circuits
Microwave generation
Radar equipment

Microwave ovens (704.2)
DT: January 1993
UF: Electric appliances—Microwave ovens*
BT: Electric ovens
Microwave devices
RT: Electric appliances

Microwave power transmission 706.1.1
DT: January 1993
UF: Electric power generation—Microwave energy*
Electric power transmission—Microwave frequencies*
BT: Electric power transmission
RT: Microwaves

Microwave radio equipment
USE: Radio equipment AND
Microwave devices

Microwave spectrographs (711) (741.3) (742.1)
SN: Microwave spectroscopes that produce photographic or fluorescent-screen spectrograms
DT: January 1993
UF: Spectrographs, Microwave*
BT: Spectrographs
RT: Microwave spectrometers
Microwave spectroscopy

Microwave spectrometers (711) (741.3)
SN: Spectroscopes capable of measuring angular deviation of microwave radiation
DT: January 1993
UF: Spectrometers, Microwave*
BT: Spectrometers
RT: Magnetic resonance spectrometers
Microwave spectrographs
Microwave spectroscopy

Microwave spectroscopy (711) (741.3) (801)
DT: January 1993
UF: Spectroscopy, Microwave*
BT: Radiofrequency spectroscopy
RT: Electron spin resonance spectroscopy
Magnetic resonance spectroscopy

* Former Ei Vocabulary term

Microwave spectroscopy *(continued)*
Microwave spectrometers
Microwave spectrographs
Paramagnetic resonance

Microwave tubes 714.1
DT: January 1993
UF: Electron tubes, Microwave*
Platinotron tubes
Resnatron tubes
BT: Electron tubes
Microwave devices
NT: Electron wave tubes
Gyrotrons

Microwaves 711
DT: January 1977
BT: Electromagnetic waves
RT: Microwave devices
Microwave generation
Microwave holography
Microwave measurement
Microwave power transmission
Millimeter waves

Mild steel
USE: Carbon steel

Military aircraft 652.1.2
DT: January 1993
UF: Aircraft, Military*
BT: Aircraft
Military equipment
NT: Antisubmarine aircraft
Bombers
Fighter aircraft
Military helicopters
Military seaplanes
Target drones
RT: Aircraft carriers
Deck landing aircraft
Military aviation

Military airports 404.1, 431.4
DT: January 1993
UF: Airports—Military*
BT: Airports
RT: Military bases
Military engineering
Military equipment

Military applications 404.1
DT: January 1993
UF: Operations research—Military purposes*
Television applications—Military*
BT: Applications
RT: Military engineering
Military operations

Military aviation 404.1, 431.1
DT: January 1993
UF: Aviation, Military*
BT: Aviation
Military operations
RT: Ground operations
Military aircraft

Military bases 404.1
DT: January 1993
BT: Facilities
NT: Missile bases
Naval bases
RT: Military airports
Military operations

Military bridges 401.1, 404.1
DT: January 1993
UF: Bridges, Military*
BT: Bridges
Military equipment

Military communications 404.1 (716)
DT: Predates 1975
BT: Military operations
RT: Command and control systems
Communication systems
Military electronic countermeasures
Military engineering
Military equipment
Military radio equipment
Military satellites
Military telephone exchanges

Military construction equipment 404.1, 405.1
DT: January 1993
UF: Construction equipment—Military*
BT: Construction equipment
Military equipment
RT: Military engineering
Military operations

Military countermeasures
USE: Military electronic countermeasures

Military data processing 723.2, 404.1
DT: January 1993
UF: Data processing—Military purposes*
BT: Data processing
Military operations

Military electronic countermeasures
404.1 (715) (716)
DT: January 1993
UF: Countermeasures (military)
ECCM
ECM
Electronic counter-countermeasures
Electronic countermeasures (military)
Military countermeasures
Military engineering—Electronic countermeasures*
BT: Electronic warfare
NT: Radar countermeasures
RT: Military communications
Military engineering

Military engineering 404.1
DT: Predates 1975
BT: Engineering
RT: Ballistics
Camouflage
Civil defense
Electronic warfare
Logistics

* Former Ei Vocabulary term

Military engineering *(continued)*
Military airports
Military applications
Military communications
Military construction equipment
Military electronic countermeasures
Military geology
Military mapping
Military operations
Military power plants
Military textiles
Naval warfare

*Military engineering—Camouflage**
USE: Camouflage

*Military engineering—Chemical warfare**
USE: Chemical warfare

*Military engineering—Electronic countermeasures**
USE: Military electronic countermeasures

*Military engineering—Electronic warfare**
USE: Electronic warfare

*Military engineering—Logistics**
USE: Logistics

*Military engineering—Photography**
USE: Military photography

Military equipment 404.1
DT: Predates 1975
BT: Equipment
NT: Arsenals
Fire control systems
Military aircraft
Military bridges
Military construction equipment
Military radar
Military radio equipment
Military rockets
Military satellites
Military seaplanes
Military telephone exchanges
Military vehicles
Missile launching systems
Ordnance
Shelters (from attack)
RT: Ballistics
Camouflage
Electromagnetic launchers
Gas masks
Military airports
Military communications
Military operations
Military power plants
Military textiles
Mobile telecommunication systems
Naval bases
Radar
Radar equipment
Remote sensing

*Military equipment—Armor**
USE: Armor

*Military equipment—Arsenals**
USE: Arsenals

*Military equipment—Shelters**
USE: Shelters (from attack)

Military geology 404.1, 481.1
DT: January 1993
UF: Geology—Military*
BT: Geology
RT: Military engineering
Military operations

Military helicopters 652.4, 404.1
DT: January 1993
UF: Helicopters, Military*
BT: Helicopters
Military aircraft

Military mapping 404.1, 405.3 (902.1)
DT: January 1993
UF: Maps and mapping—Military*
BT: Mapping
Military operations.
RT: Military engineering

Military operations 404.1
DT: January 1993
UF: Warfare
NT: Bombing
Chemical warfare
Electronic warfare
Ground operations
Logistics
Military aviation
Military communications
Military data processing
Military mapping
Military photography
Naval warfare
RT: Camouflage
Civil defense
Military applications
Military bases
Military construction equipment
Military engineering
Military equipment
Military geology
Military power plants
Military textiles
Missile bases
Naval bases
Space surveillance
Strategic materials
Warship preservation

Military photography 742.1, 404.1
DT: January 1993
UF: Military engineering—Photography*
BT: Military operations
Photography

Military power plants
404.1 (612) (613) (614) (615)
DT: January 1993
UF: Power plants—Military*
BT: Power plants

Military power plants *(continued)*
RT: Military engineering
Military equipment
Military operations

Military radar 716.2, 404.1
DT: January 1993
UF: Radar—Military*
BT: Military equipment
Radar
RT: Radar warning systems

Military radio equipment 716.3, 404.1
DT: January 1993
UF: Radio equipment—Military*
BT: Military equipment
Radio equipment
RT: Military communications

Military rockets 654.1, 404.1
DT: January 1993
UF: Defense rockets
Rockets and missiles—Military*
BT: Military equipment
Rockets
RT: Missiles
Ordnance

Military satellites 655.2, 404.1
DT: January 1993
UF: Defense satellites
Satellites—Defense applications*
BT: Military equipment
Satellites
RT: Military communications

Military seaplanes 652.1.2, 404.1
DT: January 1993
UF: Seaplanes, Military*
BT: Military aircraft
Military equipment
Seaplanes

Military telephone exchanges 718.1, 404.1
DT: January 1993
UF: Telephone exchanges—Military*
BT: Military equipment
Telephone exchanges
RT: Military communications

Military textiles 404.1, 819.2
DT: January 1993
UF: Textiles—Military*
BT: Textiles
RT: Military engineering
Military equipment
Military operations

Military vehicles 404.1 (662.1) (663.1)
SN: For use on land
DT: Predates 1975
BT: Military equipment
Vehicles
NT: Tanks (military)
RT: Automobiles

*Military vehicles—Tanks**
USE: Tanks (military)

Milking machines 821.1
DT: Predates 1975
BT: Agricultural machinery

Millimeter wave devices (714) (715)
DT: January 1995
BT: Microwave devices

Millimeter waves 711
DT: January 1977
BT: Electromagnetic waves
RT: Microwaves

Milling (comminution)
USE: Comminution

Milling (machining) 604.2
DT: January 1993
UF: Milling*
BT: Machining
RT: Gear cutting
Grinding (machining)
Metal cutting
Milling cutters
Milling machines
Thread cutting

Milling (sugar cane)
USE: Sugar cane milling

Milling (textile)
USE: Textile finishing

Milling cutters 603.1
DT: Predates 1975
BT: Metal working tools
RT: Carbide tools
Ceramic tools
Milling (machining)
Milling machines
Saws

Milling machines 603.1
DT: Predates 1975
BT: Machine tools
RT: Milling (machining)
Milling cutters
Planers

*Milling**
USE: Milling (machining)

Mills (rolling)
USE: Rolling mills

MIM devices 714.2
DT: January 1993
UF: Metal insulator metal devices
Solid state devices, MIM*
BT: Solid state devices

Mine augers 502.2
DT: January 1993
BT: Cutting machines (mining)

Mine cars 502.2
DT: Predates 1975
BT: Railroad cars
RT: Freight cars
Hopper cars

* Former Ei Vocabulary term

Mine cars *(continued)*
Materials handling equipment
Mining equipment
Underground mine transportation

Mine dust 502.1, 451.1
DT: Predates 1975
BT: Dust
RT: Air pollution
Dust collectors
Dust control
Hazards
Health hazards
Mine explosions
Mines
Mining

Mine explosions 502.1 (914.1) (914.2)
DT: January 1993
UF: Mines and mining—Explosions*
BT: Underground explosions
RT: Hazards
Mine dust
Mine fires
Mine rescue
Mines
Mining

Mine fires 914.2, 502.1
DT: January 1993
UF: Mines and mining—Fires*
BT: Fires
RT: Hazards
Mine explosions
Mine rescue
Mines
Mining

Mine flooding 914.1, 502.1
DT: January 1993
UF: Flooding (mines)
Mines and mining—Flooding*
RT: Floods
Hazards
Mine rescue
Mines
Mining

Mine hoists 502.2
DT: Predates 1975
UF: Winding engines
BT: Hoists
Mining machinery
NT: Head frames
Koepe winders
RT: Cages (mine hoists)

*Mine hoists—Cages**
USE: Cages (mine hoists)

*Mine hoists—Guides**
USE: Guides (mechanical)

*Mine hoists—Head frames**
USE: Head frames

*Mine hoists—Koepe system**
USE: Koepe winders

*Mine hoists—Safety devices**
USE: Safety devices

*Mine hoists—Wire rope**
USE: Wire rope

Mine locomotives 502.2, 682.1.2
DT: January 1993
UF: Locomotives—Mine*
BT: Industrial locomotives
RT: Mine transportation
Mining machinery

Mine power plants 502.2 (612) (614)
DT: January 1993
UF: Power plants—Mines*
BT: Power plants
RT: Mining equipment

Mine rescue 502.1, 914.1
DT: Predates 1975
UF: Rescue (mining accidents)
RT: Hazards
Mine explosions
Mine fires
Mine flooding
Mining
Protective clothing

*Mine rescue—Protective clothing**
USE: Protective clothing

Mine roof control 502.1 (914.1)
DT: January 1993
UF: Mines and mining—Roof control*
Roof control (mines)
BT: Control
Mining
RT: Mines
Roofs

Mine roof supports 502.2 (914.1)
DT: January 1993
UF: Mines and mining—Roof supports*
Roof supports (mine)
BT: Mining equipment
Supports

Mine shaft linings (502)
DT: January 1993
UF: Mine shafts—Lining*
Mines and mining—Concrete lining*
BT: Linings
Mine shafts

Mine shafts (502)
DT: Predates 1975
UF: Shafts (mine)
BT: Mines
NT: Mine shaft linings
RT: Boreholes
Raise driving
Shaft sinking

*Mine shafts—Lining**
USE: Mine shaft linings

*Mine surveying**
USE: Mine surveys

* Former Ei Vocabulary term

Mine surveys (405.3) (501.1) (911.2) (911.5)
SN: Besides alignment, traverses, etc., includes value-estimation concept
DT: January 1993
UF: Mine surveying*
BT: Surveys
RT: Mining

Mine transportation (502)
DT: January 1993
BT: Mining
Transportation
NT: Surface mine transportation
Underground mine transportation
RT: Mine locomotives

Mine trucks 502.2, 663.1
DT: January 1993
UF: Motor trucks—Mines*
BT: Industrial trucks
RT: Mining equipment

Mine ventilation 502, 643.5
DT: Predates 1975
BT: Ventilation
RT: Mines
Mining

Mineral deposits
USE: Mineral resources

Mineral dressing
USE: Beneficiation

Mineral exploration 501.1
DT: Predates 1975
BT: Natural resources exploration
RT: Minerals

Mineral industry (501)
SN: Metallic or nonmetallic. General and economic aspects
DT: January 1993
UF: Mineral industry and resources*
BT: Industry
NT: Coal industry
Petroleum industry
RT: Earth (planet)
Mineral resources
Mineralogy

*Mineral industry and resources**
USE: Mineral industry OR
Mineral resources

*Mineral industry and resources—Subaqueous**
USE: Underwater mineral resources

*Mineral oil**
USE: Mineral oils

Mineral oils 513.3 (607.1)
DT: January 1993
UF: Mineral oil*
BT: Petroleum products

Mineral resources (501)
DT: January 1993
UF: Mineral deposits
Mineral industry and resources*
BT: Natural resources
NT: Asbestos deposits
Clay deposits
Diamond deposits
Feldspar deposits
Fluorspar deposits
Fossil fuel deposits
Kyanite deposits
Mica deposits
Ore deposits
Phosphate deposits
Potash deposits
Quartz deposits
Salt deposits
Sulfur deposits
Talc deposits
Tellurium deposits
Underwater mineral resources
RT: Earth (planet)
Mineral industry
Mineralogy
Natural resources exploration
Raw materials

Mineral springs (444)
DT: Predates 1975
BT: Springs (water)
RT: Geothermal springs
Hot springs

Mineral wool 819.2 (413.2)
DT: Predates 1975
UF: Slag wool
BT: Synthetic fibers
RT: Glass fibers
Slags

Mineralogy 482
DT: Predates 1975
BT: Earth sciences
NT: Experimental mineralogy
Underwater mineralogy
RT: Clay alteration
Earth (planet)
Geochemistry
Mineral industry
Mineral resources
Minerals
Natural resources exploration

*Mineralogy—Analytical methods**
USE: Analysis

*Mineralogy—Arsenates**
USE: Arsenate minerals

*Mineralogy—Beryllium materials**
USE: Beryllium minerals

*Mineralogy—Borates**
USE: Borate minerals

*Mineralogy—Carbides**
USE: Carbide minerals

*Mineralogy—Carbonates**
USE: Carbonate minerals

*Mineralogy—Chlorides**
USE: Chloride minerals

*Mineralogy—Chlorites**
USE: Chlorite minerals

*Mineralogy—Experimental**
USE: Experimental mineralogy

*Mineralogy—Fluorides**
USE: Fluoride minerals

*Mineralogy—Germanates**
USE: Germanate minerals

*Mineralogy—Halides**
USE: Halide minerals

*Mineralogy—Molybdates**
USE: Molybdate minerals

*Mineralogy—Niobates and tantalates**
USE: Niobate minerals OR
Tantalate minerals

*Mineralogy—Organic materials**
USE: Organic minerals

*Mineralogy—Oxides**
USE: Oxide minerals

*Mineralogy—Phosphates**
USE: Phosphate minerals

*Mineralogy—Silicates**
USE: Silicate minerals

*Mineralogy—Subaqueous**
USE: Underwater mineralogy

*Mineralogy—Sulfates**
USE: Sulfate minerals

*Mineralogy—Sulfides and sulfosalts**
USE: Sulfide minerals

*Mineralogy—Titanates**
USE: Titanate minerals

*Mineralogy—Tungstates**
USE: Tungstate minerals

*Mineralogy—Uranates**
USE: Uranate minerals

*Mineralogy—Vanadates**
USE: Vanadate minerals

Minerals 482.2
DT: January 1986
UF: Biochemistry—Minerals in the body fluids*
Minerals in body fluids
BT: Chemical compounds
Materials
NT: Arsenate minerals
Beryllium minerals
Borate minerals
Calcite
Carbide minerals
Carbonate minerals
Germanate minerals
Halide minerals
Molybdate minerals
Niobate minerals
Ores
Organic minerals
Oxide minerals
Phosphate minerals
Silicate minerals
Sulfate minerals
Sulfide minerals
Tantalate minerals
Titanate minerals
Tungstate minerals
Uranate minerals
Vanadate minerals
RT: Alumina
Clay
Mineral exploration
Mineralogy
Rocks

Minerals in body fluids
USE: Body fluids AND
Minerals

Miners 502, 912.4
DT: Predates 1975
BT: Personnel
RT: Mines
Mining

Miners lamps 707.2, 502.2
DT: Predates 1975
BT: Electric lamps
RT: Mining equipment

Mines 502.1
DT: January 1993
UF: Mines and mining*
BT: Facilities
NT: Abandoned mines
Antimony mines
Asbestos mines
Bauxite mines
Bismuth mines
Chromium mines
Coal mines
Cobalt mines
Copper mines
Diamond mines
Gold mines
Iron mines
Lead mines
Magnesite mines
Manganese mines
Mercury mines
Mine shafts
Molybdenum mines
Nickel mines
Niobium mines
Phosphate mines
Platinum mines
Potash mines
Quarries
Salt mines

* Former Ei Vocabulary term

Mines *(continued)*
Silver mines
Talc mines
Tantalum mines
Tin mines
Titanium mines
Tungsten mines
Uranium mines
Zinc mines
RT: Firedamp
Land reclamation
Mine dust
Mine explosions
Mine fires
Mine flooding
Mine roof control
Mine ventilation
Miners
Mining
Mining laws and regulations
Natural resources exploration
Resource valuation
Roadway supports
Rock pressure
Subsidence
Underground structures

*Mines and mining**
USE: Mines OR
Mining

*Mines and mining—Abandoned**
USE: Abandoned mines

*Mines and mining—Augers**
USE: Augers

*Mines and mining—Compressed air**
USE: Compressed air

*Mines and mining—Concrete lining**
USE: Mine shaft linings AND
Concretes

*Mines and mining—Continuous miners**
USE: Continuous miners

*Mines and mining—Conveying**
USE: Conveying

*Mines and mining—Cutter loaders**
USE: Cutter loaders

*Mines and mining—Cutters**
USE: Cutting machines (mining)

*Mines and mining—Drills**
USE: Mining drills

*Mines and mining—Dust control**
USE: Dust control

*Mines and mining—Explosions**
USE: Mine explosions

*Mines and mining—Explosives**
USE: Explosives

*Mines and mining—Firedamp**
USE: Firedamp

*Mines and mining—Fires**
USE: Mine fires

*Mines and mining—Flooding**
USE: Mine flooding

*Mines and mining—Gas bursts**
USE: Gas bursts

*Mines and mining—Grouting**
USE: Grouting

*Mines and mining—Hydraulic equipment**
USE: Hydraulic equipment

*Mines and mining—Hydraulic process**
USE: Hydraulic mining

*Mines and mining—Lighting**
USE: Lighting

*Mines and mining—Lining**
USE: Linings

*Mines and mining—Loaders**
USE: Loaders

*Mines and mining—Longwall**
USE: Longwall mining

*Mines and mining—Noise**
USE: Acoustic noise

*Mines and mining—Open pit**
USE: Open pit mining

*Mines and mining—Pillar extraction**
USE: Pillar extraction

*Mines and mining—Placers and placering**
USE: Placer mining OR
Placers

*Mines and mining—Power supply**
USE: Electric power utilization

*Mines and mining—Radiation hazards**
USE: Radiation hazards

*Mines and mining—Raise driving**
USE: Raise driving

*Mines and mining—Roadway supports**
USE: Roadway supports

*Mines and mining—Rock bolting**
USE: Rock bolting

*Mines and mining—Rock bursts**
USE: Rock bursts

*Mines and mining—Rock mechanics**
USE: Rock mechanics

*Mines and mining—Rock pressure**
USE: Rock pressure

*Mines and mining—Roof control**
USE: Mine roof control

*Mines and mining—Roof supports**
USE: Mine roof supports

*Mines and mining—Room and pillar**
USE: Room and pillar mining

* Former Ei Vocabulary term

*Mines and mining—Sewage treatment**
USE: Sewage treatment

*Mines and mining—Solution mining**
USE: Solution mining

*Mines and mining—Stoping**
USE: Stoping

*Mines and mining—Stowage**
USE: Stowage (mines)

*Mines and mining—Subsidence**
USE: Subsidence

*Mines and mining—Surface transportation**
USE: Surface mine transportation

*Mines and mining—Time study**
USE: Time and motion study

*Mines and mining—Tunneling**
USE: Tunneling (excavation)

*Mines and mining—Underground transportation**
USE: Underground mine transportation

*Mines and mining—Valuation**
USE: Resource valuation

Miniature automobiles 662.2
SN: Includes minicars
DT: January 1993
UF: Automobiles—Miniature*
BT: Automobiles
RT: Model automobiles

Miniature batteries 702.1
DT: January 1993
UF: Electric batteries—Miniature*
BT: Electric batteries

Miniature bearings 601.2
DT: January 1993
UF: Bearings—Miniature*
BT: Bearings (machine parts)

Miniature computers
USE: Minicomputers

Miniature instruments
DT: January 1993
UF: Instruments—Miniature*
BT: Instruments

Miniature radio equipment 716.3
DT: January 1993
UF: Radio equipment—Miniature*
Radio receivers—Miniature*
Radio transmitters—Miniature*
BT: Radio equipment
RT: Radio receivers
Radio transmitters

Minicomputers 722.4
DT: January 1993
UF: Computers, Miniature*
Computers, Minicomputer*
Miniature computers
BT: Digital computers
RT: Microcomputers

Minimization of switching nets 721.1
DT: January 1993
UF: Switching theory—Minimization of switching nets*
BT: Optimization
Switching theory
RT: Logic design
Switching functions
Switching networks

Mining 502.1
DT: January 1993
UF: Mines and mining*
Photogrammetry—Mining applications*
BT: Production
NT: Hydraulic mining
Longwall mining
Mine roof control
Mine transportation
Open pit mining
Pillar extraction
Placer mining
Quarrying
Raise driving
Room and pillar mining
Solution mining
Stoping
RT: Blasting
Compressed air
Excavation
Explosives
Firedamp
Gas bursts
Grouting
Land reclamation
Landslides
Mine dust
Mine explosions
Mine fires
Mine flooding
Mine rescue
Mine surveys
Mine ventilation
Miners
Mines
Mining engineering
Mining equipment
Mining laws and regulations
Natural resources exploration
Ore treatment
Roadway supports
Rock bolting
Rock bursts
Rock mechanics
Rock pressure
Stowage (mines)
Subsidence
Tunneling (excavation)
Underground structures

Mining drills 502.2
DT: January 1993
UF: Mines and mining—Drills*

* Former Ei Vocabulary term

Mining drills *(continued)*
BT: Drills
Mining machinery
RT: Augers

Mining engineering 506
DT: Predates 1975
BT: Engineering
RT: Mining
Rock mechanics

Mining equipment 502.2
DT: January 1993
BT: Equipment
NT: Mine roof supports
Mining machinery
RT: Mine cars
Mine power plants
Mine trucks
Miners lamps
Mining

Mining laws and regulations 502, 902.3
DT: Predates 1975
BT: Laws and legislation
RT: Mines
Mining

*Mining laws and regulations—International law**
USE: International law

Mining machinery 502.2
DT: January 1986
BT: Machinery
Mining equipment
NT: Cutting machines (mining)
Mine hoists
Mining drills
RT: Mine locomotives

Mints (402.1) (535.2)
DT: Predates 1975
BT: Facilities
RT: Coinage

Mirrors 741.3
DT: Predates 1975
UF: Optical instruments—Reflectors*
Optical reflectors
Reflectors (optical)
BT: Optical devices
RT: Laser accessories
Optical beam splitters
Optical materials
Optics
Polishing
Reflection
Telescopes

MIS
USE: Management information systems

MIS devices 714.2
SN: Metal-insulator-semiconductor devices
DT: January 1993
UF: Semiconductor devices, MIS*
BT: Field effect semiconductor devices
NT: Charge transfer devices
MOS devices
RT: Field effect transistors
Semiconductor device structures

Miscibility
USE: Solubility

MISFET devices 714.2
SN: Metal-insulator-semiconductor field-effect transistors
DT: January 1993
UF: Metal insulator semiconductor field effect transistors
Semiconductor devices, MISFET*
BT: Field effect transistors

Misfit dislocations
USE: Dislocations (crystals)

Missile bases 654.1, 404.1
DT: January 1993
UF: Rockets and missiles—Bases*
BT: Military bases
NT: Missile silos
RT: Military operations

Missile guidance
USE: Electronic guidance systems

Missile launching systems 404.1, 654.1 (672.1)
DT: January 1993
UF: Warships—Missile launching systems*
BT: Military equipment
RT: Launching
Missiles
Rockets
Warships

Missile silos 654.1, 404.1 (408)
DT: January 1993
UF: Rockets and missiles—Silos*
Silos (missile)
BT: Missile bases
RT: Aerospace ground support
Launching

Missiles 654.1, 404.1
DT: January 1993
UF: Rockets and missiles*
Rockets and missiles—Military*
BT: Ordnance
NT: Marine missiles
RT: Electronic guidance systems
Fire control systems
Fuel sloshing
Heat problems
Heat shielding
Launching
Military rockets
Missile launching systems
Nose cones
Plasma sheaths
Projectiles
Propellants
Rockets
Shielding

* Former Ei Vocabulary term

Missiles *(continued)*
Tracking (position)
Trajectories

Mist
USE: Fog

Mist eliminators 451.2
DT: January 1993
UF: Industrial plants—Mist eliminators*
BT: Equipment

Mixed convection
USE: Heat convection

Mixed gas fuels 522
DT: January 1993
UF: Gas manufacture—Mixed gas*
BT: Gas fuels
RT: Gas fuel manufacture

Mixed oxide fuels 621.1.2
DT: January 1993
UF: Nuclear fuels—Mixed oxides*
BT: Nuclear fuels
RT: Ceramic materials
Oxides

Mixer circuits 713.3
DT: January 1993
UF: Electronic circuits, Mixer*
Mixers (circuits)
Mixing circuits
BT: Networks (circuits)
RT: Demodulators
Frequency converter circuits
Heterodyning
Modulators
Oscillators (electronic)
Signal processing

Mixers (circuits)
USE: Mixer circuits

Mixers (machinery) (405.1) (601.1) (802.1) (816.2)
DT: January 1993
UF: Mixers*
Roadbuilding machinery—Mixers*
BT: Machinery
NT: Concrete mixers
RT: Mixing
Plastics machinery
Roadbuilding machinery

*Mixers**
USE: Mixers (machinery)

Mixing 802.3
SN: Mechanical mixing. For signal mixing, use MIXER CIRUITS
DT: Predates 1975
UF: Flow of fluids—Mixing*
BT: Chemical operations
NT: Blending
Concrete mixing
Mechanical alloying
RT: Diffusion
Emulsification
Mixers (machinery)
Mixtures
Solubility
Turbulence
Vortex flow

Mixing circuits
USE: Mixer circuits

Mixtures (802.3)
DT: January 1987
BT: Materials
NT: Azeotropes
Binary mixtures
Coal oil mixtures
Coal water mixtures
Solutions
RT: Dispersions
Mixing

*Mixtures—Azeotrope**
USE: Azeotropes

MMIC
USE: Monolithic microwave integrated circuits

Mo
USE: Molybdenum

Mobile antennas (716)
DT: January 1993
UF: Antennas, Mobile*
BT: Antennas

Mobile communication systems
USE: Mobile telecommunication systems

Mobile electric power plants (612) (613) (614) (615) (706.1)
DT: January 1993
UF: Electric power plants, Mobile*
BT: Electric power plants
Mobile power plants

Mobile homes 402.3
DT: January 1993
UF: Houses—Mobile*
BT: Houses
RT: Light trailers

Mobile nuclear reactors 621
DT: January 1993
UF: Nuclear reactors—Mobile*
BT: Nuclear reactors

Mobile power plants (612) (613) (614) (615)
DT: January 1993
UF: Portable power plants
Power plants—Mobile*
Power plants—Portable*
BT: Power plants
NT: Mobile electric power plants

Mobile radio systems 716.3
DT: January 1993
UF: Radio systems, Mobile*
BT: Mobile telecommunication systems
Radio systems
NT: Cellular radio systems

* Former Ei Vocabulary term

Mobile radio systems *(continued)*
RT: Radio equipment
Radio links
Radio receivers
Radio transmitters
Spread spectrum communication

Mobile robots 731.5 (731.6)
DT: January 1993
UF: Robots, Industrial—Mobile*
BT: Robots
RT: Collision avoidance
Industrial robots

Mobile telecommunication systems (716) (717) (718)
DT: January 1993
UF: Mobile communication systems
Telecommunication systems, Mobile*
BT: Telecommunication systems
NT: Mobile radio systems
Mobile telephone exchanges
RT: Aircraft communication
Military equipment
Personal communication systems
Radio communication
Radio links
Radio telephone

Mobile telephone exchanges 718.1
DT: January 1993
UF: Telephone exchanges, Mobile*
BT: Mobile telecommunication systems
Telephone exchanges

Mobility aids for blind persons 461.5
DT: January 1993
UF: Human rehabilitation engineering—Mobility aids for blind*
BT: Human rehabilitation equipment
RT: Human rehabilitation engineering

Mock-ups
USE: Mockups

Mockups
DT: January 1993
UF: Mock-ups
BT: Models
NT: Aircraft mockups
Ship mockups

MOCVD
USE: Metallorganic chemical vapor deposition

Modal analysis 921
DT: January 1995
BT: Mathematical techniques

Mode locking
USE: Laser mode locking

Model aircraft
USE: Aircraft models

Model automobiles 662.1
DT: January 1993
BT: Models
RT: Automobiles
Miniature automobiles

Model buildings (402)
DT: January 1993
BT: Models
RT: Architecture
Buildings

Model structures (408)
DT: January 1993
UF: Structural design—Models*
BT: Models
RT: Structural design
Structures (built objects)

Modeling (computer)
USE: Computer simulation

Modeling (physical)
USE: Models

Models
SN: Physical models; for mathematical models, use MATHEMATICAL MODELS
DT: Predates 1975
UF: Modeling (physical)
NT: Aircraft models
Geologic models
Human form models
Hydraulic models
Mockups
Model automobiles
Model buildings
Model structures
Physiological models
Ship models
Simulators
RT: Computer simulation
Control theory
Identification (control systems)
Mathematical programming
Mechanisms
Optimal control systems
Simulation
Step response
System theory

Models (mathematical)
USE: Mathematical models

Modems (718.1) (722.3)
DT: January 1986
BT: Data communication equipment
Demodulators
Modulators

Moderators (621.1.1)
DT: January 1993
UF: Nuclear reactors—Moderators*
BT: Materials
RT: Beryllium
Beryllium alloys
Graphite
Heavy water
Nuclear fuels
Nuclear reactors
Reactor cores

* Former Ei Vocabulary term

Modernization
DT: January 1993
UF: Buildings—Modernization*
Furnaces—Modernization*
Industrial plants—Modernization*
Machinery—Modernization*
RT: Buildings
Furnaces
Industrial plants
Machinery
Maintenance
Modification
Obsolescence
Restoration
Retrofitting

Modes (laser)
USE: Laser modes

Modification
DT: January 1993
UF: Changes
NT: Chemical modification
RT: Modernization

Modula (programming language) 723.1.1
DT: January 1993
UF: Computer programming languages—Modula*
BT: High level languages

Modular construction 405.2
DT: Predates 1975
BT: Construction

Modular robots 731.5 (731.6)
DT: January 1993
UF: Robots, Industrial—Modular*
BT: Robots
RT: Industrial robots

Modulation (716) (717) (718)
DT: Predates 1975
NT: Amplitude modulation
Frequency modulation
Intermodulation
Light modulation
Phase modulation
Pulse modulation
RT: Demodulation
Electromagnetic waves
Information theory
Modulators
Phase locked loops
Radio transmission
Telecommunication
Transmitters
Velocity modulation tubes

Modulation transfer function
USE: Optical transfer function

Modulators 713.3
DT: Predates 1975
BT: Electronic equipment
NT: Light modulators
Modems
RT: Amplitude modulation
Demodulators
Frequency modulation
Mixer circuits
Modulation
Phase locked loops
Phase modulation
Pulse amplitude modulation
Pulse code modulation
Pulse modulation
Pulse time modulation
Pulse width modulation
Radio equipment
Television equipment
Transmitters

Modulus of elasticity
USE: Elastic moduli

Moire fringes 741.1
DT: January 1993
UF: Optics—Moire fringes*
RT: Electromagnetic wave interference
Light interference

Moistening
USE: Wetting

Moisture (443.1) (801.4)
DT: Predates 1975
UF: Sand, Foundry—Moisture effects*
RT: Atmospheric humidity
Moisture control
Moisture determination
Moisture meters
Regain
Weathering

Moisture control 731.3, 944.2
SN: For control of moisture in an atmosphere, use HUMIDITY CONTROL
DT: Predates 1975
BT: Control
NT: Humidity control
RT: Chemical variables control
Mechanical variables control
Moisture
Moisture determination
Tropical engineering

Moisture determination 944.2
DT: Predates 1975
BT: Materials testing
RT: Moisture
Moisture control
Moisture meters

Moisture meters 944.1
SN: Use for measurement of moisture in substances. For measurement of relative humidity, use HYGROMETERS or PSYCHROMETERS
DT: Predates 1975
BT: Instruments
RT: Hygrometers
Moisture
Moisture determination
Psychrometers

Moisture regain
USE: Regain

Molasses 822.3
DT: January 1987
BT: Food products
RT: Sugar (sucrose)

Mold
USE: Fungi

Mold release agents 803, 816.1
DT: January 1993
UF: Plastics—Mold release agents*
BT: Agents
RT: Plastics

Molding (535.2) (816.1) (818.3)
DT: January 1993
BT: Forming
NT: Compression molding
Extrusion molding
Gating and feeding
Metal molding
Patternmaking
Plastics molding
Rubber molding
RT: Casting
Foundries
Foundry practice
Foundry sand
Molds
Preforming
Tapping (furnace)

Molding compounds
USE: Sheet molding compounds

Molding machines (plastics)
USE: Plastics molding machines

Molding sand
USE: Foundry sand

Molds (535.2.1) (812.3) (816.2) (818.4)
DT: January 1993
UF: Glass manufacture—Molds*
Plastics—Molds*
Tires—Molds*
BT: Equipment
NT: Forms (concrete)
Ingot molds
Plastic molds
Rubber molds
RT: Foundry sand
Glass manufacture
Molding
Plastics molding machines
Plastics molding
Tires

Molecular alignment
USE: Molecular orientation

Molecular beam epitaxy
931.3, 933.1.2 (712.1) (714.2)
DT: January 1986
UF: MBE
BT: Vapor phase epitaxy
RT: Chemical beam epitaxy
Molecular beams

Molecular beams 931.3
DT: Predates 1975
BT: Particle beams
RT: Molecular beam epitaxy

Molecular catalysis
USE: Catalysis

Molecular crystals 933.1
DT: January 1977
BT: Crystals
RT: Crystal lattices

Molecular dynamics 801.4
DT: January 1995
BT: Physical chemistry
RT: Molecules
Reaction kinetics

Molecular orientation 931.1, 933.3
DT: January 1995
UF: Molecular alignment
RT: Crystal orientation
Liquid crystals

Molecular physics 931.3
DT: January 1993
UF: Physics—Molecular*
BT: Physics
RT: Atomic physics
Molecules
Nuclear physics

Molecular sieves 803 (802.3)
DT: January 1986
UF: Chromatographic analysis—Molecular sieve*
BT: Adsorbents

Molecular spectroscopy (741.3) (931.3)
DT: January 1993
BT: Spectroscopy
RT: Electromagnetic waves
Molecular vibrations
Molecules

Molecular structure 801.4, 931.3
DT: January 1993
BT: Structure (composition)
RT: Chemical bonds
Electronic structure
Hydrogen bonds
Interpenetrating polymer networks
Isomerization
Microstructure
Molecules

Molecular vibrations 931.3, 801.4
DT: June 1990
UF: Vibrations (molecular)
RT: Molecular spectroscopy

Molecular weight 931.3 (801)
DT: Predates 1975
UF: Weight (molecular)
RT: Molecular weight distribution

* Former Ei Vocabulary term

Molecular weight *(continued)*
Molecules
Units of measurement

Molecular weight distribution 931.3 (801) (815.1)
DT: January 1995
RT: Molecular weight
Polymers

Molecules 931.3
DT: January 1986
NT: Macromolecules
RT: Atoms
Conformations
Free radicals
Kinetic theory
Molecular dynamics
Molecular physics
Molecular spectroscopy
Molecular structure
Molecular weight
Permselective membranes

Molten materials (931.2)
DT: January 1993
UF: Molten*
BT: Liquids
RT: Melting

Molten metals
USE: Liquid metals

*Molten**
USE: Molten materials

Molybdate minerals 482.2
DT: January 1993
UF: Mineralogy—Molybdates*
BT: Minerals
Molybdenum compounds
Oxides
RT: Molybdenum

Molybdenum 543.3
DT: January 1993
UF: Mo
Molybdenum and alloys*
BT: Nonferrous metals
Refractory metals
RT: Molybdate minerals
Molybdenum alloys
Molybdenum compounds
Molybdenum deposits
Molybdenum metallurgy
Molybdenum metallography
Molybdenum mines
Molybdenum ore treatment
Molybdenum plating

Molybdenum alloys 543.3
DT: January 1993
UF: Molybdenum and alloys*
BT: Refractory alloys
RT: Molybdenum
Molybdenum compounds
Stainless steel

*Molybdenum and alloys**
USE: Molybdenum OR
Molybdenum alloys

Molybdenum compounds (804.1) (804.2)
DT: Predates 1975
BT: Refractory metal compounds
NT: Molybdate minerals
RT: Molybdenum
Molybdenum alloys

Molybdenum deposits 504.3, 543.3
DT: January 1993
BT: Ore deposits
RT: Molybdenum
Molybdenum mines

Molybdenum metallography 531.2, 543.3
DT: January 1993
BT: Metallography
RT: Molybdenum
Molybdenum metallurgy

Molybdenum metallurgy 531.1, 543.3
DT: Predates 1975
BT: Metallurgy
NT: Molybdenum powder metallurgy
RT: Molybdenum
Molybdenum metallography
Molybdenum ore treatment

Molybdenum mines 504.3, 543.3
DT: January 1993
UF: Molybdenum mines and mining*
BT: Mines
RT: Molybdenum
Molybdenum deposits

*Molybdenum mines and mining**
USE: Molybdenum mines

Molybdenum ore treatment 533.1, 543.3
DT: Predates 1975
BT: Ore treatment
NT: Copper molybdenum ore treatment
RT: Molybdenum
Molybdenum metallurgy

Molybdenum plating 539.3, 543.3
DT: Predates 1975
BT: Plating
RT: Molybdenum

Molybdenum powder metallurgy 536.1, 543.3
DT: January 1993
UF: Powder metallurgy—Molybdenum*
BT: Molybdenum metallurgy
Powder metallurgy

Monazite 482.2
DT: Predates 1975
BT: Ores
RT: Monazite deposits
Rare earth elements
Thorium

Monazite deposits 504.3
DT: Predates 1975
BT: Ore deposits

* Former Ei Vocabulary term

Monazite deposits *(continued)*
RT: Monazite
Thorium deposits

Monel metal 544.2, 548.2
DT: Predates 1975
BT: Copper alloys
Nickel alloys

Monitoring
DT: January 1993
NT: Patient monitoring
RT: Alarm systems
Detectors
Inspection
Measurements
Recording

Monitoring (patients)
USE: Patient monitoring

Monitors (computer)
USE: Computer monitors

Monitors (displays)
USE: Display devices

Monochromatic light sources
USE: Monochromators

Monochromators 741.3
DT: Predates 1975
UF: Monochromatic light sources
Optical monochromators
BT: Spectrographs
RT: Light sources

Monoclonal antibodies 461.9.1 (461.8)
DT: January 1993
UF: Immunology—Monoclonal antibody*
BT: Antibodies
RT: Biotechnology
Immunology

Monolayers (712.1) (813.2)
DT: January 1995
UF: Monomolecular films
Monomolecular layers
Unimolecular layers
RT: Langmuir Blodgett films
Semiconducting films
Ultrathin films

Monolithic integrated circuits 714.2
DT: January 1993
UF: Integrated circuits, Monolithic*
BT: Integrated circuits
NT: Application specific integrated circuits
CMOS integrated circuits
LSI circuits
Monolithic microwave integrated circuits
RT: Integrated optoelectronics

Monolithic microwave integrated circuits 714.2
DT: January 1995
UF: MMIC
BT: Microwave integrated circuits
Monolithic integrated circuits

Monomers 804 (815.1)
DT: October 1975
BT: Chemical compounds
NT: Acrylic monomers
Styrene
RT: Butadiene
Polymerization
Polymers

Monomolecular films
USE: Monolayers

Monomolecular layers
USE: Monolayers

Monorail cars 682.1.1
DT: January 1993
UF: Cars—Monorail*
BT: Railroad cars
RT: Electric railroads
Monorails

Monorail conveyors 692.1
DT: January 1993
UF: Conveyors—Monorail*
BT: Conveyors
RT: Monorails

Monorails 682.1
DT: October 1975
UF: Electric railroads—Monorail*
BT: Light rail transit
RT: Electric railroads
Guideways
Materials handling
Monorail cars
Monorail conveyors
Railroad transportation
Urban planning

*Monorails—Guideways**
USE: Guideways

Monte Carlo methods 922.2
DT: January 1993
UF: Mathematical statistics—Monte Carlo methods*
Monte Carlo simulation
BT: Numerical analysis
Statistical methods
RT: Operations research
Probability
Random processes

Monte Carlo simulation
USE: Monte Carlo methods AND Computer simulation

Moon 657.2
DT: Predates 1975
UF: Photogrammetry—Moon*
Radar—Lunar measurements*
Radio communication—Moon*
BT: Solar system
RT: Astronomy
Lunar landing

* Former Ei Vocabulary term

Moon *(continued)*
Lunar missions
Lunar surface analysis
Moon bases

Moon bases (656)
DT: January 1993
UF: Lunar bases
Moon—Bases*
BT: Facilities
RT: Extraterrestrial communication links
Lunar landing
Lunar missions
Moon
Space applications
Space flight
Space stations

Moon surface analysis
USE: Lunar surface analysis

*Moon—Bases**
USE: Moon bases

*Moon—Surface analysis**
USE: Lunar surface analysis

Mooring (672) (674.1)
DT: January 1993
UF: Ships—Mooring*
BT: Joining
NT: Single point mooring
RT: Boats
Docking
Mooring cables
Ships

Mooring cables (472) (671.2)
DT: October 1975
BT: Cables
RT: Docking
Mooring

Mopeds
USE: Motorcycles

Morphology 931.2
SN: Study of the structure and shape of materials
DT: January 1993
UF: Surface morphology
BT: Materials science
RT: Size determination

Mortar 414.3
DT: Predates 1975
BT: Masonry materials
RT: Cements
Portland cement

MOS devices 714.2
DT: January 1993
UF: Metal oxide semiconductor devices
Semiconductor devices, MOS*
BT: MIS devices

MOS transistors
USE: MOSFET devices

MOSFET devices 714.2
DT: January 1993
UF: Insulated gate field effect transistors
Metal oxide semiconductor field effect transistors
MOS transistors
Semiconductor devices, MOSFET*
BT: Field effect transistors

Mosquito control (461.7) (804.1) (804.2)
DT: Predates 1975
BT: Insect control
RT: Disease control
Malaria control

Mossbauer spectroscopy (801) (932.1) (932.2) (933.1.1)
DT: January 1993
UF: Spectroscopy, Mossbauer*
BT: Spectroscopy

Motels 402.2
DT: January 1993
BT: Housing
RT: Hotels
Reservation systems

Mothproofing (804.1) (819.5)
DT: January 1993
UF: Textiles—Mothproofing*
BT: Insect control
Textile finishing

Motion control 731.3
DT: January 1993
UF: Control, Mechanical variables—Motion*
BT: Mechanical variables control

Motion picture cameras 742.2
DT: January 1993
UF: Motion pictures—Cameras*
BT: Cameras
RT: Footage counters
Motion pictures

Motion picture editing machines 742.2
DT: January 1993
UF: Editing machines (motion pictures)
Motion pictures—Editing machines*
BT: Machinery
RT: Footage counters
Motion pictures

Motion picture film
USE: Photographic films

Motion picture screens 742.2
DT: January 1993
UF: Motion pictures—Screens*
Screens (motion picture)
BT: Equipment
RT: Motion picture theaters
Motion pictures
Projection screens

Motion picture studios (402.2) (742.1)
DT: January 1993
UF: Motion pictures—Studios*

Motion picture studios *(continued)*
BT: Studios
RT: Motion pictures

Motion picture theaters (402.2) (742.2)
DT: Predates 1975
BT: Facilities
NT: Outdoor motion picture theaters
RT: Motion picture screens
Motion pictures
Projection rooms

*Motion picture theaters—Outdoor**
USE: Outdoor motion picture theaters

*Motion picture theaters—Projection rooms**
USE: Projection rooms

Motion pictures (742.1)
DT: Predates 1975
UF: Cinematography
Made for television motion pictures
Television motion pictures
Television—Motion pictures*
NT: Color motion pictures
Educational motion pictures
RT: Animation
Light sources
Motion picture editing machines
Motion picture cameras
Motion picture screens
Motion picture theaters
Motion picture studios
Optical projectors
Photography
Recording
Sound stages

*Motion pictures, Color**
USE: Color motion pictures

*Motion pictures—Animation**
USE: Animation

*Motion pictures—Cameras**
USE: Motion picture cameras

*Motion pictures—Editing machines**
USE: Motion picture editing machines

*Motion pictures—Educational**
USE: Educational motion pictures

*Motion pictures—Footage counters**
USE: Footage counters

*Motion pictures—Industrial**
USE: Industrial applications

*Motion pictures—Light sources**
USE: Light sources

*Motion pictures—Recording and reproduction**
USE: Recording

*Motion pictures—Screens**
USE: Motion picture screens

*Motion pictures—Sound stages**
USE: Sound stages

*Motion pictures—Studios**
USE: Motion picture studios

Motion planning (431.5) (723.4) (731.5)
DT: January 1995
UF: Path planning
BT: Planning
RT: Artificial intelligence
Collision avoidance
Knowledge based systems
Manipulators
Robotics
Robots

Motivation 912.4
DT: January 1993
UF: Personnel—Motivation*
RT: Job satisfaction
Personnel

Motor boats 674.1
DT: Predates 1975
UF: Motorboats
BT: Boats
NT: Hydrofoil boats

*Motor bus drivers**
USE: Bus drivers

*Motor bus terminals**
USE: Bus terminals

*Motor bus transportation**
USE: Bus transportation

*Motor buses**
USE: Buses

Motor cycles
USE: Motorcycles

Motor generators (AC)
USE: AC generator motors

Motor generators (DC)
USE: DC generator motors

Motor scooters 662.2
DT: Predates 1975
UF: Scooters (motor)
BT: Motorcycles

*Motor ships**
USE: Ships

Motor transportation (432)
SN: Motorized land transport, as distinguished from aerospace, railway, and waterway transport
DT: Predates 1975
UF: Highway transportation
Land transportation (motor)
BT: Transportation
NT: Bus transportation
Fleet operations
Truck transportation
RT: Parking
Roads and streets
Traffic surveys
Transportation routes
Truck drivers

* Former Ei Vocabulary term

Motor transportation *(continued)*
Urban planning
Weigh stations

*Motor truck drivers**
USE: Truck drivers

*Motor truck terminals**
USE: Truck terminals

*Motor truck transportation**
USE: Truck transportation

*Motor trucks**
USE: Trucks

*Motor trucks—Cabs**
USE: Cabs (truck)

*Motor trucks—Loading**
USE: Loading

*Motor trucks—Mines**
USE: Mine trucks

*Motor trucks—Power takeoff**
USE: Power takeoffs

*Motor trucks—Public utilities**
USE: Utility trucks

*Motor trucks—Refrigerator**
USE: Refrigerator trucks

*Motor trucks—Refuse collecting**
USE: Garbage trucks

*Motor trucks—Tank**
USE: Tank trucks

*Motor trucks—Tractors**
USE: Tractors (truck)

*Motor trucks—Winches**
USE: Winches

Motorboats
USE: Motor boats

Motorcycles 662.2
SN: Includes motorized bicycles
DT: January 1986
UF: Mopeds
Motor cycles
BT: Ground vehicles
NT: Motor scooters

Motors (632.2) (632.4) (705.3)
SN: Machines utilizing external energy sources for normal operation; for machines using internal sources, use ENGINES
DT: January 1993
BT: Machinery
NT: Compressed air motors
Electric motors
Hydraulic motors
RT: Engines

Mountings 601.2
DT: January 1993
UF: Equipment mountings
Grinding wheels—Mounting*
Mountings*
BT: Components
NT: Antivibration mountings
Bearing mountings
Engine mountings

*Mountings**
USE: Mountings

Mouse (computer peripheral)
USE: Mice (computer peripherals)

Movable bridges 401.1
DT: January 1993
UF: Bridges, Movable*
Bridges—Raising*
Drawbridges
BT: Bridges
NT: Bascule bridges
Lift bridges
Swing bridges

Movable dams 441.1
DT: January 1993
UF: Dams, Movable*
BT: Dams

Moving
DT: January 1993
UF: Oil tanks—Moving*
NT: Building moving
RT: Relocation

MOVPE
USE: Metallorganic vapor phase epitaxy

Mowers
USE: Lawn mowers

MRI
USE: Magnetic resonance imaging

Mud logging 512.1.2
DT: January 1993
UF: Oil well logging—Mud*
BT: Well logging
RT: Drilling fluids
Oil well logging

Mud pumps 511.2, 618.2
DT: January 1993
UF: Oil well drilling—Mud pumps*
BT: Oil well drilling equipment
Pumps
RT: Oil well drilling

Mufflers 612.1.1
DT: January 1993
UF: Internal combustion engines—Mufflers*
BT: Internal combustion engines
NT: Automobile mufflers
RT: Acoustic noise
Exhaust systems (engine)
Noise abatement
Noise pollution control equipment

* Former Ei Vocabulary term

Multichip modules 714.2
DT: January 1995
RT: Electronics packaging
Integrated circuits

Multilayer neural networks
USE: Neural networks

Multilayers (813.2) (933.1)
DT: January 1995
RT: Films
Superlattices

Multilayers (optical)
USE: Optical multilayers

Multiphase flow 631.1
DT: January 1993
UF: Flow of fluids—Multiphase*
BT: Flow of fluids
NT: Two phase flow
RT: Unsteady flow

Multiphoton processes (741.1) (744.1)
DT: January 1993
RT: Optical pumping

Multiple fuel firing 521.1 (522) (523) (524) (614.2)
DT: January 1993
UF: Boiler firing—Multiple fuel*
BT: Boiler firing

Multiple processor systems
USE: Computer networks

Multiple quantum wells
USE: Semiconductor quantum wells

Multiple zones 512.1.1
SN: More than one oil or gas field in the same area, but at different depths
DT: January 1993
UF: Oil well completion—Multiple zone*
BT: Petroleum reservoirs

Multiplex radio telephone 716.3, 718.1
DT: January 1993
UF: Radio telephone—Multiplex systems*
BT: Radio telephone
RT: Multiplex radio transmission
Multiplexing

Multiplex radio transmission 716.3
DT: January 1993
UF: Radio transmission—Multiplex systems*
BT: Radio transmission
RT: Multiplex radio telephone
Multiplexing
Radio communication

Multiplexing (716) (717) (718)
DT: Predates 1975
NT: Frequency division multiplexing
Time division multiplexing
RT: Data communication systems
Multiplex radio transmission
Multiplex radio telephone
Multiplexing equipment
Signal processing
Telecommunication

Multiplexing equipment (716) (717) (718)
DT: Predates 1975
UF: Concentrators (telecommunications)
BT: Telecommunication equipment
NT: Line concentrators
RT: Data communication equipment
Multiplexing
Radio transmission

*Multiplexing, Frequency division**
USE: Frequency division multiplexing

*Multiplexing, Time division**
USE: Time division multiplexing

Multiplier phototubes
USE: Photomultipliers

Multipliers (electron)
USE: Electron multipliers

Multipliers (frequency)
USE: Frequency multiplying circuits

Multiplying circuits 721.3
DT: January 1993
UF: Computers—Multiplying circuits*
BT: Digital circuits
RT: Analog computers
Digital arithmetic

Multiprocessing programs 723.1
DT: January 1993
UF: Computer systems programming—Multiprocessing programs*
BT: Supervisory and executive programs
RT: Concurrency control
Multiprocessing systems
Multiprogramming
Pipeline processing systems

Multiprocessing systems 722.4
DT: January 1993
UF: Computer systems, Digital—Multiprocessing*
BT: Distributed computer systems
NT: Parallel processing systems
Pipeline processing systems
RT: Concurrency control
Interconnection networks
Multiprocessing programs

Multiprocessor interconnection networks
USE: Interconnection networks

Multiprogramming 723.1
DT: January 1993
UF: Computer systems programming—Multiprogramming*
Multitasking
BT: Computer systems programming
RT: Multiprocessing programs
Supervisory and executive programs
Time sharing programs
Time sharing systems

* Former Ei Vocabulary term

Multipurpose robots 731.5, 731.6
DT: January 1993
UF: Robots, Industrial—Multipurpose*
BT: Robots
RT: Industrial robots

Multisensor systems
USE: Sensor data fusion

Multispectral scanners (741.3)
DT: January 1993
UF: Remote sensing—Multispectral scanners*
BT: Sensors
RT: Remote sensing
Scanning

Multistable circuits (703.1) (713.5)
DT: January 1993
UF: Electronic circuits, Multistable*
BT: Networks (circuits)

Multitasking
USE: Multiprogramming

Multivariable control systems 731.1
DT: January 1993
UF: Control systems, Multivariable*
BT: Control systems
Multivariable systems
RT: Decentralized control
Hierarchical systems
Process control

Multivariable systems (731.1)
DT: January 1993
UF: Systems science and cybernetics—Multivariable systems*
BT: Systems science
NT: Multivariable control systems
RT: Hierarchical systems

Multivibrators 713.4
DT: January 1993
UF: Electronic circuits, Multivibrator*
BT: Networks (circuits)
NT: Flip flop circuits
RT: Comparator circuits
Frequency dividing circuits
Frequency multiplying circuits
Oscillators (electronic)
Pulse generators
Relaxation oscillators
Signal generators
Square wave generators
Trigger circuits

Multiwave mixing
USE: Four wave mixing

Municipal engineering 403.1
DT: January 1993
BT: Engineering
RT: Civil engineering
Sanitary engineering
Urban planning

Muriatic acid
USE: Hydrochloric acid

Muscle 461.2 (461.3)
DT: January 1993
UF: Biological materials—Muscle*
BT: Musculoskeletal system
Tissue
RT: Adenosinetriphosphate
Biomechanics
Electromyography

Muscle controlled prosthetics
USE: Myoelectrically controlled prosthetics

Musculoskeletal system 461.3
DT: January 1993
UF: Biomechanics—Musculoskeletal systems*
NT: Bone
Cartilage
Joints (anatomy)
Ligaments
Muscle
Tendons
RT: Biomechanics
Neuromuscular rehabilitation
Orthopedics

Museums 402.2
DT: Predates 1975
BT: Facilities
RT: Planetariums

Music (computer)
USE: Computer music

Musical instruments 752.4
SN: Excludes devices which are used to reproduce recorded music, for which use PHONOGRAPHS or TAPE RECORDERS
DT: Predates 1975
BT: Acoustic generators
NT: Electronic musical instruments
Organs (musical instruments)
RT: Bells

*Musical instruments, Electronic**
USE: Electronic musical instruments

Muskeg (483.1)
DT: January 1993
UF: Roads and streets—Muskeg*
Soils—Muskeg*
BT: Landforms
RT: Cold weather problems
Permafrost
Roads and streets
Soils

Mutagenesis 461.8.1 (801.2)
DT: January 1987
RT: Genetic engineering
Mutagens

Mutagens 803 (461.8.1) (801.2)
DT: January 1986
BT: Agents
RT: Air pollution
Antibiotics
Biochemistry
Carcinogens

Mutagens *(continued)*
Chemical analysis
Mutagenesis

Mycology
USE: Fungi

Myoelectrically controlled prosthetics 461.5, 462.4
DT: January 1993
UF: Muscle controlled prosthetics
Prosthetics—Myoelectric control*
BT: Prosthetics
RT: Biomaterials

N
USE: Nitrogen

N-path filters
USE: Switched filters

Na
USE: Sodium

Nails 605
DT: Predates 1975
BT: Fasteners

Nameplates (694.1)
DT: Predates 1975
BT: Labels

NAND circuits 721.2, 721.3
DT: January 1993
UF: Logic circuits, NAND*
BT: Logic circuits

Nanocomposite materials
USE: Nanostructured materials

Nanocrystalline materials
USE: Nanostructured materials

Nanofabrication
USE: Nanotechnology

Nanolithography
USE: Nanotechnology

Nanophase materials
USE: Nanostructured materials

Nanostructured materials 933.1
DT: January 1995
UF: Nanocomposite materials
Nanocrystalline materials
Nanophase materials
Nanotubes
BT: Materials
RT: Grain boundaries
Grain size and shape
Nanotechnology
Powders

Nanotechnology (714.2) (933.1)
DT: January 1995
UF: Nanofabrication
Nanolithography
RT: Integrated circuit manufacture
Nanostructured materials
Scanning tunneling microscopy

Nanotubes
USE: Nanostructured materials

Naphthalene 804.1
DT: Predates 1975
BT: Aromatic hydrocarbons

Naphthas 513.3, 804.1
DT: Predates 1975
BT: Petroleum distillates
RT: Petroleum products
Steam cracking

Natural convection 641.2 (641.3)
DT: January 1995
UF: Free convection
BT: Heat convection

Natural fibers 819.1
DT: January 1993
BT: Fibers
NT: Bast fibers
Cotton fibers
Flax
Silk
Wool fibers
RT: Textile blends

Natural frequencies (711.1) (741.1) (751.1)
DT: January 1995
UF: Resonant frequencies
BT: Frequencies
RT: Acoustic impedance
Bandwidth
Cavity resonators
Damping
Dynamics
Harmonic generation
Oscillators (electronic)
Oscillators (mechanical)
Resonance
Resonators
Tuners
Tuning

Natural gas 522
DT: Predates 1975
BT: Gas fuels
NT: Liquefied natural gas
Sour gas
RT: Butane
Enhanced recovery
Gas condensates
Gas fuel storage
Gas hydrates
Gas supply
Gases
Methane
Natural gas conditioning
Natural gas deposits
Natural gas fields
Natural gas substitutes
Natural gas transportation
Natural gas well production
Natural gas wells
Natural gasoline

* Former Ei Vocabulary term

Natural gas *(continued)*
Odorization
Public utilities
Sulfur determination

Natural gas conditioning 522
DT: January 1993
UF: Conditioning (natural gas)
Natural gas—Conditioning*
RT: Natural gas

Natural gas deposits 512.2
DT: January 1977
BT: Fossil fuel deposits
NT: Natural gas fields
RT: Acidization
Natural gas
Natural gas well production

Natural gas fields 512.2.1
DT: June 1990
BT: Natural gas deposits
RT: Natural gas
Natural gas well production

Natural gas pipelines 522, 619.1
SN: Use for transmission application. For distribution application, use GAS PIPELINES
DT: Predates 1975
BT: Pipelines
RT: Natural gas transportation
Pipeline changeover

Natural gas substitutes 522
SN: Gases derived from sources other than natural gas, and used as fuels
DT: January 1993
UF: Natural gas—Substitutes*
BT: Fuels
RT: Natural gas

Natural gas transportation 522 (619.1) (432.3) (433.3) (434.3)
DT: January 1993
UF: Natural gas—Transportation*
BT: Transportation
RT: Natural gas
Natural gas pipelines

Natural gas well completion 512.2.2
DT: January 1993
UF: Natural gas wells—Completion*
BT: Well completion
RT: Natural gas wells

Natural gas well drilling 512.2.2
DT: January 1993
UF: Natural gas wells—Drilling*
BT: Well drilling
RT: Drilling fluids
Natural gas wells
Natural gas well logging
Natural gas well production
Offshore drilling

Natural gas well logging 512.2.2
DT: January 1977
BT: Well logging
RT: Acoustic logging
Directional logging
Electric logging
Electromagnetic logging
Induced polarization logging
Induction logging
Natural gas wells
Natural gas well drilling
Neutron logging
Nuclear magnetic logging

Natural gas well production 512.2.1
DT: June 1990
BT: Production
RT: Enhanced recovery
Natural gas
Natural gas deposits
Natural gas fields
Natural gas wells
Natural gas well drilling
Oil well production
Petroleum industry
Well spacing

Natural gas wells 512.2.1
DT: Predates 1975
BT: Wells
RT: Blowouts
Gas hydrates
Natural gas
Natural gas well production
Natural gas well logging
Natural gas well completion
Natural gas well drilling
Petroleum industry
Sand consolidation
Well perforation
Well pressure
Well spacing

*Natural gas wells—Blowout**
USE: Blowouts

*Natural gas wells—Completion**
USE: Natural gas well completion

*Natural gas wells—Drilling**
USE: Natural gas well drilling

*Natural gas wells—Flow**
USE: Flow of fluids

*Natural gas wells—Hydrates**
USE: Hydrates

*Natural gas wells—Nuclear stimulation**
USE: Explosive well stimulation

*Natural gas wells—Sour gas**
USE: Sour gas

*Natural gas, Liquefied**
USE: Liquefied natural gas

*Natural gas—Conditioning**
USE: Natural gas conditioning

*Natural gas—Odorizing**
USE: Odorization

*Natural gas—Substitutes**
USE: Natural gas substitutes

*Natural gas—Transportation**
USE: Natural gas transportation

Natural gasoline 523, 513.3
DT: Predates 1975
BT: Hydrocarbons
NT: Liquefied petroleum gas
RT: Liquefied natural gas
Natural gas
Natural gasoline plants
Petroleum industry

Natural gasoline plants 513.2 (402.1)
DT: Predates 1975
BT: Industrial plants
RT: Natural gasoline

Natural gems 482.2.1
DT: January 1993
UF: Gems—Natural origin*
BT: Gems

Natural language processing systems 723.2
DT: June 1990
BT: Data processing
RT: Query languages

Natural polymers 815.1.1 (461.2) (818.1)
DT: January 1987
BT: Organic polymers
NT: Amber
Biopolymers
Lignin
Polypeptides
Polysaccharides
Rubber

Natural resources (444) (481.3.1) (501) (512)
DT: January 1993
NT: Geothermal energy
Mineral resources
Water resources
RT: Conservation
Earth (planet)
Energy resources
Geology
Natural resources exploration
Oceanography
Rocks
Soils
Watersheds

Natural resources exploration
(444) (481.3.1) (481.4) (501) (512)
SN: For purposes of subsequent exploitation
DT: January 1993
UF: Exploration (natural resources)
Exploration*
Ore prospecting
NT: Geophysical prospecting
Mineral exploration
Petroleum prospecting
Water resources exploration
RT: Earth (planet)
Exploratory geochemistry
Geological surveys
Mineral resources
Mineralogy
Mines
Mining
Natural resources
Ore deposits
Photomapping
Remote sensing
Surveying
Surveys
Underwater probing

Natural rubber
USE: Rubber

Natural sciences (931) (933) (461.9) (801)
DT: January 1993
NT: Astronomy
Biology
Chemistry
Crystallography
Earth sciences
Physics
RT: Medicine
Natural sciences computing
Research

Natural sciences computing 723.2
DT: January 1993
UF: Data processing—Natural sciences applications*
Scientific computing
BT: Computer applications
RT: Data processing
Natural sciences
Numerical analysis

Natural water geochemistry 444, 481.2
DT: January 1993
UF: Geochemistry—Natural waters*
Surface water geochemistry
BT: Geochemistry
RT: Groundwater geochemistry
Surface waters

Naval architecture 671
SN: Primarily concerned with hulls and other structural features of waterborne craft
DT: Predates 1975
RT: Boats
Marine engineering
Naval vessels
Shipbuilding
Ships

Naval bases 404.1 (672)
DT: Predates 1975
BT: Military bases
RT: Military equipment
Military operations
Naval vessels

* Former Ei Vocabulary term

Naval vessels 672
SN: Use for non-combat and mixed classes of vessels. For exclusively combat-class vessels, use WARSHIPS or other appropriate terms
DT: Predates 1975
BT: Ships
NT: Warships
RT: Naval architecture
Naval bases
Submarine tenders
Submarines

Naval warfare 672.1, 404.1
DT: Predates 1975
BT: Military operations
RT: Marine missiles
Military engineering
Underwater ballistics
Underwater launched missiles

*Naval warfare—Training**
USE: Personnel training

Navier Stokes equations 921.2 (631.1) (931.1)
DT: January 1995
BT: Equations of motion
RT: Drag
Fluid dynamics
Kinetic theory of gases
Viscous flow

Navigation (431.5) (434.4) (655.1) (716.3)
DT: Predates 1975
UF: Spacecraft—Navigation aids*
NT: Air navigation
Radio navigation
RT: Chronometers
Collision avoidance
Distance measurement
Doppler effect
Gyroscopes
Inertial navigation systems
Navigation charts
Navigation systems
Orbits
Robots
Satellites
Seakeeping
Sextants
Sonar
Space flight
Tracking (position)
Trajectories
Triangulation

Navigation charts (431.5) (434.4) (655.1)
DT: January 1993
UF: Space flight—Navigation charts*
RT: Graphic methods
Maps
Navigation
Navigation systems
Space flight

Navigation systems (431.5) (434.4)
DT: January 1993
UF: Aircraft—Navigation systems*
Ships—Navigation systems*
Vehicles—Navigation systems*
BT: Equipment
NT: Inertial navigation systems
Satellite navigation aids
RT: Aircraft
Aircraft parts and equipment
Navigation
Navigation charts
Satellites
Ships
Spacecraft
Vehicles

*Navigation—Inertial systems**
USE: Inertial navigation systems

Nb
USE: Niobium

NDRO
USE: Nondestructive readout

NDT
USE: Nondestructive examination

Ne
USE: Neon

Needles
DT: Predates 1975
BT: Equipment
RT: Syringes

Negative feedback
USE: Feedback

Negative feedback control systems
USE: Closed loop control systems

Negative impedance converters (703.1)
DT: January 1993
UF: Electric networks—Negative impedance converters*
Negative proportional transformers
BT: Active networks
Electric converters

Negative proportional transformers
USE: Negative impedance converters

Negative resistance (701.1)
DT: January 1993
UF: Electric impedance—Negative resistance*
BT: Electric resistance
RT: Oscillators (electronic)
Tunnel diodes

Negative temperature coefficient thermistors
USE: Thermistors

Nematic liquid crystals (804) (931.2)
DT: January 1993
UF: Crystals, Liquid—Nematic*
BT: Liquid crystals

Neodymium 547.2
DT: January 1993
UF: Neodymium and alloys*
BT: Rare earth elements
RT: Neodymium alloys
Neodymium compounds

Neodymium alloys 547.2
DT: January 1993
UF: Neodymium and alloys*
BT: Rare earth alloys
RT: Neodymium
Neodymium compounds

*Neodymium and alloys**
USE: Neodymium OR
Neodymium alloys

Neodymium compounds (804.1) (804.2)
DT: January 1977
BT: Rare earth compounds
RT: Neodymium
Neodymium alloys

Neodymium lasers 744.4
DT: January 1993
BT: Solid state lasers

Neon 804
DT: Predates 1975
UF: Ne
BT: Inert gases

Neonatal monitoring 461.1
DT: January 1993
UF: Biomedical engineering—Neonatal monitoring*
BT: Patient monitoring
RT: Biomedical engineering

Nepheline syenite (481.1) (482.2) (505) (812.1)
DT: Predates 1975
BT: Igneous rocks
RT: Ceramic materials

Nephelometers 941.3
SN: Photoelectric instruments for determining the amount of light transmitted or scattered by suspension of particles. For visual comparators, use TURBIDIMETERS
DT: Predates 1975
BT: Optical instruments
RT: Densitometers
Particle size analysis
Turbidimeters

Neptunium 622.1
DT: Predates 1975
UF: Np
BT: Transuranium elements

Nervous system
USE: Neurology

Network analysis
USE: Electric network analysis

Network analyzers
USE: Electric network analyzers

Network components (703.1)
DT: January 1993
BT: Components
Networks (circuits)
NT: Electric reactors
Resistors

Network interfaces
USE: Interfaces (computer)

Network protocols (723)
DT: January 1993
UF: Communication protocols
Computer networks—Protocols*
Protocols
RT: Computer networks
Data communication systems
Interfaces (computer)
Standards

Network schematics
USE: Schematic diagrams

Network theory
USE: Circuit theory

Networks (circuits) 703.1
SN: Circuits which operate principally by conduction of electrons in vacuum, gas or semiconductors. For theoretical aspects use CIRCUIT THEORY
DT: January 1993
UF: Circuits
Electric circuits
Electronic circuits*
Servomechanisms—Circuits*
NT: Active networks
Bridge circuits
Buffer circuits
Compandor circuits
Comparator circuits
Computer circuits
Coupled circuits
Delay circuits
Detector circuits
Differentiating circuits
Digital circuits
Discriminators
Distributed parameter networks
Electric attenuators
Electric filters
Electric power transmission networks
Equalizers
Equivalent circuits
Fiber optic networks
Frequency converter circuits
Gyrators
Instrument circuits
Integrated circuits
Integrating circuits
Ladder networks
Limiters
Linear networks
Local area networks
Logic circuits
Lumped parameter networks

* Former Ei Vocabulary term

Networks (circuits) *(continued)*
Magnetic circuits
Microwave circuits
Mixer circuits
Multistable circuits
Multivibrators
Network components
Nonlinear networks
Passive networks
Phase changing circuits
Phase locked loops
Phase shifters
Power supply circuits
Printed circuits
Pulse analyzing circuits
Pulse circuits
Radar circuits
Rectifying circuits
Resonant circuits
Summing circuits
Sweep circuits
Switching circuits
Switching networks
Telecommunication networks
Telegraph circuits
Telephone circuits
Time varying networks
Timing circuits
Voltage stabilizing circuits
RT: Amplifiers (electronic)
Circuit oscillations
Circuit resonance
Circuit theory
Correlation detectors
Demodulators
Electric circuit breakers
Electric network analyzers
Electric power factor
Electronic circuit tracking
Frequency response
Radio
Schematic diagrams
Signal receivers
Television
Transmitters

Networks (computer)
USE: Computer networks

Networks (television)
USE: Television networks

Neural nets
USE: Neural networks

Neural networks (461.1) (723.4)
DT: June 1990
UF: Artificial neural networks
Multilayer neural networks
Neural nets
Perceptrons
NT: Cellular neural networks
Feedforward neural networks
RT: Artificial intelligence
Backpropagation
Brain models
Cybernetics
Learning systems
Neurophysiology

Neurology 461.6
DT: January 1993
UF: Biomedical engineering—Neurology*
Nervous system
BT: Medicine
NT: Neurophysiology
RT: Biomedical engineering
Brain
Neuromuscular rehabilitation
Neurosurgery

Neuromuscular rehabilitation 461.5
DT: January 1993
UF: Human rehabilitation engineering—Neuromuscular rehabilitation*
BT: Patient rehabilitation
NT: Functional neural stimulation
RT: Human rehabilitation engineering
Musculoskeletal system
Neurology

Neurophysiology 461.6
DT: January 1993
UF: Biomedical engineering—Neurophysiology*
BT: Neurology
Physiology
RT: Biomedical engineering
Brain
Neural networks

Neurosurgery 461.6
DT: January 1993
UF: Biomedical engineering—Neurosurgery*
BT: Surgery
RT: Biomedical engineering
Brain
Neurology

Neutron absorption 932.1 (931.3)
DT: January 1993
UF: Neutrons—Absorption*
BT: Energy absorption
RT: Neutrons

Neutron activation analysis (801) (931.3) (932.1)
SN: Chemical analysis based on the identification of radionuclides, following neutron bombardment of a specimen
DT: January 1993
UF: Neutrons—Activation analysis*
BT: Activation analysis
RT: Mass spectrometers
Mass spectrometry
Microanalysis
Neutron irradiation apparatus
Neutrons

Neutron detectors 944.7 (931.3) (932.1)
DT: January 1993
UF: Neutrons—Detectors*
BT: Particle detectors
RT: Neutrons

* Former Ei Vocabulary term

Neutron diffraction (931.3) (932.1)
DT: January 1993
UF: Neutrons—Diffraction*
BT: Diffraction
RT: Neutron diffraction apparatus
Neutrons

Neutron diffraction apparatus 944.7
DT: Predates 1975
BT: Instruments
RT: Neutron diffraction

Neutron emission (931.3) (932.1)
DT: January 1993
UF: Emission (neutron)
Neutrons—Emission*
RT: Neutrons

Neutron irradiation (931.3) (932.1)
DT: January 1995
BT: Irradiation
RT: Ion bombardment

Neutron irradiation apparatus (931.3) (932.1)
DT: January 1993
UF: Irradiation apparatus (neutron)
Neutrons—Irradiation apparatus*
BT: Equipment
RT: Irradiation
Neutron activation analysis
Neutrons

Neutron logging (512.1.2) (512.2.1) (932.1)
DT: January 1993
UF: Oil well logging—Neutron*
BT: Radioactivity logging
RT: Natural gas well logging
Oil well logging

Neutron radiography (461) (932.1)
DT: January 1993
UF: Radiography—Neutron*
BT: Radiography
RT: Nondestructive examination

Neutron reflection (931.3) (932.1)
DT: January 1993
UF: Neutrons—Reflection*
BT: Reflection
RT: Neutrons

Neutron scattering (931.3) (932.1)
DT: January 1993
UF: Neutrons—Scattering*
BT: Scattering
RT: Neutrons

Neutron sources (931.3) (932.1)
DT: January 1993
UF: Neutrons—Sources*
BT: Elementary particle sources
RT: Neutrons

Neutron spectrometers (931.3) (932.1) (944.7)
DT: January 1993
UF: Spectrometers, Neutron*
BT: Particle spectrometers
RT: Neutrons

Neutrons (931.3) (932.1)
DT: Predates 1975
BT: Elementary particles
RT: Neutron absorption
Neutron activation analysis
Neutron detectors
Neutron diffraction
Neutron emission
Neutron irradiation apparatus
Neutron reflection
Neutron scattering
Neutron sources
Neutron spectrometers
Nuclear reactor reflectors

*Neutrons—Absorption**
USE: Neutron absorption

*Neutrons—Activation analysis**
USE: Neutron activation analysis

*Neutrons—Detectors**
USE: Neutron detectors

*Neutrons—Diffraction**
USE: Neutron diffraction

*Neutrons—Emission**
USE: Neutron emission

*Neutrons—Irradiation apparatus**
USE: Neutron irradiation apparatus

*Neutrons—Reflection**
USE: Neutron reflection

*Neutrons—Scattering**
USE: Neutron scattering

*Neutrons—Sources**
USE: Neutron sources

Newsprint (811.1)
DT: January 1993
UF: Paper—Newsprint*
BT: Paper
RT: Paper and pulp industry

Newtonian flow 631.1
DT: January 1993
UF: Flow of fluids—Newtonian*
BT: Flow of fluids
RT: Newtonian liquids

Newtonian liquids 931.2
DT: January 1993
UF: Liquids—Newtonian*
BT: Liquids
RT: Newtonian flow

Ni
USE: Nickel

Nicad batteries
USE: Nickel cadmium batteries

Nicarbing
USE: Carbonitriding

* Former Ei Vocabulary term

Nickel 548.1
DT: January 1993
UF: Ni
Nickel and alloys*
BT: Nonferrous metals
Transition metals
RT: Nickel alloys
Nickel compounds
Nickel deposits
Nickel metallography
Nickel metallurgy
Nickel mines
Nickel ore treatment
Nickel pipe
Nickel plating
Nickel refining
Tin nickel plating

Nickel alloys 548.2
DT: January 1993
UF: Nickel and alloys*
BT: Transition metal alloys
NT: Monel metal
Nickel silver
RT: Nickel
Nickel compounds
Stainless steel

*Nickel and alloys**
USE: Nickel OR
Nickel alloys

Nickel cadmium batteries 702.1.2 (548.1) (549.3)
DT: January 1993
UF: Nicad batteries
BT: Secondary batteries

Nickel compounds (804.1) (804.2)
DT: Predates 1975
BT: Transition metal compounds
RT: Nickel
Nickel alloys

Nickel deposits 504.3, 548.1
DT: Predates 1975
BT: Ore deposits
NT: Copper nickel deposits
RT: Nickel
Nickel mines

Nickel metallography 531.2, 548.1
DT: Predates 1975
BT: Metallography
RT: Nickel
Nickel metallurgy

Nickel metallurgy 531.1, 548.1
DT: Predates 1975
BT: Metallurgy
NT: Nickel powder metallurgy
RT: Nickel
Nickel metallography
Nickel ore treatment
Nickel refining
Nickel smelting

Nickel mines 504.3, 548.1
DT: January 1993
UF: Nickel mines and mining*
BT: Mines
RT: Nickel
Nickel deposits

*Nickel mines and mining**
USE: Nickel mines

Nickel ore treatment 533.1, 548.1
DT: Predates 1975
BT: Ore treatment
NT: Copper nickel ore treatment
Nickel smelting
RT: Nickel
Nickel metallurgy

Nickel pipe 619.1, 548.1
DT: January 1993
UF: Pipe, Nickel*
BT: Pipe
RT: Nickel

Nickel plating 539.3, 548.1
DT: Predates 1975
BT: Plating
NT: Tin nickel plating
RT: Nickel

Nickel powder metallurgy 536.1, 548.1
DT: January 1993
UF: Powder metallurgy—Cobalt nickel*
Powder metallurgy—Nickel silver*
Powder metallurgy—Nickel*
BT: Nickel metallurgy
Powder metallurgy

Nickel refining 533.2, 548.1
DT: Predates 1975
BT: Metal refining
RT: Nickel
Nickel metallurgy

Nickel silver 548.2, 544.2, 546.3
SN: Copper-nickel-zinc alloy
DT: Predates 1975
BT: Copper alloys
Nickel alloys
Zinc alloys

Nickel smelting 548.1, 533.1
DT: Predates 1975
BT: Nickel ore treatment
Smelting
RT: Nickel metallurgy

Niobate minerals 482.2
DT: January 1993
UF: Mineralogy—Niobates and tantalates*
BT: Minerals
RT: Niobium
Niobium compounds

Niobium 549.3
DT: January 1993
UF: Columbium
Nb

* Former Ei Vocabulary term

Niobium *(continued)*
Niobium and alloys*
BT: Nonferrous metals
Refractory metals
RT: Niobate minerals
Niobium alloys
Niobium compounds
Niobium deposits
Niobium metallography
Niobium metallurgy
Niobium mines
Niobium ore treatment
Niobium refining

Niobium alloys 549.3
DT: January 1993
UF: Niobium and alloys*
BT: Refractory alloys
RT: Niobium
Niobium compounds

*Niobium and alloys**
USE: Niobium OR
Niobium alloys

Niobium compounds (804.1) (804.2)
DT: Predates 1975
BT: Refractory metal compounds
RT: Niobate minerals
Niobium
Niobium alloys

Niobium deposits 504.3, 549.3
DT: Predates 1975
BT: Ore deposits
RT: Niobium
Niobium mines

Niobium metallography 531.2, 549.3
DT: Predates 1975
BT: Metallography
RT: Niobium
Niobium metallurgy

Niobium metallurgy 531.1, 549.3
DT: Predates 1975
BT: Metallurgy
NT: Niobium powder metallurgy
RT: Niobium
Niobium metallography
Niobium ore treatment
Niobium refining

Niobium mines 504.3, 549.3
DT: January 1993
UF: Niobium mines and mining*
BT: Mines
RT: Niobium
Niobium deposits

*Niobium mines and mining**
USE: Niobium mines

Niobium ore treatment 533.1, 549.3
DT: Predates 1975
BT: Ore treatment
RT: Niobium
Niobium metallurgy

Niobium powder metallurgy 536.1, 549.3
DT: January 1993
UF: Powder metallurgy—Niobium*
BT: Niobium metallurgy
Powder metallurgy

Niobium refining 533.2, 549.3
DT: Predates 1975
BT: Metal refining
RT: Niobium
Niobium metallurgy

Nitrates 804.2
DT: Predates 1975
BT: Nitrogen compounds
RT: Nitration
Nitrification

Nitration 802.2
DT: January 1993
BT: Chemical reactions
RT: Nitrates

Nitric acid 804.2
DT: Predates 1975
BT: Inorganic acids

Nitrides 804.2
DT: January 1986
BT: Nitrogen compounds
NT: Cubic boron nitride
Silicon nitride
Titanium nitride

Nitriding 537.1 (545.3)
DT: January 1993
UF: Chapmanizing
Heat treatment—Nitriding*
Malcolmizing
Steel heat treatment—Nitriding*
BT: Heat treatment
RT: Carbonitriding
Case hardening
Surface treatment

Nitrification 802.2
DT: January 1977
UF: Chemical reactions—Nitrification*
BT: Chemical reactions
RT: Bacteria
Biochemistry
Biotechnology
Nitrates
Nitrogen fixation

Nitrile resins 815.1.1
DT: January 1987
BT: Nitrogen compounds
Organic polymers

Nitrocarburizing
USE: Carbonitriding

Nitrocellulose 811.3
DT: Predates 1975
UF: Cellulose nitrate
BT: Cellulose derivatives
Esters
Nitrogen compounds

Nitrogen 804
DT: Predates 1975
UF: N
BT: Nonmetals
RT: Denitrification
Nitrogen compounds
Nitrogen fertilizers
Nitrogen fixation
Nitrogen removal

Nitrogen compounds (804.1) (804.2)
DT: Predates 1975
BT: Chemical compounds
NT: Amines
Ammonia
Ammonium compounds
Chlorophyll
Cyanides
Hydrazine
Nitrates
Nitrides
Nitrile resins
Nitrocellulose
Nitrogen oxides
Nitrosamines
Urea
RT: Nitrogen
Nitrogen fertilizers

Nitrogen fertilizers 804, 821.2
DT: January 1993
UF: Fertilizers—Nitrogen*
BT: Fertilizers
NT: Urea fertilizers
RT: Agriculture
Nitrogen
Nitrogen compounds

Nitrogen fixation 802.2 (801.2)
DT: January 1993
UF: Fixation (nitrogen)
Nitrogen—Fixation*
BT: Chemical reactions
RT: Bacteria
Bioconversion
Nitrification
Nitrogen

Nitrogen oxides 804.2
DT: January 1986
UF: Air pollution—Nitrogen oxides*
BT: Nitrogen compounds
Oxides
RT: Air pollution

Nitrogen removal (513.1) (802.2) (802.3)
DT: January 1993
UF: Petroleum refining—Nitrogen removal*
BT: Removal
RT: Nitrogen

*Nitrogen—Fixation**
USE: Nitrogen fixation

Nitrosamines 804.1 (461.7)
DT: January 1993
UF: Carcinogens—Nitrosamines*
BT: Amines
Carcinogens
Nitrogen compounds

NMR
USE: Nuclear magnetic resonance

NMR spectroscopy
USE: Nuclear magnetic resonance spectroscopy

Nobelium 622.1
DT: January 1993
BT: Transuranium elements

Noble gases
USE: Inert gases

Nodular iron 534.2, 545.1
DT: January 1993
UF: Cast iron—Nodular*
Ductile iron
Spheroidal iron
BT: Cast iron

Noise (acoustic)
USE: Acoustic noise

Noise (spurious signals)
USE: Spurious signal noise

Noise abatement 751.4
DT: Predates 1975
BT: Noise pollution control
RT: Acoustic noise
Acoustic noise measurement
Acoustic wave absorption
Environmental engineering
Human engineering
Mufflers
Noise pollution
Noise pollution control equipment
Sound insulating materials
Sound insulation

Noise control
USE: Acoustic variables control

Noise generators (715)
SN: Generators of electrical noise signals. For sound generators use ACOUSTIC GENERATORS
DT: January 1993
BT: Signal generators
RT: Signal noise measurement
Spurious signal noise

Noise pollution 751.4 (461.7)
DT: October 1975
BT: Pollution
RT: Acoustic noise
Acoustic waves

* Former Ei Vocabulary term

Noise pollution *(continued)*
Human engineering
Noise abatement

Noise pollution control 751.4 (461.7)
DT: January 1993
BT: Acoustic variables control
Pollution control
NT: Noise abatement
RT: Noise pollution control equipment

Noise pollution control equipment 751.4
DT: January 1993
BT: Pollution control equipment
RT: Mufflers
Noise abatement
Noise pollution control
Sound insulating materials
Sound insulation

*Noise, Acoustic**
USE: Acoustic noise

*Noise, Acoustic—White noise**
USE: White acoustic noise

*Noise, Spurious signal**
USE: Spurious signal noise

*Noise, Spurious signal—Microphonism**
USE: Microphonism

*Noise, Spurious signal—Shot noise**
USE: Shot noise

*Noise, Spurious signal—Signal to noise ratio**
USE: Signal to noise ratio

*Noise, Spurious signal—Thermal noise**
USE: Thermal noise

*Noise, Spurious signal—White noise**
USE: White noise

Nomenclature
USE: Terminology

Nomograms 921.4
DT: Predates 1975
UF: Nomographs
RT: Graphic methods
Numerical analysis

Nomographs
USE: Nomograms

Non Newtonian flow 631.1
DT: January 1993
UF: Flow of fluids—Non Newtonian*
BT: Flow of fluids
RT: Non Newtonian liquids
Rheology
Unsteady flow

Non Newtonian liquids 931.2
DT: January 1993
UF: Liquids—Non Newtonian*
BT: Liquids
RT: Colloids
Non Newtonian flow
Rheology
Sols
Viscosity

Nonbibliographic retrieval systems 903.3
DT: January 1993
UF: Data retrieval systems
Information retrieval systems—Nonbibliographic*
BT: Information retrieval systems
NT: Geographic information systems

Nondestructive examination (421) (422.2) (423.2)
DT: January 1977
UF: NDT
Nondestructive testing
BT: Testing
NT: Acoustic emission testing
Eddy current testing
RT: Equipment testing
Gamma radiography
High temperature testing
Industrial radiography
Inspection
Materials testing
Neutron radiography
Radiography

*Nondestructive examination—Eddy current testing**
USE: Eddy current testing

Nondestructive readout 722.1
DT: January 1993
UF: Data storage, Digital—Nondestructive*
NDRO
RT: Digital storage

Nondestructive testing
USE: Nondestructive examination

Nonelectric final control devices (732)
DT: January 1993
UF: Final control devices, Nonelectric*
BT: Final control devices
RT: Air brakes
Cams
Fluidic amplifiers
Hydraulic brakes
Mechanical control equipment
Torque converters

Nonelectric street lighting 406.2, 707
DT: January 1993
UF: Street lighting—Nonelectric lighting*
BT: Street lighting

Nonferrous metals 531.1
DT: January 1977
BT: Metals
NT: Actinides
Alkali metals
Alkaline earth metals
Gallium
Hafnium
Heavy metals
Indium

* Former Ei Vocabulary term

Nonferrous metals *(continued)*
Light metals
Manganese
Molybdenum
Nickel
Niobium
Precious metals
Rare earth elements
Scandium
Tantalum
Technetium
Yttrium
Zinc
Zirconium

Nonfibrous paper materials 811.1
DT: January 1993
UF: Papermaking—Nonfibrous materials*
BT: Materials
RT: Paper
Papermaking

Noninvasive medical procedures 461.6
DT: January 1993
UF: Biomedical engineering—Noninvasive procedures*
BT: Patient treatment
RT: Biomedical engineering
Diagnosis

Nonlinear control systems 731.1
DT: January 1993
UF: Control systems, Nonlinear*
BT: Control systems
Nonlinear systems
RT: Bang bang control systems
Control nonlinearities
On-off control systems
Phase space methods
Piecewise linear techniques
Relay control systems

Nonlinear dynamics
USE: Dynamics

Nonlinear equations (921.1)
DT: January 1993
UF: Mathematical techniques—Nonlinear equations*
BT: Mathematical techniques
RT: Bifurcation (mathematics)
Numerical analysis
Solitons

Nonlinear network analysis 703.1.1
DT: January 1993
BT: Electric network analysis
RT: Nonlinear network synthesis
Nonlinear networks

Nonlinear network synthesis 703.1.2
DT: January 1993
BT: Electric network synthesis
RT: Nonlinear network analysis
Nonlinear networks

Nonlinear networks 703.1
DT: January 1993
UF: Electric networks, Nonlinear*
BT: Networks (circuits)
RT: Nonlinear network analysis
Nonlinear network synthesis

Nonlinear optical phase conjugation
USE: Optical phase conjugation

Nonlinear optics 741.1.1
DT: January 1993
UF: Light—Nonlinear optical effects*
Optics—Nonlinear*
BT: Optics
NT: Four wave mixing
Optical frequency conversion
Optical phase conjugation
Thermal blooming
RT: Electrooptical effects
Optical bistability
Raman spectroscopy
Ultrafast phenomena

Nonlinear programming (723.1) (921.5) (922)
DT: January 1993
UF: Mathematical programming, Nonlinear*
BT: Mathematical programming
RT: Constraint theory
Mathematical models

Nonlinear systems (731.1) (921)
DT: January 1995
NT: Nonlinear control systems
RT: Bang bang control systems
Control nonlinearities
Control theory
Describing functions
Linear control systems
On-off control systems
Relay control systems

Nonlinearities (control system)
USE: Control nonlinearities

Nonmetallic bearings 601.2
DT: January 1993
UF: Bearings—Nonmetallic materials*
BT: Bearings (machine parts)
RT: Nonmetallic materials

Nonmetallic gears 601.2
DT: January 1993
UF: Gears—Nonmetallic*
BT: Gears
RT: Nonmetallic materials

Nonmetallic materials
DT: Predates 1975
BT: Materials
NT: Nonmetallic matrix composites
RT: Nonmetallic bearings
Nonmetallic gears
Nonmetallic ores

* Former Ei Vocabulary term

Nonmetallic matrix composites (415) (811) (817)
DT: January 1993
UF: Composite materials—Nonmetallic matrix composites*
BT: Composite materials
Nonmetallic materials

Nonmetallic ores 505.1
DT: January 1993
UF: Ore deposits—Nonmetallic*
Ore treatment—Nonmetallic*
BT: Ores
RT: Nonmetallic materials
Nonmetals

Nonmetals 804
DT: January 1993
BT: Chemical elements
NT: Carbon
Halogen elements
Hydrogen
Inert gases
Metalloids
Nitrogen
Oxygen
Phosphorus
Selenium
Sulfur
RT: Nonmetallic ores

Nonmotorized transportation (432.1)
DT: January 1993
UF: Transportation—Nonmotorized*
BT: Transportation
RT: Bicycles

Nontextile fibers
USE: Fibers

Nonvocal aids
USE: Communication aids for nonvocal persons

Nonvolatile storage 722.1
SN: Storage which retains its contents if the power source is removed
DT: January 1993
UF: Data storage, Digital—Nonvolatile*
BT: Digital storage
NT: Magnetic bubble memories
Magnetic disk storage
Magnetic film storage
Magnetic tape storage

Nonwoven fabrics 819.5
DT: January 1993
UF: Textiles—Nonwovens*
BT: Fabrics
NT: Felt
Synthetic leather

Nose cones 654
DT: January 1993
UF: Rockets and missiles—Nose cones*
BT: Cones
RT: Missiles
Rockets
Space shuttles

Notch ductility
USE: Ductility

Notch filters 703.2
DT: January 1993
UF: Band elimination filters
Band rejection filters
Bandstop filters
Electric filters, Notch*
BT: Electric filters
RT: Bandpass filters

Notched bar tensile testing (422.2) (531)
DT: January 1993
UF: Materials testing—Notched bar*
BT: Metal testing
Tensile testing

Notebook computers
USE: Personal computers

Nozzles (631.1)
DT: Predates 1975
BT: Components
RT: Flow of fluids
Jets
Orifices
Vents

Np
USE: Neptunium

NP-complete problems
USE: Computational complexity

NP-hard problems
USE: Computational complexity

Nuclear batteries 702.1.1
DT: January 1993
UF: Electric batteries—Nuclear*
BT: Primary batteries

Nuclear energy 932.2 (621) (622.3)
DT: Predates 1975
UF: Atomic energy
Nuclear power
RT: Fission reactions
Fusion reactions
International cooperation
Nuclear engineering
Nuclear explosions
Nuclear fuels
Nuclear physics
Nuclear power plants
Targets

*Nuclear energy—Fission reactions**
USE: Fission reactions

*Nuclear energy—Fusion reactions**
USE: Fusion reactions

*Nuclear energy—International cooperation**
USE: International cooperation

*Nuclear energy—Thermonuclear reactions**
USE: Thermonuclear reactions

Nuclear engineering (613) (621) (622) (932.2)
DT: January 1977
BT: Engineering
RT: Nuclear energy
Nuclear fuels
Nuclear industry
Nuclear power plants

Nuclear explosions (404) (621) (622.3) (914.1) (932.2)
DT: Predates 1975
BT: Explosions
RT: Civil defense
Electromagnetic pulse
Excavation
Fallout
Fission reactions
Fusion reactions
Nuclear energy
Shock waves
Thermonuclear reactions

Nuclear fission
USE: Fission reactions

Nuclear fuel accounting 622.1, 911.3
DT: January 1993
UF: Accounting (nuclear fuel)
Nuclear fuels—Accountability*
RT: Inventory control
Nuclear fuels
Nuclear materials safeguards

Nuclear fuel cladding (621.1.2) (621.2.2)
DT: January 1993
UF: Nuclear fuels—Cladding*
Nuclear reactors—Fuel element cladding*
BT: Cladding (coating)
RT: Nuclear fuels
Nuclear reactors

Nuclear fuel elements (621.1.2) (621.2.2)
DT: January 1993
UF: Fuel elements (nuclear fuel)
Nuclear reactors—Fuel elements*
Reactor fuel elements
BT: Nuclear fuels
RT: Nuclear reactors
Reactor cores

Nuclear fuel pellets (621.1.2) (621.2.2)
DT: January 1993
UF: Fuel pellets (nuclear fuel)
Nuclear fuels—Pelletizing*
BT: Nuclear fuels

Nuclear fuel reprocessing (621.1.2) (621.2.2)
DT: January 1993
UF: Fuel reprocessing
Nuclear fuels—Reprocessing*
Reprocessing (nuclear fuel)
BT: Recovery
RT: Nuclear fuels
Nuclear industry
Nuclear materials safeguards
Radioactive wastes
Waste utilization

Nuclear fuels (621.1.2) (621.2.2)
DT: Predates 1975
BT: Energy resources
Fuels
NT: Mixed oxide fuels
Nuclear fuel elements
Nuclear fuel pellets
Spent fuels
RT: Coated fuel particles
Moderators
Nuclear energy
Nuclear engineering
Nuclear fuel accounting
Nuclear fuel cladding
Nuclear fuel reprocessing
Nuclear industry
Nuclear materials safeguards
Nuclear reactors
Reactor operation
Reactor refueling
Uranium carbide

*Nuclear fuels—Accountability**
USE: Nuclear fuel accounting

*Nuclear fuels—Cladding**
USE: Nuclear fuel cladding

*Nuclear fuels—Coated particles**
USE: Coated fuel particles

*Nuclear fuels—Fission**
USE: Fission reactions

*Nuclear fuels—Fusion**
USE: Fusion reactions

*Nuclear fuels—Metallography**
USE: Metallography

*Nuclear fuels—Metallurgy**
USE: Metallurgy

*Nuclear fuels—Mixed oxides**
USE: Mixed oxide fuels

*Nuclear fuels—Pelletizing**
USE: Nuclear fuel pellets

*Nuclear fuels—Reprocessing**
USE: Nuclear fuel reprocessing

*Nuclear fuels—Safeguards**
USE: Nuclear materials safeguards

Nuclear fusion reactions
USE: Fusion reactions

Nuclear gyroscopes (932.2) (943.3)
DT: January 1993
UF: Gyroscopes—Nuclear*
BT: Gyroscopes
Nuclear instrumentation

Nuclear industry (613) (621) (622) (932.2)
DT: Predates 1975
BT: Industry
RT: Nuclear engineering
Nuclear fuel reprocessing
Nuclear fuels

* Former Ei Vocabulary term

Nuclear industry *(continued)*
Nuclear materials safeguards
Nuclear power plants
Nuclear reactors
Radioactive wastes
Radioactive waste disposal

Nuclear instrumentation (943.3) (944.7)
DT: Predates 1975
BT: Instruments
NT: Dosimeters
Nuclear gyroscopes
Particle detectors
Radiation counters
RT: Nuclear reactors

Nuclear logging
USE: Nuclear magnetic logging

Nuclear magnetic logging 512.1.2 (932.2)
DT: January 1993
UF: Nuclear logging
Oil well logging—Nuclear*
BT: Well logging
RT: Natural gas well logging
Oil well logging

Nuclear magnetic resonance (931.2) (932.2)
DT: January 1986
UF: NMR
Proton magnetic resonance
BT: Magnetic resonance
Nuclear properties
RT: Magnetic resonance imaging
Nuclear magnetic resonance spectroscopy
Radiofrequency spectroscopy

Nuclear magnetic resonance spectroscopy (931.3) (932.2)
DT: January 1995
UF: NMR spectroscopy
BT: Magnetic resonance spectroscopy
RT: Magnetic resonance imaging
Magnetic resonance spectrometers
Nuclear magnetic resonance
Radiofrequency spectroscopy

Nuclear materials safeguards 914.1 (621.1.2) (621.2.2)
DT: January 1993
UF: Nuclear fuels—Safeguards*
Nuclear power plants—Safeguard systems*
Nuclear reactors—Safeguard systems*
Safeguards (nuclear material)
RT: Hazards
Inspection
Nuclear fuel accounting
Nuclear fuel reprocessing
Nuclear fuels
Nuclear industry
Nuclear reactors
Protection
Radioactive wastes

Nuclear medicine 622.3, 461.6
DT: October 1975
BT: Medicine
RT: Biomedical engineering
Radioactive tracers
Radioisotopes
Radiotherapy

Nuclear physics 932.2
DT: January 1993
UF: Physics—Nuclear*
BT: Physics
RT: Atomic physics
Criticality (nuclear fission)
Fission reactions
Fusion reactions
Laser fusion
Molecular physics
Nuclear energy
Nuclear properties
Nuclear radiation spectroscopy
Radioactivity
Research reactors

Nuclear power
USE: Nuclear energy

Nuclear power plants 613
DT: Predates 1975
UF: Nuclear power stations
BT: Power plants
RT: Electric power plants
Nuclear energy
Nuclear engineering
Nuclear industry
Nuclear reactors
Tornado generated missiles
Wastewater disposal

*Nuclear power plants—Safeguard systems**
USE: Nuclear materials safeguards

*Nuclear power plants—Tornado generated missiles**
USE: Tornado generated missiles

Nuclear power stations
USE: Nuclear power plants

Nuclear powered vehicles 622.3 (662.1) (682.1.2)
DT: January 1993
UF: Automobiles—Nuclear power*
Locomotives—Nuclear power*
BT: Vehicles
RT: Automobile engines
Locomotives
Nuclear propulsion

Nuclear properties 932.2 (931.2)
DT: January 1993
NT: Nuclear magnetic resonance
RT: Nuclear physics

Nuclear propulsion 622.3 (653.1) (675.1)
DT: January 1993
UF: Aircraft engines, Jet and turbine—Nuclear propulsion*
Seaplanes—Nuclear propulsion*
Ship propulsion—Nuclear*
BT: Propulsion

Nuclear propulsion *(continued)*
RT: Jet engines
Nuclear powered vehicles
Nuclear reactors
Seaplanes
Ship propulsion
Submarines

Nuclear radiation spectroscopy (932.2)
DT: January 1993
UF: Spectroscopy, Nuclear radiation*
BT: Spectroscopy
RT: Nuclear physics

Nuclear reactor accidents 914.1 (621)
DT: January 1993
UF: Radiation accidents
Radiation releases
BT: Accidents
NT: Core disruptive accidents
Loss of coolant accidents
RT: Nuclear reactors
Tornado generated missiles

Nuclear reactor cores
USE: Reactor cores

Nuclear reactor licensing 902.3 (621)
DT: January 1993
UF: Licensing (reactors)
Nuclear reactors—Licensing*
RT: Laws and legislation
Nuclear reactors

Nuclear reactor operation
USE: Reactor operation

Nuclear reactor reactivity
USE: Reactivity (nuclear)

Nuclear reactor reflectors 621.1.1
DT: January 1993
UF: Nuclear reactors—Reflectors*
BT: Nuclear reactors
RT: Neutrons

Nuclear reactor simulators (621)
DT: January 1993
UF: Nuclear reactors—Simulators*
BT: Simulators
RT: Nuclear reactors

Nuclear reactors (621)
SN: Use for the general subject, and for fission reactors
DT: Predates 1975
UF: Fission reactors
Nuclear reactors—Carryover*
Nuclear reactors—Fission*
Reactors (nuclear)
BT: Facilities
NT: Breeder reactors
Breeding blankets
Containment vessels
Control rods
Educational nuclear reactors
Experimental reactors
Fast reactors
Gas cooled reactors
High temperature reactors
Liquid metal cooled reactors
Mobile nuclear reactors
Nuclear reactor reflectors
Particle injectors
Pressure tube reactors
Process heat reactors
Reactor cores
Research reactors
Small nuclear reactors
Underwater reactors
Water cooled reactors
RT: Core disruptive accidents
Core meltdown
Criticality (nuclear fission)
Decommissioning (nuclear reactors)
Fission products
Fission reactions
Fusion reactions
Loss of coolant accidents
Moderators
Nuclear fuel cladding
Nuclear fuel elements
Nuclear fuels
Nuclear industry
Nuclear instrumentation
Nuclear materials safeguards
Nuclear power plants
Nuclear propulsion
Nuclear reactor accidents
Nuclear reactor licensing
Nuclear reactor simulators
Pressure vessels
Reactivity (nuclear)
Reactor operation
Reactor refueling
Reactor shielding
Reactor shutdowns
Reactor startup
Spent fuels
Transients

*Nuclear reactors, Boiling water**
USE: Boiling water reactors

*Nuclear reactors, Breeder**
USE: Breeder reactors

*Nuclear reactors, Experimental**
USE: Experimental reactors

*Nuclear reactors, Fast**
USE: Fast reactors

*Nuclear reactors, Fusion**
USE: Fusion reactors

*Nuclear reactors, Gas cooled**
USE: Gas cooled reactors

*Nuclear reactors, Heavy water**
USE: Heavy water reactors

*Nuclear reactors, High temperature**
USE: High temperature reactors

* Former Ei Vocabulary term

*Nuclear reactors, Light water**
USE: Light water reactors

*Nuclear reactors, Liquid metal cooled**
USE: Liquid metal cooled reactors

*Nuclear reactors, Pressurized water**
USE: Pressurized water reactors

*Nuclear reactors, Process heat**
USE: Process heat reactors

*Nuclear reactors, Water cooled**
USE: Water cooled reactors

*Nuclear reactors—Auxiliary equipment**
USE: Auxiliary equipment

*Nuclear reactors—Breeding blankets**
USE: Breeding blankets

*Nuclear reactors—Carryover**
USE: Nuclear reactors

*Nuclear reactors—Containment vessels**
USE: Containment vessels

*Nuclear reactors—Control rods**
USE: Control rods

*Nuclear reactors—Core disruptive accident**
USE: Core disruptive accidents

*Nuclear reactors—Core meltdown**
USE: Core meltdown

*Nuclear reactors—Cores**
USE: Reactor cores

*Nuclear reactors—Criticality**
USE: Criticality (nuclear fission)

*Nuclear reactors—Decommissioning**
USE: Decommissioning (nuclear reactors)

*Nuclear reactors—Educational**
USE: Educational nuclear reactors

*Nuclear reactors—Fission products**
USE: Fission products

*Nuclear reactors—Fission**
USE: Nuclear reactors

*Nuclear reactors—Fuel element cladding**
USE: Nuclear fuel cladding

*Nuclear reactors—Fuel elements**
USE: Nuclear fuel elements

*Nuclear reactors—Licensing**
USE: Nuclear reactor licensing

*Nuclear reactors—Loss of coolant accident**
USE: Loss of coolant accidents

*Nuclear reactors—Mobile**
USE: Mobile nuclear reactors

*Nuclear reactors—Moderators**
USE: Moderators

*Nuclear reactors—Noise**
USE: Acoustic noise

*Nuclear reactors—Particle injectors**
USE: Particle injectors

*Nuclear reactors—Power transients**
USE: Transients

*Nuclear reactors—Pressure tubes**
USE: Pressure tube reactors

*Nuclear reactors—Reflectors**
USE: Nuclear reactor reflectors

*Nuclear reactors—Refueling**
USE: Reactor refueling

*Nuclear reactors—Research reactors**
USE: Research reactors

*Nuclear reactors—Safeguard systems**
USE: Nuclear materials safeguards

*Nuclear reactors—Shielding**
USE: Reactor shielding

*Nuclear reactors—Shutdown**
USE: Reactor shutdowns

*Nuclear reactors—Simulators**
USE: Nuclear reactor simulators

*Nuclear reactors—Small**
USE: Small nuclear reactors

*Nuclear reactors—Spent fuels**
USE: Spent fuels

*Nuclear reactors—Start up**
USE: Reactor startup

*Nuclear reactors—Underwater**
USE: Underwater reactors

Nuclear stimulation
USE: Explosive well stimulation

Nuclear wastes
USE: Radioactive wastes

Nucleate boiling 641.1
DT: January 1995
BT: Boiling liquids
RT: Heat transfer
Nucleation

Nucleation 933.1.2
DT: January 1995
RT: Cloud seeding
Condensation
Crystal growth
Crystallization
Freezing
Heat treatment
Nucleate boiling

Nucleic acid sequences 461.2, 801.2, 804.1
DT: January 1993
UF: Biochemistry—Nucleic acid sequences*
NT: DNA sequences
RT: Biochemistry
Nucleic acids

* Former Ei Vocabulary term

Nucleic acids 461.2, 801.2, 804.1
DT: Predates 1975
BT: Biological materials
Biopolymers
Macromolecules
Organic acids
NT: DNA
RNA
RT: Nucleic acid sequences

*Nuclides**
USE: Isotopes

*Nuclides—Radioactive**
USE: Radioisotopes

Number theory (921)
DT: January 1993
UF: Mathematical techniques—Number theory*
BT: Mathematical techniques
Theory
RT: Digital arithmetic
Random number generation
Root loci

Numbering systems (902.2) (921)
SN: For identification using numerals, letters, or a combination of these
DT: Predates 1975
NT: Preferred numbers
Telephone numbering systems
Textile numbering systems
RT: Labeling

*Numbering systems—Preferred numbers**
USE: Preferred numbers

Numerical analysis 921.6
DT: January 1993
UF: Mathematical techniques—Numerical analysis*
BT: Mathematical techniques
NT: Boundary element method
Chebyshev approximation
Curve fitting
Difference equations
Error analysis
Finite difference method
Finite element method
Function evaluation
Interpolation
Iterative methods
Least squares approximations
Monte Carlo methods
RT: Boundary value problems
Computational fluid dynamics
Computers
Differential equations
Differentiation (calculus)
Integral equations
Integration
Linear algebra
Mathematical models
Natural sciences computing
Nomograms
Nonlinear equations
Numerical methods

Numerical control systems 731.1
DT: January 1993
UF: Control systems, Numerical*
BT: Control systems
RT: Computer control systems
Digital control systems
Process control

Numerical methods 921.6
DT: January 1993
UF: Mathematical techniques—Numerical methods*
BT: Mathematical techniques
NT: Convergence of numerical methods
RT: Boundary value problems
Describing functions
Eigenvalues and eigenfunctions
Numerical analysis

Numerical simulation
USE: Computer simulation

Numerical stability
USE: Convergence of numerical methods

Nursing 461.7
DT: January 1993
UF: Health care—Nursing*
RT: Health care
Medicine

Nusselt number 641.2
DT: January 1995
RT: Heat conduction
Heat convection
Heat transfer
Prandtl number
Thermal conductivity

Nutrition 461.7 (822.3)
DT: January 1993
UF: Diet
Health care—Nutrition*
BT: Health care
RT: Food products
Health
Human rehabilitation engineering
Vitamins

Nuts (fasteners) (601.2) (605)
DT: January 1993
UF: Bolts and nuts*
BT: Fasteners
RT: Bolts

Nuvistors 714.1
DT: January 1993
UF: Electron tubes, Nuvistor*
BT: Electron tubes

* Former Ei Vocabulary term

Nylon polymers 815.1.1
SN: Use for nylon as a chemical substance or as a plastics product. For nylon as a textile material, use NYLON TEXTILES
DT: Predates 1975
BT: Polyamides
RT: Nylon textiles

Nylon textiles (815.1.1) (817.1) (819.2) (819.5)
SN: Use only for nylon as a textile material, i.e., for nylon fibers, filaments, yarns, yarn intermediates, or fabrics. For other applications of nylon and for treatment as a chemical substance, use NYLON POLYMERS
DT: Predates 1975
BT: Textiles
RT: Nylon polymers
Synthetic fibers
Textile fibers

Nyquist diagrams (731.1) (921.4)
DT: January 1993
UF: Control systems—Nyquist diagrams*
RT: Control system analysis
Control system synthesis
Describing functions
Frequency domain analysis
Graphic methods
System stability

Nystagmus
USE: Eye movements

O
USE: Oxygen

O rings
USE: Gaskets

Object oriented languages
USE: Object oriented programming

Object oriented programming 723.1
DT: January 1993
UF: Computer programming—Object oriented programming*
Object oriented languages
OOP
BT: Computer programming
RT: Computer aided software engineering
PROLOG (programming language)

Object recognition (716) (723.5) (741.1) (751.1)
DT: January 1995
BT: Image processing
Pattern recognition
RT: Computer graphics
Computer vision

Observability 731.1
DT: January 1993
UF: Control systems—Observability*
BT: Control theory
RT: Control systems
Stability
State space methods
System stability

Observatories (402.1) (443) (657)
DT: Predates 1975
BT: Facilities
NT: Spacecraft observatories
RT: Astronomy
Radar astronomy
Radio astronomy
Telescopes

Obsolescence
DT: January 1993
UF: Machine tools—Obsolescence*
RT: Equipment
Machine tools
Machinery
Modernization

Obstacle avoidance
USE: Collision avoidance

Obstacle detectors (681.3) (914.1)
SN: Railroad signal systems
DT: January 1993
UF: Railroad signals and signaling—Obstacle detectors*
BT: Detectors
Railroad signal systems

Obstetrics 461.6
DT: January 1993
UF: Biomedical engineering—Obstetrics*
BT: Medicine
RT: Biomedical engineering

Occupational diseases 914.3.1 (461.6) (461.7) (912.4)
DT: Predates 1975
BT: Diseases
RT: Health hazards
Industrial hygiene
Industrial poisons
Industry
Medical departments (industrial plants)
Occupational risks
Personnel

*Occupational diseases—Dermatitis**
USE: Dermatitis

*Occupational diseases—Pulmonary**
USE: Pulmonary diseases

Occupational risks (912.4) (914.1) (914.3) (914.3.1)
DT: January 1993
UF: Risk studies—Occupational risks*
BT: Risks
RT: Health risks
Industry
Occupational diseases
Personnel

Occupational therapy 461.5 (914.3.1)
DT: January 1993
UF: Human rehabilitation engineering—Occupational therapy*
BT: Patient rehabilitation
RT: Human rehabilitation engineering

* Former Ei Vocabulary term

Ocean currents 471.4
DT: January 1993
UF: Currents (ocean)
Oceanography—Currents*
RT: Flow of water
Oceanography
Tides
Water waves

Ocean dumping 453.1 (452) (472)
DT: January 1993
UF: Waste disposal—Ocean dumping*
BT: Waste disposal
RT: Marine pollution
Ocean engineering

Ocean engineering 472
DT: January 1977
UF: Undersea technology*
BT: Engineering
RT: Hydrographic surveys
Marine applications
Ocean dumping
Ocean habitats
Ocean structures
Ocean thermal energy conversion
Oceanographic equipment
Oceanography
Production platforms
Underwater construction
Underwater cutting
Underwater drilling
Underwater equipment
Underwater explosions
Underwater mineral resources
Underwater power plants
Underwater probing
Underwater structures
Underwater welding

*Ocean engineering—Habitats**
USE: Ocean habitats

*Ocean engineering—Searching**
USE: Underwater probing

*Ocean engineering—Structures**
USE: Ocean structures

Ocean floor nodules
USE: Manganese nodules

Ocean habitats 454.3 (472)
DT: January 1993
UF: Deep sea habitats
Habitats (ocean)
Ocean engineering—Habitats*
Underwater habitats
BT: Ocean structures
RT: Ocean engineering
Oceanography

Ocean platforms
USE: Production platforms

Ocean pollution
USE: Marine pollution

Ocean structures 472
DT: January 1993
UF: Ocean engineering—Structures*
BT: Offshore structures
NT: Ocean habitats
Production platforms
RT: Ocean engineering
Shore protection
Underwater structures

Ocean thermal energy conversion 615.5, 472
DT: January 1981
BT: Energy conversion
RT: Heating
Ocean engineering

Ocean waves
USE: Water waves

Oceanographic applications
USE: Marine applications

Oceanographic equipment 472 (471.2) (943.3)
DT: January 1993
BT: Equipment
NT: Oceanographic instruments
Oceanographic submarines
RT: Diving apparatus
Ocean engineering
Oceanography

Oceanographic instruments 471.2 (472)
DT: January 1993
UF: Oceanography—Instruments*
BT: Instruments
Oceanographic equipment
RT: Oceanography

Oceanographic submarines 472, 674.1
DT: January 1993
UF: Oceanography—Submarines*
BT: Oceanographic equipment
Submarines
RT: Oceanography

Oceanography 471.1
DT: Predates 1975
BT: Earth sciences
RT: Bathymetry
Beaches
Breakwaters
Coastal zones
Diving apparatus
Earth (planet)
Hydrographic surveys
Hydrology
Hydrophones
Marine applications
Marine biology
Natural resources
Ocean currents
Ocean engineering
Ocean habitats
Oceanographic equipment
Oceanographic instruments
Oceanographic submarines
Reefs

* Former Ei Vocabulary term

Oceanography *(continued)*
Remote sensing
Salinity measurement
Sea ice
Sea level
Seawater
Seaweed
Submarine geology
Submarine geophysics
Tides
Tsunamis
Underwater acoustics
Underwater mineralogy
Underwater soils
Water waves

*Oceanography—Bathymetry**
USE: Bathymetry

*Oceanography—Coastal zones**
USE: Coastal zones

*Oceanography—Currents**
USE: Ocean currents

*Oceanography—Instruments**
USE: Oceanographic instruments

*Oceanography—Radioactivity**
USE: Radioactivity

*Oceanography—Salinity measurements**
USE: Salinity measurement

*Oceanography—Sea ice**
USE: Sea ice

*Oceanography—Sea level changes**
USE: Sea level

*Oceanography—Submarines**
USE: Oceanographic submarines

OCR
USE: Optical character recognition

Octane number
USE: Antiknock rating

Odor control (445.1) (451.2) (452.2)
DT: Predates 1975
BT: Control
RT: Anaerobic digestion
Odor removal
Odors
Sewage treatment plants
Water treatment

Odor removal (445.1) (451.2) (452.2) (822.3)
DT: January 1993
UF: Oils and fats—Odor removal*
RT: Odor control
Odors
Oils and fats

Odorization 522
DT: January 1993
UF: Gas manufacture—Odorizing*
Natural gas—Odorizing*
BT: Gas fuel manufacture
RT: Natural gas

Odors (451.1) (452.1) (453.1)
DT: January 1993
RT: Odor control
Odor removal
Sensory perception

OEIC
USE: Integrated optoelectronics

Off road vehicles (662) (663)
DT: January 1993
UF: Automobiles—Off highway*
ORV
Vehicles—Off road operation*
BT: Ground vehicles

Office automation 723.5 (912.2)
DT: January 1986
UF: Electronic office
Office of the future
BT: Automation
RT: Administrative data processing
Electronic mail
Facsimile
Information technology
Management
Spreadsheets
Telecommunication systems
Teleconferencing
Telegraph
Telephone
Telephone systems
Videotex
Word processing

Office buildings 402.2
DT: Predates 1975
BT: Buildings
RT: Intelligent buildings

Office equipment (722.4) (912.2)
DT: Predates 1975
BT: Equipment
NT: Typewriters
RT: Business machines
Computers
Copying

*Office equipment—Finishing**
USE: Finishing

Office of the future
USE: Office automation

Offset printing 745.1
DT: January 1993
UF: Printing—Offset*
BT: Printing
RT: Lithography
Offset printing plates

Offset printing plates 745.1.1
DT: January 1993
UF: Printing plates—Offset*
BT: Printing plates
RT: Offset printing

* Former Ei Vocabulary term

Offshore boreholes (501.1) (512) (674.2)
DT: January 1993
UF: Boreholes—Offshore*
BT: Boreholes
RT: Offshore drilling
Offshore oil fields
Offshore oil wells
Offshore petroleum prospecting
Offshore structures
Production platforms

Offshore drilling (511.1) (512) (674.2)
DT: January 1993
UF: Oil well drilling—Offshore*
BT: Underwater drilling
RT: Marine risers
Natural gas well drilling
Offshore boreholes
Offshore oil fields
Offshore oil wells
Offshore petroleum prospecting
Oil well drilling
Production platforms

Offshore oil fields 512.1.1 (674.2)
DT: January 1993
UF: Oil fields—Offshore*
BT: Oil fields
RT: Offshore boreholes
Offshore drilling
Offshore oil wells
Offshore oil well production
Offshore petroleum prospecting
Offshore pipelines

Offshore oil well production 511.1 (674.2)
DT: January 1993
UF: Oil well completion—Offshore operations*
Oil well production—Offshore*
Oil well production—Sub-sea production system*
Sub-sea oil well production
Submarine oil well production
Subsea oil well production
BT: Oil well production
RT: Offshore oil fields
Underwater structures

Offshore oil wells 512.1.1 (674.2)
DT: January 1993
UF: Oil wells—Offshore*
BT: Oil wells
RT: Offshore boreholes
Offshore drilling
Offshore oil fields
Offshore pipelines
Offshore structures
Production platforms

Offshore petroleum prospecting 512 (674.2)
DT: January 1993
UF: Petroleum prospecting—Offshore*
BT: Petroleum prospecting
RT: Offshore boreholes
Offshore drilling
Offshore oil fields

Offshore pipelines 511.2, 619.1
DT: January 1993
UF: Pipelines—Offshore*
BT: Offshore structures
Pipelines
NT: Submarine pipelines
RT: Offshore oil fields
Offshore oil wells

Offshore power plants (402.1) (613.2) (614.2)
DT: January 1993
UF: Power plants—Offshore*
BT: Offshore structures
Power plants

Offshore structures 674.2
DT: January 1977
BT: Structures (built objects)
NT: Ocean structures
Offshore pipelines
Offshore power plants
RT: Marine insurance
Offshore boreholes
Offshore oil wells
Semisubmersibles

Ohmic contacts (704.1) (714.2)
DT: January 1993
UF: Electric contacts, Ohmic*
BT: Electric contacts

Ohmmeters 942.1
DT: Predates 1975
BT: Electric measuring instruments
RT: Ammeters
Electric resistance measurement

Oil (crude)
USE: Crude petroleum

Oil bearing formations 512.1.1 (481.1)
DT: January 1993
UF: Formations (oil bearing)
Oil fields—Formations*
RT: Acoustic logging
Geology
Landforms
Oil fields
Petroleum industry
Petroleum reservoirs

Oil booms 453.2 (407)
SN: For removing spilled oil from the water surface, or for preventing it from drifting onto beaches or into harbors
DT: Predates 1975
UF: Booms (spill retention)
Oil retention booms
Oil terminals—Spill booms*
Spill booms
BT: Water pollution control equipment
RT: Hazardous materials spills
Oil spills

Oil burners 521.3, 523 (642.2) (643.2)
DT: Predates 1975
BT: Fuel burners

* Former Ei Vocabulary term

Oil burners *(continued)*
RT: Fuel oils
Oil furnaces

*Oil burners—Noise**
USE: Acoustic noise

Oil circuit breakers 704.2 (706.2)
DT: January 1993
UF: Electric circuit breakers, Oil*
BT: Electric circuit breakers
RT: Insulating oil

Oil engines
USE: Diesel engines

Oil field development 512.1.2
DT: January 1993
UF: Oil fields—Field development*
RT: Oil fields

Oil field equipment 511.2
DT: Predates 1975
BT: Equipment
NT: Blowout preventers
Christmas trees (wellheads)
Oil well casings
Oil well drilling equipment
Oil well pumps
Packers
RT: Petroleum industry
Well equipment

*Oil field equipment—Belts**
USE: Belts

*Oil field equipment—Blowout preventers**
USE: Blowout preventers

*Oil field equipment—Christmas trees**
USE: Christmas trees (wellheads)

*Oil field equipment—Heaters**
USE: Heating equipment

*Oil field equipment—Packers**
USE: Packers

*Oil field equipment—Tubular goods**
USE: Tubes (components)

Oil fields 512.1.1 (511)
SN: Surface boundaries of areas from which petroleum may be obtained, whether corresponding to the boundaries of oil reservoirs or to political or legal limits
DT: Predates 1975
BT: Petroleum deposits
NT: Offshore oil fields
RT: Brines
Crude petroleum
Oil bearing formations
Oil field development
Petroleum industry
Petroleum reservoir evaluation
Petroleum reservoir engineering
Petroleum reservoirs
Pipeline terminals
Reserves to production ratio
Resource valuation
Wastewater disposal

*Oil fields—Brines**
USE: Brines

*Oil fields—Field development**
USE: Oil field development

*Oil fields—Formations**
USE: Oil bearing formations

*Oil fields—Offshore**
USE: Offshore oil fields

*Oil fields—Reserves to production ratio**
USE: Reserves to production ratio

*Oil fields—Reservoir evaluation**
USE: Petroleum reservoir evaluation

*Oil fields—Unit operation**
USE: Unit operations (oil wells)

*Oil fields—Valuation**
USE: Resource valuation

*Oil fields—Wastewater disposal**
USE: Wastewater disposal

Oil filled cables 706.2
DT: January 1993
UF: Electric cables—Oil filled*
Oil insulated cables
BT: Electric cables
RT: Insulating oil

Oil filled transformers 704.2
DT: January 1993
UF: Electric transformers—Oil filled*
Oil insulated transformers
BT: Electric transformers
RT: Insulating oil

Oil fired boilers 614 (521.1) (523)
DT: January 1993
BT: Boilers
RT: Fuel oils

*Oil fuel**
USE: Fuel oils

*Oil fuel—Residual**
USE: Residual fuels

Oil furnaces (521.1) (523) (532) (534) (642) (643)
DT: January 1993
UF: Furnaces—Oil fuel*
BT: Furnaces
RT: Fuel oils
Oil burners

Oil insulated cables
USE: Oil filled cables

Oil insulated transformers
USE: Oil filled transformers

Oil pollution of refrigerants 644.2 (804)
DT: January 1993
UF: Refrigerants—Oil pollution*
BT: Pollution

Oil pollution of refrigerants *(continued)*
RT: Refrigerants
Refrigeration

Oil resistance (801)
DT: January 1993
BT: Chemical resistance

Oil retention booms
USE: Oil booms

*Oil rigs, Jack-up**
USE: Jack-up rigs

Oil sand refining 513.1 (512.1)
DT: January 1993
UF: Oil sands—Refining*
BT: Petroleum refining
RT: Oil sands

Oil sands 512.1
DT: Predates 1975
BT: Fossil fuel deposits
RT: Crude petroleum
Fossil fuels
Oil sand refining
Oil shale
Petroleum industry
Thermal oil recovery

*Oil sands—Refining**
USE: Oil sand refining

*Oil sands—Thermal recovery**
USE: Thermal oil recovery

Oil shale 512.1 (481.1)
DT: Predates 1975
BT: Carbonaceous shale
Fossil fuel deposits
RT: Crude petroleum
Fossil fuels
Kerogen
Oil sands
Oil shale processing
Oil shale refining
Petroleum deposits
Petroleum industry
Shale oil

Oil shale processing 513.1
DT: June 1990
BT: Processing
NT: Oil shale refining
Oil shale retorting
RT: Oil shale
Wastewater disposal

*Oil shale processing—Retorting**
USE: Oil shale retorting

*Oil shale processing—Wastewater disposal**
USE: Wastewater disposal

Oil shale refining 513.1 (512.1)
DT: January 1993
UF: Oil shale—Refining*
BT: Hydrocarbon refining
Oil shale processing
RT: Oil shale
Petroleum refining

Oil shale retorting 513.1 (512.1)
DT: January 1993
UF: Oil shale processing—Retorting*
BT: Decomposition
Oil shale processing
RT: Pyrolysis

*Oil shale—Refining**
USE: Oil shale refining

Oil spills 453.1 (914.1)
DT: January 1993
UF: Water pollution—Oil spills*
BT: Hazardous materials spills
RT: Marine pollution
Oil booms
Petroleum industry
Soil pollution
Water pollution

Oil tankers
USE: Tankers (ships)

Oil tanks 523, 619.2
DT: Predates 1975
BT: Tanks (containers)
NT: Bleeding systems
RT: Aluminum cladding
Heating equipment
Paraffin sedimentation

*Oil tanks—Aluminum cladding**
USE: Aluminum cladding

*Oil tanks—Bleeding systems**
USE: Bleeding systems

*Oil tanks—Concrete**
USE: Concrete tanks

*Oil tanks—Gaging**
USE: Gaging

*Oil tanks—Heaters**
USE: Heating equipment

*Oil tanks—Losses**
USE: Losses

*Oil tanks—Moving**
USE: Moving

*Oil tanks—Paraffin sedimentation**
USE: Paraffin sedimentation

Oil terminals 407.1
DT: Predates 1975
UF: Terminals (oil)
BT: Facilities
RT: Service vessels

*Oil terminals—Service vessels**
USE: Service vessels

*Oil terminals—Spill booms**
USE: Oil booms

* Former Ei Vocabulary term

*Oil well casing**
USE: Oil well casings

*Oil well casing—Gun perforators**
USE: Gun perforators

*Oil well casing—Perforators**
USE: Perforators

Oil well casings 511.2
DT: January 1993
UF: Casings (oil well)
Oil well casing*
BT: Oil field equipment
Tubes (components)
RT: Fishing (oil wells)
Gun perforators
Oil wells
Perforators
Pipe

Oil well cementing 512.1.2
DT: Predates 1975
BT: Oil well completion
Well cementing
RT: Oil wells
Portland cement

Oil well completion 512.1.2
DT: Predates 1975
UF: Completion (oil well)
BT: Well completion
NT: Oil well cementing
RT: Oil wells

*Oil well completion—Multiple zone**
USE: Multiple zones

*Oil well completion—Offshore operations**
USE: Offshore oil well production

Oil well drilling 512.1.2
DT: Predates 1975
UF: Deflected oil well drilling
BT: Well drilling
NT: Deep oil well drilling
Exploratory oil well drilling
Explosive oil well drilling
RT: Blowout prevention
Circulating media
Core drilling
Drilling fluids
Horizontal wells
Mud pumps
Offshore drilling
Oil well drilling equipment
Oil well drills
Oil wells

Oil well drilling equipment 511.2
DT: January 1993
BT: Drilling equipment
Oil field equipment
NT: Drilling rigs
Mud pumps
Oil well drills
RT: Drill pipe
Fishing (oil wells)
Oil well drilling

*Oil well drilling—Bits**
USE: Bits

*Oil well drilling—Blowout prevention**
USE: Blowout prevention

*Oil well drilling—Circulating media**
USE: Circulating media

*Oil well drilling—Core**
USE: Core drilling

*Oil well drilling—Deep**
USE: Deep oil well drilling

*Oil well drilling—Deflected**
USE: Deflected well drilling

*Oil well drilling—Directional**
USE: Directional drilling

*Oil well drilling—Drill pipe**
USE: Drill pipe

*Oil well drilling—Drilling fluids**
USE: Drilling fluids

*Oil well drilling—Exploratory**
USE: Exploratory oil well drilling

*Oil well drilling—Explosive**
USE: Explosive oil well drilling

*Oil well drilling—Fishing**
USE: Fishing (oil wells)

*Oil well drilling—Mud pumps**
USE: Mud pumps

*Oil well drilling—Offshore**
USE: Offshore drilling

*Oil well drilling—Rigs**
USE: Drilling rigs

*Oil well drilling—Rotary mud**
USE: Drilling fluids

*Oil well drilling—Turbodrills**
USE: Turbodrills

Oil well drills 511.2
DT: January 1993
BT: Drills
Oil well drilling equipment
RT: Oil well drilling
Turbodrills

Oil well fires 512.1, 914.2
DT: January 1993
BT: Fires
RT: Fire hazards
Oil wells

Oil well flooding 511.1
DT: January 1993
UF: Flooding (oil wells)
Oil well production—Flooding*
RT: Oil well production
Oil wells

* Former Ei Vocabulary term

Oil well logging 512.1.2
DT: Predates 1975
BT: Well logging
RT: Acoustic logging
Directional logging
Electric logging
Electromagnetic logging
Induced polarization logging
Induction logging
Mud logging
Neutron logging
Nuclear magnetic logging
Oil wells

*Oil well logging—Acoustic**
USE: Acoustic logging

*Oil well logging—Directional**
USE: Directional logging

*Oil well logging—Electric**
USE: Electric logging

*Oil well logging—Electromagnetic**
USE: Electromagnetic logging

*Oil well logging—Geothermal**
USE: Geothermal logging

*Oil well logging—Induced polarization**
USE: Induced polarization logging

*Oil well logging—Induction**
USE: Induction logging

*Oil well logging—Mud**
USE: Mud logging

*Oil well logging—Neutron**
USE: Neutron logging

*Oil well logging—Nuclear**
USE: Nuclear magnetic logging

*Oil well logging—Radioactive**
USE: Radioactivity logging

*Oil well logging—Spontaneous potential**
USE: Spontaneous potential logging

*Oil well logging—Thermal**
USE: Thermal logging

Oil well production 511.1
DT: Predates 1975
UF: Oil well production*
BT: Production
NT: Gas lifts
Heavy oil production
Injection (oil wells)
Offshore oil well production
Oil well pumping
Repressuring
Unit operations (oil wells)
RT: Enhanced recovery
Explosive well stimulation
In situ combustion
Natural gas well production
Oil well flooding
Oil well testing
Oil wells
Paraffin problems
Petroleum industry
Sand consolidation
Well spacing

*Oil well production**
USE: Oil well production

*Oil well production—Enhanced recovery**
USE: Enhanced recovery

*Oil well production—Flooding**
USE: Oil well flooding

*Oil well production—Flow**
USE: Flow of fluids

*Oil well production—Flowlines**
USE: Flowlines

*Oil well production—Gas lift**
USE: Gas lifts

*Oil well production—Heavy oil**
USE: Heavy oil production

*Oil well production—In situ combustion**
USE: In situ combustion

*Oil well production—Offshore**
USE: Offshore oil well production

*Oil well production—Paraffin troubles**
USE: Paraffin problems

*Oil well production—Repressuring**
USE: Repressuring

*Oil well production—Secondary**
USE: Enhanced recovery

*Oil well production—Sub-sea production system**
USE: Offshore oil well production

*Oil well production—Taxation**
USE: Taxation

*Oil well production—Tertiary**
USE: Enhanced recovery

*Oil well production—Thermal**
USE: Thermal oil recovery

*Oil well production—Well testing**
USE: Oil well testing

Oil well pumping 511.1 (618.2)
DT: Predates 1975
BT: Oil well production
NT: Electric oil well pumping
RT: Oil well pumps
Oil wells

*Oil well pumping—Electric**
USE: Electric oil well pumping

*Oil well pumping—Equipment**
USE: Oil well pumps

Oil well pumps 511.2, 618.2
DT: January 1993
UF: Oil well pumping—Equipment*
BT: Oil field equipment
Well pumps
RT: Oil well pumping

*Oil well shooting**
USE: Explosive well stimulation

Oil well testing 511.1
DT: January 1993
UF: Oil well production—Well testing*
BT: Well testing
RT: Oil well production
Oil wells

Oil wells 512.1.1
DT: Predates 1975
BT: Wells
NT: Horizontal wells
Offshore oil wells
RT: Blowouts
Bottom hole pressure
Fracturing (oil wells)
Fracturing fluids
Hydraulic fracturing
Oil well casings
Oil well cementing
Oil well completion
Oil well drilling
Oil well fires
Oil well flooding
Oil well logging
Oil well production
Oil well pumping
Oil well testing
Paraffin sedimentation
Petroleum industry
Petroleum reservoir engineering
Sand consolidation
Scale (deposits)
Well perforation
Well pressure
Well spacing
Well workover

*Oil wells—Acid treatment**
USE: Acidization

*Oil wells—Bottom hole pressure**
USE: Bottom hole pressure

*Oil wells—Fracturing fluids**
USE: Fracturing fluids

*Oil wells—Fracturing**
USE: Fracturing (oil wells)

*Oil wells—Hydraulic fracturing**
USE: Hydraulic fracturing

*Oil wells—Offshore**
USE: Offshore oil wells

*Oil wells—Paraffin sedimentation**
USE: Paraffin sedimentation

*Oil wells—Perforation**
USE: Well perforation

*Oil wells—Sand consolidation**
USE: Sand consolidation

*Oil wells—Scale problems**
USE: Scale (deposits)

*Oil wells—Spacing**
USE: Well spacing

*Oil wells—Workover**
USE: Well workover

Oils and fats 804.1 (822.3)
SN: Plant and animal products.
DT: Predates 1975
UF: Edible fats
Edible oils
Fats
Food products—Fats*
BT: Organic compounds
NT: Vegetable oils
RT: Hydrogenation
Lipids
Lubricating oils
Odor removal
Textile auxiliary materials

*Oils and fats—Odor removal**
USE: Odor removal

Olefins 804.1
DT: January 1981
UF: Alkenes
BT: Hydrocarbons
NT: Butenes
Ethylene
Propylene
RT: Butadiene
Polyolefins

Oleic acid
USE: Fatty acids

Oligomers 815.1.1 (804)
DT: January 1977
BT: Organic polymers
RT: Polymerization

Olivine 482.2
DT: Predates 1975
UF: Sand, Foundry—Olivine*
BT: Silicate minerals
RT: Foundry sand

OMCVD
USE: Chemical vapor deposition

OMVPE
USE: Metallorganic vapor phase epitaxy

On line searching
USE: Online searching

On line systems
USE: Online systems

On off control systems
USE: On-off control systems

* Former Ei Vocabulary term

On-off control systems 731.1
DT: January 1993
UF: Control systems, On/off*
On off control systems
On-off systems
BT: Control systems
RT: Bang bang control systems
Describing functions
Nonlinear control systems
Nonlinear systems
Relay control systems

On-off systems
USE: On-off control systems

Oncogenic materials
USE: Carcinogens

Oncogenic viruses 461.2, 801.2
SN: Viruses that cause tumors
DT: January 1993
UF: Cancer-causing viruses
Tumor-causing viruses
Viruses—Oncogenic*
BT: Viruses
RT: Carcinogens
Oncology

Oncology 461.6
DT: January 1993
UF: Biomedical engineering—Oncology*
Cancer treatment
BT: Medicine
RT: Biomedical engineering
Carcinogens
Oncogenic viruses

Online searching 903.3 (723.5)
DT: January 1993
UF: Information retrieval systems—Online searching*
On line searching
Searching online
BT: Information retrieval
RT: Information retrieval systems
Online systems

Online systems 722.4
DT: January 1993
UF: Computer systems, Digital—On line operation*
On line systems
BT: Computer systems
RT: Interactive computer systems
Online searching
Real time systems

OOP
USE: Object oriented programming

Opacifiers 803, 812.3
DT: January 1993
UF: Glass manufacture—Opacifiers*
BT: Additives
RT: Glass manufacture
Opacity

Opacity 741.1
DT: January 1993
UF: Optical transmittance
BT: Optical properties
RT: Densitometers
Density (optical)
Light
Opacifiers
Transparency

Opacity measurement
USE: Densitometers

OPDAR
USE: Optical radar

Open channel flow 631.1
DT: January 1993
UF: Flow of fluids—Open channels*
BT: Channel flow
RT: Turbulent flow

Open hearth furnaces 532.6
DT: Predates 1975
UF: Openhearth furnaces
BT: Melting furnaces
Metallurgical furnaces
RT: Open hearth process
Regenerators

Open hearth process 532.6, 545.3
DT: January 1993
UF: Openhearth process
Steelmaking—Open hearth process*
BT: Steelmaking
RT: Open hearth furnaces

Open pit mining 502.1
DT: January 1993
UF: Mines and mining—Open pit*
Surface mining
BT: Mining

Openhearth furnaces
USE: Open hearth furnaces

Openhearth process
USE: Open hearth process

Opera houses 402.2 (751.3)
DT: Predates 1975
BT: Facilities
RT: Auditoriums
Stages
Theaters (legitimate)

*Opera houses—Stages**
USE: Stages

Operating procedures (computer)
USE: Computer operating procedures

Operating rooms 462.2 (402.2)
DT: January 1993
UF: Hospitals—Operating rooms*
BT: Hospitals
RT: Biomedical equipment
Surgery

Operating systems (computer)
USE: Computer operating systems

Operational amplifiers 713.1
DT: January 1993
UF: Amplifiers, Operational*
BT: Amplifiers (electronic)

Operations research 912.3
DT: Predates 1975
NT: PERT
RT: Constraint theory
Cost effectiveness
Dynamic programming
Electric network analysis
Heuristic programming
Industrial management
Industrial research
Information theory
Integer programming
Job analysis
Lagrange multipliers
Linear programming
Management science
Mathematical models
Monte Carlo methods
Optimization
Probability
Quality control
Queueing theory
Reliability
Resource allocation
Scheduling
Simulation
Statistical methods
Systems analysis
Systems engineering
Value engineering

*Operations research—Military purposes**
USE: Military applications

Operators (mathematics)
USE: Mathematical operators

Ophthalmology 461.6
DT: January 1993
UF: Biomedical engineering—Ophthalmology*
BT: Medicine
RT: Biomedical engineering
Vision
Vision aids

Optical aberrations
USE: Aberrations

Optical absorption
USE: Light absorption

Optical amplifiers
USE: Light amplifiers

Optical beam splitters 741.3
DT: January 1993
UF: Beam splitters (optical)
Beamsplitters (optical)
Optical devices—Optical beam splitters*
BT: Optical devices
RT: Mirrors

Optical bistability 741.1.1
DT: January 1993
UF: Bistability (optical)
RT: Electrooptical devices
Nonlinear optics
Quantum optics

Optical cables 717.2 (741.1.2)
DT: January 1993
UF: Fiber optic cables
Optical fiber cables
Telecommunication cables, Optical*
BT: Optical communication equipment
Telecommunication cables
RT: Cable jointing
Electric cable laying
Optical fibers
Optical links

Optical chaos
USE: Chaos theory

Optical character recognition 741.1 (723)
DT: January 1993
UF: Character recognition, Optical*
OCR
BT: Character recognition
RT: Image processing
Pattern recognition systems
Scanning

Optical coatings 741.3, 813.2
DT: January 1977
BT: Coatings
NT: Antireflection coatings
RT: Deposition
Optical films
Optical multilayers

Optical coherence
USE: Coherent light

Optical collimators 741.3 (941.3)
DT: January 1993
UF: Collimators (optical)
BT: Optical devices
RT: Optical instruments

Optical communication 717.1
DT: Predates 1975
BT: Telecommunication
RT: Fiber optic networks
Fiber optics
Laser applications
Laser beams
Light
Optical communication equipment
Optical interconnects
Optical links
Optics
Visual communication

Optical communication equipment 717.2
DT: Predates 1975
BT: Telecommunication equipment
NT: Optical cables
RT: Fiber optics

Optical communication equipment *(continued)*
Integrated optoelectronics
Light modulators
Optical communication
Optical fibers
Optical switches
Optical waveguides

Optical computing
USE: Optical data processing

Optical correlation 741.1
DT: January 1993
UF: Correlation (optical)
BT: Correlation methods
Optical data processing
RT: Holography

Optical data processing 723.2 (741.3)
DT: October 1975
UF: Optical computing
Optical information processing
BT: Data processing
NT: Optical correlation
RT: Computer peripheral equipment
Data communication systems
Fourier optics
Holography
Interfaces (computer)
Optical image storage
Optical transfer function

Optical data storage 722.1 (741.3)
DT: January 1993
UF: Data storage, Optical*
Data storage, Optical—Storage devices*
Optical memories
BT: Data storage equipment
NT: Optical disk storage
Optical image storage
RT: Digital image storage
Recording

Optical density
USE: Density (optical)

Optical density measuring instruments
USE: Densitometers

Optical design 741.1
DT: January 1993
BT: Design
RT: Optics

Optical devices 741.3
DT: January 1981
UF: Optical elements
BT: Equipment
NT: Binoculars
Contact lenses
Diffraction gratings
Diffusers (optical)
Electrooptical devices
Holographic optical elements
Infrared devices
Intraocular lenses
Laser windows
Magnetooptical devices
Mirrors
Optical beam splitters
Optical collimators
Optical filters
Optical instrument lenses
Optical projectors
Optical resonators
Optical sensors
Optical shutters
Optical waveguides
Optoelectronic devices
Plastic lenses
Prisms
Ultraviolet devices
RT: Choppers (circuits)
Lenses
Optical disk storage
Optical films
Optical instruments
Optical materials
Optical systems
Optics
Photographic accessories
Photographic equipment
Sol-gels

*Optical devices—Laser windows**
USE: Laser windows

*Optical devices—Optical beam splitters**
USE: Optical beam splitters

*Optical devices—Optical shutters**
USE: Optical shutters

Optical direction and ranging
USE: Optical radar

Optical disk storage 722.1, 741.3
DT: January 1993
UF: Data storage, Optical—Disk*
BT: Optical data storage
NT: CD-ROM
RT: Optical devices
Optical image storage
Videodisks

Optical elements
USE: Optical devices

Optical ellipsometry
USE: Ellipsometry

Optical engineering (741)
DT: January 1993
BT: Engineering
RT: Optics

Optical fiber cables
USE: Optical cables

Optical fiber chemical sensors
USE: Fiber optic chemical sensors

Optical fiber coupling 741.1.2 (717)
DT: January 1993
UF: Coupled fibers
Coupled optical fibers

Optical fiber coupling *(continued)*
Fiber optic coupling
Optical fibers—Coupling*
BT: Optical fibers

Optical fiber fabrication 741.1.2 (812.3)
DT: January 1993
UF: Fiber fabrication
BT: Fabrication
RT: Optical fibers

Optical fiber lasers
USE: Fiber lasers

Optical fibers 741.1.2
SN: Use for properties of fibers. For light transmission by filaments, use FIBER OPTICS
DT: January 1981
UF: Single mode fibers
BT: Synthetic fibers
NT: Optical fiber coupling
RT: Fiber lasers
Fiber optic components
Fiber optics
Glass fibers
Gradient index optics
Optical cables
Optical communication equipment
Optical fiber fabrication
Optical glass
Optical interconnects
Optical waveguides

*Optical fibers—Coupling**
USE: Optical fiber coupling

Optical films 741.3
DT: January 1993
UF: Optical materials—Films*
BT: Films
Optical materials
RT: Antireflection coatings
Dielectric films
Integrated optics
Light absorption
Light interference
Optical coatings
Optical devices
Optical filters
Optical multilayers
Optics
Reflection
Thin films

Optical filters 741.3 (717.2)
DT: Predates 1975
UF: Light filters
Optical instruments—Filters*
BT: Optical devices
Wave filters
RT: Cameras
Light absorption
Light interference
Light transmission
Optical films

Optical flows 741.1
DT: January 1993
RT: Image analysis

Optical frequency conversion 741.1.1
DT: January 1993
UF: Frequency conversion (optical)
BT: Nonlinear optics
RT: Frequency converters

Optical glass 741.3, 812.3
DT: January 1993
UF: Glass—Optical quality*
Optical materials—Glass*
BT: Glass
Optical materials
RT: Fiber optics
Glass fibers
Optical fibers
Optical properties

*Optical gratings**
USE: Diffraction gratings

Optical image storage 722.1 (741.3)
DT: January 1993
UF: Image archiving
Image storage (optical)
Image storage, Optical*
BT: Optical data storage
RT: Holograms
Holography
Image processing
Imaging techniques
Optical data processing
Optical disk storage

Optical information processing
USE: Optical data processing

Optical instrument lenses 741.3 (941.3)
DT: January 1993
UF: Optical instruments—Lenses*
BT: Lenses
Optical devices
Optical instruments
NT: Camera lenses

Optical instruments 941.3 (741.3)
DT: Predates 1975
BT: Instruments
NT: Cameras
Comparators (optical)
Densitometers
Diffractometers
Fiber optic chemical sensors
Fluorometers
Infrared instruments
Interferometers
Nephelometers
Optical instrument lenses
Optical telescopes
Optometers
Photometers
Polarimeters
Polariscopes
Sensitometers

Optical instruments *(continued)*
Spectrographs
Spectrometers
Ultraviolet instruments
RT: Aberrations
Acoustooptical devices
Alignment
Diffusers (optical)
Endoscopy
Instrument displays
Light
Light sources
Optical collimators
Optical devices
Optical materials
Optical resolving power
Optical systems
Optical testing
Optical variables control
Optics
Photographic accessories
Photographic equipment
Pollution
Reflectometers
Surveying instruments

*Optical instruments—Diffusers**
USE: Diffusers (optical)

*Optical instruments—Display systems**
USE: Instrument displays

*Optical instruments—Filters**
USE: Optical filters

*Optical instruments—Fungus protection**
USE: Fungus protection

*Optical instruments—Gratings**
USE: Diffraction gratings

*Optical instruments—Infrared**
USE: Infrared instruments

*Optical instruments—Lenses**
USE: Optical instrument lenses

*Optical instruments—Light sources**
USE: Light sources

*Optical instruments—Reflectors**
USE: Mirrors

*Optical instruments—Resolving power**
USE: Optical resolving power

*Optical instruments—Ultraviolet**
USE: Ultraviolet instruments

Optical interconnections
USE: Optical interconnects

Optical interconnects (717.1) (741.3)
DT: January 1993
UF: Interconnects (optical)
Optical interconnections
RT: Electric connectors
Holographic optical elements
Integrated circuit manufacture
Integrated optics
Integrated optoelectronics
Optical communication
Optical fibers

Optical interference
USE: Light interference

Optical Kerr effect 741.1
DT: January 1995
UF: Kerr effect (optical)
BT: Optical properties
RT: Anisotropy
Kerr electrooptical effect
Light polarization

Optical links 717.1 (741.3)
DT: January 1993
UF: Laser communication
Telecommunication links, Optical*
BT: Telecommunication links
RT: Laser applications
Light
Optical cables
Optical communication

Optical masers
USE: Lasers

Optical materials 741.3
DT: January 1986
BT: Materials
NT: Light sensitive materials
Optical films
Optical glass
RT: Gradient index optics
Lenses
Mirrors
Optical devices
Optical instruments
Optics
Polymers

*Optical materials—Films**
USE: Optical films

*Optical materials—Glass**
USE: Optical glass

*Optical materials—Light sensitive materials**
USE: Light sensitive materials

Optical memories
USE: Optical data storage

Optical microscopy 741.1 (741.3)
DT: January 1995
UF: Light microscopy
BT: Microscopic examination
RT: Metallography

Optical modulation
USE: Light modulation

Optical modulators
USE: Light modulators

Optical monochromators
USE: Monochromators

Optical multilayers 741.3
DT: January 1993
UF: Multilayers (optical)
RT: Optical coatings
Optical films

Optical phase conjugation 741.1.1
DT: January 1993
UF: Nonlinear optical phase conjugation
Phase conjugation (optical)
Time-reversal reflection
Wavefront reversal
BT: Nonlinear optics
RT: Adaptive optics

Optical polarimeters
USE: Polarimeters

Optical polarization
USE: Light polarization

Optical prisms
USE: Prisms

Optical projection systems
USE: Projection systems

Optical projectors 741.3 (742.2)
DT: Predates 1975
UF: Photography—Projectors*
Projectors (optical)
BT: Optical devices
RT: Motion pictures
Optical shutters
Photographic accessories
Photography
Projection systems

Optical propagation
USE: Light propagation

Optical properties 741.1
DT: Predates 1975
UF: Glass—Spectral properties*
BT: Physical properties
NT: Acoustooptical effects
Birefringence
Color
Density (optical)
Electrochromism
Opacity
Optical Kerr effect
Optical rotation
Photochromism
Photoreactivity
Photosensitivity
Refractive index
Transparency
RT: Atmospheric optics
Coherent light
Electrooptical effects
Light control glass
Magnetooptical effects
Optical glass
Optical variables control
Optical variables measurement
Optics
Photoelasticity
Visibility

Optical pumping (741.1)
DT: January 1986
UF: Pumping (optical)
BT: Pumping (laser)
RT: Lasers
Multiphoton processes
Optically pumped lasers

Optical radar 716.2, 741.3
DT: January 1993
UF: Lidar
OPDAR
Optical direction and ranging
Radar, Optical*
RT: Distance measurement
Infrared radiation
Laser beams
Range finders
Tracking (position)

Optical recording 741.3 (744.9)
DT: January 1993
BT: Recording
NT: Laser recording
RT: Kerr magnetooptical effect

Optical reflectors
USE: Mirrors

Optical resolving power 741.1
DT: January 1993
UF: Optical instruments—Resolving power*
Resolution (optical)
Resolving power
RT: Optical instruments
Optics

Optical resonators 741.3
DT: January 1993
UF: Light resonators
Light—Optical resonators*
BT: Optical devices
Resonators
NT: Laser resonators
RT: Light

Optical rotation 741.1
DT: January 1993
UF: Light—Optical rotation*
BT: Optical properties
RT: Birefringence
Faraday effect
Kerr magnetooptical effect
Light
Light polarization

Optical scanning
USE: Scanning

Optical sensors 741.3 (732.2) (801)
DT: January 1995
BT: Optical devices
Sensors
NT: Fiber optic sensors
RT: Image sensors

* Former Ei Vocabulary term

Optical shutters 741.3 (742.2)
DT: January 1993
UF: Optical devices—Optical shutters*
Shutters (optical)
BT: Optical devices
NT: Camera shutters
RT: Cameras
Lenses
Optical projectors

Optical sources
USE: Light sources

Optical squeezing
USE: Quantum optics

Optical switches 741.3
DT: January 1993
UF: Photonic switches
BT: Switches
RT: Optical communication equipment
Switching systems

Optical systems 741.3
DT: January 1986
NT: Projection systems
RT: Adaptive optics
Optical devices
Optical instruments
Optics

Optical telescopes 741.3, 941.3
DT: January 1993
BT: Optical instruments
Telescopes

Optical testing 741.3 (941.4)
DT: January 1993
BT: Testing
RT: Optical instruments
Optical transfer function
Optical variables measurement
Optics

Optical transfer function 741.1
DT: January 1993
UF: Modulation transfer function
BT: Transfer functions
RT: Optical data processing
Optical testing
Optics

Optical transmission
USE: Light transmission

Optical transmittance
USE: Opacity

Optical variables control 731.3 (741.3) (941.4)
DT: January 1993
UF: Control, Optical variables*
BT: Control
RT: Optical instruments
Optical properties
Optical variables measurement
Optics

Optical variables measurement 941.4
DT: Predates 1975
BT: Measurements
NT: Colorimetry
Density measurement (optical)
Ellipsometry
Light measurement
Photometry
RT: Light
Optical properties
Optical testing
Optical variables control
Optics
Reflectometers
Streak cameras

Optical waveguides 714.3, 741.3 (717.2)
DT: January 1993
UF: Light waveguides
Waveguides, Optical*
BT: Dielectric waveguides
Optical devices
RT: Integrated optics
Laser accessories
Light
Light transmission
Optical communication equipment
Optical fibers

Optically pumped lasers 744.1
DT: January 1993
UF: Lasers—Optical pumping*
BT: Lasers
RT: Optical pumping

Optics 741.1
SN: Geometrical and physical optics
DT: Predates 1975
UF: Biomedical optics
Micro-optics
Microoptics
BT: Physics
NT: Adaptive optics
Aspherics
Atmospheric optics
Chromogenics
Fiber optics
Fourier optics
Geometrical optics
Gradient index optics
Industrial optics
Integrated optics
Laser optics
Nonlinear optics
Particle optics
Quantum optics
Space optics
Statistical optics
RT: Aberrations
Acoustooptical devices
Electric lamps
Electron lenses
Focusing
Holography
Light
Light modulation
Light modulators

Optics *(continued)*
Mirrors
Optical communication
Optical design
Optical devices
Optical engineering
Optical films
Optical instruments
Optical materials
Optical properties
Optical resolving power
Optical systems
Optical testing
Optical transfer function
Optical variables measurement
Optical variables control
Photography
Photomechanics
Sol-gels
Vision

*Optics—Geometrical**
USE: Geometrical optics

*Optics—Moire fringes**
USE: Moire fringes

*Optics—Nonlinear**
USE: Nonlinear optics

Optimal control systems 731.1 (921.5)
DT: January 1993
UF: Control systems, Optimal*
BT: Control systems
Optimal systems
RT: Decision theory
Difference equations
Dynamic programming
Game theory
Identification (control systems)
Maximum principle
Models
Optimization
Predictive control systems

Optimal systems (731.1) (921.5)
DT: January 1993
UF: Systems science and cybernetics—Optimal systems*
BT: Systems science
NT: Optimal control systems
RT: Control systems
Optimization

Optimization 921.5
DT: October 1975
UF: Design optimization
NT: Mathematical programming
Minimization of switching nets
Simulated annealing
RT: Constraint theory
Correlation methods
Genetic algorithms
Lagrange multipliers
Least squares approximations
Operations research
Optimal control systems
Optimal systems
Quality control
Scheduling
Systems analysis
Systems science
Value engineering

Optoacoustic effect
USE: Photoacoustic effect

Optoacoustic spectroscopy
USE: Photoacoustic spectroscopy

Optoacoustics
USE: Photoacoustic effect

Optoelectronic devices 741.3 (714.2) (717.2)
DT: January 1983
UF: Optoelectronics
BT: Electronic equipment
Optical devices
NT: Image intensifiers (solid state)
Image sensors
Integrated optoelectronics
Light emitting diodes
Photoelectric devices
RT: Acoustooptical devices
Acoustooptical effects

Optoelectronic integrated circuits
USE: Integrated optoelectronics

Optoelectronics
USE: Optoelectronic devices

Optometers 941.3
DT: Predates 1975
BT: Biomedical equipment
Optical instruments
RT: Vision

Orbital laboratories 655.2, 656.2 (901.3)
DT: January 1993
UF: Satellites—Orbital laboratories*
BT: Laboratories
Spacecraft

Orbital transfer 655.1
DT: January 1993
UF: Spacecraft—Orbital transfer*
Transfer orbits
RT: Satellites
Space flight
Space rendezvous
Spacecraft
Trajectories

Orbiting solar power plants 615.2, 655.1 (702.3)
DT: January 1993
UF: Solar power plants—Orbiting*
BT: Satellites
Solar power plants

Orbits (654.1) (655.2)
DT: January 1993
UF: Rockets and missiles—Orbits and trajectories*
Satellites—Orbits and trajectories*
BT: Trajectories

* Former Ei Vocabulary term

Orbits *(continued)*
RT: Astrophysics
Interplanetary flight
Lunar missions
Navigation
Satellites
Space flight
Space rendezvous
Spacecraft

Orchards 821.3
DT: Predates 1975
BT: Farms
RT: Agriculture
Frost protection
Fruits

*Orchards—Frost protection**
USE: Frost protection

Order disorder transitions (531.2) (802.3) (931.3) (933)
DT: January 1993
UF: Order-disorder*
BT: Phase transitions
RT: Solid solutions
Superlattices

*Order-disorder**
USE: Order disorder transitions

Ordnance 404.1
DT: Predates 1975
UF: Weapons (military)
BT: Military equipment
NT: Ammunition
Armor
Bombs (ordnance)
Missiles
RT: Guns (armament)
Military rockets

Ore analysis 533.1, 801
DT: Predates 1975
BT: Chemical analysis
RT: Ore sampling
Ores

Ore carriers 671
DT: January 1993
UF: Ships—Ore carriers*
BT: Ships

Ore deposit geology 481.1 (504) (505)
DT: January 1993
UF: Ore deposits—Structural control*
BT: Geology

Ore deposits (504) (505)
DT: Predates 1975
UF: Ore prospecting
BT: Mineral resources
NT: Antimony deposits
Bauxite deposits
Beryllium deposits
Boron deposits
Chromite deposits
Cobalt deposits
Copper deposits
Corundum deposits
Germanium deposits
Gold deposits
Iron deposits
Lead deposits
Lithium deposits
Magnesium deposits
Manganese deposits
Mercury deposits
Molybdenum deposits
Monazite deposits
Nickel deposits
Niobium deposits
Placers
Scandium deposits
Silver deposits
Sodium deposits
Strontium deposits
Tantalum deposits
Thorium deposits
Tin deposits
Titanium deposits
Tungsten deposits
Uranium deposits
Vanadium deposits
Wall rock
Yttrium deposits
Zinc deposits
Zircon deposits
RT: Clay alteration
Natural resources exploration
Petrology
Underwater mineral resources

*Ore deposits—Clay alteration**
USE: Clay alteration

*Ore deposits—Nonmetallic**
USE: Nonmetallic ores

*Ore deposits—Structural control**
USE: Ore deposit geology

*Ore deposits—Wall rock alteration**
USE: Wall rock

*Ore handling**
USE: Materials handling AND
Ores

Ore pellets 533.1
DT: January 1993
UF: Pellets (ore)
BT: Ores
NT: Iron ore pellets
RT: Pelletizing

Ore processing
USE: Ore treatment

Ore prospecting
USE: Ore deposits AND
Natural resources exploration

Ore reduction 533.1, 802.2
DT: January 1993
UF: Ore treatment—Reduction*

* Former Ei Vocabulary term

Ore reduction *(continued)*
BT: Ore treatment
Reduction
NT: Iron ore reduction

Ore roasting 533.1 (802.3)
DT: January 1993
UF: Ore treatment—Roasting*
BT: Ore treatment
RT: Pyrometallurgy

Ore sampling 533.1 (502) (504) (505)
DT: January 1993
BT: Sampling
RT: Ore analysis
Ores

Ore sinter 533.1
DT: January 1993
UF: Sinter (ore)
BT: Ores
NT: Iron ore sinter
RT: Ore sintering

Ore sintering 533.1
DT: January 1993
BT: Ore treatment
Sintering
RT: Ore sinter

Ore tailings 452.3, 533.1 (502.1)
DT: January 1993
UF: Ore tailings disposal
BT: Tailings
RT: Ore treatment

Ore tailings disposal
USE: Tailings disposal AND
Ore tailings

Ore treatment 533.1
DT: Predates 1975
UF: Ore processing
BT: Production
NT: Antimony ore treatment
Asbestos ore treatment
Bauxite ore treatment
Beneficiation
Beryllium ore treatment
Bismuth ore treatment
Cadmium ore treatment
Cerium ore treatment
Chromium ore treatment
Cobalt ore treatment
Copper ore treatment
Electrostatic ore treatment
Fluidized bed process
Gold ore treatment
Iron ore treatment
Lead ore treatment
Magnesite ore treatment
Manganese ore treatment
Molybdenum ore treatment
Nickel ore treatment
Niobium ore treatment
Ore reduction
Ore roasting
Ore sintering
Phosphate ore treatment
Platinum ore treatment
Potash ore treatment
Silver ore treatment
Smelting
Tantalum ore treatment
Tin ore treatment
Titanium ore treatment
Tungsten ore treatment
Uranium ore treatment
Vanadium ore treatment
Zinc ore treatment
Zirconium ore treatment
RT: Classifiers
Crushing
Electrolytic reduction
Extractive metallurgy
Flotation
Fume control
Heavy media separation
Leaching
Metal refining
Metallurgical furnaces
Mining
Ore tailings
Refining

*Ore treatment—Beneficiation**
USE: Beneficiation

*Ore treatment—Classifiers**
USE: Classifiers

*Ore treatment—Electrolytic reduction**
USE: Electrolytic reduction

*Ore treatment—Electrostatic**
USE: Electrostatic ore treatment

*Ore treatment—Fluidized bed process**
USE: Fluidized bed process

*Ore treatment—Fume control**
USE: Fume control

*Ore treatment—Heavy media separation**
USE: Heavy media separation

*Ore treatment—Nonmetallic**
USE: Nonmetallic ores

*Ore treatment—Reduction**
USE: Ore reduction

*Ore treatment—Roasting**
USE: Ore roasting

*Ore treatment—Screening**
USE: Screening

*Ore treatment—Smelting**
USE: Smelting

*Ore treatment—Tailings disposal**
USE: Tailings disposal

Ores (482.2) (502) (504) (505)
DT: January 1993
UF: Ore handling*

Ores *(continued)*
BT: Minerals
NT: Barite
Chromite
Fluorspar
Ilmenite
Iron ores
Monazite
Nonmetallic ores
Ore pellets
Ore sinter
RT: Ore analysis
Ore sampling
Zircon

Organ transplantation
USE: Transplantation (surgical)

Organic acids 804.1
DT: January 1993
UF: Acids—Organic*
BT: Acids
Organic compounds
NT: Amino acids
Carboxylic acids
Nucleic acids

Organic chemicals 804.1
DT: October 1975
UF: Electric insulating materials—Organic substances*
BT: Chemicals
RT: Organic coatings
Organic compounds
Petrochemicals

Organic coatings 804.1, 813.2
DT: January 1993
UF: Protective coatings—Organic*
BT: Coatings
NT: Lacquers
Varnish
RT: Organic chemicals
Organic compounds
Paint

Organic compounds 804.1
DT: January 1977
UF: Carbon organic compounds
Geochemistry—Organic compounds*
Glass—Organic compounds*
Hydrogen compounds (organic)
Hydrogen organic compounds
BT: Chemical compounds
NT: Acetonitrile
Acrylic monomers
Alcohols
Aldehydes
Alkyd resins
Amines
Aromatic compounds
Azo dyes
Carbohydrates
Carbon disulfide
Carbon tetrachloride
Esters
Furfural
Hydrocarbons
Ketones
Lipids
Melamine
Oils and fats
Organic acids
Organic ion exchangers
Organic minerals
Organic polymers
Organic solvents
Organometallics
Phenolic resins
Polychlorinated biphenyls
Semiconducting organic compounds
Turpentine
Unsaturated compounds
Urea
Volatile organic compounds
RT: Carbon
Friedel-Crafts reaction
Luminescence of organic solids
Macromolecules
Organic chemicals
Organic coatings
Polymers

Organic conductors 708.2 (804.1)
DT: January 1995
UF: Synthetic metals
BT: Materials
RT: Semiconducting organic compounds
Superconducting materials

Organic ion exchangers 804.1 (802.1)
DT: January 1993
UF: Ion exchangers—Organic*
BT: Ion exchangers
Organic compounds

Organic minerals 482.2, 804.1
DT: January 1993
UF: Mineralogy—Organic materials*
BT: Minerals
Organic compounds

Organic polymers 815.1.1
DT: January 1993
UF: Conjugated polymers
BT: Organic compounds
Polymers
NT: Aromatic polymers
Chlorine containing polymers
Copolymers
Elastomers
Fluorine containing polymers
Ion exchange resins
Ionomers
Metallorganic polymers
Natural polymers
Nitrile resins
Oligomers
Polyacetylenes
Polyelectrolytes
Polyethers
Polyimides

* Former Ei Vocabulary term

Organic polymers *(continued)*
Polymer blends
Polyurethanes
Semiconducting polymers
Thermoplastics
Thermosets
Unsaturated polymers

Organic semiconductors
USE: Semiconducting organic compounds

Organic semiconductors (polymeric)
USE: Semiconducting polymers

Organic solvents 803, 804.1
DT: January 1995
BT: Organic compounds
Solvents
RT: Benzene
Methanol

Organizations
USE: Societies and institutions

Organometallics 804.1
DT: January 1977
UF: Metallorganic compounds
Metalorganic compounds
BT: Organic compounds
NT: Metallorganic polymers
RT: Chelation
Complexation
Metallorganic chemical vapor deposition
Metallorganic vapor phase epitaxy

Organs (musical instruments) 752.4
DT: January 1993
BT: Musical instruments

Orientation (crystal)
USE: Crystal orientation

Orifices (619.1) (631.1)
DT: January 1993
UF: Flow of fluids—Orifices*
BT: Components
RT: Ducts
Flow measurement
Flowmeters
Nozzles

Orthopedics 461.3, 461.6
DT: January 1993
UF: Biomedical engineering—Orthopedics*
BT: Medicine
RT: Biomedical engineering
Bone
Musculoskeletal system
Orthotics

Orthotics 461.5, 462.4
DT: January 1977
BT: Biomedical equipment
NT: Braces (for limbs and joints)
Dental orthoses
RT: Functional neural stimulation
Human rehabilitation equipment
Orthopedics
Prosthetics
Walking aids

*Orthotics—Braces**
USE: Braces (for limbs and joints) OR
Dental braces

ORV
USE: Off road vehicles

Oscillating conveyors
USE: Vibrating conveyors

Oscillations (931.1)
DT: January 1993
NT: Circuit oscillations
Plasma oscillations
RT: Circuit resonance
Damping
Oscillators (electronic)
Oscillators (mechanical)
Pendulums
Pulsatile flow
Resonance
Resonators
Stability
Vibrations (mechanical)
Waves

Oscillators (electronic) 713.2
DT: January 1993
UF: Oscillators*
BT: Electronic equipment
NT: Crystal oscillators
Microwave oscillators
Oscillistors
Parametric oscillators
Relaxation oscillators
Solid state oscillators
Tunnel diode oscillators
Variable frequency oscillators
RT: Active networks
Cavity resonators
Flip flop circuits
Frequency stability
Frequency standards
Frequency synthesizers
Harmonic generation
Klystrons
Magnetrons
Masers
Mixer circuits
Multivibrators
Natural frequencies
Negative resistance
Oscillations
Radio equipment
Radio receivers
Radio transmitters
Resonators
Semiconductor devices
Signal generators
Transmitters
Tuners
Tuning

* Former Ei Vocabulary term

Oscillators (mechanical) 601.1
DT: January 1993
BT: Equipment
NT: Vibrators
RT: Frequency standards
Natural frequencies
Oscillations
Pendulums

*Oscillators**
USE: Oscillators (electronic)

*Oscillators, Crystal**
USE: Crystal oscillators

*Oscillators, Gunn**
USE: Gunn oscillators

*Oscillators, Microwave**
USE: Microwave oscillators

*Oscillators, Parametric**
USE: Parametric oscillators

*Oscillators, Relaxation**
USE: Relaxation oscillators

*Oscillators, Solid state**
USE: Solid state oscillators

*Oscillators, Swept frequency**
USE: Swept frequency oscillators

*Oscillators, Tunnel diode**
USE: Tunnel diode oscillators

*Oscillators—Noise**
USE: Spurious signal noise

*Oscillators—Variable frequency**
USE: Variable frequency oscillators

Oscillistors 713.2, 713.2 (701.2)
DT: Predates 1975
BT: Oscillators (electronic)
Semiconductor devices

Oscillograms
USE: Oscillographs

Oscillographs 942.1
SN: Includes recording oscilloscopes, but not cathode-ray oscilloscopes, for which see CATHODE RAY OSCILLOSCOPES
DT: Predates 1975
UF: Oscillograms
BT: Electric measuring instruments
Recording instruments
RT: Cathode ray oscilloscopes
Time measurement

Oscilloscopes (cathode ray)
USE: Cathode ray oscilloscopes

*Oscilloscopes, Cathode ray**
USE: Cathode ray oscilloscopes

Osmium 547.1
DT: January 1993
UF: Osmium and alloys*
BT: Platinum metals
RT: Osmium alloys
Osmium compounds

Osmium alloys 547.1
DT: January 1993
UF: Osmium and alloys*
BT: Platinum metal alloys
RT: Osmium
Osmium compounds

*Osmium and alloys**
USE: Osmium OR
Osmium alloys

Osmium compounds (804.1) (804.2)
DT: Predates 1975
BT: Platinum metal compounds
RT: Osmium
Osmium alloys

Osmosis 802.3
DT: January 1977
BT: Diffusion
NT: Reverse osmosis
RT: Desalination
Diaphragms
Osmosis membranes

Osmosis membranes 802.1, 802.3
DT: January 1993
UF: Membranes—Osmosis membranes*
BT: Membranes
RT: Osmosis

*Osmosis, Reverse**
USE: Reverse osmosis

Otto cycle 641.1
DT: January 1993
UF: Thermodynamics—Otto cycle*
BT: Thermodynamics

Outages 706.1 (913.5)
DT: January 1993
UF: Power plants—Outages*
RT: Electric power plants
Maintenance
Power plants
Reliability

Outdoor electric lighting 707.1
DT: Predates 1975
UF: Electric lighting—Outdoor*
BT: Electric lighting
RT: Electric lamps
Street lighting

Outdoor motion picture theaters (402.2) (742.2)
DT: January 1993
UF: Drive-in theaters
Motion picture theaters—Outdoor*
BT: Drive-in facilities
Motion picture theaters

Outfalls 452.1, 453.1
DT: January 1993
UF: Sewers—Outfall*

Outfalls *(continued)*
RT: Drainage
Sewers
Water pollution

Outfitting (water craft) 673.1 (674.1)
DT: January 1993
UF: Barges—Constructing and outfitting*
Boat outfitting
Boats—Constructing and outfitting*
Ferry boats—Constructing and outfitting*
Fire boats—Constructing and outfitting*
Fishing vessels—Constructing and outfitting*
Ship construction and outfitting*
Ship outfitting
Ships—Constructing and outfitting*
NT: Ship conversion
RT: Drydocks
Shipbuilding
Ships
Shipways

Outgassing
USE: Degassing

Oven afterburners
USE: Afterburners (oven)

Ovens (642.2) (704.2) (822.1)
DT: January 1993
BT: Heating equipment
NT: Electric ovens
Industrial ovens
RT: Electric appliances
Electric heating
Heat treating furnaces
Heating furnaces
Industrial heating

*Ovens, Industrial**
USE: Industrial ovens

*Ovens, Industrial—Afterburners**
USE: Afterburners (oven)

*Ovens, Industrial—Exhaust gases**
USE: Exhaust gases

Overcurrent protection
(704.2) (706.1) (706.2) (914.1)
DT: Predates 1975
BT: Protection

Overhead lines 706.2
DT: January 1993
UF: Electric lines—Aerial conductors*
BT: Electric lines
RT: Telecommunication cables

Overlays (pavement)
USE: Pavement overlays

Overpasses (406.1) (433.1)
DT: January 1993
UF: Highway overpasses
Railroad overpasses
BT: Public works
RT: Highway systems
Intersections
Railroad crossings
Railroad plant and structures

Overpotential protection
USE: Overvoltage protection

Overvoltage protection
(704.2) (706.1) (706.2) (914.1)
DT: Predates 1975
UF: Overpotential protection
BT: Protection

Oxidation 802.2
DT: January 1993
BT: Chemical reactions
NT: Anodic oxidation
Internal oxidation
Thermooxidation
RT: Antioxidants
Combustion
Corrosion
Oxidation resistance
Oxides
Oxygen
Ozonization
Passivation
Redox reactions
Reduction
Weathering

Oxidation reduction reactions
USE: Redox reactions

Oxidation resistance 539.1, 802.2
DT: January 1995
BT: Corrosion resistance
RT: Corrosion prevention
Corrosion resistant alloys
Oxidation

Oxide minerals 482.2 (804.2)
DT: January 1993
UF: Mineralogy—Oxides*
BT: Minerals
Oxides
NT: Corundum
Ilmenite
Magnetite
Perovskite
Quartz
Rare earths
RT: Silica

Oxide superconductors 708.3.1
DT: January 1995
UF: Cuprate superconductors
BT: Oxides
Superconducting materials
RT: Barium compounds
Bismuth compounds
Lead compounds
Strontium compounds
Yttrium compounds

Oxides 804
DT: Predates 1975
BT: Chemical compounds

* Former Ei Vocabulary term

Oxides *(continued)*
NT: Alumina
Beryllia
Carbon dioxide
Carbon monoxide
Copper oxides
Iron oxides
Magnesia
Molybdate minerals
Nitrogen oxides
Oxide minerals
Oxide superconductors
Peroxides
Silica
Sulfur dioxide
Thoria
Titanium oxides
Uranium dioxide
Zinc oxide
Zirconia
RT: Ceramic materials
Mixed oxide fuels
Oxidation

Oximeters 462.1 (943.3)
DT: January 1993
UF: Biomedical equipment—Oximeters*
BT: Biomedical equipment
Photometers
RT: Biosensors
Hemoglobin oxygen saturation
Oxygen

Oxygen 804
DT: Predates 1975
UF: O
BT: Nonmetals
NT: Ozone
RT: Dissolved oxygen sensors
Hemoglobin oxygen saturation
Oxidation
Oximeters
Oxygen sensors
Partial pressure sensors

Oxygen blast enrichment (532.2) (533.2) (534.2)
DT: January 1993
UF: Cupola practice—Oxygen blast enrichment*
RT: Blast furnaces
Combustion
Cupola practice
Fuel injection
Temperature control

Oxygen cutting 604.1 (538) (802.2) (804)
DT: Predates 1975
UF: Gas cutting
BT: Cutting
RT: Fluxes
Metal cutting
Oxygen cutting machines
Welding

Oxygen cutting machines (538) (604.1) (804)
DT: Predates 1975
BT: Machine tools
RT: Oxygen cutting
Welding machines

*Oxygen cutting—Fire hazards**
USE: Fire hazards

Oxygen demand (biochemical)
USE: Biochemical oxygen demand

Oxygen permeable membranes 802.1 (802.3)
DT: January 1993
UF: Membranes—Oxygen permeable*
BT: Gas permeable membranes

Oxygen regulators 943.3 (454.1) (652.3)
DT: January 1993
UF: Aircraft instruments—Oxygen regulators*
BT: Instruments
RT: Aircraft instruments

Oxygen sensors (462.1) (801.2)
DT: January 1993
UF: Oxygen—Sensors*
BT: Chemical sensors
NT: Dissolved oxygen sensors
RT: Biosensors
Oxygen

Oxygen steelmaking
USE: Basic oxygen process

Oxygen supply 804 (545.3) (652.1) (672.1)
DT: January 1993
UF: Aircraft—Oxygen supply*
Iron and steel plants—Oxygen supply*
Submarines—Oxygen supply*
RT: Aircraft
Submarines

*Oxygen—Dissolved oxygen sensors**
USE: Dissolved oxygen sensors

*Oxygen—Partial pressure sensors**
USE: Partial pressure sensors

*Oxygen—Sensors**
USE: Oxygen sensors

Oxygenators (462.1) (462.2)
DT: January 1993
UF: Biomedical equipment—Oxygenators*
BT: Biomedical equipment

Ozone 804
DT: Predates 1975
BT: Oxygen
RT: Air pollution
Ozone layer
Ozone water treatment
Ozonization

Ozone layer 443.1, 804
DT: January 1993
UF: Earth atmosphere—Ozone layer*
Ozonosphere
BT: Earth atmosphere
RT: Ozone
Ultraviolet radiation
Upper atmosphere

* Former Ei Vocabulary term

Ozone resistance 802.2, 804
DT: January 1993
BT: Chemical resistance
RT: Chemical attack
Corrosion resistance

Ozone water treatment 445.1, 802.2, 804
DT: January 1993
UF: Water treatment—Ozone*
BT: Water treatment
RT: Ozone

Ozonization 802.2
DT: January 1993
BT: Chemical operations
RT: Oxidation
Ozone
Water treatment

Ozonosphere
USE: Ozone layer

P
USE: Phosphorus

PA
USE: Polyamides

PA systems
USE: Public address systems

PABX
USE: Private telephone exchanges

Pacemakers 462.4
DT: January 1993
UF: Prosthetics—Pacemakers*
BT: Prosthetics
RT: Biomaterials
Cardiology
Cardiovascular system
Cardiovascular surgery

Packaged boilers (614.2)
DT: January 1993
UF: Boilers—Packaged*
BT: Boilers

Packaging 694.1
SN: Wrapping, boxing, or crating for handling or shipping. For packaging as the design problem of grouping or combining parts into a unit, use ELECTRONICS PACKAGING, or PRODUCT DESIGN
DT: Predates 1975
UF: Boxing (packaging)
Crating
Wrapping (packaging)
NT: Tamper resistant packaging
RT: Containers
Labeling
Materials handling
Packaging machines
Packaging materials
Shock problems
Transportation

Packaging machines 694.3
DT: Predates 1975
BT: Machinery
RT: Materials handling equipment
Packaging

Packaging materials 694.2
SN: Used to wrap articles for shipment
DT: Predates 1975
BT: Materials
NT: Strapping
RT: Containers
Packaging
Twine

*Packaging materials—Aluminum**
USE: Aluminum

*Packaging materials—Paper**
USE: Paper

*Packaging materials—Permeability**
USE: Mechanical permeability

*Packaging materials—Plastics**
USE: Plastics applications

*Packaging materials—Steel**
USE: Steel

*Packaging materials—Strapping**
USE: Strapping

*Packaging materials—Twine**
USE: Twine

*Packaging—Shock problems**
USE: Shock problems

*Packaging—Tamper resistant**
USE: Tamper resistant packaging

*Packaging—Weight control**
USE: Weight control

Packed beds 802.1
DT: January 1993
UF: Chemical equipment—Packed beds*
Flow of fluids—Packed beds*
Heat transfer—Packed beds*
Mass transfer—Packed beds*
BT: Chemical equipment
RT: Flow of fluids
Fluidized beds
Heat transfer

Packers 511.2
DT: January 1993
UF: Oil field equipment—Packers*
Production packers
BT: Oil field equipment

Packet networks (716) (717) (718) (721.1)
DT: January 1995
UF: Packet switched networks
BT: Telecommunication networks
RT: Computer networks
Packet switching

Packet switched networks
USE: Packet networks

* Former Ei Vocabulary term

Packet switching (716) (717) (718)
DT: January 1993
UF: Data transmission—Packet switching*
BT: Time division multiplexing
RT: Asynchronous transfer mode
Computer networks
Data communication systems
Packet networks
Visual communication

Packing
SN: Materials or devices for minimizing leakage, usually between surfaces in relative motion. For packing, in sense of preparing for shipment, use PACKAGING
DT: Predates 1975
UF: Pipelines—Packing*
Pumps—Packing*
Stuffing
BT: Seals
RT: Bearings (machine parts)
Pipelines
Piston rings
Pumps
Valves (mechanical)

Packing plants 822.1 (402.1)
DT: Predates 1975
UF: Meat packing plants
BT: Food products plants
RT: Meats

Paddle wheel propulsion 675.1
DT: January 1993
UF: Ship propulsion—Paddle wheels*
BT: Ship propulsion

Pads (bearing)
USE: Bearing pads

Paging (virtual storage)
USE: Virtual storage

Paging systems (716.3) (718.1)
DT: January 1993
UF: Beepers
Radio telephone—Paging systems*
BT: Telecommunication systems
RT: Radio telephone

PAI
USE: Polyamideimides

Paint 813.2 (539.2.2)
DT: Predates 1975
BT: Coatings
NT: Antifouling paint
Casein paint
Chlorinated rubber paint
Luminous paint
RT: Casein
Dairies
Drying oils
Enamels
Glazes
Lacquers
Organic coatings
Paint thinners
Painting
Pigments
Varnish

Paint spraying 813.1
DT: Predates 1975
UF: Spray painting
BT: Painting
Spraying
RT: Electrostatic coatings

Paint thinners 813.2 (803)
DT: January 1993
UF: Paint—Thinners*
BT: Solvents
RT: Paint

*Paint—Antifouling**
USE: Antifouling paint

*Paint—Casein**
USE: Casein paint

*Paint—Chlorinated rubber**
USE: Chlorinated rubber paint Paint—Luminous*
USE: Luminous paint

*Paint—Thinners**
USE: Paint thinners

Painting 813.1
DT: Predates 1975
BT: Coating techniques
NT: Paint spraying
RT: Coloring
Decoration
Enameling
Lacquering
Paint
Sealing (finishing)

Palladium 547.1
DT: January 1993
UF: Palladium and alloys*
BT: Platinum metals
RT: Palladium alloys
Palladium compounds

Palladium alloys 547.1
DT: January 1993
UF: Palladium and alloys*
BT: Platinum metal alloys
RT: Palladium
Palladium compounds

*Palladium and alloys**
USE: Palladium OR
Palladium alloys

Palladium compounds (804.1) (804.2)
DT: Predates 1975
BT: Platinum metal compounds
RT: Palladium
Palladium alloys

* Former Ei Vocabulary term

Pallets 691.1
DT: January 1993
UF: Materials handling—Pallets*
BT: Materials handling equipment

PAM
USE: Pulse amplitude modulation

Panels (structural)
USE: Structural panels

Pantographs (682.1.2) (706)
DT: January 1993
UF: Electric traction—Pantographs*
BT: Electric current collectors
RT: Electric traction

Paper 811.1
DT: Predates 1975
UF: Electric insulating materials—Paper*
Packaging materials—Paper*
BT: Materials
NT: Kraft paper
Newsprint
Paper laminates
Paper sheeting
Waste paper
Waxed papers
RT: Antistatic agents
Color fastness
Deinking
Fabrics
Gloss measurement
Nonfibrous paper materials
Paper and pulp mills
Paper coating
Paper products
Paperboards
Papermaking
Papermaking machinery
Printing properties
Roughness measurement

Paper and pulp industry 811.1
DT: Predates 1975
UF: Pulp industry
BT: Industry
RT: Newsprint

Paper and pulp mills 811.1.2 (402.1)
DT: Predates 1975
UF: Pulp mills
BT: Industrial plants
NT: Woodrooms
RT: Paper

*Paper and pulp mills—Woodrooms**
USE: Woodrooms

Paper bearings
USE: Paper products AND
Bearings (machine parts)

Paper boards
USE: Paperboards

Paper capacitors 704.1 (708.1) (811.1)
DT: January 1993
UF: Capacitors, Paper*
BT: Capacitors
Paper products

Paper coating 811.1.1 (813.1)
DT: January 1993
BT: Coating techniques
Papermaking
RT: Paper

Paper containers 694.2, 811.1
DT: January 1993
UF: Containers—Paper and paperboard*
BT: Containers
Paper products

Paper laminates 811.1 (415.4)
DT: January 1993
UF: Paper—Laminates*
BT: Laminates
Paper

Paper products 811.1
DT: Predates 1975
UF: Bearings—Paper*
Paper bearings
NT: Paper capacitors
Paper containers
RT: Paper

Paper sheeting 811.1.1
DT: January 1993
UF: Paper—Sheeting*
Sheeting (paper)
BT: Paper

*Paper—Antistatic agents**
USE: Antistatic agents

*Paper—Calendering**
USE: Calendering

*Paper—Coating**
USE: Coating techniques

*Paper—Color fastness**
USE: Color fastness

*Paper—Deinking**
USE: Deinking

*Paper—Flame resistance**
USE: Flame resistance

*Paper—Gloss measurement**
USE: Gloss measurement

*Paper—Laminates**
USE: Paper laminates

*Paper—Newsprint**
USE: Newsprint

*Paper—Permeability**
USE: Mechanical permeability

*Paper—Printing properties**
USE: Printing properties

*Paper—Roughness measurement**
USE: Roughness measurement

* Former Ei Vocabulary term

*Paper—Sheeting**
USE: Paper sheeting

*Paper—Waxed papers**
USE: Waxed papers

Paperboard containers 694.2, 811.1
DT: January 1993
UF: Containers—Paper and paperboard*
BT: Containers
RT: Paperboards

Paperboards 811.1
SN: Includes cardboard, chipboard, pulpboard
DT: Predates 1975
UF: Paper boards
BT: Materials
NT: Corrugated paperboard
Kraft paperboards
RT: Paper
Paperboard containers

*Paperboards—Coating**
USE: Coating techniques

*Paperboards—Corrugated**
USE: Corrugated paperboard

*Paperboards—Kraft**
USE: Kraft paperboards

Papermaking 811.1.1
DT: Predates 1975
BT: Manufacture
NT: Paper coating
RT: Bleaching
Calendering
Nonfibrous paper materials
Paper
Papermaking machinery
Papermaking slime control
Sizing (finishing operation)
Wastewater disposal

Papermaking machinery 811.1.2
DT: Predates 1975
BT: Machinery
NT: Fourdrinier machines
Wet ends (papermaking machinery)
RT: Calenders
Dryers (equipment)
Headboxes
Paper
Papermaking
Rolls (machine components)

*Papermaking machinery—Calenders**
USE: Calenders

*Papermaking machinery—Dryers**
USE: Dryers (equipment)

*Papermaking machinery—Felts**
USE: Felts

*Papermaking machinery—Fourdrinier machines**
USE: Fourdrinier machines

*Papermaking machinery—Headbox**
USE: Headboxes

*Papermaking machinery—Rolls**
USE: Rolls (machine components)

*Papermaking machinery—Wet end**
USE: Wet ends (papermaking machinery)

Papermaking slime control 811.1.1
DT: January 1993
UF: Papermaking—Slime control*
Slime control (papermaking)
BT: Control
RT: Papermaking

*Papermaking—Bleaching**
USE: Bleaching

*Papermaking—Coating**
USE: Coating techniques

*Papermaking—Finishing**
USE: Finishing

*Papermaking—Latex applications**
USE: Latexes

*Papermaking—Nonfibrous materials**
USE: Nonfibrous paper materials

*Papermaking—Sizing**
USE: Sizing (finishing operation)

*Papermaking—Slime control**
USE: Papermaking slime control

*Papermaking—Synthetic fibers**
USE: Synthetic fibers

Parabolic antennas (716)
DT: January 1993
UF: Antennas, Parabolic*
BT: Antennas

Parachutes (404.1) (431.1) (914.1)
DT: Predates 1975
BT: Equipment
RT: Aircraft
Aircraft parts and equipment
Safety devices
Satellites
Spacecraft

Paraffin oils
USE: Kerosene

Paraffin problems (511.1) (804.1)
DT: January 1993
UF: Oil well production—Paraffin troubles*
RT: Oil well production
Paraffin sedimentation
Paraffin waxes
Paraffins

Paraffin sedimentation
(511.1) (523) (619.2) (802.3) (804.1)
DT: January 1993
UF: Oil tanks—Paraffin sedimentation*
Oil wells—Paraffin sedimentation*
BT: Sedimentation

Paraffin sedimentation *(continued)*
RT: Oil tanks
Oil wells
Paraffin problems
Paraffin waxes
Paraffins

Paraffin waxes (513.3) (804.1)
DT: Predates 1975
UF: Waxes (paraffin)
BT: Hydrocarbons
RT: Paraffin problems
Paraffin sedimentation
Paraffins

Paraffins (513.3) (804.1)
DT: Predates 1975
UF: Alkanes
BT: Hydrocarbons
NT: Butane
Ethane
RT: Kerosene
Methane
Paraffin problems
Paraffin sedimentation
Paraffin waxes
Propane

Parallel algorithms (723) (921)
DT: January 1995
BT: Algorithms
Computer software
RT: Distributed computer systems
Parallel processing systems

Parallel computing
USE: Parallel processing systems

Parallel processing systems 722.4
DT: January 1993
UF: Array processors
Computer systems, Digital—Parallel processing*
Parallel computing
BT: Multiprocessing systems
RT: Concurrency control
Interconnection networks
Parallel algorithms
Pipeline processing systems
Supercomputers
Systolic arrays
Transputers

Parallel resonator filters
USE: Bandpass filters

Paramagnetic resonance 701.2
DT: January 1993
UF: Electron paramagnetic resonance
Electron spin resonance
EPR
ESR
BT: Magnetic resonance
RT: Color centers
Electron spin resonance spectroscopy
Microwave spectroscopy
Paramagnetism

Paramagnetism 701.2
DT: January 1993
UF: Magnetic materials—Paramagnetism*
Magnetism—Paramagnetism*
BT: Magnetism
RT: Magnetic materials
Paramagnetic resonance

Parameter estimation (731.1)
DT: January 1993
BT: Identification (control systems)
RT: Control systems

Parametric amplifiers 713.1
DT: January 1993
UF: Amplifiers, Parametric*
BT: Amplifiers (electronic)
Parametric devices
RT: Time varying networks
Varactors

Parametric devices (714)
DT: Predates 1975
BT: Electronic equipment
NT: Parametric amplifiers
Parametric oscillators
RT: Electron tubes
Masers

Parametric oscillators 713.2
DT: January 1993
UF: Oscillators, Parametric*
BT: Oscillators (electronic)
Parametric devices

Parking 432.4
DT: January 1993
UF: Street traffic control—Parking*
RT: Garages (parking)
Motor transportation
Street traffic control

Parking garages
USE: Garages (parking)

Parks (403.1) (403.2)
DT: Predates 1975
UF: Regional planning—Parks*
BT: Public works
RT: Land use
Recreation centers
Regional planning
Urban planning

Parkways
USE: Highway systems

Partial differential equations 921.2
DT: January 1995
BT: Differential equations
NT: Maxwell equations
RT: Boundary value problems
Eigenvalues and eigenfunctions

Partial discharges 701.1
DT: January 1993
UF: Electric discharges—Partial discharges*

Partial discharges *(continued)*
BT: Electric discharges
RT: Electric breakdown of solids

Partial pressure sensors (804) (944.3)
DT: January 1993
UF: Carbon dioxide—Partial pressure sensors*
Oxygen—Partial pressure sensors*
BT: Sensors
RT: Blood gas analysis
Oxygen

Particle accelerator accessories 932.1.1
DT: January 1993
UF: Accelerators—Accessories*
BT: Accessories
RT: Particle accelerators

Particle accelerators 932.1.1
DT: January 1993
UF: Accelerators*
Electron accelerators
Proton accelerators
BT: Equipment
NT: Accelerator magnets
Accelerator shielding
Betatrons
Colliding beam accelerators
Cyclotrons
Electron ring accelerators
Electrostatic accelerators
Linear accelerators
Storage rings
Synchrocyclotrons
Synchrotrons
Targets
RT: Elementary particles
Focusing
High energy physics
Ion sources
Particle accelerator accessories
Particle beam dynamics
Particle beam extraction
Particle beam injection
Particle beam tracking
Particle beams
Particle optics

Particle beam dynamics 932.1
DT: January 1993
UF: Accelerators—Beam dynamics*
BT: Dynamics
RT: Particle accelerators
Particle beams

Particle beam extraction 932.1
DT: January 1993
UF: Accelerators—Beam extraction*
Extraction (beam)
RT: Particle accelerators
Particle beams

Particle beam injection 932.1
DT: January 1993
UF: Accelerators—Beam injection*
Injection (beam)
RT: Particle accelerators
Particle beams

Particle beam tracking 932.1
DT: Predates 1975
UF: Beam tracking*
Tracking (beam)
RT: Particle accelerators
Particle beams

Particle beams 932.1
DT: Predates 1975
UF: Beams (particle)
BT: Elementary particles
NT: Atomic beams
Electron beams
Ion beams
Molecular beams
RT: Beam plasma interactions
Electrostatic lenses
Particle accelerators
Particle beam dynamics
Particle beam extraction
Particle beam injection
Particle beam tracking
Particle optics

Particle board 415.4, 811.2, 817.1
DT: Predates 1975
BT: Building materials
Wood products

Particle counters
USE: Radiation counters

Particle detectors 944.7
DT: Predates 1975
BT: Nuclear instrumentation
Radiation detectors
NT: Bubble chambers
Cloud chambers
Cosmic ray detectors
Ionization chambers
Neutron detectors
Spark chambers
RT: Counting circuits
Dosimeters
Dosimetry
Particle separators

Particle injectors (621)
DT: January 1993
UF: Nuclear reactors—Particle injectors*
BT: Nuclear reactors

Particle optics 932.1 (701.1) (931.3)
DT: Predates 1975
BT: Optics
NT: Electron optics
RT: Aberrations
Atomic beams
Electric field effects
Electrodynamics
Electrostatic lenses
Elementary particles
Focusing
High energy physics

* Former Ei Vocabulary term

Particle optics *(continued)*
Magnetic field effects
Magnetic lenses
Particle accelerators
Particle beams
Particle spectrometers

Particle physics
USE: High energy physics

Particle separators (932.1)
SN: For separation of particles of physics
DT: Predates 1975
BT: Particle spectrometers
NT: Electron velocity analyzers
RT: Elementary particles
Particle detectors

Particle size analysis (943.3)
SN: Size analysis of particulate matter
DT: Predates 1975
UF: Particle size distribution
BT: Analysis
Size determination
RT: Microscopic examination
Nephelometers
Particles (particulate matter)
Screens (sizing)
Turbidimeters

Particle size distribution
USE: Particle size analysis

Particle sources (elementary particles)
USE: Elementary particle sources

Particle spectrometers (932.1)
DT: January 1993
UF: Spectrometers, Particle*
BT: Spectrometers
NT: Alpha particle spectrometers
Beta ray spectrometers
Gamma ray spectrometers
Neutron spectrometers
Particle separators
RT: Elementary particles
Particle optics
X ray spectrometers

Particles (elementary)
USE: Elementary particles

Particles (particulate matter) (451.1) (804)
DT: January 1993
RT: Aerosols
Colloids
Dust
Granular materials
Particle size analysis

Particulate emissions 451.1
DT: January 1993
UF: Air pollution—Particulate emissions*
Emissions (particulate)
BT: Wastes
RT: Air pollution
Industrial emissions
Smoke

Partitions (building) 408.2 (402)
DT: January 1993
UF: Buildings—Partitions*
BT: Building components

Parts
USE: Components

Pascal (programming language) 723.1.1
DT: January 1993
UF: Computer programming languages—Pascal*
BT: Procedure oriented languages
RT: Ada (programming language)

Passenger cars 682.1.1
SN: Railroad cars. For passenger autos, use AUTOMOBILES
DT: January 1993
UF: Cars—Passenger*
BT: Railroad cars
NT: Trolley cars

Passenger conveyors
USE: People movers

Passivation 539.2.1 (801.4.1) (802.2) (813.1)
DT: January 1993
BT: Corrosion prevention
RT: Oxidation
Protective coatings

Passive circuits
USE: Passive networks

Passive filters 703.2
DT: January 1993
UF: Cauer filters
Electric filters, Passive*
BT: Electric filters
Passive networks
RT: Acoustic surface wave filters

Passive networks 703.1
DT: January 1993
UF: Electric networks, Passive*
Passive circuits
BT: Networks (circuits)
NT: Passive filters
RT: Microwave circuits

Passive solar buildings 657.1 (402)
DT: January 1993
UF: Buildings—Solar, Passive*
BT: Solar buildings

Pastes (adhesives)
USE: Adhesives

Patents and inventions (901.3) (902.3)
DT: Predates 1975
UF: Inventions
BT: Intellectual property
RT: Engineering

Path planning
USE: Motion planning

Patient monitoring 461.6 (462.2)
DT: January 1993
UF: Biomedical engineering—Patient monitoring*

Patient monitoring *(continued)*
Monitoring (patients)
BT: Monitoring
NT: Fetal monitoring
Neonatal monitoring
RT: Biomedical engineering
Biomedical equipment
Electrocardiography
Electroencephalography
Patient treatment

Patient rehabilitation 461.5
DT: January 1993
UF: Biomedical engineering—Patient rehabilitation*
Rehabilitation
NT: Functional assessment
Functional electric stimulation
Neuromuscular rehabilitation
Occupational therapy
Physical therapy
RT: Artificial limbs
Biomedical engineering
Biomedical equipment
Handicapped persons
Human rehabilitation equipment
Human rehabilitation engineering
Medicine
Patient treatment
Sensory feedback

Patient treatment 461.6
DT: January 1993
UF: Biomedical engineering—Patient treatment*
NT: Acupuncture
Cryotherapy
Drug therapy
Electrotherapeutics
Fracture fixation
Hyperthermia therapy
Immunization
Noninvasive medical procedures
Photodynamic therapy
Radiotherapy
Respiratory therapy
Resuscitation
Sensory feedback
Surgery
RT: Biomedical engineering
Biomedical equipment
Medical departments (industrial plants)
Medical imaging
Medicine
Patient monitoring
Patient rehabilitation

Pattern classification
USE: Pattern recognition

Pattern recognition (716) (723.5) (741.1) (751.1)
DT: January 1983
UF: Pattern classification
NT: Character recognition
Computer vision
Edge detection
Feature extraction
Object recognition
Speech recognition
RT: Computer peripheral equipment
Cybernetics
Detectors
Image processing
Pattern recognition systems
System theory
Wavelet transforms

Pattern recognition systems (716) (723.5) (741.1) (751.1)
DT: Predates 1975
RT: Character recognition
Computer vision
Optical character recognition
Pattern recognition
Speech recognition

Patternmaking 534.2
DT: January 1993
UF: Foundry practice—Patternmaking*
BT: Molding
RT: Casting
Foundry practice

Pavement overlays (406.1) (406.2) (913.5)
DT: January 1993
UF: Overlays (pavement)
Pavements—Overlays*
BT: Pavements
RT: Maintenance
Repair

Pavements (406.1) (406.2)
DT: January 1977
BT: Roads and streets
NT: Asphalt pavements
Concrete pavements
Pavement overlays
RT: Asphalt
Curbs
Driveways
Highway systems
Road construction

*Pavements—Asphalt**
USE: Asphalt pavements

*Pavements—Concrete**
USE: Concrete pavements

*Pavements—Overlays**
USE: Pavement overlays

Pay per view television
USE: Subscription television

Pay telephones
USE: Coin operated telephones

Pb
USE: Lead

PBX
USE: Private telephone exchanges

PC
USE: Personal computers

* Former Ei Vocabulary term

PCB
USE: Polychlorinated biphenyls

PCM
USE: Pulse code modulation

PCM links
USE: Pulse code modulation links

PDM
USE: Pulse width modulation

PE
USE: Polyethylenes

Peak limiters
USE: Limiters

Peaking circuits
USE: Differentiating circuits

Pearlite 531.2 (545.3)
DT: January 1993
UF: Iron and steel metallography—Pearlite*
Metallography—Pearlite*
BT: Metallographic phases
RT: Pearlitic transformations
Steel metallography

Pearlitic transformations 531.2 (545.3)
DT: January 1993
BT: Phase transitions
RT: Pearlite

Peat 524 (503)
DT: Predates 1975
BT: Fossil fuels
RT: Peat cutters
Soils

Peat cutters 524 (503) (603.1) (605)
DT: January 1993
UF: Peat—Cutters*
BT: Cutting equipment
RT: Peat

*Peat—Briquetting**
USE: Briquetting

*Peat—Cutters**
USE: Peat cutters

Pedestrian bridges
USE: Footbridges

Pedestrian overpasses
USE: Footbridges

Pedestrian safety 406.2, 914.1
DT: January 1993
UF: Street traffic control—Pedestrian safety*
BT: Accident prevention
RT: Footbridges
Pedestrian tunnels
Street traffic control

Pedestrian tunnels 401.2 (406.2)
DT: January 1993
UF: Tunnels and tunneling—Pedestrian*
BT: Tunnels
RT: Pedestrian safety

Pediatrics 461.6
DT: January 1993
UF: Biomedical engineering—Pediatrics*
BT: Medicine
RT: Biomedical engineering

PEEK
USE: Polyether ether ketones

Peeling
DT: January 1993
UF: Adhesives—Peeling*
RT: Adhesion
Adhesives
Cleaning
Delamination
Stripping (removal)

Peening
USE: Shot peening

PEG
USE: Polyethylene glycols

PEI
USE: Polyetherimides

Pelleting
USE: Pelletizing

Pelletizing (533.1) (802.3)
DT: Predates 1975
UF: Pelleting
BT: Chemical operations
RT: Agglomeration
Briquetting
Compaction
Densification
Granular materials
Ore pellets
Powder metallurgy

Pellets (ore)
USE: Ore pellets

Pendulums (601.1) (931.5)
DT: Predates 1975
BT: Components
RT: Gravitation
Oscillations
Oscillators (mechanical)
Time measurement

Penitentiaries
USE: Prisons

Pens (light)
USE: Light pens

Pension plans (911.2) (912.4)
DT: Predates 1975
BT: Compensation (personnel)
RT: Management
Personnel

Penstocks (441.3) (611.1) (619.1) (631.1)
DT: Predates 1975
BT: Pipe
RT: Hydroelectric power plants

* Former Ei Vocabulary term

*Penstocks—Concrete**
USE: Concrete construction

*Penstocks—Welded steel**
USE: Welded steel structures

Pentodes 714.1
DT: January 1993
UF: Electron tubes, Pentode*
BT: Electron tubes
RT: Thermionic tubes

PEO
USE: Polyethylene oxides

People movers (431.4) (692.1)
DT: January 1993
UF: Conveyors—Passenger transportation*
Passenger conveyors
BT: Conveyors

Perception
USE: Sensory perception

Perceptrons
USE: Neural networks

Percolation (computer storage) 722.1
DT: January 1993
UF: Percolation*
RT: Data storage equipment

Percolation (fluids) (483.1) (631.1) (802.3)
DT: January 1993
UF: Liquids—Percolation*
Percolation*
BT: Flow of fluids
RT: Capillarity
Concentration (process)
Diffusion
Filtration
Hydraulic conductivity
Infiltration
Leaching
Liquids
Lysimeters
Mechanical permeability
Seepage
Separation
Solvents
Surface tension

Percolation (solid state) (933)
DT: January 1993
UF: Percolation*
RT: Alloys
Crystal lattices
Electric conductivity
Solid state devices

*Percolation**
USE: Percolation (fluids) OR
Percolation (computer storage) OR
Percolation (solid state)

Percussion welding 538.2.1
DT: January 1993
UF: Welding—Percussion*
BT: Pressure welding
Resistance welding
RT: Spot welding
Stud welding

Perforating (535.2.2)
SN: Forming of materials. For perforation of well linings, use WELL PERFORATION
DT: January 1993
UF: Sheet and strip metal—Perforating*
BT: Forming
RT: Boring
Cutting
Drilling
Metal pressing
Piercing
Punching

Perforation (well)
USE: Well perforation

Perforators 511.2
SN: Well equipment
DT: January 1993
UF: Oil well casing—Perforators*
BT: Well equipment
NT: Gun perforators
RT: Oil well casings
Well perforation

Performance
DT: January 1993
RT: Availability
Durability
Evaluation
Fuel consumption
Fuel economy
Maintainability
Maintenance
Productivity
Quality assurance
Quality control
Queueing theory
Reliability
Response time (computer systems)
Riding qualities
Service life
Specifications
Standards
Testing

Perfumes
USE: Fragrances

Periodic systems
USE: Time varying systems

Peripheral controllers
USE: Interfaces (computer)

Peripherals (computer)
USE: Computer peripheral equipment

Permafrost (483.1) (483.2)
DT: January 1993
UF: Foundations—Permafrost*
Soils—Permafrost*
BT: Frozen soils

* Former Ei Vocabulary term

Permafrost *(continued)*
RT: Arctic engineering
Muskeg

Permanent magnets 704.1 (708.4)
DT: January 1993
BT: Magnets

Permanent mold casting 534.2
DT: January 1993
UF: Foundry practice—Permanent mold*
BT: Metal casting
NT: Die casting
RT: Centrifugal casting
Continuous casting

Permanent press textiles
USE: Durable press textiles

Permeability (magnetic)
USE: Magnetic permeability

Permeability (mechanical)
USE: Mechanical permeability

Permeability coefficient
USE: Hydraulic conductivity

*Permeability, Magnetic**
USE: Magnetic permeability

*Permeability, Mechanical**
USE: Mechanical permeability

Permeameters (magnetic permeability) 942.3
SN: For measuring magnetic permeability
DT: January 1993
UF: Magnetic permeability permeameters
Permeameters, Magnetic permeability*
BT: Magnetic measuring instruments
RT: Electric measuring instruments
Magnetic permeability measurement

Permeameters (mechanical permeability) 943.1
SN: For measuring porous permeability
DT: January 1993
UF: Mechanical permeability permeameters
Permeameters, Mechanical permeability*
BT: Flow measuring instruments
RT: Flow measurement
Flowmeters
Lysimeters
Mechanical permeability
Porosimeters

*Permeameters, Magnetic permeability**
USE: Permeameters (magnetic permeability)

*Permeameters, Mechanical permeability**
USE: Permeameters (mechanical permeability)

Permittivity (701) (708.1)
DT: January 1993
UF: Dielectric constant
Dielectric dispersion
BT: Dielectric properties
RT: Capacitance
Insulating materials
Insulation
Permittivity measurement

Permittivity measurement 942.2
DT: January 1993
UF: Electric measurements—Permittivity*
BT: Electric variables measurement
RT: Permittivity

Permselective membranes (801.4) (802.2)
DT: January 1993
UF: Membranes—Permselective*
BT: Membranes
RT: Molecules

Perovskite 482.2 (804.2) (812.1)
DT: January 1993
BT: Calcium compounds
Oxide minerals
Titanium oxides

Peroxides (804.1) (804.2)
DT: January 1987
BT: Oxides
NT: Hydrogen peroxide

Personal aircraft 652.1
DT: January 1993
UF: Aircraft, Personal*
BT: Aircraft

Personal communication networks
USE: Personal communication systems

Personal communication services
USE: Personal communication systems

Personal communication systems (716) (717) (718)
DT: January 1995
UF: Personal communication networks
Personal communication services
BT: Telecommunication systems
RT: Mobile telecommunication systems
Telephone systems

Personal communications
USE: Cellular telephone systems

Personal computers 722.4
DT: January 1993
UF: Computers, Personal*
Desktop computers
Hand held computers
Laptop computers
Notebook computers
PC
BT: Microcomputers

Personnel 912.4
DT: Predates 1975
UF: Employees
Labor
NT: Apprentices
Engineers
Miners
Supervisory personnel
Transportation personnel
RT: Ability testing
Compensation (personnel)
Computer operating procedures

Personnel *(continued)*
Economic and social effects
Employment
Eye protection
Handicapped persons
Health
Industrial hygiene
Industrial plants
Industrial relations
Job analysis
Job satisfaction
Management
Motivation
Occupational diseases
Occupational risks
Pension plans
Personnel rating
Personnel selection
Personnel testing
Personnel training
Productivity
Professional aspects
Protective clothing
Wages
Work-rest schedules

Personnel rating 912.4
DT: January 1993
UF: Personnel—Rating*
Rating (personnel)
BT: Evaluation
RT: Personnel

Personnel relations
USE: Industrial relations

Personnel selection 912.4
DT: January 1993
UF: Personnel—Selection*
BT: Selection
RT: Personnel
Personnel testing

Personnel testing 912.4
DT: January 1993
UF: Personnel—Testing*
BT: Testing
NT: Ability testing
Subjective testing
RT: Human engineering
Personnel
Personnel selection
Personnel training

Personnel training 912.4 (901.2)
DT: Predates 1975
UF: Helicopters—Training applications*
Naval warfare—Training*
Training
BT: Education
NT: Driver training
RT: Apprentices
Automobile simulators
Engineering education
Industrial plants
Personnel
Personnel testing
Professional aspects
Training aircraft

*Personnel—Ability testing**
USE: Ability testing

*Personnel—Eye protection**
USE: Eye protection

*Personnel—Handicapped persons**
USE: Handicapped persons

*Personnel—Health**
USE: Health

*Personnel—Job satisfaction**
USE: Job satisfaction

*Personnel—Motivation**
USE: Motivation

*Personnel—Protective clothing**
USE: Protective clothing

*Personnel—Rating**
USE: Personnel rating

*Personnel—Selection**
USE: Personnel selection

*Personnel—Supervisory**
USE: Supervisory personnel

*Personnel—Testing**
USE: Personnel testing

*Personnel—Wages**
USE: Wages

PERT 912.2 (912.3)
DT: January 1993
UF: Data processing—PERT*
Program evaluation and review technique
BT: Management science
Operations research
RT: Critical path analysis

Perturbation techniques 921
DT: January 1993
UF: Mathematical techniques—Perturbation techniques*
BT: Mathematical techniques
RT: Asymptotic stability
Control system analysis
Control system synthesis

Pervaporation 802.3
DT: January 1995
BT: Separation
RT: Membranes
Solutions

Pest control (461.9) (803) (821.2)
DT: January 1993
BT: Control
NT: Algae control
Fumigation
Insect control
Rat control
Weed control

* Former Ei Vocabulary term

Pest control *(continued)*
RT: Marine borers
Pesticide effects
Pesticides

Pesticide effects (453.1) (454.2) (803) (821.2)
DT: January 1993
UF: Water pollution—Pesticide effects*
BT: Effects
RT: Pest control
Pesticides
Pollution
Soil pollution
Water pollution

Pesticides 803 (821.2)
DT: Predates 1975
BT: Biocides
NT: Fungicides
Herbicides
Insecticides
RT: Fumigation
Pest control
Pesticide effects

*Pesticides—Applicators**
USE: Applicators

PET
USE: Polyethylene terephthalates

PET scanning
USE: Positron emission tomography

PETP
USE: Polyethylene terephthalates

Petri nets 921.4
DT: January 1993
UF: Mathematical techniques—Petri nets*
BT: Graph theory
RT: Automata theory
Control theory
Finite automata
Programming theory

Petrochemical plants 802.1 (402.1) (513.3)
DT: October 1975
BT: Chemical plants
RT: Petrochemicals
Petroleum industry
Petroleum refineries

Petrochemicals 513.3, 804.1
DT: January 1977
BT: Petroleum products
RT: Butadiene
Organic chemicals
Petrochemical plants
Plastics

Petrography 481.1.2
DT: January 1993
UF: Coal—Petrography*
Petrology—Petrography*
BT: Geology
RT: Coal
Petrology

Petroleum
USE: Crude petroleum

Petroleum additives 513.1, 804
DT: January 1993
UF: Petroleum products—Additive compounds*
BT: Additives
RT: Petroleum products

Petroleum analysis (513.1) (801)
DT: Predates 1975
BT: Chemical analysis
RT: Crude petroleum
Petroleum chemistry
Petroleum industry
Sulfur determination

Petroleum chemistry 513.1 (801)
DT: Predates 1975
BT: Chemistry
RT: Crude petroleum
Petroleum analysis
Petroleum industry
Petroleum research

Petroleum coke 513.3 (524)
DT: January 1987
BT: Coke
RT: Petroleum products

Petroleum cracking
USE: Cracking (chemical)

Petroleum deposits 512.1
DT: June 1990
BT: Fossil fuel deposits
NT: Oil fields
Petroleum reservoirs
RT: Acidization
Crude petroleum
Oil shale
Petroleum geology
Petroleum industry

Petroleum distillates 513.3 (804.1)
DT: January 1993
UF: Fuels—Distillates*
Petroleum fractions
BT: Materials
NT: Naphthas
RT: Crude petroleum
Fuels

Petroleum engineering (511) (512) (513) (523)
DT: Predates 1975
BT: Engineering
NT: Petroleum reservoir engineering
RT: Crude petroleum
Petroleum industry

Petroleum fractions
USE: Petroleum distillates

*Petroleum gas, Liquefied**
USE: Liquefied petroleum gas

Petroleum geology 512.1, 481.1
DT: Predates 1975
BT: Geology

* Former Ei Vocabulary term

Petroleum geology *(continued)*
RT: Acoustic logging
Crude petroleum
Petroleum deposits
Petroleum industry
Petroleum prospecting
Petroleum research
Petrology

Petroleum industry (511) (512) (513)
DT: Predates 1975
UF: Helicopters—Petroleum industry*
BT: Mineral industry
RT: Chemical industry
Crude petroleum
Fuel oils
Liquefied natural gas
Liquefied petroleum gas
Natural gas wells
Natural gas well production
Natural gasoline
Oil bearing formations
Oil field equipment
Oil fields
Oil sands
Oil shale
Oil spills
Oil well production
Oil wells
Petrochemical plants
Petroleum analysis
Petroleum chemistry
Petroleum deposits
Petroleum engineering
Petroleum geology
Petroleum pipelines
Petroleum products
Petroleum prospecting
Petroleum refineries
Petroleum refining
Petroleum research
Petroleum transportation

*Petroleum industry—Accounting**
USE: Cost accounting

*Petroleum industry—Losses**
USE: Losses

Petroleum pipelines 511.1, 619.1
DT: Predates 1975
UF: Petroleum pipelines—Interface separation*
BT: Pipelines
NT: Flowlines
RT: Crude petroleum
Gas hazards
Petroleum industry
Petroleum transportation
Pipeline changeover

*Petroleum pipelines—Gas hazards**
USE: Gas hazards

*Petroleum pipelines—Interface separation**
USE: Petroleum pipelines

Petroleum products 513.3
DT: Predates 1975
BT: Materials
NT: Diesel fuels
Gas oils
Gasoline
Liquefied petroleum gas
Mineral oils
Petrochemicals
Petroleum tar
RT: Blending
Insulating oil
Lubricants
Lubricating oils
Naphthas
Petroleum additives
Petroleum coke
Petroleum industry
Petroleum refineries

*Petroleum products—Additive compounds**
USE: Petroleum additives

*Petroleum products—Blending**
USE: Blending

*Petroleum products—Tar**
USE: Petroleum tar

Petroleum prospecting 512.1.2 (481.4)
DT: Predates 1975
BT: Natural resources exploration
NT: Exploratory oil well drilling
Offshore petroleum prospecting
RT: Core analysis
Exploratory boreholes
Geophysical prospecting
Petroleum geology
Petroleum industry

*Petroleum prospecting—Core analysis**
USE: Core analysis

*Petroleum prospecting—Offshore**
USE: Offshore petroleum prospecting

Petroleum refineries 513.2 (402.1)
DT: Predates 1975
BT: Industrial plants
NT: Flare stacks
RT: Fractionating units
Heating equipment
Petrochemical plants
Petroleum industry
Petroleum products
Petroleum refining

*Petroleum refineries—Flare stacks**
USE: Flare stacks

*Petroleum refineries—Fractionating units**
USE: Fractionating units

*Petroleum refineries—Heaters**
USE: Heating equipment

*Petroleum refineries—Sewers**
USE: Sewers

* Former Ei Vocabulary term

Petroleum refining 513.1
DT: Predates 1975
BT: Hydrocarbon refining
NT: Dewaxing
Gasoline refining
Oil sand refining
RT: Alkylation
Aromatization
Catalysis
Catalytic cracking
Coking
Cracking (chemical)
Fluid catalytic cracking
Fractionation
Hydrocracking
Hydrogenation
Oil shale refining
Petroleum industry
Petroleum refineries
Reforming reactions
Salt removal

*Petroleum refining—Demulsification**
USE: Demulsification

*Petroleum refining—Dewaxing**
USE: Dewaxing

*Petroleum refining—Nitrogen removal**
USE: Nitrogen removal

*Petroleum refining—Salt removal**
USE: Salt removal

Petroleum research (511) (512) (513) (901.3)
DT: Predates 1975
BT: Research
RT: Petroleum chemistry
Petroleum geology
Petroleum industry

Petroleum reservoir engineering 512.1.2 (481.1)
DT: Predates 1975
BT: Petroleum engineering
RT: Core analysis
Oil fields
Oil wells

*Petroleum reservoir engineering—Core analysis**
USE: Core analysis

Petroleum reservoir evaluation 512.1.2 (481.4)
DT: January 1993
UF: Oil fields—Reservoir evaluation*
Reservoir evaluation
BT: Evaluation
RT: Oil fields
Petroleum reservoirs
Resource valuation

Petroleum reservoirs 512.1.1 (481.1)
DT: January 1993
UF: Reservoirs (petroleum)
BT: Petroleum deposits
NT: Multiple zones
RT: Low permeability reservoirs
Oil bearing formations
Oil fields
Petroleum reservoir evaluation

Petroleum tar 513.3
DT: January 1993
UF: Petroleum products—Tar*
BT: Petroleum products
Tar

Petroleum transportation (432.3) (433.3) (434.3)
SN: For pipeline transport use PETROLEUM PIPELINES
DT: Predates 1975
BT: Transportation
RT: Materials handling
Petroleum industry
Petroleum pipelines
Tank cars
Tank trucks
Tankers (ships)

*Petroleum, Crude**
USE: Crude petroleum

Petrology 481.1.2
DT: Predates 1975
UF: Geophysics—Rock properties*
BT: Geology
RT: Crystalline rocks
Earth (planet)
Geochemistry
Geophysics
Igneous rocks
Metamorphic rocks
Ore deposits
Petrography
Petroleum geology
Rock mechanics
Rocks
Sedimentary rocks
Sedimentology
Volcanic rocks

*Petrology—Petrography**
USE: Petrography

*Petrology—Sedimentology**
USE: Sedimentology

Pewter 544.2, 546.2, 546.4
DT: Predates 1975
BT: Copper alloys
Tin alloys
RT: Antimony alloys

pH 801.1
DT: January 1993
UF: Acidity
Alkalinity
Basicity
Causticity
Hydrogen ion concentration*
RT: pH effects
pH meters
pH sensors
Physical chemistry

* Former Ei Vocabulary term

pH effects 801.1
DT: January 1993
BT: Effects
RT: pH

pH meters 801.1, 943.3 (802.1)
DT: January 1993
UF: Hydrogen ion concentration—pH meters*
BT: Instruments
RT: pH
pH sensors

pH sensors 801.1, 943.3 (802.1)
DT: January 1993
UF: Hydrogen ion concentration—Sensors*
Hydrogen ion sensors
BT: Chemical sensors
RT: pH
pH meters

Phage
USE: Bacteriophages

Pharmaceutical plants
USE: Drug products plants

Pharmaceuticals
USE: Drug products

Pharmacodynamics 461.6
DT: January 1993
UF: Drug products—Pharmacodynamics*
NT: Drug interactions
Pharmacokinetics
RT: Drug therapy

Pharmacokinetics 461.6
DT: January 1993
UF: Drug products—Pharmacokinetics*
BT: Pharmacodynamics
RT: Drug therapy

Pharmacotherapy
USE: Drug therapy

Phase angle meters
USE: Phase meters

Phase changing circuits 703.1, 713.5
DT: January 1993
UF: Electronic circuits, Phase changer*
BT: Networks (circuits)
RT: Phase shift
Phase shifters

Phase comparators 713.5
DT: January 1993
UF: Electronic circuits, Phase comparator*
BT: Comparator circuits
RT: Phase meters

Phase composition 641.1
DT: January 1995
BT: Composition
RT: Phase separation
Volume fraction

Phase conjugation (optical)
USE: Optical phase conjugation

Phase control 701.1, 731.3
DT: January 1993
UF: Control, Electric variables—Phase*
Phase regulation
BT: Electric variables control
RT: Phase locked loops
Phase measurement
Phase shifters

Phase diagrams (531) (701) (801.4) (921.4)
DT: January 1993
RT: Binary alloys
Graphic methods
Metallurgy
Phase equilibria
Phase transitions
Solid solutions
Solubility
Systems (metallurgical)
Ternary systems

Phase equilibria (531) (641.1) (801.4) (931.2)
DT: January 1993
UF: Equilibrium (phase)
Vapor liquid equilibria
RT: Crystal microstructure
Phase diagrams
Phase transitions
Physical chemistry
Solubility
Solutions

Phase interfaces 801.4 (931.2)
DT: January 1993
UF: Colloids—Phase interfaces*
BT: Interfaces (materials)
RT: Colloids
Phase separation

Phase locked loops 713.5
DT: Predates 1975
UF: PLL
BT: Networks (circuits)
RT: Control systems
Demodulators
Modulation
Modulators
Phase control

Phase measurement 942.2
DT: January 1993
UF: Electric measurements—Phase*
BT: Electric variables measurement
RT: Phase control
Phase meters
Phase shifters
Synchronization

Phase meters 942.1
DT: Predates 1975
UF: Phase angle meters
BT: Electric measuring instruments
RT: Phase comparators
Phase measurement
Phase sequence indicators
Phase shift

* Former Ei Vocabulary term

*Phase meters—Sequence indication**
USE: Phase sequence indicators

Phase modulation (713.3) (716)
DT: Predates 1975
BT: Modulation
NT: Phase shift keying
RT: Demodulation
Demodulators
Modulators
Phase shift

*Phase modulation—Phase shift keying**
USE: Phase shift keying

Phase regulation
USE: Phase control

Phase separation 641.1, 802.3
DT: January 1995
BT: Separation
RT: Phase composition
Phase interfaces
Phase transitions

Phase sequence indicators 942.1
DT: January 1993
UF: Phase meters—Sequence indication*
Sequence indicators
BT: Electric measuring instruments
RT: Phase meters

Phase shift (703.1) (942.2)
DT: January 1993
RT: Phase changing circuits
Phase meters
Phase modulation
Phase shifters
Waves

Phase shift circuits
USE: Phase shifters

Phase shift keying (713.3) (716)
DT: January 1993
UF: Phase modulation—Phase shift keying*
PSK
BT: Phase modulation

Phase shifter circuits
USE: Phase shifters

Phase shifters 713.5 (703.1)
DT: January 1993
UF: Electric phase shift circuits
Electric phase shifters*
Phase shift circuits
Phase shifter circuits
BT: Networks (circuits)
RT: Ferrite devices
Gyrators
Phase changing circuits
Phase control
Phase measurement
Phase shift

Phase space methods 921
DT: January 1993
UF: Mathematical techniques—Phase space methods*
BT: Mathematical techniques
RT: Control system analysis
Control system synthesis
Frequency response
Nonlinear control systems

Phase transformations
USE: Phase transitions

Phase transitions 801.4 (531.2) (641.1)
DT: January 1993
UF: Changes of state
Metallography—Transformations*
Phase transformations
NT: Austenitic transformations
Bainitic transformations
Condensation
Freezing
Gasification
Glass transition
Liquefaction of gases
Martensitic transformations
Melting
Order disorder transitions
Pearlitic transformations
Solidification
Vaporization
RT: Aging of materials
Crystal microstructure
Distillation
Equations of state
Graphitization
Metallography
Phase diagrams
Phase equilibria
Phase separation
Recrystallization (metallurgy)

Phased arrays (antenna)
USE: Antenna phased arrays

Phasotrons
USE: Cyclotrons

Phenolic epoxy resins
USE: Phenolic resins

Phenolic plastics
USE: Phenolic resins

Phenolic resins 815.1.1
DT: Predates 1975
UF: Phenolic epoxy resins
Phenolic plastics
BT: Organic compounds
Thermosets
RT: Urea formaldehyde resins

Phenols 804.1
DT: Predates 1975
UF: Industrial wastes—Phenols*
BT: Aromatic compounds
RT: Alcohols
Industrial wastes

* Former Ei Vocabulary term

Phenoxy resins 815.1.1
DT: Predates 1975
BT: Polyethers

Philosophical aspects (901.1)
DT: January 1993
UF: Technology—Philosophical aspects*
RT: Engineering
Social aspects
Technology
Theory

Phonetics
USE: Speech analysis

Phonograph records 752.3.1
DT: Predates 1975
RT: Audio equipment
Audio systems
Compact disks
Phonographs
Sound recording
Sound reproduction
Stereophonic recordings

*Phonograph records—Stereophonic recordings**
USE: Stereophonic recordings

Phonographs 752.3.1
DT: Predates 1975
UF: Gramophones
Record players
Turntables
BT: Audio equipment
RT: Audio frequency amplifiers
Phonograph records
Pickups

Phonons (751.1) (931.3)
DT: January 1981
RT: Acoustic waves
Elastic waves
Electron energy levels
Elementary particles
Lattice vibrations
Solitons

Phosphate coatings 539.2.1, 813.2
DT: January 1993
UF: Protective coatings—Phosphate*
BT: Coatings
RT: Phosphates

Phosphate deposits 505.1
DT: Predates 1975
BT: Mineral resources
RT: Phosphate minerals
Phosphate mines
Phosphates
Phosphorus

Phosphate fertilizers 804.2, 821.2
DT: January 1993
UF: Fertilizers—Phosphates*
BT: Fertilizers
RT: Phosphates
Phosphorus compounds

Phosphate minerals 482.2
DT: January 1993
UF: Mineralogy—Phosphates*
BT: Minerals
Phosphates
RT: Phosphate deposits
Phosphate mines
Phosphate ore treatment
Phosphorus

Phosphate mines 505.1
DT: January 1993
UF: Phosphate mines and mining*
BT: Mines
RT: Phosphate deposits
Phosphate minerals
Phosphates
Phosphorus

*Phosphate mines and mining**
USE: Phosphate mines

Phosphate ore treatment 533.1
DT: Predates 1975
BT: Ore treatment
RT: Phosphate minerals
Phosphates
Phosphorus

Phosphates (804.1) (804.2)
DT: Predates 1975
UF: Water treatment—Phosphate*
BT: Phosphorus compounds
NT: Phosphate minerals
RT: Adenosinetriphosphate
Borophosphate glass
Phosphate coatings
Phosphate deposits
Phosphate fertilizers
Phosphate mines
Phosphate ore treatment
Phosphorus
Water pollution
Water treatment

Phosphorescence 741.1
DT: Predates 1975
BT: Luminescence
RT: Electromagnetic wave absorption
Fluorescence
Light sources
Phosphors
Scintillation

Phosphoric acid 804.2
DT: Predates 1975
BT: Inorganic acids

Phosphors 741.1, 804
DT: Predates 1975
UF: Luminophors
Luminors
Scintillators
BT: Materials
RT: Image intensifiers (electron tube)
Luminescence
Luminescent devices

* Former Ei Vocabulary term

Phosphors *(continued)*
Phosphorescence
Radiation detectors
Scintillation
Scintillation counters
Television picture tubes

Phosphorus 804
DT: Predates 1975
UF: P
BT: Nonmetals
RT: Phosphate deposits
Phosphate minerals
Phosphate mines
Phosphate ore treatment
Phosphates
Phosphorus compounds

Phosphorus compounds (804.1) (804.2)
DT: Predates 1975
BT: Chemical compounds
NT: Phosphates
RT: Phosphate fertilizers
Phosphorus

Photelectric electron emissions
USE: Photoemission

Photoacoustic effect 741.1, 751.1
DT: January 1993
UF: Optoacoustic effect
Optoacoustics
Photoacoustics
Thermoacoustic effect
RT: Acoustic generators
Acoustics
Photoacoustic spectroscopy

Photoacoustic spectroscopy 741.1, 751.1 (801)
DT: January 1993
UF: Optoacoustic spectroscopy
Spectroscopy, Photoacoustic*
BT: Acoustic spectroscopy
RT: Photoacoustic effect

Photoacoustics
USE: Photoacoustic effect

Photocathodes 714.1, 741.3
DT: Predates 1975
BT: Cathodes
Photoemissive devices
RT: Electron tube components
Image intensifiers (electron tube)
Photoemission
Photomultipliers
Phototubes
Quantum efficiency

Photocells
USE: Photoelectric cells

Photochemical forming (535.2) (741.1) (802.2)
DT: January 1993
UF: Metal forming—Photochemical*
BT: Forming
RT: Photochemical reactions

Photochemical reactions 741.1, 802.2
DT: January 1993
UF: Photochemistry
Photopolymerization
BT: Chemical reactions
NT: Photolysis
Photosynthesis
RT: Atmospheric chemistry
Chemiluminescence
Light
Photochemical forming
Photochromism
Photoelectrochemical cells
Photographic films
Photography
Photoreactivity
Photosensitive glass
Photosensitivity

Photochemistry
USE: Photochemical reactions

Photochromism 741.1
DT: January 1986
UF: Photodarkening
Plastics—Photochromism*
BT: Optical properties
RT: Color
Light sensitive materials
Photochemical reactions

Photocomposition 745.1
DT: January 1993
UF: Printing—Photocomposition*
BT: Printing

Photoconducting devices 714.2
DT: Predates 1975
BT: Photoelectric devices
NT: Photodiodes
RT: Photoconducting materials
Photoconductivity
Photoelectric cells
Photoresistors
Semiconductor devices

Photoconducting materials 708.2, 712.1
DT: Predates 1975
UF: Photoconductors
BT: Materials
RT: Photoconducting devices
Photoconductivity
Photodetectors
Photoelectricity
Semiconductor materials

Photoconductivity 701.1, 741.1
DT: January 1987
BT: Electric conductivity
Photoelectricity
RT: Photoconducting devices
Photoconducting materials
Semiconductor devices
Semiconductor materials

Photoconductors
USE: Photoconducting materials

Photocopying 745.2
DT: January 1993
UF: Photographic reproduction*
BT: Copying
NT: Color photocopying
Electrostatic reproduction
RT: Blueprints
Photography
Publishing

Photodarkening
USE: Photochromism

Photodetectors (714.1) (714.2) (741.3)
DT: Predates 1975
UF: Light detectors
BT: Photoelectric devices
Radiation detectors
NT: Infrared detectors
Ultraviolet detectors
RT: Integrated optoelectronics
Light
Photoconducting materials
Photodiodes
Photoelectric cells
Photometers
Phototransistors

Photodiodes (714.2) (741.3)
DT: January 1993
UF: Semiconductor diodes, Photodiode*
BT: Photoconducting devices
Semiconductor diodes
RT: Integrated optoelectronics
Photodetectors
Photoelectric cells
Phototubes

Photodynamic therapy 461.6 (741.1)
DT: January 1995
BT: Patient treatment
RT: Dosimetry
Endoscopy
Fiber optics
Laser applications

Photoelastic effect
USE: Photoelasticity

Photoelasticity 741.1 (408) (421)
DT: Predates 1975
UF: Photoelastic effect
BT: Mechanical properties
RT: Birefringence
Elasticity
Light polarization
Optical properties
Stress analysis

Photoelectric cells (714.2) (741.3)
DT: Predates 1975
UF: Photocells
BT: Photoelectric devices
NT: Photovoltaic cells
RT: Photoconducting devices
Photodetectors
Photodiodes
Photoelectricity
Photometers
Photomultipliers
Transducers

Photoelectric devices (714.2) (741.3)
DT: January 1993
BT: Optoelectronic devices
NT: Photoconducting devices
Photodetectors
Photoelectric cells
Photoelectric microscopes
Photoelectric relays
Phototubes
RT: Photoelectricity

Photoelectric effects
USE: Photoelectricity

Photoelectric emission
USE: Photoemission

Photoelectric microscopes 741.3
DT: January 1993
UF: Microscopes—Photoelectric*
BT: Microscopes
Photoelectric devices

Photoelectric relays 741.3 (704.2) (706.2) (714.1)
DT: January 1993
UF: Electric relays—Photoelectric*
BT: Electric relays
Photoelectric devices

Photoelectric tubes
USE: Phototubes

Photoelectricity 701.1, 741.1
DT: Predates 1975
UF: Photoelectric effects
BT: Electric properties
NT: Photoconductivity
Photoelectromagnetic effects
Photoemission
Photovoltaic effects
RT: Photoconducting materials
Photoelectric cells
Photoelectric devices
Photoemissive devices
Photoionization
Quantum efficiency

Photoelectrochemical cells 702.1 (741.3) (801.4.1)
DT: June 1990
BT: Electric batteries
RT: Photochemical reactions
Solar cells

Photoelectromagnetic effects (701.2) (712.1) (741.1)
DT: Predates 1975
BT: Magnetic field effects
Photoelectricity
RT: Electromagnetism

Photoelectron multipliers
USE: Photomultipliers

* Former Ei Vocabulary term

Photoelectron spectroscopy (741.1) (801) (931.3)
DT: January 1993
UF: Spectroscopy, Photoelectron*
BT: Electron spectroscopy
NT: X ray photoelectron spectroscopy

Photoemission 701.1, 741.1
DT: Predates 1975
UF: Photelectric electron emissions
Photoelectric emission
BT: Photoelectricity
RT: Photocathodes
Photoemissive devices

Photoemissive devices 741.3 (714)
DT: Predates 1975
BT: Equipment
NT: Photocathodes
RT: Photoelectricity
Photoemission
Phototubes

Photoglow tubes
USE: Phototubes

Photogrammetry 742.1, 405.3
DT: Predates 1975
UF: Aerial photographic surveys
NT: Underwater photogrammetry
RT: Aerial photography
Cameras
Data reduction
Geodesy
Geophysics
Infrared imaging
Mapping
Photointerpretation
Remote sensing
Space surveillance
Stockpile surveys
Surveying
Surveys

Photogrammetry interpretation
USE: Photointerpretation

*Photogrammetry—Cameras**
USE: Cameras

*Photogrammetry—Data reduction**
USE: Data reduction

*Photogrammetry—Forestry applications**
USE: Forestry

*Photogrammetry—Hydraulics applications**
USE: Hydraulics

*Photogrammetry—Infrared radiation**
USE: Infrared imaging

*Photogrammetry—Interpretation**
USE: Photointerpretation

*Photogrammetry—Mining applications**
USE: Mining

*Photogrammetry—Moon**
USE: Moon

*Photogrammetry—Stockpile surveys**
USE: Stockpile surveys

*Photogrammetry—Underwater**
USE: Underwater photogrammetry

Photograph interpretation
USE: Photointerpretation

Photographic accessories 742.2
DT: January 1993
UF: Cameras—Attachments*
Photography—Accessories*
BT: Accessories
Photographic equipment
RT: Exposure meters
Optical devices
Optical instruments
Optical projectors
Photography
Projection screens

Photographic emulsions 742.3
DT: Predates 1975
BT: Emulsions
Light sensitive materials
RT: Photographic films
Photography
Photosensitivity
Protective colloids
Sensitometers

*Photographic emulsions—Sensitivity**
USE: Photosensitivity

Photographic equipment 742.2
DT: January 1993
UF: Photography—Equipment*
BT: Equipment
NT: Cameras
Photographic accessories
RT: Optical devices
Optical instruments
Photographic films
Photography

Photographic films 742.3
DT: January 1993
UF: Camera films
Motion picture film
Photographic films and plates*
Photographic plates
BT: Films
Light sensitive materials
NT: Color films
Microforms
X ray films
RT: Densitometers
Photochemical reactions
Photographic emulsions
Photographic equipment
Photography
Photosensitivity

*Photographic films and plates**
USE: Photographic films

*Photographic films and plates, Color**
USE: Color films

*Photographic films and plates—Fire hazards**
USE: Fire hazards

*Photographic films and plates—Spools**
USE: Reels

Photographic lenses
USE: Camera lenses

Photographic plates
USE: Photographic films

*Photographic reproduction**
USE: Photocopying

*Photographic reproduction—Color**
USE: Color photocopying

*Photographic reproduction—Electrostatic**
USE: Electrostatic reproduction

*Photographic reproduction—Xerography**
USE: Xerography

Photography 742.1
DT: Predates 1975
UF: Cinematography
BT: Imaging techniques
NT: Aerial photography
Color photography
High speed photography
Military photography
Underwater photography
RT: Cameras
Color
Exposure meters
Graphic methods
Holography
Light sources
Motion pictures
Optical projectors
Optics
Photochemical reactions
Photocopying
Photographic accessories
Photographic emulsions
Photographic equipment
Photographic films
Photomapping
Photometers
Projection screens
Recording
Remote sensing
Schlieren systems
Special effects

*Photography, Color**
USE: Color photography

*Photography, High speed**
USE: High speed photography

*Photography—Accessories**
USE: Photographic accessories

*Photography—Equipment**
USE: Photographic equipment

*Photography—Exposure meters**
USE: Exposure meters

*Photography—Infrared radiation**
USE: Infrared imaging

*Photography—Projection screens**
USE: Projection screens

*Photography—Projectors**
USE: Optical projectors

*Photography—Schlieren system**
USE: Schlieren systems

*Photography—Special effects**
USE: Special effects

*Photography—Underwater**
USE: Underwater photography

Photointerpretation 742.1 (405.3)
DT: January 1993
UF: Photogrammetry interpretation
Photogrammetry—Interpretation*
Photograph interpretation
RT: Photogrammetry
Photomapping

Photoionisation
USE: Photoionization

Photoionization 802.2 (741.1)
DT: January 1993
UF: Photoionisation
BT: Ionization
RT: Atmospheric ionization
Gases
Photoelectricity

Photolithography (714.2) (745.1)
DT: January 1993
UF: Lithography—Photolithography*
BT: Lithography
NT: Photoresists
X ray lithography
RT: Integrated circuit manufacture
Masks
Printed circuits
Semiconductor device manufacture

Photoluminescence 741.1
DT: Predates 1975
BT: Luminescence
RT: Light absorption
Scintillation counters

Photolysis 802.2 (741.1)
DT: January 1993
BT: Decomposition
Photochemical reactions
RT: Radiation chemistry

Photomapping 742.1, 405.3
DT: January 1993
UF: Maps and mapping—Photomaps*
BT: Mapping
RT: Aerial photography
Geodesy
Geology

Photomapping *(continued)*
Holography
Natural resources exploration
Photography
Photointerpretation

Photomechanics 741.1, 931.1
DT: January 1993
RT: Mechanics
Optics

Photometers 941.3
DT: Predates 1975
UF: Light meters (photometers)
BT: Optical instruments
NT: Exposure meters
Illumination meters
Oximeters
Spectrophotometers
RT: Comparators (optical)
Densitometers
Photodetectors
Photoelectric cells
Photography
Photometry
Polarimeters
Radiometers
Radiometry
Reflectometers
Spectrometers
Ultraviolet detectors

*Photometers—Illumination meters**
USE: Illumination meters

Photometry 941.4
DT: Predates 1975
BT: Optical variables measurement
NT: Spectrophotometry
RT: Chemical analysis
Exposure controls
Illuminating engineering
Light
Lighting
Photometers
Radiometry
Spectrometry

Photomultipliers 714.1
DT: January 1993
UF: Electron tubes, Photomultiplier*
Multiplier phototubes
Photoelectron multipliers
BT: Electron multipliers
Phototubes
RT: Image intensifiers (electron tube)
Image intensifiers (solid state)
Photocathodes
Photoelectric cells
Scintillation counters

Photonic propulsion (654.2) (931.3) (931.4)
DT: January 1993
UF: Propulsion—Photon energy*
Rocket engines—Photon propulsion*
BT: Propulsion
RT: Rocket engines

Photonic switches
USE: Optical switches

Photons 931.3 (741.1)
DT: Predates 1975
BT: Elementary particles
RT: Gamma rays
Light
Quantum efficiency
Quantum electronics
X rays

Photopolymerization
USE: Polymerization AND
Photochemical reactions

Photoreactivity 741.1
DT: January 1993
UF: Plastics—Photoreactivity*
BT: Optical properties
RT: Photochemical reactions

Photoresistors (714.2) (941.3) (942.2)
DT: Predates 1975
UF: Light sensitive resistors
BT: Resistors
RT: Photoconducting devices

Photoresists 714.2, 813.2 (741.1) (745.1)
DT: January 1981
BT: Photolithography
RT: Integrated circuit manufacture
Printed circuits
Semiconductor device manufacture

Photosensitive glass 741.1, 812.3
DT: January 1993
UF: Glass—Photosensitive*
BT: Glass
RT: Photochemical reactions
Photosensitivity

Photosensitivity 741.1
DT: January 1993
UF: Light sensitivity
Photographic emulsions—Sensitivity*
BT: Optical properties
RT: Color films
Light sensitive materials
Photochemical reactions
Photographic emulsions
Photographic films
Photosensitive glass
Phototransistors
Sensitometers

Photosynthesis 741.1, 802.2
DT: January 1993
BT: Photochemical reactions
Synthesis (chemical)
RT: Biosynthesis
Chlorophyll
Photosynthetic membranes

Photosynthetic membranes (741.1) (802.1)
DT: January 1993
UF: Membranes—Photosynthetic membranes*
BT: Membranes
RT: Photosynthesis

Phototransistors 714.2 (741.3)
DT: January 1993
UF: Transistors, Photosensitive*
BT: Transistors
RT: Integrated optoelectronics
Photodetectors
Photosensitivity

Phototubes 714.1 (741.3)
DT: January 1993
UF: Electron tubes, Phototube*
Photoelectric tubes
Photoglow tubes
BT: Cold cathode tubes
Photoelectric devices
NT: Photomultipliers
RT: Photocathodes
Photodiodes
Photoemissive devices
Quantum efficiency

Photovoltaic cells (714.2) (741.3) (941.3)
DT: Predates 1975
BT: Photoelectric cells
RT: Photovoltaic effects
Solar cells

Photovoltaic effects (701.1) (712.1) (741.1)
DT: January 1986
BT: Photoelectricity
RT: Photovoltaic cells
Semiconductor devices
Semiconductor materials
Solar cells

Photovoltaic generators (solar)
USE: Solar cells

Physical chemistry 801.4
DT: Predates 1975
BT: Chemistry
NT: Electrochemistry
Laser chemistry
Molecular dynamics
Radiation chemistry
RT: Activation energy
Chemical analysis
Chemical engineering
Chemical equipment
Chemical reactions
Colloid chemistry
Geochemistry
Metallurgy
pH
Phase equilibria
Reaction kinetics
Thermodynamics
Van der Waals forces

*Physical chemistry—Van der Waals force**
USE: Van der Waals forces

Physical properties 931.2
DT: January 1993
NT: Acoustic properties
Anisotropy
Calorific value
Electric properties
Electronic properties
Flammability
Fluidity
High temperature properties
Low temperature properties
Magnetic properties
Optical properties
Polarization
Porosity
Surface properties
Thermoelasticity
Transport properties
RT: Composition
Inspection
Weldability

Physical therapy 461.5
DT: January 1993
UF: Human rehabilitation engineering—Physical therapy*
Hydrotherapy
Physiotherapy
BT: Patient rehabilitation
RT: Biomedical engineering
Human rehabilitation engineering

Physics (931) (932) (933)
DT: Predates 1975
BT: Natural sciences
NT: Acoustics
Astrophysics
Atomic physics
Electromagnetic field theory
Electrostatics
Geophysics
High energy physics
Laser theory
Magnetostatics
Molecular physics
Nuclear physics
Optics
Plasma theory
Quantum theory
Solid state physics
RT: Materials science
Plasmas
Radar theory
Radiation
Relativity
Research
Theory
Thermodynamics

*Physics—Atomic**
USE: Atomic physics

*Physics—High energy**
USE: High energy physics

*Physics—Molecular**
USE: Molecular physics

*Physics—Nuclear**
USE: Nuclear physics

*Physics—Solid state**
USE: Solid state physics

Physiological models (461.1)
DT: January 1993
UF: Biomedical engineering—Physiological models*
BT: Models
RT: Biomedical engineering
Human form models
Physiology

Physiology 461.9
DT: January 1993
UF: Vision—Physiology*
BT: Biology
NT: Electrophysiology
Neurophysiology
Psychophysiology
RT: Audition
Blood
Ergonomics
Hemodynamics
Medicine
Physiological models
Sleep research
Vision

Physiotherapy
USE: Physical therapy

PI
USE: Polyimides

Pick-ups
USE: Pickups

Pickling 539, 802.3
SN: Chemical cleaning
DT: Predates 1975
BT: Chemical cleaning
RT: Metal cleaning
Ultrasonic cleaning

Pickups (752.2.1) (752.3.1)
SN: Transducers that convert recorded sound into electric signals
DT: Predates 1975
UF: Pick-ups
BT: Acoustic transducers
Audio equipment
RT: Phonographs
Recording
Sound reproduction

Picture processing
USE: Image processing

Picture tubes
USE: Television picture tubes

Piecewise linear techniques 921.4 (731.1)
DT: January 1993
UF: Mathematical techniques—Piecewise linear techniques*
BT: Mathematical techniques
RT: Control system analysis
Control system synthesis
Difference equations
Nonlinear control systems

Piercing (535.2) (604.1)
DT: January 1993
UF: Puncturing
Sheet and strip metal—Piercing*
BT: Forming
RT: Boring
Broaching
Cold working
Drilling
Extrusion
Metal pressing
Perforating
Punching

Piers 407.1
SN: Port structures. For bridge supports use BRIDGE PIERS
DT: Predates 1975
BT: Port structures
RT: Docks

Piers (bridges)
USE: Bridge piers

Piezoelectric devices (704) (708.1) (714)
DT: Predates 1975
BT: Dielectric devices
NT: Crystal filters
Crystal resonators
Piezoelectric transducers
RT: Acoustic variables measurement
Mechanical testing
Piezoelectric materials
Piezoelectricity
Sonar
Ultrasonic transducers

*Piezoelectric effects**
USE: Piezoelectricity

Piezoelectric filters
USE: Crystal filters

Piezoelectric materials 708.1 (812.1)
DT: Predates 1975
UF: Ceramic materials—Piezoelectric*
BT: Dielectric materials
RT: Ceramic materials
Electrostriction
Intelligent materials
Piezoelectric devices
Piezoelectric transducers
Piezoelectricity
Semiconductor materials

Piezoelectric transducers (704) (708.1) (752.1)
DT: Predates 1975
BT: Piezoelectric devices
Transducers

Piezoelectric transducers *(continued)*
RT: Piezoelectric materials
Piezoelectricity
Pressure transducers

Piezoelectricity 701.1 (708.1)
DT: Predates 1975
UF: Piezoelectric effects*
BT: Dielectric properties
Electric field effects
RT: Ferroelectric materials
Piezoelectric devices
Piezoelectric materials
Piezoelectric transducers
Pyroelectricity
Semiconductor materials
Stresses

Pig iron 545.1
DT: Predates 1975
BT: Iron alloys
RT: Tapping (furnace)
Wrought iron

Piggyback freight handling 691.2
DT: January 1993
UF: Freight handling—Piggyback*
BT: Cargo handling

Pigments (803) (804) (813.2)
DT: Predates 1975
BT: Materials
NT: Chlorophyll
Hemoglobin
RT: Additives
Enamels
Lacquers
Luminescence
Paint
Titanium dioxide
Zinc oxide

*Pigments—Luminescence**
USE: Luminescence

*Pigments—Zinc oxide**
USE: Zinc oxide

Pile drivers 405.1
DT: Predates 1975
BT: Construction equipment
RT: Pile driving

Pile driving 405.2
DT: January 1993
UF: Piles—Driving*
BT: Construction
RT: Pile drivers
Pile foundations
Piles

Pile foundations 483.2
DT: January 1993
UF: Foundations—Piles*
BT: Foundations
RT: Pile driving
Piles

Piles 408.2
DT: Predates 1975
UF: Pilings
BT: Structural members
RT: Pile driving
Pile foundations
Poles

*Piles—Concrete**
USE: Concrete construction

*Piles—Driving**
USE: Pile driving

*Piles—Wood**
USE: Wood products

Pilings
USE: Piles

Pillar and room mining
USE: Room and pillar mining

Pillar extraction 502.1
DT: January 1993
UF: Mines and mining—Pillar extraction*
BT: Mining
RT: Room and pillar mining

Pilot boats 674.1
DT: Predates 1975
BT: Boats

Pilot lights 521.3, 522
DT: January 1993
UF: Gas appliances—Pilot lights*
BT: Gas appliances
RT: Gas stoves

Pilot plants (402.1) (912.1)
DT: January 1993
UF: Industrial plants—Pilot plants*
BT: Industrial plants

Pinch effect (701) (932.3)
DT: January 1993
UF: Plasmas—Pinch effect*
RT: Magnetic fields
Plasma confinement
Plasmas

Pipe 619.1
SN: For pipe in use, use PIPELINES or PIPING SYSTEMS
DT: Predates 1975
UF: Heat transfer—Pipes*
Materials handling—Pipe*
Rolling mill practice—Pipe*
BT: Components
NT: Aluminum pipe
Asbestos cement pipe
Cast iron pipe
Clay pipe
Concrete pipe
Copper pipe
Drill pipe
Heat pipes
Marine risers
Nickel pipe

* Former Ei Vocabulary term

Pipe *(continued)*
Penstocks
Pipe fittings
Pipe joints
Pipe linings
Plastic pipe
Steel pipe
RT: Flow of fluids
Materials handling equipment
Oil well casings
Pipe laying barges
Pipelines
Piping systems
Tubes (components)

Pipe fittings 619.1.1
DT: Predates 1975
UF: Fittings (pipe)
BT: Pipe

Pipe flow 631.1 (619.1)
DT: January 1993
UF: Flow of fluids—Pipes*
Flow of fluids—Tubes*
Pipelines—Flow*
Tube flow
BT: Confined flow
RT: Channel flow
Flow patterns
Pipeline surges
Pipelines
Pulsatile flow
Tubes (components)
Turbulence
Turbulent flow
Unsteady flow
Water hammer

Pipe joints 619.1.1
DT: Predates 1975
UF: Pipeline joints
Pipelines—Joints*
BT: Joints (structural components)
Pipe
RT: Expansion joints
Flanges
Gaskets
Pipelines
Welds

*Pipe joints—Flanges**
USE: Flanges

Pipe laying barges (619.1) (674)
DT: January 1993
UF: Barges—Pipe laying*
BT: Barges
RT: Pipe
Pipelines

Pipe linings 619.1.1
DT: January 1993
UF: Pipe—Lining*
Pipeline linings
Pipelines—Lining*
BT: Linings
Pipe
RT: Pipelines

*Pipe, Aluminum**
USE: Aluminum pipe

*Pipe, Asbestos cement**
USE: Asbestos cement pipe

*Pipe, Cast iron**
USE: Cast iron pipe

*Pipe, Clay**
USE: Clay pipe

*Pipe, Concrete**
USE: Concrete pipe

*Pipe, Copper**
USE: Copper pipe

*Pipe, Nickel**
USE: Nickel pipe

*Pipe, Plastic**
USE: Plastic pipe

*Pipe, Steel**
USE: Steel pipe

*Pipe—Handling**
USE: Materials handling

*Pipe—Lining**
USE: Pipe linings

*Pipe—Supports**
USE: Supports

Pipeline bends 619.1
DT: January 1993
UF: Bends (in pipelines)
Pipelines—Bends*
BT: Pipelines

Pipeline changeover 619.1
DT: January 1993
UF: Changeover (pipelines)
Gas pipelines—Changeover*
RT: Gas pipelines
Natural gas pipelines
Petroleum pipelines
Pipelines

Pipeline codes 619.1, 902.2
SN: Regulations
DT: January 1993
UF: Pipelines—Codes*
BT: Codes (standards)
RT: Laws and legislation

Pipeline compressor stations 618.1, 619.1
DT: January 1993
UF: Pipelines—Compressor stations*
BT: Pipelines
RT: Compressors

Pipeline joints
USE: Pipe joints AND
Pipelines

Pipeline laying 619.1 (674)
DT: January 1993
UF: Laying (pipelines)
Pipelines—Laying*
RT: Pipelines

Pipeline linings
USE: Pipelines AND
Pipe linings

Pipeline processing systems 722.4
DT: January 1993
UF: Computer systems, Digital—Pipeline processing*
BT: Multiprocessing systems
RT: Computer software
Multiprocessing programs
Parallel processing systems

Pipeline surges 631.1 (619.1)
DT: January 1993
UF: Pipelines—Surges*
BT: Surges (fluid)
RT: Pipe flow
Pipelines
Surge tanks

Pipeline terminals (402) (619.1)
DT: January 1993
UF: Pipelines—Terminals*
Terminals (pipeline)
BT: Facilities
RT: Oil fields
Pipelines

Pipelines 619.1
SN: Stationary means used to transport fluids over relatively great distances. For systems of pipes to distribute fluids throughout a building, building complex, or other structure or facility of limited extent, use PIPING SYSTEMS
DT: Predates 1975
UF: Aluminum pipelines
Bridges—Pipeline crossings*
Cast iron pipelines
Concrete pipelines
Copper pipelines
Heating—Pipelines*
Heating—Piping systems
Pipeline joints
Pipeline linings
Pipelines, Aluminum*
Pipelines, Cast iron*
Pipelines, Concrete*
Pipelines, Copper*
Pipelines, Steel*
Railroad plant and structures—Pipeline crossings*
Steel pipelines
NT: Extension pipelines
Gas pipelines
High pressure pipelines
Natural gas pipelines
Offshore pipelines
Petroleum pipelines
Pipeline bends
Pipeline compressor stations
Slurry pipelines
Steam pipelines
Temporary pipelines
Water pipelines
RT: Crossings (pipe and cable)
Leakage (fluid)
Location
Materials handling
Packing
Pipe
Pipe flow
Pipe joints
Pipe laying barges
Pipe linings
Pipeline changeover
Pipeline laying
Pipeline surges
Pipeline terminals
Piping systems
Pumping plants
Purging
Rights of way
Rupture disks
Trenching
Valves (mechanical)

*Pipelines, Aluminum**
USE: Aluminum pipe AND
Pipelines

*Pipelines, Cast iron**
USE: Pipelines AND
Cast iron pipe

*Pipelines, Concrete**
USE: Concrete pipe AND
Pipelines

*Pipelines, Copper**
USE: Pipelines AND
Copper pipe

*Pipelines, Steel**
USE: Steel pipe AND
Pipelines

*Pipelines, Submarine**
USE: Submarine pipelines

*Pipelines—Bends**
USE: Pipeline bends

*Pipelines—Bridge crossings**
USE: Crossings (pipe and cable)

*Pipelines—Codes**
USE: Pipeline codes

*Pipelines—Compressor stations**
USE: Pipeline compressor stations

*Pipelines—Expansion joints**
USE: Expansion joints

*Pipelines—Extension**
USE: Extension pipelines

*Pipelines—Flow**
USE: Pipe flow

*Pipelines—Induced currents**
USE: Induced currents

*Pipelines—Joints**
USE: Pipe joints

*Pipelines—Laying**
USE: Pipeline laying

*Pipelines—Leakage**
USE: Leakage (fluid)

*Pipelines—Lining**
USE: Pipe linings

*Pipelines—Location**
USE: Location

*Pipelines—Maps**
USE: Maps

*Pipelines—Offshore**
USE: Offshore pipelines

*Pipelines—Packing**
USE: Packing

*Pipelines—Pressure regulation**
USE: Pressure regulation

*Pipelines—Pumping stations**
USE: Pumping plants

*Pipelines—Purging**
USE: Purging

*Pipelines—Relocation**
USE: Relocation

*Pipelines—Right of way**
USE: Rights of way

*Pipelines—River crossings**
USE: Crossings (pipe and cable)

*Pipelines—Rupture disks**
USE: Rupture disks

*Pipelines—Supports**
USE: Supports

*Pipelines—Surges**
USE: Pipeline surges

*Pipelines—Temporary installations**
USE: Temporary pipelines

*Pipelines—Terminals**
USE: Pipeline terminals

*Pipelines—Thawing**
USE: Thawing

*Pipelines—Trenching**
USE: Trenching

Piping systems 619.1 (402)
SN: Systems of pipes to distribute fluids throughout a building, building complex, or other structure or facility of limited extent. For stationary means of transporting fluids over much greater distances, use PIPELINES
DT: Predates 1975
NT: Gas piping systems
High pressure piping
Refrigerating piping systems
Steam piping systems
Water piping systems
RT: Pipe
Pipelines

Pistols
USE: Guns (armament)

Piston pumps
USE: Reciprocating pumps

Piston rings 612.1.1
SN: Separable parts used to enhance the gas seal between piston and cylinder wall
DT: Predates 1975
UF: Internal combustion engines—Piston rings*
BT: Rings (components)
Seals
RT: Internal combustion engines
Packing
Pistons

Pistons 612.1.1
DT: Predates 1975
BT: Components
NT: Engine pistons
RT: Connecting rods
Piston rings
Pumps
Refrigerating machinery
Shock absorbers

Pitting 539.1
DT: January 1993
UF: Corrosion—Pitting*
BT: Corrosion
RT: Chemical attack

PL/1 (programming language) 723.1.1
DT: January 1993
UF: Computer programming languages—PL/1*
BT: High level languages

Placer mining 502.1
DT: January 1993
UF: Mines and mining—Placers and placering*
BT: Mining

Placers (482.2)
DT: January 1993
UF: Mines and mining—Placers and placering*
BT: Ore deposits

Placing (concrete)
USE: Concrete placing

Planers 603.1
DT: January 1993
BT: Machine tools
NT: Metal working planers
RT: Milling machines

*Planers, Metal working**
USE: Metal working planers

Planetariums 657.2 (402.2)
DT: Predates 1975
BT: Facilities
RT: Astronomy
Museums

Planetary landers 655.1
DT: January 1993
UF: Landers (planetary)
Spacecraft—Planetary landers*
BT: Spacecraft
RT: Interplanetary spacecraft
Interplanetary flight
Lunar landing

Planetary space flight
USE: Interplanetary flight

Planetary spacecraft
USE: Interplanetary spacecraft

Planets 657.2
DT: January 1986
BT: Solar system
RT: Extraterrestrial atmospheres
Interplanetary flight

Planimeters 943.1
DT: Predates 1975
UF: Area measuring instruments
BT: Instruments
RT: Spatial variables measurement
Surface measurement

Planning 912.2 (403)
DT: January 1993
NT: Highway planning
Motion planning
Regional planning
Strategic planning
RT: Design
Forecasting
Management
Scheduling
Statistical methods

Plant cell culture (461.8) (801.2)
DT: January 1993
UF: Cell culture—Plant*
BT: Cell culture
RT: Plants (botany)

Plant expansion (402.1) (912.1)
DT: January 1993
UF: Expansion (industrial plants)
Industrial plants—Expansion*
RT: Industrial plants

Plant layout (912.1) (913.1) (402.1)
DT: January 1993
UF: Industrial plants—Layout*
Layout (industrial plants)
RT: Industrial plants

Plant life extension (402.1) (912.1)
DT: January 1993
UF: Industrial plants—Life extension*
Life extension (industrial plants)
RT: Industrial plants
Service life

Plant location
USE: Location

Plant management 912.2
DT: January 1993
UF: Factory management
BT: Industrial management
RT: Industrial plants
Process control
Production control

Plant shutdowns (402.1) (912.1)
DT: January 1993
UF: Closings (industrial plants)
Industrial plants—Shutdown*
Shutdowns (industrial plants)
RT: Industrial plants
Reactor shutdowns

Plant site selection
USE: Site selection

Plant startup (402.1) (912.1)
DT: January 1993
UF: Industrial plants—Startup problems*
Start up (industrial plants)
Startup (industrial plants)
RT: Industrial plants
Reactor startup

Plants (botany) (461.9)
DT: January 1993
NT: Algae
Bamboo
Fungi
RT: Biology
Biomass
Plant cell culture

Plasma accelerators 932.1.1, 932.3
DT: January 1993
UF: Plasma devices—Accelerators*
BT: Plasma devices
RT: Plasma applications

Plasma applications 932.3
DT: January 1993
BT: Applications
NT: Plasma arc cutting
Plasma arc melting
Plasma diagnostics
Plasma spraying
Plasma welding
RT: Plasma accelerators
Plasma devices
Plasma diodes
Plasma display devices
Plasma filled waveguides
Plasma guns
Plasma jets
Plasma torches
Plasmas

Plasma arc cutting 604.1, 932.3
DT: January 1977
BT: Cutting
Plasma applications
RT: Metal cutting
Plasma arc melting
Plasma torches

Plasma arc melting 534.2, 932.3
DT: January 1993
UF: Metal melting—Plasma arc*
Plasma arc remelting
Steelmaking—Plasma arc remelting*
BT: Metal melting
Plasma applications
RT: Plasma arc cutting
Steelmaking

Plasma arc remelting
USE: Plasma arc melting

Plasma arc torches
USE: Plasma torches

Plasma arc welding
USE: Plasma welding

Plasma beam interactions
USE: Beam plasma interactions

Plasma collision processes 932.3
DT: January 1993
UF: Collision processes (plasmas)
Plasmas—Collision processes*
RT: Plasmas

Plasma confinement 932.3
DT: January 1993
UF: Confinement (plasmas)
Plasma containment
Plasmas—Confinement*
RT: Cyclotron resonance
Fusion reactors
Magnetic fields
Pinch effect
Plasma devices
Plasma oscillations
Plasma sheaths
Plasma shock waves
Plasma stability
Plasmas
Tokamak devices

Plasma containment
USE: Plasma confinement

Plasma density 932.3
DT: January 1993
UF: Plasmas—Density*
BT: Density of gases
RT: Plasma diagnostics
Plasmas

Plasma devices 932.3
DT: Predates 1975
BT: Equipment
NT: Plasma accelerators
Plasma diodes
Plasma display devices
Plasma filled waveguides
Plasma guns
Plasma torches
Tokamak devices
RT: Magnetohydrodynamic converters
Plasma applications
Plasma confinement
Plasma jets
Plasmas

*Plasma devices—Accelerators**
USE: Plasma accelerators

*Plasma devices—Diodes**
USE: Plasma diodes

*Plasma devices—Guns**
USE: Plasma guns

*Plasma devices—Jets**
USE: Plasma jets

*Plasma devices—Probes**
USE: Plasma probes

*Plasma devices—Torches**
USE: Plasma torches

Plasma diagnostics 932.3
DT: January 1993
UF: Diagnostics (plasma)
Plasmas—Diagnostics*
BT: Measurements
Plasma applications
RT: Plasma density
Plasma probes
Plasmas

Plasma diodes 714.1, 932.3
DT: January 1993
UF: Plasma devices—Diodes*
BT: Diodes
Direct energy converters
Plasma devices
RT: Plasma applications
Plasmas

Plasma display devices 932.3 (722.2)
DT: January 1993
UF: Display devices—Plasma*
BT: Display devices
Plasma devices
RT: Plasma applications
Plasmas

Plasma etching 802.2, 932.3 (531) (714.2)
DT: January 1995
UF: Dry plasma etching
BT: Etching
RT: Integrated circuit manufacture

Plasma filled waveguides 714.3, 932.3
DT: January 1993
UF: Plasma waveguides
Waveguides—Plasma filled*
BT: Plasma devices
Waveguides

* Former Ei Vocabulary term

Plasma filled waveguides *(continued)*
RT: Plasma applications
Plasmas

Plasma flow 932.3 (631.1.2)
DT: January 1993
UF: Flow of fluids—Plasmas*
Plasmas—Flow*
BT: Flow of fluids
RT: Plasma shock waves
Plasma turbulence
Plasmas

Plasma guns 932.3
DT: January 1993
UF: Guns (plasma)
Plasma devices—Guns*
BT: Plasma devices
RT: Magnetic lenses
Plasma applications

Plasma heating 932.3 (642)
DT: January 1993
UF: Plasmas—Heating*
BT: Heating
RT: Cyclotron resonance
Fusion reactors
Ionization
Plasmas

Plasma instability
USE: Plasma stability

Plasma interactions 932.3
DT: January 1993
NT: Beam plasma interactions
Wave plasma interactions
RT: Plasmas

Plasma jet welding
USE: Plasma welding

Plasma jets 932.3 (631.1.2)
DT: January 1993
UF: Plasma devices—Jets*
BT: Jets
RT: Plasma applications
Plasma devices

Plasma oscillations 932.3
DT: January 1993
UF: Plasmas—Oscillations*
BT: Oscillations
RT: Plasma confinement
Plasma stability
Plasma waves
Plasmas

Plasma probes 932.3
DT: January 1993
UF: Plasma devices—Probes*
Plasmas—Probes*
BT: Probes
RT: Plasma diagnostics

Plasma sheaths 932.3
DT: January 1993
UF: Plasmas—Sheaths*
Sheaths (plasma)
RT: Missiles
Plasma confinement
Plasmas
Spacecraft

Plasma shock waves 932.3
DT: January 1993
UF: Plasmas—Shock waves*
BT: Plasma waves
Shock waves
RT: Plasma confinement
Plasma flow
Plasmas

Plasma simulation 932.3
DT: January 1993
UF: Plasmas—Simulation*
BT: Simulation
RT: Plasmas

Plasma sources 932.3
DT: January 1993
UF: Plasmas—Production*
RT: Ion sources
Plasmas

Plasma spraying 932.3, 813.1
DT: January 1993
UF: Protective coatings—Plasma spraying*
BT: Coating techniques
Plasma applications
Spraying
RT: Plasmas
Protective coatings
Sprayed coatings

Plasma stability 932.3
DT: January 1993
UF: Instability (plasma)
Plasma instability
Plasmas—Stability*
BT: Stability
RT: Plasma confinement
Plasma oscillations
Plasmas

Plasma theory 932.3
DT: January 1993
UF: Plasmas—Theory*
BT: Physics
Theory
RT: Plasmas

Plasma torches 932.3
DT: January 1993
UF: Arc torches (plasma)
Plasma arc torches
Plasma devices—Torches*
Torches (plasma)
BT: Plasma devices
RT: Plasma applications
Plasma arc cutting

Plasma turbulence 932.3 (631.1.2)
DT: January 1993
UF: Plasmas—Turbulence*

* Former Ei Vocabulary term

Plasma turbulence *(continued)*
BT: Turbulence
RT: Plasma flow
Plasmas

Plasma waveguides
USE: Plasma filled waveguides

Plasma waves 932.3
DT: January 1993
UF: Plasmas—Waves*
BT: Waves
NT: Plasma shock waves
RT: Plasma oscillations
Plasmas

Plasma welding 932.3, 538.2.1
DT: January 1993
UF: Arc plasma welding
Plasma arc welding
Plasma jet welding
Welding—Plasma arc*
Welding—Plasma jet*
BT: Plasma applications
Welding

Plasmas 932.3
DT: Predates 1975
NT: Laser produced plasmas
Magnetoplasma
Semiconductor plasmas
RT: Beam plasma interactions
Electric space charge
Electromagnetic wave propagation in plasma
Physics
Pinch effect
Plasma applications
Plasma collision processes
Plasma confinement
Plasma density
Plasma devices
Plasma diagnostics
Plasma diodes
Plasma display devices
Plasma filled waveguides
Plasma flow
Plasma heating
Plasma interactions
Plasma oscillations
Plasma sheaths
Plasma shock waves
Plasma simulation
Plasma sources
Plasma spraying
Plasma stability
Plasma theory
Plasma turbulence
Plasma waves
Semiconductor materials
Wave plasma interactions

*Plasmas—Beam-plasma interactions**
USE: Beam plasma interactions

*Plasmas—Collision processes**
USE: Plasma collision processes

*Plasmas—Confinement**
USE: Plasma confinement

*Plasmas—Density**
USE: Plasma density

*Plasmas—Diagnostics**
USE: Plasma diagnostics

*Plasmas—Flow**
USE: Plasma flow

*Plasmas—Heating**
USE: Plasma heating

*Plasmas—Laser-produced**
USE: Laser produced plasmas

*Plasmas—Magnetoplasma**
USE: Magnetoplasma

*Plasmas—Oscillations**
USE: Plasma oscillations

*Plasmas—Pinch effect**
USE: Pinch effect

*Plasmas—Probes**
USE: Plasma probes

*Plasmas—Production**
USE: Plasma sources

*Plasmas—Sheaths**
USE: Plasma sheaths

*Plasmas—Shock waves**
USE: Plasma shock waves

*Plasmas—Simulation**
USE: Plasma simulation

*Plasmas—Stability**
USE: Plasma stability

*Plasmas—Theory**
USE: Plasma theory

*Plasmas—Turbulence**
USE: Plasma turbulence

*Plasmas—Waves**
USE: Plasma waves

Plaster 414
DT: Predates 1975
UF: Concrete—Plaster adherence*
Gypsum plaster
BT: Materials
RT: Building materials
Cements
Ceramic materials
Gypsum
Stucco

Plastic adhesives 817.2
DT: January 1993
UF: Joints, Adhesive—Plastics adhesives*
Plastics—Adhesives*
BT: Adhesives
Plastic products
RT: Adhesive joints

* Former Ei Vocabulary term

*Plastic and rubber molds**
USE: Plastic molds OR
Rubber molds

Plastic bottles 694.2, 817.1
DT: January 1993
UF: Bottles—Plastics*
BT: Bottles
Plastic containers

Plastic building materials 415.2, 817.2
DT: January 1993
UF: Building materials—Plastics*
Plastics—Structural application*
BT: Building materials
Plastic products

Plastic castings 817.1 (816.1)
DT: January 1993
BT: Castings
RT: Plastics casting
Plastics molding

Plastic coatings 813.2, 817.2 (539.2.2)
DT: January 1993
UF: Protective coatings—Plastics*
Wire—Plastic coating*
BT: Coatings
Plastics
RT: Plastic films
Plastic products
Powder coatings

Plastic containers 817.1 (691) (694)
DT: January 1993
UF: Containers—Plastics*
BT: Containers
Plastic products
NT: Plastic bottles
RT: Plastics

Plastic deformation (421) (931.1)
DT: January 1995
BT: Deformation
NT: Plastic flow
RT: Ductility
Plasticity

Plastic filaments 817.1, 819.2
DT: January 1993
UF: Filaments (plastic)
Plastics—Filaments*
BT: Plastics
Synthetic fibers
RT: Filament winding
Spinning (fibers)
Wire

Plastic films 817.1
DT: January 1993
UF: Plastics films*
BT: Films
Plastics
NT: Cellophane
RT: Plastic coatings
Plastic products
Plastic sheets

Plastic flow (421) (931.1) (933)
DT: January 1995
UF: Flow stress
BT: Flow of solids
Plastic deformation
RT: Internal friction
Plasticity
Rheology
Shear flow

Plastic foams
USE: Foamed plastics

Plastic laminates 817.1 (415.2)
DT: January 1993
UF: Plastics laminates*
BT: Laminates
Plastics
RT: Plastic products

Plastic lenses 741.3, 817.1
DT: January 1993
UF: Lenses—Plastics*
BT: Lenses
Optical devices
Plastic products

Plastic molds 817.1
SN: Molds made of plastic; for molding plastic objects, use PLASTICS and MOLDS
DT: January 1993
UF: Plastic and rubber molds*
BT: Molds
RT: Plastic products
Plastics

Plastic parts 817.1
DT: January 1993
UF: Conveyors—Plastics parts*
Electric appliances—Plastics parts*
Machinery—Plastics parts*
BT: Components
Plastic products
RT: Machinery

Plastic pipe 619.1, 817.1
DT: January 1993
UF: Pipe, Plastic*
BT: Pipe
Plastic products

Plastic products 817.1
DT: October 1975
UF: Plastics products*
NT: Plastic adhesives
Plastic building materials
Plastic containers
Plastic lenses
Plastic parts
Plastic pipe
Plastic tapes
RT: Plastic coatings
Plastic films
Plastic laminates
Plastic molds
Plastic sheets
Plastics

* Former Ei Vocabulary term

Plastic products *(continued)*
Plastics industry
Plastics molding

Plastic sheets 817.1
DT: January 1993
UF: Plastics sheets*
Sheet plastics
BT: Plastics
RT: Plastic films
Plastic products
Sheet molding compounds

*Plastic sheets—Sheet molding compounds**
USE: Sheet molding compounds

Plastic tapes 817.1
DT: January 1993
UF: Plastics—Tapes*
BT: Plastic products
Tapes

Plasticity
DT: Predates 1975
BT: Mechanical properties
NT: Superplasticity
RT: Brittleness
Continuum mechanics
Creep
Deformation
Ductility
Elastoplasticity
Flow of solids
Formability
Hardness
High temperature testing
Plastic deformation
Plastic flow
Plasticity testing
Rheology
Strain
Viscoplasticity

Plasticity testing (422.2)
DT: January 1993
UF: Plasticity—Testing*
BT: Mechanical testing
RT: Low temperature testing
Plasticity

*Plasticity—Testing**
USE: Plasticity testing

Plasticizers 803 (816.1)
DT: Predates 1975
BT: Agents
RT: Additives
Plastics
Plastisols
Solvents
Surface active agents

Plastics 817.1
SN: Includes filled polymers or polymers processed into semifinished or finished products. For basic polymers, use POLYMERS. For specific applications, use PLASTICS APPLICATIONS and the name of the application, e.g., ROADBUILDING MATERIALS
DT: Predates 1975
UF: Industrial wastes—Plastics*
BT: Materials
NT: Conductive plastics
Foamed plastics
Plastic coatings
Plastic filaments
Plastic films
Plastic laminates
Plastic sheets
Plastisols
Reinforced plastics
RT: Adhesives
Antistatic agents
Blending
Blowing agents
Compounding (chemical)
Dielectric materials
Disposability
Glass transition
Heat stabilizers
Industrial wastes
Latexes
Mold release agents
Petrochemicals
Plastic containers
Plastic molds
Plastic products
Plasticizers
Plastics applications
Plastics casting
Plastics fillers
Plastics forming
Plastics industry
Plastics machinery
Plastics molding
Plastics plants
Polymer blends
Polymerization
Polymers

Plastics applications 817.2
DT: January 1993
UF: Aircraft materials—Plastics*
Automobile materials—Plastics*
Biomedical equipment—Plastics*
Chemical equipment—Plastics*
Dies—Plastics*
Domes and shells—Plastics*
Electric insulating materials—Plastics*
Heat insulating materials—Plastics*
Hose—Plastics*
Lighting fixtures—Plastics*
Packaging materials—Plastics*
Roadbuilding materials—Plastics*
Rolls—Plastics*
Shipbuilding materials—Plastics*
Sporting goods—Plastics*
Springs—Plastics*
Tools, jigs and fixtures—Plastics*
Toy manufacture—Plastics*
Windows—Plastics*

* Former Ei Vocabulary term

Plastics applications *(continued)*
Wood products—Plastics encasing*
BT: Applications
RT: Plastics

*Plastics blends**
USE: Polymer blends

Plastics casting 816.1
DT: January 1993
UF: Plastics—Casting*
BT: Casting
Plastics forming
RT: Plastic castings
Plastics
Plastics molding

Plastics extruders 816.2
DT: January 1993
BT: Extruders
Plastics machinery
RT: Extrusion molding

Plastics fillers 803 (804)
DT: January 1993
UF: Plastics—Fillers*
BT: Fillers
RT: Plastics

*Plastics films**
USE: Plastic films

Plastics forming 816.1
DT: January 1993
BT: Forming
NT: Plastics casting
Plastics molding
Thermoforming
RT: Cold forming machines
Compression molding
Extrusion molding
Injection molding
Plastics
Preforming
Pultrusion
Reaction injection molding

Plastics forming dies 816.2
DT: January 1993
UF: Plastics machinery—Dies and presses*
BT: Dies
Plastics machinery

Plastics industry (816) (817.1)
DT: Predates 1975
BT: Industry
RT: Chemical industry
Plastic products
Plastics
Plastics machinery
Plastics plants

*Plastics laminates**
USE: Plastic laminates

Plastics machinery 816.2
DT: Predates 1975
UF: Plastics machinery—Mills*
BT: Machinery
NT: Cold forming machines
Plastics extruders
Plastics forming dies
Plastics molding machines
Plastics presses
RT: Calendering
Calenders
Extruders
Granulators
Laminating machinery
Mixers (machinery)
Plastics
Plastics industry
Plastics plants

*Plastics machinery—Calenders**
USE: Calenders

*Plastics machinery—Cold forming machines**
USE: Cold forming machines

*Plastics machinery—Dies and presses**
USE: Plastics forming dies OR
Plastics presses

*Plastics machinery—Granulators**
USE: Granulators

*Plastics machinery—Laminating machines**
USE: Laminating machinery

*Plastics machinery—Mills**
USE: Plastics machinery

*Plastics machinery—Molding machines**
USE: Plastics molding machines

Plastics molding 816.1
DT: January 1993
UF: Plastics—Molding*
BT: Molding
Plastics forming
NT: Blow molding
Reactive molding
Rotational molding
Transfer molding
RT: Compression molding
Extrusion molding
Injection molding
Molds
Plastic castings
Plastic products
Plastics
Plastics casting
Reaction injection molding

Plastics molding machines 816.2
DT: January 1993
UF: Molding machines (plastics)
Plastics machinery—Molding machines*
BT: Plastics machinery
RT: Molds

Plastics plants 816.2 (402.1)
DT: Predates 1975
BT: Industrial plants
RT: Plastics

* Former Ei Vocabulary term

Plastics plants *(continued)*
Plastics industry
Plastics machinery

Plastics presses 816.2
DT: January 1993
UF: Plastics machinery—Dies and presses*
BT: Plastics machinery

*Plastics products**
USE: Plastic products

*Plastics sheets**
USE: Plastic sheets

*Plastics, Foamed**
USE: Foamed plastics

*Plastics, Foamed—Blowing agents**
USE: Blowing agents

*Plastics, Foamed—Flexible**
USE: Flexible foamed plastics

*Plastics, Foamed—Rigid**
USE: Rigid foamed plastics

*Plastics, Reinforced**
USE: Reinforced plastics

*Plastics, Reinforced—Carbon fiber**
USE: Carbon fiber reinforced plastics

*Plastics, Reinforced—Glass fiber**
USE: Glass fiber reinforced plastics

*Plastics, Reinforced—Graphite fiber**
USE: Graphite fiber reinforced plastics

*Plastics, Reinforced—Wire**
USE: Wire reinforced plastics

*Plastics—Adhesives**
USE: Plastic adhesives

*Plastics—Antistatic agents**
USE: Antistatic agents

*Plastics—Blending**
USE: Blending

*Plastics—Blow molding**
USE: Blow molding

*Plastics—Blowing agents**
USE: Blowing agents

*Plastics—Calendering**
USE: Calendering

*Plastics—Casting**
USE: Plastics casting

*Plastics—Compounding**
USE: Compounding (chemical)

*Plastics—Compression molding**
USE: Compression molding

*Plastics—Conductive**
USE: Conductive plastics

*Plastics—Disposability**
USE: Disposability

*Plastics—Electric heating**
USE: Electric heating

*Plastics—Extrusion molding**
USE: Extrusion molding

*Plastics—Filament winding**
USE: Filament winding

*Plastics—Filaments**
USE: Plastic filaments

*Plastics—Fillers**
USE: Plastics fillers

*Plastics—Finishing**
USE: Finishing

*Plastics—Flame resistance**
USE: Flame resistance

*Plastics—Fungus resistance**
USE: Fungus resistance

*Plastics—Glass transition**
USE: Glass transition

*Plastics—Heat stabilizers**
USE: Heat stabilizers

*Plastics—Laminating**
USE: Laminating

*Plastics—Latex**
USE: Latexes

*Plastics—Mold release agents**
USE: Mold release agents

*Plastics—Molding**
USE: Plastics molding

*Plastics—Molds**
USE: Molds

*Plastics—Permeability**
USE: Mechanical permeability

*Plastics—Photochromism**
USE: Photochromism

*Plastics—Photoreactivity**
USE: Photoreactivity

*Plastics—Preforming**
USE: Preforming

*Plastics—Pultrusion**
USE: Pultrusion

*Plastics—Reaction injection molding**
USE: Reaction injection molding

*Plastics—Reactive molding**
USE: Reactive molding

*Plastics—Rotational molding**
USE: Rotational molding

*Plastics—Sealing**
USE: Sealing (finishing)

*Plastics—Structural application**
USE: Plastic building materials

* Former Ei Vocabulary term

*Plastics—Tapes**
USE: Plastic tapes

*Plastics—Termite resistance**
USE: Termite resistance

*Plastics—Thermoforming**
USE: Thermoforming

*Plastics—Trademarks**
USE: Trademarks

*Plastics—Transfer molding**
USE: Transfer molding

*Plastics—Transparency**
USE: Transparency

*Plastics—Tropical applications**
USE: Tropics

*Plastics—Unsaturated**
USE: Unsaturated polymers

*Plastics—Winding**
USE: Winding

Plastisols 803 (801.3) (816.1)
DT: January 1987
BT: Plastics
Sols
RT: Plasticizers
Vinyl resins

Plate girder bridges 401.1
DT: January 1993
UF: Bridges, Plate girder*
BT: Bridges
RT: Beams and girders

Plate metal (415.1) (535.1.2)
SN: Semifabricated material
DT: January 1993
UF: Plates*
Rolling mill practice—Plate*
BT: Materials
RT: Armor
Billets (metal bars)
Blooms (metal)
Metals
Rolling mill practice
Rolling mills
Sheet metal

Plate tectonics
USE: Tectonics

Plated wire storage 722.1 (708.4)
DT: January 1993
UF: Data storage, Magnetic—Plated wire*
BT: Magnetic film storage

Plates (structural components) 408.2
DT: January 1993
UF: Heat transfer—Plates*
Plates*
BT: Structural members
RT: Beams and girders
Disks (structural components)

*Plates**
USE: Plates (structural components) OR
Plate metal

Plating 539.3
DT: January 1993
UF: Metal plating
Metals and alloys—Plating*
BT: Coating techniques
NT: Aluminum plating
Bismuth plating
Brass plating
Bronze plating
Cadmium plating
Chromium plating
Cobalt plating
Copper plating
Electroless plating
Electroplating
Gold plating
Indium plating
Iridium plating
Iron plating
Lead plating
Molybdenum plating
Nickel plating
Rhenium plating
Rhodium plating
Silver plating
Tinning
Titanium plating
Tungsten plating
Zinc plating
RT: Alloys
Cladding (coating)
Electrodeposition
Hard facing
Laminates
Metallizing
Metals
Substrates
Surface treatment

Platinotron tubes
USE: Microwave tubes

Platinum 547.1
DT: January 1993
UF: Platinum and alloys*
BT: Platinum metals
RT: Platinum alloys
Platinum compounds
Platinum mines
Platinum ore treatment

Platinum alloys 547.1
DT: January 1993
UF: Platinum and alloys*
BT: Platinum metal alloys
RT: Platinum
Platinum compounds

*Platinum and alloys**
USE: Platinum OR
Platinum alloys

Platinum compounds (804.1) (804.2)
DT: Predates 1975
BT: Platinum metal compounds
RT: Intermetallics
Platinum
Platinum alloys

Platinum metal alloys 547.1
DT: January 1993
BT: Precious metal alloys
NT: Iridium alloys
Osmium alloys
Palladium alloys
Platinum alloys
Rhodium alloys
Ruthenium alloys
RT: Platinum metal compounds
Platinum metals

Platinum metal compounds (804.1) (804.2)
DT: January 1993
BT: Precious metal compounds
NT: Iridium compounds
Osmium compounds
Palladium compounds
Platinum compounds
Rhodium compounds
Ruthenium compounds
RT: Platinum metal alloys
Platinum metals

Platinum metals 547.1
DT: Predates 1975
BT: Precious metals
NT: Iridium
Osmium
Palladium
Platinum
Rhodium
Ruthenium
RT: Platinum metal alloys
Platinum metal compounds

Platinum mines 504.3, 547.1
DT: January 1993
UF: Platinum mines and mining*
BT: Mines
RT: Platinum

*Platinum mines and mining**
USE: Platinum mines

Platinum ore treatment 533.1, 547.1
DT: January 1986
BT: Ore treatment
RT: Platinum

PLC
USE: Programmable logic controllers

Plethysmography 461.6
DT: January 1993
UF: Biomedical engineering—Plethysmography*
BT: Diagnosis
RT: Cardiology
Cardiovascular system

PLL
USE: Phase locked loops

Plotters (computer) 722.2
DT: January 1993
UF: Computer peripheral equipment—Plotters*
Computer plotters
BT: Computer peripheral equipment
Recording instruments
RT: Digital to analog conversion
Printers (computer)

Plugs (electric)
USE: Electric connectors

Plumbing 619.1 (402) (446.1)
DT: Predates 1975
RT: Plumbing fixtures
Water hammer
Water pipelines
Water piping systems

Plumbing fixtures 619.1.1
DT: January 1993
UF: Fixtures (plumbing)
Plumbing—Fixtures*
BT: Equipment
RT: Plumbing

*Plumbing—Fixtures**
USE: Plumbing fixtures

Plumes (thermal)
USE: Thermal plumes

Plutonium 622.1
DT: January 1993
UF: Plutonium and alloys*
Pu
BT: Transuranium elements
RT: Plutonium alloys
Plutonium compounds

Plutonium alloys 622.1
DT: January 1993
UF: Plutonium and alloys*
BT: Alloys
RT: Plutonium
Plutonium compounds

*Plutonium and alloys**
USE: Plutonium OR
Plutonium alloys

Plutonium compounds (804.1) (804.2)
DT: Predates 1975
BT: Metallic compounds
RT: Plutonium
Plutonium alloys

Plywood 415.3, 811.2
DT: Predates 1975
BT: Wood laminates
RT: Building materials
Veneers

Pneumatic control 632.3, 731.1
DT: January 1993
UF: Industrial plants—Pneumatic control*

* Former Ei Vocabulary term

Pneumatic control *(continued)*
Ship equipment—Pneumatic control*
BT: Control
RT: Pneumatic control equipment

Pneumatic control equipment 632.4, 732.1
DT: January 1993
UF: Control equipment, Pneumatic*
Ships—Pneumatic control equipment*
BT: Mechanical control equipment
Pneumatic equipment
NT: Pneumatic servomechanisms
RT: Pneumatic control
Pneumatics

Pneumatic conveyors 632.4, 692.1
DT: January 1993
UF: Conveyors—Pneumatic*
BT: Conveyors
Pneumatic materials handling equipment
RT: Pneumatic tubes

Pneumatic drills 632.4 (405.1) (502.2)
DT: January 1993
UF: Rock drills—Pneumatic*
BT: Drills
Pneumatic equipment

*Pneumatic drive**
USE: Pneumatic drives

*Pneumatic drive—Variable speed**
USE: Variable speed drives

Pneumatic drives 632.4
DT: January 1993
UF: Pneumatic drive*
BT: Drives
Pneumatic equipment
RT: Pneumatic propulsion
Variable speed drives

Pneumatic equipment 632.4
DT: January 1993
UF: Aircraft—Pneumatic equipment*
Automobiles—Pneumatic equipment*
Breakwaters—Pneumatic*
Rockets and missiles—Pneumatic equipment*
Scales and weighing—Pneumatic*
Separators—Pneumatic*
Ship equipment—Pneumatic*
Springs—Pneumatic*
Typewriters—Pneumatic*
BT: Equipment
NT: Air brakes
Pneumatic control equipment
Pneumatic drills
Pneumatic drives
Pneumatic gages
Pneumatic materials handling equipment
Pneumatic tools
RT: Pneumatics

Pneumatic gages 632.4, 943.3
DT: January 1993
UF: Gages—Pneumatic*
BT: Gages
Pneumatic equipment

Pneumatic materials handling equipment 632.4, 691.1
DT: January 1993
UF: Materials handling—Pneumatic*
BT: Materials handling equipment
Pneumatic equipment
NT: Pneumatic conveyors
Pneumatic tubes

Pneumatic propulsion 632.4
DT: January 1993
UF: Propulsion—Pneumatic energy*
BT: Propulsion
RT: Pneumatic drives

Pneumatic servomechanisms 632.4, 732.1
DT: January 1993
UF: Servomechanisms—Pneumatic*
BT: Pneumatic control equipment
Servomechanisms

Pneumatic steelmaking 545.3, 632.3
DT: January 1993
UF: Steelmaking—Pneumatic process*
BT: Steelmaking
RT: Pneumatics

Pneumatic tools 632.4
DT: January 1993
UF: Tools, jigs and fixtures—Pneumatic*
BT: Pneumatic equipment
Tools

Pneumatic tubes 632.4
DT: Predates 1975
BT: Pneumatic materials handling equipment
Tubes (components)
RT: Pneumatic conveyors

Pneumatics 632.3
DT: October 1975
BT: Fluid mechanics
RT: Compressed air
Pneumatic control equipment
Pneumatic equipment
Pneumatic steelmaking

Po
USE: Polonium

Pocket calculators (715.1) (722.4) (921)
DT: January 1993
UF: Mathematical instruments—Pocket calculators*
BT: Mathematical instruments
RT: Computers
Microcomputers

Point contacts 704.1, 714.2
DT: January 1993
UF: Electric contacts, Point*
BT: Electric contacts
RT: Crystal whiskers

Point defects 933.1.1
DT: January 1995
BT: Crystal defects
NT: Color centers

* Former Ei Vocabulary term

Point defects *(continued)*
RT: Crystal impurities
Crystal lattices

Point groups
USE: Crystal symmetry

Point of sale terminals 722.2
DT: January 1993
UF: Business machines—Point of sale terminals*
POS terminals
Terminals (point of sale)
BT: Business machines
Computer terminals
RT: Cash registers
Marketing
Smart cards

Pointing (metal finishing)
USE: Wire pointing

Poison gas warfare
USE: Chemical warfare

Poisoning (catalysts)
USE: Catalyst poisoning

Poisons (industrial)
USE: Industrial poisons

Polarimeters 941.3
DT: Predates 1975
UF: Optical polarimeters
Polarimetry
BT: Optical instruments
RT: Chemical analysis
Ellipsometry
Light polarization
Photometers
Polariscopes
Polarization
Polarographic analysis

Polarimetry
USE: Polarimeters

Polarisation
USE: Polarization

Polariscopes 941.3
DT: Predates 1975
BT: Optical instruments
RT: Chemical analysis
Polarimeters

Polarization (701.1) (711.1) (741.1)
DT: January 1993
UF: Polarisation
BT: Physical properties
NT: Anodic polarization
Electromagnetic wave polarization
RT: Electrooptical effects
Ferroelectricity
Magnetooptical effects
Polarimeters

Polarized light
USE: Light polarization

Polarographic analysis (801)
DT: Predates 1975
UF: Polarography
BT: Chemical analysis
RT: Electrochemistry
Polarimeters
Polarographs
Volumetric analysis

Polarographs 942.1
DT: Predates 1975
BT: Electric measuring instruments
RT: Coulometers
Polarographic analysis

Polarography
USE: Polarographic analysis

Poles 408.2
SN: Supports
DT: Predates 1975
BT: Supports
RT: Columns (structural)
Electric line supports
Electric lines
Piles
Street lighting
Structural members
Towers

Poles and zeros 921 (731.1)
DT: January 1993
UF: Mathematical techniques—Poles and zeros*
BT: Mathematical techniques
RT: Polynomials
Root loci
Transfer functions

*Poles—Composite materials**
USE: Composite materials

*Poles—Concrete**
USE: Concrete construction

*Poles—Wood**
USE: Wood products

Policy
USE: Public policy

Polishing 604.2
DT: Predates 1975
BT: Finishing
Surface treatment
NT: Chemical polishing
Lapping
RT: Buffing
Descaling
Lenses
Mirrors
Polishing machines

Polishing machines 603.1
DT: Predates 1975
BT: Machine tools
RT: Buffing machines
Lapping machines
Polishing

* Former Ei Vocabulary term

*Polishing—Chemical**
USE: Chemical polishing

*Polishing—Electrolytic**
USE: Electrolytic polishing

Pollution (454.2)
SN: Prefer a term for the specific type of pollution if appropriate
DT: January 1993
NT: Air pollution
Noise pollution
Oil pollution of refrigerants
Soil pollution
Thermal pollution
Water pollution
RT: Contamination
Fiber optic sensors
Health hazards
Lasers
Optical instruments
Pesticide effects
Pollution control
Pollution detection
Pollution induced corrosion
Toxicity

Pollution control (454.2)
DT: January 1993
BT: Control
NT: Air pollution control
Noise pollution control
Soil pollution control
Water pollution control
RT: Environmental engineering
Landfill linings
Pollution
Pollution control equipment
Pollution detection
Remote sensing

Pollution control equipment (454.2)
DT: January 1993
BT: Control equipment
NT: Air pollution control equipment
Noise pollution control equipment
Water pollution control equipment
RT: Pollution control

Pollution detection (454.2)
DT: January 1993
RT: Detectors
Pollution
Pollution control

Pollution induced corrosion 539.1 (454.2)
DT: January 1993
UF: Corrosion—Pollution induced*
BT: Corrosion
RT: Air pollution
Pollution
Water pollution

Polonium 622.1
DT: Predates 1975
UF: Po
BT: Metalloids
RT: Polonium alloys
Polonium compounds

Polonium alloys 622.1
DT: January 1993
BT: Alloys
RT: Polonium
Polonium compounds

Polonium compounds (804.1) (804.2)
DT: January 1993
BT: Chemical compounds
RT: Polonium
Polonium alloys

*Polyacenes**
USE: Aromatic polymers

Polyacenic materials
USE: Aromatic polymers

Polyacetals
USE: Acetal resins

Polyacetylenes 815.1.1
DT: January 1987
BT: Organic polymers

Polyacrylates 815.1.1
DT: January 1987
BT: Acrylics

Polyacrylonitrile
USE: Polyacrylonitriles

Polyacrylonitriles 815.1.1
DT: June 1990
UF: Polyacrylonitrile
BT: Acrylics

Polyamide-imides
USE: Polyamideimides

Polyamideimides 815.1.1
DT: January 1987
UF: PAI
Polyamide-imides
BT: Aromatic polymers
RT: Polyamides
Polyimides

Polyamides 815.1.1
DT: Predates 1975
UF: PA
BT: Thermoplastics
NT: Nylon polymers
Polysulfonamides
RT: Polyamideimides

Polybutadienes 815.1.1 (818.2.1)
DT: January 1977
BT: Synthetic rubber

Polybutene
USE: Polybutenes

Polybutenes 815.1.1 (818.2.1)
DT: June 1990
UF: Polybutene
Polybutylenes
BT: Polyolefins

* Former Ei Vocabulary term

Polybutenes *(continued)*
RT: Polyethylenes
Polypropylenes

Polybutylenes
USE: Polybutenes

Polycarbonates 815.1.1
DT: Predates 1975
BT: Thermoplastics
RT: Polyesters

Polychlorinated biphenyls 804.1
DT: January 1986
UF: PCB
BT: Organic compounds
RT: Liquid insulating materials

Polycondensation 815.2 (802.2)
DT: January 1995
UF: Condensation polymerization
BT: Condensation
Polymerization
RT: Anionic polymerization
Thermoplastic elastomers

Polycrystalline materials 933.1
DT: January 1995
BT: Crystalline materials
RT: Polycrystals

Polycrystals 933.1
DT: January 1995
BT: Crystals
RT: Crystal structure
Polycrystalline materials

Polyelectrolytes 815.1.1, 817.1
DT: January 1977
BT: Electrolytes
Organic polymers
RT: Ion exchangers

Polyesters 815.1.1
DT: Predates 1975
BT: Thermoplastics
NT: Polyethylene terephthalates
RT: Polycarbonates

Polyether ether ketones 815.1.1
DT: January 1995
UF: PEEK
BT: Thermoplastics
RT: Polyethers

Polyether glycol
USE: Polyethylene glycols

Polyetherimides 815.1.1
DT: January 1987
UF: PEI
BT: Thermoplastics
RT: Polyethers
Polyimides

Polyethers 815.1.1
DT: Predates 1975
BT: Organic polymers
NT: Acetal resins
Crown ethers
Epoxy resins
Phenoxy resins
Polyethylene glycols
Polyethylene oxides
Polyphenylene oxides
Polypropylene oxides
RT: Polyether ether ketones
Polyetherimides

Polyethylene glycols 815.1.1
DT: January 1995
UF: PEG
Polyether glycol
Polyoxyethylenes
BT: Polyethers
RT: Polyethylene oxides

Polyethylene oxides 815.1.1
DT: January 1995
UF: PEO
BT: Polyethers
RT: Polyethylene glycols

Polyethylene terephthalate
USE: Polyethylene terephthalates

Polyethylene terephthalates 815.1.1
DT: January 1987
UF: PET
PETP
Polyethylene terephthalate
BT: Aromatic polymers
Polyesters

Polyethylenes 815.1.1
DT: January 1993
UF: PE
BT: Polyolefins
NT: High density polyethylenes
Low density polyethylenes
Medium density polyethylenes
Ultrahigh molecular weight polyethylenes
RT: Ethylene
Polybutenes
Polypropylenes

*Polyethylenes—High density**
USE: High density polyethylenes

*Polyethylenes—Linear low density**
USE: Linear low density polyethylenes

*Polyethylenes—Low density**
USE: Low density polyethylenes

*Polyethylenes—Medium density**
USE: Medium density polyethylenes

*Polyethylenes—Ultrahigh molecular weight**
USE: Ultrahigh molecular weight polyethylenes

Polyimides 815.1.1
DT: Predates 1975
UF: PI
BT: Organic polymers
RT: Polyamideimides
Polyetherimides

* Former Ei Vocabulary term

Polyisoprene
USE: Polyisoprenes

Polyisoprenes 815.1.1 (818.2.1)
DT: January 1987
UF: Polyisoprene
BT: Thermoplastic elastomers
RT: Synthetic rubber

Polymer blends 817.1 (815.1) (816.1)
DT: January 1993
UF: Blends (plastic)
Plastics blends*
BT: Organic polymers
NT: Interpenetrating polymer networks
RT: Blending
Graft copolymers
Plastics
Textile blends

Polymer membrane electrodes (704.1) (801.4.1) (802.1) (817.2)
DT: January 1993
UF: Electrodes, Electrochemical—Polymer membrane*
BT: Electrochemical electrodes
RT: Polymeric membranes
Sensors

Polymeric glass 815.1, 933.2
DT: January 1993
UF: Glass, Polymeric*
RT: Polymers

Polymeric materials
USE: Polymers

Polymeric membranes 817.1 (802.1)
DT: January 1993
UF: Membranes—Polymeric*
BT: Membranes
RT: Polymer membrane electrodes

Polymerization 815.2
DT: Predates 1975
UF: Photopolymerization
BT: Chemical reactions
NT: Anionic polymerization
Cationic polymerization
Copolymerization
Electropolymerization
Emulsion polymerization
Free radical polymerization
Polycondensation
Ring opening polymerization
Terpolymerization
RT: Crosslinking
Curing
Monomers
Oligomers
Plastics
Polymers

*Polymerization—Anionic polymerization**
USE: Anionic polymerization

*Polymerization—Cationic polymerization**
USE: Cationic polymerization

*Polymerization—Coordination reactions**
USE: Coordination reactions

*Polymerization—Copolymerization**
USE: Copolymerization

*Polymerization—Free radical polymerization**
USE: Free radical polymerization

*Polymerization—Ring opening**
USE: Ring opening polymerization

*Polymerization—Terpolymerization**
USE: Terpolymerization

Polymers 815.1
SN: Basic polymers only. Excludes filled polymers and polymers processed into products, for which use PLASTICS or PLASTIC PRODUCTS
DT: Predates 1975
UF: Polymeric materials
Resins
BT: Chemical compounds
Materials
NT: Inorganic polymers
Liquid crystal polymers
Organic polymers
RT: Macromolecules
Molecular weight distribution
Monomers
Optical materials
Organic compounds
Plastics
Polymeric glass
Polymerization
Vulcanization

Polymethyl methacrylate
USE: Polymethyl methacrylates

Polymethyl methacrylates 815.1.1
DT: June 1990
UF: Polymethyl methacrylate
BT: Acrylics

Polynomials 921.1
DT: January 1993
UF: Mathematical techniques—Polynomials*
BT: Algebra
RT: Poles and zeros

Polyolefins 815.1.1
DT: Predates 1975
BT: Thermoplastics
NT: Polybutenes
Polyethylenes
Polypropylenes
RT: Olefins
Synthetic fibers

Polyoxyethylenes
USE: Polyethylene glycols

Polypeptides 804.1, 815.1.1
DT: January 1977
BT: Natural polymers
RT: Amino acids
Proteins

* Former Ei Vocabulary term

Polyphenylene oxides 815.1.1
DT: Predates 1975
UF: PPO
BT: Polyethers

Polypropylene
USE: Polypropylenes

Polypropylene oxides 815.1.1
DT: June 1990
BT: Polyethers
RT: Propylene

Polypropylenes 815.1.1
DT: June 1990
UF: Polypropylene
PP
BT: Polyolefins
RT: Polybutenes
Polyethylenes
Propylene

Polysaccharides 804.1, 815.1.1
DT: January 1986
BT: Carbohydrates
Natural polymers
NT: Cellulose
Chitin
Starch

Polysilanes 815.1.2
DT: June 1990
BT: Inorganic polymers
Silicon compounds
RT: Silicones

Polysiloxanes
USE: Silicones

Polystyrenes 815.1.1
DT: Predates 1975
UF: PS
BT: Aromatic polymers
Thermoplastics
RT: Styrene

Polysulfide rubber
USE: Polysulfides

Polysulfides 815.1.1, 818.2.1
DT: Predates 1975
UF: Polysulfide rubber
BT: Synthetic rubber

Polysulfonamides 815.1.1
DT: January 1987
BT: Polyamides
RT: Polysulfones

Polysulfones 815.1.1
DT: Predates 1975
BT: Thermoplastics
RT: Polysulfonamides

Polytetrafluoroethylene
USE: Polytetrafluoroethylenes

Polytetrafluoroethylenes 815.1.1
DT: January 1987
UF: Polytetrafluoroethylene
PTFE
Teflon
BT: Fluorine containing polymers

Polyurethane resins
USE: Polyurethanes

Polyurethanes 815.1.1
DT: Predates 1975
UF: Polyurethane resins
BT: Organic polymers

Polyvinyl acetate
USE: Polyvinyl acetates

Polyvinyl acetates 815.1.1
DT: June 1990
UF: Polyvinyl acetate
PVAC
BT: Vinyl resins

Polyvinyl alcohol
USE: Polyvinyl alcohols

Polyvinyl alcohols 815.1.1
DT: January 1987
UF: Polyvinyl alcohol
PVAL
BT: Vinyl resins

Polyvinyl chloride
USE: Polyvinyl chlorides

Polyvinyl chlorides 815.1.1
DT: June 1990
UF: Polyvinyl chloride
PVC
BT: Vinyl resins

Polyvinyl resins
USE: Vinyl resins

Polyvinylidene chlorides 815.1.1
DT: January 1987
UF: PVDC
BT: Vinyl resins

Ponding (402)
SN: Accumulation of water on flat roofs
DT: January 1993
UF: Roofs—Ponding*
RT: Drainage
Leakage (fluid)
Roofs

Pontoon dredges (405) (407)
DT: January 1993
BT: Dredges
RT: Pontoons

Pontoons (401.1) (407) (434) (673)
SN: Flat-bottomed boats or floats
DT: Predates 1975
BT: Hydraulic structures
RT: Pontoon dredges
Port structures
Salvaging
Seaplanes
Waterway transportation

* Former Ei Vocabulary term

Pontryagin maximum principle
USE: Maximum principle

Population statistics (922.2)
DT: January 1993
BT: Statistics
RT: Regional planning
Social aspects

Porcelain 812.1
DT: Predates 1975
BT: Ceramic materials

Pore pressure 483.1
DT: January 1993
UF: Soils—Pore pressure*
BT: Pressure
RT: Soils

Pore size distribution
USE: Porous materials

Porosimeters (481.1) (512.1) (931.2) (943.1)
DT: Predates 1975
BT: Density measuring instruments
RT: Flowmeters
Permeameters (mechanical permeability)
Porosity

Porosity 931.2
DT: January 1993
BT: Physical properties
RT: Brittleness
Buoyancy
Capillarity
Compressibility of solids
Density (specific gravity)
Impregnation
Infiltration
Leakage (fluid)
Mechanical permeability
Microstructure
Porosimeters
Sintering
Surface phenomena
Surface properties

Porous materials
SN: Materials having an open cell structure
DT: Predates 1975
UF: Flow of fluids—Porous materials*
Heat transfer—Porous materials*
Pore size distribution
BT: Materials
NT: Foams
Porous silicon
Sponge iron
Zirconium sponge
RT: Aggregates
Impregnation
Infiltration
Mechanical permeability
Sintering

Porous silicon 549.3
DT: January 1995
BT: Porous materials
Silicon

Port structures 407.1
DT: Predates 1975
BT: Structures (built objects)
NT: Bulkheads (retaining walls)
Docks
Dolphins (structures)
Drydocks
Fenders (port structures)
Piers
Port terminals
Quay walls
RT: Coastal engineering
Hydraulic structures
Ice control
Marinas
Marine borers
Pontoons
Salt water barriers
Shore protection

*Port structures—Bulkheads**
USE: Bulkheads (retaining walls)

*Port structures—Concrete**
USE: Concrete construction

*Port structures—Dolphins**
USE: Dolphins (structures)

*Port structures—Fenders**
USE: Fenders (port structures)

*Port structures—Ice control**
USE: Ice control

*Port structures—Quay walls**
USE: Quay walls

*Port structures—Wood**
USE: Wood products

Port terminals 407.1
DT: Predates 1975
UF: Marine terminals
Terminals (port)
BT: Facilities
Port structures
RT: Ferry terminals

Portability (software)
USE: Computer software portability

Portable equipment
DT: January 1993
UF: Compressors—Portable*
Cranes—Portable*
Electric machinery—Portable*
Electric transformers—Portable*
Internal combustion engines—Portable types*
BT: Equipment

Portable industrial plants (402.1) (912.1) (913.1)
DT: January 1993
UF: Industrial plants—Portable*
BT: Industrial plants

Portable power plants
USE: Mobile power plants

* Former Ei Vocabulary term

Portland cement 412.1
DT: January 1995
BT: Cements
RT: Concretes
Mortar
Oil well cementing
Pozzolan

Ports and harbors 407.1
DT: Predates 1975
UF: Harbors
BT: Facilities
RT: Ice control
Marinas
Retaining walls
Service vessels

*Ports and harbors—Ice control**
USE: Ice control

*Ports and harbors—Service vessels**
USE: Service vessels

POS terminals
USE: Point of sale terminals

Posistors 704.1
DT: Predates 1975
UF: Positive temperature coefficient resistors
BT: Resistors

Position control 731.3
DT: January 1993
UF: Control, Mechanical variables—Position*
BT: Spatial variables control
RT: Position measurement
Servomechanisms
Servomotors
Tracking (position)

Position measurement 943.2
DT: January 1993
UF: Mechanical variables measurement—Position*
BT: Spatial variables measurement
RT: Position control
Servomechanisms
Servomotors
Tracking (position)

Position tracking
USE: Tracking (position)

Positive feedback
USE: Feedback

Positive impedance inverters
USE: Gyrators

Positive temperature coefficient resistors
USE: Posistors

Positron emission tomography (461.1)
DT: January 1993
UF: Biomedical engineering—Positron emission tomography*
PET scanning
RT: Biomedical equipment
Computerized tomography
Diagnostic radiography
Radioisotopes

Post offices 402.2
DT: Predates 1975
BT: Facilities
RT: Mail handling
Postal services

Postal cars 682.1.1 (433)
DT: January 1993
UF: Cars—Postal*
Mail cars
BT: Freight cars
RT: Mail handling
Postal services

Postal services (691)
DT: Predates 1975
RT: Mail handling
Post offices
Postal cars

Potable water 444
DT: January 1995
UF: Drinking water
BT: Water
RT: Reservoirs (water)
Water quality
Water resources
Water supply
Water treatment

Potash 804.2 (821.2)
DT: Predates 1975
UF: Potassium carbonate
Salt of tartar
BT: Carbonates
Potassium compounds
RT: Carbonate minerals
Potash deposits
Potash mines
Potash ore treatment
Potassium fertilizers

Potash deposits 505.1
DT: Predates 1975
BT: Mineral resources
RT: Potash
Potash mines
Potassium

Potash mines 505.1
DT: January 1993
UF: Potash mines and mining*
BT: Mines
RT: Potash
Potash deposits
Potassium

*Potash mines and mining**
USE: Potash mines

Potash ore treatment 533.1
DT: Predates 1975
BT: Ore treatment
RT: Potash
Potassium

* Former Ei Vocabulary term

Potassium 549.1
DT: January 1993
UF: K
Potassium and alloys*
BT: Alkali metals
RT: Potash deposits
Potash mines
Potash ore treatment
Potassium alloys
Potassium compounds
Potassium fertilizers

Potassium alloys 549.1
DT: January 1993
UF: Potassium and alloys*
BT: Alkali metal alloys
RT: Potassium
Potassium compounds

*Potassium and alloys**
USE: Potassium OR
Potassium alloys

Potassium carbonate
USE: Potash

Potassium compounds (804.1) (804.2)
DT: Predates 1975
BT: Alkali metal compounds
NT: Potash
RT: Potassium
Potassium alloys

Potassium fertilizers 804.2, 821.2
DT: January 1993
UF: Fertilizers—Potassium*
BT: Fertilizers
RT: Potash
Potassium

Potential dividers
USE: Voltage dividers

Potential flow 631.1
DT: January 1993
UF: Flow of fluids—Potential flow*
BT: Flow of fluids

Potentials (bioelectric)
USE: Bioelectric potentials

*Potentiometer instruments**
USE: Potentiometers (electric measuring instruments)

Potentiometers (electric measuring instruments) 942.1
DT: January 1993
UF: Potentiometer instruments*
BT: Voltmeters
RT: Voltage measurement

Potentiometers (resistors) 704.1
SN: Resistors with one or more sliding contacts between terminals, such as three-terminal rheostats
DT: January 1993
UF: Adjustable resistors
Potentiometers*
BT: Resistors
Voltage dividers

*Potentiometers**
USE: Potentiometers (resistors)

Potentiometric sensors (942.1)
DT: January 1993
UF: Sensors—Potentiometric measurements*
BT: Sensors

Powder coatings 813.2 (539.2.2)
DT: January 1993
UF: Protective coatings—Powder*
BT: Coatings
RT: Plastic coatings

Powder magnetic cores (708.4) (722.1)
DT: January 1993
UF: Dust cores
Magnetic cores, Powder*
BT: Magnetic cores
RT: Ferrite devices
Ferrites

Powder metal extrusion
USE: Powder metals AND
Metal extrusion

Powder metal finishing
USE: Powder metals AND
Metal finishing

Powder metal products 536.3
DT: Predates 1975
UF: Bearings—Powder metal*
Gear manufacture—Powder metals*
Magnets—Powder metal*
RT: Alumina
Powder metallurgy
Powder metals

*Powder metal products—Alumina**
USE: Alumina

*Powder metal products—Catalytic**
USE: Catalysts

*Powder metal products—Extruding**
USE: Metal extrusion

*Powder metal products—Finishing**
USE: Metal finishing

Powder metallurgy (536)
DT: Predates 1975
UF: Silicon powder metallurgy
BT: Metallurgy
NT: Aluminum powder metallurgy
Beryllium powder metallurgy
Chromium powder metallurgy
Cobalt powder metallurgy
Copper powder metallurgy
Iron powder metallurgy
Lead powder metallurgy
Magnesium powder metallurgy
Molybdenum powder metallurgy
Nickel powder metallurgy
Niobium powder metallurgy

* Former Ei Vocabulary term

Powder metallurgy *(continued)*
Silver powder metallurgy
Steel powder metallurgy
Tantalum powder metallurgy
Thorium powder metallurgy
Tin powder metallurgy
Titanium powder metallurgy
Tungsten powder metallurgy
Uranium powder metallurgy
Zinc powder metallurgy
Zirconium powder metallurgy
RT: Cermets
Copper powder
Dispersion hardening
Explosives
Fire hazards
Hot isostatic pressing
Hot pressing
Hydrostatic pressing
Iron powder
Magnesium powder
Mechanical alloying
Pelletizing
Powder metal products
Powder metals
Rapid solidification
Sintering

*Powder metallurgy—Aluminum copper**
USE: Aluminum powder metallurgy AND
Copper powder metallurgy

*Powder metallurgy—Aluminum**
USE: Aluminum powder metallurgy

*Powder metallurgy—Beryllium copper**
USE: Copper powder metallurgy AND
Beryllium powder metallurgy

*Powder metallurgy—Beryllium**
USE: Beryllium powder metallurgy

*Powder metallurgy—Chromium**
USE: Chromium powder metallurgy

*Powder metallurgy—Cobalt nickel**
USE: Nickel powder metallurgy AND
Cobalt powder metallurgy

*Powder metallurgy—Cobalt**
USE: Cobalt powder metallurgy

*Powder metallurgy—Compacting**
USE: Compaction

*Powder metallurgy—Copper tin**
USE: Copper powder metallurgy AND
Tin powder metallurgy

*Powder metallurgy—Copper**
USE: Copper powder metallurgy

*Powder metallurgy—Dispersion hardening**
USE: Dispersion hardening

*Powder metallurgy—Explosives applications**
USE: Explosives

*Powder metallurgy—Fire hazards**
USE: Fire hazards

*Powder metallurgy—Hot pressing**
USE: Hot pressing

*Powder metallurgy—Hydrostatic pressing**
USE: Hydrostatic pressing

*Powder metallurgy—Iron copper**
USE: Copper powder metallurgy AND
Iron powder metallurgy

*Powder metallurgy—Iron graphite**
USE: Iron powder metallurgy AND
Graphite

*Powder metallurgy—Iron**
USE: Iron powder metallurgy

*Powder metallurgy—Lead**
USE: Lead powder metallurgy

*Powder metallurgy—Magnesium**
USE: Magnesium powder metallurgy

*Powder metallurgy—Molybdenum**
USE: Molybdenum powder metallurgy

*Powder metallurgy—Nickel silver**
USE: Nickel powder metallurgy AND
Silver powder metallurgy

*Powder metallurgy—Nickel**
USE: Nickel powder metallurgy

*Powder metallurgy—Niobium**
USE: Niobium powder metallurgy

*Powder metallurgy—Silicon compounds**
USE: Silicon compounds

*Powder metallurgy—Silicon**
USE: Silicon

*Powder metallurgy—Silver**
USE: Silver powder metallurgy

*Powder metallurgy—Stainless steel**
USE: Steel powder metallurgy AND
Stainless steel

*Powder metallurgy—Steel**
USE: Steel powder metallurgy

*Powder metallurgy—Superalloys**
USE: Superalloys

*Powder metallurgy—Tantalum**
USE: Tantalum powder metallurgy

*Powder metallurgy—Thorium**
USE: Thorium powder metallurgy

*Powder metallurgy—Tin**
USE: Tin powder metallurgy

*Powder metallurgy—Titanium oxide**
USE: Titanium powder metallurgy AND
Titanium oxides

*Powder metallurgy—Titanium**
USE: Titanium powder metallurgy

*Powder metallurgy—Tungsten**
USE: Tungsten powder metallurgy

* Former Ei Vocabulary term

*Powder metallurgy—Uranium**
USE: Uranium powder metallurgy

*Powder metallurgy—Zinc alloys**
USE: Zinc powder metallurgy

*Powder metallurgy—Zinc**
USE: Zinc powder metallurgy

*Powder metallurgy—Zirconium**
USE: Zirconium powder metallurgy

Powder metals (536)
DT: January 1993
UF: Aircraft materials—Powder metals*
Automobile materials—Powder metals*
Metal powders
Powder metal extrusion
Powder metal finishing
Rolling mill practice—Powder metals*
BT: Powders
NT: Copper powder
Iron powder
Magnesium powder
RT: Alumina
Catalysts
Granular materials
Metals
Powder metal products
Powder metallurgy
Size separation

Powder method
USE: X ray powder diffraction

Powders (536) (804)
DT: Predates 1975
BT: Materials
NT: Powder metals
RT: Granular materials
Nanostructured materials
Size separation
X ray powder diffraction

Power amplifiers 713.1
DT: January 1993
UF: Amplifiers, Power type*
Power type amplifiers
BT: Amplifiers (electronic)
NT: Magnetic amplifiers
RT: Power electronics

Power control 731.3
DT: January 1993
UF: Control, Power*
Energy control
BT: Electric variables control

Power converters (704.2) (705.3)
DT: January 1993
UF: Electric converters, Power type*
BT: Electric converters
NT: Mercury vapor rectifiers
Rotary converters
RT: Choppers (circuits)
Power electronics
Torque converters

Power electronics (713.5) (715.2)
DT: January 1995
RT: Electric inverters
Electric power supplies to apparatus
Power amplifiers
Power converters
Power supply circuits
Rectifying circuits

Power generation
SN: Sources of power not specifically referred to generation of electric power
DT: Predates 1975
BT: Energy conversion
NT: Electric power generation
Electrogasdynamic power generation
Magnetohydrodynamic power generation
Solar power generation
RT: Fuels
Gas turbine power plants
Renewable energy resources
Solar energy
Tidal power
Water power
Wave power
Wind power

*Power generation—Electrogasdynamic**
USE: Electrogasdynamic power generation

*Power generation—Magnetohydrodynamic**
USE: Magnetohydrodynamic power generation

*Power generation—Seawater**
USE: Wave power

*Power generation—Solar energy**
USE: Solar power generation

*Power generation—Water**
USE: Water power

Power generators (electric)
USE: Electric generators

Power inductors (704.1) (714)
DT: January 1993
UF: Electric inductors, Power type*
BT: Electric inductors

Power measurement (electric)
USE: Electric power measurement

Power plant intakes (611.1) (614.2)
DT: January 1993
UF: Hydroelectric power plants—Intakes*
Steam power plants—Intakes*
BT: Power plants
RT: Hydroelectric power plants
Steam power plants

Power plant laboratories 901.3
DT: January 1993
UF: Power plants—Laboratories*
BT: Laboratories
Power plants

* Former Ei Vocabulary term

Power plants
SN: General treatments, and plants which output nonelectric power. For plants which output strictly electric power, use ELECTRIC POWER PLANTS
DT: Predates 1975
BT: Industrial plants
NT: Combined cycle power plants
Control rooms (power plants)
Diesel and gas turbine combined power plants
Diesel and hydroelectric combined power plants
Diesel power plants
Electric power plants
Floating power plants
Fossil fuel power plants
Gas turbine power plants
Geothermal power plants
Hydroelectric and gas turbine combined power plants
Hydroelectric and nuclear combined power plants
Hydroelectric and steam combined power plants
Magnetohydrodynamic power plants
Marine power plants
Military power plants
Mine power plants
Mobile power plants
Nuclear power plants
Offshore power plants
Power plant intakes
Power plant laboratories
Small power plants
Solar power plants
Steam power plants
Underground power plants
Underwater power plants
RT: Blowers
Boilers
Economizers
Feedwater heaters
Feedwater pumps
Outages
Site selection
Standby power service
Thermal effluents

*Power plants—Auxiliary equipment**
USE: Auxiliary equipment

*Power plants—Blowers**
USE: Blowers

*Power plants—Control rooms**
USE: Control rooms (power plants)

*Power plants—Depreciation**
USE: Depreciation

*Power plants—Diesel and gas turbine combined**
USE: Diesel and gas turbine combined power plants

*Power plants—Floating**
USE: Floating power plants

*Power plants—Gas and steam turbine combined**
USE: Combined cycle power plants

*Power plants—Geothermal energy**
USE: Geothermal power plants

*Power plants—Laboratories**
USE: Power plant laboratories

*Power plants—Magnetohydrodynamic**
USE: Magnetohydrodynamic power plants

*Power plants—Marine**
USE: Marine power plants

*Power plants—Military**
USE: Military power plants

*Power plants—Mines**
USE: Mine power plants

*Power plants—Mobile**
USE: Mobile power plants

*Power plants—Noise**
USE: Acoustic noise

*Power plants—Offshore**
USE: Offshore power plants

*Power plants—Outages**
USE: Outages

*Power plants—Portable**
USE: Mobile power plants

*Power plants—Power supply**
USE: Electric power utilization

*Power plants—Site selection**
USE: Site selection

*Power plants—Small capacity**
USE: Small power plants

*Power plants—Standby service**
USE: Standby power service

*Power plants—Thermal effluents**
USE: Thermal effluents

*Power plants—Underground**
USE: Underground power plants

*Power plants—Underwater**
USE: Underwater power plants

Power saws
USE: Metal working saws

Power supplies to apparatus
USE: Electric power supplies to apparatus

Power supply circuits (703.1) (713.5)
DT: January 1993
UF: Electronic circuits, Power supply*
BT: Networks (circuits)
RT: Electric power supplies to apparatus
Electric rectifiers
Electric transformers
Power electronics

* Former Ei Vocabulary term

Power supply industry
USE: Electric utilities

Power supply systems
USE: Electric power systems

Power systems (standby)
USE: Standby power systems

Power takeoffs (663.2)
DT: January 1993
UF: Motor trucks—Power takeoff*
BT: Trucks

Power transmission 602.2
SN: Transmission of power from the engines to the moving parts of vehicles or machines
DT: Predates 1975
UF: Tractors—Power transmission*
Transmission (power)
RT: Automobile transmissions
Transmissions
Vehicle transmissions

Power transmission (electric)
USE: Electric power transmission

*Power transmission—Variable speed**
USE: Variable speed transmissions

Power type amplifiers
USE: Power amplifiers

Power utilization
USE: Electric power utilization

Pozzolan 412.1
DT: January 1993
UF: Cement—Pozzolan*
BT: Cements
RT: Portland cement

PP
USE: Polypropylenes

PPM
USE: Pulse time modulation

PPO
USE: Polyphenylene oxides

Prandtl number 641.2, 931.2
DT: January 1995
RT: Diffusion
Heat convection
Nusselt number
Viscosity

Praseodymium 547.2
DT: January 1993
UF: Praseodymium and alloys*
BT: Rare earth elements
RT: Praseodymium alloys
Praseodymium compounds

Praseodymium alloys 547.2
DT: January 1993
UF: Praseodymium and alloys*
BT: Rare earth alloys
RT: Praseodymium
Praseodymium compounds

*Praseodymium and alloys**
USE: Praseodymium OR
Praseodymium alloys

Praseodymium compounds (804.1) (804.2)
DT: June 1990
BT: Rare earth compounds
RT: Praseodymium
Praseodymium alloys

Precast concrete (412)
DT: January 1993
UF: Concrete construction—Precast*
Prefabricated concrete
BT: Concretes
RT: Prefabricated construction

Precious metal alloys 547.1
DT: January 1993
BT: Transition metal alloys
NT: Gold alloys
Platinum metal alloys
Silver alloys
RT: Precious metal compounds
Precious metals

Precious metal compounds (804.1) (804.2)
DT: January 1993
BT: Transition metal compounds
NT: Gold compounds
Platinum metal compounds
Silver compounds
RT: Precious metal alloys
Precious metals

Precious metals 547.1
DT: January 1993
BT: Nonferrous metals
Transition metals
NT: Gold
Platinum metals
Silver
RT: Precious metal alloys
Precious metal compounds

Precious stones
USE: Gems

Precipitation (chemical) 802.3
DT: January 1993
UF: Precipitation*
BT: Separation
RT: Agglomeration
Coagulation
Colloids
Crystal microstructure
Crystallization
Deposition
Filtration
Sedimentation
Supersaturation

Precipitation (meteorology) 443.3
DT: January 1993
UF: Meteorology—Atmospheric precipitation*
NT: Rain
Snow

* Former Ei Vocabulary term

Precipitation (meteorology) *(continued)*
RT: Climatology
Clouds
Earth atmosphere
Extraterrestrial atmospheres
Floods
Meteorology
Storms

Precipitation hardening
USE: Age hardening

*Precipitation**
USE: Precipitation (chemical)

Precision balances 943.3
DT: January 1993
UF: Balances
Scales and weighing—Precision balances*
Wind tunnels—Balances*
BT: Scales (weighing instruments)
RT: Weighing

Precision casting 534.2
DT: January 1993
UF: Foundry practice—Precision casting*
BT: Metal casting
RT: Foundry practice
Investment casting

Precision engineering
DT: January 1981
BT: Engineering

Precoated metals 531 (539) (813)
DT: January 1993
UF: Sheet and strip metal—Precoating*
Welding—Precoated metals*
BT: Coated materials
RT: Coating techniques
Strip metal

Prediction theory
USE: Signal filtering and prediction

*Prediction**
USE: Forecasting

Predictive control systems 731.1
DT: January 1993
UF: Control systems, Predictive*
BT: Control systems
RT: Optimal control systems
Signal filtering and prediction

Prefabricated bridges 401.1
DT: January 1993
UF: Bridges—Prefabricated*
BT: Bridges
RT: Prefabricated construction

Prefabricated buildings
USE: Prefabricated construction

Prefabricated concrete
USE: Precast concrete

Prefabricated construction 405.2
DT: January 1993
UF: Buildings—Prefabricated*
Prefabricated buildings
BT: Construction
RT: Precast concrete
Prefabricated bridges
Prefabricated roofs
Prefabricated ships
Preforming

Prefabricated roofs (402) (405.2) (408.2)
DT: January 1993
UF: Roofs—Prefabricated*
BT: Roofs
RT: Prefabricated construction

Prefabricated ships (671) (672) (673.1)
DT: January 1993
UF: Ships—Prefabricated*
BT: Ships
RT: Prefabricated construction

Preferred numbers 921 (704.1) (902.2)
SN: Numbers for nominal values of resistors and capacitors, used to reduce the number of different sizes that must be kept in stock
DT: January 1993
UF: Numbering systems—Preferred numbers*
Preferred values
BT: Numbering systems
RT: Capacitors
Resistors

Preferred values
USE: Preferred numbers

Preforming (535.2.2) (816.1)
DT: January 1993
UF: Plastics—Preforming*
BT: Forming
RT: Briquetting
Compaction
Metal pressing
Molding
Plastics forming
Prefabricated construction

Preheaters (steam)
USE: Feedwater heaters

Preheating 642.1 (538.2)
DT: January 1993
UF: Welding—Preheating*
BT: Industrial heating
RT: Welding

Preignition 521.1, 612.1
SN: In internal combustion engines, ignition of the charge before ignition of the spark
DT: January 1993
UF: Automobile engines—Preignition*
BT: Ignition
RT: Internal combustion engines

Preprocessors (computer program)
USE: Program processors

Preservation (biological materials)
USE: Biological materials preservation

* Former Ei Vocabulary term

Preservation (food)
USE: Food preservation

Preservation (warships)
USE: Warship preservation

Preservation (wood)
USE: Wood preservation

Preservatives (food)
USE: Food preservatives

Press load control 603.1, 731.3
DT: January 1993
UF: Load control (presses)
Presses—Load control*
BT: Control
RT: Presses (machine tools)
Pressure control

Presses (machine tools) 603.1
DT: January 1993
UF: Industrial plants—Presses*
Presses*
Presses—Tools*
BT: Metal forming machines
RT: Dies
Extruders
Hammers
Press load control
Tools

Presses (printing)
USE: Printing presses

*Presses**
USE: Presses (machine tools)

*Presses—Hydraulic**
USE: Hydraulic machinery

*Presses—Load control**
USE: Press load control

*Presses—Tools**
USE: Presses (machine tools)

Pressing (forming) (535.2.2) (812.3)
DT: January 1993
UF: Glass—Pressing*
BT: Forming
NT: Hot pressing
Hydrostatic pressing
Metal pressing
RT: Briquetting
Calendering
Compaction
Deep drawing
Extrusion
Stamping
Straightening
Stretching
Swaging

Pressing (textiles)
USE: Textile pressing

Pressure
DT: January 1993
NT: Atmospheric pressure
Bottom hole pressure
Hydrostatic pressure
Pore pressure
Rock pressure
Uplift pressure
Vapor pressure
Well pressure
RT: Compressed air
High pressure effects
Pressure control
Pressure effects
Pressure gages
Pressure regulators
Pressure regulation
Pressure vessels
Pressurization
Surge tanks
Vacuum

Pressure control 731.3
DT: January 1987
BT: Mechanical variables control
NT: Pressure regulation
RT: Press load control
Pressure
Pressure effects
Pressurization

Pressure drop (631.1) (931.1)
DT: January 1995
RT: Flow interactions
Friction

Pressure effects
DT: January 1981
BT: Effects
RT: Pressure
Pressure control
Pressure measurement

Pressure forming
USE: Forming

Pressure gages 944.3
DT: January 1993
UF: Gages—Pressure measurement*
BT: Gages
NT: Vacuum gages
RT: Depth indicators
Pressure
Pressure measurement

Pressure measurement 944.4
DT: January 1977
BT: Mechanical variables measurement
RT: Barometers
Depth indicators
Flow measurement
Pressure effects
Pressure gages
Pressure transducers
Vacuum gages
Velocity measurement

Pressure molding
USE: Compression molding

* Former Ei Vocabulary term

Pressure pouring 534.2
DT: January 1993
UF: Foundry practice—Pressure pouring*
BT: Metal casting
RT: Foundry practice

Pressure regulation 731.3 (619.1)
DT: January 1993
UF: Pipelines—Pressure regulation*
BT: Pressure control
RT: Pressure
Pressure regulators
Pressurization
Surge tanks

Pressure regulators 732.1 (522) (619.1)
DT: Predates 1975
UF: Gas appliances—Pressure regulators*
BT: Control equipment
RT: Gas appliances
Pressure
Pressure regulation
Surge tanks

Pressure relief valves 601.2 (619.1.1)
DT: January 1993
UF: Relief valves
BT: Valves (mechanical)
NT: Rupture disks

Pressure thermit welding
USE: Thermit welding

Pressure transducers 944.3
DT: Predates 1975
BT: Transducers
NT: Barometers
Manometers
Vacuum transducers
Working fluid pressure indicators
RT: Diaphragms
Piezoelectric transducers
Pressure measurement

*Pressure transducers—Joints**
USE: Joints (structural components)

*Pressure transducers—Vacuum**
USE: Vacuum transducers

Pressure tube reactors 621.1 (619.1)
DT: January 1993
UF: Nuclear reactors—Pressure tubes*
BT: Nuclear reactors

Pressure vessel codes 619.2, 902.2
DT: January 1993
UF: Pressure vessels—Codes*
BT: Codes (standards)
RT: Laws and legislation

Pressure vessels 619.2
DT: Predates 1975
UF: Steel pressure vessels
BT: Structures (built objects)
RT: Autoclaves
Containers
Nuclear reactors
Pressure
Pressurization
Safety valves
Steam generators
Tanks (containers)

*Pressure vessels—Codes**
USE: Pressure vessel codes

*Pressure vessels—Concrete**
USE: Concrete construction

*Pressure vessels—Joints**
USE: Joints (structural components)

*Pressure vessels—Steel**
USE: Steel structures

Pressure welding 538.2.1
SN: Hot or cold
DT: January 1993
UF: Forge welding
Welding—Pressure*
BT: Welding
NT: Cold welding
Explosive welding
Flash welding
Friction welding
Percussion welding
Projection welding
Resistance welding
Stud welding
Ultrasonic welding
Vacuum welding
RT: Electric arc welding
Electric welding
Gas welding
Inert gas welding
Spot welding
Thermit welding

Pressurization (614)
DT: January 1993
UF: Boiler firing—Pressurization*
Containers—Pressurization*
Pressurizing
RT: Boiler firing
Boilers
Containers
Pressure
Pressure control
Pressure regulation
Pressure vessels

Pressurized water reactors (621.1)
DT: January 1993
UF: Nuclear reactors, Pressurized water*
PWR reactors
BT: Water cooled reactors

Pressurizing
USE: Pressurization

Prestressed beams and girders 408.2
DT: January 1993
UF: Beams and girders—Prestressed*
BT: Beams and girders

* Former Ei Vocabulary term

Prestressed beams and girders *(continued)*
RT: Prestressed materials
Prestressing

Prestressed concrete 412
DT: January 1993
UF: Concrete construction—Prestressing*
BT: Concretes
Prestressed materials
RT: Prestressing

Prestressed materials (421)
DT: January 1993
UF: Bridges—Prestressed materials*
BT: Building materials
NT: Prestressed concrete
RT: Prestressed beams and girders
Prestressing

Prestressing (405.2) (408.2) (412)
DT: January 1993
UF: Structural design—Prestressing*
BT: Construction
RT: Fatigue of materials
Prestressed beams and girders
Prestressed concrete
Prestressed materials
Structural design
Structural members

Prevention (913.5) (914.1)
DT: January 1993
NT: Accident prevention
RT: Hazards
Maintenance
Repair
Risks

Primary batteries 702.1.1
SN: Non-rechargeable
DT: January 1993
UF: Electric batteries, Primary*
BT: Electric batteries
NT: Lithium batteries
Nuclear batteries

Primary cosmic rays
USE: Cosmic rays

Printed circuit boards (714.2) (715)
DT: January 1993
UF: Boards (printed circuits)
Printed circuits—Boards*
BT: Printed circuits

Printed circuit design (714.2) (715)
DT: January 1993
UF: Printed circuits—Design*
BT: Product design
RT: Electronics packaging
Integrated circuit layout
Printed circuits
Printed circuit manufacture
Printed circuit testing

Printed circuit manufacture (714.2) (715)
DT: January 1993
UF: Printed circuits—Manufacture*
BT: Electronic equipment manufacture
RT: Assembly
Electronics packaging
Integrated circuit manufacture
Printed circuits
Printed circuit design
Printed circuit testing
Surface mount technology

Printed circuit motors 705.3 (714.2)
DT: January 1993
UF: Electric motors, Printed circuit*
BT: Electric motors
RT: Printed circuits

Printed circuit testing (714.2) (715)
DT: January 1993
UF: Printed circuits—Testing*
BT: Electron device testing
RT: Integrated circuit testing
Printed circuit design
Printed circuit manufacture
Printed circuits

Printed circuits (714.2) (715)
DT: Predates 1975
BT: Networks (circuits)
NT: Printed circuit boards
RT: Electric wiring
Electronic equipment
Electronics packaging
Etching
Lithography
Photolithography
Photoresists
Printed circuit motors
Printed circuit design
Printed circuit manufacture
Printed circuit testing
Screen printing
Surface mount technology

*Printed circuits—Boards**
USE: Printed circuit boards

*Printed circuits—Design**
USE: Printed circuit design

*Printed circuits—Manufacture**
USE: Printed circuit manufacture

*Printed circuits—Testing**
USE: Printed circuit testing

Printers (computer) 722.2 (745.1.1)
DT: January 1993
UF: Computer peripheral equipment—Printers*
Computer printers
BT: Computer peripheral equipment
RT: Plotters (computer)
Typewriters

Printing 745.1
DT: Predates 1975
NT: Color printing
Electrostatic printing
Offset printing
Photocomposition

* Former Ei Vocabulary term

Printing *(continued)*
Screen printing
Textile printing
Thermal printing
Typesetting
RT: Character sets
Graphic methods
Information dissemination
Ink
Labeling
Printing machinery
Printing plates
Printing presses
Printing properties
Word processing

Printing ink
USE: Ink

Printing machinery 745.1.1
DT: Predates 1975
BT: Machinery
NT: Printing presses
RT: Printing
Printing plates

*Printing machinery—Rolls**
USE: Rolls (machine components)

Printing plates 745.1.1
DT: Predates 1975
BT: Equipment
NT: Magnesium printing plates
Offset printing plates
RT: Printing
Printing machinery

*Printing plates—Magnesium**
USE: Magnesium printing plates

*Printing plates—Offset**
USE: Offset printing plates

Printing presses 745.1.1
DT: Predates 1975
UF: Presses (printing)
BT: Printing machinery
RT: Printing

Printing properties 745.1 (811.1.1)
DT: January 1993
UF: Paper—Printing properties*
RT: Paper
Printing

*Printing—Character sets**
USE: Character sets

*Printing—Color**
USE: Color printing

*Printing—Electrostatic**
USE: Electrostatic printing

*Printing—Offset**
USE: Offset printing

*Printing—Photocomposition**
USE: Photocomposition

*Printing—Screen**
USE: Screen printing

*Printing—Thermal**
USE: Thermal printing

Prisms 741.3
DT: October 1975
UF: Optical prisms
BT: Optical devices
RT: Light polarization

Prisons (402.2)
DT: June 1990
UF: Jails
Penitentiaries
BT: Facilities

Private telephone exchanges 718.1
DT: January 1993
UF: PABX
PBX
Telephone exchanges, Private*
BT: Telephone exchanges

Probabilistic logics 721.1 (922.1)
DT: January 1993
UF: Computer metatheory—Probabilistic logics*
BT: Formal logic
RT: Automata theory
Fuzzy sets
Logic circuits
Probability

Probability 922.1
DT: Predates 1975
UF: Electric power systems—Loss of load probability*
Error probability
Probability distributions
BT: Mathematical techniques
RT: Chaos theory
Decision theory
Game theory
Heuristic programming
Information theory
Markov processes
Monte Carlo methods
Operations research
Probabilistic logics
Quality control
Queueing theory
Random processes
Sampling
Scheduling
Statistical methods
Statistics

Probability density function 922.1 (921)
DT: January 1995
UF: Density function
Frequency function
BT: Functions

Probability distributions
USE: Probability

*Probability—Game theory**
USE: Game theory

*Probability—Queueing theory**
USE: Queueing theory

*Probability—Random processes**
USE: Random processes

Probes (941) (942) (943) (944)
DT: Predates 1975
UF: Instruments—Probes*
BT: Equipment
NT: Plasma probes
Space probes
RT: Instruments
Microelectrodes
Sensors
Underwater probing

Probing (underwater)
USE: Underwater probing

Problem oriented languages 723.1.1
DT: January 1993
UF: Computer programming languages—Problem orientation*
BT: High level languages

Problem solving (723.4) (921)
DT: January 1995
RT: Artificial intelligence
Decision theory
Theorem proving

Procedure libraries
USE: Subroutines

Procedure oriented languages 723.1.1
DT: January 1993
UF: Computer programming languages—Procedure orientation*
BT: High level languages
NT: Ada (programming language)
BASIC (programming language)
FORTRAN (programming language)
Pascal (programming language)

Process control (731)
DT: Predates 1975
BT: Control
NT: Computer control
Statistical process control
RT: Automation
Control systems
Cybernetics
Electric switches
Industrial management
Integrated control
Multivariable control systems
Numerical control systems
Plant management
Process engineering
Processing
Production control
Programmed control systems
Recording instruments
Remote control
SCADA systems
Temperature control

Process engineering 913.1 (731)
DT: June 1990
BT: Engineering
RT: Process control
Processing
Production engineering

Process heat reactors (621.1)
DT: January 1993
UF: Nuclear reactors, Process heat*
BT: Nuclear reactors

*Process heating**
USE: Industrial heating

Processing 913.4
DT: January 1993
BT: Manufacture
NT: Air entrainment
Bleaching
Comminution
Compaction
Degumming
Finishing
In situ processing
Indexing (materials working)
Metal working
Microelectronic processing
Microgravity processing
Oil shale processing
Refining
Spinning (fibers)
Tanning
Textile processing
RT: Industrial robots
Process control
Process engineering
Synthesis (chemical)

Processing fumes control
USE: Fume control

Processors (computer program)
USE: Program processors

Product design 913.1
DT: Predates 1975
UF: Industrial design
BT: Design
NT: Integrated circuit layout
Printed circuit design
RT: Product liability
Weight control

*Product design—Weight control**
USE: Weight control

Product liability 902.3 (913.1) (914.1)
DT: October 1975
UF: Liability (product)
RT: Accident prevention
Hazards
Product design

* Former Ei Vocabulary term

Production (913.1) (913.2)
DT: January 1993
NT: Acidization
Assembly
Enhanced recovery
Fracturing (oil wells)
Low temperature production
Manufacture
Mining
Natural gas well production
Oil well production
Ore treatment
Well stimulation
RT: Production control
Production engineering
Productivity
Quality control
Reliability

Production control 913.2 (731.1)
DT: Predates 1975
BT: Control
Industrial management
NT: Just in time production
RT: Industrial plants
Inventory control
Job analysis
Plant management
Process control
Production
Production engineering
Quality control
Standardization
Standards
Statistical process control

Production engineering 913.1
DT: January 1981
BT: Engineering
RT: Concurrent engineering
Process engineering
Production
Production control
Standardization

Production packers
USE: Packers

Production platforms 674.2
DT: January 1977
UF: Deep sea platforms
Marine platforms*
Ocean platforms
BT: Ocean structures
RT: Boats
Marine insurance
Marine risers
Ocean engineering
Offshore boreholes
Offshore drilling
Offshore oil wells

Productivity (913.1)
DT: January 1983
RT: Economics
Efficiency
Performance
Personnel
Production
Quality control
Reliability
Standards

Professional aspects 901.1
DT: January 1993
UF: Engineering—Professional aspects*
NT: Registration of engineers
RT: Laws and legislation
Personnel
Personnel training
Social aspects
Societies and institutions

Professional societies
USE: Societies and institutions

Program assemblers (723.1)
DT: January 1993
UF: Assemblers (program)
Assembly programs
Computer operating systems—Program assemblers*
BT: Program processors

Program compilers (723.1)
DT: January 1993
UF: Compilers (program)
Computer operating systems—Program compilers*
BT: Program processors
RT: Computational grammars
Program diagnostics

Program debugging 723.1
DT: January 1993
UF: Computer programming—Program debugging*
Debugging (computer programs)
BT: Computer programming
RT: Computer debugging
Error detection
Program diagnostics

Program diagnostics 723.1
DT: January 1993
UF: Computer programming—Program diagnostics*
Diagnostics (computer programs)
BT: Computer programming
RT: Program compilers
Program debugging

Program documentation 723.1
DT: January 1993
UF: Computer programming—Program documentation*
Documentation (computer programs)
BT: Computer programming
NT: System program documentation
RT: Decision tables

Program evaluation and review technique
USE: PERT

* Former Ei Vocabulary term

Program interpreters (723.1)
DT: January 1993
UF: Computer operating systems—Program interpreters*
Interpreters (computer program)
BT: Program processors

Program listings
USE: Computer program listings

Program preprocessors
USE: Program processors

Program processors (723.1)
DT: January 1993
UF: Computer operating systems—Program processors*
Preprocessors (computer program)
Processors (computer program)
Program preprocessors
BT: Computer software
NT: Program assemblers
Program compilers
Program interpreters
Program translators
RT: Computer systems programming
Interfaces (computer)

Program translators (723.1)
DT: January 1993
UF: Computer operating systems—Program translators*
Translators (computer program)
BT: Program processors
RT: Computer systems programming

Programmable controllers
USE: Programmable logic controllers

Programmable logic controllers 732.1
DT: January 1995
UF: PLC
Programmable controllers
BT: Control equipment
RT: Computer applications
Industrial applications

Programmable read only memory
USE: PROM

Programmable robots 731.5 (723.1)
DT: January 1993
UF: Robots, Industrial—Programmable*
BT: Robots
RT: Industrial robots
Intelligent robots

Programmed control systems 731.1 (723.1)
DT: January 1993
UF: Control systems, Programmed*
BT: Control systems
RT: Process control

Programming (computers)
USE: Computer programming

Programming (mathematical)
USE: Mathematical programming

Programming languages
USE: Computer programming languages

Programming theory 721.1, 723.1
DT: January 1993
UF: Computer metatheory—Programming theory*
BT: Computation theory
RT: Computational geometry
Computer programming
Computer programming languages
Computer software
Petri nets

Programs (computer)
USE: Computer software

Progressive wave antennas
USE: Traveling wave antennas

Project management 912.2
DT: January 1993
UF: Engineering—Project management*
BT: Management
NT: Scheduling
RT: Engineering

Projectiles (404.1) (654)
DT: January 1993
UF: Ballistics—Projectiles*
RT: Ammunition
Ballistics
Electromagnetic launchers
Missiles
Rockets

Projection rooms (402.2) (742)
DT: January 1993
UF: Motion picture theaters—Projection rooms*
BT: Facilities
RT: Motion picture theaters

Projection screens 742.2
DT: January 1993
UF: Photography—Projection screens*
Screens (projection)
BT: Equipment
RT: Motion picture screens
Photographic accessories
Photography

Projection systems 742.2 (741.3)
DT: January 1993
UF: Optical projection systems
BT: Optical systems
RT: Optical projectors

Projection welding 538.2.1
DT: January 1993
UF: Welding—Projection*
BT: Pressure welding
Resistance welding

Projectors (optical)
USE: Optical projectors

PROLOG (programming language) 723.1.1
DT: January 1993
UF: Computer programming languages—PROLOG*

* Former Ei Vocabulary term

PROLOG (programming language) *(continued)*
BT: High level languages
RT: Logic programming
Object oriented programming

PROM 722.1 (723.1)
DT: January 1993
UF: Data storage, Digital—PROM*
Programmable read only memory
BT: ROM
RT: Cellular arrays
Random access storage
Semiconductor storage

Promethium 547.2
DT: Predates 1975
BT: Rare earth elements

Propagation (crack)
USE: Crack propagation

Propagation (wave)
USE: Wave propagation

Propane (522) (804.1)
DT: Predates 1975
BT: Gas condensates
RT: Liquefied petroleum gas
Paraffins

Propellants (523) (524) (654.2) (804)
DT: January 1983
UF: Rocket engines—Propellants*
Rockets and missiles—Propellants*
BT: Materials
NT: Liquid propellants
Solid propellants
RT: Explosives
Freons
Hydrogen fuels
Missiles
Propulsion
Rocket engines
Spacecraft power supplies
Spacecraft propulsion

*Propellants—Liquid**
USE: Liquid propellants

*Propellants—Solid**
USE: Solid propellants

Propellers (652.3) (671.2)
DT: Predates 1975
BT: Engines
NT: Aircraft propellers
Ship propellers
RT: Propulsion

Propene
USE: Propylene

Proportional control systems 731.1
DT: January 1993
UF: Control systems, Proportional*
BT: Control systems
RT: Three term control systems
Two term control systems

Proportional counters 944.7
DT: Predates 1975
BT: Radiation counters
RT: Ionization chambers

Propulsion (653.1) (654.2) (661.1) (675.1) (682)
DT: Predates 1975
NT: Aircraft propulsion
Diesel propulsion
Electric propulsion
Flywheel propulsion
Land vehicle propulsion
Nuclear propulsion
Photonic propulsion
Pneumatic propulsion
Ship propulsion
Spacecraft propulsion
Steam propulsion
RT: Drives
Propellants
Propellers
Vehicles

*Propulsion—Aerospace applications**
USE: Aircraft propulsion AND
Spacecraft propulsion

*Propulsion—Electric energy**
USE: Electric propulsion

*Propulsion—Hydraulic energy**
USE: Hydraulics

*Propulsion—Ion energy**
USE: Ion propulsion

*Propulsion—Land vehicle applications**
USE: Land vehicle propulsion

*Propulsion—Photon energy**
USE: Photonic propulsion

*Propulsion—Pneumatic energy**
USE: Pneumatic propulsion

Propulsive wing aircraft 652.1
DT: January 1993
UF: Aircraft—Propulsive wing*
BT: Aircraft
RT: Aircraft propulsion

Propylene 804.1
DT: January 1981
UF: Methyl ethylene
Propene
BT: Olefins
RT: Polypropylene oxides
Polypropylenes

Prostheses
USE: Prosthetics

Prosthetic implants
USE: Implants (surgical)

Prosthetics 462.4
DT: January 1993
UF: Prostheses
BT: Biomedical equipment
NT: Artificial limbs

* Former Ei Vocabulary term

Prosthetics *(continued)*
Artificial organs
Blood vessel prostheses
Dental prostheses
Heart valve prostheses
Intraocular lenses
Joint prostheses
Maxillofacial prostheses
Myoelectrically controlled prosthetics
Pacemakers
RT: Biocontrol
Biomaterials
Biomedical engineering
Grafts
Hearing aids
Human rehabilitation equipment
Implants (surgical)
Orthotics
Sensory feedback
Transplants
Vision aids
Walking aids

*Prosthetics—Artificial limbs**
USE: Artificial limbs

*Prosthetics—Artificial organs**
USE: Artificial organs

*Prosthetics—Blood vessel prostheses**
USE: Blood vessel prostheses

*Prosthetics—Grafts**
USE: Grafts

*Prosthetics—Heart valves**
USE: Heart valve prostheses

*Prosthetics—Hip prostheses**
USE: Hip prostheses

*Prosthetics—Joint prostheses**
USE: Joint prostheses

*Prosthetics—Myoelectric control**
USE: Myoelectrically controlled prosthetics

*Prosthetics—Pacemakers**
USE: Pacemakers

*Prosthetics—Sensory feedback**
USE: Sensory feedback

*Prosthetics—Transplantation**
USE: Transplantation (surgical)

Protactinium 622.1 (547)
DT: January 1993
BT: Actinides

Protection (914.1)
DT: January 1993
NT: Bank protection
Corrosion protection
Electric equipment protection
Electric power system protection
Electric switchgear protection
Environmental protection
Explosionproofing
Eye protection
Fire protection
Frost protection
Fungus protection
Lightning protection
Overcurrent protection
Overvoltage protection
Radiation protection
Relay protection
Shore protection
Slope protection
Surge protection
Termite proofing
Waterproofing
RT: Accident prevention
Alarm systems
Earthquake resistance
Hazards
Insect control
Insulation
Nuclear materials safeguards
Protective coatings
Radiation hardening
Safety devices
Security systems
Shielding

Protective atmospheres (538.1.1) (538.2) (804.2) (914.1)
DT: Predates 1975
UF: Atmospheres*
Controlled atmospheres
RT: Air conditioning
Corrosion prevention
Foundry practice
Furnaces
Hardening
Heat treatment
Inert gases
Quenching
Welding

Protective clothing 914.1 (502.1) (912.4)
DT: January 1993
UF: Mine rescue—Protective clothing*
Personnel—Protective clothing*
BT: Equipment
RT: Ear protectors
Gas masks
Mine rescue
Personnel
Safety devices

Protective coatings (539.2.2) (813.2)
DT: Predates 1975
UF: Protective films
BT: Coatings
NT: Antifouling paint
Electric insulating coatings
Waterproof coatings
RT: Corrosion protection
Decontamination
Films
Galvanizing
Hazards
Metal finishing

* Former Ei Vocabulary term

Protective coatings *(continued)*
Passivation
Plasma spraying
Protection
Varnish

*Protective coatings—Bituminous**
USE: Bituminous coatings

*Protective coatings—Borides**
USE: Borides

*Protective coatings—Ceramics**
USE: Ceramic coatings

*Protective coatings—Chromate**
USE: Chromate coatings

*Protective coatings—Electrophoretic**
USE: Electrophoretic coatings

*Protective coatings—Electrostatic**
USE: Electrostatic coatings

*Protective coatings—Flame spraying**
USE: Flame spraying

*Protective coatings—Hazards**
USE: Hazards

*Protective coatings—Inorganic**
USE: Inorganic coatings

*Protective coatings—Organic**
USE: Organic coatings

*Protective coatings—Permeability**
USE: Mechanical permeability

*Protective coatings—Phosphate**
USE: Phosphate coatings

*Protective coatings—Plasma spraying**
USE: Plasma spraying

*Protective coatings—Plastics**
USE: Plastic coatings

*Protective coatings—Powder**
USE: Powder coatings

*Protective coatings—Rubber**
USE: Rubber coatings

*Protective coatings—Silicones**
USE: Silicone coatings

*Protective coatings—Sputtering**
USE: Sputter deposition

*Protective coatings—Stripping**
USE: Stripping (removal)

*Protective coatings—Tape**
USE: Tape coatings

*Protective coatings—Temporary**
USE: Temporary coatings

*Protective coatings—Vacuum application**
USE: Vacuum deposited coatings

*Protective coatings—Water borne**
USE: Water borne coatings

Protective colloids 801.3
DT: January 1993
UF: Colloids—Protective*
BT: Colloids
RT: Cellulose derivatives
Photographic emulsions

Protective films
USE: Protective coatings

Protective relays
USE: Relay protection

Proteins 804.1
DT: Predates 1975
BT: Biopolymers
NT: Antibodies
Casein
Collagen
Enzymes
Hemoglobin
Interferons
Zein
RT: Amino acids
Antigens
Biocatalysts
DNA
Food products
Macromolecules
Polypeptides

Protocols
USE: Network protocols

Proton accelerators
USE: Particle accelerators

Proton magnetic resonance
USE: Nuclear magnetic resonance

Protons (931.3) (932.1)
DT: Predates 1975
BT: Charged particles
RT: Hydrogen
Ions

Prototyping (software)
USE: Software prototyping

Protozoa 461.9 (801.2)
DT: January 1993
UF: Microorganisms—Protozoa*
BT: Microorganisms

Proving grounds (automobile)
USE: Automobile proving grounds

Proximity indicators 652.3 (943.3)
DT: January 1993
UF: Aircraft instruments—Proximity indicators*
BT: Aircraft instruments
Indicators (instruments)
Range finders
RT: Distance measurement
Proximity sensors

Proximity measurement
USE: Distance measurement

Proximity sensors (731.5) (943.3)
DT: January 1993
UF: Robots, Industrial—Proximity sensors*
BT: Sensors
RT: Contact sensors
Distance measurement
Industrial robots
Proximity indicators
Robots

PS
USE: Polystyrenes

PSK
USE: Phase shift keying

Psychology computing 723.5
DT: January 1993
UF: Data processing—Psychology applications*
BT: Computer applications
RT: Data processing

Psychophysiology (461.4)
DT: January 1993
UF: Human engineering—Psychophysiology*
BT: Physiology
RT: Human engineering

Psychrometers 944.1, 443.2
DT: Predates 1975
UF: Wet and dry bulb hygrometers
Wet bulb thermometers
BT: Hygrometers
RT: Atmospheric humidity
Moisture meters

PTFE
USE: Polytetrafluoroethylenes

PTM
USE: Pulse time modulation

Pu
USE: Plutonium

Public address systems 752
DT: Predates 1975
UF: PA systems
BT: Audio equipment

*Public address systems—Airborne**
USE: Aircraft parts and equipment

Public policy 901
DT: January 1993
UF: Engineering—Public policy*
Government policy
Policy
NT: Energy policy
RT: Economics
Laws and legislation
Public risks
Taxation

Public relations (901) (912.2) (911.4)
DT: January 1993
UF: Industrial plants—Public relations*
RT: Industrial management
Industrial plants

Public risks (901.4)
DT: January 1993
UF: Risk studies—Public risks*
BT: Risks
RT: Public policy

Public utilities 706
DT: Predates 1975
BT: Industry
NT: Electric utilities
RT: Natural gas
Sewers

*Public utilities—Rate making**
USE: Utility rates

Public works 403
DT: Predates 1975
NT: Highway systems
Overpasses
Parks
Roads and streets
Underpasses
RT: Facilities
Hospitals
Recreational facilities
Urban planning

Publishing 903.2
DT: January 1993
UF: Information dissemination—Publishing*
BT: Information dissemination
NT: Desktop publishing
Electronic publishing
RT: Microforms
Photocopying

Pulmonary diseases 461.6 (914.3.1)
DT: January 1993
UF: Lung diseases
Occupational diseases—Pulmonary*
Respiratory diseases
BT: Diseases
RT: Respiratory therapy

Pulp 811.1
DT: Predates 1975
BT: Materials
NT: Bleached pulp
Chemical pulp
Kraft pulp
Mechanical pulp
Pulp materials
Sulfate pulp
Sulfite pulp
Unbleached pulp
RT: Pulp manufacture

Pulp beating 811.1.1
DT: January 1993
UF: Beating (pulp)
Pulp manufacture—Beating*
BT: Pulp manufacture

Pulp cooking 811.1.1 (802.2)
DT: January 1993
UF: Cooking (pulp)

* Former Ei Vocabulary term

Pulp cooking *(continued)*
Pulp manufacture—Cooking*
BT: Pulp manufacture

Pulp digesters 811.1.2 (802.1)
DT: Predates 1975
UF: Digesters (pulp)
BT: Chemical equipment
RT: Pulp manufacture

*Pulp digesters—Lining**
USE: Linings

Pulp industry
USE: Paper and pulp industry

Pulp manufacture 811.1.1
DT: Predates 1975
BT: Manufacture
NT: Biopulping
Kraft process
Pulp beating
Pulp cooking
Pulp refining
Sulfite process
Thermomechanical pulping process
RT: Bleaching
Pulp
Pulp digesters
Waste liquor utilization

*Pulp manufacture—Beating**
USE: Pulp beating

*Pulp manufacture—Biopulping**
USE: Biopulping

*Pulp manufacture—Bleaching**
USE: Bleaching

*Pulp manufacture—Cooking**
USE: Pulp cooking

*Pulp manufacture—Flow**
USE: Flow of fluids

*Pulp manufacture—Kraft process**
USE: Kraft process

*Pulp manufacture—Refining**
USE: Pulp refining

*Pulp manufacture—Sulfite process**
USE: Sulfite process

*Pulp manufacture—Thermomechanical process**
USE: Thermomechanical pulping process

*Pulp manufacture—Washing**
USE: Washing

*Pulp manufacture—Waste liquor utilization**
USE: Waste liquor utilization

Pulp materials (811.2) (811.3)
DT: Predates 1975
BT: Pulp
RT: Bagasse
Bamboo
Cotton
Kenaf fibers
Sawdust
Straw
Waste paper
Wood

*Pulp materials—Bagasse**
USE: Bagasse

*Pulp materials—Bamboo**
USE: Bamboo

*Pulp materials—Cotton**
USE: Cotton

*Pulp materials—Kenaf**
USE: Kenaf fibers

*Pulp materials—Sawdust**
USE: Sawdust

*Pulp materials—Straw**
USE: Straw

*Pulp materials—Waste paper**
USE: Waste paper

*Pulp materials—Wood**
USE: Wood

Pulp mills
USE: Paper and pulp mills

Pulp refining 811.1.1
DT: January 1993
UF: Pulp manufacture—Refining*
BT: Pulp manufacture
Refining

*Pulp—Bleached**
USE: Bleached pulp

*Pulp—Chemical**
USE: Chemical pulp

*Pulp—Kraft**
USE: Kraft pulp

*Pulp—Mechanical**
USE: Mechanical pulp

*Pulp—Radioactivity**
USE: Radioactivity

*Pulp—Screening**
USE: Screening

*Pulp—Sulfate**
USE: Sulfate pulp

*Pulp—Sulfite**
USE: Sulfite pulp

*Pulp—Thermomechanical**
USE: Thermomechanical pulp

*Pulp—Unbleached**
USE: Unbleached pulp

*Pulp—Weight determination**
USE: Weighing

* Former Ei Vocabulary term

Pulsatile flow 631.1 (461.1)
DT: January 1993
UF: Flow of fluids—Pulsatile flow*
Pulsed flow
Pulsing flow
BT: Flow of fluids
RT: Oscillations
Pipe flow

Pulse amplifiers 713.1, 713.4
DT: January 1993
UF: Amplifiers, Pulse signal*
BT: Amplifiers (electronic)
Pulse circuits

Pulse amplitude modulation (716)
DT: Predates 1975
UF: PAM
BT: Pulse modulation
RT: Demodulators
Modulators

Pulse analyzing circuits 713.5
DT: January 1993
UF: Electronic circuits, Pulse analyzing*
BT: Networks (circuits)

Pulse circuits 713.4
DT: January 1993
UF: Electronic circuits, Pulse signal*
Pulse signal circuits
BT: Networks (circuits)
NT: Coincidence circuits
Counting circuits
Pulse amplifiers
Pulse generators
Pulse shaping circuits
RT: Comparator circuits
Digital circuits
Logic circuits
Switching circuits

Pulse code modulation (716)
DT: Predates 1975
UF: PCM
BT: Pulse modulation
RT: Compandor circuits
Delta modulation
Demodulators
Encoding (symbols)
Modulators
Pulse code modulation links
Telecommunication links
Vocoders

Pulse code modulation links (716)
DT: January 1993
UF: PCM links
Telecommunication links—Pulse code modulation*
BT: Telecommunication links
RT: Pulse code modulation

Pulse duration modulation
USE: Pulse width modulation

Pulse frequency modulation
USE: Pulse time modulation

Pulse generators 713.4
DT: Predates 1975
UF: Pulse oscillators
BT: Function generators
Pulse circuits
NT: Light pulse generators
RT: Multivibrators
Pulse shaping circuits
Square wave generators
Sweep circuits

Pulse modulation (716)
DT: Predates 1975
BT: Modulation
NT: Delta modulation
Pulse amplitude modulation
Pulse code modulation
Pulse time modulation
RT: Demodulators
Intersymbol interference
Modulators
Radar
Sampling
Telemetering
Time division multiplexing

Pulse motors
USE: Stepping motors

Pulse oscillators
USE: Pulse generators

Pulse position modulation
USE: Pulse time modulation

Pulse shaping circuits 713.4
DT: January 1993
UF: Electronic circuits, Pulse shaping*
BT: Pulse circuits
RT: Pulse generators
Signal processing

Pulse signal circuits
USE: Pulse circuits

Pulse time modulation (716)
DT: Predates 1975
UF: PPM
PTM
Pulse frequency modulation
Pulse position modulation
BT: Pulse modulation
NT: Pulse width modulation
RT: Demodulators
Modulators

Pulse transformers (704) (714) (715)
DT: January 1993
UF: Electric transformers, Pulse*
BT: Electric transformers

Pulse width modulation (716)
DT: January 1993
UF: PDM
Pulse duration modulation

* Former Ei Vocabulary term

Pulse width modulation *(continued)*
BT: Pulse time modulation
RT: Demodulators
Modulators

Pulsed flow
USE: Pulsatile flow

Pulsed laser applications 744.9
DT: January 1995
BT: Laser applications
RT: Laser pulses

Pulsed pressure welding
USE: Hydrodynamic welding

Pulses (laser)
USE: Laser pulses

Pulsing flow
USE: Pulsatile flow

Pultrusion 816.1
DT: January 1993
UF: Plastics—Pultrusion*
BT: Forming
RT: Composite materials
Extrusion
Plastics forming

Pulverization
USE: Comminution

*Pulverization**
USE: Comminution

Pulverized fuel 524 (521.3)
DT: Predates 1975
UF: Fuel burners—Pulverized coal*
BT: Fuels
RT: Pulverized fuel fired boilers

Pulverized fuel fired boilers 614 (521.1) (524)
DT: January 1993
UF: Boiler firing—Pulverized fuel*
BT: Boilers
RT: Pulverized fuel

Pulverizers
USE: Grinding mills

Pumice
USE: Abrasives

Pump linings 618.2
DT: January 1993
UF: Pumps—Lining*
BT: Linings
Pumps

Pumped storage power plants 611.1 (402.1) (706.1)
DT: January 1993
UF: Hydroelectric power plants—Pumped storage*
BT: Hydroelectric power plants
RT: AC generator motors
DC generator motors
Hydraulic turbines
Hydroelectric power
Pumping plants
Pumps

Pumping (laser) 744.1
DT: January 1993
NT: Electron beam pumping
Optical pumping
RT: Lasers

Pumping (optical)
USE: Optical pumping

Pumping plants 446 (402.1)
DT: Predates 1975
UF: Drainage—Pumping plants*
Irrigation—Pumping plants*
Pipelines—Pumping stations*
BT: Facilities
NT: Floating pumping plants
Sewage pumping plants
Underground pumping plants
RT: Drainage
Irrigation
Pipelines
Pumped storage power plants
Pumps
Water supply

*Pumping plants—Floating**
USE: Floating pumping plants

*Pumping plants—Underground**
USE: Underground pumping plants

*Pumping plants—Wind power**
USE: Wind power

Pumps 618.2
DT: Predates 1975
BT: Equipment
NT: Electromagnetic pumps
Feedwater pumps
Fuel pumps
Jet pumps
Mud pumps
Pump linings
Reciprocating pumps
Rotary pumps
Screw pumps
Sewage pumps
Submersible pumps
Sump pumps
Vacuum pumps
Well pumps
RT: Compressors
Diaphragms
Heat pump systems
Packing
Pistons
Pumped storage power plants
Pumping plants
Refrigerating machinery
Turbomachinery

*Pumps, Centrifugal**
USE: Centrifugal pumps

*Pumps, Electromagnetic**
USE: Electromagnetic pumps

* Former Ei Vocabulary term

*Pumps, Gear**
USE: Gear pumps

*Pumps, Jet**
USE: Jet pumps

*Pumps, Reciprocating**
USE: Reciprocating pumps

*Pumps, Rotary**
USE: Rotary pumps

*Pumps, Screw**
USE: Screw pumps

*Pumps, Sewage**
USE: Sewage pumps

*Pumps, Turbine**
USE: Turbine pumps

*Pumps—Ejectors**
USE: Ejectors (pumps)

*Pumps—Lining**
USE: Pump linings

*Pumps—Liquid metals**
USE: Liquid metals

*Pumps—Packing**
USE: Packing

*Pumps—Submerged motor**
USE: Submersible pumps

*Pumps—Sumps**
USE: Sump pumps

*Pumps—Vanes**
USE: Vane pumps

Punch card systems 722.2
DT: Predates 1975
UF: Punched card systems
BT: Computer peripheral equipment
RT: Data acquisition
Recording

Punch presses 603.1
DT: Predates 1975
UF: Drop presses
BT: Machine tools
RT: Punching

Punch tape systems 722.2
DT: Predates 1975
UF: Punched tape systems
BT: Computer peripheral equipment
RT: Data acquisition
Recording

Punched card systems
USE: Punch card systems

Punched tape systems
USE: Punch tape systems

Punching (535.2)
DT: January 1993
UF: Metal forming—Punching*
BT: Forming
RT: Cold working
Metal pressing
Metal stamping
Perforating
Piercing
Punch presses
Shearing
Stamping

Puncturing
USE: Piercing

Purchasing (911)
DT: January 1993
UF: Engineering—Purchasing*
RT: Management

Purging (619.1) (802.3)
DT: January 1993
UF: Pipelines—Purging*
BT: Cleaning
RT: Decontamination
Flow measurement
Pipelines
Purification

Purification (802.3)
DT: January 1993
UF: Gases—Purification*
NT: Air purification
Fuel purification
RT: Air cleaners
Air filters
Beneficiation
Cleaning
Crystallization
Decontamination
Distillation
Ion exchange
Leaching
Metal refining
Purging
Refining
Solvent extraction
Sublimation
Washing
Zone melting

Push button telephone systems 718.1
DT: January 1993
UF: Telephone—Push button systems*
BT: Telephone systems
NT: Touch tone telephone systems
RT: Telephone

Putty (412) (414)
DT: Predates 1975
BT: Cements
Sealants

PVAC
USE: Polyvinyl acetates

PVAL
USE: Polyvinyl alcohols

PVC
USE: Polyvinyl chlorides

* Former Ei Vocabulary term

PVDC
USE: Polyvinylidene chlorides

PWR reactors
USE: Pressurized water reactors

Pyrites (482.2) (545.1)
DT: Predates 1975
UF: Iron pyrites
BT: Iron compounds
Iron ores

Pyroelectricity 701.1
DT: Predates 1975
BT: Dielectric properties
Thermodynamic properties
RT: Piezoelectricity

Pyrolysis 802.2
DT: January 1993
UF: Thermal decomposition
Thermal degradation
BT: Decomposition
NT: Cracking (chemical)
RT: Calcination
Coal gasification
Degradation
Disintegration
Dissociation
Oil shale retorting

Pyrometallurgy 531.1
DT: January 1977
BT: Metallurgy
RT: Hydrometallurgy
Metal melting
Metal refining
Ore roasting
Refining
Sintering
Smelting

Pyrometers 944.5
DT: Predates 1975
BT: Temperature measuring instruments
RT: Pyrometry
Temperature measurement
Thermometers

Pyrometry 944.6
DT: January 1993
UF: Foundry practice—Pyrometry*
Iron and steel plants—Pyrometry*
BT: Temperature measurement
RT: Foundry practice
Pyrometers

Q factor measurement 942.2
DT: January 1993
UF: Electric measurements—Q factor*
BT: Electric variables measurement
RT: Q meters

Q meters 942.1
DT: Predates 1975
BT: Electric measuring instruments
RT: Q factor measurement

Q switched lasers 744.1
DT: January 1993
UF: Giant pulsed lasers
Lasers—Q switching*
BT: Lasers
RT: Electrooptical devices
Q switching

Q switching 744.8
DT: January 1993
UF: Lasers—Q switching*
Q-switching
BT: Switching
RT: Electrooptical devices
Q switched lasers

Q-switching
USE: Q switching

Quadrature mirror filters
USE: Digital filters

Quadrupole lenses
USE: Electron lenses

Quality (images)
USE: Image quality

Quality assurance 913.3
DT: October 1975
RT: Engineering
Management
Manufacture
Performance
Quality control
Reliability
Specifications
Standardization
Standards

Quality control 913.3
DT: Predates 1975
RT: Correlation methods
Decision theory
Errors
Fits and tolerances
Grading
High temperature testing
Industrial plants
Inspection
Operations research
Optimization
Performance
Probability
Production
Production control
Productivity
Quality assurance
Reliability
Sampling
Standardization
Standards
Statistical methods
Statistical process control
Statistics

* Former Ei Vocabulary term

Quality control *(continued)*
Testing
Total quality management
Value engineering

Quantum dots
USE: Semiconductor quantum dots

Quantum efficiency 931.4
DT: January 1995
UF: Efficiency (quantum)
RT: Electrons
Photocathodes
Photoelectricity
Photons
Phototubes

Quantum electronics 931.4 (744)
DT: January 1993
RT: Lasers
Masers
Photons
Quantum optics
Quantum theory

Quantum interference devices 714.2 (931.4)
DT: January 1995
NT: SQUIDs
RT: High electron mobility transistors
Semiconductor quantum wells
Tunnel diodes

Quantum mechanics
USE: Quantum theory

Quantum optics 741.1, 931.4
DT: January 1995
UF: Optical squeezing
BT: Optics
RT: Chaos theory
Coherent light
Optical bistability
Quantum electronics

Quantum theory 931.4
DT: Predates 1975
UF: Quantum mechanics
BT: Physics
Theory
RT: Elementary particles
Green's function
Mechanics
Quantum electronics
Statistical mechanics

Quantum well lasers 744.1 (931.4)
DT: January 1995
BT: Semiconductor lasers
RT: Semiconductor quantum wells

Quantum wells
USE: Semiconductor quantum wells

Quantum wires
USE: Semiconductor quantum wires

Quarries 502
DT: January 1993
UF: Quarries and quarrying*
BT: Mines
RT: Gravel
Quarrying
Rocks
Sand

*Quarries and quarrying**
USE: Quarries OR
Quarrying

*Quarries and quarrying—Screening**
USE: Screening

Quarrying 502
DT: January 1993
UF: Quarries and quarrying*
BT: Mining
RT: Gravel
Quarries
Rocks
Sand

Quartz 482.2
DT: Predates 1975
UF: Rock crystal
BT: Oxide minerals
Silica
RT: Gems
Granite
Quartz applications
Quartz deposits

Quartz applications (482.2)
DT: January 1993
BT: Applications
RT: Quartz

Quartz deposits 505.1
DT: Predates 1975
BT: Mineral resources
RT: Quartz

Quasicrystals (933)
DT: January 1995
BT: Solids
RT: Crystals

Quay walls 407.1
DT: January 1993
UF: Port structures—Quay walls*
Walls (quay)
BT: Port structures
RT: Shore protection

Quenching 537.1
DT: January 1993
UF: Heat treatment—Quenching*
Steel heat treatment—Quenching*
BT: Heat treatment
RT: Hardening
Melt spinning
Protective atmospheres
Steel heat treatment
Supersaturation

Query languages (723.1.1) (723.3)
DT: January 1993
UF: Data manipulation languages

* Former Ei Vocabulary term

Query languages *(continued)*
Database systems—Query languages*
Fourth generation languages
Information retrieval languages
Languages (query)
BT: Database systems
RT: Computer programming languages
Formal languages
Information retrieval systems
Information retrieval
Natural language processing systems

Queueing networks
USE: Computer networks

Queueing theory 922.1
DT: January 1993
UF: Probability—Queueing theory*
BT: Random processes
Theory
RT: Evaluation
Operations research
Performance
Probability
Statistical methods

Quicklime
USE: Lime

R and D management
USE: Research and development management

Ra
USE: Radium

Race automobiles
USE: Racing automobiles

Race cars
USE: Racing automobiles

Race conditions
USE: Hazards and race conditions

Racing (automobiles)
USE: Racing automobiles

Racing automobile engines 661.1
DT: January 1993
UF: Automobile engines—Racing applications*
BT: Automobile engines
RT: Racing automobiles

Racing automobiles 662.1
DT: January 1993
UF: Automobiles—Racing*
Race automobiles
Race cars
Racing (automobiles)
BT: Automobiles
RT: Racing automobile engines

Radar 716.2
DT: Predates 1975
UF: Radar—Rescue systems*
Radar—Shelters*
NT: Doppler radar
Marine radar
Meteorological radar
Military radar
Surveillance radar
Synthetic aperture radar
Tracking radar
RT: Backscattering
Doppler effect
Electronic warfare
Frequency agility
Jamming
Military equipment
Pulse modulation
Radar astronomy
Radar clutter
Radar countermeasures
Radar cross section
Radar equipment
Radar imaging
Radar interference
Radar shielding
Radar simulators
Radar stations
Radar systems
Radar theory
Radar tracking
Radar transmitters
Radar warning systems
Radio direction finding systems
Radio navigation
Tracking (position)

Radar antennas 716.2
DT: January 1993
UF: Antennas—Radar*
BT: Antennas
Radar equipment
RT: Radar cross section
Radar stations
Radomes

Radar astronomy 657.2, 716.2
DT: January 1981
BT: Astronomy
RT: Observatories
Radar

Radar circuits 716.2 (713)
DT: Predates 1975
BT: Networks (circuits)
Radar equipment

Radar clutter 716.2 (711)
DT: January 1993
UF: Clutter (radar)
Radar interference—Clutter*
BT: Radar interference
RT: Jamming
Radar
Radar displays

Radar countermeasures 716.2 (404.1)
DT: January 1993
UF: Countermeasures (radar)
Radar—Countermeasures*
BT: Military electronic countermeasures
RT: Jamming
Radar

Radar countermeasures *(continued)*
Radar interference
Signal interference

Radar cross section 716.2
DT: January 1993
UF: Radar—Cross sections*
RT: Backscattering
Electromagnetic wave scattering
Radar
Radar antennas

Radar displays 716.2
DT: January 1993
UF: Radar equipment—Displays*
BT: Instrument displays
Radar equipment
RT: Cathode ray tubes
Fluorescent screens
Radar clutter

Radar equipment 716.2 (652.3)
DT: Predates 1975
BT: Range finders
NT: Aircraft radar equipment
Radar antennas
Radar circuits
Radar displays
Radar receivers
Radar towers
Radar transmitters
Radomes
RT: Microwave filters
Microwave oscillators
Military equipment
Radar
Radar stations
Radio equipment

*Radar equipment—Displays**
USE: Radar displays

Radar imaging 716.2
DT: January 1986
BT: Imaging techniques
RT: Radar
Synthetic apertures

Radar interference 716.2 (711)
DT: Predates 1975
BT: Signal interference
NT: Radar clutter
RT: Jamming
Radar
Radar countermeasures

*Radar interference—Clutter**
USE: Radar clutter

Radar measurement 716.2 (943.2)
SN: Use of radar for measurement; for measurement of radar equipment itself, use RADAR TESTING and EQUIPMENT TESTING
DT: January 1993
UF: Radar—Lunar measurements*
Radar—Measurement application*
BT: Distance measurement

Radar receivers 716.2
DT: Predates 1975
BT: Radar equipment
Signal receivers
RT: Radio equipment
Video amplifiers

Radar reflection 716.2 (711)
DT: January 1993
UF: Radar—Reflection*
BT: Electromagnetic wave reflection

Radar shielding 716.2 (711)
DT: January 1993
UF: Radar—Shields*
BT: Electromagnetic shielding
RT: Radar

Radar simulators 716.2
DT: January 1993
UF: Radar—Simulators*
BT: Simulators
RT: Radar

Radar stations 716.2
DT: January 1993
UF: Radar systems—Stations*
Stations (radar)
BT: Radar systems
RT: Radar
Radar antennas
Radar equipment
Radar transmitters

Radar systems 716.2
DT: Predates 1975
NT: Radar stations
RT: Adaptive filtering
Radar

*Radar systems—Stations**
USE: Radar stations

Radar theory 716.2
DT: January 1993
UF: Radar—Theory*
BT: Theory
RT: Physics
Radar

Radar towers 716.2, 402.4
DT: January 1993
UF: Radar—Towers*
BT: Radar equipment
Towers

Radar tracking 716.2
DT: January 1993
UF: Radar—Tracking*
BT: Tracking (position)
RT: Doppler radar
Radar

Radar transmitters 716.2
DT: Predates 1975
BT: Radar equipment
RT: Radar
Radar stations

Radar warning systems 716.2 (404.1)
DT: January 1993
UF: Radar—Warning systems*
BT: Detectors
RT: Alarm systems
Jamming
Military radar
Radar

*Radar, Optical**
USE: Optical radar

*Radar—Countermeasures**
USE: Radar countermeasures

*Radar—Cross sections**
USE: Radar cross section

*Radar—Frequency agility**
USE: Frequency agility

*Radar—Lunar measurements**
USE: Radar measurement AND
Moon

*Radar—Marine**
USE: Marine radar

*Radar—Measurement application**
USE: Radar measurement

*Radar—Meteorological**
USE: Meteorological radar

*Radar—Military**
USE: Military radar

*Radar—Noise**
USE: Spurious signal noise

*Radar—Power supply**
USE: Electric power supplies to apparatus

*Radar—Radiation hazards**
USE: Radiation hazards

*Radar—Radomes**
USE: Radomes

*Radar—Reflection**
USE: Radar reflection

*Radar—Rescue systems**
USE: Radar

*Radar—Shelters**
USE: Radar

*Radar—Shields**
USE: Radar shielding

*Radar—Simulators**
USE: Radar simulators

*Radar—Surveillance application**
USE: Surveillance radar

*Radar—Synthetic aperture**
USE: Synthetic aperture radar

*Radar—Theory**
USE: Radar theory

*Radar—Towers**
USE: Radar towers

*Radar—Tracking**
USE: Tracking radar AND
Radar tracking

*Radar—Warning systems**
USE: Radar warning systems

Radial flow 631.1
DT: January 1993
BT: Flow of fluids
RT: Radial flow turbomachinery

Radial flow turbomachinery
(612.3) (617) (618) (631.1)
DT: January 1993
UF: Turbomachinery—Radial flow*
BT: Turbomachinery
RT: Radial flow

Radial tires (818.5)
DT: January 1993
UF: Tires—Radial*
BT: Tires

Radiant heating (641.2) (642.1) (643.1)
DT: January 1993
UF: Heating—Radiant*
BT: Heating
NT: Infrared heating
Solar heating
RT: Space heating

Radiation (657.1) (711)
DT: January 1993
UF: Ionizing radiation
NT: Antenna radiation
Cosmic rays
Electromagnetic waves
Solar radiation
RT: Irradiation
Physics
Radiation effects
Radiation hazards
Radiation protection
Radiation shielding
Radioactivity
Radioactivity measurement
Radiogenic gases
Radioisotope removal (water treatment)
Waves

Radiation (heat)
USE: Heat radiation

Radiation accidents
USE: Nuclear reactor accidents

Radiation belts (443.1) (657.1)
DT: January 1993
UF: Earth atmosphere—Radiation belts*
Van Allen belts
BT: Magnetosphere

Radiation chemistry 801.4.2
DT: October 1975
BT: Physical chemistry

* Former Ei Vocabulary term

Radiation chemistry *(continued)*
RT: Chemical reactions
Photolysis

Radiation counters 944.7
DT: Predates 1975
UF: Counters (radiation)
Ion counters
Particle counters
BT: Nuclear instrumentation
NT: Cerenkov counters
Crystal counters
Geiger counters
Proportional counters
Scintillation counters
Semiconductor counters
RT: Counting tubes
Dosimeters
Radiation detectors
Radioactivity measurement

Radiation damage (622.2) (914.1)
DT: January 1977
BT: Radiation effects
RT: Radiation hazards
Radiation protection
Radiation shielding

Radiation detectors 944.7
DT: Predates 1975
BT: Detectors
NT: Particle detectors
Photodetectors
Radiometers
RT: Dosimeters
Phosphors
Radiation counters
Radioactivity measurement

Radiation dosemeters
USE: Dosimeters

Radiation doses
USE: Dosimetry

Radiation dosimetry
USE: Dosimetry

Radiation effects (622.2) (711) (744)
DT: January 1977
BT: Effects
NT: Cathodoluminescence
Laser beam effects
Radiation damage
RT: Dosimeters
Dosimetry
Radiation
Radiation hazards
Radiation protection
Radiation shielding
Weathering

Radiation hardening (622.2) (714.2)
DT: January 1993
UF: Semiconductor devices—Radiation hardening*
RT: Protection
Semiconductor devices

Radiation hazards 914.1 (622.2) (711) (744)
DT: January 1993
UF: Aircraft—Radiation hazards*
Electromagnetic waves—Radiation hazards*
Mines and mining—Radiation hazards*
Radar—Radiation hazards*
Spacecraft—Radiation hazards*
BT: Hazards
RT: Radiation
Radiation damage
Radiation effects
Spacecraft

Radiation patterns (antenna)
USE: Directional patterns (antenna)

Radiation protection 914.1 (622.2) (711) (744)
DT: Predates 1975
BT: Protection
RT: Dosimetry
Electric equipment protection
Medicine
Radiation
Radiation damage
Radiation effects
Radiation shielding
Shielding

Radiation releases
USE: Nuclear reactor accidents

Radiation shielding (622.2) (655.1) (655.2) (711)
DT: January 1993
UF: Satellites—Radiation shielding*
Spacecraft—Radiation shielding*
BT: Shielding
NT: Accelerator shielding
Electromagnetic shielding
Reactor shielding
RT: Radiation
Radiation damage
Radiation effects
Radiation protection

Radiation therapy
USE: Radiotherapy

Radiators 616.1
DT: Predates 1975
BT: Heat exchangers
NT: Automobile radiators
RT: Cooling systems
Fans
Heat transfer

Radiators (acoustic)
USE: Acoustic radiators

Radio 716.3
SN: Very general term; prefer specific aspect
DT: Predates 1975
UF: Wireless
BT: Telecommunication
NT: Diversity reception
RT: Electromagnetic waves
Frequency allocation
High frequency telecommunication lines

* Former Ei Vocabulary term

Radio *(continued)*
Networks (circuits)
Radio astronomy
Radio broadcasting
Radio communication
Radio direction finding systems
Radio equipment
Radio interference
Radio links
Radio navigation
Radio stations
Radio studios
Radio systems
Radio telegraph
Radio telephone
Radio telescopes
Radio transmission
Reception quality
Telecommunication control

Radio altimeters 716.3, 943.3 (652.3)
SN: Radiowave-echo type
DT: Predates 1975
UF: Altimeters (radio)
BT: Radio equipment
Range finders
RT: Aircraft instruments
Surveying instruments

Radio astronomy 657.2, 716.3
DT: Predates 1975
BT: Astronomy
RT: Observatories
Radio
Radio telescopes
Radiofrequency spectroscopy
Tracking (position)

Radio balloons 716.3, 652.5
DT: January 1993
UF: Balloons—Radio communication*
BT: Balloons
Radio equipment

Radio broadcasting 716.3
DT: Predates 1975
UF: Stereophonic radio broadcasting
BT: Broadcasting
Radio communication
RT: Frequency allocation
Radio
Radio equipment
Radio links
Radio receivers
Radio stations
Radio studios
Radio transmission
Radio transmitters
Stereophonic broadcasting

*Radio broadcasting—Stereophonic signals**
USE: Stereophonic broadcasting

Radio cables
USE: Radio equipment AND
Telecommunication cables

Radio communication 716.3
DT: Predates 1975
UF: Radiocommunication
BT: Telecommunication
NT: Meteor burst communication
Radio broadcasting
Spread spectrum communication
RT: Frequency allocation
Mobile telecommunication systems
Multiplex radio transmission
Radio
Radio equipment
Radio links
Radio systems
Radio telegraph
Radio telephone
Telemetering
Vocoders

*Radio communication—Meteor bursts**
USE: Meteor burst communication

*Radio communication—Moon**
USE: Moon

Radio detectors
USE: Demodulators

Radio direction finding systems 716.3 (652.3)
DT: January 1993
UF: Direction finding systems*
Radio direction finding*
BT: Instruments
Radio equipment
NT: Adcock direction finders
RT: Aircraft instruments
Goniometers
Radar
Radio
Radio navigation
Ships

*Radio direction finding**
USE: Radio direction finding systems

Radio equipment 716.3
DT: Predates 1975
UF: Microwave radio equipment
Radio cables
Radio equipment—Embedded*
Radio shielding
Rockets and missiles—Radio equipment*
Satellites—Radio equipment*
Ships—Radio equipment*
BT: Telecommunication equipment
NT: Automobile radio equipment
Military radio equipment
Miniature radio equipment
Radio altimeters
Radio balloons
Radio direction finding systems
Radio frequency amplifiers
Radio receivers
Radio telescopes
Radio transmitters

Radio equipment *(continued)*
RT: Amplifiers (electronic)
Antennas
Capacitors
Demodulators
Electric attenuators
Electric coils
Electric connectors
Electric filters
Electric switches
Electric transformers
Loudspeakers
Microwave devices
Mobile radio systems
Modulators
Oscillators (electronic)
Radar equipment
Radar receivers
Radio
Radio broadcasting
Radio communication
Radio stations
Radio studios
Radio systems
Radio telegraph
Radio telephone
Radio towers
Rectifying circuits
Resistors
Resonators
Telecommunication cables
Telegraph equipment
Telephone equipment

*Radio equipment—Contacts**
USE: Electric contacts

*Radio equipment—Embedded**
USE: Radio equipment

*Radio equipment—High temperature**
USE: High temperature applications

*Radio equipment—Military**
USE: Military radio equipment

*Radio equipment—Miniature**
USE: Miniature radio equipment

*Radio equipment—Noise**
USE: Spurious signal noise

*Radio equipment—Power supply**
USE: Electric power supplies to apparatus

*Radio equipment—Shielding**
USE: Electromagnetic shielding

*Radio equipment—Waterproofing**
USE: Waterproofing

*Radio equipment—Wiring**
USE: Electric wiring

Radio frequency amplifiers 713.1 (716.3)
DT: January 1993
UF: Amplifiers, Radio frequency*
BT: Amplifiers (electronic)
Radio equipment

Radio industry
USE: Electronics industry

Radio interference 716.3 (711)
DT: Predates 1975
UF: Automobiles—Radio interference*
Interference (radio)
BT: Signal interference
RT: Atmospherics
Jamming
Radio
Radio transmission
White noise

Radio lines
USE: High frequency telecommunication lines

Radio links 716.3
DT: January 1993
UF: Radio relay systems
Relay systems (radio)
Telecommunication links, Radio*
BT: Radio systems
Telecommunication links
NT: Microwave links
RT: Meteor burst communication
Mobile radio systems
Mobile telecommunication systems
Radio
Radio broadcasting
Radio communication
Radio telegraph
Radio telephone
Radio transmitters
Telegraph systems
Telemetering
Telemetering systems

Radio navigation 716.3
DT: Predates 1975
BT: Navigation
RT: Global positioning system
Intelligent vehicle highway systems
Radar
Radio
Radio direction finding systems
Transponders

*Radio navigation—Global positioning system**
USE: Global positioning system

Radio propagation
USE: Radio transmission

Radio receivers 716.3
DT: Predates 1975
UF: Radios
BT: Radio equipment
Signal receivers
NT: Single sideband receivers
Transceivers
RT: Amplifiers (electronic)
Demodulation
Demodulators
Diversity reception
Electric filters

* Former Ei Vocabulary term

Radio receivers *(continued)*
Integrated circuits
Loudspeakers
Miniature radio equipment
Mobile radio systems
Oscillators (electronic)
Radio broadcasting
Radio telegraph
Radio telephone
Reception quality
Sound reproduction
Spurious signal noise
Stereophonic receivers
Superregeneration
Tuners
Tuning

*Radio receivers—Microwave frequencies**
USE: Microwave devices

*Radio receivers—Miniature**
USE: Miniature radio equipment

*Radio receivers—Noise**
USE: Spurious signal noise

*Radio receivers—Single sideband**
USE: Single sideband receivers

*Radio receivers—Stereophonic signals**
USE: Stereophonic receivers

*Radio receivers—Superregeneration**
USE: Superregeneration

Radio refractometers
USE: Refractometers

Radio relay systems
USE: Radio links

Radio shielding
USE: Electromagnetic shielding AND Radio equipment

Radio stations 716.3 (402.2)
DT: January 1993
UF: Broadcasting stations
Radio systems—Stations*
Stations (radio)
BT: Radio systems
RT: Radio
Radio broadcasting
Radio equipment
Radio studios
Radio telegraph

Radio studios 716.3 (402.2)
DT: Predates 1975
UF: Control rooms (radio)
BT: Studios
RT: Audio acoustics
Control equipment
Microphones
Radio
Radio broadcasting
Radio equipment
Radio stations
Radio systems
Sound reproduction

Radio systems 716.3
DT: Predates 1975
BT: Telecommunication systems
NT: Mobile radio systems
Radio links
Radio stations
RT: Diversity reception
High frequency telecommunication lines
Radio
Radio communication
Radio equipment
Radio studios

*Radio systems, Diversity**
USE: Diversity reception

*Radio systems, Mobile**
USE: Mobile radio systems

*Radio systems, Mobile—Cellular technology**
USE: Cellular radio systems

*Radio systems—Stations**
USE: Radio stations

Radio telegraph 716.3, 718.2
DT: Predates 1975
UF: Radiotelegraphy
RTTY
BT: Telegraph
RT: Radio
Radio communication
Radio equipment
Radio links
Radio receivers
Radio stations
Radio transmitters
Telegraph equipment
Telegraph systems

Radio telephone 716.3, 718.1
DT: Predates 1975
UF: Radio telephone—Audio induction*
Radiotelephony
BT: Telephone
NT: Multiplex radio telephone
RT: Aircraft instruments
Cellular radio systems
Cordless telephones
Crosstalk
Mobile telecommunication systems
Paging systems
Radio
Radio communication
Radio equipment
Radio links
Radio receivers
Radio transmitters
Spurious signal noise
Telephone equipment
Vocoders

*Radio telephone—Audio induction**
USE: Radio telephone

* Former Ei Vocabulary term

*Radio telephone—Multiplex systems**
USE: Multiplex radio telephone

*Radio telephone—Noise**
USE: Spurious signal noise

*Radio telephone—Paging systems**
USE: Paging systems

*Radio telephone—Power supply**
USE: Electric power supplies to apparatus

Radio telescopes 657.2, 716.3
DT: Predates 1975
UF: Radiotelescopes
BT: Radio equipment
Telescopes
RT: Antenna arrays
Antennas
Radio
Radio astronomy

Radio towers 402.4, 716.3
DT: January 1993
UF: Radio—Towers*
BT: Towers
RT: Radio equipment

Radio transmission 716.3
SN: Propagation of electromagnetic waves at radio frequencies, as distinguished from electromagnetic wave propagation in general
DT: Predates 1975
UF: Radio propagation
Radio transmission—Optical characteristics*
Radiowave propagation
NT: Fading (radio)
Multiplex radio transmission
RT: Atmospherics
Backscattering
Diversity reception
Electromagnetic compatibility
Electromagnetic wave transmission
Electromagnetic wave propagation
Ionospheric electromagnetic wave propagation
Modulation
Multiplexing equipment
Radio
Radio broadcasting
Radio interference
Radio transmitters
Radiowave propagation effects
Spread spectrum communication
Spurious signal noise

*Radio transmission—Backscattering**
USE: Electromagnetic wave backscattering

*Radio transmission—Fading**
USE: Fading (radio)

*Radio transmission—Multiplex systems**
USE: Multiplex radio transmission

*Radio transmission—Noise**
USE: Spurious signal noise

*Radio transmission—Optical characteristics**
USE: Radio transmission

*Radio transmission—Propagation effects**
USE: Radiowave propagation effects

*Radio transmission—Spread spectrum**
USE: Spread spectrum communication

Radio transmitters 716.3
DT: Predates 1975
BT: Radio equipment
Transmitters
NT: Radiosondes
Transceivers
RT: Antennas
Electromagnetic field measurement
Microphones
Miniature radio equipment
Mobile radio systems
Oscillators (electronic)
Radio broadcasting
Radio links
Radio telegraph
Radio telephone
Radio transmission

*Radio transmitters—Microwave frequencies**
USE: Microwave devices

*Radio transmitters—Miniature**
USE: Miniature radio equipment

*Radio transmitters—Power supply**
USE: Electric power supplies to apparatus

Radio waves
USE: Electromagnetic waves

*Radio—Frequency allocation**
USE: Frequency allocation

*Radio—Reception quality**
USE: Reception quality

*Radio—Towers**
USE: Radio towers

Radioactivation analysis (622.4) (801)
DT: Predates 1975
BT: Activation analysis

Radioactive elements 622.1, 804
DT: January 1993
UF: Geochemistry—Radioactive elements*
BT: Chemical elements
Radioactive materials
NT: Actinides
Astatine
Radium
Radon
Strontium

Radioactive fallout
USE: Fallout

Radioactive industrial wastes
USE: Industrial wastes AND
Radioactive wastes

Radioactive isotopes
USE: Radioisotopes

* Former Ei Vocabulary term

Radioactive materials 622.1
DT: Predates 1975
UF: Water pollution—Radioactive materials*
BT: Materials
NT: Fission products
Radioactive elements
Radioactive tracers
Radioactive wastes
Radioisotopes
Strontium alloys
Strontium compounds
Thorium alloys
Uranium alloys
Uranium compounds
RT: Hazardous materials
Health hazards
Radioactivity
Radiogenic gases
Safe handling
Water pollution

*Radioactive materials—Tracers**
USE: Radioactive tracers

Radioactive prospecting 622.4, 481.4 (501.1) (512)
DT: January 1993
UF: Geophysical prospecting—Radioactive methods*
Geophysics—Radioactivity investigations*
BT: Geophysical prospecting
RT: Radioactivity logging

Radioactive tracers 622.1, 803
DT: January 1993
UF: Radioactive materials—Tracers*
Tracers (radioactive)
BT: Radioactive materials
RT: Diagnosis
Isotopes
Nuclear medicine
Radioisotopes

Radioactive waste disposal 622.5 (452.4)
DT: January 1993
UF: Radioactive wastes—Disposal*
BT: Industrial waste disposal
NT: Radioactive waste encapsulation
Radioactive waste vitrification
RT: Geological repositories
Nuclear industry
Radioactive waste storage
Radioactive waste transportation
Radioactive wastes

Radioactive waste encapsulation 622.5 (452.4)
DT: January 1993
UF: Radioactive wastes—Encapsulation*
BT: Encapsulation
Radioactive waste disposal
RT: Radioactive wastes
Radioactive waste storage

Radioactive waste storage 622.5, 694.4 (452.4)
DT: January 1993
UF: Radioactive wastes—Storage*
BT: Storage (materials)
RT: Radioactive wastes
Radioactive waste disposal
Radioactive waste encapsulation
Radioactive waste transportation
Radioactive waste vitrification

Radioactive waste transportation 622.5 (432.3) (433.3) (434.3) (452.4)
DT: January 1993
UF: Radioactive wastes—Transportation*
BT: Transportation
RT: Radioactive wastes
Radioactive waste disposal
Radioactive waste storage

Radioactive waste vitrification 452.4, 622.5, 802.2
DT: January 1993
UF: Immobilization (radioactive wastes)
Radioactive wastes—Vitrification*
BT: Radioactive waste disposal
Vitrification
RT: Radioactive wastes
Radioactive waste storage

Radioactive wastes 622.5 (452.3)
DT: January 1986
UF: Industrial radioactive wastes
Industrial wastes—Radioactive materials*
Nuclear wastes
Radioactive industrial wastes
Sewage treatment—Radioactive materials*
BT: Radioactive materials
Wastes
RT: Fission products
Industrial wastes
Nuclear fuel reprocessing
Nuclear industry
Nuclear materials safeguards
Radioactive waste encapsulation
Radioactive waste storage
Radioactive waste transportation
Radioactive waste vitrification
Radioactive waste disposal
Radioactivity
Radioisotopes
Sewage treatment
Spent fuels
Water pollution

*Radioactive wastes—Disposal**
USE: Radioactive waste disposal

*Radioactive wastes—Encapsulation**
USE: Radioactive waste encapsulation

*Radioactive wastes—Geological repositories**
USE: Geological repositories

*Radioactive wastes—Storage**
USE: Radioactive waste storage

*Radioactive wastes—Transportation**
USE: Radioactive waste transportation

*Radioactive wastes—Vitrification**
USE: Radioactive waste vitrification

* Former Ei Vocabulary term

Radioactivity 622 (932.1)
DT: Predates 1975
UF: Ceramic products—Radioactivity*
Oceanography—Radioactivity*
Pulp—Radioactivity*
Water analysis—Radioactivity*
NT: Atmospheric radioactivity
RT: Fission products
Gamma rays
Nuclear physics
Radiation
Radioactive materials
Radioactive wastes
Radioactivity measurement
Radiogenic gases
Radioisotope removal (water treatment)
Radioisotopes

Radioactivity logging 481.4, 622.3 (512.1.2) (512.2.2)
DT: January 1993
UF: Oil well logging—Radioactive*
BT: Well logging
NT: Neutron logging
RT: Radioactive prospecting

Radioactivity measurement 622.4 (944.8)
DT: Predates 1975
BT: Measurements
RT: Dosimeters
Radiation
Radiation counters
Radiation detectors
Radioactivity

Radiocommunication
USE: Radio communication

Radiofrequency spectroscopy (801) (931.3) (932.2)
DT: January 1993
UF: RF spectroscopy
Spectroscopy, Radiofrequency*
BT: Spectroscopy
NT: Microwave spectroscopy
RT: Magnetic resonance
Nuclear magnetic resonance
Nuclear magnetic resonance spectroscopy
Radio astronomy

Radiogenic gases 622.1 (931.2)
DT: January 1993
BT: Gases
NT: Radon
RT: Radiation
Radioactive materials
Radioactivity

Radiogoniometers
USE: Goniometers

Radiography (461.1) (931.2)
DT: Predates 1975
UF: Roentgenography
X-ray photography
BT: Imaging techniques
NT: Computerized tomography
Diagnostic radiography
Gamma radiography
Industrial radiography
Neutron radiography
X ray radiography
RT: Contrast media
Crystallography
Fractography
Gamma rays
Image processing
Metallography
Nondestructive examination
Radiology
Radiotherapy
X ray analysis
X ray diffraction
X ray diffraction analysis
X rays

*Radiography—Contrast media**
USE: Contrast media

*Radiography—Diagnostic applications**
USE: Diagnostic radiography

*Radiography—Gamma ray**
USE: Gamma radiography

*Radiography—Neutron**
USE: Neutron radiography

*Radiography—X-ray**
USE: X ray radiography

Radioisotope gages 622.1.1, 943.3
DT: January 1993
UF: Gages—Radioisotope*
BT: Gages

Radioisotope removal (water treatment) 445.1, 622.1.1
DT: January 1993
UF: Water treatment—Radioisotope removal*
BT: Chemicals removal (water treatment)
RT: Radiation
Radioactivity
Radioisotopes

Radioisotopes 622.1.1
DT: January 1993
UF: Biomedical equipment—Radionuclides*
Isotopes—Radioactivity*
Nuclides—Radioactive*
Radioactive isotopes
Radionuclides
BT: Isotopes
Radioactive materials
NT: Tritium
RT: Biomedical equipment
Deuterium compounds
Heavy water
Nuclear medicine
Positron emission tomography
Radioactive tracers
Radioactive wastes

* Former Ei Vocabulary term

Radioisotopes *(continued)*
Radioactivity
Radioisotope removal (water treatment)
Radiotherapy

Radiology 622.3, 461.6
DT: January 1993
BT: Medicine
RT: Diagnostic radiography
Radiography
Radiotherapy

Radiometers 944.7
DT: Predates 1975
BT: Radiation detectors
NT: Bolometers
RT: Infrared detectors
Photometers
Radiometry
Thermal variables measurement

Radiometry 944.8
DT: January 1993
BT: Measurements
RT: Infrared detectors
Microwave measurement
Photometers
Photometry
Radiometers

Radionuclides
USE: Radioisotopes

Radiopaque agents
USE: Contrast media

Radios
USE: Radio receivers

Radiosondes 716.3, 443.2
DT: Predates 1975
BT: Radio transmitters
RT: Meteorological instruments
Sounding rockets

Radiotelegraphy
USE: Radio telegraph

Radiotelephony
USE: Radio telephone

Radiotelescopes
USE: Radio telescopes

Radiotherapy 461.6, 622.3
DT: January 1993
UF: Biomedical engineering—Radiotherapy*
Radiation therapy
BT: Patient treatment
RT: Dosimetry
Nuclear medicine
Radiography
Radioisotopes
Radiology

Radiowave propagation
USE: Radio transmission

Radiowave propagation effects 711 (716.3)
DT: January 1993
UF: Radio transmission—Propagation effects*
BT: Wave effects
RT: Backscattering
Radio transmission

Radiowaves
USE: Electromagnetic waves

Radium 622.1
DT: January 1977
UF: Ra
BT: Alkaline earth metals
Radioactive elements

Radomes 716.2 (652.3)
DT: January 1993
UF: Aircraft—Radomes*
Radar—Radomes*
BT: Antenna accessories
Radar equipment
RT: Radar antennas

Radon 622.1, 804
DT: Predates 1975
UF: Rn
BT: Inert gases
Radioactive elements
Radiogenic gases

Rail cranes
USE: Locomotive cranes

Rail guns 704.2 (404.1)
DT: January 1993
UF: Electromagnetic launchers—Rail guns*
BT: Electromagnetic launchers

Rail laying 681.1
DT: January 1993
UF: Laying (rails)
Laying (track)
Railroad track laying
Rails—Laying*
Track laying
RT: Railroad tracks

Rail motor cars 682.1.1
DT: January 1993
UF: Cars, Rail motor*
BT: Railroad cars

Rail sanders
USE: Sanding equipment

Railings 408.2 (402)
DT: January 1993
UF: Buildings—Railings*
Handrails
BT: Structural members
NT: Luminous railings
RT: Bridges
Building components

Railroad accident prevention
USE: Railroad accidents AND
Accident prevention

Railroad accidents 433, 914.1
DT: October 1975
UF: Railroad accident prevention
BT: Accidents
NT: Derailments
RT: Railroads

*Railroad accidents—Derailment**
USE: Derailments

Railroad bearings 601.2, 682.2
DT: January 1993
UF: Bearings—Railroad*
BT: Bearings (machine parts)

Railroad bridges 401.1 (681.1)
SN: Also use material or type of bridge
DT: January 1993
UF: Bridges, Railroad*
BT: Bridges
Railroad plant and structures

Railroad car buffers 682.2
DT: January 1993
UF: Buffers (railroad cars)
Cars—Buffers*
BT: Railroad car equipment

Railroad car couplings 682.2
DT: January 1993
UF: Cars—Couplings*
BT: Couplings
Railroad car equipment

Railroad car equipment 682.2
DT: January 1993
BT: Railroad cars
Vehicle parts and equipment
NT: Air cushioning
Railroad car buffers
Railroad car couplings
RT: Railroad plant and structures

Railroad cars 682.1.1
DT: January 1993
UF: Cars*
BT: Railroad rolling stock
NT: Ambulance cars
Cabooses
Dining cars
Freight cars
Mine cars
Monorail cars
Passenger cars
Rail motor cars
Railroad car equipment
Subway cars
Track test cars
RT: Axles
Riding qualities

Railroad crossings 681.1
SN: Grade-level vehicular crossings. For pipelines or cable use CROSSINGS (PIPE AND CABLE); for underpasses use RAILROAD UNDERPASSES
DT: January 1993
UF: Crossing gates (railroad)
Crossings (railroad)
Gates (railroad crossing)
Railroad plant and structures—Crossing gates*
Railroad plant and structures—Crossings*
BT: Railroad plant and structures
RT: Overpasses
Underpasses

Railroad electrification
USE: Electric railroads

Railroad overpasses
USE: Overpasses

Railroad plant and structures 681.1 (402.1)
DT: January 1993
BT: Facilities
Railroads
NT: Railroad bridges
Railroad crossings
Railroad signal systems
Railroad stations
Railroad tracks
Railroad tunnels
Railroad yards and terminals
RT: Crossings (pipe and cable)
Embankments
Maintenance of way
Overpasses
Railroad car equipment
Railroad transportation
Snow and ice removal
Underpasses

*Railroad plant and structures—Ballast**
USE: Ballast (railroad track)

*Railroad plant and structures—Crossing gates**
USE: Railroad crossings

*Railroad plant and structures—Crossings**
USE: Railroad crossings

*Railroad plant and structures—Embankments**
USE: Embankments

*Railroad plant and structures—Maintenance of way**
USE: Maintenance of way

*Railroad plant and structures—Pipeline crossings**
USE: Crossings (pipe and cable) AND
Pipelines

*Railroad plant and structures—Snow and ice removal**
USE: Snow and ice removal

*Railroad plant and structures—Stations**
USE: Railroad stations

*Railroad plant and structures—Track inspection**
USE: Inspection AND
Railroad tracks

* Former Ei Vocabulary term

*Railroad plant and structures—Track switches**
USE: Railroad track switches

*Railroad plant and structures—Track ties**
USE: Railroad ties

*Railroad plant and structures—Track**
USE: Railroad tracks

*Railroad plant and structures—Underpasses**
USE: Underpasses

*Railroad plant and structures—Weed control**
USE: Weed control

Railroad rolling stock 682.1
SN: Very general term; prefer specific types, e.g., LOCOMOTIVES, RAILROAD CARS
DT: Predates 1975
UF: Rolling stock (railroads)
BT: Ground vehicles
NT: Bogies (railroad rolling stock)
Light weight rolling stock
Locomotive cranes
Locomotives
Railroad cars
RT: Axles
Railroad transportation
Railroads
Switching (rolling stock)
Vehicle wheels

*Railroad rolling stock—Axles**
USE: Axles

*Railroad rolling stock—Bogies**
USE: Bogies (railroad rolling stock)

*Railroad rolling stock—Light weight**
USE: Light weight rolling stock

Railroad signal systems 681.3
DT: January 1993
UF: Railroad signals and signaling*
BT: Railroad plant and structures
Signal systems
NT: Interlocking signals
Obstacle detectors
Subway signal systems
RT: Centralized signal control
Electric signal systems
Signaling

*Railroad signals and signaling**
USE: Railroad signal systems

*Railroad signals and signaling—Centralized control**
USE: Centralized signal control

*Railroad signals and signaling—Interlocking**
USE: Interlocking signals

*Railroad signals and signaling—Obstacle detectors**
USE: Obstacle detectors

*Railroad signals and signaling—Power supply**
USE: Electric power supplies to apparatus

Railroad stations 681.1 (402.2)
DT: January 1993
UF: Railroad plant and structures—Stations*
Stations (railroad)
BT: Railroad plant and structures
NT: Subway stations
RT: Railroad yards and terminals

Railroad ties 681.1
DT: January 1993
UF: Railroad plant and structures—Track ties*
Ties (railroad)
BT: Railroad tracks

Railroad track laying
USE: Rail laying

Railroad track switches 681.1
DT: January 1993
UF: Railroad plant and structures—Track switches*
Track switches (railroad)
BT: Railroad tracks

Railroad tracks 681.1
DT: January 1993
UF: Railroad plant and structures—Track inspection*
Railroad plant and structures—Track*
Tracks (railroad)
BT: Railroad plant and structures
NT: Railroad ties
Railroad track switches
RT: Ballast (railroad track)
Rail laying
Rails
Track test cars

Railroad traffic control 433.4 (681.3)
DT: January 1993
UF: Railroads—Traffic control*
BT: Traffic control
RT: Railroad transportation
Railroads

Railroad transportation 433.1
DT: January 1981
UF: Land transportation (rail)
BT: Transportation
RT: Guideways
Monorails
Railroad plant and structures
Railroad rolling stock
Railroad traffic control
Railroads
Urban planning

Railroad tunnels 401.2 (681.1)
DT: January 1993
UF: Tunnels and tunneling—Railroad*
BT: Railroad plant and structures
Vehicular tunnels

Railroad underpasses
USE: Underpasses

Railroad yards and terminals 681.2 (402.1)
DT: October 1975
UF: Terminals (railroad)
Yards (railroad)
BT: Railroad plant and structures
NT: Classification yards
RT: Railroad stations

*Railroad yards and terminals—Classification yards**
USE: Classification yards

Railroads (433.1) (681) (682)
DT: Predates 1975
NT: Electric railroads
Industrial railroads
Light rail transit
Railroad plant and structures
Subways
RT: Automatic train control
Driver training
Magnetic levitation vehicles
Railroad accidents
Railroad rolling stock
Railroad traffic control
Railroad transportation
Rapid transit
Reservation systems
Rights of way
Train ferries
Urban planning

*Railroads—Automatic train control**
USE: Automatic train control

*Railroads—Industrial**
USE: Industrial railroads

*Railroads—Traffic control**
USE: Railroad traffic control

Rails 681.1
SN: Rails on which vehicles move. For railroad rails which are laid on ties, use RAILROAD TRACKS; for handrails use RAILINGS
DT: Predates 1975
BT: Equipment
NT: Crane rails
RT: Railroad tracks

*Rails—Fastening**
USE: Joining

*Rails—Joints**
USE: Joints (structural components)

*Rails—Laying**
USE: Rail laying

Rain 443.3
DT: January 1993
UF: Rain and rainfall*
Rainfall
BT: Precipitation (meteorology)
NT: Acid rain
RT: Cloud seeding
Combined sewers
Hurricanes
Meteorology
Rain gages
Storm sewers
Storms
Thunderstorms
Water
Water supply
Watersheds

*Rain and rainfall**
USE: Rain

*Rain and rainfall—Cloud seeding**
USE: Cloud seeding

*Rain and rainfall—Gages**
USE: Rain gages

Rain gages 443.2, 943.3
DT: January 1993
UF: Rain and rainfall—Gages*
BT: Liquid level indicators
RT: Rain

Rainfall
USE: Rain

Raise driving 502.1
DT: January 1993
UF: Mines and mining—Raise driving*
BT: Mining
RT: Mine shafts

RAM
USE: Random access storage

Raman effect
USE: Raman scattering

Raman scattering 741.1
DT: Predates 1975
UF: Raman effect
Raman spectra
BT: Light scattering

Raman spectra
USE: Raman scattering

Raman spectroscopy (741.1)
DT: January 1993
UF: Spectroscopy, Raman*
BT: Spectroscopy
RT: Infrared spectroscopy
Laser applications
Nonlinear optics
Rayleigh scattering

Ramjet engines 653.1
DT: January 1993
UF: Aircraft engines, Jet and turbine—Ramjet*
BT: Aircraft engines
Jet engines

Ramp function generators
USE: Ramp generators

Ramp generators 713.5
DT: Predates 1975
UF: Ramp function generators
BT: Signal generators

Random access storage 722.1
DT: January 1993
UF: Data storage, Digital—Random access*
RAM
BT: Digital storage
RT: Buffer storage
PROM
ROM
Semiconductor storage

Random electronic noise
USE: Spurious signal noise

Random number generation 922.2
DT: January 1993
UF: Mathematical statistics—Random number generation*
BT: Mathematical techniques
RT: Number theory
Random processes
Statistical methods

Random processes 922.1
DT: January 1993
UF: Probability—Random processes*
Stochastic processes
NT: Brownian movement
Markov processes
Queueing theory
RT: Chaos theory
Decision theory
Encoding (symbols)
Information theory
Monte Carlo methods
Probability
Random number generation
Signal filtering and prediction
Statistical methods

Range finders 943.1 (404.1) (941.3)
SN: General treatments of range finders, as well as optical range finders specifically. Former scope was limited to optical range finders
DT: Predates 1975
UF: Rangefinders
BT: Instruments
NT: Aneroid altimeters
Proximity indicators
Radar equipment
Radio altimeters
RT: Bombsights
Cameras
Distance measurement
Gunsights
Optical radar

Range finding
USE: Distance measurement

Rangefinders
USE: Range finders

Ranging
USE: Distance measurement

Rankine cycle 641.1
DT: January 1993
UF: Thermodynamics—Rankine cycle*
BT: Thermodynamics

Rapid prototyping 723.5
DT: January 1995
RT: Computer aided design
Computer aided manufacturing

Rapid solidification (531) (537.1) (802.3) (933)
DT: January 1995
BT: Solidification
RT: Casting
Dispersion hardening
Melt spinning
Metallic glass
Powder metallurgy

Rapid thermal annealing
USE: Annealing

Rapid transit 433.2 (403)
DT: Predates 1975
BT: Transportation
RT: Electric railroads
Light rail transit
Mass transportation
Railroads
Subways

Rare earth additions 531 (547.2)
DT: January 1993
UF: Metals and alloys—Rare earth additions*
Steelmaking—Rare earth additions*
BT: Additives
RT: Alloying
Alloying elements
Metallurgy
Rare earth alloys
Rare earth compounds
Rare earth elements
Steelmaking

Rare earth alloys 547.2
DT: January 1993
BT: Alloys
NT: Cerium alloys
Dysprosium alloys
Erbium alloys
Europium alloys
Gadolinium alloys
Holmium alloys
Lanthanum alloys
Lutetium alloys
Neodymium alloys
Praseodymium alloys
Samarium alloys
Terbium alloys
Thulium alloys
Ytterbium alloys

* Former Ei Vocabulary term

Rare earth alloys *(continued)*
RT: Rare earth additions
Rare earth compounds
Rare earth elements

Rare earth compounds (804.1) (804.2)
DT: Predates 1975
BT: Metallic compounds
NT: Cerium compounds
Dysprosium compounds
Erbium compounds
Europium compounds
Gadolinium compounds
Holmium compounds
Lanthanum compounds
Lutetium compounds
Neodymium compounds
Praseodymium compounds
Rare earths
Samarium compounds
Terbium compounds
Thulium compounds
Ytterbium compounds
RT: Rare earth additions
Rare earth alloys
Rare earth elements

Rare earth elements 547.2
DT: Predates 1975
UF: Lanthanides
BT: Nonferrous metals
NT: Cerium
Dysprosium
Erbium
Europium
Gadolinium
Holmium
Lanthanum
Lutetium
Neodymium
Praseodymium
Promethium
Samarium
Terbium
Thulium
Ytterbium
RT: Monazite
Rare earth additions
Rare earth alloys
Rare earth compounds

Rare earths 804.2 (481.2)
SN: Oxides of rare earth elements
DT: January 1993
UF: Geochemistry—Rare earths*
BT: Oxide minerals
Rare earth compounds

Rare gases
USE: Inert gases

Rasps
USE: Files and rasps

Rat control (452.1) (803) (804)
DT: January 1993
UF: Sewers—Rat eradication*
Ships—Rat eradication*
BT: Pest control
RT: Sewers
Ships

Rate constants
USE: Reaction kinetics

Rates (public utilities)
USE: Utility rates

Rating
DT: January 1993
NT: Antiknock rating
RT: Standards

Rating (personnel)
USE: Personnel rating

Raw materials
DT: January 1993
BT: Materials
NT: Feedstocks
RT: Manufacture
Mineral resources

Rayleigh fading
USE: Fading (radio)

Rayleigh scattering 711
DT: January 1995
BT: Electromagnetic wave scattering
RT: Raman spectroscopy

Rayon 819.2
DT: Predates 1975
BT: Synthetic fibers
RT: Rayon fabrics
Rayon yarn

Rayon fabrics 819.5 (819.2)
DT: Predates 1975
BT: Fabrics
RT: Rayon
Rayon yarn
Synthetic fibers

Rayon yarn 819.2
DT: Predates 1975
BT: Yarn
RT: Rayon
Rayon fabrics
Synthetic fibers

Rb
USE: Rubidium

RC engines
USE: Wankel engines

Re-entry
USE: Reentry

Reactance measurement
USE: Electric reactance measurement

Reaction injection molding 816.1
DT: January 1993
UF: Plastics—Reaction injection molding*
RIM

* Former Ei Vocabulary term

Reaction injection molding *(continued)*
BT: Injection molding
RT: Plastics forming
Plastics molding

Reaction kinetics 802.2 (801.4)
DT: January 1993
UF: Chemical kinetics
Chemical reactions—Reaction kinetics*
Kinetics of chemical reactions
Rate constants
RT: Catalysis
Catalysts
Chemical reactions
Combustion
Explosions
Molecular dynamics
Physical chemistry
Thermodynamics

Reaction mechanisms
USE: Reaction kinetics

Reaction time
USE: Human reaction time

Reactive ion etching 802.2 (531) (714.2)
DT: January 1995
UF: RIE
BT: Etching
RT: Integrated circuit manufacture
Ions

Reactive molding 816.1
DT: January 1993
UF: Plastics—Reactive molding*
BT: Plastics molding

Reactive power (703.1) (706)
DT: January 1993
UF: Electric networks—Reactive power*
Electric power systems—Reactive power*
Reactive voltamperes
Wattless power
BT: Electric properties
RT: Electric power factor
Electric power distribution
Electric power systems
Electric power transmission
Electric power transmission networks
Electric power measurement
Electric reactors

Reactive voltamperes
USE: Reactive power

Reactivity (nuclear) 622.1, 932.2.1
DT: January 1995
UF: Nuclear reactor reactivity
RT: Criticality (nuclear fission)
Nuclear reactors

Reactor cores (621.1.1)
DT: January 1993
UF: Cores (reactor)
Nuclear reactor cores
Nuclear reactors—Cores*
BT: Nuclear reactors
RT: Core disruptive accidents
Core meltdown
Moderators
Nuclear fuel elements

Reactor fuel elements
USE: Nuclear fuel elements

Reactor meltdown
USE: Core meltdown

Reactor operation 621
DT: January 1993
UF: Nuclear reactor operation
NT: Reactor refueling
Reactor startup
RT: Nuclear fuels
Nuclear reactors
Reactor shutdowns

Reactor refueling 621
DT: January 1993
UF: Nuclear reactors—Refueling*
Refueling (reactor)
BT: Reactor operation
RT: Nuclear fuels
Nuclear reactors

Reactor shielding 914.1, 621 (622.2)
DT: January 1993
UF: Nuclear reactors—Shielding*
BT: Radiation shielding
RT: Nuclear reactors
Safety devices

Reactor shutdowns 621
DT: January 1993
UF: Nuclear reactors—Shutdown*
Shutdowns (reactors)
RT: Nuclear reactors
Plant shutdowns
Reactor operation

Reactor startup 621
DT: January 1993
UF: Nuclear reactors—Start up*
Start up (reactors)
Startup (reactors)
BT: Reactor operation
RT: Nuclear reactors
Plant startup

Reactors (electric)
USE: Electric reactors

Reactors (nuclear)
USE: Nuclear reactors

Read only memory
USE: ROM

Read only storage
USE: ROM

Readout systems (941) (942) (943) (944)
DT: January 1993
UF: Instrument readouts
Instruments—Readout systems*
BT: Instrument components

* Former Ei Vocabulary term

Readout systems *(continued)*
NT: Instrument scales
Remote readouts
RT: Display devices
Indicators (instruments)
Instrument dials
Instrument displays

Real time systems 722.4
DT: January 1993
UF: Computer systems, Digital—Real time operation*
BT: Computer systems
RT: Online systems

Reamers (603.1) (605)
DT: Predates 1975
BT: Cutting tools
RT: Hand tools

Reaming 604.1
DT: Predates 1975
BT: Machining
RT: Boring
Deburring
Drilling
Metal cutting
Tapping (threads)
Trimming

Rear engine automobiles 662.1
DT: January 1993
UF: Automobiles—Rear engine*
BT: Automobiles

Reboilers
USE: Boilers

Receivers (containers) (691) (694)
DT: January 1993
UF: Compressed air—Receivers*
BT: Containers

Receivers (signal)
USE: Signal receivers

Receiving antennas (716)
DT: January 1993
UF: Antennas—Receiving*
BT: Antennas
RT: Signal receivers

Reception quality (716.3) (716.4)
DT: January 1993
UF: Radio—Reception quality*
Signal receivers—Reception quality*
Television—Reception quality*
RT: Radio
Radio receivers
Signal distortion
Signal interference
Signal processing
Signal receivers
Television
Television interference
Television picture quality

Rechargeable batteries
USE: Secondary batteries

Recharging (underground waters) 444.2
DT: January 1993
UF: Aquifers—Recharging*
Geothermal wells—Recharging*
Water, Underground—Recharging*
RT: Aquifers
Geothermal wells
Groundwater
Groundwater resources
Hydrology
Replenishment (water resources)
Water resources
Water wells

Reciprocating pumps 618.2
DT: January 1993
UF: Piston pumps
Pumps, Reciprocating*
BT: Pumps
RT: Gear pumps
Vacuum pumps

Reclamation (452.4)
DT: January 1993
UF: Cutting fluids—Reclamation*
Insulating oil—Reclamation*
Lubricants—Reclamation*
Rubber reclamation*
Sand, Foundry—Reclamation*
Shafts and shafting—Reclamation*
Sludge reclamation
Steam condensate—Oil reclamation*
Waste reclamation
Water treatment plants—Sludge reclamation*
BT: Recovery
NT: Coal reclamation
Land reclamation
Recycling
Wastewater reclamation
RT: Waste utilization

Reclamation (land)
USE: Land reclamation

Reclosing circuit breakers (704.2) (706.2)
DT: January 1993
UF: Electric circuit breakers—Reclosing*
BT: Electric circuit breakers

Recombinant DNA technology
USE: Genetic engineering

Reconnaissance aircraft (404.1) (652.1.2)
DT: January 1993
UF: Aircraft—Reconnaissance*
BT: Aircraft
RT: Jet aircraft

Reconstruction (images)
USE: Image reconstruction

Reconstruction (structural) 405.2
DT: January 1993
UF: Bridges—Reconstruction*
RT: Maintenance

* Former Ei Vocabulary term

Reconstruction (structural) *(continued)*
Repair
Restoration

Record players
USE: Phonographs

Recording (716.4) (741.3) (742.1) (752.2)
DT: Predates 1975
UF: Motion pictures—Recording and reproduction*
NT: Data recording
Image recording
Magnetic recording
Optical recording
Sound recording
Telephone traffic recording
Video recording
Well logging
RT: Digital storage
Magnetic storage
Monitoring
Motion pictures
Optical data storage
Photography
Pickups
Punch card systems
Punch tape systems
Recording instruments
Surveying
Surveys
Tape recorders
Traffic surveys

Recording instruments (941) (942) (943) (944)
DT: January 1993
UF: Data processing—Recorders*
Instruments—Recording*
BT: Instruments
NT: Carbon dioxide recorders
Heliographs
Oscillographs
Plotters (computer)
Sulfur dioxide recorders
RT: Data recording
Process control
Recording
Tape recorders
Telecontrol equipment
Telemetering equipment
Transducers

Records management (903.3) (912.2)
DT: January 1993
UF: Engineering—Record preservation*
BT: Management
RT: Information retrieval systems

Recovery (531) (622.5)
DT: January 1993
NT: Metal recovery
Nuclear fuel reprocessing
Reclamation

Recovery (waste heat)
USE: Waste heat utilization

Recreation centers 403 (402.2)
DT: Predates 1975
BT: Facilities
RT: Parks
Recreational facilities
Skating rinks
Stadiums
Swimming pools
Tennis courts
Urban planning

Recreational facilities 403 (402.2)
DT: January 1993
UF: Regional planning—Recreational facilities*
Reservoirs—Recreational facilities*
BT: Facilities
RT: Gymnasiums
Public works
Recreation centers
Regional planning
Reservoirs (water)
Skating rinks
Ski jumps
Swimming pools
Tennis courts
Urban planning

Recrystallization (metallurgy) 531.1
DT: January 1993
UF: Recrystallization*
BT: Heat treatment
RT: Epitaxial growth
Grain growth
Metallographic microstructure
Phase transitions
Strain hardening
Zone melting

Recrystallization (separation)
USE: Crystallization

*Recrystallization**
USE: Recrystallization (metallurgy)

Rectangular waveguides 714.3
DT: January 1993
UF: Waveguides, Rectangular*
BT: Waveguides

Rectifier substations 706.1.2
DT: January 1993
UF: Electric substations, Rectifier*
BT: Electric substations
RT: Electric rectifiers

Rectifier tubes
USE: Electron tube rectifiers

Rectifier valves
USE: Electron tube rectifiers

Rectifiers (electric)
USE: Electric rectifiers

Rectifying circuits (703.1) (713.5)
DT: January 1993
UF: Electronic circuits, Rectifying*
BT: Networks (circuits)

* Former Ei Vocabulary term

Rectifying circuits *(continued)*
RT: Charging (batteries)
Electric rectifiers
Power electronics
Radio equipment

Recuperative cupolas
USE: Hot blast cupolas

Recuperators 616.2, 642.2
DT: January 1993
UF: Furnaces—Recuperators*
BT: Furnaces

Recursive functions 721.1
DT: January 1993
UF: Automata theory—Recursive functions*
BT: Formal logic
Functions

Recycling 452.3
DT: January 1993
BT: Reclamation
NT: Deinking
Water recycling
RT: Energy conservation
Refining
Waste paper
Waste utilization

*Rediffusion**
USE: Broadcasting

Redox reactions 802.2
DT: January 1993
UF: Chemical reactions—Redox*
Oxidation reduction reactions
BT: Chemical reactions
RT: Oxidation
Reduction

Reduced instruction set computing (722) (723)
DT: January 1993
UF: Computer architecture—Reduced instruction set computing*
RISC
BT: Computer architecture

Reduction 802.2
SN: Chemical reaction
DT: January 1993
BT: Chemical reactions
NT: Electrolytic reduction
Ore reduction
RT: Extractive metallurgy
Oxidation
Redox reactions

Redundancy
DT: Predates 1975
RT: Control
Information theory
Majority logic
Reliability

Reed relays 704.1
DT: January 1993
UF: Electric relays, Reed*
Reed switches
BT: Electric relays

Reed switches
USE: Reed relays

Reefs 481.1 (471.1)
DT: January 1993
UF: Geology—Reefs*
BT: Landforms
RT: Oceanography
Rocks

Reeling
USE: Winding

Reels 691.2 (722.1) (742.2)
DT: January 1993
UF: Electric cables—Reels*
Magnetic tape—Spools*
Materials handling—Reels*
Photographic films and plates—Spools*
Spools
Wire—Reels*
BT: Materials handling equipment

Reentry (654.1) (656.1)
DT: January 1993
UF: Re-entry
Rockets and missiles—Re-entry*
RT: Ablation
Aerodynamic heating
Aerodynamics
Rockets
Satellites
Space flight
Spacecraft

Refineries (metal)
USE: Metal refineries

Refining (513.1) (533.2) (802.3) (811.1.1) (822.2)
DT: January 1993
BT: Processing
NT: Hydrocarbon refining
Metal refining
Pulp refining
Sugar refining
RT: Beneficiation
Cleaning
Crystallization
Desulfurization
Extractive metallurgy
Hydrometallurgy
Isomerization
Ore treatment
Purification
Pyrometallurgy
Recycling
Smelting
Zone melting

Reflectance measurement
USE: Reflectometers

Reflection (711) (741.1)
DT: January 1993
NT: Acoustic wave reflection

Reflection *(continued)*
Electromagnetic wave reflection
Electron reflection
Neutron reflection
RT: Diffusion
Energy absorption
Mirrors
Optical films
Reverberation
Scattering
Wave propagation
Wave transmission
Waves

Reflection high energy electron diffraction 931.3, 932.1
DT: January 1995
UF: RHEED
BT: High energy electron diffraction

Reflective coatings 813.2 (741.3)
DT: January 1993
UF: Coatings—Reflective*
BT: Coatings
RT: Solar control films

Reflectometers 941.3
DT: Predates 1975
UF: Light reflectometers
Reflectance measurement
Reflectometers (microwave)
Reflectometers (optical)
BT: Instruments
RT: Comparators (optical)
Microwave devices
Optical instruments
Optical variables measurement
Photometers

Reflectometers (microwave)
USE: Reflectometers

Reflectometers (optical)
USE: Reflectometers

Reflectors (optical)
USE: Mirrors

Reforming reactions 802.2 (513.1)
DT: January 1993
UF: Reforming*
BT: Chemical reactions
RT: Catalytic cracking
Petroleum refining

*Reforming**
USE: Reforming reactions

Refraction (711) (741.1)
DT: January 1993
NT: Acoustic wave refraction
Electromagnetic wave refraction
RT: Birefringence
Electromagnetic dispersion
Refractive index
Wave propagation
Waves

Refractive index 741.1
DT: January 1993
UF: Gases—Refractive index*
Refractivity
BT: Optical properties
RT: Birefringence
Dispersion (waves)
Electromagnetic wave refraction
Gases
Gradient index optics
Light refraction
Refraction
Refractometers
Ultrasonic refraction

Refractivity
USE: Refractive index

Refractometers 941.3
DT: Predates 1975
UF: Light refractometers
Radio refractometers
BT: Instruments
RT: Refractive index

Refractories
USE: Refractory materials

Refractory alloys 531 (543)
DT: January 1993
BT: Refractory materials
Transition metal alloys
NT: Chromium alloys
Molybdenum alloys
Niobium alloys
Rhenium alloys
Tantalum alloys
Tungsten alloys
Vanadium alloys
RT: Refractory metals
Refractory metal compounds

Refractory concrete 412, 812.2
DT: January 1993
UF: Refractory materials—Concrete*
BT: Concretes
Refractory materials

Refractory fibers
USE: Ceramic fibers

Refractory materials 812.2
DT: Predates 1975
UF: Refractories
BT: Materials
NT: Aluminous refractories
Carbonaceous refractories
Fireclay
Lime brick
Magnesia refractories
Magnesite refractories
Refractory alloys
Refractory concrete
Refractory metal compounds
Silica brick
Zirconia refractories
RT: Beryllia

* Former Ei Vocabulary term

Refractory materials *(continued)*
Brick
Ceramic materials
Cermets
Firing (of materials)
Heat resistance
Hot pressing
Hydrostatic pressing
Kyanite
Refractory metals
Silicon nitride
Sintering
Titanium nitride

*Refractory materials—Alumina**
USE: Aluminous refractories

*Refractory materials—Beryllia**
USE: Beryllia

*Refractory materials—Carbon**
USE: Carbonaceous refractories

*Refractory materials—Casting**
USE: Casting

*Refractory materials—Concrete**
USE: Refractory concrete

*Refractory materials—Disintegration**
USE: Disintegration

*Refractory materials—Fireclay**
USE: Fireclay

*Refractory materials—Firing**
USE: Firing (of materials)

*Refractory materials—Hot pressing**
USE: Hot pressing

*Refractory materials—Hydrostatic pressing**
USE: Hydrostatic pressing

*Refractory materials—Lime brick**
USE: Lime brick

*Refractory materials—Magnesia**
USE: Magnesia refractories

*Refractory materials—Magnesite**
USE: Magnesite refractories

*Refractory materials—Silica brick**
USE: Silica brick

*Refractory materials—Silicon nitride**
USE: Silicon nitride

*Refractory materials—Thoria**
USE: Thoria

*Refractory materials—Zirconia**
USE: Zirconia refractories

Refractory metal compounds 804.1, 812.2
DT: January 1993
BT: Refractory materials
Transition metal compounds
NT: Chromium compounds
Molybdenum compounds
Niobium compounds
Rhenium compounds
Tantalum compounds
Tungsten compounds
Vanadium compounds
RT: Refractory alloys
Refractory metals
Thoria

Refractory metals 531 (543) (812.2)
DT: Predates 1975
UF: Metallic refractories
BT: Transition metals
NT: Chromium
Molybdenum
Niobium
Rhenium
Tantalum
Tungsten
Vanadium
RT: Refractory alloys
Refractory materials
Refractory metal compounds
Superalloys

Refrigerants 644.2, 803
DT: Predates 1975
BT: Coolants
RT: Air conditioning
Ammonia
Carbon dioxide
Cooling systems
Driers (materials)
Freons
Ice
Oil pollution of refrigerants
Refrigeration
Refrigerators

*Refrigerants—Ammonia**
USE: Ammonia

*Refrigerants—Carbon dioxide**
USE: Carbon dioxide

*Refrigerants—Dryers**
USE: Driers (materials)

*Refrigerants—Flow**
USE: Flow of fluids

*Refrigerants—Freons**
USE: Freons

*Refrigerants—Oil pollution**
USE: Oil pollution of refrigerants

Refrigerated containers (644.1) (691.1) (694.4)
DT: January 1993
UF: Containers—Refrigerated*
BT: Containers
RT: Refrigerated freight handling
Refrigeration

Refrigerated freight handling (644.1) (691.1)
DT: January 1993
UF: Freight handling—Refrigerated*
BT: Cargo handling
RT: Refrigerated containers

* Former Ei Vocabulary term

Refrigerated freight handling *(continued)*
Refrigerator cars
Refrigerator trucks

Refrigerating machinery 644.3
DT: Predates 1975
BT: Cooling systems
Machinery
NT: Refrigerating piping systems
Refrigerators
Turboexpanders
RT: Air conditioning
Blowers
Capillary tubes
Compressors
Condensers (liquefiers)
Evaporators
Fans
Pistons
Pumps
Refrigeration
Refrigerator cars
Refrigerator trucks
Thermoelectric equipment

*Refrigerating machinery—Capillary tubes**
USE: Capillary tubes

*Refrigerating machinery—Condensers**
USE: Condensers (liquefiers)

*Refrigerating machinery—Evaporators**
USE: Evaporators

*Refrigerating machinery—Turboexpanders**
USE: Turboexpanders

Refrigerating piping systems 619.1, 644.3
DT: Predates 1975
BT: Piping systems
Refrigerating machinery
RT: Refrigeration

Refrigeration 644.1 (644.4)
DT: Predates 1975
BT: Cold storage
Cooling
NT: Industrial refrigeration
Solar refrigeration
Thermoelectric refrigeration
RT: Compressors
Electric power utilization
Food storage
Oil pollution of refrigerants
Refrigerants
Refrigerated containers
Refrigerating machinery
Refrigerating piping systems
Refrigerator cars
Refrigerator trucks
Refrigerators
Ventilation

*Refrigeration—Industrial**
USE: Industrial refrigeration

*Refrigeration—Solar**
USE: Solar refrigeration

*Refrigeration—Thermoelectric**
USE: Thermoelectric refrigeration

Refrigerator cars 644.3, 682.1.1
DT: January 1993
UF: Cars—Refrigerator*
BT: Freight cars
RT: Refrigerated freight handling
Refrigerating machinery
Refrigeration

Refrigerator trucks 644.3, 663.1
DT: January 1993
UF: Motor trucks—Refrigerator*
BT: Trucks
RT: Refrigerated freight handling
Refrigerating machinery
Refrigeration
Tank trucks

Refrigerators 644.3
DT: Predates 1975
BT: Refrigerating machinery
RT: Closed cycle machinery
Compressors
Defrosting
Refrigerants
Refrigeration

*Refrigerators—Closed cycle**
USE: Closed cycle machinery

*Refrigerators—Defrosting**
USE: Defrosting

*Refrigerators—Door handles**
USE: Door handles

*Refrigerators—Gas fuel**
USE: Gas appliances

*Refrigerators—Noise**
USE: Acoustic noise

Refueling (reactor)
USE: Reactor refueling

Refuse collection 452
DT: January 1993
UF: Collection (refuse)
Refuse disposal—Collection*
BT: Refuse disposal
RT: Garbage trucks

Refuse compaction 452 (802.3)
DT: January 1993
UF: Refuse disposal—Compaction*
BT: Compaction
Refuse disposal

Refuse composting 452 (801.2) (802.2)
DT: January 1993
UF: Refuse disposal—Composting*
BT: Composting
RT: Waste disposal

Refuse derived fuels 452 (522) (523) (524)
DT: January 1993
UF: Fuels—Refuse derived fuels*
BT: Fuels

* Former Ei Vocabulary term

Refuse derived fuels *(continued)*
RT: Refuse disposal
Synthetic fuels
Waste utilization

Refuse digestion 452 (801.2) (802.2)
DT: January 1993
UF: Digestion (refuse)
Refuse disposal—Digestion*
BT: Refuse disposal
Waste treatment

Refuse disposal 452
SN: Use for disposal of general refuse, as from municipalities
DT: Predates 1975
UF: Garbage disposal
Trash disposal
BT: Waste disposal
NT: Land fill
Refuse collection
Refuse compaction
Refuse digestion
Refuse incineration
RT: Composting
Environmental engineering
Refuse derived fuels
Sewage treatment
Sewers
Transfer stations
Urban planning
Waste incineration

*Refuse disposal—Collection**
USE: Refuse collection

*Refuse disposal—Compaction**
USE: Refuse compaction

*Refuse disposal—Composting**
USE: Refuse composting

*Refuse disposal—Digestion**
USE: Refuse digestion

*Refuse disposal—Incineration**
USE: Refuse incineration

*Refuse disposal—Land fill**
USE: Land fill

*Refuse disposal—Leachate treatment**
USE: Leachate treatment

*Refuse disposal—Transfer stations**
USE: Transfer stations

Refuse incineration 452
DT: January 1993
UF: Incineration (refuse)
Refuse disposal—Incineration*
BT: Refuse disposal
Waste incineration
RT: Refuse incinerators

Refuse incinerators 452, 642.2
SN: General or industrial refuse
DT: Predates 1975
UF: Buildings—Refuse incinerators*
Incinerators
BT: Furnaces
Structures (built objects)
RT: Grinding mills
Refuse incineration

*Refuse incinerators—Grinding mills**
USE: Grinding mills

Regain 819.3
DT: January 1993
UF: Moisture regain
Textiles—Regain*
RT: Diffusion
Drying
Humidity control
Moisture

Regenerators 616.1 (642.2) (653.1)
DT: January 1993
UF: Aircraft engines, Jet and turbine—Regenerators*
Furnaces—Regenerators*
Heat exchangers—Regenerators*
BT: Heat exchangers
RT: Open hearth furnaces

Regional planning 403.2
SN: Designing, developing, or otherwise controlling regions or communities on the basis of human needs and/or ecology
DT: Predates 1975
BT: Planning
RT: Air pollution control
Communication systems
Environmental engineering
Flood control
Highway planning
Land reclamation
Land use
Laws and legislation
Parks
Population statistics
Recreational facilities
Shopping centers
Transportation
Urban planning
Water conservation
Water pollution control
Water supply
Zoning

*Regional planning—Land use**
USE: Land use

*Regional planning—Parks**
USE: Parks

*Regional planning—Recreational facilities**
USE: Recreational facilities

*Regional planning—Statistical methods**
USE: Statistical methods

*Regional planning—Zoning**
USE: Zoning

Registration of engineers 901.1, 912.4
DT: January 1993
UF: Certification of engineers
Engineers—Registration*
BT: Professional aspects
RT: Engineers

Regression analysis 922.2
DT: January 1993
UF: Statistical methods—Regression analysis*
BT: Statistical methods

Regular languages
USE: Formal languages

Regulations
USE: Laws and legislation

Regulators (electric current)
USE: Electric current regulators

Rehabilitation
USE: Patient rehabilitation

Rehabilitation equipment
USE: Human rehabilitation equipment

Reheat cycle 614.2
DT: January 1993
UF: Steam power plants—Reheat cycle*
RT: Steam power plants
Steam turbines

Reinforced concrete 412
DT: January 1993
UF: Concrete construction—Reinforced concrete*
Ferrocement
BT: Concretes
RT: Concrete reinforcements

Reinforced plastics 817.1 (415.2)
DT: January 1993
UF: Plastics, Reinforced*
BT: Composite materials
Plastics
NT: Fiber reinforced plastics
Wire reinforced plastics
RT: Reinforcement

Reinforcement (412.2) (415) (816.1)
DT: January 1993
NT: Concrete reinforcements
RT: Ceramic matrix composites
Composite materials
Crystal whiskers
Reinforced plastics
Supports
Wire

Relational database systems 723.3
DT: January 1993
UF: Database systems—Relational*
BT: Database systems
RT: Data structures
Distributed database systems

Relative humidity
USE: Atmospheric humidity

Relativity 931.5
DT: Predates 1975
RT: Astronomy
Astrophysics
Physics

Relaxation oscillators 713.2
DT: January 1993
UF: Oscillators, Relaxation*
BT: Oscillators (electronic)
RT: Multivibrators

Relaxation processes 931.1
DT: January 1986
NT: Anelastic relaxation
Chemical relaxation
Dielectric relaxation
Magnetic relaxation
Stress relaxation
Ultrasonic relaxation
RT: Viscoelasticity

Relay control systems 731.1
DT: January 1993
UF: Control systems, Relay*
BT: Control systems
RT: Electric relays
Nonlinear control systems
Nonlinear systems
On-off control systems

Relay protection (706) (914.1)
SN: Protection by use of relays
DT: Predates 1975
UF: Buchholz relays
Protective relays
BT: Protection
RT: Electric grounding
Electric relays

Relay systems (radio)
USE: Radio links

Relays (electric)
USE: Electric relays

Relays (repeaters)
USE: Telecommunication repeaters

Reliability (421) (913.3) (922.2)
DT: Predates 1975
RT: Availability
Failure (mechanical)
Failure analysis
Fault tolerant computer systems
Fits and tolerances
Inspection
Instrument errors
Maintainability
Maintenance
Operations research
Outages
Performance
Production
Productivity
Quality assurance
Quality control

* Former Ei Vocabulary term

Reliability *(continued)*
Redundancy
Reliability theory
Safety factor
Sampling
Service life
Specifications
Standardization
Standards
Statistical methods
Statistical tests
Testing

Reliability theory 922.2
DT: Predates 1975
BT: Theory
RT: Failure analysis
Reliability
Statistical methods
Statistical tests

Relief valves
USE: Pressure relief valves

Religious buildings 402.2
SN: Includes all buildings designed for public worship
DT: January 1993
UF: Churches*
BT: Buildings

Relocation (619.1)
DT: January 1993
UF: Pipelines—Relocation*
BT: Location
RT: Moving
Site selection

Reluctance motors 705.3.1
DT: January 1993
UF: Electric motors, Reluctance type*
Subsynchronous reluctance motors
BT: Synchronous motors

Remanence 701.2 (708.4)
DT: January 1993
UF: Magnetic materials—Remanence*
Magnetization—Remanence*
Remanent magnetism
RT: Coercive force
Magnetic hysteresis
Magnetic materials
Magnetization

Remanent magnetism
USE: Remanence

Remelting 534.1 (545.3)
DT: January 1993
UF: Steelmaking—Remelting*
BT: Metal melting
RT: Steelmaking

Remote consoles 722.2
DT: January 1993
UF: Computer peripheral equipment—Remote consoles*
Consoles (remote)
BT: Computer terminals
RT: Control
Control equipment
Display devices

Remote control 731.1
DT: October 1975
UF: Control systems—Telecontrol*
Telecontrol
BT: Control
RT: Alarm systems
Control equipment
Process control
Remote readouts
Telecommunication equipment
Telecontrol equipment
Telemetering

Remote control equipment
USE: Telecontrol equipment

Remote metering
USE: Telemetering

Remote readouts (731.1)
DT: January 1993
UF: Instruments—Remote reading*
BT: Readout systems
RT: Remote control

Remote sensing (731.1)
DT: October 1975
RT: Geophysical prospecting
Hydrology
Infrared imaging
Military equipment
Multispectral scanners
Natural resources exploration
Oceanography
Photogrammetry
Photography
Pollution control

*Remote sensing—Environmental applications**
USE: Environmental engineering

*Remote sensing—Multispectral scanners**
USE: Multispectral scanners

Removal
DT: January 1993
NT: Chemicals removal (water treatment)
Color removal (water treatment)
Nitrogen removal

Renewable energy resources 525.1
DT: January 1993
UF: Energy resources—Renewable*
BT: Energy resources
NT: Biomass
Geothermal energy
Solar energy
Water power
Wind power
RT: Low grade fuel firing
Power generation
Wood

* Former Ei Vocabulary term

Repair 913.5
DT: January 1993
RT: Maintenance
Pavement overlays
Prevention
Reconstruction (structural)
Restoration
Retreading
Retrofitting

Repeaters
USE: Telecommunication repeaters

Replenishment (water resources) 444
DT: January 1993
UF: Water resources—Replenishment*
RT: Recharging (underground waters)

Report generators 723.1.1
DT: January 1993
UF: Computer operating systems—Report generators*
BT: Computer programming languages
RT: Computer operating systems

Repository disposal
USE: Geological repositories

Repressuring 511.1
SN: Forcing a fluid under pressure into an oil well in order to increase the recovery of oil
DT: January 1993
UF: Oil well production—Repressuring*
BT: Oil well production

Reprocessing (nuclear fuel)
USE: Nuclear fuel reprocessing

Reprocessing (scrap metal)
USE: Scrap metal reprocessing

Repulsion motors
USE: AC motors

Rescue (mining accidents)
USE: Mine rescue

Rescue services (helicopter)
USE: Helicopter rescue services

Rescue vessels 674.1, 914.1 (672.2)
DT: Predates 1975
UF: Submersibles—Rescue vessels*
BT: Water craft
RT: Submersibles

Research 901.3
DT: January 1993
NT: Behavioral research
Coal research
Engineering research
Experiments
Flame research
Industrial research
Iron research
Petroleum research
Sleep research
Space research
Steel research
RT: Engineering
Natural sciences
Physics
Research aircraft
Research laboratories
Research reactors
Societies and institutions
Theory

Research aircraft 901.3, 652.1
DT: January 1993
UF: Aircraft, Research*
BT: Aircraft
RT: Research
Supersonic aircraft

Research and development management 901.3, 912.2
DT: January 1993
UF: Industrial research management
Management—Research and development application*
R and D management
BT: Management
RT: Forecasting
Industrial research

Research laboratories 901.3 (402)
DT: Predates 1975
BT: Laboratories
RT: Research

Research reactors 621, 901.3
DT: January 1993
UF: Nuclear reactors—Research reactors*
BT: Nuclear reactors
RT: Atomic physics
Nuclear physics
Research

Reservation systems 723.2 (431) (432) (433) (434)
DT: January 1993
UF: Computer reservation systems
Transportation—Reservation systems*
RT: Administrative data processing
Air transportation
Hotels
Motels
Railroads
Transportation

Reserves to production ratio 512.1
DT: January 1993
UF: Oil fields—Reserves to production ratio*
RT: Oil fields

Reservoir evaluation
USE: Petroleum reservoir evaluation

Reservoir inlets 441.2
DT: January 1993
UF: Inlets (reservoir)
Reservoirs—Inlets*
BT: Reservoirs (water)

Reservoir linings (441.2) (813.2)
DT: January 1993

* Former Ei Vocabulary term

Reservoir linings *(continued)*
UF: Reservoirs—Lining*
BT: Linings
Reservoirs (water)

Reservoirs (petroleum)
USE: Petroleum reservoirs

Reservoirs (water) 441.2
DT: January 1993
UF: Reservoirs*
Water reservoirs
BT: Hydraulic structures
NT: Reservoir inlets
Reservoir linings
Underground reservoirs
RT: Algae
Dams
Lakes
Potable water
Recreational facilities
Rock mechanics
Seepage
Thermal stratification
Water bacteriology
Water supply
Weed control

*Reservoirs**
USE: Reservoirs (water)

*Reservoirs—Algae**
USE: Algae

*Reservoirs—Bacteriology**
USE: Water bacteriology

*Reservoirs—Inlets**
USE: Reservoir inlets

*Reservoirs—Lining**
USE: Reservoir linings

*Reservoirs—Recreational facilities**
USE: Recreational facilities

*Reservoirs—Rock mechanics**
USE: Rock mechanics

*Reservoirs—Roofs**
USE: Roofs

*Reservoirs—Seepage**
USE: Seepage

*Reservoirs—Thermal stratification**
USE: Thermal stratification

*Reservoirs—Underground**
USE: Underground reservoirs

Residual fuels 523 (513.3)
DT: January 1993
UF: Heavy fuels
Oil fuel—Residual*
BT: Fuel oils

Residual stresses (421)
DT: January 1993
UF: Internal stresses
Materials—Residual stresses*
BT: Stresses
RT: Creep
Strain hardening
Stress analysis

Resins
USE: Polymers

Resistance (electric)
USE: Electric resistance

Resistance welding 538.2.1
DT: January 1993
UF: Electric resistance welding
Welding, Electric resistance*
BT: Electric welding
Pressure welding
NT: Flash welding
Percussion welding
Projection welding
Spot welding

Resistance welding machines 538.2.2
DT: January 1993
UF: Welding machines—Resistance*
BT: Welding machines

Resistivity (electric)
USE: Electric conductivity

Resistors (704) (706) (714)
DT: Predates 1975
UF: Electric resistors
BT: Network components
NT: Photoresistors
Posistors
Potentiometers (resistors)
Thermistors
Varistors
RT: Electric conductors
Electric inductors
Preferred numbers
Radio equipment
Voltage dividers

Resnatron tubes
USE: Microwave tubes

Resolution (optical)
USE: Optical resolving power

Resolving power
USE: Optical resolving power

Resonance 931.1 (701) (751.1)
DT: January 1993
NT: Circuit resonance
Cyclotron resonance
Electron resonance
Magnetic resonance
RT: Cavity resonators
Dynamics
Natural frequencies
Oscillations
Resonators
Synchronization
Tuning
Vibrations (mechanical)

* Former Ei Vocabulary term

Resonant cavities
USE: Cavity resonators

Resonant circuits (703.1) (713)
DT: January 1993
UF: Electric networks—Resonant circuits*
BT: Networks (circuits)
RT: Circuit resonance
Resonators

Resonant frequencies
USE: Natural frequencies

Resonators (714) (741.3) (752.1)
DT: Predates 1975
BT: Equipment
NT: Cavity resonators
Crystal resonators
Optical resonators
RT: Natural frequencies
Oscillations
Oscillators (electronic)
Radio equipment
Resonance
Resonant circuits
Tuners
Tuning
Waveguide components

*Resonators, Cavity**
USE: Cavity resonators

*Resonators, Crystal**
USE: Crystal resonators

Resource allocation 912.2 (912.3)
DT: January 1995
RT: Computer operating systems
Management
Operations research
Scheduling

Resource valuation (501.1) (512)
DT: January 1993
UF: Mines and mining—Valuation*
Oil fields—Valuation*
Valuation (resources)
RT: Mines
Oil fields
Petroleum reservoir evaluation

Respirators 462.1
DT: January 1993
UF: Biomedical equipment—Respirators*
BT: Biomedical equipment
RT: Respiratory mechanics

Respiratory diseases
USE: Pulmonary diseases

Respiratory mechanics 461.3
DT: January 1993
UF: Biomechanics—Respiratory mechanics*
BT: Biomechanics
RT: Respirators
Respiratory therapy

Respiratory therapy 461.6
DT: January 1993
UF: Biomedical engineering—Respiratory therapy*
BT: Patient treatment
RT: Biomedical engineering
Pulmonary diseases
Respiratory mechanics

Response time (computer systems) (722.4)
DT: January 1993
RT: Computer systems
Performance

Rest areas (highway)
USE: Highway service areas

Restoration (402) (409) (913.5)
DT: January 1993
RT: Maintenance
Modernization
Reconstruction (structural)
Repair

Restoration (images)
USE: Image reconstruction

Resuscitation 461.6 (914.1)
DT: January 1993
UF: Electric accidents—Resuscitation*
BT: Patient treatment

Retail stores (402.2) (911.4)
DT: January 1993
UF: Department stores
Materials handling—Department stores*
Stores (retail)
NT: Shopping centers
RT: Store buildings

Retaining walls (405) (406) (407)
DT: Predates 1975
UF: Seawalls
Walls (retaining)
BT: Structures (built objects)
RT: Embankments
Levees
Ports and harbors
Road construction
Shore protection

Retreading 818.5, 913.5
SN: Of tires
DT: January 1993
UF: Tires—Retreading*
RT: Repair
Tires

Retrofitting 913.5
DT: January 1993
RT: Maintenance
Modernization
Repair

Return lines (condensate)
USE: Condensate return lines

Reusable rockets 654.1 (404.1)
DT: January 1993
UF: Rockets and missiles—Reusable construction*
BT: Rockets

* Former Ei Vocabulary term

Revegetation 442.2
DT: January 1993
UF: Land reclamation—Revegetation*
BT: Land reclamation

Reverberation (751.1) (751.3) (751.4)
DT: January 1993
UF: Acoustics—Reverberation*
RT: Acoustic wave reflection
Acoustics
Anechoic chambers
Architectural acoustics
Reflection

Reverse combustion 521.1
DT: January 1993
UF: Combustion—Reverse*
BT: Combustion
RT: In situ combustion

Reverse osmosis 802.3
DT: January 1993
UF: Osmosis, Reverse*
BT: Osmosis

Reversible lanes 406.1
DT: January 1993
UF: Highway systems—Reversible lanes*
BT: Roads and streets
RT: Highway systems

Revetments 407.2 (434.1)
DT: January 1993
UF: Inland waterways—Revetments*
BT: Structures (built objects)
RT: Embankments
Facings
Inland waterways

Reviews
SN: Do not use; use Treatment Code G
DT: January 1993

Rewinding
USE: Winding

Reynolds number 631.1 (931.2)
DT: January 1995
RT: Flow of fluids
Viscosity

RF spectroscopy
USE: Radiofrequency spectroscopy

RHEED
USE: Reflection high energy electron diffraction

Rhenium 549.3
DT: January 1993
UF: Rhenium and alloys*
BT: Refractory metals
RT: Rhenium alloys
Rhenium compounds
Rhenium plating

Rhenium alloys 549.3
DT: January 1993
UF: Rhenium and alloys*
BT: Refractory alloys
RT: Rhenium
Rhenium compounds

*Rhenium and alloys**
USE: Rhenium OR
Rhenium alloys

Rhenium compounds (804.1) (804.2)
DT: Predates 1975
BT: Refractory metal compounds
RT: Rhenium
Rhenium alloys

Rhenium plating 539.3, 549.3
DT: Predates 1975
BT: Plating
RT: Rhenium

Rheology 931.1 (631.1)
DT: Predates 1975
BT: Mechanics
RT: Capillarity
Elasticity
Flow measurement
Flow of solids
Non Newtonian flow
Non Newtonian liquids
Plastic flow
Plasticity
Rheometers
Viscoelasticity
Viscoplasticity
Viscosity

Rheometers 943.3
SN: For measuring viscous flow
DT: Predates 1975
BT: Flowmeters
RT: Rheology

Rhodium 547.1
DT: January 1993
UF: Rhodium and alloys*
BT: Platinum metals
RT: Rhodium alloys
Rhodium compounds
Rhodium metallography
Rhodium plating

Rhodium alloys 547.1
DT: January 1993
UF: Rhodium and alloys*
BT: Platinum metal alloys
RT: Rhodium
Rhodium compounds

*Rhodium and alloys**
USE: Rhodium OR
Rhodium alloys

Rhodium compounds (804.1) (804.2)
DT: Predates 1975
BT: Platinum metal compounds
RT: Rhodium
Rhodium alloys

Rhodium metallography 531.2, 547.1
DT: Predates 1975
BT: Metallography
RT: Rhodium

Rhodium plating 539.3, 547.1
DT: Predates 1975
BT: Plating
RT: Rhodium

Ribbon cables
USE: Flat cables

Ribonucleic acids
USE: RNA

Riccati equations 921.2 (731.1)
DT: January 1995
BT: Differential equations
RT: Control theory
Estimation

Riding qualities (662.1) (682.1.1)
DT: January 1993
UF: Automobiles—Riding qualities*
Cars—Riding qualities*
RT: Automobile seats
Automobiles
Performance
Railroad cars
Vehicle springs
Vehicle suspensions
Vehicles

RIE
USE: Reactive ion etching

Rifles
USE: Guns (armament)

Rights of way 432.4 (406.1) (406.2)
DT: January 1993
UF: Highway systems—Right of way*
Pipelines—Right of way*
RT: Highway administration
Highway systems
Laws and legislation
Pipelines
Railroads

Rigid foamed plastics 817.1
DT: January 1993
UF: Plastics, Foamed—Rigid*
BT: Foamed plastics

RIM
USE: Reaction injection molding

Ring gages 943.3
DT: January 1993
UF: Gages—Ring*
BT: Gages
RT: Rings (components)

Ring lasers 744
DT: January 1993
UF: Laser gyros
Lasers, Ring*
BT: Lasers
RT: Gyroscopes

Ring opening polymerization 815.2 (802.2)
DT: January 1993
UF: Polymerization—Ring opening*
BT: Polymerization
RT: Anionic polymerization
Cationic polymerization

Ring springs 601.2
DT: January 1993
UF: Springs—Ring*
BT: Springs (components)
RT: Rings (components)

Ringers
USE: Bells

Rings (components) 601.2
DT: January 1993
UF: Rings*
BT: Components
NT: Piston rings
RT: Ring gages
Ring springs

Rings (storage)
USE: Storage rings

*Rings**
USE: Rings (components)

RISC
USE: Reduced instruction set computing

Risers (marine)
USE: Marine risers

Risk assessment 914.1 (922.1)
DT: January 1993
UF: Risk studies—Assessment*
RT: Forecasting
Risks

Risk perception 914.1 (922.1)
DT: January 1993
UF: Risk studies—Perception*
RT: Risks

*Risk studies**
USE: Risks

*Risk studies—Assessment**
USE: Risk assessment

*Risk studies—Health risks**
USE: Health risks

*Risk studies—Occupational risks**
USE: Occupational risks

*Risk studies—Perception**
USE: Risk perception

*Risk studies—Public risks**
USE: Public risks

Risks 914.1 (922.1)
DT: January 1993
UF: Risk studies*
NT: Health risks

* Former Ei Vocabulary term

Risks *(continued)*
Occupational risks
Public risks
RT: Forecasting
Mathematical models
Prevention
Risk assessment
Risk perception

River basin projects 441
DT: Predates 1975
RT: Construction
Dams
Hydroelectric power plants
Lakes
River control
River diversion
Rivers

River control (407.2) (441) (444.1)
SN: For control of floods, use FLOOD CONTROL
DT: January 1993
UF: Rivers—Control*
BT: Control
RT: Bank protection
Dredging
Flood control
Ice control
River basin projects
River diversion
Rivers

River diversion 441
DT: January 1993
UF: Diversion (rivers)
Rivers—Diversion*
RT: River basin projects
River control
Rivers

Rivers (407.2) (444.1)
DT: Predates 1975
BT: Surface waters
RT: Banks (bodies of water)
Discharge (fluid mechanics)
Estuaries
Flood control
Inland waterways
River basin projects
River control
River diversion
Salinity measurement
Stream flow
Surface water resources
Water supply
Watersheds

*Rivers—Control**
USE: River control

*Rivers—Discharge**
USE: Discharge (fluid mechanics)

*Rivers—Diversion**
USE: River diversion

*Rivers—Estuaries**
USE: Estuaries

*Rivers—Salinity measurements**
USE: Salinity measurement

Riveting (405.2) (605)
DT: Predates 1975
BT: Joining
RT: Rivets
Sealing (closing)

Riveting machines (405.1) (603.1) (605)
DT: Predates 1975
BT: Machine tools
RT: Hand tools

Rivets (605)
DT: Predates 1975
BT: Fasteners
RT: Riveting

Rn
USE: Radon

RNA 461.2
DT: January 1993
UF: Biological materials—RNA*
Ribonucleic acids
BT: Nucleic acids

Road accidents
USE: Highway accidents

Road and street markings 406.2, 432
DT: January 1993
UF: Traffic signs, signals and markings*
RT: Highway markings

Road construction 405.2, 406
DT: January 1993
UF: Roadbuilding
Roads and streets—Construction*
BT: Construction
RT: Concrete pavements
Embankments
Highway systems
Pavements
Retaining walls
Roadbuilding machinery
Roads and streets
Soil surveys

Roadbuilding
USE: Road construction

Roadbuilding machinery 405.1, 406
DT: Predates 1975
BT: Construction equipment
Machinery
NT: Rollers (roadbuilding machinery)
RT: Concrete mixers
Earthmoving machinery
Mixers (machinery)
Road construction
Roads and streets
Striping machines

*Roadbuilding machinery—Mixers**
USE: Mixers (machinery)

*Roadbuilding machinery—Rollers**
USE: Rollers (roadbuilding machinery)

Roadbuilding materials 406 (411) (412) (817.2) (818.6)
DT: Predates 1975
BT: Building materials
RT: Aggregates
Bituminous paving materials
Concretes
Epoxy resins
Fly ash
Gravel
Lime
Limestone
Roads and streets
Rubber

*Roadbuilding materials—Aggregates**
USE: Aggregates

*Roadbuilding materials—Bituminous**
USE: Bituminous paving materials

*Roadbuilding materials—Concrete**
USE: Concretes

*Roadbuilding materials—Epoxy resins**
USE: Epoxy resins

*Roadbuilding materials—Fly ash**
USE: Fly ash

*Roadbuilding materials—Gravel**
USE: Gravel

*Roadbuilding materials—Lime**
USE: Lime

*Roadbuilding materials—Limestone**
USE: Limestone

*Roadbuilding materials—Plastics**
USE: Plastics applications

*Roadbuilding materials—Rubber**
USE: Rubber applications

Roads and streets 406.2
DT: Predates 1975
UF: Streets
BT: Public works
NT: Curbs
Curves (road)
Driveways
Escape ramps
Gravel roads
Guard rails
Intersections
Median dividers
Pavements
Reversible lanes
Roadsides
Rural roads
Shoulders (road)
Temporary roads
RT: Culverts
Drainage
Embankments
Frost effects
Highway administration
Highway engineering
Highway markings
Highway systems
Ice control
Motor transportation
Muskeg
Road construction
Roadbuilding materials
Roadbuilding machinery
Roughness measurement
Skid resistance
Slope stability
Snow and ice removal
Soil cement
Soil surveys
Street cleaning
Street lighting
Street traffic control
Traffic signals
Traffic signs
Traffic surveys
Transportation routes
Uplift pressure
Urban planning
Weigh stations
Widening (transportation arteries)

*Roads and streets—Bituminous**
USE: Bituminous paving materials

*Roads and streets—Concrete**
USE: Concrete pavements

*Roads and streets—Construction**
USE: Road construction

*Roads and streets—Curbs**
USE: Curbs

*Roads and streets—Curves**
USE: Curves (road)

*Roads and streets—Driveways**
USE: Driveways

*Roads and streets—Dust control**
USE: Dust control

*Roads and streets—Embankments**
USE: Embankments

*Roads and streets—Escape ramps**
USE: Escape ramps

*Roads and streets—Flood damage**
USE: Flood damage

*Roads and streets—Frost effect**
USE: Frost effects

*Roads and streets—Gravel**
USE: Gravel roads

*Roads and streets—Guard rails**
USE: Guard rails

* Former Ei Vocabulary term

*Roads and streets—Intersections**
USE: Intersections

*Roads and streets—Joints**
USE: Joints (structural components)

*Roads and streets—Muskeg**
USE: Muskeg

*Roads and streets—Roughness measurement**
USE: Roughness measurement

*Roads and streets—Rural**
USE: Rural roads

*Roads and streets—Shoulders**
USE: Shoulders (road)

*Roads and streets—Skid resistance**
USE: Skid resistance

*Roads and streets—Slope stability**
USE: Slope stability

*Roads and streets—Snow and ice removal**
USE: Snow and ice removal

*Roads and streets—Soil cement**
USE: Soil cement

*Roads and streets—Soil surveys**
USE: Soil surveys

*Roads and streets—Stabilization**
USE: Stabilization

*Roads and streets—Temporary**
USE: Temporary roads

*Roads and streets—Widening**
USE: Widening (transportation arteries)

Roadsides 406
DT: January 1993
UF: Highway systems—Adjacent land areas*
Highway systems—Roadside improvement*
BT: Roads and streets
RT: Embankments
Highway systems
Land use
Shoulders (road)
Weed control
Widening (transportation arteries)

Roadway supports 406 (502.2)
DT: January 1993
UF: Mines and mining—Roadway supports*
BT: Supports
RT: Mines
Mining

Robot applications 731.6
DT: January 1986
BT: Applications
NT: Robotic assembly
RT: Industrial robots
Robotics
Robots

Robot learning 731.5
DT: January 1993
UF: Robots, Industrial—Learning*
BT: Learning systems
RT: Industrial robots
Robots

Robot programming 723.1, 731.5
DT: January 1993
UF: Robots, Industrial—Teach programming*
Teach programming
BT: Computer programming
RT: Industrial robots
Robotics
Robots

Robot programming languages
USE: Computer programming languages

Robot software
USE: Computer software

Robot vision
USE: Computer vision

Robotic arms 731.5 (731.6)
DT: January 1993
UF: Robots, Industrial—Arms*
BT: Robots
RT: Industrial robots

Robotic assembly 601, 731.6
DT: January 1986
BT: Assembly
Robot applications
RT: Assembly machines
Industrial robots
Robotics

Robotic hands
USE: End effectors

Robotics 731.5
DT: January 1983
RT: Artificial intelligence
Automata theory
Automation
Computer vision
Cybernetics
Industrial robots
Motion planning
Robot applications
Robot programming
Robotic assembly
Robots

*Robotics—Vision systems**
USE: Computer vision

Robots 731.5 (731.6)
DT: June 1990
UF: Autonomous robots
BT: Equipment
NT: Anthropomorphic robots
End effectors
Grippers
Industrial robots
Intelligent robots

Robots *(continued)*
Mobile robots
Modular robots
Multipurpose robots
Programmable robots
Robotic arms
RT: Artificial intelligence
Automata theory
Computer vision
Contact sensors
Control equipment
Cybernetics
Motion planning
Navigation
Proximity sensors
Robot applications
Robot learning
Robot programming
Robotics
Servomechanisms
Stereo vision

*Robots, Industrial**
USE: Industrial robots

*Robots, Industrial—Anthropomorphic**
USE: Anthropomorphic robots

*Robots, Industrial—Arms**
USE: Robotic arms

*Robots, Industrial—Contact sensors**
USE: Contact sensors

*Robots, Industrial—Depth perception**
USE: Depth perception

*Robots, Industrial—Grippers**
USE: Grippers

*Robots, Industrial—Intelligent**
USE: Intelligent robots

*Robots, Industrial—Learning**
USE: Robot learning

*Robots, Industrial—Manipulators**
USE: Manipulators

*Robots, Industrial—Materials handling applications**
USE: Materials handling equipment

*Robots, Industrial—Mobile**
USE: Mobile robots

*Robots, Industrial—Modular**
USE: Modular robots

*Robots, Industrial—Multipurpose**
USE: Multipurpose robots

*Robots, Industrial—Programmable**
USE: Programmable robots

*Robots, Industrial—Proximity sensors**
USE: Proximity sensors

*Robots, Industrial—Socioeconomic aspects**
USE: Economic and social effects

*Robots, Industrial—Software**
USE: Computer software

*Robots, Industrial—Teach programming**
USE: Robot programming

*Robots, Industrial—Vision systems**
USE: Computer vision

Robust stability
USE: Robustness (control systems) AND System stability

Robustness (control systems) 731.1
DT: January 1993
UF: Control systems—Robustness*
Robust stability
BT: Control theory

Rock bolting (483.1) (484.2) (502.1)
DT: January 1993
UF: Landslides—Rock bolting*
Mines and mining—Rock bolting*
BT: Joining
RT: Landslides
Mining

Rock bursts 502.1, 914.1 (483.1)
DT: January 1993
UF: Mines and mining—Rock bursts*
Rockbursts
RT: Gas bursts
Hazards
Mining
Rock mechanics

Rock crystal
USE: Quartz

Rock drilling (405.2) (502.1)
DT: Predates 1975
BT: Drilling
RT: Circulating media
Drilling fluids
Earth boring machines
Explosive drilling
Jet drilling
Rock drills
Underwater drilling
Well drilling

*Rock drilling—Circulating media**
USE: Circulating media

*Rock drilling—Explosive**
USE: Explosive drilling

*Rock drilling—Jet**
USE: Jet drilling

*Rock drilling—Underwater**
USE: Underwater drilling

Rock drills (405.1) (502.2)
DT: Predates 1975
BT: Drills
RT: Drilling equipment
Rock drilling
Rocks
Vibrations (mechanical)

* Former Ei Vocabulary term

*Rock drills—Electric**
USE: Electric drills

*Rock drills—Hydraulic**
USE: Hydraulic drills

*Rock drills—Pneumatic**
USE: Pneumatic drills

Rock fill dams
USE: Embankment dams

Rock mechanics 483.1 (502.1) (931.1)
DT: Predates 1975
UF: Mines and mining—Rock mechanics*
Reservoirs—Rock mechanics*
Rock mechanics*
BT: Mechanics
RT: Geology
Mining
Mining engineering
Petrology
Reservoirs (water)
Rock bursts
Rocks
Soil mechanics

*Rock mechanics**
USE: Rock mechanics

Rock pressure 502.1 (483.1)
DT: January 1993
UF: Mines and mining—Rock pressure*
BT: Pressure
RT: Mines
Mining

Rock products 505
DT: Predates 1975
RT: Ballast (railroad track)
Rock products plants

Rock products plants 505 (402.1)
DT: Predates 1975
BT: Industrial plants
RT: Rock products
Sand and gravel plants

*Rock**
USE: Rocks

*Rock—Crystalline**
USE: Crystalline rocks

*Rock—Igneous**
USE: Igneous rocks

*Rock—Metamorphic**
USE: Metamorphic rocks

*Rock—Sedimentary**
USE: Sedimentary rocks

Rockbursts
USE: Rock bursts

Rocket boosters
USE: Boosters (rocket)

Rocket engines 654.2
DT: Predates 1975
BT: Engines
NT: Boosters (rocket)
Combustion chambers
Flame deflectors
RT: Electric propulsion
Electromagnetic propulsion
Exhaust systems (engine)
Fuel injection
Ion propulsion
Photonic propulsion
Propellants

*Rocket engines—Boosters**
USE: Boosters (rocket)

*Rocket engines—Combustion chambers**
USE: Combustion chambers

*Rocket engines—Electric propulsion**
USE: Electric propulsion

*Rocket engines—Electromagnetic propulsion**
USE: Electromagnetic propulsion

*Rocket engines—Exhausts**
USE: Exhaust systems (engine)

*Rocket engines—Flame deflectors**
USE: Flame deflectors

*Rocket engines—Fuel injection**
USE: Fuel injection

*Rocket engines—Ion propulsion**
USE: Ion propulsion

*Rocket engines—Photon propulsion**
USE: Photonic propulsion

*Rocket engines—Propellants**
USE: Propellants

Rocket guidance
USE: Electronic guidance systems

Rocket sondes
USE: Sounding rockets

Rockets 654.1 (404.1)
DT: January 1993
UF: Rockets and missiles*
BT: Equipment
NT: Military rockets
Reusable rockets
Sounding rockets
RT: Aerospace ground support
Deceleration
Electronic guidance systems
Firing cables
Fuel sloshing
Heat problems
Heat shielding
Launching
Missile launching systems
Missiles
Nose cones
Projectiles
Reentry
Rupture disks

Rockets *(continued)*
Rupture disks
Space flight
Spacecraft
Tracking (position)
Trajectories

*Rockets and missiles**
USE: Missiles OR
Rockets

*Rockets and missiles—Auxiliary power systems**
USE: Auxiliary power systems

*Rockets and missiles—Bases**
USE: Missile bases

*Rockets and missiles—Cameras**
USE: Cameras

*Rockets and missiles—Cryogenic equipment**
USE: Cryogenic equipment

*Rockets and missiles—Deceleration**
USE: Deceleration

*Rockets and missiles—Firing cables**
USE: Firing cables

*Rockets and missiles—Flame shields**
USE: Shielding

*Rockets and missiles—Fuel sloshing**
USE: Fuel sloshing

*Rockets and missiles—Fuel tanks**
USE: Fuel tanks

*Rockets and missiles—Guidance**
USE: Electronic guidance systems

*Rockets and missiles—Heat problems**
USE: Heat problems

*Rockets and missiles—Heat shields**
USE: Heat shielding

*Rockets and missiles—Hydraulic equipment**
USE: Hydraulic equipment

*Rockets and missiles—Launching**
USE: Launching

*Rockets and missiles—Marine**
USE: Marine missiles

*Rockets and missiles—Military**
USE: Military rockets OR
Missiles

*Rockets and missiles—Nose cones**
USE: Nose cones

*Rockets and missiles—Orbits and trajectories**
USE: Orbits OR
Trajectories

*Rockets and missiles—Pneumatic equipment**
USE: Pneumatic equipment

*Rockets and missiles—Propellants**
USE: Propellants

*Rockets and missiles—Radio equipment**
USE: Radio equipment

*Rockets and missiles—Re-entry**
USE: Reentry

*Rockets and missiles—Reusable construction**
USE: Reusable rockets

*Rockets and missiles—Shielding**
USE: Shielding

*Rockets and missiles—Shock absorbers**
USE: Shock absorbers

*Rockets and missiles—Silos**
USE: Missile silos

*Rockets and missiles—Simulators**
USE: Simulators

*Rockets and missiles—Sounding**
USE: Sounding rockets

*Rockets and missiles—Tracking**
USE: Tracking (position)

*Rockets and missiles—Underwater**
USE: Underwater launched missiles

Rocks (481.1)
DT: January 1993
UF: Rock*
BT: Materials
NT: Crystalline rocks
Sedimentary rocks
RT: Acoustic logging
Aquifers
Geology
Gravel
Lithology
Minerals
Natural resources
Petrology
Quarries
Quarrying
Reefs
Rock drills
Rock mechanics
Sand
Tectonics
Weathering

Roentgenography
USE: Radiography

Roll bonding 535.1
DT: January 1993
UF: Sheet and strip metal—Roll bonding*
BT: Bonding
RT: Laminating
Sealing (closing)
Sheet metal
Strip metal

* Former Ei Vocabulary term

Roller bearings 601.2
DT: January 1993
UF: Bearings—Roller*
BT: Antifriction bearings
RT: Ball bearings
Thrust bearings

Rollers (machine components) 601.2
DT: January 1993
UF: Metal finishing—Rollers*
BT: Machine components
RT: Metal finishing
Rolls (machine components)

Rollers (roadbuilding machinery) 405.1, 406
DT: January 1993
UF: Roadbuilding machinery—Rollers*
BT: Roadbuilding machinery

Rolling (535.1) (818.3)
DT: January 1993
UF: Rubber products—Rolling*
Sheet and strip metal—Rolling*
BT: Metal forming
NT: Thread rolling
RT: Extrusion
Forging
Metal cladding
Rolling mill practice
Rolling mills
Straightening

Rolling mill practice 535.1.2
DT: Predates 1975
NT: Cold rolling
Hot rolling
RT: Bars (metal)
Billets (metal bars)
Blooms (metal)
Descaling
Plate metal
Rolling
Rolling mills
Sheet metal
Soaking pits
Strip mills
Tube mills

*Rolling mill practice—Bars**
USE: Bars (metal)

*Rolling mill practice—Billet**
USE: Billet mills

*Rolling mill practice—Bloom**
USE: Blooming mills

*Rolling mill practice—Cold rolling**
USE: Cold rolling

*Rolling mill practice—Hot rolling**
USE: Hot rolling

*Rolling mill practice—Light metals**
USE: Light metals

*Rolling mill practice—Pipe**
USE: Pipe

*Rolling mill practice—Plate**
USE: Plate metal

*Rolling mill practice—Powder metals**
USE: Powder metals

*Rolling mill practice—Scale removal**
USE: Descaling

*Rolling mill practice—Sheet and strip**
USE: Sheet metal OR
Strip mills

*Rolling mill practice—Slab**
USE: Slab mills

*Rolling mill practice—Soaking pits**
USE: Soaking pits

*Rolling mill practice—Trimming**
USE: Trimming

*Rolling mill practice—Tube**
USE: Tube mills

Rolling mills 535.1.1
DT: Predates 1975
UF: Mills (rolling)
BT: Industrial plants
NT: Aluminum rolling mills
Cold rolling mills
Hot rolling mills
Strip mills
Tube mills
RT: Bars (metal)
Billets (metal bars)
Blooms (metal)
Iron and steel plants
Plate metal
Rolling
Rolling mill practice
Sheet metal
Strip metal

*Rolling mills—Gears**
USE: Gears

*Rolling mills—Guides**
USE: Guides (mechanical)

Rolling resistance 818.5
SN: Of tires
DT: January 1993
UF: Tires—Rolling resistance*
BT: Friction
RT: Tires
Traction (friction)

Rolling stock (railroads)
USE: Railroad rolling stock

Rolls (machine components) 601.2
DT: January 1993
UF: Papermaking machinery—Rolls*
Printing machinery—Rolls*
Textile machinery—Rolls*
BT: Machine components
RT: Papermaking machinery
Rollers (machine components)
Textile machinery

* Former Ei Vocabulary term

*Rolls—Cast iron**
USE: Cast iron

*Rolls—Plastics**
USE: Plastics applications

*Rolls—Spalling**
USE: Spalling

*Rolls—Steel**
USE: Steel

ROM 722.1
DT: January 1993
UF: Data storage, Digital—ROM*
Read only memory
Read only storage
BT: Digital storage
NT: PROM
RT: Compact disks
Random access storage
Semiconductor storage

Roof control (mines)
USE: Mine roof control

Roof coverings 402
DT: January 1993
UF: Coverings (roof)
Roofs—Coverings*
RT: Roofs

Roof supports (mine)
USE: Mine roof supports

Roofs (402) (619.2)
SN: Coverings for buildings or other structures
DT: Predates 1975
UF: Buildings—Roofs*
Inflatable roofs
Reservoirs—Roofs*
Tanks—Roofs*
BT: Structures (built objects)
NT: Cable supported roofs
Cable suspended roofs
Concrete shell roofs
Prefabricated roofs
RT: Building components
Heat losses
Mine roof control
Ponding
Roof coverings
Tanks (containers)
Wind stress

*Roofs—Cable supported**
USE: Cable supported roofs

*Roofs—Cable suspended**
USE: Cable suspended roofs

*Roofs—Concrete shell**
USE: Concrete shell roofs

*Roofs—Coverings**
USE: Roof coverings

*Roofs—Heat losses**
USE: Heat losses

*Roofs—Ponding**
USE: Ponding

*Roofs—Prefabricated**
USE: Prefabricated roofs

*Roofs—Wind stresses**
USE: Wind stress

Room and pillar mining 502.1
DT: January 1993
UF: Mines and mining—Room and pillar*
Pillar and room mining
Stall and pillar mining
BT: Mining
RT: Pillar extraction

Root loci 731.1
DT: January 1993
UF: Control systems—Root loci*
RT: Control system analysis
Control system synthesis
Control systems
Frequency domain analysis
Number theory
Poles and zeros
Stability
Stability criteria

Rope (535) (819.4)
DT: Predates 1975
UF: Cord
Cordage
BT: Materials
NT: Wire rope
RT: Cables

Rotary combustion engines
USE: Wankel engines

Rotary converters 704.1
DT: January 1993
UF: Dynamotors
Electric converters, Rotary*
BT: Power converters
RT: Electric machinery
Electric rectifiers

Rotary engines 612.1
DT: January 1993
UF: Internal combustion engines—Rotary*
BT: Engines
NT: Wankel engines

Rotary flow
USE: Rotational flow

Rotary kilns 642.2
DT: January 1993
UF: Kilns—Rotary*
BT: Kilns

Rotary machinery
USE: Rotating machinery

Rotary magnetic amplifiers
USE: Rotating magnetic amplifiers

Rotary mud
USE: Drilling fluids

* Former Ei Vocabulary term

Rotary pumps 618.2
DT: January 1993
UF: Pumps, Rotary*
BT: Pumps
NT: Centrifugal pumps
Gear pumps
Turbine pumps
Vane pumps

Rotating disks 601.2
DT: January 1993
UF: Disks—Rotating*
BT: Disks (machine components)

Rotating machinery 601.1
DT: January 1993
UF: Machinery—Rotating*
Rotary machinery
BT: Machinery
NT: Rotating magnetic amplifiers
RT: Rotors
Stators

Rotating magnetic amplifiers 713.1
DT: January 1993
UF: Amplidynes
Amplifiers, Rotary magnetic*
Rotary magnetic amplifiers
BT: Magnetic amplifiers
Rotating machinery

Rotation 931.1 (601.1)
DT: January 1993
UF: Shaft whirling
Shafts and shafting—Whirling*
Whirling
BT: Dynamics

Rotational casting
USE: Rotational molding

Rotational flow 631.1
DT: January 1993
UF: Flow of fluids—Rotating*
Rotary flow
BT: Flow of fluids

Rotational molding 816.1
DT: January 1993
UF: Plastics—Rotational molding*
Rotational casting
Rotomolding
BT: Plastics molding

Rotomolding
USE: Rotational molding

Rotors 601.2
DT: Predates 1975
UF: Turbomachine impellers
Turbomachinery—Impellers*
BT: Engines
NT: Helicopter rotors
RT: Rotating machinery
Turbomachinery
Wings

Rotors (helicopter)
USE: Helicopter rotors

Rotors (windings) 704.1
DT: January 1993
UF: Electric machinery—Rotors*
BT: Machine windings
RT: AC generators
AC machinery
AC motors
Electric equipment protection
Stators

Roughness
USE: Surface roughness

Roughness measurement (406) (811.1) (943.3)
DT: January 1993
UF: Paper—Roughness measurement*
Roads and streets—Roughness measurement*
Surfaces—Roughness measurement*
BT: Surface measurement
RT: Paper
Roads and streets
Surface roughness
Surfaces

Rowing machines
USE: Exercise equipment

RTTY
USE: Radio telegraph

Ru
USE: Ruthenium

Rubber 818.1
DT: Predates 1975
UF: Heat transfer—Rubber*
Natural rubber
Rubber reclamation*
BT: Materials
Natural polymers
RT: Blowing agents
Chlorinated rubber paint
Foamed rubber
Latexes
Roadbuilding materials
Rubber additions
Rubber applications
Rubber coatings
Rubber compounding
Rubber containers
Rubber factories
Rubber forming dies
Rubber industry
Rubber machinery
Rubber plantations
Rubber products
Rubber testing
Synthetic rubber
Vulcanization

* Former Ei Vocabulary term

Rubber additions 818.5 (411)
DT: January 1993
UF: Bituminous materials—Rubber additions*
BT: Additives
RT: Bituminous materials
Rubber
Rubber products

Rubber applications 818.6
DT: January 1993
UF: Aircraft materials—Rubber*
Automobile materials—Rubber*
Building materials—Rubber*
Electric insulating materials—Rubber*
Roadbuilding materials—Rubber*
BT: Applications
RT: Rubber
Rubber industry
Rubber products

Rubber base adhesives 818.5
DT: January 1993
BT: Adhesives
Rubber products
RT: Adhesive joints

Rubber coatings 813.2, 818.5
DT: January 1993
UF: Protective coatings—Rubber*
BT: Coatings
RT: Rubber
Rubber products

Rubber compounding 818.3
DT: Predates 1975
BT: Compounding (chemical)
RT: Rubber

Rubber containers 818.5 (691.1) (694.4)
DT: January 1993
BT: Containers
RT: Rubber

Rubber factories 818.4 (402.1)
DT: Predates 1975
BT: Industrial plants
RT: Rubber
Rubber industry
Rubber machinery
Rubber molding
Rubber molds

Rubber films 818.5
DT: January 1993
BT: Films
Rubber products

Rubber forming dies 818.4
DT: January 1993
BT: Dies
RT: Rubber

Rubber industry 818.4 (912.1)
DT: Predates 1975
BT: Industry
RT: Chemical industry
Rubber
Rubber applications
Rubber factories
Rubber machinery
Rubber plantations
Tire industry

*Rubber industry—Plantations**
USE: Rubber plantations

Rubber machinery 818.4
DT: Predates 1975
BT: Machinery
RT: Rubber
Rubber factories
Rubber industry
Rubber molds

Rubber molding 818.3
DT: January 1993
UF: Rubber products—Molding*
BT: Molding
RT: Rubber factories
Rubber molds
Rubber products

Rubber molds 818.5
SN: Molds made of rubber; for molds for rubber objects, use RUBBER PRODUCTS and MOLDS
DT: January 1993
UF: Plastic and rubber molds*
BT: Molds
Rubber products
RT: Rubber factories
Rubber machinery
Rubber molding

Rubber plantations 818.1, 821.3
DT: January 1993
UF: Rubber industry—Plantations*
BT: Farms
RT: Rubber
Rubber industry

Rubber products 818.5
DT: Predates 1975
UF: Biomedical equipment—Rubber*
Containers—Rubber*
Dies—Rubber*
Hose—Rubber*
Springs—Rubber*
Tools, jigs and fixtures—Rubber*
NT: Magnetic rubber products
Rubber base adhesives
Rubber films
Rubber molds
Rubber thread
Tires
RT: Rubber
Rubber additions
Rubber applications
Rubber coatings
Rubber molding
Windshield wipers

*Rubber products—Magnetic**
USE: Magnetic rubber products

* Former Ei Vocabulary term

*Rubber products—Molding**
USE: Rubber molding

*Rubber products—Rolling**
USE: Rolling

*Rubber products—Thread**
USE: Rubber thread

*Rubber reclamation**
USE: Rubber AND
Reclamation

Rubber testing 818 (421) (422.2) (423.2)
DT: Predates 1975
BT: Materials testing
RT: Rubber

Rubber thread 818.5
DT: January 1993
UF: Rubber products—Thread*
Thread (rubber)
BT: Rubber products
Yarn

Rubber tires
USE: Tires

*Rubber, Foamed**
USE: Foamed rubber

*Rubber, Synthetic**
USE: Synthetic rubber

Rubidium 549.1
DT: Predates 1975
UF: Rb
BT: Alkali metals
RT: Rubidium alloys
Rubidium compounds

Rubidium alloys 549.1
DT: January 1993
BT: Alkali metal alloys
RT: Rubidium
Rubidium compounds

Rubidium compounds (804.1) (804.2)
DT: Predates 1975
BT: Alkali metal compounds
RT: Rubidium
Rubidium alloys

Ruby 482.2.1
DT: January 1986
BT: Gems
RT: Corundum

Rudders 671.2
DT: January 1993
UF: Ships—Rudders*
BT: Ship steering equipment

Rule based programming
USE: Logic programming

Rule based systems
USE: Knowledge based systems

Running in 661.1
DT: January 1993
UF: Automobile engines—Running in*
RT: Automobile engines

Runoff 442.1, 444.1
DT: Predates 1975
BT: Mass transfer
NT: Agricultural runoff
RT: Combined sewers
Flood control
Storm sewers
Stream flow
Water resources
Water supply
Watersheds

Runway foaming 431.4
DT: January 1993
UF: Airport runways—Foaming*
Foaming (runways)
BT: Fire protection
RT: Airport runways

Runway markings 431.4
DT: January 1993
UF: Airport runways—Markings*
Markings (runway)
RT: Airport runways
Highway markings

Runways (airport)
USE: Airport runways

Runways (crane)
USE: Crane runways

Rupture disks (601.2) (619.1.1) (914.1)
DT: January 1993
UF: Burst disks
Bursting disks
Pipelines—Rupture disks*
BT: Disks (machine components)
Pressure relief valves
RT: Pipelines
Rockets

Rural areas (403.2) (821)
DT: January 1993
UF: Rural*
RT: Agriculture
Farms
Rural roads

Rural roads 406.2 (821)
DT: January 1993
UF: Roads and streets—Rural*
BT: Roads and streets
RT: Rural areas

*Rural**
USE: Rural areas

Ruthenium 547.1
DT: January 1993
UF: Ru
Ruthenium and alloys*
BT: Platinum metals
RT: Ruthenium alloys
Ruthenium compounds

* Former Ei Vocabulary term

Ruthenium alloys 547.1
DT: January 1993
UF: Ruthenium and alloys*
BT: Platinum metal alloys
RT: Ruthenium
Ruthenium compounds

*Ruthenium and alloys**
USE: Ruthenium OR
Ruthenium alloys

Ruthenium compounds (804.1) (804.2)
DT: Predates 1975
BT: Platinum metal compounds
RT: Ruthenium
Ruthenium alloys

Rutherford backscattering spectroscopy (801) (931.2) (932.2)
DT: January 1995
BT: Spectroscopy
RT: Backscattering
Charged particles

S
USE: Sulfur

Saccades
USE: Eye movements

Saccharification 802.2 (801.2) (811.3)
DT: January 1993
BT: Hydrolysis
RT: Cellulose
Enzymes
Glucose

Safe handling 914.1, 691.2
DT: January 1993
BT: Materials handling
RT: Accident prevention
Radioactive materials

Safeguards (nuclear material)
USE: Nuclear materials safeguards

Safety
USE: Accident prevention

*Safety codes**
USE: Codes (standards)

Safety devices 914.1
DT: January 1993
UF: Gas appliances—Safety devices*
Industrial plants—Safety devices*
Mine hoists—Safety devices*
BT: Equipment
NT: Automobile safety devices
Ear protectors
Fire detectors
Gas masks
Goggles
Guards (shields)
RT: Accident prevention
Alarm systems
Blowout preventers
Fire alarm systems
Fire extinguishers
Fire fighting equipment
Gas detectors
Guard rails
Hazards
Lifesaving equipment
Lightning arresters
Parachutes
Protection
Protective clothing
Reactor shielding
Safety glass
Safety valves
Security systems
Sprinkler systems (fire fighting)

Safety engineering
USE: Accident prevention

Safety factor 914.1 (408.1) (652.1) (662.1)
DT: January 1993
UF: Aircraft—Safety factor*
Automobiles—Safety factor*
Structural design—Safety factor*
NT: Load limits
RT: Accident prevention
Reliability
Structural design

Safety glass 914.1 (662.1) (812.3)
DT: January 1993
UF: Glass—Safety*
BT: Glass
RT: Automobile windshields
Goggles
Safety devices

Safety systems
USE: Security systems

Safety testing 914.1 (423.2)
DT: January 1993
BT: Testing
NT: Flammability testing
RT: Equipment testing
Materials testing

Safety valves 619.1.1, 914.1
DT: Predates 1975
BT: Valves (mechanical)
RT: Boilers
Pressure vessels
Safety devices
Steam pipelines

Sailboats
USE: Sailing vessels

Sailing vessels 674.1
DT: Predates 1975
UF: Sailboats
Sailships
BT: Water craft
RT: Boats
Ships
Single point mooring

Sailships
USE: Sailing vessels

* Former Ei Vocabulary term

Salaries
USE: Wages

Sales
USE: Marketing

Saline water 444
DT: January 1993
UF: Salt water
Water resources—Saline water*
BT: Water
NT: Brines
Seawater
RT: Salinity measurement
Water resources

Salinity measurement (444) (471.4) (483.1) (943.3)
DT: January 1993
UF: Oceanography—Salinity measurements*
Rivers—Salinity measurements*
Soils—Salinity measurements*
BT: Measurements
RT: Oceanography
Rivers
Saline water
Soils

Salt
USE: Sodium chloride

Salt bath furnaces 532.4
DT: January 1993
UF: Furnaces—Salt bath*
Salt pot furnaces
BT: Heat treating furnaces

Salt deposits 505.1
SN: Natural occurrences of halite or sodium chloride
DT: Predates 1975
UF: Halite deposits
BT: Mineral resources
RT: Salt mines
Sodium chloride
Sodium deposits

Salt mines 505.1
DT: January 1993
UF: Salt mines and mining*
BT: Mines
RT: Salt deposits
Sodium chloride

*Salt mines and mining**
USE: Salt mines

Salt of tartar
USE: Potash

Salt pot furnaces
USE: Salt bath furnaces

Salt removal 802.3 (513.2) (445.1)
DT: January 1993
UF: Desalting
Petroleum refining—Salt removal*
BT: Separation
NT: Desalination
RT: Petroleum refining

Salt water
USE: Saline water

Salt water barriers 407.2
DT: January 1993
UF: Inland waterways—Salt water barriers*
BT: Structures (built objects)
RT: Inland waterways
Port structures
Salt water intrusion

Salt water intrusion 444, 471.4
DT: January 1993
UF: Seawater intrusion
Water resources—Salt water intrusion*
RT: Salt water barriers
Water resources

Salts (804.1) (804.2)
DT: January 1977
BT: Chemical compounds
NT: Fused salts
Sodium chloride
RT: Ammonium compounds
Esters
Ethers
Hydrates
Ketones

*Salts—Fused**
USE: Fused salts

Salvage
USE: Salvaging

Salvaging 914.1 (652.1) (671)
DT: January 1993
UF: Aircraft—Salvaging*
Salvage
Ship salvaging*
RT: Accidents
Aircraft
Pontoons
Ships
Transportation

Samarium 547.2
DT: January 1993
UF: Samarium and alloys*
BT: Rare earth elements
RT: Samarium alloys
Samarium compounds

Samarium alloys 547.2
DT: January 1993
UF: Samarium and alloys*
BT: Rare earth alloys
RT: Samarium
Samarium compounds

*Samarium and alloys**
USE: Samarium OR
Samarium alloys

Samarium compounds (804.1) (804.2)
DT: Predates 1975
BT: Rare earth compounds

* Former Ei Vocabulary term

Samarium compounds *(continued)*
NT: Semiconducting samarium compounds
RT: Samarium
Samarium alloys

Sampled data control systems 731.1
DT: January 1993
UF: Control systems, Sampled data*
BT: Discrete time control systems
RT: Sampling

Sampling 801 (913.3) (922.1)
DT: Predates 1975
NT: Ore sampling
RT: Analysis
Assays
Chemical analysis
Core samples
Inspection
Probability
Pulse modulation
Quality control
Reliability
Sampled data control systems
Standards
Statistical methods

Sand 483.1 (505)
DT: January 1993
UF: Sand and gravel*
BT: Materials
NT: Foundry sand
Silica sand
RT: Granular materials
Gravel
Quarries
Quarrying
Rocks
Sand and gravel plants
Sandstone
Silica
Soils

Sand and gravel plants 483 (402.1) (505)
DT: Predates 1975
BT: Industrial plants
RT: Gravel
Rock products plants
Sand

*Sand and gravel**
USE: Gravel OR
Sand

Sand consolidation 511.1
DT: January 1993
UF: Oil wells—Sand consolidation*
RT: Geology
Natural gas wells
Oil well production
Oil wells
Well completion

*Sand, Foundry**
USE: Foundry sand

*Sand, Foundry—Binders**
USE: Binders

*Sand, Foundry—Moisture effects**
USE: Moisture

*Sand, Foundry—Olivine**
USE: Olivine

*Sand, Foundry—Reclamation**
USE: Reclamation

*Sand, Silica**
USE: Silica sand

Sanders 603.1, 811.2
DT: January 1993
UF: Woodworking machinery—Sanders*
BT: Woodworking machinery
RT: Grinding machines
Grinding wheels
Hand tools

Sanding equipment (406.1) (681.1)
SN: For application of sand to improve friction on rails, highways, etc.
DT: January 1993
UF: Locomotives—Sanding equipment*
Rail sanders
BT: Equipment
RT: Ice control
Snow and ice removal
Traction (friction)

Sandstone 482.2
DT: January 1986
BT: Sedimentary rocks
RT: Sand

Sandwich structures (415)
DT: Predates 1975
UF: Aircraft manufacture—Sandwich construction*
BT: Structures (built objects)
RT: Composite materials
Honeycomb structures
Walls (structural partitions)

Sanitary engineering 409 (454.1) (901)
DT: Predates 1975
BT: Civil engineering
RT: Environmental engineering
Municipal engineering

Sanitary landfills
USE: Land fill

Sanitary sewers 452.1
DT: January 1993
UF: Sewers—Sanitary*
BT: Sewers
RT: Combined sewers

Sanitation 454.1
DT: January 1993
UF: Beaches—Sanitation*
Industrial plants—Sanitation*
RT: Cleaning

Saponification 802.2
DT: January 1993
BT: Chemical reactions
RT: Hydrolysis
Metallic soaps

Sappare
USE: Kyanite

Sapphire 482.2.1
DT: Predates 1975
BT: Gems
RT: Corundum

Satellite communication systems 655.2.1 (716) (717) (718)
SN: For communication to and/or from satellites, as for control or telemetering
DT: January 1993
UF: Satellites—Communication systems*
BT: Telecommunication systems
RT: Satellite ground stations
Satellites

Satellite ground stations 655.2.1 (716) (717) (718)
DT: January 1993
UF: Satellites—Ground stations*
BT: Aerospace ground support
RT: Satellite communication systems
Satellite relay systems
Satellites
Telecommunication equipment

Satellite interception 655.2
DT: January 1993
UF: Interception (satellite)
Satellites—Interception*
RT: Satellites

Satellite links 655.2.1 (716) (717) (718)
DT: January 1993
UF: Telecommunication links, Satellite*
BT: Extraterrestrial communication links
RT: Satellite relay systems
Satellites
Transponders

Satellite navigation aids 655.2
SN: Use of satellites as aids to navigation
DT: January 1993
UF: Satellites—Navigation aids application*
BT: Navigation systems
NT: Global positioning system
RT: Satellites

Satellite observatories 655.2 (443) (657)
DT: January 1993
UF: Satellites—Observatories*
BT: Satellites
Spacecraft observatories
NT: Astronomical satellites
RT: Airborne telescopes

Satellite relay systems 655.2.1 (716) (717) (718)
DT: January 1993
UF: Telecommunication systems, Satellite relay*
BT: Telecommunication systems
RT: Communication satellites
Global positioning system
Satellite ground stations
Satellite links

Satellite simulators (655.2) (656.2)
DT: January 1993
UF: Satellites—Simulators*
BT: Simulators
RT: Environmental chambers
Satellites

Satellite television services
USE: Subscription television

Satellites 655.2
DT: Predates 1975
UF: Artificial satellites
Inflatable satellites
BT: Spacecraft
NT: Communication satellites
Geodetic satellites
Geostationary satellites
Military satellites
Orbiting solar power plants
Satellite observatories
Weather satellites
RT: Heat problems
Interplanetary spacecraft
Launching
Meteor impacts
Microgravity processing
Navigation
Navigation systems
Orbital transfer
Orbits
Parachutes
Reentry
Satellite communication systems
Satellite ground stations
Satellite interception
Satellite links
Satellite navigation aids
Satellite simulators
Telemetering
Telemetering equipment
Tracking (position)
Trajectories

*Satellites—Astronomical**
USE: Astronomical satellites

*Satellites—Communication systems**
USE: Satellite communication systems

*Satellites—Computers**
USE: Computers

*Satellites—Defense applications**
USE: Military satellites

*Satellites—Geodetic**
USE: Geodetic satellites

*Satellites—Geostationary**
USE: Geostationary satellites

* Former Ei Vocabulary term

*Satellites—Ground stations**
USE: Satellite ground stations

*Satellites—Heat problems**
USE: Heat problems

*Satellites—Inflatable**
USE: Inflatable equipment

*Satellites—Interception**
USE: Satellite interception

*Satellites—Launching**
USE: Launching

*Satellites—Meteor impact**
USE: Meteor impacts

*Satellites—Microgravity materials processing**
USE: Microgravity processing

*Satellites—Navigation aids application**
USE: Satellite navigation aids

*Satellites—Observatories**
USE: Satellite observatories

*Satellites—Orbital laboratories**
USE: Orbital laboratories

*Satellites—Orbits and trajectories**
USE: Orbits OR
Trajectories

*Satellites—Power supply**
USE: Spacecraft power supplies

*Satellites—Radiation shielding**
USE: Radiation shielding

*Satellites—Radio equipment**
USE: Radio equipment

*Satellites—Simulators**
USE: Satellite simulators

*Satellites—Television equipment**
USE: Television equipment

*Satellites—Temperature**
USE: Temperature

*Satellites—Tracking**
USE: Tracking (position)

*Satellites—Wakes**
USE: Wakes

*Satellites—Weather**
USE: Weather satellites

Saturable core reactors 704.1
DT: Predates 1975
UF: Electric reactors, Saturable core*
Saturable reactors
Transductors
BT: Electric reactors
RT: Electric coils
Magnetic amplifiers
Magnetic cores

Saturable reactors
USE: Saturable core reactors

Saturation (materials composition) 801.4
DT: January 1995
NT: Supersaturation
RT: Solubility
Solutions
Wetting

Sawdust 811.2
DT: January 1993
UF: Concrete aggregates—Sawdust*
Pulp materials—Sawdust*
Wood waste—Sawdust*
BT: Wood wastes
RT: Concrete aggregates
Pulp materials

Sawing (604.1) (811.2)
DT: January 1993
BT: Cutting
RT: Cold working
Sawmills
Saws
Shearing
Slitting
Trimming

Sawmills 821.6 (402.1) (811.2)
DT: Predates 1975
UF: Lumber mills
BT: Industrial plants
RT: Logging (forestry)
Lumber
Sawing

Saws (603.1) (605)
DT: Predates 1975
UF: Woodworking saws
BT: Cutting tools
NT: Metal working saws
RT: Diamond cutting tools
Hand tools
Machine tools
Milling cutters
Sawing
Woodworking tools

*Saws—Diamond**
USE: Diamond cutting tools

*Saws—Metal working**
USE: Metal working saws

*Saws—Woodworking**
USE: Woodworking tools

Sb
USE: Antimony

SCADA systems 731.1
DT: January 1995
UF: Supervisory control and data acquisition systems
BT: Control systems
Data acquisition
RT: Process control

* Former Ei Vocabulary term

Scaffolds 405.1
DT: Predates 1975
BT: Construction equipment

Scale (deposits) (511.1) (539.1) (615.1) (616.1)
DT: January 1993
UF: Geothermal wells—Scale problems*
Heat exchangers—Scale formation*
Oil wells—Scale problems*
BT: Deposits
RT: Corrosion
Descaling
Geothermal wells
Oil wells

Scale removal
USE: Descaling

Scales (readouts)
USE: Instrument scales

Scales (weighing instruments) 943.3
DT: January 1993
UF: Scales and weighing*
BT: Instruments
NT: Electronic scales
Magnetic scales
Precision balances
RT: Weigh stations
Weighing

*Scales and weighing**
USE: Scales (weighing instruments) OR
Weighing

*Scales and weighing—Electronic**
USE: Electronic scales

*Scales and weighing—Magnetic**
USE: Magnetic scales

*Scales and weighing—Pneumatic**
USE: Pneumatic equipment

*Scales and weighing—Precision balances**
USE: Precision balances

*Scales and weighing—Weigh stations**
USE: Weigh stations

Scaling tubes
USE: Counting tubes

Scandium 549.3
DT: January 1993
UF: Scandium and alloys*
BT: Nonferrous metals
Transition metals
RT: Scandium alloys
Scandium compounds
Scandium deposits

Scandium alloys 549.3
DT: January 1993
UF: Scandium and alloys*
BT: Transition metal alloys
RT: Scandium
Scandium compounds

*Scandium and alloys**
USE: Scandium OR
Scandium alloys

Scandium compounds (804.1) (804.2)
DT: June 1990
BT: Transition metal compounds
RT: Scandium
Scandium alloys

Scandium deposits 504.3, 549.3
DT: Predates 1975
BT: Ore deposits
RT: Scandium

Scanning (741.3)
DT: January 1993
UF: Optical scanning
RT: Character recognition
Facsimile
Multispectral scanners
Optical character recognition

Scanning antennas (716)
DT: January 1993
UF: Antennas, Scanning*
BT: Antennas

Scanning electron microscopy (741.1) (931.2)
DT: January 1993
UF: Microscopic examination—Scanning electron microscopy*
SEM
BT: Electron microscopy
RT: Atomic force microscopy
Electron microscopes
Specimen preparation

Scanning tunneling microscopy (931.2) (933)
DT: January 1995
BT: Microscopic examination
RT: Atomic force microscopy
Electron microscopy
Electron tunneling
Nanotechnology
Surface structure

Scanning yokes
USE: Deflection yokes

Scarfing 537.1, 545.3
DT: January 1993
UF: Steel heat treatment—Scarfing*
BT: Cutting
RT: Deburring
Trimming

Scattering (711) (741.1)
DT: January 1993
NT: Acoustic wave scattering
Backscattering
Electromagnetic wave scattering
Electron scattering
Neutron scattering
RT: Attenuation
Reflection
Targets

* Former Ei Vocabulary term

Scavenging 612.1
DT: January 1993
UF: Internal combustion engines—Scavenging*
RT: Cleaning
Engine cylinders
Internal combustion engines

Scheduling 912.2
DT: January 1977
BT: Project management
RT: Critical path analysis
Decision theory
Forecasting
Game theory
Operations research
Optimization
Planning
Probability
Resource allocation
Statistical methods
Systems engineering
Turnaround time
Work-rest schedules

Schematic diagrams (703.1)
DT: January 1993
UF: Circuit schematics
Electric network schematics
Electric networks—Schematic diagrams*
Electronic circuit schematics
Network schematics
Schematics
RT: Graphic methods
Masks
Networks (circuits)

Schematics
USE: Schematic diagrams

Schlieren systems 742.2
DT: January 1993
UF: Photography—Schlieren system*
BT: Equipment
RT: Interferometry
Photography

School buildings 402.2
DT: Predates 1975
BT: Buildings
NT: College buildings

Schottky barrier diodes 714.2
DT: January 1993
UF: Semiconductor devices, Schottky barrier*
BT: Semiconductor devices
RT: Semiconductor metal boundaries

Schottky gate field effect transistors
USE: MESFET devices

Schottky noise
USE: Shot noise

Scientific computing
USE: Natural sciences computing

Scintillation 741.1
DT: Predates 1975
RT: Phosphorescence
Phosphors
Scintillation counters

Scintillation counters 944.7 (741.3)
DT: Predates 1975
UF: Scintillation detectors
Scintillation spectrometers
BT: Luminescent devices
Radiation counters
RT: Phosphors
Photoluminescence
Photomultipliers
Scintillation

Scintillation detectors
USE: Scintillation counters

Scintillation spectrometers
USE: Scintillation counters

Scintillators
USE: Phosphors

Scooters (motor)
USE: Motor scooters

Scour (401.1) (421)
DT: January 1993
UF: Bridge piers—Scour*
BT: Wear of materials
RT: Erosion
Floods

Scouring (textiles)
USE: Textile scouring

SCR
USE: Thyristors

Scrap metal 452.3, 531
DT: Predates 1975
UF: Materials handling—Scrap metal*
BT: Materials
NT: Copper scrap
Iron scrap
Steel scrap
Zinc scrap
RT: Industrial wastes
Metals
Scrap metal reprocessing

Scrap metal reprocessing 452.4, 534.2
DT: January 1993
UF: Foundries—Scrap reclamation*
Reprocessing (scrap metal)
Scrap metal—Reprocessing*
BT: Metal recovery
RT: Scrap metal

*Scrap metal—Reprocessing**
USE: Scrap metal reprocessing

Screen printing 745.1 (714.2)
DT: January 1993
UF: Printing—Screen*
BT: Printing
RT: Printed circuits

Screening 802.3 (524) (533.1) (811.1.1)
DT: January 1993
UF: Clay—Screening*
Coal preparation—Screening*
Ore treatment—Screening*
Pulp—Screening*
Quarries and quarrying—Screening*
BT: Size separation

Screens (display)
USE: Fluorescent screens

Screens (motion picture)
USE: Motion picture screens

Screens (projection)
USE: Projection screens

Screens (sizing) (405.1) (631.3) (802.1)
DT: January 1993
UF: Flow of fluids—Screens*
Screens and sieves*
BT: Classifiers
NT: Vibrating screens
RT: Particle size analysis
Sieves
Wire products
Wire screen cloth

Screens (window)
USE: Window screens

Screens (X-ray)
USE: X ray screens

*Screens and sieves**
USE: Screens (sizing) OR
Sieves

*Screens and sieves—Vibrating**
USE: Vibrating screens

Screw conveyors 692.1
DT: January 1993
UF: Auger conveyors
Conveyors—Screw*
BT: Conveyors

Screw pumps 618.2
DT: January 1993
UF: Pumps, Screw*
BT: Pumps

Screw thread cutting
USE: Thread cutting

Screw thread cutting dies
USE: Thread cutting dies

Screw thread gages 943.3
DT: January 1993
UF: Gages—Screw thread*
Thread gages
BT: Gages

Screw thread grinders
USE: Thread grinders

Screw thread rolling
USE: Thread rolling

Screw threads 601.2
DT: Predates 1975
BT: Screws
RT: Tapping (threads)
Thread cutting
Thread grinders
Thread rolling

*Screw threads—Finishing**
USE: Metal finishing

*Screw threads—Inserts**
USE: Die casting inserts

*Screw threads—Rolling**
USE: Thread rolling

*Screw threads—Tapping**
USE: Tapping (threads)

Screws 605
DT: Predates 1975
BT: Fasteners
NT: Screw threads
Self drilling screws
RT: Tapping (threads)
Thread cutting
Thread grinders
Thread rolling
Washers

*Screws—Self drilling**
USE: Self drilling screws

*Screws—Thread cutting**
USE: Thread cutting

Scrubbers 802.1 (451.2)
SN: For removal of undesired materials from process gas streams
DT: January 1977
UF: Gas purification—Scrubbers*
Separators—Gas scrubbers*
BT: Separators
RT: Air pollution control equipment
Gas fuel purification

Sea floor nodules
USE: Manganese nodules

Sea ice 471.1 (471.4)
DT: January 1993
UF: Oceanography—Sea ice*
BT: Ice
RT: Freezing
Oceanography
Seawater

Sea level 471.1
DT: January 1993
UF: Oceanography—Sea level changes*
RT: Oceanography

Seakeeping 671
SN: Ability of ships to navigate safely during storms
DT: January 1993
UF: Ships—Seakeeping*

* Former Ei Vocabulary term

Seakeeping *(continued)*
RT: Navigation
Ships

Sealants (652.2) (411) (803) (817) (818)
DT: January 1993
UF: Aircraft materials—Sealants*
Building materials—Sealants*
Joints—Sealants*
Sealing materials
Seals*
BT: Materials
NT: Putty
RT: Adhesives
Building materials
Joints (structural components)
Sealing (closing)
Seals
Tapes

Sealing (closing)
DT: January 1993
UF: Windows—Sealing*
NT: Grouting
RT: Adhesion
Bonding
Cements
Encapsulation
Gluing
Linings
Riveting
Roll bonding
Sealants
Seals
Soldering
Waterproofing

Sealing (finishing) (531) (534.2) (813.1)
DT: January 1993
UF: Foundry practice—Sealing*
Metals and alloys—Sealing*
Plastics—Sealing*
BT: Finishing
RT: Coating techniques
Foundry practice
Painting

Sealing materials
USE: Sealants

Seals 619.1.1
SN: Stoppers and other devices for minimizing leakage between stationary and/or moving surfaces. For materials designed for this purpose, use SEALANTS
DT: Predates 1975
BT: Components
NT: Gaskets
Packing
Piston rings
RT: Adhesives
Diaphragms
Joints (structural components)
Sealants
Sealing (closing)
Tapes
Valves (mechanical)

*Seals**
USE: Sealants

Seam welding 538.2.1
DT: January 1993
UF: Welding—Seam*
BT: Electric welding

Seaplanes 652.1
DT: Predates 1975
BT: Aircraft
NT: Military seaplanes
Submersible seaplanes
RT: Hydrofoils
Nuclear propulsion
Pontoons

*Seaplanes, Military**
USE: Military seaplanes

*Seaplanes—Nuclear propulsion**
USE: Nuclear propulsion

*Seaplanes—Submersible**
USE: Submersible seaplanes

Searching (underwater)
USE: Underwater probing

Searching online
USE: Online searching

Searchlights 707.2
DT: Predates 1975
BT: Electric lamps

Seashores
USE: Coastal zones

Seat belts (automobile)
USE: Automobile seat belts

Seating for disabled persons 461.5
DT: January 1993
UF: Human rehabilitation engineering—Seating for disabled*
BT: Seats
RT: Human rehabilitation engineering
Human rehabilitation equipment

Seats (461.5) (652.2) (662.1)
DT: January 1993
BT: Equipment
NT: Aircraft seats
Automobile seats
Seating for disabled persons

Seawalls
USE: Retaining walls

Seawater 471.4
DT: Predates 1975
UF: Seawater, Thermal gradients*
BT: Saline water
RT: Desalination
Marine pollution
Oceanography
Sea ice
Seawater corrosion

* Former Ei Vocabulary term

Seawater *(continued)*
Seawater effects
Wave power

Seawater corrosion 539.1, 471.4
DT: January 1993
UF: Corrosion—Seawater*
BT: Corrosion
Seawater effects
RT: Seawater

Seawater effects 471.4
DT: January 1993
UF: Concrete—Seawater effects*
BT: Effects
NT: Seawater corrosion
RT: Seawater

Seawater intrusion
USE: Salt water intrusion

*Seawater, Thermal gradients**
USE: Thermal gradients AND
Seawater

*Seawater—Salt removal**
USE: Desalination

Seaweed 471.1 (471.5)
DT: Predates 1975
BT: Algae
RT: Oceanography

Second harmonic generation (713.2) (741.1) (744.1)
DT: January 1993
BT: Harmonic generation
RT: Lasers

Secondary batteries 702.1.2
SN: Rechargeable
DT: January 1993
UF: Electric batteries, Secondary*
Rechargeable batteries
BT: Electric batteries
NT: Nickel cadmium batteries
RT: Charging (batteries)
Electrolysis
Storage battery vehicles

Secondary ion mass spectrometry (801) (943.3)
DT: January 1995
UF: SIMS
BT: Mass spectrometry
RT: Mass spectrometers

Secondary recovery
USE: Enhanced recovery

Security of data 723.2
DT: January 1993
UF: Computer security
Data processing—Security of data*
Data security
Information security
Security of information
RT: Computer crime
Cryptography
Data handling
Data processing
Security systems

Security of information
USE: Security of data

Security systems 914.1
DT: January 1986
UF: Safety systems
BT: Equipment
NT: Alarm systems
RT: Computer crime
Electric power system protection
Enclosures
Protection
Safety devices
Security of data

Sediment transport (483.1) (631.3)
DT: January 1993
UF: Flow of fluids—Sediment transport*
RT: Flow of fluids
Flow of solids
Suspensions (fluids)

Sediment traps 943.3 (444)
SN: Devices for measuring the rate of sediment accumulation on the floor of bodies of water
DT: January 1993
BT: Instruments
RT: Sedimentation

Sedimentary rocks 482.2
DT: January 1993
UF: Rock—Sedimentary*
BT: Rocks
NT: Limestone
Sandstone
Shale
RT: Petrology

Sedimentation 802.3
DT: Predates 1975
UF: Sewage treatment—Sedimentation*
BT: Deposition
NT: Paraffin sedimentation
RT: Agglomeration
Clarification
Coagulation
Crystallization
Effluents
Flocculation
Flotation
Precipitation (chemical)
Sediment traps
Sediments
Settling tanks
Sewage treatment
Sols

Sedimentation basins
USE: Settling tanks

Sedimentation tanks
USE: Settling tanks

Sedimentology 481.1
DT: January 1993
UF: Geology—Sedimentology*
Petrology—Sedimentology*
BT: Geology
RT: Petrology
Sediments

Sediments 483
DT: January 1993
UF: Soils—Sediments*
BT: Materials
RT: Sedimentation
Sedimentology
Soils

Seed 821.4
DT: January 1993
UF: Agricultural products—Seed*
BT: Agricultural products

Seeding (clouds)
USE: Cloud seeding

Seepage (407) (441)
DT: January 1993
UF: Dams—Seepage*
Inland waterways—Seepage*
Reservoirs—Seepage*
RT: Dams
Flow of fluids
Grouting
Infiltration
Inland waterways
Leakage (fluid)
Percolation (fluids)
Reservoirs (water)

Segmentation (image)
USE: Image segmentation

Segregation (metallography) 531.2
DT: January 1993
UF: Metallography—Segregation*
BT: Metallography
RT: Beneficiation
Inclusions
Smelting

Seismic prospecting 481.4, 484.1 (501.1) (512)
DT: January 1993
UF: Geophysical prospecting—Seismic*
BT: Geophysical prospecting
RT: Acoustic variables measurement
Seismology

Seismic water waves
USE: Tsunamis

Seismic waves 484
DT: Predates 1975
BT: Elastic waves
RT: Acoustic wave transmission
Earthquakes
Seismographs
Seismology
Tsunamis

Seismographs 484.1, 943.3
SN: Includes seismometers
DT: Predates 1975
BT: Instruments
RT: Acoustic variables measurement
Earthquakes
Seismic waves
Seismology

Seismology 484.1 (481.1)
DT: Predates 1975
UF: Spectrum analysis—Seismological applications*
BT: Geophysics
RT: Earth (planet)
Earthquakes
Seismic prospecting
Seismic waves
Seismographs
Shock waves
Spectrum analysis

Seizing (601.2) (931.1)
DT: January 1993
UF: Bearings—Seizure*
BT: Wear of materials
RT: Adhesion
Friction

Selection
DT: January 1993
NT: Personnel selection
RT: Computer selection and evaluation
Computer software selection and evaluation

Selectivity (catalyst)
USE: Catalyst selectivity

Selenium 549.3 (712.1.1)
DT: January 1983
BT: Nonmetals
NT: Semiconducting selenium
RT: Selenium compounds

Selenium compounds (712.1.2) (804.1) (804.2)
DT: Predates 1975
BT: Chemical compounds
NT: Semiconducting selenium compounds
RT: Selenium

Self adjusting control systems 731.1
DT: January 1993
UF: Control systems, Self adjusting*
BT: Adaptive control systems

Self drilling screws 605.2
DT: January 1993
UF: Screws—Self drilling*
BT: Screws

Self erecting cranes 693.1
DT: January 1993
UF: Cranes—Self erecting*
BT: Cranes

* Former Ei Vocabulary term

Self lubricating bearings 601.2 (607.2)
DT: January 1993
UF: Bearings—Self lubricating*
BT: Antifriction bearings

Self lubricating composites (415) (607.2) (817.1)
DT: January 1993
UF: Composite materials—Self lubricating*
BT: Composite materials
RT: Lubrication

Self organizing storage 722.1
DT: January 1993
UF: Data storage, Digital—Self organizing*
BT: Digital storage
RT: Associative storage

Self potential logging
USE: Spontaneous potential logging

Self reproducing automata 721.1
DT: January 1993
UF: Automata theory—Self reproducing automata*
BT: Automata theory

Self tuning control systems 731.1
DT: January 1993
UF: Control systems, Self tuning*
BT: Adaptive control systems

Selling
USE: Marketing

SEM
USE: Scanning electron microscopy

Semi trailers
USE: Truck trailers

Semi-submersibles
USE: Semisubmersibles

Semiconducting aluminum compounds 712.1.2
DT: Predates 1975
BT: Aluminum compounds
Semiconductor materials

Semiconducting antimony 712.1.1
DT: January 1977
BT: Antimony
Semiconductor materials

Semiconducting antimony compounds 712.1.2
DT: Predates 1975
BT: Antimony compounds
Semiconductor materials

Semiconducting bismuth compounds 712.1.2
DT: Predates 1975
BT: Bismuth compounds
Semiconductor materials

Semiconducting boron 712.1.1
DT: Predates 1975
BT: Boron
Semiconductor materials

Semiconducting cadmium compounds 712.1.2
DT: Predates 1975
BT: Cadmium compounds
Semiconductor materials
RT: Cadmium sulfide solar cells

Semiconducting diamonds 712.1.1
DT: January 1993
BT: Diamonds
Semiconductor materials

Semiconducting films 712.1
DT: Predates 1975
BT: Films
Semiconductor materials
RT: Metallorganic chemical vapor deposition
Metallorganic vapor phase epitaxy
Monolayers
Thin films

Semiconducting gallium 712.1.1
DT: Predates 1975
BT: Gallium
Semiconductor materials

Semiconducting gallium arsenide 712.1.2
DT: January 1981
UF: Gallium arsenide materials
BT: Semiconducting gallium compounds

Semiconducting gallium compounds 712.1.2
DT: Predates 1975
BT: Gallium compounds
Semiconductor materials
NT: Semiconducting gallium arsenide

Semiconducting germanium 712.1.1
DT: Predates 1975
BT: Germanium
Semiconductor materials

Semiconducting germanium compounds 712.1.2
DT: January 1983
BT: Germanium compounds
Semiconductor materials

Semiconducting glass 712.1 (812.3)
DT: Predates 1975
BT: Glass
Semiconductor materials
RT: Semiconducting silicon compounds

Semiconducting indium 712.1.1
DT: Predates 1975
BT: Indium
Semiconductor materials

Semiconducting indium compounds 712.1.2
DT: Predates 1975
BT: Indium compounds
Semiconductor materials
NT: Semiconducting indium phosphide

Semiconducting indium phosphide 712.1.2
DT: June 1990
BT: Semiconducting indium compounds

Semiconducting intermetallics 712.1
DT: Predates 1975
BT: Intermetallics
Semiconductor materials

* Former Ei Vocabulary term

Semiconducting lead compounds 712.1.2
DT: Predates 1975
BT: Lead compounds
Semiconductor materials

Semiconducting liquids 712.1
DT: Predates 1975
BT: Liquids
Semiconductor materials
RT: Electric conductivity of liquids

Semiconducting manganese compounds 712.1.2
DT: Predates 1975
BT: Manganese compounds
Semiconductor materials

Semiconducting organic compounds 712.1.2
DT: Predates 1975
UF: Organic semiconductors
BT: Organic compounds
Semiconductor materials
NT: Semiconducting polymers
RT: Organic conductors

Semiconducting plastics
USE: Semiconducting polymers

Semiconducting polymers 712.1 (815.1) (817.1)
DT: Predates 1975
UF: Organic semiconductors (polymeric)
Semiconducting plastics
BT: Organic polymers
Semiconducting organic compounds

Semiconducting samarium compounds 712.1.2
DT: Predates 1975
BT: Samarium compounds
Semiconductor materials

Semiconducting selenium 712.1.1
DT: Predates 1975
BT: Selenium
Semiconductor materials

Semiconducting selenium compounds 712.1.2
DT: January 1983
BT: Selenium compounds
Semiconductor materials

Semiconducting silicon 712.1.1
DT: Predates 1975
BT: Semiconductor materials
Silicon
NT: Silicon wafers

Semiconducting silicon compounds 712.1.2
DT: Predates 1975
BT: Semiconductor materials
Silicon compounds
RT: Semiconducting glass

Semiconducting silver compounds 712.1.2
DT: Predates 1975
BT: Semiconductor materials
Silver compounds

Semiconducting tellurium 712.1.1
DT: January 1977
BT: Semiconductor materials
Tellurium

Semiconducting tellurium compounds 712.1.2
DT: Predates 1975
BT: Semiconductor materials
Tellurium compounds

Semiconducting tin compounds 712.1.2
DT: Predates 1975
BT: Semiconductor materials
Tin compounds

Semiconducting zinc compounds 712.1.2
DT: Predates 1975
BT: Semiconductor materials
Zinc compounds

Semiconductor counters 714.2, 944.7 (712.1)
DT: Predates 1975
BT: Radiation counters
Semiconductor devices

Semiconductor defects
USE: Crystal defects

Semiconductor device manufacture 714.2
DT: Predates 1975
BT: Electron device manufacture
NT: Silicon on insulator technology
Silicon on sapphire technology
RT: Clean rooms
Etching
Ion beam lithography
Microelectronic processing
Photolithography
Photoresists
Semiconductor devices
Silicon wafers
X ray lithography

*Semiconductor device manufacture—Silicon on insulator technology**
USE: Silicon on insulator technology

*Semiconductor device manufacture—Silicon on sapphire technology**
USE: Silicon on sapphire technology

Semiconductor device models 714.2 (921)
DT: January 1993
UF: Semiconductor devices—Modeling*
Semiconductor models
BT: Mathematical models
RT: Equivalent circuits
Semiconductor devices

Semiconductor device structures 714.2
DT: January 1993
UF: Semiconductor devices—Structures*
BT: Solid state device structures
RT: MIS devices
Semiconductor devices
Semiconductor insulator boundaries
Semiconductor metal boundaries

* Former Ei Vocabulary term

Semiconductor device testing 714.2
DT: Predates 1975
BT: Electron device testing
RT: Semiconductor devices

Semiconductor devices 714.2
DT: Predates 1975
BT: Solid state devices
NT: Beam lead devices
Bipolar semiconductor devices
Field effect semiconductor devices
Gunn devices
Integrated circuits
Oscillistors
Schottky barrier diodes
Semiconductor counters
Semiconductor diodes
Semiconductor lasers
Semiconductor relays
Semiconductor switches
Spacistors
Thermistors
Thyristors
Transistors
Transit time devices
Varistors
RT: Detectors
Dry etching
Electric rectifiers
Hall effect devices
Heterojunctions
Lithography
Masks
Metal insulator boundaries
Oscillators (electronic)
Photoconducting devices
Photoconductivity
Photovoltaic effects
Radiation hardening
Semiconductor device manufacture
Semiconductor device testing
Semiconductor doping
Semiconductor device models
Semiconductor device structures
Semiconductor growth
Semiconductor insulator boundaries
Semiconductor junctions
Semiconductor materials
Semiconductor metal boundaries
Semiconductor storage
Solid state rectifiers
Spurious signal noise

*Semiconductor devices, Bipolar**
USE: Bipolar semiconductor devices

*Semiconductor devices, Charge coupled**
USE: Charge coupled devices

*Semiconductor devices, Charge transfer**
USE: Charge transfer devices

*Semiconductor devices, Field effect**
USE: Field effect semiconductor devices

*Semiconductor devices, Gunn effect**
USE: Gunn devices

*Semiconductor devices, MESFET**
USE: MESFET devices

*Semiconductor devices, MIS**
USE: MIS devices

*Semiconductor devices, MISFET**
USE: MISFET devices

*Semiconductor devices, MOS**
USE: MOS devices

*Semiconductor devices, MOSFET**
USE: MOSFET devices

*Semiconductor devices, Schottky barrier**
USE: Schottky barrier diodes

*Semiconductor devices, Transit time**
USE: Transit time devices

*Semiconductor devices—Contacts**
USE: Electric contacts

*Semiconductor devices—Heterojunctions**
USE: Heterojunctions

*Semiconductor devices—Junctions**
USE: Semiconductor junctions

*Semiconductor devices—Modeling**
USE: Semiconductor device models

*Semiconductor devices—Noise**
USE: Spurious signal noise

*Semiconductor devices—Radiation hardening**
USE: Radiation hardening

*Semiconductor devices—Semiconductor insulator boundaries**
USE: Semiconductor insulator boundaries

*Semiconductor devices—Semiconductor metal boundaries**
USE: Semiconductor metal boundaries

*Semiconductor devices—Structures**
USE: Semiconductor device structures

*Semiconductor devices—Tunneling**
USE: Electron tunneling

Semiconductor diode light emitters
USE: Light emitting diodes

Semiconductor diodes 714.2
DT: Predates 1975
UF: Crystal diodes
Crystal rectifiers
BT: Diodes
Semiconductor devices
NT: Avalanche diodes
Gunn diodes
Light emitting diodes
Photodiodes
Tunnel diodes
Varactors

* Former Ei Vocabulary term

Semiconductor diodes *(continued)*
Zener diodes
RT: Limited space charge accumulation
Solid state rectifiers

*Semiconductor diodes, Avalanche**
USE: Avalanche diodes

*Semiconductor diodes, Gunn**
USE: Gunn diodes

*Semiconductor diodes, IMPATT**
USE: IMPATT diodes

*Semiconductor diodes, Light emitting**
USE: Light emitting diodes

*Semiconductor diodes, Photodiode**
USE: Photodiodes

*Semiconductor diodes, Tunnel**
USE: Tunnel diodes

*Semiconductor diodes, Zener**
USE: Zener diodes

*Semiconductor diodes—Limited space charge accumulation**
USE: Limited space charge accumulation

Semiconductor doping 712.1 (714.2)
DT: January 1993
UF: Semiconductor materials—Doping*
BT: Doping (additives)
RT: Ion implantation
Semiconductor devices
Semiconductor materials

Semiconductor growth 712.1 (714.2) (933.1.2)
DT: January 1993
UF: Semiconductor materials—Growth*
BT: Growth (materials)
RT: Crystal growth
Metallorganic chemical vapor deposition
Metallorganic vapor phase epitaxy
Semiconductor devices
Semiconductor materials

Semiconductor insulator boundaries 714.2 (712.1)
DT: January 1993
UF: Boundaries (semiconductor)
Insulator semiconductor boundaries
Semiconductor devices—Semiconductor insulator boundaries*
RT: Insulating materials
Semiconductor devices
Semiconductor device structures
Semiconductor materials

Semiconductor junction lasers
USE: Semiconductor lasers

Semiconductor junctions 714.2 (712.1)
DT: January 1993
UF: Junctions (semiconductor)
Semiconductor devices—Junctions*
NT: Heterojunctions
Semiconductor quantum wells
RT: Semiconductor devices
Semiconductor quantum dots
Semiconductor quantum wires
Tunnel junctions

Semiconductor lasers 744.4.1 (712.1) (714.2)
DT: January 1993
UF: Diode lasers
Junction lasers
Laser diodes
Lasers, Semiconductor*
Semiconductor junction lasers
BT: Semiconductor devices
NT: Injection lasers
Quantum well lasers
RT: Distributed feedback lasers
Semiconductor quantum wells
Semiconductor quantum dots
Semiconductor quantum wires

Semiconductor materials 712.1
DT: Predates 1975
UF: Semiconductors
BT: Materials
NT: Charge carriers
Magnetic semiconductors
Semiconducting aluminum compounds
Semiconducting antimony compounds
Semiconducting antimony
Semiconducting bismuth compounds
Semiconducting boron
Semiconducting cadmium compounds
Semiconducting diamonds
Semiconducting films
Semiconducting gallium compounds
Semiconducting gallium
Semiconducting germanium
Semiconducting germanium compounds
Semiconducting glass
Semiconducting indium compounds
Semiconducting indium
Semiconducting intermetallics
Semiconducting lead compounds
Semiconducting liquids
Semiconducting manganese compounds
Semiconducting organic compounds
Semiconducting samarium compounds
Semiconducting selenium compounds
Semiconducting selenium
Semiconducting silicon compounds
Semiconducting silicon
Semiconducting silver compounds
Semiconducting tellurium compounds
Semiconducting tellurium
Semiconducting tin compounds
Semiconducting zinc compounds
Semiconductor plasmas
RT: Acoustoelectric effects
Band structure
Carrier concentration
Crystal growth from melt
Crystals
Current voltage characteristics
Electric conductivity
Electron energy levels
Electron tunneling

* Former Ei Vocabulary term

Semiconductor materials *(continued)*
Excitons
Gunn effect
Interdiffusion (solids)
Ion implantation
Photoconducting materials
Photoconductivity
Photovoltaic effects
Piezoelectric materials
Piezoelectricity
Plasmas
Semiconductor devices
Semiconductor doping
Semiconductor growth
Semiconductor insulator boundaries
Semiconductor metal boundaries
Zener effect

*Semiconductor materials—Charge carriers**
USE: Charge carriers

*Semiconductor materials—Doping**
USE: Semiconductor doping

*Semiconductor materials—Energy gap**
USE: Energy gap

*Semiconductor materials—Growth**
USE: Semiconductor growth

*Semiconductor materials—Gunn effect**
USE: Gunn effect

*Semiconductor materials—Hot carriers**
USE: Hot carriers

*Semiconductor materials—Plasmas**
USE: Semiconductor plasmas

*Semiconductor materials—Thermoelectric effects**
USE: Thermoelectricity

*Semiconductor materials—Zener effect**
USE: Zener effect

Semiconductor metal boundaries 714.2 (531) (712.1)
DT: January 1993
UF: Boundaries (semiconductor metal)
Metal semiconductor boundaries
Semiconductor devices—Semiconductor metal boundaries*
RT: Metals
Schottky barrier diodes
Semiconductor device structures
Semiconductor devices
Semiconductor materials

Semiconductor models
USE: Semiconductor device models

Semiconductor plasmas 712.1, 932.3
DT: January 1993
UF: Semiconductor materials—Plasmas*
BT: Plasmas
Semiconductor materials

Semiconductor quantum dots 714.2
DT: January 1995
UF: Quantum dots
RT: Semiconductor junctions
Semiconductor lasers
Semiconductor quantum wells
Semiconductor quantum wires

Semiconductor quantum wells 714.2 (712.1)
DT: January 1993
UF: Multiple quantum wells
Quantum wells
BT: Semiconductor junctions
RT: Quantum interference devices
Quantum well lasers
Semiconductor lasers
Semiconductor quantum dots
Semiconductor quantum wires
Semiconductor superlattices

Semiconductor quantum wires 714.2
DT: January 1995
UF: Quantum wires
RT: Semiconductor junctions
Semiconductor lasers
Semiconductor quantum dots
Semiconductor quantum wells

Semiconductor relays 714.2 (712.1)
DT: January 1993
UF: Electric relays, Semiconductor*
Solid state relays
BT: Electric relays
Semiconductor devices

Semiconductor storage 722.1 (712.1) (714.2)
DT: January 1993
UF: Data storage, Semiconductor*
Data storage, Semiconductor—Storage devices*
Memory expansion boards
BT: Data storage equipment
RT: PROM
Random access storage
ROM
Semiconductor devices

Semiconductor superlattices 714.2 (712.1) (933.1)
DT: January 1993
UF: Superlattices—Semiconductor*
BT: Superlattices
RT: Heterojunction bipolar transistors
High electron mobility transistors
Semiconductor quantum wells

Semiconductor switches 714.2
DT: January 1993
UF: Electric switches, Semiconductor*
BT: Electric switches
Semiconductor devices
RT: Thyristors

Semiconductors
USE: Semiconductor materials

Semicustom integrated circuits
USE: Application specific integrated circuits

* Former Ei Vocabulary term

Semimetals
USE: Metalloids

Seminars
USE: Technical presentations

Semisubmersibles 674.1
DT: January 1981
UF: Semi-submersibles
RT: Amphibious vehicles
Offshore structures
Submersibles

Sensitivity analysis 921 (731.1)
DT: January 1993
UF: Mathematical techniques—Sensitivity analysis*
BT: Mathematical techniques
RT: Control system analysis
Control system synthesis
Dynamic response
Frequency response
Transfer functions

Sensitometers 941.3
SN: Instruments which apply to photographic emulsion a graduated series of exposures to radiation of controlled spectral quality, intensity and duration
DT: Predates 1975
BT: Optical instruments
RT: Densitometers
Photographic emulsions
Photosensitivity

Sensor data fusion 723.2 (732.2)
DT: January 1993
UF: Multisensor systems
RT: Data processing
Sensors
Signal processing

Sensors (732.2) (801)
DT: October 1975
BT: Detectors
NT: Amperometric sensors
Biosensors
Chemical sensors
Contact sensors
Hybrid sensors
Image sensors
Multispectral scanners
Optical sensors
Partial pressure sensors
Potentiometric sensors
Proximity sensors
Silicon sensors
RT: Arrays
Blood gas analysis
Chemically sensitive field effect transistors
Ion selective electrodes
Ion sensitive field effect transistors
Liquid membrane electrodes
Polymer membrane electrodes
Probes
Sensor data fusion
Transducers
Urea electrodes

*Sensors—Amperometric measurements**
USE: Amperometric sensors

*Sensors—Calcium ion sensors**
USE: Calcium ion sensors

*Sensors—Electrochemical**
USE: Electrochemical sensors

*Sensors—Fiber optic chemical sensors**
USE: Fiber optic chemical sensors

*Sensors—Glucose sensors**
USE: Glucose sensors

*Sensors—Hybrid construction**
USE: Hybrid sensors

*Sensors—Potentiometric measurements**
USE: Potentiometric sensors

*Sensors—Silicon sensors**
USE: Silicon sensors

*Sensors—Urea electrodes**
USE: Urea electrodes

Sensory aids 461.5, 462.1
DT: Predates 1975
BT: Biomedical equipment
NT: Hearing aids
Vision aids
RT: Human rehabilitation engineering
Sensory feedback
Sensory perception
Tactile reading aids

Sensory feedback 461.5
DT: January 1993
UF: Human rehabilitation engineering—Sensory feedback techniques*
Prosthetics—Sensory feedback*
BT: Patient treatment
RT: Human rehabilitation engineering
Patient rehabilitation
Prosthetics
Sensory aids
Sensory perception

Sensory perception 461.4
SN: Includes perception of pain, proprioception, smell, taste, temperature sense, and touch
DT: January 1993
UF: Human engineering—Sensory perception*
Perception
NT: Audition
Vision
RT: Arousal
Biofeedback
Human engineering
Odors
Sensory aids
Sensory feedback

Separation 802.3
SN: Of materials. Very general term; prefer a term for the specific method of separation

Separation *(continued)*
DT: Predates 1975
BT: Chemical operations
NT: Centrifugation
Chemicals removal (water treatment)
Clarification
Concentration (process)
Degassing
Demulsification
Dewatering
Dewaxing
Dialysis
Distillation
Extraction
Filtration
Flotation
Fractionation
Heavy media separation
Magnetic separation
Pervaporation
Phase separation
Precipitation (chemical)
Salt removal
Size separation
RT: Chemical reactions
Chromatographic analysis
Chromatography
Crystallization
Ion exchange
Leaching
Percolation (fluids)
Separators
Sublimation
Vaporization

Separators 802.1
DT: January 1993
UF: Ship equipment—Separators*
BT: Equipment
NT: Centrifuges
Clarifiers
Classifiers
Cyclone separators
Dryers (equipment)
Dust collectors
Electrostatic separators
Filters (for fluids)
Fractionating units
Ion exchangers
Magnetic separators
Scrubbers
Settling tanks
Sieves
Steam separators and traps
RT: Crystallizers
Separation

*Separators—Centrifugal**
USE: Centrifuges

*Separators—Electrostatic**
USE: Electrostatic separators

*Separators—Gas scrubbers**
USE: Scrubbers

*Separators—Hydraulic**
USE: Hydraulic equipment

*Separators—Magnetic**
USE: Magnetic separators

*Separators—Pneumatic**
USE: Pneumatic equipment

Septic tanks 452.2 (619.2)
DT: January 1993
UF: Sewage treatment—Septic tanks*
BT: Tanks (containers)
RT: Sewage tanks
Sewage treatment
Sewage treatment plants
Sewers
Waste disposal

Sequence indicators
USE: Phase sequence indicators

Sequential circuits 721.2 (721.1)
DT: January 1993
UF: Logic circuits, Sequential*
BT: Logic circuits
NT: Asynchronous sequential logic
RT: Combinatorial circuits
Sequential switching

Sequential machines 721.1
DT: January 1993
UF: Automata theory—Sequential machines*
BT: Automata theory

Sequential switching 721.1
DT: January 1993
UF: Switching theory—Sequential switching*
BT: Switching
RT: Asynchronous sequential logic
Finite automata
Sequential circuits
Switching theory

Serpentine 482.2
DT: Predates 1975
BT: Silicate minerals

Service areas (highway)
USE: Highway service areas

Service life
DT: January 1993
UF: Life in service
RT: Performance
Plant life extension
Reliability
Weibull distribution

Service stations
USE: Filling stations

Service vessels 674.1
SN: Vessels for service in port terminals and harbors
DT: January 1993
UF: Oil terminals—Service vessels*
Ports and harbors—Service vessels*
BT: Ships

* Former Ei Vocabulary term

Service vessels *(continued)*
RT: Oil terminals
Ports and harbors

Servo amplifiers 713.1, 732.1
DT: January 1993
UF: Amplifiers, Servo*
BT: Amplifiers (electronic)
Servomechanisms

Servo systems
USE: Servomechanisms

Servomechanisms 705, 732.1
DT: Predates 1975
UF: Servo systems
BT: Control equipment
NT: Hydraulic servomechanisms
Pneumatic servomechanisms
Servo amplifiers
Servomotors
RT: Actuators
Position control
Position measurement
Robots

*Servomechanisms—Circuits**
USE: Networks (circuits)

*Servomechanisms—Hydraulic**
USE: Hydraulic servomechanisms

*Servomechanisms—Pneumatic**
USE: Pneumatic servomechanisms

*Servomechanisms—Symbols**
USE: Codes (symbols)

Servomotors 705.3, 732.1
DT: Predates 1975
UF: Servos
BT: Servomechanisms
NT: Synchros
RT: Actuators
Position control
Position measurement

Servos
USE: Servomotors

Set theory 921.4
DT: January 1993
UF: Mathematical techniques—Set theory*
Sets (mathematics)
Subsets (mathematics)
BT: Mathematical techniques
Theory
NT: Fuzzy sets
RT: Boolean algebra
Combinatorial mathematics

Sets (mathematics)
USE: Set theory

Setting 412.1 (802.3)
DT: January 1993
UF: Cement—Setting*
BT: Solidification
RT: Cements
Concretes

Settlement of structures 483.1 (402) (483.2)
DT: January 1993
UF: Buildings—Settlement*
Foundations—Settlement*
Settling of structures
RT: Buildings
Foundations
Soil liquefaction
Soil structure interactions
Structures (built objects)

Settling basins
USE: Settling tanks

Settling of structures
USE: Settlement of structures

Settling tanks (445.1) (452.2) (802.1)
DT: Predates 1975
UF: Basins (settling)
Sedimentation basins
Sedimentation tanks
Settling basins
BT: Separators
Tanks (containers)
NT: Sewage settling tanks
RT: Sedimentation

Sewage 452.1 (453.1)
DT: January 1987
BT: Wastes
NT: Sewage sludge
RT: Composting
Effluents
Industrial wastes
Sewage aeration
Sewage analysis
Sewage bacteriology
Sewage lagoons
Sewage pumping plants
Sewage tanks
Sewage treatment

Sewage aeration 452.2
DT: January 1993
UF: Aeration (sewage)
Sewage treatment—Aeration*
BT: Sewage treatment
RT: Sewage

Sewage analysis 452.1, 801
DT: Predates 1975
BT: Chemical analysis
RT: Sewage
Sewage bacteriology
Sewage treatment

Sewage bacteriology 452.1 (461.9) (801.2)
DT: Predates 1975
BT: Bacteriology
RT: Sewage
Sewage analysis
Sewage treatment

Sewage disposal
USE: Sewage treatment

* Former Ei Vocabulary term

Sewage disposal charges 452.2
DT: January 1993
UF: Sewage treatment plants—Disposal charges*
BT: Utility rates
RT: Sewage treatment plants

Sewage lagoons 452.2
DT: January 1993
UF: Lagoons (sewage)
Sewage treatment—Lagoons*
BT: Sewage treatment plants
RT: Sewage
Sewage tanks
Sewage treatment

Sewage pumping plants 452.2 (402.1) (618.2)
DT: Predates 1975
BT: Pumping plants
RT: Sewage
Sewage pumps
Sewage treatment
Sewage treatment plants

*Sewage pumping plants—Pumps**
USE: Sewage pumps

*Sewage pumping plants—Underground**
USE: Underground pumping plants

Sewage pumps 618.2, 452.1
DT: January 1993
UF: Pumps, Sewage*
Sewage pumping plants—Pumps*
BT: Pumps
RT: Sewage pumping plants

Sewage settling tanks 452.2, 802.1
DT: January 1993
UF: Sewage tanks—Settling tanks*
BT: Settling tanks
Sewage tanks

Sewage sludge 452.2 (453.1)
DT: June 1990
UF: Sludge (sewage)
BT: Sewage
RT: Sludge digestion
Sludge disposal

Sewage tanks 452.2 (619.2)
DT: Predates 1975
BT: Tanks (containers)
NT: Sewage settling tanks
RT: Septic tanks
Sewage
Sewage lagoons
Sewage treatment
Stabilization ponds

*Sewage tanks—Concrete**
USE: Concrete tanks

*Sewage tanks—Settling tanks**
USE: Sewage settling tanks

Sewage treatment 452.2
DT: Predates 1975
UF: Mines and mining—Sewage treatment*
Sewage disposal
Ships—Sewage treatment*
BT: Waste treatment
NT: Activated sludge process
Biological sewage treatment
Sewage aeration
RT: Biochemical oxygen demand
Chemical oxygen demand
Chlorination
Flocculation
Flotation
Radioactive wastes
Refuse disposal
Sedimentation
Septic tanks
Sewage
Sewage analysis
Sewage bacteriology
Sewage lagoons
Sewage pumping plants
Sewage tanks
Sewage treatment plants
Sewage treatment charges
Sludge disposal
Stabilization ponds
Trickling filtration
Wastewater reclamation
Water pollution

Sewage treatment charges 452.2
DT: January 1993
UF: Sewage treatment—Service charges*
BT: Utility rates
RT: Sewage treatment
Sewage treatment plants

Sewage treatment plants 452.2 (402.1)
DT: Predates 1975
BT: Industrial plants
NT: Grit chambers
Sewage lagoons
Stabilization ponds
RT: Clarifiers
Odor control
Septic tanks
Sewage disposal charges
Sewage pumping plants
Sewage treatment
Sewage treatment charges

*Sewage treatment plants—Clarifiers**
USE: Clarifiers

*Sewage treatment plants—Disposal charges**
USE: Sewage disposal charges

*Sewage treatment plants—Fly control**
USE: Insect control

*Sewage treatment plants—Grit chambers**
USE: Grit chambers

*Sewage treatment—Activated sludge**
USE: Activated sludge process

*Sewage treatment—Aeration**
USE: Sewage aeration

* Former Ei Vocabulary term

*Sewage treatment—Biochemical oxygen demand**
USE: Biochemical oxygen demand

*Sewage treatment—Biological treatment**
USE: Biological sewage treatment

*Sewage treatment—Chemical oxygen demand**
USE: Chemical oxygen demand

*Sewage treatment—Detergents effects**
USE: Detergents

*Sewage treatment—Lagoons**
USE: Sewage lagoons

*Sewage treatment—Radioactive materials**
USE: Radioactive wastes

*Sewage treatment—Sedimentation**
USE: Sedimentation

*Sewage treatment—Septic tanks**
USE: Septic tanks

*Sewage treatment—Service charges**
USE: Sewage treatment charges

*Sewage treatment—Sludge digestion**
USE: Sludge digestion

*Sewage treatment—Sludge disposal**
USE: Sludge disposal

*Sewage treatment—Stabilization ponds**
USE: Stabilization ponds

*Sewage treatment—Trickling filtration**
USE: Trickling filtration

*Sewage treatment—Water reclamation**
USE: Wastewater reclamation

Sewer linings 452
DT: January 1993
UF: Sewers—Lining*
BT: Linings
Sewers

Sewer tunnels 401.2, 452.1
DT: January 1993
UF: Sewers—Tunnels*
BT: Sewers
Tunnels

Sewers 452.1 (403.1)
DT: Predates 1975
UF: Petroleum refineries—Sewers*
BT: Facilities
NT: Combined sewers
Sanitary sewers
Sewer linings
Sewer tunnels
Storm sewers
RT: Drainage
Outfalls
Public utilities
Rat control
Refuse disposal
Septic tanks
Urban planning
Waste disposal

*Sewers—Bituminized fiber**
USE: Bituminized fibers

*Sewers—Combined**
USE: Combined sewers

*Sewers—Corrugated metal**
USE: Corrugated metal

*Sewers—Flow**
USE: Flow of fluids

*Sewers—Frozen ground**
USE: Frozen soils

*Sewers—Infiltration**
USE: Infiltration

*Sewers—Joints**
USE: Joints (structural components)

*Sewers—Leakage**
USE: Leakage (fluid)

*Sewers—Lining**
USE: Sewer linings

*Sewers—Maps**
USE: Maps

*Sewers—Outfall**
USE: Outfalls

*Sewers—Rat eradication**
USE: Rat control

*Sewers—Sanitary**
USE: Sanitary sewers

*Sewers—Storm drainage**
USE: Storm sewers

*Sewers—Tunnels**
USE: Sewer tunnels

Sewing machines 819.6
DT: Predates 1975
BT: Textile machinery

Sextants 943.3 (652.3) (671.2)
DT: January 1993
UF: Aircraft instruments—Sextants*
BT: Instruments
RT: Aircraft instruments
Navigation
Ship instruments

Sferics
USE: Atmospherics

Shaft displacement 602.1
DT: January 1993
UF: Shafts and shafting—Displacement*
RT: Alignment
Shafts (machine components)

Shaft sinking (405.2) (502.1)
DT: Predates 1975
BT: Excavation
NT: Cementing (shafts)

* Former Ei Vocabulary term

Shaft sinking *(continued)*
RT: Drilling
Grouting
Linings
Mine shafts

*Shaft sinking—Cementing**
USE: Cementing (shafts)

*Shaft sinking—Freezing**
USE: Freezing

*Shaft sinking—Grouting**
USE: Grouting

*Shaft sinking—Lining**
USE: Linings

Shaft whirling
USE: Rotation

Shaftless motors 705.3
DT: January 1993
UF: Electric motors—Shaftless*
BT: Electric motors

Shafts (machine components) 601.2
DT: January 1993
UF: Electric machinery—Shafts*
Shafts and shafting*
BT: Machine components
NT: Crankshafts
Lead screws
RT: Bushings
Shaft displacement
Splines
Straightening

Shafts (mine)
USE: Mine shafts

*Shafts and shafting**
USE: Shafts (machine components)

*Shafts and shafting—Alignment**
USE: Alignment

*Shafts and shafting—Displacement**
USE: Shaft displacement

*Shafts and shafting—Reclamation**
USE: Reclamation

*Shafts and shafting—Straightening**
USE: Straightening

*Shafts and shafting—Whirling**
USE: Rotation

Shale (481.1.2) (482.2) (483.1) (505)
DT: Predates 1975
BT: Sedimentary rocks
NT: Carbonaceous shale
RT: Slate

Shale oil 523 (513.3)
DT: January 1987
BT: Fossil fuels
RT: Oil shale

*Shale—Carbonaceous**
USE: Carbonaceous shale

Shape memory effect 931.2
DT: January 1993
UF: Materials with memory*
Memory (shape)
BT: Mechanical properties
RT: Intelligent materials
Thermoelasticity

Shaped charges (404.1) (405.2) (502.2) (804)
DT: January 1993
UF: Explosives—Shaped charges*
BT: Explosives

Shapers (metal working)
USE: Metal working shapers

*Shapers, Metal working**
USE: Metal working shapers

Shaw process 534.2
DT: January 1993
UF: Foundry practice—Shaw process*
BT: Metal casting
RT: Foundry practice

Shear deformation (421) (422)
DT: January 1995
BT: Deformation
RT: Shear stress

Shear flow 631.1 (421)
DT: January 1995
RT: Creep
Flow of fluids
Flow of solids
Plastic flow

Shear strength (421) (422)
DT: January 1993
UF: Soils—Shear strength*
BT: Strength of materials
RT: Fatigue of materials
Shear stress
Soils

Shear stress (421)
DT: January 1993
UF: Stresses—Shear*
BT: Stresses
RT: Fatigue of materials
Shear deformation
Shear strength
Shear walls
Stress analysis
Torsional stress

Shear walls 402, 484.3
SN: Building walls designed to resist the shear forces of earthquakes
DT: January 1993
UF: Buildings—Shear walls*
BT: Walls (structural partitions)
RT: Earthquake resistance
Shear stress

Shear waves
USE: Elastic waves

Shearing 604.1 (535.1)
DT: January 1993
UF: Sheet and strip metal—Shearing*
BT: Cutting
RT: Metal cutting
Punching
Sawing
Shearing machines
Sheet metal
Slitting
Slotting
Strip metal

Shearing machines 603.1 (535.1)
DT: Predates 1975
BT: Cutting tools
Machine tools
RT: Shearing
Slitting machines

Sheathing (cable)
USE: Cable sheathing

Sheaths (plasma)
USE: Plasma sheaths

Shedding (electric loads)
USE: Electric load shedding

*Sheet and strip metal**
USE: Sheet metal OR
Strip metal

*Sheet and strip metal—Coating**
USE: Coating techniques

*Sheet and strip metal—Finishing**
USE: Metal finishing

*Sheet and strip metal—Perforating**
USE: Perforating

*Sheet and strip metal—Piercing**
USE: Piercing

*Sheet and strip metal—Precoating**
USE: Precoated metals

*Sheet and strip metal—Roll bonding**
USE: Roll bonding

*Sheet and strip metal—Rolling**
USE: Rolling

*Sheet and strip metal—Shearing**
USE: Shearing

*Sheet and strip metal—Slitting**
USE: Slitting

*Sheet and strip metal—Slotting**
USE: Slotting

*Sheet and strip metal—Stitching**
USE: Stitching (metal joining)

*Sheet and strip metal—Straightening**
USE: Straightening

*Sheet and strip metal—Stretching**
USE: Stretching

*Sheet and strip metal—Thickness control**
USE: Thickness control

Sheet metal (535.1)
DT: January 1993
UF: Materials handling—Sheet metal*
Metal sheet
Rolling mill practice—Sheet and strip*
Sheet and strip metal*
Welding—Sheet metal*
BT: Materials
NT: Aluminum sheet
Corrugated metal
Metal foil
Steel sheet
Tin plate
Titanium sheet
Tungsten sheet
RT: Metals
Plate metal
Roll bonding
Rolling mill practice
Rolling mills
Shearing

Sheet molding compounds 816.1, 817.1
DT: January 1993
UF: Molding compounds
Plastic sheets—Sheet molding compounds*
BT: Materials
RT: Plastic sheets

Sheet plastics
USE: Plastic sheets

Sheeting (paper)
USE: Paper sheeting

Shell casting
USE: Shell mold casting

Shell mold casting 534.2
DT: January 1993
UF: Foundry practice—Shell process*
Shell casting
BT: Metal casting
RT: Centrifugal casting
Foundry practice

Shells (structures) 408.2
DT: January 1993
UF: Domes and shells*
Inflatable shells
BT: Structures (built objects)
RT: Walls (structural partitions)

Shelters (civil defense) 404.2 (402)
DT: January 1993
UF: Civil defense—Shelters*
BT: Facilities
RT: Civil defense
Shelters (from attack)

* Former Ei Vocabulary term

Shelters (from attack) 404.1 (404.2)
DT: January 1993
UF: Military equipment—Shelters*
BT: Military equipment
RT: Shelters (civil defense)

Shielding
(622) (654.1) (655.1) (701.1) (711.1) (914.1)
DT: January 1993
UF: Rockets and missiles—Flame shields*
Rockets and missiles—Shielding*
BT: Equipment
NT: Electric shielding
Heat shielding
Magnetic shielding
Radiation shielding
RT: Missiles
Protection
Radiation protection
Spacecraft

Shift registers 721.3
DT: January 1993
UF: Computers—Shift registers*
BT: Computer hardware
RT: Adders
Logic circuits

Shims 601.2
DT: Predates 1975
UF: Bearings—Shims*
BT: Components

Ship construction
USE: Shipbuilding

*Ship construction and outfitting**
USE: Outfitting (water craft)

Ship conversion 673.1
DT: January 1993
UF: Conversion (ships)
Ships—Conversion*
BT: Outfitting (water craft)

Ship cranes 671.2, 693.1
DT: January 1993
UF: Ship derricks
Ship equipment—Cranes*
Ship equipment—Derricks*
BT: Cranes
Ship equipment

Ship crew accommodations
USE: Crew accommodations

Ship derricks
USE: Ship cranes

Ship equipment 671.2
DT: Predates 1975
BT: Equipment
Ships
Water craft parts and equipment
NT: Anchor cables
Anchors
Ballast tanks
Davits
Electric ship equipment
Electronic ship equipment
Hatch covers
Hatches
Lifesaving equipment
Ship cranes
Ship instruments
Ship propellers
Ship smokestacks
Ship steering equipment
Windlasses
RT: Boat equipment

*Ship equipment—Anchor cables**
USE: Anchor cables

*Ship equipment—Anchors**
USE: Anchors

*Ship equipment—Cranes**
USE: Ship cranes

*Ship equipment—Davits**
USE: Davits

*Ship equipment—Derricks**
USE: Ship cranes

*Ship equipment—Electric control**
USE: Electric control equipment

*Ship equipment—Electric**
USE: Electric ship equipment

*Ship equipment—Electronic**
USE: Electronic ship equipment

*Ship equipment—Evaporators**
USE: Evaporators

*Ship equipment—Hydraulic control**
USE: Hydraulic control equipment

*Ship equipment—Hydraulic**
USE: Hydraulic equipment

*Ship equipment—Lifesaving**
USE: Lifesaving equipment

*Ship equipment—Pneumatic control**
USE: Pneumatic control

*Ship equipment—Pneumatic**
USE: Pneumatic equipment

*Ship equipment—Separators**
USE: Separators

*Ship equipment—Winches**
USE: Winches

*Ship equipment—Windlasses**
USE: Windlasses

Ship fenders
USE: Fenders (boat)

Ship fueling 671 (522) (523) (524)
DT: January 1993
UF: Ships—Fueling*
BT: Fueling
RT: Ships

* Former Ei Vocabulary term

Ship instruments 671.2
DT: January 1993
UF: Ships—Instruments*
BT: Ship equipment
RT: Electric ship equipment
Electronic ship equipment
Sextants

Ship launched missiles
USE: Marine missiles

Ship mockups 671.3
DT: January 1993
UF: Ships—Models*
BT: Mockups
RT: Ship models
Ship simulators

Ship model tanks 671.3
SN: For activities such as towing tests, buoyancy studies and wave effects
DT: January 1993
UF: Ship models—Tanks*
Towing tanks
BT: Test facilities
RT: Laboratories
Ship models
Ships
Simulators
Water waves

Ship models 671.3
SN: Scale models, used for purposes such as recognition training or towing-tank tests. For actual-size mockups, use SHIP MOCKUPS
DT: Predates 1975
BT: Models
RT: Ship mockups
Ship model tanks
Ship simulators
Ship testing
Ships

*Ship models—Tanks**
USE: Ship model tanks

Ship outfitting
USE: Outfitting (water craft)

Ship propellers 671.2 (675.1)
DT: January 1993
UF: Ships—Propellers*
BT: Propellers
Ship equipment
RT: Ship propulsion

Ship propulsion 675.1
DT: Predates 1975
UF: Ships—Propulsion*
BT: Propulsion
NT: Dynamic positioning
Hydraulic jet propulsion
Hydrofoil propulsion
Paddle wheel propulsion
RT: Diesel propulsion
Electric propulsion
Marine engines
Nuclear propulsion
Ship propellers
Ships
Steam propulsion

*Ship propulsion—Diesel**
USE: Diesel propulsion

*Ship propulsion—Electric**
USE: Electric propulsion

*Ship propulsion—Gas and steam turbine combined**
USE: Gas turbines AND
Steam turbines

*Ship propulsion—Gas turbine**
USE: Gas turbines

*Ship propulsion—Hydraulic jet**
USE: Hydraulic jet propulsion

*Ship propulsion—Nuclear**
USE: Nuclear propulsion

*Ship propulsion—Paddle wheels**
USE: Paddle wheel propulsion

*Ship propulsion—Steam**
USE: Steam propulsion

*Ship salvaging**
USE: Salvaging

Ship simulators 671.3
DT: January 1993
UF: Ships—Simulators*
BT: Simulators
RT: Ship mockups
Ship models
Ship testing
Ships

Ship smokestacks 671.2
DT: January 1993
UF: Ships—Smokestacks*
Smokestacks (ship)
BT: Ship equipment

Ship steering equipment 671.2
DT: January 1993
UF: Ships—Steering equipment*
Steering equipment (ships)
BT: Ship equipment
NT: Bow thrusters
Rudders
RT: Steering

Ship testing 671
DT: January 1993
UF: Ships—Testing*
BT: Equipment testing
RT: Ship models
Ship simulators
Ships

Shipbuilding 673.1
DT: January 1993
UF: Barges—Constructing and outfitting*
Boat building

* Former Ei Vocabulary term

Shipbuilding *(continued)*
Boat construction
Boatbuilding
Boats—Constructing and outfitting*
Ferry boats—Constructing and outfitting*
Fire boats—Constructing and outfitting*
Fishing vessels—Constructing and outfitting*
Ship construction
Ships—Constructing and outfitting*
BT: Construction
RT: Drydocks
Naval architecture
Outfitting (water craft)
Shipbuilding materials
Ships
Shipways
Shipyards
Water craft

Shipbuilding materials 673.2
DT: Predates 1975
BT: Building materials
RT: Shipbuilding
Ships
Shipyards

*Shipbuilding materials—Bronze**
USE: Bronze

*Shipbuilding materials—Cast iron**
USE: Cast iron

*Shipbuilding materials—Composite materials**
USE: Composite materials

*Shipbuilding materials—Plastics**
USE: Plastics applications

*Shipbuilding materials—Steel**
USE: Steel

*Shipbuilding—Drafting**
USE: Drafting practice

Shipping
USE: Freight transportation

Ships (671) (672) (674)
SN: Very general term; prefer specific type of vessel
DT: Predates 1975
UF: Motor ships*
BT: Water craft
NT: Cable ships
Crew accommodations
Decks (ship)
Drillships
Hulls (ship)
Icebreakers
Naval vessels
Ore carriers
Prefabricated ships
Service vessels
Ship equipment
Steamships
Submarine tenders
Submarines
Tankers (ships)
Whaling vessels
RT: Boats
Docking
Drydocks
Fueling
Ice problems
Launching
Loading
Maneuverability
Marine borers
Marine engineering
Marine insurance
Merchant marine
Mooring
Naval architecture
Navigation systems
Outfitting (water craft)
Radio direction finding systems
Rat control
Sailing vessels
Salvaging
Seakeeping
Ship fueling
Ship model tanks
Ship models
Ship propulsion
Ship simulators
Ship testing
Shipbuilding
Shipbuilding materials
Shipyards
Shock testing
Single point mooring
Slamming (ships)
Unloading
Waterway transportation

*Ships—Auxiliary machinery**
USE: Auxiliary equipment

*Ships—Ballast tanks**
USE: Ballast tanks

*Ships—Bow thrusters**
USE: Bow thrusters

*Ships—Cable laying application**
USE: Electric cable laying

*Ships—Constructing and outfitting**
USE: Outfitting (water craft) OR
Shipbuilding

*Ships—Conversion**
USE: Ship conversion

*Ships—Crew accommodations**
USE: Crew accommodations

*Ships—Dynamic positioning**
USE: Dynamic positioning

*Ships—Elevators**
USE: Elevators

*Ships—Fire prevention**
USE: Fire protection

* Former Ei Vocabulary term

*Ships—Fueling**
USE: Ship fueling

*Ships—Hatch covers**
USE: Hatch covers

*Ships—Hatches**
USE: Hatches

*Ships—Hulls**
USE: Hulls (ship)

*Ships—Icebreakers**
USE: Icebreakers

*Ships—Icing**
USE: Ice problems

*Ships—Instruments**
USE: Ship instruments

*Ships—Launching**
USE: Launching

*Ships—Lighting**
USE: Lighting

*Ships—Loading**
USE: Loading

*Ships—Losses**
USE: Losses

*Ships—Maneuverability**
USE: Maneuverability

*Ships—Models**
USE: Ship mockups

*Ships—Mooring**
USE: Mooring

*Ships—Navigation systems**
USE: Navigation systems

*Ships—Ore carriers**
USE: Ore carriers

*Ships—Pneumatic control equipment**
USE: Pneumatic control equipment

*Ships—Prefabricated**
USE: Prefabricated ships

*Ships—Propellers**
USE: Ship propellers

*Ships—Propulsion**
USE: Ship propulsion

*Ships—Radio equipment**
USE: Radio equipment

*Ships—Rat eradication**
USE: Rat control

*Ships—Resistance**
USE: Drag

*Ships—Rudders**
USE: Rudders

*Ships—Seakeeping**
USE: Seakeeping

*Ships—Sewage treatment**
USE: Sewage treatment

*Ships—Shock testing**
USE: Shock testing

*Ships—Simulators**
USE: Ship simulators

*Ships—Slamming**
USE: Slamming (ships)

*Ships—Smokestacks**
USE: Ship smokestacks

*Ships—Speed**
USE: Speed

*Ships—Stabilizers**
USE: Stabilizers (marine vessel)

*Ships—Steering equipment**
USE: Ship steering equipment

*Ships—Testing**
USE: Ship testing

*Ships—Turnaround time**
USE: Turnaround time

*Ships—Unloading**
USE: Unloading

Shipways 673.1
SN: The supports on which ships are constructed, repaired, or launched
DT: Predates 1975
UF: Slipways
BT: Equipment
RT: Drydocks
Outfitting (water craft)
Shipbuilding
Shipyards

Shipyards 673.3
DT: Predates 1975
BT: Industrial plants
RT: Drydocks
Shipbuilding
Shipbuilding materials
Ships
Shipways

Shock absorbers
(601.2) (632) (654.1) (662.4) (682.1.1) (931.1)
DT: Predates 1975
UF: Cars—Shock absorbers*
Rockets and missiles—Shock absorbers*
BT: Equipment
NT: Automobile shock absorbers
RT: Antivibration mountings
Damping
Pistons
Vibrations (mechanical)

Shock problems (421) (422) (691) (694)
SN: Physical shock; for electric shock use ELECTRIC ACCIDENTS
DT: January 1993
UF: Freight handling—Shock problems*
Materials handling—Shock problems
Packaging—Shock problems*
RT: Materials handling
Packaging
Shock testing

Shock testing (421) (422)
DT: Predates 1975
UF: Ships—Shock testing*
BT: Mechanical testing
RT: Equipment testing
Impact testing
Materials testing
Ships
Shock problems

Shock tubes (422.1) (631.1) (632) (651.2) (715)
DT: Predates 1975
UF: Electromagnetic shock tubes
Wave superheaters
BT: Tubes (components)
RT: Aerodynamics
Hypersonic flow
Shock waves
Wind tunnels

Shock waves 931
DT: Predates 1975
BT: Elastic waves
NT: Plasma shock waves
RT: Acoustic noise
Acoustic wave propagation
Acoustic wave velocity
Acoustic waves
Acoustics
Aerodynamics
Detonation
Explosions
Fluid dynamics
Mechanics
Nuclear explosions
Seismology
Shock tubes
Solitons
Transonic flow
Ultrasonics

Shoe manufacture
(811) (814) (817.2) (818.6) (819.5) (913.4)
DT: Predates 1975
BT: Manufacture
RT: Garment manufacture

Shooting (well)
USE: Explosive well stimulation

Shopping centers (402.2) (403) (911.4)
DT: Predates 1975
UF: Malls
Shopping malls
BT: Retail stores
RT: Regional planning
Store buildings

Shopping malls
USE: Shopping centers

Shore protection 407.1
SN: Seacoast protection. For lake, river, and inland waterway shores, use BANK PROTECTION
DT: Predates 1975
BT: Protection
RT: Bank protection
Beaches
Breakwaters
Coastal engineering
Coastal zones
Environmental engineering
Erosion
Jetties
Levees
Ocean structures
Port structures
Quay walls
Retaining walls
Soil conservation

Short circuit currents 701.1 (914.1)
DT: January 1993
UF: Electric fault currents—Short circuit currents*
Short circuits
BT: Electric fault currents
RT: Electric arcs
Electric discharges
Electric grounding

Short circuits
USE: Short circuit currents

Short takeoff and landing aircraft
USE: VTOL/STOL aircraft

Shot effect
USE: Shot noise

Shot firing cables
USE: Firing cables

Shot noise (701.1) (716)
DT: January 1993
UF: Noise, Spurious signal—Shot noise*
Schottky noise
Shot effect
BT: Spurious signal noise

Shot peening 539
DT: January 1993
UF: Metal cleaning—Shot peening*
Peening
BT: Finishing
RT: Metal cleaning

Shotcreting 412
DT: January 1993
UF: Concrete construction—Shotcreting*
BT: Concrete construction

Shoulders (road) 406.2
DT: January 1993
UF: Roads and streets—Shoulders*
BT: Roads and streets
RT: Roadsides

* Former Ei Vocabulary term

Shovels (405.1) (605.2)
DT: Predates 1975
BT: Construction equipment
RT: Hand tools
Tools

Shrink fitting
USE: Shrinkfitting

Shrink resistant textiles
USE: Shrinkproofing (textiles)

Shrinkage (537.1)
DT: January 1993
UF: Contraction
Heat treatment—Shrinking*
Shrinking
RT: Dimensional stability
Growth (materials)
Shrinkfitting
Textiles

Shrinkfitting 604.2
DT: Predates 1975
UF: Shrink fitting
RT: Joining
Shrinkage

Shrinking
USE: Shrinkage

Shrinkproofing (textiles) 819.5
DT: January 1993
UF: Shrink resistant textiles
Textiles—Shrinkproofing*
BT: Textile finishing

Shubnikov-de Haas effect
USE: Magnetoresistance

Shutdowns (industrial plants)
USE: Plant shutdowns

Shutdowns (reactors)
USE: Reactor shutdowns

Shutters (camera)
USE: Camera shutters

Shutters (optical)
USE: Optical shutters

Shuttles (space)
USE: Space shuttles

Si
USE: Silicon

Sieves (405.1) (502.2) (605) (802.1)
DT: January 1993
UF: Screens and sieves*
BT: Separators
RT: Screens (sizing)

Signal delay lines
USE: Electric delay lines

Signal detection 716.1
DT: January 1981
UF: Information theory—Signal acquisition*
BT: Signal processing
RT: Demodulation
Information theory
Signal receivers

Signal distortion 716.1
SN: Of acoustical, electrical or optical signals
DT: Predates 1975
UF: Electron tubes—Distortion*
BT: Distortion (waves)
RT: Amplifiers (electronic)
Fading (radio)
Intermodulation
Reception quality
Signal interference
Signal processing
Spurious signal noise
Telephone

Signal encoding 716.1
DT: January 1993
UF: Coding (signals)
Encoding (signal)
Signal processing—Signal encoding*
BT: Signal processing
RT: Vocoders

Signal filtering and prediction 716.1 (731.1)
DT: January 1977
UF: Filtering and prediction theory
Prediction theory
BT: Signal processing
NT: Adaptive filtering
Kalman filtering
RT: Information theory
Predictive control systems
Random processes
Walsh transforms

*Signal filtering and prediction—Kalman filtering**
USE: Kalman filtering

Signal generators (713.5) (715)
SN: Electronic equipment for generation of special electrical signals
DT: Predates 1975
BT: Electronic equipment
NT: Ancillary signal generators
Frequency synthesizers
Function generators
Noise generators
Ramp generators
RT: Acoustic generators
Multivibrators
Oscillators (electronic)
Ultrasonic equipment
Video signal processing

*Signal generators, Ancillary**
USE: Ancillary signal generators

Signal interference 716.1
SN: Electrical interference with a telecommunication by another, external signal, as distinguished from INTERMODULATION and from SPURIOUS SIGNAL NOISE
DT: Predates 1975

Signal interference *(continued)*
UF: Interference (signal)
Telecommunication interference
NT: Crosstalk
Intersymbol interference
Jamming
Radar interference
Radio interference
Telegraph interference
Telephone interference
Television interference
RT: Atmospherics
Echo suppression
Electric shielding
Electromagnetic compatibility
Electromagnetic wave interference
Frequency allocation
Interference suppression
Radar countermeasures
Reception quality
Signal distortion
Spurious signal noise
Telecommunication
Wave interference

*Signal interference—Crosstalk**
USE: Crosstalk

*Signal interference—Jamming**
USE: Jamming

*Signal interference—Suppression**
USE: Interference suppression

Signal lights (Aircraft)
USE: Aircraft signal lights

Signal noise measurement (716.1) (942.2)
DT: January 1993
BT: Electric variables measurement
RT: Noise generators
Spurious signal noise

Signal processing 716.1
DT: Predates 1975
NT: Acoustic signal processing
Data compression
Digital signal processing
Heterodyning
Image processing
Interference suppression
Signal detection
Signal encoding
Signal filtering and prediction
Video signal processing
RT: Correlation detectors
Decision theory
Digital filters
Electric filters
Encoding (symbols)
Frequency multiplying circuits
Information theory
Integrating circuits
Mixer circuits
Multiplexing
Pulse shaping circuits
Reception quality
Sensor data fusion
Signal distortion
Signal theory
Spectrum analysis
Speech processing
Telecommunication equipment
Walsh transforms
Wavelet transforms
Z transforms

*Signal processing—Correlation detectors**
USE: Correlation detectors

*Signal processing—Digital techniques**
USE: Digital signal processing

*Signal processing—Signal encoding**
USE: Signal encoding

*Signal processing—Video signals**
USE: Video signal processing

Signal receivers (716) (717) (718)
DT: Predates 1975
UF: Receivers (signal)
BT: Telecommunication equipment
NT: Acoustic receivers
Radar receivers
Radio receivers
Stereophonic receivers
Television receivers
Transponders
RT: Amplifiers (electronic)
Antennas
Demodulation
Demodulators
Detectors
Gain control
Limiters
Networks (circuits)
Receiving antennas
Reception quality
Signal detection
Telecommunication repeaters
Teleprinters
Transmitters
Tuners
Tuning

*Signal receivers—Reception quality**
USE: Reception quality

Signal systems (716) (717) (718) (914.1)
DT: January 1993
BT: Communication systems
NT: Alarm systems
Building signal systems
Electric signal systems
Marine signal systems
Railroad signal systems
RT: Centralized signal control
Signaling
Signaling equipment
Signs
Sirens
Telecommunication systems

* Former Ei Vocabulary term

Signal theory 716.1
DT: January 1977
RT: Information theory
Signal processing
Signal to noise ratio
Spurious signal noise

Signal to noise ratio 716.1
DT: January 1993
UF: Noise, Spurious signal—Signal to noise ratio*
RT: Signal theory
Spurious signal noise

Signaling (404) (914.1)
SN: By methods other than telecommunication
DT: Predates 1975
BT: Communication
RT: Alarm systems
Electric signal systems
Marine signal systems
Railroad signal systems
Signal systems
Signaling equipment
Subway signal systems

Signaling equipment (402) (404) (914.1)
DT: January 1993
BT: Equipment
RT: Signal systems
Signaling

Signs 432
SN: Restricted to the sense of signboards or billboards
DT: Predates 1975
NT: Electric signs
Traffic signs
RT: Labels
Signal systems

Silanes 804.2
DT: January 1987
UF: Silicon hydride
BT: Hydrides
Silicon compounds

Silencers (intake)
USE: Intake silencers

Silica (482.2) (804.2) (812)
DT: Predates 1975
UF: Silicon dioxide
BT: Oxides
Silicon compounds
NT: Fused silica
Quartz
Silica gel
RT: Oxide minerals
Sand
Silica brick
Silica sand

Silica brick 812.2
DT: January 1993
UF: Refractory materials—Silica brick*
BT: Brick
Refractory materials
RT: Lime brick
Silica

Silica gel 804 (804.2) (801.3)
DT: Predates 1975
BT: Gels
Silica
RT: Adsorbents

Silica glass
USE: Fused silica

Silica sand (483) (482.2) (812)
DT: January 1993
UF: Sand, Silica*
BT: Sand
RT: Silica

*Silica—Fused**
USE: Fused silica

Silicate minerals 482.2
DT: January 1993
UF: Mineralogy—Silicates*
BT: Minerals
Silicates
NT: Bentonite
Chlorite minerals
Clay minerals
Feldspar
Kyanite
Mica
Olivine
Serpentine
Topaz
Zircon

Silicates (414) (482.2) (812)
DT: Predates 1975
BT: Silicon compounds
NT: Garnets
Silicate minerals
Zeolites
RT: Borosilicate glass

Silicon 549.3 (712.1.1)
DT: January 1993
UF: Cast iron—Silicon content*
Powder metallurgy—Silicon*
Si
Silicon and alloys*
Silicon powder metallurgy
BT: Metalloids
NT: Amorphous silicon
Porous silicon
Semiconducting silicon
RT: Silicon alloys
Silicon compounds
Silicon sensors

Silicon alloys 549.3
DT: January 1993
UF: Silicon and alloys*
BT: Alloys
RT: Silicon
Silicon compounds

*Silicon and alloys**
USE: Silicon OR
Silicon alloys

Silicon batteries 702.1
DT: January 1993
UF: Electric batteries—Silicon*
BT: Electric batteries
RT: Silicon solar cells

Silicon carbide 804.2 (812.2)
DT: Predates 1975
BT: Carbides
Silicon compounds

Silicon compounds (712.1.2) (804.1) (804.2)
DT: Predates 1975
UF: Powder metallurgy—Silicon compounds*
BT: Chemical compounds
NT: Polysilanes
Semiconducting silicon compounds
Silanes
Silica
Silicates
Silicon carbide
Silicon nitride
Silicones
RT: Silicon
Silicon alloys

Silicon controlled rectifiers
USE: Thyristors

Silicon dioxide
USE: Silica

Silicon hydride
USE: Silanes

Silicon iron
USE: Silicon steel

Silicon nitride 804.2 (812.1)
DT: Predates 1975
UF: Ceramic materials—Silicon nitride*
Refractory materials—Silicon nitride*
BT: Nitrides
Silicon compounds
RT: Ceramic materials
Ceramic matrix composites
Refractory materials

Silicon on insulator technology 714.2
DT: January 1993
UF: Semiconductor device manufacture—Silicon on insulator technology*
BT: Semiconductor device manufacture

Silicon on sapphire technology 714.2
DT: January 1993
UF: Semiconductor device manufacture—Silicon on sapphire technology*
BT: Semiconductor device manufacture

Silicon powder metallurgy
USE: Silicon AND
Powder metallurgy

Silicon sensors 714.2 (732.2) (801)
SN: Fabricated from silicon
DT: January 1993
UF: Sensors—Silicon sensors*
BT: Sensors
RT: Silicon

Silicon solar cells 702.3
DT: January 1993
UF: Solar cells—Silicon*
BT: Solar cells
RT: Silicon batteries

Silicon steel 545.3, 549.3
DT: Predates 1975
UF: Silicon iron
BT: Steel

Silicon wafers 714.2 (712.1.1)
DT: January 1995
BT: Semiconducting silicon
RT: Semiconductor device manufacture
Thin films

Silicone coatings 813.2 (539.2) (817.1) (818.5)
DT: January 1993
UF: Protective coatings—Silicones*
BT: Coatings
RT: Silicones

Silicone resins
USE: Silicones

Silicones 815.1.1 (816) (817) (818.2)
DT: Predates 1975
UF: Automobile materials—Silicones*
Electric insulating materials—Silicones*
Polysiloxanes
Silicone resins
BT: Inorganic polymers
Silicon compounds
RT: Polysilanes
Silicone coatings
Thermosets

Silk 819.1
SN: Includes raw, boiled-off, thrown and spun silk
DT: Predates 1975
BT: Natural fibers

Silos (agricultural) 821.6 (402.1)
DT: January 1993
UF: Silos*
Steel silos
BT: Farm buildings

Silos (missile)
USE: Missile silos

*Silos**
USE: Silos (agricultural)

*Silos—Steel**
USE: Steel structures

*Silos—Unloading**
USE: Unloading

* Former Ei Vocabulary term

Silt 483.1
DT: January 1993
UF: Soils—Silt*
BT: Soils

Silver 547.1
DT: January 1993
UF: Silver and alloys*
BT: Precious metals
RT: Silver alloys
Silver compounds
Silver deposits
Silver metallography
Silver metallurgy
Silver mines
Silver ore treatment
Silver plating
Silver refining
Silverware

Silver alloys 547.1
DT: January 1993
UF: Brazing—Silver alloy fillers*
Silver and alloys*
Solders—Silver solders*
BT: Precious metal alloys
RT: Silver
Silver compounds

*Silver and alloys**
USE: Silver OR
Silver alloys

Silver compounds (804.1) (804.2)
DT: Predates 1975
BT: Precious metal compounds
NT: Semiconducting silver compounds
RT: Silver
Silver alloys

Silver deposits 504.3, 547.1
DT: Predates 1975
BT: Ore deposits
NT: Copper silver deposits
RT: Silver
Silver gold mines
Silver mines

Silver gold mines 504.3, 547.1
DT: January 1993
UF: Silver gold mines and mining*
BT: Gold mines
Silver mines
RT: Gold deposits
Silver deposits

*Silver gold mines and mining**
USE: Silver gold mines

Silver metallography 531.2, 547.1
DT: Predates 1975
BT: Metallography
RT: Silver
Silver metallurgy

Silver metallurgy 531.1, 547.1
DT: Predates 1975
BT: Metallurgy
NT: Silver powder metallurgy
RT: Silver
Silver metallography
Silver ore treatment

Silver mines 504.3, 547.1
DT: January 1993
UF: Silver mines and mining*
BT: Mines
NT: Silver gold mines
RT: Copper silver deposits
Silver
Silver deposits

*Silver mines and mining**
USE: Silver mines

Silver ore treatment 533.1, 547.1
DT: Predates 1975
BT: Ore treatment
RT: Silver
Silver metallurgy

Silver plating 539.3, 547.1
DT: Predates 1975
BT: Plating
RT: Silver
Silverware

Silver powder metallurgy 536.1, 547.1
DT: January 1993
UF: Powder metallurgy—Nickel silver*
Powder metallurgy—Silver*
BT: Powder metallurgy
Silver metallurgy

Silver refining 533.2, 547.1
DT: Predates 1975
BT: Metal refining
RT: Silver

Silverware 547.1 (539.3)
DT: Predates 1975
RT: Silver
Silver plating

SIMS
USE: Secondary ion mass spectrometry

Simulated annealing 537.1 (921)
DT: January 1995
UF: Annealing (simulated)
BT: Optimization
RT: Genetic algorithms

Simulation
SN: Use for conceptual, electric, hydraulic, or mechanical simulations, as distinguished from COMPUTER SIMULATION, MATHEMATICAL MODELS, or actual physical MODELS.
DT: January 1993
UF: Analogies
Electric analogies
NT: Computer simulation
Plasma simulation
RT: Computer aided design
Computer graphics
Control theory

Simulation *(continued)*
Identification (control systems)
Mathematical models
Models
Operations research
Simulators
System theory
Systems analysis
Systems science

Simulation languages
USE: Computer simulation languages

Simulation modeling
USE: Computer simulation

Simulators
(621) (654.1) (655.2) (657.1) (662.1) (671)
DT: January 1993
UF: Rockets and missiles—Simulators*
Solar radiation—Simulators*
BT: Models
NT: Automobile driver simulators
Automobile simulators
Direct analogs
Environmental chambers
Flight simulators
Nuclear reactor simulators
Radar simulators
Satellite simulators
Ship simulators
RT: Computer simulation
Mathematical models
Ship model tanks
Simulation
Wind tunnels

Simultaneous engineering
USE: Concurrent engineering

Single crystals 933.1
DT: January 1995
BT: Crystals
NT: Crystal whiskers
RT: Crystal microstructure
Diamonds

Single mode fibers
USE: Optical fibers

Single point mooring 674
DT: October 1975
BT: Mooring
RT: Boats
Sailing vessels
Ships

Single sideband receivers 716.3
DT: January 1993
UF: Radio receivers—Single sideband*
BT: Radio receivers

Sinter (ore)
USE: Ore sinter

Sintered alumina 804.2, 812
DT: January 1993
UF: Alumina—Sintered*
BT: Alumina
RT: Briquets
Sintering

Sintered carbides 804.2, 812.2
DT: January 1993
UF: Carbides—Sintered*
Tantalum carbide—Sintered*
BT: Carbides
RT: Sintering

Sintering (533.1) (536.1) (802.3) (812.2)
DT: January 1993
BT: Heat treatment
NT: Ore sintering
RT: Hot isostatic pressing
Porosity
Porous materials
Powder metallurgy
Pyrometallurgy
Refractory materials
Sintered alumina
Sintered carbides

Sirens 752.4 (914.1)
DT: Predates 1975
BT: Acoustic generators
RT: Alarm systems
Signal systems

Site selection (402) (403)
DT: January 1993
UF: Industrial plants—Site selection*
Plant site selection
Power plants—Site selection*
RT: Buildings
Facilities
Industrial plants
Location
Power plants
Relocation
Structures (built objects)

Size determination (943.3)
DT: January 1993
UF: Containers—Size determination*
Electric conductors—Size determination*
Granular materials—Size determination*
BT: Spatial variables measurement
NT: Particle size analysis
RT: Classifiers
Dilatometers
Dimensional stability
Grading
Morphology
Size separation

Size exclusion chromatography 801 (802.3)
DT: January 1993
UF: Chromatographic analysis—Size exclusion*
BT: Chromatography

Size separation 802.3
DT: January 1993
BT: Separation
NT: Screening
RT: Beneficiation

* Former Ei Vocabulary term

Size separation *(continued)*
Classifiers
Filtration
Flotation
Powder metals
Powders
Size determination

Sizing (finishing operation) (811.1.1) (819.5)
DT: January 1993
UF: Papermaking—Sizing*
Textiles—Sizing*
BT: Finishing
RT: Binders
Fillers
Papermaking
Textile finishing
Textile processing

Skating rinks 403.1 (402.2) (644.3)
DT: Predates 1975
BT: Facilities
RT: Recreation centers
Recreational facilities

Ski jumps 403.2
DT: Predates 1975
BT: Facilities
RT: Recreational facilities

Skid resistance (406.2) (818.5) (914.1)
DT: January 1993
UF: Containers—Skid resistant*
Roads and streets—Skid resistance*
Tires—Skid resistance*
RT: Accident prevention
Highway accidents
Roads and streets
Skidding
Tires

Skidding 662.1 (406.2) (818.5) (914.1)
DT: January 1993
UF: Automobiles—Skidding*
RT: Automobiles
Highway accidents
Skid resistance

Skids 691.1
DT: January 1993
UF: Materials handling—Skids*
BT: Materials handling equipment

Skin 461.2
DT: January 1993
UF: Biological materials—Skin*
BT: Biological materials
RT: Dermatitis

Skin effect 701.1 (708.2)
SN: The tendency of alternating current to flow near the surface of a conductor
DT: January 1993
UF: Conductor skin effect
Electric conductors—Skin effect*
Kelvin skin effect
RT: Bimetals
Conductive materials
Electric conductors
Electric currents
Electrodynamics

Skyscrapers
USE: Tall buildings

Slab mills 535.1.1
DT: January 1993
UF: Rolling mill practice—Slab*
BT: Hot rolling mills
RT: Billet mills
Blooming mills
Strip mills

Slag cement 412.1
DT: January 1993
UF: Cement—Slag*
Lime—Slag admixture*
BT: Cements
RT: Lime
Slags

Slag reaction process
USE: Steelmaking

Slag wool
USE: Mineral wool

Slags (406) (412) (413) (452.3) (532.2) (545.3) (614.2) (804.2) (821.2)
DT: Predates 1975
UF: Blast furnace slag
Boiler corrosion and deposits—Slag*
Boilers—Slag tap*
Steelmaking—Slags*
BT: Industrial wastes
RT: Blast furnace practice
Boiler corrosion
Metal recovery
Mineral wool
Slag cement
Steelmaking
Waste utilization

*Slags—Aluminum recovery**
USE: Metal recovery AND Aluminum

*Slags—Iron recovery**
USE: Metal recovery AND Iron

*Slags—Manganese recovery**
USE: Metal recovery AND Manganese

*Slags—Vanadium recovery**
USE: Metal recovery AND Vanadium

*Slags—Zinc recovery**
USE: Metal recovery AND Zinc

Slaked lime
USE: Hydrated lime

* Former Ei Vocabulary term

Slamming (ships) 671
DT: January 1993
UF: Ships—Slamming*
RT: Hydrodynamics
Ships

Slate 482.2 (414) (505)
DT: Predates 1975
BT: Metamorphic rocks
RT: Building materials
Shale

Sleep research 461.4
DT: January 1993
UF: Human engineering—Sleep studies*
BT: Research
RT: Human engineering
Physiology

*Slide rules**
USE: Mathematical instruments

Slideways 603.2
SN: Guideways used to position workpieces prior to machine tool operation
DT: January 1993
UF: Guideways (machine tools)
Machine tools—Slideways*
BT: Machine tools

Slime control (papermaking)
USE: Papermaking slime control

Slings 691.1
DT: January 1993
UF: Materials handling—Slings*
BT: Materials handling equipment

Slip forming 412 (405.2)
DT: January 1993
UF: Concrete construction—Slip forming*
BT: Concrete construction
RT: Forms (concrete)

Slipways
USE: Shipways

Slitting 604.1 (535)
DT: January 1993
UF: Sheet and strip metal—Slitting*
BT: Cutting
RT: Sawing
Shearing
Slitting machines
Slotting
Trimming

Slitting machines 603.1 (535)
DT: January 1993
BT: Cutting tools
Machine tools
RT: Shearing machines
Slitting
Slotting machines

Slope protection (407.1) (441.1) (483.1)
DT: January 1993
UF: Dams—Slope protection*
BT: Protection
RT: Bank protection
Slope stability
Soil conservation

Slope stability 406.2 (483.1)
DT: January 1993
UF: Roads and streets—Slope stability*
BT: Stability
RT: Roads and streets
Slope protection
Soil liquefaction

Sloshing (fuel)
USE: Fuel sloshing

Slot antennas (716)
DT: January 1993
UF: Antennas, Slot*
BT: Antennas

Slotting 604.1 (535)
DT: January 1993
UF: Sheet and strip metal—Slotting*
BT: Cutting
RT: Metal cutting
Shearing
Slitting
Slotting machines

Slotting machines 603.1 (535) (604.1)
DT: Predates 1975
BT: Cutting tools
Machine tools
RT: Slitting machines
Slotting

Sludge (sewage)
USE: Sewage sludge

Sludge digestion 452.2
DT: January 1993
UF: Digestion (sludge)
Sewage treatment—Sludge digestion*
BT: Waste treatment
RT: Sewage sludge
Sludge disposal

Sludge disposal 452.2
DT: January 1993
UF: Sewage treatment—Sludge disposal*
BT: Waste disposal
RT: Sewage sludge
Sewage treatment
Sludge digestion

Sludge reclamation
USE: Reclamation

Slurries 801.3 (524) (533) (631.1) (804)
DT: June 1990
BT: Suspensions (fluids)
NT: Coal slurries
RT: Slurry pipelines

Slurry pipelines 619.1 (801.3)
DT: January 1977
BT: Pipelines
RT: Coal slurries

* Former Ei Vocabulary term

Slurry pipelines *(continued)*
Materials handling
Slurries

Small automobile engines 661.1 (662.2)
DT: January 1993
UF: Automobile engines—Small*
BT: Automobile engines

Small nuclear reactors 621
DT: January 1993
UF: Nuclear reactors—Small*
BT: Nuclear reactors

Small power plants (611) (612) (613) (614) (615)
DT: January 1993
UF: Power plants—Small capacity*
BT: Power plants

Small turbomachinery (612.3) (617.1) (617.2) (618) (632)
DT: January 1993
UF: Turbomachinery—Small*
BT: Turbomachinery

Smart buildings
USE: Intelligent buildings

Smart cards 722.4
DT: January 1993
UF: Microprocessor chips—Smart cards*
BT: Digital devices
RT: Microprocessor chips
Point of sale terminals

Smart houses
USE: Intelligent buildings

Smart materials
USE: Intelligent materials

Smart structures
USE: Intelligent structures

SMD
USE: Surface mount technology

Smectic liquid crystals 804 (931.2)
DT: January 1993
UF: Crystals, Liquid—Smectic*
BT: Liquid crystals

Smelting 533.2
DT: January 1993
UF: Ore treatment—Smelting*
BT: Ore treatment
NT: Copper smelting
Lead smelting
Nickel smelting
Tin smelting
Zinc smelting
RT: Desulfurization
Metallurgy
Pyrometallurgy
Refining
Segregation (metallography)

Smoke (451.1) (801.3) (804)
DT: Predates 1975
BT: Dispersions
RT: Aerosols
Air pollution
Cigarette filters
Dust
Exhaust gases
Fume control
Particulate emissions
Smoke abatement
Soot
Vapors

Smoke abatement 451.2
DT: Predates 1975
BT: Air pollution control
RT: Electrostatic separators
Environmental engineering
Fume control
Smoke

Smoke detectors 914.2
DT: January 1987
BT: Fire detectors
RT: Alarm systems
Fire protection

Smoke filters (cigarette)
USE: Cigarette filters

Smokestacks (petroleum refinery)
USE: Flare stacks

Smokestacks (ship)
USE: Ship smokestacks

SMT
USE: Surface mount technology

Sn
USE: Tin

Snow 443.3
DT: January 1993
UF: Snow and snowfall*
Snowfall
BT: Precipitation (meteorology)
RT: Avalanches (snowslides)
Cold weather problems
Meteorology
Snow and ice removal
Snow gages
Snow making
Snow melting systems
Snow plows
Snowfall measurement
Storms
Trafficability
Water
Water supply

Snow and ice removal (406.2) (431.4) (443.3) (681.1)
DT: January 1993
UF: Airport runways—Snow and ice removal*
Airports—Snow and ice removal*
Ice removal
Railroad plant and structures—Snow and ice removal*
Roads and streets—Snow and ice removal*

* Former Ei Vocabulary term

Snow and ice removal *(continued)*
BT: Cleaning
RT: Airport runways
Cold weather problems
Highway systems
Ice
Ice control
Ice problems
Maintenance of way
Railroad plant and structures
Roads and streets
Sanding equipment
Snow
Snow melting systems
Snow plows
Street cleaning

*Snow and snowfall**
USE: Snow

*Snow and snowfall—Avalanches and slides**
USE: Avalanches (snowslides)

*Snow and snowfall—Manufacture**
USE: Snow making

*Snow and snowfall—Measurements**
USE: Snowfall measurement

*Snow and snowfall—Trafficability**
USE: Trafficability

Snow gages 443.2, 443.3 (943.3)
DT: Predates 1975
BT: Gages
RT: Snow
Snowfall measurement

Snow making 644 (443.3)
DT: January 1993
UF: Snow and snowfall—Manufacture*
Snowmaking
RT: Snow

Snow melting systems (406.2) (431.4) (443.3) (681.1)
DT: Predates 1975
BT: Equipment
RT: Snow
Snow and ice removal

Snow plows 406
DT: Predates 1975
UF: Snowplows
BT: Equipment
RT: Snow
Snow and ice removal
Trucks
Vehicle parts and equipment

Snow surveys
USE: Snowfall measurement

Snowfall
USE: Snow

Snowfall measurement (443.3) (943.3)
DT: January 1993
UF: Snow and snowfall—Measurements*
Snow surveys
BT: Measurements
RT: Meteorology
Snow
Snow gages
Water resources
Water supply

Snowmaking
USE: Snow making

Snowplows
USE: Snow plows

Snowslides
USE: Avalanches (snowslides)

Soaking pits 535.1.1
DT: January 1993
UF: Rolling mill practice—Soaking pits*
BT: Furnaces
RT: Rolling mill practice

*Soap**
USE: Soaps (detergents)

*Soap—Metallic**
USE: Metallic soaps

Soaps (detergents) 804.1
DT: January 1993
UF: Soap*
BT: Detergents
RT: Laundering

Soaps (metallic)
USE: Metallic soaps

Soapstone (steatite)
USE: Steatite

Social aspects 901.4
DT: January 1993
UF: Engineering—Social aspects*
Sociological aspects
RT: Economic and social effects
Philosophical aspects
Population statistics
Professional aspects
Social sciences

Social effects
USE: Economic and social effects

Social sciences (461.4) (901.4) (911.2) (912.2)
DT: January 1993
NT: Behavioral research
Economics
Management science
RT: Economic and social effects
Social aspects
Social sciences computing

Social sciences computing 723.2 (461.4) (901.4) (911.2) (912.2)
DT: January 1993
UF: Behavioral science computing
Data processing—Social and behavioral sciences applications*
BT: Computer applications

Social sciences computing *(continued)*
RT: Data processing
Social sciences

Societies and institutions 901.1.1
DT: Predates 1975
UF: Institutions
Organizations
Professional societies
RT: Professional aspects
Research

Socioeconomic effects
USE: Economic and social effects

Sociological aspects
USE: Social aspects

Sociological effects
USE: Economic and social effects

Sockets (electric)
USE: Electric connectors

Sodium 549.1
DT: January 1993
UF: Na
Sodium and alloys*
BT: Alkali metals
RT: Sodium alloys
Sodium compounds
Sodium deposits
Sodium metallography

Sodium alloys 549.1
DT: January 1993
UF: Sodium and alloys*
BT: Alkali metal alloys
RT: Sodium
Sodium compounds

*Sodium and alloys**
USE: Sodium OR
Sodium alloys

Sodium chloride (482.2) (804.2)
SN: Use for the chemical compound. For natural occurrences of halite or sodium chloride use SALT DEPOSITS, and for mines use SALT MINES
DT: Predates 1975
UF: Halite
Salt
BT: Chlorine compounds
Salts
Sodium compounds
RT: Desalination
Salt deposits
Salt mines

Sodium compounds (804.1) (804.2)
DT: Predates 1975
BT: Alkali metal compounds
NT: Caustic soda
Sodium chloride
RT: Sodium
Sodium alloys

Sodium deposits 504.1, 549.1
DT: Predates 1975
BT: Ore deposits
RT: Salt deposits
Sodium

Sodium metallography 531.2, 549.1
DT: Predates 1975
BT: Metallography
RT: Sodium

Sodium silicate process 534.2
DT: January 1993
UF: Foundry practice—Sodium silicate process*
BT: Metal casting
RT: Foundry practice

Soft coal
USE: Bituminous coal

Software (computer)
USE: Computer software

Software development
USE: Software engineering

Software engineering 723.1
DT: January 1993
UF: Software development
BT: Computer applications
NT: Computer aided software engineering
Software prototyping

Software portability
USE: Computer software portability

Software prototyping 723.1
DT: January 1995
UF: Prototyping (software)
BT: Software engineering

Software tools
USE: Computer aided software engineering

Soil cement 406.2, 412 (483.1)
DT: January 1993
UF: Roads and streets—Soil cement*
BT: Cements
RT: Roads and streets
Soils

Soil conditioners 483.1
DT: January 1993
UF: Conditioners (soil)
Soils—Conditioners*
BT: Materials
RT: Soils

Soil conservation 483.1 (821.3)
DT: January 1993
UF: Soils—Conservation*
BT: Conservation
RT: Agriculture
Bank protection
Erosion
Land reclamation
Shore protection
Slope protection

* Former Ei Vocabulary term

Soil conservation *(continued)*
Soil mechanics
Soils

Soil liquefaction 483.1
DT: January 1993
UF: Soils—Liquefaction*
BT: Liquefaction
RT: Landslides
Settlement of structures
Slope stability
Soils

Soil load testing
USE: Soil mechanics AND
Load testing

Soil mechanics 483.1
DT: Predates 1975
UF: Soil load testing
BT: Mechanics
RT: Load testing
Rock mechanics
Soil conservation
Soil structure interactions
Soils

Soil pollution 483.1 (454.2)
DT: October 1975
BT: Pollution
RT: Agricultural runoff
Hazardous materials spills
Oil spills
Pesticide effects
Soil pollution control
Soils
Water pollution

Soil pollution control 454.2, 483.1
DT: January 1993
UF: Soil pollution—Control*
BT: Pollution control
RT: Soil pollution

*Soil pollution—Control**
USE: Soil pollution control

Soil resistance (textiles) 819.3
DT: January 1993
UF: Textiles—Soil resistance*
RT: Cleaning
Textiles

Soil structure interactions 483.2
DT: January 1993
UF: Foundations—Soil structure interaction*
RT: Dynamic loads
Earthquakes
Engineering geology
Foundations
Settlement of structures
Soil mechanics

Soil surveys 483.1 (406.2)
DT: January 1993
UF: Roads and streets—Soil surveys*
Soils—Surveys*
BT: Surveys
RT: Geological surveys
Road construction
Roads and streets
Soils

Soil testing 483.1
DT: January 1993
UF: Soils—Testing*
BT: Mechanical testing

Soils 483.1
DT: Predates 1975
UF: Tractors—Soil factors*
Vehicles—Soil factors*
NT: Clay
Frozen soils
Silt
Underwater soils
RT: Consolidation
Earth (planet)
Erosion
Frost effects
Fullers earth
Geology
Granular materials
Gravel
Infiltration
Landslides
Muskeg
Natural resources
Peat
Pore pressure
Salinity measurement
Sand
Sediments
Shear strength
Soil cement
Soil conditioners
Soil conservation
Soil liquefaction
Soil mechanics
Soil pollution
Soil surveys
Subsidence
Trafficability

*Soils—Conditioners**
USE: Soil conditioners

*Soils—Conservation**
USE: Soil conservation

*Soils—Consolidation**
USE: Consolidation

*Soils—Erosion**
USE: Erosion

*Soils—Frost effect**
USE: Frost effects

*Soils—Frozen**
USE: Frozen soils

*Soils—Infiltration**
USE: Infiltration

* Former Ei Vocabulary term

*Soils—Liquefaction**
USE: Soil liquefaction

*Soils—Load testing**
USE: Load testing

*Soils—Muskeg**
USE: Muskeg

*Soils—Permafrost**
USE: Permafrost

*Soils—Pore pressure**
USE: Pore pressure

*Soils—Salinity measurements**
USE: Salinity measurement

*Soils—Sediments**
USE: Sediments

*Soils—Shear strength**
USE: Shear strength

*Soils—Silt**
USE: Silt

*Soils—Stabilization**
USE: Stabilization

*Soils—Subsidence**
USE: Subsidence

*Soils—Surveys**
USE: Soil surveys

*Soils—Testing**
USE: Soil testing

*Soils—Trafficability**
USE: Trafficability

*Soils—Underwater**
USE: Underwater soils

Sol-gels 804 (801.3)
DT: January 1993
BT: Colloids
RT: Gels
Glass
Optical devices
Optics
Sols

Solar absorbers 657.1, 813.2 (615.2) (702.3)
DT: January 1993
BT: Materials
RT: Black coatings
Coatings
Energy absorption
Solar collectors

Solar batteries
USE: Solar cells

Solar buildings 402, 657.1
DT: January 1993
BT: Buildings
NT: Active solar buildings
Passive solar buildings

Solar cell arrays 702.3 (615.2)
DT: January 1993
UF: Solar cells—Arrays*
BT: Arrays
Solar equipment
RT: Solar cells

Solar cells 702.3 (615.2)
DT: Predates 1975
UF: Batteries (solar)
Photovoltaic generators (solar)
Solar batteries
BT: Direct energy converters
Solar equipment
NT: Cadmium sulfide solar cells
Silicon solar cells
RT: Electric batteries
Electric power generation
Photoelectrochemical cells
Photovoltaic cells
Photovoltaic effects
Solar cell arrays
Solar energy
Solar power generation
Solar power plants

*Solar cells—Arrays**
USE: Solar cell arrays

*Solar cells—Cadmium sulfide**
USE: Cadmium sulfide solar cells

*Solar cells—Silicon**
USE: Silicon solar cells

Solar collectors 657.1 (615.2) (702.3)
DT: January 1993
UF: Collectors (solar)
Solar radiation—Collectors*
BT: Solar equipment
NT: Solar ponds
RT: Solar absorbers
Solar radiation

Solar concentrators 657.1 (615.2) (702.3)
DT: January 1993
UF: Concentrators (solar)
Heliostats (solar concentrators)
Solar energy concentrators
Solar radiation—Concentrators*
BT: Solar equipment
RT: Solar radiation

Solar control films 657.1, 813.2 (615.2) (702.3)
DT: January 1993
UF: Coatings—Solar control films*
Light control films
BT: Films
RT: Coatings
Reflective coatings
Windows

Solar dryers 657.1, 802.1
DT: January 1993
UF: Dryers—Solar*
BT: Dryers (equipment)
Solar equipment

* Former Ei Vocabulary term

Solar energy 657.1 (615.2)
DT: July 1974
UF: Solar power
BT: Renewable energy resources
RT: Power generation
Solar cells
Solar ponds
Solar power generation
Solar power plants
Solar radiation
Solar refrigeration
Sun

Solar energy concentrators
USE: Solar concentrators

Solar equipment 657.1 (615.2)
DT: January 1993
UF: Air conditioning—Solar energy systems*
Calorimeters—Solar*
Furnaces—Solar*
Solar powered equipment
Solar stoves
Stoves—Solar*
BT: Equipment
NT: Solar cell arrays
Solar cells
Solar collectors
Solar concentrators
Solar dryers
Solar water heaters
RT: Solar heating
Solar power generation
Solar power plants
Solar refrigeration
Stoves

Solar heating 657.1 (642.1) (643.1)
DT: January 1993
UF: Heating—Solar*
BT: Radiant heating
RT: Solar equipment
Solar ponds

Solar ponds 657.1 (615.2)
DT: January 1993
UF: Solar radiation—Solar ponds*
BT: Solar collectors
Surface waters
RT: Solar energy
Solar heating
Solar radiation

Solar power
USE: Solar energy

Solar power generation 615.2 (657.1) (702.3)
DT: January 1993
UF: Power generation—Solar energy*
BT: Power generation
RT: Solar cells
Solar energy
Solar equipment
Solar power plants

Solar power plants 615.2 (402.1) (657.1) (702.3)
DT: January 1977
BT: Power plants
NT: Orbiting solar power plants
RT: Electric power plants
Solar cells
Solar energy
Solar equipment
Solar power generation

*Solar power plants—Orbiting**
USE: Orbiting solar power plants

Solar powered equipment
USE: Solar equipment

Solar radiation 657.1
DT: Predates 1975
BT: Radiation
RT: Cosmic rays
Daylighting
Heliostats (instruments)
Infrared radiation
Solar collectors
Solar concentrators
Solar energy
Solar ponds

*Solar radiation—Collectors**
USE: Solar collectors

*Solar radiation—Concentrators**
USE: Solar concentrators

*Solar radiation—Heliostats**
USE: Heliostats (instruments)

*Solar radiation—Simulators**
USE: Simulators

*Solar radiation—Solar ponds**
USE: Solar ponds

Solar refrigeration 644.3, 657.1
DT: January 1993
UF: Refrigeration—Solar*
BT: Refrigeration
RT: Solar energy
Solar equipment

Solar stoves
USE: Solar equipment AND
Stoves

Solar system 657.2
DT: January 1993
NT: Earth (planet)
Moon
Planets
Sun
RT: Astronomy
Interplanetary flight
Space research

Solar water heaters 657.1 (642.1) (643.2)
DT: January 1993
UF: Water heaters—Solar*
BT: Solar equipment
Water heaters

* Former Ei Vocabulary term

Soldered joints 538.1.1
DT: January 1993
UF: Joints—Soldered*
BT: Joints (structural components)
RT: Soldering
Welds

Soldering 538.1.1
DT: Predates 1975
BT: Bonding
NT: Brazing
RT: Fluxes
Metal working
Sealing (closing)
Soldered joints
Soldering alloys
Tinning
Welding

Soldering alloys 538.1.1
DT: January 1993
UF: Soldering—Filler metals*
Solders*
BT: Filler metals
RT: Copper alloys
Lead alloys
Soldering
Tin alloys
Zinc alloys

*Soldering—Filler metals**
USE: Soldering alloys

*Solders**
USE: Soldering alloys

*Solders—Silver solders**
USE: Silver alloys

Solenoids 704.1
DT: January 1977
BT: Electric coils
RT: Electric relays
Electromagnets

Solid electrolytes 803 (702.1) (802.2) (804)
DT: January 1993
UF: Electrolytes, Solid*
BT: Electrolytes

Solid film lubricants
USE: Solid lubricants

Solid flow
USE: Flow of solids

Solid lasers
USE: Solid state lasers

Solid lubricants 607.1
DT: January 1993
UF: Lubricants—Solid films*
Solid film lubricants
BT: Lubricants

Solid propellants 524, 804
DT: January 1993
UF: Propellants—Solid*
BT: Propellants

Solid solutions 933
DT: January 1986
BT: Solids
RT: Aging of materials
Alloying
Alloys
Binary alloys
Eutectics
Order disorder transitions
Phase diagrams
Solidification
Superlattices
Supersaturation

Solid state device structures 714.2
DT: January 1993
NT: Semiconductor device structures
RT: Solid state devices

Solid state devices 714.2
DT: Predates 1975
BT: Electron devices
NT: MIM devices
Semiconductor devices
Solid state lasers
Solid state oscillators
Solid state rectifiers
Thick film devices
Thin film devices
RT: Magnetic devices
Metal insulator boundaries
Percolation (solid state)
Solid state device structures
Solid state physics

*Solid state devices, MIM**
USE: MIM devices

*Solid state devices, Thick film**
USE: Thick film devices

*Solid state devices, Thin film**
USE: Thin film devices

*Solid state devices—Metal/insulator boundaries**
USE: Metal insulator boundaries

Solid state lasers 744.4 (714.2)
DT: January 1993
UF: Lasers, Solid state*
Solid lasers
BT: Lasers
Solid state devices
NT: Fiber lasers
Neodymium lasers

Solid state oscillators 713.2
DT: January 1993
UF: Oscillators, Solid state*
BT: Oscillators (electronic)
Solid state devices

Solid state physics 933
DT: January 1993
UF: Physics—Solid state*
BT: Physics

Solid state physics *(continued)*
RT: Solid state devices
Solids

Solid state rectifiers 714.2
DT: January 1993
UF: Electric rectifiers, Solid state*
BT: Electric rectifiers
Solid state devices
RT: Semiconductor diodes
Semiconductor devices

Solid state relays
USE: Semiconductor relays

Solidification 802.3 (933)
DT: January 1993
UF: Directional solidification
BT: Phase transitions
NT: Rapid solidification
Setting
RT: Freezing
Solid solutions
Ternary systems

Solids 933 (931.2)
DT: Predates 1975
BT: Materials
NT: Quasicrystals
Solid solutions
RT: Band structure
Compressibility of solids
Creep
Crystal atomic structure
Dielectric properties of solids
Diffusion in solids
Electric breakdown of solids
Electric conductivity of solids
Equations of state of solids
Flow of solids
Ionic conduction in solids
Ionization of solids
Luminescence of solids
Solid state physics
Specific heat of solids
Thermal conductivity of solids
Thermal diffusion in solids
Volume fraction

*Solids—Dielectric properties**
USE: Dielectric properties of solids

*Solids—Diffusion**
USE: Diffusion in solids

*Solids—Electric breakdown**
USE: Electric breakdown of solids

*Solids—Electric conductivity**
USE: Electric conductivity of solids

*Solids—High pressure effects**
USE: High pressure effects in solids

*Solids—Ionic conduction**
USE: Ionic conduction in solids

*Solids—Ionization**
USE: Ionization of solids

*Solids—Thermal conductivity**
USE: Thermal conductivity of solids

Solitons (921.1) (931.3)
DT: January 1995
RT: Nonlinear equations
Phonons
Shock waves

Sols 804 (801.3)
DT: June 1990
BT: Colloids
NT: Plastisols
RT: Aerosols
Non Newtonian liquids
Sedimentation
Sol-gels

Solubility 801.4
DT: January 1993
UF: Immiscibility
Insolubility
Miscibility
RT: Mixing
Phase diagrams
Phase equilibria
Saturation (materials composition)
Solutions
Solvents

Solution mining 502.1
DT: January 1993
UF: Mines and mining—Solution mining*
BT: Mining
RT: Leaching

Solutions (801) (803) (804)
SN: Physical-chemical sense only
DT: Predates 1975
UF: Aqueous solutions
BT: Liquids
Mixtures
NT: Antifreeze solutions
Electroplating solutions
RT: Dielectric properties of liquids
Eutectics
Extraction
Ion exchange
Leaching
Luminescence of liquids and solutions
Pervaporation
Phase equilibria
Saturation (materials composition)
Solubility
Solvents
Supersaturation
Washing

Solvent extraction 802.3
DT: January 1993
BT: Extraction
NT: Supercritical fluid extraction
RT: Leaching
Purification
Solvents

Solvents 803 (804.1) (813.2)
DT: Predates 1975
BT: Chemicals
NT: Organic solvents
Paint thinners
RT: Acetone
Additives
Agents
Benzene
Carbon disulfide
Carbon tetrachloride
Coatings
Drying oils
Ethanol
Ethers
Extraction
Lacquers
Leaching
Methanol
Percolation (fluids)
Plasticizers
Solubility
Solutions
Solvent extraction
Supercritical fluid extraction
Toluene
Turpentine

Sonar 752.1 (751.1)
DT: Predates 1975
UF: Asdic
Sound ranging
BT: Acoustic equipment
Detectors
RT: Acoustic wave transmission
Distance measurement
Hydrophones
Magnetostrictive devices
Navigation
Piezoelectric devices
Sounding apparatus
Tracking (position)
Underwater acoustics

Sonic delay lines
USE: Acoustic delay lines

Sonoluminescence 741.1, 751.1
DT: Predates 1975
UF: Acoustoluminescence
BT: Acoustic wave effects
Luminescence

Soot 804 (451.1) (521)
DT: January 1987
BT: Wastes
RT: Air pollution
Carbon
Carbon black
Coal
Combustion
Smoke
Soot blowers

Soot blowers 618.3 (614.2)
DT: Predates 1975
BT: Blowers
RT: Boilers
Soot

Sorption 802.3
DT: June 1990
BT: Chemical operations
NT: Absorption
Adsorption
Chemisorption
Desorption
RT: Chromatographic analysis
Surface properties

Sorting (723.1) (723.2)
DT: January 1993
UF: Computer systems programming—Sorting*
BT: Data handling
RT: Merging

Sound
USE: Acoustic waves

Sound generators
USE: Acoustic generators

Sound insulating materials 413.3 (751.3) (752.1)
DT: Predates 1975
UF: Acoustic insulating materials
BT: Insulating materials
RT: Acoustic wave absorption
Noise abatement
Noise pollution control equipment
Sound insulation

Sound insulation 413.3 (751.3)
DT: January 1993
UF: Acoustic insulation
Buildings—Sound insulation*
BT: Insulation
RT: Acoustic wave absorption
Noise abatement
Noise pollution control equipment
Sound insulating materials

Sound intensity
USE: Acoustic intensity

Sound measurement
USE: Acoustic variables measurement

*Sound measuring instruments**
USE: Acoustic measuring instruments

Sound ranging
USE: Sonar

Sound recording 752.2 (402.2)
DT: Predates 1975
BT: Recording
RT: Audio acoustics
Audio systems
Compact disks
Microphones
Phonograph records
Sound stages
Stereophonic recordings
Tape recorders

* Former Ei Vocabulary term

*Sound recording—Compact disks**
USE: Compact disks

Sound reproduction 752.3
DT: Predates 1975
RT: Acoustic devices
Acoustic equipment
Acoustics
Audio acoustics
Audio systems
Compact disks
Hearing aids
Loudspeakers
Microphones
Phonograph records
Pickups
Radio receivers
Radio studios
Tape recorders

*Sound reproduction—Compact disks**
USE: Compact disks

Sound stages 752.2 (402.2)
DT: January 1993
UF: Motion pictures—Sound stages*
BT: Stages
RT: Motion pictures
Sound recording
Studios

Sound velocity
USE: Acoustic wave velocity

Sound waves
USE: Acoustic waves

Sounding apparatus 752.1
DT: Predates 1975
BT: Acoustic equipment
RT: Sonar

Sounding rockets 654.1, 443.2
DT: January 1993
UF: Meteorological rockets
Rocket sondes
Rockets and missiles—Sounding*
BT: Rockets
RT: Meteorological instruments
Radiosondes
Weather satellites

Sour gas 512.2
DT: January 1993
UF: Natural gas wells—Sour gas*
BT: Natural gas
RT: Hydrogen sulfide
Sulfur compounds

Sources (light)
USE: Light sources

Sources (particle)
USE: Elementary particle sources

SP logging
USE: Spontaneous potential logging

Space applications 656
DT: January 1993
BT: Applications
RT: Aerospace applications
Moon bases
Space flight
Space platforms
Space research
Space stations
Spacecraft

Space based space surveillance
USE: Space surveillance

Space charge
USE: Electric space charge

Space debris 656.1 (454.2) (655.1)
DT: January 1993
UF: Space flight—Debris*
BT: Wastes
RT: Space flight
Spacecraft

Space engineering
USE: Aerospace engineering

Space flight 656.1
DT: Predates 1975
UF: Astrodynamics
Astronautics
BT: Transportation
NT: Interplanetary flight
Manned space flight
RT: Extraterrestrial communication links
Human engineering
Lunar missions
Moon bases
Navigation
Navigation charts
Orbital transfer
Orbits
Reentry
Rockets
Space applications
Space debris
Space optics
Space platforms
Space rendezvous
Space research
Space stations
Spacecraft
Spacecraft propulsion
Trajectories
Weightlessness

*Space flight—Debris**
USE: Space debris

*Space flight—Interplanetary flight**
USE: Interplanetary flight

*Space flight—Manned flight**
USE: Manned space flight

*Space flight—Navigation charts**
USE: Navigation charts

* Former Ei Vocabulary term

*Space flight—Weightlessness**
USE: Weightlessness

Space heating 643.1
DT: January 1993
BT: Climate control
Heating
NT: District heating
Electric space heating
RT: Environmental engineering
Gas heating
Hot air heating
Hot water heating
Radiant heating

Space heating furnaces 643.2
DT: January 1993
UF: Domestic furnaces
Furnaces, Space heating*
BT: Furnaces

Space optics 656.1, 741.1
DT: January 1993
BT: Optics
RT: Airborne telescopes
Space flight
Space research

Space platforms 655.1
DT: January 1983
BT: Spacecraft
RT: Interplanetary flight
Space applications
Space flight
Space stations

Space probes 655.3
DT: Predates 1975
BT: Probes
Spacecraft
RT: Interplanetary flight
Spacecraft observatories

Space rendezvous 656.1 (655.1)
DT: January 1993
UF: Spacecraft rendezvous
Spacecraft—Rendezvous*
RT: Docking
Orbital transfer
Orbits
Space flight
Spacecraft
Trajectories

Space research 656.2
DT: Predates 1975
BT: Research
RT: Astronomy
Astrophysics
Extraterrestrial communication links
Solar system
Space applications
Space flight
Space optics
Space simulators
Space stations
Space surveillance

*Space research—Simulation chambers**
USE: Space simulators

*Space research—Surveillance**
USE: Space surveillance

Space shuttles 656.1
DT: Predates 1975
UF: Shuttles (space)
BT: Spacecraft
RT: Manned space flight
Nose cones

Space simulators 656.2
DT: January 1993
UF: Space research—Simulation chambers*
BT: Environmental chambers
RT: Centrifuges
Flight simulators
Space research
Spacecraft

Space stations 656.1
DT: June 1990
BT: Spacecraft
RT: Inflatable structures
Interplanetary flight
Manned space flight
Moon bases
Space applications
Space flight
Space platforms
Space research

Space surveillance 656.2 (404.1) (405.3)
DT: January 1993
UF: Ground based space surveillance
Space based space surveillance
Space research—Surveillance*
Surveillance (space)
RT: Military operations
Photogrammetry
Space research
Spacecraft

Spacecraft 655.1
DT: January 1977
BT: Vehicles
NT: Interplanetary spacecraft
Orbital laboratories
Planetary landers
Satellites
Space platforms
Space probes
Space shuttles
Space stations
Spacecraft equipment
Spacecraft observatories
RT: Aerospace ground support
Aerospace vehicles
Docking
Electronic guidance systems
Launching
Lunar landing
Lunar missions
Maneuverability

* Former Ei Vocabulary term

Spacecraft *(continued)*
Meteor impacts
Meteorological problems
Navigation systems
Orbital transfer
Orbits
Parachutes
Plasma sheaths
Radiation hazards
Reentry
Rockets
Shielding
Space applications
Space debris
Space flight
Space rendezvous
Space simulators
Space surveillance
Spacecraft landing
Spacecraft propulsion
Trajectories
Water landing (spacecraft)

Spacecraft equipment 655.1
DT: January 1993
UF: Spacecraft—Equipment*
Spacecraft—Furniture*
BT: Equipment
Spacecraft
Vehicle parts and equipment
NT: Life support systems (spacecraft)
Spacecraft escape systems
Spacecraft instruments
Spacecraft power supplies
Tetherlines

Spacecraft escape systems 655.1, 914.1
DT: January 1993
UF: Escape systems (spacecraft)
Spacecraft—Escape systems*
BT: Spacecraft equipment

Spacecraft guidance
USE: Electronic guidance systems

Spacecraft instruments 655.1
DT: January 1993
UF: Spacecraft—Instruments*
BT: Instruments
Spacecraft equipment

Spacecraft landing 655.1
DT: January 1993
UF: Spacecraft—Landing*
BT: Landing
NT: Lunar landing
Water landing (spacecraft)
RT: Spacecraft

Spacecraft observatories 655.1 (443.2) (657.2)
DT: January 1993
UF: Spacecraft—Observatories*
BT: Observatories
Spacecraft
NT: Satellite observatories
RT: Space probes

Spacecraft power supplies 655.1 (702)
DT: January 1993
UF: Satellites—Power supply*
Spacecraft—Power supply*
BT: Spacecraft equipment
RT: Direct energy conversion
Electric batteries
Electric power supplies to apparatus
Fuel cells
Propellants

Spacecraft propulsion 656.1 (654) (655.1)
DT: January 1993
UF: Propulsion—Aerospace applications*
BT: Propulsion
RT: Ion propulsion
Propellants
Space flight
Spacecraft

Spacecraft rendezvous
USE: Space rendezvous

*Spacecraft—Computers**
USE: Computers

*Spacecraft—Docking**
USE: Docking

*Spacecraft—Equipment**
USE: Spacecraft equipment

*Spacecraft—Escape systems**
USE: Spacecraft escape systems

*Spacecraft—Flexible structures**
USE: Flexible structures

*Spacecraft—Fuel cells**
USE: Fuel cells

*Spacecraft—Furniture**
USE: Spacecraft equipment

*Spacecraft—Inflatable equipment**
USE: Inflatable equipment

*Spacecraft—Instruments**
USE: Spacecraft instruments

*Spacecraft—Interplanetary**
USE: Interplanetary spacecraft

*Spacecraft—Landing**
USE: Spacecraft landing

*Spacecraft—Life support systems**
USE: Life support systems (spacecraft)

*Spacecraft—Lunar landing**
USE: Lunar landing

*Spacecraft—Lunar missions**
USE: Lunar missions

*Spacecraft—Magnetic effects**
USE: Magnetic field effects

*Spacecraft—Meteor impact**
USE: Meteor impacts

* Former Ei Vocabulary term

*Spacecraft—Meteorological problems**
USE: Meteorological problems

*Spacecraft—Navigation aids**
USE: Navigation

*Spacecraft—Observatories**
USE: Spacecraft observatories

*Spacecraft—Orbital transfer**
USE: Orbital transfer

*Spacecraft—Planetary landers**
USE: Planetary landers

*Spacecraft—Power supply**
USE: Spacecraft power supplies

*Spacecraft—Radiation hazards**
USE: Radiation hazards

*Spacecraft—Radiation shielding**
USE: Radiation shielding

*Spacecraft—Rendezvous**
USE: Space rendezvous

*Spacecraft—Tethers**
USE: Tetherlines

*Spacecraft—Water impact**
USE: Water landing (spacecraft)

Spacers (electric lines)
USE: Electric lines AND
Electric insulators

Spacistors 714.2
DT: Predates 1975
BT: Semiconductor devices
RT: Transistors

Spalling (535.1.1)
DT: January 1993
UF: Rolls—Spalling*
RT: Defects
Fatigue of materials
Wear of materials

Spark chambers 944.7 (932.2)
DT: Predates 1975
BT: Particle detectors
RT: Bubble chambers
Ionization chambers

Spark cutting 604.1
DT: January 1993
UF: Electric spark cutting
Metal cutting—Electric spark*
BT: Cutting

Spark hardening 537.1
DT: January 1993
UF: Steel heat treatment—Spark hardening*
BT: Hardening
Heat treatment
RT: Boriding
Case hardening
Flame hardening
Steel heat treatment

Spark plugs 612.1.1 (661.1)
DT: Predates 1975
BT: Internal combustion engines

Sparking
USE: Electric sparks

Sparkover
USE: Electric sparks

Sparks
USE: Electric sparks

Spatial light modulators
USE: Light modulation

Spatial variables control 731.3
DT: January 1993
BT: Control
NT: Level control
Position control
Thickness control
Volume control (spatial)
RT: Dimensional stability
Mechanical variables control
Spatial variables measurement

Spatial variables measurement 943.2
DT: January 1993
BT: Measurements
NT: Angle measurement
Distance measurement
Level measurement
Position measurement
Size determination
Surface measurement
Thickness measurement
Volume measurement
RT: Aspect ratio
Mechanical variables measurement
Micrometers
Planimeters
Spatial variables control

Special effects (716.4) (742.1)
DT: January 1993
UF: Photography—Special effects*
Stereoscopic photography
Television broadcasting—Special effects*
Time lapse photography
RT: Computer applications
Computer graphics
Computer music
Photography
Television broadcasting

Special purpose machine tools 603.1
DT: January 1993
UF: Machine tools—Special purpose*
BT: Machine tools

Specific density
USE: Density (specific gravity)

Specific gravity
USE: Density (specific gravity)

Specific gravity measuring instruments
USE: Gravitometers

* Former Ei Vocabulary term

Specific heat 641.1
DT: Predates 1975
UF: Heat capacity
Thermal capacity
BT: Thermodynamic properties
NT: Specific heat of gases
Specific heat of liquids
Specific heat of solids
RT: Differential scanning calorimetry
Thermal conductivity

Specific heat of gases 641.1
DT: January 1993
UF: Specific heat—Gases*
BT: Specific heat
RT: Gases

Specific heat of liquids 641.1
DT: January 1993
UF: Specific heat—Liquids*
BT: Specific heat
RT: Liquids

Specific heat of solids 641.1
DT: January 1993
UF: Specific heat—Solids*
BT: Specific heat
RT: Solids

*Specific heat—Gases**
USE: Specific heat of gases

*Specific heat—Liquids**
USE: Specific heat of liquids

*Specific heat—Solids**
USE: Specific heat of solids

Specific volume
USE: Density (specific gravity)

Specific weight
USE: Density (specific gravity)

Specification languages
USE: Computer hardware description languages

Specification writing
USE: Technical writing

Specifications 902.2
DT: January 1995
RT: Evaluation
Inspection
Maintenance
Performance
Quality assurance
Reliability
Standardization
Standards

Specimen preparation
DT: January 1993
UF: Metallography—Specimen preparation*
Microscopic examination—Specimen preparation*
RT: Electron microscopy
Fractography
Metallography
Microscopic examination
Scanning electron microscopy

Speckle 741.1
DT: January 1993
UF: Light—Speckle*
BT: Light scattering
RT: Coherent light
Holographic interferometry
Light interference

Spectra (atmospheric)
USE: Atmospheric spectra

Spectral analysers
USE: Spectrum analyzers

Spectral analysis
USE: Spectrum analysis

Spectrographs 741.3 (801)
SN: Spectroscopes that produce photographic or fluorescent-screen spectrograms
DT: Predates 1975
UF: Spectroscopes (recording)
BT: Optical instruments
NT: Infrared spectrographs
Microwave spectrographs
Monochromators
Ultraviolet spectrographs
X ray spectrographs
RT: Spectroscopic analysis
Spectrum analysis

*Spectrographs, Infrared**
USE: Infrared spectrographs

*Spectrographs, Microwave**
USE: Microwave spectrographs

*Spectrographs, Ultraviolet**
USE: Ultraviolet spectrographs

*Spectrographs, X-ray**
USE: X ray spectrographs

Spectrometers 741.3 (801) (941.3)
SN: Spectroscopes capable of measuring angular deviation of radiation of different wavelengths
DT: Predates 1975
UF: Spectroscopes (measuring instruments)
BT: Optical instruments
NT: Infrared spectrometers
Magnetic resonance spectrometers
Mass spectrometers
Microwave spectrometers
Particle spectrometers
Spectrophotometers
Ultraviolet spectrometers
X ray spectrometers
RT: Diffraction gratings
Photometers
Spectrometry
Spectroscopic analysis
Spectrum analysis

*Spectrometers, Alpha particle**
USE: Alpha particle spectrometers

*Spectrometers, Beta ray**
USE: Beta ray spectrometers

*Spectrometers, Gamma ray**
USE: Gamma ray spectrometers

*Spectrometers, Infrared**
USE: Infrared spectrometers

*Spectrometers, Magnetic resonance**
USE: Magnetic resonance spectrometers

*Spectrometers, Microwave**
USE: Microwave spectrometers

*Spectrometers, Neutron**
USE: Neutron spectrometers

*Spectrometers, Particle**
USE: Particle spectrometers

*Spectrometers, Ultraviolet**
USE: Ultraviolet spectrometers

*Spectrometers, X-ray**
USE: X ray spectrometers

Spectrometry 941.4 (741.3) (801)
DT: June 1990
BT: Measurements
NT: Mass spectrometry
RT: Photometry
Spectrometers
Spectrophotometers
Spectroscopy

Spectrophotometers 741.3 (801) (941.3)
SN: Spectrometers for measuring spectral energy distributions
DT: Predates 1975
BT: Photometers
Spectrometers
NT: Infrared spectrophotometers
Ultraviolet spectrophotometers
X ray spectrophotometers
RT: Color
Colorimeters
Colorimetric analysis
Colorimetry
Spectrometry
Spectrophotometry
Spectroscopic analysis
Spectrum analysis

*Spectrophotometers, Infrared**
USE: Infrared spectrophotometers

*Spectrophotometers, Ultraviolet**
USE: Ultraviolet spectrophotometers

*Spectrophotometers, X-ray**
USE: X ray spectrophotometers

Spectrophotometry 941 (741.3) (801)
DT: June 1990
BT: Photometry
RT: Colorimetric analysis
Infrared spectrophotometers
Spectrophotometers
Spectroscopic analysis
Ultraviolet spectrophotometers
X ray spectrophotometers

Spectroscopes (measuring instruments)
USE: Spectrometers

Spectroscopes (recording)
USE: Spectrographs

Spectroscopic analysis 801
SN: Application of spectroscopy to chemical analysis
DT: Predates 1975
BT: Chemical analysis
RT: Spectrographs
Spectrometers
Spectrophotometers
Spectrophotometry
Spectroscopy

Spectroscopy (801) (931.3) (932.2)
SN: Theory and interpretation of spectral phenomena from materials or material particles when subjected to some form of radiant energy. For interpretation of spectra of waveforms as distinguished from substances, use SPECTRUM ANALYSIS
DT: Predates 1975
UF: Electrochemical impedance spectroscopy
NT: Absorption spectroscopy
Acoustic spectroscopy
Atomic spectroscopy
Deep level transient spectroscopy
Electron spectroscopy
Emission spectroscopy
Infrared spectroscopy
Magnetic resonance spectroscopy
Molecular spectroscopy
Mossbauer spectroscopy
Nuclear radiation spectroscopy
Radiofrequency spectroscopy
Raman spectroscopy
Rutherford backscattering spectroscopy
Ultraviolet spectroscopy
X ray spectroscopy
RT: Spectrometry
Spectroscopic analysis
Ultrafast phenomena

*Spectroscopy, Absorption**
USE: Absorption spectroscopy

*Spectroscopy, Acoustic**
USE: Acoustic spectroscopy

*Spectroscopy, Auger electron**
USE: Auger electron spectroscopy

*Spectroscopy, Electron**
USE: Electron spectroscopy

*Spectroscopy, Emission**
USE: Emission spectroscopy

*Spectroscopy, Infrared**
USE: Infrared spectroscopy

*Spectroscopy, Microwave**
USE: Microwave spectroscopy

* Former Ei Vocabulary term

*Spectroscopy, Mossbauer**
USE: Mossbauer spectroscopy

*Spectroscopy, Nuclear radiation**
USE: Nuclear radiation spectroscopy

*Spectroscopy, Photoacoustic**
USE: Photoacoustic spectroscopy

*Spectroscopy, Photoelectron**
USE: Photoelectron spectroscopy

*Spectroscopy, Radiofrequency**
USE: Radiofrequency spectroscopy

*Spectroscopy, Raman**
USE: Raman spectroscopy

*Spectroscopy, Ultraviolet**
USE: Ultraviolet spectroscopy

*Spectroscopy, X-ray**
USE: X ray spectroscopy

Spectrum analysis (921) (922.2) (941) (942) (943) (944)
SN: Mathematical or physical determinations of the distribution of frequencies, and of their amplitudes and energies in spectra of waveforms; and of spectra of electromagnetic or mechanical vibrations or waves. For investigation of substances use SPECTROSCOPIC ANALYSIS or SPECTROSCOPY
DT: Predates 1975
UF: Glass—Spectral properties*
Spectral analysis
BT: Waveform analysis
NT: Speech analysis
RT: Data reduction
Electric waveforms
Light
Seismology
Signal processing
Spectrographs
Spectrometers
Spectrophotometers
Spectrum analyzers
Speech recognition

*Spectrum analysis—Electric waveforms**
USE: Electric waveforms

*Spectrum analysis—Instrumentation**
USE: Spectrum analyzers

*Spectrum analysis—Seismological applications**
USE: Seismology

Spectrum analyzers (941) (942) (943) (944)
DT: Predates 1975
UF: Spectral analysers
Spectrum analysis—Instrumentation*
BT: Equipment
RT: Spectrum analysis

Speech 751.5
DT: Predates 1975
UF: Articulation (speech)
Voice
RT: Acoustic wave propagation
Audition
Biocommunications
Linguistics
Speech analysis
Speech coding
Speech communication
Speech intelligibility
Speech processing
Speech recognition
Speech synthesis

Speech aids
USE: Communication aids for nonvocal persons

Speech analysis 751.5
DT: January 1993
UF: Phonetics
Speech—Analysis*
BT: Spectrum analysis
RT: Acoustic signal processing
Speech
Speech coding
Speech processing
Speech recognition
Speech synthesis

Speech coding 751.5 (723.5)
DT: January 1993
UF: Coding (speech)
Speech—Coding*
RT: Encoding (symbols)
Speech
Speech analysis
Speech processing

Speech communication 751.5
SN: Directly to hearer or other receiver. Otherwise use SPEECH TRANSMISSION
DT: January 1993
UF: Information dissemination—Speech communication*
Speech—Transmission*
Voice communication
BT: Communication
RT: Information dissemination
Speech
Speech transmission

Speech intelligibility 751.5 (461.4)
DT: January 1993
UF: Acoustic masking
Intelligibility (speech)
Masking (acoustic)
Speech—Intelligibility*
RT: Audition
Human engineering
Information theory
Speech
Speech recognition

Speech processing 751.5 (723.5)
DT: January 1993
UF: Speech—Processing*
RT: Signal processing

* Former Ei Vocabulary term

Speech processing *(continued)*
Speech
Speech analysis
Speech coding
Speech recognition
Speech synthesis

Speech production aids 461.5
DT: January 1993
UF: Human rehabilitation engineering—Speech production aids*
BT: Human rehabilitation equipment
RT: Human rehabilitation engineering

Speech recognition 751.5 (723.5)
DT: January 1993
UF: Speech—Recognition*
Voice input
Voice recognition
BT: Pattern recognition
RT: Acoustics
Audition
Pattern recognition systems
Spectrum analysis
Speech
Speech analysis
Speech intelligibility
Speech processing
Speech synthesis

Speech synthesis 751.5 (723.5)
DT: January 1993
UF: Speech—Synthesis*
Text to speech synthesis
Voice synthesis
RT: Computer peripheral equipment
Speech
Speech analysis
Speech processing
Speech recognition
Vocoders

Speech transmission 751.5
SN: Telecommunication. For communication directly to hearer or other receiver, use SPEECH COMMUNICATION
DT: January 1993
UF: Telecommunication—Speech transmission*
Voice transmission
BT: Telecommunication
RT: Information dissemination
Speech communication
Vocoders
Voice/data communication systems

*Speech—Analysis**
USE: Speech analysis

*Speech—Coding**
USE: Speech coding

*Speech—Intelligibility**
USE: Speech intelligibility

*Speech—Processing**
USE: Speech processing

*Speech—Recognition**
USE: Speech recognition

*Speech—Synthesis**
USE: Speech synthesis

*Speech—Transmission**
USE: Speech communication

Speed
DT: January 1993
UF: Ships—Speed*
RT: Doppler effect
Speed control
Velocity

Speed control 731.3
DT: January 1993
BT: Mechanical variables control
RT: Speed

Speed indicators 943.3
DT: Predates 1975
BT: Indicators (instruments)
NT: Anemometers
Tachometers
RT: Governors
Speed regulators
Stroboscopes
Velocimeters

Speed of light
USE: Light velocity

Speed of sound
USE: Acoustic wave velocity

Speed reducers 602
DT: Predates 1975
BT: Gears

Speed regulators 732.1 (731.3)
DT: Predates 1975
NT: Governors
RT: Final control devices
Speed indicators
Velocity control

Spent fuels 621 (452.3) (452.4)
DT: January 1993
UF: Nuclear reactors—Spent fuels*
BT: Nuclear fuels
RT: Fission products
Nuclear reactors
Radioactive wastes

Spheres
DT: January 1986
UF: Flow of fluids—Spheres*
BT: Bodies of revolution
RT: Flow of fluids

Spheroidal iron
USE: Nodular iron

Spill booms
USE: Oil booms

Spills
USE: Hazardous materials spills

* Former Ei Vocabulary term

Spillway gates 441.3
DT: January 1993
UF: Gates (spillway)
Spillways—Gates*
BT: Spillways
RT: Hydraulic gates

Spillways 441.3
DT: Predates 1975
BT: Hydraulic structures
NT: Spillway gates
RT: Dams
Hydroelectric power plants
Waterworks

*Spillways—Gates**
USE: Spillway gates

Spinning (fibers) 819.3
DT: January 1993
UF: Textiles—Spinning*
BT: Processing
RT: Fibers
Plastic filaments
Spinning machines
Textile fibers
Textile processing
Yarn

Spinning (metal)
USE: Metal spinning

Spinning machines 819.6
SN: For conversion of fibers or filaments into thread or yarn
DT: January 1993
UF: Textile machinery—Spinning machines*
BT: Machinery
RT: Spinning (fibers)
Textile machinery

Splicing (cable)
USE: Cable jointing

Splines 601.2
DT: Predates 1975
BT: Machine components
RT: Couplings
Fasteners
Shafts (machine components)

Sponge iron 545.1
DT: Predates 1975
BT: Iron alloys
Porous materials

Spontaneous combustion 521.1
DT: January 1993
UF: Combustion—Spontaneous*
BT: Combustion
RT: Explosions
Fire hazards
Fire protection
Fires
Flammability
Ignition

Spontaneous potential logging 512.1.2
DT: January 1993
UF: Oil well logging—Spontaneous potential*
Self potential logging
SP logging
BT: Electric logging

Spooling
USE: Winding

Spools
USE: Reels

Sporadic E layer
USE: E region

Sporting goods (817.2)
DT: Predates 1975

*Sporting goods—Plastics**
USE: Plastics applications

Sports medicine 461.6 (461.3)
DT: January 1993
UF: Biomechanics—Sports*
BT: Medicine
RT: Biomechanics

Spot welding 538.2.1
DT: January 1993
UF: Welding—Spot*
BT: Resistance welding
NT: Stitching (metal joining)
RT: Electric arc welding
Inert gas welding
Percussion welding
Pressure welding
Submerged arc welding
Ultrasonic welding
Welding machines

Spray guns (813.1)
DT: Predates 1975
BT: Applicators
RT: Hand tools
Spraying

Spray painting
USE: Paint spraying

Spray steelmaking 545.3
DT: January 1993
UF: Steelmaking—Spray process*
BT: Basic oxygen process

Sprayed coatings 813.2
DT: January 1993
UF: Coatings—Sprayed*
BT: Coatings
RT: Plasma spraying
Spraying
Varnish

Spraying (813.1)
DT: January 1993
UF: Insecticides—Spraying*
NT: Paint spraying
Plasma spraying
RT: Spray guns

* Former Ei Vocabulary term

Spraying *(continued)*
Sprayed coatings
Vaporization
Wetting

Spread spectrum communication 716.3
DT: January 1993
UF: Radio transmission—Spread spectrum*
BT: Radio communication
RT: Frequency hopping
Mobile radio systems
Radio transmission

Spreaders (802.1) (821.1)
DT: January 1993
UF: Adhesives—Spreaders*
Fertilizers—Spreaders*
BT: Applicators

Spreadsheets 723.1 (723.2)
DT: January 1993
UF: Computer programming—Spreadsheet*
RT: Administrative data processing
Financial data processing
Graphic methods
Office automation

Springs (components) 601.2
DT: January 1993
UF: Springs*
BT: Components
NT: Helical springs
Ring springs
Vehicle springs
RT: Wire products

Springs (water) 444 (481.3.1)
DT: January 1993
UF: Water springs
NT: Geothermal springs
Hot springs
Mineral springs
RT: Groundwater
Surface waters
Water resources

*Springs**
USE: Springs (components)

*Springs—Copper**
USE: Copper

*Springs—Helical**
USE: Helical springs

*Springs—Plastics**
USE: Plastics applications

*Springs—Pneumatic**
USE: Pneumatic equipment

*Springs—Ring**
USE: Ring springs

*Springs—Rubber**
USE: Rubber products

*Springs—Steel**
USE: Steel

Sprinkler systems (fire fighting) 914.2
DT: January 1993
UF: Fire fighting equipment—Sprinkler systems*
Fire protection—Sprinkler systems*
BT: Fire fighting equipment
RT: Fire protection
Safety devices

Sprinkler systems (irrigation) 821.1
DT: January 1993
UF: Irrigation—Sprinkler systems*
RT: Irrigation

Sprockets 601.2
DT: Predates 1975
BT: Components
RT: Gears
Transmissions

Spur gears 601.2
DT: January 1995
BT: Gears
RT: Helical gears

Spurious signal noise (701.1) (713) (716)
SN: Limited to unwanted, random signals originating within electronic equipment. Distinguish from SIGNAL INTERFERENCE
DT: January 1993
UF: Electric transformers—Noise*
Electron device noise
Electron tubes—Noise*
Gaussian noise (electronic)
Masers—Noise*
Noise (spurious signals)
Noise, Spurious signal*
Oscillators—Noise*
Radar—Noise*
Radio equipment—Noise*
Radio receivers—Noise*
Radio telephone—Noise*
Radio transmission—Noise*
Random electronic noise
Semiconductor devices—Noise*
Telephone—Noise*
Television receivers—Noise*
NT: Microphonism
Shot noise
Thermal noise
RT: Electron tubes
Interference suppression
Noise generators
Radio receivers
Radio telephone
Radio transmission
Semiconductor devices
Signal distortion
Signal interference
Signal noise measurement
Signal theory
Signal to noise ratio
Telecommunication
White noise

Sputter deposition (539.3) (813.1) (932.1)
DT: January 1993
UF: Metals and alloys—Sputtering*
Protective coatings—Sputtering*
BT: Deposition
Sputtering
RT: Coating techniques
Metallizing
Thin films

Sputtering (539.3) (813.1) (932.1)
DT: Predates 1975
UF: Cathodic sputtering
NT: Magnetron sputtering
Sputter deposition
RT: Ion beams

Square wave generators 715.1
DT: Predates 1975
BT: Function generators
RT: Multivibrators
Pulse generators

SQUIDs (704.2) (708.3) (931.4)
DT: January 1995
UF: Superconducting quantum interference devices
BT: Quantum interference devices
Superconducting devices
RT: Magnetometers

Squirrel cage motors 705.3.1
DT: January 1993
UF: Electric motors—Squirrel cage*
BT: Induction motors

Sr
USE: Strontium

Stability
DT: January 1993
UF: Instability
NT: Asymptotic stability
Dimensional stability
Plasma stability
Slope stability
System stability
Thermodynamic stability
RT: Controllability
Damping
Durability
Equations of motion
Fuel sloshing
Observability
Oscillations
Root loci
Stabilization

Stability (numerical methods)
USE: Convergence of numerical methods

Stability criteria 731.4
DT: January 1993
UF: System stability—Criteria*
RT: Root loci
System stability

Stability matrix
USE: Stiffness matrix

Stabilization
DT: January 1993
UF: Airport runways—Stabilization*
Roads and streets—Stabilization*
Soils—Stabilization*
Textiles—Stabilization*
RT: Stability
Stress relief

Stabilization ponds 452.2
DT: January 1993
UF: Sewage treatment—Stabilization ponds*
BT: Sewage treatment plants
RT: Sewage tanks
Sewage treatment

Stabilizers (agents) 803
DT: January 1993
UF: Stabilizers*
BT: Agents
NT: Heat stabilizers
RT: Antioxidants
Viscosity

Stabilizers (marine vessel) 671.2
DT: January 1993
UF: Boat stabilizers
Boats—Stabilizers*
Ships—Stabilizers*
BT: Water craft parts and equipment

*Stabilizers**
USE: Stabilizers (agents)

Stacking faults 933.1.1
DT: January 1995
BT: Crystal defects
RT: Dislocations (crystals)

Stadiums 402.2
DT: Predates 1975
UF: Grandstands
BT: Facilities
RT: Recreation centers

Stages 402.2
SN: For performances
DT: January 1993
UF: Opera houses—Stages*
BT: Facilities
NT: Sound stages
RT: Auditoriums
Opera houses
Theaters (legitimate)

Stained glass 812.3 (402)
DT: January 1993
UF: Windows—Stained glass*
BT: Glass

Stainless steel 545.3
DT: Predates 1975
UF: Powder metallurgy—Stainless steel*
Stainless steel powder metallurgy
Welding—Stainless steel*

* Former Ei Vocabulary term

Stainless steel *(continued)*
BT: Steel
RT: Chromium alloys
Maraging steel
Molybdenum alloys
Nickel alloys

Stainless steel powder metallurgy
USE: Steel powder metallurgy AND
Stainless steel

Stairs 402
DT: Predates 1975
UF: Buildings—Stairs*
Stairways
BT: Building components
RT: Escalators
Ladders

Stairways
USE: Stairs

Stall and pillar mining
USE: Room and pillar mining

Stall indicators 652.3
DT: January 1993
UF: Aircraft instruments—Stall indicators*
BT: Aircraft instruments

Stamping 535.2
DT: January 1993
BT: Forming
NT: Metal stamping
RT: Drawing (forming)
Pressing (forming)
Punching

Standardization 902.2
DT: Predates 1975
NT: Calibration
RT: Production control
Production engineering
Quality assurance
Quality control
Reliability
Specifications
Standards
Testing

Standards 902.2
DT: Predates 1975
NT: Codes (standards)
Frequency standards
Television standards
Temperature scales
RT: Calibration
Computer software portability
Grading
Inspection
Laws and legislation
Network protocols
Performance
Production control
Productivity
Quality assurance
Quality control
Rating
Reliability
Sampling
Specifications
Standardization
Terminology
Testing
Units of measurement
Value engineering

Standby power service
(706.1) (611) (613) (614) (615)
DT: January 1993
UF: Power plants—Standby service*
RT: Electric power plants
Power plants

Standby power systems
(706.1) (611) (613) (614) (615)
DT: January 1993
UF: Buildings—Standby power systems*
Power systems (standby)
BT: Electric power systems
RT: Electric power supplies to apparatus

Standing wave detectors
USE: Standing wave meters

Standing wave indicators
USE: Standing wave meters

Standing wave meters 942.1
DT: Predates 1975
UF: Standing wave detectors
Standing wave indicators
BT: Electric measuring instruments
Indicators (instruments)
RT: Bolometers
Impedance matching (electric)

Stands
USE: Supports

Starch (804.1) (822.3)
DT: Predates 1975
BT: Polysaccharides
RT: Food products

Start up (industrial plants)
USE: Plant startup

Start up (reactors)
USE: Reactor startup

Starters (612.1.1) (617) (661.2) (705.1)
DT: January 1993
UF: Electric machinery—Starters*
BT: Machine components
RT: Actuators
Electric exciters
Ignition systems
Internal combustion engines
Starting

Starting (612) (661.1) (617) (705.1)
SN: Of machinery
DT: January 1993
UF: Automobile engines—Starting*
Diesel engines—Starting*

* Former Ei Vocabulary term

Starting *(continued)*
Electric machinery—Starting*
Turbomachinery—Starting*
RT: Actuators
Ignition
Ignition systems
Internal combustion engines
Launching
Starters
Vapor lock

Startup (industrial plants)
USE: Plant startup

Startup (reactors)
USE: Reactor startup

State assignment 721.1
DT: January 1993
UF: Switching theory—State assignment*
BT: Switching functions

State estimation 731.1
DT: January 1993
BT: Identification (control systems)
RT: Control systems

State feedback
USE: Feedback

State space methods 921 (703.1) (731.1)
DT: January 1993
UF: Mathematical techniques—State space methods*
BT: Mathematical techniques
RT: Control system analysis
Control system synthesis
Controllability
Electric network analysis
Electric network synthesis
Observability

State taxes
USE: Taxation

Static (atmospherics)
USE: Atmospherics

Static converters 704.1
DT: January 1993
UF: Electric converters, Static*
BT: Electric converters

Static electricity 701.1
DT: January 1993
BT: Electricity
RT: Electrostatics

Stations (radar)
USE: Radar stations

Stations (radio)
USE: Radio stations

Stations (railroad)
USE: Railroad stations

Stations (subway)
USE: Subway stations

Stations (telegraph)
USE: Telegraph stations

Stations (television)
USE: Television stations

Statistical analysis
USE: Statistical methods

Statistical linguistics
USE: Computational linguistics

Statistical mechanics 922.2, 931.1
DT: January 1987
UF: Statistical thermodynamics
BT: Mechanics
RT: Kinetic theory
Quantum theory
Statistical methods
Thermodynamics

Statistical methods 922.2
DT: Predates 1975
UF: Mathematical statistics*
Regional planning—Statistical methods*
Statistical analysis
BT: Mathematical techniques
NT: Correlation methods
Game theory
Monte Carlo methods
Regression analysis
Statistical tests
Time series analysis
RT: Chaos theory
Correlation theory
Forecasting
Measurement errors
Operations research
Planning
Probability
Quality control
Queueing theory
Random number generation
Random processes
Reliability
Reliability theory
Sampling
Scheduling
Statistical mechanics
Statistical optics
Statistical process control
Statistics
Surveying
Surveys
Total quality management
Weibull distribution

*Statistical methods—Regression analysis**
USE: Regression analysis

*Statistical methods—Statistical tests**
USE: Statistical tests

*Statistical methods—Time series analysis**
USE: Time series analysis

Statistical optics 741.1, 922.2
DT: January 1993
BT: Optics
RT: Statistical methods

Statistical process control 731.1 (913.3) (922.2)
DT: January 1995
BT: Process control
RT: Computer control
Manufacturing data processing
Production control
Quality control
Statistical methods

Statistical tests 922.2
DT: January 1993
UF: Statistical methods—Statistical tests*
BT: Statistical methods
Testing
RT: Correlation methods
Measurement errors
Reliability
Reliability theory
Statistics

Statistical thermodynamics
USE: Statistical mechanics

Statistics 922.2
DT: January 1993
NT: Error statistics
Population statistics
RT: Measurement errors
Probability
Quality control
Statistical methods
Statistical tests

Stators 705.1
DT: January 1993
UF: Electric machinery—Stators*
BT: Machine windings
RT: AC generators
AC machinery
AC motors
Electric equipment protection
Rotating machinery
Rotors (windings)

Steam (614) (617.2) (617.3) (641.1)
DT: Predates 1975
BT: Vapors
RT: Boilers
Feedwater heaters
Fog
Steam accumulators
Steam condensate
Steam condensers
Steam cracking
Steam engineering
Steam engines
Steam generators
Steam pipelines
Steam piping systems
Steam power plants
Steam separators and traps
Steam tables and charts
Steam turbines
Steamships
Superheaters
Thermodynamics
Water

Steam accumulators 614.2
DT: Predates 1975
BT: Boilers
RT: Heat storage
Steam

Steam boats
USE: Steamships

Steam charts
USE: Steam tables and charts

Steam condensate (614.2) (617.2) (617.3) (641.1)
DT: Predates 1975
BT: Materials
RT: Boilers
Condensate return lines
Condensate treatment
Steam
Steam condensers
Steam piping systems
Vapors
Water

Steam condensate treatment
USE: Condensate treatment

*Steam condensate—Oil reclamation**
USE: Reclamation

*Steam condensate—Treatment**
USE: Condensate treatment

Steam condensers 616.1 (614.2)
DT: Predates 1975
BT: Condensers (liquefiers)
NT: Condenser tubes
RT: Boilers
Condensate return lines
Heat exchangers
Heat transfer
Heating equipment
Steam
Steam condensate
Steam separators and traps

*Steam condensers—Tubes**
USE: Condenser tubes

Steam cracking (513.1) (802.2)
DT: January 1993
BT: Cracking (chemical)
RT: Gas oils
Naphthas
Steam

Steam engineering 641.1, 901
DT: Predates 1975
BT: Engineering
RT: Steam
Steam tables and charts

Steam engines 617.3
DT: Predates 1975
BT: Heat engines
NT: Steam turbines
RT: Boilers
Compound engines
Steam
Steam locomotives
Steamships
Stokers

Steam generators 614.2
DT: October 1975
BT: Heating equipment
NT: Boilers
Superheaters
RT: Economizers
Heat exchangers
Heat transfer
Pressure vessels
Steam

Steam lines
USE: Steam piping systems

Steam locomotives 682.1.2 (641.1)
DT: January 1993
UF: Locomotives—Steam power*
BT: Locomotives
RT: Steam engines

Steam pipelines 619.1 (641.1) (642.2) (643.2)
SN: Stationary means to transport steam over distances. For systems of pipes to distribute steam throughout a building or facility, use STEAM PIPING SYSTEMS
DT: Predates 1975
BT: Pipelines
NT: Condensate return lines
RT: High pressure pipelines
Safety valves
Steam
Steam piping systems
Steam separators and traps

*Steam pipelines—Condensate return**
USE: Condensate return lines

*Steam pipelines—High pressure**
USE: High pressure pipelines

Steam piping systems
619.1 (642.2) (643.2) (802.1)
SN: Systems of pipes used to distribute steam throughout a building, building complex, or other structure or facility of limited extent. For transportation of steam over greater distances use PIPELINES
DT: Predates 1975
UF: Steam lines
BT: Piping systems
RT: Condensate return lines
Steam
Steam condensate
Steam pipelines
Steam separators and traps

Steam power plants 614 (402.1)
DT: Predates 1975
UF: Coal handling—Steam power plants*
Cranes—Steam power plants*
BT: Power plants
RT: Electric power plants
Feedwater treatment
Power plant intakes
Reheat cycle
Steam
Steam turbines

*Steam power plants—Intakes**
USE: Power plant intakes

*Steam power plants—Reheat cycle**
USE: Reheat cycle

Steam preheaters
USE: Feedwater heaters

Steam propulsion 675.1 (641.1)
DT: January 1993
UF: Ship propulsion—Steam*
BT: Propulsion
RT: Ship propulsion
Steamships

Steam separators and traps 619.1.1 (641.1)
DT: Predates 1975
UF: Steam traps
BT: Separators
RT: Condensate return lines
Steam
Steam condensers
Steam pipelines
Steam piping systems

Steam ships
USE: Steamships

Steam tables and charts 641.1
DT: Predates 1975
UF: Steam charts
RT: Handbooks
Steam
Steam engineering

Steam traps
USE: Steam separators and traps

Steam turbines 617.2
DT: Predates 1975
UF: Ship propulsion—Gas and steam turbine combined*
BT: Steam engines
Turbines
RT: Boilers
Combined cycle power plants
Gas turbines
Hydroelectric and steam combined power plants
Reheat cycle
Steam
Steam power plants
Steamships
Turbogenerators

* Former Ei Vocabulary term

Steamboats
USE: Steamships

Steamships 671
SN: Passenger and freight ships using steam as a power source
DT: Predates 1975
UF: Boats—Steam*
Steam boats
Steam ships
Steamboats
BT: Ships
RT: Steam
Steam engines
Steam propulsion
Steam turbines

Steatite 482.2
DT: Predates 1975
UF: Soapstone (steatite)
BT: Metamorphic rocks
RT: Talc
Talc deposits

Steel 545.3
DT: Predates 1975
UF: Aircraft materials—Steel*
Automobile materials—Steel*
Packaging materials—Steel*
Rolls—Steel*
Shipbuilding materials—Steel*
Springs—Steel*
Welding—Steel*
Windows—Steel*
Wire screen cloth—Steel*
BT: Iron alloys
NT: Aluminum coated steel
Carbon steel
Copper clad steel
Maraging steel
Silicon steel
Stainless steel
Tool steel
RT: Armor
Cold heading
Descaling
Iron
Iron and steel plants
Iron and steel industry
Steel analysis
Steel beams and girders
Steel bridges
Steel castings
Steel construction
Steel containers
Steel corrosion
Steel foundry practice
Steel ingots
Steel metallography
Steel metallurgy
Steel pipe
Steel research
Steel scrap
Steel sheet
Steel structures
Steel testing
Steelmaking

Steel analysis 545.3, 801
DT: January 1993
UF: Iron and steel analysis*
BT: Metal analysis
RT: Iron analysis
Steel

Steel beams and girders 545.3, 408.2
DT: January 1993
UF: Beams and girders—Steel*
BT: Beams and girders
RT: Steel
Steel bridges
Steel construction
Steel structures

Steel bridges 545.3, 401.1
DT: January 1993
UF: Bridges, Steel*
BT: Bridges
RT: Steel
Steel beams and girders
Steel construction

Steel castings 534.2, 545.3
DT: January 1986
UF: Welding—Steel castings*
BT: Metal castings
NT: Steel ingots
RT: Steel

Steel construction 545.3, 405.2 (415.1)
DT: June 1990
BT: Construction
RT: Steel
Steel beams and girders
Steel bridges
Steel structures

Steel containers 545.3 (691.1) (694.4)
DT: January 1993
UF: Containers—Steel*
BT: Containers
RT: Steel

Steel corrosion 539.1, 545.3
DT: Predates 1975
UF: Steel—Staining*
BT: Corrosion
RT: Steel

Steel foundry practice 534.2, 545.3
DT: Predates 1975
BT: Foundry practice
Steel metallurgy
RT: Steel
Steelmaking

Steel heat treatment 537.1, 545.3
DT: Predates 1975
BT: Heat treatment
RT: Case hardening
Quenching
Spark hardening

* Former Ei Vocabulary term

*Steel heat treatment—Carburizing**
USE: Carburizing

*Steel heat treatment—Case hardening**
USE: Case hardening

*Steel heat treatment—Flame hardening**
USE: Flame hardening

*Steel heat treatment—Nitriding**
USE: Nitriding

*Steel heat treatment—Quenching**
USE: Quenching

*Steel heat treatment—Scarfing**
USE: Scarfing

*Steel heat treatment—Spark hardening**
USE: Spark hardening

Steel industry
USE: Iron and steel industry

Steel ingots 545.3 (534.2)
DT: Predates 1975
BT: Ingots
Steel castings
RT: Hot topping
Steel
Steelmaking

*Steel ingots—Hot topping**
USE: Hot topping

*Steel ingots—Weighing**
USE: Weighing

Steel metallography 531.2, 545.3
DT: January 1993
UF: Iron and steel metallography*
BT: Metallography
RT: Austenite
Bainite
Ferrite
Graphitization
Iron metallography
Martensite
Pearlite
Steel
Steel metallurgy

Steel metallurgy 531.1, 545.5
DT: January 1993
UF: Iron and steel metallurgy*
BT: Metallurgy
NT: Steel foundry practice
Steel powder metallurgy
Steelmaking
RT: Iron and steel plants
Iron metallurgy
Steel
Steel metallography

Steel mills
USE: Iron and steel plants

Steel pipe 545.3, 619.1
DT: January 1993
UF: Pipe, Steel*
Pipelines, Steel*
Steel pipelines
BT: Pipe
RT: Steel

Steel pipelines
USE: Pipelines AND
Steel pipe

Steel plants
USE: Iron and steel plants

Steel powder metallurgy 536.1, 545.3
DT: January 1993
UF: Powder metallurgy—Stainless steel*
Powder metallurgy—Steel*
Stainless steel powder metallurgy
BT: Powder metallurgy
Steel metallurgy
RT: Iron powder metallurgy

Steel pressure vessels
USE: Steel structures AND
Pressure vessels

Steel research 901.3, 545.3
DT: January 1993
UF: Iron and steel research*
BT: Research
RT: Iron and steel industry
Iron research
Materials science
Steel
Steel testing
Steelmaking

Steel scrap 452.3, 545.3
DT: January 1993
UF: Iron and steel scrap*
BT: Scrap metal
RT: Iron scrap
Steel

Steel sheet 535.1, 545.3
DT: January 1977
BT: Sheet metal
RT: Steel

Steel silos
USE: Steel structures AND
Silos (agricultural)

Steel structures 408.2, 545.3
DT: Predates 1975
UF: Pressure vessels—Steel*
Silos—Steel*
Steel pressure vessels
Steel silos
Steel water tanks
Steel water towers
Water tanks and towers—Steel*
BT: Structures (built objects)
NT: Tubular steel structures
Welded steel structures
RT: Joints (structural components)
Steel
Steel beams and girders
Steel construction

* Former Ei Vocabulary term

*Steel structures—Connections**
USE: Joints (structural components)

*Steel structures—Tubular**
USE: Tubular steel structures

*Steel structures—Weight control**
USE: Weight control

Steel tanks 545.3, 619.2
DT: January 1993
UF: Tanks—Steel*
BT: Tanks (containers)

Steel testing 545.3 (421) (422.2)
DT: Predates 1975
BT: Metal testing
RT: Steel
Steel research

Steel water tanks
USE: Steel structures AND
Water tanks

Steel water towers
USE: Water towers AND
Steel structures

*Steel—Aluminum coating**
USE: Aluminum coated steel

*Steel—Cold heading**
USE: Cold heading

*Steel—Cold working**
USE: Cold working

*Steel—Copper clad**
USE: Copper clad steel

*Steel—Scale removal**
USE: Descaling

*Steel—Staining**
USE: Steel corrosion

*Steel—Symbols**
USE: Codes (symbols)

Steelmaking 545.3
DT: Predates 1975
UF: Slag reaction process
Steelmaking—Slag reaction process*
BT: Steel metallurgy
NT: Basic oxygen process
Bessemer process
Direct reduction process
Electric furnace process
Electroslag process
Open hearth process
Pneumatic steelmaking
RT: Deoxidants
Desulfurization
Direct process refining
Iron and steel industry
Iron and steel plants
Ladle metallurgy
Plasma arc melting
Rare earth additions
Remelting
Slags
Steel
Steel foundry practice
Steel ingots
Steel research

Steelmaking furnaces 532, 545.3
DT: January 1993
UF: Furnaces—Steelmaking*
BT: Metallurgical furnaces

*Steelmaking—Basic oxygen process**
USE: Basic oxygen process

*Steelmaking—Bessemer process**
USE: Bessemer process

*Steelmaking—Deoxidants**
USE: Deoxidants

*Steelmaking—Direct process**
USE: Direct process refining

*Steelmaking—Direct reduction process**
USE: Direct reduction process

*Steelmaking—Electric furnace process**
USE: Electric furnace process

*Steelmaking—Electroslag process**
USE: Electroslag process

*Steelmaking—Ladle process**
USE: Ladle metallurgy

*Steelmaking—Open hearth process**
USE: Open hearth process

*Steelmaking—Plasma arc remelting**
USE: Plasma arc melting

*Steelmaking—Pneumatic process**
USE: Pneumatic steelmaking

*Steelmaking—Rare earth additions**
USE: Rare earth additions

*Steelmaking—Remelting**
USE: Remelting

*Steelmaking—Slag reaction process**
USE: Steelmaking

*Steelmaking—Slags**
USE: Slags

*Steelmaking—Spray process**
USE: Spray steelmaking

Steering (662.4) (671.2) (663.2)
DT: January 1993
UF: Vehicles—Steering*
RT: Automobile steering equipment
Maneuverability
Ship steering equipment

Steering equipment (automobile)
USE: Automobile steering equipment

Steering equipment (ships)
USE: Ship steering equipment

* Former Ei Vocabulary term

Stellite 543.1, 543.5, 549.3
DT: Predates 1975
BT: Chromium alloys
Cobalt alloys
Superalloys
Tungsten alloys

Step motors
USE: Stepping motors

Step response (731.1)
DT: January 1993
UF: Control systems—Step response*
BT: Control theory
RT: Control system analysis
Control system synthesis
Identification (control systems)
Models

Stepped motors
USE: Stepping motors

Stepping motors 705.3
DT: January 1993
UF: Electric motors, Stepping type*
Pulse motors
Step motors
Stepped motors
BT: Electric motors

Stereo broadcasting
USE: Stereophonic broadcasting

Stereo receivers
USE: Stereophonic receivers

Stereo recordings
USE: Stereophonic recordings

Stereo vision 723.5, 741.2 (731.6)
DT: January 1995
RT: Computer vision
Robots
Three dimensional computer graphics

Stereophonic broadcasting (716.3) (716.4)
DT: January 1993
UF: Radio broadcasting—Stereophonic signals*
Stereo broadcasting
Stereophonic radio broadcasting
Stereophonic television broadcasting
Television broadcasting—Stereophonic*
BT: Broadcasting
RT: Radio broadcasting
Stereophonic receivers
Television broadcasting

Stereophonic radio broadcasting
USE: Stereophonic broadcasting AND
Radio broadcasting

Stereophonic receivers (716.3) (716.4)
DT: January 1993
UF: Radio receivers—Stereophonic signals*
Stereo receivers
BT: Signal receivers
RT: Radio receivers
Stereophonic broadcasting
Television receivers

Stereophonic recordings 752.2
DT: January 1993
UF: Audio recordings
Audiotape recordings
Phonograph records—Stereophonic recordings*
Stereo recordings
Tape recordings
RT: Audio equipment
Audio systems
Compact disks
Phonograph records
Sound recording
Video recording

Stereophonic television broadcasting
USE: Television broadcasting AND
Stereophonic broadcasting

Stereoscopic photography
USE: Special effects

Sterilization (cleaning) (462.1) (802.3) (822.2)
SN: Process which destroys all microbes; distinguish from DISINFECTION, which is designed to destroy pathogens
DT: January 1993
UF: Sterilization*
BT: Cleaning
RT: Biomedical engineering
Decontamination
Disinfection
Food processing
Fumigation
Sterilizers

*Sterilization**
USE: Sterilization (cleaning)

Sterilizers (462.1) (802.1) (822.1)
DT: Predates 1975
BT: Biomedical equipment
RT: Autoclaves
Hospitals
Laboratories
Sterilization (cleaning)

Stiction (603.1) (931.1)
DT: January 1993
UF: Machine tools—Stiction*
BT: Friction
RT: Adhesion
Machine tools

Stiffness (421) (422)
DT: January 1995
BT: Mechanical properties
RT: Deformation
Stiffness matrix
Tensile testing

Stiffness matrix 921.1 (408.1) (421)
DT: January 1995
UF: Stability matrix
BT: Matrix algebra
RT: Stiffness
Structural analysis

Stilling basins 441.3, 525.4
SN: Hydraulic structures for energy dissipation
DT: Predates 1975
UF: Stilling boxes
BT: Energy dissipators
Hydraulic structures

Stilling boxes
USE: Stilling basins

Stirling cycle 641.1
DT: January 1993
UF: Thermodynamics—Stirling cycle*
BT: Thermodynamics
RT: Air engines
Carnot cycle
Heat engines

Stirling cycle engines
USE: Heat engines

Stitching (metal joining) 538.2.1 (535)
DT: January 1993
UF: Sheet and strip metal—Stitching*
BT: Spot welding

Stochastic control systems 731.1
DT: January 1993
UF: Control systems, Stochastic*
BT: Control systems

Stochastic processes
USE: Random processes

Stock control
USE: Inventory control

Stockpile surveys 742.1 (405.3)
SN: Surveying of reserve supplies using photogrammetric techniques
DT: January 1993
UF: Photogrammetry—Stockpile surveys*
BT: Surveys
RT: Photogrammetry

Stoichiometry 801.4 (802.2)
DT: January 1995
RT: Chemical reactions
Composition

Stokers (614.2) (617.3) (643.2)
DT: Predates 1975
BT: Heating equipment
RT: Boiler firing
Boilers
Furnaces
Steam engines

STOL aircraft
USE: VTOL/STOL aircraft

Stop watches 943.3
DT: Predates 1975
UF: Stopwatches
BT: Watches
RT: Time measurement

Stoping 502.1
DT: January 1993
UF: Mines and mining—Stoping*
BT: Mining

Stoppers (container)
USE: Container closures

Stopwatches
USE: Stop watches

Storage (data)
USE: Data storage equipment

Storage (energy)
USE: Energy storage

Storage (materials) 694.4
DT: January 1993
UF: Gas storage*
Storage*
NT: Cold storage
Food storage
Fuel storage
Radioactive waste storage
RT: Materials handling
Warehouses

Storage allocation (computer) 722.1
DT: January 1993
UF: Computer operating systems—Storage allocation*
Memory allocation
RT: Computer operating systems
Computer systems programming
Data storage equipment
Supervisory and executive programs

Storage battery vehicles 662.1, 702.1.2
DT: Predates 1975
BT: Electric vehicles
RT: Electric propulsion
Secondary batteries

Storage devices (data)
USE: Data storage equipment

Storage rings 932.1.1
DT: January 1993
UF: Accelerators—Storage rings*
Rings (storage)
BT: Particle accelerators

Storage tubes 714.1
DT: January 1993
UF: Electron tubes, Storage*
BT: Electron tubes
NT: Image storage tubes
RT: Electric energy storage

*Storage**
USE: Storage (materials)

Store buildings 402.2
DT: Predates 1975
BT: Buildings
RT: Retail stores
Shopping centers

Stores (retail)
USE: Retail stores

* Former Ei Vocabulary term

Stores control
USE: Inventory control

Storm sewers 452.1 (442.1) (443.3)
DT: January 1993
UF: Sewers—Storm drainage*
BT: Sewers
RT: Combined sewers
Drainage
Rain
Runoff
Storms

Storms 443.3
DT: January 1993
UF: Cyclones
Meteorology—Storms*
NT: Hurricanes
Thunderstorms
Tornadoes
RT: Clouds
Combined sewers
Earth atmosphere
Extraterrestrial atmospheres
Meteorology
Precipitation (meteorology)
Rain
Snow
Storm sewers
Turbulence
Wind effects

Stoves (642.2) (643.2)
DT: Predates 1975
UF: Blast furnaces—Stoves*
Solar stoves
BT: Heating equipment
NT: Electric stoves
Gas stoves
Industrial stoves
RT: Electric appliances
Solar equipment

*Stoves—Electric**
USE: Electric stoves

*Stoves—Gas**
USE: Gas stoves

*Stoves—Solar**
USE: Solar equipment

Stowage (mines) 502.1
DT: January 1993
UF: Mines and mining—Stowage*
RT: Mining

Straightening 535.2 (602.2)
DT: January 1993
UF: Shafts and shafting—Straightening*
Sheet and strip metal—Straightening*
Wire—Straightening*
BT: Forming
RT: Alignment
Bending (forming)
Cold working
Hot working
Metal working
Pressing (forming)
Rolling
Shafts (machine components)
Straightening machines
Stretching
Wire

Straightening machines 535.2.1
DT: Predates 1975
BT: Metal forming machines
RT: Straightening

Strain (408.1) (421)
DT: October 1975
BT: Mechanical properties
RT: Bending (deformation)
Buckling
Deformation
Elasticity
Plasticity
Strain control
Strain gages
Strain measurement
Strain rate
Stresses
Structural analysis
Torsional stress

Strain control 731.3 (421)
DT: January 1993
UF: Control, Mechanical variables—Strain*
BT: Mechanical variables control
RT: Strain
Strain measurement

Strain gages 943.1 (422.1)
DT: Predates 1975
UF: Strain gauges
BT: Gages
RT: Dilatometers
Mechanical testing
Strain
Strain measurement
Stress analysis
Transducers

Strain gauges
USE: Strain gages

Strain hardening 537.1
DT: January 1995
UF: Work hardening
BT: Hardening
RT: Age hardening
Cold working
Recrystallization (metallurgy)
Residual stresses

Strain measurement 943.2
DT: January 1993
UF: Mechanical variables measurement—Strain*
BT: Mechanical variables measurement
RT: Mechanical testing
Strain
Strain control
Strain gages

Strain rate (421) (422)
DT: January 1995
RT: Impact testing
Strain
Tensile testing

Strapping 694.2
DT: January 1993
UF: Packaging materials—Strapping*
BT: Packaging materials
RT: Fasteners

Strategic materials (404.1) (911.2)
DT: January 1993
UF: Critical materials
Industrial economics—Strategic materials*
BT: Materials
RT: Chromium
Cobalt
Manganese
Military operations

Strategic planning 912.2
DT: January 1986
BT: Planning
RT: Management

Stratification (thermal)
USE: Thermal stratification

Stratified charge engines 612.1
DT: January 1993
UF: Internal combustion engines—Stratified charge*
BT: Internal combustion engines
RT: Automobiles
Automotive fuels
Combustion
Fuel injection

Stratigraphy 481.1
DT: January 1993
UF: Geology—Stratigraphy*
BT: Geology
RT: Geomorphology
Landforms

Straw 821.5
DT: January 1993
UF: Agricultural wastes—Straw*
Pulp materials—Straw*
BT: Agricultural wastes
RT: Pulp materials

Stray light 741.1
DT: January 1993
BT: Light

Strays (atmospheric)
USE: Atmospherics

Streak cameras 742.2
DT: January 1993
UF: Streak photography
BT: Cameras
RT: Optical variables measurement
Time measurement

Streak photography
USE: Streak cameras

Stream flow 631.1, 407.2
DT: Predates 1975
BT: Flow of water
RT: Discharge (fluid mechanics)
Rivers
Runoff

Streaming (acoustic)
USE: Acoustic streaming

Street cleaning 406.2, 452
DT: Predates 1975
BT: Cleaning
RT: Roads and streets
Snow and ice removal

Street lighting 406.2, 707.1
SN: Includes artificial illumination of roads and highways
DT: January 1993
BT: Lighting
NT: Nonelectric street lighting
RT: Electric lighting
Fluorescent lamps
Glare
Highway systems
Outdoor electric lighting
Poles
Roads and streets
Urban planning

*Street lighting—Fluorescent**
USE: Fluorescent lamps

*Street lighting—Glare**
USE: Glare

*Street lighting—Nonelectric lighting**
USE: Nonelectric street lighting

Street railroad cars
USE: Trolley cars

Street traffic control 406.2, 432.4
DT: Predates 1975
BT: Traffic control
NT: Emergency traffic control
RT: Highway markings
Highway traffic control
Parking
Pedestrian safety
Roads and streets
Traffic signals
Traffic signs
Traffic surveys

*Street traffic control—Emergency measures**
USE: Emergency traffic control

*Street traffic control—Parking**
USE: Parking

*Street traffic control—Pedestrian safety**
USE: Pedestrian safety

Streetcars
USE: Trolley cars

* Former Ei Vocabulary term

Streets
USE: Roads and streets

Strength of materials (421) (422)
DT: Predates 1975
UF: Fatigue strength
Mechanical strength
BT: Mechanical properties
NT: Bending strength
Bond strength (materials)
Compressive strength
Shear strength
RT: Delamination
Failure (mechanical)
Stresses

Strengthening (metal) 531.1 (537.1)
SN: By or including control of constituents
DT: January 1993
UF: Metals and alloys—Strengthening*
BT: Metallurgy
RT: Hardening
Heat treatment
Toughening

Stress analysis (421)
DT: January 1993
UF: Stresses—Analysis*
BT: Analysis
RT: Bending (deformation)
Inclusions
Mechanical engineering
Photoelasticity
Residual stresses
Shear stress
Strain gages
Stresses
Structural analysis
Thermal stress
Torsional stress
Wind stress
Yield stress

Stress concentration (421) (422)
DT: January 1995
UF: Stress distribution
RT: Fatigue testing
Fracture testing
Impact testing
Stress intensity factors
Stresses

Stress corrosion cracking 539.1 (421)
DT: January 1993
UF: Corrosion cracking
Corrosion—Stress corrosion cracking*
Metals and alloys—Stress corrosion cracking*
BT: Corrosion
RT: Crack propagation
Cracks
Fatigue of materials
Fracture
Weathering

Stress distribution
USE: Stress concentration

Stress intensity factors (421) (422)
DT: January 1995
RT: Stress concentration
Stresses

Stress relaxation 931 (421)
DT: January 1993
BT: Relaxation processes
RT: Anelastic relaxation
Annealing
Creep
Heat treatment
Mechanical properties
Viscoelasticity

Stress relief (421) (537.1) (538.2)
DT: January 1993
UF: Grinding—Stress relief*
Heat treatment—Stress relief*
Stress relieving
Welds—Stress relief*
RT: Annealing
Grinding (machining)
Heat treatment
Stabilization
Tempering
Welds

Stress relieving
USE: Stress relief

Stress wave emissions
USE: Acoustic emissions

Stresses (408.1) (421)
DT: Predates 1975
NT: Residual stresses
Shear stress
Thermal stress
Torsional stress
Wind stress
RT: Cavitation
Dynamic loads
Elastoplasticity
Loads (forces)
Mechanical testing
Piezoelectricity
Strain
Strength of materials
Stress analysis
Stress concentration
Stress intensity factors
Structural design

*Stresses—Analysis**
USE: Stress analysis

*Stresses—Shear**
USE: Shear stress

*Stresses—Thermal**
USE: Thermal stress

*Stresses—Torsional**
USE: Torsional stress

*Stresses—Wind origin**
USE: Wind stress

* Former Ei Vocabulary term

Stretchers 462.1
DT: Predates 1975
BT: Biomedical equipment

Stretching 535.2
DT: January 1993
UF: Metal forming—Stretching*
Sheet and strip metal—Stretching*
BT: Forming
RT: Cold working
Deep drawing
Drawing (forming)
Hot working
Pressing (forming)
Straightening
Winding

Strip lines
USE: Strip telecommunication lines

Strip lines (telecommunication)
USE: Strip telecommunication lines

Strip metal 535.1
DT: January 1993
UF: Metal strip
Sheet and strip metal*
BT: Materials
RT: Metals
Precoated metals
Roll bonding
Rolling mills
Shearing

Strip mills 535.1.2 (402.2)
DT: January 1993
UF: Rolling mill practice—Sheet and strip*
BT: Rolling mills
RT: Cold rolling mills
Hot rolling mills
Metal working
Rolling mill practice
Slab mills

Strip telecommunication lines (716)
DT: January 1993
UF: Strip lines
Strip lines (telecommunication)
Striplines (telecommunication)
Telecommunication lines, Strip*
BT: Telecommunication lines
NT: Microstrip lines
RT: Microwave integrated circuits

Striping machines 405.1, 406
SN: For placing stripes on roads and highways
DT: January 1993
UF: Traffic signs, signals and markings—Striping machines*
BT: Machinery
RT: Highway markings
Roadbuilding machinery

Striplines (telecommunication)
USE: Strip telecommunication lines

Stripping (bark)
USE: Bark stripping

Stripping (dyes) 802.3 (803)
DT: January 1993
UF: Dye stripping
Dyes and dyeing—Stripping*
BT: Stripping (removal)
RT: Dyeing
Dyes

Stripping (removal) (811.2) (813.1)
DT: January 1993
UF: Protective coatings—Stripping*
NT: Bark stripping
Stripping (dyes)
RT: Coating techniques
Coatings
Peeling

Stroboscopes 943.3
DT: Predates 1975
BT: Instruments
RT: Speed indicators
Synchronization
Velocimeters
Vibration measurement

Strontium 549.2
DT: January 1993
UF: Sr
Strontium and alloys*
BT: Alkaline earth metals
Radioactive elements
RT: Strontium alloys
Strontium compounds
Strontium deposits

Strontium alloys 549.2
DT: January 1993
UF: Strontium and alloys*
BT: Alkaline earth metal alloys
Radioactive materials
RT: Strontium
Strontium compounds

*Strontium and alloys**
USE: Strontium OR
Strontium alloys

Strontium compounds (804.1) (804.2)
DT: Predates 1975
BT: Alkaline earth metal compounds
Radioactive materials
RT: Oxide superconductors
Strontium
Strontium alloys

Strontium deposits 504.3, 549.2
DT: Predates 1975
BT: Ore deposits
RT: Strontium

Structural analysis 408.1
DT: October 1975
BT: Structural design
RT: Dynamic response
Mathematical techniques
Stiffness matrix
Strain
Stress analysis

* Former Ei Vocabulary term

*Structural analysis—Dynamic response**
USE: Dynamic response

Structural ceramics 415.4, 812.1
DT: January 1995
BT: Building materials
Ceramic materials

Structural design 408.1
DT: Predates 1975
UF: Structural engineering
BT: Design
NT: Structural analysis
RT: Anchorages (foundations)
Architectural design
Blast resistance
Building materials
Buildings
Civil engineering
Deflection (structures)
Earthquake resistance
Engineering
Failure (mechanical)
Fire resistance
Hurricane effects
Hurricane resistance
Impact resistance
Light weight structures
Mechanical engineering
Mechanics
Model structures
Prestressing
Safety factor
Stresses
Structural loads
Structures (built objects)
Struts
Underground structures
Undersnow structures
Underwater structures
Vibrations (mechanical)
Wind effects

*Structural design—Anchorages**
USE: Anchorages (foundations)

*Structural design—Blast resistance**
USE: Blast resistance

*Structural design—Impact resistance**
USE: Impact resistance

*Structural design—Light weight**
USE: Light weight structures

*Structural design—Loads**
USE: Structural loads

*Structural design—Models**
USE: Model structures

*Structural design—Prestressing**
USE: Prestressing

*Structural design—Safety factor**
USE: Safety factor

*Structural design—Underground**
USE: Underground structures

*Structural design—Undersnow**
USE: Undersnow structures

*Structural design—Underwater**
USE: Underwater structures

Structural engineering
USE: Structural design

Structural frames 408.2
DT: January 1981
UF: Frames (structural)
BT: Structures (built objects)
RT: Beams and girders
Structural panels
Supports

Structural geology
USE: Tectonics

Structural loads 408.1
DT: January 1993
UF: Loads (structural)
Structural design—Loads*
BT: Loads (forces)
RT: Bearing capacity
Load limits
Structural design
Structures (built objects)

Structural materials
USE: Building materials

Structural members 408.2
DT: January 1993
BT: Components
Structures (built objects)
NT: Arches
Beams and girders
Bearings (structural)
Columns (structural)
Disks (structural components)
Floors
Joints (structural components)
Piles
Plates (structural components)
Railings
Structural panels
Struts
Studs (structural members)
Underpinnings
RT: Building components
Poles
Prestressing

Structural metals 531, 415.1
DT: January 1993
UF: Metals and alloys—Structural*
BT: Building materials
Materials
RT: Metals

Structural panels 408.2
DT: October 1975
UF: Panels (structural)
BT: Structural members
RT: Structural frames

* Former Ei Vocabulary term

Structure (composition)
SN: The construction or makeup of substances and equipment
DT: January 1993
UF: Chemical structure
Structure*
NT: Crystal structure
Electronic structure
Molecular structure

*Structure**
USE: Structure (composition)

Structured programming 723.1
DT: January 1993
UF: Computer programming—Structured programming*
BT: Computer programming

Structures (built objects) (408)
SN: Things that are built and designed to sustain a load
DT: January 1993
NT: Bridges
Buildings
Causeways
Composite structures
Crossings (pipe and cable)
Domes
Embankments
Energy dissipators
Flexible structures
Foundations
Fountains
Honeycomb structures
Hydraulic structures
Inflatable structures
Intelligent structures
Light weight structures
Offshore structures
Port structures
Pressure vessels
Refuse incinerators
Retaining walls
Revetments
Roofs
Salt water barriers
Sandwich structures
Shells (structures)
Steel structures
Structural frames
Structural members
Towers
Trusses
Tunnels
Underground structures
Undersnow structures
Underwater structures
RT: Architecture
Building materials
Civil engineering
Facilities
Hurricane effects
Hurricane resistance
Industrial plants
Model structures
Settlement of structures
Site selection
Structural design
Structural loads
Wind stress

Struts 408.2
DT: Predates 1975
BT: Structural members
RT: Beams and girders
Structural design
Supports

Stucco (412.1) (413) (414)
DT: Predates 1975
BT: Building materials
RT: Cements
Plaster

Stud welding 538.2.1
DT: January 1993
UF: Welding—Stud*
BT: Pressure welding
RT: Inert gas welding
Percussion welding

Studios (716.3) (716.4) (402.1)
DT: January 1993
BT: Facilities
NT: Audio studios
Motion picture studios
Radio studios
Television studios
RT: Sound stages

Studs (fasteners) 605.2
DT: January 1993
UF: Studs*
BT: Fasteners

Studs (structural members) 408.2
DT: January 1993
UF: Studs*
BT: Structural members
RT: Columns (structural)
Walls (structural partitions)

*Studs**
USE: Studs (structural members) OR Studs (fasteners)

Stuffing
USE: Packing

Styrene 804.1
DT: Predates 1975
BT: Aromatic compounds
Monomers
RT: Polystyrenes

Sub-sea oil well production
USE: Offshore oil well production

Subcrustal geology
USE: Geology

Subcrustal geophysics
USE: Geophysics

* Former Ei Vocabulary term

Subjective testing 461.4 (912.4)
DT: January 1993
UF: Human engineering—Subjective tests*
BT: Personnel testing
RT: Human engineering

Sublimation 802.3
DT: January 1993
BT: Vaporization
RT: Ablation
Condensation
Evaporation
Purification
Separation

Submarine cables (706.2) (718.1)
DT: January 1993
UF: Electric cables—Submarine*
BT: Electric cables
RT: Cable ships
Electric cable laying
Telephone lines

Submarine geology 471.1, 481.1
DT: January 1993
UF: Geology—Subaqueous*
BT: Geology
RT: Oceanography
Submarine geophysics

Submarine geophysics 471.1, 481.3
DT: January 1993
UF: Geophysics—Subaqueous*
BT: Geophysics
RT: Oceanography
Submarine geology

Submarine oil well production
USE: Offshore oil well production

Submarine pipelines 619.1 (471.1)
DT: January 1993
UF: Pipelines, Submarine*
Underwater pipelines
BT: Offshore pipelines
RT: Underwater structures

Submarine tenders 672.2
DT: January 1993
UF: Submarines—Tenders*
Tenders (submarine)
BT: Ships
RT: Naval vessels
Submarines

Submarines 672.1
DT: Predates 1975
BT: Ships
NT: Oceanographic submarines
RT: Air cleaners
Air purification
Amphibious vehicles
Naval vessels
Nuclear propulsion
Oxygen supply
Submarine tenders
Submersibles

*Submarines—Air purification**
USE: Air purification

*Submarines—Air purifiers**
USE: Air cleaners

*Submarines—Oxygen supply**
USE: Oxygen supply

*Submarines—Tenders**
USE: Submarine tenders

Submerged arc welding 538.2.1
DT: January 1993
UF: Welding—Submerged arc*
BT: Electric arc welding
RT: Spot welding

Submerged melt welding 538.2.1
DT: January 1993
UF: Welding—Submerged melt*
BT: Welding

Submerged pumps
USE: Submersible pumps

Submersible motors 705.3
DT: January 1993
UF: Electric motors—Submerged*
Underwater motors
BT: Electric motors
RT: Submersible pumps

Submersible pumps 618.2
DT: January 1993
UF: Pumps—Submerged motor*
Submerged pumps
BT: Pumps
RT: Submersible motors
Sump pumps
Water well pumps
Well pumps

Submersible seaplanes 652.1
DT: January 1993
UF: Seaplanes—Submersible*
BT: Seaplanes
RT: Submersibles

Submersibles 674.1
SN: Craft designed primarily for undersea exploration or rescue, such as bathyscaphes or diving bells. For vessels designed for other work, use SUBMARINES
DT: Predates 1975
UF: Bathyscaphes
Deep submergence vehicles
Diving bells
BT: Boats
RT: Amphibious vehicles
Diving apparatus
Rescue vessels
Semisubmersibles
Submarines
Submersible seaplanes

*Submersibles—Rescue vessels**
USE: Rescue vessels

Subroutines 723.1
DT: January 1993
UF: Computer programming—Subroutines*
Computer subroutines
Procedure libraries
BT: Computer software
RT: Algorithms
Computer programming
Computer program listings
Macros

Subscription television 716.4
DT: January 1993
UF: Pay per view television
Satellite television services
Television—Subscriber systems*
BT: Television
RT: Cable television systems

Subsea oil well production
USE: Offshore oil well production

Subsets (mathematics)
USE: Set theory

Subsidence 483.1 (405) (483.2) (502.1)
SN: Ground subsidence
DT: Predates 1975
UF: Foundations—Subsidence*
Ground subsidence
Mines and mining—Subsidence*
Soils—Subsidence*
RT: Caves
Erosion
Foundations
Mines
Mining
Soils

Subsonic aerodynamics 651.1
DT: January 1993
UF: Aerodynamics—Subsonic*
BT: Aerodynamics
RT: Subsonic flow

Subsonic flow 631.1
DT: January 1993
UF: Flow of fluids—Subsonic*
BT: Flow of fluids
RT: Aerodynamics
Subsonic aerodynamics
Transonic flow

Substations (electric)
USE: Electric substations

Substitution reactions 802.2
DT: January 1993
UF: Substitution*
BT: Chemical reactions
NT: Friedel-Crafts reaction

*Substitution**
USE: Substitution reactions

Substrates (712.1) (714.2) (461) (801) (813.1)
DT: January 1986
RT: Coatings
Epitaxial growth
Films
Integrated circuit manufacture
Integrated circuits
Laminates
Materials
Metallizing
Plating
Thick films
Thin films
Vapor deposition

Subsynchronous reluctance motors
USE: Reluctance motors

Subway cars 682.1.1
DT: January 1993
UF: Cars—Subway*
BT: Railroad cars
RT: Subways

Subway signal systems 433.4, 681.3
DT: January 1993
UF: Subways—Signal systems*
BT: Railroad signal systems
Subways
RT: Electric signal systems
Signaling

Subway stations 433.2, 681.1 (402.2)
DT: January 1993
UF: Stations (subway)
Subways—Stations*
BT: Railroad stations
Subways

Subways 681.1, 433.2 (403.1)
DT: Predates 1975
BT: Railroads
NT: Subway signal systems
Subway stations
RT: Electric railroads
Light rail transit
Mass transportation
Rapid transit
Subway cars
Urban planning

*Subways—Signal systems**
USE: Subway signal systems

*Subways—Stations**
USE: Subway stations

Sucrose
USE: Sugar (sucrose)

Sugar (sucrose) 822.3
DT: January 1993
UF: Cane sugar*
Sucrose
Sugar handling*
BT: Food products
Sugars
NT: Liquid sugar
RT: Molasses
Sugar beets
Sugar cane

Sugar (sucrose) *(continued)*
Sugar factories
Sugar industry
Sugar manufacture
Sugar substitutes

Sugar beets 821.4
DT: Predates 1975
BT: Agricultural products
RT: Sugar (sucrose)
Sugar industry

*Sugar beets—Growing**
USE: Cultivation

Sugar cane 821.4
DT: Predates 1975
BT: Agricultural products
RT: Sugar (sucrose)
Sugar industry

Sugar cane milling 822.2
DT: January 1993
UF: Milling (sugar cane)
Sugar cane—Milling*
BT: Sugar manufacture

*Sugar cane—Harvesting**
USE: Harvesting

*Sugar cane—Milling**
USE: Sugar cane milling

Sugar factories 822.1 (402.1)
DT: Predates 1975
BT: Food products plants
RT: Sugar (sucrose)
Sugar industry
Sugar manufacture

*Sugar factories—Accounting**
USE: Cost accounting

*Sugar factories—Diffusers**
USE: Diffusers (fluid)

*Sugar factories—Dust control**
USE: Dust control

*Sugar factories—Evaporators**
USE: Evaporators

*Sugar handling**
USE: Sugar (sucrose) AND
Materials handling

Sugar industry 822 (912.1)
DT: January 1993
BT: Industry
RT: Sugar (sucrose)
Sugar beets
Sugar cane
Sugar factories
Sugar manufacture

Sugar manufacture 822.2
DT: Predates 1975
BT: Manufacture
NT: Sugar cane milling
Sugar refining
RT: Clarification
Food processing
Sugar (sucrose)
Sugar factories
Sugar industry

*Sugar manufacture—Clarification**
USE: Clarification

*Sugar manufacture—Refining**
USE: Sugar refining

Sugar refining 822.2
DT: January 1993
UF: Sugar manufacture—Refining*
BT: Refining
Sugar manufacture

Sugar substitutes 822.3 (804)
DT: January 1993
UF: Sugar—Substitutes*
BT: Synthetic foods
RT: Food additives
Sugar (sucrose)

*Sugar—Liquid**
USE: Liquid sugar

*Sugar—Substitutes**
USE: Sugar substitutes

Sugars 804.1
DT: January 1987
BT: Carbohydrates
NT: Dextrose
Fructose
Glucose
Maltose
Sugar (sucrose)
Xylose

Sulfamic acid 804.2
DT: Predates 1975
BT: Inorganic acids

Sulfate minerals 482.2
DT: January 1993
UF: Mineralogy—Sulfates*
BT: Minerals
Sulfur compounds
NT: Gypsum

Sulfate pulp 811.1
DT: January 1993
UF: Pulp—Sulfate*
BT: Pulp
RT: Kraft process

Sulfate pulping
USE: Kraft process

Sulfide corrosion cracking 539.1 (421) (802.2)
DT: January 1993
UF: Corrosion—Sulfide corrosion cracking*
BT: Corrosion
RT: Cracks

Sulfide minerals 482.2
DT: January 1993
UF: Mineralogy—Sulfides and sulfosalts*

* Former Ei Vocabulary term

Sulfide minerals *(continued)*
Sulfosalt minerals
BT: Minerals
Sulfur compounds
RT: Metallic compounds
Sulfur deposits

Sulfite process 811.1.1
DT: January 1993
UF: Pulp manufacture—Sulfite process*
BT: Pulp manufacture
RT: Sulfite pulp

Sulfite pulp 811.1
DT: January 1993
UF: Pulp—Sulfite*
BT: Pulp
RT: Sulfite process

Sulfonation 802.2
DT: January 1993
BT: Chemical reactions

Sulfosalt minerals
USE: Sulfide minerals

Sulfur 804
DT: Predates 1975
UF: Cast iron—Sulfur content*
S
Sulphur
BT: Nonmetals
RT: Desulfurization
Sulfur deposits

Sulfur compounds (804.1) (804.2)
DT: Predates 1975
UF: Gasoline refining—Sulfur compounds*
BT: Chemical compounds
NT: Carbon disulfide
Hydrogen sulfide
Sulfate minerals
Sulfide minerals
Sulfur dioxide
Zinc sulfide
RT: Sour gas
Sulfur hexafluoride circuit breakers

Sulfur deposits 505.1
DT: Predates 1975
BT: Mineral resources
RT: Sulfide minerals
Sulfur

Sulfur determination 801
DT: January 1993
UF: Fuels—Sulfur determination*
BT: Chemical analysis
RT: Coal
Desulfurization
Fuels
Natural gas
Petroleum analysis

Sulfur dioxide 804.2
DT: Predates 1975
BT: Oxides
Sulfur compounds
RT: Sulfur dioxide recorders

Sulfur dioxide recorders 801, 804.2
DT: January 1993
UF: Sulfur dioxide—Recorders*
BT: Recording instruments
RT: Sulfur dioxide

*Sulfur dioxide—Recorders**
USE: Sulfur dioxide recorders

Sulfur hexafluoride circuit breakers 704.2 (804.2)
DT: January 1993
UF: Electric circuit breakers—Sulfur hexafluoride*
BT: Electric circuit breakers
RT: Sulfur compounds

Sulfur removal
USE: Desulfurization

Sulfuric acid 804.2
DT: Predates 1975
BT: Inorganic acids

*Sulfuric acid—Corrosive properties**
USE: Corrosive effects

Sulphur
USE: Sulfur

Summing circuits 721.3
DT: January 1993
UF: Computers—Summing circuits*
BT: Networks (circuits)
RT: Adders
Digital circuits

Sump pumps 618.2
DT: January 1993
UF: Pumps—Sumps*
BT: Pumps
RT: Submersible pumps

Sun 657.2
DT: January 1986
BT: Solar system
RT: Solar energy

Sun hoods 402
DT: January 1993
UF: Awnings
Buildings—Sun hoods*
Sun screens
Sunscreens
BT: Building components

Sun screens
USE: Sun hoods

Sunscreens
USE: Sun hoods

Superalloys 531
DT: January 1977
UF: Heat resistant alloys
Powder metallurgy—Superalloys*
BT: Alloys
NT: Stellite

Superalloys *(continued)*
RT: Heat resistance
Refractory metals

Superchargers (612.1.1) (618.1)
DT: January 1993
UF: Superchargers and supercharging*
Supercharging
Supercharging*
BT: Compressors
RT: Blowers
Intake systems
Internal combustion engines
Turbomachinery

*Superchargers and supercharging**
USE: Superchargers

Supercharging
USE: Superchargers

*Supercharging**
USE: Superchargers

Supercomputers 722.4
DT: January 1993
UF: Computers, Supercomputer*
BT: Digital computers
RT: Parallel processing systems

Superconducting cables 706.2, 708.3
SN: For cables in use, use SUPERCONDUCTING ELECTRIC LINES
DT: January 1977
UF: Cryogenic cables
Electric cables—Superconducting*
BT: Electric cables
Superconducting devices
RT: Superconducting electric lines

Superconducting devices (704.2) (708.3)
DT: Predates 1975
BT: Equipment
NT: Cryotrons
Josephson junction devices
SQUIDs
Superconducting cables
Superconducting electric lines
Superconducting magnets
RT: Cryoelectric storage
Electric coils
Superconducting materials
Superconductivity

*Superconducting devices—Josephson junctions**
USE: Josephson junction devices

Superconducting electric lines 706.2, 708.3
DT: January 1993
UF: Electric lines—Superconducting*
BT: Electric lines
Superconducting devices
RT: Superconducting cables

Superconducting films 708.3
DT: June 1990
UF: Superconducting thin films
BT: Films
Superconducting materials
RT: Superconducting transition temperature
Thin films

*Superconducting films—Transition temperature**
USE: Superconducting transition temperature

Superconducting junction devices
USE: Josephson junction devices

Superconducting magnets 708.3, 708.4
DT: Predates 1975
BT: Electromagnets
Superconducting devices

Superconducting materials 708.3
DT: Predates 1975
UF: Superconductors
BT: Materials
NT: High temperature superconductors
Oxide superconductors
Superconducting films
RT: Critical current density (superconductivity)
Critical currents
Electric conductors
Fullerenes
Organic conductors
Superconducting devices
Superconducting transition temperature
Superconducting wire
Superconductivity

*Superconducting materials—Transition temperature**
USE: Superconducting transition temperature

Superconducting quantum interference devices
USE: SQUIDs

Superconducting thin films
USE: Superconducting films

Superconducting transition temperature 701.1, 708.3
DT: January 1993
UF: High temperature superconductors—Transition temperature*
Superconducting films—Transition temperature*
Superconducting materials—Transition temperature*
Transition temperature
BT: Temperature
RT: High temperature superconductors
Superconducting materials
Superconducting films
Superconductivity

Superconducting wire 708.3
DT: January 1995
BT: Electric wire
RT: Superconducting materials

* Former Ei Vocabulary term

Superconductivity 701.1, 708.3
DT: Predates 1975
BT: Electric conductivity
RT: Critical currents
Cryogenics
Electric conductivity of solids
Magnetic field effects
Superconducting devices
Superconducting materials
Superconducting transition temperature
Superfluid helium

Superconductors
USE: Superconducting materials

Supercritical fluid extraction 802.3
DT: January 1993
UF: Extraction—Supercritical fluid*
BT: Solvent extraction
RT: Solvents

Superfluid helium 804 (708.3) (931.2)
DT: January 1995
UF: Superfluidity
BT: Helium
Liquefied gases
RT: Cryogenics
Cryostats
Superconductivity

Superfluidity
USE: Superfluid helium

Superheater tubes 616.1 (614.2)
DT: January 1993
UF: Superheaters—Tubes*
BT: Superheaters
Tubes (components)

Superheaters 616.1 (614.2)
SN: For superheating of steam. For wave superheaters, use SHOCK TUBES
DT: Predates 1975
BT: Steam generators
NT: Superheater tubes
RT: Boilers
Heat exchangers
Steam

*Superheaters—Tubes**
USE: Superheater tubes

Superlattices 933.1 (531) (712.1)
DT: June 1990
BT: Crystal lattices
NT: Metallic superlattices
Semiconductor superlattices
RT: Crystal atomic structure
Crystal microstructure
Multilayers
Order disorder transitions
Solid solutions

*Superlattices—Metallic**
USE: Metallic superlattices

*Superlattices—Semiconductor**
USE: Semiconductor superlattices

Superplasticity (421) (931.2)
DT: January 1993
BT: Plasticity
RT: Deformation

Superregeneration 716.3
DT: January 1993
UF: Radio receivers—Superregeneration*
RT: Detector circuits
Radio receivers

Supersaturation 801.4
DT: January 1995
BT: Saturation (materials composition)
RT: Crystallization
Heat treatment
Precipitation (chemical)
Quenching
Solid solutions
Solutions

Supersonic aerodynamics 651.1
DT: January 1993
UF: Aerodynamics—Supersonic*
BT: Aerodynamics
RT: Hypersonic aerodynamics
Supersonic aircraft
Transonic aerodynamics

Supersonic aircraft 652.1
DT: January 1993
UF: Aircraft—Supersonic speeds*
Transonic aircraft
BT: Aircraft
RT: Fighter aircraft
Jet aircraft
Research aircraft
Supersonic aerodynamics
Supersonic flow

Supersonic flow 631.1
DT: January 1993
UF: Flow of fluids—Supersonic*
Wind tunnels—Supersonic flow*
BT: Flow of fluids
RT: Aerodynamics
Hypersonic flow
Supersonic aircraft
Transonic flow
Wind tunnels

Supervisors
USE: Supervisory personnel

Supervisory and executive programs 723.1
DT: January 1993
UF: Computer systems programming—Supervisory and executive programs*
Executive programs
Supervisory programs
BT: Computer software
NT: Multiprocessing programs
Time sharing programs
RT: Computer operating systems
Computer systems programming
Multiprogramming
Storage allocation (computer)

* Former Ei Vocabulary term

Supervisory control and data acquisition systems
USE: SCADA systems

Supervisory personnel 912.4
DT: January 1993
UF: Foremen
Managers
Personnel—Supervisory*
Supervisors
BT: Personnel
RT: Management

Supervisory programs
USE: Supervisory and executive programs

Supply systems (electric)
USE: Electric power systems

Supports (401.1) (408.2) (619.1) (693.1) (706.2)
DT: January 1993
UF: Bridges—Supports*
Cranes—Supports*
Electric lines—Supports*
Pipe—Supports*
Pipelines—Supports*
Stands
BT: Components
NT: Electric line supports
Ground supports
Mine roof supports
Poles
Roadway supports
RT: Axles
Bearings (structural)
Bridge bearings
Bridge piers
Bushings
Columns (structural)
Reinforcement
Structural frames
Struts

Supports (catalyst)
USE: Catalyst supports

Surface acoustic wave devices
USE: Acoustic surface wave devices

Surface acoustic wave filters
USE: Acoustic surface wave filters

Surface active agents 803 (804.1) (804.2)
DT: January 1993
UF: Surfactants
BT: Agents
NT: Detergents
RT: Plasticizers

Surface analysis (moon)
USE: Lunar surface analysis

Surface cleaning (539) (802)
DT: January 1993
UF: Surface treatment—Cleaning*
BT: Cleaning
NT: Metal cleaning
Ultrasonic cleaning
RT: Finishing

Surface composition
USE: Surface structure

Surface discharges 701.1
DT: January 1993
UF: Electric discharges—Surface phenomena*
Tracking (insulation)
BT: Electric discharges
Surface phenomena
NT: Electric corona
RT: Electric insulating materials
Electric insulation
Electric insulators
Flashover
Voltage distribution measurement

Surface energy
USE: Interfacial energy

Surface measurement 943.2
DT: January 1993
UF: Granular materials—Surface measurement*
BT: Spatial variables measurement
NT: Contour measurement
Roughness measurement
RT: Granular materials
Planimeters
Surface properties
Surfaces

Surface mine transportation 502.1
DT: January 1993
UF: Mines and mining—Surface transportation*
BT: Mine transportation

Surface mining
USE: Open pit mining

Surface modification
USE: Surface treatment

Surface morphology
USE: Surfaces AND
Morphology

Surface mount technology 714.2
DT: January 1993
UF: Electronics packaging—Surface mount technology*
SMD
SMT
Surface mounted devices
BT: Electronic equipment manufacture
RT: Assembly
Electronics packaging
Integrated circuit manufacture
Printed circuit manufacture
Printed circuits

Surface mounted devices
USE: Surface mount technology

Surface phenomena 931
DT: Predates 1975
NT: Capillarity
Crazing
Interfacial energy

Surface phenomena *(continued)*
Surface discharges
Surface tension
Surface waves
Wetting
RT: Adhesion
Corrosion
Electron emission
Films
Friction
Interfaces (materials)
Porosity
Surface properties
Surface structure
Surface treatment
Surfaces
Tribology

Surface properties 931.2
DT: January 1993
BT: Physical properties
RT: Color
Corrosion
Diffusion
Electromigration
Friction
Interfaces (materials)
Mechanical properties
Porosity
Sorption
Surface measurement
Surface phenomena
Surfaces
Textures
Tribology

Surface roughness 931.2
DT: January 1995
UF: Roughness
BT: Surface structure
RT: Roughness measurement

Surface structure 931.2
DT: January 1995
UF: Surface composition
NT: Surface roughness
RT: Atomic force microscopy
Low energy electron diffraction
Scanning tunneling microscopy
Surface phenomena

Surface tension 931.2
DT: January 1993
UF: Glass—Surface tension*
Interface tension
Interfacial tension
Liquids—Surface tension*
BT: Surface phenomena
RT: Capillarity
Contact angle
Glass
Interfaces (materials)
Interfacial energy
Liquids
Percolation (fluids)

Surface testing (423.2)
DT: January 1993
UF: Materials testing—Surface*
BT: Materials testing
RT: Surfaces

Surface treatment (539) (604.2) (802) (931)
DT: Predates 1975
UF: Surface modification
NT: Etching
Galvanizing
Polishing
RT: Cleaning
Coating techniques
Coatings
Finishing
Metal finishing
Metallurgy
Nitriding
Plating
Surface phenomena

*Surface treatment—Cleaning**
USE: Surface cleaning

Surface water geochemistry
USE: Natural water geochemistry

Surface water resources 444.1
DT: January 1993
UF: Estuarine water resources
Water resources—Surface water*
BT: Water resources
RT: Estuaries
Lakes
Rivers
Surface waters

Surface waters 444.1
DT: January 1993
NT: Estuaries
Lakes
Rivers
Solar ponds
RT: Banks (bodies of water)
Evapotranspiration
Fish ponds
Floods
Groundwater
Hydrology
Inland waterways
Irrigation
Natural water geochemistry
Springs (water)
Surface water resources
Water
Watersheds
Waterway transportation

Surface waves (711) (751) (931)
DT: Predates 1975
BT: Surface phenomena
Waves
RT: Mechanical waves

Surfaces (604.2) (801.4) (931)
DT: Predates 1975
UF: Surface morphology
RT: Contour measurement
Geometry
Interfaces (materials)
Low energy electron diffraction
Roughness measurement
Surface measurement
Surface phenomena
Surface properties
Surface testing

*Surfaces—Contour measurement**
USE: Contour measurement

*Surfaces—Roughness measurement**
USE: Roughness measurement

Surfacing
USE: Hard facing

Surfactants
USE: Surface active agents

Surge protection 704, 914.1
DT: Predates 1975
BT: Protection
RT: Electric surges
Lightning protection

Surge tanks 619.2 (632.2) (914.1)
DT: Predates 1975
BT: Hydraulic structures
Tanks (containers)
RT: Hydraulic accumulators
Pipeline surges
Pressure
Pressure regulation
Pressure regulators

Surgery 461.6
DT: January 1993
UF: Biomedical engineering—Surgery applications*
BT: Patient treatment
NT: Cardiovascular surgery
Cryosurgery
Electrosurgery
Laser surgery
Neurosurgery
Transplantation (surgical)
RT: Biomedical engineering
Implants (surgical)
Medicine
Operating rooms

Surges (electric)
USE: Electric surges

Surges (fluid) 631.1
DT: January 1993
BT: Fluid dynamics
NT: Pipeline surges

Surgical equipment
USE: Biomedical equipment

Surgical implants
USE: Implants (surgical)

Surveillance (space)
USE: Space surveillance

Surveillance radar 716.2
DT: January 1993
UF: Radar—Surveillance application*
BT: Radar
RT: Doppler radar
Synthetic aperture radar

Surveying 405.3
SN: Geographic locations
DT: January 1993
RT: Construction
Data acquisition
Geodesy
Geometry
Geophysics
Mapping
Maps
Measurements
Natural resources exploration
Photogrammetry
Recording
Statistical methods
Surveying instruments
Surveys
Triangulation

Surveying instruments 405.3 (941) (943.3)
DT: Predates 1975
BT: Instruments
RT: Aneroid altimeters
Compasses (magnetic)
Optical instruments
Radio altimeters
Surveying
Surveys
Triangulation

*Surveying instruments—Electronics**
USE: Electronic equipment

*Surveying—Triangulation**
USE: Triangulation

Surveys
SN: Very general term; prefer specific type of survey
DT: January 1993
NT: Geological surveys
Hydrographic surveys
Mine surveys
Soil surveys
Stockpile surveys
Traffic surveys
RT: Construction
Data acquisition
Geodesy
Geometry
Geophysics
Mapping
Maps
Measurements

* Former Ei Vocabulary term

Surveys*(continued)*
Natural resources exploration
Photogrammetry
Recording
Statistical methods
Surveying
Surveying instruments
Triangulation

Suspension bridges 401.1
DT: January 1993
UF: Bridges, Suspension*
BT: Bridges

Suspensions (components) 601.2
DT: January 1993
BT: Components
NT: Vehicle suspensions

Suspensions (fluids) 804 (801.3)
DT: January 1993
UF: Suspensions*
BT: Dispersions
NT: Slurries
RT: Brownian movement
Sediment transport

*Suspensions**
USE: Suspensions (fluids)

Swaging 535.2
DT: Predates 1975
BT: Metal forming
RT: Cold working
Forging
Hot working
Pressing (forming)
Swaging machines

Swaging machines 535.2.1
DT: Predates 1975
BT: Metal forming machines
RT: Swaging

Sweep circuits 713.4
DT: January 1993
UF: Electronic circuits, Sweep*
BT: Networks (circuits)
RT: Pulse generators

Swelling (421)
DT: January 1993
RT: Dimensional stability
Expansion
Growth (materials)
Thermal expansion
Weathering

Swept frequency oscillators 713.2
DT: January 1993
UF: Oscillators, Swept frequency*
BT: Variable frequency oscillators

Swimming pools 403.2 (402.2)
DT: Predates 1975
BT: Facilities
RT: Recreation centers
Recreational facilities

Swing bridges 401.1
DT: January 1993
UF: Bridges, Swing*
BT: Movable bridges

Switchboards (electric)
USE: Electric switchboards

Switchboards (telephone)
USE: Telephone switchboards

Switchboxes
USE: Electric switchgear

Switched filters 703.2
DT: January 1993
UF: Electric filters, Switched*
N-path filters
BT: Electric filters
Switching networks

Switches (704.1) (741.3)
DT: January 1993
BT: Equipment
NT: Electric switches
Optical switches
RT: Final control devices
Switching networks
Switching systems
Switching theory

Switchgear
USE: Electric switchgear

Switching (721.1)
DT: January 1993
NT: Combinatorial switching
Q switching
Sequential switching
Switching (rolling stock)
RT: Asynchronous sequential logic
Electric switchgear
Switching circuits
Switching functions
Switching systems
Switching theory

Switching (rolling stock) 682.1
DT: January 1993
BT: Switching
RT: Locomotives
Railroad rolling stock

Switching circuits 713.4 (721.3)
DT: January 1993
UF: Electronic circuits, Switching*
BT: Networks (circuits)
NT: Choppers (circuits)
Trigger circuits
RT: Digital circuits
Diode transistor logic circuits
Electric circuit breakers
Electric relays
Electric switches
Emitter coupled logic circuits
Logic circuits
Pulse circuits

* Former Ei Vocabulary term

Switching circuits *(continued)*
Switching
Switching networks
Switching systems
Switching theory
Telephone switching equipment

Switching functions 721.1, 921
DT: January 1993
BT: Functions
Switching theory
NT: State assignment
RT: Boolean functions
Minimization of switching nets
Switching

Switching networks (703.1) (721)
DT: January 1993
UF: Electric networks, Switching*
BT: Networks (circuits)
NT: Interconnection networks
Iterated switching networks
Switched filters
RT: Minimization of switching nets
Switches
Switching circuits
Switching systems
Switching theory
Telephone switching equipment

Switching systems (716) (717) (718) (721.1)
DT: Predates 1975
NT: Line concentrators
RT: Automatic telephone systems
Intelligent networks
Optical switches
Switches
Switching
Switching circuits
Switching networks
Switching theory
Telecommunication
Telecommunication systems
Telephone switching equipment
Telephone systems

Switching theory 721.1
DT: Predates 1975
BT: Circuit theory
NT: Hazards and race conditions
Minimization of switching nets
Switching functions
RT: Asynchronous sequential logic
Combinatorial switching
Computation theory
Electric network synthesis
Formal logic
Iterated switching networks
Logic circuits
Logic design
Majority logic
Sequential switching
Switches
Switching
Switching circuits
Switching networks
Switching systems
Topology

*Switching theory—Asynchronous sequential logic**
USE: Asynchronous sequential logic

*Switching theory—Combinatorial switching**
USE: Combinatorial switching

*Switching theory—Hazards and race conditions**
USE: Hazards and race conditions

*Switching theory—Iterated switching networks**
USE: Iterated switching networks

*Switching theory—Minimization of switching nets**
USE: Minimization of switching nets

*Switching theory—Sequential switching**
USE: Sequential switching

*Switching theory—State assignment**
USE: State assignment

Symbolic codes
USE: Codes (symbols)

Symbols (codes)
USE: Codes (symbols)

Symmetry (crystal)
USE: Crystal symmetry

Synchrocyclotrons 932.1.1
DT: January 1993
UF: Accelerators, Synchrocyclotron*
BT: Particle accelerators

Synchronisation
USE: Synchronization

Synchronism
USE: Synchronization

Synchronization (731.1)
DT: January 1993
UF: Synchronisation
Synchronism
RT: Coincidence circuits
Concurrency control
Control
Control systems
Phase measurement
Resonance
Stroboscopes
Time measurement
Tuning
Video signal processing

Synchronizing reactors
USE: Current limiting reactors

Synchronous generators 705.2.1
DT: January 1993
UF: Alternators (generators)
Electric generators, Synchronous*

Synchronous generators *(continued)*
BT: AC generators
Synchronous machinery

Synchronous machinery 705.1
DT: January 1993
UF: Electric machinery, Synchronous*
BT: AC machinery
NT: Synchronous generators
Synchronous motors

Synchronous motors 705.3.1
DT: January 1993
UF: Electric motors, Synchronous*
BT: AC motors
Synchronous machinery
NT: Hysteresis motors
Reluctance motors

Synchros 705.3, 732.1
DT: Predates 1975
BT: Servomotors

Synchrotron radiation 932.1.1
DT: June 1990
BT: Electromagnetic waves
RT: Synchrotrons

Synchrotrons 932.1.1
DT: January 1993
UF: Accelerators, Synchrotron*
BT: Particle accelerators
RT: Synchrotron radiation

Synthesis (chemical) 802.2
DT: January 1993
UF: Mica—Synthesizing*
Synthesis*
BT: Chemical reactions
NT: Biosynthesis
Fischer-Tropsch synthesis
Photosynthesis
RT: Analysis
Manufacture
Processing

Synthesis gas (522) (803) (804)
DT: January 1986
BT: Gases
RT: Carbon monoxide
Methanation
Synthesis gas manufacture

Synthesis gas manufacture 802.2 (522)
DT: January 1993
UF: Gas manufacture—Synthesis gas*
BT: Gas fuel manufacture
RT: Methanation
Synthesis gas

*Synthesis**
USE: Synthesis (chemical)

Synthesized music
USE: Computer music

Synthetic aperture radar 716.2
DT: January 1993
UF: Radar—Synthetic aperture*
BT: Radar
RT: Surveillance radar
Synthetic apertures

Synthetic apertures 716.2
DT: January 1993
RT: Holography
Imaging techniques
Radar imaging
Synthetic aperture radar

Synthetic blood
USE: Blood substitutes

Synthetic blood vessels
USE: Blood vessel prostheses

Synthetic diamonds 804 (482.2.1)
DT: Predates 1975
UF: Diamonds—Synthetic*
BT: Diamonds

Synthetic fibers 819.2
DT: January 1977
UF: Dyes and dyeing—Synthetic fibers*
Papermaking—Synthetic fibers*
BT: Fibers
NT: Bituminized fibers
Carbon fibers
Ceramic fibers
Glass fibers
Mineral wool
Optical fibers
Plastic filaments
Rayon
Synthetic textile fibers
RT: Acrylics
Nylon textiles
Polyolefins
Rayon fabrics
Rayon yarn
Textile blends
Textile fibers

Synthetic foods 822.3 (804.1)
DT: January 1993
UF: Artificial foods
Food products—Synthetics*
BT: Food products
NT: Sugar substitutes

Synthetic fuels (522) (523) (524)
DT: January 1977
UF: Alternate fuels
Alternative fuels
BT: Fuels
NT: Alcohol fuels
Hydrogen fuels
RT: Biogas
Coal gas
Coal gasification
Coal liquefaction
Refuse derived fuels

* Former Ei Vocabulary term

Synthetic gems 482.2.1
DT: January 1993
UF: Gems—Synthetic origin*
BT: Gems

Synthetic graphite
USE: Artificial graphite

Synthetic leather (814.1) (817.1)
DT: January 1993
UF: Leather—Synthetic*
BT: Nonwoven fabrics
RT: Leather

Synthetic lubricants 607.1
DT: January 1993
UF: Lubricants—Synthetic products*
BT: Lubricants

Synthetic marble 414
DT: January 1993
UF: Marble—Synthetic*
BT: Materials
RT: Building materials
Marble

Synthetic metals
USE: Organic conductors

Synthetic rubber 818.2.1
DT: January 1993
UF: Rubber, Synthetic*
BT: Elastomers
NT: Polybutadienes
Polysulfides
RT: Butadiene
Foamed rubber
Latexes
Polyisoprenes
Rubber

Synthetic textile fibers 819.2
DT: January 1993
UF: Synthetic textiles
Textile fibers—Synthetic*
Textiles—Synthetic materials*
BT: Synthetic fibers
Textile fibers
RT: Textiles

Synthetic textiles
USE: Synthetic textile fibers

Syringes 462.1
DT: January 1993
UF: Biomedical equipment—Syringes*
BT: Biomedical equipment
RT: Disposable biomedical equipment
Needles

System analysis
USE: Systems analysis

System design
USE: Systems analysis

System engineering
USE: Systems engineering

System identification
USE: Identification (control systems)

System program documentation 723.1
DT: January 1993
UF: Computer systems programming—Documentation*
Documentation (systems programs)
Systems documentation
BT: Program documentation
RT: Computer systems programming
Systems analysis

System programming
USE: Computer systems programming

System science
USE: Systems science

System stability 731.4
DT: Predates 1975
UF: Robust stability
Systems stability
BT: Stability
NT: Frequency stability
RT: Bode diagrams
Control systems
Control theory
Controllability
Damping
Lyapunov methods
Nyquist diagrams
Observability
Stability criteria

*System stability—Criteria**
USE: Stability criteria

*System stability—Frequency stability**
USE: Frequency stability

*System stability—Lyapunov methods**
USE: Lyapunov methods

System theory (461.1) (731.4) (912.3)
DT: January 1993
UF: Systems science and cybernetics—System theory*
Systems theory
BT: Systems science
Theory
RT: Artificial intelligence
Identification (control systems)
Information theory
Models
Pattern recognition
Simulation

Systematic errors
USE: Measurement errors

Systems (computer)
USE: Computer systems

Systems (metallurgical) (531.1.) (531.2)
DT: January 1995
NT: Ternary systems
RT: Binary alloys
Phase diagrams

* Former Ei Vocabulary term

Systems analysis 912.3 (731.1)
DT: January 1983
UF: System analysis
System design
Systems design
NT: Flowcharting
RT: Computer systems programming
Critical path analysis
Decision tables
Decision theory
Information theory
Mathematical models
Operations research
Optimization
Simulation
System program documentation
Systems engineering
Value engineering

Systems design
USE: Systems analysis

Systems documentation
USE: System program documentation

Systems engineering (912)
DT: January 1977
UF: System engineering
BT: Engineering
RT: Computers
Concurrent engineering
Decision making
Ergonomics
Man machine systems
Management
Mathematical models
Mechanization
Operations research
Scheduling
Systems analysis

Systems programming
USE: Computer systems programming

Systems science (461.1) (731.1) (912.3)
DT: January 1993
UF: Bioengineering (systems science)
System science
Systems science and cybernetics*
NT: Constraint theory
Cybernetics
Decision theory
Hierarchical systems
Large scale systems
Learning systems
Man machine systems
Multivariable systems
Optimal systems
System theory
Time varying systems
RT: Computer programming
Heuristic programming
Identification (control systems)
Optimization
Simulation

*Systems science and cybernetics**
USE: Systems science

*Systems science and cybernetics—Adaptive systems**
USE: Adaptive systems

*Systems science and cybernetics—Biocommunications**
USE: Biocommunications

*Systems science and cybernetics—Biocontrol**
USE: Biocontrol

*Systems science and cybernetics—Bionics**
USE: Bionics

*Systems science and cybernetics—Brain models**
USE: Brain models

*Systems science and cybernetics—Cognitive systems**
USE: Cognitive systems

*Systems science and cybernetics—Constraint theory**
USE: Constraint theory

*Systems science and cybernetics—Economic cybernetics**
USE: Economic cybernetics

*Systems science and cybernetics—Heuristic programming**
USE: Heuristic programming

*Systems science and cybernetics—Hierarchical systems**
USE: Hierarchical systems

*Systems science and cybernetics—Identification**
USE: Identification (control systems)

*Systems science and cybernetics—Large scale systems**
USE: Large scale systems

*Systems science and cybernetics—Man machine systems**
USE: Man machine systems

*Systems science and cybernetics—Multivariable systems**
USE: Multivariable systems

*Systems science and cybernetics—Optimal systems**
USE: Optimal systems

*Systems science and cybernetics—System theory**
USE: System theory

*Systems science and cybernetics—Time varying systems**
USE: Time varying systems

Systems stability
USE: System stability

* Former Ei Vocabulary term

Systems theory
USE: System theory

Systolic algorithms
USE: Systolic arrays

Systolic arrays (722.1) (921)
DT: January 1995
UF: Systolic algorithms
BT: Cellular arrays
RT: Digital signal processing
Parallel processing systems

Ta
USE: Tantalum

Table lookup 723.1
DT: January 1993
UF: Computer systems programming—Table lookup*
Lookup tables
BT: Data handling

Tables (decision)
USE: Decision tables

Tachometers 943.3 (652.3)
DT: Predates 1975
BT: Speed indicators
RT: Aircraft instruments
Electric frequency measurement
Velocimeters

Tackiness
USE: Adhesion

Tactile reading aids 461.5
DT: January 1993
UF: Human rehabilitation engineering—Tactile reading aids*
BT: Human rehabilitation equipment
RT: Human rehabilitation engineering
Sensory aids

Tail assemblies (aircraft)
USE: Empennages

Tailings 452.3 (502.1) (533.1)
DT: January 1993
BT: Industrial wastes
NT: Coal tailings
Ore tailings
RT: Tailings disposal

Tailings disposal 452.4 (502.1) (533.1)
DT: January 1993
UF: Coal preparation—Tailings disposal*
Coal tailings disposal
Ore tailings disposal
Ore treatment—Tailings disposal*
BT: Industrial waste disposal
RT: Tailings

Takeoff 652.1
SN: Of aircraft
DT: January 1993
UF: Aircraft—Takeoff*
RT: Air traffic control
Aircraft
Airports
Aviation
Launching
Takeoff indicators

Takeoff indicators 652.3
DT: January 1993
UF: Aircraft instruments—Takeoff indicators*
BT: Aircraft instruments
RT: Takeoff

Talc 482.2
DT: Predates 1975
UF: Talcs
BT: Materials
RT: Fillers
Steatite
Talc deposits
Talc mines

Talc deposits 505.1
DT: Predates 1975
BT: Mineral resources
RT: Steatite
Talc
Talc mines

Talc mines 505.1
DT: January 1993
UF: Talc mines and mining*
BT: Mines
RT: Talc
Talc deposits

*Talc mines and mining**
USE: Talc mines

Talcs
USE: Talc

Tall buildings 402
DT: January 1993
UF: Buildings—Tall*
Skyscrapers
BT: Buildings

Tall oil 804.1, 811.2
DT: Predates 1975
UF: Liquid rosin
Tallol
BT: Materials
RT: Waste liquor utilization

Tallol
USE: Tall oil

Tamper resistant packaging 694.1 (694.2) (914.1)
DT: January 1993
UF: Packaging—Tamper resistant*
Tamperproof packaging
BT: Packaging

Tamperproof packaging
USE: Tamper resistant packaging

Tandem connections
USE: Cascade connections

* Former Ei Vocabulary term

Tank cars 682.1.1 (433.3)
DT: January 1993
UF: Cars—Tank*
BT: Freight cars
RT: Gas fuel storage
Petroleum transportation

Tank linings 619.2
DT: January 1993
UF: Tanks—Lining*
BT: Linings
Tanks (containers)

Tank trucks 663.1 (432.3)
DT: January 1993
UF: Motor trucks—Tank*
Tankers (trucks)
BT: Trucks
RT: Gas fuel storage
Petroleum transportation
Refrigerator trucks

Tankers (ships) 671 (434.3) (672.2)
SN: Ships. For trucks, use TANK TRUCKS
DT: January 1993
UF: Oil tankers
Tankers*
BT: Ships
RT: Petroleum transportation

Tankers (trucks)
USE: Tank trucks

*Tankers**
USE: Tankers (ships)

Tanks (containers) 619.2
DT: January 1993
BT: Containers
NT: Ballast tanks
Concrete tanks
Fuel tanks
Oil tanks
Septic tanks
Settling tanks
Sewage tanks
Steel tanks
Surge tanks
Tank linings
Water tanks
RT: Heat losses
Hydraulic structures
Materials handling
Pressure vessels
Roofs
Vents

Tanks (military) 663.1, 404.1
DT: January 1993
UF: Military vehicles—Tanks*
BT: Ground vehicles
Military vehicles

*Tanks—Aluminum**
USE: Aluminum containers

*Tanks—Concrete**
USE: Concrete tanks

*Tanks—Heat losses**
USE: Heat losses

*Tanks—Lining**
USE: Tank linings

*Tanks—Roofs**
USE: Roofs

*Tanks—Steel**
USE: Steel tanks

*Tanks—Vents**
USE: Vents

*Tanks—Welded steel**
USE: Welded steel structures

Tanning 814.2 (802)
DT: Predates 1975
UF: Leather—Tanning*
BT: Processing
RT: Leather

Tantalate minerals 482.2
DT: January 1993
UF: Mineralogy—Niobates and tantalates*
BT: Minerals
Tantalum compounds
RT: Tantalum

Tantalum 543.4
DT: January 1993
UF: Ta
Tantalum and alloys*
BT: Nonferrous metals
Refractory metals
RT: Tantalate minerals
Tantalum alloys
Tantalum carbide
Tantalum compounds
Tantalum deposits
Tantalum metallography
Tantalum mines
Tantalum ore treatment
Tantalum powder metallurgy

Tantalum alloys 543.4
DT: January 1993
UF: Tantalum and alloys*
BT: Refractory alloys
RT: Tantalum
Tantalum compounds

*Tantalum and alloys**
USE: Tantalum OR
Tantalum alloys

Tantalum carbide 804.2 (812.1)
DT: Predates 1975
BT: Carbides
Tantalum compounds
RT: Tantalum

*Tantalum carbide—Sintered**
USE: Sintered carbides

Tantalum compounds (804.1) (804.2)
DT: Predates 1975
BT: Refractory metal compounds

Tantalum compounds *(continued)*
NT: Tantalate minerals
Tantalum carbide
RT: Tantalum
Tantalum alloys

Tantalum deposits 504.3, 543.4
DT: Predates 1975
BT: Ore deposits
RT: Tantalum
Tantalum mines

Tantalum metallography 531.2, 543.4
DT: Predates 1975
BT: Metallography
RT: Tantalum

Tantalum mines 504.3, 543.4
DT: January 1993
UF: Tantalum mines and mining*
BT: Mines
RT: Tantalum
Tantalum deposits

*Tantalum mines and mining**
USE: Tantalum mines

Tantalum ore treatment 533.1, 543.4
DT: Predates 1975
BT: Ore treatment
RT: Tantalum

Tantalum powder metallurgy 536.1, 543.4
DT: January 1993
UF: Powder metallurgy—Tantalum*
BT: Powder metallurgy
RT: Tantalum

Tape coatings 813.2 (539.2.2)
DT: January 1993
UF: Protective coatings—Tape*
BT: Coatings
RT: Tapes

Tape drives 722.2
DT: January 1993
UF: Computer peripheral equipment—Tape drives*
BT: Computer peripheral equipment
RT: Magnetic tape storage

Tape recorders 752.2.1
DT: Predates 1975
BT: Audio equipment
NT: Videocassette recorders
RT: Magnetic recording
Magnetic tape storage
Recording
Recording instruments
Sound recording
Sound reproduction
Video recording

Tape recordings
USE: Stereophonic recordings

Tape storage (of data)
USE: Magnetic tape storage

Tapes (708.2) (708.4) (817.1)
DT: January 1993
BT: Materials
NT: Electric tapes
Magnetic tape
Plastic tapes
RT: Sealants
Seals
Tape coatings

Taping (cable)
USE: Cable taping

Tapping (furnace) 534.2
DT: January 1993
UF: Cupola practice—Tapping*
BT: Metallurgy
RT: Blast furnace practice
Casting
Cupola practice
Melting
Molding
Pig iron

Tapping (threads) 604.1
SN: Cutting internal threads
DT: January 1993
UF: Screw threads—Tapping*
Tapping*
Water pipelines—Tapping*
BT: Thread cutting
RT: Drilling
Metal cutting
Reaming
Screw threads
Screws

*Tapping**
USE: Tapping (threads)

Taps 605.2
DT: January 1993
UF: Taps and dies*
Thread cutting taps
BT: Cutting tools
RT: Carbide tools
Hand tools
Thread cutting
Thread cutting dies

*Taps and dies**
USE: Taps OR
Thread cutting dies

Tar 804 (411) (513) (524)
DT: January 1987
BT: Materials
NT: Coal tar
Petroleum tar
RT: Asphalt
Bituminous materials

Target drones 652.1.2 (404.1)
DT: January 1993
UF: Aircraft—Target drones*
Drones (target)
BT: Military aircraft

* Former Ei Vocabulary term

Targets 932.1.1
DT: January 1993
UF: Accelerators—Targets*
BT: Particle accelerators
RT: Nuclear energy
Scattering

Taste control (water treatment) 445.1.1
DT: January 1993
UF: Water treatment—Taste control*
BT: Water treatment
RT: Chemicals removal (water treatment)

Taxation (902.3) (911.2)
DT: January 1993
UF: Corporate taxes
Electric utilities—Taxation*
Federal taxes
Oil well production—Taxation*
State taxes
Taxes
BT: Finance
RT: Fees and charges
Public policy

Taxes
USE: Taxation

Taxi meters 943.3 (432.2)
DT: January 1993
UF: Taxicabs—Meters*
BT: Instruments
Taxicabs

Taxicabs 662.1 (432.2)
DT: Predates 1975
BT: Automobiles
NT: Taxi meters

*Taxicabs—Meters**
USE: Taxi meters

Tb
USE: Terbium

TBE
USE: Binding energy

TDM
USE: Time division multiplexing

Te
USE: Tellurium

Teach programming
USE: Robot programming

Teaching (901.2)
DT: January 1993
UF: Education—Teaching*
BT: Education
RT: Information dissemination
Technical presentations

*Teaching machines**
USE: Computer aided instruction

Technetium 549.3 (622.1)
DT: Predates 1975
BT: Nonferrous metals
Transition metals
RT: Technetium compounds

Technetium compounds 804.1, 804.2
DT: June 1990
BT: Metallic compounds
RT: Technetium

Technical communication
USE: Technical writing

Technical presentations (901.2) (903.2)
DT: January 1993
UF: Education—Technical presentation*
Seminars
Workshops (seminars)
RT: Demonstrations
Education
Information dissemination
Teaching

Technical writing 903.2
DT: January 1993
UF: Engineering writing*
Information dissemination—Technical writing*
Specification writing
Technical communication
Writing (technical)
BT: Information dissemination
RT: Engineering

Technological forecasting (901.4)
DT: Predates 1975
BT: Forecasting
RT: Economic and social effects
Employment
Technology

Technology (901)
SN: All means used to provide society with material objects, treated in general terms. Prefer the specific technology or technological activity, if appropriate
DT: Predates 1975
NT: Biotechnology
Vacuum technology
RT: Economic and social effects
Engineering
History
Philosophical aspects
Technological forecasting
Technology transfer

Technology transfer (901.4) (911.2)
DT: January 1993
RT: Economic and social effects
Technology

*Technology—Economic and sociological effects**
USE: Economic and social effects

*Technology—Philosophical aspects**
USE: Philosophical aspects

* Former Ei Vocabulary term

Tectonics 481.1 (484.1)
DT: January 1993
UF: Geology—Tectonics*
Plate tectonics
Structural geology
BT: Geology
RT: Earth (planet)
Earthquakes
Geophysics
Landforms
Rocks

Teeth (gears)
USE: Gear teeth

Teflon
USE: Polytetrafluoroethylenes

Telecommunication (716) (718)
SN: Very general term; prefer specific subject
DT: Predates 1975
UF: Electric communication
Telecommunications
BT: Communication
NT: Aircraft communication
Bandwidth compression
Carrier communication
Optical communication
Radio
Radio communication
Speech transmission
Telecommunication services
Teleconferencing
Telegraph
Telemetering
Telephone
Television
Visual communication
RT: Channel capacity
Code converters
Electromagnetic waves
Electronics engineering
Electronics industry
Information technology
Modulation
Multiplexing
Signal interference
Spurious signal noise
Switching systems
Telecommunication equipment
Telecommunication systems
Telecommunication links
Telecommunication control
Telecommunication traffic

Telecommunication cables (716) (717) (718)
SN: For cables in use, use
TELECOMMUNICATION LINES
DT: Predates 1975
UF: Communication cables
Radio cables
Telephone cables
Television cables
BT: Electric cables
Telecommunication equipment
NT: Coaxial cables
Optical cables
RT: Electric cable laying
Overhead lines
Radio equipment
Telecommunication links
Telecommunication lines
Telegraph equipment
Telephone equipment
Television equipment

*Telecommunication cables, Coaxial**
USE: Coaxial cables

*Telecommunication cables, Optical**
USE: Optical cables

Telecommunication control (716) (717) (718)
SN: Functional control. For legislation use
TELECOMMUNICATION and LAWS AND
LEGISLATION
DT: January 1993
UF: Telecommunication—Control*
BT: Control
RT: Radio
Telecommunication
Telegraph
Telephone
Television

Telecommunication equipment (716) (717) (718)
DT: October 1975
UF: Boats—Communication equipment*
BT: Electronic equipment
NT: Antennas
Communication satellites
Compandor circuits
Data communication equipment
Facsimile equipment
Multiplexing equipment
Optical communication equipment
Radio equipment
Signal receivers
Telecommunication cables
Telecommunication lines
Telecommunication repeaters
Telecontrol equipment
Telegraph equipment
Telemetering equipment
Telephone apparatus
Telephone equipment
Television equipment
Transmitters
Tuners
RT: Information technology
Remote control
Satellite ground stations
Signal processing
Telecommunication
Telecommunication systems

Telecommunication interference
USE: Signal interference

Telecommunication lines (716) (717) (718)
DT: Predates 1975
BT: Electric lines

Telecommunication lines *(continued)*
Telecommunication equipment
NT: High frequency telecommunication lines
Strip telecommunication lines
Telephone lines
RT: Telecommunication links
Telecommunication cables

*Telecommunication lines, High frequency**
USE: High frequency telecommunication lines

*Telecommunication lines, Strip**
USE: Strip telecommunication lines

Telecommunication links (716) (717) (718)
DT: Predates 1975
UF: Communication links
Links (telecommunication)
Television relay systems
BT: Telecommunication systems
NT: Extraterrestrial communication links
Optical links
Pulse code modulation links
Radio links
RT: Interfaces (computer)
Pulse code modulation
Telecommunication
Telecommunication cables
Telecommunication lines
Waveguides

*Telecommunication links, Extraterrestrial**
USE: Extraterrestrial communication links

*Telecommunication links, Microwave**
USE: Microwave links

*Telecommunication links, Optical**
USE: Optical links

*Telecommunication links, Radio**
USE: Radio links

*Telecommunication links, Satellite**
USE: Satellite links

*Telecommunication links—Pulse code modulation**
USE: Pulse code modulation links

Telecommunication networks (716) (717) (718)
DT: January 1993
UF: Communication networks
Electric networks, Communication*
BT: Networks (circuits)
NT: Broadband networks
Intelligent networks
Packet networks
RT: Congestion control (communication)
Telecommunication systems

Telecommunication repeaters (718.1) (718.2)
DT: Predates 1975
UF: Relays (repeaters)
Repeaters
BT: Telecommunication equipment
RT: Amplifiers (electronic)
Light amplifiers
Signal receivers
Telegraph equipment
Telephone equipment
Transmitters

Telecommunication services (716) (717) (718)
DT: January 1993
BT: Telecommunication
NT: Broadcasting
Electronic mail
Facsimile
Videotex
RT: Information services
Intelligent networks
Telecommunication systems
Telemetering

Telecommunication systems (716) (717) (718)
DT: Predates 1975
BT: Communication systems
NT: Data communication systems
Digital communication systems
Mobile telecommunication systems
Paging systems
Personal communication systems
Radio systems
Satellite communication systems
Satellite relay systems
Telecommunication links
Telegraph systems
Telemetering systems
Telephone systems
Television systems
RT: Electromagnetic compatibility
Interference suppression
Office automation
Signal systems
Switching systems
Telecommunication
Telecommunication equipment
Telecommunication networks
Telecommunication services
Telecommunication traffic

*Telecommunication systems, Mobile**
USE: Mobile telecommunication systems

*Telecommunication systems, Satellite relay**
USE: Satellite relay systems

Telecommunication traffic (716) (717) (718)
DT: January 1993
UF: Communication traffic
Telecommunication—Traffic*
Traffic (telecommunication)
RT: Telecommunication
Telecommunication systems
Telephone traffic recording

*Telecommunication—Control**
USE: Telecommunication control

*Telecommunication—Speech transmission**
USE: Speech transmission

*Telecommunication—Traffic**
USE: Telecommunication traffic

* Former Ei Vocabulary term

Telecommunications
USE: Telecommunication

Teleconferencing 718.1
DT: January 1986
UF: Video conferencing
BT: Telecommunication
RT: Administrative data processing
Closed circuit television systems
Conference calls
Electronic mail
Office automation
Telephone
Television applications
Video telephone equipment
Visual communication

Telecontrol
USE: Remote control

Telecontrol equipment 732.1
DT: January 1993
UF: Remote control equipment
BT: Control equipment
Telecommunication equipment
RT: Recording instruments
Remote control
Telemetering equipment
Transducers

Telegraph 718.2
DT: Predates 1975
BT: Telecommunication
NT: Carrier telegraph
Radio telegraph
RT: Data communication systems
Diplex transmission
Office automation
Telecommunication control
Telegraph circuits
Telegraph codes
Telegraph equipment
Telegraph interference
Telegraph stations
Telegraph systems

Telegraph circuits 718.2
DT: Predates 1975
BT: Networks (circuits)
RT: Telegraph
Telegraph equipment
Telegraph systems

Telegraph codes 718.2
DT: January 1993
UF: Telegraph—Codes*
BT: Codes (symbols)
RT: Telegraph

Telegraph equipment 718.2
DT: Predates 1975
BT: Telecommunication equipment
NT: Teleprinters
RT: Data communication equipment
Line concentrators
Radio equipment
Radio telegraph
Telecommunication repeaters
Telecommunication cables
Telegraph
Telegraph circuits
Telegraph systems

Telegraph interference 718.2 (711)
DT: Predates 1975
BT: Signal interference
RT: Telegraph

Telegraph stations 718.2 (402.1)
DT: January 1993
UF: Stations (telegraph)
Telegraph systems—Stations*
BT: Telegraph systems
RT: Telegraph

Telegraph systems 718.2
DT: Predates 1975
UF: Telex
BT: Telecommunication systems
NT: Telegraph stations
RT: Coaxial cables
Digital communication systems
Radio links
Radio telegraph
Telegraph
Telegraph circuits
Telegraph equipment

*Telegraph systems—Stations**
USE: Telegraph stations

*Telegraph, Carrier**
USE: Carrier telegraph

*Telegraph—Codes**
USE: Telegraph codes

*Telegraph—Diplex transmission**
USE: Diplex transmission

Telemetering (716) (717) (718) (731.2)
DT: Predates 1975
UF: Remote metering
Telemetry
BT: Telecommunication
RT: Aerospace ground support
Pulse modulation
Radio communication
Radio links
Remote control
Satellites
Telecommunication services
Telemetering equipment
Telemetering systems
Time division multiplexing

Telemetering equipment
(716) (717) (718) (731.2) (732.2)
DT: Predates 1975
UF: Telemeters
BT: Telecommunication equipment
RT: Aerospace ground support
Biomedical equipment
Data communication equipment
Recording instruments

* Former Ei Vocabulary term

Telemetering equipment *(continued)*
Satellites
Telecontrol equipment
Telemetering
Telemetering systems
Transducers

Telemetering systems (716) (717) (718) (731.2) (732.1)
DT: Predates 1975
BT: Telecommunication systems
RT: Aerospace ground support
Radio links
Telemetering
Telemetering equipment

Telemeters
USE: Telemetering equipment

Telemetry
USE: Telemetering

Telephone 718.1
DT: Predates 1975
BT: Telecommunication
NT: Carrier telephone
Conference calls
Radio telephone
RT: Echo suppression
Intercom systems
Office automation
Push button telephone systems
Signal distortion
Telecommunication control
Teleconferencing
Telephone accounting systems
Telephone apparatus
Telephone circuits
Telephone equipment
Telephone exchanges
Telephone interference
Telephone lines
Telephone numbering systems
Telephone switching equipment
Telephone systems
Telephone traffic recording
Telephone traffic analysis
Touch tone telephone systems
Weather information services

Telephone accounting systems 718.1, 911.1
DT: January 1993
UF: Telephone—Accounting systems*
BT: Telephone systems
RT: Telephone
Telephone equipment

Telephone apparatus 718.1
SN: Devices or equipment located at subscriber station, or similar apparatus located elsewhere
DT: Predates 1975
UF: Telephone station equipment
BT: Telecommunication equipment
NT: Coin operated telephones
Cordless telephones
Telephone cords
Telephone hearing aids
Telephone sets
RT: Earphones
Facsimile equipment
Headphones
Telephone

*Telephone apparatus—Coin operation**
USE: Coin operated telephones

*Telephone apparatus—Cordless**
USE: Cordless telephones

*Telephone apparatus—Cords**
USE: Telephone cords

*Telephone apparatus—Hearing aids**
USE: Telephone hearing aids

*Telephone apparatus—Telephone sets**
USE: Telephone sets

Telephone cables
USE: Telephone equipment AND
Telecommunication cables

Telephone circuits 718.1
DT: Predates 1975
BT: Networks (circuits)
RT: Telephone
Telephone equipment
Telephone systems

Telephone cords 718.1
DT: January 1993
UF: Telephone apparatus—Cords*
BT: Telephone apparatus
Wire products

Telephone equipment 718.1
SN: Devices or equipment located outside the subscriber station. For subscriber-station equipment, use TELEPHONE APPARATUS
DT: Predates 1975
UF: Telephone cables
BT: Telecommunication equipment
NT: Telephone lines
Telephone switching equipment
Telephone traffic analyzers
Video telephone equipment
Vocoders
RT: Amplifiers (electronic)
Automatic telephone systems
Compandor circuits
Echo suppression
Equalizers
Radio equipment
Radio telephone
Telecommunication repeaters
Telecommunication cables
Telephone
Telephone accounting systems
Telephone circuits
Telephone exchanges
Telephone systems

*Telephone equipment—Depreciation**
USE: Depreciation

*Telephone equipment—Power supply**
USE: Electric power supplies to apparatus

*Telephone equipment—Traffic analyzers**
USE: Telephone traffic analyzers

*Telephone equipment—Video telephone**
USE: Video telephone equipment

*Telephone equipment—Wiring**
USE: Electric wiring

Telephone exchanges 718.1
DT: Predates 1975
UF: Exchanges (telephone)
BT: Telephone systems
NT: Automatic telephone exchanges
Military telephone exchanges
Mobile telephone exchanges
Private telephone exchanges
RT: Telephone
Telephone equipment
Telephone lines
Telephone switching equipment
Telephone traffic recording

*Telephone exchanges, Automatic**
USE: Automatic telephone exchanges

*Telephone exchanges, Mobile**
USE: Mobile telephone exchanges

*Telephone exchanges, Private**
USE: Private telephone exchanges

*Telephone exchanges—Military**
USE: Military telephone exchanges

Telephone filters
USE: Electric filters

Telephone hearing aids
461.5, 462.1, 718.1 (752.1)
DT: January 1993
UF: Telephone apparatus—Hearing aids*
BT: Hearing aids
Telephone apparatus

Telephone interference 718.1 (711)
DT: Predates 1975
BT: Signal interference
RT: Compandor circuits
Crosstalk
Telephone

Telephone lines 718.1
DT: Predates 1975
UF: Lines (telephone)
BT: Telecommunication lines
Telephone equipment
RT: Carrier telephone
Coaxial cables
Comparator circuits
Drop wires
Line concentrators
Submarine cables
Telephone
Telephone exchanges
Telephone systems
Waveguides

*Telephone lines—Drop wires**
USE: Drop wires

Telephone numbering systems 718.1
DT: January 1993
UF: Telephone—Numbering systems*
BT: Numbering systems
Telephone systems
RT: Telephone

Telephone relays
USE: Electric relays

Telephone sets 718.1
SN: For operators or subscribers
DT: January 1993
UF: Handsets (telephone)
Telephone apparatus—Telephone sets*
Telephones
BT: Telephone apparatus
RT: Coin operated telephones
Earphones
Electroacoustic transducers

Telephone station equipment
USE: Telephone apparatus

Telephone switchboards 718.1
DT: January 1993
UF: Switchboards (telephone)
Telephone switching equipment—Switchboards*
BT: Electric switchboards
Telephone switching equipment

Telephone switching equipment 718.1
DT: Predates 1975
BT: Electric switches
Telephone equipment
NT: Crossbar equipment
Telephone switchboards
RT: Electric contacts
Line concentrators
Switching circuits
Switching networks
Switching systems
Telephone
Telephone exchanges

*Telephone switching equipment—Contacts**
USE: Electric contacts

*Telephone switching equipment—Crossbar equipment**
USE: Crossbar equipment

*Telephone switching equipment—Switchboards**
USE: Telephone switchboards

Telephone systems 718.1
DT: Predates 1975
BT: Telecommunication systems
NT: Automatic telephone systems
Cellular telephone systems
Intercom systems
Long distance telephone systems

Telephone systems *(continued)*
Push button telephone systems
Telephone accounting systems
Telephone exchanges
Telephone numbering systems
RT: Digital communication systems
Office automation
Personal communication systems
Switching systems
Telephone
Telephone circuits
Telephone equipment
Telephone lines
Telephone traffic recording
Telephone traffic analysis
Video telephone equipment
Videotex
Weather information services

*Telephone systems, Automatic**
USE: Automatic telephone systems

*Telephone systems—Long distance**
USE: Long distance telephone systems

*Telephone systems—Traffic analysis**
USE: Telephone traffic analysis

*Telephone systems—Traffic recording**
USE: Telephone traffic recording

Telephone traffic analysis 718.1
DT: January 1993
UF: Telephone systems—Traffic analysis*
BT: Analysis
RT: Telephone
Telephone systems
Telephone traffic analyzers
Telephone traffic recording

Telephone traffic analyzers 718.1
DT: January 1993
UF: Telephone equipment—Traffic analyzers*
BT: Telephone equipment
RT: Telephone traffic analysis
Telephone traffic recording

Telephone traffic recording 718.1
DT: January 1993
UF: Telephone systems—Traffic recording*
BT: Recording
RT: Telecommunication traffic
Telephone
Telephone exchanges
Telephone systems
Telephone traffic analysis
Telephone traffic analyzers

*Telephone, Carrier**
USE: Carrier telephone

*Telephone—Accounting systems**
USE: Telephone accounting systems

*Telephone—Conference calls**
USE: Conference calls

*Telephone—Intercommunication**
USE: Intercom systems

*Telephone—Noise**
USE: Spurious signal noise

*Telephone—Numbering systems**
USE: Telephone numbering systems

*Telephone—Personal signaling**
USE: Cellular telephone systems

*Telephone—Push button systems**
USE: Push button telephone systems

*Telephone—Touch tone system**
USE: Touch tone telephone systems

*Telephone—Weather information service**
USE: Weather information services

Telephones
USE: Telephone sets

Teleprinters 718.2
DT: Predates 1975
UF: Teletypes
BT: Telegraph equipment
Typewriters
RT: Signal receivers
Transmitters

Telescopes (716.3) (741.3) (657)
DT: Predates 1975
BT: Instruments
NT: Airborne telescopes
Optical telescopes
Radio telescopes
RT: Antennas
Astronomy
Image intensifiers (electron tube)
Image intensifiers (solid state)
Lenses
Mirrors
Observatories

*Telescopes—Airborne installations**
USE: Airborne telescopes

Teletext
USE: Videotex

Teletypes
USE: Teleprinters

Television 716.4
DT: Predates 1975
UF: TV
BT: Telecommunication
NT: Color television
High definition television
Infrared television
Subscription television
RT: Cable television systems
Closed circuit television systems
Networks (circuits)
Reception quality
Telecommunication control
Television applications
Television broadcasting
Television equipment
Television interference

Television *(continued)*
Television networks
Television picture quality
Television receivers
Television standards
Television stations
Television studios
Television systems
Television transmitters
Television transmission
Video cameras
Video recording
Visual communication

Television amplifiers
USE: Amplifiers (electronic)

Television antennas 716.4
DT: January 1993
UF: Antennas—Television*
BT: Antennas
Television equipment
RT: Television receivers
Television stations
Television transmitters

Television applications 716.4
DT: Predates 1975
BT: Applications
RT: Teleconferencing
Television
Video telephone equipment
Videotex

*Television applications—Aerospace**
USE: Aerospace applications

*Television applications—Education**
USE: Education

*Television applications—Industry**
USE: Industrial applications

*Television applications—Medicine**
USE: Medicine

*Television applications—Military**
USE: Military applications

*Television applications—Underground**
USE: Underground equipment

*Television applications—Underwater**
USE: Underwater equipment

Television broadcasting 716.4
DT: Predates 1975
UF: Stereophonic television broadcasting
BT: Broadcasting
RT: Color television
High definition television
Special effects
Stereophonic broadcasting
Television
Television equipment
Television networks
Television stations
Television studios
Television systems

*Television broadcasting—Special effects**
USE: Special effects

*Television broadcasting—Stereophonic**
USE: Stereophonic broadcasting

Television cables
USE: Television equipment AND
Telecommunication cables

Television camera tubes 714.1, 716.4
DT: January 1993
UF: Camera tubes (television)
Electron tubes, Television camera*
BT: Cathode ray tubes
Television equipment
RT: Image storage tubes

Television cameras
USE: Video cameras

Television equipment 716.4
DT: Predates 1975
UF: Satellites—Television equipment*
Television cables
Video equipment
BT: Telecommunication equipment
NT: Automobile television equipment
Television antennas
Television camera tubes
Television picture tubes
Television receivers
Television transmitters
Video amplifiers
Video cameras
Videocassette recorders
Videodisks
RT: Antennas
Cable television systems
Cathode ray tubes
Closed circuit television systems
Compact disk players
High definition television
Modulators
Telecommunication cables
Television
Television broadcasting
Television networks
Television picture quality
Television stations
Television systems
Television towers
Video telephone equipment

*Television equipment—Cameras**
USE: Video cameras

Television interference 716.4 (711)
DT: Predates 1975
BT: Signal interference
RT: Reception quality
Television
Television picture quality

Television motion pictures
USE: Motion pictures

* Former Ei Vocabulary term

Television networks 716.4
DT: Predates 1975
UF: Networks (television)
BT: Television systems
RT: Cable television systems
Television
Television broadcasting
Television equipment
Television stations

Television picture quality 716.4
DT: January 1993
UF: Television—Picture quality*
RT: Reception quality
Television
Television equipment
Television interference
Television picture tubes
Television receivers
Television systems
Television transmission

Television picture tubes 714.1, 716.4
DT: January 1993
UF: Display tubes (television)
Electron tubes, Television picture*
Picture tubes
BT: Cathode ray tubes
Television equipment
RT: Display devices
Phosphors
Television picture quality

Television receivers 716.4
DT: Predates 1975
BT: Signal receivers
Television equipment
NT: Color television receivers
Large screen projection television
RT: Deflection yokes
Fluorescent screens
Integrated circuits
Microphones
Stereophonic receivers
Television
Television antennas
Television picture quality
Tuners
Tuning

*Television receivers, Color**
USE: Color television receivers

*Television receivers—Deflection yokes**
USE: Deflection yokes

*Television receivers—Noise**
USE: Spurious signal noise

*Television receivers—Power supply**
USE: Electric power supplies to apparatus

*Television receivers—Tuners**
USE: Tuners

Television recording
USE: Video recording

Television relay systems
USE: Telecommunication links

Television standards 716.4, 902.2
DT: January 1993
UF: Television standards conversion
Television—Standards conversion*
BT: Standards
RT: High definition television
Television

Television standards conversion
USE: Television standards

Television stations 716.4 (402.1)
DT: Predates 1975
UF: Stations (television)
BT: Television systems
RT: Television
Television antennas
Television broadcasting
Television equipment
Television networks
Television transmitters

Television studios 716.4 (402.1)
DT: Predates 1975
BT: Studios
RT: Television
Television broadcasting
Video cameras

Television systems 716.4
DT: Predates 1975
BT: Telecommunication systems
NT: Cable television systems
Closed circuit television systems
Television networks
Television stations
RT: Bandwidth compression
Coaxial cables
High definition television
Television
Television broadcasting
Television equipment
Television picture quality
Video telephone equipment

*Television systems, Cable**
USE: Cable television systems

*Television systems, Closed circuit**
USE: Closed circuit television systems

Television towers 716.4, 402.4
DT: January 1993
BT: Towers
RT: Television equipment

Television transmission 716.4
DT: January 1977
BT: Electromagnetic wave transmission
RT: Electromagnetic wave propagation
Television
Television picture quality
Television transmitters

* Former Ei Vocabulary term

Television transmitters 716.4
DT: Predates 1975
BT: Television equipment
Transmitters
NT: Airborne television transmitters
Color television transmitters
RT: Television
Television antennas
Television stations
Television transmission

*Television transmitters, Color**
USE: Color television transmitters

*Television transmitters—Airborne**
USE: Airborne television transmitters

*Television, Color**
USE: Color television

*Television—High definition**
USE: High definition television

*Television—Infrared**
USE: Infrared television

*Television—Large screen projection**
USE: Large screen projection television

*Television—Motion pictures**
USE: Motion pictures

*Television—Picture quality**
USE: Television picture quality

*Television—Reception quality**
USE: Reception quality

*Television—Recording**
USE: Video recording

*Television—Standards conversion**
USE: Television standards

*Television—Subscriber systems**
USE: Subscription television

Telex
USE: Telegraph systems

Telluric prospecting 481.4 (501.1) (512)
DT: January 1993
UF: Geophysical prospecting—Telluric*
BT: Electric prospecting

Tellurium 549.3
DT: Predates 1975
UF: Te
BT: Metalloids
NT: Semiconducting tellurium
RT: Tellurium compounds
Tellurium deposits

Tellurium compounds (804.1) (804.2)
DT: Predates 1975
BT: Chemical compounds
NT: Semiconducting tellurium compounds
RT: Tellurium

Tellurium deposits 504.3, 549.3
DT: Predates 1975
BT: Mineral resources
RT: Tellurium

TEM
USE: Transmission electron microscopy

Temperature 641.1
DT: January 1993
UF: Absolute temperature
Ambient temperature
Bearings—Temperature*
Electric transformers—Temperature*
Glass transition temperature
Satellites—Temperature*
Turbomachinery—Temperature*
NT: Atmospheric temperature
Superconducting transition temperature
RT: Air conditioning
Cold effects
Cooling
Heating
High temperature effects
High temperature operations
Isotherms
Temperature control
Temperature distribution
Temperature indicating cameras
Temperature measurement
Temperature scales
Thermodynamic properties

Temperature control 731.3
DT: Predates 1975
BT: Thermal variables control
RT: Climate control
Combustion
Cooling
Cryostats
Environmental engineering
Environmental testing
Heating
Oxygen blast enrichment
Process control
Temperature
Temperature measurement
Thermal insulation
Thermostats
Ventilation

Temperature distribution 641.1
DT: Predates 1975
UF: Thermoclines
RT: Environmental engineering
Environmental testing
Temperature
Temperature measurement
Thermal gradients
Thermal stratification
Thermography (temperature measurement)

Temperature effects
USE: Thermal effects

Temperature indicating cameras 742.2, 944.5
DT: January 1993
UF: Cameras—Temperature indicating*

Temperature indicating cameras *(continued)*
BT: Cameras
RT: Temperature
Temperature measurement

Temperature measurement 944.6
DT: Predates 1975
BT: Thermal variables measurement
NT: Pyrometry
Thermography (temperature measurement)
Underground temperature measurement
Underwater temperature measurement
RT: Bolometers
Pyrometers
Temperature
Temperature control
Temperature distribution
Temperature indicating cameras
Temperature scales
Thermometers

*Temperature measurement—Underground**
USE: Underground temperature measurement

*Temperature measurement—Underwater**
USE: Underwater temperature measurement

Temperature measuring instruments 944.5
DT: Predates 1975
BT: Gages
NT: Pyrometers
Thermometers
RT: Bolometers
Calorimeters
Thermocouples

Temperature programmed desorption 802.3
DT: January 1995
UF: TPD
RT: Desorption
Materials testing

Temperature scales 944.6 (902.2)
DT: Predates 1975
BT: Standards
RT: Temperature
Temperature measurement

Tempering 537.1
DT: January 1993
UF: Heat treatment—Tempering*
BT: Heat treatment
RT: Hardening
Stress relief

Temporary bridges 401.1
DT: January 1993
UF: Bridges—Temporary*
BT: Bridges

Temporary coatings 813.2 (539.2.2)
DT: January 1993
UF: Protective coatings—Temporary*
BT: Coatings

Temporary pipelines 619.1
DT: January 1993
UF: Pipelines—Temporary installations*
BT: Pipelines

Temporary roads 406.2
DT: January 1993
UF: Roads and streets—Temporary*
BT: Roads and streets

Temporary runways 431.4
DT: January 1993
UF: Airport runways—Temporary*
BT: Airport runways
RT: Landing mats

Tenacity
USE: Tensile strength

Tenders (submarine)
USE: Submarine tenders

Tendons 461.2
DT: January 1993
UF: Biological materials—Tendons*
BT: Musculoskeletal system
Tissue

Tennis courts 403.2 (402)
DT: Predates 1975
BT: Facilities
RT: Recreation centers
Recreational facilities

Tensile modulus
USE: Elastic moduli

Tensile properties (421) (422)
DT: January 1995
BT: Mechanical properties
NT: Tensile strength
RT: Creep
Fatigue of materials
Hysteresis
Tensile testing

Tensile strength (421) (422)
DT: January 1995
UF: Tenacity
BT: Tensile properties
RT: Impact testing
Loads (forces)
Tensile testing

Tensile testing (422.2)
DT: January 1993
UF: Materials testing—Tensile tests*
Tension testing
BT: Mechanical testing
NT: Notched bar tensile testing
RT: Compression testing
Elastic moduli
Elasticity testing
Fatigue testing
Fracture testing
Hardness testing
High temperature testing
Low temperature testing
Stiffness
Strain rate
Tensile properties

* Former Ei Vocabulary term

Tensile testing *(continued)*
Tensile strength
Torsion testing
Yield stress

Tension testing
USE: Tensile testing

Tensors 921.1
DT: January 1993
UF: Mathematical techniques—Tensors*
BT: Linear algebra
RT: Matrix algebra

Terbium 547.2
DT: January 1981
UF: Tb
BT: Rare earth elements
RT: Terbium alloys
Terbium compounds

Terbium alloys 547.2
DT: January 1993
BT: Rare earth alloys
RT: Terbium
Terbium compounds

Terbium compounds (804.1) (804.2)
DT: January 1993
BT: Rare earth compounds
RT: Terbium
Terbium alloys

Terminals (bus)
USE: Bus terminals

Terminals (computer)
USE: Computer terminals

Terminals (electric) 704.1
DT: January 1993
UF: Electric machinery—Terminals*
Electric terminals
BT: Electric equipment

Terminals (ferry)
USE: Ferry terminals

Terminals (oil)
USE: Oil terminals

Terminals (pipeline)
USE: Pipeline terminals

Terminals (point of sale)
USE: Point of sale terminals

Terminals (port)
USE: Port terminals

Terminals (railroad)
USE: Railroad yards and terminals

Terminals (truck)
USE: Truck terminals

Terminology (902.2) (903.2)
SN: Use for wordlists that include definitions and/or translations of words, as well as for information about the terminology of a field
DT: January 1993
UF: Color—Terminology*
Dictionaries
Glossaries*
Instruments—Terminology*
Nomenclature
RT: Standards
Units of measurement

Termite proofing
(412) (415.3) (803) (811.2) (817.1)
DT: January 1993
UF: Concrete—Termite proofing*
Wood—Termite proofing*
BT: Insect control
Protection
RT: Termite resistance
Wood

Termite resistance (415.3) (803) (811.2) (817.1)
DT: January 1993
UF: Plastics—Termite resistance*
RT: Insect control
Termite proofing

Ternary systems (531.1) (531.2)
DT: January 1995
BT: Systems (metallurgical)
RT: Eutectics
Phase diagrams
Solidification

Terpolymerization 815.2 (802.2)
DT: January 1993
UF: Polymerization—Terpolymerization*
BT: Polymerization
RT: Copolymerization
Terpolymers

Terpolymers 815.1
DT: Predates 1975
BT: Copolymers
NT: ABS resins
RT: Block copolymers
Graft copolymers
Terpolymerization

Terrestrial atmosphere
USE: Earth atmosphere

Terrestrial magnetism
USE: Geomagnetism

Tertiary recovery
USE: Enhanced recovery

Test facilities (402.1) (422.1) (423.1)
DT: January 1993
BT: Facilities
NT: Ship model tanks
Wind tunnels
RT: Anechoic chambers
Industrial laboratories
Laboratories
Testing

Testing (422.2) (423.2)
DT: January 1993
NT: Automatic testing

* Former Ei Vocabulary term

Testing *(continued)*
Environmental testing
Equipment testing
High temperature testing
Impact testing
Laser diagnostics
Low temperature testing
Materials testing
Mechanical testing
Nondestructive examination
Optical testing
Personnel testing
Safety testing
Statistical tests
Well testing
RT: Analysis
Errors
Evaluation
Inspection
Measurements
Performance
Quality control
Reliability
Standardization
Standards
Test facilities

Tetherlines 655.1
SN: Lines which connect astronauts to the spacecraft during extra-vehicular activity, or which connect equipment to the spacecraft
DT: January 1993
UF: Spacecraft—Tethers*
Tethers (spacecraft)
BT: Cables
Spacecraft equipment

Tethers (spacecraft)
USE: Tetherlines

Tetrodes 714.1
DT: January 1993
UF: Electron tubes, Tetrode*
BT: Electron tubes
RT: Thermionic tubes

Text to speech synthesis
USE: Speech synthesis

Textbooks 903.2 (901.2)
DT: January 1993
UF: Engineering—Textbooks*
RT: Handbooks
Information dissemination

Textile auxiliary materials 803, 819.5
DT: Predates 1975
BT: Materials
RT: Glycerol
Oils and fats
Textiles

Textile blends 819.5
DT: January 1993
UF: Blends (fibers)
Dyes and dyeing—Mixed fibers*
Textile fibers—Mixed*
Textiles—Mixed*
BT: Textiles
RT: Blending
Natural fibers
Polymer blends
Synthetic fibers

Textile carding 819.5
DT: January 1993
UF: Carding (textiles)
Textiles—Carding*
BT: Textile processing

Textile chemical treatment 802, 819.5 (803)
DT: January 1993
UF: Textile finishing—Chemical treatment*
BT: Chemical finishing
Textile finishing
RT: Chemical operations

Textile classing 819.5
DT: January 1993
UF: Textiles—Classing*
BT: Textile processing

Textile crimping 819.5
DT: January 1993
UF: Textiles—Crimping*
BT: Texturing

Textile fibers (819.1) (819.2)
SN: Includes natural and manmade fibers, chopped fibers, and filaments
DT: Predates 1975
BT: Fibers
NT: High modulus textile fibers
Synthetic textile fibers
RT: Asbestos
Bast fibers
Cellulose derivatives
Hemp
Hemp fibers
Jute fibers
Kenaf fibers
Nylon textiles
Spinning (fibers)
Synthetic fibers
Textiles
Yarn

*Textile fibers—High modulus**
USE: High modulus textile fibers

*Textile fibers—Mixed**
USE: Textile blends

*Textile fibers—Synthetic**
USE: Synthetic textile fibers

Textile finishing 802, 819.5
DT: Predates 1975
UF: Milling (textile)
BT: Finishing
Textile processing
NT: Mercerization
Mothproofing
Shrinkproofing (textiles)
Textile chemical treatment

* Former Ei Vocabulary term

Textile finishing *(continued)*
Textile pressing
Textile printing
Textile scouring
Texturing
RT: Bleaching
Sizing (finishing operation)
Textiles
Waterproofing

*Textile finishing—Chemical treatment**
USE: Textile chemical treatment

*Textile finishing—Mercerization**
USE: Mercerization

*Textile finishing—Texturizing**
USE: Texturing

Textile industry 819.6
DT: Predates 1975
BT: Industry
RT: Garment manufacture
Hosiery mills
Textile machinery
Textile mills
Textile processing
Textiles
Woolen and worsted mills

Textile machinery 819.6
DT: Predates 1975
BT: Machinery
NT: Bobbins
Knitting machinery
Looms
Sewing machines
RT: Dryers (equipment)
Rolls (machine components)
Spinning machines
Textile industry
Textile processing
Textiles
Winding machines

*Textile machinery—Bobbins**
USE: Bobbins

*Textile machinery—Dryers**
USE: Dryers (equipment)

*Textile machinery—Looms**
USE: Looms

*Textile machinery—Rolls**
USE: Rolls (machine components)

*Textile machinery—Spinning machines**
USE: Spinning machines

Textile manufacture
USE: Textile processing

*Textile measuring instruments**
USE: Instruments AND
Textiles

Textile mills 819.6 (402.1)
DT: Predates 1975
BT: Industrial plants
NT: Hosiery mills
Woolen and worsted mills
RT: Garment manufacture
Textile industry
Textile processing

Textile numbering systems 819.5
DT: January 1993
UF: Textiles—Numbering systems*
BT: Numbering systems
RT: Textiles

Textile pilling 819.5 (421)
DT: January 1993
UF: Textiles—Pilling*
RT: Knit fabrics
Laundering
Textiles
Wear of materials

Textile pressing 819.5
DT: January 1993
UF: Pressing (textiles)
Textiles—Pressing*
BT: Textile finishing

Textile printing 819.5 (745.1) (802.3)
DT: January 1993
UF: Textiles—Printing*
BT: Printing
Textile finishing

Textile processing 819.5
DT: January 1993
UF: Textile manufacture
Textiles—Processing*
BT: Processing
NT: Textile carding
Textile classing
Textile finishing
Weaving
RT: Bleaching
Carpet manufacture
Dyeing
Garment manufacture
Labeling
Sizing (finishing operation)
Spinning (fibers)
Textile industry
Textile machinery
Textile mills
Wrinkle recovery

Textile scouring 819.5
DT: January 1993
UF: Scouring (textiles)
Textiles—Scouring*
BT: Cleaning
Textile finishing
RT: Dyeing

Textiles 819
DT: Predates 1975
UF: Heat transfer—Textiles*
Textile measuring instruments*
BT: Materials
NT: Cotton fabrics

* Former Ei Vocabulary term

Textiles *(continued)*
Durable press textiles
Fabrics
Geotextiles
Metallic textiles
Military textiles
Nylon textiles
Textile blends
Wire screen cloth
RT: Antistatic agents
Color fastness
Cotton
Dry cleaning
Dyeing
Laundering
Leather
Shrinkage
Soil resistance (textiles)
Synthetic textile fibers
Textile auxiliary materials
Textile fibers
Textile finishing
Textile industry
Textile machinery
Textile numbering systems
Textile pilling
Wrinkle recovery
Yarn

*Textiles—Air permeability**
USE: Air permeability

*Textiles—Antistatic agents**
USE: Antistatic agents

*Textiles—Bleaching**
USE: Bleaching

*Textiles—Carding**
USE: Textile carding

*Textiles—Classing**
USE: Textile classing

*Textiles—Color fastness**
USE: Color fastness

*Textiles—Crimping**
USE: Textile crimping

*Textiles—Dry cleaning**
USE: Dry cleaning

*Textiles—Durable press**
USE: Durable press textiles

*Textiles—Flame resistance**
USE: Flame resistance

*Textiles—Heat resistance**
USE: Heat resistance

*Textiles—Laundering**
USE: Laundering

*Textiles—Metallic materials**
USE: Metallic textiles

*Textiles—Military**
USE: Military textiles

*Textiles—Mixed**
USE: Textile blends

*Textiles—Mothproofing**
USE: Mothproofing

*Textiles—Nonwovens**
USE: Nonwoven fabrics

*Textiles—Numbering systems**
USE: Textile numbering systems

*Textiles—Pilling**
USE: Textile pilling

*Textiles—Pressing**
USE: Textile pressing

*Textiles—Printing**
USE: Textile printing

*Textiles—Processing**
USE: Textile processing

*Textiles—Regain**
USE: Regain

*Textiles—Scouring**
USE: Textile scouring

*Textiles—Shrinkproofing**
USE: Shrinkproofing (textiles)

*Textiles—Sizing**
USE: Sizing (finishing operation)

*Textiles—Soil resistance**
USE: Soil resistance (textiles)

*Textiles—Spinning**
USE: Spinning (fibers)

*Textiles—Stabilization**
USE: Stabilization

*Textiles—Synthetic materials**
USE: Synthetic textile fibers

*Textiles—Waterproofing**
USE: Waterproofing

*Textiles—Weaving**
USE: Weaving

*Textiles—Winding**
USE: Winding

*Textiles—Wrinkle recovery**
USE: Wrinkle recovery

Textures (933)
DT: January 1993
BT: Microstructure
RT: Anisotropy
Crystallography
Mechanical properties
Surface properties

Texturing 819.5
DT: January 1993
UF: Textile finishing—Texturizing*
Texturizing
BT: Textile finishing
NT: Textile crimping

* Former Ei Vocabulary term

Texturizing
USE: Texturing

Th
USE: Thorium

Thallium 549.3
DT: January 1993
UF: Thallium and alloys*
Tl
BT: Heavy metals
RT: Getters
Thallium alloys
Thallium compounds
Thallium metallography

Thallium alloys 549.3
DT: January 1993
UF: Thallium and alloys*
BT: Heavy metal alloys
RT: Thallium
Thallium compounds

*Thallium and alloys**
USE: Thallium OR
Thallium alloys

Thallium compounds (804.1) (804.2)
DT: Predates 1975
BT: Heavy metal compounds
RT: Thallium
Thallium alloys

Thallium metallography 531.2, 549.3
DT: Predates 1975
BT: Metallography
RT: Thallium

Thawing (619.1) (822.2)
DT: January 1993
UF: Food products—Thawing*
Pipelines—Thawing*
BT: Melting

Theaters (legitimate) 402.2 (751.3)
DT: January 1993
UF: Theaters*
BT: Facilities
RT: Auditoriums
Opera houses
Stages

*Theaters**
USE: Theaters (legitimate)

Theorem proving (721.1) (723.4) (921)
DT: January 1993
UF: Automata theory—Theorem proving*
BT: Computation theory
RT: Artificial intelligence
Formal logic
Problem solving

Theory
SN: Very general term; prefer specific type of theory. Also use Treatment Code T
DT: January 1993
NT: Approximation theory
Chaos theory
Circuit theory
Computation theory
Electric machine theory
Electromagnetic field theory
Laser theory
Measurement theory
Number theory
Plasma theory
Quantum theory
Queueing theory
Radar theory
Reliability theory
Set theory
System theory
RT: Engineering
Philosophical aspects
Physics
Research

Thermal analysis
USE: Thermoanalysis

Thermal blooming 741.1.1 (744.8)
DT: January 1993
BT: Nonlinear optics
RT: Focusing
Laser beams

Thermal capacity
USE: Specific heat

Thermal conductivity 641.1 (641.2) (931.2)
DT: Predates 1975
UF: Heat conductivity
Thermal resistivity
BT: Thermodynamic properties
Transport properties
NT: Thermal conductivity of gases
Thermal conductivity of liquids
Thermal conductivity of solids
RT: Heat conduction
Heat transfer
Low temperature testing
Nusselt number
Specific heat
Thermal effects
Thermal insulation
Thermal insulating materials
Thermoelasticity

Thermal conductivity of gases
641.1 (641.2) (931.2)
DT: January 1993
UF: Gases—Thermal conductivity*
BT: Thermal conductivity
RT: Gases

Thermal conductivity of liquids
641.2 (641.2) (931.2)
DT: January 1993
UF: Liquids—Thermal conductivity*
BT: Thermal conductivity
RT: Liquids

* Former Ei Vocabulary term

Thermal conductivity of solids
641.1 (641.2) (931.2) (933)
DT: January 1993
UF: Solids—Thermal conductivity*
BT: Thermal conductivity
RT: Solids

Thermal cycling (421) (537.1)
DT: January 1995
UF: Cyclical heating
Thermocycling
RT: Cooling
Heating
Thermal stress

Thermal decomposition
USE: Pyrolysis

Thermal degradation
USE: Pyrolysis

Thermal diffusion 641.2 (931.2)
DT: Predates 1975
BT: Diffusion
NT: Thermal diffusion in gases
Thermal diffusion in liquids
Thermal diffusion in solids
RT: Heat transfer
Thermal effects
Thermal gradients
Thermal insulation

Thermal diffusion in gases 641.2 (931.2)
DT: January 1993
UF: Thermal diffusion—Gases*
BT: Diffusion in gases
Thermal diffusion
RT: Gases

Thermal diffusion in liquids 641.2 (931.2)
DT: January 1993
UF: Thermal diffusion—Liquids*
BT: Diffusion in liquids
Thermal diffusion
RT: Liquids

Thermal diffusion in solids 641.2 (931.2) (933)
DT: January 1993
UF: Thermal diffusion—Solids*
BT: Diffusion in solids
Thermal diffusion
RT: Solids

*Thermal diffusion—Gases**
USE: Thermal diffusion in gases

*Thermal diffusion—Liquids**
USE: Thermal diffusion in liquids

*Thermal diffusion—Solids**
USE: Thermal diffusion in solids

Thermal effects (421) (531)
DT: Predates 1975
UF: Heat effects
Temperature effects
BT: Effects
NT: High temperature effects
RT: Heat problems
Hot weather problems
Thermal conductivity
Thermal diffusion
Thermal expansion
Thermal noise
Thermal stress
Thermal variables control
Thermal variables measurement
Thermoanalysis
Thermooxidation

Thermal effluents
(452.3) (453.1) (613.2) (614.2) (615)
DT: January 1993
UF: Power plants—Thermal effluents*
BT: Effluents
RT: Power plants
Thermal plumes

Thermal expansion 641.1 (421)
DT: January 1993
BT: Expansion
Thermodynamic properties
RT: Ice
Swelling
Thermal effects

Thermal gradients 641.1
DT: January 1993
UF: Seawater, Thermal gradients*
RT: Temperature distribution
Thermal diffusion
Thermal stratification
Thermodynamic properties

Thermal imaging
USE: Infrared imaging

Thermal insulating materials 413.2
DT: January 1993
UF: Heat insulating materials*
BT: Insulating materials
RT: Asbestos
Foamed products
Thermal conductivity
Thermal insulation

Thermal insulation 413.2
DT: January 1981
UF: Heat insulation
BT: Insulation
RT: Cooling
Heat losses
Heat shielding
Heat sinks
Heat transfer
Heating
Temperature control
Thermal conductivity
Thermal diffusion
Thermal insulating materials

Thermal load (643.1) (643.3)
DT: January 1993
UF: Air conditioning—Thermal load*
Heat load
RT: Air conditioning
Heating

* Former Ei Vocabulary term

Thermal logging 512.1.2
DT: January 1993
UF: Oil well logging—Thermal*
BT: Well logging

Thermal noise (701.1) (706.1) (713)
DT: January 1993
UF: Noise, Spurious signal—Thermal noise*
BT: Spurious signal noise
RT: Thermal effects

Thermal oil recovery 511.1
DT: January 1993
UF: Oil sands—Thermal recovery*
Oil well production—Thermal*
BT: Enhanced recovery
RT: In situ combustion
Oil sands

Thermal plumes (451.1) (453.1)
DT: January 1977
UF: Plumes (thermal)
RT: Air pollution
Thermal effluents
Thermal pollution
Waste heat
Water pollution

Thermal pollution (451.1) (453.1)
DT: January 1993
UF: Water pollution—Waste heat effects*
BT: Pollution
RT: Air pollution
Marine biology
Marine pollution
Thermal plumes
Waste heat
Water pollution

Thermal printing 745.1
DT: January 1993
UF: Printing—Thermal*
Thermographic printing
Thermography*
BT: Printing

Thermal problems
USE: Heat problems

*Thermal properties**
USE: Thermodynamic properties

Thermal radiation
USE: Heat radiation

Thermal resistance
USE: Heat resistance

Thermal resistivity
USE: Thermal conductivity

Thermal shielding
USE: Heat shielding

Thermal springs
USE: Geothermal springs

Thermal stability
USE: Thermodynamic stability

Thermal stratification (443.1) (441.2)
DT: January 1993
UF: Reservoirs—Thermal stratification*
Stratification (thermal)
RT: Atmospheric temperature
Earth atmosphere
Heat transfer
Lakes
Reservoirs (water)
Temperature distribution
Thermal gradients
Thermodynamic properties
Water
Wave power

Thermal stress 931.2 (421)
DT: January 1993
UF: Stresses—Thermal*
BT: Stresses
RT: Stress analysis
Thermal cycling
Thermal effects
Thermodynamic properties
Thermoelasticity

Thermal treatment
USE: Heat treatment

Thermal variables control 731.3
DT: January 1993
UF: Control, Thermal variables*
BT: Control
NT: Temperature control
RT: Heating
Thermal effects
Thermodynamic properties

Thermal variables measurement 944.6
DT: Predates 1975
UF: Heat measurement
BT: Measurements
NT: Temperature measurement
RT: Bolometers
Heating
Infrared detectors
Radiometers
Thermal effects
Thermodynamic properties

Thermionic cathodes 714.1 (712.2)
DT: January 1993
UF: Cathodes, Thermionic*
BT: Cathodes
RT: Thermionic tubes

Thermionic conversion
USE: Thermionic power generation

Thermionic power generation 615.7 (712.2)
DT: Predates 1975
UF: Thermionic conversion
BT: Direct energy conversion
RT: Thermoelectric energy conversion

Thermionic tubes 714.1 (712.2)
DT: January 1993
UF: Electron tubes, Thermionic*

Thermionic tubes *(continued)*
Thermionic valves
BT: Electron tubes
NT: Cathode ray tubes
Electron wave tubes
Thyratrons
X ray tubes
RT: Pentodes
Tetrodes
Thermionic cathodes
Triodes

Thermionic valves
USE: Thermionic tubes

Thermistors 714.2
DT: Predates 1975
UF: Negative temperature coefficient thermistors
BT: Resistors
Semiconductor devices
RT: Thermostats

Thermit welding 538.2.1
DT: January 1993
UF: Aluminothermic welding
Pressure thermit welding
Welding—Thermit*
BT: Welding
RT: Pressure welding

Thermoacoustic effect
USE: Photoacoustic effect

Thermoanalysis 801
DT: Predates 1975
UF: Thermal analysis
BT: Chemical analysis
NT: Differential thermal analysis
Thermogravimetric analysis
RT: Calorimetry
Thermal effects

Thermoclines
USE: Temperature distribution

Thermocouples 944.5
DT: Predates 1975
RT: Temperature measuring instruments
Thermoelectricity
Thermometers
Thermopiles

*Thermocouples—Thermopiles**
USE: Thermopiles

Thermocycling
USE: Thermal cycling

Thermodynamic properties 641.1
DT: January 1993
UF: Activity (thermodynamics)
Thermal properties*
Thermophysical properties
NT: Calorific value
Enthalpy
Entropy
Free energy
Gibbs free energy
Interfacial energy
Pyroelectricity
Specific heat
Thermal conductivity
Thermal expansion
Thermodynamic stability
Vapor pressure
RT: Heating
High temperature testing
Low temperature testing
Temperature
Thermal gradients
Thermal stratification
Thermal stress
Thermal variables control
Thermal variables measurement
Thermodynamics
Thermoluminescence

Thermodynamic stability 641.1
DT: January 1993
UF: Colloids—Thermodynamic stability*
Thermal stability
BT: Stability
Thermodynamic properties

Thermodynamics 641.1
DT: Predates 1975
NT: Atmospheric thermodynamics
Brayton cycle
Carnot cycle
Otto cycle
Rankine cycle
Stirling cycle
RT: Adiabatic engines
Catalysis
Chemical activation
Chemical reactions
Engines
Enthalpy
Equations of state
Gases
Heating
Physical chemistry
Physics
Reaction kinetics
Statistical mechanics
Steam
Thermodynamic properties

*Thermodynamics—Brayton cycle**
USE: Brayton cycle

*Thermodynamics—Carnot cycle**
USE: Carnot cycle

*Thermodynamics—Entropy**
USE: Entropy

*Thermodynamics—Otto cycle**
USE: Otto cycle

*Thermodynamics—Rankine cycle**
USE: Rankine cycle

*Thermodynamics—Stirling cycle**
USE: Stirling cycle

* Former Ei Vocabulary term

Thermoelasticity 931.2 (421)
DT: Predates 1975
BT: Physical properties
RT: Elasticity
Shape memory effect
Thermal conductivity
Thermal stress

Thermoelectric effects
USE: Thermoelectricity

Thermoelectric energy conversion 615.4
DT: Predates 1975
BT: Direct energy conversion
RT: Thermionic power generation
Thermoelectric power plants
Thermoelectric equipment
Thermoelectricity

Thermoelectric equipment 615.4 (704.2)
SN: Also use the type of equipment, e.g., AIR CONDITIONING
DT: January 1993
UF: Air conditioning—Thermoelectric systems*
BT: Electric equipment
RT: Air conditioning
Refrigerating machinery
Thermoelectric power plants
Thermoelectric energy conversion
Thermoelectric refrigeration
Thermoelectricity

Thermoelectric power plants 615.4 (402.1)
DT: January 1993
UF: Electric power plants—Thermoelectric*
BT: Electric power plants
RT: Thermoelectric energy conversion
Thermoelectric equipment

Thermoelectric refrigeration 615.4, 644.3
DT: January 1993
UF: Refrigeration—Thermoelectric*
BT: Refrigeration
RT: Thermoelectric equipment

Thermoelectricity 701.1 (615.4)
DT: Predates 1975
UF: Semiconductor materials—Thermoelectric effects*
Thermoelectric effects
BT: Electricity
RT: Electric properties
Thermocouples
Thermoelectric equipment
Thermoelectric energy conversion
Thermopiles

Thermoforming 816.1
DT: January 1993
UF: Plastics—Thermoforming*
Thermoplastic forming
BT: Plastics forming
RT: Thermoplastics

Thermographic printing
USE: Thermal printing

Thermography (imaging) 742.1
DT: January 1993
UF: Thermography*
BT: Infrared imaging
RT: Biomedical engineering
Diagnosis
Thermography (temperature measurement)

Thermography (temperature measurement) 944.6 (461.1) (462.1)
DT: January 1993
UF: Thermography*
BT: Temperature measurement
RT: Temperature distribution
Thermography (imaging)
Thermoluminescence

*Thermography**
USE: Thermal printing OR
Thermography (imaging) OR
Thermography (temperature measurement)

Thermogravimetric analysis 801
DT: January 1995
UF: Thermogravimetry
BT: Gravimetric analysis
Thermoanalysis

Thermogravimetry
USE: Thermogravimetric analysis

Thermoluminescence 741.1 (931.2)
DT: Predates 1975
BT: Luminescence
RT: Thermodynamic properties
Thermography (temperature measurement)

Thermomechanical pulp 811.1
DT: January 1993
UF: Pulp—Thermomechanical*
BT: Mechanical pulp
RT: Thermomechanical pulping process

Thermomechanical pulping process 811.1
DT: January 1993
UF: Pulp manufacture—Thermomechanical process*
BT: Pulp manufacture
RT: Thermomechanical pulp

Thermomechanical treatment (535.2) (537.1)
DT: January 1993
BT: Forming
Heat treatment

Thermometers 944.5
DT: Predates 1975
BT: Temperature measuring instruments
RT: Bimetals
Pyrometers
Temperature measurement
Thermocouples
Thermopiles

Thermonuclear reactions 621.2 (932.2.1)
DT: January 1993
UF: Nuclear energy—Thermonuclear reactions*

* Former Ei Vocabulary term

Thermonuclear reactions *(continued)*
BT: Fusion reactions
RT: Nuclear explosions

Thermonuclear reactors
USE: Fusion reactors

Thermooxidation 802.2
DT: January 1993
BT: Oxidation
RT: Thermal effects

Thermophysical properties
USE: Thermodynamic properties

Thermopiles 944.5
DT: January 1993
UF: Thermocouples—Thermopiles*
RT: Thermocouples
Thermoelectricity
Thermometers

Thermoplastic elastomers 818.2 (815.1.1)
DT: January 1977
UF: TPE
BT: Elastomers
Thermoplastics
NT: Polyisoprenes
RT: Polycondensation

Thermoplastic forming
USE: Thermoforming

Thermoplastic resins
USE: Thermoplastics

Thermoplastics 815.1.1 (816.1) (817.1)
DT: January 1977
UF: Thermoplastic resins
BT: Organic polymers
NT: Acrylics
Polyamides
Polycarbonates
Polyesters
Polyether ether ketones
Polyetherimides
Polyolefins
Polystyrenes
Polysulfones
Thermoplastic elastomers
Vinyl resins
RT: Thermoforming

Thermosets 815.1.1 (816.1) (817.1)
DT: January 1977
UF: Thermosetting resins
BT: Organic polymers
NT: Alkyd resins
Allyl resins
Furan resins
Melamine formaldehyde resins
Phenolic resins
Urea formaldehyde resins
RT: Silicones
Transfer molding

Thermosetting resins
USE: Thermosets

Thermostats 732.1
DT: Predates 1975
BT: Electric switches
RT: Bimetals
Cryostats
Temperature control
Thermistors

Thermotropic liquid crystals 804
DT: January 1993
UF: Crystals, Liquid—Thermotropic*
BT: Liquid crystals

Thesauri
USE: Vocabulary control

Thick film circuits 714.2
DT: January 1993
UF: Integrated circuits, Thick film*
BT: Integrated circuits
RT: Thick film devices
Thick films

Thick film devices 714.2
DT: January 1993
UF: Solid state devices, Thick film*
BT: Solid state devices
RT: Thick film circuits
Thick films

Thick films (714.2)
DT: January 1993
BT: Films
NT: Magnetic thick films
RT: Metallic films
Substrates
Thick film circuits
Thick film devices

Thickness control 731.3
DT: January 1993
UF: Control, Mechanical variables—Thickness*
Sheet and strip metal—Thickness control*
BT: Spatial variables control
RT: Thickness measurement

Thickness gages 943.3 (423.1)
DT: January 1993
UF: Gages—Thickness measurement*
BT: Gages
RT: Thickness measurement

Thickness measurement 943.2 (423.2) (943.3)
DT: Predates 1975
BT: Spatial variables measurement
RT: Micrometers
Thickness control
Thickness gages

Thin film circuits 714.2
DT: January 1993
UF: Integrated circuits, Thin film*
BT: Integrated circuits
RT: Cryotrons
Flip chip devices
Thin film devices
Thin films

* Former Ei Vocabulary term

Thin film devices 714.2
DT: January 1993
UF: Solid state devices, Thin film*
BT: Solid state devices
NT: Magnetic thin film devices
Microstrip devices
Thin film transistors
RT: Thin film circuits
Thin films

Thin film storage
USE: Magnetic film storage

Thin film transistors 714.2
DT: January 1995
BT: Thin film devices
Transistors

Thin films (714.2)
DT: January 1993
BT: Films
NT: Magnetic thin films
Ultrathin films
RT: Coatings
Dielectric films
Epitaxial growth
Metallic films
Microstrip lines
Optical films
Semiconducting films
Silicon wafers
Sputter deposition
Substrates
Superconducting films
Thin film circuits
Thin film devices
Vapor deposition

Thin layer chromatography 801 (802.3)
DT: January 1993
UF: Chromatographic analysis—Thin layer*
BT: Chromatography

Thoria 804.2 (812.2)
DT: January 1993
UF: Refractory materials—Thoria*
Thorium dioxide
BT: Oxides
Thorium compounds
RT: Refractory metal compounds

Thorium 549.3 (622.1)
DT: January 1993
UF: Th
Thorium and alloys*
BT: Actinides
RT: Monazite
Thorium alloys
Thorium compounds
Thorium deposits
Thorium metallography

Thorium alloys 622.1 (549.3)
DT: January 1993
UF: Thorium and alloys*
BT: Alloys
Radioactive materials
RT: Thorium
Thorium compounds

*Thorium and alloys**
USE: Thorium OR
Thorium alloys

Thorium compounds (804.1) (804.2)
DT: Predates 1975
BT: Metallic compounds
NT: Thoria
RT: Thorium
Thorium alloys

Thorium deposits 504.3, 549.3 (622.1)
DT: Predates 1975
BT: Ore deposits
RT: Monazite deposits
Thorium

Thorium dioxide
USE: Thoria

Thorium metallography 531.2, 549.3
DT: January 1983
BT: Metallography
RT: Thorium

Thorium powder metallurgy 536.1, 549.3 (622.1)
DT: January 1993
UF: Powder metallurgy—Thorium*
BT: Powder metallurgy

Thread (rubber)
USE: Rubber thread

Thread cutting 604.1
DT: January 1993
UF: Screw thread cutting
Screws—Thread cutting*
BT: Machining
NT: Tapping (threads)
RT: Metal cutting
Milling (machining)
Screw threads
Screws
Taps
Thread cutting dies
Thread grinders
Thread rolling
Turning

Thread cutting dies (603) (605)
DT: January 1993
UF: Screw thread cutting dies
Taps and dies*
BT: Cutting tools
Dies
RT: Taps
Thread cutting

Thread cutting taps
USE: Taps

Thread gages
USE: Screw thread gages

* Former Ei Vocabulary term

Thread grinders 603.1
DT: January 1993
UF: Grinding machines—Thread grinding*
Screw thread grinders
Thread grinding machines
BT: Grinding machines
RT: Screw threads
Screws
Thread cutting

Thread grinding machines
USE: Thread grinders

Thread rolling 535.1
DT: January 1993
UF: Screw thread rolling
Screw threads—Rolling*
BT: Rolling
RT: Cold rolling
Machining
Screw threads
Screws
Thread cutting
Turning

Three D graphics
USE: Three dimensional computer graphics

Three dimensional
DT: January 1993
RT: Display devices
Geometry
Holograms
Three dimensional computer graphics

Three dimensional computer graphics 723.5
DT: January 1993
UF: Computer graphics—Three dimensional graphics*
Three D graphics
BT: Computer graphics
RT: Stereo vision
Three dimensional

Three term control systems 731.1
SN: Proportional - Integral - Derivative
DT: January 1993
UF: Control systems, Three term*
BT: Control systems
RT: Proportional control systems
Two term control systems

Threshold elements 721.2
DT: January 1993
UF: Logic devices—Threshold elements*
BT: Logic devices
RT: Threshold logic

Threshold logic 721.1 (721.2) (721.3)
DT: January 1993
UF: Computer metatheory—Threshold logic*
BT: Formal logic
RT: Logic circuits
Threshold elements

Thrust bearings 601.2
DT: January 1993
UF: Bearings—Thrust*
BT: Bearings (machine parts)
RT: Antifriction bearings
Ball bearings
Roller bearings

Thrust reversal (612.3) (617.1) (617.2) (618) (632)
DT: January 1993
UF: Turbomachinery—Thrust reversal*
RT: Braking
Deceleration
Turbomachinery

Thruways
USE: Highway systems

Thulium 547.2
DT: January 1981
UF: Tm
BT: Rare earth elements
RT: Thulium alloys
Thulium compounds

Thulium alloys 547.2
DT: January 1993
BT: Rare earth alloys
RT: Thulium
Thulium compounds

Thulium compounds (804.1) (804.2)
DT: January 1993
BT: Rare earth compounds
RT: Thulium
Thulium alloys

Thunderstorms 443.1
DT: January 1993
UF: Meteorology—Thunderstorms*
BT: Storms
RT: Atmospheric electricity
Clouds
Hurricanes
Lightning
Meteorology
Rain
Tornadoes

Thyratrons 714.1
DT: January 1993
UF: Electron tubes, Thyratron*
BT: Gas discharge tubes
Thermionic tubes
RT: Electric rectifiers
Thyristors

Thyristors 714.2
DT: Predates 1975
UF: Gate turn-off devices
GTO devices
SCR
Silicon controlled rectifiers
Triacs
BT: Semiconductor devices
RT: Electric rectifiers
Semiconductor switches
Thyratrons

Ti
USE: Titanium

Ticket issuing machines 601.1 (911.4)
DT: Predates 1975
UF: Ticket machines
BT: Vending machines

Ticket machines
USE: Ticket issuing machines

Tidal power 615.6 (611.2)
DT: Predates 1975
BT: Water power
RT: Power generation
Tidal power plants
Tides
Wave power

Tidal power plants 615.6 (402.1)
DT: Predates 1975
BT: Hydroelectric power plants
RT: Tidal power
Tides

Tidal waves
USE: Tsunamis

Tide gages 471.2, 943.3 (471.4)
DT: Predates 1975
BT: Gages
RT: Tides

Tides 471.4
SN: Hydrospheric. Excludes atmospheric tides, for which use ATMOSPHERIC MOVEMENTS
DT: Predates 1975
RT: Floods
Ocean currents
Oceanography
Tidal power
Tidal power plants
Tide gages
Water waves

Ties (railroad)
USE: Railroad ties

Tightening (bolts)
USE: Bolt tightening

Tile (414) (415) (812.1)
DT: Predates 1975
BT: Building materials
RT: Ceramic materials
Ceramic products

Time and motion study 461.4, 912.2
DT: Predates 1975
UF: Industrial plants—Time and motion study*
Mines and mining—Time study*
Time study
BT: Management
RT: Human engineering
Industrial engineering
Industrial plants
Job analysis
Time measurement
Work simplification

Time delay circuits
USE: Delay circuits

Time division multiplexing (716) (717) (718)
DT: January 1993
UF: Multiplexing, Time division*
TDM
BT: Multiplexing
NT: Asynchronous transfer mode
Packet switching
RT: Pulse modulation
Telemetering

Time domain analysis 921 (703.1.1) (731.1)
DT: January 1993
UF: Mathematical techniques—Time domain analysis*
BT: Mathematical techniques
RT: Control system analysis
Electric network analysis

Time domain networks
USE: Time varying networks

Time lapse photography
USE: Special effects

Time measurement 943.3
DT: Predates 1975
BT: Measurements
RT: Atomic clocks
Chronometers
Clocks
Electronic timing devices
Frequency meters
Masers
Oscillographs
Pendulums
Stop watches
Streak cameras
Synchronization
Time and motion study
Timing circuits
Timing devices
Units of measurement
Watches

Time series analysis 922.2
DT: January 1993
UF: Statistical methods—Time series analysis*
BT: Statistical methods

Time sharing programs 723.1
DT: January 1993
UF: Computer systems programming—Time sharing programs*
BT: Supervisory and executive programs
RT: Multiprogramming
Time sharing systems

Time sharing systems 722.4
DT: January 1993
UF: Computer systems, Digital—Time sharing*
BT: Computer systems
RT: Multiprogramming
Time sharing programs

* Former Ei Vocabulary term

Time study
USE: Time and motion study

Time switches (704.1) (715)
DT: January 1993
UF: Electric switches, Time*
BT: Electric switches

Time varying circuits
USE: Time varying networks

Time varying control systems 731.1
DT: January 1993
UF: Control systems, Time varying*
BT: Control systems
Time varying systems

Time varying networks 703.1
DT: January 1993
UF: Electric networks, Time varying*
Time domain networks
Time varying circuits
BT: Networks (circuits)
RT: Circuit theory
Parametric amplifiers
Time varying systems
Varactors

Time varying systems 731.1
DT: January 1993
UF: Periodic systems
Systems science and cybernetics—Time varying systems*
BT: Systems science
NT: Time varying control systems
RT: Time varying networks

Time-reversal reflection
USE: Optical phase conjugation

Timing circuits 713.4
DT: January 1993
UF: Electronic circuits, Timing*
BT: Networks (circuits)
RT: Computer circuits
Electric clocks
Time measurement
Timing devices

Timing devices 943.3
DT: Predates 1975
BT: Instruments
NT: Clocks
Electronic timing devices
Watches
RT: Time measurement
Timing circuits

*Timing devices—Electronic**
USE: Electronic timing devices

Tin 546.2
DT: January 1993
UF: Cast iron—Tin content*
Sn
Tin and alloys*
BT: Heavy metals
RT: Tin alloys
Tin compounds
Tin deposits
Tin foil
Tin metallography
Tin metallurgy
Tin mines
Tin nickel plating
Tin ore treatment
Tin plate
Tin refining
Tinning

Tin alloys 546.2
DT: January 1993
UF: Tin and alloys*
BT: Heavy metal alloys
NT: Bronze
Pewter
RT: Soldering alloys
Tin
Tin compounds

*Tin and alloys**
USE: Tin OR
Tin alloys

Tin compounds (804.1) (804.2)
DT: Predates 1975
BT: Heavy metal compounds
NT: Semiconducting tin compounds
RT: Tin
Tin alloys

Tin deposits 504.3, 546.2
DT: Predates 1975
BT: Ore deposits
RT: Tin
Tin mines

Tin foil 535.1, 546.2
DT: Predates 1975
BT: Metal foil
RT: Tin

Tin metallography 531.2, 546.2
DT: Predates 1975
BT: Metallography
RT: Tin
Tin metallurgy

Tin metallurgy 531.1, 546.2
DT: Predates 1975
BT: Metallurgy
NT: Tin powder metallurgy
RT: Tin
Tin metallography
Tin ore treatment
Tin smelting

Tin mines 504.3, 546.2
DT: January 1993
UF: Tin mines and mining*
BT: Mines
RT: Tin
Tin deposits

*Tin mines and mining**
USE: Tin mines

Tin nickel plating 539.3, 546.2, 548.2
DT: Predates 1975
BT: Nickel plating
Tinning
RT: Nickel
Tin
Tin plate

Tin ore treatment 533.1, 546.2
DT: Predates 1975
BT: Ore treatment
NT: Tin smelting
RT: Tin
Tin metallurgy

Tin plate 546.2, 545.3 (539.3)
SN: Thin sheet, of iron or steel, tincoated by dipping or electrolysis. For coating with tin, use TINNING
DT: Predates 1975
UF: Welding—Tin plate*
BT: Sheet metal
RT: Tin
Tin nickel plating
Tinning

Tin plating
USE: Tinning

Tin powder metallurgy 536.1, 546.2
DT: January 1993
UF: Powder metallurgy—Copper tin*
Powder metallurgy—Tin*
BT: Powder metallurgy
Tin metallurgy

Tin refining 533.2, 546.2
DT: Predates 1975
BT: Metal refining
RT: Tin

Tin smelting 533.2, 546.2
DT: Predates 1975
BT: Smelting
Tin ore treatment
RT: Tin metallurgy

Tinning 539.3, 546.2
SN: Coating with tin, whether by hot-dipping or by electroplating
DT: Predates 1975
UF: Tin plating
BT: Plating
NT: Tin nickel plating
RT: Soldering
Tin
Tin plate

Tire cords 819.4 (818.5)
DT: January 1993
UF: Cords (tire)
Tires—Cords*
BT: Tires
Yarn

Tire industry 818
DT: January 1977
BT: Industry
RT: Rubber industry
Tires

Tires 818.5
DT: Predates 1975
UF: Automobile tires
Rubber tires
BT: Rubber products
Vehicle parts and equipment
NT: Radial tires
Tire cords
Tubeless tires
RT: Molds
Retreading
Rolling resistance
Skid resistance
Tire industry
Traction (friction)
Wheels

*Tires—Cords**
USE: Tire cords

*Tires—Molds**
USE: Molds

*Tires—Noise**
USE: Acoustic noise

*Tires—Radial**
USE: Radial tires

*Tires—Retreading**
USE: Retreading

*Tires—Rolling resistance**
USE: Rolling resistance

*Tires—Skid resistance**
USE: Skid resistance

*Tires—Traction**
USE: Traction (friction)

*Tires—Tubeless**
USE: Tubeless tires

Tissue 461.2
DT: January 1993
UF: Biological materials—Tissue*
BT: Biological materials
NT: Bone
Cartilage
Chitin
Ligaments
Muscle
Tendons
RT: Tissue culture

Tissue culture (461.2) (801.2)
DT: January 1993
UF: Cell culture—Tissue*
BT: Cell culture
RT: Tissue

Tissue transplantation
USE: Transplantation (surgical)

Titanate minerals 482.2
DT: January 1993
UF: Mineralogy—Titanates*
BT: Minerals
Titanium compounds
RT: Titanium

Titanium 542.3
DT: January 1993
UF: Cast iron—Titanium content*
Ti
Titanium and alloys*
BT: Light metals
Transition metals
RT: Titanate minerals
Titanium alloys
Titanium castings
Titanium compounds
Titanium deposits
Titanium foundry practice
Titanium metallography
Titanium metallurgy
Titanium mines
Titanium ore treatment
Titanium plating
Titanium sheet

Titanium alloys 542.3
DT: January 1993
UF: Titanium and alloys*
BT: Transition metal alloys
RT: Titanium
Titanium compounds

*Titanium and alloys**
USE: Titanium OR
Titanium alloys

Titanium carbide 804.2 (812.2)
DT: Predates 1975
BT: Carbides
Titanium compounds
RT: Ceramic materials
Ceramic matrix composites

Titanium castings 534.2, 542.3
DT: January 1987
BT: Metal castings
RT: Titanium

Titanium compounds (804.1) (804.2)
DT: Predates 1975
BT: Transition metal compounds
NT: Barium titanate
Titanate minerals
Titanium carbide
Titanium nitride
Titanium oxides
RT: Titanium
Titanium alloys

Titanium deposits 504.1, 542.3
DT: Predates 1975
BT: Ore deposits
RT: Titanium
Titanium mines

Titanium dioxide 804.2
DT: January 1993
BT: Titanium oxides
RT: Pigments

Titanium foundry practice 534.2, 542.3
DT: Predates 1975
BT: Foundry practice
RT: Titanium

Titanium metallography 531.2, 542.3
DT: Predates 1975
BT: Metallography
RT: Titanium
Titanium metallurgy

Titanium metallurgy 531.1, 542.3
DT: Predates 1975
BT: Metallurgy
NT: Titanium powder metallurgy
RT: Titanium
Titanium metallography
Titanium ore treatment

Titanium mines 504.1, 542.3
DT: January 1993
UF: Titanium mines and mining*
BT: Mines
RT: Titanium
Titanium deposits

*Titanium mines and mining**
USE: Titanium mines

Titanium nitride 804.2 (812.2)
DT: January 1995
BT: Nitrides
Titanium compounds
RT: Cermets
Refractory materials

Titanium ore treatment 533.1, 542.3
DT: Predates 1975
BT: Ore treatment
NT: Ilmenite ore treatment
RT: Titanium
Titanium metallurgy

*Titanium oxide**
USE: Titanium oxides

Titanium oxides 804.2
DT: Predates 1975
UF: Powder metallurgy—Titanium oxide*
Titanium oxide*
BT: Oxides
Titanium compounds
NT: Perovskite
Titanium dioxide

Titanium plating 539.3, 542.3
DT: Predates 1975
BT: Plating
RT: Titanium

Titanium powder metallurgy 536.3, 542.3
DT: January 1993
UF: Powder metallurgy—Titanium oxide*
Powder metallurgy—Titanium*

* Former Ei Vocabulary term

Titanium powder metallurgy *(continued)*
BT: Powder metallurgy
Titanium metallurgy

Titanium sheet 535.1, 542.3
DT: Predates 1975
BT: Sheet metal
RT: Titanium

Titration 801
DT: January 1993
UF: Chemical analysis—Titration*
BT: Volumetric analysis

Tl
USE: Thallium

Tm
USE: Thulium

Tobacco 821.4
DT: Predates 1975
BT: Agricultural products
RT: Cigarette manufacture

*Tobacco—Grading**
USE: Grading

Tokamak devices 621.2, 932.3
DT: January 1977
BT: Fusion reactors
Plasma devices
RT: Plasma confinement

Tolerances
USE: Fits and tolerances

Toll bridges 401.1 (406.1) (911.2)
DT: January 1993
BT: Bridges
RT: Fees and charges
Toll highways

Toll highways 406.1 (911.2)
DT: January 1993
UF: Highway systems—Toll systems*
Toll roads
BT: Highway systems
RT: Fees and charges
Toll bridges

Toll roads
USE: Toll highways

Toluene 804.1
DT: Predates 1975
BT: Aromatic hydrocarbons
RT: Solvents

Tool steel 545.3
DT: Predates 1975
UF: Drill steel
BT: Steel

Tools (603) (605)
DT: January 1993
UF: Industrial plants—Tools, jigs and fixtures*
Tools, jigs and fixtures*
BT: Equipment
NT: Bending tools
Brushes
Carbide tools
Ceramic tools
Chucks
Clamping devices
Cutting tools
Electric tools
Explosive actuated tools
Hand tools
Hydraulic tools
Machine tools
Pneumatic tools
Woodworking tools
RT: Fixtures (tooling)
Hardware
Indexing (materials working)
Mechanization
Presses (machine tools)
Shovels

*Tools, Hand**
USE: Hand tools

*Tools, jigs and fixtures**
USE: Tools OR
Jigs OR
Fixtures (tooling)

*Tools, jigs and fixtures—Bending**
USE: Bending tools

*Tools, jigs and fixtures—Carbide**
USE: Carbide tools

*Tools, jigs and fixtures—Ceramic**
USE: Ceramic tools

*Tools, jigs and fixtures—Electric**
USE: Electric tools

*Tools, jigs and fixtures—Explosive**
USE: Explosive actuated tools

*Tools, jigs and fixtures—Hydraulic**
USE: Hydraulic tools

*Tools, jigs and fixtures—Indexing**
USE: Indexing (materials working)

*Tools, jigs and fixtures—Magnesium**
USE: Magnesium

*Tools, jigs and fixtures—Plastics**
USE: Plastics applications

*Tools, jigs and fixtures—Pneumatic**
USE: Pneumatic tools

*Tools, jigs and fixtures—Rubber**
USE: Rubber products

*Tools, jigs and fixtures—Welding**
USE: Welding

Tooth enamel 461.2
DT: January 1993
UF: Biological materials—Tooth enamel*
BT: Biological materials
RT: Dentistry

* Former Ei Vocabulary term

Topaz 482.2.1
DT: Predates 1975
BT: Gems
Silicate minerals
RT: Industrial gems

Topology 921.4 (703.1)
DT: January 1993
UF: Mathematical techniques—Topology*
BT: Mathematical techniques
NT: Electric network topology
Graph theory
RT: Combinatorial mathematics
Distance measurement
Electric network analysis
Electric network synthesis
Fractals
Switching theory
Trees (mathematics)
Vectors

Topping (hot)
USE: Hot topping

Topping cycle systems 614
DT: January 1993
UF: Cogeneration plants—Topping systems*
BT: Cogeneration plants

Torches (plasma)
USE: Plasma torches

Tornado generated missiles 443.1, 914.1
SN: High-speed flying objects set in motion by tornadoes
DT: January 1993
UF: Nuclear power plants—Tornado generated missiles*
RT: Accidents
Nuclear power plants
Nuclear reactor accidents
Tornadoes

Tornadoes 443.1
DT: January 1981
BT: Storms
RT: Meteorology
Thunderstorms
Tornado generated missiles
Turbulence

Torpedoes 404.1, 654.1
DT: Predates 1975
BT: Underwater launched missiles
RT: Launching
Underwater ballistics

*Torpedoes—Launching**
USE: Launching

Torque 931.2 (421)
DT: January 1993
RT: Mechanical properties
Torque control
Torque converters
Torque measurement

Torque control 731.3
DT: January 1993
UF: Control, Mechanical variables—Torques*
BT: Mechanical variables control
RT: Torque
Torque converters
Torque measurement

Torque converters 602.1 (662.4)
DT: Predates 1975
UF: Converters (torque)
BT: Drives
NT: Hydraulic torque converters
Mechanical torque converters
RT: Nonelectric final control devices
Power converters
Torque
Torque control
Transmissions
Vehicle transmissions

*Torque converters, Hydraulic**
USE: Hydraulic torque converters

*Torque converters, Mechanical**
USE: Mechanical torque converters

Torque converting turbomachinery
USE: Hydraulic torque converters

Torque measurement 943.2
DT: January 1993
UF: Mechanical variables measurement—Torques*
BT: Mechanical variables measurement
RT: Dynamometers
Torque
Torque control
Torque meters

Torque meters 943.1
DT: Predates 1975
UF: Torquemeters
BT: Instruments
RT: Dynamometers
Torque measurement

Torquemeters
USE: Torque meters

Torsion
USE: Torsional stress

Torsion loading
USE: Torsional stress

Torsion meters 943.1
DT: Predates 1975
RT: Torsion testing

Torsion testing (422.2)
DT: January 1993
UF: Materials testing—Torsion tests*
BT: Mechanical testing
RT: Elastic moduli
Tensile testing
Torsion meters
Torsional stress
Yield stress

Torsional stress (421)
DT: January 1993
UF: Stresses—Torsional*
Torsion
Torsion loading
BT: Stresses
RT: Bending (deformation)
Buckling
Deformation
Shear stress
Strain
Stress analysis
Torsion testing

Torsional waves
USE: Elastic waves

Total binding energy
USE: Binding energy

Total heat
USE: Enthalpy

Total quality management 913.3 (912.2) (922.2)
DT: January 1995
UF: TQM
BT: Management
RT: Quality control
Statistical methods

Touch tone telephone systems 718.1
DT: January 1993
UF: Telephone—Touch tone system*
BT: Push button telephone systems
RT: Electric signal systems
Telephone

Toughening (421) (537.1)
DT: January 1993
RT: Strengthening (metal)
Toughness

Toughness (421) (422)
DT: January 1995
BT: Mechanical properties
NT: Fracture toughness
RT: Brittleness
Toughening

Towboats
USE: Tugboats

Tower cranes 693.1
DT: January 1993
UF: Cranes—Tower*
BT: Cranes
RT: Towers

Towers 402.4
SN: Structures
DT: Predates 1975
BT: Structures (built objects)
NT: Coke quenching towers
Control towers
Cooling towers
Electric towers
Radar towers
Radio towers
Television towers
Water towers
RT: Poles
Tower cranes

Towing tanks
USE: Ship model tanks

Toxic materials
USE: Hazardous materials

Toxic wastes
USE: Hazardous materials

Toxicity 461.7 (454.2) (804) (914.1)
DT: January 1993
RT: Accidents
Contamination
Health hazards
Industrial poisons
Pollution

Toy manufacture (601.1) (913.4)
DT: Predates 1975
BT: Manufacture

*Toy manufacture—Plastics**
USE: Plastics applications

TPD
USE: Temperature programmed desorption

TPE
USE: Thermoplastic elastomers

TQM
USE: Total quality management

Trace analysis 801
DT: January 1981
BT: Chemical analysis

Trace elements (481.2) (801) (804)
DT: January 1993
UF: Geochemistry—Trace elements*
BT: Chemical elements

Tracers (radioactive)
USE: Radioactive tracers

Track laying
USE: Rail laying

Track switches (railroad)
USE: Railroad track switches

Track test cars 681.1, 682.1.1
SN: Cars used to test railroad tracks
DT: January 1993
UF: Cars—Track testing*
BT: Materials testing apparatus
Railroad cars
RT: Equipment testing
Railroad tracks

Tracking (beam)
USE: Particle beam tracking

Tracking (circuit)
USE: Electronic circuit tracking

Tracking (insulation)
USE: Surface discharges

Tracking (position) (654.1) (656.1) (716.2)
DT: January 1993
UF: Aerospace vehicle tracking*
Position tracking
Rockets and missiles—Tracking*
Satellites—Tracking*
Vehicle tracking*
NT: Radar tracking
RT: Aerospace ground support
Aerospace vehicles
Missiles
Navigation
Optical radar
Position control
Position measurement
Radar
Radio astronomy
Rockets
Satellites
Sonar
Tracking radar
Vehicle locating systems

Tracking radar 716.2
DT: January 1993
UF: Radar—Tracking*
BT: Radar
RT: Tracking (position)

Trackless trolleys 663.1 (403.1) (432.2)
DT: Predates 1975
UF: Trolleys (trackless)
BT: Electric vehicles
RT: Buses
Light rail transit

Tracks (railroad)
USE: Railroad tracks

Traction (electric)
USE: Electric traction

Traction (friction) 931.1 (682.1.1) (682.1.2)
DT: January 1993
UF: Cars—Traction*
Locomotives—Traction*
Tires—Traction*
BT: Friction
RT: Locomotives
Rolling resistance
Sanding equipment
Tires
Vehicles

Traction motors (612.2) (705.3)
DT: January 1993
UF: Diesel engines—Traction applications*
Electric motors—Traction*
BT: Electric motors
RT: Electric railroads
Electric traction

Tractors (agricultural) 663.1, 821.1
DT: January 1993
UF: Tractors—Agricultural*
BT: Agricultural machinery
Ground vehicles
RT: Agricultural implements

Tractors (truck) 663.1
DT: January 1993
UF: Motor trucks—Tractors*
BT: Trucks

*Tractors—Agricultural**
USE: Tractors (agricultural)

*Tractors—Antarctic expedition**
USE: Arctic vehicles

*Tractors—Gears**
USE: Gears

*Tractors—Implements**
USE: Agricultural implements

*Tractors—Power transmission**
USE: Power transmission

*Tractors—Soil factors**
USE: Soils

*Trade marks**
USE: Trademarks

Trade names
USE: Trademarks

Trademarks 902.3 (911.2)
DT: January 1993
UF: Plastics—Trademarks*
Trade marks*
Trade names
BT: Intellectual property

Traffic (telecommunication)
USE: Telecommunication traffic

Traffic accidents
USE: Highway accidents

Traffic control (431.5) (432.4) (433.4)
SN: Vehicular traffic
DT: January 1993
UF: Transportation—Traffic control*
BT: Control
NT: Air traffic control
Highway traffic control
Railroad traffic control
Street traffic control
RT: Transportation
Vehicle locating systems

Traffic signals 432.4 (406)
DT: January 1993
UF: Highway signals
Highway signs, signals and markings*
Traffic signs, signals and markings*
BT: Electric signal systems
NT: Vehicle actuated signals
RT: Highway markings
Highway systems
Highway traffic control
Roads and streets

Traffic signals *(continued)*
Street traffic control
Traffic signs

Traffic signs 432.4 (406)
DT: January 1993
UF: Highway signs
Highway signs, signals and markings*
Traffic signs, signals and markings*
BT: Signs
RT: Highway systems
Highway traffic control
Roads and streets
Street traffic control
Traffic signals

*Traffic signs, signals and markings**
USE: Traffic signals OR
Highway markings OR
Traffic signs OR
Road and street markings

*Traffic signs, signals and markings—Striping machines**
USE: Striping machines

*Traffic signs, signals and markings—Vehicle actuation**
USE: Vehicle actuated signals

Traffic surveys 432.4 (406)
SN: Vehicular traffic
DT: Predates 1975
BT: Surveys
RT: Highway systems
Highway traffic control
Motor transportation
Recording
Roads and streets
Street traffic control
Urban planning

Trafficability 432.4 (406) (443.3) (483.1)
DT: January 1993
UF: Snow and snowfall—Trafficability*
Soils—Trafficability*
RT: Snow
Soils

*Trailers**
USE: Light trailers

*Trailers—Motor truck**
USE: Truck trailers

Train ferries 433.1, 674.1 (681.1) (682)
DT: Predates 1975
BT: Ferry boats
RT: Railroads

Training
USE: Personnel training

Training aircraft 652.1 (912.4)
DT: January 1993
UF: Aircraft, Training*
BT: Aircraft
RT: Aviators
Personnel training

Trajectories (404.1) (654.1) (655.2)
DT: January 1993
UF: Rockets and missiles—Orbits and trajectories*
Satellites—Orbits and trajectories*
NT: Orbits
RT: Astrophysics
Interplanetary flight
Lunar missions
Missiles
Navigation
Orbital transfer
Rockets
Satellites
Space flight
Space rendezvous
Spacecraft

Transceivers 716.3
DT: June 1990
UF: Transmitter receivers
BT: Radio receivers
Radio transmitters

Transconductance 701.1 (714.1)
DT: January 1995
RT: Electron tubes

Transducers (704) (715) (732.2) (752.1)
SN: Simple devices which, for energy input, supply related-energy output of the same or a different kind
DT: Predates 1975
BT: Equipment
NT: Acoustic transducers
Dilatometers
Hall effect transducers
Piezoelectric transducers
Pressure transducers
RT: Actuators
Control equipment
Direct energy conversion
Force measurement
Magnetostrictive devices
Mechanical testing
Photoelectric cells
Recording instruments
Sensors
Strain gages
Telecontrol equipment
Telemetering equipment

Transductors
USE: Saturable core reactors

Transfer cases (vehicles) (691.1) (694.1)
DT: January 1993
UF: Vehicle transfer cases
Vehicles—Transfer cases*
BT: Containers

Transfer dies 603.2
DT: January 1993
UF: Dies—Transfer*
BT: Dies

* Former Ei Vocabulary term

Transfer functions 921 (703.1) (731.1) (741.1)
DT: January 1993
UF: Mathematical techniques—Transfer functions*
BT: Functions
Mathematical techniques
NT: Optical transfer function
RT: Control system analysis
Control system synthesis
Damping
Electric network analysis
Frequency response
Poles and zeros
Sensitivity analysis

Transfer molding 816.1
DT: January 1993
UF: Plastics—Transfer molding*
BT: Plastics molding
RT: Thermosets

Transfer orbits
USE: Orbital transfer

Transfer stations 452
SN: For refuse disposal
DT: January 1993
UF: Refuse disposal—Transfer stations*
BT: Facilities
RT: Land fill
Refuse disposal

Transfluxors
USE: Magnetic cores

Transformations (mathematical)
USE: Mathematical transformations

Transformer magnetic circuits 704.1 (708.4)
DT: January 1993
UF: Electric transformers—Magnetic circuit*
BT: Electric transformers
Magnetic circuits

Transformer oil
USE: Insulating oil

Transformer protection 704.1, 914.1
DT: January 1993
BT: Electric equipment protection
RT: Electric transformers

Transformer substations 706.1.2 (402.1) (704.1)
DT: January 1993
UF: Electric substations, Transformer*
BT: Electric substations
RT: Electric transformers

Transformer windings 704.1
DT: January 1993
UF: Electric windings, Transformer*
BT: Electric transformers
Electric windings

Transformers (electric)
USE: Electric transformers

Transforms
USE: Mathematical transformations

Transient capacitance spectroscopy
USE: Deep level transient spectroscopy

Transients (621) (703.1) (731.1)
DT: January 1993
UF: Nuclear reactors—Power transients*
RT: Electric fault currents
Electric surges
Nuclear reactors

Transistor transistor logic circuits 721.2, 721.3 (714.2)
DT: January 1993
UF: Logic circuits, Transistor transistor*
TTL
BT: Logic circuits
RT: Digital integrated circuits

Transistors 714.2
DT: Predates 1975
BT: Semiconductor devices
NT: Bipolar transistors
Field effect transistors
Phototransistors
Thin film transistors
RT: Spacistors

*Transistors, Bipolar**
USE: Bipolar transistors

*Transistors, Field effect**
USE: Field effect transistors

*Transistors, Field effect—Chemically sensitive**
USE: Chemically sensitive field effect transistors

*Transistors, Field effect—Gates**
USE: Gates (transistor)

*Transistors, Field effect—Ion sensitive**
USE: Ion sensitive field effect transistors

*Transistors, Field effect—Junction gate**
USE: Junction gate field effect transistors

*Transistors, High electron mobility**
USE: High electron mobility transistors

*Transistors, Photosensitive**
USE: Phototransistors

Transit time devices 714.2
DT: January 1993
UF: Semiconductor devices, Transit time*
BT: Semiconductor devices
NT: IMPATT diodes
RT: Avalanche diodes

Transition alloys
USE: Transition metal alloys

Transition flow 631.1
DT: January 1993
UF: Flow of fluids—Transition flow*
BT: Flow of fluids
RT: Laminar flow
Turbulence
Turbulent flow

* Former Ei Vocabulary term

Transition metal alloys 531
DT: January 1993
UF: Transition alloys
BT: Alloys
NT: Cadmium alloys
Cobalt alloys
Copper alloys
Hafnium alloys
Iron alloys
Manganese alloys
Mercury amalgams
Nickel alloys
Precious metal alloys
Refractory alloys
Scandium alloys
Titanium alloys
Yttrium alloys
Zinc alloys
Zirconium alloys

Transition metal compounds (804.1) (804.2)
DT: January 1993
BT: Metallic compounds
NT: Cadmium compounds
Cobalt compounds
Copper compounds
Hafnium compounds
Iron compounds
Manganese compounds
Mercury compounds
Nickel compounds
Precious metal compounds
Refractory metal compounds
Scandium compounds
Titanium compounds
Yttrium compounds
Zinc compounds
Zirconium compounds

Transition metals 531
DT: January 1986
BT: Metals
NT: Cadmium
Cobalt
Copper
Hafnium
Iron
Manganese
Mercury (metal)
Nickel
Precious metals
Refractory metals
Scandium
Technetium
Titanium
Yttrium
Zinc
Zirconium

Transition temperature
USE: Superconducting transition temperature

Transitions (electron)
USE: Electron transitions

*Translating machines**
USE: Computer aided language translation

Translation (languages) 903.1 (723.5)
DT: January 1993
UF: Information science—Language translation and linguistics*
BT: Information analysis
NT: Computer aided language translation

Translators (computer program)
USE: Program translators

Transmission (carrier)
USE: Carrier communication

Transmission (diplex)
USE: Diplex transmission

Transmission (power)
USE: Power transmission

Transmission (waves)
USE: Wave transmission

Transmission electron microscopy (741.3)
DT: January 1993
UF: Microscopic examination—Transmission electron microscopy*
TEM
BT: Electron microscopy
RT: Atomic force microscopy
Electron microscopes

Transmission equipment (for data)
USE: Data communication equipment

Transmission line theory 703.1 (706.1.1)
DT: January 1993
UF: Electric networks—Transmission line theory*
BT: Circuit theory
RT: Distributed parameter networks
Electric lines

Transmission lines
USE: Electric lines

Transmission network calculations 703.1 (706.1.1) (921)
DT: January 1993
UF: Electric power transmission—Network calculations*
BT: Calculations
RT: Electric load flow
Electric network analyzers
Electric network analysis
Electric power transmission networks

Transmission networks
USE: Electric power transmission networks

Transmissions 602.2
SN: Gearing systems by which power is transmitted from the engines to the moving parts of vehicles or machines.
DT: January 1993
UF: Agricultural machinery—Transmissions*
Construction equipment—Transmissions*
BT: Mechanical drives
NT: Variable speed transmissions

Transmissions *(continued)*
Vehicle transmissions
RT: Gears
Power transmission
Sprockets
Torque converters

Transmitter receivers
USE: Transceivers

Transmitters (716)
DT: Predates 1975
BT: Telecommunication equipment
NT: Radio transmitters
Television transmitters
Transponders
RT: Acoustic generators
Antennas
Electric exciters
Electromagnetic field measurement
Gain control
Microphones
Modulation
Modulators
Networks (circuits)
Oscillators (electronic)
Signal receivers
Telecommunication repeaters
Teleprinters

Transonic aerodynamics 651.1
DT: January 1993
UF: Aerodynamics—Transonic*
BT: Aerodynamics
RT: Supersonic aerodynamics
Transonic flow
Wind tunnels

Transonic aircraft
USE: Supersonic aircraft

Transonic flow 631.1
DT: January 1993
UF: Flow of fluids—Transonic*
Wind tunnels—Transonic flow*
BT: Flow of fluids
RT: Aerodynamics
Shock waves
Subsonic flow
Supersonic flow
Transonic aerodynamics
Wind tunnels

Transparency 741.1
SN: Of materials
DT: January 1993
UF: Plastics—Transparency*
BT: Optical properties
RT: Electromagnetic wave absorption
Opacity
Turbidimeters

Transplantation (surgical) 462.4 (461.6)
DT: January 1993
UF: Organ transplantation
Prosthetics—Transplantation*
Tissue transplantation
BT: Surgery
RT: Biomedical engineering
Graft vs. host reactions
Implants (surgical)
Transplants

Transplants 462.4
DT: January 1993
NT: Grafts
RT: Graft vs. host reactions
Prosthetics
Transplantation (surgical)

Transponders (716) (752.1)
DT: January 1986
BT: Signal receivers
Transmitters
RT: Amplifiers (electronic)
Radio navigation
Satellite links

Transport aircraft 652.1 (431.3)
DT: January 1993
UF: Aircraft, Transport*
BT: Aircraft
RT: Cargo aircraft

Transport properties 931.2
DT: January 1993
BT: Physical properties
NT: Thermal conductivity
Viscosity
RT: Diffusion
Electric properties
Kinetic theory
Mass transfer

Transport properties (electron)
USE: Electron transport properties

Transportation (431) (432) (433) (434)
DT: Predates 1975
NT: Air transportation
Airport passenger transportation
Airport vehicular traffic
Freight transportation
Mass transportation
Mine transportation
Motor transportation
Natural gas transportation
Nonmotorized transportation
Petroleum transportation
Radioactive waste transportation
Railroad transportation
Rapid transit
Space flight
Waterway transportation
RT: Bridges
Communication systems
Loading
Magnetic levitation vehicles
Materials handling
Packaging
Regional planning
Reservation systems
Salvaging

Transportation *(continued)*
Traffic control
Transportation routes
Transportation charges
Transportation personnel
Urban planning
Vehicles

Transportation charges (431) (432) (433) (434) (911.1)
DT: January 1993
UF: Fares (transportation)
Transportation—Rates*
BT: Fees and charges
RT: Transportation

Transportation personnel 912.4 (431) (432) (433) (434)
DT: January 1993
UF: Transportation—Personnel*
BT: Personnel
NT: Automobile drivers
Aviators
Bus drivers
Truck drivers
RT: Transportation

Transportation routes (431) (432) (433) (434)
DT: January 1993
UF: Transportation—Route analysis*
RT: Air transportation
Highway systems
Inland waterways
Motor transportation
Roads and streets
Transportation

*Transportation—Accounting**
USE: Cost accounting

*Transportation—Freight**
USE: Freight transportation

*Transportation—Guideways**
USE: Guideways

*Transportation—Noise**
USE: Acoustic noise

*Transportation—Nonmotorized**
USE: Nonmotorized transportation

*Transportation—Personnel**
USE: Transportation personnel

*Transportation—Rates**
USE: Transportation charges

*Transportation—Reservation systems**
USE: Reservation systems

*Transportation—Route analysis**
USE: Analysis AND
Transportation routes

*Transportation—Traffic control**
USE: Traffic control

Transputers (714.2) (721) (722)
DT: January 1995
BT: Microprocessor chips
RT: Microcomputers
Parallel processing systems

Transuranium elements 622.1
DT: January 1993
BT: Actinides
NT: Americium
Berkelium
Californium
Curium
Einsteinium
Fermium
Kurchatovium
Lawrencium
Mendelevium
Neptunium
Nobelium
Plutonium

Trash disposal
USE: Refuse disposal

Traveling cranes
USE: Gantry cranes

Traveling wave antennas (716)
DT: January 1993
UF: Antennas—Traveling wave*
Progressive wave antennas
Travelling wave antennas
BT: Antennas

Traveling wave tubes 714.1
DT: January 1993
UF: Electron tubes, Traveling wave*
Travelling wave tubes
BT: Electron wave tubes
NT: Backward wave tubes

Travelling wave antennas
USE: Traveling wave antennas

Travelling wave tubes
USE: Traveling wave tubes

Treadmills
USE: Exercise equipment

Treaties
USE: International cooperation

Trees (mathematics) 921.4
DT: January 1993
UF: Mathematical techniques—Trees*
BT: Graph theory
RT: Topology

Trellis codes 723.2
DT: January 1995
BT: Codes (symbols)
RT: Coding errors
Encoding (symbols)
Error correction
Error detection

* Former Ei Vocabulary term

Trenching 619.1 (405.2)
DT: January 1993
UF: Pipelines—Trenching*
BT: Excavation
RT: Blasting
Pipelines

Triacs
USE: Thyristors

Triangulation 405.3
DT: January 1993
UF: Surveying—Triangulation*
RT: Navigation
Surveying
Surveying instruments
Surveys

Triboelectric emission
USE: Triboelectricity

Triboelectricity 701.1 (931)
DT: Predates 1975
UF: Frictional electricity
Triboelectric emission
Triboelectricity—Triboelectric emission*
BT: Electricity
RT: Electrostatics
Friction
Tribology

*Triboelectricity—Triboelectric emission**
USE: Triboelectricity

Tribology 931 (931.2)
DT: Predates 1975
UF: Friction theory
RT: Friction
Friction materials
Friction meters
Lubrication
Surface phenomena
Surface properties
Triboelectricity
Triboluminescence
Wear of materials

Triboluminescence 741.1 (931.2)
DT: Predates 1975
UF: Frictional light
Mechanoluminescence
BT: Luminescence
RT: Friction
Tribology

Trickling filtration 452.2, 802.3
DT: January 1993
UF: Sewage treatment—Trickling filtration*
BT: Filtration
RT: Sewage treatment
Waste disposal

Trigger circuits 713.5
DT: January 1993
UF: Electronic circuits, Trigger*
BT: Switching circuits
RT: Flip flop circuits
Logic circuits
Multivibrators

Trimming 535.1.2
DT: January 1993
UF: Rolling mill practice—Trimming*
BT: Cutting
RT: Cold working
Deburring
Grinding (machining)
Machining
Reaming
Sawing
Scarfing
Slitting
Turning

Triodes 714.1
DT: January 1993
UF: Electron tubes, Triode*
BT: Electron tubes
RT: Thermionic tubes

Tritium 622.1.1 (804)
SN: Radioactive hydrogen isotope, mass-number 3
DT: Predates 1975
BT: Radioisotopes
RT: Hydrogen

Troffers 707.1
DT: January 1993
UF: Lighting fixtures—Troffers*
BT: Lighting fixtures

Trolley cars 682.1.1 (433.2)
SN: Cars which run on rails
DT: January 1993
UF: Cars—Street railroad*
Street railroad cars
Streetcars
BT: Passenger cars
RT: Light rail transit

Trolleys (trackless)
USE: Trackless trolleys

Tropical buildings 402 (443)
DT: January 1993
UF: Buildings—Tropics*
BT: Buildings
RT: Air conditioning
Climate control
Hot weather problems
Tropical engineering
Tropics

Tropical engineering (409) (443)
DT: Predates 1975
BT: Engineering
RT: Air conditioning
Hot weather problems
Humidity control
Moisture control
Tropical buildings
Tropics

Tropics 443
DT: January 1993
UF: Corrosion—Tropics*

* Former Ei Vocabulary term

Tropics *(continued)*
Plastics—Tropical applications*
Water treatment—Tropics*
BT: Geographical regions
RT: Hot weather problems
Malaria control
Tropical buildings
Tropical engineering

Troposphere 443.1
DT: January 1993
UF: Electromagnetic waves—Propagation in troposphere*
BT: Earth atmosphere

Truck drivers 912.4 (663.1)
DT: January 1993
UF: Drivers (truck)
Motor truck drivers*
BT: Transportation personnel
RT: Driver training
Fleet operations
Motor transportation
Truck transportation

Truck fleet operation
USE: Truck transportation

Truck terminals 432.3 (402.1)
DT: January 1993
UF: Motor truck terminals*
Terminals (truck)
BT: Facilities
RT: Truck transportation
Trucks

Truck trailers 663.2
DT: January 1993
UF: Semi trailers
Trailers—Motor truck*
BT: Trucks

Truck transportation 432.3
DT: January 1993
UF: Motor truck transportation*
Truck fleet operation
BT: Motor transportation
RT: Truck drivers
Truck terminals
Trucks

Trucks 663.1
DT: January 1993
UF: Motor trucks*
BT: Ground vehicles
NT: Cabs (truck)
Garbage trucks
Industrial trucks
Power takeoffs
Refrigerator trucks
Tank trucks
Tractors (truck)
Truck trailers
Utility trucks
RT: Loading
Materials handling equipment
Snow plows
Truck terminals
Truck transportation
Unloading
Weigh stations
Winches

Trusses 408.2
DT: Predates 1975
BT: Structures (built objects)
RT: Beams and girders

Tsunamis 471.4 (484)
DT: January 1993
UF: Seismic water waves
Tidal waves
Water waves—Tsunamis*
BT: Water waves
RT: Earthquakes
Oceanography
Seismic waves

TTL
USE: Transistor transistor logic circuits

Tube flow
USE: Pipe flow

Tube mills 535.1.1 (402.1)
DT: January 1993
UF: Rolling mill practice—Tube*
BT: Rolling mills
RT: Rolling mill practice

Tubeless tires 818.5
DT: January 1993
UF: Tires—Tubeless*
BT: Tires

Tubes (components) 619.1 (511.2) (616.1)
DT: January 1993
UF: Heat exchangers—Finned tubes*
Heat exchangers—Tubes*
Heat transfer—Tubes*
Oil field equipment—Tubular goods*
BT: Components
NT: Capillary tubes
Collapsible tubes
Condenser tubes
Expandable tubes
Extendable tubes
Oil well casings
Pneumatic tubes
Shock tubes
Superheater tubes
RT: Bending (forming)
Heat exchangers
Hose
Pipe
Pipe flow
Tubing
Tubular steel structures
Tubular turbines

Tubes (electron)
USE: Electron tubes

*Tubes—Bending**
USE: Bending (forming)

* Former Ei Vocabulary term

*Tubes—Collapsible**
USE: Collapsible tubes

*Tubes—Expandable**
USE: Expandable tubes

*Tubes—Extendable**
USE: Extendable tubes

Tubing 619.1 (652.2) (662.3)
DT: January 1993
UF: Aircraft materials—Tubing*
Automobile materials—Tubing*
BT: Materials
RT: Automobile materials
Hose
Tubes (components)

Tubular steel structures 408.2, 545.3
DT: January 1993
UF: Steel structures—Tubular*
BT: Steel structures
RT: Tubes (components)

Tubular turbines 617.3
DT: January 1993
UF: Hydraulic turbines—Tubular*
BT: Hydraulic turbines
RT: Tubes (components)

Tugboats 674.1 (672.2)
DT: Predates 1975
UF: Towboats
BT: Boats

Tumbling
USE: Barreling

Tumor-causing viruses
USE: Oncogenic viruses

Tuners (716.3) (716.4)
DT: January 1993
UF: Television receivers—Tuners*
BT: Telecommunication equipment
RT: Natural frequencies
Oscillators (electronic)
Radio receivers
Resonators
Signal receivers
Television receivers
Tuning

Tungstate minerals 482.2
DT: January 1993
UF: Mineralogy—Tungstates*
BT: Minerals
Tungsten compounds
RT: Tungsten

Tungsten 543.5
DT: January 1993
UF: Cast iron—Tungsten content*
Tungsten and alloys*
W
Wolfram
BT: Refractory metals
RT: Tungstate minerals
Tungsten alloys
Tungsten compounds
Tungsten deposits
Tungsten metallography
Tungsten metallurgy
Tungsten mines
Tungsten ore treatment
Tungsten plating
Tungsten sheet

Tungsten alloys 543.5
DT: January 1993
UF: Tungsten and alloys*
BT: Refractory alloys
NT: Stellite
RT: Tungsten
Tungsten compounds

*Tungsten and alloys**
USE: Tungsten OR
Tungsten alloys

Tungsten carbide 804.2 (812.1)
DT: Predates 1975
BT: Carbides
Tungsten compounds

Tungsten compounds (804.1) (804.2)
DT: Predates 1975
BT: Refractory metal compounds
NT: Tungstate minerals
Tungsten carbide
RT: Tungsten
Tungsten alloys

Tungsten deposits 504.3, 543.5
DT: Predates 1975
BT: Ore deposits
RT: Tungsten
Tungsten mines

Tungsten metallography 531.2, 543.5
DT: Predates 1975
BT: Metallography
RT: Tungsten
Tungsten metallurgy

Tungsten metallurgy 531.1, 543.5
DT: Predates 1975
BT: Metallurgy
NT: Tungsten powder metallurgy
RT: Tungsten
Tungsten metallography
Tungsten ore treatment

Tungsten mines 504.3, 543.5
DT: January 1993
UF: Tungsten mines and mining*
BT: Mines
RT: Tungsten
Tungsten deposits

*Tungsten mines and mining**
USE: Tungsten mines

Tungsten ore treatment 533.1, 543.5
DT: Predates 1975
BT: Ore treatment

* Former Ei Vocabulary term

Tungsten ore treatment *(continued)*
RT: Tungsten
Tungsten metallurgy

Tungsten plating 539.3, 543.5
DT: Predates 1975
BT: Plating
RT: Tungsten

Tungsten powder metallurgy 536.1, 543.5
DT: January 1993
UF: Powder metallurgy—Tungsten*
BT: Powder metallurgy
Tungsten metallurgy

Tungsten sheet 535.1, 543.5
DT: Predates 1975
BT: Sheet metal
RT: Tungsten

Tuning (713) (716.3) (716.4) (744.1)
DT: January 1993
UF: Electronic circuits—Tuning*
NT: Laser tuning
RT: Electronic circuit tracking
Gain control
Natural frequencies
Oscillators (electronic)
Radio receivers
Resonance
Resonators
Signal receivers
Synchronization
Television receivers
Tuners

Tunnel diode amplifiers 713.1, 714.2
DT: January 1993
UF: Amplifiers, Tunnel diode*
BT: Amplifiers (electronic)
RT: Tunnel diodes

Tunnel diode oscillators 713.2, 714.2
DT: January 1993
UF: Oscillators, Tunnel diode*
BT: Oscillators (electronic)
RT: Tunnel diodes

Tunnel diodes 714.2
DT: January 1993
UF: Semiconductor diodes, Tunnel*
BT: Semiconductor diodes
RT: Electron tunneling
Negative resistance
Quantum interference devices
Tunnel diode amplifiers
Tunnel diode oscillators

Tunnel effect
USE: Electron tunneling

Tunnel junctions (714.2) (931.3) (932)
DT: January 1995
RT: Electron tunneling
Josephson junction devices
Semiconductor junctions

Tunnel linings 401.2
DT: January 1993
UF: Tunnels and tunneling—Lining*
Water supply tunnels—Lining*
BT: Linings
Tunnels

Tunneling (electron)
USE: Electron tunneling

Tunneling (excavation) 401.2 (502.1)
DT: January 1993
UF: Mines and mining—Tunneling*
Tunnelling (excavation)
Tunnels and tunneling*
BT: Excavation
RT: Construction
Drainage
Mining
Tunneling machines
Tunnels
Waterproofing

Tunneling machines 401.2, 405.1 (502.2)
DT: January 1977
BT: Machinery
RT: Tunneling (excavation)

Tunnelling (electron)
USE: Electron tunneling

Tunnelling (excavation)
USE: Tunneling (excavation)

Tunnels 401.2
DT: January 1993
UF: Tunnels and tunneling*
BT: Structures (built objects)
NT: Ground supports
Pedestrian tunnels
Sewer tunnels
Tunnel linings
Vehicular tunnels
Water supply tunnels
RT: Caissons
Tunneling (excavation)

Tunnels (wind)
USE: Wind tunnels

*Tunnels and tunneling**
USE: Tunnels OR
Tunneling (excavation)

*Tunnels and tunneling—Ground supports**
USE: Ground supports

*Tunnels and tunneling—Lining**
USE: Tunnel linings

*Tunnels and tunneling—Pedestrian**
USE: Pedestrian tunnels

*Tunnels and tunneling—Railroad**
USE: Railroad tunnels

*Tunnels and tunneling—Vehicular**
USE: Vehicular tunnels

*Tunnels and tunneling—Washing machines**
USE: Washing machines

*Tunnels and tunneling—Waterproofing**
USE: Waterproofing

Turbidimeters 941.3
SN: For direct visual determination of reduction in light transmission due to solid particles in a solution, as compared with a standard. For photoelectric instruments, use NEPHELOMETERS
DT: Predates 1975
BT: Density measuring instruments
RT: Aerosols
Chemical analysis
Light scattering
Nephelometers
Particle size analysis
Transparency

Turbine engines
USE: Turbines

Turbine generators
USE: Turbogenerators

Turbine pumps 618.2
DT: January 1993
UF: Pumps, Turbine*
BT: Rotary pumps
Turbomachinery
RT: Centrifugal pumps
Turbines

Turbine rings (653.2)
DT: January 1993
UF: Aircraft engines, Jet and turbine—Turbine rings*
BT: Turbines

Turbines (612.3) (617.1) (617.2)
DT: January 1993
UF: Turbine engines
BT: Turbomachinery
NT: Gas turbines
Hydraulic turbines
Steam turbines
Turbine rings
Wind turbines
RT: Hydroelectric power plants
Turbine pumps
Turbodrills
Turbogenerators

Turboalternators
USE: Turbogenerators

Turbodrills 511.2
DT: January 1993
UF: Oil well drilling—Turbodrills*
BT: Drills
Turbomachinery
RT: Drilling
Oil well drills
Turbines

Turboexpanders 644.3
DT: January 1993
UF: Refrigerating machinery—Turboexpanders*
BT: Refrigerating machinery
Turbomachinery

Turbofan engines 653.1
DT: January 1993
UF: Aircraft engines, Jet and turbine—Turbofan*
BT: Turbojet engines
RT: Ducted fan engines
Turboprop engines

Turbogenerators 705.2
DT: Predates 1975
UF: Turbine generators
Turboalternators
Turbogenerators, Direct current*
BT: Electric generators
Turbomachinery
RT: Gas turbines
Hydroelectric generators
Steam turbines
Turbines

*Turbogenerators, Direct current**
USE: DC machinery AND Turbogenerators

Turbojet engines 653.1
DT: January 1993
UF: Aircraft engines, Jet and turbine*
BT: Jet engines
NT: Turbofan engines
Turboprop engines
RT: Gas turbines

Turbomachine blades
(612.3) (617.1) (617.2) (618) (631.1)
DT: January 1993
UF: Blades (turbomachinery)
Turbomachinery—Blades*
BT: Turbomachinery

Turbomachine impellers
USE: Rotors

Turbomachinery
(612.3) (617.1) (617.2) (618) (632.1)
SN: Turbines, and compressors and pumps which operate on the same principle
DT: Predates 1975
BT: Machinery
NT: Axial flow turbomachinery
High pressure turbomachinery
Radial flow turbomachinery
Small turbomachinery
Turbine pumps
Turbines
Turbodrills
Turboexpanders
Turbogenerators
Turbomachine blades
RT: Blowers
Cascades (fluid mechanics)
Cavitation
Closed cycle machinery

* Former Ei Vocabulary term

Turbomachinery *(continued)*
Compressed air motors
Compressors
Ejectors (pumps)
Hydraulic torque converters
Intake systems
Pumps
Rotors
Superchargers
Thrust reversal

*Turbomachinery—Automotive applications**
USE: Automobile engines

*Turbomachinery—Axial flow**
USE: Axial flow turbomachinery

*Turbomachinery—Blades**
USE: Turbomachine blades

*Turbomachinery—Cascades**
USE: Cascades (fluid mechanics)

*Turbomachinery—Closed cycle**
USE: Closed cycle machinery

*Turbomachinery—Deposits**
USE: Deposits

*Turbomachinery—Diffusers**
USE: Diffusers (fluid)

*Turbomachinery—Ejectors**
USE: Ejectors (pumps)

*Turbomachinery—Exhausts**
USE: Exhaust systems (engine)

*Turbomachinery—High pressure**
USE: High pressure turbomachinery

*Turbomachinery—Impellers**
USE: Rotors

*Turbomachinery—Inlets**
USE: Intake systems

*Turbomachinery—Radial flow**
USE: Radial flow turbomachinery

*Turbomachinery—Small**
USE: Small turbomachinery

*Turbomachinery—Starting**
USE: Starting

*Turbomachinery—Temperature**
USE: Temperature

*Turbomachinery—Thrust reversal**
USE: Thrust reversal

Turboprop engines 653.1
DT: January 1993
UF: Aircraft engines, Jet and turbine—Turboprop*
BT: Turbojet engines
RT: Turbofan engines

Turbulence (443.1) (631.1) (932.3)
DT: January 1993
BT: Fluid dynamics
NT: Atmospheric turbulence
Plasma turbulence
RT: Aerodynamics
Cavitation
Diffusion
Drag
Flow of fluids
Mixing
Pipe flow
Storms
Tornadoes
Transition flow
Turbulent flow
Two phase flow
Unsteady flow
Vortex flow
Wakes
Wall flow
Wind

Turbulent flow 631.1
DT: January 1993
UF: Flow of fluids—Turbulent*
BT: Flow of fluids
NT: Wakes
RT: Aerodynamics
Cavitation
Drag
Laminar flow
Open channel flow
Pipe flow
Transition flow
Turbulence
Two phase flow
Unsteady flow
Viscous flow
Vortex flow
Wall flow

Turing machines 721.1
DT: January 1993
UF: Automata theory—Turing machines*
BT: Automata theory
RT: Digital computers
Formal logic

Turnaround time 912.2 (434.1)
DT: January 1993
UF: Ships—Turnaround time*
RT: Loading
Management
Scheduling
Unloading

Turning 604.2
DT: January 1993
BT: Machining
RT: Boring
Lathes
Metal cutting
Thread cutting
Thread rolling
Trimming

Turntables
USE: Phonographs

Turpentine 804.1, 811.2 (813.2)
DT: Predates 1975
BT: Organic compounds
Wood products
RT: Solvents

Turret lathes 603.1
SN: Includes hand and automatic screw machines
DT: January 1993
UF: Lathes, Turret*
BT: Lathes
RT: Chucking machines

TV
USE: Television

Twine 694.2, 819.4
DT: January 1993
UF: Packaging materials—Twine*
BT: Materials
RT: Packaging materials
Yarn

Twinning 933.1.2
DT: January 1993
UF: Crystal twinning
Crystals—Twinning*
BT: Crystal growth
RT: Crystal structure
Crystals

Two fuel engines
USE: Dual fuel engines

Two fuel oil burners
USE: Dual fuel burners

Two phase flow 631.1
DT: January 1993
UF: Flow of fluids—Two phase flow*
BT: Multiphase flow
RT: Flow of solids
Flow patterns
Heat transfer
Laminar flow
Turbulence
Turbulent flow

Two term control systems 731.1
SN: Proportional - integral systems
DT: January 1993
UF: Control systems, Two term*
BT: Control systems
RT: Proportional control systems
Three term control systems

Typesetting 745.1
DT: Predates 1975
BT: Printing

Typewriter keyboards 745.1.1
DT: January 1993
UF: Keyboards (typewriter)
Typewriters—Keyboards*
BT: Typewriters
RT: Computer keyboards
Electric typewriters

Typewriters 745.1.1
DT: Predates 1975
UF: Typewriters—Phonetic actuation*
BT: Business machines
Office equipment
NT: Electric typewriters
Teleprinters
Typewriter keyboards
RT: Printers (computer)
Word processing

*Typewriters—Electric**
USE: Electric typewriters

*Typewriters—Keyboards**
USE: Typewriter keyboards

*Typewriters—Phonetic actuation**
USE: Typewriters

*Typewriters—Pneumatic**
USE: Pneumatic equipment

U
USE: Uranium

UHMWPE
USE: Ultrahigh molecular weight polyethylenes

UHV power transmission 706.1.1
SN: Greater than 800 kv
DT: January 1993
UF: Electric power transmission, UHV*
Ultrahigh voltage transmission
BT: EHV power transmission

ULSI circuits 714.2
DT: January 1993
UF: Integrated circuits, ULSI*
Ultra large scale integration
BT: VLSI circuits

Ultra large scale integration
USE: ULSI circuits

Ultrafast phenomena (741.1.1) (744.8)
DT: January 1993
RT: Laser pulses
Nonlinear optics
Spectroscopy

Ultrafiltration 802.3
DT: January 1987
BT: Filtration
RT: Filters (for fluids)

Ultrahigh frequency amplifiers 713.1
DT: January 1993
UF: Amplifiers, Ultrahigh frequency*
BT: Amplifiers (electronic)

Ultrahigh molecular weight polyethylenes 815.1.1
DT: January 1993
UF: Polyethylenes—Ultrahigh molecular weight*
UHMWPE
BT: Polyethylenes

Ultrahigh voltage transmission
USE: UHV power transmission

* Former Ei Vocabulary term

Ultrasonic absorption 753.1
DT: January 1993
UF: Ultrasonic waves—Absorption*
BT: Acoustic wave absorption
RT: Ultrasonic transmission
Ultrasonic waves

Ultrasonic applications 753.3
DT: Predates 1975
BT: Applications
NT: Ultrasonic cleaning
Ultrasonic cutting
Ultrasonic imaging
Ultrasonic measurement
Ultrasonic welding
RT: Ultrasonic devices
Ultrasonic equipment
Ultrasonic waves
Ultrasonics

Ultrasonic cameras 753.2
DT: January 1993
UF: Cameras—Ultrasonic*
BT: Cameras
Ultrasonic equipment

Ultrasonic cleaning 753.3 (539)
DT: January 1993
UF: Metal cleaning—Ultrasonic*
BT: Surface cleaning
Ultrasonic applications
RT: Chemical cleaning
Descaling
Pickling
Vibratory finishing

Ultrasonic cutting 753.3, 604.1
DT: January 1993
UF: Metal cutting—Ultrasonic*
BT: Cutting
Ultrasonic applications

Ultrasonic delay lines 753.2
DT: January 1993
BT: Acoustic delay lines
Ultrasonic devices

Ultrasonic devices 753.2
DT: Predates 1975
BT: Acoustic devices
Ultrasonic equipment
NT: Ultrasonic delay lines
Ultrasonic transducers
RT: Bragg cells
Ultrasonic applications
Ultrasonic effects
Ultrasonics

Ultrasonic diffraction 753.1
DT: January 1993
UF: Ultrasonic waves—Diffraction*
BT: Acoustic wave diffraction
RT: Ultrasonic dispersion
Ultrasonic propagation
Ultrasonic waves

Ultrasonic dispersion 753.1
DT: January 1993
UF: Ultrasonic waves—Dispersion*
BT: Acoustic dispersion
RT: Ultrasonic diffraction
Ultrasonic propagation
Ultrasonic velocity
Ultrasonic waves

Ultrasonic distortion 753.1
DT: January 1993
UF: Ultrasonic waves—Distortion*
BT: Acoustic distortion
RT: Ultrasonic waves

Ultrasonic Doppler velocimeters 753.3, 943.3
DT: January 1993
UF: Velocimeters—Ultrasonic Doppler*
BT: Ultrasonic equipment
Velocimeters
RT: Ultrasonic velocity measurement

Ultrasonic effects 753.1
DT: Predates 1975
BT: Acoustic wave effects
RT: Ultrasonic devices
Ultrasonic equipment
Ultrasonic waves
Ultrasonics

Ultrasonic equipment 753.2
SN: Equipment which utilizes or produces ultrasonic waves
DT: Predates 1975
BT: Acoustic equipment
NT: Ultrasonic cameras
Ultrasonic devices
Ultrasonic Doppler velocimeters
Ultrasonic flowmeters
Ultrasonic machine tools
RT: Signal generators
Ultrasonic applications
Ultrasonic effects
Ultrasonics

Ultrasonic flowmeters 753.2, 943.3
DT: January 1993
UF: Flowmeters—Ultrasonic*
BT: Flowmeters
Ultrasonic equipment

Ultrasonic imaging (753.1) (753.3)
DT: January 1993
BT: Acoustic imaging
Ultrasonic applications
NT: Echocardiography

Ultrasonic machine tools 753.2, 603.1
DT: January 1993
UF: Machine tools—Ultrasonic*
BT: Machine tools
Ultrasonic equipment

Ultrasonic measurement 753.1, 941.2
DT: January 1993
UF: Ultrasonics—Measurements*
BT: Acoustic variables measurement

* Former Ei Vocabulary term

Ultrasonic measurement *(continued)*
Ultrasonic applications
NT: Ultrasonic velocity measurement

Ultrasonic propagation 753.1
SN: In gases. For propagation in liquids and solids use ULTRASONIC TRANSMISSION
DT: January 1993
UF: Ultrasonic waves—Propagation*
BT: Acoustic wave propagation
RT: Ultrasonic diffraction
Ultrasonic dispersion
Ultrasonic reflection
Ultrasonic refraction
Ultrasonic relaxation
Ultrasonic scattering
Ultrasonic transmission
Ultrasonic waves

Ultrasonic reflection 753.1
DT: January 1993
UF: Ultrasonic waves—Reflection*
BT: Acoustic wave reflection
RT: Ultrasonic propagation
Ultrasonic waves

Ultrasonic refraction 753.1
DT: January 1993
UF: Ultrasonic waves—Refraction*
BT: Acoustic wave refraction
RT: Refractive index
Ultrasonic propagation
Ultrasonic waves

Ultrasonic relaxation 753.1
DT: January 1993
BT: Relaxation processes
RT: Ultrasonic propagation

Ultrasonic scattering 753.1
DT: January 1993
UF: Ultrasonic waves—Scattering*
BT: Acoustic wave scattering
RT: Ultrasonic propagation
Ultrasonic waves

Ultrasonic transducers 753.2
DT: Predates 1975
BT: Acoustic transducers
Ultrasonic devices
RT: Microwave acoustic devices
Piezoelectric devices

Ultrasonic transmission 753.1
SN: In liquids and solids. For transmission in gases use ULTRASONIC PROPAGATION
DT: January 1993
UF: Ultrasonic waves—Transmission*
BT: Acoustic wave transmission
RT: Ultrasonic absorption
Ultrasonic propagation
Ultrasonic waves

Ultrasonic velocity 753.1
DT: January 1993
UF: Ultrasonic waves—Velocity*
BT: Acoustic wave velocity
RT: Ultrasonic dispersion
Ultrasonic velocity measurement
Ultrasonic waves

Ultrasonic velocity measurement 753.1, 941.2
DT: January 1993
UF: Ultrasonics—Velocity measurement*
BT: Ultrasonic measurement
Velocity measurement
RT: Ultrasonic Doppler velocimeters
Ultrasonic velocity

Ultrasonic waves 753.1
DT: Predates 1975
BT: Acoustic waves
RT: Ultrasonic absorption
Ultrasonic applications
Ultrasonic diffraction
Ultrasonic dispersion
Ultrasonic distortion
Ultrasonic effects
Ultrasonic propagation
Ultrasonic reflection
Ultrasonic refraction
Ultrasonic scattering
Ultrasonic transmission
Ultrasonic velocity
Ultrasonics

*Ultrasonic waves—Absorption**
USE: Ultrasonic absorption

*Ultrasonic waves—Diffraction**
USE: Ultrasonic diffraction

*Ultrasonic waves—Dispersion**
USE: Ultrasonic dispersion

*Ultrasonic waves—Distortion**
USE: Ultrasonic distortion

*Ultrasonic waves—Propagation**
USE: Ultrasonic propagation

*Ultrasonic waves—Reflection**
USE: Ultrasonic reflection

*Ultrasonic waves—Refraction**
USE: Ultrasonic refraction

*Ultrasonic waves—Scattering**
USE: Ultrasonic scattering

*Ultrasonic waves—Transmission**
USE: Ultrasonic transmission

*Ultrasonic waves—Velocity**
USE: Ultrasonic velocity

Ultrasonic welding 753.3, 538.2.1
DT: January 1993
UF: Welding—Ultrasonic*
BT: Pressure welding
Ultrasonic applications
RT: Spot welding

Ultrasonics 753.1
DT: Predates 1975
UF: Ultrasound
BT: Acoustics

* Former Ei Vocabulary term

Ultrasonics *(continued)*
RT: Acoustooptical effects
Shock waves
Ultrasonic applications
Ultrasonic devices
Ultrasonic effects
Ultrasonic equipment
Ultrasonic waves

*Ultrasonics—Measurements**
USE: Ultrasonic measurement

*Ultrasonics—Velocity measurement**
USE: Ultrasonic velocity measurement

Ultrasound
USE: Ultrasonics

Ultrathin films (714.2)
DT: January 1995
BT: Thin films
RT: Monolayers

Ultraviolet detectors 741.3
DT: Predates 1975
BT: Photodetectors
Ultraviolet devices
RT: Photometers
Ultraviolet radiation

Ultraviolet devices 741.3
DT: June 1990
BT: Optical devices
NT: Ultraviolet detectors
RT: Ultraviolet instruments
Ultraviolet lamps
Ultraviolet radiation

Ultraviolet instruments 741.3, 941.3
DT: January 1993
UF: Optical instruments—Ultraviolet*
BT: Optical instruments
NT: Ultraviolet spectrographs
Ultraviolet spectrometers
RT: Ultraviolet devices
Ultraviolet radiation

Ultraviolet lamps 707.2, 741.3
DT: January 1993
UF: Electric lamps, Ultraviolet*
BT: Electric lamps
RT: Ultraviolet devices
Ultraviolet radiation

Ultraviolet radiation 741.1
DT: Predates 1975
UF: UV light
UV radiation
BT: Light
RT: Ozone layer
Ultraviolet detectors
Ultraviolet devices
Ultraviolet instruments
Ultraviolet lamps

Ultraviolet spectrographs 741.3 (801)
SN: Ultraviolet spectroscopes that produce photographic or fluorescent-screen spectrograms
DT: January 1993
UF: Spectrographs, Ultraviolet*
BT: Spectrographs
Ultraviolet instruments
RT: Ultraviolet spectrometers
Ultraviolet spectroscopy

Ultraviolet spectrometers 741.3 (801)
SN: Spectroscopes capable of measuring angular deviation of ultraviolet radiation
DT: January 1993
UF: Spectrometers, Ultraviolet*
BT: Spectrometers
Ultraviolet instruments
NT: Ultraviolet spectrophotometers
RT: Ultraviolet spectrographs
Ultraviolet spectroscopy

Ultraviolet spectrophotometers
741.3 (801) (941.3)
SN: Spectrometers for measuring ultraviolet spectral energy distributions
DT: January 1993
UF: Spectrophotometers, Ultraviolet*
BT: Spectrophotometers
Ultraviolet spectrometers
RT: Spectrophotometry
Ultraviolet spectroscopy

Ultraviolet spectroscopy (801) (931.3) (932.2)
DT: January 1993
UF: Spectroscopy, Ultraviolet*
UV spectroscopy
BT: Spectroscopy
RT: Ultraviolet spectrometers
Ultraviolet spectrographs
Ultraviolet spectrophotometers

Unbleached pulp 811.1
DT: January 1993
UF: Pulp—Unbleached*
BT: Pulp

Underground air conditioning systems
USE: Air conditioning

Underground buildings 402
DT: January 1993
UF: Buildings—Underground*
BT: Buildings
Underground structures
RT: Underground equipment
Underground heat transfer
Underground temperature measurement

Underground cables 706.2
DT: January 1993
UF: Electric cables—Underground*
BT: Electric cables
Underground equipment
RT: Underground electric power distribution

* Former Ei Vocabulary term

Underground cameras 742.2
DT: January 1993
UF: Cameras—Underground*
BT: Cameras
Underground equipment

Underground corrosion 539.1 (483.1)
DT: January 1993
UF: Corrosion—Underground*
BT: Corrosion
RT: Underground equipment
Underground structures

Underground electric power distribution 706.1.2
DT: January 1993
UF: Electric power distribution—Underground installation*
BT: Electric power distribution
RT: Underground cables
Underground equipment
Underground structures

Underground equipment (643.4) (716.4)
DT: January 1993
UF: Air conditioning—Underground*
Television applications—Underground*
BT: Equipment
NT: Underground cables
Underground cameras
RT: Underground buildings
Underground corrosion
Underground electric power distribution
Underground gas storage
Underground gas manufacture
Underground mine transportation
Underground structures

Underground explosions (502.2) (914.1)
DT: January 1993
UF: Explosions—Underground*
BT: Explosions
NT: Mine explosions

Underground gas manufacture 522
DT: January 1993
UF: Gas manufacture—Underground*
BT: Gas fuel manufacture
RT: Underground equipment
Underground structures

Underground gas storage 522, 694.4
DT: January 1993
UF: Gas storage—Underground*
BT: Gas fuel storage
RT: Underground equipment
Underground structures

Underground heat transfer 641.2 (483.1)
DT: January 1993
UF: Heat transfer—Underground*
BT: Heat transfer
RT: Underground buildings
Underground structures
Underground temperature measurement

Underground mine transportation (502.2)
DT: January 1993
UF: Mines and mining—Underground transportation*
BT: Mine transportation
RT: Mine cars
Underground equipment

Underground power plants (402.1) (611.1) (613) (614) (615.1)
DT: January 1993
UF: Power plants—Underground*
BT: Power plants
Underground structures

Underground pumping plants 618.2
DT: January 1993
UF: Pumping plants—Underground*
Sewage pumping plants—Underground*
BT: Pumping plants
Underground structures

Underground reservoirs 441.2
DT: January 1993
UF: Reservoirs—Underground*
BT: Reservoirs (water)
Underground structures
RT: Low permeability reservoirs

Underground structures 408.1
DT: January 1993
UF: Structural design—Underground*
BT: Structures (built objects)
NT: Underground buildings
Underground power plants
Underground pumping plants
Underground reservoirs
RT: Mines
Mining
Structural design
Underground corrosion
Underground electric power distribution
Underground equipment
Underground gas storage
Underground gas manufacture
Underground heat transfer
Underground temperature measurement

Underground temperature measurement 944.6 (483.1)
DT: January 1993
UF: Temperature measurement—Underground*
BT: Temperature measurement
RT: Underground buildings
Underground heat transfer
Underground structures

Underground water
USE: Groundwater

Underpasses 406.2
DT: January 1993
UF: Highway underpasses
Railroad plant and structures—Underpasses*
Railroad underpasses
BT: Public works
RT: Highway systems

* Former Ei Vocabulary term

Underpasses *(continued)*
Railroad crossings
Railroad plant and structures
Vehicular tunnels

Underpinnings 408.2 (402) (483.2)
DT: January 1993
UF: Buildings—Underpinning*
Foundations—Underpinning*
BT: Structural members
RT: Buildings
Foundations

Undersea manganese nodules
USE: Manganese nodules

*Undersea technology**
USE: Ocean engineering

Undersnow structures 408.1 (443.3)
DT: January 1993
UF: Structural design—Undersnow*
BT: Structures (built objects)
RT: Arctic buildings
Arctic engineering
Arctic vehicles
Cold weather problems
Structural design

Underwater acoustics 751.1
DT: January 1993
UF: Acoustics, Underwater*
BT: Acoustics
RT: Acoustic wave transmission
Oceanography
Sonar

Underwater audition 751.1
DT: January 1993
UF: Audition—Underwater*
BT: Audition

Underwater ballistics 931 (404.1)
DT: January 1993
UF: Ballistics—Underwater*
BT: Ballistics
RT: Naval warfare
Torpedoes
Underwater launched missiles

Underwater cameras 742.2
DT: January 1993
UF: Cameras—Underwater*
BT: Cameras
Underwater equipment

Underwater construction 405.2, 472
DT: January 1993
UF: Concrete construction—Underwater*
BT: Construction
RT: Caissons
Diving apparatus
Ocean engineering
Underwater cutting
Underwater equipment
Underwater explosions
Underwater structures

Underwater cutting 604.1
DT: January 1993
UF: Metal cutting—Underwater*
BT: Cutting
RT: Ocean engineering
Underwater construction
Underwater equipment
Underwater structures

Underwater drilling 472 (405.2) (501.1) (511.1)
DT: January 1993
UF: Rock drilling—Underwater*
BT: Drilling
NT: Offshore drilling
RT: Ocean engineering
Rock drilling
Underwater equipment
Underwater mineral resources
Underwater structures

Underwater equipment 472
DT: January 1993
UF: Television applications—Underwater*
BT: Equipment
NT: Underwater cameras
Underwater lamps
RT: Ocean engineering
Underwater construction
Underwater cutting
Underwater drilling
Underwater photography
Underwater power plants
Underwater probing
Underwater reactors
Underwater structures
Underwater temperature measurement
Underwater welding

Underwater explosions 472 (405.2)
DT: January 1993
UF: Explosions—Underwater*
BT: Explosions
RT: Ocean engineering
Underwater construction

Underwater foundations 472, 483.2
DT: January 1993
UF: Foundations—Underwater*
BT: Foundations
Underwater structures

Underwater habitats
USE: Ocean habitats

Underwater heat transfer 641.2 (472)
DT: January 1993
UF: Heat transfer—Underwater*
BT: Heat transfer
RT: Underwater structures
Underwater temperature measurement

Underwater lamps 707.2
DT: January 1993
UF: Electric lamps—Underwater*
BT: Electric lamps
Underwater equipment

* Former Ei Vocabulary term

Underwater launched missiles 654.1 (404.1)
DT: January 1993
UF: Rockets and missiles—Underwater*
BT: Marine missiles
NT: Torpedoes
RT: Naval warfare
Underwater ballistics

Underwater mineral resources 501.1, 471.5
DT: January 1993
UF: Mineral industry and resources—Subaqueous*
BT: Mineral resources
NT: Manganese nodules
RT: Ocean engineering
Ore deposits
Underwater drilling
Underwater mineralogy

Underwater mineralogy 471.1, 482.1
DT: January 1993
UF: Mineralogy—Subaqueous*
BT: Mineralogy
RT: Oceanography
Underwater mineral resources

Underwater motors
USE: Submersible motors

Underwater photogrammetry 742.1, 405.3, 472
DT: January 1993
UF: Photogrammetry—Underwater*
BT: Photogrammetry

Underwater photography 742.1 (472)
DT: January 1993
UF: Photography—Underwater*
BT: Photography
RT: Underwater equipment

Underwater pipelines
USE: Submarine pipelines

Underwater power plants
(402.1) (611) (612) (613) (614.1) (615.1)
DT: January 1993
UF: Power plants—Underwater*
BT: Power plants
Underwater structures
RT: Ocean engineering
Underwater equipment

Underwater probing 472
DT: January 1993
UF: Ocean engineering—Searching*
Probing (underwater)
Searching (underwater)
RT: Natural resources exploration
Ocean engineering
Probes
Underwater equipment
Underwater visibility

Underwater reactors 621 (472)
DT: January 1993
UF: Nuclear reactors—Underwater*
BT: Nuclear reactors
Underwater structures
RT: Underwater equipment

Underwater soils 471.1, 483.1
DT: January 1993
UF: Soils—Underwater*
BT: Soils
RT: Oceanography

Underwater structures 408.1, 472
DT: January 1993
UF: Structural design—Underwater*
BT: Structures (built objects)
NT: Underwater foundations
Underwater power plants
Underwater reactors
RT: Ocean engineering
Ocean structures
Offshore oil well production
Structural design
Submarine pipelines
Underwater construction
Underwater cutting
Underwater drilling
Underwater equipment
Underwater heat transfer
Underwater temperature measurement
Underwater welding

Underwater temperature measurement
944.6, 471.3
DT: January 1993
UF: Temperature measurement—Underwater*
BT: Temperature measurement
RT: Underwater equipment
Underwater heat transfer
Underwater structures

Underwater visibility 471.4, 741.2 (472)
DT: January 1993
UF: Visibility—Underwater*
BT: Visibility
RT: Underwater probing

Underwater welding 538.2.1 (472)
DT: January 1993
UF: Welding—Underwater*
BT: Welding
RT: Ocean engineering
Underwater equipment
Underwater structures

Unimolecular layers
USE: Monolayers

Uninterruptible power supplies
USE: Electric power supplies to apparatus

Unit operations (chemical)
USE: Chemical operations

Unit operations (oil wells) 511.1
DT: January 1993
UF: Oil fields—Unit operation*
BT: Oil well production

Unit processes
USE: Chemical reactions

Units of measurement 902.2
DT: Predates 1975
UF: Electric units
Measurement units
Weights and measures
NT: Metric system
RT: Measurements
Molecular weight
Standards
Terminology
Time measurement
Weighing

Universal joints 601.2
DT: Predates 1975
UF: Joints (universal)
BT: Machine components

University buildings
USE: College buildings

UNIX (722) (723)
DT: January 1993
UF: Computer operating systems—UNIX*
BT: Computer operating systems
RT: C (programming language)

Unloading 691.2 (671.2) (674.1)
DT: January 1993
UF: Boats—Unloading*
Containers—Unloading*
Materials handling—Unloading*
Ships—Unloading*
Silos—Unloading*
BT: Materials handling
RT: Loading
Ships
Trucks
Turnaround time

Unsaturated compounds 804.1
DT: June 1990
BT: Organic compounds
NT: Unsaturated polymers

Unsaturated polymers 815.1.1
DT: January 1993
UF: Plastics—Unsaturated*
BT: Organic polymers
Unsaturated compounds

Unsteady flow 631.1
DT: January 1993
UF: Flow of fluids—Unsteady flow*
BT: Flow of fluids
RT: Aerodynamics
Flow of solids
Heat transfer
Laminar flow
Multiphase flow
Non Newtonian flow
Pipe flow
Turbulence
Turbulent flow

Unwinding
USE: Winding

Uplift pressure (441.1) (402) (406.2)
DT: January 1993
UF: Dams—Uplift pressure*
BT: Pressure
RT: Dams
Floors
Roads and streets

Upper atmosphere 443.1
SN: Above about 30 km altitude
DT: January 1993
UF: Earth atmosphere—Upper atmosphere*
BT: Earth atmosphere
NT: Ionosphere
Magnetosphere
RT: Ozone layer

UPS
USE: Electric power supplies to apparatus

Upsetting (forming) 535.2
DT: January 1993
UF: Metal forming—Upsetting*
BT: Forging
RT: Cold heading
Cold working
Hot working

Uranate minerals 482.2
DT: January 1993
UF: Mineralogy—Uranates*
BT: Minerals
Uranium compounds
RT: Uranium

Uranium 549.3, 622.1
DT: January 1993
UF: U
Uranium and alloys*
BT: Actinides
RT: Uranate minerals
Uranium alloys
Uranium compounds
Uranium deposits
Uranium metallography
Uranium metallurgy
Uranium mines
Uranium ore treatment

Uranium alloys 549.3, 622.1
DT: January 1993
UF: Uranium and alloys*
BT: Alloys
Radioactive materials
RT: Uranium
Uranium compounds

*Uranium and alloys**
USE: Uranium OR
Uranium alloys

*Uranium and alloys—Fission**
USE: Fission reactions

Uranium carbide (621.1.2) (622.1) (804.2)
DT: Predates 1975
BT: Carbides

Uranium carbide *(continued)*
Uranium compounds
RT: Nuclear fuels

Uranium compounds (622.1) (804.1) (804.2)
DT: Predates 1975
BT: Metallic compounds
Radioactive materials
NT: Uranate minerals
Uranium carbide
Uranium dioxide
RT: Uranium
Uranium alloys

Uranium deposits 504.5, 549.3, 622.1
DT: Predates 1975
BT: Ore deposits
RT: Uranium
Uranium mines

Uranium dioxide 622.1, 804.2
DT: Predates 1975
BT: Oxides
Uranium compounds

Uranium metallography 531.2, 549.3
DT: January 1993
BT: Metallography
RT: Uranium
Uranium metallurgy

Uranium metallurgy 531.1, 549.3
DT: Predates 1975
BT: Metallurgy
NT: Uranium powder metallurgy
RT: Uranium
Uranium metallography
Uranium ore treatment

Uranium mines 504.5, 549.3, 622.1
DT: January 1993
UF: Uranium mines and mining*
BT: Mines
RT: Uranium
Uranium deposits

*Uranium mines and mining**
USE: Uranium mines

Uranium ore treatment 533.1, 549.3, 622.1
DT: Predates 1975
BT: Ore treatment
RT: Uranium
Uranium metallurgy

Uranium powder metallurgy 536.1, 549.3, 622.1
DT: January 1993
UF: Powder metallurgy—Uranium*
BT: Powder metallurgy
Uranium metallurgy

Urban planning 403.1
DT: Predates 1975
UF: City planning
RT: Bridges
Building codes
Civil engineering
Highway systems
Housing
Land reclamation
Land use
Light rail transit
Monorails
Motor transportation
Municipal engineering
Parks
Public works
Railroad transportation
Railroads
Recreation centers
Recreational facilities
Refuse disposal
Regional planning
Roads and streets
Sewers
Street lighting
Subways
Traffic surveys
Transportation
Waste disposal
Water supply
Zoning

*Urban planning—Land use**
USE: Land use

*Urban planning—Zoning**
USE: Zoning

Urea 804.1 (815.1.1)
DT: Predates 1975
UF: Carbamide
BT: Nitrogen compounds
Organic compounds
RT: Urea electrodes
Urea fertilizers
Urea formaldehyde resins

Urea electrodes (801) (801.4.1) (804.1)
DT: January 1993
UF: Sensors—Urea electrodes*
BT: Electrochemical electrodes
RT: Sensors
Urea

Urea fertilizers 804.1, 821.2
DT: January 1993
UF: Fertilizers—Urea*
BT: Nitrogen fertilizers
RT: Urea

Urea formaldehyde resins 815.1.1
DT: January 1987
UF: Amino resins (urea)
BT: Thermosets
RT: Formaldehyde
Melamine formaldehyde resins
Phenolic resins
Urea

Urology 461.6
DT: January 1993
UF: Biomedical engineering—Urology*
BT: Medicine
RT: Biomedical engineering

* Former Ei Vocabulary term

User interfaces 722.2
DT: January 1993
BT: Interfaces (computer)
NT: Graphical user interfaces
RT: Human computer interaction
Interactive computer systems
Interactive devices
Man machine systems
Virtual reality

Utility programs 723.1
DT: January 1993
UF: Computer systems programming—Utility programs*
BT: Computer software

Utility rates 706.1, 911.1
DT: January 1993
UF: Public utilities—Rate making*
Rates (public utilities)
BT: Fees and charges
NT: Electric rates
Sewage disposal charges
Sewage treatment charges
RT: Water meters

Utility trucks 663.1
SN: Public utility vehicles
DT: January 1993
UF: Motor trucks—Public utilities*
BT: Trucks

Utilization (power)
USE: Electric power utilization

Utilization (waste heat)
USE: Waste heat utilization

Utilization (wastes)
USE: Waste utilization

UV light
USE: Ultraviolet radiation

UV radiation
USE: Ultraviolet radiation

UV spectroscopy
USE: Ultraviolet spectroscopy

Vaccines 461.6 (461.9.1)
DT: January 1993
UF: Drug products—Vaccines*
BT: Drug products
RT: Biocompatibility
Immunization
Immunology

Vacuum 633
DT: January 1993
RT: Pressure
Vacuum applications

Vacuum applications 633.1
DT: January 1993
UF: Materials handling—Vacuum*
BT: Applications
NT: Vacuum brazing
Vacuum die casting
Vacuum welding
RT: Electron tubes
Getters
Vacuum
Vacuum cleaners
Vacuum deposited coatings
Vacuum furnaces
Vacuum gages
Vacuum pumps
Vacuum transducers

Vacuum brazing 538.1.1, 633.1
DT: January 1993
UF: Brazing—Vacuum*
BT: Brazing
Vacuum applications

Vacuum cleaners 704.2 (633.1)
DT: Predates 1975
BT: Electric appliances
RT: Cleaning
Vacuum applications

Vacuum deposited coatings 813.2 (539.2.1) (633.1)
DT: January 1993
UF: Protective coatings—Vacuum application*
BT: Coatings
RT: Vacuum applications

Vacuum die casting 534.2 (633.1)
DT: January 1993
UF: Die casting—Vacuum*
BT: Die casting
Vacuum applications

Vacuum furnaces (532) (537.2) (642.2) (633.2)
DT: January 1993
UF: Furnaces—Vacuum*
BT: Electric furnaces
NT: Electron beam furnaces
RT: Heat treating furnaces
Vacuum applications

Vacuum gages 944.3, 633.2
DT: January 1993
UF: Vacuum technology—Gages*
BT: Pressure gages
RT: Pressure measurement
Vacuum applications
Vacuum technology

Vacuum plating
USE: Vapor deposition

Vacuum pumps 618.2, 633.2
DT: Predates 1975
BT: Pumps
RT: Compressors
Ejectors (pumps)
Reciprocating pumps
Vacuum applications
Vacuum technology

Vacuum technology 633
DT: Predates 1975
BT: Technology
RT: Electron tubes

* Former Ei Vocabulary term

Vacuum technology *(continued)*
Vacuum gages
Vacuum pumps

*Vacuum technology—Gages**
USE: Vacuum gages

*Vacuum technology—Getters**
USE: Getters

Vacuum transducers 633.2, 944.3
DT: January 1993
UF: Pressure transducers—Vacuum*
BT: Pressure transducers
RT: Vacuum applications

Vacuum tubes
USE: Electron tubes

Vacuum welding 538.2.1, 633.2
DT: January 1993
UF: Welding—Vacuum*
BT: Pressure welding
Vacuum applications
RT: Electron beam welding
Laser beam welding

Valuation (resources)
USE: Resource valuation

Value analysis
USE: Value engineering

Value engineering 911.5
DT: Predates 1975
UF: Value analysis
BT: Engineering
RT: Operations research
Optimization
Quality control
Standards
Systems analysis

Valves (electron)
USE: Electron tubes

Valves (mechanical) 601.2 (619.1.1) (619.2.1)
SN: Includes valve gear
DT: January 1993
UF: Locomotives—Valve gear*
Valves and valve gear*
Valves*
BT: Mechanical control equipment
NT: Pressure relief valves
Safety valves
RT: Actuators
Diaphragms
Hydraulic equipment
Packing
Pipelines
Seals
Water hammer

*Valves and valve gear**
USE: Valves (mechanical)

*Valves**
USE: Valves (mechanical)

Van Allen belts
USE: Radiation belts

Van de Graaff accelerators 932.1.1
DT: January 1993
UF: Accelerators, Van de Graaff*
BT: Electrostatic accelerators

Van de Graaff generators 932.1.1 (701.1)
DT: January 1993
UF: Electrostatic generators, Van de Graaff*
BT: Electrostatic generators

Van der Waals forces 801.4, 931.3
DT: January 1993
UF: Physical chemistry—Van der Waals force*
BT: Mechanics
RT: Physical chemistry

Vanadate minerals 482.2
DT: January 1993
UF: Mineralogy—Vanadates*
BT: Minerals
Vanadium compounds
RT: Vanadium

Vanadium 543.6
DT: January 1993
UF: Cast iron—Vanadium content*
Slags—Vanadium recovery*
Vanadium and alloys*
BT: Refractory metals
RT: Vanadate minerals
Vanadium alloys
Vanadium compounds
Vanadium deposits
Vanadium metallurgy
Vanadium metallography
Vanadium ore treatment

Vanadium alloys 543.6
DT: January 1993
UF: Vanadium and alloys*
BT: Refractory alloys
RT: Vanadium
Vanadium compounds

*Vanadium and alloys**
USE: Vanadium OR
Vanadium alloys

Vanadium compounds (804.1) (804.2)
DT: Predates 1975
BT: Refractory metal compounds
NT: Vanadate minerals
RT: Vanadium
Vanadium alloys

Vanadium deposits 504.3, 543.6
DT: Predates 1975
BT: Ore deposits
RT: Vanadium

Vanadium metallography 531.2, 543.6
DT: Predates 1975
BT: Metallography
RT: Vanadium
Vanadium metallurgy

Vanadium metallurgy 531.1, 543.6
DT: Predates 1975
BT: Metallurgy
RT: Vanadium
Vanadium metallography
Vanadium ore treatment

Vanadium ore treatment 533.1, 543.6
DT: Predates 1975
BT: Ore treatment
RT: Vanadium
Vanadium metallurgy

Vane pumps 618.2
DT: January 1993
UF: Pumps—Vanes*
BT: Rotary pumps
RT: Centrifugal pumps
Gear pumps

Vapor density
USE: Density of gases

Vapor deposition (539.3) (633.2) (802.2) (813.1)
DT: January 1993
UF: Vacuum plating
BT: Deposition
NT: Chemical vapor deposition
RT: Coating techniques
Epitaxial growth
Gas alloying
Metallizing
Substrates
Thin films
Vaporization

Vapor liquid equilibria
USE: Phase equilibria

Vapor lock 523 (612.1)
DT: January 1993
UF: Fuels—Vapor lock*
RT: Fuel systems
Internal combustion engines
Starting

Vapor phase epitaxial growth
USE: Vapor phase epitaxy

Vapor phase epitaxy 933.1.2 (802.2)
DT: January 1995
UF: CVD epitaxy
Vapor phase epitaxial growth
VPE
BT: Epitaxial growth
NT: Chemical beam epitaxy
Metallorganic vapor phase epitaxy
Molecular beam epitaxy
RT: Chemical vapor deposition

Vapor pressure 641.1 (804)
DT: January 1993
UF: Chemicals—Vapor pressure*
Hydrocarbons—Vapor pressure*
BT: Pressure
Thermodynamic properties

Vaporisation
USE: Vaporization

Vaporization 802.3
DT: January 1993
UF: Vaporisation
Vaporizing
BT: Phase transitions
NT: Evaporation
Sublimation
RT: Ablation
Boiling liquids
Concentration (process)
Desalination
Distillation
Gasification
Heating
Separation
Spraying
Vapor deposition
Vapors

Vaporizing
USE: Vaporization

Vapors (641.1) (804)
DT: Predates 1975
UF: Heat transfer—Vapors*
Vapours
BT: Gases
NT: Steam
RT: Condensation
Exhaust gases
Heat transfer
Smoke
Steam condensate
Vaporization

Vapours
USE: Vapors

Varactors 714.2
DT: Predates 1975
UF: Voltage sensitive capacitors
BT: Capacitors
Semiconductor diodes
RT: Frequency multiplying circuits
Harmonic generation
Parametric amplifiers
Time varying networks

Variable frequency oscillators 713.2
DT: January 1993
UF: Oscillators—Variable frequency*
VCO
Voltage controlled oscillators
BT: Oscillators (electronic)
NT: Swept frequency oscillators

Variable speed drives
602.1 (705.1) (632.2) (632.4)
DT: January 1993
UF: Drives—Variable speed*
Electric drive—Variable speed*
Hydraulic drive—Variable speed*
Mechanical drive—Variable speed*
Pneumatic drive—Variable speed*
BT: Drives

* Former Ei Vocabulary term

Variable speed drives *(continued)*
RT: AC motors
Electric drives
Hydraulic drives
Mechanical drives
Pneumatic drives

Variable speed transmissions 602.2
DT: January 1993
UF: Power transmission—Variable speed*
BT: Transmissions

Variational calculus
USE: Variational techniques

Variational techniques 921.2
DT: January 1993
UF: Mathematical techniques—Variational techniques*
Variational calculus
BT: Mathematical techniques

Variometers
USE: Electric inductors

Varistors 714.2
DT: Predates 1975
UF: Voltage sensitive resistors
BT: Resistors
Semiconductor devices

Varnish 813.2 (413.1) (539.2.2)
DT: Predates 1975
UF: Electric insulating materials—Varnish*
Varnishes
BT: Organic coatings
RT: Drying oils
Electric insulating materials
Enamels
Glazes
Lacquers
Paint
Protective coatings
Sprayed coatings

Varnishes
USE: Varnish

Vat dyes 803 (804.1)
DT: January 1993
UF: Dyes and dyeing—Vat dyes*
BT: Dyes

VCO
USE: Variable frequency oscillators

VCR
USE: Videocassette recorders

VDT
USE: Computer terminals

VDU
USE: Computer terminals

Vector quantization 921.1 (741) (723.2)
DT: January 1995
BT: Image compression
RT: Image coding

Vectors 921.1
DT: January 1993
UF: Mathematical techniques—Vectors*
BT: Linear algebra
RT: Eigenvalues and eigenfunctions
Electric network analysis
Electric network synthesis
Topology

Vegetable oils (804.1) (822.3)
DT: Predates 1975
BT: Oils and fats
NT: Cottonseed oil
Drying oils
Essential oils
RT: Degumming
Food products

*Vegetable oils—Cottonseed**
USE: Cottonseed oil

*Vegetable oils—Degumming**
USE: Degumming

Vehicle actuated signals 432.4 (406)
DT: January 1993
UF: Traffic signs, signals and markings—Vehicle actuation*
BT: Traffic signals

Vehicle locating systems 432.4 (406)
DT: October 1975
RT: Tracking (position)
Traffic control

Vehicle parts and equipment
(652.3) (662.1) (663.1) (671.2) (682.1)
DT: January 1993
BT: Equipment
Vehicles
NT: Aircraft parts and equipment
Boat equipment
Ground vehicle parts and equipment
Railroad car equipment
Spacecraft equipment
Tires
Vehicle springs
Vehicle suspensions
Vehicle transmissions
Vehicle wheels
Water craft parts and equipment
Windshield wipers
Windshields
RT: Snow plows

Vehicle springs
601.2 (652.3) (662.4) (663.2) (671.2) (682.2)
DT: January 1993
UF: Cars—Springs and suspensions*
Locomotives—Springs and suspensions*
Vehicles—Springs and suspensions*
BT: Springs (components)
Vehicle parts and equipment
NT: Automobile springs
RT: Riding qualities

Vehicle suspensions 601.2
DT: January 1993
UF: Cars—Springs and suspensions*
Locomotives—Springs and suspensions*
Vehicles—Springs and suspensions*
BT: Suspensions (components)
Vehicle parts and equipment
NT: Automobile suspensions
RT: Riding qualities

*Vehicle tracking**
USE: Tracking (position)

Vehicle transfer cases
USE: Transfer cases (vehicles)

Vehicle transmissions 602.2
DT: January 1993
UF: Aircraft engines, Jet and turbine—Transmissions*
Earthmoving machinery—Transmissions*
Helicopters—Transmissions*
Locomotives—Transmissions*
Vehicles—Transmission*
BT: Transmissions
Vehicle parts and equipment
NT: Automobile transmissions
RT: Power transmission
Torque converters

Vehicle wheels 601.2
DT: January 1993
UF: Vehicles—Wheels*
BT: Vehicle parts and equipment
Wheels
RT: Axles
Railroad rolling stock

Vehicles (432)
DT: Predates 1975
NT: Air cushion vehicles
Amphibious vehicles
Arctic vehicles
Cycloidal propulsion vehicles
Electric vehicles
Emergency vehicles
Ground vehicles
Light weight vehicles
Military vehicles
Nuclear powered vehicles
Spacecraft
Vehicle parts and equipment
Water craft
RT: Fueling
Guideways
Intelligent vehicle highway systems
Navigation systems
Propulsion
Riding qualities
Traction (friction)
Transportation

*Vehicles—All-wheel drive**
USE: All wheel drive vehicles

*Vehicles—Amphibious**
USE: Amphibious vehicles

*Vehicles—Antarctic expedition**
USE: Arctic vehicles

*Vehicles—Cycloidal propulsion**
USE: Cycloidal propulsion vehicles

*Vehicles—Ground effect**
USE: Air cushion vehicles

*Vehicles—Guideways**
USE: Guideways

*Vehicles—Light weight**
USE: Light weight vehicles

*Vehicles—Magnetic levitation**
USE: Magnetic levitation vehicles

*Vehicles—Navigation systems**
USE: Navigation systems

*Vehicles—Noise**
USE: Acoustic noise

*Vehicles—Off road operation**
USE: Off road vehicles

*Vehicles—Soil factors**
USE: Soils

*Vehicles—Springs and suspensions**
USE: Vehicle springs OR
Vehicle suspensions

*Vehicles—Steering**
USE: Steering

*Vehicles—Transfer cases**
USE: Transfer cases (vehicles)

*Vehicles—Transmission**
USE: Vehicle transmissions

*Vehicles—Wheels**
USE: Vehicle wheels

*Vehicles—Windshields**
USE: Windshields

Vehicular traffic (airports)
USE: Airport vehicular traffic

Vehicular tunnels 401.2 (432)
DT: January 1993
UF: Tunnels and tunneling—Vehicular*
BT: Highway systems
Tunnels
NT: Railroad tunnels
RT: Underpasses

Velocimeters 943.3
DT: January 1983
BT: Instruments
NT: Laser Doppler velocimeters
Ultrasonic Doppler velocimeters
RT: Accelerometers
Anemometers
Flowmeters
Speed indicators
Stroboscopes
Tachometers
Velocity

Velocimeters *(continued)*
Velocity control
Velocity measurement

*Velocimeters—Laser Doppler**
USE: Laser Doppler velocimeters

*Velocimeters—Ultrasonic Doppler**
USE: Ultrasonic Doppler velocimeters

Velocimetry
USE: Velocity measurement

Velocity
DT: January 1993
NT: Acoustic wave velocity
Light velocity
RT: Acceleration
Deceleration
Doppler effect
Electron velocity analyzers
Mechanical properties
Speed
Velocimeters
Velocity control
Velocity measurement

Velocity control 731.3
DT: January 1993
UF: Control, Mechanical variables—Velocity*
BT: Mechanical variables control
RT: Speed regulators
Velocimeters
Velocity
Velocity measurement

Velocity measurement 943.3
DT: January 1993
UF: Mechanical variables measurement—Velocity*
Velocimetry
BT: Mechanical variables measurement
NT: Ultrasonic velocity measurement
RT: Electron velocity analyzers
Pressure measurement
Velocimeters
Velocity
Velocity control

Velocity modulation tubes 714.1
DT: January 1993
UF: Electron tubes, Velocity modulation*
BT: Electron wave tubes
RT: Modulation

Velocity of light
USE: Light velocity

Velocity of sound
USE: Acoustic wave velocity

Vending machines 601.1
DT: Predates 1975
BT: Coin operated equipment
Machinery
NT: Ticket issuing machines

*Veneer**
USE: Veneers

Veneers 811.2
DT: January 1993
UF: Veneer*
RT: Coatings
Facings
Furniture manufacture
Plywood
Wood laminates
Wood products

Ventilation 643.5
DT: Predates 1975
BT: Climate control
NT: Mine ventilation
RT: Air conditioning
Air filters
Air purification
Blowers
Cooling
Domestic appliances
Environmental engineering
Environmental testing
Fans
Life support systems (spacecraft)
Refrigeration
Temperature control
Ventilation codes
Ventilation ducts
Ventilation exhausts
Vents

Ventilation codes 643.5, 902.2
DT: January 1993
UF: Ventilation—Codes*
BT: Codes (standards)
RT: Building codes
Laws and legislation
Ventilation

Ventilation ducts 619.1, 643.5
DT: January 1993
UF: Ventilation—Ducts*
BT: Ducts
RT: Ventilation
Vents

Ventilation exhausts 619.1, 643.5 (451.1)
DT: January 1993
UF: Exhausts (ventilation)
Ventilation—Exhausts*
BT: Equipment
RT: Ventilation
Vents

*Ventilation—Codes**
USE: Ventilation codes

*Ventilation—Ducts**
USE: Ventilation ducts

*Ventilation—Exhausts**
USE: Ventilation exhausts

*Ventilation—Noise**
USE: Acoustic noise

* Former Ei Vocabulary term

Vents (522) (619.1.1) (619.2.1)
DT: January 1993
UF: Gas appliances—Vents*
Tanks—Vents*
BT: Components
RT: Gas appliances
Nozzles
Tanks (containers)
Ventilation
Ventilation ducts
Ventilation exhausts

Vertical takeoff and landing aircraft
USE: VTOL/STOL aircraft

Very large scale integration
USE: VLSI circuits

Vibrating (concrete)
USE: Concrete vibrating

Vibrating conveyors 692.1
DT: January 1993
UF: Conveyors—Shaking*
Conveyors—Vibrating*
Oscillating conveyors
BT: Conveyors
RT: Vibrators

Vibrating screens 605.1
DT: January 1993
UF: Screens and sieves—Vibrating*
BT: Screens (sizing)
RT: Vibrators

Vibration control 731.3
DT: January 1993
UF: Vibrations—Control*
BT: Mechanical variables control
RT: Balancing
Damping
Vibration measurement
Vibrations (mechanical)

Vibration measurement 943.2
DT: January 1993
UF: Vibrations—Measurements*
BT: Mechanical variables measurement
RT: Acoustic variables measurement
Stroboscopes
Vibration control
Vibrations (mechanical)

Vibrations (lattice)
USE: Lattice vibrations

Vibrations (mechanical) 931.1
SN: Periodic or random, free or forced
DT: January 1993
BT: Dynamics
NT: Machine vibrations
RT: Acoustic noise
Acoustic waves
Acoustics
Balancing
Compaction
Concrete vibrating
Damping
Drilling
Elastic waves
Oscillations
Resonance
Rock drills
Shock absorbers
Structural design
Vibration control
Vibration measurement
Vibrators

Vibrations (molecular)
USE: Molecular vibrations

*Vibrations—Control**
USE: Vibration control

*Vibrations—Damping**
USE: Damping

*Vibrations—Measurements**
USE: Vibration measurement

Vibrators 601.1 (691.1) (752.1)
SN: Devices for producing mechanical vibrations
DT: Predates 1975
UF: Materials handling—Vibrators*
BT: Oscillators (mechanical)
RT: Acoustic generators
Materials handling equipment
Vibrating conveyors
Vibrating screens
Vibrations (mechanical)
Vibrotactile aids

Vibratory finishing (539) (604.2) (606.2) (753.3)
DT: January 1993
UF: Metal finishing—Vibration*
BT: Finishing
RT: Ultrasonic cleaning

Vibrotactile aids 461.5
SN: For blind and/or deaf persons
DT: January 1993
UF: Human rehabilitation engineering—Vibrotactile aids*
BT: Human rehabilitation equipment
RT: Human rehabilitation engineering
Vibrators

Video amplifiers 713.1, 716.4
DT: January 1993
UF: Amplifiers, Video*
BT: Amplifiers (electronic)
Television equipment
RT: Radar receivers

Video cameras 716.4, 742.2
DT: January 1993
UF: Television cameras
Television equipment—Cameras*
BT: Cameras
Television equipment
RT: Television
Television studios

Video cassette recorders
USE: Videocassette recorders

* Former Ei Vocabulary term

Video coding
USE: Image coding

Video compression
USE: Image compression

Video conferencing
USE: Teleconferencing

Video display screens
USE: Fluorescent screens

Video equipment
USE: Television equipment

Video recording 716.4
SN: Recording of signals in the video-frequency range
DT: Predates 1975
UF: Television recording
Television—Recording*
BT: Recording
RT: Stereophonic recordings
Tape recorders
Television
Videocassette recorders

*Video recording—Cassette recorders**
USE: Videocassette recorders

*Video recording—Laser disk**
USE: Videodisks

Video signal processing 716.4
DT: January 1993
UF: Signal processing—Video signals*
BT: Signal processing
RT: Image processing
Signal generators
Synchronization
Visual communication

Video telephone equipment 716.4, 718.1
DT: January 1993
UF: Telephone equipment—Video telephone*
Videophone equipment
BT: Telephone equipment
RT: Teleconferencing
Telephone systems
Television applications
Television equipment
Television systems
Visual communication

Videocassette recorders 716.4, 752.2.1
DT: January 1993
UF: VCR
Video cassette recorders
Video recording—Cassette recorders*
BT: Tape recorders
Television equipment
RT: Video recording

Videodisks 716.4, 741.3 (744.9)
DT: January 1993
UF: Laser disks
Video recording—Laser disk*
BT: Television equipment
RT: Laser applications
Optical disk storage

Videophone equipment
USE: Video telephone equipment

Videotex 716.4, 718.1, 903.3
DT: January 1993
UF: Information retrieval systems—Teletext and videotex*
Teletext
Videotext
Viewdata
BT: Telecommunication services
RT: Data communication systems
Information dissemination
Information services
Office automation
Telephone systems
Television applications

Videotext
USE: Videotex

Viewdata
USE: Videotex

Vinyl resins 815.1.1
DT: Predates 1975
UF: Polyvinyl resins
BT: Thermoplastics
NT: Polyvinyl acetates
Polyvinyl alcohols
Polyvinyl chlorides
Polyvinylidene chlorides
RT: Plastisols

Virtual environment
USE: Virtual reality

Virtual reality 723
DT: January 1995
UF: Artificial reality
Virtual environment
RT: Computer graphics
Flight simulators
User interfaces

Virtual storage 722.1
DT: January 1993
UF: Data storage, Digital—Virtual*
Paging (virtual storage)
VS
BT: Digital storage
RT: Computer operating systems
File organization

Viruses 461.9 (801.2)
DT: January 1986
BT: Microorganisms
NT: Bacteriophages
Oncogenic viruses
RT: Biological materials
Interferons
Microbiology

Viruses (computer crime)
USE: Computer viruses

* Former Ei Vocabulary term

*Viruses—Bacterial**
USE: Bacteriophages

*Viruses—Bacteriophage**
USE: Bacteriophages

*Viruses—Oncogenic**
USE: Oncogenic viruses

Viscoelasticity 931.2 (421)
DT: October 1975
BT: Mechanical properties
RT: Creep
Elasticity
High temperature testing
Internal friction
Low temperature testing
Relaxation processes
Rheology
Stress relaxation
Viscoplasticity
Viscosity

Viscometers 943.3
DT: Predates 1975
BT: Instruments
RT: Viscosity
Viscosity measurement

Viscometry
USE: Viscosity measurement

Viscoplasticity 931.2 (421)
DT: January 1986
BT: Mechanical properties
RT: Plasticity
Rheology
Viscoelasticity
Viscosity

Viscosity 631.1, 931.2
DT: Predates 1975
BT: Transport properties
NT: Viscosity of gases
Viscosity of liquids
RT: Fluidity
Hydrogels
Internal friction
Non Newtonian liquids
Prandtl number
Reynolds number
Rheology
Stabilizers (agents)
Viscoelasticity
Viscometers
Viscoplasticity
Viscosity measurement
Viscous flow

Viscosity measurement 943.3
DT: January 1993
UF: Viscometry
Viscosity—Measurement*
BT: Measurements
RT: Viscometers
Viscosity
Viscous flow

Viscosity of gases 931.2, 631.1
DT: January 1993
UF: Gases—Viscosity*
BT: Viscosity
RT: Fluid dynamics
Gases

Viscosity of liquids 931.2, 631.1
DT: January 1993
UF: Liquids—Viscosity*
BT: Viscosity
RT: Hydrodynamics
Liquids
Lubrication

*Viscosity—Measurement**
USE: Viscosity measurement

Viscous flow 631.1
DT: January 1993
UF: Flow of fluids—Viscous*
BT: Flow of fluids
RT: Aerodynamics
Computational fluid dynamics
Laminar flow
Navier Stokes equations
Turbulent flow
Viscosity
Viscosity measurement

Visibility 741.2
DT: Predates 1975
NT: Glare effects
Underwater visibility
RT: Atmospheric optics
Color
Flickering
Fog dispersal
Illuminating engineering
Light
Light transmission
Optical properties
Vision

*Visibility—Flicker effects**
USE: Flickering

*Visibility—Glare effects**
USE: Glare effects

*Visibility—Underwater**
USE: Underwater visibility

Vision (461.4) (741.2)
DT: Predates 1975
UF: Vision—Anatomy*
BT: Sensory perception
NT: Binocular vision
Color vision
Depth perception
RT: Biology
Computer vision
Eye movements
Eye protection
Illuminating engineering
Ophthalmology
Optics

* Former Ei Vocabulary term

Vision *(continued)*
Optometers
Physiology
Visibility
Vision aids

Vision (artificial)
USE: Computer vision

Vision aids 741.3, 461.5
DT: January 1993
UF: Correction of vision
Vision correction
Vision—Sensory aids*
BT: Sensory aids
NT: Contact lenses
Eyeglasses
Intraocular lenses
RT: Ophthalmology
Prosthetics
Vision

Vision correction
USE: Vision aids

*Vision, Color**
USE: Color vision

*Vision—Anatomy**
USE: Vision

*Vision—Binocular effect**
USE: Binocular vision

*Vision—Contact lenses**
USE: Contact lenses

*Vision—Eye movements**
USE: Eye movements

*Vision—Eyeglasses**
USE: Eyeglasses

*Vision—Intraocular lenses**
USE: Intraocular lenses

*Vision—Physiology**
USE: Physiology

*Vision—Sensory aids**
USE: Vision aids

Visual communication 717.1 (716.4)
DT: January 1993
BT: Telecommunication
RT: Broadband networks
Optical communication
Packet switching
Teleconferencing
Television
Video signal processing
Video telephone equipment

Visualisation
USE: Visualization

Visualization
DT: January 1993
UF: Visualisation
NT: Flow visualization

Visualization (flow)
USE: Flow visualization

Vitamins (461.7) (801.2) (804.1)
DT: January 1987
BT: Drug products
RT: Food products
Nutrition

Vitreous fibers
USE: Ceramic fibers

Vitreous silica
USE: Fused silica

Vitrification 802.2 (622.5) (812.3)
DT: January 1993
BT: Chemical reactions
NT: Radioactive waste vitrification
RT: Amorphization
Glass
Glass transition

VLSI circuits 714.2
DT: January 1993
UF: Integrated circuits, VLSI*
Very large scale integration
BT: LSI circuits
NT: ULSI circuits
WSI circuits
RT: Application specific integrated circuits

VOC
USE: Volatile organic compounds

Vocabulary control 903.1
DT: January 1993
UF: Controlled vocabularies
Information science—Vocabulary control*
Thesauri
BT: Information analysis
RT: Indexing (of information)

Vocoders 752.2 (716.3) (716.4) (718.1) (751.5)
DT: Predates 1975
UF: Voice coders
BT: Frequency synthesizers
Telephone equipment
RT: Frequency modulation
Pulse code modulation
Radio communication
Radio telephone
Signal encoding
Speech synthesis
Speech transmission

Voice
USE: Speech

Voice activated input devices
752.2.1 (461.5) (462.1) (722.2) (723.5) (751.5)
DT: January 1993
BT: Interactive devices
RT: Computer keyboard substitutes (handicapped aid)

Voice coders
USE: Vocoders

* Former Ei Vocabulary term

Voice communication
USE: Speech communication

Voice input
USE: Speech recognition

Voice recognition
USE: Speech recognition

Voice synthesis
USE: Speech synthesis

Voice transmission
USE: Speech transmission

Voice/data communication systems (716.3) (716.4) (718.1) (751.5)
SN: Integrated systems for communicating both voice and digital data
DT: January 1993
UF: Broadband ISDN
Data and voice communications
Digital communication systems—Voice/data integrated services*
Integrated services digital network
Integrated voice/data services
ISDN
Voice/data integrated services
BT: Digital communication systems
RT: Asynchronous transfer mode
Speech transmission

Voice/data integrated services
USE: Voice/data communication systems

Volatile organic compounds 801.4 (451.1)
DT: January 1995
UF: VOC
BT: Organic compounds
RT: Air pollution
Industrial emissions
Land fill
Water pollution

Volcanic rocks 482.2 (481.1.2)
DT: January 1993
UF: Geochemistry—Volcanic rocks*
BT: Igneous rocks
NT: Basalt
RT: Geochemistry
Petrology
Volcanoes

Volcanoes 484
DT: Predates 1975
BT: Landforms
RT: Earthquakes
Geology
Geothermal energy
Volcanic rocks

Voltage comparators
USE: Comparator circuits

Voltage control 731.3
DT: January 1993
UF: Control, Electric variables—Voltage*
Electric lines—Voltage regulation*
Voltage regulation
BT: Electric variables control
RT: Electric equipment protection
Electric transformers
Voltage dividers
Voltage measurement
Voltage regulators

Voltage controlled oscillators
USE: Variable frequency oscillators

Voltage distribution measurement 942.2
DT: January 1993
UF: Electric measurements—Voltage distribution*
BT: Electric variables measurement
RT: Electric breakdown
Surface discharges
Voltage measurement

Voltage dividers 704.2
DT: June 1990
UF: Dividers (voltage)
Potential dividers
BT: Electric equipment
NT: Electric autotransformers
Potentiometers (resistors)
RT: Resistors
Voltage control

Voltage measurement 942.2
DT: January 1993
UF: Electric measurements—Voltage*
BT: Electric variables measurement
RT: Electrometers
Potentiometers (electric measuring instruments)
Voltage control
Voltage distribution measurement
Voltmeters

Voltage regulation
USE: Voltage control

Voltage regulators 732.1
DT: June 1990
BT: Electric control equipment
NT: Voltage stabilizing circuits
RT: Voltage control

Voltage sensitive capacitors
USE: Varactors

Voltage sensitive resistors
USE: Varistors

Voltage stabilizing circuits 713.5
DT: January 1993
UF: Electronic circuits, Voltage stabilizing*
BT: Networks (circuits)
Voltage regulators

Voltmeters 942.1
DT: Predates 1975
UF: Electric field meters
BT: Electric measuring instruments
NT: Digital voltmeters
Electronic voltmeters
Potentiometers (electric measuring instruments)

* Former Ei Vocabulary term

Voltmeters *(continued)*
RT: Ammeters
Voltage measurement

*Voltmeters, Digital**
USE: Digital voltmeters

*Voltmeters—Electronic**
USE: Electronic voltmeters

Volume acoustic wave devices
USE: Acoustic bulk wave devices

Volume control (gain)
USE: Gain control

Volume control (spatial) 731.3
DT: January 1993
UF: Control, Mechanical variables—Volumes*
BT: Spatial variables control
RT: Density control (specific gravity)
Volume measurement

Volume fraction 641.1
DT: January 1995
RT: Fractionation
Gases
Liquids
Phase composition
Solids

Volume measurement 943.2
DT: January 1993
UF: Mechanical variables measurement—Volumes*
BT: Spatial variables measurement
RT: Density measurement (specific gravity)
Volume control (spatial)

Volumetric analysis 801
DT: Predates 1975
BT: Chemical analysis
NT: Titration
RT: Gas fuel analysis
Gravimetric analysis
Measurements
Microanalysis
Polarographic analysis

Vortex flow 631.1
DT: January 1993
UF: Flow of fluids—Vortex flow*
Vortex tubes*
Vortexes
Vortices
BT: Flow of fluids
RT: Cavitation
Mixing
Turbulence
Turbulent flow
Vortex shedding
Wakes

Vortex shedding 631.1 (931.1)
DT: January 1995
RT: Fluid structure interaction
Vortex flow

*Vortex tubes**
USE: Vortex flow

Vortexes
USE: Vortex flow

Vortices
USE: Vortex flow

Voting machines 601.1
DT: Predates 1975
BT: Machinery

VPE
USE: Vapor phase epitaxy

VS
USE: Virtual storage

VTOL/STOL aircraft 652.1
DT: January 1993
UF: Aircraft, VTOL/STOL*
Short takeoff and landing aircraft
STOL aircraft
Vertical takeoff and landing aircraft
BT: Aircraft
RT: Compound helicopters
VTOL/STOL services

VTOL/STOL services 431.4 (652.1) (652.4)
DT: January 1993
UF: Airports—VTOL/STOL services*
BT: Air transportation
RT: VTOL/STOL aircraft

Vulcanization 802.2, 818.3
DT: Predates 1975
BT: Chemical reactions
RT: Curing
Elastomers
Polymers
Rubber
Vulcanization agents

Vulcanization agents 803 (818.3.1)
DT: January 1993
UF: Vulcanization—Agents*
BT: Agents
RT: Vulcanization

*Vulcanization—Agents**
USE: Vulcanization agents

W
USE: Tungsten

Wafer scale integration
USE: WSI circuits

*Wage payment plans**
USE: Wages

Wages 912.4 (911.1)
DT: January 1993
UF: Personnel—Wages*
Salaries
Wage payment plans*
BT: Compensation (personnel)
RT: Personnel

Wakes 631.1 (655.2) (656.1)
DT: January 1993
UF: Flow of fluids—Wakes*
Satellites—Wakes*
BT: Turbulent flow
RT: Cavitation
Drag
Ground effect
Turbulence
Vortex flow

Walkers
USE: Walking aids

Walking aids 461.5, 462.1
SN: Includes canes, crutches, and walkers
DT: January 1993
UF: Crutches
Human rehabilitation engineering—Walking aids*
Walkers
BT: Human rehabilitation equipment
RT: Human rehabilitation engineering
Orthotics
Prosthetics

Wall flow 631.1 (651.2)
DT: January 1993
UF: Wind tunnels—Wall interference*
BT: Flow of fluids
RT: Boundary layer flow
Heat transfer
Laminar flow
Turbulence
Turbulent flow
Wind tunnels

Wall rock 482.2 (504) (505.1)
DT: January 1993
UF: Ore deposits—Wall rock alteration*
BT: Ore deposits

Walls (quay)
USE: Quay walls

Walls (retaining)
USE: Retaining walls

Walls (structural partitions) 408.2 (402)
DT: January 1993
UF: Buildings—Walls*
Heat transfer—Walls*
BT: Building components
NT: Shear walls
RT: Enclosures
Floors
Sandwich structures
Shells (structures)
Studs (structural members)

Walsh transforms 921.3
DT: January 1993
UF: Mathematical transformations—Walsh transforms*
BT: Mathematical transformations
RT: Signal filtering and prediction
Signal processing

WAN
USE: Wide area networks

Wankel engines 612.1
DT: January 1993
UF: Internal combustion engines—Wankel*
RC engines
Rotary combustion engines
BT: Internal combustion engines
Rotary engines

Warehouses 694.4 (402.1)
DT: Predates 1975
UF: Materials handling—Warehouses*
BT: Facilities
NT: Collapsible warehouses
RT: Materials handling
Materials handling equipment
Storage (materials)

*Warehouses—Collapsible**
USE: Collapsible warehouses

Warfare
USE: Military operations

Warm springs
USE: Geothermal springs

Warning systems
USE: Alarm systems

Warship preservation 672.1 (404.1)
DT: January 1993
UF: Preservation (warships)
Warships—Preservation*
RT: Military operations
Warships

Warships 672.1 (404.1)
SN: Combat-class naval vessels
DT: Predates 1975
BT: Naval vessels
NT: Aircraft carriers
RT: Degaussing
Missile launching systems
Warship preservation

*Warships—Ammunition handling**
USE: Ammunition AND
Materials handling

*Warships—Degaussing**
USE: Degaussing

*Warships—Missile launching systems**
USE: Missile launching systems

*Warships—Preservation**
USE: Warship preservation

Wash boring
USE: Jet drilling

Washers 601.2
SN: Fasteners; for cleaning machines use WASHING MACHINES
DT: Predates 1975
BT: Fasteners

* Former Ei Vocabulary term

Washers *(continued)*
RT: Bolts
Screws

Washing (503.2) (802.3) (811.1.1) (819.5)
DT: January 1993
UF: Coal washing
Granular materials—Washing*
Pulp manufacture—Washing*
BT: Cleaning
NT: Laundering
RT: Leaching
Purification
Solutions
Washing machines

Washing machines 601.1
DT: Predates 1975
UF: Tunnels and tunneling—Washing machines*
BT: Machinery
NT: Bottle washing machines
RT: Cleaning
Domestic appliances
Electric appliances
Washing

Waste disposal 452.4
SN: General disposal of material waste. Formerly included industrial-process wastes.
DT: Predates 1975
UF: Disposal (waste)
NT: Deep well disposal
Industrial waste disposal
Ocean dumping
Refuse disposal
Sludge disposal
Wastewater disposal
RT: Compaction
Composting
Disposability
Disposable biomedical equipment
Industrial wastes
Industrial waste treatment
Refuse composting
Septic tanks
Sewers
Trickling filtration
Urban planning
Waste treatment
Waste utilization
Wastes

*Waste disposal—Compaction**
USE: Compaction

*Waste disposal—Composting**
USE: Composting

*Waste disposal—Deepwell disposal**
USE: Deep well disposal

*Waste disposal—Incineration**
USE: Waste incineration

*Waste disposal—Ocean dumping**
USE: Ocean dumping

Waste heat 525.4 (454.2) (614.1) (642.1) (643.1)
DT: January 1993
UF: Boilers—Waste heat*
BT: Wastes
RT: Heating
Thermal plumes
Thermal pollution

Waste heat recovery
USE: Waste heat utilization

Waste heat utilization
525.3 (454.2) (614.1) (642.1) (643.1)
DT: Predates 1975
UF: Electric lighting—Heat utilization*
Heat recovery
Recovery (waste heat)
Utilization (waste heat)
Waste heat recovery
BT: Waste utilization
RT: Electric lighting
Heat exchangers
Heating

Waste incineration 452.4
DT: January 1993
UF: Incineration (waste)
Waste disposal—Incineration*
BT: Combustion
NT: Refuse incineration
RT: Refuse disposal

Waste liquor utilization 452.4 (522) (811.1.1)
DT: January 1993
UF: Gas manufacture—Waste liquor utilization*
Pulp manufacture—Waste liquor utilization*
BT: Waste utilization
RT: Gas fuel manufacture
Pulp manufacture
Tall oil

Waste materials
USE: Wastes

Waste paper 452.3, 811.1
DT: January 1993
UF: Pulp materials—Waste paper*
BT: Paper
Wastes
RT: Pulp materials
Recycling
Waste utilization

Waste reclamation
USE: Reclamation

Waste treatment 452.4
DT: January 1993
NT: Anaerobic digestion
Industrial waste treatment
Refuse digestion
Sewage treatment
Sludge digestion
Wastewater treatment
RT: Waste disposal
Wastes

Waste utilization (452.3)
DT: Predates 1975
UF: Utilization (wastes)
NT: Composting
Waste heat utilization
Waste liquor utilization
RT: Industrial wastes
Metal recovery
Nuclear fuel reprocessing
Reclamation
Recycling
Refuse derived fuels
Slags
Waste disposal
Waste paper
Wastes

*Waste utilization—Wastewater reclamation**
USE: Wastewater reclamation

Waste water
USE: Wastewater

Wastes (452.3)
DT: January 1993
UF: Waste materials
NT: Agricultural wastes
Effluents
Exhaust gases
Industrial wastes
Particulate emissions
Radioactive wastes
Sewage
Soot
Space debris
Waste heat
Waste paper
Wastewater
Wood wastes
RT: Ash handling
Low grade fuel firing
Waste disposal
Waste treatment
Waste utilization

Wastewater 452.3
DT: January 1993
UF: Waste water
BT: Wastes
Water
RT: Wastewater disposal
Wastewater reclamation
Wastewater treatment
Water pollution

Wastewater disposal 452.4 (454.2)
DT: January 1993
UF: Oil fields—Wastewater disposal*
Oil shale processing—Wastewater disposal*
BT: Waste disposal
RT: Nuclear power plants
Oil fields
Oil shale processing
Papermaking
Wastewater
Wastewater reclamation
Wastewater treatment
Water pollution

Wastewater reclamation 452.4 (454.2)
DT: January 1993
UF: Industrial wastes—Water reclamation*
Sewage treatment—Water reclamation*
Waste utilization—Wastewater reclamation*
Water reclamation
BT: Reclamation
RT: Industrial wastes
Sewage treatment
Wastewater
Wastewater disposal
Wastewater treatment
Water conservation
Water pollution
Water recycling
Water treatment

Wastewater treatment 452.4
DT: June 1990
BT: Waste treatment
Water treatment
RT: Industrial waste treatment
Wastewater
Wastewater disposal
Wastewater reclamation
Water conservation
Water pollution
Water recycling

Watches 943.3
DT: Predates 1975
BT: Timing devices
NT: Electric watches
Stop watches
RT: Clocks
Time measurement

*Watches—Electric operation**
USE: Electric watches

Water (444) (804.2)
DT: Predates 1975
BT: Hydrogen inorganic compounds
NT: Cooling water
Groundwater
Heavy water
Potable water
Saline water
Wastewater
RT: Erosion
Flow of water
Glaciers
Ice
Liquids
Rain
Snow
Steam
Steam condensate
Surface waters
Thermal stratification
Water absorption
Water aeration
Water bacteriology

Water *(continued)*
Water conservation
Water filtration
Water levels
Water pollution
Water pollution control
Water power
Water resources
Water supply
Water treatment
Water waves
Water wells

Water absorption 802.3
DT: January 1993
UF: Leather—Water absorption*
BT: Absorption
RT: Water

Water aeration 445.1 (802.3)
DT: January 1993
UF: Aeration (water)
Water treatment—Aeration*
BT: Water treatment
RT: Water

Water analysis 445.2, 801
DT: Predates 1975
BT: Chemical analysis
NT: Feedwater analysis
RT: Water bacteriology
Water pollution
Water supply
Water treatment

*Water analysis—Radioactivity**
USE: Radioactivity

Water bacteriology 445.2, 801.2 (914.3) (461.7)
DT: Predates 1975
UF: Reservoirs—Bacteriology*
BT: Bacteriology
RT: Reservoirs (water)
Water
Water analysis
Water pollution
Water supply

Water boilers
USE: Boilers

Water borne coatings 813.2
DT: January 1993
UF: Protective coatings—Water borne*
BT: Coatings

Water conservation 444 (403.1)
DT: January 1993
UF: Water resources—Conservation*
Water—Conservation*
BT: Conservation
RT: Regional planning
Wastewater reclamation
Wastewater treatment
Water
Water quality
Water recycling
Water resources

Water cooled reactors (621.1) (621.2)
DT: January 1993
UF: Nuclear reactors, Water cooled*
BT: Nuclear reactors
NT: Boiling water reactors
Heavy water reactors
Light water reactors
Pressurized water reactors

Water cooling systems 616.1 (802.1)
SN: Systems for cooling by means of water
DT: Predates 1975
UF: Industrial plants—Water cooling systems*
BT: Cooling systems
NT: Water cooling towers

Water cooling towers 616.1 (802.1)
DT: Predates 1975
BT: Cooling towers
Water cooling systems

Water craft 674.1
DT: January 1993
BT: Vehicles
NT: Boats
Rescue vessels
Sailing vessels
Ships
Water craft parts and equipment
RT: Shipbuilding

Water craft parts and equipment 671.2
DT: January 1993
BT: Vehicle parts and equipment
Water craft
NT: Marine boilers
Marine engines
Ship equipment
Stabilizers (marine vessel)
RT: Marine couplings
Marine signal systems

Water distribution systems 446.1
DT: Predates 1975
BT: Water supply systems
NT: Cross connections (water distribution)
Hot water distribution systems
Water piping systems
RT: Water meters
Water supply
Water tanks
Water towers

*Water distribution systems—Cross connections**
USE: Cross connections (water distribution)

*Water distribution systems—Waterworks**
USE: Waterworks

Water filtration 445.1, 802.3
DT: Predates 1975
BT: Filtration
RT: Water
Water pollution control
Water supply
Water treatment

Water hammer 632.2 (631.1)
DT: Predates 1975
RT: Hydraulic equipment
Hydrodynamics
Pipe flow
Plumbing
Valves (mechanical)
Water pipelines
Water piping systems

Water heaters (616.1) (643)
DT: Predates 1975
BT: Heating equipment
NT: Feedwater heaters
Solar water heaters
RT: Domestic appliances
Electric appliances
Heat exchangers
Hot water heating

*Water heaters—Gas**
USE: Gas appliances

*Water heaters—Solar**
USE: Solar water heaters

Water injection 612.1
SN: Introduction of water to an internal combustion engine to enhance combustion power for quick takeoff
DT: January 1993
UF: Internal combustion engines—Water injection*
RT: Combustion
Internal combustion engines

Water landing (spacecraft) 655.1
DT: January 1993
UF: Spacecraft—Water impact*
BT: Spacecraft landing
RT: Spacecraft

Water levels (407.2) (614.2)
DT: January 1993
UF: Boiler control—Water level*
Inland waterways—Water level*
RT: Boilers
Inland waterways
Water

Water lines
USE: Water piping systems

Water meters 943.3 (446.1)
DT: Predates 1975
BT: Flowmeters
RT: Utility rates
Water distribution systems
Water supply

Water pipelines 446.1, 619.1
SN: Stationary means of transporting water over distances. For systems of pipes used to distribute water within a building or facility, use WATER PIPING SYSTEMS
DT: Predates 1975
BT: Pipelines
Water supply systems
RT: Irrigation
Plumbing
Water hammer
Water piping systems
Water supply

*Water pipelines—Cement asbestos**
USE: Asbestos cement

*Water pipelines—Tapping**
USE: Tapping (threads)

Water piping systems 446.1, 619.1 (643.2) (802.1)
SN: Systems of pipes to distribute water throughout a building, building complex, or other structure or facility of limited extent. For transportation of water over greater distances use WATER PIPELINES
DT: Predates 1975
UF: Water lines
BT: Piping systems
Water distribution systems
RT: Irrigation
Plumbing
Water hammer
Water pipelines
Water supply

Water pollution 453
DT: Predates 1975
BT: Pollution
NT: Groundwater pollution
Marine pollution
RT: Agricultural runoff
Environmental engineering
Fertilizers
Hazardous materials spills
Industrial wastes
Inland waterways
Landfill linings
Oil spills
Outfalls
Pesticide effects
Phosphates
Pollution induced corrosion
Radioactive materials
Radioactive wastes
Sewage treatment
Soil pollution
Thermal plumes
Thermal pollution
Volatile organic compounds
Wastewater
Wastewater disposal
Wastewater reclamation
Wastewater treatment
Water
Water analysis
Water bacteriology
Water pollution control
Water quality
Water supply

Water pollution control 453.2
DT: January 1993
UF: Water pollution—Control*
BT: Pollution control
RT: Regional planning
Water
Water filtration
Water pollution
Water pollution control equipment
Water supply

Water pollution control equipment 453.2
DT: January 1993
BT: Pollution control equipment
NT: Oil booms
RT: Water pollution control

*Water pollution—Agricultural runoffs**
USE: Agricultural runoff

*Water pollution—Control**
USE: Water pollution control

*Water pollution—Detergents effects**
USE: Detergents

*Water pollution—Marine pollution**
USE: Marine pollution

*Water pollution—Oil spills**
USE: Oil spills

*Water pollution—Pesticide effects**
USE: Pesticide effects

*Water pollution—Radioactive materials**
USE: Radioactive materials

*Water pollution—Underground**
USE: Groundwater pollution

*Water pollution—Waste heat effects**
USE: Thermal pollution

*Water pollution—Water quality**
USE: Water quality

Water power 611
DT: January 1993
UF: Power generation—Water*
BT: Renewable energy resources
NT: Hydroelectric power
Tidal power
Wave power
RT: Power generation
Water

Water quality 445.2 (453.2)
DT: January 1993
UF: Water pollution—Water quality*
Water supply—Water quality*
RT: Environmental engineering
Potable water
Water conservation
Water pollution
Water supply
Water treatment

Water reclamation
USE: Wastewater reclamation

Water recycling (444) (445.1) (453.2)
DT: January 1993
UF: Industrial plants—Water recycling*
BT: Recycling
RT: Wastewater reclamation
Wastewater treatment
Water conservation
Water resources
Water supply
Water treatment

Water reservoirs
USE: Reservoirs (water)

Water resources 444
DT: Predates 1975
BT: Natural resources
NT: Geothermal water resources
Groundwater resources
Surface water resources
RT: Arid regions
Drought
Earth (planet)
Evapotranspiration
Hydrology
Potable water
Recharging (underground waters)
Runoff
Saline water
Salt water intrusion
Snowfall measurement
Springs (water)
Water
Water conservation
Water recycling
Water resources exploration
Water supply
Watersheds
Waterway transportation

Water resources exploration 444
DT: January 1993
UF: Water resources—Exploration*
BT: Natural resources exploration
RT: Arid regions
Water resources
Water supply

*Water resources—Arid regions**
USE: Arid regions

*Water resources—Conservation**
USE: Water conservation

*Water resources—Drought**
USE: Drought

*Water resources—Estuarine**
USE: Estuaries

*Water resources—Evapotranspiration**
USE: Evapotranspiration

*Water resources—Exploration**
USE: Water resources exploration

*Water resources—Groundwater**
USE: Groundwater resources

* Former Ei Vocabulary term

*Water resources—Replenishment**
USE: Replenishment (water resources)

*Water resources—Saline water**
USE: Saline water

*Water resources—Salt water intrusion**
USE: Salt water intrusion

*Water resources—Surface water**
USE: Surface water resources

*Water resources—Thermal**
USE: Geothermal water resources

*Water resources—Underground**
USE: Groundwater resources

Water springs
USE: Springs (water)

Water supply 446.1
DT: Predates 1975
RT: Aquifers
Cooling towers
Dams
Flood control
Hydraulic structures
Hydrology
Irrigation
Lakes
Potable water
Pumping plants
Rain
Regional planning
Reservoirs (water)
Rivers
Runoff
Snow
Snowfall measurement
Urban planning
Water
Water analysis
Water bacteriology
Water distribution systems
Water filtration
Water meters
Water pipelines
Water piping systems
Water pollution control
Water pollution
Water quality
Water recycling
Water resources exploration
Water resources
Water supply systems
Water supply tunnels
Water tanks
Water towers
Water treatment
Water treatment plants
Water wells
Watersheds

Water supply systems 446.1
DT: January 1993
BT: Waterworks
NT: Water distribution systems
Water pipelines
RT: Water supply

Water supply tunnels 401.2, 446.1
DT: Predates 1975
BT: Tunnels
Waterworks
RT: Water supply

*Water supply tunnels—Lining**
USE: Tunnel linings

*Water supply—Water quality**
USE: Water quality

Water tanks 446.1, 619.2
DT: January 1993
UF: Steel water tanks
Water tanks and towers*
BT: Tanks (containers)
RT: Water distribution systems
Water supply
Water towers

*Water tanks and towers**
USE: Water tanks OR
Water towers

*Water tanks and towers—Aluminum**
USE: Aluminum

*Water tanks and towers—Steel**
USE: Steel structures

Water towers 446.1
DT: January 1993
UF: Steel water towers
Water tanks and towers*
BT: Towers
RT: Water distribution systems
Water supply
Water tanks

Water treatment 445.1
SN: General water treatment, and treatment of potable water
DT: Predates 1975
NT: Activated carbon treatment
Activated silica treatment
Biological water treatment
Chemical water treatment
Chemicals removal (water treatment)
Color removal (water treatment)
Fluoridation
Industrial water treatment
Ozone water treatment
Taste control (water treatment)
Wastewater treatment
Water aeration
RT: Algae control
Chlorination
Coagulation
Condensate treatment
Denitrification
Desalination
Disinfection
Flocculation

Water treatment *(continued)*
Flotation
Ion exchange
Lime
Odor control
Ozonization
Phosphates
Potable water
Wastewater reclamation
Water
Water analysis
Water filtration
Water quality
Water recycling
Water supply
Water treatment plants

Water treatment plants 445.1 (402.1) (446.1)
DT: Predates 1975
BT: Facilities
RT: Water supply
Water treatment

*Water treatment plants—Fly control**
USE: Insect control

*Water treatment plants—Lime handling**
USE: Lime

*Water treatment plants—Sludge reclamation**
USE: Reclamation

*Water treatment, Industrial**
USE: Industrial water treatment

*Water treatment—Activated carbon**
USE: Activated carbon treatment

*Water treatment—Activated silica**
USE: Activated silica treatment

*Water treatment—Aeration**
USE: Water aeration

*Water treatment—Algae control**
USE: Algae control

*Water treatment—Biological treatment**
USE: Biological water treatment

*Water treatment—Carbonization**
USE: Carbonization

*Water treatment—Chemicals removal**
USE: Chemicals removal (water treatment)

*Water treatment—Chemicals**
USE: Chemical water treatment

*Water treatment—Chlorination**
USE: Chlorination

*Water treatment—Color removal**
USE: Color removal (water treatment)

*Water treatment—Detergents effects**
USE: Detergents

*Water treatment—Fluoridation**
USE: Fluoridation

*Water treatment—Hydrogen sulfide removal**
USE: Hydrogen sulfide removal (water treatment)

*Water treatment—Iron removal**
USE: Iron removal (water treatment)

*Water treatment—Lead removal**
USE: Lead removal (water treatment)

*Water treatment—Lime**
USE: Lime

*Water treatment—Manganese removal**
USE: Manganese removal (water treatment)

*Water treatment—Ozone**
USE: Ozone water treatment

*Water treatment—Phosphate**
USE: Phosphates

*Water treatment—Radioisotope removal**
USE: Radioisotope removal (water treatment)

*Water treatment—Salt removal**
USE: Desalination

*Water treatment—Taste control**
USE: Taste control (water treatment)

*Water treatment—Tropics**
USE: Tropics

Water tube boilers 614.2
DT: January 1993
UF: Boilers—Water tube*
BT: Boilers

Water wave effects 471.4 (631.1) (671.1)
DT: January 1993
BT: Wave effects
RT: Hurricane effects
Water waves

Water waves 471.4 (631.1)
DT: Predates 1975
UF: Ocean waves
BT: Liquid waves
NT: Tsunamis
RT: Hurricanes
Hydrodynamics
Ocean currents
Oceanography
Ship model tanks
Tides
Water
Water wave effects

*Water waves—Tsunamis**
USE: Tsunamis

*Water waves—Wave energy conversion**
USE: Wave energy conversion

Water well logging 444.2
DT: January 1993
UF: Water wells—Logging*
BT: Well logging
RT: Water wells

Water well pumps 446.2, 618.2
DT: January 1993
BT: Well pumps
RT: Submersible pumps
Water wells

Water wells 444.2
DT: Predates 1975
BT: Wells
NT: Geothermal wells
RT: Brines
Directional drilling
Groundwater resources
Hydraulic fracturing
Recharging (underground waters)
Water
Water supply
Water well logging
Water well pumps

*Water wells—Hydraulic fracturing**
USE: Hydraulic fracturing

*Water wells—Logging**
USE: Water well logging

*Water, Underground**
USE: Groundwater

*Water, Underground—Recharging**
USE: Recharging (underground waters)

*Water—Conservation**
USE: Water conservation

Waterflooding
USE: Well flooding

Waterproof cement 412.1
DT: January 1993
UF: Cement—Waterproof*
BT: Cements
RT: Waterproofing

Waterproof coatings 813.2
DT: January 1993
UF: Coatings—Waterproofing*
BT: Protective coatings
RT: Waterproofing

Waterproofing (802.3)
DT: January 1993
UF: Brick—Waterproofing*
Buildings—Waterproofing*
Concrete construction—Waterproofing*
Concrete—Waterproofing*
Instruments—Waterproofing*
Leather—Waterproofing*
Radio equipment—Waterproofing*
Textiles—Waterproofing*
Tunnels and tunneling—Waterproofing*
BT: Protection
RT: Sealing (closing)
Textile finishing
Tunneling (excavation)
Waterproof cement
Waterproof coatings

Watersheds 444.1
DT: Predates 1975
RT: Lakes
Land use
Natural resources
Rain
Rivers
Runoff
Surface waters
Water resources
Water supply

*Watersheds—Mapping**
USE: Mapping

Waterway transportation 434.1
DT: Predates 1975
BT: Transportation
RT: Boats
Canals
Inland waterways
Merchant marine
Pontoons
Ships
Surface waters
Water resources

Waterworks 446
DT: January 1993
UF: Water distribution systems—Waterworks*
BT: Facilities
NT: Water supply systems
Water supply tunnels
RT: Spillways

Watt hour meters 942.1
DT: Predates 1975
UF: Watthour meters
BT: Electric measuring instruments
RT: Electric power measurement

Watthour meters
USE: Watt hour meters

Wattless power
USE: Reactive power

Wattmeters 942.1
DT: Predates 1975
BT: Electric measuring instruments
RT: Electric power measurement

Wave effects (711) (751.1)
DT: January 1993
BT: Effects
NT: Acoustic wave effects
Radiowave propagation effects
Water wave effects
RT: Waves

Wave energy conversion 615.6 (471.4)
SN: Water waves
DT: January 1993
UF: Water waves—Wave energy conversion*
BT: Energy conversion
RT: Wave power

* Former Ei Vocabulary term

Wave filters 703.2
DT: January 1993
UF: Filters (wave)
BT: Electronic equipment
NT: Acoustic surface wave filters
Electric filters
Optical filters
RT: Electromagnetic waves

Wave interference (711) (751.1)
DT: January 1993
UF: Interference (wave)
NT: Acoustic wave interference
Electromagnetic wave interference
RT: Signal interference
Waves

Wave plasma interactions 932.3
DT: January 1993
BT: Plasma interactions
RT: Electromagnetic waves
Plasmas

Wave power 615.6
SN: Excludes tidal power
DT: January 1993
UF: Power generation—Seawater*
BT: Water power
RT: Power generation
Seawater
Thermal stratification
Tidal power
Wave energy conversion

Wave propagation (711) (751.1)
SN: In gases. For propagation in liquids and solids, use WAVE TRANSMISSION
DT: January 1993
UF: Propagation (wave)
NT: Acoustic wave propagation
Electromagnetic wave propagation
RT: Attenuation
Dispersion (waves)
Distortion (waves)
Ion acoustic waves
Reflection
Refraction
Wave transmission
Waves

Wave superheaters
USE: Shock tubes

Wave transmission (711) (751.1)
SN: In solids and liquids. For transmission in gases, use WAVE PROPAGATION
DT: January 1993
UF: Transmission (waves)
NT: Acoustic wave transmission
Electromagnetic wave transmission
RT: Attenuation
Reflection
Wave propagation
Waves

Waveform analysis (921)
DT: Predates 1975
BT: Analysis
NT: Spectrum analysis
RT: Harmonic analysis

Wavefront reversal
USE: Optical phase conjugation

Wavefronts (711) (741.1) (751.1)
DT: January 1993
BT: Waves
RT: Light

Waveguide attenuators 714.3
DT: Predates 1975
UF: Attenuators (waveguide)
BT: Electric attenuators
Waveguide components

Waveguide circulators 714.3
DT: January 1993
UF: Circulators (waveguide)
Waveguide components—Circulators*
BT: Waveguide components
RT: Microwave circulators

Waveguide components 714.3
DT: Predates 1975
BT: Components
Waveguides
NT: Waveguide attenuators
Waveguide circulators
Waveguide couplers
Waveguide isolators
Waveguide junctions
Waveguide transformers
RT: Ferrite devices
Microwave filters
Microwave isolators
Microwave limiters
Resonators

*Waveguide components—Circulators**
USE: Waveguide circulators

*Waveguide components—Couplers**
USE: Waveguide couplers

*Waveguide components—Isolators**
USE: Waveguide isolators

*Waveguide components—Junctions**
USE: Waveguide junctions

*Waveguide components—Transformers**
USE: Waveguide transformers

Waveguide connections
USE: Waveguide couplers

Waveguide couplers 714.3
DT: January 1993
UF: Connections (waveguide)
Couplers (waveguide)
Waveguide components—Couplers*
Waveguide connections
BT: Waveguide components
RT: Coupled circuits

* Former Ei Vocabulary term

Waveguide isolators 714.3
DT: January 1993
UF: Waveguide components—Isolators*
BT: Waveguide components
RT: Microwave isolators

Waveguide junctions 714.3
DT: January 1993
UF: Junctions (waveguide)
Waveguide components—Junctions*
BT: Waveguide components

Waveguide transformers 714.3
DT: January 1993
UF: Waveguide components—Transformers*
BT: Waveguide components

Waveguides 714.3
DT: Predates 1975
UF: Coplanar waveguides
BT: Electronic equipment
NT: Circular waveguides
Dielectric waveguides
Helical waveguides
Plasma filled waveguides
Rectangular waveguides
Waveguide components
RT: Antenna feeders
Antennas
Coaxial cables
Guided electromagnetic wave propagation
Gyrators
Microwave devices
Telecommunication links
Telephone lines

*Waveguides, Circular**
USE: Circular waveguides

*Waveguides, Dielectric**
USE: Dielectric waveguides

*Waveguides, Helical**
USE: Helical waveguides

*Waveguides, Optical**
USE: Optical waveguides

*Waveguides, Rectangular**
USE: Rectangular waveguides

*Waveguides—Losses**
USE: Electric losses

*Waveguides—Plasma filled**
USE: Plasma filled waveguides

Wavelength division multiplexing
USE: Frequency division multiplexing

Wavelet transforms 921.3
DT: January 1995
BT: Mathematical transformations
RT: Pattern recognition
Signal processing

Wavemeters 942.1
DT: Predates 1975
BT: Electric measuring instruments
RT: Frequency meters
Microwave measurement

Waves (711) (741.1) (751.1)
DT: January 1983
UF: Acceleration waves*
NT: Electromagnetic waves
Mechanical waves
Plasma waves
Surface waves
Wavefronts
RT: Attenuation
Dispersion (waves)
Distortion (waves)
Doppler effect
Oscillations
Phase shift
Radiation
Reflection
Refraction
Wave effects
Wave interference
Wave propagation
Wave transmission

Waxed papers 811.1
DT: January 1993
UF: Paper—Waxed papers*
BT: Paper
RT: Waxes

Waxes 804.1, 821.4
SN: Plant or insect derivatives. For petroleum waxes, use PARAFFIN WAXES
DT: Predates 1975
BT: Lipids
RT: Waxed papers

Waxes (paraffin)
USE: Paraffin waxes

WDM
USE: Frequency division multiplexing

Weapons (guns)
USE: Guns (armament)

Weapons (military)
USE: Ordnance

Wear of materials 931.2 (421)
DT: Predates 1975
UF: Abrasion
Mechanical wear
Wear*
NT: Fatigue of materials
Scour
Seizing
RT: Ablation
Corrosion
Deformation
Durability
Erosion
Failure (mechanical)
Friction
Grinding (machining)
Hardness
Lubrication

Wear of materials *(continued)*
Mechanical properties
Spalling
Textile pilling
Tribology
Wear resistance

Wear resistance 931.2 (421)
DT: January 1993
UF: Abrasion resistance*
Wear resisting*
BT: Mechanical properties
RT: Hardness
Wear of materials

*Wear resisting**
USE: Wear resistance

*Wear**
USE: Wear of materials

Weather balloons
USE: Meteorological balloons

Weather control
USE: Weather modification

Weather forecasting 443
DT: January 1993
UF: Electric utilities—Weather forecasting*
Meteorology—Weather forecasting*
BT: Forecasting
RT: Meteorological balloons
Meteorological problems
Meteorology
Weather information services
Weather satellites

Weather information services
443 (716.3) (716.4) (718.1)
DT: January 1993
UF: Telephone—Weather information service*
BT: Information services
RT: Telephone
Telephone systems
Weather forecasting

Weather modification 443
DT: January 1993
UF: Meteorology—Weather modification*
Weather control
NT: Cloud seeding
RT: Control
Meteorological problems
Meteorology

Weather problems
USE: Meteorological problems

Weather radar
USE: Meteorological radar

Weather satellites 443.2, 655.2
DT: January 1993
UF: Meteorological satellites
Satellites—Weather*
BT: Satellites
RT: Meteorological instruments
Sounding rockets
Weather forecasting

Weathering (802.2) (421)
DT: January 1993
UF: Environmental degradation
BT: Degradation
RT: Aging of materials
Corrosion
Decomposition
Disintegration
Environmental testing
Erosion
Geology
Leaching
Moisture
Oxidation
Radiation effects
Rocks
Stress corrosion cracking
Swelling
Wetting

Weaving 819.5
DT: January 1993
UF: Textiles—Weaving*
BT: Textile processing
RT: Filament winding
Looms

Weed control
(406.2) (407.2) (681.1) (803) (821.2)
DT: Predates 1975
UF: Inland waterways—Weed control*
Railroad plant and structures—Weed control*
BT: Pest control
RT: Herbicides
Highway systems
Inland waterways
Maintenance of way
Reservoirs (water)
Roadsides

Weibull distribution 922.2
DT: January 1995
RT: Components
Service life
Statistical methods

Weigh stations (902.1) (943.3)
DT: January 1993
UF: Scales and weighing—Weigh stations*
BT: Facilities
RT: Highway systems
Laws and legislation
Motor transportation
Roads and streets
Scales (weighing instruments)
Trucks
Weighing

Weighing 943.3
DT: January 1993
UF: Pulp—Weight determination*
Scales and weighing*
Steel ingots—Weighing*
Weight measurement
BT: Measurements

* Former Ei Vocabulary term

Weighing *(continued)*
RT: Precision balances
Scales (weighing instruments)
Units of measurement
Weigh stations

Weight (molecular)
USE: Molecular weight

Weight control (662.1) (694.1)
SN: Of products or packaging
DT: January 1993
UF: Automobiles—Weight control*
Packaging—Weight control*
Product design—Weight control*
Steel structures—Weight control*
BT: Control
RT: Product design

Weight machines
USE: Exercise equipment

Weight measurement
USE: Weighing

Weightlessness 656.1 (461.4)
DT: January 1993
UF: Space flight—Weightlessness*
RT: Aviation medicine
Gravitation
Human engineering
Life support systems (spacecraft)
Microgravity processing
Space flight

Weights and measures
USE: Units of measurement

Weirs 441.1
DT: Predates 1975
BT: Dams

Weldability 531, 538.2
DT: January 1993
UF: Metals and alloys—Weldability*
BT: Formability
RT: Ductility
Physical properties
Welding

Welded steel structures
538.2, 545.3 (415.1) (619.2)
DT: Predates 1975
UF: Buildings—Welded steel*
Footbridges—Welded steel*
Gas holders—Welded steel*
Penstocks—Welded steel*
Tanks—Welded steel*
BT: Steel structures
RT: Welding
Welds

Welding 538.2
DT: Predates 1975
UF: Tools, jigs and fixtures—Welding*
NT: Butt welding
Electric welding
Electron beam welding
Gas welding
Hydrodynamic welding
Laser beam welding
Plasma welding
Pressure welding
Submerged melt welding
Thermit welding
Underwater welding
RT: Alloys
Brazing
Fasteners
Filler metals
Fluxes
Hard facing
Heat affected zone
Metal working
Metals
Oxygen cutting
Preheating
Protective atmospheres
Soldering
Weldability
Welded steel structures
Welding codes
Welding machines
Welding shops
Welds

Welding codes 538.2, 902.2 (902.3)
DT: Predates 1975
BT: Codes (standards)
RT: Building codes
Laws and legislation
Welding

Welding electrodes 538.2.2
DT: January 1993
UF: Welding—Electrodes*
BT: Electrodes
NT: Welding rods
RT: Coated wire electrodes
Electric welding
Filler metals

Welding machines 538.2.2
DT: Predates 1975
BT: Machine tools
NT: Laser welding machines
Resistance welding machines
RT: Oxygen cutting machines
Spot welding
Welding

*Welding machines—Lasers**
USE: Laser welding machines

*Welding machines—Resistance**
USE: Resistance welding machines

*Welding machines—Transformers**
USE: Electric transformers

Welding rods 538.2.2
DT: Predates 1975
BT: Welding electrodes
RT: Coated wire electrodes
Electric welding

* Former Ei Vocabulary term

Welding rods *(continued)*
Filler metals
Wire

Welding shops 538.2 (402.1)
DT: Predates 1975
BT: Industrial plants
RT: Welding

*Welding, Electric arc**
USE: Electric arc welding

*Welding, Electric resistance**
USE: Resistance welding

*Welding, Electric**
USE: Electric welding

*Welding, Gas**
USE: Gas welding

*Welding—Brass**
USE: Brass

*Welding—Butt**
USE: Butt welding

*Welding—Carbon dioxide**
USE: Carbon dioxide arc welding

*Welding—Clad metals**
USE: Clad metals

*Welding—Coated metals**
USE: Coated materials

*Welding—Cold method**
USE: Cold welding

*Welding—Corrosion resisting materials**
USE: Corrosion resistant alloys

*Welding—Dissimilar metals**
USE: Dissimilar metals

*Welding—Electrodes**
USE: Welding electrodes

*Welding—Electron beam**
USE: Electron beam welding

*Welding—Electroslag**
USE: Electroslag welding

*Welding—Explosive**
USE: Explosive welding

*Welding—Filler metals**
USE: Filler metals

*Welding—Flash**
USE: Flash welding

*Welding—Fluxed core**
USE: Flux core wire welding

*Welding—Fluxes**
USE: Fluxes

*Welding—Friction**
USE: Friction welding

*Welding—Galvanized metal**
USE: Galvanized metal

*Welding—Hydrodynamic**
USE: Hydrodynamic welding

*Welding—Inert gas**
USE: Inert gas welding

*Welding—Iron castings**
USE: Cast iron

*Welding—Laser**
USE: Laser beam welding

*Welding—Light metals**
USE: Light metals

*Welding—Maraging steel**
USE: Maraging steel

*Welding—Percussion**
USE: Percussion welding

*Welding—Plasma arc**
USE: Plasma welding

*Welding—Plasma jet**
USE: Plasma welding

*Welding—Power supply**
USE: Electric power supplies to apparatus

*Welding—Precoated metals**
USE: Precoated metals

*Welding—Preheating**
USE: Preheating

*Welding—Pressure**
USE: Pressure welding

*Welding—Projection**
USE: Projection welding

*Welding—Seam**
USE: Seam welding

*Welding—Sheet metal**
USE: Sheet metal

*Welding—Spot**
USE: Spot welding

*Welding—Stainless steel**
USE: Stainless steel

*Welding—Steel castings**
USE: Steel castings

*Welding—Steel**
USE: Steel

*Welding—Stud**
USE: Stud welding

*Welding—Submerged arc**
USE: Submerged arc welding

*Welding—Submerged melt**
USE: Submerged melt welding

*Welding—Thermit**
USE: Thermit welding

*Welding—Tin plate**
USE: Tin plate

* Former Ei Vocabulary term

*Welding—Ultrasonic**
USE: Ultrasonic welding

*Welding—Underwater**
USE: Underwater welding

*Welding—Vacuum**
USE: Vacuum welding

*Welding—Wire products**
USE: Wire products

*Welding—Wire screen**
USE: Wire screen cloth

*Welding—Wrought iron**
USE: Wrought iron

Welds 538.2
DT: Predates 1975
UF: Joints—Welded*
BT: Joints (structural components)
RT: Heat affected zone
Heat treatment
Metallurgy
Pipe joints
Soldered joints
Stress relief
Welded steel structures
Welding

*Welds—Finishing**
USE: Metal finishing

*Welds—Heat affected zone**
USE: Heat affected zone

*Welds—Metallurgy**
USE: Metallurgy

*Welds—Stress relief**
USE: Stress relief

Well cementing (446.1) (512.1.2) (512.2.2)
DT: January 1993
BT: Cementing (shafts)
NT: Oil well cementing
RT: Bonding
Cements
Grouting
Well completion

Well completion (446.1) (512.1.2) (512.2.2)
SN: Final sealing off of a drilled well with devices to permit control and use
DT: January 1993
UF: Completion (well)
NT: Natural gas well completion
Oil well completion
Well perforation
RT: Sand consolidation
Well cementing
Well drilling
Wells

Well drilling (512.1.2) (512.2.2) (446.1)
DT: January 1993
BT: Drilling
NT: Deflected well drilling
Natural gas well drilling
Oil well drilling
RT: Boreholes
Drilling equipment
Drilling fluids
Drills
Drillships
Earth boring machines
Geothermal wells
Jet drilling
Rock drilling
Well completion

Well equipment (446.1) (511.2)
DT: January 1993
BT: Equipment
NT: Perforators
RT: Oil field equipment

Well flooding 511.1
DT: January 1993
UF: Waterflooding
BT: Enhanced recovery

Well logging (446.1) (512.1.1) (512.2.2)
DT: January 1993
UF: Boreholes—Logging*
Logging (wells)
BT: Recording
NT: Acoustic logging
Directional logging
Electric logging
Electromagnetic logging
Geothermal logging
Mud logging
Natural gas well logging
Nuclear magnetic logging
Oil well logging
Radioactivity logging
Thermal logging
Water well logging
RT: Boreholes
Geophysical prospecting
Wells

Well perforation (446.1) (512.1.2) (512.2.2)
DT: January 1993
UF: Oil wells—Perforation*
Perforation (well)
BT: Well completion
RT: Gun perforators
Natural gas wells
Oil wells
Perforators

Well pressure (446.1) (512.1.2) (512.2.2)
DT: January 1993
UF: Geothermal wells—Well pressure*
BT: Pressure
RT: Geothermal wells
Natural gas wells
Oil wells

Well pumps 618.2 (446.1) (512.1.2) (512.2.2)
SN: Scope formerly limited to water well pumps
DT: Predates 1975
UF: Deep well pumps

* Former Ei Vocabulary term

Well pumps *(continued)*
BT: Pumps
NT: Oil well pumps
Water well pumps
RT: Submersible pumps
Wells

Well shooting
USE: Explosive well stimulation

Well spacing (446.1) (512.1.1) (512.2.1)
DT: January 1993
UF: Oil wells—Spacing*
RT: Natural gas wells
Natural gas well production
Oil well production
Oil wells
Wells

Well stimulation (512.1.2) (512.2.2)
SN: Technique to increase oil and gas reservoir production
DT: January 1993
BT: Production
NT: Explosive well stimulation
RT: Acidization

Well testing (512.1.2) (512.2.2) (446.1)
DT: January 1993
UF: Wells—Testing*
BT: Testing
NT: Oil well testing
RT: Wells

Well workover (446.1) (511)
DT: January 1993
UF: Oil wells—Workover*
RT: Enhanced recovery
Oil wells

Wells (446.1) (512.1) (512.2)
DT: January 1993
NT: Natural gas wells
Oil wells
Water wells
RT: Boreholes
Well completion
Well logging
Well pumps
Well spacing
Well testing

*Wells—Testing**
USE: Well testing

Wet and dry bulb hygrometers
USE: Psychrometers

Wet bulb thermometers
USE: Psychrometers

Wet ends (papermaking machinery) 811.1.2
DT: January 1993
UF: Papermaking machinery—Wet end*
BT: Papermaking machinery

Wetting (802.3) (931.2)
DT: January 1993
UF: Moistening
BT: Surface phenomena
RT: Adhesion
Adsorption
Capillarity
Contact angle
Cooling
Saturation (materials composition)
Spraying
Weathering

Whaling vessels 671 (471.5)
DT: Predates 1975
BT: Ships

Wheel dressing 606.2
DT: January 1993
UF: Dressing (wheels)
Grinding wheels—Dressing*
BT: Maintenance
RT: Grinding wheels

Wheelchairs 461.5, 462.1
DT: January 1993
UF: Biomedical equipment—Wheelchairs*
BT: Human rehabilitation equipment
RT: Human rehabilitation engineering

Wheels 601.2
DT: Predates 1975
BT: Equipment
NT: Grinding wheels
Vehicle wheels
RT: Axles
Bogies (railroad rolling stock)
Brakes
Gears
Mechanical drives
Tires

Whirling
USE: Rotation

Whiskers (crystal)
USE: Crystal whiskers

White acoustic noise 751.4
DT: January 1993
UF: Noise, Acoustic—White noise*
BT: Acoustic noise

White clay
USE: Kaolin

White noise (711) (716.1)
DT: January 1993
UF: Noise, Spurious signal—White noise*
RT: Jamming
Radio interference
Spurious signal noise

Wide area networks (722.3)
DT: January 1993
UF: Computer networks—Wide area networks*
WAN
BT: Computer networks

Wideband networks
USE: Broadband networks

* Former Ei Vocabulary term

Widening (transportation arteries) (401.1) (406.2) (407.2)
DT: January 1993
UF: Bridges—Widening*
Inland waterways—Widening*
Roads and streets—Widening*
BT: Improvement
RT: Bridges
Inland waterways
Roads and streets
Roadsides

Winches 605 (663.2) (671.2)
DT: January 1993
UF: Boats—Winches*
Motor trucks—Winches*
Ship equipment—Winches*
BT: Materials handling equipment
RT: Cranes
Elevators
Hoists
Trucks

Wind 443.1
DT: January 1993
BT: Atmospheric movements
RT: Anemometers
Turbulence
Wind effects

Wind effects 443.1 (408.1)
DT: Predates 1975
UF: Gust loads
BT: Effects
NT: Wind stress
RT: Erosion
Hurricane effects
Hurricanes
Storms
Structural design
Wind

Wind power 615.8
DT: Predates 1975
UF: Pumping plants—Wind power*
BT: Renewable energy resources
RT: Electric power plants
Power generation
Wind turbines

Wind stress 443.1 (402) (421)
DT: January 1993
UF: Buildings—Wind stresses*
Roofs—Wind stresses*
Stresses—Wind origin*
BT: Stresses
Wind effects
RT: Buildings
Hurricane effects
Load limits
Roofs
Stress analysis
Structures (built objects)

Wind tunnels 651.2 (422.1)
DT: Predates 1975
UF: Tunnels (wind)
BT: Test facilities
RT: Aerodynamics
Blowers
Flow visualization
Heating equipment
Hypersonic flow
Laboratories
Shock tubes
Simulators
Supersonic flow
Transonic aerodynamics
Transonic flow
Wall flow

*Wind tunnels—Air cleaners**
USE: Air cleaners

*Wind tunnels—Balances**
USE: Precision balances

*Wind tunnels—Blowers**
USE: Blowers

*Wind tunnels—Heaters**
USE: Heating equipment

*Wind tunnels—Hypersonic flow**
USE: Hypersonic flow

*Wind tunnels—Supersonic flow**
USE: Supersonic flow

*Wind tunnels—Transonic flow**
USE: Transonic flow

*Wind tunnels—Visualization**
USE: Flow visualization

*Wind tunnels—Wall interference**
USE: Wall flow

Wind turbines 615.8
DT: January 1983
BT: Turbines
RT: Wind power

Winding 691.2 (816.1) (819.3)
DT: January 1993
UF: Plastics—Winding*
Reeling
Rewinding
Spooling
Textiles—Winding*
Unwinding
BT: Materials handling
NT: Filament winding
RT: Stretching

Winding engines
USE: Mine hoists

Winding machines 691.1 (816.1) (819.3)
DT: Predates 1975
BT: Machinery
RT: Electric coils
Textile machinery
Wire

Windings (electric)
USE: Electric windings

Windlasses 671.2
DT: January 1993
UF: Ship equipment—Windlasses*
BT: Ship equipment

Window screens (402)
DT: Predates 1975
UF: Screens (window)
BT: Building components
RT: Windows
Wire screen cloth

Windows 402 (812.3)
DT: Predates 1975
BT: Building components
RT: Daylighting
Solar control films
Window screens

*Windows—Aluminum**
USE: Aluminum

*Windows—Plastics**
USE: Plastics applications

*Windows—Sealing**
USE: Sealing (closing)

*Windows—Stained glass**
USE: Stained glass

*Windows—Steel**
USE: Steel

Windscreens
USE: Windshields

Windshield wipers 662.4
DT: January 1993
UF: Automobiles—Windshield wipers*
BT: Vehicle parts and equipment
RT: Aircraft parts and equipment
Aircraft windshields
Automobile parts and equipment
Automobile windshields
Rubber products
Windshields

Windshields (652.1) (662.4) (812.3)
DT: January 1993
UF: Vehicles—Windshields*
Windscreens
BT: Vehicle parts and equipment
NT: Aircraft windshields
Automobile windshields
RT: Windshield wipers

Wine 822.3
DT: Predates 1975
BT: Beverages
RT: Fermentation

Wings (651.1) (652.1)
DT: January 1993
UF: Aerodynamics—Wings and airfoils*
Aircraft wings
Aircraft—Wings*
BT: Aircraft parts and equipment
Airfoils
RT: Aerodynamics
Airframes
Rotors

Wire 535.2
DT: Predates 1975
UF: Clad wire
BT: Materials
NT: Electric wire
Exploding wires
RT: Cables
Cladding (coating)
Glass bonding
Plastic filaments
Reinforcement
Straightening
Welding rods
Winding machines
Wire belts
Wire drawing
Wire flattening
Wire mills
Wire pointing
Wire products
Wire reinforced plastics

Wire belts 602.2
DT: January 1993
UF: Belts—Wire*
BT: Belts
RT: Wire

Wire drawing 535.2.2
DT: Predates 1975
BT: Metal drawing
RT: Wire
Wire drawing machines

Wire drawing machines 535.2.1
DT: Predates 1975
BT: Wire forming machines
RT: Wire drawing
Wire mills

*Wire drawing—Dies**
USE: Drawing dies

Wire flattening 535.2.2
DT: January 1993
UF: Wire—Flattening*
BT: Metal working
RT: Wire

Wire forming machines 535.2.1
DT: Predates 1975
BT: Metal forming machines
NT: Wire drawing machines
RT: Wire mills

Wire insulating extruders 413.1, 535.2.1
DT: Predates 1975
BT: Extruders
RT: Insulated wire
Insulation

*Wire insulating extruders—Dies**
USE: Extrusion dies

Wire mills 535.2.1 (402.1)
DT: Predates 1975
BT: Industrial plants
RT: Wire
Wire drawing machines
Wire forming machines
Wire products

Wire pointing 535.2.2
DT: January 1993
UF: Pointing (metal finishing)
Wire—Pointing*
BT: Metal finishing
RT: Wire

Wire products 535.2
DT: Predates 1975
UF: Welding—Wire products*
NT: Cables
Telephone cords
Wire reinforced plastics
Wire screen cloth
RT: Screens (sizing)
Springs (components)
Wire
Wire mills

Wire reinforced plastics 817.1 (415.2) (535.2)
DT: January 1993
UF: Plastics, Reinforced—Wire*
BT: Reinforced plastics
Wire products
RT: Wire

Wire rope 535.2
DT: Predates 1975
UF: Mine hoists—Wire rope*
BT: Cables
Rope
RT: Cable cores

*Wire rope—Connectors**
USE: Connectors (structural)

*Wire rope—Cores**
USE: Cable cores

Wire screen cloth 535.2
DT: Predates 1975
UF: Welding—Wire screen*
BT: Textiles
Wire products
RT: Screens (sizing)
Window screens

*Wire screen cloth—Steel**
USE: Steel

*Wire—Cladding**
USE: Metal cladding

*Wire—Enameling**
USE: Enameling

*Wire—Finishing**
USE: Finishing

*Wire—Flattening**
USE: Wire flattening

*Wire—Gaging**
USE: Gaging

*Wire—Glass bonding**
USE: Glass bonding

*Wire—Plastic coating**
USE: Plastic coatings

*Wire—Pointing**
USE: Wire pointing

*Wire—Reels**
USE: Reels

*Wire—Scale removal**
USE: Descaling

*Wire—Straightening**
USE: Straightening

Wireless
USE: Radio

Wiring
USE: Electric wiring

Wolfram
USE: Tungsten

Wood 811.2
DT: Predates 1975
UF: Pulp materials—Wood*
BT: Materials
RT: Bark stripping
Biomass
Decay (organic)
Fuels
Fungus attack
Impregnation
Lumber
Pulp materials
Renewable energy resources
Termite proofing
Wood chemicals
Wood fuels
Wood laminates
Wood preservation
Wood products
Wooden construction
Xylose

Wood alcohol
USE: Methanol

Wood chemicals 804.1, 811.2
DT: January 1987
BT: Chemicals
RT: Wood
Wood products

Wood fuels 524 (811.2)
DT: January 1993
BT: Energy resources
Fuels
RT: Biomass

* Former Ei Vocabulary term

Wood fuels *(continued)*
Charcoal
Wood

Wood laminates 811.2 (415.3)
DT: January 1993
UF: Wood products—Laminates*
BT: Laminates
Wood products
NT: Plywood
RT: Building materials
Veneers
Wood

Wood preservation 811.2 (803)
DT: Predates 1975
UF: Preservation (wood)
RT: Wood
Wood products

Wood products 811.2
DT: Predates 1975
UF: Joints, Adhesive—Wood products*
Piles—Wood*
Poles—Wood*
Port structures—Wood*
NT: Charcoal
Lumber
Particle board
Turpentine
Wood laminates
Wooden containers
Wooden fences
Wooden floors
RT: Adhesive joints
Furniture manufacture
Gluing
Veneers
Wood
Wood chemicals
Wood preservation
Wood wastes
Woodworking

*Wood products—Finishing**
USE: Finishing

*Wood products—Gluing**
USE: Gluing

*Wood products—Laminates**
USE: Wood laminates

*Wood products—Plastics encasing**
USE: Plastics applications AND Encapsulation

*Wood products—Wastes**
USE: Wood wastes

Wood sugar
USE: Xylose

*Wood waste**
USE: Wood wastes

*Wood waste—Sawdust**
USE: Sawdust

Wood wastes 452.3 (811.2) (821.5)
DT: January 1993
UF: Wood products—Wastes*
Wood waste*
BT: Wastes
NT: Sawdust
RT: Wood products

*Wood—Bark stripping**
USE: Bark stripping

*Wood—Decay**
USE: Decay (organic)

*Wood—Finishing**
USE: Finishing

*Wood—Fungus attack**
USE: Fungus attack

*Wood—Grading**
USE: Grading

*Wood—Permeability**
USE: Mechanical permeability

*Wood—Termite proofing**
USE: Termite proofing

Wooden beams and girders 408.2, 415.3 (811.2)
DT: January 1993
UF: Beams and girders—Wood*
BT: Beams and girders
RT: Wooden construction

Wooden bridges 401.1, 415.3 (811.2)
DT: January 1993
UF: Bridges, Wood*
BT: Bridges
RT: Wooden construction

Wooden buildings 402, 415.3 (811.2)
DT: January 1993
UF: Buildings—Wood*
BT: Buildings
RT: Wooden construction

Wooden construction 405.2 (415.3) (811.2)
DT: Predates 1975
BT: Construction
RT: Adhesive joints
Wood
Wooden beams and girders
Wooden bridges
Wooden buildings
Wooden containers
Wooden fences
Wooden floors

*Wooden construction—Connections**
USE: Joints (structural components)

Wooden containers 811.2 (691.1) (694.4)
DT: January 1993
UF: Containers—Wood*
BT: Containers
Wood products
RT: Wooden construction

* Former Ei Vocabulary term

Wooden fences 415.3 (811.2)
DT: January 1993
UF: Fences—Wood*
BT: Fences
Wood products
RT: Wooden construction

Wooden floors 415.3 (811.2)
DT: January 1993
UF: Floors—Wood*
BT: Floors
Wood products
RT: Wooden construction

Woodrooms 811.1.2 (402.1) (811.2)
DT: January 1993
UF: Paper and pulp mills—Woodrooms*
BT: Paper and pulp mills

Woodworking 811.2
DT: Predates 1975
UF: Cabinetry
BT: Manufacture
RT: Furniture manufacture
Wood products
Woodworking machinery
Woodworking tools

Woodworking machinery 603.1, 811.2
DT: Predates 1975
BT: Machine tools
NT: Sanders
RT: Woodworking
Woodworking tools

*Woodworking machinery—Sanders**
USE: Sanders

Woodworking saws
USE: Saws AND
Woodworking tools

Woodworking tools 811.2 (603) (605)
DT: January 1993
UF: Cutting tools—Woodworking*
Saws—Woodworking*
Woodworking saws
BT: Tools
RT: Cutting tools
Saws
Woodworking
Woodworking machinery

*Woodworking—Finishing**
USE: Finishing

Wool 821.4 (819.1)
DT: Predates 1975
BT: Agricultural products
RT: Felt
Wool fibers
Woolen and worsted fabrics
Woolen and worsted yarn

Wool fibers 819.1
DT: Predates 1975
BT: Natural fibers
RT: Wool
Woolen and worsted fabrics
Woolen and worsted yarn

Woolen and worsted fabrics 819.1, 819.5
DT: Predates 1975
BT: Fabrics
RT: Wool
Wool fibers
Woolen and worsted mills
Woolen and worsted yarn

Woolen and worsted mills 819.6 (402.1) (819.1)
DT: Predates 1975
BT: Textile mills
RT: Textile industry
Woolen and worsted fabrics
Woolen and worsted yarn

Woolen and worsted yarn 819.1, 819.4
DT: Predates 1975
BT: Yarn
RT: Wool
Wool fibers
Woolen and worsted fabrics
Woolen and worsted mills

Word processing 723.2
DT: January 1993
UF: Data processing—Word processing*
BT: Data processing
RT: Administrative data processing
Computer workstations
Desktop publishing
Office automation
Printing
Typewriters

Work hardening
USE: Strain hardening

Work simplification 912.1
DT: Predates 1975
RT: Industrial engineering
Job analysis
Time and motion study

Work-rest cycles
USE: Work-rest schedules

Work-rest schedules 461.4, 912.4
DT: January 1993
UF: Human engineering—Work-rest schedules*
Work-rest cycles
RT: Human engineering
Personnel
Scheduling

Working fluid pressure indicators 944.3
DT: January 1993
UF: Engine indicators
Indicators—Working fluid pressure*
BT: Indicators (instruments)
Pressure transducers

Workshops (seminars)
USE: Technical presentations

Workstations (computer)
USE: Computer workstations

* Former Ei Vocabulary term

Worm gears 601.2 (602)
DT: January 1993
UF: Gears—Worm*
BT: Gears
RT: Helical gears

Worms (computer crime)
USE: Computer worms

Wrapping (packaging)
USE: Packaging

Wrenches
USE: Hand tools

Wrinkle recovery 819.5
SN: Of textiles
DT: January 1993
UF: Textiles—Wrinkle recovery*
RT: Durable press textiles
Textile processing
Textiles

Writing (technical)
USE: Technical writing

Wrought iron 545.1
DT: Predates 1975
UF: Welding—Wrought iron*
BT: Iron alloys
RT: Pig iron

WSI circuits 714.2
DT: January 1993
UF: Integrated circuits, WSI*
Wafer scale integration
BT: VLSI circuits

X ray analysis (421) (801)
DT: January 1993
UF: X-ray analysis*
NT: X ray diffraction analysis
X ray spectroscopy
RT: Chemical analysis
Radiography
X ray apparatus
X ray crystallography
X ray laboratories
X ray tubes
X rays

X ray apparatus (422.1) (462.1) (801)
DT: January 1993
UF: X-ray apparatus*
BT: Equipment
NT: X ray cameras
X ray films
X ray microscopes
X ray screens
X ray spectrographs
X ray spectrometers
X ray tubes
RT: Biomedical equipment
Image intensifiers (electron tube)
X ray analysis
X ray laboratories
X ray lasers
X ray lithography
X ray radiography
X ray spectroscopy
X rays

X ray cameras 742.2 (422.1) (462.1)
DT: January 1993
UF: X-ray apparatus—Cameras*
BT: Cameras
X ray apparatus

X ray crystallography 933.1.1 (931.2)
DT: January 1995
UF: X-ray crystallography
BT: Crystallography
RT: Crystal defects
Crystal structure
X ray analysis
X ray diffraction
X ray powder diffraction
X ray spectrometers

X ray diffraction (931.3) (933.1.1)
SN: For studies of X ray diffraction of crystals use X RAY CRYSTALLOGRAPHY.
DT: January 1995
UF: X-ray diffraction
BT: Electromagnetic wave diffraction
RT: Metallography
Radiography
X ray crystallography
X ray diffraction analysis
X ray spectroscopy

X ray diffraction analysis (801) (931.3) (933.1.1)
DT: January 1995
UF: X ray diffractometry
X-ray diffraction analysis
X-ray diffractometry
BT: X ray analysis
NT: X ray powder diffraction
RT: Chemical analysis
Radiography
X ray diffraction
X rays

X ray diffractometry
USE: X ray diffraction analysis

X ray films 742.3
DT: January 1993
UF: X-ray films*
BT: Photographic films
X ray apparatus

X ray laboratories (402.1) (422.1) (462.1) (901.3)
DT: January 1993
UF: X-ray laboratories*
BT: Laboratories
RT: Biomedical engineering
X ray analysis
X ray apparatus
X ray radiography

X ray lasers 744.6
DT: January 1993
UF: Lasers, X-ray*
BT: Lasers
RT: X ray apparatus

X ray lithography (745.1)
DT: January 1993
UF: Lithography—X-ray*
BT: Photolithography
RT: Integrated circuit manufacture
Masks
Semiconductor device manufacture
X ray apparatus
X rays

X ray microscopes 741.3
DT: January 1993
UF: Microscopes—X-ray*
BT: Microscopes
X ray apparatus

X ray photoelectron spectroscopy (801) (931.3)
DT: January 1995
UF: X-ray photoelectron spectroscopy
XPS
BT: Photoelectron spectroscopy
RT: X rays

X ray powder analysis
USE: X ray powder diffraction

X ray powder diffraction (931.3) (933.1.1)
DT: January 1995
UF: Powder method
X ray powder analysis
X-ray powder analysis
X-ray powder diffraction
BT: X ray diffraction analysis
RT: Powders
X ray crystallography

X ray production 932.1
DT: January 1993
UF: X-ray and gamma ray production*
RT: X ray radiography
X ray tubes
X rays

X ray radiography (422.2) (461.1)
DT: January 1993
UF: Radiography—X-ray*
BT: Radiography
RT: X ray apparatus
X ray laboratories
X ray production
X ray tubes

X ray screens (422.1) (462.1)
DT: January 1993
UF: Screens (X-ray)
X-ray apparatus—Screens*
BT: X ray apparatus

X ray spectrographs (741.3) (801)
DT: January 1993
UF: Spectrographs, X-ray*
BT: Spectrographs
X ray apparatus
RT: X ray spectrometers
X ray spectroscopy

X ray spectrometers 741.3 (801) (941.3)
DT: January 1993
UF: Spectrometers, X-ray*
BT: Spectrometers
X ray apparatus
NT: X ray spectrophotometers
RT: Particle spectrometers
X ray crystallography
X ray spectrographs
X ray spectroscopy

X ray spectrophotometers 741.3 (801) (941.3)
DT: January 1993
UF: Spectrophotometers, X-ray*
BT: Spectrophotometers
X ray spectrometers
RT: Spectrophotometry
X ray spectroscopy

X ray spectroscopy (801) (931.3) (932.2)
DT: January 1993
UF: Spectroscopy, X-ray*
BT: Spectroscopy
X ray analysis
RT: X ray apparatus
X ray diffraction
X ray spectrographs
X ray spectrometers
X ray spectrophotometers
X rays

X ray tubes (422.1) (462.1)
DT: January 1993
UF: X-ray tubes*
BT: Thermionic tubes
X ray apparatus
RT: X ray analysis
X ray production
X ray radiography

X rays 932.1
DT: January 1993
UF: X-rays*
BT: Electromagnetic waves
RT: Photons
Radiography
X ray analysis
X ray apparatus
X ray diffraction analysis
X ray lithography
X ray photoelectron spectroscopy
X ray production
X ray spectroscopy

*X-ray analysis**
USE: X ray analysis

*X-ray and gamma ray production**
USE: Gamma ray production OR
X ray production

*X-ray apparatus**
USE: X ray apparatus

*X-ray apparatus—Cameras**
USE: X ray cameras

*X-ray apparatus—Screens**
USE: X ray screens

X-ray crystallography
USE: X ray crystallography

X-ray diffraction
USE: X ray diffraction

X-ray diffraction analysis
USE: X ray diffraction analysis

X-ray diffractometry
USE: X ray diffraction analysis

*X-ray films**
USE: X ray films

*X-ray laboratories**
USE: X ray laboratories

X-ray photoelectron spectroscopy
USE: X ray photoelectron spectroscopy

X-ray photography
USE: Radiography

X-ray powder analysis
USE: X ray powder diffraction

X-ray powder diffraction
USE: X ray powder diffraction

*X-ray tubes**
USE: X ray tubes

*X-rays**
USE: X rays

Xe
USE: Xenon

Xenon 804
DT: Predates 1975
UF: Xe
BT: Inert gases

Xerography 745.2
DT: January 1993
UF: Photographic reproduction—Xerography*
BT: Electrostatic reproduction

XPS
USE: X ray photoelectron spectroscopy

Xylene 804.1
DT: Predates 1975
BT: Aromatic hydrocarbons

Xylose 804.1
DT: January 1987
UF: Wood sugar
BT: Aldehydes
Sugars
RT: Wood

Yachts 674.1
DT: Predates 1975
BT: Boats

Yards (railroad)
USE: Railroad yards and terminals

Yarn 819.4
SN: Continuous strands of textile fibers, of one or more filaments, or other material suitable for intertwining into a textile fabric. Includes cords and threads
DT: Predates 1975
BT: Materials
NT: Cotton yarn
Rayon yarn
Rubber thread
Tire cords
Woolen and worsted yarn
RT: Fabrics
Spinning (fibers)
Textile fibers
Textiles
Twine

Yeast 822.3 (801.2)
DT: Predates 1975
BT: Fungi
RT: Fermentation
Microorganisms

Yield strength
USE: Yield stress

Yield stress (421) (422)
DT: January 1995
UF: Yield strength
BT: Mechanical properties
RT: Bearing capacity
Compression testing
Stress analysis
Tensile testing
Torsion testing

Yokes
USE: Deflection yokes

Young's modulus
USE: Elastic moduli

Ytterbium 547.2
DT: Predates 1975
BT: Rare earth elements
RT: Ytterbium alloys
Ytterbium compounds

Ytterbium alloys 547.2
DT: January 1993
BT: Rare earth alloys
RT: Ytterbium
Ytterbium compounds

Ytterbium compounds (804.1) (804.2)
DT: Predates 1975
BT: Rare earth compounds
RT: Ytterbium
Ytterbium alloys

* Former Ei Vocabulary term

Yttrium 549.3
DT: January 1993
UF: Yttrium and alloys*
BT: Nonferrous metals
Transition metals
RT: Yttrium alloys
Yttrium compounds
Yttrium deposits
Yttrium metallography

Yttrium alloys 549.3
DT: January 1993
UF: Yttrium and alloys*
BT: Transition metal alloys
RT: Yttrium
Yttrium compounds

*Yttrium and alloys**
USE: Yttrium OR
Yttrium alloys

Yttrium compounds (804.1) (804.2)
DT: Predates 1975
BT: Transition metal compounds
RT: Oxide superconductors
Yttrium
Yttrium alloys

Yttrium deposits 504.3, 549.3
DT: Predates 1975
BT: Ore deposits
RT: Yttrium

Yttrium metallography 531.2, 549.3
DT: Predates 1975
BT: Metallography
RT: Yttrium

Z transforms 921.3
DT: January 1993
UF: Mathematical transformations—Z transforms*
BT: Mathematical transformations
RT: Circuit theory
Difference equations
Electric network analysis
Signal processing

Zein 804.1 (813.2) (819.1) (821.4)
DT: Predates 1975
BT: Proteins
RT: Fibers

Zener breakdown
USE: Zener effect

Zener diodes 714.2
DT: January 1993
UF: Semiconductor diodes, Zener*
BT: Semiconductor diodes
RT: Zener effect

Zener effect 701.1 (712.1)
DT: January 1993
UF: Semiconductor materials—Zener effect*
Zener breakdown
BT: Electric breakdown of solids
Electric properties
RT: Electron tunneling
Semiconductor materials
Zener diodes

Zeolites 804.2
DT: January 1977
BT: Aluminum compounds
Silicates
RT: Adsorbents
Ion exchangers

Zero gravity casting 534.2
DT: January 1993
UF: Foundry practice—Zero gravity*
BT: Metal casting
RT: Foundry practice

Zinc 546.3
DT: January 1993
UF: Slags—Zinc recovery*
Zinc and alloys*
Zn
BT: Nonferrous metals
Transition metals
RT: Galvanizing
Zinc alloys
Zinc castings
Zinc compounds
Zinc deposits
Zinc foundry practice
Zinc metallography
Zinc mines
Zinc ore treatment
Zinc plating
Zinc refining
Zinc scrap

Zinc alloys 546.3
DT: January 1993
UF: Zinc and alloys*
BT: Transition metal alloys
NT: Brass
Nickel silver
RT: Soldering alloys
Zinc
Zinc compounds

*Zinc and alloys**
USE: Zinc OR
Zinc alloys

Zinc castings 534.2, 546.3
DT: January 1986
BT: Metal castings
RT: Zinc

Zinc compounds (804.1) (804.2)
DT: Predates 1975
BT: Transition metal compounds
NT: Semiconducting zinc compounds
Zinc oxide
Zinc sulfide
RT: Ferrites
Intermetallics
Zinc
Zinc alloys

* Former Ei Vocabulary term

Zinc deposits 504.3, 546.3
DT: Predates 1975
BT: Ore deposits
NT: Copper zinc deposits
Lead zinc deposits
RT: Zinc
Zinc mines

Zinc foundry practice 534.2, 546.3
DT: Predates 1975
BT: Foundry practice
Zinc metallurgy
RT: Zinc

Zinc metallography 531.2, 546.3
DT: Predates 1975
BT: Metallography
RT: Zinc
Zinc metallurgy

Zinc metallurgy 531.1, 546.3
DT: Predates 1975
BT: Metallurgy
NT: Zinc foundry practice
Zinc powder metallurgy
RT: Zinc metallography
Zinc ore treatment
Zinc smelting

Zinc mines 504.3, 546.3
DT: January 1993
UF: Zinc mines and mining*
BT: Mines
NT: Copper zinc mines
RT: Lead zinc deposits
Zinc
Zinc deposits

Zinc mines and mining*
USE: Zinc mines

Zinc ore treatment 533.1, 546.3
DT: Predates 1975
BT: Ore treatment
NT: Copper lead zinc ore treatment
Copper zinc ore treatment
Zinc smelting
RT: Zinc
Zinc metallurgy

Zinc oxide 804.2
DT: January 1993
UF: Pigments—Zinc oxide*
BT: Oxides
Zinc compounds
RT: Pigments

Zinc plating 539.3, 546.3
DT: Predates 1975
BT: Plating
NT: Galvanizing
RT: Zinc

Zinc powder metallurgy 536.1, 546.3
DT: January 1993
UF: Powder metallurgy—Zinc alloys*
Powder metallurgy—Zinc*
BT: Powder metallurgy
Zinc metallurgy

Zinc refining 533.2, 546.3
DT: Predates 1975
BT: Metal refining
RT: Zinc

Zinc scrap 452.3, 546.2
DT: Predates 1975
BT: Scrap metal
RT: Zinc

Zinc smelting 533.2, 546.3
DT: Predates 1975
BT: Smelting
Zinc ore treatment
RT: Zinc metallurgy

Zinc sulfide 804.2
DT: Predates 1975
BT: Sulfur compounds
Zinc compounds

Zippers
USE: Fasteners

Zircon 482.2.1 (804.2) (812.1)
SN: Zirconium silicate
DT: Predates 1975
BT: Silicate minerals
Zirconium compounds
RT: Gems
Ores
Zircon deposits

Zircon deposits 505.1
DT: January 1993
UF: Zircon—Deposits*
Zirconium deposits*
BT: Ore deposits
RT: Zircon
Zirconium

Zircon—Deposits*
USE: Zircon deposits

Zirconia 804.2 (812.2)
DT: June 1990
UF: Zirconium oxide
BT: Oxides
Zirconium compounds
RT: Zirconia refractories

Zirconia refractories 812.2 (804.2)
DT: January 1993
UF: Refractory materials—Zirconia*
BT: Refractory materials
RT: Zirconia

Zirconium 549.3
DT: January 1993
UF: Zirconium and alloys*
Zr
BT: Nonferrous metals
Transition metals
RT: Zircon deposits
Zirconium alloys
Zirconium compounds
Zirconium metallurgy

* Former Ei Vocabulary term

Zirconium *(continued)*
Zirconium metallography
Zirconium ore treatment
Zirconium sponge

Zirconium alloys 549.3
DT: January 1993
UF: Zirconium and alloys*
BT: Transition metal alloys
RT: Zirconium
Zirconium compounds

*Zirconium and alloys**
USE: Zirconium OR
Zirconium alloys

Zirconium compounds (804.1) (804.2)
DT: Predates 1975
BT: Transition metal compounds
NT: Zircon
Zirconia
RT: Zirconium
Zirconium alloys

*Zirconium deposits**
USE: Zircon deposits

Zirconium metallography 531.2, 549.3
DT: Predates 1975
BT: Metallography
RT: Zirconium
Zirconium metallurgy

Zirconium metallurgy 531.1, 549.3
DT: Predates 1975
BT: Metallurgy
NT: Zirconium powder metallurgy
RT: Zirconium
Zirconium metallography
Zirconium ore treatment

Zirconium ore treatment 533.1, 549.3
DT: Predates 1975
BT: Ore treatment
RT: Zirconium
Zirconium metallurgy

Zirconium oxide
USE: Zirconia

Zirconium powder metallurgy 536.1, 549.3
DT: January 1993
UF: Powder metallurgy—Zirconium*
BT: Powder metallurgy
Zirconium metallurgy

Zirconium sponge 549.3
DT: Predates 1975
BT: Porous materials
RT: Zirconium

Zn
USE: Zinc

Zone melting 533.2
DT: January 1993
UF: Metals and alloys—Zone melting*
Zone refining
BT: Metal melting
RT: Freezing
Levitation melting
Purification
Recrystallization (metallurgy)
Refining

Zone refining
USE: Zone melting

Zoning (403.1) (403.2)
DT: January 1993
UF: Regional planning—Zoning*
Urban planning—Zoning*
RT: Land use
Laws and legislation
Regional planning
Urban planning

Zr
USE: Zirconium

* Former Ei Vocabulary term

CLASSIFICATION CODES

Table of Contents

ENGINEERING INFORMATION

CLASSIFICATION CODES

400 Series — Civil Engineering, General

401 Bridges and Tunnels
- 401.1 Bridges
- 401.2 Tunnels and Tunneling

402 Buildings and Towers
- 402.1 Industrial and Agricultural Buildings
- 402.2 Public Buildings
- 402.3 Residences
- 402.4 Towers

403 Urban and Regional Planning and Development
- 403.1 Urban Planning and Development
- 403.2 Regional Planning and Development

404 Civil Defense and Military Engineering
- 404.1 Military Engineering
- 404.2 Civil Defense

405 Construction Equipment and Methods; Surveying
- 405.1 Construction Equipment
- 405.2 Construction Methods
- 405.3 Surveying

406 Highway Engineering
- 406.1 Highway Systems
- 406.2 Roads and Streets

407 Maritime and Port Structures; Rivers and other Waterways
- 407.1 Maritime Structures
- 407.2 Waterways
- 407.3 Coastal Engineering

408 Structural Design
- 408.1 Structural Design (General)
- 408.2 Structural Members and Shapes

409 Civil Engineering, General

410 Series — Construction Materials

411 Bituminous Materials
- 411.1 Asphalt
- 411.2 Coal Tar

412 Concrete
- 412.1 Cement
- 412.2 Concrete Reinforcements

413 Insulating Materials
- 413.1 Electric Insulating Materials
- 413.2 Heat Insulating Materials
- 413.3 Sound Insulating Materials

414 Masonry Materials
- 414.1 Brickmaking
- 414.2 Brick Materials
- 414.3 Mortar (Before 1993, use code 412)

415 Metals, Plastics, Wood and Other Structural Materials
- 415.1 Metal Structural Materials
- 415.2 Plastics Structural Materials
- 415.3 Wood Structural Materials
- 415.4 Other Structural Materials

420 Series Building Materials Properties and Testing

421 Strength of Building Materials; Mechanical Properties

422 Strength of Building Materials; Test Equipment and Methods
- 422.1 Test Equipment
- 422.2 Test Methods

423 Non Mechanical Properties and Tests of Building Materials
- 423.1 Test Equipment
- 423.2 Test Methods

430 Series Transportation

431 Air Transportation
- 431.1 Air Transportation (General)
- 431.2 Passenger Air Transportation
- 431.3 Cargo Air Transportation
- 431.4 Airports
- 431.5 Air Navigation and Traffic Control

432 Highway Transportation
- 432.1 Highway Transportation (General)
- 432.2 Passenger Highway Transportation
- 432.3 Cargo Highway Transporation
- 432.4 Highway Traffic Control

433 Railroad Transportation
- 433.1 Railroad Transportation (General)
- 433.2 Passenger Railroad Transportation
- 433.3 Freight Railroad Transportation
- 433.4 Railroad Traffic Control

434 Waterway Transportation
- 434.1 Waterway Transportation (General)
- 434.2 Passenger Waterway Transportation
- 434.3 Cargo Waterway Transportation
- 434.4 Waterway Navigation

440 Series Water and Waterworks Engineering

441 Dams and Reservoirs; Hydro Development
- 441.1 Dams
- 441.2 Reservoirs
- 441.3 Related Hydraulic Structures

442 Flood Control; Land Reclamation
- 442.1 Flood Control
- 442.2 Land Reclamation

443 Meteorology
- 443.1 Atmospheric Properties
- 443.2 Meteorological Instrumentation
- 443.3 Precipitation

444 Water Resources
- 444.1 Surface Water
- 444.2 Groundwater

445 Water Treatment
- 445.1 Water Treatment Techniques
  - 445.1.1 Potable Water Treatment Techniques
  - 445.1.2 Water Treatment Techniques for Industrial Use
- 445.2 Water Analysis

446 Waterworks
- 446.1 Water Supply Systems
- 446.2 Related Hydraulic Structures

450 Series Pollution, Sanitary Engineering, Wastes

451 Air Pollution
- 451.1 Air Pollution Sources
- 451.2 Air Pollution Control

452 Sewage and Industrial Wastes Treatment
- 452.1 Sewage
- 452.2 Sewage Treatment
- 452.3 Industrial Wastes
- 452.4 Industrial Wastes Treatment

453 Water Pollution
- 453.1 Water Pollution Sources
- 453.2 Water Pollution Control

454 Environmental Engineering
- 454.1 Environmental Engineering (General)
- 454.2 Environmental Impact and Protection
- 454.3 Ecology and Ecosystems

460 Series Bioengineering

461 Bioengineering
- 461.1 Biomedical Engineering
- 461.2 Biological Materials
- 461.3 Biomechanics
- 461.4 Human Engineering
- 461.5 Human Rehabilitation Engineering
- 461.6 Medicine
- 461.7 Health Care
- 461.8 Biotechnology
  - 461.8.1 Genetic Engineering
- 461.9 Biology
  - 461.9.1 Immunology

462 Biomedical Equipment
- 462.1 Biomedical Equipment (General)
- 462.2 Hospitals, Equipment and Supplies
- 462.3 Dental Equipment and Supplies
- 462.4 Prosthetics
- 462.5 Biomaterials

470 Series Ocean and Underwater Technology

471 Marine Science and Oceanography
- 471.1 Oceanography (General)
- 471.2 Oceanographic Research Instruments
- 471.3 Oceanographic Techniques
- 471.4 Seawater, Tides and Waves
- 471.5 Sea as Source of Minerals and Food

472 Ocean Engineering

480 Series Engineering Geology

481 Geology and Geophysics
- 481.1 Geology
  - 481.1.1 Geomorphlogy
  - 481.1.2 Petrology (Before 1993, use code 482)
- 481.2 Geochemistry
- 481.3 Geophysics
  - 481.3.1 Geothermal Phenomena
  - 481.3.2 Earth Magnetism and Terrestrial Electricity
- 481.4 Geophysical Prospecting

482 Minerology
- 482.1 Mineralogical Techniques
- 482.2 Minerals
  - 482.2.1 Gems

483 Soil Mechanics and Foundations
- 483.1 Soils and Soil Mechanics
- 483.2 Foundations

484 Seismology
- 484.1 Earthquake Measurements and Analysis
- 484.2 Secondary Earthquake Effects
- 484.3 Earthquake Resistance

500 Series Mining Engineering, General

501 Exploration and Prospecting
- 501.1 Exploration and Prospecting Methods
- 501.2 Exploration and Prospecting Instrumentation

502 Mines and Quarry Equipment and Operations
- 502.1 Mine and Quarry Operations
- 502.2 Mine and Quarry Equipment

503 Mines and Mining, Coal
- 503.1 Coal Mines
- 503.2 Coal Mining Operations
- 503.3 Coal Mining Equipment

504 Mines and Mining, Metal
- 504.1 Light Metal Mines
- 504.2 Light Metal Mining
- 504.3 Heavy Metal Mines
- 504.4 Heavy Metal Mining
- 504.5 Uranium Mines
- 504.6 Uranium Mining

505 Mines and Mining, Nonmetallic
- 505.1 Nonmetallic Mines
- 505.2 Nonmetallic Mining Operations
- 505.3 Nonmetallic Mining Equipment

506 Mining Engineering, General

510 Series Petroleum Engineering

511 Oil Field Equipment and Production Operations
- 511.1 Oil Field Production Operations
- 511.2 Oil Field Equipment

512 Petroleum and Related Deposits
- 512.1 Petroleum Deposits
  - 512.1.1 Oil Fields
  - 512.1.2 Development Operations
- 512.2 Natural Gas Deposits
  - 512.2.1 Natural Gas Fields
  - 512.2.2 Development Operations

513 Petroleum Refining
- 513.1 Petroleum Refining (General)
- 513.2 Petroleum Refineries
- 513.3 Petroleum Products

520 Series Fuel Technology

521 Fuel Combustion and Flame Research
- 521.1 Fuel Combustion
- 521.2 Combustors
- 521.3 Fuel Burners
- 521.4 Flame Research

522 Gas Fuels

523 Liquid Fuels

524 Solid Fuels

525 Energy Management
- 525.1 Energy Resources
- 525.2 Energy Conservation
- 525.3 Energy Utilization
- 525.4 Energy Losses/Dissipation
- 525.5 Energy Conversion
- 525.6 Energy Policy
- 525.7 Energy Storage

530 Series Metallurgical Engineering, General

531 Metallurgy and Metallography
- 531.1 Metallurgy
- 531.2 Metallography

532 Metallurgical Furnaces
- 532.1 Basic Oxygen Converters
- 532.2 Blast Furnaces
- 532.3 Electric Metallurgical Furnaces
- 532.4 Heat Treating Furnaces
- 532.5 Melting Furnaces
- 532.6 Open Hearth Furnaces

533 Ore Treatment and Metal Refining
- 533.1 Ore Treatment
- 533.2 Metal Refining

534 Foundries and Foundry Practice
- 534.1 Foundries
- 534.2 Foundry Practice

535 Rolling, Forging and Forming
- 535.1 Metal Rolling
  - 535.1.1 Rolling Mills
  - 535.1.2 Rolling Mill Practice
- 535.2 Metal Forming
  - 535.2.1 Metal Forming Machines
  - 535.2.2 Metal Forming Practice

536 Powder Metallurgy
- 536.1 Powder Metallurgy Operations
- 536.2 Powder Metallurgy Equipment
- 536.3 Powder Metal Products

537 Heat Treatment
- 537.1 Heat Treatment Processes

537.2 Heat Treating Equipment

538 Welding and Bonding
- 538.1 Metal Bonding
  - 538.1.1 Soldering
- 538.2 Welding
  - 538.2.1 Welding Processes
  - 538.2.2 Welding Equipment

539 Metals Corrosion and Protection; Metal Plating
- 539.1 Metals Corrosion
- 539.2 Corrosion Protection
  - 539.2.1 Protection Methods
  - 539.2.2 Protecting Materials
- 539.3 Metal Plating
  - 539.3.1 Electroplating
  - 539.3.2 Electroless Plating

540 Series Metallurgical Engineering, Metal Groups

541 Aluminum and Alloys
- 541.1 Aluminum
- 541.2 Aluminum Alloys

542 Beryllium, Magnesium, Titanium and Other Light Metals and Alloys
- 542.1 Beryllium and Alloys
- 542.2 Magnesium and Alloys
- 542.3 Titanium and Alloys
- 542.4 Lithium and Alloys

543 Chromium, Manganese, Molybdenum, Tantalum, Tungsten, Vanadium and Alloys
- 543.1 Chromium and Alloys
- 543.2 Manganese and Alloys
- 543.3 Molybdenum and Alloys
- 543.4 Tantalum and Alloys
- 543.5 Tungsten and Alloys
- 543.6 Vanadium and Alloys

544 Copper and Alloys
- 544.1 Copper
- 544.2 Copper Alloys

545 Iron and Steel
- 545.1 Iron
- 545.2 Iron Alloys
- 545.3 Steel

546 Lead, Tin, Zinc, Antimony and Alloys
- 546.1 Lead and Alloys
- 546.2 Tin and Alloys
- 546.3 Zinc and Alloys
- 546.4 Antimony and Alloys

547 Minor, Precious and Rare Earth Metals and Alloys
- 547.1 Precious Metals
- 547.2 Rare Earth Metals

548 Nickel and Alloys
- 548.1 Nickel
- 548.2 Nickel Alloys

549 Nonferrous Metals and Alloys
- 549.1 Alkali Metals
- 549.2 Alkaline Earth Metals
- 549.3 Others (including Bismuth, Boron, Cadmium, Cobalt, Mercury, Niobium, Selenium, Silicon, Tellurium and Zirconium)

600 Series Mechanical Engineering, General

601 Mechanical Design
- 601.1 Machanical Devices
- 601.2 Machine Components
- 601.3 Mechanisms

602 Mechanical Drives and Transmissions
- 602.1 Mechanical Drives
- 602.2 Mechanical Transmissions

603 Machine Tools
- 603.1 Machine Tools (General)
- 603.2 Machine Tool Accessories

604 Metal Cutting and Machining
- 604.1 Metal Cutting
- 604.2 Machining Operations

605 Small Tools and Hardware
- 605.1 Small Tools, Powered
- 605.2 Small Tools, Unpowered

606 Abrasives
- 606.1 Abrasive Materials
- 606.2 Abrasive Devices and Processes

607 Lubricants and Lubrication
- 607.1 Lubricants
- 607.2 Lubrication

608 Mechanical Engineering, General

610 Series Mechanical Engineering, Plant and Power

611 Hydro and Tidal Power Plants
- 611.1 Hydroelectric Power Plants
- 611.2 Tidal Power Plants

612 Internal Combustion Engines
- 612.1 Internal Combustion Engines (General)
  - 612.1.1 Internal Combustion Engine Components
- 612.2 Diesel Engines
- 612.3 Gas Turbines and Engines

613 Nuclear Power Plants
- 613.1 Nuclear Power Plant Design and Construction
- 613.2 Nuclear Power Plant Equipment and Operation

614 Steam Power Plants
- 614.1 Steam Power Plant Design and Construction
- 614.2 Steam Power Plant Equipment and Operation

615 Thermoelectric, Magnetohydrodynamic and Other Power Generators
- 615.1 Geothermal Energy
- 615.2 Solar Power
- 615.3 Magnetohydrodynamics
- 615.4 Thermoelectric Energy
- 615.5 Ocean Thermal Energy
- 615.6 Wave Energy
- 615.7 Thermionic Power
- 615.8 Wind Power (Before 1993, use code 611)

616 Heat Exchangers
- 616.1 Heat Exchange Equipment and Components
- 616.2 Heat Exchanger Applications

617 Turbines and Steam Turbines
- 617.1 Hydraulic Turbines
- 617.2 Steam Turbines
- 617.3 Steam Engines

618 Compressors and Pumps
- 618.1 Compressors
- 618.2 Pumps
- 618.3 Blowers and Fans

619 Pipes, Tanks and Accessories; Plant Engineering Generally
- 619.1 Pipe, Piping and Pipelines
  - 619.1.1 Pipe Accessories
  - 619.1.2 Pipe Materials
- 619.2 Tanks
  - 619.2.1 Tank Accessories
  - 619.2.2 Tank Materials

620 Series Nuclear Technology

621 Nuclear Reactors
- 621.1 Fission Reactors
  - 621.1.1 Fission Reactor Equipment and Components
  - 621.1.2 Nuclear Fuels for Fission Reactors
- 621.2 Fusion Reactors
  - 621.2.1 Fusion Reactor Equipment and Components
  - 621.2.2 Fusion Fuels

622 Radioactive Materials
- 622.1 Radioactive Materials (General)
  - 622.1.1 Radioisotopes
- 622.2 Radiation Effects
- 622.3 Radioactive Material Applications
- 622.4 Radioactivity Measurements
- 622.5 Radioactive Wastes

630 Series Fluid Flow; Hydraulics, Pneumatics and Vacuum

631 Fluid Flow
- 631.1 Fluid Flow (General)
- 631.1.1 Liquid Dynamics
- 631.1.2 Gas Dynamics
- 631.2 Hydrodynamics
- 631.3 Flow of Fluid-Like Materials

632 Hydraulics, Pneumatics and Related Equipment
- 632.1 Hydraulics
- 632.2 Hydraulic Equipment and Machinery
- 632.3 Pneumatics
- 632.4 Pneumatic Equipment and Machinery

633 Vacuum Technology
- 633.1 Vacuum Applications
- 633.2 Vacuum Equipment

640 Series Heat and Thermodynamics

641 Heat and Mass Transfer; Thermodynamics
- 641.1 Thermodynamics
- 641.2 Heat Transfer
- 641.3 Mass Transfer

642 Industrial Furnaces and Process Heating
- 642.1 Process Heating
- 642.2 Industrial Furnaces and Components

643 Space Heating and Air Conditioning
- 643.1 Space Heating
- 643.2 Space Heating Equipment and Components
- 643.3 Air Conditioning
- 643.4 Air Conditioning Equipment and Components
- 643.5 Ventilation

644 Refrigeration and Cryogenics
- 644.1 Refrigeration Methods
- 644.2 Refrigerants
- 644.3 Refrigeration Equipment and Components
- 644.4 Cryogenics
- 644.5 Cryogenic Equipment and Components

650 Series Aerospace Engineering

651 Aerodynamics
- 651.1 Aerodynamics (General)
- 651.2 Wind Tunnels

652 Aircraft
- 652.1 Aircraft (General)
- 652.1.1 Commerical Aircraft
- 652.1.2 Military Aircraft
- 652.2 Aircraft Materials
- 652.3 Aircraft Instruments and Equipment
- 652.4 Helicopters
- 652.5 Balloons and Gliders

653 Aircraft Engines
- 653.1 Aircraft Engines (General)
- 653.2 Aircraft Engine Components

654 Rockets and Rocket Propulsion
- 654.1 Rockets and Missiles
- 654.2 Rocket Engines

655 Spacecraft
- 655.1 Spacecraft (General)
- 655.2 Satellites
  - 655.2.1 Communication Satellites
- 655.3 Space Probes

656 Space Flight
- 656.1 Space Flight
- 656.2 Space Research

657 Space Physics
- 657.1 Solar Energy and Phenomena
- 657.2 Extraterrestrial Physics and Stellar Phenomena

658 Aerospace Engineering (General)

660 Series Automotive Engineering

661 Automotive Engines and Related Equipment
- 661.1 Automotive Engines
- 661.2 Automotive Engine Components

662 Automobiles and Smaller Vehicles
- 662.1 Automobiles
- 662.2 Smaller Vehicles
- 662.3 Automobile and Smaller Vehicle Materials
- 662.4 Automobile and Smaller Vehicle Components

663 Buses, Tractors and Trucks
- 663.1 Heavy Duty Motor Vehicles
- 663.2 Heavy Duty Motor Vehicle Components
- 663.3 Heavy Duty Motor Vehicle Applications

664 Automotive Engineering, General

670 Series Naval Architecture and Marine Engineering

671 Naval Architecture
- 671.1 Ship Design
- 671.2 Ship Equipment
- 671.3 Ship Models

672 Naval Vessels
- 672.1 Combat Naval Vessels
- 672.2 Noncombat Naval Vessels

673 Shipbuilding and Shipyards
- 673.1 Shipbuilding
- 673.2 Shipbuilding Materials
- 673.3 Shipyards

674 Small Craft and Other Marine Craft
- 674.1 Small Marine Craft
- 674.2 Marine Drilling Rigs and Platforms

675 Marine Engineering
- 675.1 Ship Propulsion (Before 1993, use code 671)

680 Series Railroad Engineering

681 Railroad Plant and Structures
- 681.1 Railroad Plant and Structures (General)
- 681.2 Railroad Yards and Terminals
- 681.3 Railroad Signals and Signaling

682 Railroad Rolling Stock
- 682.1 Railroad Rolling Stock (General)
  - 682.1.1 Railroad Cars
  - 682.1.2 Locomotives
- 682.2 Rolling Stock Equipment and Components

690 Series Materials Handling

691 Bulk Handling and Unit Loads
- 691.1 Materials Handling Equipment
- 691.2 Materials Handling Methods

692 Conveyors and Elevators
- 692.1 Conveyors
- 692.2 Elevators

693 Cranes and Derricks
- 693.1 Cranes
- 693.2 Derricks

694 Packaging
- 694.1 Packaging
- 694.2 Packaging Materials
- 694.3 Packaging Equipment
- 694.4 Storage

700 Series Electrical Engineering, General

701 Electricity and Magnetism
- 701.1 Electricity: Basic Concepts and Phenomena
- 701.2 Magnetism: Basic Concepts and Phenomena

702 Electric Batteries and Fuel Cells
- 702.1 Electric Batteries
  - 702.1.1 Primary Batteries
  - 702.1.2 Secondary Batteries
- 702.2 Fuel Cells
- 702.3 Solar Cells
- 702.4 Other Direct Energy Converters

703 Electric Circuits
- 703.1 Electric Networks
  - 703.1.1 Electric Network Analysis
  - 703.1.2 Electric Network Synthesis
- 703.2 Electric Filters
  - 703.2.1 Electric Filter Analysis
  - 703.2.2 Electric Filter Synthesis

704 Electric Components and Equipment
- 704.1 Electric Components
- 704.2 Electric Equipment

705 Electric Generators and Motors
- 705.1 Electric Machinery (General)
- 705.2 Electric Generators
  - 705.2.1 AC Generators
  - 705.2.2 DC Generators
- 705.3 Electric Motors
  - 705.3.1 AC Motors
  - 705.3.2 DC Motors

706 Electric Transmission and Distribution
- 706.1 Electric Power Systems
  - 706.1.1 Electric Power Transmission
  - 706.1.2 Electric Power Distribution
- 706.2 Electric Power Lines and Equipment

707 Illuminating Engineering
- 707.1 Electric Lighting
- 707.2 Electric Lamps

708 Electric and Magnetic Materials
- 708.1 Dielectric Materials
- 708.2 Conducting Materials
- 708.3 Superconducting Materials
  - 708.3.1 High Temperature Superconducting Materials
- 708.4 Magnetic Materials

709 Electrical Engineering, General

710 Series Electronics and Communication Engineering

711 Electromagnetic Waves
- 711.1 Electromagnetic Waves in Different Media
- 711.2 Electromagnetic Waves in Relation to Various Structures

712 Electronic and Thermionic Materials
- 712.1 Semiconducting Materials
  - 712.1.1 Single Element Semiconducting Materials
  - 712.1.2 Compound Semiconducting Materials
- 712.2 Thermionic Materials

713 Electronic Circuits
- 713.1 Amplifiers
- 713.2 Oscillators
- 713.3 Modulators, Demodulators, Limiters, Discriminators, Mixers
- 713.4 Pulse Circuits
- 713.5 Other Electronic Circuits

714 Electronic Components and Tubes
- 714.1 Electron Tubes
- 714.2 Semiconductor Devices and Integrated Circuits
- 714.3 Waveguides

715 Electronic Equipment, General Purpose and Industrial
- 715.1 Electronic Equipment (non-communication)
- 715.2 Industrial Electronic Equipment

716 Electronic Equipment, Radar, Radio and Television
- 716.1 Information and Communication Theory
- 716.2 Radar Systems and Equipment
- 716.3 Radio Systems and Equipment
- 716.4 Television Systems and Equipment

717 Electro-Optical Communication
- 717.1 Optical Communication Systems
- 717.2 Optical Communication Equipment

718 Telephone and Other Line Communications
- 718.1 Telephone Systems and Equipment
- 718.2 Telegraph Systems and Equipment
- 718.3 Facsimile Systems and Equipment

720 Series Computers and Data Processing

721 Computer Circuits and Logic Elements
- 721.1 Computer Theory (Includes Formal Logic, Automata Theory, Switching Theory, Programming Theory)
- 721.2 Logic Elements
- 721.3 Computer Circuits

722 Computer Hardware
- 722.1 Data Storage (Equipment and Techniques)
- 722.2 Computer Peripheral Equipment
- 722.3 Data Communication (Equipment and Techniques)
- 722.4 Digital Computers and Systems
- 722.5 Analog and Hybrid Computers

723 Computer Software, Data Handling and Applications
- 723.1 Computer Programming
  - 723.1.1 Computer Programming Languages
- 723.2 Data Processing
- 723.3 Database Systems
- 723.4 Artificial Intelligence
  - 723.4.1 Expert Systems
- 723.5 Computer Applications

730 Series Control Engineering

731 Automatic Control Principles and Applications
- 731.1 Control Systems
- 731.2 Control System Applications
- 731.3 Specific Variables Control
- 731.4 System Stability
- 731.5 Robotics
- 731.6 Robot Applications

732 Control Devices
- 732.1 Control Equipment
- 732.2 Control Instrumentation

740 Series Light and Optical Technology

741 Light, Optics and Optical Devices
- 741.1 Light/Optics
  - 741.1.1 Nonlinear Optics
  - 741.1.2 Fiber Optics
- 741.2 Vision
- 741.3 Optical Devices and Systems

742 Cameras and Photography
- 742.1 Photography
- 742.2 Photographic Equipment
- 742.3 Photographic Materials and Chemicals

743 Holography
- 743.1 Holographic Techniques
  - 743.1.1 Optical Holography
  - 743.1.2 Acoustic Holography
  - 743.1.3 Microwave Holography
- 743.2 Holographic Applications

744 Lasers
- 744.1 Lasers (General)
- 744.2 Gas Lasers
- 744.3 Liquid Lasers
- 744.4 Solid State Lasers
  - 744.4.1 Semiconductor Lasers
- 744.5 Free Electron Lasers
- 744.6 Other Lasers
- 744.7 Laser Components
- 744.8 Laser Beam Interactions
- 744.9 Laser Applications

745 Printing and Reprography
- 745.1 Printing
  - 745.1.1 Printing Equipment
- 745.2 Reproduction (Copying)
  - 745.2.1 Reproduction Equipment

750 Series Sound and Acoustical Technology

751 Acoustics, Noise, Sound
- 751.1 Acoustic Waves
- 751.2 Acoustic Properties of Materials
- 751.3 Architectural Acoustics
- 751.4 Acoustic Noise
- 751.5 Speech

752 Sound Devices, Equipment and Systems
- 752.1 Acoustic Devices
- 752.2 Sound Recording
  - 752.2.1 Sound Recording Equipment
- 752.3 Sound Reproduction
  - 752.3.1 Sound Reproduction Equipment
- 752.4 Acoustic Generators

753 Sound Technology and Ultrasonics
- 753.1 Ultrasonic Waves
- 753.2 Ultrasonic Devices
- 753.3 Ultrasonic Applications

800 Series Chemical Engineering, General

801 Chemistry
- 801.1 Chemistry (General)
- 801.2 Biochemistry
- 801.3 Colloid Chemistry
- 801.4 Physical Chemistry
  - 801.4.1 Electrochemistry
  - 801.4.2 Radiation Chemistry

802 Chemical Apparatus and Plants; Unit Operations; Unit Processes
- 802.1 Chemical Plants and Equipment
- 802.2 Chemical Reactions
- 802.3 Chemical Operations

803 Chemical Agents

804 Chemical Products Generally
- 804.1 Organic Compounds
- 804.2 Inorganic Compounds

805 Chemical Engineering, General
- 805.1 Chemical Engineering
  - 805.1.1 Biochemical Engineering

810 Series Chemical Engineering, Process Industries

811 Cellulose, Paper and Wood Products
- 811.1 Pulp and Paper
  - 811.1.1 Papermaking Processes
  - 811.1.2 Papermaking Equipment
- 811.2 Wood and Wood Products
- 811.3 Cellulose and Derivatives

812 Ceramics, Refractories and Glass
- 812.1 Ceramics
- 812.2 Refractories
- 812.3 Glass

813 Coatings and Finishes
- 813.1 Coating Techniques
- 813.2 Coating Materials

814 Leather and Tanning
- 814.1 Leather
- 814.2 Tanning

815 Polymers and Polymer Science
- 815.1 Polymeric Materials
  - 815.1.1 Organic Polymers
  - 815.1.2 Inorganic Polymers
- 815.2 Polymerization

816 Plastics Processing and Machinery
- 816.1 Plastics Processing
- 816.2 Plastics Plants and Machinery

817 Plastics, Products and Applications
- 817.1 Plastics Products
- 817.2 Plastics Applications

818 Rubber and Elastomers
- 818.1 Natural Rubber
- 818.2 Elastomers
  - 818.2.1 Synthetic Rubber
- 818.3 Rubber and Elastomer Processing
  - 818.3.1 Processing Agents
- 818.4 Rubber Factories and Machinery
- 818.5 Rubber Products
- 818.6 Rubber Applications

819 Synthetic and Natural Fibers; Textile Technology
- 819.1 Natural Fibers
- 819.2 Synthetic Fibers
- 819.3 Fiber Chemistry and Processing
- 819.4 Fiber Products
- 819.5 Textile Products and Processing
- 819.6 Textile Mills, Machinery and Equipment

820 Series Agricultural Engineering and Food Technology

821 Agricultural Equipment and Methods
- 821.1 Agricultural Machinery and Equipment
- 821.2 Agricultural Chemicals
- 821.3 Agricultural Methods
- 821.4 Agricultural Products
- 821.5 Agricultural Wastes
- 821.6 Farm Buildings and Other Structures

822 Food Technology
- 822.1 Food Products Plants and Equipment
- 822.2 Food Processing Operations
- 822.3 Food Products

900 Series Engineering, General

901 Engineering Profession
- 901.1 Engineering Professional Aspects
- 901.1.1 Societies and Institutions
- 901.2 Education
- 901.3 Engineering Research
- 901.4 Impact of Technology on Society

902 Engineering Graphics; Engineering Standards; Patents
- 902.1 Engineering Graphics
- 902.2 Codes and Standards
- 902.3 Legal Aspects

903 Information Science
- 903.1 Information Sources and Analysis
- 903.2 Information Dissemination
- 903.3 Information Retrieval and Use
- 903.4 Information Services
- 903.4.1 Libraries

910 Series Engineering Management

911 Cost and Value Engineering; Industrial Economics
- 911.1 Cost Accounting
- 911.2 Industrial Economics
- 911.3 Inventory Control
- 911.4 Marketing
- 911.5 Value Engineering

912 Industrial Engineering and Management
- 912.1 Industrial Engineering
- 912.2 Management
- 912.3 Operations Research
- 912.4 Personnel

913 Production Planning and Control; Manufacturing
- 913.1 Production Engineering
- 913.2 Production Control
- 913.3 Quality Assurance and Control
- 913.3.1 Inspection
- 913.4 Manufacturing
- 913.4.1 Flexible Manufacturing Systems
- 913.4.2 Computer Aided Manufacturing
- 913.4.3 Cellular Manufacturing
- 913.5 Maintenance
- 913.6 Concurrent Engineering

914 Safety Engineering
- 914.1 Accidents and Accident Prevention
- 914.2 Fires and Fire Protection
- 914.3 Industrial Hygiene
- 914.3.1 Occupational Diseases

920 Series Engineering Mathematics

921 Applied Mathematics
- 921.1 Algebra
- 921.2 Calculus
- 921.3 Mathematical Transformations
- 921.4 Combinatorial Mathematics (Includes Graph Theory, Set Theory)
- 921.5 Optimization Techniques
- 921.6 Numerical Methods

922 Statistical Methods
- 922.1 Probability Theory
- 922.2 Mathematical Statistics

930 Series Engineering Physics

931 Applied Physics Generally
- 931.1 Mechanics
- 931.2 Physical Properties of Gases, Liquids and Solids
- 931.3 Atomic and Molecular Physics
- 931.4 Quantum Theory
- 931.5 Gravitation and Relativity

932 High Energy Physics; Nuclear Physics; Plasma Physics
- 932.1 High Energy Physics
  - 932.1.1 Particle Accelerators
- 932.2 Nuclear Physics
  - 932.2.1 Fission and Fusion Reactions
- 932.3 Plasma Physics

933 Solid State Physics
- 933.1 Crystalline Solids
  - 933.1.1 Crystal Lattice
  - 933.1.2 Crystal Growth
- 933.2 Amorphous Solids
- 933.3 Electronic Structure of Solids

940 Series Instruments and Measurement

941 Acoustical and Optical Measuring Instruments
- 941.1 Acoustical Instruments
- 941.2 Acoustic Variables Measurements
- 941.3 Optical Instruments
- 941.4 Optical Variables Measurements

942 Electric and Electronic Measuring Instruments
- 942.1 Electric and Electronic Instruments
- 942.2 Electric Variables Measurements
- 942.3 Magnetic Instruments
- 942.4 Magnetic Variables Measurements

943 Mechanical and Miscellaneous Measuring Instruments
- 943.1 Mechanical Instruments
- 943.2 Mechanical Variables Measurements
- 943.3 Special Purpose Instruments

944 Moisture, Pressure and Temperature, and Radiation Measuring Instruments

944.1	Moisture Measuring Instruments
944.2	Moisture Measurements
944.3	Pressure Measuring Instruments
944.4	Pressure Measurements
944.5	Temperature Measuring Instruments
944.6	Temperature Measurements
944.7	Radiation Measuring Instruments
944.8	Radiation Measurements

ENGINEERING INFORMATION'S PRODUCTS AND SERVICES

PRINT MEDIA

The Engineering Index® Monthly—A monthly compilation of abstracts with an author index and subject index covering the journal and conference literature of engineering and related technical disciplines. Each issue contains indexed and classified abstracts of approximately 14,000 articles from some 3,000 journals representing the world's professional periodical and conference literature and technical reports.

The Engineering Index® Annual—A casebound cumulative edition of each year's The Engineering Index Monthly issues, with subject, author and affiliation indexes and other features, containing over 180,000 abstracts in recent editions. Back issues are available in print from 1984 and in microfilm from 1884. Over 4 million records have been published in *The Engineering Index* since 1884.

Energy Abstracts—A specialized monthly publication with an author and subject index, drawn from the Ei Monthly, covering the literature of conventional and alternative energy sources. A cumulative Annual Subject Index is also provided.

PIE: Publications In Engineering—Lists all serial and other source publications reviewed and indexed during the year. Published annually.

Ei Thesaurus—Complete hierarchical indexing guide to *The Engineering Index Monthly* and related products from 1993 forward, with cross references to indexing terms from prior years. Second edition, 1995, includes Ei classification (CAL) codes linked with descriptors.

Newsletters published jointly with SPIE, The International Society for Optical Engineering:

Ei/SPIE Page One: Optics — A bi-weekly newsletter listing tables of contents. Available in print or on diskette.

Ei/SPIE Critical Papers — Five bi-weekly newsletters in optical engineering including abstracts.

Engineering specialty monthlies published jointly by Ei and Cambridge Scientific Abstracts (CSA) in the following areas:

- Bioengineering and Biotechnology
- Computer and Information Systems
- Electronics and Communications
- Mechanical Engineering
- Solid State and Superconductivity

Ei Spotlights — newsletters for engineers. Over 155 topics in all branches of engineering are available monthly with abstracts. Customized topics also available.

ELECTRONIC MEDIA

Ei Compendex*Plus® is the machine-readable version of *The Engineering Index*. It contains both journal and conference literature and is currently available on CDROM, by direct magnetic tape subscription, in site licenses or by online searching and retrieval through several public computer utility systems worldwide. Also available through Ei Connexion.

The following public computer systems offer online and/or batch searches of Ei's databases. Contact the individual vendor for prices and terms.

- CISTI (Canada)
- CEDOCAR (Center de Documentation de L'Armement) (France)
- DATA-STAR (USA & UK)
- DIALOG Online (USA & UK)
- ESA-IRS (European Space Agency Information Retrieval Service) (Italy)
- FIZ Technik (Germany)
- ORBIT-Questel
- STN International (The Scientific & Technical Network) (USA, Germany & Japan)
- OCLC (USA)

Engineering specialty CDROMs containing subsets of Ei Compendex*Plus include:

- Ei ChemDisc ®, the database for chemical engineering.
- Ei EEDisc™ , the database for computers and electrical engineering.
- Ei Energy/Environment Disc ™ .
- Ei Materials™
- Ei Manufacturing™
- Ei Civil Disc™

Ei Page One on CDROM, on magnetic tape or in site licensing is a tables-of-contents alerting service.

Ei Industry Monitors — a table of contents service organized by industry. Available on diskette or CDROM.

ORDERING INFORMATION

All orders and inquiries should be sent to:

Engineering Information Inc.
Sales Department
Castle Point on the Hudson
Hoboken, NJ 07030, USA

Telephone: 201-216-8500 or 800-221-1044
FAX: 201-216-8532
E-mail: ei@einet.org
Telex 4990438

In Western Europe, please contact the following sales offices :

Ei Europe
London Road, Sunningdale
Berks SL5 OEP, ENGLAND.
Tel: +44 (0) 1344-291072. Fax: +44 (0) 344-26120

Ei Europe
Am Holzhausenpark
Justinianstrasse 22
60322 Frankfurt am Main, GERMANY.
Tel: +49 (0) 69-95515-219. Fax: +49 (0) 69-595770

Ei Europe
Immeuble le Luzard 2 Noisiel
B.P.217
77441 Marne la Vallee Cedex 2, FRANCE
Tel: +33 (1) 64-11-41-85. Fax: +33 (1) 64-11-41-42

In Eastern Europe, please contact:

INFO TECHNOLOGY SUPPLY LTD.
204-226 Imperial Drive
Harrow, Middlesex HA2 7HH, ENGLAND.
Tel: +44 (0) 181-429-3970,-4455. Fax: +44 (0) 181-429-3642 or +44 (0) 181-868-2330

In Asia, please contact:

Kinokuniya Company Ltd.
38-1 Sakuragaoka 5-chome
Setagaya-ku, Tokyo 156, JAPAN.
Tel: (03) 3439-0123. Fax: (03) 3439-1094.

INQUIRY FORM

Please send me current prices and information for the following publications:

___Engineering Index Monthly
___Engineering Index Annual
___Combination Subscription
(current year Monthly, prior year Annual)
___Ei Industry Monitors
___Ei Spotlights
___Page One: Optics
___Ei/SPIE Critical Papers

User Aids
___PIE: Publications In Engineering
___Ei Thesaurus

Ordered by ______________________
Title/ Dept. ______________________
Organization ______________________
Street ______________________
City ______________________
State/Province ____________ Zip ________
Country ______________________
Telephone ____________ Fax ________

Ei Document Delivery Service

Full-text copies at your fingertips

The full text of the articles and conference papers covered in *The Engineering Index®*, *Ei Compendex*Plus*™ and *Ei Page One*™ can be ordered directly from EiDDS . Ordering is easily done by toll-free phone, mail, fax or electronically through major online ordering services such as Dialog, ORBIT-Questel, STN International, ESA/IRS, RLG/RLIN and OCLC. All documents are provided as published in the original language.

Other features include:

- **Copyright** — Ei takes care of your copyright headaches and hassles by paying all royalties either directly to publishers or through the Copyright Clearance Center.
- **Delivery** — Documents are delivered by first-class mail for fast turnaround.
- **Rerouting** — If the document requested is not in Ei's collection, we try other sources. It takes a little longer, but it saves you the trouble.
- **Billing** — Invoices are shipped monthly. VISA, Mastercard and American Express are accepted.
- **Deposit Accounts** — Deposit accounts are available. Monthly statements are sent to detail all account activity.
- **Contract Services** — Volume discounts and customized delivery services are available for large volume users.

Engineering Information
EiDDS
Castle Point on the Hudson
Hoboken, NJ 07030
FAX: 201-216-8532
E-mail: DDS@einet.ei.org